Analysis, Synthesis, and Design of Chemical Processes

Third Edition

Analysis, Synthesis, and Design of Chemical Processes

Third Edition

Richard Turton

Richard C. Bailie

Wallace B. Whiting

Joseph A. Shaeiwitz

Upper Saddle River, New Jersey • Boston • Indianapolis • San Francisco
New York • Toronto • Montreal • London • Munich • Paris • Madrid
Capetown • Sydney • Tokyo • Singapore • Mexico City

The publisher offers excellent discounts on this book when ordered in quantity for bulk purchases or special sales, which may include electronic versions and/or custom covers and content particular to your business, training goals, marketing focus, and branding interests. For more information, please contact:

U.S. Corporate and Government Sales
(800) 382-3419
corpsales@pearsontechgroup.com

For sales outside the United States please contact:

International Sales
international@pearson.com

Visit us on the Web: informit.com/ph

Library of Congress Cataloging-in-Publication Data
Analysis, synthesis, and design of chemical processes / Richard Turton . . . [et al.]—3rd ed.
p. cm.
Originally published: Upper Saddle River, N.J. : Prentice Hall PTR, c1998.
Includes bibliographical references and index.
ISBN 0-13-512966-4 (hardback : alk. paper)
1. Chemical processes. I. Turton, Richard, 1955-
TP155.7.A53 2009
660'.2812—dc22 2008040404

ISBN-13: 978-0-13-512966-1
ISBN-10: 0-13-512966-4

Text printed in the United States on recycled paper at Courier Westford in Westford, Massachusetts.

4th Printing November 2010

Contents

Material on the CD-ROM

Preface

This book represents the culmination of many years of teaching experience in the senior design course at West Virginia University (WVU) and University of Nevada, Reno. Although the program at WVU has evolved over the past 30 years and is still evolving, it is fair to say that the current program has gelled over the past 20 years as a concerted effort by the authors to integrate design throughout the undergraduate curriculum in chemical engineering.

We view design as the focal point of chemical engineering practice. Far more than the development of a set of specifications for a new chemical plant, design is the creative activity through which engineers continuously improve the operations of facilities to create products that enhance the quality of life. Whether developing the grassroots plant, proposing and guiding process modifications, or troubleshooting and implementing operational strategies for existing equipment, engineering design requires a broad spectrum of knowledge and intellectual skills to be able to analyze the big picture and the minute details and, most important, to know when to concentrate on each.

Our vehicle for helping students develop and hone their design skills is process design rather than plant design, covering synthesis of the entire chemical process through topics relating to the preliminary sizing of equipment, flowsheet optimization, economic evaluation of projects, and the operation of chemical processes. The purpose of this text is to assist chemical engineering students in making the transition from solving well-posed problems in a specific subject to integrating all the knowledge that they have gained in their undergraduate education and applying this information to solving open-ended process problems. Many of the nuts-and-bolts issues regarding plant design (for example, what schedule pipe to use for a given stream or what corrosion allowance to use for a vessel in a certain service) are not covered. Although such issues are clearly important to the practicing engineer, several excellent handbooks and textbooks are available to address such problems, and these are cited in the text where applicable.

In the third edition, we have rearranged some of the material from previous editions, added a new chapter on batch processing and a section on optimization of batch processes, and supplied new problems for all of the quantitative chapters. We continue to emphasize the importance of understanding, analyzing, and synthesizing chemical processes and process flow diagrams. To this end, we have expanded Appendix B to include an additional seven preliminary designs of chemical processes. The CAPCOST program for preliminary evaluation of fixed capital investment and profitability analysis has been expanded to include more equipment. Finally, the chapters on outcomes assessment, written and oral communications, and a written report case study have been moved to the CD accompanying the text.

The arrangement of chapters into the six sections of the book is similar to that adopted in the second edition. These sections are as follows.

Section 1—Conceptualization and Analysis of Chemical Processes
Section 2—Engineering Economic Analysis of Chemical Processes
Section 3—Synthesis and Optimization of Chemical Processes
Section 4—Analysis of Process Performance
Section 5—The Impact of Chemical Engineering Design on Society
Section 6— Interpersonal and Communication Skills

In Section 1, the student is introduced first to the principal diagrams that are used to describe a chemical process. Next, the evolution and generation of different process configurations are covered. Key concepts used in evaluating batch processes are included in the new Chapter 3, and the chapter on product design has been moved to Chapter 4. Finally, the analysis of existing processes is covered.

In Section 2, the information needed to assess the economic feasibility of a process is covered. This includes the estimation of fixed capital investment and manufacturing costs, the concepts of the time value of money and financial calculations, and finally the combination of these costs into profitability measures for the process.

Section 3 covers the synthesis of a chemical process. The minimum information required to simulate a process is given, as are the basics of using a process simulator. The choice of the appropriate thermodynamic model to use in a simulation is covered, and the choice of separation operations is covered. In addition, process optimization (including an introduction to optimization of batch processes) and heat integration techniques are covered in this section.

In Section 4, the analysis of the performance of existing processes and equipment is covered. The material in Section 4 is substantially different from that found in most textbooks. We consider equipment that is already built and operating and analyze how the operation can be changed, how an operating problem may be solved, and how to analyze what has occurred in the process to cause an observed change.

In Section 5, the impact of chemical engineering design on society is covered. The role of the professional engineer in society is addressed. Separate chapters addressing ethics and professionalism, health, safety, and the environment, and green engineering are included.

In Section 6, the interpersonal skills required by the engineer to function as part of a team and to communicate both orally and in written form are covered (both in the text and on the CD). An entire chapter (on the CD) is devoted to addressing some of the common mistakes that students make in written reports.

Finally, three appendices are included. Appendix A gives a series of cost charts for equipment. This information is embedded in the CAPCOST program for evaluating fixed capital investments and process economics. Appendix B gives the preliminary design information for 11 chemical processes: dimethyl ether, ethylbenzene, styrene, drying oil, maleic anhydride, ethylene oxide, formalin, batch manufacture of amino acids, acrylic acid, acetone, and heptenes production. This information is used in many of the end-of-chapter problems in the book. These processes can also be used as the starting point for more detailed analyses—for example, optimization studies. Other projects, detailed in Appendix C, are included on the CD accompanying this book. The reader (faculty and students) is also referred to our Web site at che.cemr.wvu.edu/publications/projects/, where a variety of design projects for sophomore- through senior-level chemical engineering courses is provided. There is also a link to another Web site that contains environmentally related design projects.

For a one-semester design course, we recommend including the following core:

- Section 1—Chapters 1 through 6
- Section 3—Chapters 11, 12, and 13
- Section 5—Chapters 23 and 24

For programs in which engineering economics is not a prerequisite to the design course, Section 2 (Chapters 7–10) should also be included. If students have previously covered engineering economics, Chapters 14 and 15 covering optimization and pinch technology could be substituted.

For the second term of a two-term sequence, we recommend Chapters 16 through 20 (and Chapters 14 and 15 if not included in the first design course) plus design projects. If time permits, we also recommend Chapter 21 (Regulating Process Conditions) and Chapter 22 (Process Troubleshooting) because these tend to solidify as well as extend the concepts of Chapters 16 through 20, that is, what an entry-level process engineer will encounter in the first few years of employment at a chemical process facility. For an environmental emphasis, Chapter 25 could be substituted for Chapters 21 and 22; however, it is recommended that supplementary material be included.

We have found that the most effective way both to enhance and to examine student progress is through oral presentations in addition to the submission of written reports. During these oral presentations, individual students or a student group defends its results to a faculty panel, much as a graduate student defends a thesis or dissertation.

Because design is at its essence a creative, dynamic, challenging, and iterative activity, we welcome feedback on and encourage experimentation with this design textbook. We hope that students and faculty will find the excitement in teaching and learning engineering design that has sustained us over the years.

Finally, we would like to thank those people who have been instrumental to the successful completion of this book. Many thanks are given to all undergraduate chemical engineering students at West Virginia University over the years, particularly the period 1992–2008. In particular, we would like to thank Joe Stoffa, who was responsible for developing the spreadsheet version of CAPCOST, and Mary Metzger and John Ramsey, who were responsible for collecting and correlating equipment cost information for this edition. We also acknowledge the many faculty who have provided, both formally and informally, feedback about this text. Finally, RT would like to thank his wife Becky for her continued support, love, and patience during the preparation of this third edition.

R.T.
R.C.B.
W.B.W.
J.A.S.

About the Authors

Richard Turton, P.E., has taught the senior design course at West Virginia University for the past 22 years. Prior to this, he spent five years in the design and construction industry. His main interests are in design education, particulate processing, and process modeling.

Richard C. Bailie has more than ten years of experience in process evaluation, pilot plant operation, plant start-up, and industrial consulting. He also ran his own chemical company. He is professor emeritus at WVU, having taught chemical engineering for more than 20 years.

Wallace B. Whiting, P.E., is professor emeritus at University of Nevada, Reno. He has been involved in the practice and teaching of chemical process design for more than 24 years.

Joseph A. Shaeiwitz has been involved in the senior design sequence and unique sophomore- and junior-level integrated design projects at WVU for 20 years. His interests include design education and outcomes assessment.

List of Nomenclature

Symbol	Definition	SI Units
A	Equipment Cost Attribute	
A	Area	m^2
A	Absorption Factor	
A	Annuity Value	$/time
A/F, i, n	Sinking Fund Factor	
A/P, i, n	Capital Recovery Factor	
BV	Book Value	$
C	Equipment Cost	$
C or *c*	Molar Concentration	$kmol/m^3$
CA	Corrosion Allowance	m
CBM	Bare Module Cost	$
COM	Cost of Manufacture	$/time
cop	Coefficient of Performance	
C_p	Heat Capacity	kJ/kg°C or kJ/kmol°C
CCP	Cumulative Cash Position	$
CCR	Cumulative Cash Ratio	
D	Diameter	m
D	Amount Allowed for Depreciation	$
D	Distillate Product Flowrate	kmol/time
d	Yearly Depreciation Allowance	$/yr
DCFROR	Discounted Cash Flow Rate of Return	
DMC	Direct Manufacturing Cost	$/time
DPBP	Discounted Payback Period	years
E	Money Earned	$
E	Weld Efficiency	
E_{act} or *E*	Activation Energy	kJ/kmol
EAOC	Equivalent Annual Operating Cost	$/yr
ECC	Equivalent Capitalized Cost	$

f_q	Quantity Factors for Trays	
F	Future Value	$
F	Molar Flowrate	kmol/s
F	Equipment Module Cost Factor	
F	Correction for Multipass Heat Exchangers	
F	Future Value	$
F_d	Drag Force	N/m^2 or kPa
f	Friction Factor	
f	Rate of Inflation	
$F/A, i, n$	Uniform Series Compound Amount Factor	
FCI	Fixed Capital Investment	$
$F/P, i, n$	Single Payment Compound Amount Factor	
FMC	Fixed Manufacturing Costs	$/time
F_{Lang}	Lang Factor	
G	Gas Flowrate	kg/s, kmol/s
GE	General Expenses	$/time
h	Individual Heat Transfer Coefficient	W/m^{2}K
H	Enthalpy or Specific Enthalpy	kJ or kJ/kg
H	Height	m
I	Cost Index	
i	Compound Interest	
i'	Effective Interest Rate Including Inflation	
$INPV$	Incremental Net Present Value	$
$IPBP$	Incremental Payback Period	years
k	Thermal Conductivity	W/m K
k_o	Preexponential Factor for Reaction Rate Constant	Depends on molecularity of reaction
K_p	Equilibrium Constant	Depends on reaction stoichiometry
k_{reac} or k_i	Reaction Rate Constant	Depends on molecularity of reaction
L	Lean Stream Flowrate	kg/s
L	Liquid Flowrate	kg/s or kmol/s
$\dot{m}$	Flowrate	kg/s
m	Partition Coefficient (y/x)	
n	Life of Equipment	years
n	Years of Investment	years
n	Number of Batches	
n_c	Number of Campaigns	
N	Number of Streams	
N	Number of Trays, Stages, or Shells	
N	Molar Flowrate	kmol/s
$NPSH$	Net Positive Suction Head	m of liquid
NPV	Net Present Value	$

N_{toG}	Number of Transfer Units	
OBJ, OF	Objective Function	usually $ or $/time
p	Price	$
P	Dimensionless Temperature Approach	
P	Pressure	bar or kPa
P	Present Value	$
P^*	Vapor Pressure	bar or kPa
$P/A, i, n$	Uniform Series Present Worth Factor	
PBP	Payback Period	year
PC	Project Cost	$
$P/F, i, n$	Single Payment Present Worth Factor	
PVR	Present Value Ratio	
$P(x)$	Probability Density Function of x	
Q or q	Rate of Heat Transfer	W or MJ/h
Q	Quantity	
r	Reaction Rate	$kmol/m^3$ or $kmol/kg$ cat s
r	Rate of Production	kg/h
R	Gas Constant	kJ/kmol K
R	Ratio of Heat Capabilities	
R	Residual Funds Needed	$
R	Reflux Ratio	
Re	Reynolds Number	
R	Rich Stream Flowrate	kg/s
Rand	Random Number	
ROROI	Rate of Return on Investment	
ROROII	Rate of Return on Incremental Investment	
S	Salvage Value	$
S	Maximum Allowable Working Pressure	bar
S	Salt Concentration Factor	
S	Sensitivity	
SF	Stream Factor	
t	Thickness of Wall	m
t	Time	s, min, h, yr
T	Total Time for a Batch	s, min, h, yr
T	Temperature	K, R, °C, or °F
u	Flow Velocity	m/s
U	Overall Heat Transfer Coefficient	W/m^2K
V	Volume	m^3
V	Vapor Flow Rate	kmol/h
v_{react}	Specific Volume of Reactor	m^3/kg of product
v_p	Velocity	m/s
$\dot{v}$	Volumetric Flowrate	m^3/s
W	Weight	kg

W	Total Moles of a Component	kmol
W or WS	Work	kJ/kg
WC	Working Capital	$
X	Conversion	
X	Base-Case Ratio	
x	Mole or Mass Fraction	
y	Mole or Mass Fraction	
YOC	Yearly Operating Cost	$/yr
YS	Yearly Cash Flow (Savings)	$/yr
z	Distance	m

Greek Symbols

α	Multiplication Cost Factor	
α	Relative Volatility	
ε	Void Fraction	
ε	Pump Efficiency	
ϕ	Fugacity Coefficient	
γ	Activity Coefficient	
η	Selectivity	
λ	Heat of Vaporization	kJ/kg
μ	Viscosity	kg/m s
ξ	Selectivity	
ρ	Density	kg/m^3
θ	Rates of Species Concentration to that of Limiting Reactant	s\
Θ	Cycle Time	s
τ	Space Time	s

Subscripts

1	Base Time
2	Desired Time
a	Required Attribute
ACT	Actual
Aux	Auxiliary Buildings
b	Base Attribute
BM	Bare Module
$clean$	Cleaning
$Cont$	Contingency
$cycle$	Cycle
d	Without Depreciation
D, d	Demand
E	Contractor Engineering Expenses
eff	Effective Interest
eq	Equivalent

Fee	Contractor Fee
FTT	Transportation, etc.
GR	Grass Roots
k	Year
L	Installation Labor
L	Lean Streams
L	Without Land Cost
m	Number of Years
M	Materials for Installation
M	Material Cost Factor
max	Maximum
MC	Matching Costs
min	Minimum
nom	Nominal Interest
O or *OH*	Construction Overhead
Off	Offsites and Utilities
OL	Operating Labor
opt	Optimum
p	Production
P	Equipment at Manufacturer's Site (Purchased)
P	Pressure Cost Factor
P&I	Piping and Instrumentation
R	Rich Stream
RM	Raw Materials
rev	Reversible
rxn, *r*	Reaction
s	Simple Interest
S	Supply
Site	Site Development
TM	Total Module
UT	Utilities
WT	Waste Treatment

Superscripts

DB	Double Declining Balance Depreciation
o	Cost for Ambient Pressure Using Carbon Steel
SL	Straight Line Depreciation
SOYD	Sum of the Years Depreciation
′	Includes Effect of Inflation on Interest

Additional Nomenclature

Table 1.2	Convention for Specifying Process Equipment
Table 1.3	Convention for Specifying Process Streams
Table 1.7	Abbreviations for Equipment and Materials of Construction
Table 1.10	Convention for Specifying Instrumentation and Control Systems

SECTION

1

Conceptualization and Analysis of Chemical Processes

The purpose of this section of the book is to introduce the tools necessary to understand, interpret, synthesize, and create chemical processes. The basis of interpreting chemical processes lies with understanding the principal diagrams that are routinely used to describe chemical processes, most important of which is the process flow diagram (PFD). Although PFDs are unique for each chemical product, they possess many of the same characteristics and attributes. Moreover, the conditions (pressure, temperature, and concentration) at which different equipment operate are unique to the chemical product and processing route chosen. In order for process engineers to understand a given process or to be able to synthesize and optimize a new process, they must be able to apply the principles outlined in this section.

Chapter 1: Diagrams for Understanding Chemical Processes
The technical diagrams commonly used by chemical engineers are presented. These diagrams include the block flow diagram (BFD), the process flow diagram (PFD), and the piping and instrumentation diagram (P&ID). A standard method for presenting a PFD is given and illustrated using a process to produce benzene via the catalytic hydrodealkylation of toluene. The 3-D topology of chemical processes is introduced, and some basic information on the spacing and elevation of equipment is presented. These concepts are further illustrated in the Virtual Plant Tour AVI file on the CD accompanying the textbook.

Chapter 2: The Structure and Synthesis of Process Flow Diagrams
The evolutionary process of design is investigated. This evolution begins with the process concept diagram that shows the *input/output* structure of all

processes. From this simple starting point, the engineer can estimate the gross profit margins of competing processes and of processes that use different chemical synthesis routes to produce the same product. In this chapter, it is shown that all processes have a similar input/output structure whereby raw materials enter a process and are reacted to form products and by-products. These products are separated from unreacted feed, which is usually recycled. The product streams are then purified to yield products that are acceptable to the market place. All equipment in a process can be categorized into one of the six elements of the generic block flow process diagram. The process of process design continues by building preliminary flowsheets from these basic functional elements that are common to all processes.

Chapter 3: Batch Processing

In this chapter, key issues relating to the production of chemical products using batch processes are explored. The major difference between continuous and batch processes is that unsteady state operations are normal to batch plants whereas steady state is the norm for continuous processes. The chapter starts with an example illustrating typical calculations required to design a sequence of batch operations to produce a given product. The remainder of the chapter is devoted to how best to sequence the different operations required to produce multiple chemical products using a fixed amount of equipment. The concepts of Gantt charts, cycle times, batch campaigning, intermediate and final product storage, and parallel operations are covered.

Chapter 4: Chemical Product Design

Chemical product design is defined to include application of chemical engineering principles to the development of new devices, development of new chemicals, development of new processes to produce these new chemicals, and development of marketable technology. The design hierarchy for chemical product design is presented. The necessity of considering customer needs in chemical product design and the need to develop interdisciplinary teams are discussed.

Chapter 5: Tracing Chemicals through the Process Flow Diagram

In order to gain a better understanding of a PFD, it is often necessary to follow the flow of key chemical components through the diagram. This chapter presents two different methods to accomplish this. The tracing of chemicals through the process reinforces our understanding of the role that each piece of equipment plays. In most cases, the major chemical species can be followed throughout the flow diagram using simple logic without referring to the flow summary table.

Chapter 6: Understanding Process Conditions
Once the connectivity or topology of the PFD has been understood, it is necessary to understand why a piece of equipment is operated at a given pressure and temperature. The idea of conditions of special concern is introduced. These conditions are either expensive to implement (due to special materials of construction and/or the use of thick-walled vessels) or use expensive utilities. The reasons for using these conditions are introduced and explained.

CHAPTER

1

Diagrams for Understanding Chemical Processes

The chemical process industry (CPI) is involved in the production of a wide variety of products that improve the quality of our lives and generate income for companies and their stockholders. In general, chemical processes are complex, and chemical engineers in industry encounter a variety of chemical process flow diagrams. These processes often involve substances of high chemical reactivity, high toxicity, and high corrosivity operating at high pressures and temperatures. These characteristics can lead to a variety of potentially serious consequences, including explosions, environmental damage, and threats to people's health. It is essential that errors or omissions resulting from missed communication between persons and/or groups involved in the design and operation do not occur when dealing with chemical processes. Visual information is the clearest way to present material and is least likely to be misinterpreted. For these reasons, it is essential that chemical engineers be able to formulate appropriate process diagrams and be skilled in analyzing and interpreting diagrams prepared by others.

The most effective way of communicating information about a process is through the use of flow diagrams.

This chapter presents and discusses the more common flow diagrams encountered in the chemical process industry. These diagrams evolve from the time a process is conceived in the laboratory through the design, construction, and the many years of plant operation. The most important of these diagrams are described and discussed in this chapter.

The following narrative is taken from Kauffman [1] and describes a representative case history related to the development of a new chemical process. It shows how teams of engineers work together to provide a plant design and introduces the types of diagrams that will be explored in this chapter.

> *The research and development group at ABC Chemicals Company worked out a way to produce alpha-beta souptol (ABS). Process engineers assigned to work with the development group have pieced together a continuous process for making ABS in commercial quantities and have tested key parts of it. This work involved hundreds of* ***block flow diagrams,*** *some more complex than others. Based on information derived from these block flow diagrams, a decision was made to proceed with this process.*
>
> *A process engineering team from ABC's central office carries out the detailed process calculations, material and energy balances, equipment sizing, etc. Working with their drafting department, they produced a series of* ***PFDs (Process Flow Diagrams)*** *for the process. As problems arise and are solved, the team may revise and redraw the PFDs. Often the work requires several rounds of drawing, checking, and revising.*
>
> *Specialists in distillation, process control, kinetics, and heat transfer are brought in to help the process team in key areas. Some are company employees and others are consultants.*
>
> *Since ABC is only a moderate-sized company, it does not have sufficient staff to prepare the 120* ***P&IDs (Piping and Instrumentation Diagrams)*** *needed for the new ABS plant. ABC hires a well-known engineering and construction firm* ***(E&C Company),*** *DEFCo, to do this work for them. The company assigns two of the ABC process teams to work at DEFCo to coordinate the job. DEFCo's process engineers, specialists, and drafting department prepare the P&IDs. They do much of the detailed engineering (pipe sizes, valve specifications, etc.) as well as the actual drawing. The job may take two to six months. Every drawing is reviewed by DEFCo's project team and by ABC's team. If there are disagreements, the engineers and specialists from the companies must resolve them.*
>
> *Finally, all the PFDs and the P&IDs are completed and approved. ABC can now go ahead with the construction. They may extend their contract with DEFCo to include this phase, or they may go out for construction bids from a number of sources.*

This narrative describes a typical sequence of events taking a project from its initial stages through plant construction. If DEFCo had carried out the construction, ABC could go ahead and take over the plant or DEFCo could be contracted to carry out the start-up and to commission the plant. Once satisfactory performance specifications have been met, ABC would take over the operation of the plant and commercial production would begin.

From conception of the process to the time the plant starts up, two or more years will have elapsed and millions of dollars will have been spent with no revenue from the plant. The plant must operate successfully for many years to produce sufficient income to pay for all plant operations and to repay the costs associated with designing and building the plant. During this operating period, many unforeseen changes are likely to take place. The quality of the raw materials used by the plant may change, product specifications may be raised, production rates may need to be increased, the equipment performance will decrease because of wear, the development of new and better catalysts will occur, the costs of utilities will change, new environmental regulations may be introduced, or improved equipment may appear on the market.

As a result of these unplanned changes, plant operations must be modified. Although the operating information on the original process diagrams remains informative, the actual performance taken from the operating plant will be different. The current operating conditions will appear on updated versions of the various process diagrams, which will act as a primary basis for understanding the changes taking place in the plant. These process diagrams are essential to an engineer who has been asked to diagnose operating problems, solve problems in operations, debottleneck systems for increased capacity, and predict the effects of making changes in operating conditions. All these activities are essential in order to maintain profitable plant operation.

In this chapter, we concentrate on three diagrams that are important to chemical engineers: block flow, process flow, and piping and instrumentation diagrams. Of these three diagrams, we will find that the most useful to chemical engineers is the PFD. The understanding of the PFD represents a central goal of this textbook.

1.1 BLOCK FLOW DIAGRAMS (BFDs)

Block flow diagrams were introduced early in the chemical engineering curriculum. In the first course in material and energy balances, often an initial step was to convert a word problem into a simple block diagram. This diagram consisted of a series of blocks representing different equipment or unit operations that were connected by input and output streams. Important information such as operating temperatures, pressures, conversions, and yield was included on the diagram along with flowrates and some chemical compositions. However, the diagram did not include any details of equipment within any of the blocks.

The block flow diagram can take one of two forms. First, a block flow diagram may be drawn for a single process. Alternatively, a block flow diagram may be drawn for a complete chemical complex involving many different chemical processes. We differentiate between these two types of diagram by calling the first a block flow process diagram and the second a block flow plant diagram.

1.1.1 Block Flow Process Diagram

An example of a block flow process diagram is shown in Figure 1.1, and the process illustrated is described below.

> *Toluene and hydrogen are converted in a reactor to produce benzene and methane. The reaction does not go to completion, and excess toluene is required. The noncondensable gases are separated and discharged. The benzene product and the unreacted toluene are then separated by distillation. The toluene is then recycled back to the reactor and the benzene removed in the product stream.*

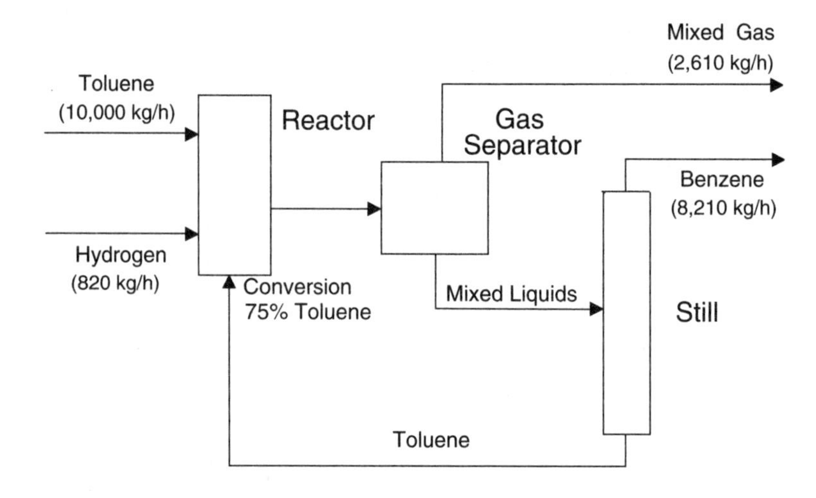

Figure 1.1 Block Flow Process Diagram for the Production of Benzene

This block flow diagram gives a clear overview of the production of benzene, unobstructed by the many details related to the process. Each block in the diagram represents a process function and may, in reality, consist of several pieces of equipment. The general format and conventions used in preparing block flow process diagrams are presented in Table 1.1.

Although much information is missing from Figure 1.1, it is clear that such a diagram is very useful for "getting a feel" for the process. Block flow process diagrams often form the starting point for developing a PFD. They are also very helpful in conceptualizing new processes and explaining the main features of the process without getting bogged down in the details.

Table 1.1 Conventions and Format Recommended for Laying Out a Block Flow Process Diagram

1. Operations shown by blocks.
2. Major flow lines shown with arrows giving direction of flow.
3. Flow goes from left to right whenever possible.
4. Light stream (gases) toward top with heavy stream (liquids and solids) toward bottom.
5. Critical information unique to process supplied.
6. If lines cross, then the horizontal line is continuous and the vertical line is broken (hierarchy for all drawings in this book).
7. Simplified material balance provided.

1.1.2 Block Flow Plant Diagram

An example of a block flow plant diagram for a complete chemical complex is illustrated in Figure 1.2. This block flow plant diagram is for a coal to higher alcohol fuels plant. Clearly, this is a complicated process in which there are a number of alcohol fuel products produced from a feedstock of coal. Each block in this diagram represents a complete chemical process (compressors and turbines are also shown as trapezoids), and we could, if we wished, draw a block flow process diagram for each block in Figure 1.2. The advantage of a diagram such as Figure 1.2 is that it allows us to get a complete picture of what this plant does and how all the different processes interact. On the other hand, in order to keep the diagram relatively uncluttered, only limited information is available about each process unit. The conventions for drawing block flow plant diagrams are similar to Table 1.1.

Both types of block flow diagrams are useful for explaining the overall operation of chemical plants. For example, consider that you have just joined a large chemical manufacturing company that produces a wide range of chemical products from the site to which you have been assigned. You would most likely be given a *block flow plant diagram* to orient you to the products and important areas of operation. Once assigned to one of these areas, you would again likely be provided with a *block flow process diagram* describing the operations in your particular area.

In addition to the orientation function described earlier, block flow diagrams are used to sketch out and screen potential process alternatives. Thus, they are used to convey information necessary to make early comparisons and eliminate competing alternatives without having to make detailed and costly comparisons.

1.2 PROCESS FLOW DIAGRAM (PFD)

The process flow diagram (PFD) represents a quantum step up from the BFD in terms of the amount of information that it contains. The PFD contains the bulk of the chemical engineering data necessary for the design of a chemical process. For all of the diagrams discussed in this chapter, there are no universally accepted standards. The PFD from one company will probably contain slightly different information than the PFD for the same process from another company. Having made this point, it is fair to say that most PFDs convey very similar information. A typical commercial PFD will contain the following information.

1. All the major pieces of equipment in the process will be represented on the diagram along with a description of the equipment. Each piece of equipment will have assigned a unique equipment number and a descriptive name.

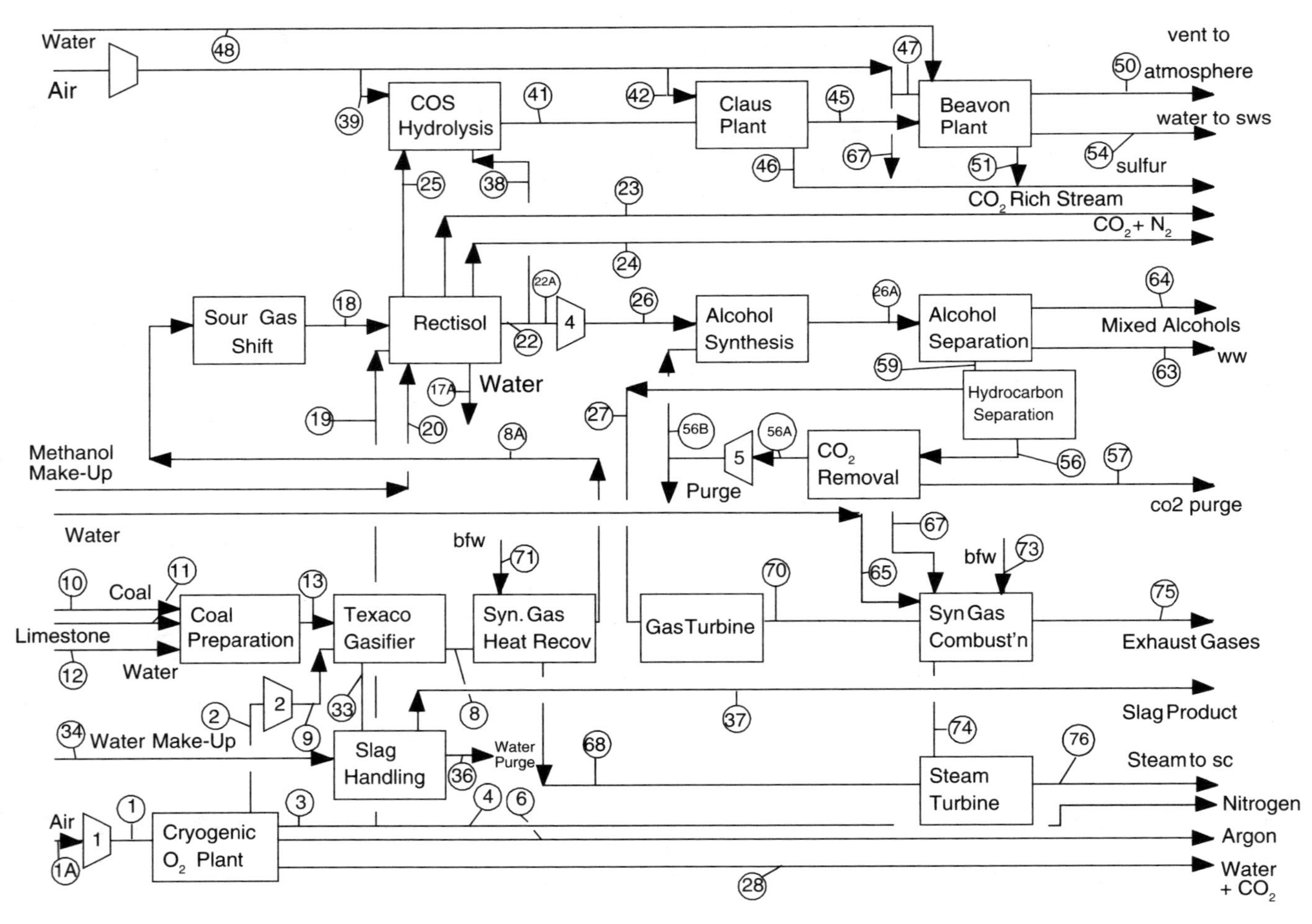

Figure 1.2 Block Flow Plant Diagram of a Coal to Higher Alcohol Fuels Process

2. All process flow streams will be shown and identified by a number. A description of the process conditions and chemical composition of each stream will be included. These data will be either displayed directly on the PFD or included in an accompanying flow summary table.
3. All utility streams supplied to major equipment that provides a process function will be shown.
4. Basic control loops, illustrating the control strategy used to operate the process during normal operations, will be shown.

It is clear that the PFD is a complex diagram requiring a substantial effort to prepare. It is essential that it should remain uncluttered and be easy to follow, to avoid errors in presentation and interpretation. Often PFDs are drawn on large sheets of paper (for example, size D: 24″ × 36″), and several connected sheets may be required for a complex process. Because of the page size limitations associated with this text, complete PFDs cannot be presented here. Consequently, certain liberties have been taken in the presentation of the PFDs in this text. Specifically, certain information will be presented in accompanying tables, and only the essential process information will be included on the PFD. The resulting PFDs will retain clarity of presentation, but the reader must refer to the flow summary and equipment summary tables in order to extract all the required information about the process.

Before we discuss the various aspects of the PFD, it should be noted that the PFD and the process that we describe in this chapter will be used throughout the book. The process is the hydrodealkylation of toluene to produce benzene. This is a well-studied and well-understood commercial process still used today. The PFD we present in this chapter for this process is technically feasible but is in no way optimized. In fact, there are many improvements to the process technology and economic performance that can be made. Many of these improvements will become evident when the appropriate material is presented. This allows the techniques provided throughout this text to be applied both to identify technical and economic problems in the process and to make the necessary process improvements. Therefore, as we proceed through the text, we will identify weak spots in the design, make improvements, and move toward an optimized process flow diagram.

The basic information provided by a PFD can be categorized into one of the following:

1. Process topology
2. Stream information
3. Equipment information

We will look at each aspect of the PFD separately. After we have addressed each of the three topics, we will bring all the information together and present the PFD for the benzene process.

1.2.1 Process Topology

Figure 1.3 is a skeleton process flow diagram for the production of benzene (see also the block flow process diagram in Figure 1.1). This skeleton diagram illustrates the location of the major pieces of equipment and the connections that the process streams make between equipment. The location of and interaction between equipment and process streams are referred to as the process topology.

Equipment is represented symbolically by "icons" that identify specific unit operations. Although the American Society of Mechanical Engineers (ASME) [2] publishes a set of symbols to use in preparing flowsheets, it is not uncommon for companies to use in-house symbols. A comprehensive set of symbols is also given by Austin [3]. Whatever set of symbols is used, there is seldom a problem in identifying the operation represented by each icon. Figure 1.4 contains a list of the symbols used in process diagrams presented in this text. This list covers more than 90% of those needed in fluid (gas or liquid) processes.

Figure 1.3 shows that each major piece of process equipment is identified by a number on the diagram. A list of the equipment numbers along with a brief descriptive name for the equipment is printed along the top of the diagram. The location of these equipment numbers and names roughly corresponds to the horizontal location of the corresponding piece of equipment. The convention for formatting and identifying the process equipment is given in Table 1.2.

Table 1.2 provides the information necessary for the identification of the process equipment icons shown in a PFD. As an example of how to use this information, consider the unit operation P-101A/B and what each number or letter means.

P-101A/B identifies the equipment as a pump.

P-**1**01A/B indicates that the pump is located in area 100 of the plant.

P-1**01**A/B indicates that this specific pump is number 01 in unit 100.

P-101**A/B** indicates that a backup pump is installed. Thus, there are two identical pumps P-101A and P-101B. One pump will be operating while the other is idle.

The 100 area designation will be used for the benzene process throughout this text. Other processes presented in the text will carry other area designations. Along the top of the PFD, each piece of process equipment is assigned a descriptive name. From Figure 1.3 it can be seen that Pump P-101 is called the "toluene feed pump." This name will be commonly used in discussions about the process and is synonymous with P-101.

During the life of the plant, many modifications will be made to the process; often it will be necessary to replace or eliminate process equipment. When a piece of equipment wears out and is replaced by a new unit that provides essentially the same process function as the old unit, then it is not uncommon for the new piece of equipment to inherit the old equipment's name and number (often an

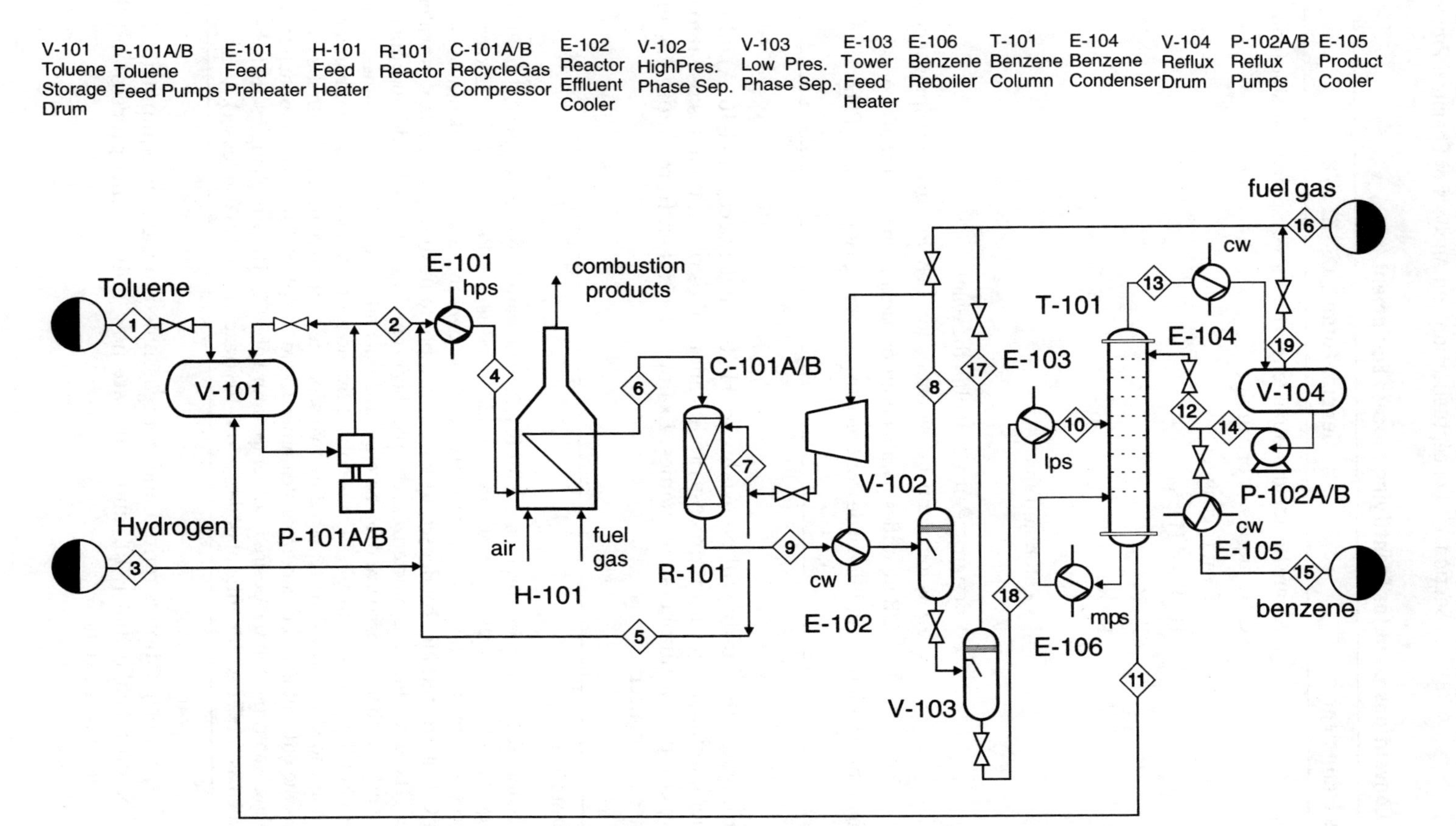

Figure 1.3 Skeleton Process Flow Diagram (PFD) for the Production of Benzene via the Hydrodealkylation of Toluene

Table 1.2 Conventions Used for Identifying Process Equipment

Process Equipment	General Format XX-YZZ A/B
	XX are the identification letters for the equipment classification
	C - Compressor or Turbine
	E - Heat Exchanger
	H - Fired Heater
	P - Pump
	R - Reactor
	T - Tower
	TK - Storage Tank
	V - Vessel
	Y designates an area within the plant
	ZZ is the number designation for each item in an equipment class
	A/B identifies parallel units or backup units not shown on a PFD
Supplemental Information	Additional description of equipment given on top of PFD

additional letter suffix will be used, e.g., H-101 might become H-101A). On the other hand, if a significant process modification takes place, then it is usual to use new equipment numbers and names. Example 1.1, taken from Figure 1.3, illustrates this concept.

Example 1.1

Operators report frequent problems with E-102, which are to be investigated. The PFD for the plant's 100 area is reviewed, and E-102 is identified as the "Reactor Effluent Cooler." The process stream entering the cooler is a mixture of condensable and noncondensable gases at 654°C that are partially condensed to form a two-phase mixture. The coolant is water at 30°C. These conditions characterize a complex heat transfer problem. In addition, operators have noticed that the pressure drop across E-102 fluctuates wildly at certain times, making control of the process difficult. Because of the frequent problems with this exchanger, it is recommended that E-102 be replaced by two separate heat exchangers. The first exchanger cools the effluent gas and generates steam needed in the plant. The second exchanger uses cooling water to reach the desired exit temperature of 38°C. These exchangers are to be designated as E-107 (reactor effluent boiler) and E-108 (reactor effluent condenser).

The E-102 designation is retired and not reassigned to the new equipment. There can be no mistake that E-107 and E-108 are new units in this process and that E-102 no longer exists.

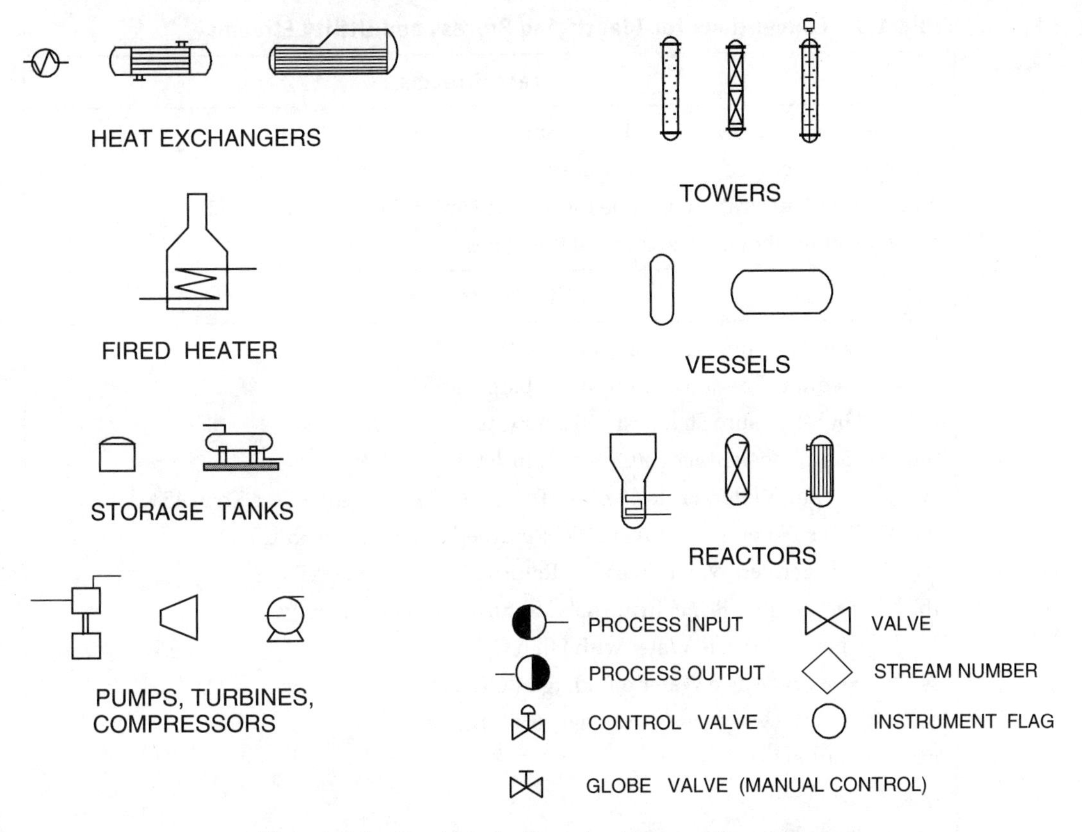

Figure 1.4 Symbols for Drawing Process Flow Diagrams

1.2.2 Stream Information

Referring back to Figure 1.3, it can be seen that each of the process streams is identified by a number in a diamond box located on the stream. The direction of the stream is identified by one or more arrowheads. The process stream numbers are used to identify streams on the PFD, and the type of information that is typically given for each stream is discussed in the next section.

Also identified in Figure 1.3 are utility streams. Utilities are needed services that are available at the plant. Chemical plants are provided with a range of central utilities that include electricity, compressed air, cooling water, refrigerated water, steam, condensate return, inert gas for blanketing, chemical sewer, waste water treatment, and flares. A list of the common services is given in Table 1.3, which also provides a guide for the identification of process streams.

Table 1.3 Conventions for Identifying Process and Utility Streams

Process Streams	
All conventions shown in Table 1.1 apply.	
Diamond symbol located in flow lines.	
Numerical identification (unique for that stream) inserted in diamond.	
Flow direction shown by arrows on flow lines.	
Utility Streams	
lps	Low-Pressure Steam: 3–5 barg (sat) ‡
mps	Medium-Pressure Steam: 10–15 barg (sat) ‡
hps	High-Pressure Steam: 40–50 barg (sat) ‡
htm	Heat Transfer Media (Organic): to 400°C
cw	Cooling Water: From Cooling Tower 30°C Returned at Less Than 45°C†
wr	River Water: From River 25°C Returned at Less Than 35°C
rw	Refrigerated Water: In at 5°C Returned at Less Than 15°C
rb	Refrigerated Brine: In at –45°C Returned at Less Than 0°C
cs	Chemical Waste Water with High COD
ss	Sanitary Waste Water with High BOD, etc.
el	Electric Heat (Specify 220, 440, 660V Service)
ng	Natural Gas
fg	Fuel Gas
fo	Fuel Oil
fw	Fire Water

‡These pressures are set during the preliminary design stages and typical values vary within the ranges shown.

†Above 45°C, significant scaling occurs.

Each utility is identified by the initials provided in Table 1.3. As an example, let us locate E-102 in Figure 1.3. The notation, cw, associated with the nonprocess stream flowing into E-102 indicates that cooling water is used as a coolant.

Electricity used to power motors and generators is an additional utility that is not identified directly on the PFD or in Table 1.3 but is treated separately. Most of the utilities shown are related to equipment that adds or removes heat within the process in order to control temperatures. This is common for most chemical processes.

From the PFD in Figure 1.3, the identification of the process streams is clear. For small diagrams containing only a few operations, the characteristics of the streams such as temperatures, pressures, compositions, and flowrates can be

Table 1.4 Information Provided in a Flow Summary

Required Information
Stream Number
Temperature (°C)
Pressure (bar)
Vapor Fraction
Total Mass Flowrate (kg/h)
Total Mole Flowrate (kmol/h)
Individual Component Flowrates (kmol/h)
Optional Information
Component Mole Fractions
Component Mass Fractions
Individual Component Flowrates (kg/h)
Volumetric Flowrates (m^3/h)
Significant Physical Properties
Density
Viscosity
Other
Thermodynamic Data
Heat Capacity
Stream Enthalpy
K-values
Stream Name

shown directly on the figure, adjacent to the stream. This is not practical for a more complex diagram. In this case, only the stream number is provided on the diagram. This indexes the stream to information on a flow summary or stream table, which is often provided below the process flow diagram. In this text the flow summary table is provided as a separate attachment to the PFD.

The stream information that is normally given in a flow summary table is given in Table 1.4. It is divided into two groups—required information and optional information—that may be important to specific processes. The flow summary table, Figure 1.3, is given in Table 1.5 and contains all the required information listed in Table 1.4.

With information from the PFD (Figure 1.3) and the flow summary table (Table 1.5), problems regarding material balances and other problems are easily analyzed. Example 1.2 and Example 1.3 are provided to offer experience in working with information from the PFD.

Table 1.5 Flow Summary Table for the Benzene Process Shown in Figure 1.3 (and Figure 1.5)

Stream Number	1	2	3	4	5	6	7	8
Temperature (°C)	25	59	25	225	41	600	41	38
Pressure (bar)	1.90	25.8	25.5	25.2	25.5	25.0	25.5	23.9
Vapor Fraction	0.0	0.0	1.00	1.0	1.0	1.0	1.0	1.0
Mass Flow (tonne/h)	10.0	13.3	0.82	20.5	6.41	20.5	0.36	9.2
Mole Flow (kmol/h)	108.7	144.2	301.0	1204.4	758.8	1204.4	42.6	1100.8
Component Mole Flow (kmol/h)								
Hydrogen	0.0	0.0	286.0	735.4	449.4	735.4	25.2	651.9
Methane	0.0	0.0	15.0	317.3	302.2	317.3	16.95	438.3
Benzene	0.0	1.0	0.0	7.6	6.6	7.6	0.37	9.55
Toluene	108.7	143.2	0.0	144.0	0.7	144.0	0.04	1.05

Example 1.2

Check the overall material balance for the benzene process shown in Figure 1.3. From the figure, we identify the input streams as Stream 1 (toluene feed) and Stream 3 (hydrogen feed) and the output streams as Stream 15 (product benzene) and Stream 16 (fuel gas). From the flow summary table, these flows are listed as (units are in (10^3 kg)/h):

Input:		Output:	
Stream 3	0.82	Stream 15	8.21
Stream 1	10.00	Stream 16	2.61
Total	10.82×10^3 kg/h	Total	10.82×10^3 kg/h

Balance is achieved since Output = Input.

Example 1.3

Determine the conversion per pass of toluene to benzene in R-101 in Figure 1.3. Conversion is defined as

$$\varepsilon = (\text{benzene produced})/(\text{total toluene introduced})$$

From the PFD, the input streams to R-101 are shown as Stream 6 (reactor feed) and Stream 7 (recycle gas quench), and the output stream is Stream 9 (reactor effluent stream). From the information in Table 1.5 (units are kmol/h):

toluene introduced = 144 (Stream 6) + 0.04 (Stream 7) = 144.04 kmol/h

benzene produced = 116 (Stream 9) – 7.6 (Stream 6) – 0.37 (Stream 7)
= 108.03 kmol/h

$$\varepsilon = 108.03/144.04 = 0.75$$

9	10	11	12	13	14	15	16	17	18	19
654	90	147	112	112	112	38	38	38	38	112
24.0	2.6	2.8	3.3	2.5	3.3	2.3	2.5	2.8	2.9	2.5
1.0	0.0	0.0	0.0	0.0	0.0	0.0	1.0	1.0	0.0	1.0
20.9	11.6	3.27	14.0	22.7	22.7	8.21	2.61	0.07	11.5	0.01
1247.0	142.2	35.7	185.2	290.7	290.7	105.6	304.2	4.06	142.2	0.90
652.6	0.02	0.0	0.0	0.02	0.0	0.0	178.0	0.67	0.02	0.02
442.3	0.88	0.0	0.0	0.88	0.0	0.0	123.05	3.10	0.88	0.88
116.0	106.3	1.1	184.3	289.46	289.46	105.2	2.85	0.26	106.3	0.0
36.0	35.0	34.6	0.88	1.22	1.22	0.4	0.31	0.03	35.0	0.0

Alternatively, we can write

moles of benzene produced = toluene in – toluene out = 144.04 – 36.00
= 108.04 kmol/h

$$\varepsilon = 108.04/144.04 = 0.75$$

1.2.3 Equipment Information

The final element of the PFD is the equipment summary. This summary provides the information necessary to estimate the costs of equipment and furnish the basis for the detailed design of equipment. Table 1.6 provides the information needed for the equipment summary for most of the equipment encountered in fluid processes.

The information presented in Table 1.6 is used in preparing the equipment summary portion of the PFD for the benzene process. The equipment summary for the benzene process is presented in Table 1.7, and details of how we estimate and choose various equipment parameters are discussed in Chapter 11.

1.2.4 Combining Topology, Stream Data, and Control Strategy to Give a PFD

Up to this point, we have kept the amount of process information displayed on the PFD to a minimum. A more representative example of a PFD for the benzene process is shown in Figure 1.5. This diagram includes all of the elements found in Figure 1.3, some of the information found in Table 1.5, plus additional information on the major control loops used in the process.

Table 1.6 Equipment Descriptions for PFD and PIDs

Equipment Type
Description of Equipment
Towers
Size (height and diameter), Pressure, Temperature Number and Type of Trays Height and Type of Packing Materials of Construction
Heat Exchangers
Type: Gas-Gas, Gas-Liquid, Liquid-Liquid, Condenser, Vaporizer Process: Duty, Area, Temperature, and Pressure for both streams Number of Shell and Tube Passes Materials of Construction: Tubes and Shell
Tanks and Vessels
Height, Diameter, Orientation, Pressure, Temperature, Materials of Construction
Pumps
Flow, Discharge Pressure, Temperature, ΔP, Driver Type, Shaft Power, Materials of Construction
Compressors
Actual Inlet Flowrate, Temperature, Pressure, Driver Type, Shaft Power, Materials of Construction
Heaters (Fired)
Type, Tube Pressure, Tube Temperature, Duty, Fuel, Material of Construction
Other
Provide Critical Information

Table 1.7 Equipment Summary for Toluene Hydrodealkylation PFD

Heat Exchangers	**E-101**	**E-102**	**E-103**	**E-104**	**E-105**	**E-106**
Type	Fl.H.	Fl.H.	MDP	Fl.H.	MDP	Fl.H.
Area (m^2)	36	763	11	35	12	80
Duty (MJ/h)	15,190	46,660	1055	8335	1085	9045
Shell						
Temp. (°C)	225	654	160	112	112	185
Pres. (bar)	26	24	6	3	3	11
Phase	Vap.	Par. Cond.	Cond.	Cond.	l	Cond.
MOC	316SS	316SS	CS	CS	CS	CS
Tube						
Temp. (°C)	258	40	90	40	40	147
Pres. (bar)	42	3	3	3	3	3
Phase	Cond.	l	l	l	l	Vap.
MOC	316SS	316SS	CS	CS	CS	CS
Vessels/Tower/ Reactors	**V-101**	**V-102**	**V-103**	**V-104**	**T-101**	**R-101**
Temperature (°C)	55	38	38	112	147	660
Pressure (bar)	2.0	24	3.0	2.5	3.0	25
Orientation	Horizn'l	Vertical	Vertical	Horizn'l	Vertical	Vertical
MOC	CS	CS	CS	CS	CS	316SS
Size						
Height/Length (m)	5.9	3.5	3.5	3.9	29	14.2
Diameter (m)	1.9	1.1	1.1	1.3	1.5	2.3
Internals		s.p.	s.p.		42 sieve trays 316SS	catalyst packed bed-10m
Pumps/Compressors	**P-101 (A/B)**	**P-102 (A/B)**	**C-101 (A/B)**	**Heater**		**H-101**
Flow (kg/h)	13,000	22,700	6770	Type		Fired
Fluid Density (kg/m^3)	870	880	8.02	MOC		316SS
Power (shaft) (kW)	14.2	3.2	49.1	Duty (MJ/h)		27,040

(continued)

Table 1.7 Equipment Summary for Toluene Hydrodealkylation PFD (*continued*)

Pumps/Compressors	P-101 (A/B)	P-102 (A/B)	C-101 (A/B)	Heater	H-101
Type/Drive	Recip./ Electric	Centrf./ Electric	Centrf./ Electric	Radiant Area (m^2)	106.8
Efficiency (Fluid Power/Shaft Power)	0.75	0.50	0.75	Convective Area (m^2)	320.2
MOC	CS	CS	CS	Tube P (bar)	26.0
Temp. (in) (°C)	55	112	38		
Pres. (in) (bar)	1.2	2.2	23.9		
Pres. (out) (bar)	27.0	4.4	25.5		

Key:

MOC	Materials of construction	Par	Partial
316SS	Stainless steel type 316	F.H.	Fixed head
CS	Carbon steel	Fl.H.	Floating head
Vap	Stream being vaporized	Rbl	Reboiler
Cond	Stream being condensed	s.p.	Splash plate
Recipr.	Reciprocating	l	Liquid
Centrf.	Centrifugal	MDP	Multiple double pipe

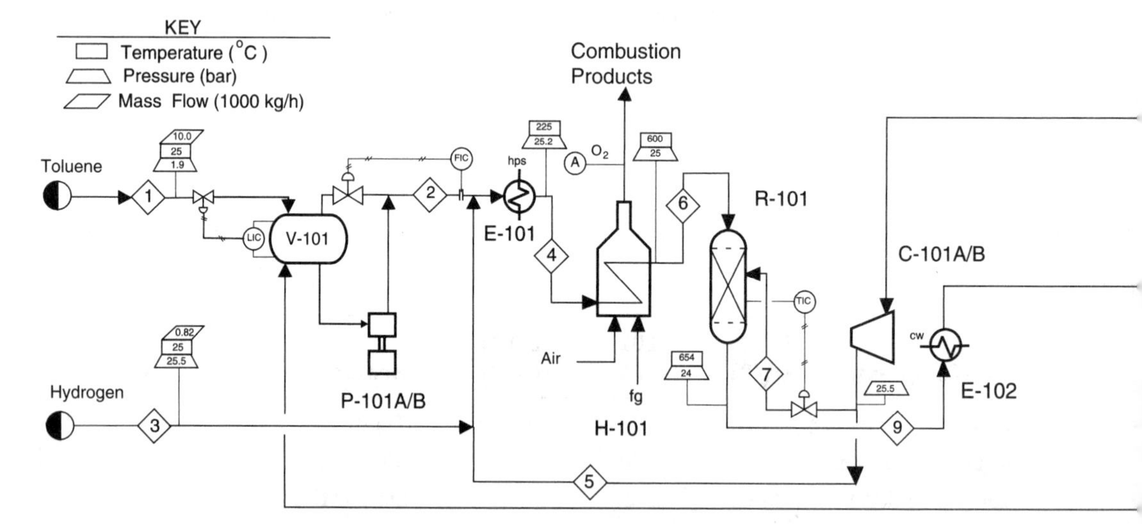

Figure 1.5 Benzene Process Flow Diagram (PFD) for the Production of Benzene via the Hydrodealkylation of Toluene

Stream information is added to the diagram by attaching "information flags." The shape of the flags indicates the specific information provided on the flag. Figure 1.6 illustrates all the flags used in this text. These information flags play a dual role. They provide information needed in the plant design leading to plant construction and in the analysis of operating problems during the life of the plant. Flags are mounted on a staff connected to the appropriate process stream. More than one flag may be mounted on a staff. Example 1.4 illustrates the different information displayed on the PFD.

Example 1.4

We locate Stream 1 in Figure 1.5 and note that immediately following the stream identification diamond a staff is affixed. This staff carries three flags containing the following stream data:

1. Temperature of 25°C
2. Pressure of 1.9 bar
3. Mass flow rate of 10.0 x 10^3 kg/h

The units for each process variable are indicated in the key provided at the left-hand side of Figure 1.5.

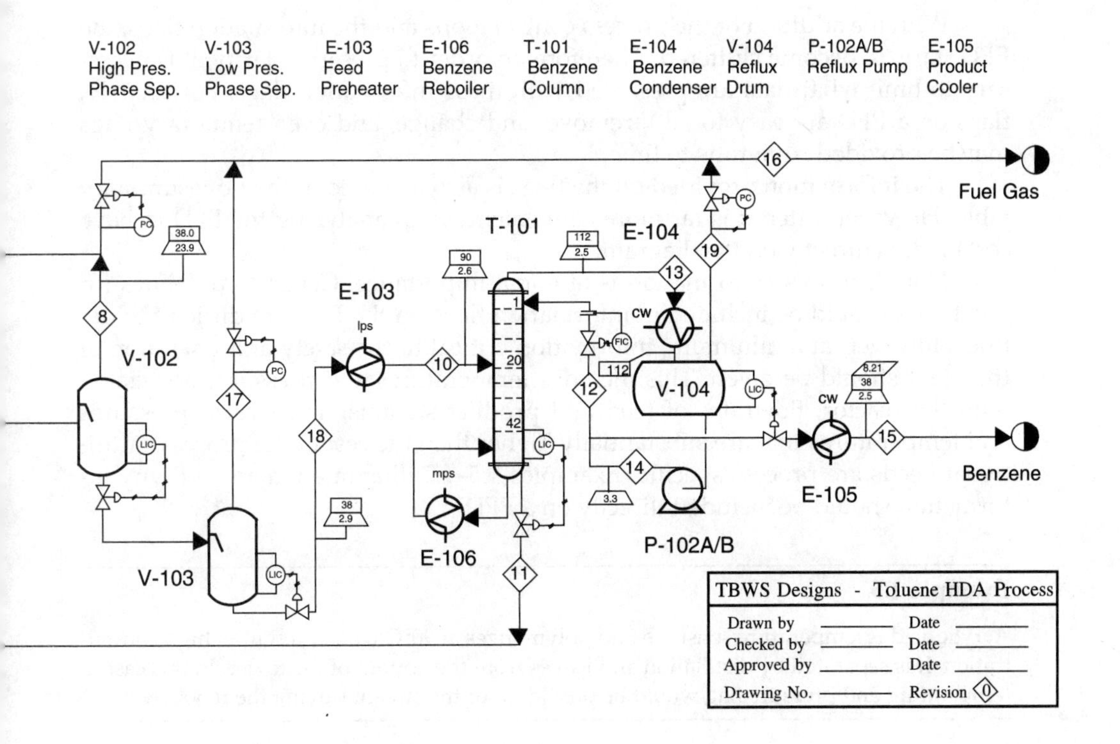

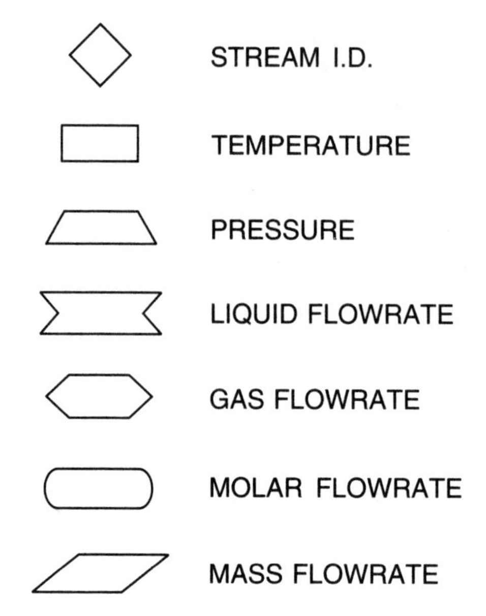

Figure 1.6 Symbols for Stream Identification

With the addition of the process control loops and the information flags, the PFD starts to become cluttered. Therefore, in order to preserve clarity, it is necessary to limit what data are presented with these information flags. Fortunately, flags on a PFD are easy to add, remove, and change, and even temporary flags may be provided from time to time.

The information provided on the flags is also included in the flow summary table. However, often it is far more convenient when analyzing the PFD to have certain data directly on the diagram.

Not all process information is of equal importance. General guidelines for what data should be included in information flags on the PFD are difficult to define. However, at a minimum, information critical to the safety and operation of the plant should be given. This includes temperatures and pressures associated with the reactor, flowrates of feed and product streams, and stream pressures and temperatures that are substantially higher than the rest of the process. Additional needs are process specific. Examples 1.5–1.7 illustrate where and why information should be included directly on a PFD.

Example 1.5

Acrylic acid is temperature sensitive and polymerizes at 90°C when present in high concentration. It is separated by distillation and leaves from the bottom of the tower. In this case, a temperature and pressure flag would be provided for the stream leaving the reboiler.

Example 1.6

In the benzene process, the feed to the reactor is substantially hotter than the rest of the process and is crucial to the operation of the process. In addition, the reaction is exothermic, and the reactor effluent temperature must be carefully monitored. For this reason Stream 6 (entering) and Stream 9 (leaving) have temperature flags.

Example 1.7

The pressures of the streams to and from R-101 in the benzene process are also important. The difference in pressure between the two streams gives the pressure drop across the reactor. This, in turn, gives an indication of any maldistribution of gas through the catalyst beds. For this reason, pressure flags are also included on Streams 6 and 9.

Of secondary importance is the fact that flags are useful in reducing the size of the flow summary table. For pumps, compressors, and heat exchangers, the mass flows are the same for the input and output streams, and complete entries in the stream table are not necessary. If the input (or output) stream is included in the stream table, and a flag is added to provide the temperature (in the case of a heat exchanger) or the pressure (in the case of a pump) for the other stream, then there is no need to present this stream in the flow summary table. Example 1.8 illustrates this point.

Example 1.8

Follow Stream 13 leaving the top of the benzene column in the benzene PFD given in Figure 1.5 and in Table 1.5. This stream passes through the benzene condenser, E-104, into the reflux drum, V-104. The majority of this stream then flows into the reflux pump, P-102, and leaves as Stream 14, while the remaining noncondensables leave the reflux drum in Stream 19. The mass flowrate and component flowrates of all these streams are given in Table 1.5. The stream leaving E-104 is not included in the stream table. Instead, a flag giving the temperature (112°C) was provided on the diagram (indicating condensation without subcooling). An additional flag, showing the pressure following the pump, is also shown. In this case the entry for Stream 14 could be omitted from the stream table, because it is simply the sum of Streams 12 and 15, and no information would be lost.

More information could be included in Figure 1.5 had space for the diagram not been limited by text format. It is most important that the PFD remain uncluttered and easy to follow in order to avoid errors and misunderstandings. Adding additional material to Figure 1.5 risks sacrificing clarity.

The flow table presented in Table 1.5, the equipment summary presented in Table 1.7, and Figure 1.5 taken together constitute all the information contained on a commercially produced PFD.

The PFD is the first comprehensive diagram drawn for any new plant or process. It provides all of the information needed to understand the chemical process. In addition, sufficient information is given on the equipment, energy, and material balances to establish process control protocol and to prepare cost estimates to determine the economic viability of the process.

Many additional drawings are needed to build the plant. All the process information required can be taken from this PFD. As described in the narrative at the beginning of this chapter, the development of the PFD is most often carried out by the operating company. Subsequent activities in the design of the plant are often contracted out.

The value of the PFD does not end with the construction of the plant. It remains the document that best describes the process, and it is used in the training of operators and new engineers. It is consulted regularly to diagnose operating problems that arise and to predict the effects of changes on the process.

1.3 PIPING AND INSTRUMENTATION DIAGRAM (P&ID)

The piping and instrumentation diagram (P&ID), also known as mechanical flow diagram (MFD), provides information needed by engineers to begin planning for the construction of the plant. The P&ID includes every mechanical aspect of the plant except the information given in Table 1.8. The general conventions used in drawing P&IDs are given in Table 1.9.

Each PFD will require many P&IDs to provide the necessary data. Figure 1.7 is a representative P&ID for the distillation section of the benzene process shown in Figure 1.5. The P&ID presented in Figure 1.7 provides information on the piping, and this is included as part of the diagram. As an alternative, each pipe can be numbered, and the specifics of every line can be provided in a separate table accompanying this diagram. When possible, the physical size of the larger-sized unit operations is reflected by the size of the symbol in the diagram.

Utility connections are identified by a numbered box in the P&ID. The number within the box identifies the specific utility. The key identifying the utility connections is shown in a table on the P&ID.

Table 1.8 Exclusions from Piping and Instrumentation Diagram

1. Operating Conditions T, P
2. Stream Flows
3. Equipment Locations
4. Pipe Routing
   a. Pipe Lengths
   b. Pipe Fittings
5. Supports, Structures, and Foundations

Table 1.9 Conventions in Constructing Piping and Instrumentation Diagrams

For Equipment—Show Every Piece Including
Spare Units Parallel Units Summary Details of Each Unit
For Piping—Include All Lines Including Drains and Sample Connections, and Specify
Size (Use Standard Sizes) Schedule (Thickness) Materials of Construction Insulation (Thickness and Type)
For Instruments—Identify
Indicators Recorders Controllers Show Instrument Lines
For Utilities—Identify
Entrance Utilities Exit Utilities Exit to Waste Treatment Facilities

All process information that can be measured in the plant is shown on the P&ID by circular flags. This includes the information to be recorded and used in process control loops. The circular flags on the diagram indicate where the information is obtained in the process and identify the measurements taken and how the information is dealt with. Table 1.10 summarizes the conventions used to identify information related to instrumentation and control. Example 1.9 illustrates the interpretation of instrumentation and control symbols.

Example 1.9

Consider the benzene product line leaving the right-hand side of the P&ID in Figure 1.7. The flowrate of this stream is controlled by a control valve that receives a signal from a level measuring element placed on V-104. The sequence of instrumentation is as follows:

A level sensing element (LE) is located on the reflux drum V-104. A level transmitter (LT) also located on V-104 sends an electrical signal (designated by a dashed line) to a level indicator and controller (LIC). This LIC is located in the control room on the control panel or console (as indicated by the horizontal line under LIC) and can be observed by the operators. From the LIC, an electrical signal is sent to an instrument (LY) that computes the

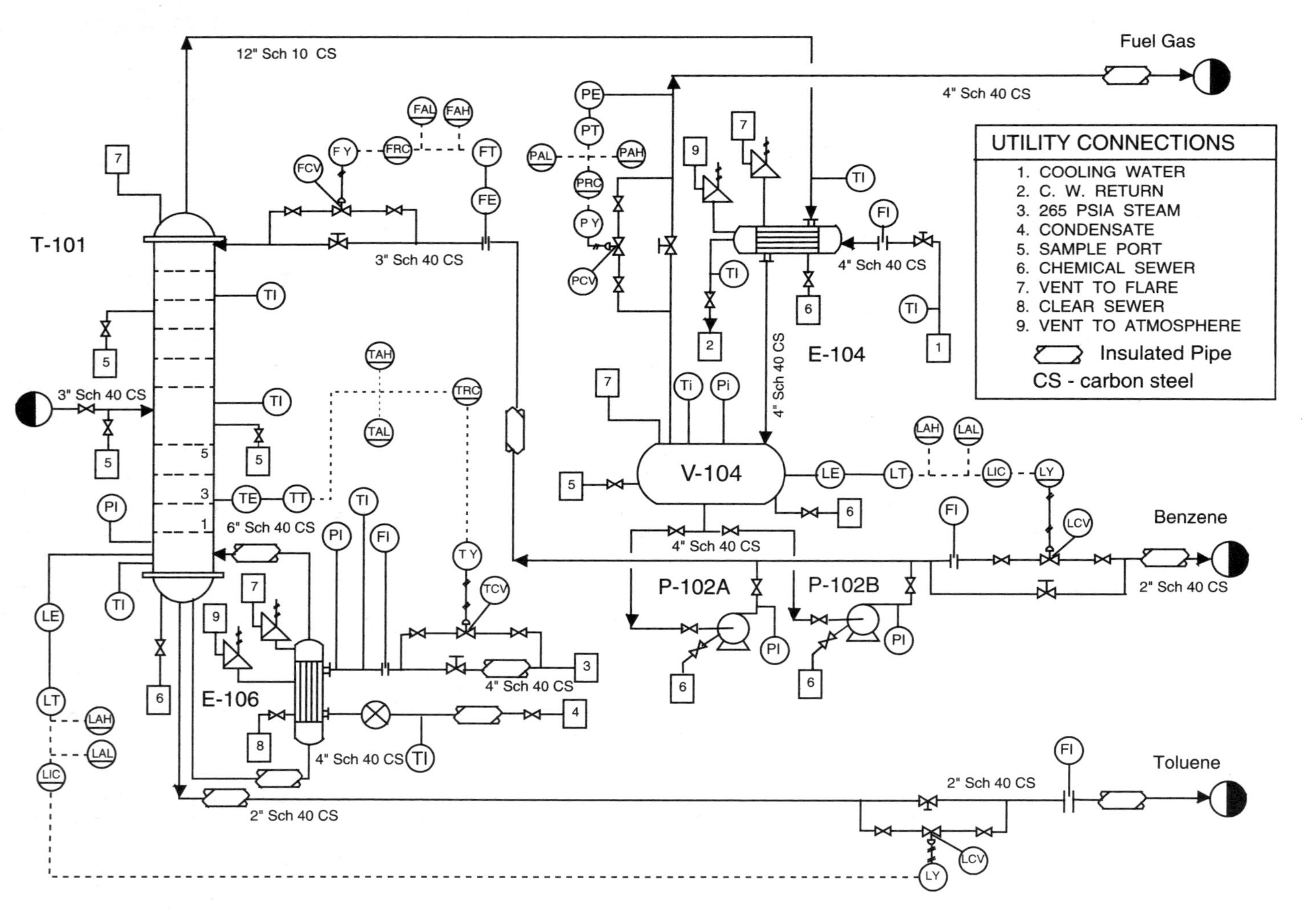

Figure 1.7 Piping and Instrumentation Diagram for Benzene Distillation (adapted from Kauffman, D, "Flow Sheets and Diagrams," AIChE Modular Instruction, Series G: Design of Equipment, series editor J. Beckman, AIChE, New York, 1986, vol 1, Chapter G.1.5, AIChE copyright © 1986 AIChE, all rights reserved)

Table 1.10 Conventions Used for Identifying Instrumentation on P&IDs (ISA standard ISA-S5-1, [4])

Location of Instrumentation	
○	Instrument Located in Plant
⊖	Instrument Located on Front of Panel in Control Room
○ (dotted line)	Instrument Located on Back of Panel in Control Room

Meanings of Identification Letters (XYY)

	First Letter (X)	*Second or Third Letter (Y)*
A	Analysis	Alarm
B	Burner Flame	
C	Conductivity	Control
D	Density or Specific Gravity	
E	Voltage	Element
F	Flowrate	
H	Hand (Manually Initiated)	High
I	Current	Indicate
J	Power	
K	Time or Time Schedule	Control Station
L	Level	Light or Low
M	Moisture or Humidity	Middle or Intermediate
O		Orifice
P	Pressure or Vacuum	Point
Q	Quantity or Event	
R	Radioactivity or Ratio	Record or print
S	Speed or Frequency	Switch
T	Temperature	Transmit
V	Viscosity	Valve, Damper, or Louver
W	Weight	Well
Y		Relay or Compute
Z	Position	Drive

Identification of Instrument Connections

Line	Connection
———— (solid line)	Capillary
—//—— (line with double slash)	Pneumatic
.... (dotted line)	Electrical

correct valve position and in turn sends a pneumatic signal (designated by a solid line with cross hatching) to activate the control valve (LCV). In order to warn operators of potential problems, two alarms are placed in the control room. These are a high-level alarm (LAH) and a low-level alarm (LAL), and they receive the same signal from the level transmitter as does the controller.

This control loop is also indicated on the PFD of Figure 1.5. However, the details of all the instrumentation are condensed into a single symbol (LIC), which adequately describes the essential process control function being performed. The control action that takes place is not described explicitly in either drawing. However, it is a simple matter to infer that if there is an increase in the level of liquid in V-104, the control valve will open slightly and the flow of benzene product will increase, tending to lower the level in V-104. For a decrease in the level of liquid, the valve will close slightly.

The details of the other control loops in Figures 1.5 and 1.7 are left to problems at the end of this chapter. It is worth mentioning that in virtually all cases of process control in chemical processes, the final control element is a valve. Thus, all control logic is based on the effect that a change in a given flowrate has on a given variable. The key to understanding the control logic is to identify which flowrate is being manipulated to control which variable. Once this has been done, it is a relatively simple matter to see in which direction the valve should change in order to make the desired change in the control variable. The response time of the system and type of control action used—for example, proportional, integral, or differential—are left to the instrument engineers and are not covered in this text.

The final control element in nearly all chemical process control loops is a valve.

The P&ID is the last stage of process design and serves as a guide for those who will be responsible for the final design and construction. Based on this diagram,

1. Mechanical engineers and civil engineers will design and install pieces of equipment.
2. Instrument engineers will specify, install, and check control systems.
3. Piping engineers will develop plant layout and elevation drawings.
4. Project engineers will develop plant and construction schedules.

Before final acceptance, the P&IDs serve as a checklist against which each item in the plant is checked.

The P&ID is also used to train operators. Once the plant is built and is operational, there are limits to what operators can do. About all that can be done to correct or alter performance of the plant is to open, close, or change the position of a valve. Part of the training would pose situations and require the operators to be able to describe what specific valve should be changed, how it should be changed, and what to observe in order to monitor the effects of the change. Plant simulators (similar to flight simulators) are sometimes involved in operator training. These programs are sophisticated, real-time process simulators that show a trainee operator how quickly changes in controlled variables propagate through the process. It is also possible for such programs to display scenarios of process upsets so that operators can get training in recognizing and correcting such situations. These types of programs are very useful and cost-effective in initial operator training. However, the use of P&IDs is still very important in this regard.

The P&ID is particularly important for the development of start-up procedures where the plant is not under the influence of the installed process control systems. An example of a start-up procedure is given in Example 1.10.

Example 1.10

Consider the start-up of the distillation column shown in Figure 1.7. What sequence would be followed? The procedure is beyond the scope of this text, but it would be developed from a series of questions such as

a. What valve should be opened first?
b. What should be done when the temperature of . . . reaches . . . ?
c. To what value should the controller be set?
d. When can the system be put on automatic control?

These last three sections have followed the development of a process from a simple BFD through the PFD and finally to the P&ID. Each step showed additional information. This can be seen by following the progress of the distillation unit as it moves through the three diagrams described.

1. **Block Flow Diagram (BFD) (see Figure 1.1):** The column was shown as a part of one of the three process blocks.
2. **Process Flow Diagram (PFD) (see Figure 1.5):** The column was shown as the following set of individual equipment: a tower, condenser, reflux drum, reboiler, reflux pumps, and associated process controls.
3. **Piping and Instrumentation Diagram (P&ID) (see Figure 1.7):** The column was shown as a comprehensive diagram that includes additional details such as pipe sizes, utility streams, sample taps, numerous indicators, and so on. It is the only unit operation on the diagram.

The value of these diagrams does not end with the start-up of the plant. The design values on the diagram are changed to represent the actual values determined under normal operating conditions. These conditions form a "base case" and are used to compare operations throughout the life of the plant.

1.4 ADDITIONAL DIAGRAMS

During the planning and construction phases of a new project, many additional diagrams are needed. Although these diagrams do not possess additional process information, they are essential to the successful completion of the project. Computers are being used more and more to do the tedious work associated with all of these drawing details. The creative work comes in the development of the concepts provided in the BFD and the process development required to produce the PFD. The computer can help with the drawings but cannot create a new process. Computers are valuable in many aspects of the design process where the size of equipment to do a specific task is to be determined. Computers may also be used when considering performance problems that deal with the operation of existing equipment. However, they are severely limited in dealing with diagnostic problems that are required throughout the life of the plant.

The diagrams presented here are in both American Engineering and SI units. The most noticeable exception is in the sizing of piping, where pipes are specified in inches and pipe schedule. This remains the way they are produced and purchased in the United States. A process engineer today must be comfortable with SI, conventional metric, and American (formerly British, who now use SI exclusively) Engineering units.

We discuss these additional diagrams briefly below.

A **utility flowsheet** may be provided that shows all the headers for utility inputs and outputs available along with the connections needed to the process. It provides information on the flows and characteristics of the utilities used by the plant.

Vessel sketches, logic ladder diagrams, wiring diagrams, site plans, structural support diagrams, and many other drawings are routinely used but add little to our understanding of the basic chemical processes that take place.

Additional drawings are necessary to locate all of the equipment in the plant. **Plot plans** and **elevation diagrams** are provided that locate the placement and elevation of all of the major pieces of equipment such as towers, vessels, pumps, heat exchangers, and so on. When constructing these drawings, it is necessary to consider and to provide for access for repairing equipment, removing tube bundles from heat exchangers, replacement of units, and so on. What remains to be shown is the addition of the structural support and piping.

Piping isometrics are drawn for every piece of pipe required in the plant. These drawings are 3-D sketches of the pipe run, indicating the elevations and orientation of each section of pipe. In the past, it was also common for comprehensive plants to build a **scale model** so the system could be viewed in three dimensions and modified to remove any potential problems. Over the past twenty years, scale models have been replaced by three-dimensional **computer aided design (CAD)** programs that are capable of representing the plant as-built in three dimensions. They provide an opportunity to view the local equipment topology from any angle at any location inside the plant. One can actually "walk through" the plant and preview what will be seen when the plant is built. The ability to "view" the plant before construction will be made even more realistic with the help of **virtual reality** software. With this new tool, it is possible not only to walk through the plant but also to "touch" the equipment, turn valves, climb to the top of distillation columns, and so on. In the next section, the information needed to complete a preliminary plant layout design is reviewed, and the logic used to locate the process units in the plant and how the elevations of different equipment are determined are briefly explained.

1.5 THREE-DIMENSIONAL REPRESENTATION OF A PROCESS

As mentioned earlier, the major design work products, both chemical and mechanical, are recorded on two-dimensional diagrams (PFD, P&ID, etc.). However, when it comes to the construction of the plant, there are many issues that require a three-dimensional representation of the process. For example, the location of shell and tube exchangers must allow for tube bundle removal for cleaning and repair. Locations of pumps must allow for access for maintenance and replacement. For compressors, this access may also require that a crane be able to remove and replace a damaged drive. Control valves must be located at elevations that allow operator access. Sample ports and instrumentation must also be located conveniently. For anyone who has toured a moderate-to-large chemical facility, the complexity of the piping and equipment layout is immediately apparent. Even for experienced engineers, the review of equipment and piping topology is far easier to accomplish in 3-D than 2-D. Due to the rapid increase in computer power and advanced software, such representations are now done routinely using the computer. In order to "build" an electronic representation of the plant in 3-D, all the information in the previously mentioned diagrams must be accessed and synthesized. This in itself is a daunting task, and a complete accounting of this process is well beyond the scope of this text. However, in order to give the reader a flavor of what can now be accomplished using such software, a brief review of the principles of plant layout design will be given. A more detailed account involving a virtual plant tour of the dimethyl ether (DME) plant (Appendix B.1) is given on the CD accompanying this book.

For a complete, detailed analysis of the plant layout, all equipment sizes, piping sizes, PFDs, P&IDs, and all other information should be known. However, for this description, a preliminary plant layout based on information given in the PFD of Figure B.1.1 is considered. Using this figure and the accompanying stream tables and equipment summary table (Tables B.1.1 and B.1.3), the following steps are followed.

1. *The PFD is divided into logical subsystems.* For the DME process, there are three logical subsections, namely, the feed and reactor section, the DME

purification section, and the methanol separation and recycle section. These sections are shown as dotted lines on Figure 1.8.

2. *For each subsystem, a preliminary plot plan is created.* The topology of the plot plan depends on many factors, the most important of which are discussed below.

   In general, the layout of the plot plan can take one of two basic configurations: the grade-level, horizontal, in-line arrangement and the structure-mounted vertical arrangement [5]. The grade-level, horizontal, in-line arrangement will be used for the DME facility. In this arrangement, the process equipment units are aligned on either side of a pipe rack that runs through the middle of the process unit. The purpose of the pipe rack is to carry piping for utilities, product, and feed to and from the process unit. Equipment is located on either side of the pipe rack, which allows for easy access. In addition, vertical mounting of equipment is usually limited to a single level. This arrangement generally requires a larger "footprint" and, hence, more land than does the structure-mounted vertical arrangement. The general arrangement for these layout types is shown in Figure 1.9.

   The minimum spacing between equipment should be set early on in the design. These distances are set for safety purposes and should be set with both local and national codes in mind. A comprehensive list of the recommended minimum distances between process equipment is given by Bausbacher and Hunt [5]. The values for some basic process equipment are listed in Table 1.11.

   The sizing of process equipment should be completed and the approximate location on the plot plan determined. Referring to Table B.1.3 for equipment specifications gives some idea of key equipment sizes. For example, the data given for the reflux drums V-202 and V-203, reactor R-201, and towers T-201 and T-202 are sufficient to sketch these units on the plot plan. However, pump sizes must be obtained from vendors or previous jobs, and additional calculations for heat exchangers must be done to estimate their required footprint on the plot plan. Calculations to illustrate the estimation of equipment footprints are given in Example 1.11.

Example 1.11

Estimate the footprint for E-202 in the DME process.

From Table B.1.3 we have the following information:

Floating Head Shell-and-Tube design

Area = 171 m^2

Hot Side—Temperatures: in at 364°C and out at 281°C

Cold Side—Temperatures: in at 154°C and out at 250°C

Choose a two-shell pass and four-tube pass exchanger

Area per shell = 171/2 = 85.5 m^2

Using 12 ft, 1-inch OD tubes, 293 tubes per shell are needed

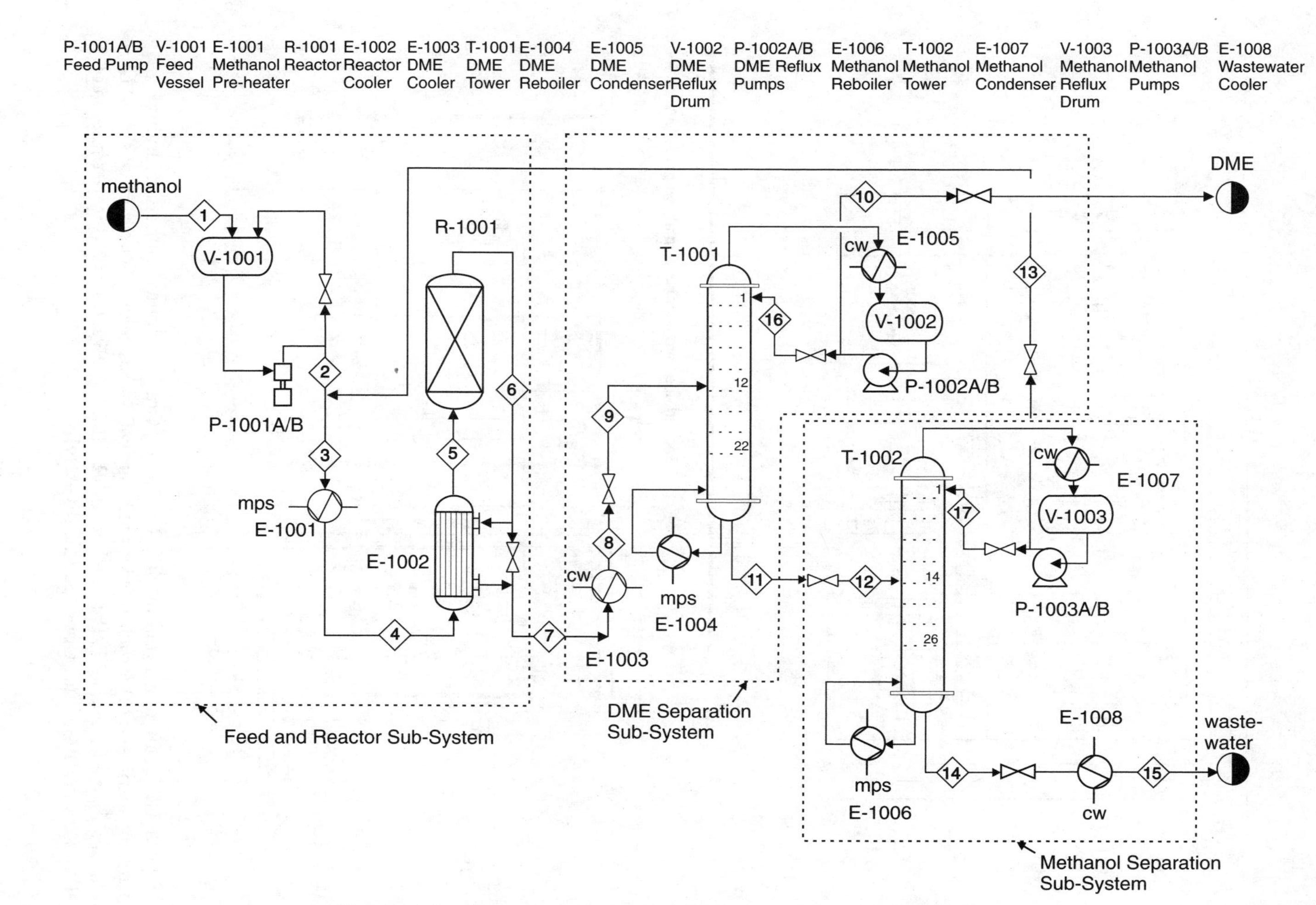

Figure 1.8 Subsystems for Preliminary Plan Layout for DME Process

(a)

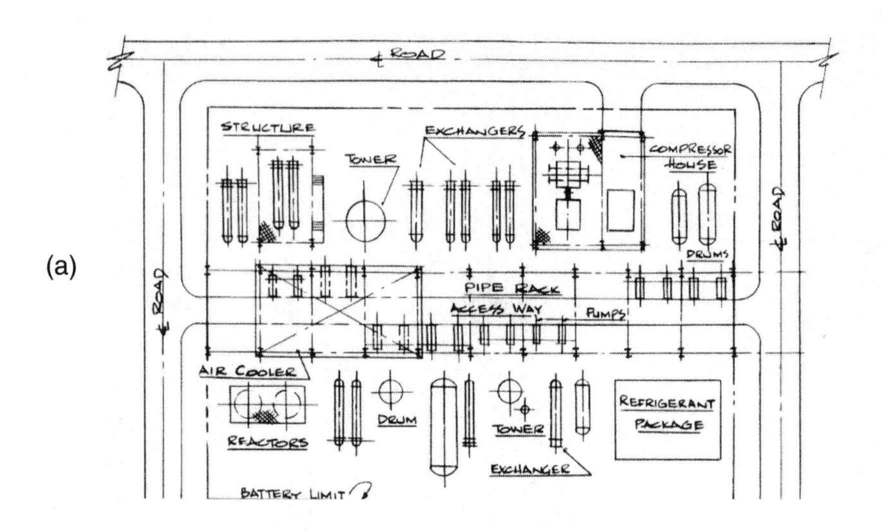

(b)

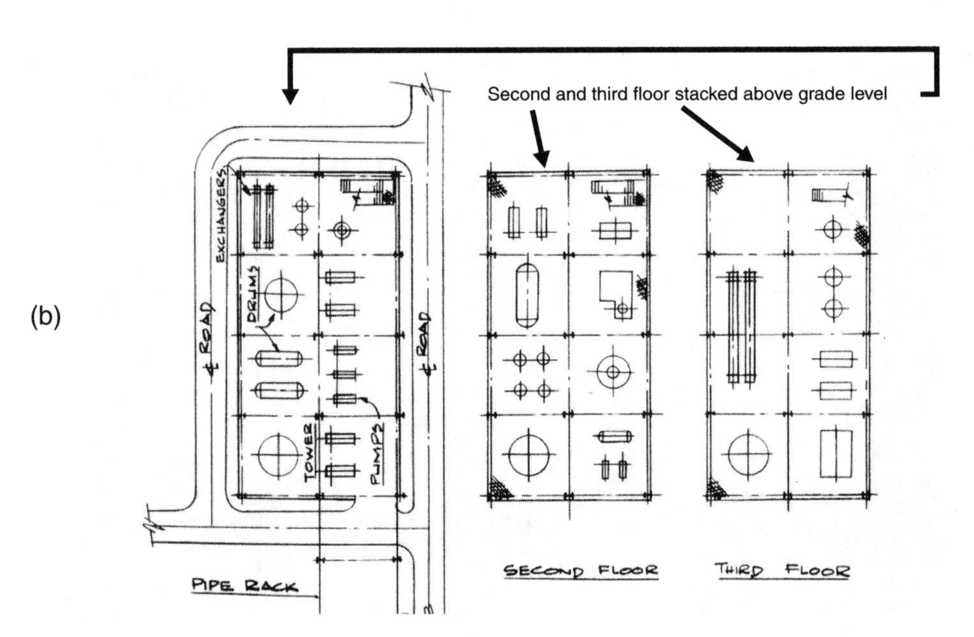

Figure 1.9 Different Types of Plant Layout: (a) Grade-Mounted Horizontal Inline Arrangement, and (b) Structure-Mounted Vertical Arrangement (*Source: Process Plant Layout and Piping Design*, by E. Bausbacher and R. Hunt, © 1994, reprinted by permission of Pearson Education, Inc., Upper Saddle River, NJ)

Table 1.11 Recommended Minimum Spacing (in Feet) between Process Equipment for Refinery, Chemical, and Petrochemical Plants

	Pumps	Compressors	Reactors	Towers and Vessels	Exchangers
Pumps	M	25	M	M	M
Compressors		M	30	M	M
Reactors			M	15	M
Towers				M	M
Exchangers					M

M = minimum for maintenance access

Source: Process Plant Layout and Piping Design, by E. Bausbacher and R. Hunt, © 1994, reprinted by permission of Pearson Education, Inc., Upper Saddle River, NJ.

Assuming the tubes are laid out on a 1¼-inch square pitch, a 27-inch ID shell is required.

Assume that the front and rear heads (where the tube fluid turns at the end of the exchanger) are 30 inches in diameter and require 2 feet each (including flanges), and that the two shells are stacked on top of each other. The footprint of the exchanger is given in Figure E1.11.

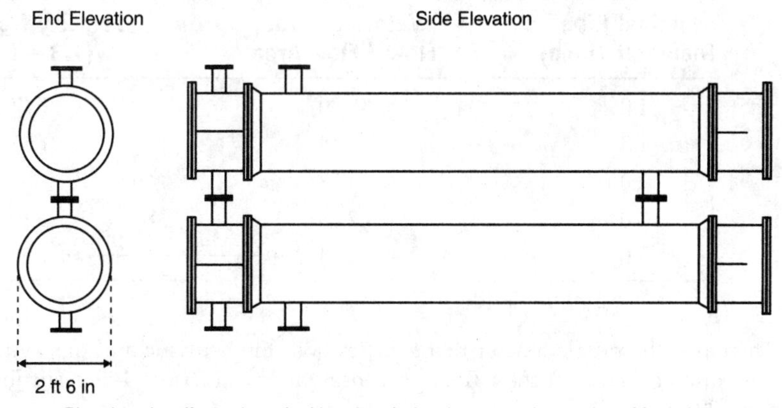

Plot plan view (from above looking down) showing approximate size of footprint

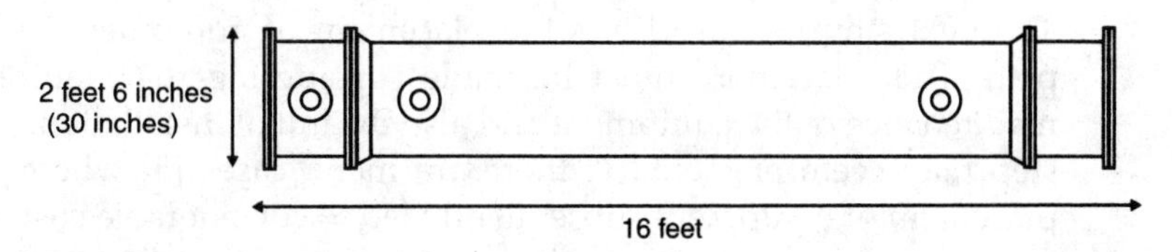

Figure E1.11 Approximate Dimensions and Footprint of Exchanger E-202

Next, the size of the major process lines must be determined. In order to estimate these pipe sizes, it is necessary to make use of some heuristics. A heuristic is a simple algorithm or hint that allows an approximate answer to be calculated. The preliminary design of a piece of equipment might well use many such heuristics, and some of these might conflict with each other. Like any simplifying procedure, the result from a heuristic must be reviewed carefully. For preliminary purposes, the heuristics from Chapter 11 can be used to estimate approximate pipe sizes. Example 1.12 illustrates the heuristic for calculating pipe size.

Example 1.12

Consider the suction line to P-202 A/B, what should be the pipe diameter?
From Table 11.8, 1(b) for liquid pump suction, the recommended liquid velocity and pipe diameter are related by $u = (1.3 + D\text{ (inch)}/6)$ ft/s.

From Table B.1.1, the mass flowrate of the stream entering P-202, $\dot{m}$ = Stream 16 + Stream 10 = 2170 + 5970 = 8140 kg/h and the density is found to be 800 kg/m^3.
The volumetric flowrate is 8140/800 = 10.2 m^3/h = 0.00283 m^3/s= 0.0998 ft^3/s.

The procedure is to calculate the velocity in the suction line and compare it to the heuristic. Using this approach, the following table is constructed.

Nominal Pipe Diameter (inch)	Velocity = Vol Flow / Flow Area	Velocity from $u = (1.3 + D/6)$
1.0	18.30	1.47
1.5	8.13	1.55
2.0	4.58	1.63
3.0	2.03	1.80
4.0	1.14 ←→	1.97

Therefore, the pipe diameter that satisfies both the heuristic and the continuity equation lies between 3 and 4 inches. Taking a conservative estimate, a 4-inch suction line is chosen for P-202.

The next step to consider is the placement of equipment within the plot plan. This placement must be made considering the required access for maintenance of the equipment and also the initial installation. Although this step may seem elementary, there are many cases [5] where the incorrect placement of equipment subsequently led to considerable cost overruns and major problems both during the construction of the plant and during maintenance operations. Consider the example shown in Figure 1.10(a), where

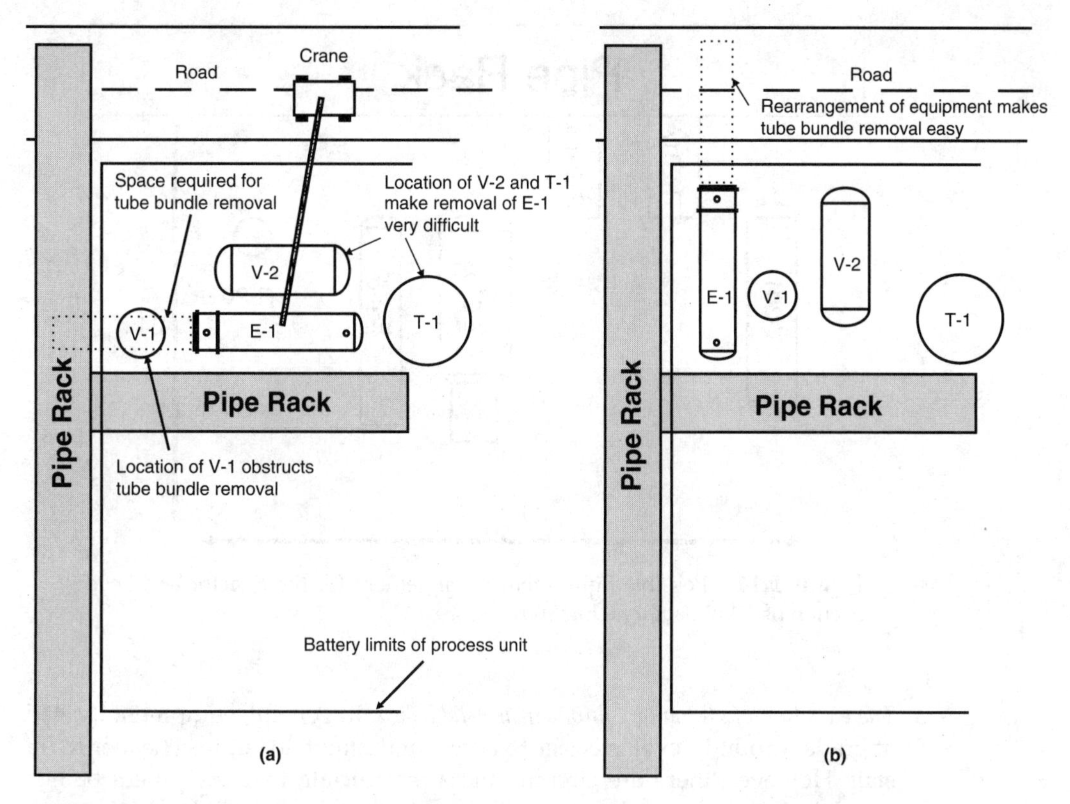

Figure 1.10 The Effect of Equipment Location on the Ease of Access for Maintenance, Installation, and Removal

two vessels, a tower, and a heat exchanger are shown in the plot plan. Clearly, V-1 blocks the access to the exchanger's tube bundle, which often requires removal to change leaking tubes or to remove scale on the outside of the tubes. With this arrangement, the exchanger would have to be lifted up vertically and placed somewhere where there was enough clearance so that the tube bundle could be removed. However, the second vessel, V-2, and the tower T-1 are located such that crane access is severely limited and a very tall (and expensive) crane would be required. The relocation of these same pieces of equipment, as shown in Figure 1.10(b), alleviates both these problems. There are too many considerations of this type to cover in detail in this text, and the reader is referred to Bausbacher and Hunt [5] for a more in-depth coverage of these types of problems. Considering the DME facility, a possible arrangement for the feed and reactor subsection is shown in Figure 1.11.

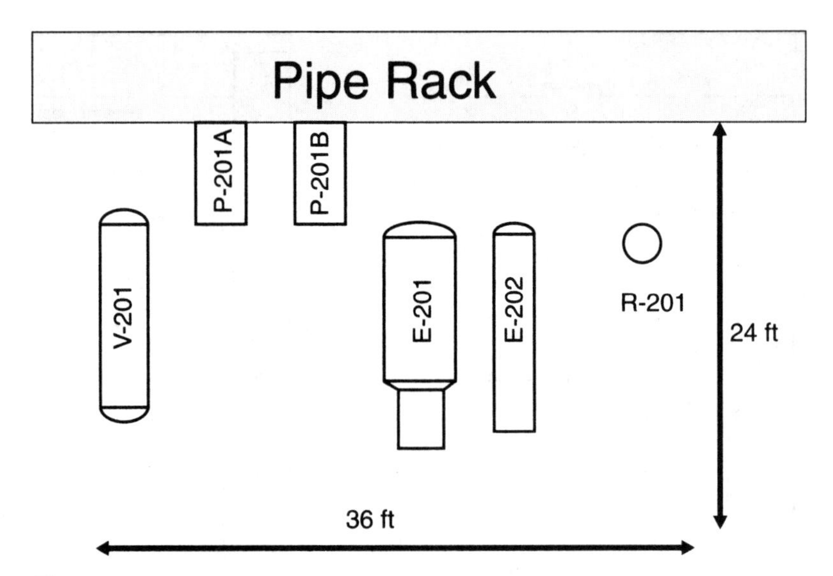

Figure 1.11 Possible Equipment Arrangement for the Reactor and Feed Section of DME Facility, Unit 200

3. *The elevation of all major equipment is established.* In general, equipment located at grade (ground) level is easier to access and maintain, and is cheaper to install. However, there are circumstances that dictate that equipment be elevated in order to provide acceptable operation. For example, the bottoms product of a distillation column is a liquid at its bubble point. If this liquid is fed to a pump, then, as the pressure drops in the suction line due to friction, the liquid boils and causes the pumps to cavitate. To alleviate this problem, it is necessary to elevate the bottom of the column relative to the pump inlet, in order to increase the Net Positive Suction Head Available (for more detail about $NPSH_A$ see Chapter 18). This can be done by digging a pit below grade for the pump or by elevating the tower. Pump pits have a tendency to accumulate denser-than-air gases, and maintenance of equipment in such pits is dangerous due to the possibility of suffocation and poisoning (if the gas is poisonous). For this reason, towers are generally elevated between 3 to 5 m (10 and 15 feet) above ground level by using a "skirt." This is illustrated in Figure 1.12. Another reason for elevating a distillation column is also illustrated in Figure 1.12. Often a thermosiphon reboiler is used. These reboilers use the difference in density between the liquid fed to the reboiler and the two-phase mixture (saturated liquid-vapor) that leaves the reboiler to "drive" the circulation of bottoms liquid through the reboiler. In order to obtain an acceptable driving force for this circulation, the static head of the liquid must be substantial, and a 3–5 m height differential between the liquid level in the column and the liquid inlet to the reboiler is typically sufficient. Examples showing when equipment elevation is required are given in Table 1.12.

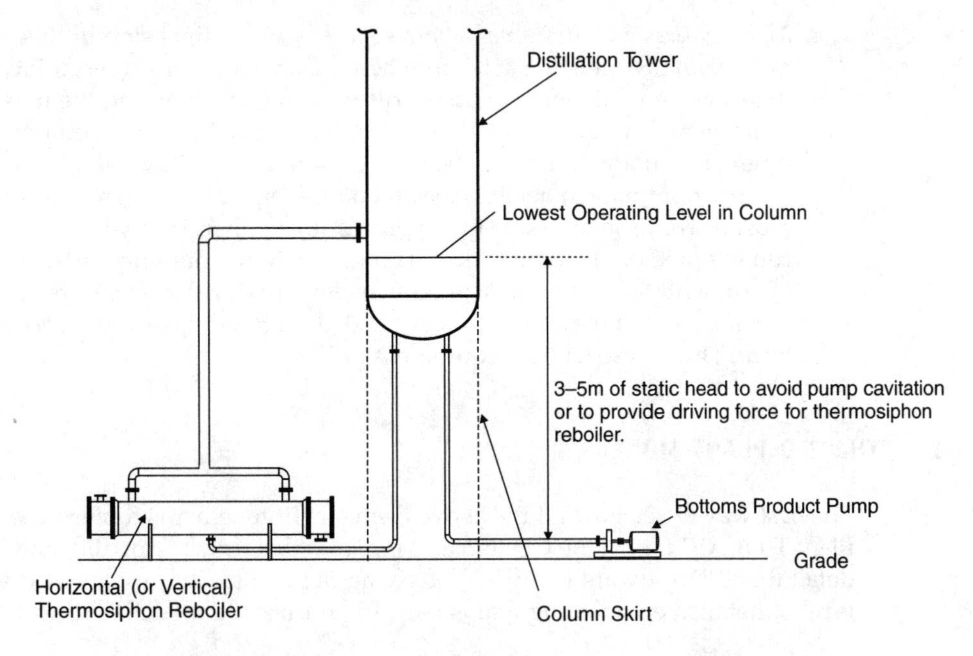

Figure 1.12 Sketch Illustrating Reasons for Elevating Distilling Column

Table 1.12 Reasons for Elevating Equipment

Equipment to Be Elevated	Reason for Elevation
Columns or vessels	When the NPSH available is too low to avoid cavitation in the discharge pump, equipment must be elevated.
Columns	To provide driving head for thermosiphon reboilers.
Any equipment containing suspended solids or slurries	To provide gravity flow of liquids containing solids that avoids the use of problematic slurry pumps.
Contact barometric condensers	This equipment is used to produce vacuum by expanding high-pressure steam through an ejector. The condensables in the vapor are removed by direct contact with a cold-water spray. The tail pipe of such a condenser is sealed with a 34-foot leg of water.
Critical fire-water tank (or cooling water holding tank)	In some instances, flow of water is absolutely critical, for example, in firefighting or critical cooling operations. The main water supply tank for these operations may be elevated to provide enough water pressure to eliminate the need for feed pumps.

4. *Major process and utility piping are sketched in.* The final step in this preliminary plant layout is to sketch in where the major process (and utility) pipes (lines) go. Again, there are no set rules to do this. However, the most direct route between equipment that avoids clashes with other equipment and piping is usually desirable. It should be noted that utility lines originate and usually terminate in headers located on the pipe rack. When process piping must be run from one side to the process to another, it may be convenient to run the pipe on the pipe rack. All control valves, sampling ports, and major instrumentation must be located conveniently for the operators. This usually means that they should be located close to grade or a steel access platform. This is also true for equipment isolation valves.

1.6 THE 3-D PLANT MODEL

The best way to see how all the above elements fit together is to view the Virtual Plant Tour AVI file on the CD that accompanies this text. The quality and level of detail that 3-D software is capable of giving depend on the system used and the level of detailed engineering that is used to produce the model. Figures 1.13–1.15

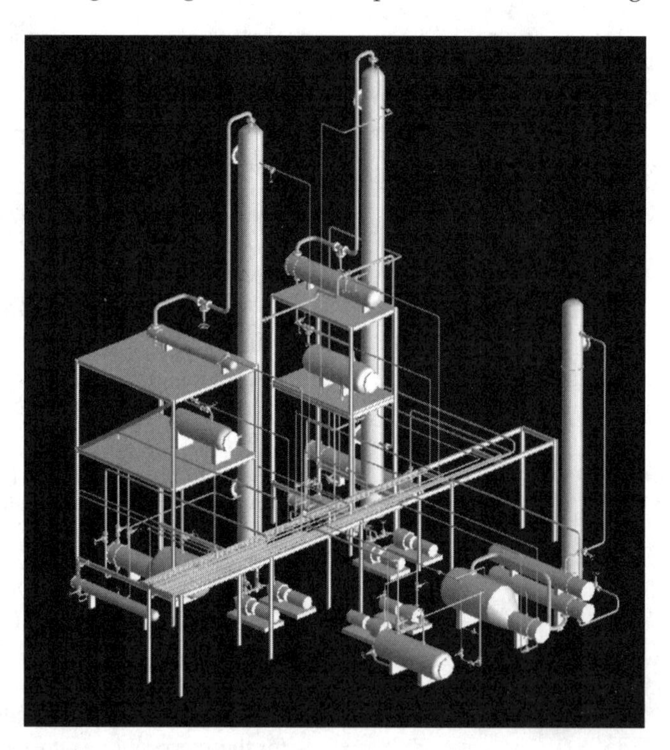

Figure 1.13 Isometric View of Preliminary 3-D Plant Layout Model for DME Process (Reproduced by Permission of Cadcentre, an Aveva Group Company, from their Vantage/PDMS Software)

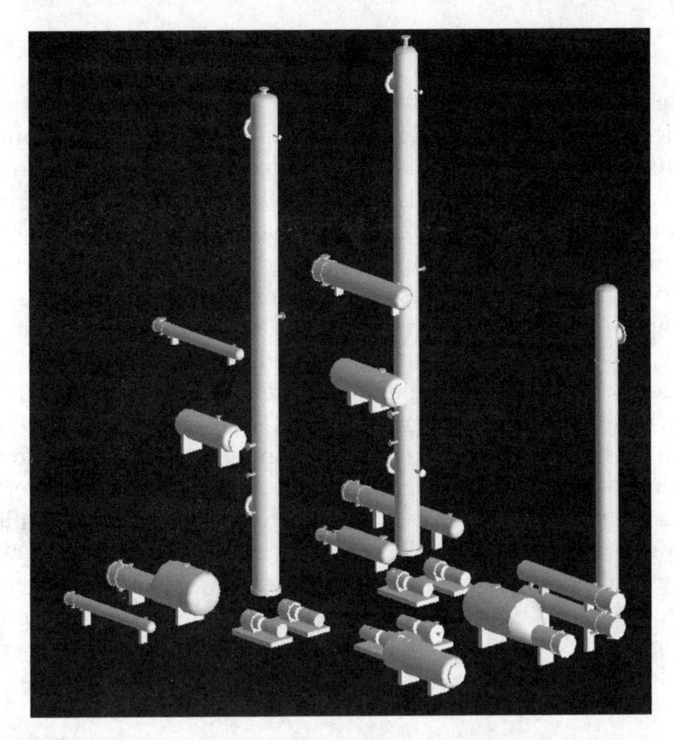

Figure 1.14 3-D Representation of Preliminary Equipment Layout for the DME Process (Reproduced by Permission of Cadcentre, an Aveva Group Company, from their Vantage/PDMS Software)

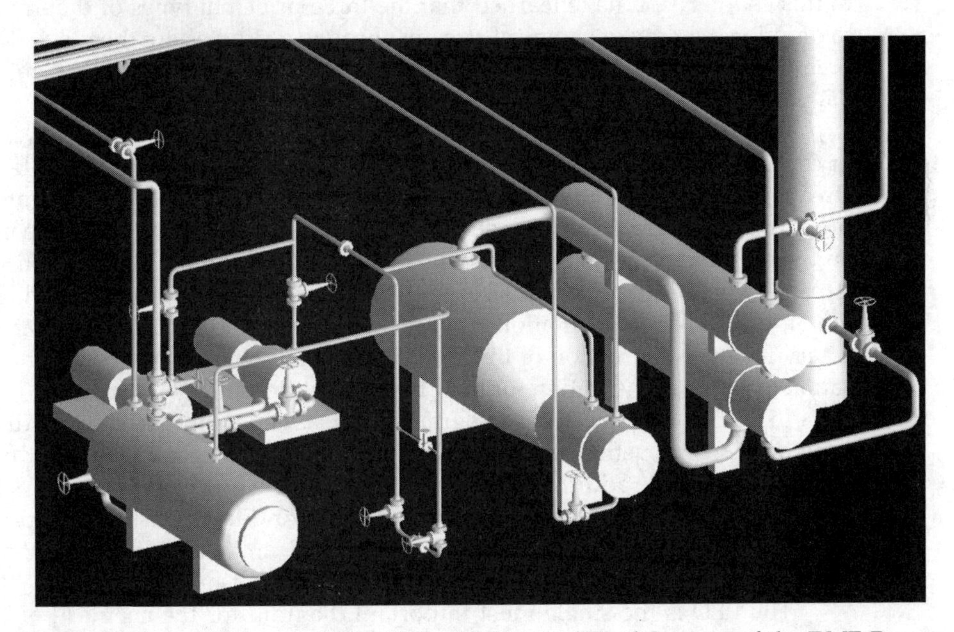

Figure 1.15 3-D Representation of the Reactor and Feed Sections of the DME Process Model (Reproduced by Permission of Cadcentre, an Aveva Group Company, from their Vantage/PDMS Software)

were generated for the DME facility using the PDMS software package from Cadcentre, Inc. (These figures and the Virtual_Plant_Tour.AVI file are presented here with permission of Cadcentre, Inc.) In Figure 1.13, an isometric view of the DME facility is shown. All major process equipment, major process and utility piping, and basic steel structures are shown. The pipe rack is shown running through the center of the process, and steel platforms are shown where support of elevated process equipment is required. The distillation sections are shown to the rear of the figure on the far side of the pipe rack. The reactor and feed section is shown on the near side of the pipe rack. The elevation of the process equipment is better illustrated in Figure 1.14, where the piping and structural steel have been removed. The only elevated equipment apparent from this figure are the overhead condensers and reflux drums for the distillation columns. The overhead condensers are located vertically above their respective reflux drums to allow for the gravity flow of condensate from the exchangers to the drums. Figure 1.15 shows the arrangement of process equipment and piping for the feed and reactor sections. The layout of equipment corresponds to that shown in Figure 1.11. It should be noted that the control valve on the discharge of the methanol feed pumps is located close to grade level for easy access.

1.7 SUMMARY

In this chapter, you have learned that the three principal types of diagrams used to describe the flow of chemical streams through a process are the block flow diagram (BFD), the process flow diagram (PFD), and the piping and instrumentation diagram (P&ID). These diagrams describe a process in increasing detail.

Each diagram serves a different purpose. The block flow diagram is useful in conceptualizing a process or a number of processes in a large complex. Little stream information is given, but a clear overview of the process is presented. The process flow diagram contains all the necessary information to complete material and energy balances on the process. In addition, important information such as stream pressures, equipment sizes, and major control loops is included. Finally, the piping and instrumentation diagram contains all the process information necessary for the construction of the plant. These data include pipe sizes and the location of all instrumentation for both the process and utility streams.

In addition to the three diagrams, there are a number of other diagrams used in the construction and engineering phase of a project. However, these diagrams contain little additional information about the process.

Finally, the logic for equipment placement and layout within the process is presented. The reasons for elevating equipment and providing access are discussed, and a 3-D representation of a DME plant is presented.

The PFD is the single most important diagram for the chemical or process engineer and will form the basis of much of the discussion covered in this book.

REFERENCES

1. Kauffman, D., "Flow Sheets and Diagrams," *AIChE Modular Instruction, Series G: Design of Equipment,* series editor J. Beckman, American Institute of Chemical Engineers, New York, 1986, vol. 1, Chapter G.1.5. Reproduced by permission of the American Institute of Chemical Engineers, AIChE copyright © 1986, all rights reserved.
2. *Graphical Symbols for Process Flow Diagrams,* ASA Y32.11 (New York: American Society of Mechanical Engineers, 1961).
3. Austin, D. G., *Chemical Engineering Drawing Symbols* (London: George Godwin, 1979).
4. *Instrument Symbols and Identification,* Research Triangle Park, NC: Instrument Society of America, Standard ISA-S5-1, 1975.
5. Bausbacher, E. and R. Hunt, *Process Plant Layout and Piping Design* (Upper Saddle River, NJ: Prentice Hall PTR, 1998).

SHORT ANSWER QUESTIONS

1. What are the three principal types of diagrams used by process engineers to describe the flow of chemicals in a process? On which of these diagrams would you expect to see the following items:
   a. the temperature and pressure of a process stream
   b. an overview of a multiple-unit process
   c. a major control loop
   d. a pressure indicator
   e. a pressure-relief valve
2. A problem has occurred in the measuring element of a level-indicating controller in a batch reactor. To what principal diagram should you refer to in order to troubleshoot the problem?
3. Why is it important for a process engineer to be able to review a three-dimensional model (actual or virtual/electronic) of the plant prior to the construction phase of a project?
4. Name five things that would affect the locations of different pieces of equipment when determining the layout of equipment in a process unit.
5. Why are accurate plant models (made of plastic parts) no longer made as part of the design process? What function did these models play and how is this function now achieved?

PROBLEMS

6. There are two common reasons for elevating the bottom of a tower by means of a "skirt." One reason is to provide enough $NPSH_A$ for bottoms product pumps to avoid cavitation. What is the other reason?
7. Which of the principal diagrams should be used to do the following:
   a. Determine the number of trays is a distillation column?
   b. Determine the top and bottom temperatures in a distillation column?
   c. Validate the overall material balance for a process?
   d. Check the instrumentation for a given piece of equipment in a "pre-start-up" review?
   e. Determine the overall material balance for a whole chemical plant?
8. What is the purpose(s) of a pipe rack in a chemical process?
9. When would a structure-mounted vertical plant layout arrangement be preferred over a grade-mounted horizontal in-line arrangement?
10. A process that is being considered for construction has been through several technical reviews; block flow, process flow, and piping and instrumentation diagrams are available for the process. Explain the changes that would have to be made to the three principal diagrams if during a final preconstruction review, the following changes were made:
    a. The efficiency of a fired heater had been specified incorrectly as 92% instead of 82%.
    b. A waste process stream flowrate (sent to a sludge pond) was calculated incorrectly and is now 30% greater than before.
    c. It has been decided to add a second (backup) drive for an existing compressor.
    d. The locations of several control valves have changed to allow for better operator access.
11. During a retrofit of an existing process, a vessel used to supply the feed pump to a batch reactor has been replaced because of excessive corrosion. The vessel is essentially identical to the original one, except it is now grounded differently to reduce the corrosion. If the function of the vessel (namely to supply liquid to a pump) has not changed, answer the following questions:
    a. Should the new vessel have a new equipment number, or should the old vessel number be used again? Explain your answer.
    b. On which diagram or diagrams (BFD, PFD, or P&D) should the change in the grounding setup be noted?
12. Draw a section of a P&ID diagram for a vessel receiving a process liquid through an insulated 4″ sch 40 pipe. The purpose of the vessel is to store approximately 5 minutes of liquid volume and to provide "capacity" for a feed pump connected to the bottom of the pump using a 6″ sch 40 pipe. The diagram should include the following features:

a. The vessel is numbered V-1402 and the pump(s) are P-1407 A/B.
b. The discharge side of the pump is made of 4″ sch 40 carbon steel pipe and all pipe is insulated.
c. A control valve is located in the discharge line of the pump, and a double block and bleed arrangement is used (see Problem 1.13 for more information).
d. Both pumps and vessel have isolation (gate) valves.
e. The pumps should be equipped with drain lines that discharge to a chemical sewer.
f. The vessel is equipped with local pressure and temperature indicators.
g. The vessel has a pressure relief valve set to 50 psig that discharges to a flare system.
h. The tank has a drain valve and a sampling valve, both of which are connected to the tank through separate 2″ sch 40 CS lines.
i. The tank level is used to control the flow of liquid out of the tank by adjusting the setting of the control valve on the discharge side of the pump. The instrumentation is similar to that shown for V-104 in Figure 1.7.

13. A standard method for instrumenting a control valve is termed the "double block and bleed," which is illustrated in Figure P1.13.

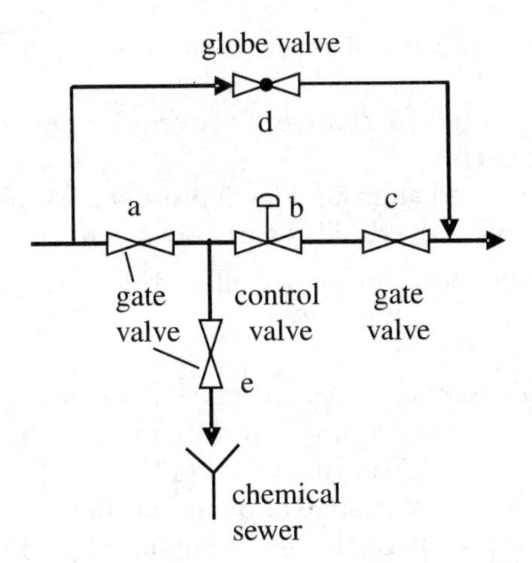

Figure P1.13 Double Block and Bleed Arrangement for Problem 13

Under normal conditions, valves a to c are open and valves d and e are closed. Answer the following:

a. Explain, carefully, the sequence of opening and closing valves required in order to change out the valve stem on the control valve (valve b).
b. What changes, if any, would you make to Figure P1.13 if the process stream did not contain a process chemical but contained process water?

c. It has been suggested that the bypass valve (valve d) be replaced with another gate valve to save money. Gate valves are cheap but essentially function as on-off valves. What do you recommend?
d. What would be the consequence of eliminating the bypass valve (valve d)?

14. Often, during the distillation of liquid mixtures, some noncondensable gases are dissolved in the feed to the tower. These noncondensables come out of solution when heated in the tower and may accumulate in the overhead reflux drum. In order for the column to operate satisfactorily, these vapors must be periodically vented to a flare or stack. One method to achieve this venting process is to implement a control scheme in which a process control valve is placed on the vent line from the reflux drum. A pressure signal from the drum is used to trigger the opening or closing of the vent line valve. Sketch the basic control loop needed for this venting process on a process flow diagram representing the top portion of the tower.
15. Repeat Problem 14, but create the sketch as a PI&D to show all the instrumentation needed for this control loop.
16. Explain how each of the following statements might affect the layout of process equipment:
    a. A specific pump requires a large NPSH.
    b. The flow of liquid from an overhead condenser to the reflux drum is gravity driven.
    c. Pumps and control valves should be located for easy access and maintenance.
    d. Shell and tube exchanges may require periodic cleaning and tube bundle replacement.
    e. Pipes located at ground level present a tripping hazard.
    f. The prevailing wind is nearly always from the west.
17. Estimate the footprint for a shell-and-tube heat exchanger from the following design data:
    - Area = 145 m^2
    - Hot side temperatures: in at 300°C out at 195°C
    - Cold side temperature: bfw at 105°C mps at 184°C
    - Use 12 ft, 1" OD tubes on a 1-1/4" square pitch, use a single shell and tube pass because of change of phase on shell side
    - Use a vapor space above boiling liquid = 3 times liquid volume
18. Make a sketch of a layout (plot plan only) of a process unit containing the following process equipment:
    - 3 reactors (vertical – diameter 1.3 m each)
    - 2 towers (1.3 and 2.1 m in diameter, respectively)
    - 4 pumps (each mounting pad is 1 m by 1.8 m)
    - 4 exchangers (footprints of 4 m by 1m, 3.5 m by 1.2 m, 3 m by 0.5 m, and 3.5 m by 1.1 m)

The two columns and the 3 reactors should all be aligned with suitable spacing and all the exchangers should have clearance for tube bundle removal.

19. Using the data from Table 1.7 estimate the footprints of all the equipment in the toluene HDA process.
    - For the shell and tube exchangers, assume 12 ft, 1.25″ tubes on a 1.5″ square pitch and assume 2 ft additional length at either end of the exchanger for tube return and feed header.
    - For double pipe exchangers, assume an 8″ schedule 20 OD and a 6″ schedule 40 ID pipe with a length of 12 ft including u-bend.
    - For the footprints of pumps, compressors, and fired heater, assume the following:
        - P-101 use 2m by 1m, P-102 use 2m by 1m
        - C-101 (+D-101) use 4m by 2m
        - H-101 use 5m by 5m
20. With the information from Problem 19 and the topology given in Figure 1.5, accurately sketch a plant layout (plot plan) of the toluene HDA process using a grade-mounted horizontal inline arrangement similar to the one shown in Figure 1.9. You should assume that the area of land available for this process unit is surrounded on three sides by an access road and that a pipe rack runs along the fourth side. Use the information in Table 1.11 as a guide to placing equipment.
21. What do the following symbols (as seen on a P&ID) indicate?

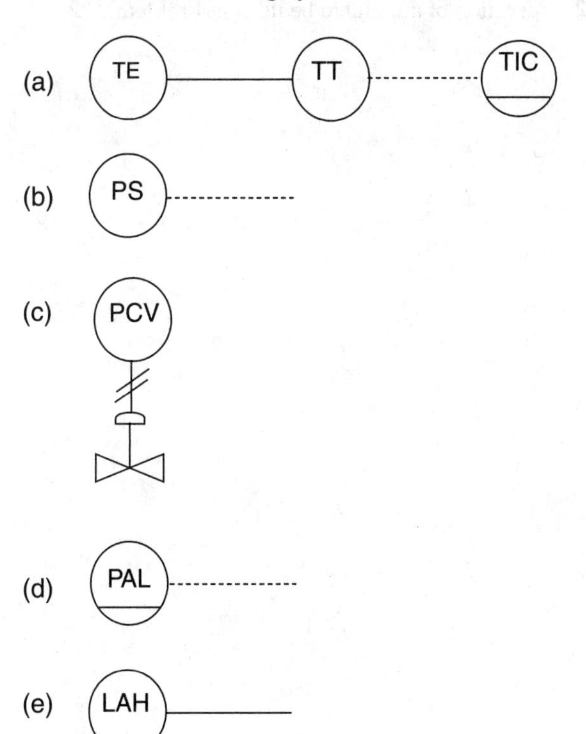

22. Determine all the errors in the following section of a P&ID.

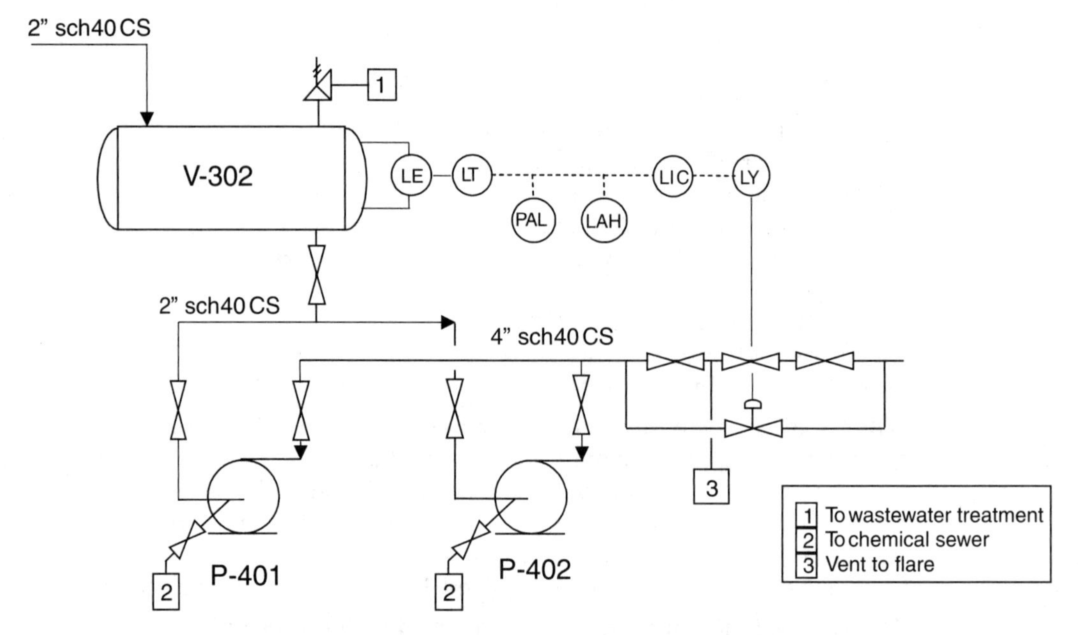

Figure P1.22 A section of a P&ID to be used in Problem 1.22

CHAPTER

2

The Structure and Synthesis of Process Flow Diagrams

When looking at a process flow diagram (PFD) for the first time, it is easy to be confused or overwhelmed by the complexity of the diagram. The purpose of this chapter is to show that the evolution of every process follows a similar path. The resulting processes will often be quite different, but the series of steps that have been followed to produce the final processes are similar. Once the path or evolution of the structure of processes has been explained and understood, the procedure for understanding existing PFDs is also made simpler. Another important benefit of this chapter is to provide a framework to generate alternative PFDs for a given process.

2.1 HIERARCHY OF PROCESS DESIGN

Before discussing the steps involved in the conceptual design of a process, it should be noted that often the most important decision in the evolution of a process is the choice of which chemical syntheses or routes should be investigated to produce a desired product. The identification of alternative process chemistries should be done at the very beginning of any conceptual design. The conceptual design and subsequent optimization of a process are "necessary conditions" for any successful new process. However, the greatest improvements (savings) associated with chemical processes are most often due to changes, sometimes radical changes, to the chemical pathway used to produce the product. Most often, there are at least two viable ways to produce a given chemical. These alternative routes may require different raw materials and may produce different by-products. The cost of the raw materials, the value of the by-products, the complexity of the synthesis, and the environmental impact of any waste materials and pollutants produced must be taken into account when evaluating alternative synthesis routes.

Douglas [1,2], among others, has proposed a hierarchical approach to conceptual process design. In this approach, the design process follows a series of decisions and steps. The order in which these decisions are made forms the hierarchy of the design process. These decisions are listed as follows.

1. Decide whether the process will be batch or continuous.
2. Identify the input/output structure of the process.
3. Identify and define the recycle structure of the process.
4. Identify and design the general structure of the separation system.
5. Identify and design the heat-exchanger network or process energy recovery system.

In designing a new process, we follow steps 1 through 5 in that order. Alternatively, by looking at an existing process, we can work backward from step 5 and eliminate or greatly simplify the PFD and, hence, reveal much about the structure of the underlying process.

Let us start with this five-step design algorithm and see how it can be applied to a chemical process. Each of the steps is discussed in some detail, and the general philosophy about the decision-making process will be covered. However, because steps 4 and 5 require extensive discussion, these will be covered in separate chapters (Chapter 12 for separations, and Chapter 15 for energy recovery).

2.2 STEP 1—BATCH VERSUS CONTINUOUS PROCESS

It should be pointed out that there is a difference between a batch process and a batch (unit) operation. Indeed, there are very few, if any, processes that use only continuous operations. For example, most chemical processes described as continuous receive their raw material feeds and ship their products to and from the plant in rail cars, tanker trucks, or barges. The unloading and loading of these materials are done in a batch manner. Indeed, the demarcation between continuous and batch processes is further complicated by situations when plants operate continuously but feed or receive material from other process units within the plant that operate in a batch mode. Such processes are often referred to as semibatch. A **batch process** is one in which a finite quantity (batch) of product is made during a period of a few hours or days. The batch process most often consists of metering feed(s) into a vessel followed by a series of unit operations (mixing, heating, reaction, distillation, etc.) taking place at discrete scheduled intervals. This is then followed by the removal and storage of the products, by-product, and waste streams. The equipment is then cleaned and made ready for the next process. Production of up to 100 different products from the same facility has been reported [3]. This type of operation is in contrast to **continuous processes,** in which feed is sent continuously to a series of equipment, with each piece usually performing a single unit operation. Products, by-products, and waste steams leave the process continuously and are sent to storage or for further processing.

There are a number of considerations to weigh when deciding between batch and continuous processes, and some of the more important of these are listed in Table 2.1. As this table indicates, there are many things to consider when making the decision regarding batch versus continuous operation. Probably the most important of these are size and flexibility. If it is desired to produce relatively small quantities, less than 500 tonne/y [1], of a variety of different products using a variety of different feed materials, then batch processing is probably the correct choice. For large quantities, greater than 5000 tonne/y of product [1], using a single or only a few raw materials, then a continuous process is probably the best choice. There are many trade-offs between the two types of processes. However, like most things, it boils down to cost. For a batch process compared to the equivalent continuous process, the capital investment is usually much lower because the same equipment can be used for multiple unit operations and can be reconfigured easily for a wide variety of feeds and products. On the other hand, operating labor costs and utility costs tend to be much higher. Recent developments in batch processing have led to the concept of the "pipeless batch process" [4]. In this type of operation, equipment is automatically moved to different workstations at which different processes are performed. For example, a reactor may be filled with raw materials and mixed at station 1, moved to station 2 for heating and reaction, to station 3 for product separation, and finally to station 4 for product removal. The workstations contain a variety of equipment to perform functions such as mixing, weighing, heating/cooling, filtration, and so on. This modular approach to the sequencing of batch operations greatly improves productivity and eases the scheduling of different events in the overall process.

Finally, it is important to recognize the role of pilot plants in the development of processes. It has been long understood that what works well in the laboratory often does not work as well on the large scale. Of course, much of the important preliminary work associated with catalyst development and phase equilibrium is most efficiently and inexpensively completed in the laboratory. However, problems associated with trace quantities of unwanted side products, difficult material handling problems, and multiple reaction steps are not easily scaled up from laboratory-scale experiments. In such cases, specific unit operations or the entire process may be "piloted" to gain better insight into the proposed full-scale operation. Often, this pilot plant work is carried out in batch equipment in order to reduce the inventory of raw materials. Sometimes, the pilot plant serves the dual purpose of testing the process at an intermediate scale and producing enough material for customers and other interested parties to test. The role and importance of pilot plants are covered in detail by Lowenstein [5].

Table 2.1 Some Factors to Consider When Deciding between Batch and Continuous Processes

Factor	Advantages/Disadvantages for Batch Processes	Advantages/Disadvantages for Continuous Processes
Size	Smaller throughput favors batch operations. As throughput increases, the required size of the process equipment increases, and the technical difficulties of moving large amounts of chemicals from equipment to equipment rapidly increase.	Economies of scale favor continuous processes for large throughput.
Batch Accountability/ Product Quality	When the product quality of each batch of material must be verified and certified, batch operations are preferred. This is especially true for pharmaceutical and food products. The manufacture of these products is strictly monitored by the Food and Drug Administration (FDA). If reworking (reprocessing) of off-specification product is usually not permitted, small batches are favored.	Continuous or periodic testing of product quality is carried out, but some potentially large quantities of off-specification product can be produced. If off-specification material may be blended or stored in dump/slop tanks and reworked through the process when the schedule permits, continuous processes are favored.
Operational Flexibility	Often the same equipment can be used for multiple operations—for example, a stirred tank can be used as a mixer, then a reactor, and then as a stage of a mixer-settler for liquid-liquid extraction.	Operational flexibility can be built in to continuous processes but often leads to inefficient use of capital. Equipment not required for one process but needed for another may sit idle for months. Often continuous processes are designed to produce a fixed suite of products from a well-defined feed material. If market forces change the feed/product availability or demand, then the plant will often be retrofitted to accommodate the change.
Standardized Equipment—Multiple Products	Often batch processes can be easily modified to produce several different products using essentially the same equipment. Examples of batch plants that can produce 100 different	The product suite or slate produced from continuous processes is usually fixed. Equipment tends to be designed and optimized for a single or small number of operating conditions.

Table 2.1 Some Factors to Consider When Deciding between Batch and Continuous Processes (continued)

Factor	Advantages/Disadvantages for Batch Processes	Advantages/Disadvantages for Continuous Processes
	products are known [3]. For such processes the optimal control and sequencing of operations are critical to the success of such a plant.	
Processing Efficiency	Operation of batch processes requires strict scheduling and control. Because different products are scheduled back-to-back, changes in schedules have a ripple effect and may cause serious problems with product availability for customers. If the same equipment is used to produce many different products, then this equipment will not be optimized for any one product. Energy integration is usually not possible, so utility usage tends to be higher than for continuous processes. Separation and reuse of raw materials are more difficult than for continuous processes.	Generally, as throughput increases, continuous processes become more efficient. For example, fugitive energy losses are reduced, and rotating equipment (pumps, compressors, etc.) operates with higher efficiency. Recycle of unused reactants and the integration of energy within the process or plant are standard practices and relatively easy to achieve.
Maintenance and Operating Labor	There are higher operating labor costs in standard batch plants due to equipment cleaning and preparation time. These costs have been shown to be reduced for the so-called pipeless batch plants [4].	For the same process, operating labor will be lower for continuous processes.
Feedstock Availability	Batch operations are favored when feedstock availability is limited, for example, seasonally. Canneries and wineries are examples of batch processing facilities that often operate for only part of the year.	Continuous plants tend to be large and need to operate throughout the year to be profitable. The only way that seasonal variations in feeds can be accommodated is through the use of massive storage facilities that are very expensive.
Product Demand	Seasonal demand for products such as fertilizers, gas-line antifreeze, deicing chips for roads and pavements, and so on, can be easily accommodated. Because batch plants are flexible, other products can be made during the off-season.	Difficult to make other products during the off-season. However, similar but different products—for example, a family of solvents—can be produced using the same processes through a series of campaigns at different times during the year. Each campaign may last several months.

(cont.)

Table 2.1 Some Factors to Consider When Deciding between Batch and Continuous Processes (continued)

Factor	Advantages/Disadvantages for Batch Processes	Advantages/Disadvantages for Continuous Processes
Rate of Reaction to Produce Products	Batch operations favor processes that have very slow reaction rates and subsequently require long residence times. Examples include fermentation, aerobic and anaerobic waste water treatment, and many other biological reactions.	Very slow reactions require very large equipment. The flow through this equipment will be slow, and dispersion can be a problem if very high conversion is desired and plug flow is required.
Equipment Fouling	When there is significant equipment fouling, batch operations are favored because cleaning of equipment is always a standard operating procedure in a batch process and can be accommodated easily in the scheduling of the process.	Significant fouling in continuous operations is a serious problem and is difficult to handle. Operating identical units in parallel, one on-line and the other off-line for cleaning, can solve this problem. However, capital investment is higher, additional labor is required, and safety problems are more likely.
Safety	Generally, worker exposure to chemicals and operator error will be higher (per pound of product) than for continuous processes. Operator training in chemical exposure and equipment operation is critical.	Large chemical plants operating continuously have excellent safety records [6], and safety procedures are well established. Operator training is still of great importance, but many of the risks associated with opening equipment containing chemicals are eliminated.
Controllability	This problem arises because batch processes often use the same equipment for different unit operations and sometimes to produce different products. The efficient scheduling of equipment becomes very important. The control used for this scheduling is complicated [3].	Generally, continuous processes are easier to control. Also, more work and research have been done for these processes. For complicated and highly integrated (energy and/or raw materials) plants, the control becomes complex, and operational flexiblity is greatly reduced.

2.3 STEP 2—THE INPUT/OUTPUT STRUCTURE OF THE PROCESS

Although all processes are different, there are common features of each. The purpose of this section is to investigate the input/output structure of the process. The inputs represent feed streams and the outputs are product streams, which may be desired or waste streams.

2.3.1 Process Concept Diagram

The first step in evaluating a process route is to construct a process concept diagram. Such a diagram uses the stoichiometry of the main reaction pathway to identify the feed and product chemicals. The first step to construct such a diagram is to identify the chemical reaction or reactions taking place within the process. The balanced chemical reaction(s) form the basis for the overall process concept diagram. Figure 2.1 shows this diagram for the toluene hydrodealkylation process discussed in Chapter 1. It should be noted that only chemicals taking place in the reaction are identified on this diagram. The steps used to create this diagram are as follows.

1. A single "cloud" is drawn to represent the concept of the process. Within this cloud the stoichiometry for all reactions that take place in the process is written. The normal convention of the reactants on the left and products on the right is used.
2. The reactant chemicals are drawn as streams entering from the left. The number of streams corresponds to the number of reactants (two). Each stream is labeled with the name of the reactant (toluene and hydrogen).
3. Product chemicals are drawn as streams leaving to the right. The number of streams corresponds to the number of products (two). Each stream is labeled with the name of the product (benzene and methane).
4. Seldom does a single reaction occur, and unwanted side reactions must be considered. All reactions that take place and the reaction stoichiometry must be included. The unwanted products are treated as by-products and must leave along with the product streams shown on the right of the diagram.

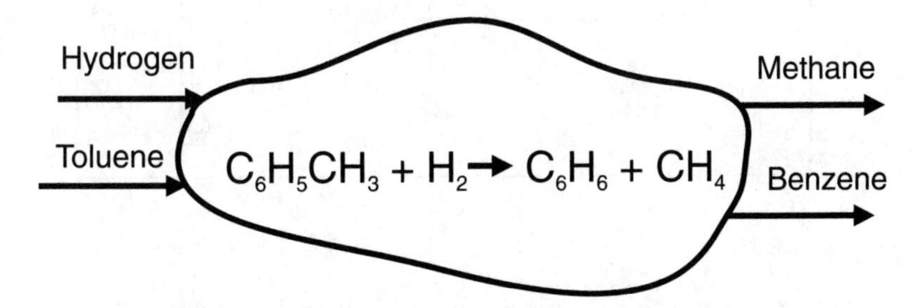

Figure 2.1 Input/Output Structure of the Process Concept Diagram for the Toluene Hydrodealkylation Process

2.3.2 The Input/Output Structure of the Process Flow Diagram

If the process concept diagram represents the most basic or rudimentary representation of a process, then the process flow diagram (PFD) represents the other extreme. However, the same input/output structure is seen in both diagrams. The PFD, by convention, shows the process feed stream(s) entering from the left and the process product stream(s) leaving to the right.

There are other auxiliary streams shown on the PFD, such as utility streams that are necessary for the process to operate but that are not part of the basic input/output structure. Ambiguities between process streams and utility streams may be eliminated by starting the process analysis with an overall input/output concept diagram.

Figure 2.2 shows the basic input/output structure for the PFD (see Figure 1.3). The input and output streams for the toluene HDA PFD are shown in bold. Both Figures 2.1 and 2.2 have the same overall input/output structure. The input streams labeled toluene and hydrogen shown on the left in Figure 2.1 appear in the streams on the left of the PFD in Figure 2.2. In Figure 2.2, these streams contain the reactant chemicals plus other chemicals that are present in the raw feed materials. These streams are identified as Streams 1 and 3, respectively. Likewise, the output streams, which contain benzene and methane, must appear on the right on the PFD. The benzene leaving the process, Stream 15, is clearly labeled, but there is no clear identification for the methane. However, by

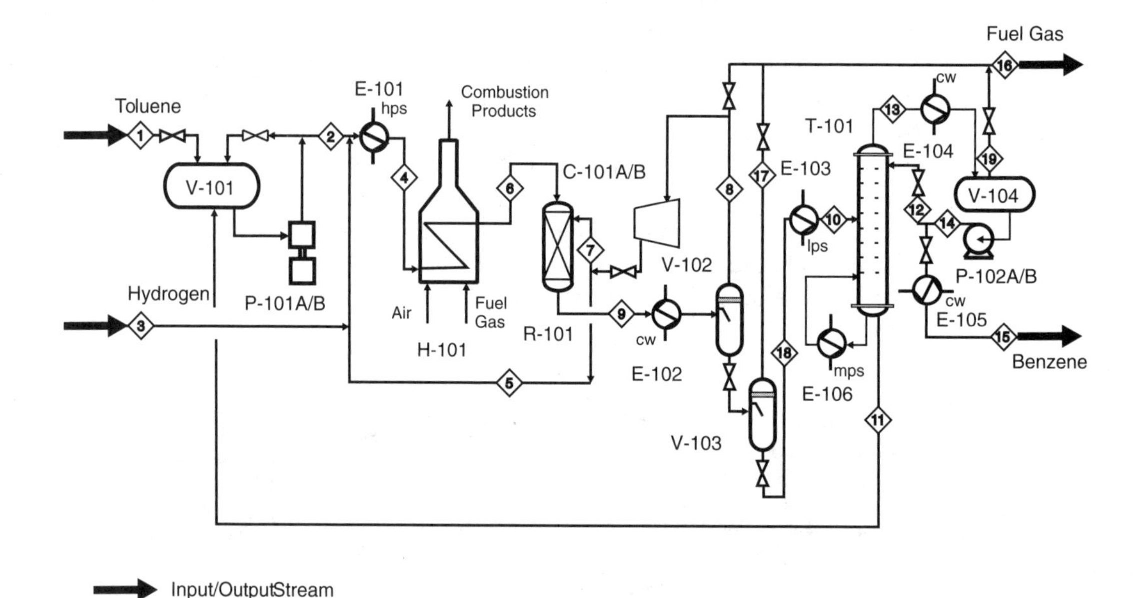

Figure 2.2 Input and Output Streams on Toluene Hydrodealkylation PFD

referring to Table 1.5 and looking at the entry for Stream 16, it can be seen that this stream contains a considerable amount of methane. From the stoichiometry of the reaction, the amount of methane and benzene produced in the process should be equal (on a mole basis). This is easily checked from the data for Streams 1, 3, 15, and 16 (Table 1.5) as follows:

Benzene produced in process = benzene leaving – benzene entering
= 105.2 (Stream 15) + 2.85 (Stream 16) – 0 = 108.05 kmol/h

Amount of methane produced = methane leaving – methane entering
= 123.05 (Stream 16) – 15.0 (Stream 3) = 108.05 kmol/h

At times, it will be necessary to use the process conditions or the flow table associated with the PFD to determine where a chemical is to be found.

There are several important factors to consider in analyzing the overall input/output structure of a PFD. We list some of these factors below.

1. Chemicals entering the PFD from the left that are not consumed in the chemical reactor are either required to operate a piece of equipment or are inert material that simply passes through the process. Examples of chemicals required but not consumed include catalyst make-up, solvent make-up, and inhibitors. In addition, feed materials that are not pure may contain inert chemicals. Alternatively, they may be added in order to control reaction rates, to keep the reactor feed outside of the explosive limits, or to act as a heat sink or heat source to control temperatures.
2. Any chemical leaving a process must either have entered in one of the feed streams or have been produced by a chemical reaction within the process.
3. Utility streams are treated differently from process streams. Utility streams, such as cooling water, steam, fuel, and electricity, rarely directly contact the process streams. They usually provide or remove thermal energy or work.

Figure 2.3 identifies, with bold lines, the utility streams in the benzene process. It can be seen that two streams—fuel gas and air—enter the fired heater. These are burned to provide heat to the process, but never come in direct contact (that is, mix) with the process streams. Other streams such as cooling water and steam are also highlighted in Figure 2.3. All these streams are utility streams and are not extended to the left or right boundaries of the diagram, as were the process streams. Other utility streams are also provided but are not shown in the PFD. The most important of these is electrical power, which is most often used to run rotating equipment such as pumps and compressors. Other utilities, such as plant air, instrument air, nitrogen for blanketing of tanks, process water, and so on, are also consumed.

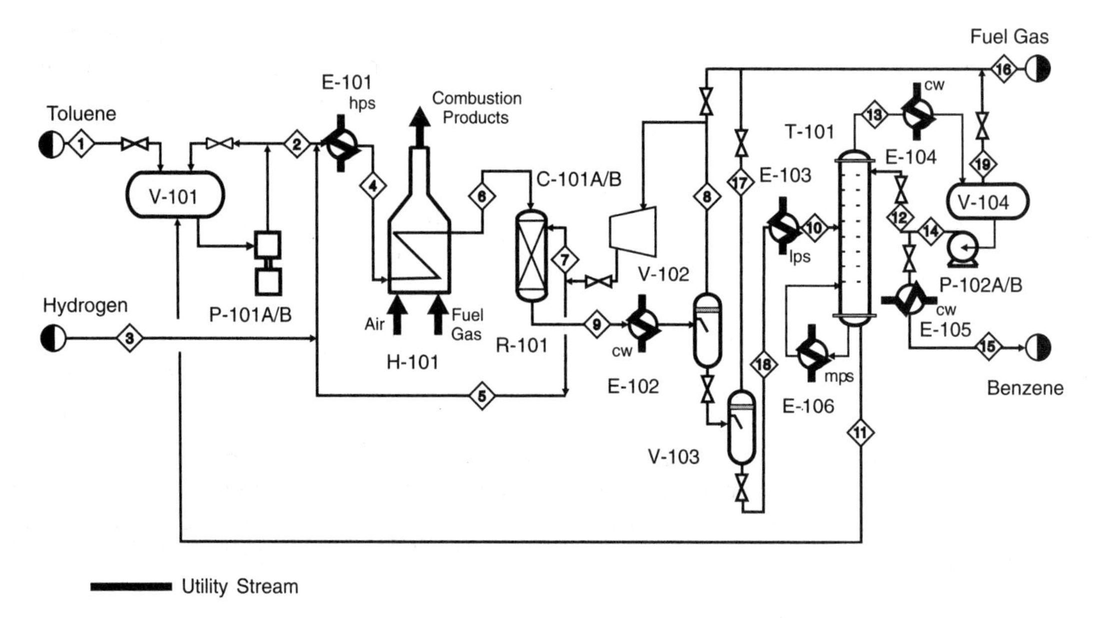

Figure 2.3 Identification of Utility Streams on the Toluene Hydrodealkylation PFD

2.3.3 The Input/Output Structure and Other Features of the Generic Block Flow Process Diagram

The **generic block flow diagram** is intermediate between the process concept diagram and the PFD. This diagram illustrates features, in addition to the basic input/output structure, that are common to all chemical processes. Moreover, in discussing the elements of new processes it is convenient to refer to this diagram because it contains the logical building blocks for all processes. Figure 2.4(a) provides a generic block flow process diagram that shows a chemical process broken down into six basic areas or blocks. Each block provides a function necessary for the operation of the process. These six blocks are as follows:

1. Reactor feed preparation
2. Reactor
3. Separator feed preparation
4. Separator
5. Recycle
6. Environmental control

An explanation of the function of each block in Figure 2.4(a) is given below.

1. **Reactor Feed Preparation Block:** In most cases, the feed chemicals entering a process come from storage. These chemicals are most often not at a suit-

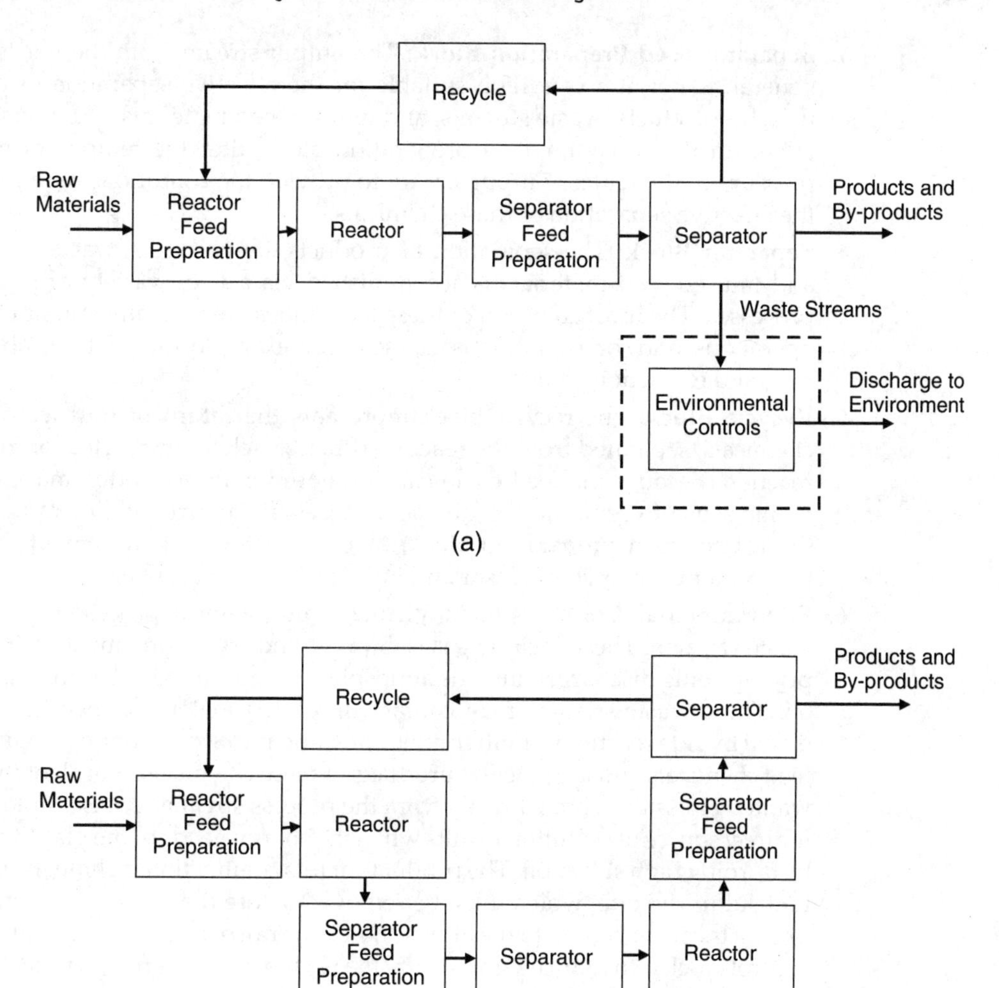

Figure 2.4 (a) The Six Elements of the Generic Block Flow Process Diagram; (b) A Process Requiring Multiple Process Blocks

able concentration, temperature, and pressure for optimal performance in the reactor. The purpose of the reactor feed preparation section is to change the conditions of these process feed streams as required in the reactor.

2. **Reactor Block:** All chemical reactions take place in this block. The streams leaving this block contain the desired product(s), any unused reactants, and a variety of undesired by-products produced by competing reactions.

3. **Separator Feed Preparation Block:** The output stream from the reactor, in general, is not at a condition suitable for the effective separation of products, by-products, waste streams, and unused feed materials. The units contained in the separator feed preparation block alter the temperature and pressure of the reactor output stream to provide the conditions required for the effective separation of these chemicals.
4. **Separator Block:** The separation of products, by-products, waste streams, and unused feed materials is accomplished via a wide variety of physical processes. The most common of these techniques are typically taught in unit operations and/or separations classes—for example, distillation, absorption, and extraction.
5. **Recycle Block:** The recycle block represents the return of unreacted feed chemicals, separated from the reactor effluent, back to the reactor for further reaction. Because the feed chemicals are not free, it most often makes economic sense to separate the unreacted reactants and recycle them back to the reactor feed preparation block. Normally, the only equipment in this block is a pump or compressor and perhaps a heat exchanger.
6. **Environmental Control Block:** Virtually all chemical processes produce waste streams. These include gases, liquids, and solids that must be treated prior to being discharged into the atmosphere, sequestered in landfills, and so on. These waste streams may contain unreacted materials, chemicals produced by side reactions, fugitive emissions, and impurities coming in with the feed chemicals and the reaction products of these chemicals. Not all of the unwanted emissions come directly from the process streams. An example of an indirect source of pollution results when the energy needs of the plant are met by burning high sulfur oil. The products of this combustion include the pollutant sulfur dioxide, which must be removed before the gaseous combustion products can be vented to the atmosphere. The purpose of the environmental control block is to reduce significantly the waste emissions from a process and to render all nonproduct streams harmless to the environment.

It can be seen that a dashed line has been drawn around the block containing the environmental control operations. This identifies the unique role of environmental control operations in a chemical plant complex. A single environmental control unit may treat the waste from several processes. For example, the waste water treatment facility for an oil refinery might treat the waste water from as many as 20 separate processes. In addition, the refinery may contain a single stack and incinerator to deal with gaseous wastes from these processes. Often, this common environmental control equipment is not shown in the PFD for an individual process, but is shown on a separate PFD as part of the "off site" section of the plant. Just because the environmental units do not appear on the PFD does not indicate that they do not exist or that they are unimportant.

Each of the process blocks may contain several unit operations. Moreover, several process blocks may be required in a given process. An example of multiple process blocks in a single process is shown in Figure 2.4(b). In this process, an intermediate product is produced in the first reactor and is subsequently separated and sent to storage. The remainder of the reaction mixture is sent to a second stage reactor in which product is formed. This product is subsequently separated and sent to storage, and unused reactant is also separated and recycled to the front end of the process. Based upon the reason for including the unit, each unit operation found on a PFD can be placed into one of these blocks. Although each process may not include all the blocks, all processes will have some of these blocks.

In Example 2.6, at the end of this chapter, different configurations will be investigated for a given process. It will be seen that these configurations are most conveniently represented using the building blocks of the generic block flow diagram.

2.3.4 Other Considerations for the Input/Output Structure of the Process Flowsheet

The effects of feed impurities and additional flows that are required to carry out specific unit operations may have a significant impact on the structure of the PFD. These issues are covered in the following section.

Feed Purity and Trace Components. In general, the feed streams entering a process do not contain pure chemicals. The option always exists to purify further the feed to the process. The question of whether this purification step should be performed can be answered only by a detailed economic analysis. However, some commonsense heuristics may be used to choose a good base case or starting point. The following heuristics are modified from Douglas [1].

- If the impurities are not present in large quantities (say, <10–20%) and these impurities do not react to form by-products, then *do not separate* them prior to feeding to the process. For example, the hydrogen fed to the toluene HDA process contains a small amount of methane (5 mol%—see Stream 3 in Table 1.5). Because the methane does not react (it is inert) and it is present as a small quantity, it is probably not worth considering separating it from the hydrogen.
- If the separation of the impurities is difficult (for example, an impurity forms an azeotrope with the feed or the feed is a gas at the feed conditions), then *do not separate* them prior to feeding to the process. For example, again consider the methane in Stream 3. The separation of methane and hydrogen is relatively expensive (see Example 2.3) because it involves low temperature and/or high pressure. This fact, coupled with the reasons given above, means that separation of the feed would not normally be attempted.
- If the impurities foul or poison the catalyst, then *purify the feed.* For example, one of the most common catalyst poisons is sulfur. This is especially true for

catalysts containing Group VIII metals such as iron, cobalt, nickel, palladium, and platinum [7]. In the steam reformation of natural gas (methane) to produce hydrogen, the catalyst is rapidly poisoned by the small amounts of sulfur in the feed. A guard bed of activated carbon (or zinc oxide) is placed upstream of the reactor to reduce the sulfur level in the natural gas to the low ppm level.

- If the impurity reacts to form difficult-to-separate or hazardous products, then *purify the feed*. For example, in the manufacture of isocyanates for use in the production of polyurethanes, the most common synthesis path involves the reaction of phosgene with the appropriate amine [8]. Because phosgene is a highly toxic chemical, all phosgene is manufactured on-site via the reaction of chlorine and carbon monoxide.

$$CO + Cl_2 \rightarrow \underset{phosgene}{COCl_2}$$

If carbon monoxide is not readily available (by pipeline), then it must be manufactured via the steam reformation of natural gas. The following equation shows the overall main reaction (carbon dioxide may also be formed in the process, but it is not considered here):

$$CH_4 + H_2O \rightarrow CO + 3H_2$$

The question to ask is, At what purity must the carbon monoxide be fed to the phosgene unit? The answer depends on what happens to the impurities in the CO. The main impurity is hydrogen. The hydrogen reacts with the chlorine to form hydrogen chloride, which is difficult to remove from the phosgene, is highly corrosive, and is detrimental to the isocyanate product. With this information, it makes more sense to remove the hydrogen to the desired level in the carbon monoxide stream rather than send it through with the CO and cause more separation problems in the phosgene unit and further downstream. Acceptable hydrogen levels in carbon monoxide feeds to phosgene units are less than 1%.

- If the impurity is present in large quantities, then *purify the feed*. This heuristic is fairly obvious because significant additional work and heating/cooling duties are required to process the large amount of impurity. Nevertheless, if the separation is difficult and the impurity acts as an inert, then separation may still not be warranted. An obvious example is the use of air, rather than pure oxygen, as a reactant. Because nitrogen often acts as an inert compound, the extra cost of purifying the air is not justified compared with the lesser expense of processing the nitrogen through the process. An added advantage of using air, as opposed to pure oxygen, is the heat absorbing capacity of nitrogen, which helps moderate the temperature rise of many highly exothermic oxidation reactions.

Addition of Feeds Required to Stabilize Products or Enable Separations. Generally, product specifications are given as a series of characteristics that the product stream must meet or exceed. Clearly, the purity of the main chemical in the product is the major concern. However, other specifications such as color, density or specific gravity, turbidity, and so on, may also be specified. Often many of these specifications can be met in a single piece or train of separation equipment. However, if the product stream is, for example, reactive or unstable, then additional stabilizing chemicals may need to be added to the product before it goes to storage. These stabilizing chemicals are additional feed streams to the process. The same argument can be made for other chemicals such as solvent or catalyst that are effectively consumed in the process. If a solvent such as water or an organic chemical is required to make a separation take place—for example, absorption of a solvent-soluble chemical from a gas stream—then this solvent is an additional feed to the process (see Appendix B, Problem 5—the production of maleic anhydride via the partial oxidation of propylene). Accounting for these chemicals both in feed costs and in the overall material balance (in the product streams) is very important.

Inert Feed Material to Control Exothermic Reactions. In some cases, it may be necessary to add additional inert feed streams to the process in order to control the reactions taking place. Common examples of this are partial oxidation reactions of hydrocarbons. For example, consider the partial oxidation of propylene to give acrylic acid, an important chemical in the production of acrylic polymers. The feeds consist of nearly pure propylene, air, and steam. The basic reactions that take place are as follows:

$$C_3H_6 + \frac{3}{2}O_2 \rightarrow C_3H_4O_2 + H_2O \quad \text{Reaction 1}$$

$$C_3H_6 + \frac{5}{2}O_2 \rightarrow C_2H_4O_2 + H_2O + CO_2 \quad \text{Reaction 2}$$

$$C_3H_6 + \frac{9}{2}O_2 \rightarrow 3H_2O + 3CO_2 \quad \text{Reaction 3}$$

All these reactions are highly exothermic, not limited by equilibrium, and potentially explosive. In order to eliminate or reduce the potential for explosion, steam is fed to the reactor to dilute the feed and provide thermal ballast to absorb the heat of reaction and make control easier. In some processes, enough steam (or other inert stream) is added to move the reaction mixture out of the flammability limits, thus eliminating the potential for explosion. The steam (or other inert stream) is considered a feed to the process, must be separated, and leaves as a product, by-product, or waste stream.

Addition of Inert Feed Material to Control Equilibrium Reactions. Sometimes it is necessary to add an inert material to shift the equilibrium of the desired reaction. Consider the production of styrene via the catalytic dehydrogenation of ethyl benzene:

$$\underset{\textit{ethyl benzene}}{C_6H_5CH_2CH_3} \rightleftharpoons \underset{\textit{styrene}}{C_6H_5CH=CH_2} + H_2$$

This reaction takes place at high temperature (600–750°C) and low pressure (<1 bar) and is limited by equilibrium. The ethyl benzene is co-fed to the reactor with superheated steam. The steam acts as an inert in the reaction and both provides the thermal energy required to preheat the ethyl benzene and dilutes the feed. As the steam-to-ethyl benzene ratio increases, the equilibrium shifts to the right (LeChatelier's principle) and the single-pass conversion increases. The optimum steam-to-ethyl benzene feed ratio is based on the overall process economics.

2.3.5 What Information Can Be Determined Using the Input/Output Diagram for a Process?

The following basic information, obtained from the input/output diagram, is limited but nevertheless very important.

- Basic economic analysis on profit margin
- What chemical components must enter with the feed and leave as products
- All the reactions, both desired and undesired, that take place

The potential profitability of a proposed process can be evaluated and a decision whether to pursue the process can be made. As an example, consider the profit margin for the toluene HDA process given in Figure 2.1.

The profit margin will be formally introduced in Chapter 10, but it is defined as the difference between the value of the products and the cost of the raw materials. To keep things simple we use the stoichiometry of the reaction as our basis. If the profit margin is a negative number, then there is no potential to make money. The profit margin for the HDA process is given in Example 2.1.

Example 2.1

Evaluate the profit margin for the HDA process.

From Tables 8.3 and 8.4, we get the following prices for raw materials and products:

Benzene = \$0.657/kg

Toluene = \$0.648/kg

Natural gas (methane and ethane, MW = 18) = \$11.10/GJ = \$11.89/1000 std. ft^3 = \$0.302/kg

Hydrogen = \$1.000/kg (based on the same equivalent energy cost as natural gas)

Using 1 kmol of toluene feed as a basis

Cost of Raw Materials

92 kg of Toluene = (92 kg)($ 0.648/kg) = $ 59.62
2 kg of Hydrogen = (2 kg)($ 1.000/kg) = $ 2.00

Value of Products

78 kg of Benzene = (78 kg)($ 0.657/kg) = $ 51.25
16 kg of Methane = (16 kg)($ 0.302/kg) = $ 4.83

Profit Margin

Profit Margin = (51.25 + 4.83) – (59.62 + 2.00) = –$ 5.54 or –$ 0.060/kg toluene

Based on this result, we conclude that further investigation of this process is definitely not warranted.

Despite the results illustrated in Example 2.1, benzene has been produced for the last 50 years and is a viable starting material for a host of petrochemical products. Therefore, how is this possible? We must conclude that benzene can be produced via at least one other route, which is less sensitive to changes in the price of toluene, benzene, and natural gas. One such commercial process is the disproportionation or transalkylation of toluene to produce benzene and a mixture of para-, ortho-, and meta-xylene by the following reaction.

$$\underset{toluene}{2C_7H_8} \rightarrow \underset{benzene}{C_6H_6} + \underset{xylene}{C_8H_{10}}$$

The profit margin for this process is given in Example 2.2.

Example 2.2

Evaluate the profit margin for the toluene disproportionation process.

From Table 6.4, we have
Mixed Xylenes = 0.608 $/kg

Using 2 kmols of toluene feed as a basis

Cost of Raw Materials

184 kg of Toluene = (184 kg)($0.648/kg) = $ 119.23

Value of Products

78 kg of Benzene = (78 kg)($0.657/kg) = $ 51.25
106 kg of xylene = (106 kg)($0.608/kg) = $ 64.45

Margin

Profit Margin = 64.45 + 51.25 – 119.23 = –\$3.53 or –\$0.019/kg toluene feed

Based on the results of Example 2.2, the production of benzene via the disproportionation of toluene is better than the toluene HDA process but is still unprofitable. However, a closer look at the cost of purified xylenes (from Table 8.4) shows that these purified xylenes are considerably more valuable (ranging from \$0.805 to \$2.91/kg) than the mixed xylene stream (\$0.608/kg). Therefore, the addition of a xylene purification section to the disproportionation process might well yield a potentially profitable process—namely, a process that is worth further, more-detailed analysis. Historically, the prices of toluene and benzene fluctuate in phase with each other, but the toluene price (\$0.648/kg) is currently elevated relative to that of benzene (\$0.657/kg). In general, toluene disproportionation has been the preferred process for benzene production over the last two decades.

Examples 2.1 and 2.2 make it apparent that a better approach to evaluating the margin for a process would be to find cost data for the feed and product chemicals over a period of several years to get average values and then use these to evaluate the margin. Another important point to note is that there are often two or more different chemical paths to produce a given product. These paths may all be technically feasible; that is, catalysts for the reactions and separation processes to isolate and purify the products probably exist. However, it is the costs of the raw materials that usually play the major role when deciding which process to choose.

2.4 STEP 3—THE RECYCLE STRUCTURE OF THE PROCESS

The remaining three steps in building the process flow diagram basically involve the recovery of materials and energy from the process. It may be instructive to break down the operating costs for a typical chemical process. This analysis for the toluene process is given in Chapter 8, Example 8.10. From the results of Example 8.10, we can see that raw material costs (toluene and hydrogen) account for $(60.549)/(87.3) \times 100 = 69\%$ of the total manufacturing costs. This value is typical for chemical processes. Peters and Timmerhaus [9] suggest that raw materials make up between 10% and 50% of the total operating costs for processing plants; however, due to increasing conservation and waste minimization techniques this estimate may be low, and an upper limit of 75% is more realistic. Because these raw materials are so valuable, it is imperative that we be able to separate and recycle unused reactants. Indeed, high efficiency for raw material usage is a requirement of the vast majority of chemical processes. This is why the generic block flow process diagram (Figure 2.4) has a recycle stream shown. However, the extent of recycling of unused reactants depends largely on the ease with which these unreacted raw materials can be separated (and purified) from the products that are formed within the reactor.

2.4.1 Efficiency of Raw Material Usage

It is important to understand the difference between single-pass conversion in the reactor, the overall conversion in the process, and the yield.

$$\textit{Single-pass Conversion} = \frac{\textit{reactant consumed in reaction}}{\textit{reactant fed to the reactor}} \tag{2.1}$$

$$\textit{Overall Conversion} = \frac{\textit{reactant consumed in process}}{\textit{reactant fed to the process}} \tag{2.2}$$

$$\textit{Yield} = \frac{\textit{moles of reactant to produce desired product}}{\textit{moles of limiting reactant reacted}} \tag{2.3}$$

For the hydrodealkylation process introduced in Chapter 1, the following values are obtained for the most costly reactant (toluene) from Table 1.5:

$$\textit{Single-pass Conversion} = \frac{(144.0 - 36.0)}{144.0} = 0.75 \textit{ or } 75\%$$

$$\textit{Overall Conversion} = \frac{(108.7 - 0.4 - 0.31)}{108.7} = 0.993 \textit{ or } 99.3\%$$

$$\textit{Yield} = \frac{(105.2 + 2.85)}{(108.7 - 0.4 - 0.31)} = 0.9995 \textit{ or } 99.95\%$$

The single-pass conversion tells us how much of the toluene that enters the reactor is converted to benzene. The lower the single-pass conversion, the greater the recycle must be, assuming that the unreacted toluene can be separated and recycled. In terms of the overall economics of the process, the single-pass conversion will affect equipment size and utility flows, because both of these are directly affected by the amount of recycle. However, the raw material costs are not changed significantly, assuming that the unreacted toluene is separated and recycled.

The overall conversion tells us what fraction of the toluene in the feed to the process (Stream 1) is converted to products. For the hydrodealkylation process, it is seen that this fraction is high (99.3%). This high overall conversion is typical for chemical processes and shows that unreacted raw materials are not being lost from the process.

Finally the yield tells us what fraction of the reacted toluene ends up in our desired product: benzene. For this case, the yield is unity (within round-off error), and this is to be expected because we have not considered any competing or side reactions. In reality, there is at least one other significant reaction that can take place, and this may reduce the yield of toluene. This case is considered in Problem 2.1 at the end of the chapter. Nevertheless, yields for this process are generally very high. For example, Lummus [10] quotes yields from 98% to 99% for their DETOL, hydrodealkylation process.

We can also look at the conversion of the other reactant, hydrogen. From the figures in Table 1.5 we get the following:

$$\textit{Single-pass Conversion} = \frac{(735.4 - 652.6)}{735.4} = 0.113 \textit{ or } 11.3\%$$

$$\textit{Overall Conversion} = \frac{(286.0 - 178.0)}{286.0} = 0.378 \textit{ or } 37.8\%$$

Clearly these conversions are much lower than for toluene. The single-pass conversion is kept low because a high hydrogen-to-hydrocarbon ratio is desired everywhere in the reactor so as to avoid or reduce coking of the catalyst. However, the low overall conversion of hydrogen indicates poor raw material usage. Therefore, the questions to ask are, Why is the material usage for toluene so much better than that of hydrogen? and, How can the hydrogen usage be improved? These questions can be answered by looking at the ease of separation of hydrogen and toluene from their respective streams and leads us to investigate the *recycle structure* of the process.

2.4.2 Identification and Definition of the Recycle Structure of the Process

There are basically three ways that unreacted raw materials can be recycled in continuous processes.

1. Separate and purify unreacted feed material from products and then recycle.
2. Recycle feed and product together and use a purge stream.
3. Recycle feed and product together and do not use a purge stream.

Separate and Purify. Through the ingenuity of chemical engineers and chemists, technically feasible separation paths exist for mixtures of nearly all commercially desired chemicals. Therefore, the decision on whether to separate the unreacted raw materials must be made purely from economic considerations. In general, the ease with which a given separation can be made is dependent on two principles.

- First, for the separation process (unit operation) being considered, what conditions (temperature and pressure) are necessary to operate the process?
- Second, for the chemical species requiring separation, are the differences in physical or chemical properties for the species, on which the separation is based, large or small? Examples that illustrate these principles are given below.

For the hydrodealkylation process, the reactor effluent, Stream 9, is cooled and separated in a two-stage flash operation. The liquid, Stream 18, contains essentially benzene and toluene. The combined vapor stream, Streams 8 and 17, contain essentially methane and hydrogen. In Example 2.3, methods to separate the hydrogen in these two streams are considered and are used to screen potential changes in the recycle structure of the HDA process.

Example 2.3

For the separation of methane and hydrogen, first look at distillation:

Normal boiling point of methane = –161°C
Normal boiling point of hydrogen = –252°C

Separation should be easy using distillation due to the large difference in boiling points of the two components. However, in order to get a liquid phase, we will have to use a combination of high pressure and low temperature. This will be very costly and suggests that distillation is not the best operation for this separation.

Absorption

It might be possible to absorb or scrub the methane from Streams 8 and 17 into a hydrocarbon liquid. In order to determine which liquids, if any, are suitable for this process, we must compare the solubility parameters for both methane and hydrogen in the different liquids. This information is available in Walas [11]. Because of the low boiling point of methane, it would require a low temperature and high pressure for effective absorption.

Pressure-Swing Adsorption

The affinity of a molecule to adhere (either chemically or physically) to a solid material is the basis of adsorption. In pressure-swing adsorption, the preferential adsorption of one species from the gas phase occurs at a given pressure, and the desorption of the adsorbed species is facilitated by reducing the pressure and allowing the solid to "de-gas." Two (or more) beds operate in parallel, with one bed adsorbing and the other desorbing. The separation and purification of hydrogen contained in gaseous hydrocarbon streams could be carried out using pressure-swing adsorption. In this case, the methane would be preferentially adsorbed on to the surface of a sorbent, and the stream leaving the unit would contain a higher proportion of hydrogen than the feed. This separation could be applied to the HDA process.

Membrane Separation

Commercial membrane processes are available to purify hydrogen from hydrocarbon streams. This separation is facilitated because hydrogen passes more readily through certain membranes than does methane. This process occurs at moderate pressures, consistent with the operation of the HDA process. However, the hydrogen is recovered at a fairly low pressure and would have to be recompressed prior to recycling. This separation could be applied to the HDA process.

From Example 2.3, we see that pressure-swing adsorption and membrane separation of the gas stream should be considered as viable process alternatives, but for the preliminary PFD for this process, no separation of hydrogen was attempted. In Example 2.4, the separation of toluene from a mixture of benzene and toluene is considered.

Example 2.4

What process should be used in the separation of toluene and benzene?

Distillation

Normal boiling point of benzene = 79.8°C
Normal boiling point of toluene = 110°C

Separation should be easy using distillation, and neither excessive temperatures nor pressures will be needed. This is a viable operation for this separation of benzene and toluene in the HDA process.

Economic considerations often make distillation the separation method of choice. The separation of benzene and toluene is routinely practiced through distillation and is the preferred method in the preliminary PFD for this process.

Recycle Feed and Product Together with a Purge Stream. If separation of unreacted feed and products is not accomplished easily, then recycling both feed and product should be considered. In the HDA process, the methane product will act as an inert because it will not react with toluene. In addition, we are not limited by equilibrium considerations; therefore, the reaction of methane and benzene to give toluene and hydrogen (the undesired path for this reaction), under the conditions used in this process, is not significant. It should be noted that for the case when a product is recycled with an unused reactant and the product does not react further, then a purge stream must be used to avoid the accumulation of product in the process. For the HDA process, the purge is the fuel gas containing the methane product and unused hydrogen, Stream 16, leaving the process. The recycle structure for the hydrogen and methane in the HDA process is illustrated in Figure 2.5.

Recycle Feed and Product Together without a Purge Stream. This recycle scheme is feasible only when the product can react further in the reactor and therefore there is no need to purge it from the process. If the product does not react and it does not leave the system with the other products, then it would accumulate in the process, and steady state operations could not be achieved. In the previous case, with hydrogen and methane, we saw that the methane did not react further and that we had to purge some of the methane and hydrogen in Stream 16 in order to prevent accumulation of methane in the system.

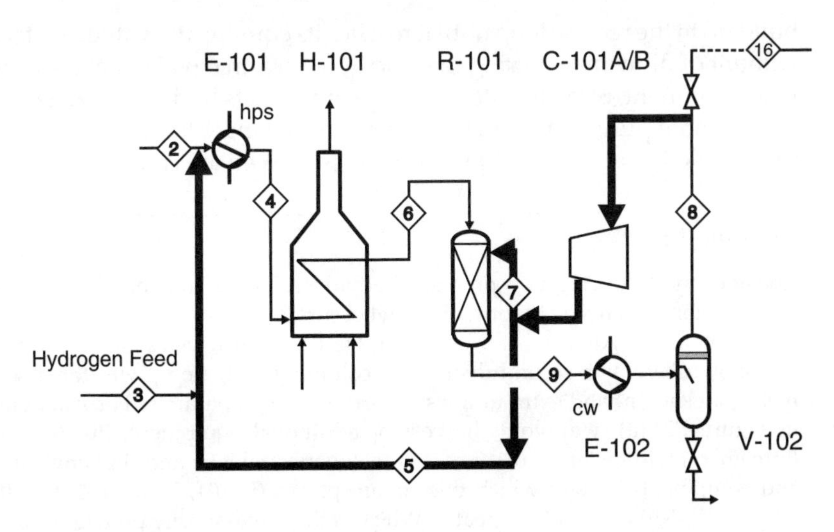

Figure 2.5 Recycle Structure of Hydrogen Stream in Toluene Hydrodealkylation Process. Methane Is Purged from the System via Stream 16.

An example where this strategy could be considered is again given in the toluene HDA process. Up to this point, we have only considered the main reaction between toluene and hydrogen:

$$\underset{toluene}{C_7H_8} + H_2 \rightarrow \underset{benzene}{C_6H_6} + CH_4$$

However, even when using a catalyst that is very specific to the production of benzene, some amount of side reaction will occur. For this process, the yield of toluene for commercial processes is on the order of 98% to 99%. Although this is high, it is still lower than the 100% that was originally assumed. A very small amount of toluene may react with the hydrogen to form small-molecule, saturated hydrocarbons, such as ethane, propane, and butane. More important, a proportion of the benzene reacts to give a two ring aromatic, diphenyl:

$$\underset{benzene}{2C_6H_6} \rightleftharpoons \underset{diphenyl}{C_{12}H_{10}} + H_2$$

The primary separation between the benzene and toluene in T-101 (see Figure 2.1) will remain essentially unchanged, because the light ends (hydrogen, methane, and trace amounts of $C_2 - C_4$ hydrocarbons) will leave in the flash separators (V-102 and V-103) or from the overhead reflux drum (V-104). However, the bottoms product from T-101 will now contain toluene and essentially all the diphenyl produced in the reactor, because it has a much higher boiling point than toluene. It is known that the benzene/diphenyl reaction is equilibrium limited at the conditions used in the reactor. Therefore, if the diphenyl is recycled with the toluene, it will simply build up in the recycle loop until it reaches its equilibrium value. At steady state, the amount of diphenyl entering the reactor in Stream 6 will equal the diphenyl in the reactor effluent, Stream 9. Because diphenyl reacts back to benzene, it can be recycled without purging it from the system. The changes to the structure of the process that would be required if diphenyl were produced are considered in Example 2.5.

Example 2.5

Consider the following two process alternatives for the toluene HDA process when the side reaction of benzene to form diphenyl occurs.

Clearly for Alternative B, shown in Figure E2.5(b), we require an additional separator, shown here as a second distillation column T-102, along with the associated equipment (not shown) and extra utilities to carry out the separation. For Alternative A, shown in Figure E2.5(a), we avoid the cost of additional equipment, but the recycle stream (Stream 11) will be larger because it now contains toluene and diphenyl, and the utilities and equipment through which this stream passes (H-101, E-101, R-101, E-102, V-102, V-103, T-101, E-106) will all be greater. Which is the economically preferable alternative?

The answer depends upon the value of the equilibrium constant for the benzene-diphenyl reaction. If the equilibrium conversion of benzene is high, then there will be a large amount of diphenyl in the recycle and the costs to recycle this material will be high, and vice versa. The equilibrium constant for this reaction is given as

$$\ln K_{eq} = 1.788 - \frac{4135.2}{T(K)}$$

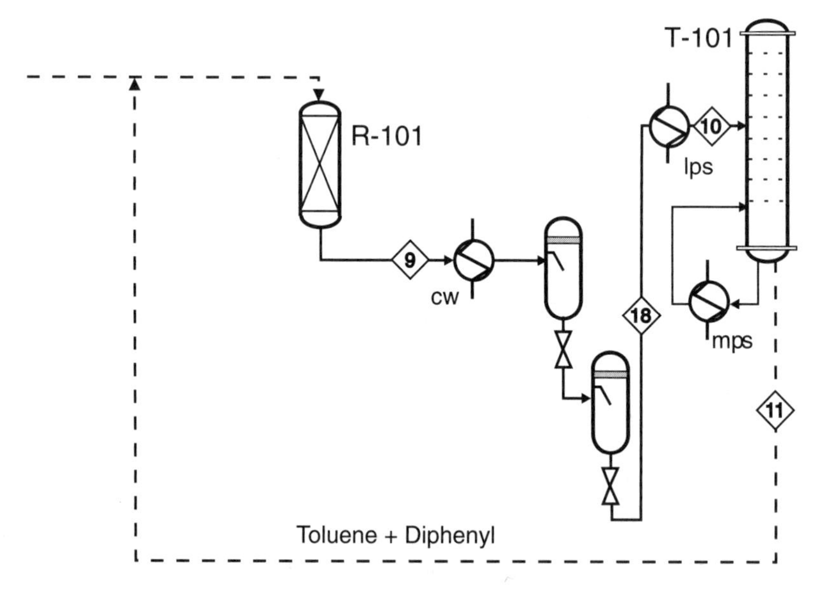

Figure E2.5(a) PFD for Alternative A in Example 2.5—Recycle of Diphenyl without Separation (E-101 and H-101 Not Shown)

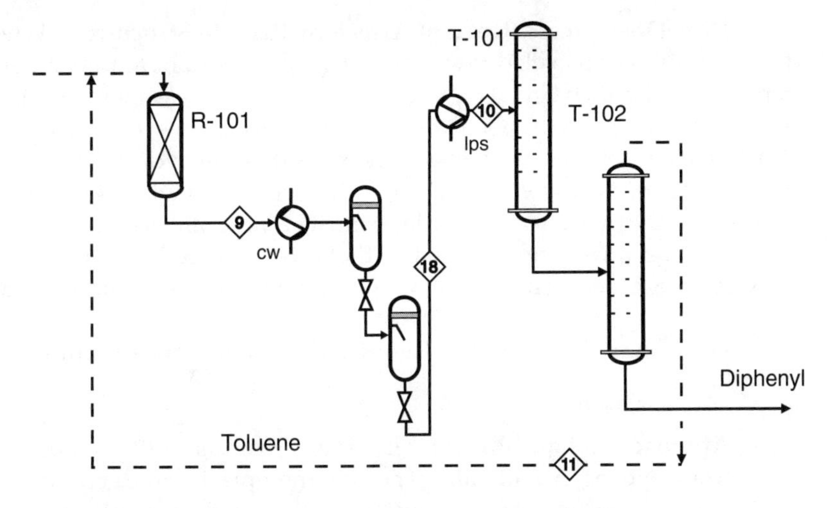

Figure E2.5(b) PFD for Alternative B in Example 2.5—Recycle of Diphenyl with Separation (E-101 and H-101 Not Shown)

The exit conditions of the reactor can be estimated by assuming that the benzene-diphenyl reaction has reached equilibrium, a conservative assumption. Using this assumption and data from Table 1.5 for Stream 9, if x kmol/h of diphenyl is present in the reactor effluent then:

$$K_{eq} = \frac{[C_{10}H_{12}][H_2]}{[C_6H_6]^2} \Rightarrow \exp\left[1.788 - \frac{4135.2}{(654 + 273)}\right] = \frac{(x)(652.6 + x)}{(116 - 2x)^2}$$

Solving for the only unknown gives $x = 1.36$ kmol/h. Thus, the toluene recycle, Stream 11, will be increased from 35.7 to 37.06 kmol/h, an increase of 4%, while the increases in Streams 4 and 6 will be approximately 0.1%. Based on this result, Alternative A will probably be less expensive than Alternative B.

2.4.3 Other Issues Affecting the Recycle Structure That Lead to Process Alternatives

There are many other issues that affect the recycle structure of the PFD. The use of excess reactant, the recycling of inert materials, and the control of an equilibrium reaction are some examples that are addressed in this section.

How Many Potential Recycle Streams Are There? Consider first the reacting species that are of value. These are essentially all reactants except air and maybe water. Each reacting species that does not have a single-pass conversion > 99% should be considered as a potential recycle stream. The value of 99% is an arbitrarily high number, and it could be anywhere from 90 to > 99%, depending on the cost of raw materials, the cost to separate and recycle unused raw materials, and the cost of disposing of any waste streams containing these chemicals.

How Does Excess Reactant Affect the Recycle Structure? When designing the separation of recycled raw materials, it is important to remember which reactant, if any, should be in excess and how much this excess should be. For the toluene HDA process, the hydrogen is required to be in excess in order to suppress coking reactions that foul the catalyst. The result is that the hydrogen:toluene ratio at the inlet of the reactor (from Table 1.5) is 735.4:144, or slightly greater than 5:1. This means that the hydrogen recycle loop must be large, and a large recycle compressor is required. If it were not for the fact that this ratio needs to be high, the hydrogen recycle stream, and hence the recycle compressor, could be eliminated.

How Many Reactors Are Required? The reasons for multiple reactors are as follows.

- **Approach to Equilibrium:** The classic example is the synthesis of ammonia from hydrogen and nitrogen. As ammonia is produced in a packed bed reactor, the heat of reaction heats the products and moves the reaction closer to equilibrium. By adding additional reactants between staged packed beds arranged in series, the concentration of the reactants is increased, and the temperature is decreased. Both these factors move the reaction away from equilibrium and allow the reaction to proceed further to produce the desired product, ammonia.
- **Temperature Control:** If the reaction is mildly exothermic or endothermic, then internal heat transfer may not be warranted, and temperature control for gas-phase reactions can be achieved by adding a "cold (or hot) shot" between staged adiabatic packed beds of catalyst. This is similar to the ammonia converter described earlier.
- **Concentration Control:** If one reactant tends to form by-products, then it may be advantageous to keep this reactant at a low concentration. Multiple side feeds to a series of staged beds or reactors may be considered. See Chapter 20 for more details.
- **Optimization of Conditions for Multiple Reactions:** When several series reactions ($A \rightarrow R \rightarrow S \rightarrow T$) must take place to produce the desired product (T) and these reactions require different catalysts and/or different operating conditions, then operating a series of staged reactors at different conditions may be warranted.

Do Unreacted Raw Material Streams Need to Be Purified Prior to Recycling? The next issue is whether the components need to be separated prior to recycle. For example, if distillation is used to separate products from unused reactants, and if two of the reactants lie next to each other in a list of relative volatility, then no separation of these products is necessary. They can be simply recycled as a mixed stream.

Is Recycling of an Inert Warranted? We next consider components in the feed streams that do not react, that is, are inert. Depending on the process, it may be worth recycling these streams. For example, consider the water feed to the absorber, Stream 8, in the acetone production process (Appendix B, Figure B.10.1). This water stream is used to absorb trace amounts of isopropyl alcohol and acetone from the hydrogen vent, Stream 5. After purification, the water leaves the process as a waste water stream, Stream 15. This water has been purified in column T-1103 and contains only trace amounts of organics. An alternative process configuration would be to recycle this water back to the absorber. This type of pollution prevention strategy is discussed further in Chapter 25.

Can Recycling an Unwanted Product or an Inert Shift the Reaction Equilibrium to Produce Less of an Unwanted Product? Another example of recycling an inert or unwanted product is to use that material to change the conversion and selectivity of an equilibrium reaction. For example, consider the production of synthesis gas (H_2 and CO) via the partial oxidation (gasification) of coal:

$$C_mH_n + \left(\frac{m}{2} + \frac{n}{4}\right)O_2 \rightarrow mCO + \frac{n}{2}H_2O \qquad \text{partial oxidation}$$

$$C_mH_n + \left(m + \frac{n}{4}\right)O_2 \rightarrow mCO_2 + \frac{n}{2}H_2O \qquad \text{complete oxidation}$$

$$CO + H_2O \rightleftharpoons CO_2 + H_2 \qquad \text{water-gas shift}$$

Coal, shown here simply as a mixture of carbon and hydrogen, is reacted with a substoichiometric amount of pure oxygen in a gasifier, and steam is added to moderate the temperature. The resulting mixture of product gases forms the basis of the synthesis gas. The carbon dioxide is an unwanted by-product of the reaction and must be removed from the product stream, usually by a physical or chemiphysical absorption process. A viable process alternative is recycling a portion of the separated carbon dioxide stream back to the reactor. This has the effect of pushing the equilibrium of the water-gas shift reaction to the left, thus favoring the production of carbon monoxide.

Is Recycling of an Unwanted Product or an Inert Warranted for the Control of Reactor Operation? As we have mentioned previously, for highly exothermic reactions such as the partial oxidation of organic molecules, it is sometimes necessary to add an inert material to the reactor feed to moderate the temperature rise in the reactor and/or to move the reacting components outside of the explosive (flammability) limits. The most often used material for this purpose is steam, but any inert material that is available may be considered. For example, in the coal gasification example given earlier, steam is used to moderate the temperature rise in the reactor. For the case of recycling carbon dioxide to affect the water-gas shift reaction, there is another potential benefit. The recycling of carbon dioxide reduces the amount of steam needed in the feed to the reactor, because the carbon dioxide can absorb heat and reduce the temperature rise in the reactor.

What Phase is the Recycle Stream? The phase of the stream to be recycled plays an important role in determining the separation and recycle structure of the process. For liquids, there are concerns about azeotropes that complicate the separations scheme. For gases, there are concerns about whether high pressures and/or low temperatures must be used to enable the desired separation to take place. In either case gas compression is required, and, generally, this is an expensive operation. For example, the use of membrane separators or pressure-swing adsorption requires that the gas be fed at an elevated pressure to these units. If separation of a gas (vapor) is to be achieved using distillation, then a portion of the gas must be condensed, which usually requires cooling the gas significantly below ambient temperatures. This cooling process generally requires the use of compressors in the refrigeration cycle; the lower the desired temperature, the more expensive the refrigeration. Some typical refrigerants and their range of temperature are given in Table 2.2. Because separations of gases require expensive, low-temperature refrigeration, they are avoided unless absolutely necessary.

Only refrigerants with critical temperatures above the typical cooling water condenser temperature of 45°C can be used in single-stage, noncascaded refrigeration systems. Therefore, such systems are usually limited to the range of –45 to –60°C (for example, propylene, propane, and methyl chloride). For lower temper-

Table 2.2 Common Refrigerants and Their Ranges of Cooling (Data from References [12] and [13])

Refrigerant	Typical Operating Temperature Range (°C)	Vapor Pressure at 45°C (bar)	Critical Pressure and Temperature (bar)	and (°C)
Methane	–129 to –184	749	46.0	–82.5
Ethane	–59 to –115	1453	48.8	32.3
Ethylene	–59 to –115	2164	50.3	9.3
Propane	4 to –46	15.3	42.5	96.7
Propylene	4 to –46	18.45	46.1	91.6
N-Butane	16 to –12	4.35	38.0	152.0
Ammonia	27 to –32	17.8	112.8	132.5
Carbon Dioxide	4 to –50	787	73.8	31.1
Methylene Chloride	4 to –12	1.21	60.8	236.9
Methyl Chloride	4 to –62	9.84	66.8	143.1
R-134a (1,1,1,2-tetrafluoro-ethane)	4 to –50	11.6	40.6	101.0
R-152a (1,1-difluoro-ethane)	4 to –50	10.4	45.0	113.5

atures, refrigeration systems with two different refrigerants are required, with the lower-temperature refrigerant rejecting heat to the higher-temperature refrigerant, which in turn rejects heat to the cooling water. Costs of refrigeration are given in Chapter 8, and these costs increase drastically as the temperature decreases. For this reason, separations of gases requiring very low temperatures are avoided unless absolutely necessary.

As a review of the concepts covered in this chapter, Example 2.6 is presented to illustrate the approach to formulating a preliminary process flow diagram.

Example 2.6

Illustrative Example Showing the Input/Output and Recycle Structure Decisions Leading to the Generation of Flowsheet Alternatives for a Process

Consider the conversion of a mixed feed stream of methanol (88 mol%), ethanol (11 mol%), and water (1 mol%) via the following dehydration reactions:

$$2CH_3OH \rightleftharpoons (CH_3)_2O + H_2O \qquad \Delta H_{reac} = -11{,}770\ kJ/kmol$$

methanol dimethyl ether (desired product)

$$2C_2H_5OH \rightleftharpoons (C_2H_5)_2O + H_2O \qquad \Delta H_{reac} = -11{,}670\ kJ/kmol$$

ethanol diethyl ether (valuable by-product)

$$C_2H_5OH \rightarrow C_2H_4 + H_2O \qquad \Delta H_{reac} = -1{,}570\ kJ/kmol$$

ethanol ethylene (less valuable by-product)

The reactions take place in the gas phase, over an alumina catalyst [14, 15], and are mildly exothermic but do not require additional diluents to control reaction temperature. The stream leaving the reactor (reactor effluent) contains the following components, listed in order of decreasing volatility (increasing boiling point):

1. Ethylene (C_2H_4)
2. Dimethyl Ether (DME)
3. Diethyl Ether (DEE)
4. Methanol (MeOH)
5. Ethanol (EtOH)
6. Water (H_2O)

Moreover, because these are all polar compounds, with varying degrees of hydrogen bonding, it is not surprising that these compounds are highly non-ideal and form a variety of azeotropes with each other. These azeotropes are as follows:

- DME – H_2O (but no azeotrope with significant presence of alcohol)
- DME – EtOH
- DEE – EtOH
- DEE – H_2O
- EtOH – H_2O

For this problem, we assume that the mixed alcohol stream is available at a relatively low price from a local source ($0.25/kg). However, pure methanol ($0.22/kg) and/or ethanol ($0.60/kg) streams may be purchased if necessary. The selling price for DME, DEE, and ethylene are $0.95/kg, $1.27/kg, and $0.57/kg, respectively. Preliminary market surveys indicate that we can sell up to 15,000 tonne/y of DEE and up to 10,000 tonne/y of ethylene.

For a proposed process to produce 50,000 tonnes/y of DME, determine the viable process alternatives.

Step 1: Batch versus Continuous

For a plant of this magnitude, a continuous process would probably be chosen. However, we will return to this issue after considering some process alternatives and see that a hybrid batch/continuous process should also be considered.

Step 2: Define the Input/Output Structure of the Process

The basic input/output diagram of the process is shown in the process concept diagram of Figure E2.6(a).

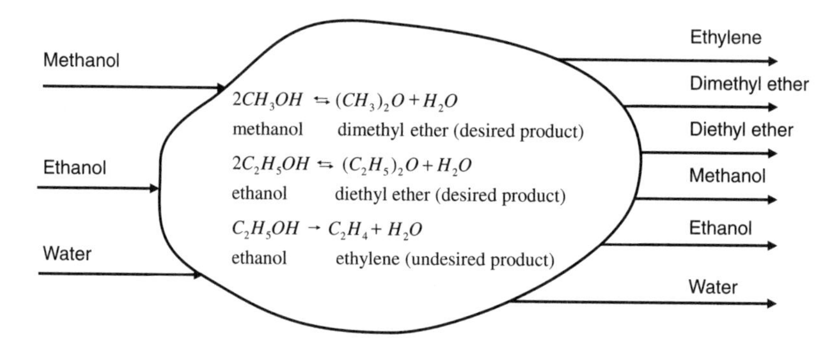

Figure E2.6(a) Process Concept Diagram for the Mixed Ethers Process of Example 2.6

First, consider a material balance for the process and estimate the profit margin:

$$\text{Desired DME production} = 50{,}000{,}000 \text{ kg/y} = \frac{50 \times 10^6}{46} = 1.087 \times 10^6 \text{ kmol/y}$$

$$\text{Required MeOH feed} = (2)(1.087 \times 10^6) = 2.174 \times 10^6 \text{ kmol/y}$$

$$\text{EtOH feed entering with methanol} = \frac{2.174 \times 10^6}{88}(11) = 0.2718 \times 10^6 \text{ kmol/y}$$

$$\text{Maximum DEE production} = \frac{0.2718 \times 10^6}{2} = 0.1309 \times 10^6 \text{ kmol/y or } 9.69 \times 10^3 \text{ tonne/y}$$

$$\text{Maximum ethylene production} = 0.2718 \times 10^6 \text{ kmol/y or } 7.61 \times 10^3 \text{ tonne/y}$$

$$\text{Cost of Feed} = \left((2.174 \times 10^6)(30) + (0.2718 \times 10^6)(46) + \frac{(2.174 \times 10^6)}{88}(18)\right)$$

$$(0.25) = \$19.54 \times 10^6$$

Value of DME = $(50 \times 10^6)(0.95) = \$47.5 \times 10^6/y$

Value of DEE (maximum production) = $(0.1309 \times 10^6)(74)(1.27) = \$12.30 \times 10^6/y$

Value of ethylene (maximum production) = $(0.2718 \times 10^6)(28)(0.57) = \$4.34 \times 10^6/y$

Margin will vary between (47.5 + 12.3 – 19.54) = \$40.26 million and (47.5 + 4.34 – 19.54) = \$32.30 million per year.

Important Points

From this margin analysis, it is clear that the amount of DEE produced should be optimized, because making ethylene is far less profitable. In addition, the maximum amount of DEE that the market can support is not currently being produced. Therefore, supplementing the feed with ethanol should be considered.

Because the main feed stream contains both reactants and an impurity (water), separation or purification of the feed prior to processing should be considered.

In order to minimize the production of by-products (ethylene), the selectivity of the DEE reaction should be optimized.

Alternative 1

In this option, shown in Figure E2.6(b), we do not separate the mixed alcohol feed, but we supplement the feed with ethanol. We use one reactor for both reactions. The disadvantages

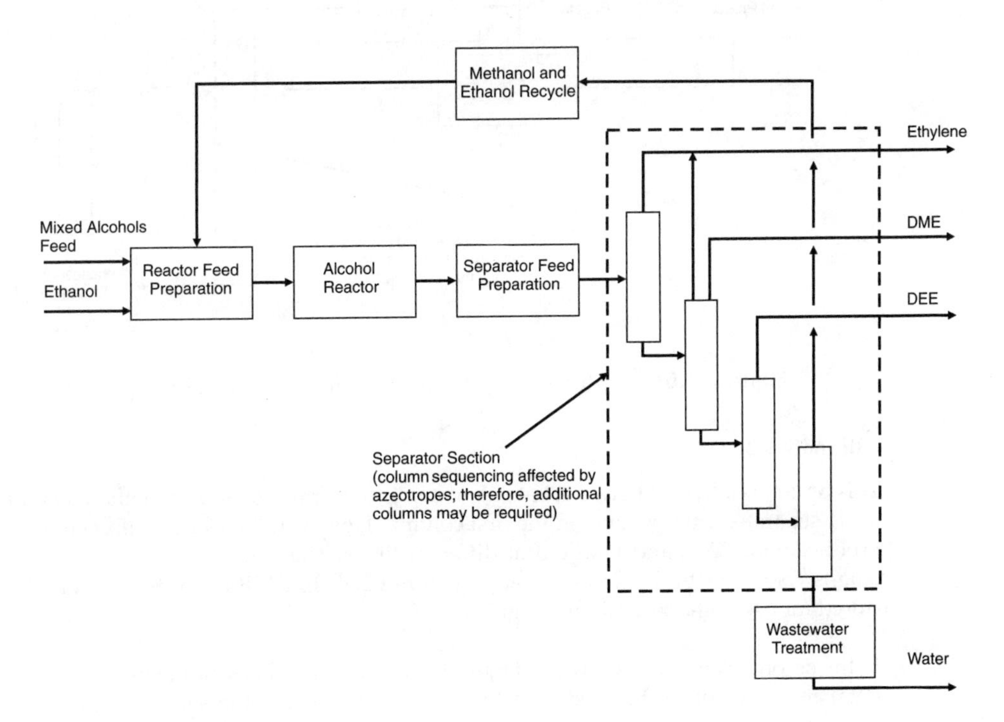

Figure E2.6(b) Structure of Process for Alternative 1 in Example 2.6

of this case are that the separations are complicated and we cannot optimize the reactor for both DME and DEE production.

Alternative 2

In this option, shown in Figure E2.6(c), feed is supplemented with ethanol and is separated into separate methanol and ethanol streams. Two reaction trains are used: one for DME and the other for DEE production. This allows the production of DME and DEE to be optimized separately and eliminates problems associated with the DME-ethanol azeotrope. However, there are two reactors and at least one more separation (column).

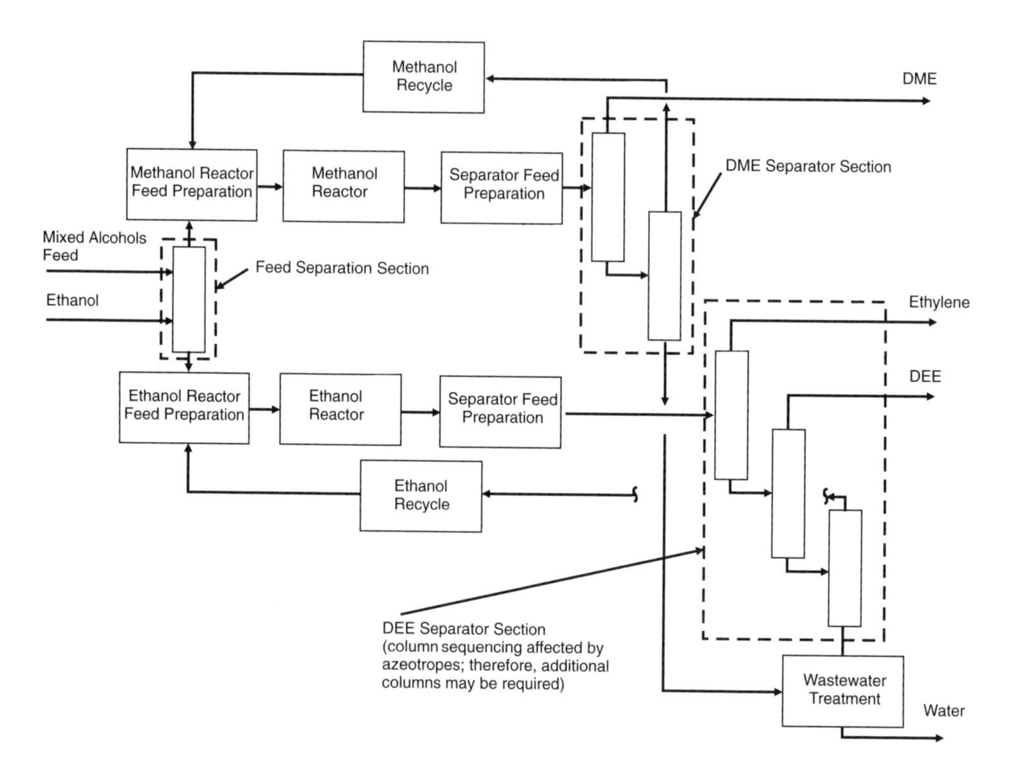

Figure E2.6(c) Structure of Process for Alternative 2 in Example 2.6

Alternative 3

This option is a hybrid between batch and continuous processes. The methanol is continuously separated from ethanol in the first column. However, the same equipment is used to produce both DME and DEE but at different times. The equipment is run in two "campaigns" per year. In the first campaign (Figure E2.6[d]), DME is produced and ethanol is stored for use in the second campaign.

In the second campaign, shown in Figure E2.6(e), methanol is sent to storage, and ethanol is taken from storage to produce DEE and ethylene using the same equipment as was

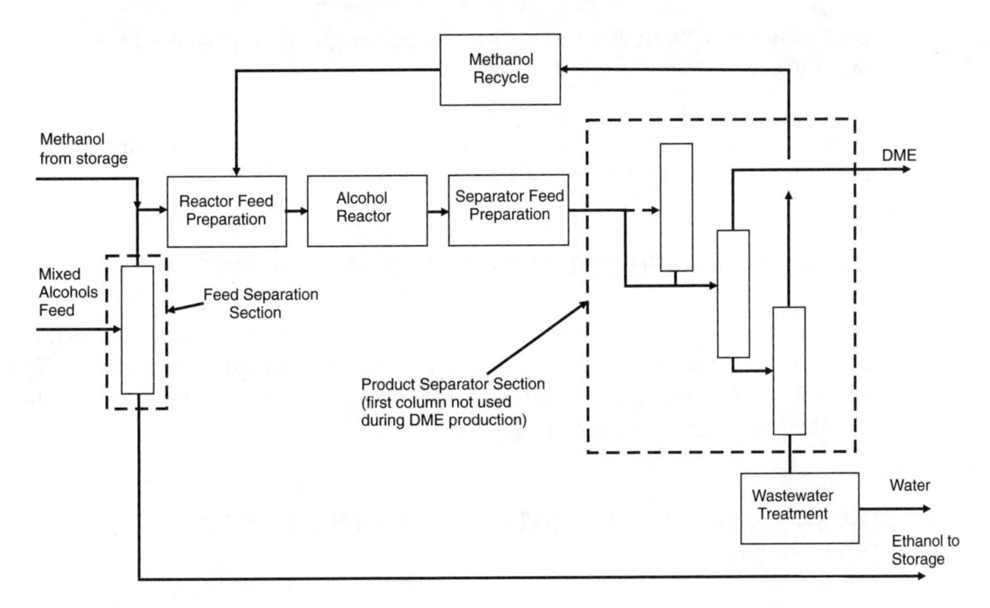

Figure E2.6(d) Structure of Process for Alternative 3—DME Campaign in Example 2.6

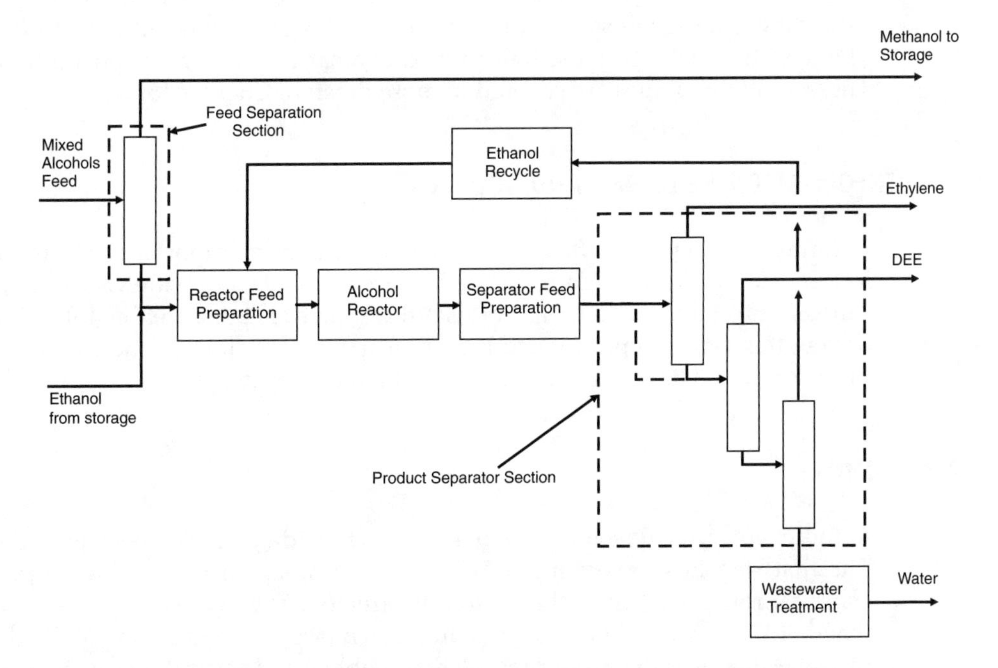

Figure E2.6(e) Structure of Process for Alternative 3—DEE Campaign in Example 2.6

used to produce DME. For this part of the campaign, the first column is used to remove the ethylene.

For this option, there is significantly less equipment to buy. However, the design and optimization of the process are more complicated because the equipment must be designed to perform two separate and quite different functions.

2.5 STEP 4—GENERAL STRUCTURE OF THE SEPARATION SYSTEM

As pointed out previously, the structure of the separation sequence is covered in detail in Chapter 12. In that chapter, considerable emphasis is placed on the sequencing of distillation columns, and some of the problems associated with azeotropic systems are covered.

2.6 STEP 5—HEAT-EXCHANGER NETWORK OR PROCESS ENERGY RECOVERY SYSTEM

The main objective of process energy recovery is to optimize the energy that a process exchanges with the utilities. At the expense of capital investment, the utility usage can be decreased by exchanging energy between process streams. The amount of energy integration is a function of the relative costs of the utilities. In addition, the process becomes more complex and more difficult to control. This loss in flexibility must be weighed against the savings in operating costs. These and other issues are covered in more detail in Chapter 15.

2.7 INFORMATION REQUIRED AND SOURCES

In formulating a process flow diagram, one of the most important tasks is the collection and synthesis of data. These data are available in a wide variety of publications. As a guide, a summary of useful resources is presented in Table 2.3. The data in this table are partitioned into information pertaining to new and existing processes and data on new and existing chemical pathways.

2.8 SUMMARY

In this chapter, the development of a process flow diagram has been investigated. The first step in synthesizing a PFD was to establish and examine all possible chemical routes that form the desired product(s). The next step was to establish whether the process should operate in a batch or continuous manner. Guidelines to make this decision were presented in Table 2.1. The next step was to establish the input/output structure of the process. The process concept diagram was

Table 2.3 Summary of Resources for Obtaining Information on Chemical Processes

Resource	Information Available
Existing Processes	
Shreve's Chemical Process Industries [16]	Gives a good review of basic processes to produce a wide variety of chemicals. Both organic and inorganic chemicals are covered.
Refinery Processes Handbook [17]	Published every other year in *Hydrocarbon Processing*. Gives basic block flow diagrams and operating cost and capital investment data for a wide range of refinery operations.
Gas Processes Handbook [18]	Published every other year in *Hydrocarbon Processing*. Gives basic block flow diagrams and operating cost and capital investment data for a wide range of gas processing operations.
Petrochemical Processes Handbook [19]	Published every other year in *Hydrocarbon Processing*. Gives basic block flow diagrams and operating cost and capital investment data for a wide range of gas petrochemical operations.
Kirk-Othmer Encyclopedia of Chemical Technology [20]	Comprehensive 25-volume encyclopedia has background information and PFDs for a wide variety of organic and inorganic chemical processes.
Encyclopedia of Chemical Processing and Design [21]	Comprehensive 20-volume encyclopedia contains background information on a variety of chemical processes. Many solutions to previous AIChE student contest problems are published as case studies.
Reactions and Kinetics	
Chemical Reactor Design for Process Plants [22]	Vol. 2 has several excellent case studies for processes, including reaction kinetics and reactor designs.
Industrial and Engineering Chemistry Research	This journal is published monthly by the American Chemical Society and contains numerous research articles containing information about processes and reaction kinetics.
Journal of Catalysis—Academic Press *Applied Catalysis*—Elsevier *Catalysis Today*—Elsevier	These (and other) journals concentrate on research conducted into the field of heterogeneous catalysis. Kinetic expressions and activity data are given for many processes of industrial importance.

Table 2.3 Summary of Resources for Obtaining Information on Chemical Processes (*continued*)

Resource	Information Available
	Reactions and Kinetics
Patents	The patent literature contains a wealth of information about new processes. Typically, single-pass conversions and catalyst activities are given. However, reaction kinetics are generally not provided and may not be derived easily from patent data. An excellent on-line patent search engine is at http://www.delphion.com.
SRI Reports	Excellent source of background information on all aspects of processes. Unfortunately, this information is available only to industrial clients of this service.

introduced that only required the identification of the raw materials, products, and stoichiometry of all the reactions that take place. At the process level, it was shown that all processes possess the same basic structure given in the generic block flow diagram.

The recycle structure of the PFD was introduced, and the three basic methods of recycle were discussed. Reasons and examples were provided to illustrate why inert material or products are sometimes recycled with unreacted raw materials. Difficulties in separating streams of products and reactants were given, and these were shown to influence the recycle structure and type of separation used.

The separation of products and unreacted raw materials and the integration of energy were covered briefly and are covered in greater depth in Chapters 12 and 15, respectively. An example showing how process alternatives are generated using the methods outlined in this chapter was provided, and several process alternatives were illustrated for this example using generic block flow diagrams. Finally, a list of resources was presented to help guide the reader to obtain basic data on chemical reactions and processes.

REFERENCES

1. Douglas, J. M., *Conceptual Design of Chemical Processes* (New York: McGraw-Hill, 1989).
2. Douglas, J. M., "A Hierarchical Design Procedure for Process Synthesis," *AIChE Journal*, 31 (1985): 353.

3. "Batch Plants Adapt to CPI's Flexible Gameplans, Newsfront," *Chemical Engineering* 95 (February 1988): 31–35.
4. Fruci, L., "Pipeless Plants Boost Batch Processing," *Chemical Engineering* 100 (June 1993): 102–110.
5. Lowenstein, J. G., "The Pilot Plant," *Chemical Engineering* 92 (December 1985): 62–79.
6. Crowl, D., and J. Louvar, *Chemical Process Safety* (Upper Saddle River, NJ: Prentice-Hall, 1990).
7. Rase, H. F., *Chemical Reactor Design for Process Plants,* Vol. 1: *Principles and Techniques* (New York: John Wiley and Sons, 1977).
8. Oertel, G., *Polyurethane Handbook,* 2nd ed. (Munich: Hanser Publ., 1993).
9. Peters, M. S., and K. D. Timmerhaus, *Plant Design and Economics for Chemical Engineers*, 4th ed. (New York: McGraw-Hill, 1991.)
10. *Technical Information for Houdry® Hydrodealkylation Detol®, Litol®, and Pyrotol® for High Purity Benzene*, ABB Lummus Global, BG503-0026, May 1997.
11. Walas, S. M., *Phase Equilibria in Chemical Engineering* (Stoneham, MA: Butterworth, 1985).
12. Mehra, Y. R., "Refrigeration Systems for Low-Temperature Processes," *Chemical Engineering,* July 12, 1982, p. 95.
13. Perry, R. H., D. W. Green, and J. O. Maloney, *Chemical Engineers' Handbook,* 7th ed. (New York: McGraw-Hill, 1991).
14. Butt, J. B., H. Bliss, and C. A. Walker, "Rates of Reaction in a Recycling System—Dehydration of Ethanol and Diethyl Ether over Alumina," *AIChE Journal* 8, no. 1 (1962): 42–47.
15. Berčič, G., and J. Lavec, "Intrinsic and Global Reaction Rates of Methanol Dehydration over γ-Al_2O_3 Pellets," *Industrial and Engineering Chemistry Research* 31 (1992): 1035–1040.
16. Austin, G. T., *Shreve's Chemical Process Industries*, 5th ed. (New York: McGraw-Hill, 1984).
17. Refinery Processes Handbook '00, in *Hydrocarbon Processing* (Houston, TX: Gulf Publishing Co., 2000).
18. Gas Processes Handbook '00, in *Hydrocarbon Processing* (Houston, TX: Gulf Publishing Co., 2000).
19. Petrochemical Processes Handbook '01, in *Hydrocarbon Processing* (Houston, TX: Gulf Publishing Co., 2001).
20. *Kirk-Othmer Encyclopedia of Chemical Technology,* 4th ed. (New York: John Wiley and Sons, 1991–1998).
21. McKetta, J. J., and W. A. Cunningham, *Encyclopedia of Chemical Processing and Design* (New York: Marcel Dekker, 1976).
22. Rase, H. F., *Chemical Reactor Design for Process Plants*, vol. 2. (New York: John Wiley and Sons, 1977).

SHORT ANSWER QUESTIONS

1. What are the five elements of the hierarchy of process design?
2. What are the three types of recycle structures possible in a chemical process? Explain when each is used.
3. Give three criteria for choosing a batch process as opposed to a continuous process.
4. When would one purposely add an inert material to a feed stream? Illustrate this strategy with an example, and explain the advantages (and disadvantages) of doing this.
5. In general, when would one purify a material prior to feeding it to a process unit? Give at least one example for each case you state.

PROBLEMS

6. In modern integrated gasification combined cycle (IGCC) coal-fed power plants, oxygen is produced via cryogenic separation of air and is fed to the IGCC plant along with coal. The separation of oxygen from air is expensive; what reason(s) can you give for doing this?
7. The production of ethylbenzene is described in Appendix B, project B.2. From the PFD (Figure B.2.1) and accompanying stream table (Table B.2.1), determine the following:
   a. The single-pass conversion of benzene
   b. The single-pass conversion of ethylene
   c. Overall conversion of benzene
   d. Overall conversion of ethylene

   Suggest two strategies to increase the overall conversion of ethylene and discuss their merits.
8. Consider the following statement: "If a reactant (G) in a process is a gas at the feed conditions, subsequent separation from the reactor effluent is difficult, hence unused G cannot be recycled." Do you agree with this statement? Give your reasoning why you agree or disagree.
9. Most pharmaceutical products are manufactured using batch processes. Give at least three reasons why this is so.
10. The formation of styrene via the dehydrogenation of ethylbenzene is a highly endothermic reaction. In addition, ethylbenzene may decompose to benzene and toluene and also may react with hydrogen to form toluene and methane:

$$\underset{ethylbenzene}{C_6H_5C_2H_5} \overset{1}{\underset{2}{\rightleftharpoons}} \underset{styrene}{C_6H_5C_2H_3} + \underset{hydrogen}{H_2} \tag{B.2.1}$$

$$\underset{ethylbenzene}{C_6H_5C_2H_5} \xrightarrow{3} \underset{benzene}{C_6H_6} + \underset{ethylene}{C_2H_4} \quad (B.2.2)$$

$$\underset{ethylbenzene}{C_6H_5C_2H_5} + \underset{hydrogen}{H_2} \xrightarrow{4} \underset{toluene}{C_6H_6CH_3} + \underset{methane}{CH_4} \quad (B.2.3)$$

This process is presented in Appendix B as project B.3. From the information given in Appendix B, determine the following:

a. the single-pass conversion of ethylbenzene.
b. the overall conversion of ethylbenzene.
c. the yield of styrene.

Suggest one strategy to increase the yield of styrene, and sketch any changes to the PFD that this strategy would require.

11. There are two technically viable routes to the production of a hydrocarbon solvent, *S*, starting with feed material *A*. Route 1 uses a disproportionation reaction, in which feed material *A* is converted to the desired solvent *S* and another solvent *R*, both of which are marketable products. Route 2 starts with the same chemical *A*, but uses a hydrodealkylation reaction to produce the desired solvent. The reaction schemes for each process are shown below.

Route 1 $2A \rightarrow S + R$
Route 2 $A + H_2 \rightarrow S + CH_4$

Assuming that pure *A* is fed to the process, the solvents *S* and *R* are separable by simple distillation, and both are much less volatile than either methane or hydrogen, sketch PFDs for Routes 1 and 2. Which process do you think will be more profitable? Explain your reasoning and assumptions.

12. When considering the evolution of a process flowsheet, it was noted that there are three forms of recycle structure for unused reactants, given as a–c below. For each case, carefully explain under what conditions you would consider or implement each strategy.
a. Separate, purify, and recycle.
b. Recycle without separation and use a purge.
c. Recycle without separation and do not use a purge.

13. Acetaldehyde is a colorless liquid with a pungent, fruity odor. It is primarily used as a chemical intermediate, principally for the production of acetic acid, pyridine and pyridine bases, peracetic acid, pentaeythritol, butylene glycol, and chloral. Acetaldehyde is a volatile and flammable liquid that is miscible in water, alcohol, ether, benzene, gasoline, and other common organic solvents. In this problem, the synthesis of acetaldehyde via the dehydrogenation of ethanol is to be considered. The following reactions occur during the dehydrogenation of ethanol:

$$CH_3CH_2OH \rightarrow \underset{acetaldehyde}{CH_3CHO} + H_2 \quad (1)$$

$$2CH_3CH_2OH \rightarrow \underset{ethyl\ acetate}{CH_3COOC_2H_5} + 2H_2 \quad (2)$$

$$2CH_3CH_2OH \rightarrow \underset{butanol}{CH_3(CH_2)_3OH} + H_2O \tag{3}$$

$$CH_3CH_2OH + H_2O \rightarrow \underset{acetic\ acid}{CH_3COOH} + 2H_2 \tag{4}$$

The conversion of ethanol is typically 60%. The yields for each reaction are approximately:
(1) acetaldehyde 92%
(2) ethyl acetate 4%
(3) butanol 2%
(4) acetic acid 2%

a. For this process, generate a process concept diagram showing all the input and output chemicals.
b. Develop two alternative preliminary process flow diagrams for this process.

14. Consider the following process in which liquid feed material *A* (normal BP of 110°C) is reacted with gaseous feed material *G* to produce main product *C* and by-products *R* and *S* via the following reactions:

$$A + G \rightarrow C + S$$

$$G + C \rightarrow R$$

Both feeds enter the process at ambient temperature and pressure. Both reactions occur in the gas phase at moderate temperature and pressure (250°C and 10 bar). The normal boiling points of *G*, *S*, and *C* are less than –120°C. By-product *R* has a normal boiling point of 75°C and is highly soluble in water. Product *C* is very soluble in water but *G* and *S* are insoluble. The single-pass conversion through the reactor is low for feed *A*, and the ratio of *G* to *A* in the feed to the reactor should be maintained in excess of 4 to minimize the chance of other unwanted side reactions. Using this information, and assuming that both *A* and *G* are expensive, do the following.

a. Draw a preliminary process flow diagram identifying the main unit operations (reactors, compressors, pumps, heat exchangers, and separators), and identify the recycle structure of the process.
b. Justify the methods used to recycle *A* and *G*.
c. What unit operations do you suggest for your separators? Justify your choices.
d. How would your PFD change if the price of feed material *G* were very low?

15. How is Scotch whisky made?

The following descriptions of malt and grain whisky manufacturing are given here courtesy of the University of Edinburgh at http://www.dcs.ed.ac.uk/home/jhb/whisky/swa/chap3.html. For each of the two processes, sketch a process flow diagram.

There are two kinds of Scotch whisky: malt whisky, which is made by the pot still process, and grain whisky, which is made by the patent still (or Coffey still) process. Malt whisky is made from malted barley only, whereas grain whisky is made from malted barley together with unmalted barley and other cereals.

Malt Whisky

The pot still process by which malt whisky is made may be divided into four main stages: malting, mashing, fermentation, and distillation.

(1) Malting

The barley is first screened to remove any foreign matter and then soaked for two or three days in tanks of water known as steeps. After this it is spread out on a concrete floor known as the malting floor and allowed to germinate. Germination may take from 8 to 12 days depending on the season of the year, the quality of the barley used, and other factors. During germination the barley secretes the enzyme diastase, which makes the starch in the barley soluble, thus preparing it for conversion into sugar. Throughout this period the barley must be turned at regular intervals to control the temperature and rate of germination.

At the appropriate moment germination is stopped by drying the malted barley or green malt in the malt kiln. More usually nowadays malting is carried out in Saladin boxes or in drum maltings, in both of which the process is controlled mechanically. Instead of germinating on the distillery floor, the grain is contained in large rectangular boxes (Saladin) or in large cylindrical drums. Temperature is controlled by blowing air at selected temperatures upward through the germinating grain, which is turned mechanically. A recent development caused by the rapid expansion of the Scotch whisky industry is for distilleries to obtain their malt from centralized maltings that supply a number of distilleries, thereby enabling the malting process to be carried out more economically.

(2) Mashing

The dried malt is ground in a mill, and the grist, as it is now called, is mixed with hot water in a large circular vessel called a mash tun. The soluble starch is thus converted into a sugary liquid known as wort. This is drawn off from the mash tun, and the solids remaining are removed for use as cattle food.

(3) Fermentation

After cooling, the wort is passed into large vessels holding anything from 9,000 to 45,000 liters of liquid, where it is fermented by the addition of yeast. The living yeast attacks the sugar in the wort and converts it into crude alcohol. Fermentation takes about 48 hours and produces a liquid known as wash, containing alcohol of low strength, some unfermentable matter, and certain by-products of fermentation.

(4) Distillation

Malt whisky is distilled twice in large copper pot stills. The liquid wash is heated to a point at which the alcohol becomes vapor. This rises up the still and is passed into the cooling plant, where it is condensed into liquid state. The cooling plant may take the form of a coiled copper tube or worm that is kept in continuously running cold water, or it may be another type of condenser.

The first distillation separates the alcohol from the fermented liquid and eliminates the residue of the yeast and unfermentable matter. This distillate, known as low wines, is then passed into another still, where it is distilled a second time. The first runnings from this second distillation are not considered potable, and it is only when the spirit reaches an acceptable standard that it is collected in the spirit receiver. Again, toward the end of the distillation, the spirit begins to fall off in strength and quality. It is then no longer collected as spirit but drawn off and kept, together with the first running, for redistillation with the next low wines.

Pot Still distillation is a batch process.

Grain Whisky

The patent still process by which grain whisky is made is continuous in operation and differs from the pot still process in four other ways.

a. The mash consists of a proportion of malted barley together with unmalted cereals.
b. Any unmalted cereals used are cooked under steam pressure in converters for about 3½ hours. During this time the mixture of grain and water is agitated by stirrers inside the cooker.
c. The starch cells in the grain burst, and when this liquid is transferred to the mash tun, with the malted barley, the diastase in the latter converts the starch into sugar.
d. The wort is collected at a specific gravity lower than in the case of the pot still process.
e. Distillation is carried out in a patent or Coffey still, and the spirit collected at a much higher strength.

Storage and aging of the whisky are also an important part of the overall process but need not be considered for this problem. Storage occurs in oak barrels that previously stored either sherry or bourbon (or both, in the case of double-aged whisky). The length of storage in the barrel determines the vintage of the whisky. Unlike wine, the time after bottling does not count, and so a 15-year-old scotch that was bought in 1960 is today still a 15-year-old scotch.

CHAPTER

3

Batch Processing

Some key reasons for choosing to manufacture a product using a batch process were discussed in Chapter 2. These include small production volume, seasonal variations in product demand, a need to document the production history of each batch, and so on. When designing a batch plant, there are many other factors an engineer must consider. The types of design calculations are very different for batch compared with continuous processes. Batch calculations involve transient balances, which are different from the steady-state design calculations taught in much of the traditional chemical engineering curriculum. Batch **sequencing**—the order and timing of the processing steps—is probably the most important factor to be considered. Determining the optimal batch sequence depends on a variety of factors. For example, will there be more than one product made using the same equipment? What is the optimal size of the equipment? How long must the equipment run to make each different product? What is the trade-off between economics and operability of the plant? In this chapter, these questions will be addressed, and an introduction to other problems that arise when considering the design and operation of batch processes will be provided.

3.1 DESIGN CALCULATIONS FOR BATCH PROCESSES

Design calculations for batch processes are different from the steady-state design calculations taught in most unit operations classes. The batch nature of the process makes all design calculations unsteady state. This is best demonstrated by example; Example 3.1 illustrates the types of design calculations required for batch processing.

Example 3.1

In the production of an API (active pharmaceutical ingredient), the following batch recipe is used.

Step 1: 500 kg of reactant A (MW = 100 kg/kmol) is added to 5000 kg of a mixture of organic solvent (MW = 200 kg/kmol) containing 60% excess of a second reactant B (MW = 125 kg/kmol) in a jacketed reaction vessel (R-301), the reactor is sealed, and the mixture is stirred and heated (using steam in the jacket) until the temperature has risen to 95°C. The density of the reacting mixture is 875 kg/m^3 (time taken = 1.5 h).

Step 2: Once the reaction mixture has reached 95°C, a solid catalyst is added, and reaction takes place while the batch of reactants is stirred. The required conversion is 94% (time taken = 2.0 h).

Step 3: The reaction mixture is drained from the reactor and passed through a filter screen (Sc-301) that removes the catalyst and stops any further reaction (time taken = 0.5 h).

Step 4: The reaction mixture (containing API, solvent, and unused reactants) is transferred to a distillation column, T-301, where it is distilled under vacuum. Virtually all of the unused reactants and approximately 50% of the solvent are removed as overhead product (time taken = 3.5 h). The end-point for the distillation is when the solution remaining in the still contains less than 1 mol% of reactant B. This ensures that the crystallized API, produced in Step 5, meets specification.

Step 5: The material remaining in the still is pumped to a crystallizer, CR-301, where the mixture is cooled under vacuum and approximately 60% of the API from Step 2 crystallizes out (time taken = 2.0 h).

Step 6: The API is filtered from the crystallizer and placed in a tray dryer, TD-301, where any entrapped solvent is removed (time taken = 4 h).

Step 7: The dried API is sealed and packaged in a packing machine, PK-301, and sent to a warehouse for shipment to the customer (time taken = 1.0 h).

Perform a preliminary design on the required equipment items for this batch process.

Solution

The equipment items will be designed in sequence.

Step 1: Reaction Vessel—Preheat

The reaction vessel, which is used to preheat the reactants and subsequently run the reaction, is designed first. For the batch size specified, the volume of the liquid in the tank, V, and the volume required for the reaction vessel, V_{tank}, are given by Equations (E3.1a) and (E3.1b), in which it is assumed that the vessel is approximately 60% full during operation.

$$V = \frac{5500[kg]}{875[kg/m^3]} = 6.286m^3 \tag{E3.1a}$$

$$V_{required} = \frac{5500[kg]}{875[kg/m^3]}\frac{1}{0.6} = 10.48m^3 = 2768\,gal \tag{E3.1b}$$

Because reactors of this sort come in standard sizes, a 3000-gallon (V_{tank}) reactor is selected.

The heat-transfer characteristics of this vessel are then checked. For a jacketed vessel, the unsteady-state design equation is

$$\rho V C_p \frac{dT}{dt} = UA(T_s - T) \quad \text{(E3.1c)}$$

where ρ is the liquid density, C_p is the liquid heat capacity, T is the temperature of the liquid in the tank (95°C is desired value in 1.5 h), U is the overall heat transfer coefficient from the jacket to the liquid in the tank, A is the heat transfer area of the jacket (cylinder surface), and T_s is the temperature of the condensing steam. (Normally, there is also a jacketed bottom to such a vessel, but this added heat-transfer area is ignored in this example for simplification.) Integration of this equation yields

$$ln \frac{(T_s - T_{final})}{(T_s - T_o)} = -\frac{UA\Delta t}{\rho V C_p} \quad \text{(E3.1d)}$$

where T_o is the initial temperature in the tank (assumed to be 25°C). The following "typical" values are assumed for this design:

C_p = 2000 J/kg°C
T_s = 120°C (200 kPa Saturated Steam)
U = 300 W/m^2°C
Tank Height to Diameter Ratio = 3/1 (so $H = 3D$)

Assuming the tank to be cylindrical and ignoring the volume of the bottom elliptical head, the tank volume is $V_{tank} = \pi D^2 H/4 = 3\pi D^3/4$. Thus, the tank diameter, D, is 1.689 m. The height of fill is $H_{fill} = 4V/(\pi D^2)$ = 2.806 m. The area for heat transfer is $A = \pi D H_{fill}$ = 14.89 m^2, because we assume negligible heat transfer to the vapor space. When these values are used in Equation (E3.1d), it is found that the time required for preheating the reactor, Δt, is 3288 s (55 min). Thus, the step time requirement of 1.5 h for this step is met. The additional time is required for filling, sealing, and inspecting the vessel prior to heating. It should be noted that there may be process issues that require a slower temperature ramp, which can be accomplished by controlling the steam pressure. Note also that it is assumed that the time requirement for cleaning the vessels in this example is included in the step times given in the problem statement.

Step 2: Reaction Vessel—Reaction

It is assumed that the reaction of one mole each of A and B to form one mole of the product is second order (first order in each reactant) and that the rate constant is 7.09×10^{-4} m^3/kmol s. The relationship for a batch reactor is

$$\frac{dC_A}{dt} = -kC_A C_B \quad \text{(E3.1e)}$$

where A and B are the two reactants, and A is the limiting reactant. The standard analysis for conversion in a reactor yields

$$C_A = C_{Ao}(1 - X) \quad \text{(E3.1f)}$$

$$C_B = C_{Ao}(\Theta - X) \tag{E3.1g}$$

$$\frac{dX}{dt} = kC_{Ao}(1-X)(\Theta - X) \tag{E3.1h}$$

where C_{Ao} = (500 kg/100 kg/kmol)(875 kg/m^3)/5500 kg = 0.796 kmol/m^3. Because reactant B is present in 60% excess, Θ = 1.6. The desired conversion, X_{final}, is 0.94. Integration of Equation (E3.1h) with an initial condition of zero conversion at time zero yields

$$\frac{1}{\Theta-1}\left[ln\frac{\Theta - X_{final}}{1-X_{final}}\right] = kC_{Ao}\Delta t \tag{E3.1i}$$

When all of the values are inserted into Equation (E3.1i), the time (Δt) is found to be 7082 s, or 118 min, which is just less than the desired two hours allotted for the reaction. For simplicity, the additional reaction time that occurs after the mixture leaves the reactor until the catalyst is removed from the reacting mixture has been ignored.

Step 3: Draining Reaction Vessel and Catalyst Filtration

This step will be modeled as a draining tank, which may significantly underestimate the actual required time for draining and filtering. In reality, experimental data on the filter medium and inclusion of the exit pipe frictional resistance would have to be included to determine the actual time for a specific tank. Generally, the filter is the bottleneck in such a step. Here, a 2-in schedule 40 exit pipe, with a cross-sectional area of 0.00216 m^2, is assumed.

For a draining tank, the model is

$$\frac{dm}{dt} = \frac{d(\rho A_t H)}{dt} = -\dot{m} = -\rho A_p v_p \tag{E3.1j}$$

where ρ is the density of the liquid in the tank, A_t is the cross-sectional area of the tank, H is the height of liquid in the tank, A_p is the cross-sectional area of the exit pipe, and v_p is the velocity of liquid in the exit pipe, which, from Bernoulli's equation (turbulent flow), is $(2gH)^{1/2}$, where g is the gravitational acceleration. Therefore, Equation (E3.1j) becomes

$$\frac{dH}{dt} = -\frac{(2g)^{1/2}A_p}{A_t}H^{1/2} \tag{E3.1k}$$

Integrating from H = 2.806 m at t = 0 to find the time when H = 0 yields

$$-2H^{0.5}\Big|_{2.806\,m}^{0} = \frac{2^{0.5}(9.81[\text{m/s}^2])^{0.5}(0.00216[\text{m}^2])}{\frac{\pi(1.689\,[\text{m}])^2}{4}}\Delta t \tag{E3.1l}$$

which gives Δt = 785 s = 13 min, which is rounded up to 30 minutes for this step. Note that this time can be further reduced by pressurizing the vessel with an inert gas.

Step 4: Distillation of Reaction Products

A material balance on the reactor at the end of Step 2 yields the following:

Component, i	kmoles	x_i	MW	mass (kg)
Reactant A	= (1 – 0.94)(5.0) = 0.3	0.0106	100	30.0
Reactant B	= (1.6)(5) – (5.0 – 0.3) = 3.3	0.1166	125	412.5
Solvent S	20.0	0.7067	200	4000.0
Product P	= (0.94)(5.0) = 4.7	0.1661	225	1057.5
Total	**28.3**	**1.0000**		**5500.0**

Initially, the reaction mixture is heated to its boiling point of 115°C at the operating pressure. This is done by condensing steam in a heat exchanger located in the still of the column. The time to heat the mixture from 95°C (the temperature leaving the reactor, assuming no heat loss in the filter) to 115°C is given by Equation (E3.1d) with the following variable values:

$T_s = 120°\text{C}$
$\rho = 875 \text{ kg/m}^3$
$C_p = 2000 \text{ J/kg°C}$
$U = 420 \text{ W/m}^2\text{°C}$
$A = 10 \text{ m}^2$

Solving for the unknown time, we get $t = 4215$ s = 70.3 min.

The distillation is performed using a still with three theoretical stages ($N = 3$), a boil-up rate, $V = 30$ kmol/h, and a reflux ratio, $R = 4.5$. The volatilities of each component relative to the product are given as follows:

$\alpha_{AP} = 3.375$
$\alpha_{BP} = 2.700$
$\alpha_{SP} = 1.350$
$\alpha_{PP} = 1.000$

The solution methodology involves a numerical integration using the method of Sundaram and Evans [1]. The overall material and component balances are given by

$$-\frac{dW}{dt} = D = \frac{V}{1 + R}$$

or in finite difference form,

$$W^{(k+1)} = W^{(k)} - \left(\frac{V}{1 + R}\right)\Delta t \qquad \text{(E3.1m)}$$

$$\frac{d(Wx_{W_i})}{dt} = x_{D_i}\frac{dW}{dt}$$

or in finite difference form,

$$x_{W_i}^{(k+1)} = x_{W_i}^{(k)} + (x_{D_i}^{(k)} - x_{W_i}^{(k)})\frac{W^{(k+1)} - W^{(k)}}{W^{(k)}} \qquad \text{(E3.1n)}$$

where W is the total moles in the still; x_{Di} and x_{Wi} are the mole fractions of component i, at any time t, in the overhead product and in the still, respectively; k is the index for time in the finite difference representation; and Δt is the time step. These equations are solved in conjunction with the sum of the gas phase mole fraction equaling unity and the Fenske-Underwood-Gilliland method for multicomponent distillation. This leads to the following additional equations:

$$x_{D_r} = \frac{x_{W_r}}{\sum_{i=1}^{C} x_{W_i} \alpha_{1,C}^{N_{min}}} \text{ and } x_{D_i} = x_{W_i}\left(\frac{x_{D_i}}{x_{w_i}}\right)\alpha_{i,r}^{N_{min}} \tag{E3.1o}$$

$$R_{min} = \frac{\alpha_{1,C}^{N_{min}} - \alpha_{1,c}}{(\alpha_{1,C}-1)\sum_{i=1}^{C} x_{w_i}\alpha_{1,C}^{N_{min}}} \text{ and } \frac{N-N_{min}}{N+1} = 0.75\left[1-\left(\frac{R-R_{min}}{R+1}\right)^{0.5668}\right] \tag{E3.1p}$$

where R_{min} and N_{min} are the minimum values for the reflux ratio and the number of theoretical stages, respectively. The solution of these equations is explained in detail by Seader and Henly [2], and the results for this example are shown in Figures E3.1(a) and E.3.1(b).

From Figures E3.1(a) and (b), the mole fraction of reactant B is seen to drop to less than the specification of 0.01 (1 mol%) at a time of approximately 2.3 h. This time, coupled with the heating time of 70.2 minutes, gives a total of 3.5 h. However, note that only about 75% of the product remains in the still to be recovered in the next step.

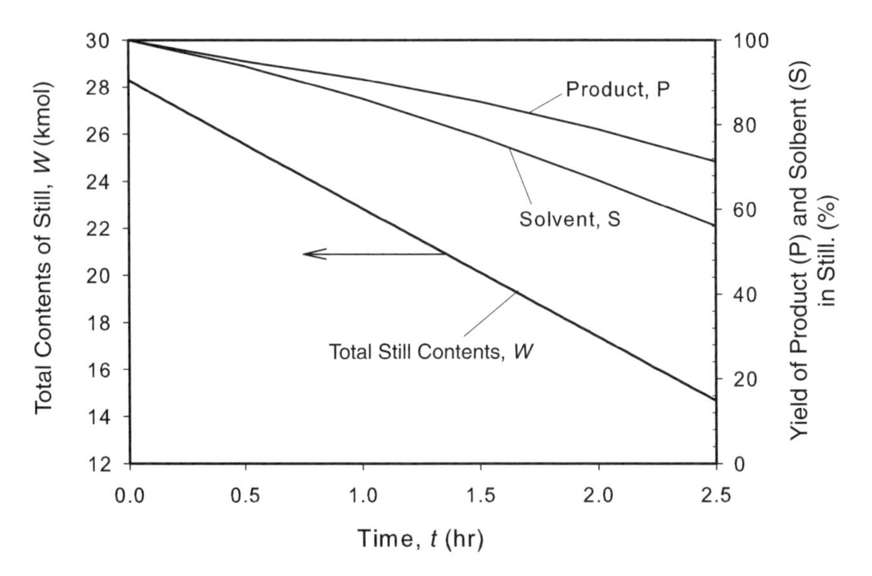

Figure E3.1(a) Change of Still Contents and Yields of P and S with Time

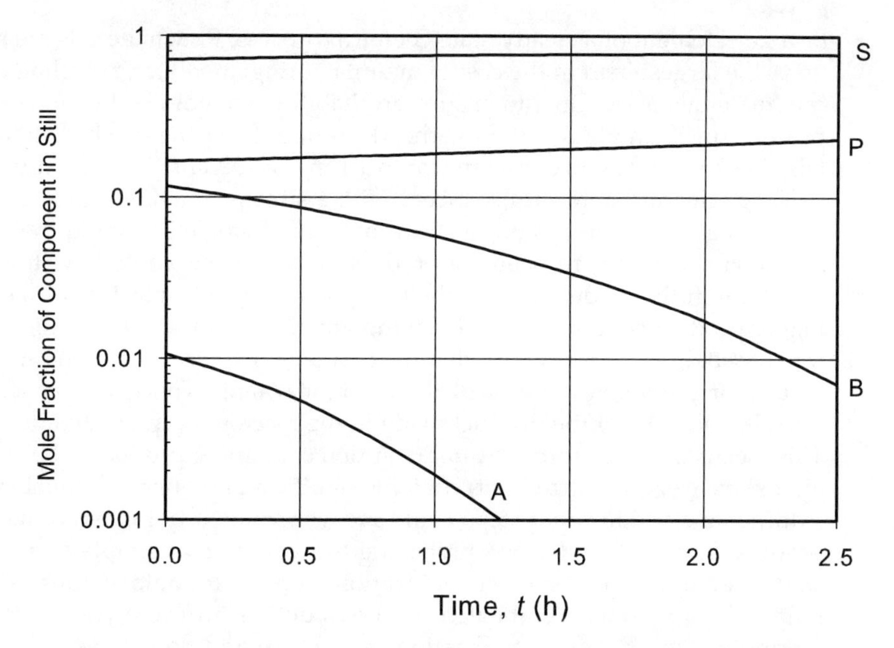

Figure E3.1(b) Change in Composition of Still Material with Time

Step 5: Cooling and Crystallization of Product

The analysis of the crystallization, filtration, drying, and packaging steps is beyond the scope of this analysis. Therefore, it is assumed that the times for each of these steps have been determined through laboratory-scale experiments, and those times are simply stated here. The amount of product crystallized is 80% of the product recovered from the still, or 60% of the 1057.5 kg produced in the reactor (634.5 kg.) The time required to cool and crystallize is 2 h.

Step 6: Filtration and Drying

The time required for filtration and drying is 4 h.

Step 7: Packaging

The time required for packaging is 1 h.

There are several unique features of batch operations observed in Example 3.1. First, the heating, reaction, and separations steps are unsteady state, which is different from the typical steady-state analysis with which most undergraduate chemical engineers are familiar. Secondly, it is observed that no provision was made to recycle the unreacted raw materials. In Chapter 2, recycle was shown to be a key element of a steady-state chemical process. Raw materials are almost always the largest item in the cost of manufacturing; therefore, recycling unreacted raw materials is essential to ensure profitability. So, how is this done in a batch process? In Example 3.1, the overhead product from the batch distillation contains unreacted raw material and product in the solvent. This could be sent to a holding tank and periodically mixed with a stream containing pure solvent and just enough reactants A and/or B to make up a single charge to the process in Step 1. However, the recycling of product to the reactor would have to be investigated carefully to determine whether unwanted side reactions take place at higher product concentrations. Even though an additional tank would need to be purchased, it is almost certain that the cost benefit of recycling the raw materials would far outweigh the cost of the additional tank. Third, it is observed that, overall, only 60% of the product made in the reactor is crystallized out in Step 5. This means that the **mother liquor** (solution containing product to be crystallized after some has crystallized out) contains significant amounts of valuable product. Additional crystallization steps could recover some, if not most, of the valuable product. The strategy for accomplishing this could be as simple as scheduling a second or third cooling or crystallization step, or it could involve storing the mother liquor from several batch processes until a sufficient volume is available for another cooling or crystallization step. These additional crystallization steps are tantamount to adding additional separation stages.

3.2 GANTT CHARTS AND SCHEDULING

Gantt charts (see, for example, Dewar [3]) are tabular representations used to illustrate a series of tasks (rows) that occur over a period of time (columns). These charts graphically represent completion dates, milestone achievements, current progress, and so on [3] and are discussed further in Chapter 26 as a planning tool for completing large design projects. In this chapter, a simplified Gantt chart is used to represent the scheduling of equipment needed to produce a given batch product. Example 3.2 illustrates the use of Gantt charts to show the movement of material as it passes through several pieces of equipment during a batch operation.

Example 3.2

Draw a Gantt chart that illustrates the sequence of events in the production of the API in Example 3.1.

Solution

Gantt charts for this process are shown in Figure E3.2. Note that in both charts, Steps 1 and 2 have been consolidated into one operation because they occur sequentially in the same piece of equipment. The top chart shows the row names as tasks, and the bottom figure

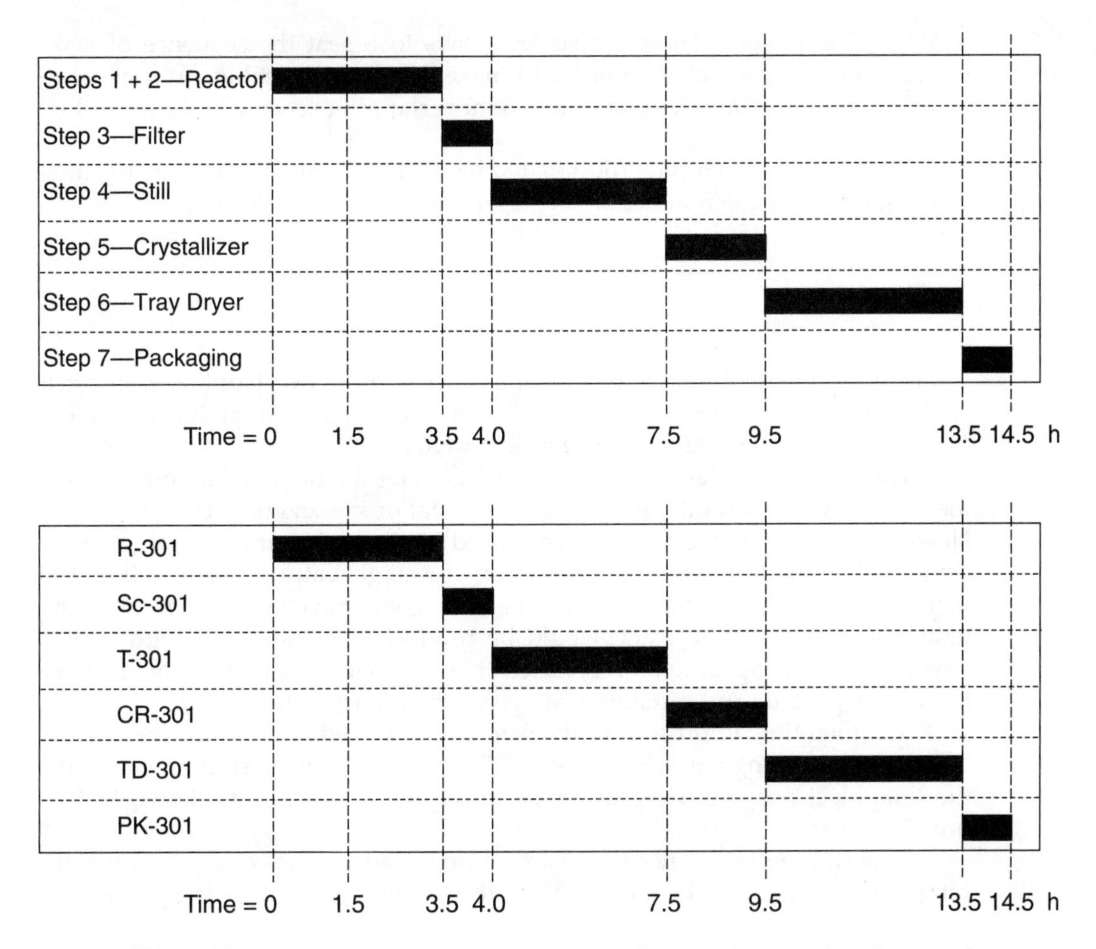

Figure E3.2 Gantt Chart Showing Sequence of Events for the Manufacture of API in Example 3.1

simply identifies each row with the equipment number. In general, the notation used in the bottom figure will be adopted.

3.3 NONOVERLAPPING OPERATIONS, OVERLAPPING OPERATIONS, AND CYCLE TIMES

In general, product is produced throughout an extended period of time by using a repeating sequence of operations. For example, the batch process described in Example 3.1 produces a certain amount of crystallized API, namely 634.5 kg. If it is desired to produce 5000 kg, then the sequence of steps must be repeated

5000/634.5 ≅ 8 times. There are several ways to repeat the sequence of tasks needed to make one batch, in order to make the desired total amount of product (5000 kg). An example of one such **nonoverlapping** scheme is shown in Figure 3.1.

For the nonoverlapping (designated by the subscript *NO*) scheme, the total processing time is the number of batches multiplied by the time to process a single batch.

$$T_{NO} = n\sum_{i=1}^{m} t_i \tag{3.1}$$

where T_{NO} is the total time to process n batches without overlapping, each batch having m steps of duration $t_1, t_2, \ldots, t_m$. For this example, the total time is equal to (8)(3.5 + 0.5 + 3.5 + 2 + 4.0 + 1.0) = (8)(14.5) = 116.0 h.

For the process described in Figure 3.1, using the nonoverlapping scheme, the equipment is used infrequently, and the total processing time is unduly long. However, such a scheme might be employed in plants that operate only a single shift per day. In such cases, the production of a single batch might be tailored to fit an 8 or 10 h shift (for this example, the shift would have to be 14.5 h), with the limitation that only one batch would be produced per day. Although such a scheme does not appear to be very efficient, it eliminates prolonged storage of intermediate product and certainly makes the scheduling problem easy.

The total time to process all the batches can be reduced by starting a batch before the preceding batch has finished. This is equivalent to shifting backward the time blocks representing the steps in the batch process, as shown in Figure 3.2.

This shifting of batches backward in time leads to the concept of **overlapping** sequencing of batches. The limit of this shifting or overlapping process oc-

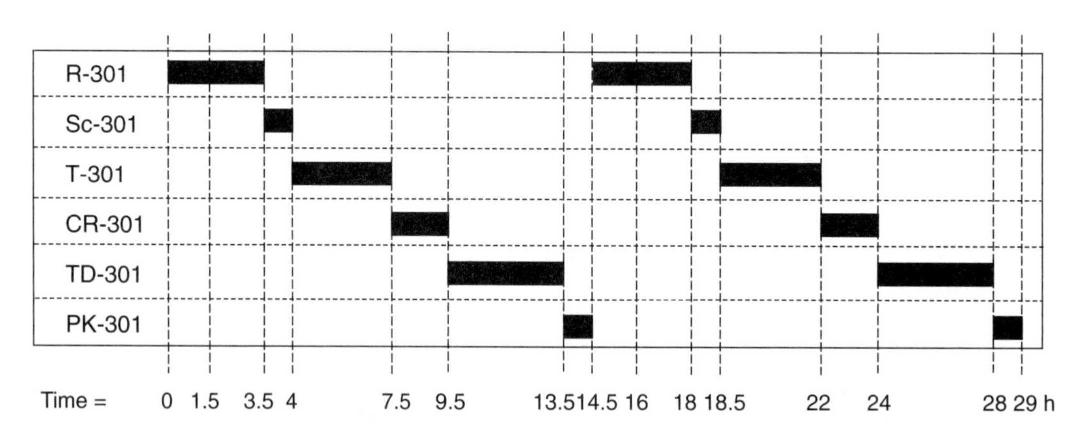

Figure 3.1 Example of a Nonoverlapping Sequence of Batch Operations

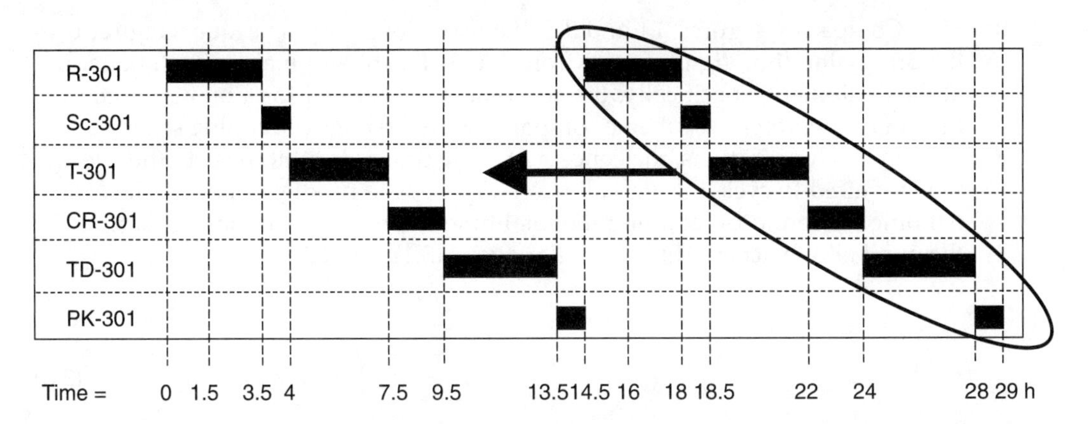

Figure 3.2 Backward Shifting of Batches, Giving Rise to Overlapping Sequencing

curs when two time blocks in consecutive batches just touch each other (assuming that cleaning, inspection, and charging times are included). This situation is shown in Figure 3.3.

From Figure 3.3, it can be seen that the limiting case for overlapping occurs when the step taking the longest time (here, the tray drying step in TD-301, which takes 4 h to complete) repeats itself without a waiting time between batches. The time to complete n batches using this limiting overlapping scheme is given by

$$T_O = T = (n-1) \max_{i=1,\ldots,m} (t_i) + \sum_{i=1}^{m} t_i \tag{3.2}$$

where T_O is the minimum total (overlapping) time, and [max (t_i)] is the maximum individual time step for the batch process. The subscript O denoting overlapping will be dropped, and T will be used as the total processing time from this point on. For the example, $T = (8–1)(4.0) + (14.5) = 42.5$ h.

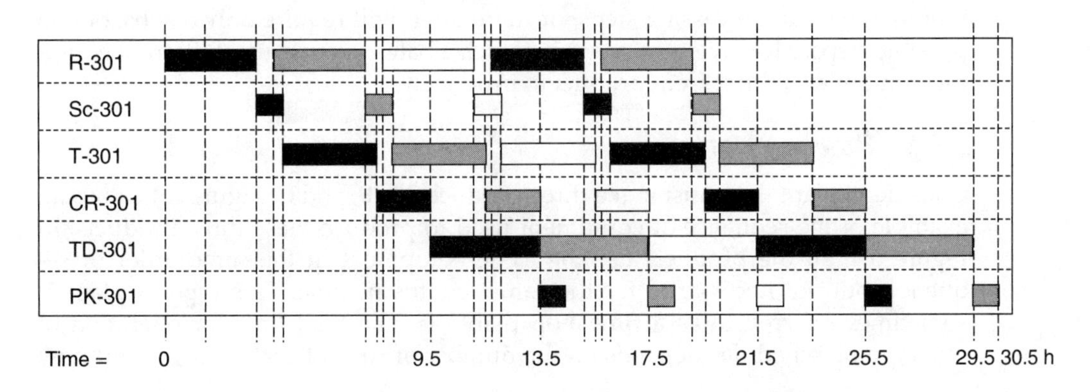

Figure 3.3 The Limiting Case for Overlapping Batch Sequencing

Comparing Figures 3.1 and 3.3, the use of overlapping sequencing reduces the processing time significantly (from 116 to 42.5 h) and makes much better use of the equipment; specifically, the equipment is operated for a higher fraction of time in the overlapping scheme compared with the nonoverlapping scheme.

In batch operations, the concept of **cycle time** is used to refer to the average time required to cycle through all necessary steps to produce a batch. The formal definition is found by dividing the total time to produce a number of batches by the number of batches. Thus, from Equations (3.1) and (3.2),

$$t_{cycle,NO} = \frac{T_{NO}}{n} = \frac{n\sum_{i=1}^{m} t_i}{n} = \sum_{i=1}^{m} t_i \tag{3.3}$$

$$t_{cycle,O} = t_{cycle} = \frac{T}{n} = \frac{(n-1)\max\limits_{i=1,\ldots,m}(t_i) + \sum_{i=1}^{m} t_i}{n} \tag{3.4}$$

For the overlapping scheme, when the number of batches (n) to be produced is large, the cycle time is approximated by

$$t_{cycle} \cong \max_{i=1,\ldots,m}\{t_i\} \tag{3.5}$$

Therefore, for Example 3.1, the nonoverlapping and overlapping cycle times are 14.5 h and 4 h, respectively.

3.4 FLOWSHOP AND JOBSHOP PLANTS

Thus far, the discussion has focused on the production of only a single product. However, most batch plants produce multiple products. All these products may require the same processing steps, or more often will require only a subset of all possible steps. Moreover, the order in which a batch process uses different equipment might also differ from product to product.

3.4.1 Flowshop Plants

Consider a plant that must make three products: A, B, and C. Figure 3.4 shows an example of the sequence of equipment used to produce these three products. In Figure 3.4, all the products use the same equipment in the same order or sequence, but not necessarily for the same lengths of time. This type of plant is sometimes referred to as a **flowshop** plant [4]. The total time for operation of overlapping schedules depends on the number of runs of each product and how

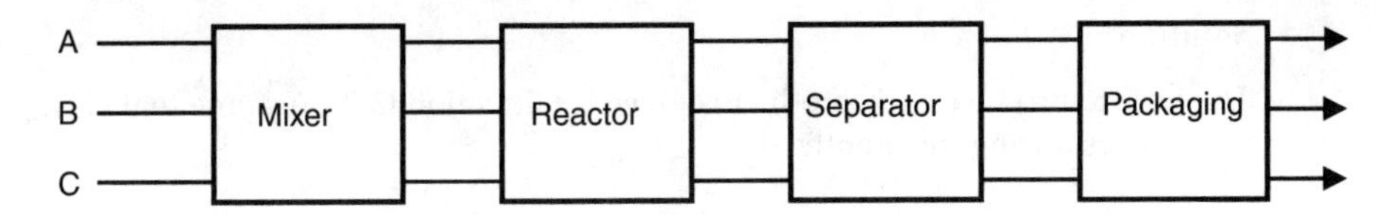

Figure 3.4 An Example of a Flowshop Plant for Three Products A, B, and C

these runs are scheduled. One approach to scheduling multiple products is to run each product in a campaign during which only that product is made. Then the plant is set up to run the next product in a campaign, and so on. The case when multiple products, using the same equipment in the same order, are to be produced in separate campaigns is considered first. If the corresponding numbers of batches for products A, B, C in a campaign are n_A, n_B, and n_C, respectively, then the total processing time, or production cycle time, can be found by adding the operation times for each product. If the number of batches per campaign is large (for example, >10), then the production cycle time can be approximated by

$$T = \sum_{j=A}^{C} n_j \{t_{cycle}\}_j \cong \sum_{j=A}^{C} n_j \left\{ \max_{i=1,\ldots,m} \{t_i\} \right\}_j \tag{3.6}$$

An illustration of a multiple-product process is given in Example 3.3.

Example 3.3

Consider three batch processes, producing products A, B, and C, as illustrated in Table E3.3. Each process uses the four pieces of equipment in the same sequencing order but for different times.

Market demand dictates that equal numbers of batches of the three products be produced over a prolonged period of time.

Determine the total number of batches that can be produced in a production cycle equal to one month of operation of the plant using separate campaigns for each product, assuming that a month of operation is equivalent to 500 h (based on 1/12 of a 6000 h year for a three-shift plant operating five days per week).

Table E3.3 Equipment Times (in Hours) Needed to Produce A, B, and C

Product	Time in Mixer	Time in Reactor	Time in Separator	Time in Packaging	Total Time
A	1.5	1.5	2.5	2.5	8.0
B	1.0	2.5	4.5	1.5	9.5
C	1.0	4.5	3.5	2.0	11.0

Solution

The time to produce each product is given by Equation (3.2). Assume that each product is run x times during the month:

$$T = 500 = \sum_{j=A}^{C}\left\{(n-1)\max_{i=1,\ldots,m}\{t_i\} + \sum_{i=1}^{m} t_i\right\}_j$$

$$500 = [(x-1)(2.5) + 8] + [(x-1)(4.5) + 9.5] + [(x-1)(4.5) + 11]$$

$$500 = (x-1)(11.5) + 28.5 \Rightarrow x = \frac{(500-28.5)}{11.5} + 1 = 42$$

Thus, 42 batches each of A, B, and C can be run as campaigns in a 500 h period. The cycle times are $t_{cycle,A} = [\,(41)(2.5) + 8/(42)] = 2.631$ h, $t_{cycle,B} = 4.619$ h, and $t_{cycle,C} = 4.655$ h.

Using Equation (3.6) with the approximations $t_{cycle,A} = 2.5$, $t_{cycle,B} = 4.5$, and $t_{cycle,C} = 4.5$,

$$T = 500 = x(2.5 + 4.5 + 4.5) \Rightarrow x = \frac{500}{11.5} = 43$$

Equation (3.6) slightly overestimates the number of batches that can be run in the 500 h period but is a very good approximation. In general, Equation (3.6) will be used to estimate cycle times and other calculations for single-product campaigns for multiproduct plants.

Running campaigns for the production of the same product is efficient and makes scheduling relatively easy. However, this strategy suffers from a drawback: The longer the production cycle, the greater the amount of product that must be stored. The concept of product storage is addressed in the following section. However, the bottom line is that storage requires additional equipment or warehouse floor space that must be purchased or rented. On the other hand, a strategy of single-product campaigns may decrease cleaning times and costs, which generally are greater when switching from one process to another. Therefore, the implementation of a batch sequencing strategy that uses sequences of single-product campaigns involves additional costs that should be included in any design and optimization. The extreme case for single-product batch campaigning occurs for seasonal produce (a certain vegetable oil, for example), where the feed material is available only for a short period of time and must be processed quickly, but the demand for the product lasts the whole year.

An alternative to running single-product campaigns (AAA..., BBB..., CCC...) over the production cycle is to run multiproduct campaigns—for example, ABCABCABC, ACBACBACB, AACBAACBAACB, and so on. In this strategy, products are run in a set sequence and the sequence is repeated. This approach is illustrated in Example 3.4.

Example 3.4

Consider the same processes given in Example 3.3. Determine the number of batches that can be produced in a month (500 h) using a multiproduct campaign strategy with the sequence ABCABCABC....

The Gantt chart for this sequence is shown in Figure E3.4.

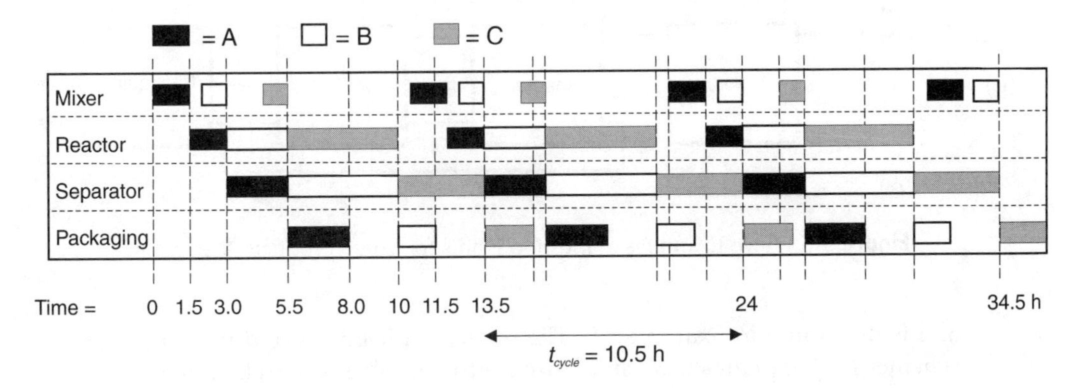

Figure E3.4 Gantt Chart Showing the Multiproduct Sequence ABCABCABC...

From Figure E3.4, it can be seen that the limiting equipment for this sequence is the separator. This means that the separator is used without downtime for the duration of the 500 h. If x batches are produced during the 500 h period, then

$$T = 500 = (3 + x(2.5 + 4.5 + 3.5) + 2) \Rightarrow x = \frac{(500-5)}{10.5} = 47$$

Therefore, an additional five batches of each product can be produced using this sequence compared with the single-product campaign discussed in Example 3.3, assuming no additional cleaning time. It should be noted that other sequences, such as BACBACBAC, could be used, and these may give more or fewer batches than the sequence used here.

3.4.2 Jobshop Plants

The flowshop plant discussed previously is one example of a batch plant that processes multiple products. When not all products use the same equipment or the sequence of using the equipment is different for different products, then the plant is referred to as a **jobshop** plant [4]. Figure 3.5 gives two examples of such plants. In Figure 3.5(a), all the products move from the left to the right—that is, they move in the same direction through the plant, but not all of them use the same equipment. In Figure 3.5(b), products A and B move from left to right, but product C uses the equipment in a different order from the other two products. The sequencing of multiproduct campaigns for this type of plant is more complex

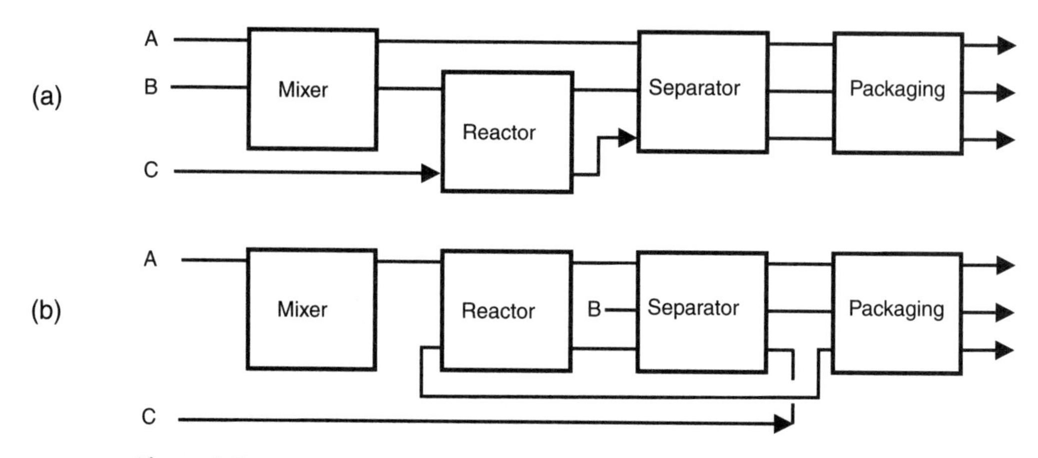

Figure 3.5 Two Examples of Jobshop Plants for Three Products A, B, and C

and is illustrated in Example 3.5. The relative efficiencies of different processing schemes for the plant shown in Figure 3.5(b) are calculated in Example 3.6.

Example 3.5

Consider the jobshop plant following the sequence shown in Figure 3.5(b) and described in Table E3.5. Construct the Gantt charts for overlapping single-product campaigns for products A, B, and C and for the multiproduct campaign with sequence ABCABCABC....

Table E3.5 Equipment Processing Times (in Hours) for Processes A, B, and C

Process	Mixer	Reactor	Separator	Packaging
A	1.0	5.0	4.0	1.5
B	—	—	4.5	1.0
C	—	3.0	5.0	1.5

Solution

The Gantt charts for the three processes are shown in Figure E3.5. The top chart shows the timing sequences for each batch, and the next three charts show overlapping campaigns for products A, B, and C, respectively. It can be seen that the rules and equations for overlapping campaigns given previously still apply. The bottom chart shows the overlapping multiproduct campaign using the sequence ABCABCABC.... Note that there are many time gaps separating the use of the different pieces of equipment, and no one piece of equipment is used all the time. This situation is common in jobshop plants, and strategies to increase equipment usage become increasingly important and complicated as the number of products increases.

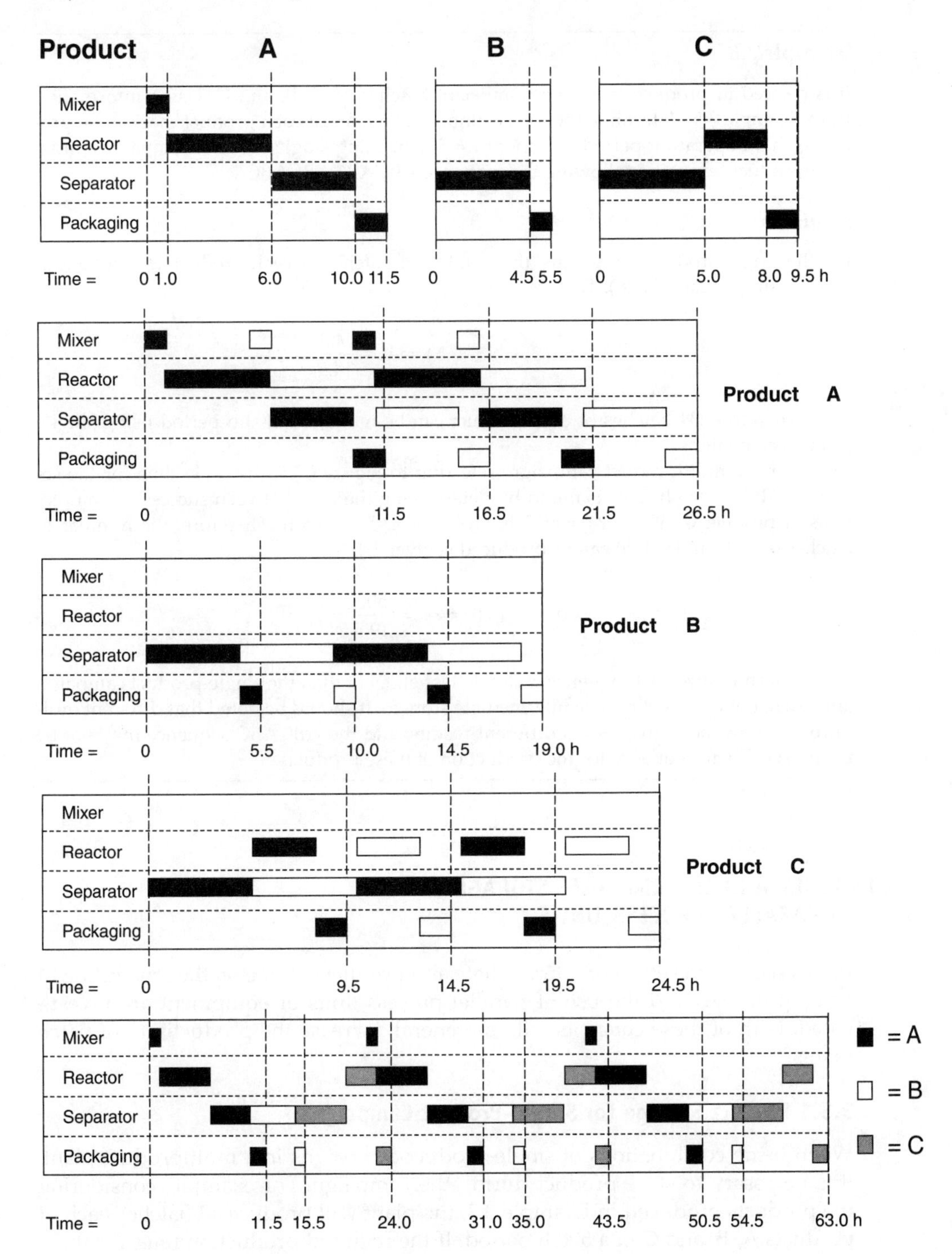

Figure E3.5 Gantt Charts for Single-Product and Multiproduct Campaigns

Example 3.6

It is desired to produce the same number of batches of A, B, and C. Using information from Example 3.5, determine the total number of batches of each product that can be produced in an operating period of 1 month = 500 h, using single-product campaigns and a multiproduct campaign following the sequence ABCABCABCABC....

Solution

For the single-product campaigns, the number of batches of each product, x, can be estimated using Equation (3.6). Thus

$$500 = x(5 + 4.5 + 5) \Rightarrow x = \frac{500}{14.5} = 34$$

Therefore, 34 batches of each product can be made in a 500 h period using single-product campaigns.

For the multiproduct campaign, referring to Figure E3.5, the cycle time for the sequence ABC is 19.5 h. This is found by determining the time between successive completions of product C: 43.5 – 24 = 19.5 h, and 63 – 43.5 = 19.5 h. Therefore, the number of batches of A, B, and C that can be produced is given by

$$500 = x(19.5) \Rightarrow x = \frac{500}{19.5} = 25$$

This multiproduct sequence is clearly less efficient than the single-product campaign approach, but it does eliminate intermediate storage. It should be noted that different multiproduct sequences give rise to different results, and the ABCABC sequence may not be the most efficient sequence for the production of these products.

3.5 PRODUCT AND INTERMEDIATE STORAGE AND PARALLEL PROCESS UNITS

In this section, the effect of intermediate and product storage on the scheduling of batch processes and the use of parallel process units or equipment are investigated. Both of these concepts will, in general, increase the productivity of batch plants.

3.5.1 Product Storage for Single-Product Campaigns

When using combinations of single-product campaigns in a multiproduct plant, it is necessary to store product during the campaign. For example, considering the products produced in Example 3.3, the plant will produce 43 batches each of products A, B, and C in a 500 h period. If the required production rates for these three products are 10,000, 15,000, and 12,000 kg/month, respectively, then what is the amount of storage required? In practice, it is the volume, and not the weight, of each product that determines the required storage capacity. For this example, it is assumed that the densities of each product are the same and equal to 1000 kg/m^3. Considering product A first and assuming that demand is steady, the demand rate (r_d) is equal to 10,000/500 = 20 kg/h = 0.020 m^3/h. Note that the demand rate is calculated on the basis of plant operating hours, and not on the basis of a 24-hour day. During the campaign, 10,000 kg of A must be made in 43 batch runs, with each run taking $t_{cycle,\,A}$ = 2.5 h. Thus, during production, the production rate (r_p) of A is equal to 10,000/(43)(2.5) = 93.0 kg/h = 0.0930 m^3/h. Results for all the products are given in Table 3.1.

When a campaign for a product is running, the rate of production is greater than the demand rate. When the campaign has stopped, the demand rate is greater than the production rate of zero. Therefore, the accumulation and depletion of product over the monthly period are similar to those shown in Figure 3.6. The changing inventory of material is represented on this figure by the bottom diagram. The maximum inventory, V_s, is the minimum storage capacity that is required for the product using this single-product campaign strategy. The expression for calculating V_s is

$$V_s = (r_p - r_d)t_{camp} \tag{3.7}$$

where t_{camp} is the campaign time. This assumes that the shipping rate of product from the plant is constant during plant operating hours. Because shipping is usually itself a batch process, the actual minimum storage capacity could be more or less than that calculated in Equation (3.7). The strategies for matching shipping schedules to minimize cost (including storage costs and missed-delivery risks) are known as logistics and are not covered here.

Determination of the minimum storage capacities for all products in Example 3.3 is given in Example 3.7.

Table 3.1 Production and Demand Rates for Products A, B, and C in Example 3.3

Rate	Product A	Product B	Product C
Volume (m^3) of product required per month	10.0	15.0	12.0
Cycle time (h)	2.5*	4.5*	4.5*
Production rate, r_p (m^3/h)	(10)/[(43)(2.5)] = 0.0930	0.07752	0.06202
Demand rate, r_d (m^3/h)	(10)/(500) = 0.020	0.030	0.024

*These are approximate cycle times based on Equation (3.5).

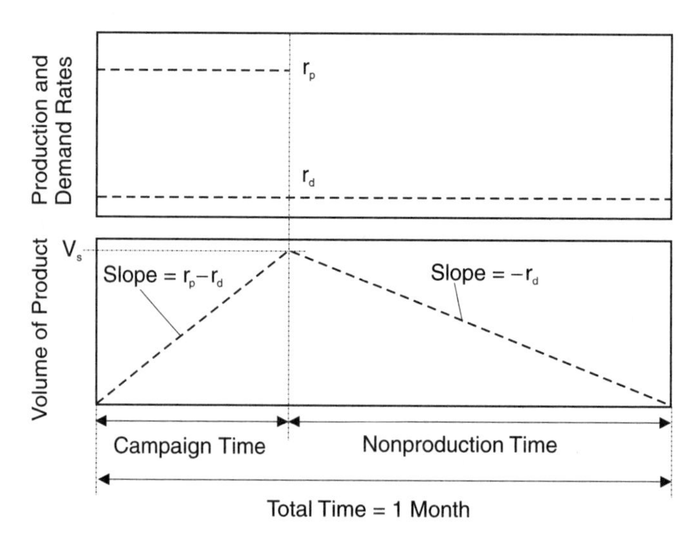

Figure 3.6 Changing Inventory of Product during Single-Product Campaign Run within a Multiproduct Process

Example 3.7

For the products A, B, and C in Example 3.3, determine the minimum storage capacities for the single-product campaign strategy outlined in Example 3.3.

Table E3.7 Results for the Estimation of Minimum Storage Volume from Equation (3.7)

Product	Campaign Time, t_{camp} (h)	$r_p - r_d$ (m^3/h)	V_s (m^3)
A	(43)(2.5) = 107.5	0.09302 – 0.020 = 0.07302	(0.07302)(107.5) = 7.85
B	(43)(4.5) = 193.5	0.07752 – 0.030 = 0.04752	(0.04752)(193.5) = 9.20
C	(43)(4.5) = 193.5	0.06202 – 0.024 = 0.03802	(0.03802)(193.5) = 7.36

Solution

Table E3.7 shows the results using data given in Example 3.3 and Table 3.1.

It should be noted that the production cycle time is equal to the sum of the campaign times, or (107.5 + 193.5 + 193.5) = 494.5 h, which is slightly less than 500 h. This discrepancy reflects the approximation of cycle times given by Equation (3.6). The actual cycle times for A, B, and C are found from Example 3.3 and are equal to 2.63, 4.62, and 4.65 h, respectively. The corresponding values of V_s are 7.79, 9.18, and 7.31 m^3. Clearly, these differences are small, and the approach using Equation (3.6) is acceptable when the number of production runs per campaign is 10 or more.

3.5.2 Intermediate Storage

Up to this point, it has been assumed that there is no intermediate product storage available. This type of process is also known as a **zero wait,** or a **zw-process** [4]. Specifically, as soon as a unit operation is completed, the products are transferred to the next unit operation in the sequence, or they go to final product storage. The concept of storing the final product to match the supply with the demand was demonstrated in Example 3.7. However, it may also be beneficial to store the output from a given piece of equipment for a period of time to increase the overall efficiency of a process. It may be possible to store product in the equipment that has just been used. For example, if two feed streams are mixed in a vessel, the mixture could be stored until the next process unit in the production sequence becomes available. In this case, the storage time is limited based on the scheduling of equipment. This **holding-in-place** method may not work for some unit operations. For example, in a reactor, a side reaction may take place, and unless the reaction can be quenched, the product yield and selectivity will suffer. The upper limit of the intermediate storage concept occurs when there is **unlimited intermediate storage (uis)** available, and this is referred to as a **uis-process** [4]. In general, cycle times can be shortened when intermediate product storage is available. This concept is illustrated in Figure 3.7, which is based on the information given in Table 3.2.

Without intermediate product storage, the shortest multiproduct campaign, as given by Equation (3.6), is 14 h, as shown in Figure 3.7. However, if the materials leaving the reactor and crystallizer are placed in storage prior to transfer to the crystallizer and dryer, respectively, then this time is reduced to 11 h. The limiting cycle time for a uis-process is given by

$$t_{cycle,uis} = \max_{j=1,m} \sum_{i=1}^{N} nc_i t_{i,j} \tag{3.8}$$

where m is the number of unit operations, N is the number of products, and nc_i is the number of campaigns of product i produced in a single multiproduct sequence. For the case shown in Table 3.2 and Figure 3.7, n =1 (because only one campaign for each product (A, B, and C) is used in the multiproduct sequence), and Equation (3.8) is the maximum value given in the last row of Table 3.2, or 11.0 h.

Table 3.2 Equipment Times (in Hours) Required for Products A, B, and C

Product	Reactor	Crystallizer	Dryer	Total
A	2.0	5.0	2.0	9.0
B	6.0	4.0	4.0	14.0
C	2.0	2.0	3.0	7.0
Total Time per Equipment	10.0	11.0	9.0	

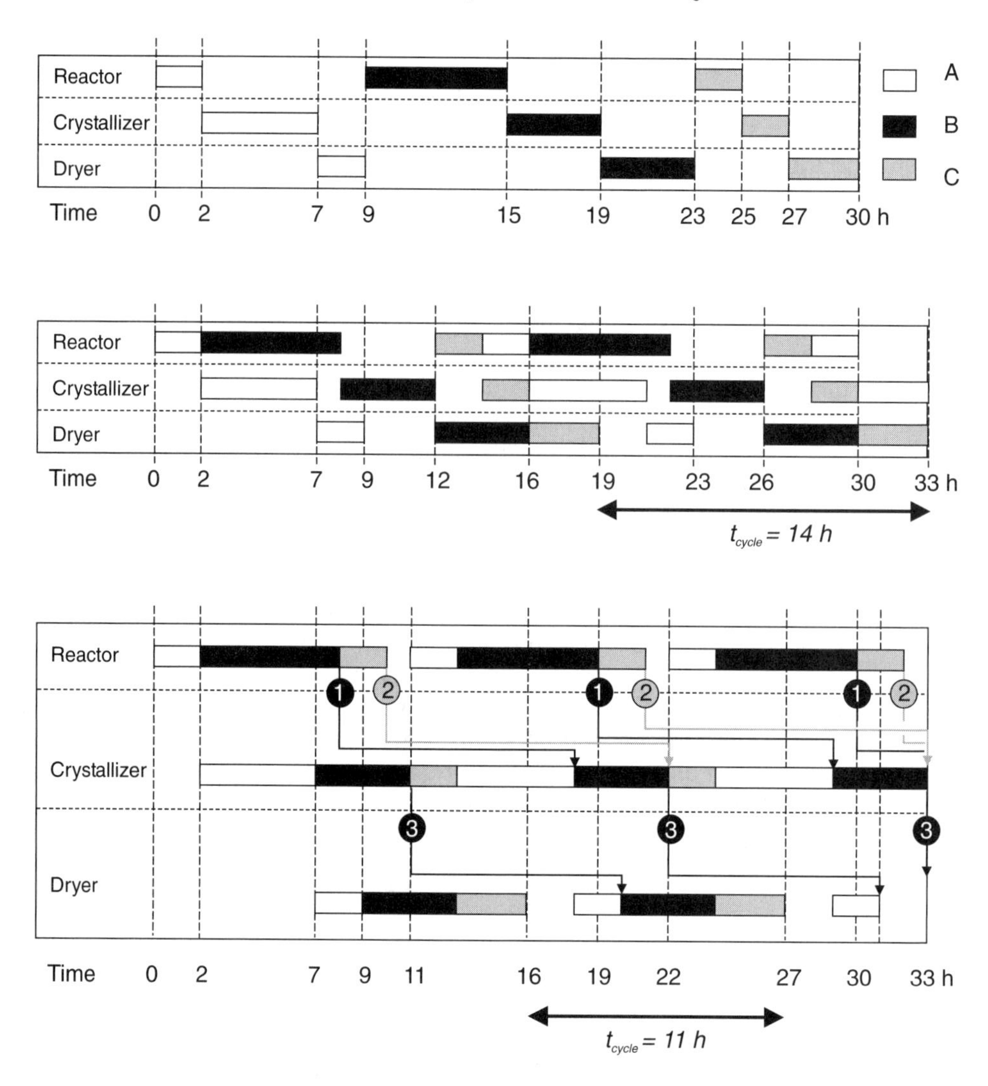

Figure 3.7 Multiproduct Sequence (ABC) for Products Given in Table 3.2 Showing Effect of Intermediate Storage (Storage Shown as Circles; Number Identifies Individual Tanks for Each Intermediate Product)

The total amount of storage required for this example is fairly small, because only three storage vessels are required, each dedicated to one intermediate product. The downside of this approach is that there are many more material transfers required, and the potential for product contamination and operator error increases significantly.

3.5.3 Parallel Process Units

Another strategy that can be employed to increase production is to use duplicate equipment. This strategy is most beneficial when there is a bottleneck involving a single piece of equipment that can be relieved by adding a second (or more) units in parallel. This strategy can be extended to a limiting case in which parallel trains of equipment are used for each product. This strategy eliminates the dependence of scheduling between the different products but is more expensive, because the number of pieces of equipment increases *m*-fold, where *m* is the number of products. An example of using parallel equipment is shown in Figure 3.8 based on the data in Table 3.3.

From the top chart in Figure 3.8, the limiting piece of equipment is seen to be the crystallizer. The bottom chart shows the effect of adding a second crystallizer that processes product C. The effect is to reduce the cycle time from 21 h to 13 h, a considerable improvement in throughput. The determination of whether to make this change must be made using an appropriate economic criterion, such as net present value (NPV) or equivalent annual operating cost (EAOC), which

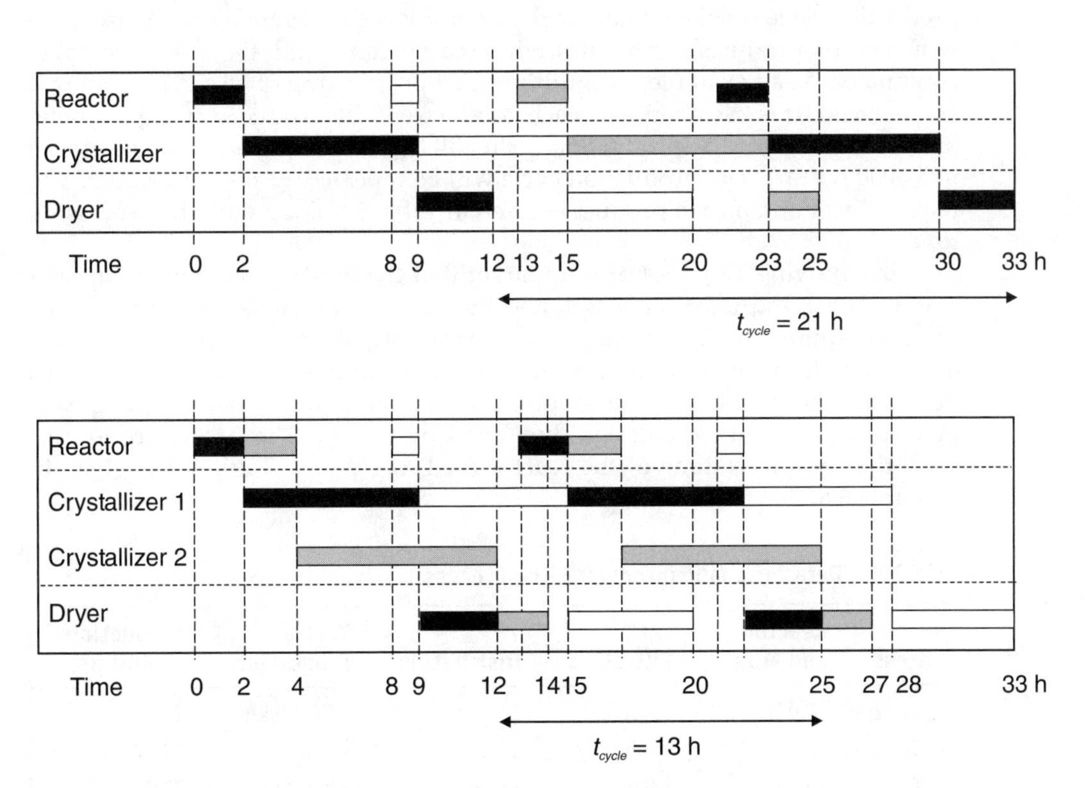

Figure 3.8 The Effect of Adding an Additional Crystallizer to the Process Given in Table 3.2

Table 3.3 Data (Times in Hours) for Multiproduct Batch Process Shown in Figure 3.8

Process	Reactor	Crystallizer	Dryer
A	2.0	7.0	3.0
B	1.0	6.0	5.0
C	2.0	8.0	2.0

are discussed in Chapter 10. The resulting trade-off is between increased product revenues versus the cost of purchasing a second crystallizer plus additional operators to run the extra equipment.

3.6 DESIGN OF EQUIPMENT FOR MULTIPRODUCT BATCH PROCESSES

The design of equipment sizes for multiproduct batch processes depends on the production cycle time, whether single- or multiproduct campaigns are used, the sequence of products for multiproduct campaigns, and the use of parallel equipment. As an example, the multiproduct process described in Table 3.4 will be analyzed. It is assumed that each product will be produced using a single-product campaign. The production cycle will be 500 h (equivalent to one month in a 6000 h year). The production cycle will be repeated 12 times in a year. The required amount of each product is given in Table 3.4 along with the processing times.

By studying Table 3.4, it is apparent that the limiting piece of equipment is the mixing and reaction vessel, and the cycle times can be found from this piece of equipment. To estimate equipment volume, it is necessary to determine the volume of each piece of equipment per unit of product produced. To determine these quantities, descriptions of the method (recipe) for using each piece of equipment for each product must be known. The procedure to estimate the specific volume of the reactor for Process A in Table 3.4 is given in Example 3.8.

Table 3.4 Data for a Multiproduct Batch Process

Process	Reactor and Mixer	Filtration	Distillation	Yearly Production	Production in 500 h
A	7.0 h	1.0 h	2.0 h	120,000 kg	10,000 kg
B	9.0 h	1.0 h	1.5 h	180,000 kg	15,000 kg
C	10.0 h	1.0 h	3.0 h	420,000 kg	35,000 kg

Example 3.8

The following is a description of the reaction in process A, based on a laboratory-scale experiment. First, 10 kg of liquid reactant (density = 980 kg/m^3) is added to 50 kg of a liquid mixture of organic solvent containing excess of a second reactant (density of mixture = 1050 kg/m^3) in a jacketed reaction vessel, the reactor is sealed, and the mixture is stirred and heated. Once the reaction mixture has reached 95°C, a solid catalyst (negligible volume) is added, and reaction takes place while the batch of reactants continues to be stirred. The required conversion is 94%, 17.5 kg of product is produced, and the time taken is 7.0 h. The reactor is filled to 60% of maximum capacity to allow for expansion and to provide appropriate vapor space above the liquid surface. Determine the volume of reaction vessel required to produce 1 kg of product.

Solution

$$\text{volume of reactor} = \left(\frac{10[\text{kg}]}{980[\text{kg/m}^3]} + \frac{50[\text{kg}]}{1050[\text{kg/m}^3]}\right)\frac{100}{60} = 0.09637\ \text{m}^3$$

$$v_{react} = \frac{\text{volume of reactor}}{\text{mass of product}} = \frac{0.09637}{17.5} = 0.005507\ \text{m}^3/\text{kg-product}$$

Similar calculations can be made for the reactor/mixer for processes B and C in Table 3.4, and these results are given in Table 3.5 along with the cycle times for each process. It should be noted that even for a preliminary design and cost estimate, other attributes of the equipment should also be considered. For example, in order to specify fully the reactor and estimate its cost, the heating duty and the size of the motor for the mixer impeller must be calculated. To simplify the current example, only the volumes of the reactor are considered, but it should be understood that other relevant equipment properties must also be considered before a final design can be completed. This procedure should be applied to all the equipment in the process.

Let the single-product campaign times for the three products be t_A, t_B, and t_C, respectively. Applying Equation (3.6), the following relationship is obtained:

$$t_A + t_B + t_C = 500 \tag{3.9}$$

Table 3.5 Specific Reactor/Mixer Volumes for Processes A, B, and C

Process	A	B	C
v_{react} (m^3/kg-product)	0.005507	0.007860	0.006103
t_{cycle} (h)	7.0	9.0	10.0

The number of batches per campaign for each product is then given by $t_x/t_{cycle,x}$ and

$$\text{batch size (kg/batch)} = \frac{\text{production of x}}{t_x/t_{cycle,x}} \tag{3.10}$$

Furthermore, the volume of a batch is found by multiplying Equation (3.10) by $v_{reac,x}$, and equating batch volumes for the different products yields

$$\text{Volume of batch} = \frac{(\text{production of x})(v_{react,x})}{t_x/t_{cycle,x}} \tag{3.11}$$

$$\frac{(10{,}000)(0.005507)}{t_A/7.0} = \frac{(15{,}000)(0.007860)}{t_B/9.0} = \frac{(35{,}000)(0.006103)}{t_B/10.0} \tag{3.12}$$

Solving Equations (3.9) and (3.12) yields

$$t_A = 53.8\text{h}$$
$$t_B = 148.1\text{h}$$
$$t_C = 298.1\text{h}$$

$$v_{react,A} = v_{react,B} = v_{react,C} = 7.17 \text{ m}^3$$

Number of batches per campaign for product A = 7.7
Number of batches per campaign for product B = 16.5
Number of batches per campaign for product C = 29.8

Clearly the number of batches should be an integer value. Rounding these numbers yields

For product A
Number of batches = 8
$t_A = (8)(7) = 56$ h
$V_A = (10{,}000)(0.005507)(7)/(56) = 6.88 \text{ m}^3$

For product B
Number of batches = 16
$t_B = (16)(9) = 144$ h
$V_B = (15{,}000)(0.007860)(9)/(144) = 7.37 \text{ m}^3$

For product C
Number of batches = 30
$t_C = (30)(10) = 300$ h
$V_C = (35{,}000)(0.006103)(10)/(300) = 7.12 \text{ m}^3$

Total time for production cycle = 500 h
Volume of reactor = 7.37 m^3 (limiting condition for Product B)
= (7.37)(264.2) =1947 gallons

The closest standard size, 2,000 gallons, is chosen.

3.7 SUMMARY

In this chapter, concepts important to the design of batch processes were introduced. Gantt charts were used to illustrate the timing and movement of product streams through batch processes. The concepts of nonoverlapping and overlapping sequences were discussed for single-product and multiproduct processes. The differences between flowshop (multipurpose) and jobshop (multiproduct) plants were introduced, and the strategies for developing single-product and multiproduct campaigns for each type of process were discussed. The role of intermediate and final product storage and the methods to estimate the minimum product storage for single-product campaigns were illustrated. The addition of parallel equipment was shown to reduce product cycle time. Finally, an example of estimating the size of vessels required in a multiproduct process was given.

REFERENCES

1. Sundaram, S., and L. B. Evans, "Shortcut Procedure for Simulating Batch Distillation Operations," *Ind. & Eng. Chem. Res.* 32 (1993): 511–518.
2. Seader, J. D., and E. J. Henley, *Separation Processes Principles* (New York: John Wiley & Sons, 1998).
3. Dewar, J. D., "If You Don't Know Where You're Going, How Will You Know When You Get There?" *CHEMTECH* 19, no. 4 (1989): 214–217.
4. Biegler, L. T., I. E. Grossman, and A. W. Westerberg, *Systematic Methods of Chemical Process Design* (Upper Saddle River, NJ: Prentice-Hall, 1997).

SHORT ANSWER QUESTIONS

1. What is a flowshop plant?
2. What is a jobshop plant?
3. What are the two main methods for sequencing multiproduct processes?
4. Give one advantage and one disadvantage of using single-product campaigns in a multiproduct plant.
5. What is the difference between a zero-wait and a uis-process?

PROBLEMS

6. Consider the processes given in Example 3.3. Determine the number of batches that can be produced in a month (500 h) using a series of single-product campaigns when the required number of batches for product A is twice that of either product B or product C.
7. Consider the processes given in Examples 3.3 and 3.4. Determine the number of batches that can be produced in a month (500 h) using a multiproduct campaign strategy with the sequence ACBACBACB. Are there any other sequences for this problem other than the one used in Example 3.4 and the one used here?
8. Consider the multiproduct batch plant described in Table P3.8.

Table P3.8 Equipment Processing Times for Processes A, B, and C

Process	Mixer	Reactor	Separator
A	2.0 h	5.0 h	4.0 h
B	3.0 h	4.0 h	3.5 h
C	1.0 h	3.0 h	4.5 h

It is required to produce the same number of batches of each product. Determine the number of batches that can be produced in a 500 h operating period using the following strategies:
a. Single-product campaigns for each product
b. A multiproduct campaign using the sequence ABCABCABC…
c. A multiproduct campaign using the sequence CBACBACBA…

9. Consider the process given in Problem 3.8. Assuming that a single-product campaign strategy is used over a 500 h operating period and further assuming that the production rates (for a year = 6,000 h) for products A, B, C are 18,000 kg/y, 24,000 kg/y, and 30,000 kg/y, respectively, determine the minimum volume of product storage required. Assume that the product densities of A, B, and C are 1100, 1200, and 1000 kg/m^3, respectively.
10. Using the data from Tables P3.10(a) and (b), and following the methodology given in Section 3.6, determine the number of batches and limiting reactor size for each product.

Table P3.10(a) Production Rates for A, B, and C

Product	Yearly Production	Production in 500 h
A	150,000 kg	12,500 kg
B	210,000 kg	17,500 kg
C	360,000 kg	30,000 kg

Table P3.10(b) Specific Reactor/Mixer Volumes for Processes A, B, and C

Process	A	B	C
v_{react} (m^3/kg-product)	0.0073	0.0095	0.0047
t_{cycle} (h)	6.0	9.5	18.5

11. Referring to the batch production of amino acids described in Project 8 in Appendix B, the batch reaction times and product filtration times are given in Table P3.11.

Table P3.11 Batch Step Times (in Hours) for Reactor and Bacteria Filter for Project 8 in Appendix B

Product	Reactor*	Precoating of Bacteria Filter	Filtration of Bacteria
L-aspartic acid	35	25	5
L-phenylalanine	65	25	5

*Includes 5 h for filling, cleaning, and heating.

The capacity of the reactors chosen for both products is 10,000 gallons each. The precoating of the filters may occur while the batch reactions are taking place, and hence the critical time for the filtration is 5 h. It should be noted that intermediate storage is not used between the reactor and filter, and hence the batch times for the reactors must be extended an additional 5 h while the contents are fed through the filter to storage tank V-901. It is desired to produce L-aspartic acid and L-phenylalanine in the ratio of 1 to 1.25 by mass. Using a campaign period of 1 year = 8000 h and assuming that there is a single reactor and filter available, determine the following.

a. The number of batches of each product that can be produced in a year, maintaining the desired ratio of the two products, if one single-product campaign is used for each amino acid per year.
b. The amount of final product storage for each product assuming a constant demand of each product over the year. Express this amount of storage as a volume of final solid crystal product (bulk density of each amino acid is 1200 kg/m^3). You may assume that the recovery of each amino acid is 90% of that produced in the reactor.
c. By how much would the answer to part (b) change if the single-product campaigns for each amino acid were repeated every month rather than every year?

12. Referring to Problem 3.11, by how much would yearly production change if the following applied?
    a. The reaction times for L-aspartic acid and L-phenylalanine were reduced by 5 hours each. Use the same scenario as described in Problem 3.11, part (a).
    b. The reaction times for L-aspartic acid and L-phenylalanine were increased by 5 hours each. Use the same scenario as described in Problem 3.11, part (a).

CHAPTER

4

Chemical Product Design

The subject of most of this book is chemical process design, the traditional capstone experience in chemical engineering curricula. For most of the history of chemical engineering, graduates have gone to work in chemical plants that manufacture commodity chemicals. **Commodity chemicals** are those manufactured by many companies in large quantities, usually in continuous processes like those illustrated in this textbook. There is little or no difference between a commodity chemical produced by different companies. The price for which a commodity chemical can be sold is essentially the same for all producers, and, because most raw materials are also commodity chemicals, the price of raw materials is also the same for all producers. For the most part, innovations regarding manufacture of commodity chemicals have occurred a long time in the past. Therefore, the only real way to be more profitable than a competitor is to have lower ancillary costs, such as a favorable union contract, a better deal on the costs of different sources of energy, superior automation, a better catalyst, and so on.

Before a chemical became a commodity, it was probably a specialty. A **specialty chemical** is one made in smaller quantities, often in batch processes, usually by the company that invented the chemical. Perhaps the best examples of specialty chemicals evolving into commodity chemicals are polymers. Polymers such as nylon, Teflon, and polyethylene were specialties when they were invented in the first half of the twentieth century, and they were seen only in selected applications. Now, they are ubiquitous commodities.

The chemical industry has also become more global. At one time, chemicals and products from chemicals for the entire world were manufactured in the centers of the chemical industry: the United States and Western Europe. This meant that a large, rapidly growing chemical industry in the United States and Western

Europe was needed to serve the needs of developing countries. Now chemicals are manufactured all over the world, closer to where they are used, as are the raw materials for these chemicals. Therefore, the traditional commodity chemical industry is not in a growth phase in places such as the United States and Western Europe. Existing chemical processes continue to operate, and chemical engineers trained to understand and work with continuous, commodity chemical processes are still needed.

It has been suggested that the future of chemical engineering—that is, the place where chemical engineers can innovate—is in chemical product design [1,2]. This is also the place where more and more chemical engineers are being employed [2]. It could be argued that the future is identical to the past. Since the 1970s, most large, commodity chemical companies have trimmed their long-range research, focusing instead on support for the global growth of commodity chemical production. They believed this to be a necessary shift of emphasis as chemicals that were once specialties evolved into commodities.

What is a chemical product? One possibility is a new specialty chemical. A new drug is a chemical product. A new catalyst or solvent for use in the commodity chemical industry is a chemical product. Post-it Notes are a chemical product. A fuel cell is a chemical product. A device for indoor air purification is a chemical product. Technologies employing chemical engineering principles could be considered to be chemical products. Even the ChemE Cars that many students build as part of the AIChE competition could be considered chemical products.

In this chapter, an introduction to procedures used for chemical product design is provided. It will be seen that there are similarities to chemical process design; however, the focus of this chapter is on the differences between process and product design.

4.1 STRATEGIES FOR CHEMICAL PRODUCT DESIGN

A strategy for chemical product design has been suggested by Cussler and Moggridge [1]. It has four steps:

1. Needs
2. Ideas
3. Selection
4. Manufacture

Needs means that a need for a product must be identified. This involves dealing with industrial customers and/or the public. If the business end of a commodity chemical industry determines there is a market for additional benzene, a process is constructed to make the same benzene product that all other benzene producers make, probably using the same process technology. If there is a market for the additional benzene, there will be customers. In contrast, in chemical product design, once the need is established, then the search for the best product begins.

Ideas means that the search for the best product has begun. This is similar to the brainstorming stage of the problem-solving strategy to be discussed in Chapter 22. Different ideas are identified as to the best possible product to serve the need.

Selection involves screening the ideas for those believed to be the best. There are quantitative methods for this step, and they are discussed later.

Manufacture involves determining how to manufacture the product in sufficient quantities. Unlike commodity chemicals, this usually involves batch rather than continuous processes.

The four-step process is a simplification that is most applicable to the design of chemical products that are actually chemicals. For the design of devices, there are additional steps needed. Two such product-design strategies, suggested by Dym and Little [3] and Ulrich and Eppinger [4] are illustrated in Table 4.1 and compared with the strategy of Cussler and Moggridge [1]. The most significant difference is the inclusion of different stages of device design not apparent for design of a chemical, although it could be considered that these steps are all part of the selection and/or manufacture step.

There are clear parallels between these three product design strategies. It is observed that the first step in each strategy is the identification of customer needs. The ideas and selection steps of Cussler and Moggridge [1] are identical to the generate product concepts and select product concepts steps of Ulrich and Eppinger [4], respectively. The strategies of Dym and Little [3] and Ulrich and Eppinger [4] all include several design and product-testing steps. Although these are not explicitly included in the strategy of Cussler and Moggridge [1], they will have to be part of any product design strategy. For example, no one would begin to manufacture a product without first making a small amount in the lab and testing it.

It is instructive to observe the parallel between the strategy of Dym and Little [3] and the increasing levels of capital cost estimates in process design

Table 4.1 Comparison of Product Design Strategies

Strategy	Cussler and Moggridge [1]	Dym and Little [3]	Ulrich and Eppinger [4]
Steps	Needs	Need	Identify customer needs
	Ideas	Problem definition	Establish target specifications
	Selection	Conceptual design	Generate product concepts
	Manufacture	Preliminary design	Select product concepts
		Detailed design	Test product concepts
		Design communication	Set final specifications
		Final design	Plan downstream development

illustrated in Table 7.1. Moving from a feasibility estimate through a detailed estimate parallels moving from a conceptual design to a detailed design. Similarly, the evolution of a detailed P&ID from a PFD also parallels the evolution from a conceptual design to a detailed design.

In the following sections, the strategy of Cussler and Moggridge [1] is illustrated using several examples.

4.2 NEEDS

A new chemical product is sought in response to a need. The need might be those of individual customers, those of groups, or those of society. Consider the case of Freon refrigerants. In the 1980s, Freons were identified as having high ozone-depleting potential because of their chlorine content. Therefore, a need for an environmentally friendly chemical with the appropriate properties of a refrigerant was established. This led to the development of fluorocarbon refrigerants (e.g., R-134a) and methods for their synthesis. However, this did not solve the problem entirely. It was then determined that the new refrigerants were incompatible with typical compressor lubricants. This created the need for a new lubricant that was compatible with the new refrigerant. Subsequently, this new lubricant was developed, and the new refrigerant began to be phased in as Freons were phased out.

Chemical companies devoted to product design (e.g., food products, personal care) deal with customers all the time. Customers are interviewed, often in focus groups, and the results of these interviews must be interpreted and made into product specifications. This is an inexact "science." Care must be taken to define the correct need.

As an example, consider the needs of vessels used for space travel (e.g., the space shuttle) as they reenter Earth's atmosphere [5]. The customer, NASA, initially sought the development of a material that would withstand the temperatures of reentry. Such a material was never developed. Once the problem was redefined, a more appropriate need was defined. The real need was not to have a material capable of withstanding the temperatures of reentry; the real need was to protect those inside the space vessel from the high temperatures generated by friction with the edge of Earth's atmosphere. This led to the development of the sacrificial tiles used in the space shuttle. The energy generated by friction is dissipated by vaporizing these sacrificial tiles, thereby protecting those inside the vessel from the heat. Only after the correct need was identified was the problem solved.

Examples 4.1 through 4.4 [6–8] illustrate definition of needs.

Example 4.1

Zebra mussels are mollusks that have been known to infest the water intake pipes of water treatment and electric power plants. Entire towns have been shut down because the infestation of zebra mussels has halted the supply of water to purification plants. The initial solution to this problem was to remove the infested zebra mussels manually. Identify the need(s) to alleviate the infestation problem.

The need is for a method to prevent the infestation, because it is undesirable to shut down water treatment facilities for manual cleaning. If this method is to involve a chemical, it is important to specify the desired features of this chemical. For example, it should be inexpensive, it must prevent infestation, it should not harm other wildlife, and it should be removable in the water treatment facility.

Example 4.2

Maintaining a swimming pool, either at home or in a public facility, is both expensive and time consuming. The water must be tested often, particularly for chlorine. The chlorine additive to a swimming pool emits a characteristic odor, irritates the eyes, and can fade colors on swimsuits due to its bleaching effect. Identify a product need.

There might be a need for a method to disinfect the pool water other than adding a chlorine-containing compound. Suppose a continuous-flow device could be developed that disinfected the pool water as it passed through the filter system. Is there a need for such a product? (This is one possible alternative to chlorine. There are others.) Such a device would undoubtedly increase the capital cost of installing a pool, even though it would save time and the cost of constantly adding chlorine. The unanswered question is whether pool purchasers would be willing to pay the incremental capital cost. Even though a net present value or an equivalent annual operating cost calculation, such as that illustrated in Chapter 10, might prove that the incremental cost of such a device is justified by the savings, it is still unclear whether people would purchase such a device. Most buyers will not sit down and do an incremental economic analysis. It is difficult to put a dollar value on the savings in time created by such a device. Clearly, it would be necessary to get feedback from potential customers before proceeding with development of such a device.

Example 4.3

Research is under way to develop a magnetic refrigerator [9]. This refrigerator operates by using magnetocaloric materials, which are materials that heat up when in a magnetic field. A magnetic refrigerator operates without a compressor and therefore does not need a refrigerant like Freon, which vaporizes and condenses in the vapor-compression cycle. Is there a need for such a refrigerator?

What are the advantages of such a refrigerator? There are two obvious ones. One is that a refrigerant like Freon is not needed. This may have spurred the initial research effort; however, the development of new refrigerants with more favorable environmental properties may have diminished this advantage. The other is the lack of a compressor, probably the most costly component of a vapor-compression refrigerator, both in capital cost and in operating cost. There are energy costs associated with the magnetic refrigerator, including a pump to circulate the cooling fluid and a motor to cycle the magnetocaloric material into and out of the magnetic field. Therefore, the savings created by compressor removal may be small. Based on these two factors, it is unclear whether there is a need for a magnetic refrigerator.

Example 4.4

With the advent of portable electric devices such as laptop computers, cellular phones, personal digital assistants, MP3 players, and so on, the length of time they can run before recharging and/or replacement of their power source is becoming an issue. Is there a product need here?

Anyone who has ever used a laptop computer where there is no source of power has probably, at one time or another, been frustrated by a battery that has run out before the desired work was completed. However, does this mean there is a need for a longer-lasting power source? Or will this be a high-end, niche market? Consider the situations when one uses a laptop computer for long periods of time away from a power source. One of the most common situations is on an airplane. However, newer aircraft now have fitted power connections at every seat. Some older aircraft have already been retrofitted with such power connections. As older aircraft are replaced or as they are modernized, will all aircraft used for longer flights have power available? If so, this could diminish the need for a longer-lasting power source, especially one that might require new technology and be costly.

4.3 IDEAS

The generation of ideas is tantamount to brainstorming. Just as in brainstorming, when ideas are being generated, there are no bad ideas. They will be screened in the next step, selection. Ideas can be sought from a variety of sources, including, but not limited to, members of the product development team, potential customers, and published literature. If there is a time for "pie in the sky," it is at this step.

It is important to remember not to "get married" to an idea at this stage. The chances of the first idea generated being the best one are slim or none. As many ideas as can be imagined should be generated before moving on to the selection step.

Examples 4.5 through 4.7 illustrate generation of ideas.

Example 4.5

Suggest some ideas for the zebra mussel problem in Example 4.1.

One possible idea is to invent a chemical or determine whether there is a chemical that can kill existing infestations. It would be nice if this were combined with a chemical that can prevent the infestations. If either of these methods were to be used, some type of delivery system would be needed. Another possibility is to place some type of filter at the water intake to keep the zebra mussels out of the water intake. Because zebra mussels are attracted to the warmer water near power plants and water treatment facilities, another possibility is to find a way to cool this water. There are undoubtedly other possibilities. Can you think of any?

Example 4.6

Suggest some ideas for the chlorine problem in Example 4.2.

An alternative disinfectant to chlorine would be one possibility. A chemical would need to be developed that has similar disinfectant properties to chlorine, but without the smell and irritation. This would address those problems but not the problem of the time and effort needed to add disinfectant. Another possibility is some type of automatic dispenser for chlorine or a replacement disinfectant, so that a disinfectant reservoir would only have to be changed periodically and would not require daily attention. Another possibility is a device that makes chlorine from a less toxic chemical, in situ, somewhere in the filtration system. This would keep chlorine out of the pool, confining it to the region where water is circulated through the filter. There are undoubtedly other possibilities. Can you think of any?

Example 4.7

Suggest some ideas for a power source for laptop computers, as discussed in Example 4.4.

One possibility is to develop better batteries. This is currently being done. Early laptop computers used NiCd batteries. The next generation was Ni-metal-hydride batteries, and the current generation is Li-ion batteries. There will probably be another generation forthcoming. Another possibility is to develop fuel cells that can be used to power a laptop computer. There may be other possibilities. Can you think of any?

4.4 SELECTION

Once a sufficient number of ideas has been generated, it is necessary to screen the ideas and select a few for more detailed investigation. Scientific principles can be applied. If it is thermodynamically impossible to manufacture an alternative, that idea can be eliminated. If it is determined that the kinetics of a desired reaction are unfavorable, that idea might be eliminated or downgraded, although this might stimulate development of a new catalyst to improve selectivity. If it is possible to determine at this stage that an alternative will be far too expensive relative to another idea, that idea might be eliminated or downgraded. However, when in doubt, it is probably best not to reject any idea too soon.

There are more quantitative methods for screening alternatives. One set of methods known as concept screening and concept scoring [4] will be briefly summarized here. More details can be found in Ulrich and Eppinger [4]. These methods are useful in that they allow subjective assessments to be quantified systematically for comparison purposes.

In **concept screening,** a selection matrix is prepared by listing a set of criteria to be used to evaluate the alternatives. Then one alternative is chosen as a reference alternative. This should be an alternative with which the team doing the evaluation is most familiar, perhaps an industry standard. All criteria for the reference standard are assigned a value of zero, meaning "same as." The criteria for all other alternatives are assigned values of +, meaning "better than"; zero; or −, meaning "worse than." Then the number of "worse thans" is subtracted from the number of "better thans." The net score for each alternative provides a relative ranking. Some type of reflection is needed at this stage to determine whether the results make sense and whether each criterion was assigned a reasonable value. The number of alternatives is now reduced, though it is up to those involved to determine how many alternatives survive to the next step. An example of concept screening is shown in Table 4.2. Here, alternative 5 is chosen as the reference alternative. It is observed that alternatives with equal scores are assigned the same rank. To proceed to the next step, we will assume that only four alternatives—those with positive scores—remain in the selection process.

In **concept scoring,** the same matrix is used, but only on those alternatives that have survived the concept screening process. The results are now more quantitative. Each criterion is now assigned a relative weight, which reflects the team's judgment as to its relative importance. A reference alternative is chosen. Then, for each alternative, each criterion is assigned a value from 1 to 5, where 1 = much worse than reference, 2 = worse than reference, 3 = same as reference, 4 = better than reference, and 5 = much better than reference. The score is calculated for each alternative by weighting the evaluations using the relative weights. Once again, some degree of reflection on the result is needed because this is a subjective process, particularly the assigning of relative weights. The best alternative is the one with the highest score. Because there is a large degree of subjectivity here, care should be exercised when differentiating between alternatives with close scores. Table 4.3 illustrates concept scoring for the four alternatives chosen during concept screening. Based on this method, alternative 1 is chosen for further study, although alternative 7 is close. Once again, small differences in total score may not be significant. Also, information obtained during product develop-

Table 4.2 Example of Concept Screening

	Alternative								
Criterion	**1**	**2**	**3**	**4**	**5**	**6**	**7**	**8**	**9**
1	+	−	0	+	0	+	0	−	0
2	0	+	+	0	0	0	+	0	+
3	+	0	0	0	0	+	+	−	0
4	0	0	−	0	0	−	+	0	0
5	0	−	+	−	0	0	−	+	−
Total Score	**2**	**−1**	**1**	**0**	**0**	**1**	**2**	**−1**	**0**
Rank	1	8	3	5	5	3	1	8	5

Table 4.3 Example of Concept Scoring

Criterion	Weight	Alternative 1	Alternative 3	Alternative 6	Alternative 7
1	25%	5	3	4	3
2	5%	3	4	3	4
3	15%	5	3	5	5
4	35%	3	1	2	5
5	20%	3	4	3	1
Total Score		**3.80**	**2.55**	**3.20**	**3.65**
Rank		**1**	**3**	**4**	**2**

ment may change the relative weights and/or individual scores sufficiently so that the total score changes enough and alternative 7 is actually the best choice.

Caution should be exercised before using this method. There are subtleties associated with it, which are explained in more detail elsewhere [4].

4.5 MANUFACTURE

This final step in the chemical product design structure is the most detailed. It includes determining whether the product can be manufactured, developing detailed product specifications, determining how the product is to be manufactured, and estimating the cost of manufacturing. It also includes sample or prototype testing, which may result in changes in the selection process and undoubtedly will result in modifications in every step of the manufacturing process until the optimal product and manufacturing process is obtained.

These feedback loops in the manufacturing process exist for all the manufacture of any product, even a commodity chemical. Before a multimillion-dollar plant is constructed, a pilot plant is usually constructed. Before a new chemical product is manufactured, small quantities are made in the laboratory to determine whether the product satisfies the need for which it was designed. Similarly, before a device is manufactured, a prototype is built and tested.

One lesson is that device manufacture is likely to be a very interdisciplinary effort. In the magnetic refrigerator example, Example 4.3, mechanical engineers would be needed for the pulley system, and electrical engineers might be needed for the control systems. Industrial engineers may be needed to determine the most efficient manufacturing procedure and to help determine the unit cost in mass production, because the cost of a prototype always exceeds the unit cost in mass production. When interdisciplinary efforts are needed, it is recommended that the interdisciplinary team be involved from the beginning, if possible. Example 4.8 illustrates the type of product that might be manufactured.

Example 4.8

Suppose the following scenario has evolved for the zebra mussel problem discussed in Examples 4.1 and 4.5. It is not possible to use a filter to prevent zebra mussel infestation because in the veliger (infancy) stage, zebra mussels are microscopic. They attach to the wall of the intake pipe, where they grow into maturity. Once the walls are saturated, they stack on each other, eventually occluding the pipe. Therefore, some type of chemical treatment is desirable. It has been determined from experimentation that alkylbenzyldimethylammonium chloride (alkyl chains between 12 and 16 carbons) will kill existing infestations, and dead mussels detach from the wall [10]. Also, assume that it has also been determined that 0.3 wt % hydrogen peroxide will inhibit veliger attachment. Describe the manufacturing stage.

A delivery system is needed, both for the initial kill and for the hydrogen peroxide to prevent infestation. One possible solution is to design and market a technology for delivery of these chemicals. For example, suppose that a grating for the intake pipe containing flow channels with holes discharging into the intake pipe were designed. A pumping system would be needed to deliver the chemicals through the holes in the grating. If the fluid mechanics of the discharge into the intake pipe were studied to optimize hole placement, the hole placement could be optimized to ensure that the chemicals covered the entire cross section of the pipe at the desired concentration. This technology could then be marketed to water treatment facilities and power plants to prevent zebra mussel infestation.

4.6 BATCH PROCESSING

In the manufacture of a chemical product that is actually a chemical, batch operations are often employed. This is because specialty chemical products are usually produced in small batches. In the Douglas hierarchy discussed in Chapter 2, the first decision to be made in designing a chemical process is batch versus continuous. For production of a commodity chemical in the quantities reflected in the examples on the CD-ROM accompanying this book, the choice will always be a continuous process. Similarly, for production of a specialty chemical, the choice will almost always be a batch process.

The issues involved in batch processing were discussed in Chapter 3.

4.7 ECONOMIC CONSIDERATIONS

When a new process is constructed for a commodity chemical, the sale price for the chemical is largely determined by the price competitors charge for the same chemical. However, the law of supply and demand does affect the price. If new capacity exists without additional demand, the value of the chemical may drop; if new capacity is created in response to a demand, the value of the chemical can probably be estimated from its value before the demand increased. Either way, the value of the chemical can probably be bracketed reasonably easily.

However, when a new product enters the market, the initial price usually reflects the value of its uniqueness. We see this every day. When new electronic devices enter the market (CD players, DVD players, projection TVs, HDTV), they usually carry a high price tag. In part, this is because they are not being produced in large quantities, and in part it is because there are customers who will pay a huge premium to be the first to have one. Eventually, prices decrease to attract new customers and then decrease significantly if the product becomes a commodity. Pharmaceuticals, an example of a chemical product, also carry a high price tag when they are new. Pharmaceutical companies must recover the extremely high costs of product research and development and the regulatory process before their patents expire and low-cost, generic alternatives become available, or before a competitor invents a superior alternative.

Therefore, although the profitability criteria that will be discussed in Chapter 10 can be used to evaluate the economics of chemical products, the details of the analysis may change. Years of research and development costs are included as capital costs. However, remember that there may be ten to fifteen years of such costs, and the time value of money requires that the price charged for the product must be high to obtain a favorable rate of return. Furthermore, there is risk with developing new products. One way to include risk in the profitability calculations that will be discussed in Chapter 10 is to increase the desired rate of return, which also increases the price of the product. (This is similar to the practice of lending institutions charging more for a loan to consumers with weaker credit histories, because they are poorer credit risks.)

4.8 SUMMARY

The challenges of chemical product design are different from those of chemical process design. These challenges include dealing with customer needs, screening alternatives, batch processing and scheduling, and the need for interdisciplinary teams more than in chemical process design. This chapter has been only a brief introduction to chemical product design. The major issues have been introduced, and examples have been presented to illustrate these principles. The readers interested in a more detailed treatment of product design should consult references [1], [3], and [4].

REFERENCES

1. Cussler, E. L., and G. D. Moggridge, *Chemical Product Design* (New York: Cambridge, 2001).
2. Cussler, E. L., "Do Changes in the Chemical Industry Imply Changes in Curriculum?" *Chem. Engr. Educ.* 33, no. 1 (1999) 12–17.
3. Dym, C. L., and P. Little, *Engineering Design: A Project-Based Introduction* (New York: Wiley, 2000).

4. Ulrich, K. T., and S. D. Eppinger, *Product Design and Development*, 4th ed. (New York: McGraw-Hill, 2008).
5. Fogler, H. S., and S. E. LeBlanc, *Strategies for Creative Problem Solving* (Upper Saddle River, NJ: Prentice Hall, 1995), 49.
6. Shaeiwitz, J. A., and R. Turton, "Chemical Product Design," Topical Conference Proceedings, *Chemical Engineering in the New Millennium—A First-Time Conference on Chemical Engineering Education*, 2000, 461–468.
7. Shaeiwitz, J. A., and R. Turton, "Chemical Product Design," *Chem. Engr. Educ.* 35, no. 4 (2001): 280–285.
8. http://www.cemr.wvu.edu/~wwwche/publications/projects/index.html.
9. Gschneider, K., and V. Pecharsky, "The Giant Magnetocaloric Effect in $Gd_5(Si_xGe_{1-x})_4$ Materials for Magnetic Refrigeration," *Advances in Cryogenic Engineering* (New York: Plenum, 1998), 1729.
10. Welker, B., "Development of an Environmentally Benign, Species-Specific Control Measure for *Corbicula fluminea*," *Science 21*, 2, no. 1 (1997): 11, 25–27.

CHAPTER

5

Tracing Chemicals through the Process Flow Diagram

In Chapter 2, we classified the unit operations from a PFD into one of the six blocks of a generic block flow process diagram. In this chapter, you gain a deeper understanding of a chemical process by learning how to trace the paths taken by chemical species through a chemical process.

5.1 GUIDELINES AND TACTICS FOR TRACING CHEMICALS

In this chapter, guidelines and some useful tactics are provided to help you trace chemicals through a process. Two important operations for tracing chemical pathways in PFDs are the adiabatic mixer and adiabatic splitter.

> **Mixer:** Two or more input streams are combined to form a single stream. This single output stream has a well-defined composition, phase(s), pressure, and temperature.
>
> **Splitter:** A single input stream is split into two or more output streams with the same temperature, pressure, and composition as the input stream. All streams involved differ only in flowrate.

These operations are found where streams meet or a stream divides on a PFD. They are little more than tees in pipelines in the plant. These operations involve little design and minimal cost. Hence, they are not important in estimating the capital cost of a plant and would not appear on a list of major equipment. However, you will find in Chapter 13 that these units are included in the design of flowsheets for implementing and using chemical process simulators.

We have highlighted the mixers and splitters as shaded boxes on the flow diagrams presented in this chapter. They carry an "m" and "s" designation, respectively.

5.2 TRACING PRIMARY PATHS TAKEN BY CHEMICALS IN A CHEMICAL PROCESS

Chemical species identified in the overall block flow process diagram (those associated with chemical reactions) are termed **primary chemicals.** The paths followed by primary chemicals between the reactor and the boundaries of the process are termed **primary flow paths.** Two general guidelines should be followed when tracing these primary chemicals.

1. **Reactants:** Start with the feed (left-hand side of the PFD) and trace chemicals forward toward the reactor.
2. **Products:** Start with the product (right-hand side of the PFD) and trace chemicals backward toward the reactor.

The following tactics for tracing chemicals apply to all unit operations *except for* chemical reactors.

Tactic 1: Any unit operation, or group of operations, that has a single or multiple input streams and a single output stream is traced in a forward direction. If chemical A is present in any input stream, it must appear in the single output stream (see Figure 5.1[a]).

Tactic 2: Any unit operation, or group of operations, that has a single input stream and single or multiple output streams is traced in a backward direction. If chemical A is present in any output stream, it must appear in the single input stream (see Figure 5.1[b]).

Tactic 3: Systems such as distillation columns are composed of multiple unit operations with a single input or output stream. It is sometimes necessary to consider such equipment combinations as blocks before implementing Tactics 1 and 2.

When tracing chemicals through a PFD, it is important to remember the following:

Only in reactors are feed chemicals transformed into product chemicals.

You may occasionally encounter situations where both reactions and physical separations take place in a single piece of equipment. In most cases, this is undesirable but unavoidable. In such situations, it will be necessary to divide the

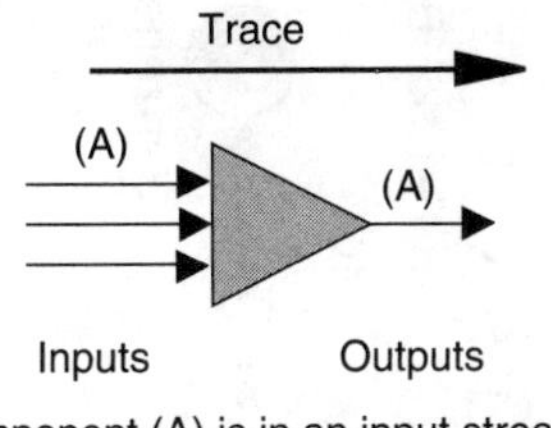

If component (A) is in an input stream, it will also be in the output stream.

(a)

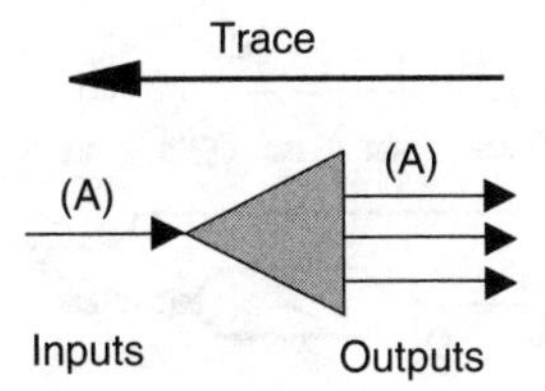

If component (A) is in an output stream, it will also be in the input stream.

(b)

Figure 5.1 Tactics for Tracing Chemical Species

unit into two imaginary, or phantom, units. The chemical reactions take place in one phantom unit, and the separation in the second phantom unit. These phantom units are never shown on the PFD, but we will see that such units are useful when building a flowsheet for a chemical process simulator (see Chapter 13).

We demonstrate these guidelines in Example 5.1, by determining the paths of the primary chemicals in the toluene hydrodealkylation process. The only information used is that provided in the skeleton process flow diagram given in Figure 1.3.

Example 5.1

For the toluene hydrodealkylation process, establish the primary flow pathway for

a. Toluene between the feed (Stream 1) and the reactor

b. Benzene between the reactor and the product (Stream 15)

Hint: Consider only one unit of the system at a time. Refer to Figure E5.1.

Toluene Feed: Tactic 1 is applied to each unit operation in succession.

a. Toluene feed Stream 1 mixes with Stream 11 in V-101. A single unidentified stream leaves tank V-101 and goes to pump P-101. All the toluene feed is in this stream.

b. Stream 2 leaves pump P-101 and goes to mixer m-102. All the feed toluene is in this stream.

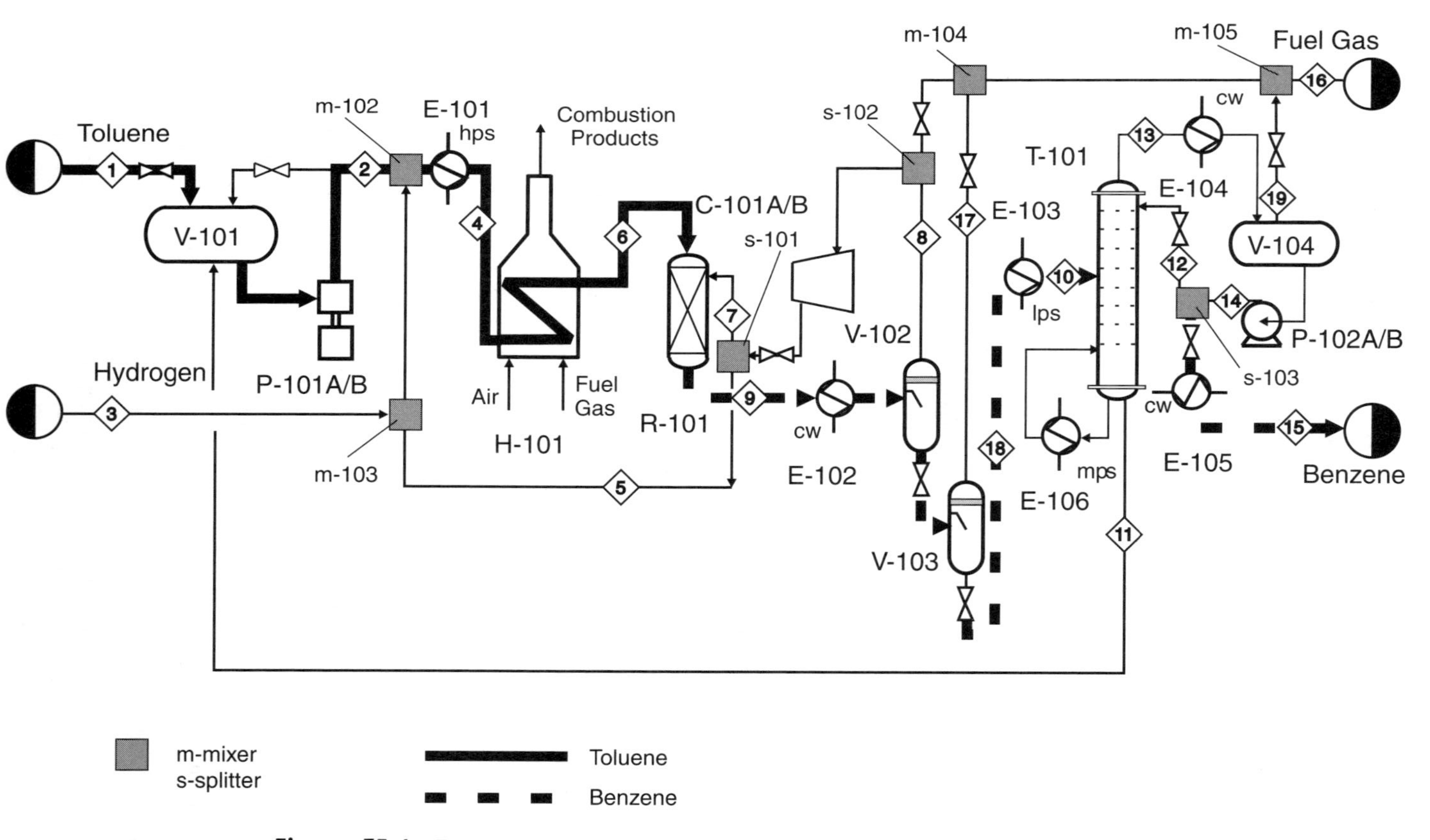

Figure E5.1 Primary Chemical Pathways for Benzene and Toluene in the Toluene Hydrodealkylation Process (Figure 1.3)

c. A single unidentified stream leaves mixer m-102 and goes to exchanger E-101. All the feed toluene is in this stream.

d. Stream 4 leaves exchanger E-101 and goes to heater H-101. All the feed toluene is in this stream.

e. Stream 6 leaves heater H-101 and goes to reactor R-101. All the feed toluene is in this stream.

Benzene Product: Tactic 2 is applied to each unit operation in succession.

a. Product Stream 15 leaves exchanger E-105.

b. Entering exchanger E-105 is an undesignated stream from s-103 of the distillation system. It contains all of the benzene product.

c. Apply Tactic 3 and treat the tower T-101, pump P-102, exchangers E-104 and E-106, vessel V-104, and splitter s-103 as a system.

d. Entering this distillation unit system is Stream 10 from exchanger E-103. It contains all the benzene product.

e. Entering exchanger E-103 is Stream 18 from vessel V-103. It contains all the benzene product.

f. Entering vessel V-103 is an undesignated stream from vessel V-102. It contains all the benzene product.

g. Entering vessel V-102 is an undesignated stream from exchanger E-102. It contains all the benzene product.

h. Entering exchanger E-102 is Stream 9 from reactor R-101. It contains all the benzene product.

The path for toluene was identified as an enhanced solid line in Example 5.1. For this case, it was not necessary to apply any additional information about the unit operations to establish this path. The two streams that joined the toluene path did not change the fact that all the feed toluene remained as part of the stream. All the toluene fed to the process in Stream 1 entered the reactor, and this path represents the primary path for toluene.

The path for benzene was identified as an enhanced dotted line in Example 5.1. The equipment that makes up the distillation system was considered as an operating system and treated as a single unit operation. The fact that, within this group of process units, some streams were split with some of the flow returning upstream did not change the fact that the product benzene always remained in the part of the stream that continued to flow toward the product discharge. All the benzene product followed this path, and it represents the primary path for the benzene. The flow path taken for the benzene through the distillation column section is shown in more detail in Figure 5.2. The concept of drawing envelopes around groups of equipment in order to carry out material and energy balances is introduced early into the chemical engineering curriculum. This concept is essentially the same as the one used here to trace the path of benzene through the distillation column. The only information needed about unit operations used in this

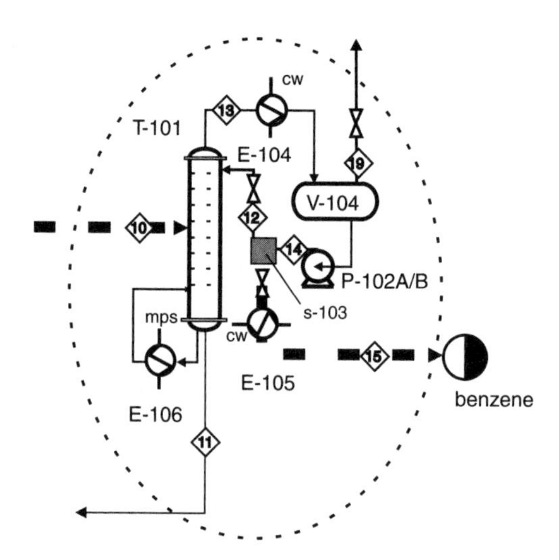

Figure 5.2 Envelope around Tower T-101 Showing Alternative Method for Tracing Benzene Stream

analysis was the identification of the multiple units that made up the distillation system. This procedure can be used to trace chemicals throughout the PFD and forms an alternative tracing method that is illustrated in Example 5.2.

Example 5.2

Establish the primary flow pathway for

a. Hydrogen between its introduction as a feed and the reactor
b. Methane between its generation in the reactor and the discharge from the process as a product

In order to determine the primary flow paths, we develop systems (by drawing envelopes around equipment) that progressively include additional unit operations. Reference should be made to Figure E5.2(a) for viewing and identifying systems for tracing hydrogen.

Hydrogen Feed: Tactic 1 is applied to each system in a forward progression. Each system includes the hydrogen feed Stream 3, and the next piece of equipment to the right.

System -a-: This system illustrates the first step in our analysis. The system includes the first unit into which the hydrogen feed stream flows. The unidentified stream leaving mixer m-103 contains the feed hydrogen.
System -b-: Includes mixers m-103 and m-102. The exit stream for this system includes the feed hydrogen.
System -c-: Includes mixers m-103, m-102, and exchanger E-101. The exit stream for this system, Stream 4, includes the feed hydrogen.

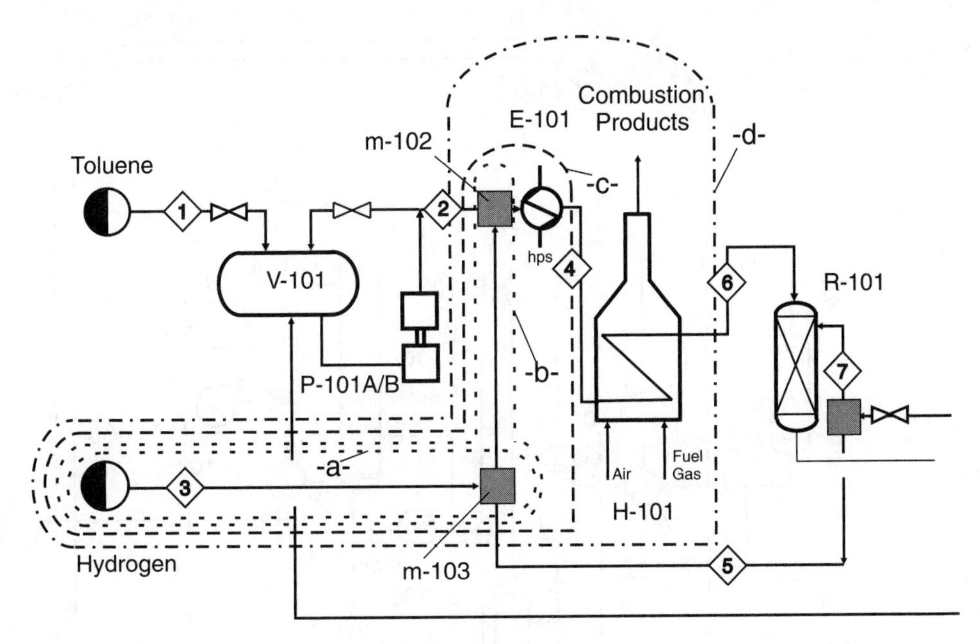

Figure E5.2(a) Tracing Primary Chemical Pathways Using the Envelope Method

System -d-: Includes mixers m-103, m-102, exchanger E-101, and heater 101. The exit stream for this system, Stream 6, includes the feed hydrogen. Stream 6 goes to the reactor.

The four steps described above are illustrated in Figure E5.2(a). A similar analysis is possible for tracing methane, and the steps necessary to do this are illustrated in Figure E5.2(b). These steps are discussed briefly below.

Methane Product: The methane produced in the process leaves in the fuel gas, Stream 16. Tactic 2 is applied to each system containing the fuel gas product, in backward progression.

System -m-: Consists of m-105, m-104, E-105, T-101, V-104, P-102, s-103, E-104, E-106, E-103, V-102, V-103, and s-102. This is the smallest system that can be found that contains the fuel gas product stream and has a single input.

System -n-: Includes the system identified above plus exchanger E-102 and compressor C-101. The inlet to E-102 contains all the methane in the fuel stream. This is Stream 9, which leaves the reactor.

In the first step of tracing methane, including only m-105 was attempted. This unit had two input streams, and it was not possible to determine which of these streams carried the methane that made up the product stream. Thus, Tactic 2 could not be used. In order to move ahead, additional units were added to m-105

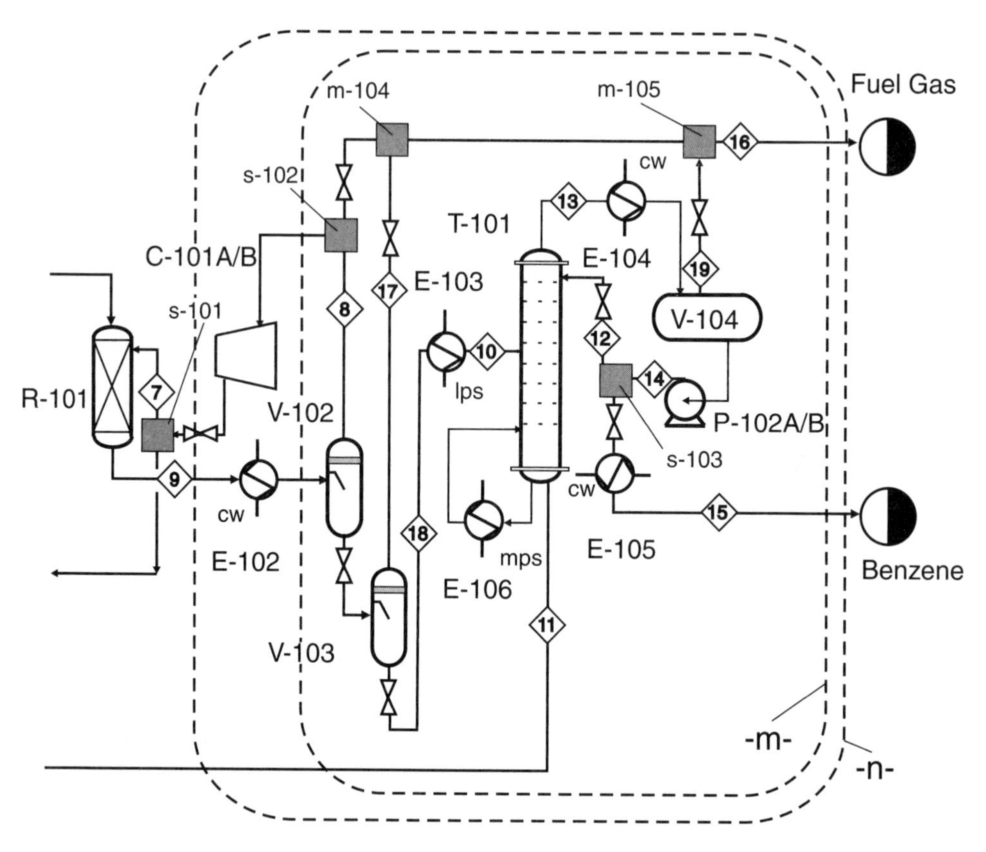

Figure E5.2(b) Tracing the Primary Flow Path for Methane in Toluene Hydrodealkylation PFD

to create a system that had a single input stream. The resulting system, System -m-, has a single input, with the unidentified stream coming from exchanger E-102. An identical problem would arise if the procedure used in Example 5.1 were implemented. Figure 5.3 shows the primary paths for the hydrogen and methane.

5.3 RECYCLE AND BYPASS STREAMS

It is important to be able to recognize recycle and bypass streams in chemical processes. When identifying recycle and bypass streams, we look for flow loops in the PFD. Any time we can identify a flow loop, we have either a recycle or a bypass stream. The direction of the streams, as indicated by the direction of the arrow heads, determines whether the loop contains a recycle or a bypass. The following tactics are applied to flow loops:

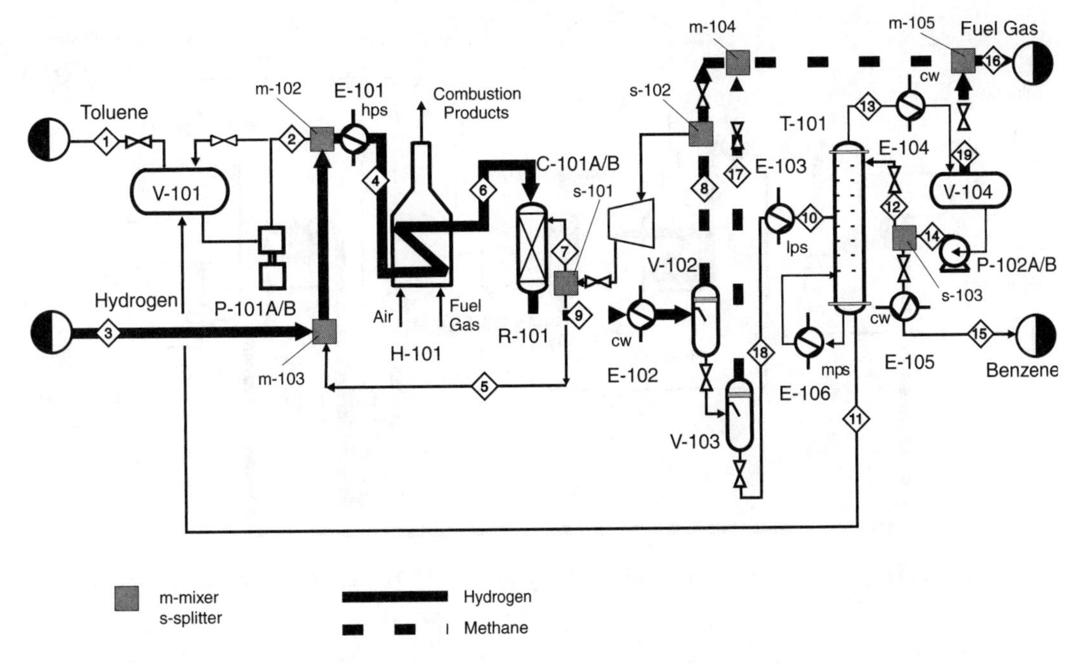

Figure 5.3 Primary Chemical Pathways for Methane and Hydrogen in the Toluene Hydrodealkylation PFD

Tactic 4: If the streams in a loop flow so that the flow path forms a complete circuit back to the point of origin, then it is a **recycle loop.**

Tactic 5: If the streams in a loop flow so that the flow path does not form a complete circuit back to the place of origin, then it is a **bypass stream.**

It is worth noting that certain pieces of equipment normally contain recycle streams. In particular, distillation columns very often have top and bottoms product reflux streams, which are essentially recycle loops. When identifying recycle loops, we can easily determine which loops contain reflux streams and which do not. Example 5.3 illustrates the procedure for identifying recycle and bypass streams in the toluene hydrodealkylation PFD.

Example 5.3

For the toluene hydrodealkylation PFD given in Figure E5.1, identify all recycle and bypass streams.

The recycle loops are identified in Figures E5.3(a) and E5.3(b). The main toluene recycle loop is highlighted in Figure E5.3(a), and the hydrogen recycle loops are shown in Figure E5.3(b)(a) and E5.3(b)(b). There are two reflux loops associated with T-101, and

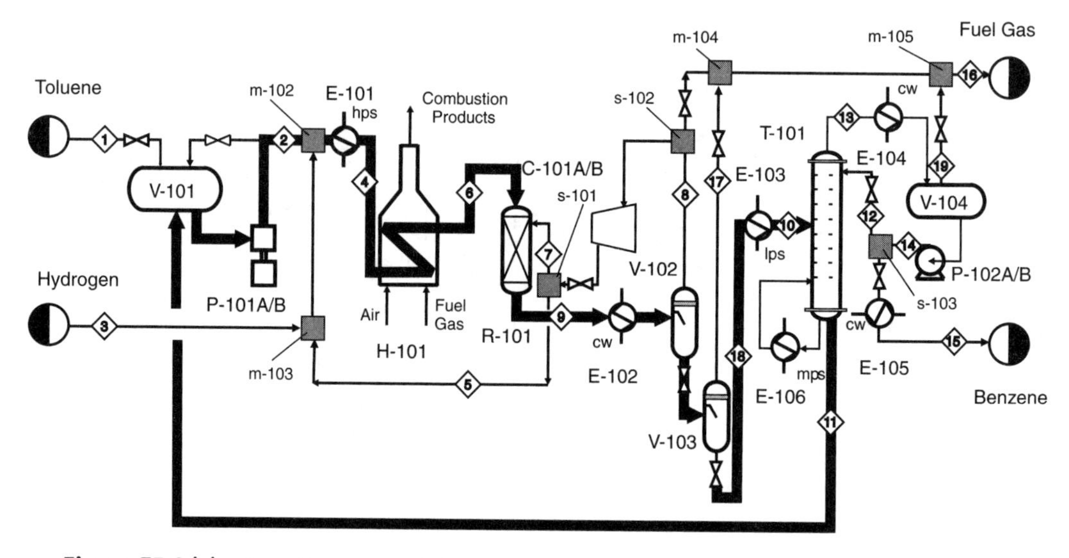

Figure E5.3(a) Identification of Toluene Recycle Loop in Toluene Hydrodealkylation PFD

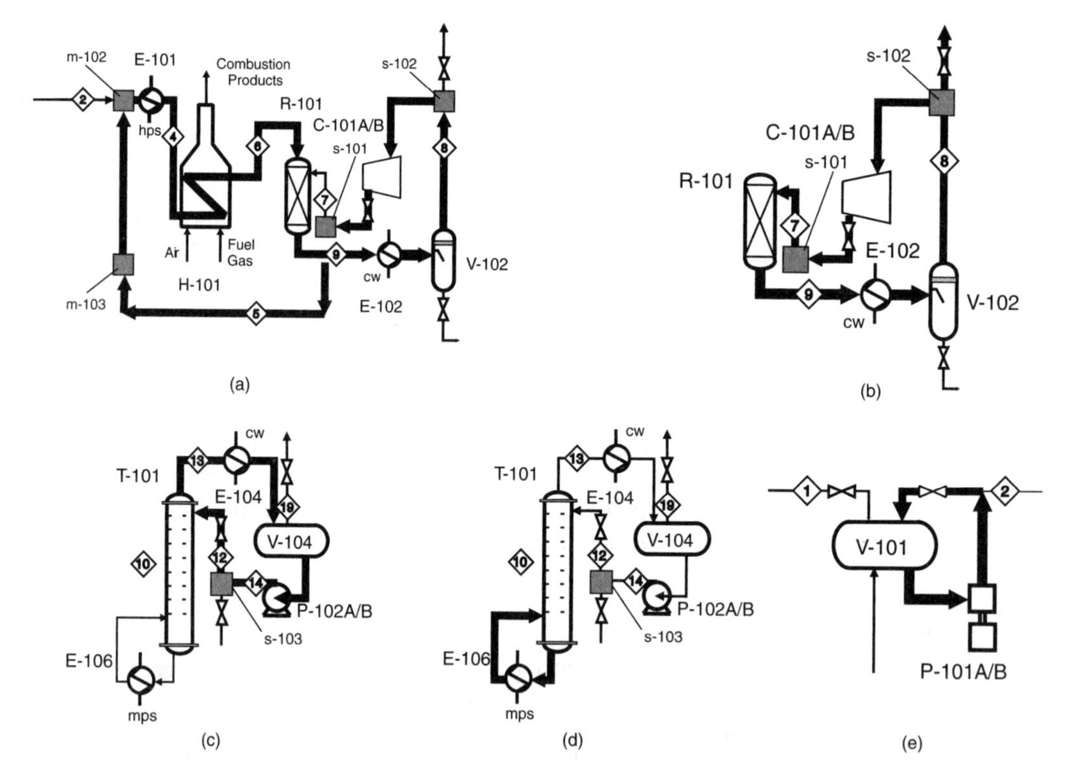

Figure E5.3(b) Identification of Other Recycle Loops in Toluene Hydrodealkylation PFD

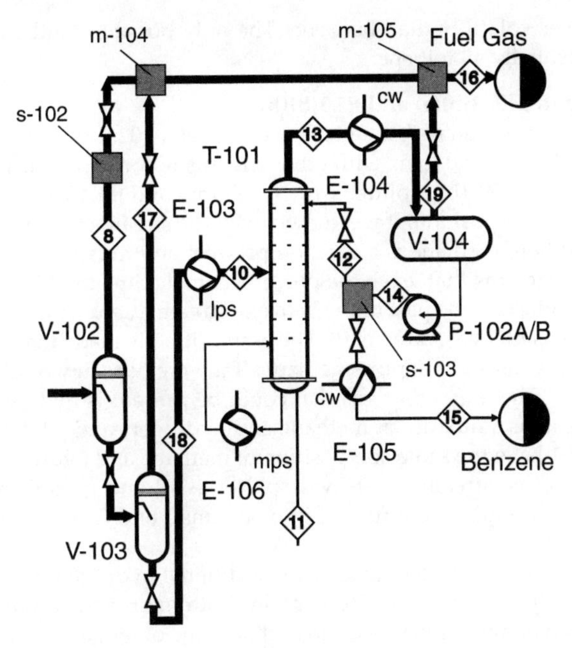

Figure E5.3(c) Identification of Bypass Streams in Toluene Hydrodealkylation PFD

these are shown in Figures E5.3(b)(c) and E5.3(b)(d). Finally, there is a second toluene recycle loop identified in Figure E5.3(b)(e). This recycle loop is used for control purposes (see Chapter 21) and is not discussed further here. The logic used to deduce what chemical is being recycled in each loop is discussed in the next example.

The bypass streams are identified in Figure E5.3(c). These bypass streams contain mostly hydrogen and methane and are combined to form the fuel gas stream, Stream 16.

It is important to remember that flow diagrams represent the most meaningful and useful documents to describe and understand a process. Although PFDs contain a lot of process information, it is sometimes necessary to apply additional knowledge about a unit operation to determine which chemicals are contained in a recycle stream. This idea is demonstrated in Example 5.4.

Example 5.4

Provide preliminary identification of the important chemical species in each of the three recycle streams identified in Example 5.3. See Figures E5.3(a), E5.3(b)(a), and E5.3(b)(b).

Figure E5.3(a)

Stream 11: This is the bottoms product stream out of the distillation tower that provides the product benzene as distillate. The bottoms product stream must have a lower volatility than benzene. The only possible candidate is toluene. Stream 11 is essentially all toluene.

Figures E5.3(b)(a) and E5.3(b)(b)
Two undesignated streams leave splitter s-102: One stream leaves as part of a product stream and joins with other streams to form Stream 16. The other stream passes through C-101 to splitter s-101. The input and the two streams leaving s-101 have the same composition. If we know any of the stream compositions, we know them all. In addition, methane is a reaction product and must leave the process. There are only two streams that leave the process, namely Streams 15 and 16. Because the methane is unlikely to be part of the benzene stream, it must therefore be in the stream identified as fuel gas, Stream 16. The assumption is made that the product stream leaving is gaseous and not pure methane. If it were pure, it would be labeled methane.

The only other gas that could be present is hydrogen. Therefore, the fuel gas stream is a mixture of methane and hydrogen, and all three streams associated with s-101 have the same composition of methane and hydrogen.

The stream that leaves splitter s-102 and goes through compressor C-101 to splitter s-101 is split further into Streams 5 and 7. All streams have the same composition.

Stream 5 then mixes with additional hydrogen from Stream 3 in mixer m-103. The stream leaving m-103 contains both hydrogen and methane, but with a composition of hydrogen greater than that in the other gas streams discussed.

Finally, Stream 7, which also leaves splitter s-101, flows back to the reactor and forms the third recycle stream.

Before the analysis in Example 5.4 can be accepted, it is necessary to check out the assumption used to develop the analysis. Up to this point, we have used the skeleton flow diagram that did not provide the important temperatures, pressures, and flowrates that are seen in the completed PFD (Figure 1.5). Figure 1.5 gives the following information for the flowrates of reactants:

Hydrogen (Stream 3): 572 kg/h (286.0 kmol/h)
Toluene (Stream 1): 10,000 kg/h (108.7 kmol/h)

Based on the information given in Table 1.5, only 108 kmol/h of hydrogen reacts to form benzene, and 178 kmol/h is excess reactant that leaves in the fuel gas. The fuel gas content is about 40 mole % methane and 60 mole % hydrogen. This confirms the assumption made in Example 5.4.

5.4 TRACING NONREACTING CHEMICALS

Chemical processes often contain nonreacting, or inert, compounds. These chemicals must appear in both the input and output streams and are neither created nor destroyed in the process. Unlike the reactants, it makes no difference in what direction we choose to trace these nonreacting chemicals. You can trace in the forward direction, the backward direction, or start in the middle and trace in both directions. Other than this additional flexibility, the tactics provided above can be applied to all nonreacting chemicals.

5.5 LIMITATIONS

When the tracing procedure resorts to combining several unit operations into a single system that provides a single stream, the path is incomplete. This can be seen in the paths of both product streams, methane and benzene, in Figure E5.1.

Benzene: The benzene flows into and out of the distillation system as the figure shows. There is no indication how it moves through the internal units consisting of V-104, s-103, E-104, E-106, and T-101.

Methane: The methane flows into and out of a system composed of V-102, V-103, s-102, and m-104. Again, there is no indication of the methane path.

In order to determine the performance and the flows through these compound systems, you need more information than provided in the skeleton PFD, and you must know the function of each of the units.

The development given in the previous sections used only the information provided on the skeleton PFD, without the description of the unit operation, and did not include the important flows, temperatures, and pressures that were given in the full PFD (Figure 1.5) and the flow table (Table 1.5). With this additional information and knowledge of the unit operations, you will be able to fill in some of the paths that are yet unknown.

Each step in tracing the flow paths increases our understanding of the process for the production of benzene represented in the PFD. As a last resort, reference should be made to the flow table to determine the composition of the streams, but this fails to develop analytical skills that are essential to understand the process.

5.6 WRITTEN PROCESS DESCRIPTION

A process description, like a flow table, is often included with a PFD. When a description is not included, it is necessary to provide a description based upon the PFD. Based on the techniques developed in this and Chapter 1, you should be able to write a detailed description of the toluene hydrodealkylation process. Table 5.1 provides such a description. You should read this description carefully and make sure you understand it fully. It would be useful, if not essential, to refer to the PFD in Figure 1.5 during your review. It is a good idea to have the PFD in front of you while you follow the process description.

Table 5.1 Process Description of the Toluene Hydrodealkylation Process (Refer to Figures 5.3 and 1.5)

Fresh toluene, Stream 1, is combined with recycled toluene, Stream 11, in the storage tank, V-101. Toluene from the storage tank is pumped, via P-101, up to a pressure of 25.8 bar and combined with the recycled and fresh hydrogen streams, Streams 3 and 5. This two-phase mixture is then fed through the feed preheater exchanger, E-101, where its temperature is raised to 225°C, and the toluene is completely vaporized. Further heating is accomplished in the heater, H-101, where the temperature of the stream is raised to 600°C. The stream leaving the heater, Stream 6, enters the reactor, R-101, at 600°C and 25.0 bar. The reactor consists of a vertical packed bed of catalyst, down through which the hot gas stream flows. The hydrogen and toluene react catalytically to produce benzene and methane according to the following exothermic reaction:

$$\underset{toluene}{C_7H_8} + H_2 \rightarrow \underset{benzene}{C_6H_6} + CH_4$$

The reactor effluent, Stream 9, consisting of benzene and methane produced from the reaction, along with the unreacted toluene and hydrogen, is quenched in exchanger E-102, where the temperature is reduced to 38°C using cooling water. Most of the benzene and toluene condenses in E-102, and the two-phase mixture leaving this exchanger is then fed to the high-pressure phase separator, V-102, where the liquid and vapor streams are allowed to disengage.

The liquid stream leaving V-102 is flashed to a pressure of 2.8 bar and is then fed to the low-pressure phase separator, V-103. The liquid leaving V-103, Stream 18, contains toluene and benzene with only trace amounts of dissolved methane and hydrogen. This stream is heated in exchanger E-103 to a temperature of 90°C prior to being fed to the benzene purification column, T-101. The benzene column, T-101, contains 42 sieve trays and operates at approximately 2.5 bar. The overhead vapor, Stream 13, from the column is condensed using cooling water in E-104, and the condensate is collected in the reflux drum, V-104. Any methane and hydrogen in the column feed accumulates in V-104, and these noncondensables, Stream 19, are sent to fuel gas. The condensed overhead vapor stream is fed from V-104 to the reflux pump P-102. The liquid stream leaving P-102, Stream 14, is split into two, one portion of which, Stream 12, is returned to the column to provide reflux. The other portion of the condensed liquid is cooled to 38°C in E-105, prior to being sent to storage as benzene product, Stream 15. The bottoms product from T-101, Stream 11, contains virtually all of the toluene fed to the column and is recycled back to V-101 for further processing.

The vapor stream leaving V-102 contains most of the methane and hydrogen in the reactor effluent stream plus small quantities of benzene and toluene. This stream is split into two, with one portion being fed to the recycle gas compressor, C-101. The stream leaving C-101 is again split into two. The major portion is contained in Stream 5, which is recycled back to the front end of the process, where it is combined with fresh hydrogen feed, Stream 3, prior to being mixed with the toluene feed upstream of E-101. The remaining gas leaving C-101, Stream 7, is used for temperature control in the reactor, R-101. The second portion of the vapor leaving V-102 constitutes the major portion of the fuel gas stream. This stream is first reduced in pressure and then combined with the flashed vapor from V-103, Stream 17, and with the noncondensables from the overhead reflux drum, Stream 19. The combination of these three streams is the total fuel gas product from the process, Stream 16.

The process description should capture all the knowledge that you have developed in the last two chapters and represents a culmination of our understanding of the process up to this point.

5.7 SUMMARY

This chapter showed how to trace many of the chemical species through a PFD, based solely upon the information shown on the skeleton PFD. It introduced operations involving splitting and mixing, not explicitly shown on the PFD, which were helpful in tracing these streams.

For situations where there was no single input or output stream, systems containing multiple unit operations were created. The tracing techniques for these compound systems did not provide the information needed to determine the internal flows for these systems. In order to determine reflux ratios for columns, for example, the process flow table must be consulted.

With the information provided, an authoritative description of the process can be prepared.

PROBLEMS

Identify the main reactant and product process streams for the following:

1. The ethylbenzene process shown in Figure B.2.1, Appendix B.
2. The styrene production facility shown in Figure B.3.1, Appendix B.
3. The drying oil production facility shown in Figure B.4.1, Appendix B.
4. The maleic anhydride production process shown in Figure B.5.1, Appendix B.
5. The ethylene oxide anhydride production process shown in Figure B.6.1, Appendix B.
6. The formalin production process shown in Figure B.7.1, Appendix B.

Identify the main recycle and bypass streams for the following:

7. The styrene production facility shown in Figure B.3.1, Appendix B.
8. The drying oil production facility shown in Figure B.4.1, Appendix B.
9. The maleic anhydride production process shown in Figure B.5.1, Appendix B.

Write a process description for the following:

10. The ethylbenzene process shown in Figure B.2.1, Appendix B.
11. The drying oil production facility shown in Figure B.4.1, Appendix B.
12. The ethylene oxide production facility shown in Figure B.6.1, Appendix B.

CHAPTER 6

Understanding Process Conditions

In previous chapters, process flow diagrams (PFDs) were accepted without evaluating the technical features of the process. The process topology and process operating conditions were provided but were not examined. Economic evaluations were carried out, but without confirming that the process would operate as indicated by the flow diagram.

It is not uncommon to investigate process economics based upon assumed process performance. For example, in order to justify spending the capital to develop a new catalyst, the economics of a process using a hypothetical catalyst with assumed characteristics, such as no unwanted side reactions, might be calculated.

The ability to make an economic analysis of a chemical process based on a PFD is not proof that the process will actually work.

In this chapter, you will learn how to analyze the reasons why the specific temperatures, pressures, and compositions selected for important streams and unit operations have been chosen. Stream specifications and process conditions are influenced by physical processes as well as economic considerations and are not chosen arbitrarily. The conditions used in a process most often represent an economic compromise between process performance and the capital and operating costs of the process equipment. Final selection of operating conditions should not be made prior to the analysis of the process economics. In this chapter, we concentrate on analyzing process conditions that require special consideration. As an example, we do not address why a reactor is run at 600°C instead of 580°C, but rather concentrate on the reasons why the reactor is not run at a much lower temperature, for example, 200°C. This type of analysis leads us to question how process conditions are chosen and makes us consider the consequences of changing these conditions.

6.1 CONDITIONS OF SPECIAL CONCERN FOR THE OPERATION OF SEPARATION AND REACTOR SYSTEMS

Process streams are rarely available at conditions most suitable for reactor and separation units. Temperatures, pressures, and stream compositions must be adjusted to provide conditions that allow effective process performance. This is discussed in Chapter 2, where the generic BFD was introduced (see Figure 2.4[a]). This figure showed two feed preparation blocks: one associated with the reactor and the second with the separation section.

We provide two generalizations to assist you in analyzing and understanding the selection of process conditions.

- It is usually easier to adjust the temperature and/or pressure of a stream than it is to change its composition. In fact, often the concentration of a compound in a stream (for a gas) is a dependent variable and is controlled by the temperature and pressure of the stream.
- In general, pressures between 1 and 10 bar and temperatures between 40°C and 260°C do not cause severe processing difficulties.

The rationale for the conditions given in the second generalization are explained below.

6.1.1 Pressure

There are economic advantages associated with operating equipment at greater than ambient pressure when gases are present. These result from the increase in gas density and a decrease in gas volume with increasing pressure. All other things being equal, in order to maintain the same gas residence time in a piece of equipment, the size of the equipment through which the gas stream flows need not be as large when the pressure is increased.

Most chemical processing equipment can withstand pressures up to 10 bar without much additional capital investment (see the cost curves in Appendix A). At pressures greater than 10 bar, thicker walled, more expensive equipment is necessary. Likewise, operating at less than ambient pressure (vacuum conditions) tends to make equipment large and may require special construction techniques, thus increasing the cost of equipment.

A decision to operate outside the pressure range of 1 to 10 bar must be justified.

6.1.2 Temperature

There are several critical temperature limits that apply to chemical processes. At elevated temperatures, common construction materials (primarily carbon steel) suffer a significant drop in physical strength and must be replaced by more costly materials. This drop in strength with temperature is illustrated in Example 6.1.

Example 6.1

The maximum allowable tensile strengths for a typical carbon steel and stainless steel, at ambient temperature, 400°C, and 550°C are provided below (Walas [1]).

	Tensile Strength of Material at Temperature Indicated (bar)		
Temperature	**Ambient**	**400°C**	**550°C**
Carbon Steel (grade 70)	1190	970	170
Stainless Steel (Type 302)	1290	1290	430

Determine the fractional decrease in the maximum allowable tensile strength (relative to the strength at ambient conditions) for the temperature intervals (a) ambient to 400°C and (b) 400°C to 550°C.

a. Interval: ambient to 400°C:
   Carbon Steel: (1190–970)/1190 = 0.18
   Stainless Steel: (1290–1290)/1290 = 0.0

b. Interval: 400°C to 550°C:
   Carbon Steel: (970–170)/1190 = 0.67
   Stainless Steel: (1290–430)/1290 = 0.67

Example 6.1 has shown that carbon steel suffers a loss of 18%, and stainless steel suffers no loss in tensile strength, when heated to 400°C. With an additional temperature increase of 150°C to 550°C, stainless steel suffers a 67% loss while carbon steel suffers an additional 67% loss in strength. At operating temperatures of 550°C, carbon steel has a maximum allowable tensile strength of about 15% of its value at ambient conditions. For stainless steel, the maximum allowable strength at 550°C is about 33% of its ambient value. For this example, it is clear that carbon steel is unacceptable for service temperatures greater than 400°C, and that the use of stainless steel is severely limited. For higher service temperatures, more exotic (and expensive) alloys are required and/or equipment may have to be refractory lined.

A decision to operate at greater than 400°C must be justified.

Thus, if we specify higher temperatures, we must be able to justify the economic penalty associated with more complicated processing equipment, such as refractory-lined vessels or exotic materials of construction. In addition to the critical temperature of 400°C, there are temperature limits associated with the availability of common utilities for heating and cooling a process stream.

Steam: High-pressure steam between 40 and 50 bar is commonly available and provides heat at 250 to 265°C. Above this temperature additional costs are involved.

Water: Water from a cooling tower is commonly available at about 30°C (and is returned to the cooling tower at around 40°C). For utilities below this temperature, costs increase due to refrigeration. As the temperature decreases, the costs increase dramatically (see Table 8.3).

If cryogenic conditions are necessary, there may be an additional need for expensive materials of construction.

A decision to operate outside the range of 40°C to 260°C, thus requiring special heating/cooling media, must be justified.

6.2 REASONS FOR OPERATING AT CONDITIONS OF SPECIAL CONCERN

When you review the PFD for different processes, you are likely to encounter conditions in reactors and separators that lie outside the temperature and pressure ranges presented in Section 6.1. This does not mean to say that these are "bad" processes, but rather that these conditions had to be used, despite the additional costs involved, to have the process operate effectively. These conditions, outside the favored temperature and pressure ranges, are identified as **conditions of special concern.**

When you encounter these conditions, you should seek a rational explanation for their selection. If no explanation can be identified, the condition used may be unnecessary. In this situation, the condition may be changed to a less severe one that provides an economic advantage.

A list of possible justifications for using temperature and pressure conditions outside the ranges given above are identified in Tables 6.1 through 6.3. The material provided in Tables 6.1 to 6.3 is based upon elementary concepts presented in undergraduate texts covering thermodynamics and reactor design.

We describe below the rationale used to justify operating at temperatures that are of special concern presented in Table 6.1. The justification for entries in Tables 6.2 and 6.3 has a similar rationale.

For chemical reactors, we considered the following items in our justification:

1. **Favorable Equilibrium Conversion:** If the reaction is endothermic and approaches equilibrium, it benefits from operating at high temperatures. We recall Le Chatelier's principle, which states, "For a reacting system at equilibrium, the extent of the reaction will change so as to oppose any changes in temperature or pressure." For an endothermic reaction, an increase in temperature tends to push the reaction equilibrium to the right (toward products). Conversely, low temperatures decrease the equilibrium conversion.
2. **Increase Reaction Rates:** All chemical reaction rates are strongly dependent upon temperature through an Arrhenius type equation:

$$k_{reaction} = k_0 e^{-\frac{E_{act}}{RT}} \tag{6.1}$$

As temperature increases, so does the reaction rate constant, $k_{reaction}$, for both catalytic and noncatalytic reactions. Therefore, temperatures greater than 250°C may be required to obtain a high enough reaction rate in order to keep the size of the reaction vessel reasonable.
3. **Maintain a Gas Phase:** Many catalytic chemical reactions used in processes today require both reactants and products to be in the gas phase. For high-boiling-point materials or operations where high pressure is used, a temperature in excess of 400°C may be required in the reactor in order to maintain all the species in the vapor phase.
4. **Improve Selectivity:** If competing reactions (series, parallel, or a combination of both) occur and the different reactions have different activation energies, then the production of the desired product may be favored by using a high temperature. Schemes for competing reactions are covered in greater detail in many of the well-known texts on chemical reaction engineering, as well as in Chapter 20.

Table 6.1 Possible Reasons for Operating Reactors and Separators Outside the Temperature Ranges of Special Concern

Stream Condition	Process Justification for Operating at This Condition	Penalty for Operating at This Condition
High Temperature [T > 250°C]	**Reactors** • Favorable equilibrium conversion for endothermic reactions • Increase reaction rates • Maintain a gas phase • Improve selectivity • •	• Use of special process heaters • T > 400°C requires special materials of construction • • •
	Separators • Obtain a gas phase required for vapor-liquid equilibrium • •	
Low Temperature [T < 40°C]	**Reactors** • Favorable equilibrium conversion for exothermic reactions • Temperature-sensitive materials • Improve selectivity • Maintain a liquid phase • •	• Uses expensive refrigerant • May require special materials of construction for very low temperatures • • •
	Separators • Obtain a liquid phase required for vapor-liquid or liquid-liquid equilibrium • Obtain a solid phase for crystallization • Temperature-sensitive materials • •	

Table 6.2 Possible Reasons for Operating Reactors and Separators Outside the Pressure Range of Special Concern

Stream Condition	Process Justification for Operating at This Condition	Penalty for Operating at This Condition
High Pressure (P > 10 bar)	**Reactors** • Favorable equilibrium conversion • Increase reaction rates for gas phase reactions (due to higher concentration) • Maintain a liquid phase • •	• Requires thicker-walled equipment • Requires expensive compressors if gas streams must be compressed • • •
	Separators • Obtain a liquid phase for vapor-liquid or liquid-liquid equilibrium • •	
Low Pressure (P < 1 bar)	**Reactors** • Favorable equilibrium conversion • Maintain a gas phase • •	• Requires large equipment • Special design for vacuum operation • Air leaks into equipment that may be dangerous and expensive to prevent • • •
	Separators • Obtain a gas phase for vapor-liquid equilibrium • Temperature-sensitive materials • •	

Table 6.3 Possible Reasons for Nonstoichiometric Reactor Feed Compositions of Special Concern

Stream Condition	Process Justification for Operating at This Condition	Penalty for Operating at This Condition
Inert Material in Feed to Reactor	• Acts as a diluent to control the rate of reaction and/or to ensure that the reaction mixture is outside the explosive limits (exothermic reactions) • Inhibits unwanted side reactions • •	• Causes reactor and downstream equipment to be larger since inert takes up space • Requires separation equipment to remove inert material • May cause side reactions (material is no longer inert) • Decreases equilibrium conversion • •
Excess Reactant	• Increases the equilibrium conversion of the limiting reactant • Inhibits unwanted side reactions • •	• Requires separation equipment to remove excess reactant • Requires recycle • Added feed material costs (due to losses in separation and/or no recycle) • •
Product Present in Feed to Reactor	• Product cannot easily be separated from recycled feed material. • Recycled product retards the formation of unwanted by-products formed from side reactions. • Product acts as a diluent to control the rate of reaction and/or to ensure that the reaction mixture is outside the explosive limits, for exothermic reactions. • •	• Causes reactor and downstream equipment to be larger • Requires larger recycle loop • Decreases equilibrium conversion • •

For separators, we considered the following item of justification:

1. **Obtain a Vapor Phase for Vapor-Liquid Equilibrium:** This situation arises quite frequently when high-boiling-point materials need to be distilled. An example is the distillation of crude oil in which the bottom of the atmospheric column is typically operated in the region of 310°C to 340°C (590°F to 645°F).

You would benefit by spending time to acquaint yourself with the information presented in Tables 6.1 to 6.3 and to convince yourself that you understand the justifications given in these tables. These tables should not be considered an exhaustive list of possible reasons for operating in the ranges of special concern. Instead, they represent a starting point in analyzing process conditions. As you discover other explanations for reasons to operate equipment in the ranges of special concern you may wish to add them to Tables 6.1–6.3. Additional blank entries are provided for this purpose.

6.3 CONDITIONS OF SPECIAL CONCERN FOR THE OPERATION OF OTHER EQUIPMENT

Additional equipment (such as pumps, compressors, heaters, exchangers, and valves) produces the temperature and pressure required by the feed streams entering the reactor and separation sections. When initially choosing the stream conditions for the reactor and separator sections, it is worthwhile using certain guidelines or heuristics. These technical heuristics are useful guidelines for doing design. Comprehensive lists of heuristics are described and applied in Chapter 11. In this chapter, we present some of the more general guidelines that apply to streams passing through process equipment. These are presented in Table 6.4. Some of these guidelines are explored in Example 6.2.

Table 6.4 Changes in Process Conditions That Are of Special Concern for a Stream Passing through a Single Piece of Equipment

Type of Equipment	Change in Stream Condition Causing Concern	Justification or Remedy	Penalty for Operating Equipment in This Manner
1. Compressors	$P_{out}/P_{in} > 3$	**Remedy:** Use multiple stages and intercoolers.	High theoretical work requirement due to large temperature rise of gas stream.
	High-temperature inlet gas	**Remedy:** Cool the gas before compression.	High theoretical work requirement and special construction materials required.
2. Heat Exchangers	$\Delta T_{lm} > 100°C$	**Remedy:** Integrate heat better within process (see Chapter 13).	Large temperature driving force means we are wasting valuable high-temperature energy.
		Justification: Heat integration not possible or not profitable.	

(continued)

Table 6.4 Changes in Process Conditions That Are of Special Concern for a Stream Passing through a Single Piece of Equipment (*Continued*)

Type of Equipment	Change in Stream Condition Causing Concern	Justification or Remedy	Penalty for Operating Equipment in This Manner
3. Process Heaters	$T_{out} < T_{steam\ available}$	**Remedy:** Use high-pressure steam to heat process stream.	Process heaters are expensive and unnecessary if heating can be accomplished by using an available utility.
		Justification: Heater may be needed during start-up.	
4. Valves	Large ΔP across valve	**Remedy:** For gas streams install a turbine to recover lost work.	Wasteful expenditure of energy due to throttling.
		Justification: (a) Valve used for control purposes. (b) Installation of turbine not profitable. (c) Liquid is being throttled.	
5. Mixers (Streams Mixing)	Streams of greatly differing temperatures mix	**Remedy:** Bring temperatures of streams closer together using heat integration.	Wasteful expenditure of high-temperature energy.
	Streams of greatly differing composition mix	**Justification:** (a) Quenching of reaction products. (b) Provides driving force for mass transfer.	Causes extra separation equipment and cost.

Example 6.2

It is necessary to provide a nitrogen stream at 80°C and a pressure of 6 bar. The source of the nitrogen is at 200°C and 1.2 bar. Determine the work and cooling duty required for three alternatives.

a. Compress in a single compression stage and cool the compressed gas.
b. Cool the feed gas to 80°C and then repeat part a.
c. Repeat part b, except use two stages of compression with an intercooler.
d. Identify any conditions of special concern that occur.

Nitrogen can be treated as an ideal diatomic gas for this comparison. Use as a basis 1 kmol of nitrogen and assume that the efficiency, ε, of each stage of compression is 70%.

For ideal diatomic gas: $C_p = 3.5R$, $C_v = 2.5R$, $\gamma = C_p/C_v = 1.4$, $R = 8.314$ kJ/kmol K, and assuming an efficiency, $\varepsilon = 0.70$

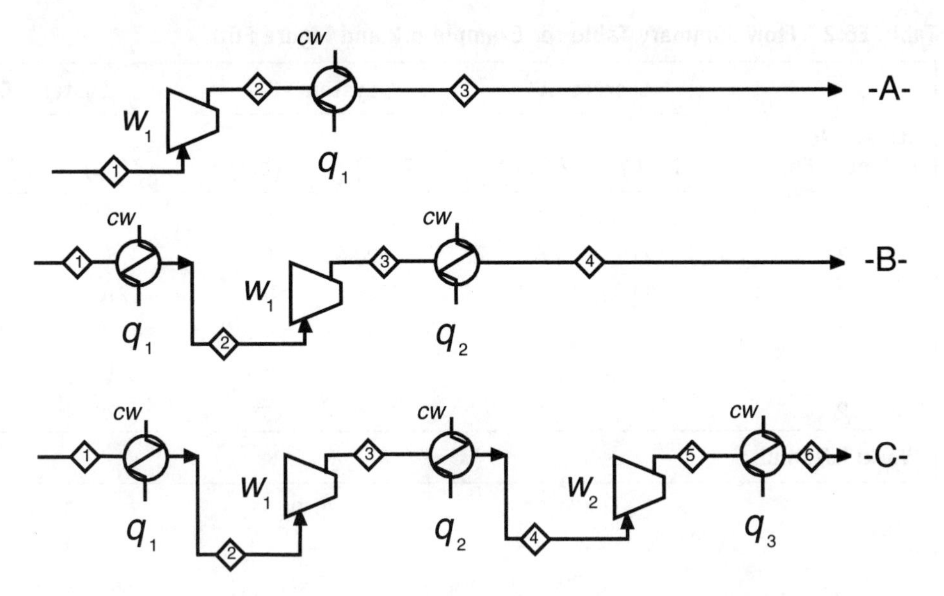

Figure E6.2 Alternative Process Schemes for Compression of Nitrogen

Equations used: $q = C_p\Delta T$, $w = RT_{in}\gamma/(\gamma - 1)[(P_{out}/P_{in})^{(\gamma-1)/\gamma} - 1]/\varepsilon$,

$$T_{out} = T_{in}\left(1 + \frac{1}{\varepsilon}[(P_{out}/P_{in})^{(\gamma-1)/\gamma} - 1]\right)$$

Figure E6.2 gives the process flow diagrams for the three alternatives and identifies stream numbers and utilities.

The results of the calculations for parts a, b, and c are provided in Table E6.2, which shows stream conditions and utility requirements. To keep the calculations simple, the pressure drops across and between equipment have been ignored.

Part d: Alternative -A- requires a compressor exit temperature of 595°C, which is a condition of special concern. Note also that although the intermediate temperature of the gas (stream) in Alternative -B- was 374°C, because this stream is to be cooled there are no concerns about utility requirements.

Example 6.2 showed three alternatives of differing complexity for achieving the same final conditions. The amount of work (w) and cooling utilities (q) required for each alternative were calculated. Based solely upon process complexity, Alternative -A- is the most desirable. However, this alternative has several disincentives that should be considered before final selection:

1. The highest electric utility demand and cost (assuming that the compressor is electrically driven).
2. The highest cooling utility demand and cost.
3. A condition of special concern, that is, T > 400°C (see Table 6.1).

   Note: Compressors are high-speed rotating devices where the loss of material strength and thermal expansion is critical. It would be expected that the

Table E6.2 Flow Summary Table for Example 6.2 and Figure E6.2

Stream No. in Figure E6.2	System -A-		System -B-		System -C-	
	T(°C)	P(bar)	T(°C)	P(bar)	T(°C)	P(bar)
1	200	1.2	200	1.2	200	1.2
2	595	6.0	80	1.2	80	1.2
3	80	6.0	374	6.0	210	2.68
4	—	—	80	6.0	80	2.68
5	—	—	—	—	210	6.0
6	—	—	—	—	80	6.0
Work: kJ/kmol						
w_1	11,470		8560		3780	
w_2	—		—		3780	
w_{total}	11,470		8560		7560	
Heat: kJ/kmol						
q_1	14,970		3490		3490	
q_2	—		8550		3780	
q_3	—		—		3780	
q_{total}	14,970		12,040		11,050	

purchase cost of the compressor would undergo a quantum jump for high-temperature operations.

4. Exceeds the 3:1 pressure ratio provided as a guideline (see Table 6.4).

Alternative -B- is more complex than Alternative -A- because it requires an additional heat exchanger, but it avoids the condition of special concern in Item 3. The result of using this extra exchanger is a significant decrease in utilities over Alternative -A-. As a result, it is likely that Alternative -B- would be preferred to Alternative -A-.

Alternative -C- requires an extra stage of compression and an additional cooler before the second compressor, something that is not required by Alternative -B-. However, Alternative -C- results in an additional savings in utilities over Alternative -B-.

The qualitative analysis given above suggests that both Alternatives -B- and -C- are superior to Alternative -A-. This conclusion is consistent with the two heuristics for compressors in Table 6.4: It is better to cool a hot gas prior to compressing it, and it is usually desirable to keep the compression ratio less than 3:1. Before a final selection is made, an economic analysis, which must include both the capital investment and the operating costs, should be carried out on each of the competing schemes. The equivalent annual operating cost (EAOC), described in Chapter 10, would be a suitable criterion to make such a comparison.

You should review the information given in Table 6.4 and convince yourself that you understand the rules, along with the penalties, remedies, and justifications, for operating equipment under these conditions. You may also be able to provide additional reasons why operating the equipment in this way would be justified. You should add these reasons to the list provided along with additional heuristics that are uncovered as you work problems and gain experience.

6.4 ANALYSIS OF IMPORTANT PROCESS CONDITIONS

In this section, we begin to analyze and to justify the conditions of special concern found in a process flow diagram. To help with this analysis, it is beneficial to prepare a process conditions matrix (PCM). In the PCM, all the equipment is listed vertically and the conditions of special concern are listed horizontally. Each unit is reviewed for conditions of special concern, and a check mark is used to identify which pieces of equipment have been identified. The PCM for the toluene hydrodealkylation process is shown in Table 6.5. The information for this PCM was obtained from Chapter 1, and you should verify that none of the areas of special concern have been missed.

We now consider and justify all of the special conditions identified in Table 6.5.

6.4.1 Evaluation of Reactor R-101

Three conditions of concern have been identified for the reactor. They are high temperature, high pressure, and non-stoichiometric feed conditions.

In order to understand why these conditions are needed, additional information about the toluene hydrodealkylation reaction is required. Table 6.6 provides additional but limited information. This information is divided into two groups.

Thermodynamic Information: This is information found in most chemical engineering thermodynamic textbooks:

a. Information required to perform energy balances, including heats of reaction and phase change, heat capacities, and so on

b. Information required to determine equilibrium conversion, including heats of formation, free energy of formation, and so on

Table 6.5 Process Conditions Matrix for the PFD of the Toluene Hydrodealklyation Process Shown in Figure 1.5

	Reactors and Separators Tables 6.1–6.3					Other Equipment Table 6.4				
Equipment	*High Temp*	*Low Temp*	*High Pres.*	*Low Pres.*	*Non-Stoich. Feed*	*Comp*	*Exch.*	*Htr.*	*Valve*	*Mix*
R-101	X		X		X					
V-101										
V-102			X							
V-103										
V-104										
T-101										
H-101										
E-101							X			
E-102							X			
E-103										
E-104										
E-105										
E-106										
C-101										
P-101										
P-102										
PCV on Stream 8									X	
PCV on Stream from V-101 to V-103									X	

Figure 6.1 is a plot of the heat of reaction and the equilibrium constant as a function of temperature, evaluated from the information provided in Table 6.6. From these plots it is evident that the chemical reaction is slightly exothermic, causing the equilibrium constant to decrease with temperature.

Reaction Kinetics Information: This information is reaction specific and must be obtained experimentally. The overall kinetics may involve homogeneous and heterogeneous reactions both catalytic and noncatalytic. The expressions are often complex.

Table 6.6 Equilibrium and Reaction Kinetics Data for the Toluene Hydrodealkylation Process

Reaction Stoichiometry

$$\underset{\text{toluene}}{C_6H_5CH_3} + H_2 = \underset{\text{benzene}}{C_6H_6} + CH_4$$

Equilibrium Constant (T is in units of K)

$$\ln(K_p) = 13.51 + \frac{5037}{T} - 2.073\ln(T) + 3.499 \times 10^{-4}T + 4.173 \times 10^{-8}T^2 + \frac{3017}{T^2}$$

Heat of Reaction

$$\Delta H_{reaction} = -37{,}190 - 17.24T + 29.09 \times 10^{-4}T^2 + 0.6939 \times 10^{-6}T^3 + \frac{50{,}160}{T} \quad \frac{\text{kJ}}{\text{kmol}}$$

At the Reaction Conditions of 600°C (873 K)

Equilibrium Constant, $K_p = 265$

Heat of Reaction, $\Delta H_{reaction} = -49{,}500 \ \frac{\text{kJ}}{\text{kmol}}$

Information on Reaction Kinetics

No side reactions

Reaction is kinetically controlled

Before a process is commercialized, reaction kinetics information, such as space velocity and residence times, must be obtained for different temperatures and pressures from pilot plant studies. Such data are necessary to design the reactor. At this point, we are not interested in the reactor design, and hence specific kinetics expressions have not been included in Table 6.6 and are not necessary for the following analysis.

The analysis of the reactor takes place in two parts.

a. Evaluation of the special conditions from the thermodynamic point of view. This assumes that chemical equilibrium is reached and provides a limiting case.

b. Evaluation of the special conditions from the kinetics point of view. This accounts for the limitations imposed by reaction kinetics, mass transfer, and heat transfer.

If a process is unattractive under equilibrium (thermodynamic) conditions, analysis of the kinetics is not necessary. For processes in which the equilibrium conditions give favorable results, further study is necessary. The reason for this is that conditions that favor high equilibrium conversion may be unfavorable from the standpoint of reaction kinetics.

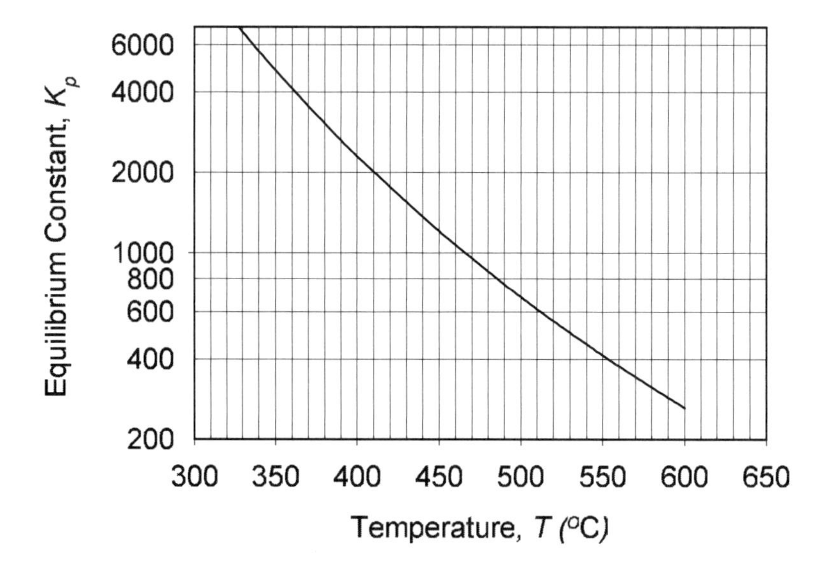

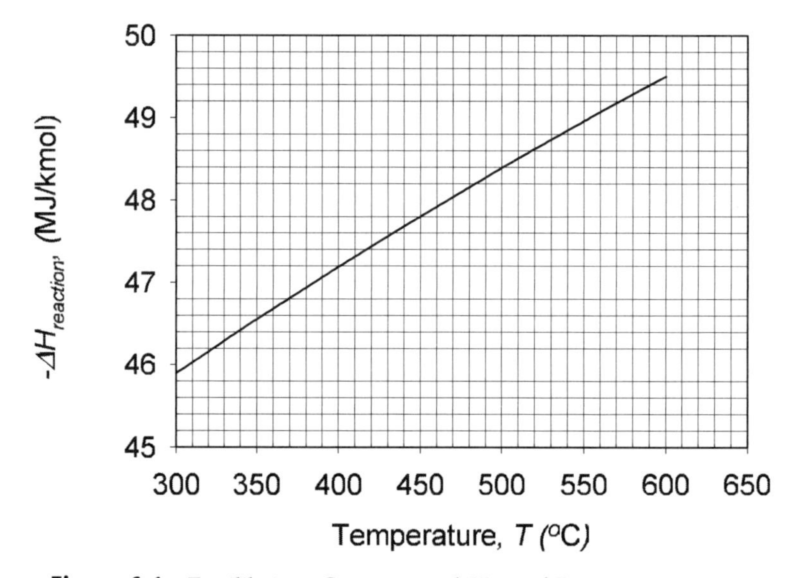

Figure 6.1 Equilibrium Constant and Heat of Reaction as a Function of Temperature for the Toluene Hydrodealkylation Reaction

Thermodynamic Considerations. We consider the use of high-temperature, high-pressure, and non-stoichiometric feed conditions separately.

High-Temperature Concern (see Table 6.1). Figure 6.1 provided the important information that the reaction is exothermic. Table 6.1 notes that for an exothermic reaction, the result of increasing temperature is a reduction in equilibrium conversion. This is confirmed by the plot of the equilibrium constant versus temperature given in Figure 6.1. The decrease in the equilibrium conversion is undesirable. The actual conversion for the HDA process is compared with the equilibrium conversion in Example 6.3.

Example 6.3

For the PFD presented in Figure 1.5,

a. Calculate the actual conversion.

b. Evaluate the equilibrium conversion at 600°C.

Assuming ideal gas behavior: $K_p = (N_{benzene} N_{methane})/(N_{toluene} N_{hydrogen})$
where N represents the moles of each species at equilibrium.

Information on the feed stream to the reactor from Table 1.5 (Stream 6 on Figure 1.5):

Hydrogen	735.4 kmol/h
Methane	317.3
Benzene	7.6
Toluene	144.0
Total	1204.3

c. Actual Conversion: Toluene in exit stream (Stream 9) = 36 kmol/h

Conversion = (144 – 36)/144 = 0.75 (75%)

d. Equilibrium Conversion at 600°C. From Table 6.6 @600°C $K_p = 265$

Let N = kmol/h of benzene formed
$265 = [(N + 7.6)(N + 317.3)]/[(735.4 - N)(144 - N)]$
$N = 143.6$
Equilibrium Conversion = 143.6/144 = 0.997 (99.7%)

The equilibrium conversion for the hydrodealkylation reaction remained high in spite of the high temperature. Although there is no real problem with using the elevated temperature in the reactor, it cannot be justified from a thermodynamic point of view.

High-Pressure Concern (see Table 6.2). From the reaction stoichiometry, we see that there are equal numbers of reactant and product moles in the hydrodealkylation reaction. For this case, there is no effect of pressure on equilibrium conversion. From a thermodynamic point of view there is no reason for the high pressure in the reactor.

Non-stoichiometric Feed (see Table 6.3). The component feed rates to the reactor (see Example 6.3) show that

1. Toluene is the limiting reactant.
2. Hydrogen is an excess reactant (more than 400% excess).
3. Methane, a reaction product, is present in significant amounts.

Reaction Products (Methane) in Feed. The presence of reaction product in the feed results in a reduction in the equilibrium conversion (see Table 6.3). However, Example 6.3 shows that at the conditions selected for the reactor, the equilibrium conversion remained high despite the presence of the methane in the feed.

Excess Reactant (Hydrogen) in Feed. The presence of excess reactants in the feed results in an increase in equilibrium conversion (see Table 6.3). Example 6.4 explores the effect of this excess hydrogen on conversion.

Example 6.4

(Reference Example 6.3). Reduce the amount of hydrogen in the feed to the reactor to the stoichiometric amount—that is, 144 kmol/h—and determine the effect on the equilibrium conversion at 600°C.

The calculations are not shown. They are similar to those in Example 6.3(b). The total moles of hydrogen in the feed were changed from 735.4 kmol/h to the stoichiometric value of 144 kmol/h.

The results obtained were $N = 128.8$ kmol/h, equilibrium conversion = 0.895 (89.5%).

Example 6.4 reveals that the presence of the large excess of hydrogen had a noticeable effect on the equilibrium conversion.

We conclude that thermodynamic considerations do not explain the selection of the high temperature, the high pressure, and the presence of reaction products in the feed. The presence of a large excess of hydrogen is the only positive effect predicted by thermodynamics.

Consideration of Reaction Kinetics. The information on reaction kinetics is limited in this chapter. We will present a more detailed description of the kinetics rate expression in a case study in Chapter 20, and we investigate the predictions made in this chapter with this limited information. However, you will find that a great deal of understanding can be extracted from the limited information presented here.

From the information provided in Table 6.6 and Chapter 1 we know that

1. The reaction takes place in the gas phase.
2. The reaction is kinetically controlled.
3. There are no significant side reactions.

High-Temperature Concern (see Table 6.1). In a region where the reaction kinetics control, the reaction rate increases rapidly with temperature, as Example 6.5 illustrates.

Example 6.5

The activation energy for the rate of reaction for the hydrodealkylation of toluene is equal to 148.1 kJ/mol (Tarhan [2]). What is the reaction rate at 600°C relative to that at 400°C?

$$\text{Ratio of Reaction Rates} = \exp[-E/R\{1/T_2 - 1/T_1\}]$$
$$= \exp[148100/8.314\{1/673 - 1/873\}] = 430$$

The size of a reactor would increase by nearly three orders of magnitude if the reaction were carried out at 400°C (the critical temperature for materials selection, Table 6.1) rather than 600°C. Clearly the effect of temperature is significant.

Most reactions are not kinetically controlled as is the case here. In most cases the rate is controlled by heat or mass transfer considerations. These are not as sensitive to temperature changes as chemical reaction rates. For more detail, see Chapter 20.

High-Pressure Concern (see Table 6.2). For gas phase reactions, the concentration of reactants is proportional to the pressure. For a situation where the reaction rate is directly proportional to the concentration, operation at 25 bar rather than at 1 bar would increase the reaction rate by a factor of 25 (assuming ideal gas behavior). Although we do not know that the rate is directly proportional to the concentration, we can predict that the effect of pressure is likely to be substantial, and the reactor size will be substantially reduced.

Non-stoichiometric Feed (see Table 6.3). The reactor feed contains both excess hydrogen and the reaction product methane.

Methane in the Feed. The effect of methane is to reduce the reactant concentrations. This decreases the reaction rate and represents a negative impact. The methane could possibly reduce the formation of side products, but we have no information to suggest that this is the case.

Excess Hydrogen in the Feed. The large amount of excess hydrogen in the feed ensures that the concentration of hydrogen will remain large throughout the reactor. This increases the reaction rate. Although there is no information provided regarding the decision to maintain the high hydrogen levels, it may be linked to reducing the formation of side products.

With the exception of the presence of methane product in the feed, the high-temperature operation, the excess hydrogen, and the elevated pressure all support an increase in reaction rate and a reduction in reactor volume. This suggests that the catalyst is not "hot"—that is, the catalyst is still operating in the reaction-controlled regime and mass transfer effects have not started to intrude. For these conditions, the manipulation of temperatures and pressures is essential to limit the reactor size.

There is a significant economic penalty for using more than 400% excess hydrogen in the reactor feed. The raw material cost of hydrogen would be reduced significantly if excess hydrogen were not used. The fact that this large excess is used in spite of the economic penalty involved suggests that the hydrogen plays an important role in the prevention of side products. The concept of selectivity is discussed further in Chapter 20.

The presence of methane in the feed has not yet been resolved. At best it behaves as an inert and occupies volume that must be handled downstream of the reactor, thus making all the equipment larger and more expensive. This question is considered in more detail in Example 6.6.

Example 6.6

It has been proposed that we handle the hydrogen/methane stream in the same manner that we handled the toluene/benzene stream. We recall that the unreacted toluene was separated from the benzene product and then recycled. It is proposed that the methane be separated from the hydrogen. The methane would then become a process by-product and the hydrogen would be recycled. Discuss this proposal using the arguments provided in Tables 6.1 and 6.2.

To use distillation for the separation of methane from hydrogen, as was used with the toluene/benzene, requires a liquid phase. For methane/hydrogen systems, this requires extremely high pressures together with cryogenic temperatures.

If the hydrogen could be separated from the methane and recycled, then the reactor feed would not contain significant quantities of methane, and the large excess of hydrogen could be maintained without the steep cost of excess hydrogen feed. Note that the overall conversion of hydrogen in the process is only 37%, whereas for toluene it is 99%.

Alternative separation schemes that do not require a liquid phase (e.g., a membrane separator) should be considered. The use of alternative separation technologies is addressed further in Chapter 12.

6.4.2 Evaluation of High-Pressure Phase Separator V-102

This vessel separates toluene and benzene as a liquid from the noncondensable gases hydrogen and methane. The reactor product is cooled and forms a vapor and a liquid stream that are in equilibrium. The vapor-liquid equilibrium is that at the temperature and pressure of the stream entering V-102. From Tables 6.1 and 6.2, we conclude that the lower temperature (38°C) was provided to obtain a liquid phase for the vapor-liquid equilibrium. The pressure was maintained to support the formation of the liquid phase. Because the separation can be affected relatively easily at high pressure, it is worthwhile maintaining V-102 at this high pressure.

6.4.3 Evaluation of Large Temperature Driving Force in Exchanger E-101

There is a large temperature driving force in this exchanger, because the heating medium is at a temperature of approximately 250°C, and the inlet to the exchanger is only 30°C. This is greater than the 100°C suggested in Table 6.4. This is an example of poor heat integration, and we will take a closer look at improving this in Chapter 15 (also see the case study presented in Chapter 28).

6.4.4 Evaluation of Exchanger E-102

Stream 9 is cooled from 654°C to 40°C using cooling water at approximately 35°C. Again this is greater than the 100°C suggested in Table 6.4, and the process stream has a lot of valuable energy that is being wasted. Again, we can save a lot of money by using heat integration (see Chapter 15).

6.4.5 Pressure Control Valve on Stream 8

The purpose of this control valve is to reduce the pressure of the stream entering the fuel gas line from 23.9 bar to 2.5 bar. This reduction in pressure represents a potential loss of useful work due to the throttling action of the valve. Referring to Table 6.4, we can see that when we throttle a gas, we can recover work by using a turbine, although this may not be economically attractive. The operation of this valve is justified because of its control function.

6.4.6 Pressure Control Valve on Stream from V-102 to V-103

The purpose of this valve is to reduce the pressure of the liquid leaving V-102. This reduction in pressure causes some additional flashing and recovery of dissolved methane and hydrogen from the toluene/benzene mixture. The flashed gas is separated in V-103 and sent to the fuel gas line. The purpose of this valve is to control the pressure of the material fed to the distillation column T-101. Because the stream passing through the valve is essentially all liquid, little useful work could be recovered from this stream.

This completes our review of the conditions of special concern for the toluene hydrodealkylation process.

6.5 SUMMARY

In this chapter, you learned to identify process conditions that are of special interest or concern in the analysis of the PFD. A series of tables was presented in which justifications for using process conditions of special concern were given. We introduced the process conditions matrix (PCM) for the toluene hydrodealkylation process and identified all the equipment in which process conditions of special concern existed. Finally, by comparing the process conditions from the

PFD to those given in the tables, we learned to analyze why these conditions were selected for the process and where improvements may be made.

REFERENCES

1. Walas, S. M., *Chemical Process Equipment: Selection and Design* (Stoneham, MA: Butterworth, 1988).
2. Tarhan, M. O., *Catalytic Reactor Design* (New York: McGraw-Hill, 1983).

SHORT ANSWER QUESTIONS

1. State two common criteria for setting the pressure of a distillation column.
2. Suggest two reasons each why distillation columns are run *above* or *below* ambient pressure. Be sure to state clearly which explanation is for above and which is for below ambient pressure.
3. Suggest two reasons why reactors are run at elevated pressures and/or temperatures. Be sure to state clearly which explanation is for elevated pressure and which is for elevated temperature.
4. Give two reasons why operation of a process greater than 250°C is undesirable. Give one reason each why one would operate a distillation column and a reactor at temperature greater than 250°C.
5. Define a "condition of special concern." Define two such conditions, and state one possible justification for each.
6. In the food and drug industries, many processes used to produce new active ingredients (drugs) or to separate and purify drugs and foods occur at vacuum conditions and often at low temperatures (less than room temperature). What is it about these types of products that requires that these conditions of special concern be used?

PROBLEMS

7. For the separation of a binary mixture in a distillation column, what will be the effect of an increase in column pressure on the following variables?
   a. Tendency to flood at a fixed reflux ratio
   b. Reflux ratio for a given top and bottom purity at a constant number of stages
   c. Number of stages required for a given top and bottom purity at constant reflux ratio
   d. Overhead condenser temperature

8. In a new chemical process, a reboiler for a tower requires a heating medium at 290°C. Two possible solutions have been suggested: (a) use high-pressure steam superheated to 320°C, and (b) use saturated steam at 320°C. Suggest one disadvantage for each suggestion.
9. As the ambient temperature and humidity increase, the temperature at which cooling water (cw) can be supplied to any piece of equipment increases. For example, in the winter, cw may be available at 27°C whereas in midsummer it may rise to 34°C. How, if at all, does this affect the pressure at which a distillation column operates (assuming that the overhead condenser uses cooling water as the heat exchange utility)?
10. It is desired to produce a hot vapor stream of benzene to feed a reactor for a certain petrochemical process. The benzene is available from an off-site storage facility at 1 atm pressure and ambient temperature (assume 25°C), and the reactor requires the benzene to be at 250°C at 10 atm. Two possible process schemes are being considered to heat and pressurize the feed: (1) pump the liquid benzene to pressure and then vaporize it in a heat exchanger, and (2) vaporize the benzene first and then compress it to the desired pressure. Answer the following.
    a. Discuss qualitatively which scheme (if either) is better.
    b. Confirm your answer to part (a) by comparing the costs using both schemes to feed 1000 kg/h of benzene to the reactor. (Assume that the cost of heating is $15/GJ and that electricity costs $0.06/kWh.)
11. One way to produce very pure oxygen and nitrogen is to separate air using a distillation process. For such a separation determine the following.
    a. Find the normal boiling point (at 1 atm pressure) of nitrogen and oxygen.
    b. For a distillation column operating at 1 atm pressure, what would be the top and bottom temperatures and top and bottom compositions of a distillation column that separates air into nitrogen and oxygen? (For this problem, you may assume that air contains only nitrogen and oxygen and that pure components leave at the top and bottom of the column.)
    c. At what pressure can oxygen and nitrogen be liquefied at ambient temperature (say 40°C)?
    d. What does the answer to part (c) tell you about the potential to distill air at ambient conditions?
12. The production of ammonia (a key ingredient for fertilizer) using the Haber process takes place at temperatures of around 500°C and pressures of 250 atm using a porous iron catalyst according the following highly exothermic synthesis reaction:

$$N_2(g) + 3H_2(g) \rightleftharpoons 2NH_3(g) \qquad \Delta H = -92.4 \text{ kJ/mol}$$

Give possible reasons for the high temperature and pressure used for this reaction.

13. Consider the ammonia process in Problem 12. For the given conditions, the maximum single-pass conversion obtained in the reactor is about 15–20%. Explain how the temperature and pressure should be adjusted to increase this conversion and the penalties for making these changes.
14. For the production of drying oil shown as Project B.4 in Appendix B, do the following.
    a. Construct a process conditions matrix (PCM) for the process, and determine all conditions of special concern.
    b. For each condition of special concern identified in part (a), suggest at least one reason why such a condition was used.
    c. For each condition of special concern identified in part (a), suggest at least one process alternative to eliminate the condition.
15. For the styrene production process given in Project B.3 in Appendix B, do the following.
    a. Construct a process conditions matrix (PCM) for the process, and determine all conditions of special concern.
    b. Explain the reasons for using the conditions of special concern in the reactor.
    c. Suggest any process alternatives for part (b).

SECTION

2

Engineering Economic Analysis of Chemical Processes

In this section, we concentrate on the evaluation of the economics of a chemical process. In order for a chemical engineer or cost engineer to evaluate the economic impact of a new (or existing) chemical process, certain technical information must be available. Although this information is gleaned from a variety of sources, it is generally presented in the form of the technical diagrams discussed in Chapter 1.

In the chapters of this section, methods to evaluate the economics of a chemical process are covered. The term *economics* refers to the evaluation of capital costs and operating costs associated with the construction and operation of a chemical process. The methods by which the one-time costs associated with the construction of the plant and the continuing costs associated with the daily operation of the process are combined into meaningful economic criteria are provided.

This material is treated in the following chapters.

Chapter 7: Estimation of Capital Costs

The common types of estimates are presented along with the basic relationships for scaling costs with equipment size. The concept of cost inflation is presented, and some common cost indexes are presented. The concept of total fixed capital investment to construct a new process is discussed, and the cost module approach to estimating is given. Finally, the software program (CAPCOST) to evaluate fixed capital costs (and other financial calculations) is described.

Chapter 8: Estimation of Manufacturing Costs

The basic components of the manufacturing costs of a process are presented. A method to relate the total cost of manufacturing (COM) to five elements—fixed capital investment, cost of operating labor, cost of raw materials, cost of utilities, and cost of waste treatment—is given. Examples of how utility costs can be calculated from the basic costs of fuel, power, and water are discussed. The estimation of labor costs based on the size and complexity of the process are also given.

Chapter 9: Engineering Economic Analysis

The concept of the time value of money is discussed. The following topics are presented: simple and compound interest, effective and nominal interest rates, annuities, cash flow diagrams, and discount factors. In addition, the concepts of depreciation, inflation, and taxation are covered.

Chapter 10: Profitability Analysis

The ideas discussed in Chapter 9 are extended to evaluate the profitability of chemical processes. Profitability criteria using nondiscounted and discounted bases are presented and include net present value (NPV), discounted cash flow rate of return (DCFROR), and payback period (PBP). A discussion of evaluating equipment alternatives using equivalent annual operating costs (EAOC) and other methods is presented. Finally, the concept of evaluating risk is covered and an introduction to the Monte Carlo method is presented.

CHAPTER

7

Estimation of Capital Costs

In Chapter 1, the information provided on a process flow diagram, including a stream table and an equipment summary table, was presented. In the next four chapters, this information will be used as a basis for estimating

1. How much money (capital cost) it takes to build a new chemical plant
2. How much money (operating cost) it takes to operate a chemical plant
3. How to combine items 1 and 2 to provide several distinct types of composite values reflecting process profitability
4. How to select a "best process" from competing alternatives
5. How to estimate the economic value of making process changes and modifications to an existing processes
6. How to quantify uncertainty when evaluating the economic potential of a process

In this chapter, we concentrate on the estimation of capital costs. **Capital cost** pertains to the costs associated with construction of a new plant or modifications to an existing chemical manufacturing plant.

7.1 CLASSIFICATIONS OF CAPITAL COST ESTIMATES

There are five generally accepted classifications of capital cost estimates that are most likely to be encountered in the process industries [1,2,3]:

1. Detailed estimate
2. Definitive estimate

3. Preliminary estimate
4. Study estimate
5. Order-of-magnitude estimate

The information required to perform each of these estimates is provided in Table 7.1.

Table 7.1 Summary of Capital Cost Estimating Classifications (References [1], [2], and [3])

Order-of-Magnitude (also known as Ratio or Feasibility) Estimate **Data:** This type of estimate typically relies on cost information for a complete process taken from previously built plants. This cost information is then adjusted using appropriate scaling factors, for capacity, and for inflation, to provide the estimated capital cost. **Diagrams:** Normally requires only a block flow diagram.
Study (also known as Major Equipment or Factored) Estimate **Data:** This type of estimate utilizes a list of the major equipment found in the process. This includes all pumps, compressors and turbines, columns and vessels, fired heaters, and exchangers. Each piece of equipment is roughly sized and the approximate cost determined. The total cost of equipment is then factored to give the estimated capital cost. **Diagrams:** Based on PFD as described in Chapter 1. Costs from generalized charts. **Note:** Most individual student designs are in this category.
Preliminary Design (also known as Scope) Estimate **Data:** This type of estimate requires more accurate sizing of equipment than used in the study estimate. In addition, approximate layout of equipment is made along with estimates of piping, instrumentation, and electrical requirements. Utilities are estimated. **Diagrams:** Based on PFD as described in Chapter 1. Includes vessel sketches for major equipment, preliminary plot plan, and elevation diagram. **Note:** Most large student group designs are in this category.
Definitive (also known as Project Control) Estimate **Data:** This type of estimate requires preliminary specifications for all the equipment, utilities, instrumentation, electrical, and off-sites. **Diagrams:** Final PFD, vessel sketches, plot plan, and elevation diagrams, utility balances, and a preliminary P&ID.
Detailed (also known as Firm or Contractor's) Estimate **Data:** This type of estimate requires complete engineering of the process and all related off-sites and utilities. Vendor quotes for all expensive items will have been obtained. At the end of a detailed estimate, the plant is ready to go to the construction stage. **Diagrams:** Final PFD and P&ID, vessel sketches, utility balances, plot plan and elevation diagrams, and piping isometrics. All diagrams are required to complete the construction of the plant if it is built.

Table 7.2 Classification of Cost Estimates

Class of Estimate	Level of Project Definition (as % of Complete Definition)	Typical Purpose of Estimate	Methodology (Estimating Method)	Expected Accuracy Range (+/- Range Relative to Best Index of 1)	Preparation Effort (Relative to Lowest Cost Index of 1)
Class 5	0% to 2%	Screening or Feasibility	Stochastic or Judgment	4 to 20	1
Class 4	1% to 15%	Concept Study or Feasibility	Primarily Stochastic	3 to 12	2 to 4
Class 3	10% to 40%	Budget, Authorization, or Control	Mixed but Primarily Stochastic	2 to 6	3 to 10
Class 2	30% to 70%	Control or Bid/Tender	Primarily Deterministic	1 to 3	5 to 20
Class 1	50% to 100%	Check Estimate or Bid/Tender	Deterministic	1	10 to 100

(From AACE Recommended Practice No. 17R-97 [4], reprinted with permission of AACE International, 209 Prairie Ave., Morgantown, WV; http://www.aacei.org)

The five classifications given in Table 7.1 roughly correspond to the five classes of estimate defined in the AACE Recommended Practice No. 17R-97 [4]. The accuracy range and the approximate cost for performing each class of estimate are given in Table 7.2.

In Table 7.2, the accuracy range associated with each class of estimate and the costs associated with carrying out the estimate are ranked relative to the most accurate class of estimate (Class 1). In order to use the information in Table 7.2, it is necessary to know the accuracy of a Class 1 estimate. For the cost estimation of a chemical plant, a Class 1 estimate (detailed estimate) is typically +6% to –4% accurate. This means that by doing such an estimate, the true cost of building the plant would likely be in the range of 6% higher than and 4% lower than the estimated price. Likewise, the effort to prepare a Class 5 estimate for a chemical process is typically in the range of 0.015% to 0.30% of the total installed cost of the plant [1,2].

The use of the information in Table 7.2, to estimate the accuracy and costs of performing estimates, is illustrated in Examples 7.1 and 7.2.

Example 7.1

The estimated capital cost for a chemical plant using the study estimate method (Class 4) was calculated to be $2 million. If the plant were to be built, over what range would you expect the actual capital estimate to vary?

For a Class 4 estimate, from Table 7.2, the expected accuracy range is between 3 and 12 times that of a Class 1 estimate. As noted in the text, a Class 1 estimate can be expected to vary from +6% to –4%. We can evaluate the narrowest and broadest expected capital cost ranges as follows.

Lowest Expected Cost Range

High value for actual plant cost $(\$2.0 \times 10^6)[1 + (0.06)(3)] = \$2.36 \times 10^6$

Low value for actual plant cost $(\$2.0 \times 10^6)[1 - (0.04)(3)] = \$1.76 \times 10^6$

Highest Expected Cost Range

High value for actual plant cost $(\$2.0 \times 10^6)[1 + (0.06)(12)] = \$3.44 \times 10^6$

Low value for actual plant cost $(\$2.0 \times 10^6)[1 - (0.04)(12)] = \$1.04 \times 10^6$

The actual expected range would depend on the level of project definition and effort. If the effort and definition are at the high end, then the expected cost range would be between \$1.76 and \$2.36 million. If the effort and definition are at the low end, then the expected cost range would be between \$1.04 and \$3.44 million.

The primary reason that capital costs are underestimated stems from the failure to include all of the equipment needed in the process. Typically, as a design progresses, the need for additional equipment is uncovered, and the estimate accuracy improves. The different ranges of cost estimates are illustrated in Example 7.2.

Example 7.2

Compare the costs for performing an order-of-magnitude estimate and a detailed estimate for a plant that cost $\$5.0 \times 10^6$ to build.

For the order-of-magnitude estimate, the cost of the estimate is in the range of 0.015% to 0.3% of the final cost of the plant:

Highest Expected Value: $(\$5.0 \times 10^6)(0.003) = \$15{,}000$

Lowest Expected Value: $(\$5.0 \times 10^6)(0.00015) = \$750$

For the detailed estimate, the cost of the estimate is in the range of 10 to 100 times that of the order-of-magnitude estimate.

For the lowest expected cost range:

Highest Expected Value: $(\$5.0 \times 10^6)(0.03) = \$150{,}000$

Lowest Expected Value: $(\$5.0 \times 10^6)(0.0015) = \$7500$

For the highest expected cost range:

Highest Expected Value: $(\$5.0 \times 10^6)(0.3) = \$1{,}500{,}000$

Lowest Expected Value: $(\$5.0 \times 10^6)(0.015) = \$75{,}000$

Capital cost estimates are essentially paper-and-pencil studies. The cost of making an estimate indicates the personnel hours required in order to complete the estimate. From Table 7.2 and Examples 7.1 and 7.2, the trend between the accuracy of an estimate and the cost of the estimate is clear. If greater accuracy is required in the capital cost estimate, then more time and money must be expended in conducting the estimate. This is the direct result of the greater detail required for the more accurate estimating techniques.

What cost estimation technique is appropriate? At the beginning of Chapter 1, a short narrative was given that introduced the evolution of a chemical process leading to the final design and construction of a chemical plant. Cost estimates are performed at each stage of this evolution.

There are many tens to hundreds of process systems examined at the block diagram level for each process that makes it to the construction stage. Most of the processes initially considered are screened out before any detailed cost estimates are made. Two major areas dominate this screening process. To continue process development, the process must be both technically sound and economically attractive.

A typical series of cost estimates that would be carried out in the narrative presented in Chapter 1 is as follows.

- Preliminary feasibility estimates (order-of-magnitude or study estimates) are made to compare many process alternatives.
- More accurate estimates (preliminary or definitive estimates) are made for the most profitable processes identified in the feasibility study.
- Detailed estimates are then made for the more promising alternatives that remain after the preliminary estimates.
- Based on the results from the detailed estimate, a final decision is made whether to go ahead with the construction of a plant.

This text focuses on the preliminary and study estimation classification based on a PFD as presented in Chapter 1. This approach will provide estimates accurate in the range of +40% to –25%.

In this chapter, it is assumed that all processes considered are technically sound and attention is focused on the economic estimation of capital costs. The technical aspects of processes will be considered in later chapters.

7.2 ESTIMATION OF PURCHASED EQUIPMENT COSTS

To obtain an estimate of the capital cost of a chemical plant, the costs associated with major plant equipment must be known. For the presentation in this chapter, it is assumed that a PFD for the process is available. This PFD is similar to the one discussed in detail in Chapter 1, which included material and energy balances with each major piece of equipment identified, materials of construction selected, and the size/capacity roughly estimated from conditions on the PFD. Additional PFDs and equipment summary tables are given for several processes in Appendix B.

The most accurate estimate of the purchased cost of a piece of major equipment is provided by a current price quote from a suitable vendor (a seller of equipment). The next best alternative is to use cost data on previously purchased equipment of the same type. Another technique, sufficiently accurate for study and preliminary cost estimates, utilizes summary graphs available for various types of common equipment. This last technique is used for study estimates emphasized in this text and is discussed in detail in Section 7.3. Any cost data must be adjusted for any difference in unit capacity (see Section 7.2.1) and also for any elapsed time since the cost data were generated (see Section 7.2.2).

7.2.1 Effect of Capacity on Purchased Equipment Cost

The most common simple relationship between the purchased cost and an attribute of the equipment related to units of capacity is given by Equation 7.1.

$$\frac{C_a}{C_b} = \left(\frac{A_a}{A_b}\right)^n \tag{7.1}$$

where A = Equipment cost attribute
C = Purchased cost
n = Cost exponent

Subscripts: a refers to equipment with the required attribute
b refers to equipment with the base attribute

The **equipment cost attribute** is the equipment parameter that is used to correlate capital costs. The equipment cost attribute is most often related to the unit capacity, and the term *capacity* is commonly used to describe and identify this attribute. Some typical values of cost exponents and unit capacities are given in Table 7.3. From Table 7.3, it can be seen that the following information is given:

1. A description of the type of equipment used
2. The units in which the capacity is measured
3. The range of capacity over which the correlation is valid
4. The cost exponent (values shown for n vary between 0.30 and 0.84)

Equation 7.1 can be rearranged to give

$$C_a = KA_a^n \tag{7.2}$$

where $K = C_b/A_b^n$

Equation 7.2 is a straight line with a slope of n when the log of C_a is plotted versus the log of A_a. To illustrate this relationship, the typical cost of a single-stage blower versus the capacity of the blower, given as the volumetric flowrate, is plotted in Figure 7.1. The value for the cost exponent, n, from this curve is 0.60.

Table 7.3 Typical Values of Cost Exponents for a Selection of Process Equipment

Equipment Type	Range of Correlation	Units of Capacity	Cost Exponent n
Reciprocating compressor with motor drive	0.75 to 1490	kW	0.84
Heat exchanger shell and tube carbon steel	1.9 to 1860	m^2	0.59
Vertical tank carbon steel	0.4 to 76	m^3	0.30
Centrifugal blower	0.24 – 71	std m^3/s	0.60
Jacketed kettle glass lined	0.2 to 3.8	m^3	0.48

(All data from Table 9-50, *Chemical Engineer's Handbook,* Perry, R.H., Green, D.W., and Maloney, J.O. (eds.), 7th ed, 1997. Reproduced by permission of The McGraw-Hill Companies, Inc., New York, NY.)

The value of the cost exponent, n, used in Equations 7.1 and 7.2, varies depending on the class of equipment being represented. See Table 7.3. The value of n for different items of equipment is often around 0.6. Replacing n in Equation 7.1 and/or 5.2 by 0.6 provides the relationship referred to as the **six-tenths rule.** A problem using the six-tenths rule is given in Example 7.3.

Example 7.3

Use the six-tenths rule to estimate the percentage increase in purchased cost when the capacity of a piece of equipment is doubled.

Using Equation 7.1 with $n = 0.6$,

$$C_a/C_b = (2/1)^{0.6} = 1.52$$

$$\% \text{ increase} = ((1.52 - 1.00)/1.00)(100) = 52\%$$

This simple example illustrates a concept referred to as the **economy of scale.** Even though the equipment capacity was doubled, the purchased cost of the equipment increased by only 52%. This leads to the following generalization.

> **The larger the equipment, the lower the cost of equipment per unit of capacity.**

Special care must be taken in using the six-tenths rule for a single piece of equipment. The cost exponent may vary considerably from 0.6, as illustrated in Example 7.4. The use of this rule for a total chemical process is more reliable and is discussed in Section 7.3.

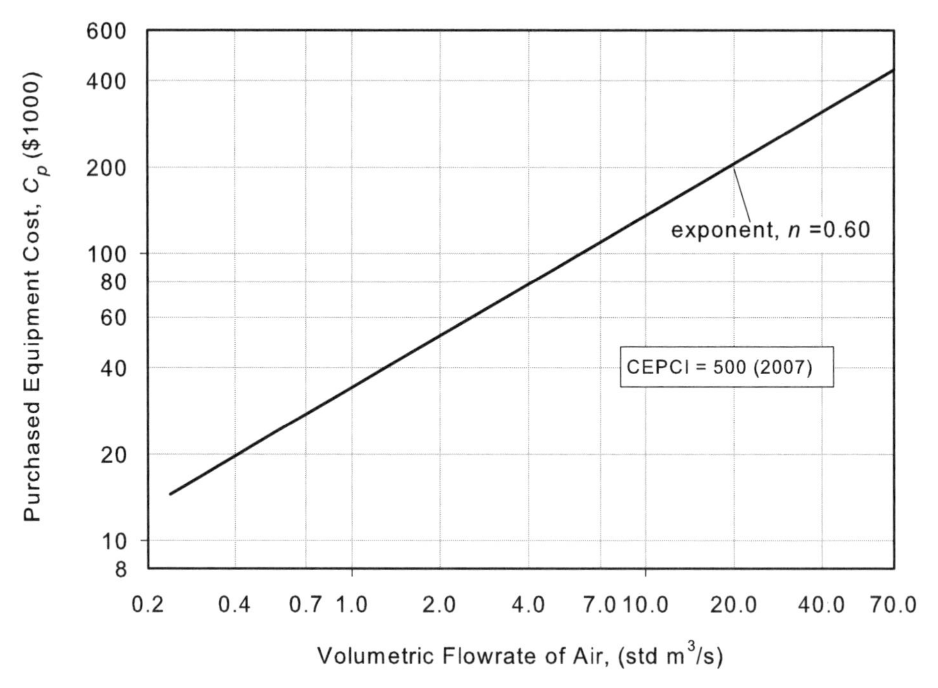

Figure 7.1 Purchased Cost of a Centrifugal Air Blower (Data adapted from Reference [3])

Example 7.4

Compare the error for the scale-up of a reciprocating compressor by a factor of 5 using the six-tenths rule in place of the cost exponent given in Table 7.3.

Using Equation 7.1,

Cost ratio using six-tenths rule (i.e., $n = 0.60$) = $5.0^{0.60} = 2.63$

Cost ratio using ($n = 0.84$) from Table 7.3 = $5.0^{0.84} = 3.86$

% Error = ((2.63 – 3.86)/3.86)(100) = –32 %

Another way to think of the economy of scale is to consider the purchased cost of equipment per unit capacity. Equation 7.2 can be rearranged to give the following relationship:

$$\frac{C}{A} = KA^{n-1} \tag{7.3}$$

If Equation 7.3 is plotted on log-log coordinates, the resulting curve will have a negative slope, as shown in Figure 7.2. The meaning of the negative slope is that as the capacity of a piece of equipment increases, the cost per unit of capacity decreases. This, of course, is a consequence of $n < 1$ but also shows clearly how

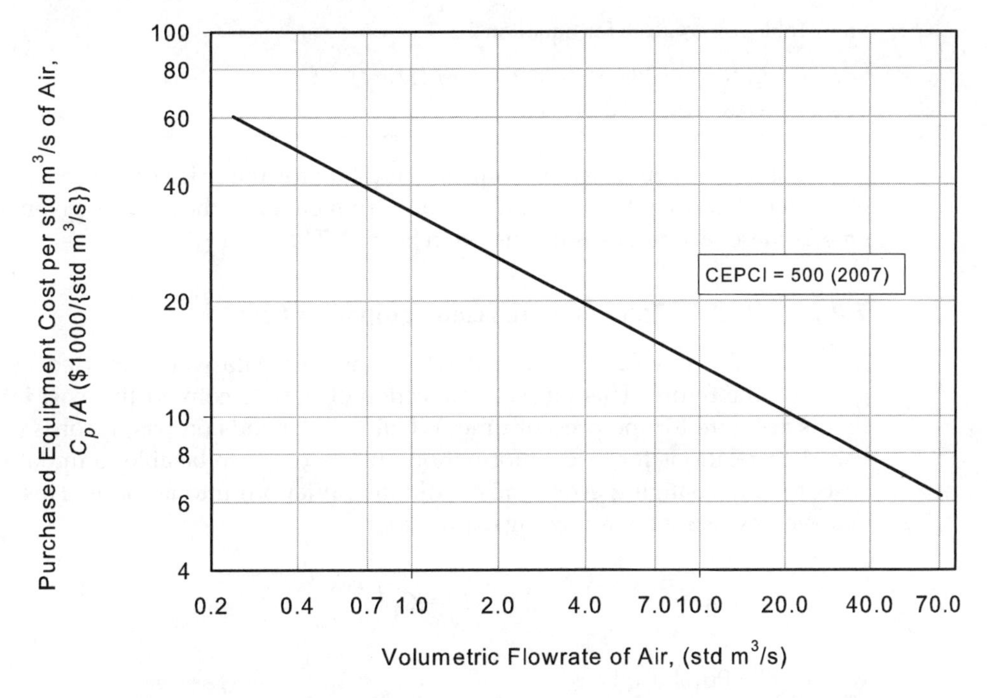

Figure 7.2 Purchased Cost per Unit of Flowrate of a Centrifugal Air Blower (Adapted from Reference [3])

the economy of scale works. As cost curves for equipment are introduced in the text, they will be presented in terms of cost per unit capacity as a function of capacity to illustrate better the idea of economy of scale. For many equipment types, the simple relationship in Equation 7.1 is not very accurate, and an equation that is second order in the attribute is used.

In the last two examples, the relative costs of equipment of differing size were calculated. It is necessary to have cost information on the equipment at some "base case" in order to be able to determine the cost of other similar equipment. This base-case information must allow for the constant, *K*, in Equation 7.2, to be evaluated, as shown in Example 7.5. This base case cost information may be obtained from a current bid provided by a manufacturer for the needed equipment or from company records of prices paid for similar equipment.

Example 7.5

The purchased cost of a recently acquired heat exchanger with an area of 100 m^2 was $10,000.

Determine

a. The constant *K* in Equation 7.2
b. The cost of a new heat exchanger with area equal to 180 m^2

From Table 7.3: $n = 0.59$: for Equation 7.2:

a. $K = C_b/(A_b)^n = 10{,}000/(100)^{0.59} = 661\ \{\$/(\text{m}^2)^{0.59}\}$

b. $C_a = (661)(180)^{0.59} = \$14{,}100$

There are additional techniques that allow for the price of equipment to be estimated that do not require information from either of the sources given above. One of these techniques is discussed in Section 7.3.

7.2.2 Effect of Time on Purchased Equipment Cost

In Figures 7.1 and 7.2, the time at which the cost data were reported (2006) is given on the figure. This raises the question of how to convert this cost into one that is accurate for the present time. When one depends on past records or published correlations for price information, it is essential to be able to update these costs to take changing economic conditions (inflation) into account. This can be achieved by using the following expression:

$$C_2 = C_1 \left(\frac{I_2}{I_1} \right) \tag{7.4}$$

where C = Purchased cost
I = Cost index

Subscripts: 1 refers to base time when cost is known
2 refers to time when cost is desired

There are several cost indices used by the chemical industry to adjust for the effects of inflation. Several of these cost indices are plotted in Figure 7.3.

All indices in Figure 7.3 show similar inflationary trends with time. The indices most generally accepted in the chemical industry and reported in the back page of every issue of *Chemical Engineering* are the Marshall and Swift Equipment Cost Index and the Chemical Engineering Plant Cost Index.

Table 7.4 provides values for both the Marshall and Swift Equipment Cost Index and the Chemical Engineering Plant Cost Index from 1991 to 2006.

Unless otherwise stated, the Chemical Engineering Plant Cost Index (CEPCI) will be used in this text to account for inflation. This is a composite index, and the items that are included in the index are listed in Table 7.5. A comparison between these two indices is given in Example 7.6.

Example 7.6

The purchased cost of a heat exchanger of 500 m^2 area in 1992 was \$25,000.

a. Estimate the cost of the same heat exchanger in 2006 using the two indices introduced above.

b. Compare the results.

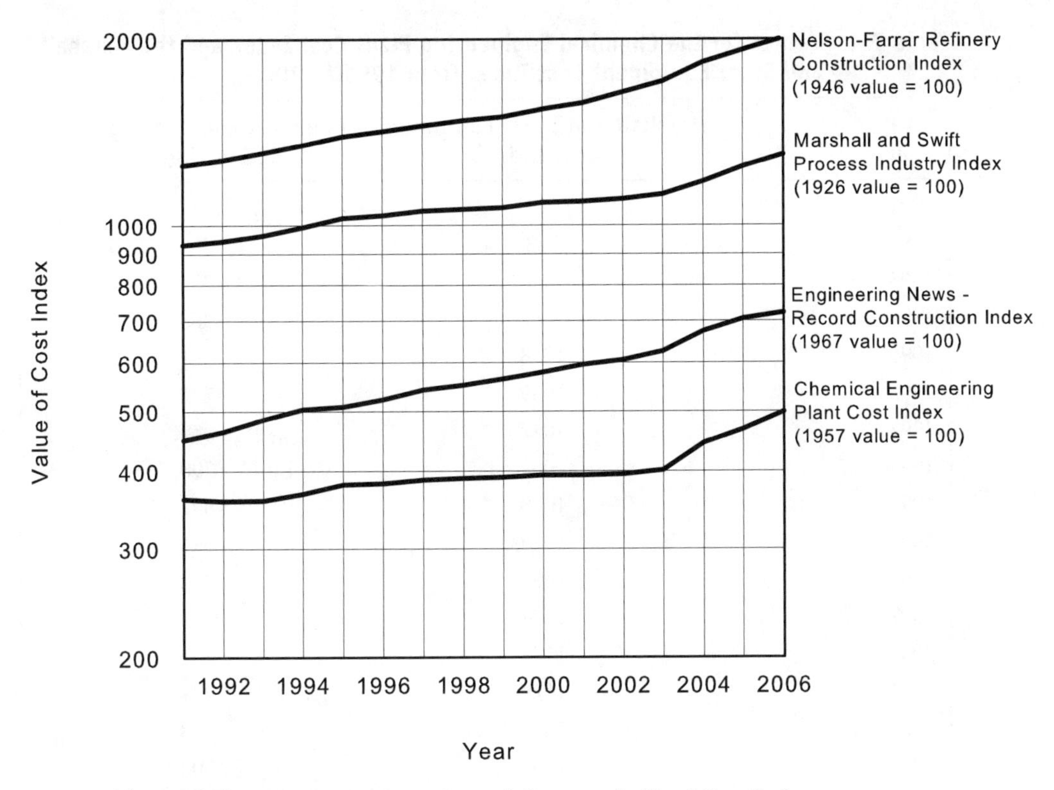

Figure 7.3 The Variations in Several Commonly Used Cost Indexes Over 15 Years (1992–2006)

From Table 7.4	**1992**	**2006**
Marshal and Swift Index	943	1302
Chemical Engineering Plant Cost Index	358	500

a. Marshal and Swift: Cost = (\$25,000)(1302/943) = \$34,518

Chemical Engineering: Cost = (\$25,000)(500/358) = \$34,916

b. Average Difference: ((\$34,518 – 34,916)/((\$34,518 + 34,916)/2)(100) = –1.1%

7.3 ESTIMATING THE TOTAL CAPITAL COST OF A PLANT

The capital cost for a chemical plant must take into consideration many costs other than the purchased cost of the equipment. As an analogy, consider the costs associated with building a new home.

The purchased cost of all the materials that are needed to build a home does not represent the cost of the home. The final cost reflects the cost of property, the cost for delivering materials, the cost of construction, the cost of a driveway, the cost for hooking up utilities, and so on.

Table 7.4 Values for the Chemical Engineering Plant Cost Index and the Marshall and Swift Equipment Cost Index from 1991 to 2006

Year	Marshall and Swift Equipment Cost Index	Chemical Engineering Plant Cost Index
1991	931	361
1992	943	358
1993	964	359
1994	993	368
1995	1028	381
1996	1039	382
1997	1057	387
1998	1062	390
1999	1068	391
2000	1089	394
2001	1094	394
2002	1104	396
2003	1124	402
2004	1179	444
2005	1245	468
2006	1302	500

Table 7.5 The Basis for the Chemical Engineering Plant Cost Index

Components of Index	Weighting of Component (%)	
Equipment, Machinery, and Supports		
(a) Fabricated equipment	37	
(b) Process machinery	14	
(c) Pipe, valves, and fittings	20	
(d) Process instruments and controls	7	
(e) Pumps and compressors	7	
(f) Electrical equipment and materials	5	
(g) Structural supports, insulation, and paint	10	
	100	61% of total
Erection and installation labor		22
Buildings, materials, and labor		7
Engineering and supervision		10
Total		100

Table 7.6 presents a summary of the costs that must be considered in the evaluation of the total capital cost of a chemical plant.

The estimating procedures to obtain the full capital cost of the plant are described in this section. If an estimate of the capital cost for a process plant is needed and access to a previous estimate for a similar plant with a different capacity is available, then the principles already introduced for the scaling of purchased costs of equipment can be used.

1. The six-tenths rule (Equation 7.1 with n set to 0.6) can be used to scale up or down to a new capacity.

Table 7.6 Factors Affecting the Costs Associated with Evaluation of Capital Cost of Chemical Plants (from References [2] and [5])

Factor Associated with the Installation of Equipment	Symbol	Comments
1. **Direct Project Expenses**		
a. Equipment f.o.b. cost (f.o.b. = free on board)	C_P	Purchased cost of equipment at manufacturer's site.
b. Materials required for installation	C_M	Includes all piping, insulation and fireproofing, foundations and structural supports, instrumentation and electrical, and painting associated with the equipment.
c. Labor to install equipment and material	C_L	Includes all labor associated with installing the equipment and materials mentioned in (a) and (b).
2. **Indirect Project Expenses**		
a. Freight, insurance, and taxes	C_{FIT}	Includes all transportation costs for shipping equipment and materials to the plant site, all insurance on the items shipped, and any purchase taxes that may be applicable.
b. Construction overhead	C_O	Includes all fringe benefits such as vacation, sick leave, retirement benefits, etc.; labor burden such as social security and unemployment insurance, etc.; and salaries and overhead for supervisory personnel.
c. Contractor engineering expenses	C_E	Includes salaries and overhead for the engineering, drafting, and project management personnel on the project.

(continued on next page)

Table 7.6 Factors Affecting the Costs Associated with Evaluation of Capital Cost of Chemical Plants (from References [2] and [5]) (*continued*)

Factor Associated with the Installation of Equipment	Symbol	Comments
3. **Contingency and Fee**		
a. Contingency	C_{Cont}	A factor to cover unforeseen circumstances. These may include loss of time due to storms and strikes, small changes in the design, and unpredicted price increases.
b. Contractor fee	C_{Fee}	This fee varies depending on the type of plant and a variety of other factors.
4. **Auxiliary Facilities**		
a. Site development	C_{Site}	Includes the purchase of land; grading and excavation of the site; installation and hookup of electrical, water, and sewer systems; and construction of all internal roads, walkways, and parking lots.
b. Auxiliary buildings	C_{Aux}	Includes administration offices, maintenance shop and control rooms, warehouses, and service buildings (e.g., cafeteria, dressing rooms, and medical facility).
c. Off-sites and utilities	C_{Off}	Includes raw material and final product storage; raw material and final product loading and unloading facilities; all equipment necessary to supply required process utilities (e.g., cooling water, steam generation, fuel distribution systems, etc.); central environmental control facilities (e.g., waste water treatment, incinerators, flares, etc.); and fire protection systems.

2. The Chemical Engineering Plant Cost Index should be used to update the capital costs (Equation 7.4).

The six-tenths rule is more accurate in this application than it is for estimating the cost of a single piece of equipment. The increased accuracy results from the fact that multiple units are required in a processing plant. Some of the process units will have cost coefficients, n, less than 0.6. For this equipment the six-tenths rule overestimates the costs of these units. In a similar way, costs for process units having coefficients greater than 0.6 are underestimated. When the sum of the costs is determined, these differences tend to cancel each other out.

The Chemical Engineering Plant Cost Index (CEPCI) can be used to account for changes that result from inflation. The CEPCI values provided in Table 7.4 are composite values that reflect the inflation of a mix of goods and services associated with the chemical process industries (CPI).

You may be familiar with the more common consumer price index issued by the government. This represents a composite cost index that reflects the effect of inflation on the cost of living. This index considers the changing cost of a "basket" of goods composed of items used by the "average" person. For example, the price of housing, cost of basic foods, cost of clothes and transportation, and so on, are included and weighted appropriately to give a single number reflecting the average cost of these goods. By comparing this number over time, it is possible to get an indication of the rate of inflation as it affects the average person.

In a similar manner, the CEPCI represents a "basket" of items directly related to the costs associated with the construction of chemical plants. A breakdown of the items included in this index was given in Table 7.5. The index is directly related to the effect of inflation on the cost of an "average" chemical plant, as shown in Example 7.7.

Example 7.7

The capital cost of a 30,000 metric ton/year isopropanol plant in 1992 was estimated to be \$23 million. Estimate the capital cost of a new plant with a production rate of 50,000 metric tons/year in 2007 (assume CEPCI = 500).

Cost in 2007 = (Cost in 1992)(Capacity Correction)(Inflation Correction)

$$= (\$23{,}000{,}000)(50{,}000/30{,}000)^{0.6}(500/358)$$

$$= (\$23{,}000{,}000)(1.359)(1.397) = \$43{,}644{,}000$$

In most situations, cost information will not be available for the same process configuration; therefore, other estimating techniques must be used.

7.3.1 Lang Factor Technique

A simple technique to estimate the capital cost of a chemical plant is the Lang Factor method, due to Lang [6, 7, 8]. The cost determined from the Lang Factor represents the cost to build a major expansion to an existing chemical plant. The total cost is determined by multiplying the total purchased cost for all the major items of equipment by a constant. The major items of equipment are those shown in the process flow diagram. The constant multiplier is called the Lang Factor. Values for Lang Factors, F_{Lang}, are given in Table 7.7.

Table 7.7 Lang Factors for the Estimation of Capital Cost for Chemical Plant (from References [6, 7, 8])

Capital Cost = (Lang Factor)(Sum of Purchased Costs of All Major Equipment)	
Type of Chemical Plant	**Lang Factor = F_{Lang}**
Fluid processing plant	4.74
Solid-fluid processing plant	3.63
Solid processing plant	3.10

The capital cost calculation is determined using Equation 7.5.

$$C_{TM} = F_{Lang} \sum_{i=1}^{n} C_{p,i} \tag{7.5}$$

where C_{TM} is the capital cost (total module) of the plant
$C_{p,i}$ is the purchased cost for the major equipment units
n is the total number of individual units
F_{Lang} is the Lang Factor (from Table 7.7)

Plants processing only fluids have the largest Lang Factor, 4.74, and plants processing only solids have a factor of 3.10. Combination fluid-solid systems fall between these two values. The greater the Lang Factor, the less the purchased costs contribute to the plant costs. For all cases, the purchased cost of the equipment is less than one-third of the capital cost of the plant. The use of the Lang Factor is illustrated in Example 7.8.

Example 7.8

Determine the capital cost for a major expansion to a fluid processing plant that has a total purchased equipment cost of \$6,800,000.

Capital Costs = (\$6,800,000)(4.74) = \$32,232,000

This estimating technique is insensitive to changes in process configuration, especially between processes in the same broad categories shown in Table 7.7. It cannot accurately account for the common problems of special materials of construction and high operating pressures. A number of alternative techniques are available. All require more detailed calculations using specific price information for the individual units/equipment.

7.3.2 Module Costing Technique

The equipment module costing technique is a common technique to estimate the cost of a new chemical plant. It is generally accepted as the best for making preliminary cost estimates and is used extensively in this text. This approach, introduced by Guthrie [9, 10] in the late 1960s and early 1970s, forms the basis of many of the equipment module techniques in use today. This costing technique relates all costs back to the purchased cost of equipment evaluated for some base conditions. Deviations from these base conditions are handled by using multiplying factors that depend on the following:

1. The specific equipment type
2. The specific system pressure
3. The specific materials of construction

The material provided in the next section is based upon information in Guthrie [9, 10], Ulrich [5], and Navarrete [11]. The reader is encouraged to review these references for further information.

Equation 7.6 is used to calculate the bare module cost for each piece of equipment. The bare module cost is the sum of the direct and indirect costs shown in Table 7.6.

$$C_{BM} = C_p^o F_{BM} \tag{7.6}$$

where C_{BM} = bare module equipment cost: direct and indirect costs for each unit
F_{BM} = bare module cost factor: multiplication factor to account for the items in Table 7.6 plus the specific materials of construction and operating pressure
C_p^o = purchased cost for base conditions: equipment made of the most common material, usually carbon steel and operating at near ambient pressures

Because of the importance of this cost estimating technique, it is described below in detail.

7.3.3 Bare Module Cost for Equipment at Base Conditions

The bare module equipment cost represents the sum of direct and indirect costs shown in Table 7.6. The conditions specified for the base case are

1. Unit fabricated from most common material, usually carbon steel (CS)
2. Unit operated at near-ambient pressure

Equation 7.6 is used to obtain the bare module cost for the base conditions. For these base conditions, a superscript zero (0) is added to the bare module cost factor and the bare module equipment cost. Thus C_{BM}^o and F_{BM}^o refer to the base conditions.

Table 7.8 supplements Table 7.6 and provides the relationships and equations for the direct, indirect, contingency, and fee costs based on the purchased cost of the equipment. These equations are used to evaluate the bare module factor. The entries in Table 7.8 are described on page 202.

Table 7.8 Equations for Evaluating Direct, Indirect, Contingency, and Fee Costs

Factor	Basic Equation	Multiplying Factor to Be Used with Purchased Cost, C_p^o
1. Direct		
a. Equipment	$C_p^o = C_p^o$	1.0
b. Materials	$C_M = \alpha_M C_p^o$	α_M
c. Labor	$C_L = \alpha_L(C_p^o + C_M)$	$(1.0 + \alpha_M)\alpha_L$
Total Direct	$C_{DE} = C_p^o + C_M + C_L$	$(1.0 + \alpha_M)(1.0 + \alpha_L)$
2. Indirect		
a. Freight, etc.	$C_{FIT} = \alpha_{FIT}(C_p^o + C_M)$	$(1.0 + \alpha_M)\alpha_{FIT}$
b. Overhead	$C_O = \alpha_O C_L$	$(1.0 + \alpha_M)\alpha_L\alpha_O$
c. Engineering	$C_E = \alpha_E(C_p^o + C_M)$	$(1.0 + \alpha_M)\alpha_E$
Total Indirect	$C_{IDE} = C_{FIT} + C_O + C_E$	$(1.0 + \alpha_M)(\alpha_{FIT} + \alpha_L\alpha_O + \alpha_E)$
Bare Module	$C_{BM}^o = C_{IDE} + C_{DE}$	$(1.0 + \alpha_M)(1.0 + \alpha_L + \alpha_{FIT} + \alpha_L\alpha_O + \alpha_E)$
3. Contingency and Fee		
a. Contingency	$C_{Cont} = \alpha_{Cont} C_{BM}^o$	$(1.0 + \alpha_M)(1.0 + \alpha_L + \alpha_{FIT} + \alpha_L\alpha_O + \alpha_E)\alpha_{Cont}$
b. Fee	$C_{Fee} = \alpha_{Fee} C_{BM}^o$	$(1.0 + \alpha_M)(1.0 + \alpha_L + \alpha_{FIT} + \alpha_L\alpha_O + \alpha_E)\alpha_{Fee}$
Total Module	$C_{TM} = C_{BM}^o + C_{Cont} + C_{Fee}$	$(1.0 + \alpha_M)(1.0 + \alpha_L + \alpha_{FIT} + \alpha_L\alpha_O + \alpha_E)(1.0 + \alpha_{Cont} + \alpha_{Fee})$

Column 1: Lists the factors given in Table 7.6.

Column 2: Lists equations used to evaluate each of the costs. These equations introduce multiplication cost factors, α_i. Each cost item, other than the purchased equipment cost, introduces a separate factor.

Column 3: For each factor, the cost is related to the purchased cost C_p^o by an equation of the form.

$$C_{XX} = C_p^o f(\alpha_{i,j,k...}) \tag{7.7}$$

The function, $f(\alpha_{i,j,k...})$, is given in column 3 of Table 5.8.

From Table 7.8 and Equations 7.6 and 7.7, it can be seen that the bare module factor is given by

$$F_{BM}^o = [1 + \alpha_L + \alpha_{FIT} + \alpha_L\alpha_o + \alpha_E][1 + \alpha_M] \tag{7.8}$$

The values for the bare module cost multiplying factors vary between equipment modules. The calculations for the bare module factor and bare module cost for a carbon steel heat exchanger are given in Example 7.9.

Example 7.9

The purchased cost for a carbon steel heat exchanger operating at ambient pressure is $10,000. For a heat exchanger module, Guthrie [9, 10] provides the following cost information.

Item	% of Purchased Equipment Cost
Equipment	100.0
Materials	71.4
Labor	63.0
Freight	8.0
Overhead	63.4
Engineering	23.3

Using the information given above, determine the equivalent cost multipliers given in Table 7.8 and the following:

a. Bare module cost factor, F^o_{BM}

b. Bare module cost, C^o_{BM}

Item	% of Purchased Equipment Cost	Cost Multiplier (Table 7.8)	Value of Multiplier
Equipment	100.0	1.0	
Materials	71.4	α_M	0.714
Labor	63.0	α_L	0.63/(1 + 0.714) = 0.368
Freight	8.0	α_{FIT}	0.08/(1 + 0.714) = 0.047
Overhead	63.4	α_O	0.634/0.368/(1 + 0.714) = 1.005
Engineering	23.3	α_E	0.233/(1 + 0.714) = 0.136
Bare Module	**329.1**		

a. Using Equation 7.8,

$$F^o_{BM} = (1 + 0.368 + 0.047 + (1.005)(0.368) + 0.136)(1 + 0.714) = 3.291$$

b. From Equation 7.6,

$$C^o_{BM} = (3.291)(\$10{,}000) = \$32{,}910$$

Fortunately, we do not have to repeat the procedure illustrated in Example 7.9 in order to estimate F^o_{BM} for every piece of equipment. This has already been done for a large number of equipment modules, and the results are given in Appendix A.

In order to estimate bare module costs for equipment, purchased costs for the equipment at base case conditions (ambient pressure using carbon steel) must be available along with the corresponding bare module factor and factors to account for different operating pressures and materials of construction. These data are made available for a variety of common gas/liquid processing equipment in Appendix A. These data were compiled during the summer of 2001 from information obtained from manufacturers and also from the R-Books software marketed by Richardson Engineering Services [12]. The method by which material and pressure factors are accounted for depends on the equipment type, and these are covered in the next section. The estimation of the bare module cost for a floating-head shell-and-tube heat exchanger is illustrated in Example 7.10 and in subsequent examples in this chapter.

Example 7.10

Find the bare module cost of a floating-head shell-and-tube heat exchanger with a heat transfer area of 100 m^2 at the end of 2006. The operating pressure of the equipment is 1.0 bar, with both shell-and-tube sides constructed of carbon steel. The cost curve for this heat exchanger is given in Appendix A, Figure A.5, and is repeated as Figure 7.4. It should be noted that unlike the examples shown in Figures 7.1 and 7.2, the log-log plot of cost per

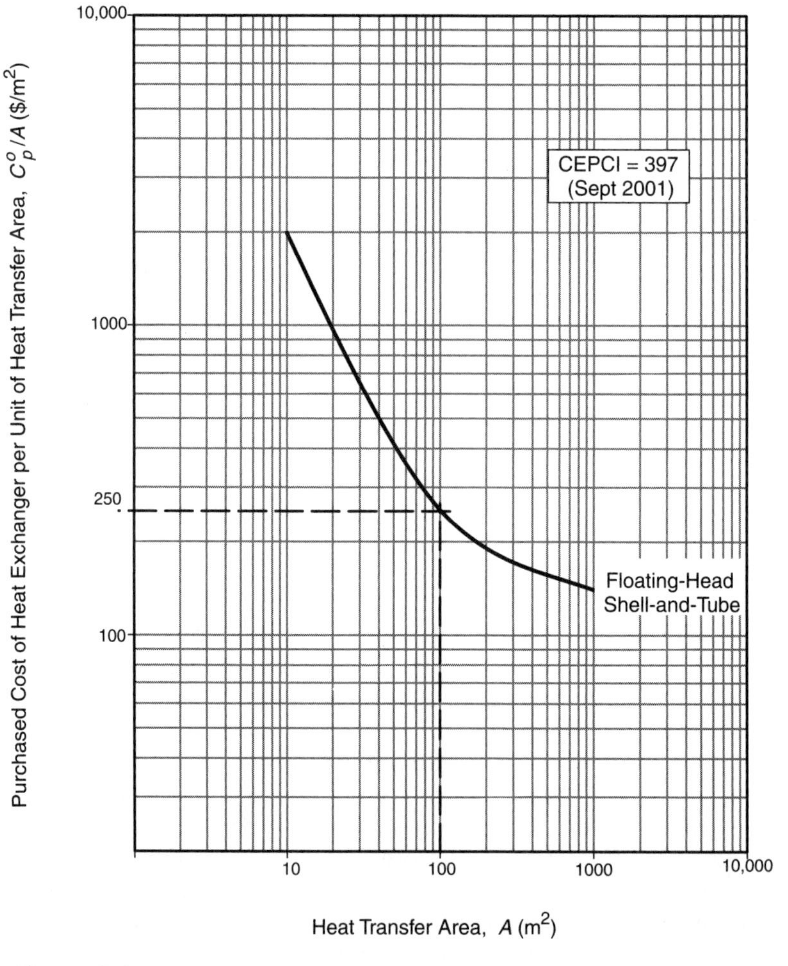

Figure 7.4 Purchased Costs for Floating-Head Shell-and-Tube Heat Exchangers

unit area versus area is nonlinear. In general this will be the case, and a second order polynomial is normally used to describe this relationship.

From Figure 7.4, $C_p^o(2001)$ = (\$ 250)(100) = \$25,000 (the evaluation path is shown on Figure 7.4).

The bare module cost for shell-and-tube heat exchangers is given by Equation A.4.

$$C_{BM} = C_p^o\,[B_1 + B_2F_pF_M] \tag{A.4}$$

The values of B_1 and B_2 for floating-head heat exchangers from Table A.4 are 1.63 and 1.66, respectively.

The pressure factor is obtained from Equation A.3.

$$\log_{10} F_p = C_1 + C_2 \log_{10} P + C_3(\log_{10}P)^2 \tag{A.3}$$

From Table A.2, for pressures <5 barg, $C_1 = C_2 = C_3 = 0$, and from Equation A.3, $F_p = 1$. Using data in Table A.3 for shell-and-tube heat exchangers with both shell and tubes made of carbon steel (Identification Number = 1) and Figure A.8, $F_M = 1$. Substituting this data into Equation A.4 gives

$$C_{BM}^o\,(2001) = C_p^o\,(2001)[1.63 + 1.66(F_p = 1)(F_M = 1)] = 3.29C_p^o = (3.29)(\$25{,}000) = \$82{,}300$$
$$C_{BM}^o\,(2006) = C_{BM}^o\,(2001)\,(500/394) = \$82{,}300\,(500/397) = \$103{,}590$$

A comparison of the value of bare module cost factor for Example 7.10 shows that it is the same as the value of 3.29 evaluated using the individual values for α_i, given in Example 7.9.

7.3.4 Bare Module Cost for Nonbase Case Conditions

For equipment made from other materials of construction and/or operating at nonambient pressure, the values for F_M and F_P are greater than 1.0. In the equipment module technique, these additional costs are incorporated into the bare module cost factor, F_{BM}. The bare module factor used for the base case, F_{BM}^o, is replaced with an actual bare module cost factor, F_{BM}, in Equation 7.6. The information needed to determine this actual bare module factor is provided in Appendix A. The effect of pressure on the cost of equipment is considered first.

Pressure Factors. As the pressure at which a piece of equipment operates increases, the thickness of the walls of the equipment will also increase. For example, consider the design of a process vessel. Such vessels, when subjected to internal pressure (or external pressure when operating at vacuum) are subject to rigorous mechanical design procedures. For the simple case of a cylindrical vessel operating at greater than ambient pressure, the relationship between design pressure and wall thickness required to withstand the radial stress in the cylindrical portion of the vessel, as recommended by the ASME [13], is given as

$$t = \frac{PD}{2SE - 1.2P} + CA \tag{7.9}$$

where t is the wall thickness in meters, P is the design pressure (bar), D is the diameter of the vessel (m), S is the maximum allowable working pressure (maximum allowable stress) of material (bar), E is a weld efficiency, and CA is the corrosion allowance (m). The weld efficiency is dependent on the type of weld and the degree of examination of the weld. Typical values are from 1.0 to 0.6. The corrosion allowance depends on the service, and typical values are from 3.15 to 6.3 mm (0.125 to 0.25 inches). However, for very aggressive environments, inert linings such as glass and graphite are often used to protect the structural metal. Finally, the maximum working pressure of the material of construction, S, is dependent not only on the material but also on the operating temperature. Some typical values of S are given for common materials of construction in Figure 7.5. From this figure, it is clear that for typical carbon steel the maximum allowable stress drops off rapidly after 350°C. However, for stainless steels (ASME SA-240) the decrease in maximum allowable stress with temperature is less steep, and operation up to 600–650°C is possible for some grades. For even higher temperatures and very corrosive environments, when the lining of vessels is not practical, more exotic alloys such as titanium and titanium-based alloys and nickel-based alloys may be used. For example, Hastelloy B has excellent resistance to alkali environments up to 850°C. Inconel 600, whose main constituents are Ni 72%, Cr 15%, and Fe 8%, has excellent corrosion resistance to oxidizing environments such as acids and can be used from cryogenic temperatures up to 1100°C. The maximum allowable working pressure for Incoloy 800HT, which also has excellent corrosion resistance in acidic environments, is shown as a function of temperature in Figure 7.5.

The relationship between cost of a vessel and its operating pressure is a complex one. However, with all other things being constant, the cost of the vessel is approximately proportional to the weight of the vessel, which in turn is proportional to the vessel thickness. From Equation 7.9, it is clear that as the operating pressure approaches $1.67SE$, the required wall thickness, and hence cost, becomes infinite. Moreover, the thickness of the vessel for a given pressure will increase as the vessel diameter increases. The effect of pressure on the weight (and ultimately cost) of carbon steel vessel shells as a function of vessel diameter is shown in Figure 7.6. The y-axis of the figure shows the ratio of the vessel thickness at the design pressure to that at ambient pressure, and the x-axis is the design pressure. A corrosion allowance of 3.15 mm (1/8 inch) and a value of S = 944 bar (13,700 psi) are assumed. It is also assumed that the vessel is designed with a minimum wall thickness of 6.3 mm (1/4 inch). A minimum wall thickness is often required to ensure that the vessel does not buckle under its own weight or when being transported. In addition to these factors, the costs for the vessel supports, manholes, nozzles, instrument wells, the vessel head, and so on, all add to the overall weight and cost of the vessel. For the sake of simplification, it is assumed that the pressure factor (F_P) for vertical and horizontal process vessels is equal to the value given on the y-axis of Figure 7.6. This, clearly, is a simplification but should be valid for the expected accuracy of this technique. Hence, the equation for F_P for process vessels is given by Equation 7.10.

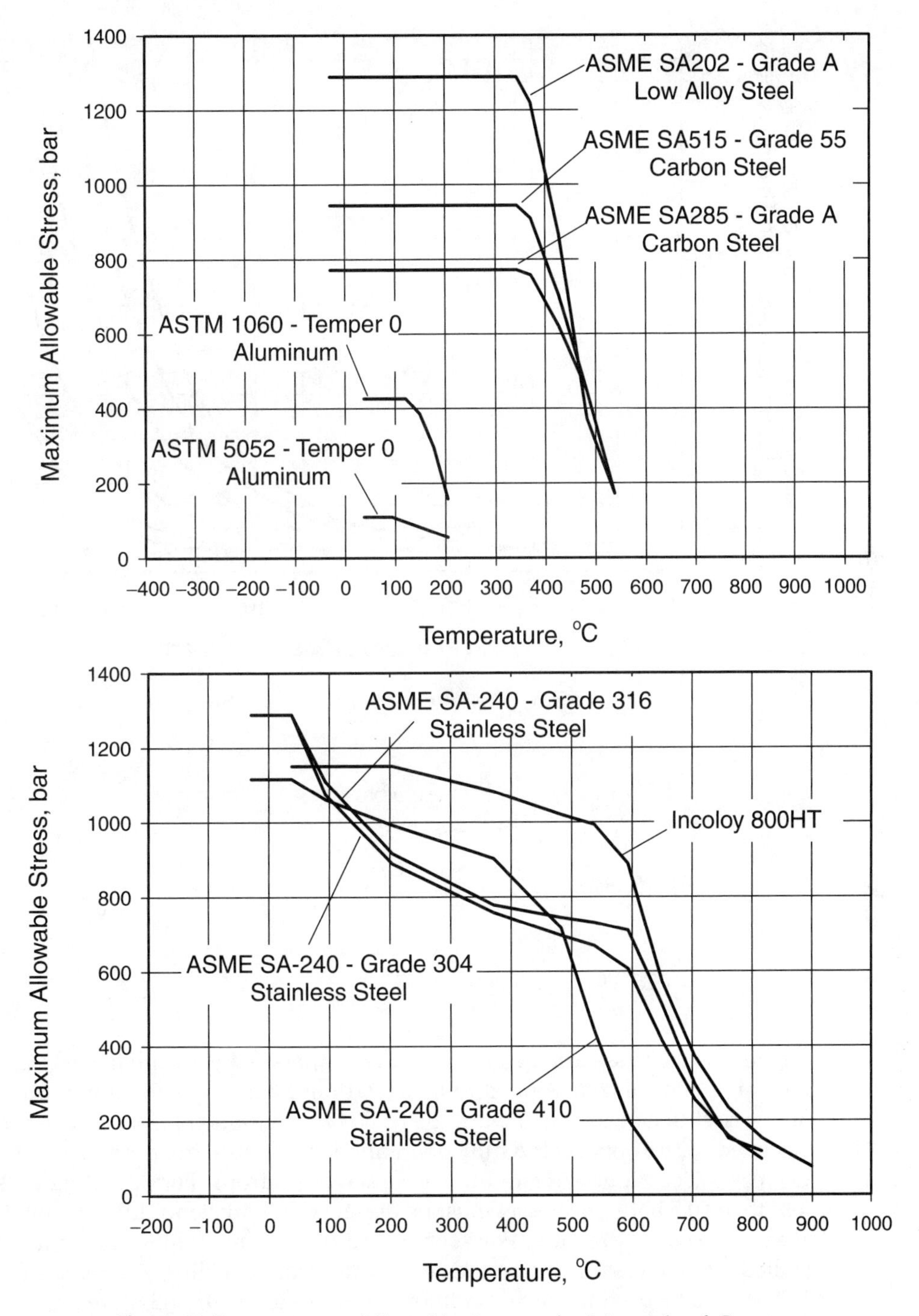

Figure 7.5 Maximum Allowable Stresses for Materials of Construction as a Function of Operating Temperature (Data from Perry et al. [3], Chapter 10, and Ref [15])

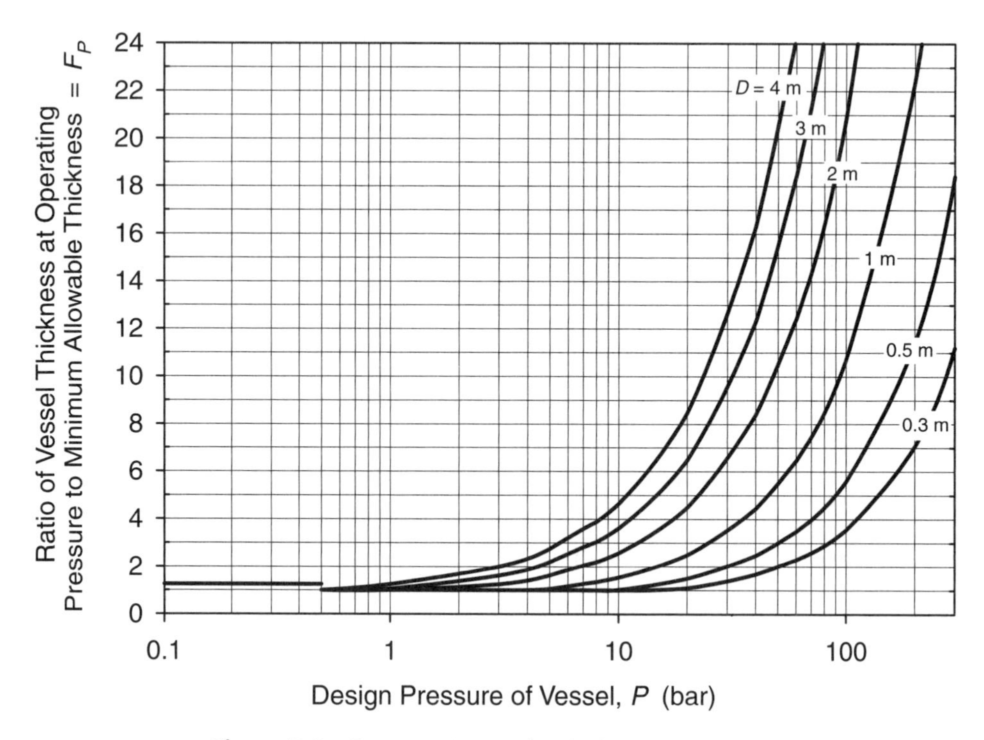

Figure 7.6 Pressure Factors for Carbon Steel Vessels

$$F_{P,vessel} \begin{cases} = 1 \quad \text{for } t < t_{\min} \text{ and } P > -0.5 \text{ barg} \\ = \dfrac{\dfrac{(P+1)D}{(2)(944)(0.9) - 1.2(P+1)} + CA}{t_{\min}} \quad \text{for } t > t_{\min} \text{ and } P > -0.5 \text{ barg} \\ = 1.25 \quad \text{for } P < -0.5 \text{ barg} \end{cases} \tag{7.10}$$

where D is the vessel diameter in m, P is the operating pressure in barg, CA is the corrosion allowance (assumed to be 0.00315 m), and t_{min} is the minimum allowable vessel thickness (assumed to be 0.0063 m). A value of S = 944 bar has been assumed for carbon steel. As the operating temperature increases, the value of S decreases (see Figure 7.5) and the accuracy of F_p drops. For operating pressures less than –0.5 barg, the vessel must be designed to withstand full vacuum, that is, 1 bar of external pressure. For such operations, strengthening rings must be installed into the vessels to stop the vessel walls from buckling. A pressure factor of 1.25 should be used for such conditions, and this is shown in Figure 7.6.

Pressure factors for different equipment are given in Appendix A, Equation A.3, and Table A.2. These pressure factors are presented in the general form given by Equation A.3:

$$\log_{10} F_p = C_1 + C_2 \log_{10} P + C_3 (\log_{10} P)^2 \quad \text{(A.3)}$$

This equation is clearly different from Equation 7.10 for process vessels. Moreover, the value predicted by this equation (using the appropriate constants) gives values of F_P much smaller than those for vessels at the same pressure. This difference arises from the fact that for other equipment, the internals of the equipment make up the major portion of the cost. Therefore, the cost of a thicker outer shell is a much smaller fraction of the equipment cost than for a process vessel, which is strongly dependent on the weight of the metal. Example 7.11 considers the effect of pressure on a shell-and-tube heat exchanger.

Example 7.11

a. Repeat Example 7.10 except consider the case when the operating pressures in both the shell- and the tube-side are 100 barg.

b. Explain why the pressure factor for the heat exchanger is much smaller than for any of the process vessels shown in Figure 7.6.

Solution

a. From Example 7.10, C_p^o (2001) = \$25,000, $F_P = 1$
From Table A.2, for $5<P<140$ barg, $C_1 = 0.03881$, $C_2 = -0.11272$, $C_3 = 0.08183$
Using Equation A.3 and substituting for P = 100 barg and the above constants,

$$\log_{10} F_p = 0.03881 - 0.11272\log_{10}(100) + 0.08183[\log_{10}(100)]^2 = 0.1407$$

$$F_P = 10^{0.1407} = 1.383$$

From Equation A.4:

$$C_{BM}(2001) = C_p^o(2001)[B_1 + B_2F_PF_M] = \$25{,}000[1.63 + 1.66(1.383)(1.0)] = \$98{,}100$$
$$C_{BM}(2006) = \$98{,}100\,(500/397) = \$123{,}590$$

b. Compared with Figure 7.6, this pressure factor (1.383) is much less than any of the vessels at P = 100 barg. Why?

The answer lies in the fact that much of the cost of a shell-and-tube heat exchanger is associated with the cost of the tubes that constitute the heat exchange surface area. Tubing is sold in standard sizes based on the BWG (Birmingham wire gauge) standard. Tubes for heat exchangers are typically between 19.1 and 31.8 mm (3/4 and 1-1/4 inch) in diameter and between 2.1 and 0.9 mm (0.083 and 0.035 inch) thick, corresponding to BWGs of 14 to 20, respectively. Using Equation 7.9, the maximum operating pressure of a 25.4 mm (1 inch) carbon steel tube can be estimated (assume that CA is zero), the results are as follows.

BWG	Thickness (t) (mm)	P (from Eq. 7.9) (barg)
20	0.889	59.1
18	1.244	81.8
16	1.651	106.9
14	2.108	134.1

From the table, it is evident that even the thinnest tube normally used for heat exchangers is capable of withstanding pressures much greater than atmospheric. Therefore, the most costly portion of a shell-and-tube heat exchanger (the cost of the tubes) is relatively insensitive to pressure. Hence, it makes sense that the pressure factors for this type of equipment are much smaller than those for process vessels at the same pressure.

The purchased cost of the equipment for the heat exchanger in Example 7.11 would be $C_P(2006)$ = (\$25,000)(1.383) (500/394) = \$43,880. If this equipment cost were multiplied by the bare module factor for the base case, the cost would become C_{BM} = (\$43,880)(3.29) = \$144,360. This is 16% greater than the \$124,490 calculated in Example 7.11. The difference between these two costs results from assuming, in the latter case, that all costs increase in direct proportion to the increase in material cost. This is far from the truth. Some costs, such as insulation, show small changes with the cost of materials, whereas other costs, such as installation materials, freight, labor, and so on, are impacted to varying extents. The method of equipment module costing accounts for these variations in the bare module factor.

Finally, some equipment is unaffected by pressure. Examples are tower trays and packing. This "equipment" is not subjected to significant differential pressure because it is surrounded by process fluid. Therefore, in Equation A.3, use $C_1 = C_2 = C_3 = 0$. Some other equipment also has zero for these constants. For example, compressor drives are not exposed to the process fluid and so are not significantly affected by operating pressure. Other equipment, such as compressors, do not have pressure corrections because such data were not available. Use of these cost correlations for equipment outside the pressure range shown in Table A.2 should be done with extreme caution.

Materials of Construction (MOCs). The choice of what MOC to use depends on the chemicals that will contact the walls of the equipment. As a guide, Table 7.9, excerpted from Sandler and Luckiewicz [14], may be used for preliminary MOC selection. However, the interaction between process streams and MOCs can be very complex and the compatibility of the MOC with the process stream must be investigated fully before the final design is completed.

Many *polymeric* compounds are nonreactive in both acidic and alkaline environments. However, polymers generally lack the structural strength and resilience of metals. Nevertheless, for operations at less than about 120°C in corrosive environments the use of polymers as liners for steel equipment or incorporated

Table 7.9 Corrosion Characteristics for Some Materials of Construction

Chemical Component	Carbon Steel	304 Stainless Steel	316 Stainless Steel	Aluminum	Copper	Brass	Monel	Hastelloy C	Titanium	TFE	Graphite
Acetaldehyde	N		A			C		A	A	A	A
Acetic acid, glacial	N		A	A	A	C	B	A	A	A	A
Acetic acid, 20%	N	A	A	A	A	C	B	A	A	A	A
Acetic anhydride	N	A	B	A	A	C		A	A	A	A
Acetone	A	A	A	A	A	A	A	A	A	A	
Ammonia, 10%	C	A	A	C	N	N	N	A	A	A	A
Aniline	A	A	A	N	N	N	A	A	A	A	A
Aqua regia	N	N	N	N	N	N	N	C	A	A	
Benzaldehyde		A	A	A	A	A	A	A	A	A	A
Benzene	A	A	A	A	A	A	A	A	A	A	A
Benzoic acid		C	A					A	B	A	A
Furfural	A	C	C	A	A	A	A	A	A	A	A
Gasoline	C	A	A	A	A	A	A	A	N	A	A
Heptane	A	A	A	A	A	A	A	A	A	A	A
Hexane		A	A	A			A	A	A	A	A
HCl, 0–25%	N	N	N	N	C	N	C	C	C	A	A
HCl, 25–37%	N	N	N	N	C	N	C	C	C	A	A
HF, 30%	N	B	B	N	N	N	A	A	N	A	A
HF, 60%	N	B	B	N	N	N	A	A	N	A	A
H_2O_2, 30%	C	C	A	C	C	N	C	A	A	A	A
H_2O_2, 90%	C	C	A	C	C	N		A	A	A	A
H_2S, aqueous	C	C	A	A	N	N	N	A	A	A	A
Maleic acid		A	A			A		A	A	A	A
Methanol		A	A	A	A		A	A	A	A	A
Methyl chloride		A	A	N			A	A	A	A	A

(continued to next page)

Table 7.9 Corrosion Characteristics for Some Materials of Construction (*continued*)

Chemical Component	Carbon Steel	304 Stainless Steel	316 Stainless Steel	Aluminum	Copper	Brass	Monel	Hastelloy C	Titanium	TFE	Graphite
Methyl ethyl ketone	A	A	A	A	A		A	A	A	A	A
Methylene chloride		A	A		N		N	A	A	A	A
Napthalene		A	A	A			A	A	A	A	A
Nitric acid, 10%	N	A	A	B	N		N	A	A	A	A
Nitric acid, 50%	N	C	C	B	N		N	A	A	A	N
Oleic acid	C	A	A	A	C		A	A	A	A	A
Oxalic acid	C	C	B	C	C		A	A	A	A	A
Phenol	N	C	C	B	N		A	A	A	A	A
Phosphoric acid, 0–50%	C	C	C	N	C		C	A	B	A	A
Phosphoric acid, 51–100%	C	C	C	N	C		C	A	B	A	A
Propyl alcohol		A	A	A	A			A	A	A	
Sodium hydroxide, 20%	A	A	A	N	C	N	A	A	A	A	A
Sodium hydroxide, 50%	A	A	A	N	C	N	A	A	A	A	A
Stearic acid		A	A	A	A		B	A	A	A	A
Sulfuric acid, 0–10%	N	N	N	N	N		C	A	B	A	A
Sulfuric acid, 10–75%	N	N	N	N	N		C	A	C	A	A
Sulfuric acid, 75–100%	N	N	N	N	N		C	C	N	A	A
Tataric acid		A	A	A	A		C	A	A	A	A
Toluene	A	A	A	A	A		A	A	A	A	
Urea		A	A	A				A	A	A	A
Xylene		A	A					A	A	A	A

A = acceptable; B = acceptable up to 30°C; C = caution, use under limited conditions; N = not recommended; no entry = information is not available.
(Reproduced from Sandler and Luckiewicz, *Practical Process Engineering, a Working Approach to Plant Design*, with permission of XIMIX, Inc., Philadelphia, 1987.)

into fiberglass structures (at moderate operating pressures) often gives the most economical solution. The most common MOCs are still *ferrous* alloys, in particular carbon steel. *Carbon steels* are distinguished from other ferrous alloys such as wrought and cast iron by the amount of carbon in them. Carbon steel has less than 1.5 wt% carbon, can be given varying amounts of hardness or ductility, is easy to weld, and is cheap. It is still the material of choice in the CPI when corrosion is not a concern.

- *Low-alloy steels* are produced in the same way as carbon steel except that amounts of chromium and molybdenum are added (chromium between 4 and 9 wt%). The molybdenum increases the strength of the steel at high temperatures, and the addition of chromium makes the steel resistant to mildly acidic and oxidizing atmospheres and to sulfur-containing streams.
- *Stainless steels* are so-called high-alloy steels containing greater than 12 wt% chromium and possessing a corrosion-resistant surface coating, also known as a passive coating. At these chromium levels, the corrosion of steel to rusting is reduced by more than a factor of 10. Chemical resistance is also increased dramatically.
- *Nonferrous alloys* are characterized by higher cost and difficulty in machining. Nevertheless, they possess improved corrosion resistance.

  Aluminum and its alloys have a high strength-to-weight ratio and are easy to machine and cast, but in some cases are difficult to weld. The addition of small amounts of other metals—for example, magnesium, zinc, silicon, and copper—can improve the weldability of aluminum. Generally, corrosion resistance is very good due to the formation of a passive oxide layer, and aluminum has been used extensively in cryogenic (low-temperature) operations.

  Copper and its alloys are often used when high thermal conductivity is required. Resistance to seawater and nonoxidizing acids such as acetic acid is very good, but copper alloys should not be used for services that contact ammonium ions (NH_4^-) or oxidizing acids. Common alloys of copper include *brasses* (containing 5–45 wt% zinc) and *bronzes* (containing tin, aluminum, and/or silicon).
- *Nickel and its alloys* are alloys in which nickel is the major component.

  Nickel-copper alloys are known by the name Monel, a trademark of the International Nickel Corp. These alloys have excellent resistance to sulfuric and hydrochloric acids, salt water, and some caustic environments.

  Nickel-chromium alloys are known by the name Inconel, a trademark of the International Nickel Corp. These alloys have excellent chemical resistance at high temperatures. They are also capable of withstanding attack from hot concentrated aqueous solutions containing chloride ions.

  Nickel-chromium-iron alloys are known by the name Incoloy, a trademark of the International Nickel Corp. These alloys have characteristics similar to Inconel but with slightly less resistance to oxidizing agents.

Nickel-molybdenum alloys are known by the name Hastelloy, a trademark of the Cabot Corp. These alloys have very good resistance to concentrated oxidizing agents.

- *Titanium and its alloys* have good strength-to-weight ratios and very good corrosion resistance to oxidizing agents. However, it is attacked by reducing agents, it is relatively expensive, and it is difficult to weld.

As previously shown, the combination of operating temperature and operating pressure will also affect the choice of MOC. From Table 7.9, it is evident that the number of MOCs available is very large and that the correct choice of materials requires input from a trained metallurgist.

Moreover, information about the cost of materials presented in this text is limited to a few different MOCs. The approximate relative cost of some common metals is given in Table 7.10. As a very approximate rule, if the metal of interest does not appear in Appendix A, then Table 7.10 can be used to find a metal that has approximately the same cost. As the metallurgy becomes more "exotic," the margin for error becomes larger, and the data provided in this text will lead to larger errors in estimating the plant cost than for a plant constructed of carbon steel or stainless steel.

To account for the cost of different materials of construction, it is necessary to use the appropriate material factor, F_M, in the bare module factor. This material factor is *not* simply the relative cost of the material of interest to that of carbon steel. The reason is that the cost to produce a piece of equipment is not directly proportional to the cost of the raw materials. For example, consider the cost of a process vessel as discussed in the previous section. Just as the bare module cost was broken down into factors relating to the purchased cost of the equipment (Tables 7.6 and 7.8), the purchased cost (or at least the manufacturing cost) can be broken down into factors relating to the cost of manufacturing the equipment. Many of these costs will be related to the size of the vessel that is in turn related to the vessel's weight, W_{vessel}. An example of these costs is given in Table 7.11.

Table 7.10 Relative Costs of Metals Using Carbon Steel as the Base Case

Material	Relative Cost
Carbon steel	Base case (lowest)
Low-alloy steel	Low to moderate
Stainless steel	Moderate
Aluminum and aluminum alloys	Moderate
Copper and copper alloys	Moderate
Titanium and titanium-based alloys	High
Nickel and nickel-based alloys	High

Table 7.11 Costs Associated with the Manufacture of a Process Vessel

Factors Associated with the Manufacturing Cost of a Vessel	Relationship relating Cost to vessel weight, W_{vessel}
Direct Expenses	
Cost of raw materials	$\beta_{RM} W_{vessel}$
Machining costs	$\beta_{MC} W_{vessel}$
Labor costs	$\beta_{L} W_{vessel}$
Indirect Costs	
Overhead	$\beta_{OH} \beta_{L} W_{vessel}$
Engineering expenses	$\beta_{E} (\beta_{RM} + \beta_{MC}) W_{vessel}$
Contingencies	$\beta_{Cont} W_{vessel}$
Total manufacturing cost	$[\beta_{RM} + \beta_{MC} + \beta_{L} + \beta_{OH} \beta_{L} + \beta_{E} (\beta_{RM} + \beta_{MC}) + \beta_{Cont}] W_{vessel}$

From Table 7.11, it is clear that the cost of the vessel is proportional to its weight. Therefore, the cost will be proportional to the vessel thickness, and thus the pressure factor derived in the previous section is valid (or at least is a reasonably good approximation). The effect of different MOCs is connected to the factor β_{RM}. Clearly, as the raw material costs increase, the total manufacturing costs will not increase proportionally to β_{RM}. In other words, if material X is 10 times as expensive as carbon steel, a vessel made from material X will be less than 10 times the cost of a similar vessel made from carbon steel. For example, over the last 15 years, the cost of stainless steel has varied between 4.7 and 7.0 times the cost of carbon steel [16]. However, the cost of a stainless steel process vessel has varied in the approximate range of 2.3 to 3.5 times the cost of a carbon steel vessel for similar service.

Materials factors for the process equipment considered in this text are given in Appendix A, Tables A.3–A.6, and Figures A.18 and A.19. These figures are constructed using averaged data from the following sources: Peters and Timmerhaus [2], Guthrie [9, 10], Ulrich [5], Navarrete [11], and Perry et al. [3]. Example 7.12 illustrates the use of these figures and tables.

Example 7.12

Find the bare module cost of a floating-head shell-and-tube heat exchanger with a heat transfer area of 100 m^2 for the following cases.

a. The operating pressure of the equipment is 1 barg on both shell and tube sides, and the MOC of the shell and tubes is stainless steel.

b. The operating pressure of the equipment is 100 barg on both shell and tube sides, and the MOC of the shell and tubes is stainless steel.

From Example 7.10, $C_p^o(2001)$ = \$25,000 and $C_p^o(2006)$ = \$25,000 (500/397) = \$31,490.

a. From Example 7.10, at 1 barg, $F_P = 1$

From Table A.3 for a shell-and-tube heat exchanger made of SS, Identification No. = 5 and using Figure A.8, $F_M = 2.73$

From Equation A.4,

$$C_{BM}(2006) = C_p^o[B_1 + B_2F_PF_M] = \$31{,}490[1.63 + 1.66(1.0)(2.73)] = \$194{,}000$$

b. From Example 7.11 for $P = 100$ barg, $F_P = 1.383$

From (a) above, $F_M = 2.73$

Substituting these values in to Equation A.4,

$$C_{BM}(2006) = C_p^o[B_1 + B_2F_PF_M] = \$31{,}490[1.63 + 1.66(1.383)(2.73)] = \$248{,}600$$

The last three examples all considered the same size heat exchanger made with different materials of construction and operating pressure. The results are summarized below.

Example	Pressure	Materials	F_{BM}	Cost
7.10	ambient	CS tubes/shell	3.29	\$103,590
7.11	100 barg	CS tubes/shell	3.93	\$123,590
7.12a	ambient	SS tubes/shell	6.16	\$194,000
7.12b	100 barg	SS tubes/shell	7.90	\$248,600

These results reemphasize the point that the cost of the equipment is strongly dependent on the materials of construction and the pressure of operation.

7.3.5 Combination of Pressure and MOC Information to Give the Bare Module Factor, F_{BM}, and Bare Module Cost, C_{BM}

In Examples 7.10–7.12, the bare module factors and costs were calculated using Equation A.4. The form of this equation is not obvious, and its derivation is based on the approach used by Ulrich [5]:

$$\text{Cost of Equipment} = C_p^oF_PF_M \tag{7.11}$$

This is the equipment cost at operating conditions:

$$\text{Cost for Equipment Installation (for base conditions)} = C_p^o(F_{BM}^o - 1) \tag{7.12}$$

This cost is calculated by taking the bare module cost, at base conditions, and subtracting the cost of the equipment at the base conditions.

The incremental cost of equipment installation due to nonbase case conditions is

$$= C_p^o(F_PF_M - 1)f_{P\&I} \tag{7.13}$$

This cost is based on the incremental cost of equipment due to nonbase conditions multiplied by a factor, ($f_{P\&I}$), that accounts for the fraction of the installation cost that is associated with piping and instrumentation. The values of $f_{P\&I}$ are modified from Guthrie [9, 10] to account for an increase in the level and cost of instrumentation that modern chemical plants enjoy compared with that at the time of Guthrie's work.

Equations 7.11 through 7.13 can be combined to give the following relationship:

$$\text{Bare module cost, } C_{BM} = C_p^o F_P F_M + C_p^o(F_{BM}^o - 1) + C_p^o(F_P F_M - 1) f_{P\&I}$$

$$= C_p^o[F_P F_M(1 + f_{P\&I}) + F_{BM}^o - 1 - f_{P\&I}] = C_p^o[B_1 + B_2 F_P F_M] \quad (7.14)$$

Equation 7.13 is the same as Equation A.4, with $B_1 = F_{BM}^o - 1 - f_{P\&I}$ and $B_2 = 1 + f_{P\&I}$.

7.3.6 Algorithm for Calculating Bare Module Costs

The following six-step algorithm is used to estimate actual bare module costs for equipment from the figures in Appendix A.

1. Using the correct figure in Appendix A (Figures A.1–A.17), or the data in Table A.1, obtain C_p^o for the desired piece of equipment. This is the purchased equipment cost for the base case (carbon steel construction and near ambient pressure).
2. Find the correct relationship for the bare module factor. For exchangers, pumps, and vessels, use Equation A.4 and the data in Table A.4. For other equipment, the form of the equation is given in Table A.5.
3. For exchangers, pumps, and vessels, find the pressure factor, F_P, Table A.2 and Equation A.2 or A.3, and the material of construction factor, F_M, Equation A.4, Table A.3, and Figure A.18. Use Equation A.4 to calculate the bare module factor, F_{BM}.
4. For other equipment find the bare module factor, F_{BM}, using Table A.6 and Figure A.19.
5. Calculate the bare module cost of equipment, C_{BM}, from Equation 7.6.
6. Update the cost from 2001 (CEPCI – 397) to the present by using Equation 7.4.

Example 7.13 illustrates the six-step algorithm for the case of a distillation column with associated trays.

Example 7.13

Find the bare module cost (in 2006) of a stainless steel tower 3 m in diameter and 30 m tall. The tower has 40 stainless steel sieve trays and operates at 20 barg.

The costs of the tower and trays are calculated separately and then added together to obtain the total cost.

For the tower,

a. Volume = $\pi D^2 L/4 = (3.14159)(3)^2(30)/4 = 212.1\ m^3$

From Equation A.1,

$$\log_{10} C_p^o(2001) = 3.4974 + 0.4485\log_{10}(212.1) + 0.1074\{\log_{10}(212.1)\}^2 = 5.1222$$

$$C_p^o(2001) = 10^{5.1222} = \$132{,}500$$
$$C_p^o(2006) = \$132{,}500(500/397) = \$166{,}880$$

b. From Equation A.3 and Table A.4, $F_{BM} = 2.25 + 1.82\ F_M F_P$

c. From Equation 7.10 with $P = 20$ barg and $D = 3$ m,

$$F_{P,vessels} = \frac{\dfrac{(20 + 1)3}{(2)[(944)(0.9) - 0.6(20 + 1)]} + 0.00315}{0.0063} = 6.47$$

From Table A.3, identification number for stainless steel vertical vessel = 20; from Figure A.8, $F_M = 3.11$

$$F_{BM} = 2.25 + 1.82(6.47)(3.11) = 38.87$$

d. $C_{BM}(2006) = (166{,}880)(38.87) = \$6{,}486{,}000$

For the trays,

a. Tray (tower) area = $\pi D^2/4 = 7.0686$

From Equation A.1,

$$\log_{10} C_p^o(2001) = 2.9949 + 0.4465\log_{10}(7.0686) + 0.3961\{\log_{10}(7.0686)\}^2 = 3.6599$$

$$C_p^o(2001) = 10^{3.6599} = \$4570$$
$$C_p^o(2006) = \$4{,}570(500/397) = \$5{,}756$$

From Table A.5,

$$C_{BM} = C_p N F_{BM} f_q$$

$N = 40$

$f_q = 1.0$ (since number of trays > 20, Table A.5)

From Table A.6, SS sieve trays identification number = 61; from Figure A.9, $F_{BM} = 1.83$

$$C_{BM,trays}(2006) = (\$5{,}756)(40)(1.83)(1.0) = \$421{,}300$$

For the tower plus trays,

$$C_{BM,tower+trays}(2006) = \$6{,}486{,}000 + \$\ 421{,}300 = \$\ 6{,}908{,}300$$

7.3.7 Grass Roots and Total Module Costs

The term *grass roots* refers to a completely new facility in which we start the construction on essentially undeveloped land, a grass field. The term *total module cost* refers to the cost of making small-to-moderate expansions or alterations to an existing facility.

To estimate these costs, it is necessary to account for other costs in addition to the direct and indirect costs. These additional costs were presented in Table 7.6 and can be divided into two groups.

Group 1: Contingency and Fee Costs: The contingency cost varies depending on the reliability of the cost data and completeness of the process flowsheet available. This factor is included in the evaluation of the cost as a protection against oversights and faulty information. Unless otherwise stated, values of 15% and 3% of the bare module cost are assumed for contingency costs and fees, respectively. These are appropriate for systems that are well understood. Adding these costs to the bare module cost provides the *total module cost.*

Group 2: Auxiliary Facilities Costs: These include costs for site development, auxiliary buildings, and off-sites and utilities. These terms are generally unaffected by the materials of construction or the operating pressure of the process. A review of costs for these auxiliary facilities by Miller [17] gives a range of approximately 20% to more than 100% of the bare module cost. Unless otherwise stated, these costs are assumed to be equal to 50% of the bare module costs for the base case conditions. Adding these costs to the total module cost provides the *grassroots cost.*

The total module cost can be evaluated from

$$C_{TM} = \sum_{i=1}^{n} C_{TM,i} = 1.18 \sum_{i=1}^{n} C_{Bm,i} \tag{7.15}$$

and the grassroots cost can be evaluated from

$$C_{GR} = C_{TM} + 0.50 \sum_{i=1}^{n} C^{o}_{BM,i} \tag{7.16}$$

where n represents the total number of pieces of equipment. The use of these equations is shown in Example 7.14.

Example 7.14

A small expansion to an existing chemical facility is being investigated, and a preliminary PFD of the process is shown in Figure E7.14.

The expansion involves the installation of a new distillation column with a reboiler, condenser, pumps, and other associated equipment. A list of the equipment, sizes, materials of construction, and operating pressures is given in Table E7.14(a). Using the information in Appendix A, calculate the total module cost for this expansion in 2006.

The same algorithm presented above is used to estimate bare module costs for all equipment. This information is listed in Table E7.14(b), along with purchased equipment cost, pressure factors, material factors, and bare module factors.

The substitutions from Table E7.14(b) are made into Equations 7.15 and 7.16 to determine the total module cost and the grassroots cost.

$$\text{total module cost } (C_{TM}) = 1.18\sum_{i=1}^{n} C_{BM,i} = 1.18(\$797{,}000)(500/397) = \$1{,}184{,}000$$

$$\text{grassroots cost } (C_{GR}) = C_{TM} + 0.50\sum_{i=1}^{n} C^o_{BM,i} = \$1{,}184{,}00 + 0.50(\$597{,}800)(500/397)$$
$$= \$1{,}561{,}000$$

Although the grassroots cost is not appropriate here (because we have only a small expansion to an existing facility), it is shown for completeness.

7.3.8 A Computer Program (CAPCOST) for Capital Cost Estimation Using the Equipment Module Approach

For processes involving only a few pieces of equipment, estimating the capital cost of the plant by hand is relatively easy. For complex processes with many pieces of equipment, these calculations become tedious. To make this process easier, a computer program has been developed that allows the user to enter data interactively and obtain cost estimates in a fraction of the time required by hand calculations with less chance for error. The program (CAPCOST_2008.xls) is programmed in Microsoft Excel, and a template copy of the program is supplied on the CD that accompanies this book.

The program is written in the Microsoft Windows programming environment. The program requires the user to input information about the equipment—for example, the capacity, operating pressure, and materials of construction. The

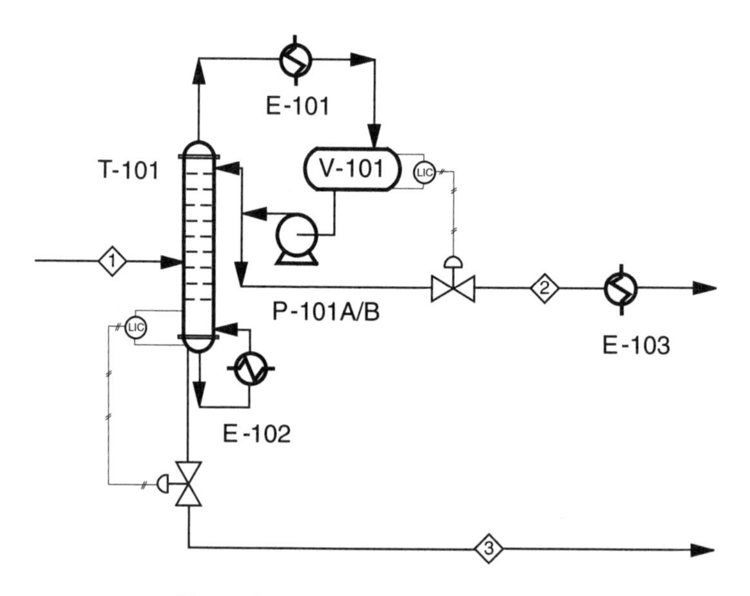

Figure E7.14 PFD for Example 7.14

Table E7.14(a) Information on Equipment Required for the Plant Expansion Described in Example 7.14

Equipment No.	Capacity / Size	Material of Construction*	Operating Pressure (barg†)
E-101 Overhead condenser	Area = 170 m^2 Shell and tube (floating-head)	Tube - CS Shell - CS	Tube = 5.0 Shell = 5.0
E-102 Reboiler	Area = 205 m^2 Shell and tube (floating-head)	Tube - SS Shell - CS	Tube = 18.0 Shell = 6.0
E-103 Product cooler	Area = 10 m^2 (double pipe)	All CS construction	Inner = 5.0 Outer = 5.0
P-101A/B Reflux pumps	$Power_{shaft}$ = 5 kW Centrifugal	CS	Discharge = 5.0
T-101 Aromatics column	Diameter = 2.1 m Height = 23 m	Vessel - CS	Column = 5.0
	32 sieve trays	Trays - SS	
V-101 Reflux drum	Diameter = 1.8 m Length = 6 m Horizontal	Vessel - CS	Vessel = 5.0

*CS = Carbon steel; SS = Stainless steel
†barg = bar gauge, thus 0.0 barg = 1.0 bar

Table E7.14(b) Results of Capital Cost Estimate for Example 7.14

Equipment	F_P	F_M	F_{BM}	C_p^o (2001)($)	C_{BM} (2001)($)	C_{BM}^o (2001)($)
E-101	1.0	1.0	3.29	33,000	108,500	108,500
E-102	1.023	1.81	4.70	36,900	177,900	121,300
E-103	1.0	1.0	3.29	3700	12,300	12,300
P-101A/B	1.0	1.55	3.98	(2)(3200)	(2)(12,600)	(2)(10,300)
T-101	1.681	1.0	5.31	54,700	290,700	222,800
32 trays		1.83	1.83	(32)(2200)	131,200	71,700
V-101	1.513	1.0	3.79	13,500	51,200	40,600
Totals				219,900	797,000	597,800
CEPCI = 397						

cost data can be adjusted for inflation by entering the current value of the CEPCI. Other information such as output file names and the number of the unit (100, 200, etc.) is also required.

The equipment options available to the user are given below.

- Blenders
- Centrifuges
- Compressors and blowers without drives
- Conveyors
- Crystallizers
- Drives for compressors, blowers, and pumps
- Dryers
- Dust collectors
- Evaporators and vaporizers
- Fans with electric drives
- Filters
- Fired heaters, thermal fluid heaters, and packaged steam boilers
- Furnaces
- Heat exchangers
- Mixers
- Process vessels with/without internals
- Power recovery equipment
- Pumps with electric drives
- Reactors
- Screens
- Storage vessels (fixed roof and floating roof)
- Towers
- User-added modules

The type of equipment required can be entered by using the mouse-activated buttons provided on the first worksheet. The user will then be asked a series of questions that appear on the screen. The user will be required to identify or enter the same information as would be needed to do the calculations by hand—that is, operating pressure, materials of construction, and the size of the equipment. The same information as contained in the cost charts and tables in Appendix A is embedded in the program, and the program should give the same results as hand calculations using these charts.

When the data for equipment are entered, a list of the costs on the first worksheet is updated. The use of the spreadsheet is explained in the CAP-COST.avi help files contained on the CD, and the reader is encouraged to view the file prior to using the software. You are strongly advised to verify the results of Example E7.14 for yourself prior to using the program to solve problems in the back of this chapter.

7.4 SUMMARY

In this chapter, the different types of capital cost estimating techniques that are available were reviewed. The accuracy of the different estimates was shown to increase significantly with the time involved in completion and the amount of data required. The information required to make an equipment module estimate based on data from the major process equipment was also covered. The effects of operating pressure and materials of construction on the bare module cost of equipment were reviewed. Several examples were given to show how the installed cost of equipment is significantly greater than the purchased cost and how the installed cost increases with increased pressure and materials of construction. The use of cost indices to adjust for the effects of inflation on equipment costs was considered, and the Chemical Engineering Plant Cost Index (CEPCI) was adopted for all inflation adjustments. The concepts of grass roots and total module costs were introduced in order to make estimates of the total capital required to build a brand new plant or make an expansion to an existing facility. To ease the calculation of the various costs, a computer program for cost estimation was introduced. This chapter contains the basic approach to estimating capital costs for new chemical plants and expansions to existing plants, and mastery of this material is assumed in the remaining chapters.

REFERENCES

1. Pikulik, A., and H. E. Diaz, "Cost Estimating for Major Process Equipment," *Chem. Eng.* 84, no. 21 (1977): 106.
2. Peters, M. S., and K. D. Timmerhaus, *Plant Design and Economics for Chemical Engineers*, 4th ed. (New York: McGraw-Hill, 1991).
3. Perry, R. H., D. W. Green, and J. O. Maloney, eds., *Chemical Engineers Handbook*, 7th ed. (New York: McGraw-Hill, 1997).
4. Cost Estimate Classification System, AACE International Recommended Practice No. 17R-97, 1997.
5. Ulrich, G. D., *A Guide to Chemical Engineering Process Design and Economics* (New York: John Wiley and Sons, 1984).
6. Lang, H. J., "Engineering Approach to Preliminary Cost Estimates," *Chem. Eng.* 54, no. 9 (1947): 130.
7. Lang, H. J., "Cost Relationships in Preliminary Cost Estimates," *Chem. Eng.* 54, no. 10 (1947): 117.

8. Lang, H. J., "Simplified Approach to Preliminary Cost Estimates," *Chem. Eng.* 55, no. 6 (1948): 112.
9. Guthrie, K. M., "Capital Cost Estimating," *Chem. Eng.* 76, no. 3 (1969): 114.
10. Guthrie, K. M., *Process Plant Estimating, Evaluation and Control* (Solana Beach, CA: Solana, 1974).
11. Navarrete, P. F., *Planning, Estimating, and Control of Chemical Construction Projects* (New York: Marcel Dekker, Inc., 1995).
12. R-Books Software, Richardson Engineering Services, Inc., 2001.
13. Section VIII, *ASME Boiler and Pressure Vessel Code,* ASME Boiler and Pressure Vessel Committee (New York: ASME, 2000).
14. Sandler, H. J., and E. T. Luckiewicz, *Practical Process Engineering, a Working Approach to Plant Design* (Philadelphia: XIMIX, Inc., 1987).
15. Incoloy Alloys 800 and 800HT, Table 22, Inco Alloys International Publication, IAI-20 4M US/1M UK (1986).
16. Construction Economics Section, *Engineering News Record,* December 24, 2001, 26.
17. Miller, C. A., "Factor Estimating Refined for Appropriation of Funds," *Amer. Assoc. Cost Engin. Bull.,* September 1965, 92.

SHORT ANSWER QUESTIONS

1. What are the three main factors that determine the capital cost of a piece of equipment such as a heat exchanger at a given time?
2. What is the Chemical Engineering Plant Cost Index (*CEPCI*) used for, and what does it measure?
3. What is the difference between the total module cost and the grassroots cost of a chemical process?
4. When would you use a cost exponent of 0.6?
5. What is meant by the economy of scale?
6. What is a Lang Factor?
7. The pressure factor F_p for a shell-and-tube heat exchanger is significantly smaller than for a vessel over the same pressure range. Why is this so?

PROBLEMS

8. The cost* of a recent plant in Alberta, Canada, to produce 1.27 million tonne/y of polyethylene was $540 million. Estimate what the range of cost estimates would likely have been for a Class 5, a Class 3, and a Class 1 estimate.

*http://www.chemicals-technology.com/projects/joffre/

9. In Appendix A, Figures A.1–A.17, the purchased costs for various types of equipment are given. The *y*-axis is given as the cost of the equipment per unit of capacity, and the *x*-axis is given as the capacity. The capacity is simply the relevant sizing parameter for the equipment. Identify all equipment that does not conform to the principle of the economy of scale.
10. A process vessel was purchased in the United Kingdom for our plant in the United States in 1993. A similar vessel, but of different capacity, was purchased in 1998. From the data given below, estimate the cost in U.S.$ of a vessel of 120 m^3 capacity purchased today (assume the current CEPCI = 500).

Date	Vessel Capacity (m^3)	Purchased Cost (Pounds Sterling = £)	Exchange Rate
1993	75	£ 7,800	$1.40/£
1998	155	£ 13,800	$1.65/£
2007	120		$2.00/£

11. You have been hired as a consultant to a legal firm. Part of your assignment is to determine the size of a storage tank purchased in 1978 (CEPCI = 219), before computerization of records. Many records from this era were destroyed in a fire (not in the plant, but in a distant office building). The tank was replaced every 10 years, and the sizes have changed due to plant capacity changes. You have the information in the table below. Estimate the original capacity of this vessel.

Date	Tank Capacity (1000 gal)	Purchased Cost
1978	?	$35,400
1988	105	$45,300
1998	85	$45,500

12. In your role as a consultant to a legal firm, you have been requested to determine whether calculations submitted in a legal action are valid. The problem is to determine what year a compressor was placed into service. The information in the table is available. It is claimed that the compressor was placed into service in 1976. History suggests that during the period from 1976 to 1985 there was significant inflation. Do you believe the information submitted is correct? If not, what year do you believe the compressor to be placed into service? Use *CEPCI* = 500 for 2006.

Date	Compressor Power (kW)	Total Module Cost (in $10^3$$)
???	1000	645.93
2000	500	500.00
2006	775	811.68

Note: CEPCI (1986) = 318, CEPCI (1981) = 297, CEPCI (1976) = 192.

13. When designing equipment for high-temperature and high-pressure service, the maximum allowable stress as a function of temperature of the material of construction is of great importance. Consider a cylindrical vessel shell that is to be designed for pressure of 150 bar (design pressure). The diameter of the vessel is 3.2 m, it is 15 m long, and a corrosion allowance of 6.35 mm (1/4") is to be used. Construct a table that shows the thickness of the vessel walls in the temperature range of 300 to 500°C (in 20°C increments) if the materials of construction are (a) ASME SA515-grade carbon steel and (b) ASME SA-240-grade 316 stainless steel.
14. Using the results of Problem 13, determine the relative costs of the vessel using the two materials of construction (CS and 316 SS) over the temperature range. You may assume that the cost of the vessel is directly proportional to the weight of the vessel and that the 316 SS costs 3.0 times that of CS. Based on these results, which material of construction is favored over the temperature range 300–500°C for this vessel?

The following problems may be solved either by using hand calculations or by using CAPCOST (use a value of CEPCI = 500).

15. Determine the bare module cost of a 1-shell pass, 2-tube pass (1-2) heat exchanger designed for the following operating conditions:
    Maximum operating pressure (tube side) = 30 barg
    Maximum operating pressure (shell side) = 5 barg
    Process fluid in tubes requires stainless steel MOC
    Shell-side utility (cooling water) requires carbon steel MOC
    Heat exchange area = 160 m^2
16. Repeat Problem 15, except reverse the shell-side and tube-side fluids. Are your results consistent with the heuristics for heat exchangers given in Chapter 11? Which heuristic is relevant?
17. In Chapter 15, the concepts of heat-exchanger networks and pinch technology are discussed. When designing these networks to recover process heat, it is often necessary to have a close temperature approach between process streams, which leads to large heat exchangers with multiple shells. Multiple-shell heat exchangers are often constructed from sets of 1-2 shell and tube exchangers stacked together. For costing considerations, the cost of the multiple-shell heat exchanger is best estimated as a number of smaller 1-2 exchangers. Consider a heat exchanger constructed of carbon steel and designed to withstand a pressure of 20 barg in both the shell and tube sides. This equipment has a heat exchange area of 400 m^2. Do the following.
    a. Determine the bare module cost of this 4-shell and 8-tube pass heat exchanger as four, 1-2 exchangers, each with a heat-exchange area of 100 m^2.
    b. Determine the bare module cost of the same exchanger as if it had a single shell.

c. What is the name of the principle given in this chapter that explains the difference between the two answers in (a) and (b)?

18. A distillation column is initially designed to separate a mixture of toluene and xylene at around ambient temperature (say, 100°C) and pressure (say, 1 barg). The column has 20 stainless steel valve trays and is 2 m in diameter and 14 m tall. Determine the purchased cost and the bare module cost using a $CEPCI = 500$.

19. A column with similar dimensions, number of trays, and operating at the same conditions as given in Problem 18 is to be used to separate a mixture containing the following chemicals. For each case determine the bare module cost using a $CEPCI = 500$.
    a. 10% nitric acid solution
    b. 50% sodium hydroxide solution
    c. 10% sulfuric acid solution
    d. 98% sulfuric acid solution

    Hint: you may need to look for the relevant MOC for part (d) on the Internet or another resource.

It is recommended that the following problems be solved using CAPCOST (use a value of CEPCI = 500).

Determine the bare module, total module, and grassroots cost of the following.

20. Toluene hydrodealkylation plant described in Chapter 1 (see Figures 1.3 and 1.5 and Tables 1.5 and 1.7).

21. Ethylbenzene plant described in Appendix B, project B.2.

22. Styrene plant described in Appendix B, project B.3.

23. Drying oil plant described in Appendix B, project B.4.

24. Maleic anhydride plant described in Appendix B, project B.5.

25. Ethylene oxide plant described in Appendix B, project B.6.

26. Formalin plant described in Appendix B, project B.7.

CHAPTER

8

Estimation of Manufacturing Costs

The costs associated with the day-to-day operation of a chemical plant must be estimated before the economic feasibility of a proposed process can be assessed. This chapter introduces the important factors affecting the manufacturing cost and provides methods to estimate each factor. In order to estimate the manufacturing cost, we need process information provided on the PFD, an estimate of the fixed capital investment, and an estimate of the number of operators required to operate the plant. The fixed capital investment is the same as either the total module cost or the grassroots cost defined in Chapter 7. Manufacturing costs are expressed in units of dollars per unit time, in contrast to the capital costs, which are expressed in dollars. How we treat these two costs, expressed in different units, to judge the economic merit of a process is covered in Chapters 9 and 10.

8.1 FACTORS AFFECTING THE COST OF MANUFACTURING A CHEMICAL PRODUCT

There are many elements that influence the cost of manufacturing chemicals. A list of the important costs involved, including a brief explanation of each cost, is given in Table 8.1.

The cost information provided in Table 8.1 is divided into three categories:

1. **Direct manufacturing costs:** These costs represent operating expenses that vary with production rate. When product demand drops, production rate is reduced to less than the design capacity. At this lower rate, we would expect a reduction in the factors making up the direct manufacturing costs. These

Table 8.1 Factors Affecting the Cost of Manufacturing (*COM*) for a Chemical Product (from References [1, 2, and 3])

Factor	Description of Factor
1. Direct Costs	**Factors that vary with the rate of production**
A. Raw materials	Costs of chemical feed stocks required by the process. Flowrates obtained from the PFD.
B. Waste treatment	Costs of waste treatment to protect environment.
C. Utilities	Costs of utility streams required by process. Includes but not limited to a. Fuel gas, oil, and/or coal b. Electric power c. Steam (all pressures) d. Cooling water e. Process water f. Boiler feed water g. Instrument air h. Inert gas (nitrogen, etc.) i. Refrigeration Flowrates for utilities found on the PFD/PIDs.
D. Operating labor	Costs of personnel required for plant operations.
E. Direct supervisory and clerical labor	Cost of administrative, engineering, and support personnel.
F. Maintenance and repairs	Costs of labor and materials associated with maintenance.
G. Operating supplies	Costs of miscellaneous supplies that support daily operation not considered to be raw materials. Examples include chart paper, lubricants, miscellaneous chemicals, filters, respirators and protective clothing for operators, etc.
H. Laboratory charges	Costs of routine and special laboratory tests required for product quality control and troubleshooting.
I. Patents and royalties	Cost of using patented or licensed technology.
2. Fixed Costs	**Factors not affected by the level of production**
A. Depreciation	Costs associated with the physical plant (buildings, equipment, etc.). Legal operating expense for tax purposes.
B. Local taxes and insurance	Costs associated with property taxes and liability insurance. Based on plant location and severity of the process.

Factor	Description of Factor
C. Plant overhead costs (sometimes referred to as factory expenses)	Catch-all costs associated with operations of auxiliary facilities supporting the manufacturing process. Costs involve payroll and accounting services, fire protection and safety services, medical services, cafeteria and any recreation facilities, payroll overhead and employee benefits, general engineering, etc.
3. General Expenses	**Costs associated with management level and administrative activities not directly related to the manufacturing process**
A. Administration costs	Costs for administration. Includes salaries, other administration, buildings, and other related activities.
B. Distribution and selling costs	Costs of sales and marketing required to sell chemical products. Includes salaries and other miscellaneous costs.
C. Research and development	Costs of research activities related to the process and product. Includes salaries and funds for research-related equipment and supplies, etc.

reductions may be directly proportional to the production rate, as for raw materials, or might be reduced slightly—for example, maintenance costs or operating labor.

2. **Fixed manufacturing costs:** These costs are independent of changes in production rate. They include property taxes, insurance, and depreciation, which are charged at constant rates even when the plant is not in operation.
3. **General expenses:** These costs represent an overhead burden that is necessary to carry out business functions. They include management, sales, financing, and research functions. General expenses seldom vary with production level. However, items such as research and development and distribution and selling costs may decrease if extended periods of low production levels occur.

The equation used to evaluate the cost of manufacture using these costs becomes:

$$\text{Cost of Manufacture } (COM) = \text{Direct Manufacturing Costs } (DMC) + \text{Fixed Manufacturing Costs } (FMC) + \text{General Expenses } (GE)$$

The approach we provide in this chapter is similar to that presented in other chemical engineering design texts [1, 2, 3].

The cost of manufacturing, *COM*, can be determined when the following costs are known or can be estimated:

1. Fixed capital investment (*FCI*): (C_{TM} or C_{GR})
2. Cost of operating labor (C_{OL})
3. Cost of utilities (C_{UT})
4. Cost of waste treatment (C_{WT})
5. Cost of raw materials (C_{RM})

Table 8.2 gives data to estimate the individual cost items identified in Table 8.1 (both tables carry the same identification of individual cost terms). With the exception of the cost of raw materials, waste treatment, utilities, and operating labor (all parts of the direct manufacturing costs), Table 8.2 presents equations that can be used to estimate each individual item. With each equation, a typical range for the constants (multiplication factors) to estimate an individual cost item is presented. If no other information is available, the midpoint value for each of these ranges is used to estimate the costs involved. It should be noted that the best information that is available should always be used to establish these constants. The method presented here should be used only when no other information on these costs is available.

By using the midpoint values given in Table 8.2, column 2, the resulting equations for the individual items are calculated in column 3. The cost items for each of the three categories are added together to provide the total cost for each category. The equations for estimating the costs for each of the categories are as follows:

$$DMC = C_{RM} + C_{WT} + C_{UT} + 1.33C_{OL} + 0.069FCI + 0.03COM$$

$$FMC = 0.708C_{OL} + 0.068FCI + \text{depreciation}$$

$$GE = 0.177C_{OL} + 0.009FCI + 0.16COM$$

We can obtain the total manufacturing cost by adding these three cost categories together and solving for the total manufacturing cost, *COM*. The result is

$$COM = 0.280FCI + 2.73C_{OL} + 1.23(C_{UT} + C_{WT} + C_{RM}) \tag{8.1}$$

In Equation (8.1), the depreciation allowance of 0.10*FCI* is added separately.

The cost of manufacture without depreciation, COM_d, is

$$COM_d = 0.180FCI + 2.73C_{OL} + 1.23(C_{UT} + C_{WT} + C_{RM}) \tag{8.2}$$

The calculation of manufacturing costs and expenses is given in Example 8.1.

Table 8.2 Multiplication Factors for Estimating Manufacturing Cost* (See Also Table 8.1)

Cost Item from Table 8.1	Typical Range of Multiplying Factors	Value Used in Text
1. Direct Manufacturing Costs		
a. Raw materials	C_{RM}†	
b. Waste treatment	C_{WT}†	
c. Utilities	C_{UT}†	
d. Operating labor	C_{OL}	C_{OL}
e. Direct supervisory and clerical labor	$(0.1 - 0.25)C_{OL}$	$0.18C_{OL}$
f. Maintenance and repairs	$(0.02 - 0.1)FCI$	$0.06FCI$
g. Operating supplies	(0.1 – 0.2)(Line 1.F)	$0.009FCI$
h. Laboratory charges	$(0.1 - 0.2)C_{OL}$	$0.15C_{OL}$
i. Patents and royalties	$(0 - 0.06)COM$	$0.03COM$
Total Direct Manufacturing Costs	$\mathbf{C_{RM} + C_{WT} + C_{UT} + 1.33C_{OL} + 0.03COM + 0.069FCI}$	
2. Fixed Manufacturing Costs		
a. Depreciation	$0.1FCI$‡	$0.1FCI$‡
b. Local taxes and insurance	$(0.014 - 0.05)FCI$	$0.032FCI$
c. Plant overhead costs	(0.50 – 0.7)(Line 1.D. + Line 1.E + Line 1.F)	$0.708C_{OL} + 0.036FCI$
Total Fixed Manufacturing Costs	$\mathbf{0.708C_{OL} + 0.068FCI}$ **+ depreciation**	
3. General Manufacturing Expenses		
a. Administration costs	0.15(Line 1.D. + Line 1.E.+ Line 1.F.)	$0.177C_{OL} + 0.009FCI$
b. Distribution and selling costs	$(0.02 - 0.2)COM$	$0.11COM$
c. Research and development	$0.05COM$	$0.05COM$
Total General Manufacturing Costs	$\mathbf{0.177C_{OL} + 0.009FCI + 0.16COM}$	
Total Costs	$\mathbf{C_{RM} + C_{WT} + C_{UT} + 2.215C_{OL} + 0.190COM + 0.146FCI}$ **+ depreciation**	

*Costs are given in dollars per unit time (usually per year).
†Costs are evaluated from information given on the PFD and the unit cost.
‡Depreciation costs are covered separately in Chapter 9. The use of 10% of *FCI* is a crude approximation at best.
From references [1], [2], and [3].

Example 8.1

The following cost information was obtained from a design for a 92,000 tonne/year nitric acid plant.

Fixed Capital Investment	\$11,000,000
Raw Materials Cost	\$ 7,950,000/yr
Waste Treatment Cost	\$ 1,000,000/yr
Utilities	\$ 356,000/yr
Direct Labor Cost	\$ 300,000/yr
Fixed Costs	\$ 1,500,000/yr

Determine

a. The manufacturing cost in \$/yr and \$/tonne of nitric acid

b. The percentage of manufacturing costs resulting from each cost category given in Tables 8.1 and 8.2

Using Equation 8.2,

$$COM_d = (0.180)(\$11{,}000{,}000) + (2.73)(\$300{,}000) + (1.23)(\$356{,}000 + \$1{,}000{,}000 + \$7{,}950{,}000) = \$14{,}245{,}000/\text{yr}$$

$$(\$14{,}245{,}000/\text{yr})/(92{,}000\ \text{tonne/yr}) = \$155/\text{tonne}$$

From the relationships given in Table 8.2,

Direct Manufacturing Costs = \$7,950,000 + \$1,000,000 + \$356,000 + (1.33)(\$300,000) + (0.069)(\$11,000,000) + (0.03)(\$14,245,000) = \$10,891,000

Percentage of manufacturing cost = (100)(10.891)/14.25 = 76%

Fixed Manufacturing Costs = (0.708)(\$300,000) + (0.068)(\$11,000,000) = \$960,000

Percentage of manufacturing cost = (100)(0.960)/14.25 = 7%

General Expenses = (0.177)(\$300,000) + (0.009)(\$11,000,000) + (0.16)(\$14,245,000) = \$2,431,000

Percentage of manufacturing cost = (100)(2.431)/14.25 = 17%

In Example 8.1, the direct costs were shown to dominate the manufacturing costs, accounting for about 76% of the manufacturing costs. Of these direct costs, the raw materials cost, the waste treatment cost, and the cost of utilities accounted for more than \$9 million of the \$10.9 million direct costs. These three cost contributions are not dependent on any of the estimating factors provided in Table 8.2. Therefore, the manufacturing cost is generally insensitive to the estimating factors provided in Table 8.2. The use of the midrange values is acceptable for this situation.

8.2 COST OF OPERATING LABOR

The technique used to estimate operating labor requirements is based on data obtained from five chemical companies and correlated by Alkayat and Gerrard [4]. According to this method, the operating labor requirement for chemical processing plants is given by Equation 8.3:

$$N_{OL} = (6.29 + 31.7P^2 + 0.23N_{np})^{0.5} \quad (8.3)$$

where N_{OL} is the number of operators per shift, P is the number of processing steps involving the handling of particulate solids—for example, transportation and distribution, particulate size control, and particulate removal. N_{np} is the number of nonparticulate processing steps and includes compression, heating and cooling, mixing, and reaction. In general, for the processes considered in this text, the value of P is zero, and the value of N_{np} is given by

$$N_{np} = \sum Equipment \quad (8.4)$$

compressors
towers
reactors
heaters
exchangers

Equation 8.3 was derived for processes with, at most, two solid handling steps. For processes with a greater number of solid handling operations, this equation should not be used.

The value of N_{OL} in Equation 8.3 is the number of operators required to run the process unit per shift. A single operator works on the average 49 weeks a year (3 weeks' time off for vacation and sick leave), five 8-hour shifts a week. This amounts to (49 weeks/year × 5 shifts/week) 245 shifts per operator per year. A chemical plant normally operates 24 hours/day. This requires (365 days/year × 3 shifts/day) 1095 operating shifts per year. The number of operators needed to provide this number of shifts is [(1095 shifts/yr)/(245 shifts/operator/yr)] or approximately 4.5 operators. Four and one-half operators are hired for each operator needed in the plant at any time. This provides the needed operating labor but does not include any support or supervisory staff.

To estimate the cost of operating labor, the average hourly wage of an operator is required. Chemical plant operators are relatively highly paid, and data from the Bureau of Labor and Statistics [5] give the hourly rate for miscellaneous plant and system operators in the Gulf Coast region at $26.48 in May 2006. This corresponds to $52,900 for a 2000-hour year. The cost of labor depends considerably on the location of the plant, and significant variations from the above figure may be expected. Historically, wage levels for chemical plant operators have grown slightly faster than the other cost indexes for process plant equipment given in Chapter 7. *The Oil and Gas Journal* and *Engineering News Record* provide appropriate indices to correct labor costs for inflation, or reference [5] can be consulted. The estimation of operating costs is illustrated in Example 8.2.

Example 8.2

Estimate the operating labor requirement and costs for the toluene hydrodealkylation facility shown in Figures 1.3 and 1.5.

From the PFD in Figure 1.5, the number and type of equipment are determined.

Table E8.2 Results for the Estimation of Operating Labor Requirements for the Toluene Hydrodealkylation Process Using the Equipment Module Approach

Equipment Type	Number of Equipment	N_{np}
Compressors	1	1
Exchangers	7	7
Heaters/Furnaces	1	1
Pumps*	2	—
Reactors	1	1
Towers	1	1
Vessels*	4	—
	Total	11

*Pumps and vessels are not counted in evaluating N_{np} in Equation 8.4.

Using Equation (8.4), an estimate of the number of operators required per shift is made. This information is shown in Table E8.2.

$$N_{OL} = [6.29 + (0)^{0.1} + (0.23)(11)]^{0.5} = [8.82]^{0.5} = 2.97$$

The number of operators required per shift = 2.97.

Operating Labor = (4.5)(2.97) = 13.4 (rounding up to the nearest integer yields 14 operators)

Labor Costs (2001) = 14 × $52,900 = $740,600/yr

8.3 UTILITY COSTS

The costs of utilities are directly influenced by the cost of fuel. Specific difficulties emerge when estimating the cost of fuel, which directly impact the price of utilities such as electricity, steam, and thermal fluids. Figure 8.1 shows the general trends for fossil fuel costs from 1991 to 2006. The costs presented represent average values and are not site specific. These costs do not reflect the wide variability of cost and availability of various fuels throughout the United States.

8.3.1 Background Information on Utilities

As seen from Figure 8.1, coal represents the lowest-cost fossil fuel on an energy basis. Most coal is consumed near the "mine mouth" in large power plants to produce electricity. The electricity is transported by power lines to the consumer. At locations remote from mines, both the availability and cost of transportation reduce and/or eliminate much of the cost advantage of coal. Coal suffers further from its negative environmental impact—for example, relatively high sulfur content and relatively high ratio of CO_2 produced per unit of energy.

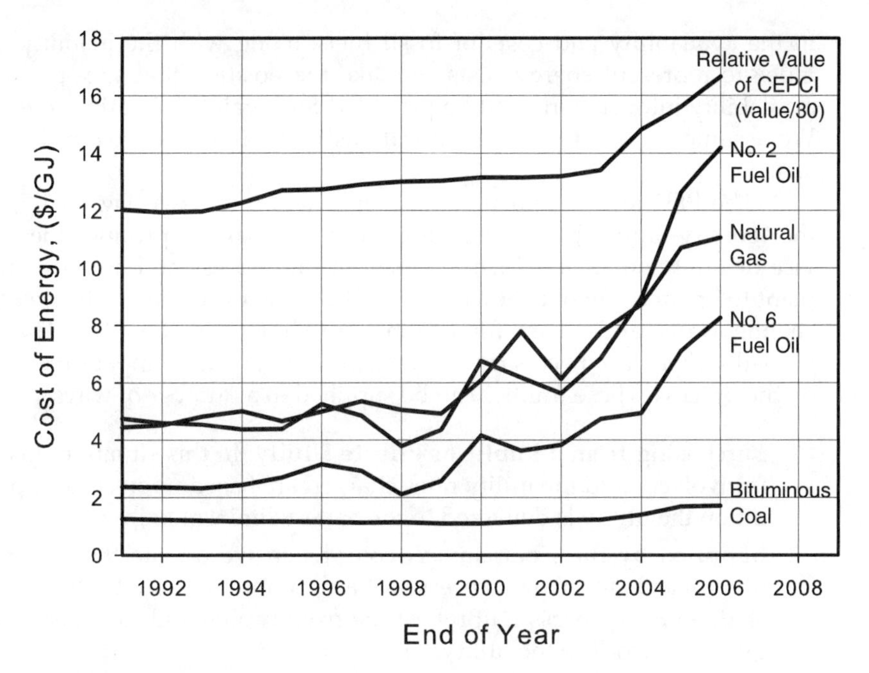

Figure 8.1 Changes in Fuel Prices from 1991 to 2006 (Information taken from Energy Information Administration [6])

After no. 6 fuel oil (a heavy oil with a relatively high sulfur content), the next lowest cost fuel source shown in Figure 8.1 is natural gas. Natural gas fuel is the least damaging fossil fuel energy supply with respect to the environment. It is transported by pipelines throughout much of the country. The cost is more uniform than coal throughout different regions of the country. There remain, however, regions in the country that are not yet serviced by the natural gas distribution system. In these regions, the use of natural gas is not an option that can be considered. Although natural gas is a mixture of several light hydrocarbons, it consists predominantly of methane. For the calculations used in this text, it is assumed that methane and natural gas are equivalent.

No. 2 fuel oil is the final fossil fuel that is commonly used as an energy source in the chemical industry. Until recently, it has been the highest-cost fossil fuel source. It is most readily available near coastal regions where oil enters the country and refining takes place. The United States has become increasingly dependent on imported oil, which may be subject to large upsets in cost and domestic availability. Uncertainties in the availability of supplies, high storage costs, and large fluctuations in cost make this source of energy least attractive in many situations. However, recently the cost of natural gas has increased substantially to the point that No. 2 fuel oil is now a viable alternative to natural gas in many plants.

Figure 8.1 shows that fuel costs have increased somewhat more rapidly and in a much more chaotic fashion than the cost index (CEPCI) that we have used previously to correct costs for inflation. As a result of the regional variations in the availability and costs of fossil fuels, along with the inability of the cost index to represent energy costs, we take the position that site-specific cost and availability information must be provided for a valid estimation of energy costs. We assume, in this text, that natural gas is the fuel of choice unless otherwise stated.

The PFD for the toluene hydrodealkylation process (Figure 1.5) represents the "battery-limits" plant. The equipment necessary to produce the various service or utility streams, which are used in the process and are necessary for the plant to operate, are not shown on the PFD. However, the utility streams such as cooling water and steam for heating are shown on the PFD. These streams, termed *utilities,* are necessary for the control of stream temperatures as required by the process. These utilities can be supplied in a number of ways.

1. **Purchasing from a Public or Private Utility:** In this situation no capital cost is involved, and the utility rates charged are based upon consumption. In addition the utility is delivered to the battery limits at known conditions.
2. **Supplied by the Company:** A comprehensive off-site facility provides the utility needs for many processes at a common location. In this case, the rates charged to a process unit reflect the fixed capital and the operating costs required to produce the utility.
3. **Self-Generated and Used by a Single Process Unit:** In this situation the capital cost for purchase and installation becomes part of the fixed capital cost of the process unit. Likewise the related operating costs for producing that particular utility are directly charged to the process unit.

Utilities that would likely be provided in a comprehensive chemical plant complex are shown in Table 8.3.

8.3.2 Calculation of Utility Costs

The calculation of utility costs can be quite complicated, and the true cost of such streams is often difficult to estimate in a large facility. For estimating operating costs associated with supplying utilities to different processes, the approach taken here is to assume that the capital investment required to build a facility to supply the utility—for example, a cooling tower, a steam boiler, and so forth—has already been made. This would be the case when a grassroots cost has been used for the fixed capital investment. The costs associated with supplying a given utility are then obtained by calculating the operating costs to generate the utility. These are the costs that have been presented in Table 8.3, and the following sections show how these cost estimates were obtained for the major utilities given in the table.

Cooling Tower Water. In most large chemical, petrochemical, and refinery plants, cooling water is supplied to process units from a central facility. This facility consists of a cooling tower (or many towers), water makeup, chemical

Table 8.3 Utilities Provided by Off-Sites for a Plant with Multiple Process Units (Costs Represent Charges for Utilities Delivered to the Battery Limit of a Process)

Utility	Description	Cost $/GJ	Cost $/Common Unit
Air Supply	Pressurized and dried air (add 20% for instrument air)		
	a. 6 barg (90 psig)		$0.49/100 std m^{3}*
	b. 3.3 barg (50 psig)		$0.35/100 std m^{3}*
Steam from Boilers	Process steam: latent heat only		
	a. Low pressure (5 barg, 160°C) from HP steam		
	With credit for power	13.28	$27.70/1000 kg
	Without credit for power	14.05	$29.29/1000 kg
	b. Medium pressure (10 barg, 184°C) from HP steam		
	With credit for power	14.19	$28.31/1000 kg
	Without credit for power	14.83	$29.59/1000 kg
	c. High pressure (41 barg, 254°C)	17.70	$29.97/1000 kg
Steam Generated from Process	Estimate savings as avoided cost of burning natural gas in boiler	12.33	
Cooling Tower Water	Processes cooling water: 30°C to 40°C or 45°C	0.354	$14.8/1000 m^{3}†
Other Water	High-purity water for		
	a. Process use		$0.067/1000 kg
	b. Boiler feed water (available at 115°C)‡		$2.45/1000 kg
	c. Potable (drinking)		$0.26/1000 kg
	d. Deionized water		$1.00/1000 kg
Electrical Substation	Electric Distribution	16.8	$0.06/kWh
	a. 110 V		
	b. 220 V		
	c. 440 V		
Fuels	a. Fuel oil (no. 2)	14.2	$549/m^{3}
	b. Natural gas	11.1§	$0.42/std m^{3}*
	c. Coal (f.o.b. mine mouth)	1.72	$41.4/tonne
Refrigeration	a. Moderately low temperature		
	Refrigerated water in at T = 5°C and returned at 15°C	4.43	$0.185/1000kg
	b. Low temperature		
	Refrigerant available at T = –20°C	7.89	
	c. Very low temperature		
	Refrigerant available at T = –50°C	13.11	Based on process cooling duty

(*continued*)

Table 8.3 Utilities Provided by Off-Sites for a Plant with Multiple Process Units (Costs Represent Charges for Utilities Delivered to the Battery Limit of a Process) (*continued*)

Utility	Description	Cost $/GJ	Cost $/Common Unit
Thermal Systems	Cost based on thermal efficiency of fired heater using natural gas		
	a. 90% efficient	12.33	Based on process
	b. 80% efficient	13.88	heating duty
Waste Disposal (Solid and Liquid)	a. Nonhazardous		$36/tonne
	b. Hazardous		$200–2000/tonne°
Waste Water Treatment	a. Primary (filtration)		$41/1000 m^3
	b. Secondary (filtration + activated sludge)		$43/1000 m^3
	c. Tertiary (filtration, activated sludge, and chemical processing)		$56/1000 m^3

*Standard conditions are 1.013 bar and 15°C.

†Based on $\Delta T_{cooling\ water}$ = 10°C. Cooling water return temperatures should not exceed 45°C due to excess scaling at higher temperatures.

‡Approximately equal credit is given for condensate returned from exchangers using steam.

§Based on lower heating value of natural gas.

°For hazardous waste, the cost of disposal varies widely. Chemical analyses are required for all materials that cannot be thoroughly identified. This does not include radioactive waste.

injection, and the cooling water feed pumps. A typical cooling water facility is shown in Figure 8.2.

The cooling of the water occurs in the cooling tower where some of the water is evaporated. Adding makeup water to the circulating cooling water stream makes up this loss. Because essentially pure water is evaporated, there is a tendency for inorganic material to accumulate in the circulating loop; therefore, there is a water purge or blowdown from the system. The makeup water stream also accounts for windage or spray losses from the tower and also the water purge. Chemicals are added to reduce the tendency of the water to foul heat-exchanger surfaces within the processes. For a detailed discussion and further information regarding the conditioning of water for cooling towers, the reader is referred to Hile et al. [7] and Gibson [8]. From Figure 8.2, we can estimate the cost to supply process users with cooling water if the following are known:

- Total heat load and circulation rate required for process users
- Composition and saturation compositions of inorganic chemicals in the feed water
- Required chemical addition rate
- Desired supply and return temperatures (shown earlier to be 30°C and 40°C, respectively)

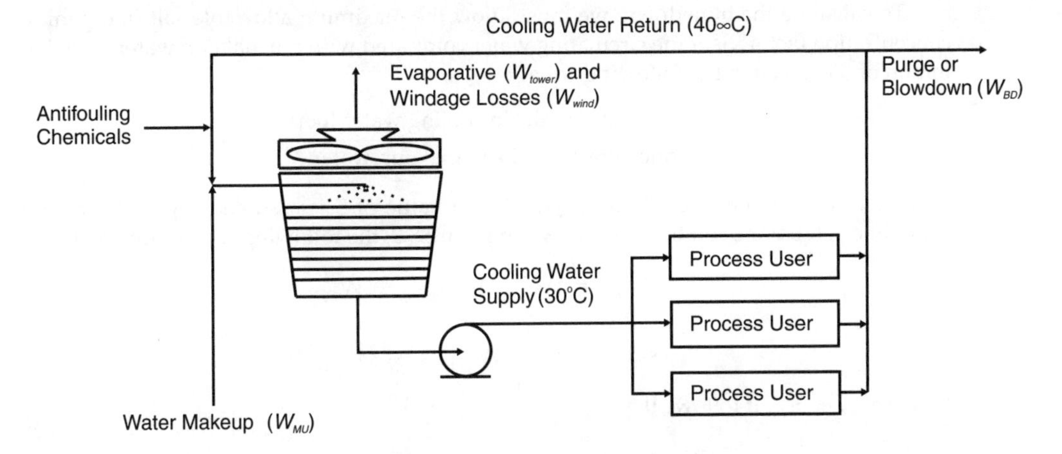

Figure 8.2 Schematic Diagram of Cooling Water Loop

- Cost of cooling tower and cooling water pumps
- Costs of supply chemicals, electricity for pumps and cooling tower fans, and makeup water

The estimation of operating costs associated with a typical cooling water system is illustrated in Example 8.3.

Example 8.3

Estimate the utility cost for producing a circulating cooling water stream using a mechanical draft cooling tower. Consider a basis of 1 GJ/h of energy removal from the process units. Flow of cooling water required to remove this energy = $\dot{m}$ kg/h.

An energy balance gives

$$\dot{m}c_p\Delta T = 1 \times 10^9 \Rightarrow (\dot{m})(4180)(40 - 30) = 41{,}800\,\dot{m} = 1 \times 10^9\,\text{J/h}$$

Therefore, $\dot{m} = \dfrac{1 \times 10^9}{41{,}800} = 23{,}923$ kg/h

Latent heat of water at average temperature of 35°C = 2417 kJ/kg
Amount of water evaporated from tower, W_{tower}

$$W_{tower} = \frac{\textit{Heat Load}}{\Delta H_{vap}} = \frac{1 \times 10^9}{2417 \times 10^3} = 413.7\text{ kg/h}$$

This is (413.7)(100)/(23,923) = 1.73% of the circulating water flowrate.

Typical windage losses from mechanical draft towers are between 0.1 and 0.3% [9, 10]; use 0.3%.

To calculate the blowdown, we must know the maximum allowable salt (inorganics) concentration factor, S, of the circulating water compared with the makeup water. The definition of S is given in the following equation:

$$S = \frac{\text{concentration salts in cooling water loop}}{\text{concentration salts in makeup water}} = \frac{S_{loop}}{S_{in}}$$

Typical values are between 3 and 7 [9]. Here a value of 5 is assumed. By performing a water and salt balance on the loop shown in Figure 8.2, the following results are obtained:

$$W_{MU} = W_{tower} + W_{wind} + W_{BD}$$

$$s_{in}W_{MU} = s_{loop}W_{wind} + s_{loop}W_{BD}$$

Because $s_{loop} = 5s_{in}$, it follows that

$$s_{in}(W_{tower} + W_{wind} + W_{BD}) = s_{loop}W_{wind} + s_{loop}W_{BD}$$

$$W_{BD} = \frac{s_{in}W_{tower} + W_{wind}(s_{in} - s_{loop})}{s_{loop} - s_{in}} = \frac{s_{in}W_{tower}}{s_{loop} - s_{in}} - W_{wind} = \frac{W_{tower}}{4} - W_{wind} = \frac{1.73\%}{4} - 0.3\% = 0.133\%$$

$$W_{MU} = 1.73 + 0.3 + 0.133 = 2.163\% = 517 \text{kg/h}$$

Pressure drop around the cooling water loop is estimated as follows: ΔP_{loop} = 15 psi (pipe losses) + 5 psi (exchanger losses) + 10 psi (control valve loss) + 8.7 psi of static head (because water must be pumped to top of cooling water tower, estimated to be 20 ft above pump inlet) = 38.7 psi = 266.7 kPa.

Power required for cooling water pumps with a volumetric flow rate $\dot{V}$, assuming an overall efficiency of 75%, is

$$\text{Pump Power} = \frac{1}{\varepsilon}\dot{V}\Delta P = \frac{1}{(0.75)}\frac{(23{,}923)}{(1000)(3600)}(266.7) = 2.36 \text{ kW}$$

Power required for fans: From reference [11], the required surface area in the tower = 0.5 ft^2/gpm (this assumes that the design wet-bulb air temperature is 26.7°C [80°F]). From the same reference, the fan horsepower per square foot of tower area is 0.041 hp/ft^2.

$$\text{Power for fan} = \frac{(23{,}923)(2.2048)}{(60)(8.337)}(0.5)(0.041) = (2.16)(0.746) = 1.61 \text{ kW}$$

From a survey of vendors, the cost of chemicals is \$0.156/1000 kg of makeup water.

Using an electricity cost of \$0.06/kWh and a process water cost of \$0.067/1000 kg, the overall cost of the cooling water is given by

Cost of cooling water = cost of electricity + cost of chemicals for makeup water + cost of makeup water

Using the cost values for electricity and process water given in Table 8.3,

$$\text{Cooling water cost} = (0.06)(2.36 + 1.61) + \frac{(517.3)(0.156)}{1000} + \frac{(517.3)(0.067)}{1000}$$

$$= \$0.354/\text{hr} = \$0.354/\text{GJ}$$

Clearly, this cost will change depending on the cost of electricity, the cost of chemicals, and the cost of process water.

Refrigeration. The basic refrigeration cycle consists of circulating a working fluid around a loop consisting of a compressor, evaporator, expansion valve or turbine, and condenser. This cycle is shown in Figure 8.3. The phases of the working fluid (L-liquid and V-vapor) are shown on the diagram.

The Carnot efficiency of a mechanical refrigeration system can be expressed by the reversible coefficient of performance, COP_{REV}:

$$COP_{REV} = \frac{\text{evaporator temperature } (T_1)}{\text{temperature difference between condenser and evaporator } (T_2 - T_1)}$$

$$COP \cong \frac{\text{evaporator heat load}}{\text{work required}} \quad \text{or} \quad \text{work required} = \frac{\text{evaporator heat load}}{\text{COP}}$$

Because all the processes for a Carnot engine must be reversible, the COP_{REV} gives the best theoretical performance of a refrigeration system. Thus the net required power (compressor-expansion turbine) will always be greater than that predicted by the equation above using COP_{REV}. Nevertheless, it is clear that as the temperature difference between the evaporator and condenser increases then the

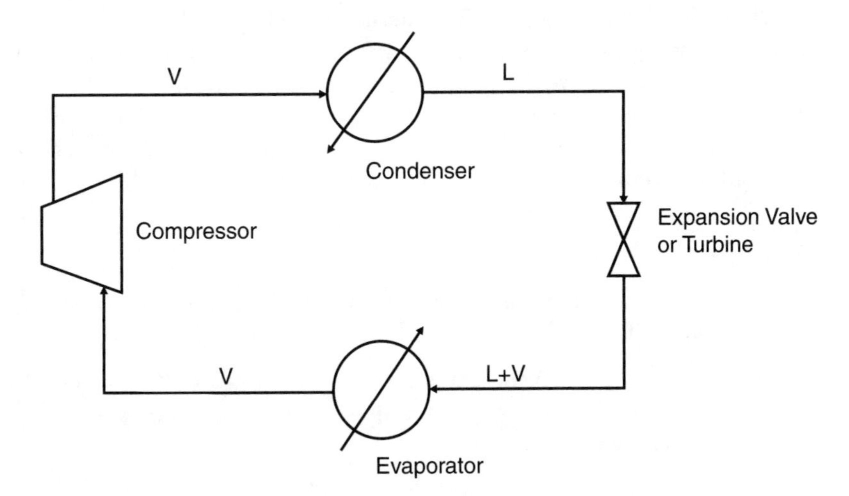

Figure 8.3 Process Flow Diagram for a Simple Refrigeration Cycle

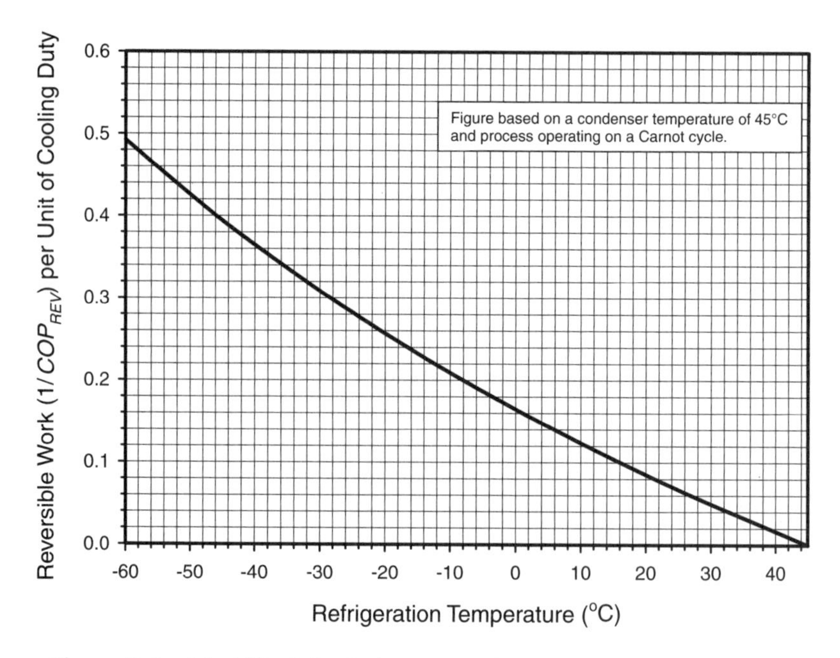

Figure 8.4 Ideal Work for Refrigeration Cycles as a Function of Refrigeration Temperature

work required per unit of energy removed in the evaporator (refrigerator) increases. Therefore, the operating costs for refrigeration will increase as the temperature at which the refrigeration is required decreases. The condensation of the working fluid will most often be achieved using cooling water, so a reasonable condensing temperature would be 45°C (giving a 5°C approach in the condensing exchanger). Figure 8.4 illustrates the effect of the evaporator temperature on the reversible work required for a given cooling load. This figure gives an approximate guide to the relative cost of refrigeration. The relative costs of refrigeration at different temperatures are explored in Example 8.4.

Example 8.4

Using Figure 8.4, calculate the relative costs of providing refrigeration at 5°C, –20°C, and –50°C. From the figure, the ordinate values are given as follows.

Temperature	$1/COP_{REV}$
5°C	0.144
–20°C	0.257
–50°C	0.426

Therefore, compared with cooling at 5°C, cooling to –20°C is 0.257/0.144 times as expensive, and cooling to –50°C is 0.426/0.144 times as expensive. This analysis assumes that the two refrigeration systems operate equally efficiently with respect to the reversible limit and that the major cost is the power to run the compressors.

In Example 8.5, a real refrigeration system is considered and operating costs are estimated.

Example 8.5

Obtain a cost estimate for a refrigerated cooling utility operating at 5°C.

Consider a single-stage refrigeration system to provide refrigeration at 5°C, using 1,1 difluoroethane (R-152a) as the refrigerant. The process flow diagram and operating conditions are given in Figure E8.5 and Table E8.5, respectively.

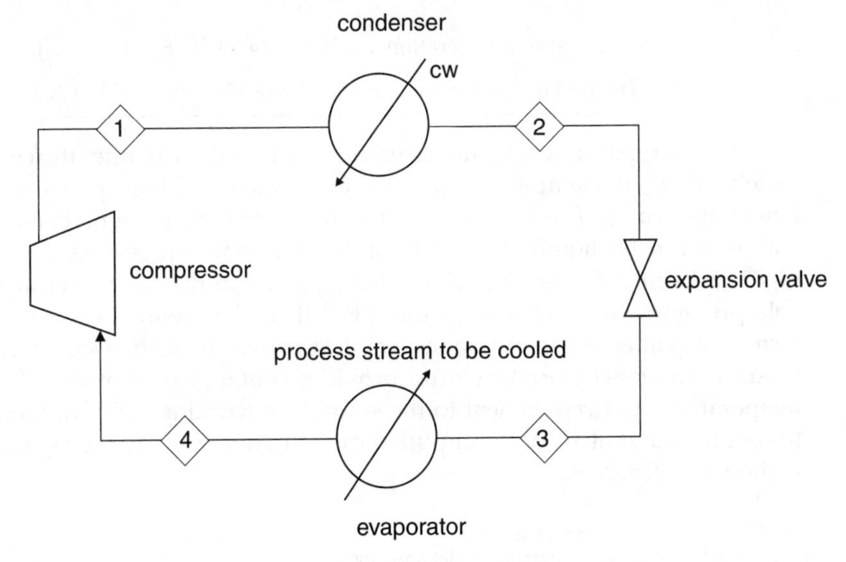

Figure E8.5 Process Flow Diagram for Simple Refrigeration Cycle of Example 8.5

Table E8.5 Stream Conditions for Figure E8.5

	Stream Number			
Condition	**1**	**2**	**3**	**4**
Pressure (bar)	10.9	10.9	3.21	3.21
Temperature (°C)	68.7	45.0	5.0	5.0
Vapor Fraction	1.0	0.0	0.2492	1.0

For the simulation shown, pressure drops across piping and heat exchangers have not been considered. When the circulation rate of R-152a is 65.3 kmol/h, the duty of the evaporator is 1 GJ/h. The compressor is assumed to be 75% efficient and the loads on the equipment are as follows:

Compressor Power = 66.5 kW (at 75% efficiency)

Condenser Duty = 1.24 GJ/h

Evaporator Duty = 1.00 GJ/h

$$\text{Compressor work per unit of cooling} = (66.5)/(1{,}000{,}000/3600) = 0.2394$$

This value compares with 0.144 for the Carnot cycle. The main differences are due to the inefficiencies in the compressor and the use of a throttling valve instead of a turbine.

$$\begin{aligned}\text{The cost of refrigeration at } 5^\circ\text{C} &= (66.5)(0.06) + (1.24)(0.354) = 3.99 + 0.44\\ &= 4.43\ \$/\text{h} = 4.43\ \$/\text{GJ}\end{aligned}$$

Using the results of Example 6.4, we can predict the cost of refrigeration at –20°C and –50°C as

$$\text{The cost of refrigeration at } -20^\circ\text{C} = (4.43)(1.78) = \$7.89/\text{GJ}$$

$$\text{The cost of refrigeration at } -50^\circ\text{C} = (4.43)(2.96) = \$13.11/\text{GJ}$$

For refrigeration systems operating at less than temperatures of approximately –60°C, the simple refrigeration cycle shown in Figures 8.4 and E8.5 is no longer applicable. The main reason for this is that there are no common refrigerants that can be liquified at 45°C under reasonable pressures (not excessively high) and still give the desired low temperature in the condenser also at reasonable pressures (not excessively low). For these low-temperature systems, some form of cascaded refrigeration system is required. In such systems, two working fluids are used. The primary fluid provides cooling to the process (at the lowest temperature) and rejects heat to the secondary working fluid that rejects its heat to cooling water at 45°C. A simplified diagram of a cascaded refrigeration system is shown in Figure 8.5.

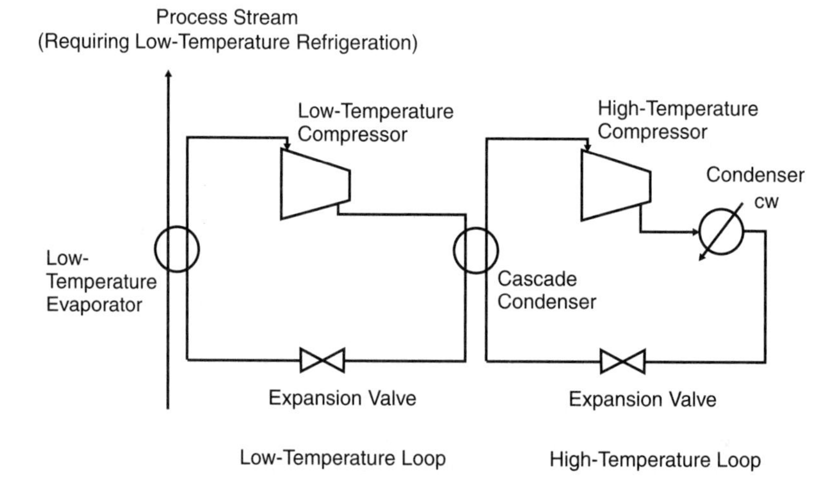

Figure 8.5 Schematic Diagram of a Simple Cascaded Refrigeration System

Steam Production. Steam is produced by the evaporation and superheating of specially treated water. The fuel that is used to supply the energy to produce steam is by far the major operating expense. However, water treatment costs can be substantial depending on the supply water composition and the degree of recovery of condensed steam in process heat exchangers. As shown in Table 8.3, for large chemical plants, steam is often required at several different pressure levels. However, it is often generated at the highest level and then let down to the lower pressure levels through turbines. These turbines produce electricity used in the plant. A typical steam generating facility is shown in Figure 8.6. Because there are losses of steam in the system due to leaks and, more important, due to process users not returning condensate, there is a need to add makeup water. This water is filtered to remove particulates and then treated to reduce the hardness. The latter can be achieved by the addition of chemicals to precipitate magnesium and calcium salts followed by filtration. These salts have **reverse solubility** characteristics and therefore precipitate at high temperatures. Alternatively, an ion-exchange system can be employed. The solids-free, "soft" water is now fed to the steam generating system. The thorough treatment of the water is necessary, because any contaminants entering with the water will ultimately deposit on heat-exchanger surfaces and boiler tubes and cause fouling and other damage. Another important issue is the dissolved oxygen and carbon dioxide

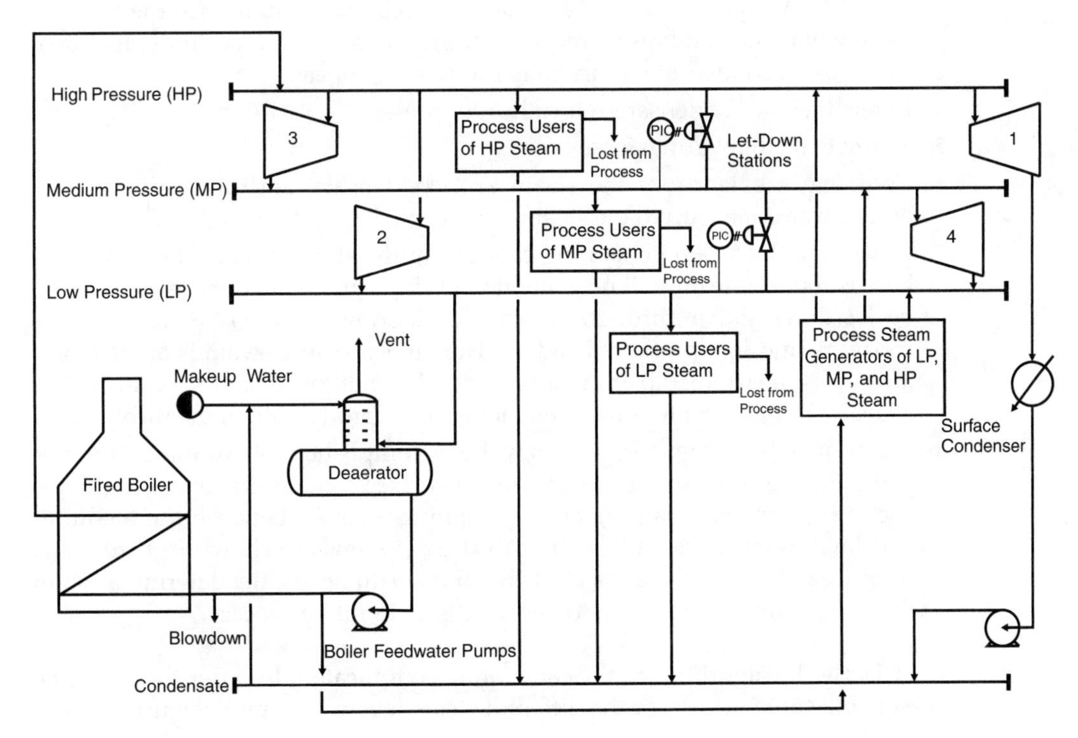

Figure 8.6 Typical Steam Producing System for a Large Chemical Facility

that enter with the makeup water. These dissolved gases must be removed in order to eliminate (reduce) corrosion of metal surfaces in the plant. The removal occurs in the deaerator, in which the makeup water is scrubbed with steam to degas the water. Oxygen scavengers are also added to the circulating condensate to remove any trace amounts of oxygen in the system. Amines may also be added to the water in order to neutralize any residual carbonic acid formed from dissolved carbon dioxide. Finally, blowdown of water from the water storage tank (situated near the boiler) is necessary to remove any heavy sludge and light solids that are picked up as steam and condensate circulate through the system [12]. The problems associated with the build-up of chemicals become even more troublesome in high-pressure (>66 bar) boilers, and several solutions are discussed by Wolfe [13].

In order to estimate accurately steam generation costs, it is necessary to complete a steam balance on the plant. An algorithm for carrying out a steam balance for a new facility is listed below.

1. Determine the pressure levels for the steam in the plant. These are usually set at around 41.0 barg (600 psig), between 10.0 barg (150 psig) and 15.5 barg (225 psig), and between 3.4 barg (50 psig) and 6.1 barg (90 psig).
2. Determine the total number of process users of the different levels of steam. These numbers become the basis for the steam balance.
3. Determine which of the above users will return condensate to the boiler feed water (BFW) system. Note: If live steam injection is required for the process, there will be no condensate returned from this service. In addition, for some small users, condensate return may not be economical.
4. Determine the condensate-return header pressure.
5. Estimate the blowdown losses.
6. Complete a balance on the steam and condensate, and determine the required water makeup to the steam system.
7. Determine the steam generating capacity of the steam boiler. The logic used here is that all steam will be generated at the highest-pressure level and will be let down either through turbines or let-down stations (valves) to the medium- and low-pressure headers. The high-pressure steam is often generated at 44.3 barg (650 psig) to allow for frictional losses and superheated to 400°C (752°F) to produce more efficient power production in the turbines.
8. Additional power generation may be accomplished by running turbines using the high-pressure steam, by using surface condensers (operating at the cooling water temperature), and by running turbines between the medium- and low-pressure steam headers. All these options are shown in Figure 8.6. In order to balance a plant's electrical and steam needs, the determination of the correct amount of steam to generate is an iterative process.

Clearly the algorithm can become quite complicated. In order to determine a reasonable value or cost for the different steam levels, the approach used here is to assume that all the steam will be generated at the highest pressure level and then let down to the appropriate pressure level through turbines or let-down stations (valves). In the former case, credit will be taken for generating power; in the latter case, credit will not be taken. The procedure for calculating the cost of steam at different pressure levels is given in Example 8.6.

Example 8.6

Determine the cost of producing high-, medium-, and low-pressure steam using a natural gas fuel source. For medium- and low-pressure steam production, assume that steam is produced at the highest pressure level, and consider both the case when this steam is sent through a turbine to make electricity and when it is simply throttled through a valve.

Again the approach taken here is to assume that the fixed capital investment associated with the initial purchase of the steam generation facilities has been accounted for elsewhere. The analysis given below accounts only for the operating costs associated with steam (and power) production. The source of fuel is assumed to be natural gas that costs \$11.10/GJ. See Table 8.3.

High-Pressure Steam (41.0 barg)

Basis is 1000 kg of HP steam generated at 45.3 bar and 400°C $\Rightarrow h_{44.3\ barg,\ 400°C} = 3204.3$ kJ/kg.

Conditions at the header are 41 bar saturated ($T_{sat} = 254°C$). Note that the steam is generated at a higher pressure and superheated for more efficient expansion, but that desuperheating will be assumed at the process user.

Assume boiler feed water comes from a deaerator that operates at exhaust steam header pressure of 0.7 barg and $T_{sat} = 115°C$ (10 psig) $\Rightarrow h_{BFW} = 483.0$ kJ/kg.

$$\Delta H_{BFW\text{-}HP\ Steam} = (3204.3 - 483.0) = 2721.3 \text{ kJ/kg}$$

$$\text{Energy required to produce HP steam} = (2721.3)(1000) = 2.721 \text{ GJ}$$

Because this HP steam is superheated, we can produce more than 1000 kg of saturated steam from it. In order to desuperheat this steam, BFW is added to produce saturated steam at 41.0 barg ($h = 2797.6$ kJ/kg). See Figure E8.6.

Let x be the amount of saturated HP steam produced from superheated HP steam; an enthalpy balance gives $(1000)(3204.3) + (x - 1000)(483) = (x)(2797.6) \Rightarrow x = 1175.7$ kg.

The cost of natural gas to produce 1000 kg of sat HP steam (assuming a 90% boiler efficiency) is given by

$$\text{Cost} = \frac{(2.721)}{(0.9)}\frac{(1000)(11.1)}{(1175.7)} = \$28.54$$

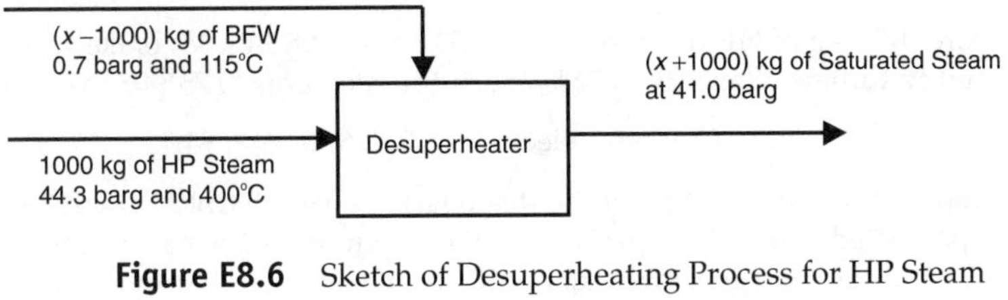

Figure E8.6 Sketch of Desuperheating Process for HP Steam

Treatment costs for circulating boiler feed water = \$0.15/1000 kg for oxygen scavengers, and so on (average from several vendors).

Boiler feed water cost is based on the assumption that 10% makeup is required.

Cost of electricity to power air blowers supplying combustion air to boiler:

Natural gas usage = 2.721/0.9/1.1757 = 2.572 GJ = (2.572)(6)/(0.23)= 67.1 std m^3 = 67.1/22.4 = 2.99 kmol

Oxygen usage (based on 3% excess over stoichiometric) = (2.99)(2)(1.03) = 6.17 kmol oxygen

This comes from (6.17)/(0.21) = 29.38 kmol of air.

Assume that this air must be raised 0.5 bar to overcome frictional losses in boiler and stack, and assuming that the blower is 60% efficient. Therefore,

The electrical usage for blower is 14 kWh/1000 kg of steam produced, giving an electricity cost = (14)(0.06) = \$0.84.

The cost of BFW is based on the water makeup, treatment chemicals, and the thermal energy in the stream.

For a basis of 1000 kg of BFW,

Cost of makeup water = \$0.067

Cost of chemicals for treatment = \$0.15

Energy in BFW = $\dot{m}c_p\Delta T$ = (1000)(4.18)(115 – 25) = 0.376 GJ

Value of energy = (\$11.10)(0.376) = \$ 4.17

BFW cost = 4.17 + 0.067 + 0.15 = \$4.39/1000 kg

Cost of BFW makeup = (0.1)(4.39) = \$0.439

Total cost of HP steam = \$28.54 + \$0.15 + \$0.84 + \$0.439 = \$29.97/1000 kg

Medium-Pressure Steam (10.0 barg)

It is assumed that letting steam down through a turbine from the high-pressure header to the medium-pressure header will generate electrical power.

The theoretical steam requirement (kg steam/kWh) for this situation is found by assuming an isentropic expansion of the steam from the HP condition to the medium-pressure level. From the steam tables, we have the following information:

$$h_{44.3\ barg,\ 400°C} = 3204.3 \text{kJ/kg and } s_{44.3\ barg,\ 400°C} = 6.6953 \text{ kJ/kg K}$$

By interpolating at constant specific entropy, we get that the outlet temperature is 212°C and the outlet enthalpy = 2851.0 kJ/kg.

$$\Delta h = (3204.3 - 2851.0) = 353.3 \text{ kJ/kg = theoretical work}$$

Therefore, 1000 kg of HP steam produces 353.3 MJ or 98.14 kWh of electricity.
Assuming a turbine efficiency of 75%, the output power is (0.75)(98.14) = 73.6 kWh.

$$\text{Credit for electricity} = (73.6)(0.06) = \$4.42$$

The actual outlet enthalpy of the steam is 3,204.3 – (353.3)(0.75) = 2939.3 kJ/kg. This is still superheated steam. Desuperheating the steam to 10.3 barg and saturated conditions (h = 2779.1 kJ/kg) will generate x kg of MP steam from the 1000 kg of HP steam, where

$$(1000)(2939.3) + (x - 1000)(483.0) = (x)(2779.1) \Rightarrow x = 1069.8 \text{ kg}$$

Therefore, the cost of natural gas to produce 1,000 kg of MP steam (assuming a 90% boiler efficiency) is

$$\text{Cost} = \frac{(2.721)}{(0.9)}\frac{(1000)(11.10)}{(1069.8)} = \$31.37$$

We can find the cost of electricity for the blower by using the ratio of the natural gas usage from the high-pressure steam case. Therefore,

$$\text{cost of electricity} = \frac{(31.37)}{(28.54)}(0.84) = \$0.92$$

Total cost of MP steam (with power production) = \$31.37 – \$4.42 + \$0.92 + \$0.439 = \$28.31/1000 kg

For the case when power production is not implemented, the HP steam is throttled to the pressure of the MP header through a let-down station, which is essentially an irreversible, isentropic process through a valve. The superheated steam is then desuperheated at the process user.

Enthalpy of HP steam (at 44.8 barg and 400°C) = 3204.3 kJ/kg
Enthalpy of saturated MP steam = 2779.1 kJ/kg
Enthalpy of BFW = 483.0 kJ/kg

If x is the amount of saturated MP steam obtained by desuperheating, then an enthalpy balance gives

$$(1000)(3204.3) + (x - 1000)(483) = (x)(2779.1) \Rightarrow x = 1185.2 \text{ kg}$$

Cost of natural gas to produce 1000 kg of sat HP steam (assuming a 90% boiler efficiency) is

$$\text{Cost} = \frac{(2.721)}{(0.9)}\frac{(1000)(11.10)}{(1185.2)} = \$28.32$$

$$\text{Cost of electricity for the air blower} = \frac{(28.32)}{(28.54)}(0.84) = \$0.83$$

Total cost of MP steam (without power production) = \$28.32 + \$0.83 + \$0.439
= \$29.59/1000 kg

Note: This is almost identical to the cost for HP steam.

Low-Pressure Steam (5.2 barg)

The calculation procedures for evaluating the cost of low-pressure steam are identical to those given above for medium-pressure steam and the results are given below.

Total cost of LP steam (with power production) = \$32.25 – \$5.94 + \$0.95 + \$0.439
= \$27.70/1000 kg

Total cost of LP steam (without power production) = \$28.03 + \$0.82 + \$0.439 = \$29.29/1000 kg

Waste Heat Boilers

When steam is generated from within the process—in a waste heat boiler, for example—the savings to the process are usually calculated from the avoided cost of using an equivalent amount of natural gas in the boiler system. If we assume that the boiler efficiency is 90%, then for every GJ of energy saved by producing steam within a process unit, the boiler facility saves (\$11.1)/(0.9) = \$12.33 in natural gas costs.

Hot Circulating Heat-Transfer Fluids. Again, the greatest cost for these systems is the fuel that is burned to heat the circulating heat-transfer fluid. Typical efficiencies (based on the lower heating value, LHV, of the fuel) for these heaters range from 60% to 82% [1]. With air preheating economizers, the efficiency can be as high as 90%. Example 8.7 illustrates the use of efficiencies in fired heaters.

Example 8.7

Estimate the utility cost of a heat-transfer medium heated in a fired heater using natural gas as the fuel.

Assuming that the heat-transfer medium is heated in a process heater that is 80% efficient and uses natural gas at \$11.10/GJ as the fuel source, we get

$$\text{Cost of 1 GJ of energy} = (1)(11.10)/(0.80) = \$\ 13.88/\text{GJ}$$

Assuming a 90% efficient heater, we get

$$\text{Cost of 1 GJ of energy} = (1)(11.10)/(0.9) = \$\ 12.33/\text{GJ}$$

8.4 RAW MATERIAL COSTS

The cost of raw materials can be estimated by using the current price listed in such publications as the *Chemical Market Reporter* (*CMR*) [14]. A list of common chemicals and their selling price, as of August 2006, are given in Table 8.4. Current raw material and product chemical prices may be obtained from the current issue of the *CMR* [14]. To locate costs for individual items, it is not sufficient to look solely at the current issue, because not all chemicals are listed in each issue. It is necessary to explore several of the most recent issues. In addition, for certain chemicals large seasonal price fluctuations may exist, and it may be advisable to look at the average price over a period of several months.

Another factor that is sometimes overlooked is that often companies will lock onto a selling or purchase price through a short- or long-term contract. Such contracts will often yield prices that are significantly lower than those given in the *CMR*. In addition, in doing economic evaluations for different chemical processes, the purchase and selling price for chemicals will not always be available from the *CMR*. For example, in January 2001, *CMR* stopped publishing the price of dimethyl ether. Likewise, prices for allyl alcohol have not been published for several years. The prices shown in Table 8.4 were obtained from manufacturers' quotes. When doing economic evaluations for new, existing, or future plants, it is advisable to establish the true selling or purchase price for all raw materials

Table 8.4: Costs of Some Common Chemicals*

Chemical	Cost ($/kg)	Typical Shipping Capacity or Basis for Price
Acetaldehyde	1.003	Railroad Tank Cars
Acetic Acid	1.090 (2004)	Railroad Tank Cars
Acetone (MMA grade)	0.948	Railroad Tank Cars
Acrylic Acid	1.929	Railroad Tank Cars
Allyl Chloride	1.80**	F.O.B. Gulf Coast
Benzene	0.657	Barge, Gulf Coast
Chlorine	0.375	Railroad Tank Car
Dimethyl Ether	0.948† (Jan. 2000)	Railroad Tank Car
Ethanol (190 Proof)	0.937	Railroad Tank Car
Ethylbenzene	1.069	Railroad Tank Car, Gulf Coast
Ethylene	1.202	Contract
Ethylene Oxide	1.764	Railroad Tank Car
Formaladehyde/Formalin (37 wt%)		
No-inhibitor	0.838	Railroad Tank Car, Gulf Coast
7% Methanol Inhibitor	0.463	Railroad Tank Car, Gulf Coast
Hydrochloric Acid (23ϒBe)	0.095	Railroad Tank Car, Works
Iso-Butylene	0.706	F.O.B. Works
Iso-Propanol (99%)	1.378	Railroad Tank Car
Maleic Anhydride	1.543	Railroad Tank Car
Methanol	0.294	F.O.B. Gulf Coast
Methyl Ethyl Ketone	1.598	Railroad Tank Car
MTBE	0.687	Barge, Gulf Coast
Propylene		
(Polymer Grade)	1.014	F.O.B. Gulf Coast
(Chemical Grade)	0.981	F.O.B. Gulf Coast
Styrene	1.543	F.O.B. Works
Sulfur (Crude)	0.043	Railroad Car
Sulfuric Acid (virgin)	0.090	Railroad Tank Car, Gulf Coast
Toluene	0.648	Barge, Gulf Coast
Mixed Xylenes	0.608	Barge, Gulf Coast
Ortho-Xylene	0.805	Railroad Tank Cars
Para-Xylene	1.135	Railroad Tank Cars
Meta-Xylene	2.910	Railroad Tank Cars

*Unless stated otherwise these are average values from ICIS.com, http://www.icis.com/StaticPages/a-e.htm#top, August 2006.
**Vendor quote.
†From *CMR*, January 2000.

and products. Because the largest operating cost is nearly always the cost of raw materials, it is important to obtain accurate prices if realistic economic evaluations are to be obtained.

8.5 YEARLY COSTS AND STREAM FACTORS

Manufacturing and associated costs are most often reported in terms of \$/yr. Information on a PFD is most often reported in terms of kg or kmol per hour or per second. In order to calculate the yearly cost of raw materials or utilities, the fraction of time that the plant is operating in a year must be known. This fraction is known as the **stream factor** (SF), where

$$\text{Stream Factor (SF)} = \frac{\text{Number of Days Plant Operates per Year}}{365} \tag{8.5}$$

Typical values of the stream factor are in the range of 0.96 to 0.90. Even the most reliable and well-managed plants will typically shut down for two weeks a year for scheduled maintenance, giving an SF = 0.96. Less reliable processes may require more downtime and hence lower SF values. The stream factor represents the fraction of time that the process unit is on-line and operating at design capacity. When estimating the size of equipment, care must be taken to use the design flowrate for a typical stream day and not a calendar day. Example 8.8 illustrates the use of the stream factor.

Example 8.8

a. Determine the yearly cost of toluene for the process given in Chapter 1.
b. What is the yearly consumption of toluene?
c. What is the yearly revenue from the sale of benzene?

Assume a stream factor of 0.95, and note that the flowrates given on the PFD are in kilograms per stream hour.

From Table 1.5, flowrate of toluene = 10,000 kg/h (Stream 1)
From Table 1.5, flowrate of benzene = 8210 kg/h (Stream 15)
From Table 8.5, cost of toluene = \$0.648/kg
From Table 8.5, cost of benzene = \$0.657/kg

a. Yearly cost of toluene = (24)(365)(10,000)(0.648)(0.95) = \$53,927,000/yr.
b. Yearly consumption of toluene = (24)(365)(10,000)(0.95) / 1000 = 83,200 tonne/yr.
c. Yearly revenue from benzene sales = (24)(365)(8210)(0.657)(0.95) = \$44,889,000/yr.

Comparing the results from Parts (a) and (c), we can see that with the current prices for these two chemicals it is not economical to produce benzene from toluene. Historically, the price differential between benzene and toluene has been greater than the $0.009/kg shown in Table 8.4, and this is the reason why this process has been used, and is currently being used, to produce benzene. Clearly, if this low price differential were to exist for a long period of time, this process might have to be shut down.

8.6 ESTIMATING UTILITY COSTS FROM THE PFD

Most often, utilities do not directly contact process streams. Instead, they exchange heat energy (fuel gas, steam, cooling water, and boiler feed water) in equipment such as heat exchangers and process heaters, or they supply work (electric power or steam) to pumps, compressors, and other rotating equipment. In most cases, the flowrate can be found either by inspection or by doing a simple heat balance around the equipment.

Steam can be used to drive a piece of rotating equipment such as a compressor. In this case, both the theoretical steam requirement and efficiency are required. Table 8.5 provides the theoretical steam requirements as a function of the steam inlet pressure and the exhaust pressure for steam turbine drives. The mechanical efficiencies of different drives are shown in Figure 8.7, using data from Walas [9].

To illustrate the techniques used to estimate the utility flowrates and utility costs for various types of equipment, see Example 8.9.

Table 8.5 Theoretical Steam Requirements (kg steam/kWh)

	Inlet Pressure of Steam (barg) (Superheat in °C)							
	10.0 (sat'd)	13.8 (sat'd)	17.2 50	27.6 170	41.4 145	41.4 185	58.6 165	58.6 205
Exhaust Pressure								
2″ Hg abs	4.77	4.54	4.11	3.34	3.22	3.07	2.98	2.85
4″ Hg abs	5.33	5.04	4.54	3.62	3.47	3.30	3.20	3.05
0 barg	8.79	7.94	6.88	5.08	4.72	4.45	4.22	4.00
0.69 barg	10.87	9.57	8.11	5.77	5.28	4.97	4.67	4.40
2.07 barg	15.24	12.72	10.40	6.91	6.18	5.78	5.35	5.02
3.45 barg	20.86	16.32	12.79	7.97	6.97	6.49	5.93	5.54
4.14 barg	24.45	18.32	14.11	8.50	7.34	6.83	6.20	5.78
4.82 barg	28.80	20.68	15.47	9.05	7.71	7.16	6.45	6.01

From Perry, R. H., and D. W. Green, *Perry's Chemical Engineering Handbook*, 6th ed., McGraw-Hill, New York, NY, 1984. Reprinted by permission of The McGraw-Hill Companies.

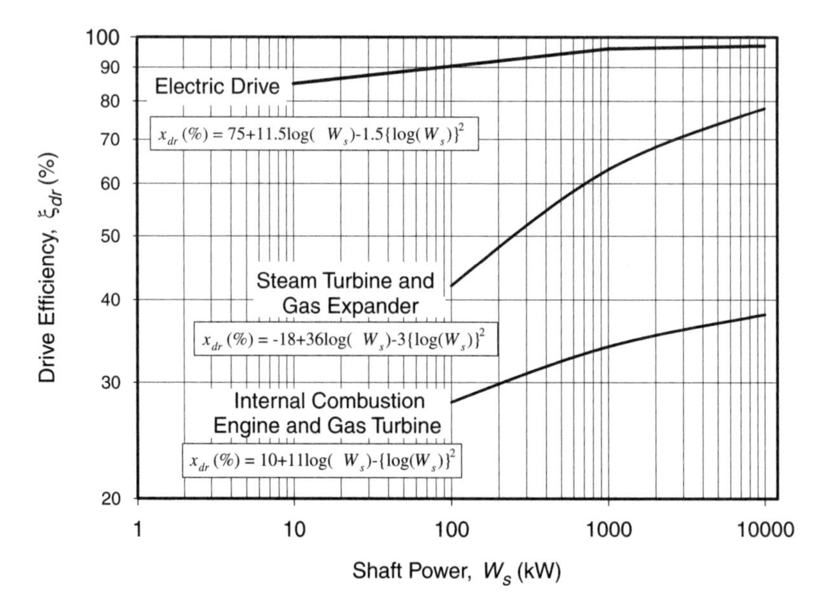

Figure 8.7 Efficiencies for Pumps and Compressor Drives (Data from Walas [9], Chapter 4)

Example 8.9

Estimate the quantities and yearly costs of the appropriate utilities for the following pieces of equipment on the toluene hydrodealkylation PFD (Figure 1.5). It is assumed that the stream factor is 0.95 and that all the numbers on the PFD are on a stream time basis. The duty on all of the units can be found in Table 1.7.

a. E-101, Feed Preheater
b. E-102, Reactor Effluent Cooler
c. H-101, Heater
d. C-101, Recycle Gas Compressor, assuming electric drive
e. C-101, Recycle Gas Compressor, assuming steam drive using 10 barg steam discharging to atmospheric pressure.
f. P-101, Toluene Feed Pump

Solution

a. E-101: Duty is 15.19 GJ/h. From Table 8.3, Cost of High-Pressure Steam = \$17.70/GJ

$$\text{Energy Balance: } Q = 15.19 \text{ GJ/h} = (\dot{m}_{steam})(\Delta H_{vap}) = (\dot{m}_{steam})(1699.3) \text{ kJ/kg}$$

$$\dot{m}_{steam} = 8939 \text{ kg/h} = 2.48 \text{ kg/s}$$

$$\text{Yearly Cost} = (Q)(C_{steam})(t) = (15.19\text{ GJ/h})(\$17.70/\text{GJ})(24)(365)(0.95) = \$\,2{,}237{,}000/\text{yr}$$

Alternatively, Yearly Cost = (Yearly flowrate)(Cost per unit mass)

$$\text{Yearly Cost} = (2.48)(3600)(24)(365)(0.95)(29.97/1000) = \$2{,}227{,}000/\text{yr}$$

(same as above within round-off error)

b. E-102: Duty is 46.66 GJ/h. From Table 8.3, Cost of Cooling Water = \$0.354/GJ

$$Q = 46.66\text{ GJ/h} = (\dot{m}_{cw})(C_{pcw})(\Delta T_{cw}) = (\dot{m}_{cw})(4.18)(10) = 41.8\,\dot{m}_{cw}$$

$$\dot{m}_{cw} = (46.66)(10^9/41.8)(10^3) = 1{,}116{,}270\text{ kg/h} = 310\text{ kg/s}$$

$$\text{Yearly Cost} = (46.66\text{ GJ/h})(24)(365)(0.95)(\$0.354/\text{GJ}) = \$137{,}000/\text{yr}$$

c. H-101: Duty is 27 GJ/h (7510 kW). Assume that an indirect, nonreactive process heater has a thermal efficiency (ξ_{th}) of 90%. From Table 8.3, natural gas costs \$11.10/GJ, and the heating value is 0.0377 GJ/m^3.

$$Q = 27\text{ GJ/h} = (\dot{v}_{gas})(\Delta H_{natural\ gas})(\text{efficiency}) = (\dot{v}_{gas})(0.0377)(0.9)$$

$$\dot{v}_{gas} = 796\text{ std m}^3/\text{h } (0.22\text{ std m}^3/\text{sec})$$

$$\text{Yearly Cost} = (27)(11.10)(24)(365)(0.95)/(0.90) = \$2{,}771{,}000/\text{yr}$$

d. C-101: Shaft power is 49.1 kW, and from Figure 8.7 the efficiency of an electric drive (ξ_{dr}) is 90%.

$$\text{Electric Power} = P_{dr} = \text{Output power}/\xi_{dr} = (49.1)/(0.90) = 54.6\text{ kW}$$

$$\text{Yearly Cost} = (54.6)(0.06)(24)(365)(0.95) = \$27{,}200/\text{yr}$$

e. Same as Part (d) with steam driven compressor. For 10 barg steam with exhaust at 0 barg, Table 8.5 provides a steam requirement of 8.79 kg steam/kWh of power. The shaft efficiency is about 35% (extrapolating from Figure 8.7).

$$\text{Steam required by drive} = (49.1)(8.79/0.35) = 1233\text{ kg/h } (0.34\text{ kg/s})$$

$$\text{Cost of Steam} = (1233)(24)(365)(0.95)(28.32\times10^{-3}) = \$290{,}600/\text{yr}$$

f. P-101: Shaft power is 14.2 kW. From Figure 8.7 the efficiency of an electric drive is about 86%.

$$\text{Electric Power} = 14.2/0.86 = 16.5\text{ kW}$$

$$\text{Yearly Cost} = (16.5)(0.06)(24)(365)(0.95) = \$8240/\text{yr}$$

Note: The cost of using steam to power the compressor is much greater than the cost of electricity even though the cost per unit energy is much lower for the steam. The reasons for this are (1) the thermodynamic efficiency is low, and (2) the efficiency of the drive is low for a small compressor. Usually steam drives are used only for compressor duties greater than 100 kW.

8.7 COST OF TREATING LIQUID AND SOLID WASTE STREAMS

As environmental regulations continue to tighten, the problems and costs associated with the treatment of waste chemical streams will increase. In recent years there has been a trend to try to reduce or eliminate the volume of these streams through waste minimization strategies. Such strategies involve utilizing alternative process technology or using additional recovery steps in order to reduce or eliminate waste streams. Although waste minimization will become increasingly important in the future, the need to treat waste streams will continue. Some typical costs associated with this treatment are given in Table 8.3, and flowrates can be obtained from the PFD. It is worth noting that the costs associated with the disposal of solid waste streams, especially hazardous wastes, have grown immensely in the past few years, and the values given in Table 8.3 are only approximate average numbers. Escalation of these costs should be done with extreme caution.

8.8 EVALUATION OF COST OF MANUFACTURE FOR THE PRODUCTION OF BENZENE VIA THE HYDRODEALKYLATION OF TOLUENE

The cost of manufacture for the production of benzene via the toluene HDA process is given in Example 8.10.

Example 8.10

Calculate the cost of manufacture without depreciation (COM_d) for the toluene hydrodealkylation process using the PFD in Figure 1.5 and the flow table given in Table 1.5.

A utility summary for all the equipment is given in Table E8.10, from which we find the total yearly utility costs for this process:

Steam = \$ 3,412,000/yr
Cooling Water = \$ 165,000/yr
Fuel Gas = \$2,771,000/yr
Electricity = \$37,400/yr

Total Utilities = \$6,385,000/yr

Raw Material Costs from the PFD, Table 8.4, and Example 8.8 are

Toluene = \$53,927,000/yr
Hydrogen = \$6,622,000/yr (based on a value of \$0.118/std m^3)

Total Raw Materials = \$60,549,000/yr

There are no waste streams shown on the PFD, so

Waste Treatment = \$0.0/yr

From Example 8.2 the cost of operating labor is
C_{OL} **= (14)(52,900) = \$741,000/yr**

Table E8.10 Summary of Utility Requirements for the Equipment in the Toluene Hydrodealkylation Process

Equipment	Electric Power (kW)	Steam High-Pressure (kg/s)	Steam Med-Pressure (kg/s)	Steam Low-Pressure (kg/s)	Cooling Water (m^3/s)	Fuel Gas (std m^3/s)
E-101	—	2.48	—	—	—	—
E-102	—	—	—	—	0.31	—
E-103	—	—	—	0.14	—	—
E-104	—	—	—	—	0.055	—
E-105	—	—	—	—	0.007	—
E-106	—	—	1.26	—	—	—
H-101	—	—	—	—	—	0.22
C-101	54.5	—	—	—	—	—
P-101	16.5	—	—	—	—	—
P-102	4.0	—	—	—	—	—
Totals	75.0	2.48	1.26	0.14	0.372	0.22
Total yearly cost $/yr	37,400	2,227,000	1,069,000	116,000	165,000	2,771,000

Data from Figure 1.5, Table 1.7, and Example 8.9.

From Problem 7.21 (using CAPCOST), we find that the fixed capital investment (C_{GR}) for the process is $ 11.7 × 10^6.

***FCI* = $ 11.7 × 10^6**

Finally, using Equation 8.2, the total manufacturing cost is estimated to be

$$COM_d = 0.180FCI_L + 2.73C_{OL} + 1.23(\text{Utilities} + \text{Raw Materials} + \text{Waste Treatment})$$

$$COM_d = (0.180)(11.7 \times 10^6) + 2.73\,(741{,}000) + 1.23\,(6{,}385{,}000 + 60{,}549{,}000 + 0)$$

$$COM_d = \$\,86.46 \times 10^6/\text{yr}$$

8.9 SUMMARY

In this chapter, the cost of manufacturing for a chemical process was shown to depend on the fixed capital investment, the cost of operating labor, the cost of utilities, the cost of waste treatment, and the cost of raw materials. In most cases, the cost of raw materials is the biggest cost. Methods to evaluate these different costs were discussed. Specifically, the amount of the raw materials and utilities can be obtained directly from the PFD. The cost of operating labor can be estimated from the number of pieces of equipment given on the PFD. Finally, the fixed capital investment can again be estimated from the PFD using the techniques given in Chapter 7.

REFERENCES

1. Ulrich, G. D., *A Guide to Chemical Engineering Process Design and Economics* (New York: John Wiley and Sons, 1984).
2. Peters, M. S., and K. D. Timmerhaus, *Plant Design and Economics for Chemical Engineers,* 4th ed. (New York: McGraw-Hill, 1990).
3. Valle-Riestra, J. F., *Project Evaluation in the Chemical Process Industries* (New York: McGraw-Hill, 1983).
4. Alkhayat, W. A., and A. M. Gerrard, *Estimating Manning Levels for Process Plants,* AACE Transactions, I.2.1–I.2.4, 1984.
5. Bureau of Labor and Statistics, U.S. Department of Labor, http://www.data.bls.gov.
6. Energy Information Administration, http://www.eia.doe.gov/overview_hd.html.
7. Hile, A. C., L. Lytton, K. Kolmetz, and J. S. Walker, *Monitor Cooling Towers for Environmental Compliance,* CEP, 37-41, March 2001.
8. Gibson, W. D., "Recycling Cooling and Boiler Water," *Chemical Engineering* (January 1999): 47–51.
9. Walas, S. M., *Chemical Process Equipment: Selection and Design* (Boston: Butterworths Publ., 1988).
10. *Engineering Data Book,* 9th ed. (Tulsa, OK: Gas Processors Suppliers Association, 1972).
11. Perry, R. H., and C. H. Chilton, *Chemical Engineers' Handbook,* 5th ed. (New York: McGraw-Hill, 1973), Figures 12-14 and 12-15.
12. Dyer, D. F., *Boiler Efficiency Improvement* (Auburn, AL: Boiler Efficiency Institute, 1981).
13. Wolfe, T. W., "Boiler-water Treatment at High Pressures, the Rules Change," *Chemical Engineering* (October 2000): 82–88.
14. *Chemical Market Reporter* (now incorporated into ICIS Chemical Business, additional chemical prices are available at http://www.icis.com/StaticPages/a-e.htm#top).

SHORT ANSWER QUESTIONS

1. In the general equation for determining the cost of manufacturing (COM_d), Equation (8.2), one of the terms is 0.180*FCI,* where *FCI* is the fixed capital investment of the plant. "This term is included to cover the interest payment

on the loan for the plant." Is this statement true or false? Explain your answer.

2. Why is the number of operators per shift multiplied by approximately 4.5 to obtain the total number of operators required to run the plant?
3. What is a stream factor?
4. When estimating the cost of manufacturing (COM_d) for a chemical process, the overall COM_d may be estimated using only five individual costs. Lists these five costs.
5. Cooling water is priced on an energy basis: $/GJ. We usually assume the temperature rise to be 10°C. Does the cooling water cost change if the return temperature changes? Are there any limitations to the return temperature? Explain.
6. In Equation (8.2), the cost of raw materials, C_{RM}, is multiplied by a factor of 1.23. The reason for this is that, in general, we expect that the estimated cost of raw materials will be about 20% low and we add a correction factor of 1.23 to adjust for this. Do you agree with this explanation? If you do not, give another reason for using the factor of 1.23.
7. In Equation (8.2), the cost of operating labor, C_{OL}, is multiplied by a factor of 2.73. One reason for this is that the value of C_{OL} includes only plant operators and not supervisory and clerical labor costs. Is this statement true or false? What other factors (if any) account for the multiplication factor of 2.73?
8. Explain the difference between direct costs, fixed costs, and general expenses. Give two examples of each.

PROBLEMS

9. You are employed at a chemical company and have recently been transferred from a plant that manufactures synthetic dyes to a new facility that makes specialty additives for the polymer resin industry.
   a. You have been asked to estimate the cost of manufacturing at this new facility. Would you
      i. Use Equation (8.2) to estimate COM_d?
      ii. Use data from the old plant where you worked, because you are very familiar with all the aspects of manufacturing for that process?
      iii. Dig up information on the new process and use these figures?
   b. When would you use a relationship such as Equation (8.2)?
10. When a chemical plant needs steam at multiple pressure levels, it is often economical to generate all the steam at a high pressure and then to let the steam down through pressure-reducing turbines to the desired pressure. This principle is illustrated in Figure 8.6. The downside of this approach is that as the exhaust pressure of the turbine increases, the theoretical (and

actual) steam requirements increase, meaning that less energy is extracted. To illustrate this point, do the following.

a. Estimate the amount of energy extracted from 10,000 kg/h of 58.6 barg steam superheated by 165°C when connected to the following turbines (each 80% efficient).
   i. Exhaust pressure is 4" Hg absolute.
   ii. Exhaust pressure is 4.82 barg.
b. Estimate the amount of energy extracted from 10,000 kg/h of saturated, 10.0 barg steam when connected to a turbine (80% efficient) exhausting at 4.82 barg.
c. Identify the locations of each of the three turbines described above on Figure 8.6.

11. What are the operating costs associated with a typical cooling water system? Based on the example given in this chapter, answer the following.
    a. What percentage of the operating costs is the makeup water?
    b. By how much would the cost of cooling water increase if the cost of power (electricity) were to double?
    c. By how much would the cost of cooling water increase if the cost of make-up water were to double?
12. Determine the cost of producing a refrigerant stream at –50°C using propane as the working fluid in a noncascaded system. You may wish to refer to Example 8.5 to do this problem. The steps you should follow are as follows.
    a. Determine the pressure at which propane can be condensed at 45°C, which assumes that cooling water with a 5°C temperature approach will be used as the condensing medium.
    b. Determine the pressure to which the propane must be throttled in order to liquefy it at –50°C.
    c. Use the results of Parts (a) and (b) to set the approximate pressure levels in the condenser and evaporator in the refrigeration system.
    d. Determine the amount of propane necessary to extract 1 GJ of heat in the evaporator.
    e. Assuming a 5kPa pressure drop in both heat exchangers and that a single-stage compressor is used with an efficiency of 75%, determine the cost of electricity to run the compressor, determine the cooling water cost, and from this determine the cost of providing refrigeration at –50°C using propane as the working fluid.
13. Repeat the process described in Problem 12 using a simple refrigeration loop to determine the cost of providing 1 GJ of refrigeration at –50°C using the following working fluids:
    a. Propylene
    b. Ethane
    c. Ammonia

Determine whether any of the working fluids given above cannot be used in a simple (noncascaded) refrigeration loop. For these fluids, would using a cascaded refrigeration system to provide –50°C refrigerant make sense? Explain carefully your answers to these questions.

14. Estimate the cost of operating labor (C_{OL}), the cost of utilities (C_{UT}), and the cost of manufacturing (COM_d) for the ethylbenzene process given in Project B.2 of Appendix B. You must do Problem 7.21 in order to estimate COM_d.
15. Estimate the cost of operating labor (C_{OL}), the cost of utilities (C_{UT}), and the cost of manufacturing (COM_d) for the styrene process given in Project B.3 of Appendix B. You must do Problem 7.22 in order to estimate COM_d.
16. Estimate the cost of operating labor (C_{OL}), the cost of utilities (C_{UT}), and the cost of manufacturing (COM_d) for the drying oil process given in Project B.4 of Appendix B. You must do Problem 7.23 in order to estimate COM_d.
17. Estimate the cost of operating labor (C_{OL}), the cost of utilities (C_{UT}), and the cost of manufacturing (COM_d) for the maleic anhydride process given in Project B.5 of Appendix B. You must do Problem 7.24 in order to estimate COM_d.
18. Estimate the cost of operating labor (C_{OL}), the cost of utilities (C_{UT}), and the cost of manufacturing (COM_d) for the ethylene oxide process given in Project B.6 of Appendix B. You must do Problem 7.25 in order to estimate COM_d.
19. Estimate the cost of operating labor (C_{OL}), the cost of utilities (C_{UT}), and the cost of manufacturing (COM_d) for the formalin process given in Project B.7 of Appendix B. You must do Problem 7.26 in order to estimate COM_d.

CHAPTER

9

Engineering Economic Analysis

The goal of any manufacturing company is to make money. This is realized by producing products with a high market value from raw materials with a low market value. The companies in the chemical process industry produce high-value chemicals from low-value raw materials.

In the previous chapters, a process flow diagram (Chapter 1), an estimate of the capital cost (Chapter 7), and an estimate of operating costs (Chapter 8) were provided for the production of benzene. From this material, an economic evaluation can be carried out to determine

1. Whether the process generates money
2. Whether the process is attractive compared with other processes (such as those for the production of ethylbenzene, ethylene oxide, formalin, and so on, given in Appendix B)

In the next two chapters, the necessary background to perform this economic analysis is provided.

The principles of economic analysis are covered in this chapter. The material presented covers all of the major topics required for completion of the Fundamentals of Engineering (FE) examination. This is the first requirement for becoming a registered professional engineer in the United States.

It is important for you, the graduating student, to understand the principles presented in this chapter at the beginning of your professional career in order to manage your money skillfully. As a result, we have elected to integrate discussions and examples of personal money management throughout the chapter.

The evaluation of profitability and comparison of alternatives for proposed projects are covered in Chapter 10.

9.1 INVESTMENTS AND THE TIME VALUE OF MONEY

The ability to profit from investing money is the key to our economic system. In this text, we introduce investment in terms of personal financing and then apply the concepts to chemical process economics.

There are various ways to distribute one's personal income. The first priority is to maintain a basic (no-frills) standard of living. This includes necessary food, clothing, housing, transportation, and expenses such as taxes imposed by the government. The remaining money, termed *discretionary money,* can then be distributed. It is wise to distribute this money in a manner that will realize both your short-term and long-term goals.

Generally, there are two classifications for spending discretionary money.

1. Consume money as received. This provides immediate personal gratification and/or satisfaction. We experience this use for money early in life.
2. Retain money for future consumption. This is money put aside to meet future needs. These may result from hard-to-predict causes such as sickness and job layoffs or from a more predictable need for long-term retirement income. It is unlikely that you have considered these types of financial needs and you probably have little experience in investing to secure a comfortable lifestyle after you stop working.

There are two approaches to setting money aside for use at a later date:

- **Simple savings:** Put money in a safety deposit box, sugar bowl, or other such container.
- **Investments:** Put money into an investment.

These two approaches are considered in Example 9.1.

Example 9.1

Upon graduation, you start your first job at $50,000/yr. You decide to set aside 10%, or $5000/yr, for retirement in 40 years' time, and you assume that you will live 20 years after retiring. You have been offered an investment that will pay you $67,468/yr during your retirement years for the money you invest.

a. How much money would you have per year in retirement if you had saved the money, but not invested it, until retirement?
b. How does this compare with the investment plan offered?
c. How much money was produced from the investment?

Solution

a. **Money saved:** ($5000)(40) = $200,000

Income during retirement: $200,000/20 = $10,000/yr

b. **Comparison:** (Income from savings)/(Income from investments) = $10,000/$67,468 = 0.15

c. Money Produced = Money Received – Money Invested = ($67,468)(20) – $200,000 = $1,149,360

The value of the investment is clear. The income in retirement from savings amounts to only 15% of the investment income. The amount of money provided during retirement, by setting $200,000 aside, was more than a million dollars. We will show later that this high return on investment resulted from two factors: the long time period for the investment and the interest rate earned on the savings.

Money, when invested, makes money.

The term *investment* will now be defined.

An **investment** is an agreement between two parties, whereby one party, the **investor,** provides money, P, to a second party, the **producer,** with the expectation that the producer will return money, F, to the investor at some future specified date, where $F > P$. The terms used in describing the investment are

P: Principal or Present Value

F: Future Value

n: Years between F and P

The amount of money earned from the investment is

$$E = F - P \tag{9.1}$$

The yearly earnings rate is

$$i_s = \frac{E}{Pn} = \frac{(F - P)}{Pn} \tag{9.2}$$

where i_s is termed the simple interest rate.

From Equation (9.2), we have

$$\frac{F}{P} = (1 + ni_s) \text{ or, in general, } \frac{F}{P} = f(n,i) \tag{9.3}$$

Example 9.2 illustrates this concept.

Example 9.2

You decide to put $1000 into a bank that offers a special rate if left in for two years. After two years you will be able to withdraw $1150.

a. Who is the producer?
b. Who is the investor?
c. What are the values of P, F, i_s, and n?

Solution

a. **Producer:** The bank has to produce $150.00 after two years.
b. **Investor:** You invest $1000 in an account at the beginning of the two-year period.
c. $P = \$1000$ (given)
   $F = \$1150$ (given)
   $n = 2$ years (given)
   From Equation (9.2),
   $i_s = (\$1150 - \$1000)/(\$1000)/(2) = 0.075$ or 7.5% per year

In Example 9.2, you were the investor and invested in the bank. The bank was the "money producer" and had to return to you more dollars ($1150) than you invested ($1000). This bank transaction is an investment commonly termed as *savings.* In the reverse situation, termed *loan,* the bank becomes the investor. You must produce money during the time of the investment.

Equations (9.1) through (9.3) apply to a single transaction between the investor and the producer that covers n years and uses simple interest. There are other investment schedules and interest formulations in practice; these will be covered later in this chapter.

Figure 9.1 illustrates a possible arrangement to provide the funds necessary to build a new chemical plant such as the one introduced in the narrative in Chapter 1.

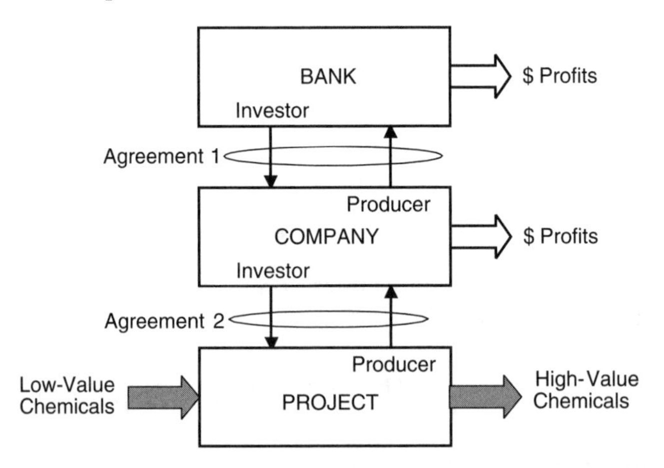

Figure 9.1 A Typical Financing Scheme for a Chemical Plant

In this arrangement, a bank invests in a company, which in turn invests in a project to produce a chemical. There are two agreements in this project (see Figure 9.1).

1. The bank is an investor and the company is the producer.
2. The company is an investor and the project is the producer.

In this illustration, all the money produced in the project is sent to the company. The company pays its investor, the bank, and draws off the rest as profit. The bank also makes a profit from its investment loan to the company. The project is the source of money to provide profits to both the company and the bank. The project converts a low-value, raw chemical into chemicals of higher value. Without the investor, the plant would not be built, and without the plant, there would be no profits for either the company or the bank.

Money is a measure of the value of products and services. The value of a chemical material is the price it can be exchanged for in dollars. Investments may be made in units other than dollars, such as stocks, bonds, grain, oil, or gold. We will often refer to value, or value added, in describing investments. The term *value* is a general one and, in our case, may be assigned a dollar figure for economic calculations.

Figure 9.1 shows that all profits were produced from an operating plant. The role of engineers in our economy should be clear. This is to ensure efficient production of high-value products, including current as well as new and improved products.

In almost all cases, the economic analysis of processes will be made from the point of view of the company as the investor in a project. The project may be the construction of a new plant or a modification to an existing plant.

Consider the decisions involved in the investment in a new plant (the project) from the point of view of the company. The company must invest the money to build the plant before any income resulting from production can begin. Once the plant has been built and is operational, it is expected to operate for many years. During this time, the plant produces a profit and the company receives income from its investment. It is necessary to be able to determine whether this future income is sufficiently attractive to make the investment worthwhile.

The **time value of money** refers to a concept that is fundamental to evaluating an investment. This is illustrated in Example 9.3.

Example 9.3

You estimate that in two years' time you will need $1150 in order to replace the linoleum in your kitchen. Consider two choices.

1. Wait two years to take action.
2. Invest $1000 now (assume that interest is offered by the bank at the same rate as given in Example 9.1).

What would you do (explain your answer)?

Solution

Consider investing the $1000 today because it will provide $1150 in two years. The key is that the dollar I have today is worth 15% more than a dollar I will have in two years' time.

From Example 9.3, we conclude that today's dollar is worth more than tomorrow's dollar because it can be invested to earn more dollars. This must not be confused with inflation, which erodes purchasing power and is discussed in Section 9.6.

Money today is worth more than money in the future.

In the upcoming sections, it will be found that when comparing capital investments made at different times, the timing of each investment must be considered.

9.2 DIFFERENT TYPES OF INTEREST

Two types of interest are used when calculating the future value of an investment. They are referred to as *simple* and *compound interest*. Simple interest calculations are rarely used today. Unless specifically noted, all interest calculations will be carried out using compound interest methods.

9.2.1 Simple Interest

In **simple interest** calculations, the amount of interest paid is based solely on the initial investment.

Interest paid in any year = Pi_s

For an investment period of n years, the total interest paid = $Pi_s n$

Total value of investment in n years = $F_n = P + Pi_s n$

$$F_n = P(1 + i_s n) \tag{9.4}$$

If, instead of setting the earned interest aside, it were reinvested, the total amount of interest earned would be greater. When earned interest is reinvested, the interest is referred to as **compound interest.**

9.2.2 Compound Interest

Let us determine the future value of an investment, F_n, after n years at an interest rate of i per year for an initial investment of P when the interest earned is reinvested each year.

a. At the start, we have our initial investment = P.

b. In year 1, we earn Pi in interest.
For year 2, we invest $P + Pi$ or $P(1 + i)$.

c. In year 2, we earn $P(1 + i)i$ in interest.
For year 3, we invest $P(1 + i) + P(1 + i)i$, or $P(1 + i)^2$.

d. In year 3, we earn $P(1 + i)^2 i$ in interest.
For year 4, we invest $P(1 + i)^2 + P(1 + i)^2 i$, or $P(1 + i)^3$.

e. By induction we find that after n years the value of our investment is $P(1 + i)^n$.

Thus, for compound interest we can write

$$F_n = P(1 + i)^n \tag{9.5}$$

We can reverse the process and ask, How much would I have to invest now, P, in order to receive a certain sum, F_n, in n years' time? The solution to this problem is found by rearranging Equation (9.5):

$$P = \frac{F_n}{(1 + i)^n} \tag{9.6}$$

We illustrate the use of these equations in Examples 9.4, 9.5, and 9.6. The letters *p.a.* following the interest refers to per year (per annum).

Example 9.4

For an investment of \$500 at an interest rate of 8% p.a. for 4 years, what would be the future value of this investment, assuming compound interest?

From Equation (9.5) for $P = 500$, $i = 0.08$, and $n = 4$ we obtain

$F_4 = P(1 + i)^n = 500(1 + 0.08)^4 = \$680.24$

Note: Simple interest would have yielded $F_4 = 500(1 + (4)(0.08)) = \$660$ (\$20.24 less).

Example 9.5

How much would I need to invest in a savings account yielding 6% interest p.a. to have \$5000 in five years' time?

From Equation (9.6) using $F_5 = \$\ 5000$, $i = 0.06$, and $n = 5$ we get

$P = F_n/(1 + i)^n = 5000/(1.06)^5 = \$3736.29$

If we invest \$3736.29 into the savings account today, we will have \$5,000 in five years' time.

Example 9.6

I need to borrow a sum of money (P) and have two loan alternatives.

a. I borrow from my local bank, which will lend me money at an interest rate of 7% p.a. and pay compound interest.

b. I borrow from "Honest Sam," who offers to lend me money at 7.3% p.a. using simple interest.

In both cases, I need the money for three years. How much money would I need in three years to pay off this loan? Consider each option separately.

Bank: From Equation (9.5) for $n = 3$ and $i = 0.07$ we get

$F_3 = (P)(1 + 0.07)^3 = 1.225\, P$

Sam: From Equation (9.4) for $n = 3$ and $i = 7.3$ we get

$F_3 = (P)(1 + (3)(0.073)) = 1.219\, P$

Sam stated a higher interest rate, and yet it is still preferable to borrow the money from Sam because $1.219P < 1.225P$. This is because Sam used simple interest, and the bank used compound interest.

9.2.3 Interest Rates Changing with Time

If we have an investment over a period of years and the interest rate changes each year, then the appropriate calculation for compound interest is given by

$$F_n = P\prod_{j=1}^{n}(1 + i_j) = P(1 + i_1)(1 + i_2)\cdots(1 + i_n) \tag{9.7}$$

9.3 TIME BASIS FOR COMPOUND INTEREST CALCULATIONS

In industrial practice, the length of time assumed when expressing interest rates is one year. However, we are sometimes confronted with terms such as 6% p.a. compounded monthly. In this case, the 6% is referred to as a **nominal annual interest rate,** i_{nom}, and the number of compounding periods per year is m (12 in this case). The nominal rate is not used directly in any calculations. The **actual rate** is the interest rate per compounding period, r. The relationship needed to evaluate r is

$$r = \frac{i_{nom}}{m} \tag{9.8}$$

This is illustrated in Example 9.7.

Example 9.7

For the case of 12% p.a. compounded monthly, what are m, r, and i_{nom}?

Given: $m = 12$ (months in a year), $i_{nom} = 12\% = 0.12$

From Equation (9.8),

$r = 0.12/12 = 0.01$ (or 1% per month)

9.3.1 Effective Annual Interest Rate

We can use an **effective annual interest rate,** i_{eff}, which will allow us to make interest calculations on an annual basis and obtain the same result as using the actual compounding periods. If we look at the value of an investment after one year, we can write

$$F_1 = P(1 + i_{eff}) = P\left(1 + \frac{i_{nom}}{m}\right)^m$$

which, upon rearrangement, gives

$$i_{eff} = \left(1 + \frac{i_{nom}}{m}\right)^m - 1 \qquad (9.9)$$

Effective annual interest rate is illustrated in Example 9.8.

Example 9.8

What is the effective annual interest rate for a nominal rate of 8% p.a. when compounded monthly?

From Equation (9.9), for $i_{nom} = 0.08$ and $m = 12$, we obtain

$i_{eff} = (1 + 0.08/12)^{12} - 1 = 0.083$ (or 8.3 % p.a.)

The effective annual interest rate is greater than the nominal annual rate. This indicates that the effective interest rate will continue to increase as the number of compounding periods per year increases. For the limiting case, interest is compounded continuously.

9.3.2 Continuously Compounded Interest

For the case of continuously compounded interest, we must look at what happens to Equation (9.9) as $m \to \infty$:

$$i_{eff} = \lim_{m\to\infty}\left[\left(1 + \frac{i_{nom}}{m}\right)^m - 1\right]$$

Rewriting the left-hand side as $\lim_{m\to\infty}\left[\left\{\left(1 + \frac{i_{nom}}{m}\right)^{\frac{m}{i_{nom}}}\right\}^{i_{nom}} - 1\right]$

and noting that $\lim_{n\to\infty}\left[1 + \frac{x}{n}\right]^{\frac{n}{x}} = e$

we find that for continuous compounding,

$$i_{eff} = e^{i_{nom}} - 1 \tag{9.10}$$

Equation (9.10) represents the maximum effective annual interest rate for a given nominal rate.

The method for calculating continuously compounded interest is illustrated in Example 9.9.

Example 9.9

What is the effective annual interest rate for an investment made at a nominal rate of 8% p.a. compounded continuously?

From Equation (9.10) for $i_{nom} = 0.08$ we obtain

$i_{eff} = e^{0.08} - 1 = 0.0833$, or 8.33% p.a.

Note: We can see, by comparison with Example 9.8, that by compounding continuously little was gained over monthly compounding.

In comparing alternatives, the effective annual rate, and not the nominal annual rate of interest, must be used.

9.4 CASH FLOW DIAGRAMS

To this point, we have considered only an investment made at a single point in time at a known interest rate, and we learned to evaluate the future value of this investment. More complicated transactions involve several investments and/or payments of differing amounts made at different times. For more complicated investment schemes, we must keep careful track of the amount and time of each transaction. An effective way to track these transactions is to utilize a **cash flow diagram**, or CFD. Such a diagram offers a visual representation of each investment. Figure 9.2 is the cash flow diagram for Example 9.10, which is used to introduce the basic elements of the **discrete CFD**.

Figure 9.2 shows that cash transactions were made periodically. The values given represent payments made at the end of the year. Figure 9.2 shows that \$1000, \$1200, and \$1500 were received at the end of the first, second, and third year, respectively. In the fifth and seventh year, \$2000 and \$$X$ were paid out. There were no transactions in the fourth and sixth years.

Each cash flow is represented by a vertical line, with length proportional to the cash value of the transaction. The sign convention uses a downward-pointing arrow when cash flows outward and an upward-pointing arrow representing inward cash flows. When a company invests money in a project, the company CFD shows a negative cash flow (outward flow), and the project CFD shows a positive cash flow (inward flow). Lines are placed periodically in the horizontal direction

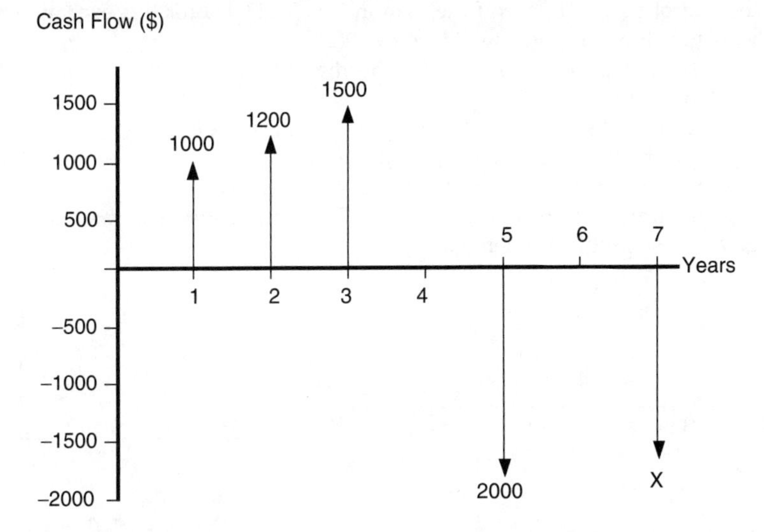

Figure 9.2 An Example of a Representative Discrete Cash Flow Diagram (CFD)

to represent the time axis. Most frequently, we perform our analysis from the point of view of the investor.

The cash flow diagram shown in Figure 9.2 can be presented in a simplified format, using the following simplifications.

1. The y-axis is not shown.
2. Units of monetary transactions are not given for every event.

In addition to the discrete CFD described above, we can show the same information in a **cumulative CFD.** This type of CFD presents the accumulated cash flow at the end of each period.

9.4.1 Discrete Cash Flow Diagram

The discrete CFD provides a clear, unambiguous pictorial record of the value, type, and timing of each transaction occurring during the life of a project. In order to avoid making mistakes and save time, it is recommended that prior to doing any calculations, you sketch a cash flow diagram. Examples 9.10 and 9.11 illustrate the use of discrete cash flow diagrams.

Example 9.10

I borrow $1000, $1200, and $1500 from a bank (at 8% p.a. effective interest rate) at the end of years 1, 2, and 3, respectively. At the end of year 5, I make a payment of $2000, and at the end of year 7, I pay off the loan in full. The CFD for this exchange from my point of view (producer) is given in Figure E9.10(a).

Note: Figure E9.10(a) is the shorthand version of the one presented in Figure 9.2 used to introduce the CFD.

Draw a discrete cash flow diagram for the investor.

The bank represents the investor. From the investor's point of view, the initial three transactions are negative and the last two are positive.

Figure E9.10(b) represents the CFD for the bank. It is the mirror image of the one given in the problem statement.

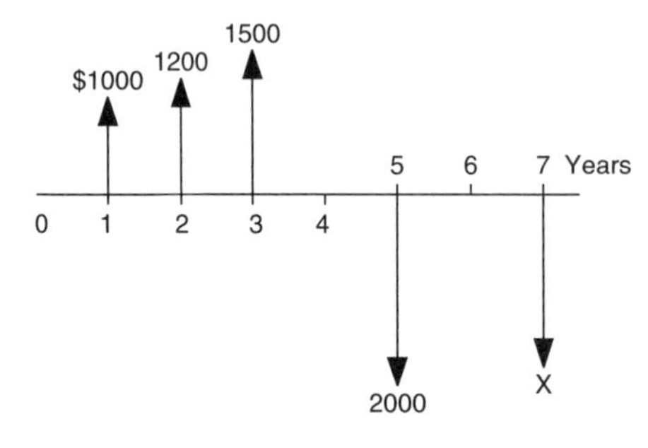

Figure E9.10a CFD for Example 9.10 from My Perspective

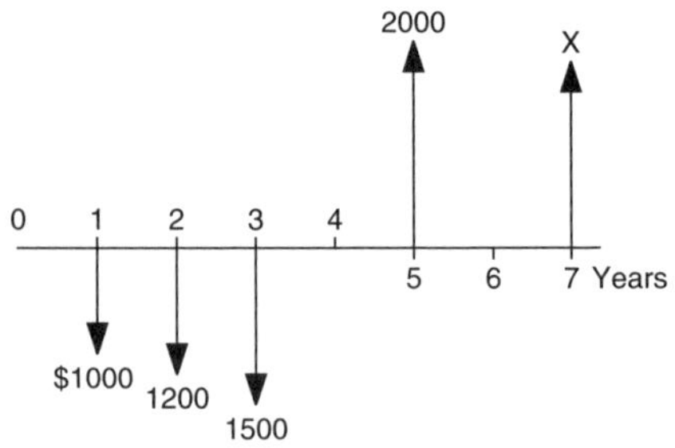

Figure E9.10b CFD for Example 9.10 from the Bank's Perspective

---

The value of X in Example 9.10 depends on the interest rate. Its value is a direct result of the time value of money. The effect of interest rate and the calculation of the value of X (in Example 9.13) are determined in the next section.

Example 9.11

You borrow $10,000 from a bank to buy a new car and agree to make 36 equal monthly payments of $320 each to repay the loan. Draw the discrete CFD for the investor in this agreement.

The bank is the investor. The discrete CFD for this investment is shown in Figure E9.11.

Notes:

1. There is a break in both the time scale and in the investment at time = 0 (the initial investment).
2. From your point of view, the cash flow diagram would be the mirror image of the one shown.

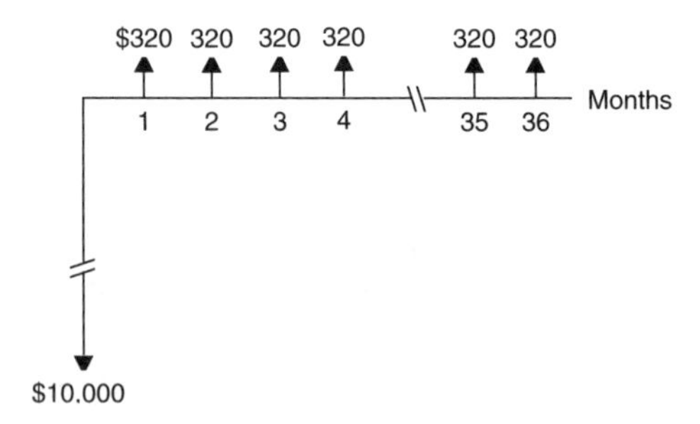

Figure E9.11 CFD for Car Loan Described in Example 9.11

The cash flow diagram constructed in Example 9.11 is typical of those you will encounter throughout this text. The investment (negative cash flow) is made early in the project during design and construction before there is an opportunity for the plant to produce product and generate money to repay the investor. In Example 9.11, payback was made in a series of equal payments over three years to repay the initial investment by the bank. In Section 9.5, you will learn how to calculate the interest rate charged by the bank in this example.

9.4.2 Cumulative Cash Flow Diagram

As the name suggests, the cumulative CFD keeps a running total of the cash flows occurring in a project. To illustrate how to construct a cumulative CFD, consider Example 9.12, which illustrates the cash flows associated with the construction and operation of a new chemical plant.

Example 9.12

The yearly cash flows estimated for a project involving the construction and operation of a chemical plant producing a new product are provided in the discrete CFD in Figure E9.12a. Using this information, construct a cumulative CFD.

The numbers shown in Table E9.12 were obtained from this diagram.

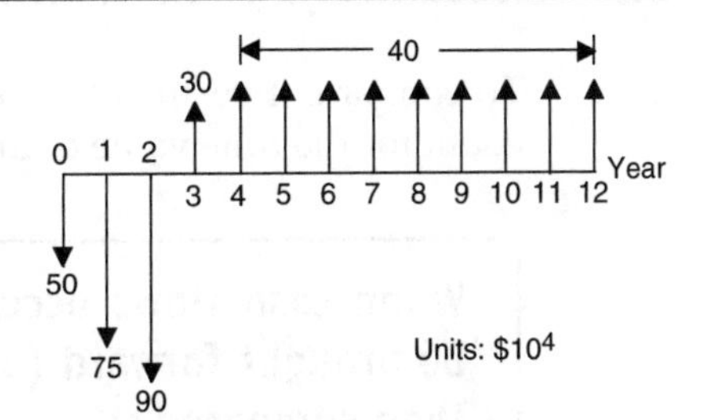

Figure E9.12a Discrete CFD for Chemical Plant Described in Example 9.12

Table E9.12 Summary of Discrete and Cumulative Cash Flows in Example 9.12

Year	Cash Flow ($) (from Discrete CFD)	Cumulative Cash Flow (Calculated)
0	–500,000	–500,000
1	–750,000	–1,250,000
2	–900,000	–2,150,000
3	300,000	–1,850,000
4	400,000	–1,450,000
5	400,000	–1,050,000
6	400,000	–650,000
7	400,000	–250,000
8	400,000	150,000
9	400,000	550,000
10	400,000	950,000
11	400,000	1,350,000
12	400,000	1,750,000

The cumulative cash flow diagram is plotted in figure E9.12b.

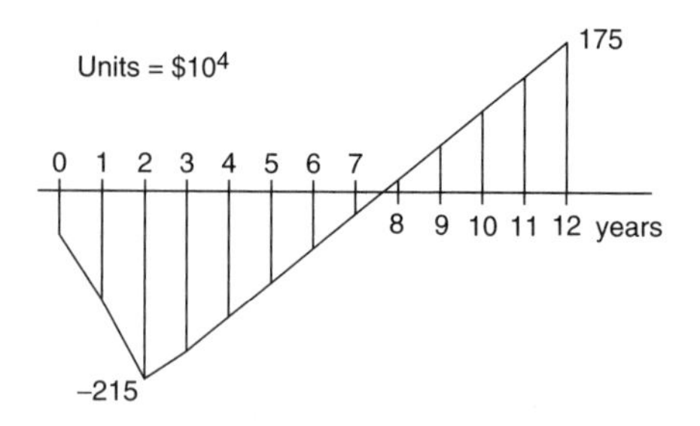

Figure 9.12b Cumulative CFD for Chemical Plant Described in Example 9.12

9.5 CALCULATIONS FROM CASH FLOW DIAGRAMS

To compare investments that take place at different times, it is necessary to account for the time value of money.

> **When cash flows occur at different times, each cash flow must be brought forward (or backward) to the same point in time and then compared.**

The point in time that is chosen is arbitrary. This is illustrated in Example 9.13.

Example 9.13

The CFD obtained from Example 9.10 (for the borrower) is repeated in Figure E9.13. The annual interest rate paid on the loan is 8% p.a.

In year 7, the remaining money owed on the loan is paid off.

a. Determine the amount, X, of the final payment.

b. Compare the value of X with the value that would be owed if there were no interest paid on the loan.

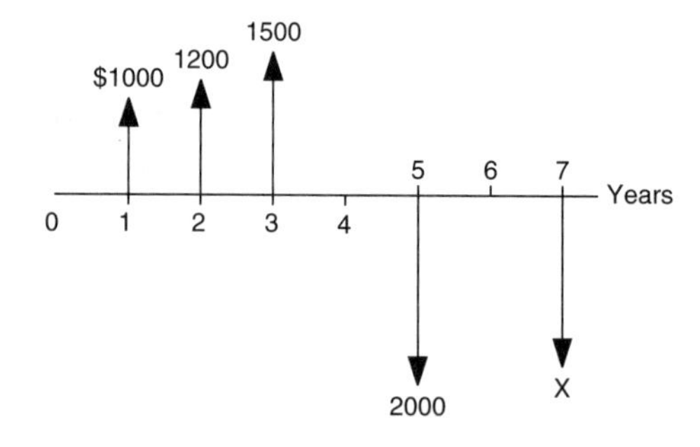

Figure E9.13 CFD for Example 9.13

With the final payment at the end of year 7, no money is owed on the loan. If we sum all the positive and negative cash flows adjusted for the time of the transactions, this adjusted sum must equal zero.

We select as the base time the date of the final payment.

a. From Equation (9.5) for $i = 0.08$ we obtain the following.

For withdrawals:

\$1000 end of year 1: $F_6 = (\$1000)(1 + 0.08)^6 = \$1586.87$
\$1200 end of year 2: $F_5 = (\$1200)(1 + 0.08)^5 = \$1763.19$
\$1500 end of year 3: $F_4 = (\$1500)(1 + 0.08)^4 = \$2040.73$

Total withdrawals = \$5390.79

For repayments:

\$2000 end of year 5: $F_2 = -(\$2000)(1 + 0.08)^2 = -\$2332.80$
\$X end of year 7: $F_0 = -(\$X)(1 + 0.08)^0 = -\$X$
Total repayments = $-\$(2332.80 + X)$

Summing the cash flows and solving for X yields

$$0 = \$5390.79 - \$(2332.80 + X)$$
$$X = \$3057.99 \approx \$3058$$

b. For $i = 0.00$

Withdrawals = \$1000 + \$1200 + \$1500 = \$3700
Repayments = $-\$(2000 + X)$

$$0 = \$3700 - \$(2000 + X)$$
$$X = \$1700$$

Note: Because of the interest paid to the bank, the borrower repaid a total of \$1358 (\$3058 – \$1700) more than was borrowed from the bank seven years earlier.

To demonstrate that any point in time could be used, as a basis, compare the amount repaid based on the end of year 1. Equation (9.6) is used, and all cash flows are moved backward in time (exponents become negative). This gives

$$0 = 1000 + \frac{1200}{1.08} + \frac{1500}{1.08^2} - \frac{2000}{1.08^4} - \frac{X}{1.08^6}$$

and solving for X yields

$$X = (1.08)^6\left[1000 + \frac{1200}{1.08} + \frac{1500}{1.08^2} - \frac{2000}{1.08^4}\right] = \$3058 \text{ (the same answer as before!)}$$

Usually, the desire is to compute investments at the start or at the end of a project, but the conclusions drawn are independent of where that comparison is made.

9.5.1 Annuities—A Uniform Series of Cash Transactions

Problems are often encountered involving a series of uniform cash transactions, each of value A, taking place at the end of each year for n consecutive years. This

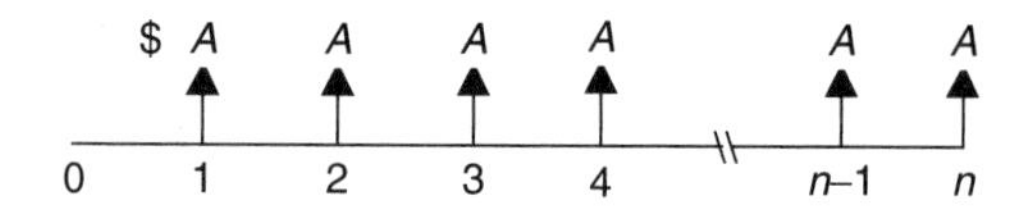

Figure 9.3 A Cash Flow Diagram for an Annuity Transaction

pattern is called an **annuity**, and the discrete CFD for an annuity is shown in Figure 9.3.

To avoid the need to do a year-by-year analysis like the one in Example 9.13, an equation can be developed to provide the future value of an annuity.

The future value of an annuity at the end of time period n is found by bringing each of the investments forward to time n, as we did in Example 9.13.

$$F_n = A(1 + i)^{n-1} + A(1 + i)^{n-2} + A(1 + i)^{n-3} + \cdots + A$$

This equation is a geometric series of the form $a, ar, ar^2, \ldots, ar^{n-1}$ with sum $S_n = F_n$.

$$S_n = a\left[\frac{r^n - 1}{r - 1}\right]$$

For the present case, $a = A$; $r = 1 + i$; $n = n$. Therefore,

$$F_n = A\left[\frac{(1 + i)^n - 1}{i}\right] \tag{9.11}$$

It is important to notice that Equation (9.11) is correct when the annuity starts at the end of the *first* time period and not at time zero. In the next section, a shorthand notation is provided that will be useful in CFD calculations.

9.5.2 Discount Factors

The shorthand notation for the future value of an annuity starts with Equation (9.11). The term F_n is shortened by simply calling it F, and then dividing through by A yields

$$F/A = [(1 + i)^n - 1]/i = f(i, n)$$

This ratio of F/A is a function of i and n—that is, $f(i,n)$. It can be evaluated when both the interest rate, i, and the time duration, n, are known. The value of $f(i,n)$ is referred to as a **discount factor.** If either A or F is known, the remaining unknown can be evaluated.

In general terms, a discount factor is designated as

$$\text{Discount factor for } X/Y = (X/Y, i, n) = f(i, n)$$

Discount factors represent simple ratios and can be multiplied or divided by each other to give additional discount factors. For example, assume that we need to know the present worth, P, of an annuity, A—that is, the discount factor for

P/A—but do not have the needed equation. The only available formula containing the annuity term, A, is the one for F/A derived above. We can eliminate the future value, F, and introduce the present value, P, by multiplying by the ratio of P/F, from Equation 9.6.

$$\text{Discount factor for } P/A = (P/A, i, n)$$
$$= (F/A, i, n)(P/F, i, n)$$

Substituting for F/A and P/F gives

$$P/A = \frac{(1+i)^n - 1}{i} \frac{1}{(1+i)^n}$$
$$= \frac{(1+i)^n - 1}{1} \frac{1}{i(1+i)^n} = (P/A, i, n)$$

Table 9.1 lists the most frequently used discount factors in this text with their common names and corresponding formulae.

Table 9.1 Commonly Used Factors for Cash Flow Diagram Calculations

Conversion	Symbol	Common Name	Eq. No.	Formula
P to F	$(F/P, i, n)$	Single Payment Compound Amount Factor	(9.5)	$(1+i)^n$
F to P	$(P/F, i, n)$	Single Payment Present Worth Factor	(9.6)	$\frac{1}{(1+i)^n}$
A to F	$(F/A, i, n)$	Uniform Series Compound Amount Factor, Future Worth of Annuity	(9.11)	$\frac{(1+i)^n - 1}{i}$
F to A	$(A/F, i, n)$	Sinking Fund Factor	(9.12)	$\frac{i}{(1+i)^n - 1}$
P to A	$(A/P, i, n)$	Capital Recovery Factor	(9.13)	$\frac{i(1+i)^n}{(1+i)^n - 1}$
A to P	$(P/A, i, n)$	Uniform Series Present Worth Factor, Present Worth of Annuity	(9.14)	$\frac{(1+i)^n - 1}{i(1+i)^n}$

The key to performing any economic analysis is the ability to evaluate and compare equivalent investments. In order to understand that the equations presented in Table 9.1 provide a comparison of alternatives, it is suggested to replace the equal sign with the words "is equivalent to." As an example, consider the equation given for the value of an annuity, A, needed to provide a specific future worth, F. From Table 9.1, Equation (9.11) can be expressed as

$$F \text{ (Future value) is equivalent to } \{f(i,n)\ A(\text{Annuity value})\}$$

where

$$f(i,n) = (F/A, i, n)$$

Example 9.14 illustrates a future value calculation.

Example 9.14

You have just won $2,000,000 in the Texas Lottery as one of seven winners splitting up a jackpot of $14,000,000. It has been announced that each winner will receive $100,000/year for the next 20 years. What is the equivalent present value of your winnings if you have a secure investment opportunity providing 9.5% p.a.?

From Table 9.1, Equation (9.14), for $n = 20$ and $i = 0.075$,

$$P = (\$100{,}000)[(1 + 0.075)^{20} - 1]/[(0.075)(1 + 0.075)^{20}]$$

$$P = \$1{,}019{,}000$$

A present value of $1,019,000 is equivalent to a 20-year annuity of $100,000/yr when the effective interest rate is 9.5%.

Examples 9.15 through 9.17 illustrate how to use these discount factors and how to approach problems involving discrete CFDs.

Example 9.15

Consider Example 9.11, involving a car loan. The discrete CFD from the bank's point of view was shown.

What interest rate is the bank charging for this loan?

You have agreed to make 36 monthly payments of $320. The time selected for evaluation is the time at which the final payment is made. At this time, the loan will be fully paid off. This means that the future value of the $10,000 borrowed is equivalent to a $320 annuity over 36 payments.

$$(\$10{,}000)(F/P, i, n) = (\$320)(F/A, i, n)$$

Substituting the equations for the discount factors given in Table 9.1, with $n = 36$ months, yields

$$0 = -(10{,}000)(1 + i)^{36} + (320)[(1 + i)^{36} - 1]/i$$

This equation cannot be solved explicitly for i. It is solved by plotting the value of the right-hand side of the equation shown above for various interest rates. This equation could also be solved using a numerical technique. From Figure E9.15, the interest rate that gives a value of zero represents the answer. From Figure E9.15 the rate of interest is $i = 0.0079$.

The nominal annual interest rate is $(12)(0.00786) = 0.095$ (9.5%).

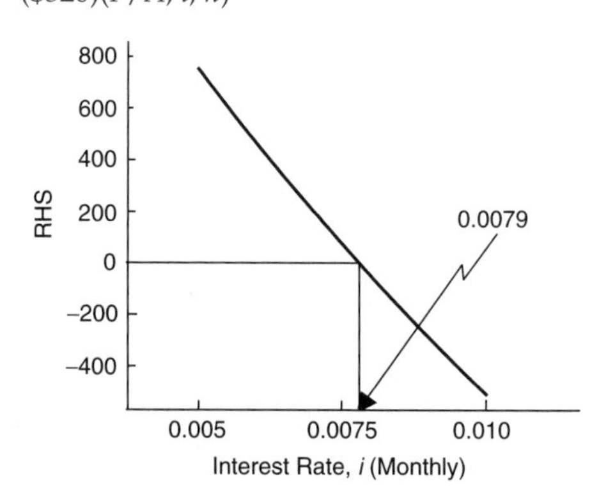

Figure E9.15 Determination of Interest Rate for Example 9.15

Example 9.16

I invest money in a savings account that pays a nominal interest rate of 6% p.a. compounded monthly. I open the account with a deposit of \$1000 and then deposit \$50 at the end of each month for a period of two years, followed by a monthly deposit of \$100 for the following three years. What will the value of my savings account be at the end of the five-year period?

First, draw a discrete CFD (Figure E9.16).

Although this CFD looks rather complicated, it can be broken down into three easy sub-problems:

1. The initial investment
2. The 24 monthly investments of \$50
3. The 36 monthly investments of \$100

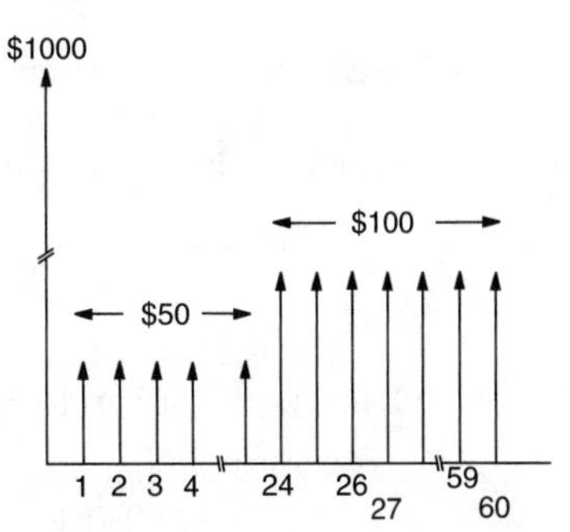

Figure E9.16 Cash Flow Diagram for Example 9.16

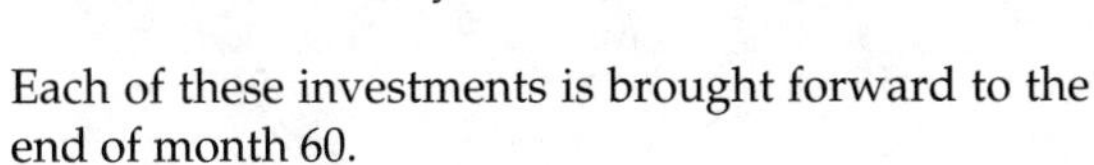

Each of these investments is brought forward to the end of month 60.

$F = (\$1000)(F/P, 0.005, 60) + (\$50)(F/A, 0.005, 24)(F/P, 0.005, 36) + (\$100)(F/A, 0.005, 36)$

Note: The effective monthly interest rate is $0.06/12 = 0.005$.

$$F = (\$1000)(1.005)^{60} + (\$50)\frac{(1.005^{24} - 1)}{0.005}(1.005)^{36} + (\$100)\frac{(1.005^{36} - 1)}{0.005} = \$6804.16$$

There are many ways to solve most complex problems. No one method is more or less correct than another. For example, the discrete CFD could be considered to be made up of a single investment of \$1000 at the start, a \$50 monthly annuity for the next 60 months, and another \$50 annuity for the last 36 months. Evaluating the future worth of the investment gives

$$F = (\$1000)(F/P, 0.005, 60) + (\$50)(F/A, 0.005, 60) + (\$50)(F/A, 0.005, 36)$$

$$F = (\$1000)(1.005)^{60} + (\$50)\frac{(1.005^{60} - 1)}{0.005} + (\$50)\frac{(1.005^{36} - 1)}{0.005} = \$6804.16$$

This is the same result as before.

Example 9.17

In Example 9.1, an investment plan for retirement was introduced. It involved investing \$5000/year for 40 years leading to retirement. The plan then provided \$67,468/year for 20 years of retirement income.

a. What yearly interest rate was used in this evaluation?
b. How much money was invested in the retirement plan before withdrawals began?

Solution

a. The evaluation is performed in two steps.

Step 1: Find the value of the \$5000 annuity investment at the end of the 40 years.

Step 2: Evaluate the interest rate of an annuity that will pay out this amount in 20 years at \$67,468/ year.

Step 1: From Equation (9.11), Table 9.1, for $A = \$5000$ and $n = 40$,

$$F_{40} = (A)(F/A, n, i) = (\$5000)[(1+i)^{40}-1]/i$$

Step 2: From Equation (9.14), Table 9.1, for $A = \$67{,}468$ and $n = 20$,

$$P = (A)(P/A, n, i) = (\$67{,}468)[(1+i)^{20} - 1]/[(i)(1+i)^{20}]$$

Set $F_{40} = P$ and solve for i. From Figure E9.17, $i = 0.060$

b. With $i = 0.060$, from Figure E9.17, F_{40} = \$774,000

Note: The interest rate of 6.0% p.a. represents a relatively low interest rate, involving small risk.

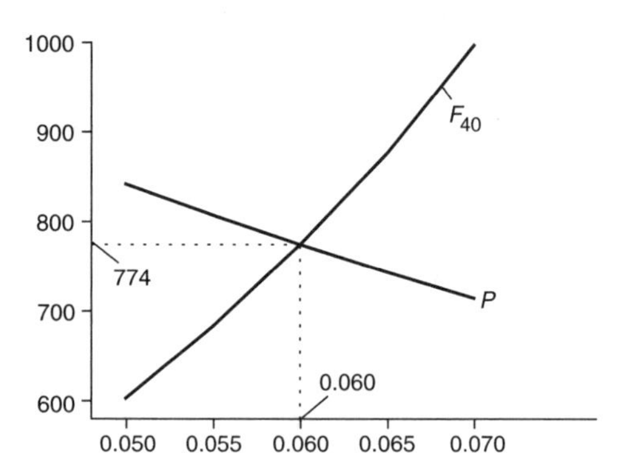

Figure E9.17 Determination of Interest Rate for Example 9.17

9.6 INFLATION

As a result of inflation, a dollar set aside (not invested) will purchase fewer goods and services in the future than the same dollar would today. In Chapters 7 and 8, it was seen that inflation of equipment, labor, and fuel costs could be tracked by the use of cost indexes. It is sometimes desirable to express these trends in terms of rates of inflation (f). This can be done using the cost indexes as follows:

$$CEPCI(j + n) = (1 + f)^n \, CEPCI(j) \tag{9.15}$$

where n = time span in years

f = average inflation rate over the time span

j = arbitrary year

The use of Equation (9.15) to estimate the inflation rate is illustrated in Example 9.18.

Example 9.18

What was the average rate of inflation for the costs associated with building a chemical plant over the following periods?

a. 1995 through 2001
b. 2001 through 2007

From Table 7.4, the values of the Chemical Engineering Plant Cost Index (CEPCI) are

CEPCI (1995) = 381
CEPCI (2001) = 394
CEPCI (2007) = 500

Equation (9.15) yields

a. $394 = (381)(1 + f)^6$,
$f = 0.006$ (0.6% p.a.)
b. $500 = 394\ (1 + f)^6$
$f = 0.049$ (4.9% p.a.)

To understand inflation, it is necessary to distinguish between cash and the purchasing power (for the purchase of goods and services) of cash. Inflation decreases this purchasing power with time. All the previous discussions on A, P, and F are given in terms of cash and not in terms of the relative purchasing power of this cash. The term F' is introduced, which represents the purchasing power of future cash. This purchasing power can be estimated using Equation (9.16).

$$F' = \frac{F}{(1 + f)^n} \tag{9.16}$$

Substituting the equation for F in terms of P, from Equation (9.5), gives

$$F' = P\frac{(1 + i)^n}{(1 + f)^n} = P\left[\frac{1 + i}{1 + f}\right]^n \tag{9.17}$$

Equation (9.17) is now written in terms of an effective interest rate, i', which includes the effect of inflation.

$$F' = P(1 + i')^n \tag{9.18}$$

By comparing Equation (9.18) with Equation (9.17), it is seen that i' is given by

$$i' = \frac{1 + i}{1 + f} - 1 = \frac{i - f}{1 + f} \tag{9.19}$$

For small values of $f < 0.05$, Equation (9.19) can be approximated by

$$i' \approx (i - f) \tag{9.20}$$

Example 9.19 demonstrates the incorporation of inflation into a calculation.

Example 9.19

In this example, consider the effect of inflation on the purchasing power of the money set aside for retirement in Example 9.17. Previously, the amount of cash available at the time of retirement in 40 years was calculated to be $774,000. This provided an income of $67,468 for 20 years.

a. Assuming an annual inflation rate of 2%, what is the purchasing power of the cash available at retirement?

b. What is the purchasing power of the retirement income in the first and twentieth years of retirement?

c. How does Part (a) compare with the total annuity payments of $5000/yr for 40 years?

Solution

a. Using Equation (9.16) for $f = 0.02$, $n = 40$, and $F = \$774{,}000$,

$$F' = \$774{,}000/(1 + 0.02)^{40} = \$351{,}000$$

b. At the end of the forty-first year (first year of retirement),
Purchasing Power $= \$67{,}468/(1 + 0.02)^{41} = \$29{,}956/\text{yr}$

At the end of the sixtieth year (twentieth year of retirement),
Purchasing Power $= \$67{,}468/(1 + 0.02)^{60} = \$20{,}563/\text{yr}$

c. Amount invested = ($5000/yr)(40 yr) = $200,000, compared with $351,000

Example 9.19 reveals the consequences of inflation. It showed that the actual income received in retirement of $67,468/yr had a purchasing power equivalent to between $29,956 and $20,563 at the time the initial investment was made. This does not come close to the $50,000/yr base salary at that time. To increase this value, you would have to increase one or more of the following:

a. The amount invested

b. The interest rate for the investment

c. The time over which the investment was made

The effects of inflation should not be overlooked in any decisions involving investments. Because inflation is influenced by politics, future world events, and so on, it is hard to predict. In this book, inflation will not be considered directly, and cash flows will be considered to be in uninflated dollars.

9.7 DEPRECIATION OF CAPITAL INVESTMENT

When a company builds and operates a chemical process plant, the physical plant (equipment and buildings) associated with the process has a finite life. The value or worth of this physical plant decreases with time. Some of the equipment wears out and has to be replaced during the life of the plant. Even if the equipment is seldom used and is well maintained, it becomes obsolete and of little value. When the plant is closed, the plant equipment can be salvaged and sold for only a fraction of the original cost.

The cash flows associated with the purchase and installation of equipment are expenses that occur before the plant is operational. This results in a negative cash flow on a discrete CFD. When the plant is closed, equipment is salvaged, and this results in a positive cash flow at that time. The difference between these costs represents capital depreciation.

For tax purposes, the government does not allow companies to charge the full costs of the plant as a one-time expense when the plant is built. Instead, it allows only a fraction of the capital depreciation to be charged as an operating expense each year until the total capital depreciation has been charged.

The amount and rate at which equipment may be depreciated are set by the federal government (Internal Revenue Service of the U.S. Treasury Department). The regulations that cover the capital depreciation change often. Both the current method of depreciation suggested by the IRS and several of the techniques that have been used in the past to depreciate capital investment are presented. Example 9.20 illustrates the need for depreciation of capital.

Example 9.20

Consider a person who owns a business with the following annual revenue and expenses:

Revenue from sales	$ 356,000
Rent	($ 22,000)
Employee salaries	($ 100,000)
Employee benefits	($ 32,000)
Utilities	($ 7000)
Miscellaneous expenses	($ 5000)
Overhead expenses	($ 40,000)
Before-tax profit	$ 150,000

The owner of the business decides that, in order to improve the manufacturing operation, she must buy a new packing and labeling machine for $100,000, which has a useful operating life of four years and can be sold for $2000 scrap value at that time. This, she estimates, will increase her sales by 5% per year. The only additional cost is an extra $1000/yr in utilities. The new, before-tax profit is estimated to be

Before-Tax Profit = $150,000 + 17,800 – 1000 = $ 166,800/yr, or an increase of $16,800/yr

Using a before-tax basis, it can be seen that her \$100,000 investment yields \$16,800/yr. The alternative to buying the new machine is to invest money in a mutual fund that yields 10% per year before tax. At face value, the investment in the new machine looks like a winner. However, take a close look at the cash flows for each case (Table E9.20).

Table E9.20 Cash Flows for Both Investment Opportunities

Year	Cash Flow for Investment in Machine (All \$ Figures in 1000)	Cash Flow for Investment in Mutual Fund (All \$ Figures in 1000)
0	–100	–100
1	16.8	10
2	16.8	11
3	16.8	12.1
4	16.8 + 2.0	13.31 + 100
Total	**–30.8**	**46.41**

Although the yearly return for buying the machine looks much better than that for the investment in the mutual fund, the big difference is that at the end of the fourth year the owner can recover her initial investment from the mutual fund, but the machine is worth only \$2000. From this example, it can be seen that the \$100,000 – \$2000 = \$98,000 investment in the machine is really a long-term expense, and the owner should be able to deduct it as a legitimate operating expense. Depreciation is the method that the government allows for businesses to obtain operating expense credits for capital investments.

9.7.1 Fixed Capital, Working Capital, and Land

When the depreciation of capital investment is discussed, care must be taken to distinguish between what can and cannot be depreciated. In general, the total capital investment in a chemical process is made up of two components:

$$\text{Total Capital Investment} = \text{Fixed Capital} + \text{Working Capital} \qquad (9.21)$$

Fixed capital is all the costs associated with building the plant and was covered in Chapter 7 (either total module cost or grassroots cost). The only part of the fixed capital investment that cannot be depreciated is the **land,** which usually represents only a small fraction of the total.

Working capital is the amount of capital required to start up the plant and finance the first few months of operation before revenues from the process start. Typically, this money is used to cover salaries, raw material inventories, and any contingencies. The working capital will be recovered at the end of the project and represents a float of money to get the project started. This concept is similar to that of paying the first and last month's rent on an apartment. The last month's rent is fully recoverable at the end of the lease but must be paid at the beginning. Because the working capital is fully recoverable, it cannot be depreciated. Typical values for the working capital are between 15% and 20% of the fixed capital investment.

9.7.2 Different Types of Depreciation

First the terms that are used to evaluate depreciation are introduced and defined.

Fixed Capital Investment, FCI_L: This represents the fixed capital investment to build the plant minus the cost of land and represents the depreciable capital investment.

Salvage Value, S: This represents the fixed capital investment of the plant, minus the value of the land, evaluated at the end of the plant life. Usually, the equipment salvage (scrap) value represents a small fraction of the initial fixed capital investment. Often the salvage value of the equipment is assumed to be zero.

Life of the Equipment, n: This is specified by the U.S. Internal Revenue Service (IRS). It does not reflect the actual working life of the equipment but rather the time allowed by the IRS for equipment depreciation. Chemical process equipment currently has a depreciation class life of 9.5 years [1].

Total Capital for Depreciation: The total amount of depreciation allowed is the difference between the fixed capital investment and the salvage value.

$$D = FCI_L - S$$

Yearly Depreciation: The amount of depreciation varies from year to year. The amount allowed in the kth year is denoted d_k.

Book Value: The amount of the depreciable capital that has not yet been depreciated.

$$BV_k = FCI_L - \sum_{1}^{k} d_j$$

A discussion of three representative depreciation methods that have been widely used to determine the depreciation allowed each year is provided. Currently, only the straight line and double declining balance methods are approved by the IRS. The sum of the years digits method has been used previously and is included here for completeness.

Straight-Line Depreciation Method, SL: An equal amount of depreciation is charged each year over the depreciation period allowed. This is shown as

$$d_k^{SL} = \frac{[FCI_L - S]}{n} \tag{9.22}$$

Sum of the Years Digits Depreciation Method, *SOYD*: The formula for calculating the depreciation allowance is as follows:

$$d_k^{soyd} = \frac{[n + 1 - k][FCI_L - S]}{\frac{n}{2}[n + 1]} \tag{9.23}$$

The method gets its name from the denominator of Equation (9.23), which is equal to the sum of the number of years over which the depreciation is allowed:

$$1 + 2 + 3... + n = (n)(n + 1)/2$$

For example, if $n = 7$, then the denominator equals 28.

Double Declining Balance Depreciation Method, *DDB*: The formula for calculating the depreciation allowance is as follows:

$$d_k^{DDB} = \frac{2}{n}\left[FCI_L - \sum_{j=0}^{j=k-1} d_j\right] \tag{9.24}$$

In the declining balance method, the amount of depreciation each year is a constant fraction of the book value, BV_{k-1}. The word *double* in *DDB* refers to the factor 2 in Equation (9.24). Values other than 2 are sometimes used; for example, for the 150% declining balance method, 1.5 is substituted for the 2 in Equation (9.24).

In this method, the salvage value does not enter into the calculations. It is not possible, however, to depreciate more than the value of *D*. To avoid this problem, the final year's depreciation is reduced to obtain this limiting value.

Example 9.21 illustrates the use of each of the above formulas to calculate the yearly depreciation allowances.

Example 9.21

The fixed capital investment (excluding the cost of land) of a new project is estimated to be $150.0 million, and the salvage value of the plant is $10.0 million. Assuming a seven-year equipment life, estimate the yearly depreciation allowances using the following:

a. The straight-line method
b. The sum of the years digits method
c. The double declining balance method

We have $FCI_L = \$150 \times 10^6$, $S = \$10.0 \times 10^6$, and $n = 7$ years.

Sample calculations for year 2 give the following:

For straight-line depreciation, using Equation (9.22),

$d_2 = (\$150 \times 10^6 - \$10 \times 10^6)/7 = \$20 \times 10^6$

For SOYD depreciation, using Equation (9.23),

$d_2 = (7 + 1 - 2)\ (\$150 \times 10^6 - \$10 \times 10^6)/28 = \$30 \times 10^6$

For double declining balance depreciation, using Equation (9.24),

$d_2 = (2/7)\ (\$150 \times 10^6 - \$42.86 \times 10^6) = \$30.6 \times 10^6$

A summary of all the calculations is given in Table E9.21 and presented graphically in Figure E9.21.

Table E9.21 Calculations and Results for Example 9.21: The Depreciation of Capital Investment for a New Chemical Plant (All Values in $\$10^7$)

Year (k)	d_k^{SL}	d_k^{SOYD}	d_k^{DDB}	Book Value $FCI_L - Sd_k^{DDB}$
0				$(15 - 0) = 15$
1	$\frac{(15-1)}{7} = 2$	$\frac{(7+1-1)(15-1)}{28^*} = 3.5$	$\frac{(2)(15)}{7} = 4.29$	$(15 - 4.29) = 10.71$
2	$\frac{(15-1)}{7} = 2$	$\frac{(7+1-2)(15-1)}{28^*} = 3.0$	$\frac{(2)(10.71)}{7} = 3.06$	$(10.71 - 3.06) = 7.65$
3	$\frac{(15-1)}{7} = 2$	$\frac{(7+1-3)(15-1)}{28^*} = 2.5$	$\frac{(2)(7.65)}{7} = 2.19$	$(7.65 - 2.19) = 5.46$
4	$\frac{(15-1)}{7} = 2$	$\frac{(7+1-4)(15-1)}{28^*} = 2.0$	$\frac{(2)(5.46)}{7} = 1.56$	$(5.46 - 1.56) = 3.90$
5	$\frac{(15-1)}{7} = 2$	$\frac{(7+1-5)(15-1)}{28^*} = 1.5$	$\frac{(2)(3.90)}{7} = 1.11$	$(3.90 - 1.11) = 2.79$
6	$\frac{(15-1)}{7} = 2$	$\frac{(7+1-6)(15-1)}{28^*} = 1.0$	$\frac{(2)(2.79)}{7} = 0.80$	$(2.79 - 0.80) = 1.99$
7	$\frac{(15-1)}{7} = 2$	$\frac{(7+1-7)(15-1)}{28^*} = 0.5$	$1.99 - 1.0 = 0.99^\dagger$	$(1.99 - 0.99) = 1.00$
Total	14.0	14.0	14.0	$1.0 =$ Salvage Value†

*Sum of digits: $[n + 1]n/2 = [7 + 1]\ 7/2 = 28$

†The depreciation allowance in the final year of the double declining balance method is adjusted to give a final book value equal to the salvage value.

From Figure E9.21 it is seen as follows.

1. The depreciation values obtained from the sum of the years' digits are similar to those obtained from the double declining balance methods.

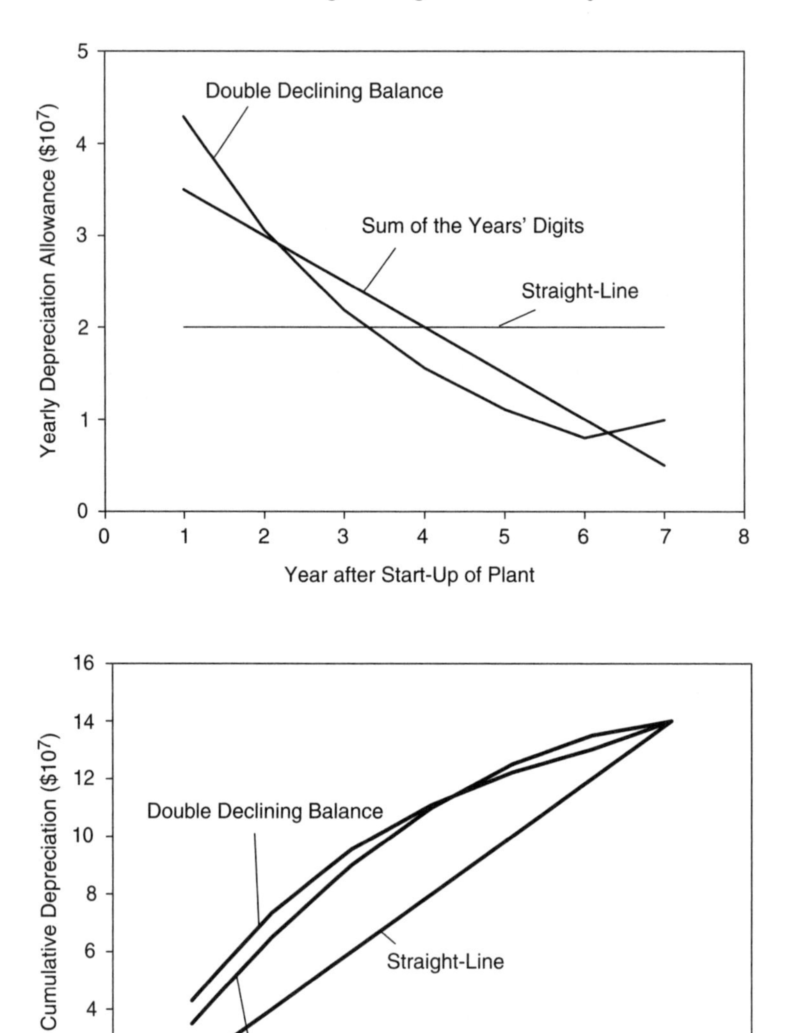

Figure E9.21 Yearly Depreciation Allowances and Cumulative Depreciation Amounts for Example 9.21

2. The double declining balance method has the largest depreciation in the early years.
3. The straight-line method represents the slowest depreciation in the early years.

The SOYD and the DDB methods are examples of accelerated depreciation schemes (relative to the straight line). It is shown in Section 9.8 that accelerated depreciation has significant economic advantages over the straight-line method.

Capital investment can be depreciated only in accordance with current tax regulations.

9.7.3 CURRENT DEPRECIATION METHOD: MODIFIED ACCELERATED COST RECOVERY SYSTEM (MACRS)

The current federal tax law is based on MACRS, using a half-year convention. All equipment is assigned a class life, which is the period over which the depreciable portion of the investment may be discounted. Most equipment in a chemical plant has a class life of 9.5 years [1] with no salvage value. This means that the capital investment may be depreciated using a straight-line method over 9.5 years. Alternatively, a MACRS method over a shorter period of time may be used, which is five years for this class life. In general, it is better to depreciate an investment as soon as possible. This is because the more the depreciation is in a given year, the less taxes paid. As shown earlier in this chapter, "money now is worth more than the same amount in the future"; therefore, it is better to pay less in taxes at the beginning of a project than at the end.

The MACRS method uses a double declining balance method and switches to a straight-line method when the straight-line method yields a greater depreciation allowance for that year. The straight-line method is applied to the remaining depreciable capital over the remaining time allowed for depreciation. The half-year convention assumes that the equipment is bought midway through the first year for which depreciation is allowed. In the first year, the depreciation is only half of that for a full year. Likewise in the sixth (and last) year, the depreciation is again for one-half year. The depreciation schedule for equipment with a 9.5-year class life and 5-year recovery period, using the MACRS method, is shown in Table 9.2.

Example 9.22 illustrates the method by which the MACRS depreciation allowances in Table 9.2 are obtained.

Example 9.22

The basic approach is to use the double declining balance (DDB) method and compare the result with the straight-line (SL) method for the remaining depreciable capital over the remaining period of time. The MACRS method requires depreciation of the total FCI_L, without regard for the salvage value. Calculations are given below, using a basis of $100:

$$\text{For DDB, we have } d_k^{DDB} = \frac{2}{5}\left[100 - \sum_{j=0}^{k-1} d_j^{DDB}\right]$$

$$\text{For SL, we have } d_k^{SL} = \frac{\text{undepreciated capital}}{\text{remaining time for depreciation}}$$

k	d_k^{DDB} (½ yr convention)		d_k^{SL} (time remaining)
1	0.4(100)(0.5) = **20**		
2	0.4(100 – 20) = **32**		(100 – 20)/4.5 = 17.78
3	0.4(100 – 52) = **19.2**	switch to SL	(100 – 52)/3.5 = 13.71
4	0.4(100 – 71.2) = **11.52**	⟶	(100 – 71.2)/2.5 = **11.52**
5	0.4(100 – 82.72) = 6.91		(100 – 82.72)/1.5 = **11.52**
6		½ yr convention ⟶	(0.5)(100 – 94.24)/0.5 = **5.76**

9.8 TAXATION, CASH FLOW, AND PROFIT

Taxation has a direct impact on the profits realized from building and operating a plant. Tax regulations are complex, and companies have tax accountants and attorneys to ensure compliance and to maximize the benefit from these laws. When we are considering individual projects or comparing similar projects, we must account for the effect of taxes. Taxation rates for companies and the laws governing taxation change frequently. The current tax rate schedule (as of May 2001) is given in Table 9.3.

For most large corporations, the basic federal taxation rate is 35%. In addition, corporations must also pay state, city, and other local taxes. The overall taxation rate is often in the range of 40% to 50%. The taxation rate used in the problems at the back of this chapter will vary and may be as low as 30%. For the

Table 9.2 Depreciation Schedule for MACRS Method for Equipment with a 9.5-Year Class Life and a 5-Year Recovery Period [1]

Year	Depreciation Allowance (% of Capital Investment)
1	20.00
2	32.00
3	19.20
4	11.52
5	11.52
6	5.76

Table 9.3 Federal Tax Rate Schedule for Corporations [2]

Range of Net Taxable Income	Taxation Rate
> \$0 and ≤ \$50,000	15%
> \$50,000 and ≤ \$75,000	\$7500 + 25% of amount over \$50,000
> \$75,000 and ≤ \$100,000	\$13,750 + 34% of amount over \$75,000
> \$100,000 and ≤ \$335,000	\$22,250 + 39% of amount over \$100,000
> \$335,000 and ≤ \$10 million	\$113,900 + 34% of amount over \$335,000
> \$10 million and ≤ \$15 million	\$3,4000,000 + 35% of amount over \$10 million
> \$15 million and ≤ \$18.333 million	\$5,150,000 + 38% of amount over \$15 million
> \$18.333 million	35%

economic analysis of a proposed (current) process, clearly it is important to use the correct taxation rate, which will, in turn, depend on the location of the proposed process.

Table 9.4 provides the definition of important terms and equations used to evaluate the cash flow and the profits produced from a project.

The equations from Table 9.4 are used in Example 9.23.

Table 9.4 Evaluation of Cash Flows* and Profits* in Terms of Revenue (*R*), Cost of Manufacturing (*COM*), Depreciation (*d*), and Tax Rate (*t*)

	Description	Formula	Equation
Expenses	= Manufacturing Costs + Depreciation	$= COM_d + d$	(9.250
Income Tax	= (Revenue − Expenses)(Tax Rate)	$= (R - COM_d - d)(t)$	(9.26)
After-Tax (Net) Profit	= Revenue − Expenses − Income Tax	$= (R - COM_d - d)(1 - t)$	(9.27)
After-Tax Cash Flow	= Net Profit + Depreciation	$= (R - COM_d - d)(1 - t) + d$	(9.28)
Variables:			
t	Tax Rate		Constant
COM_d	Cost of Manufacture Excluding Depreciation		(8.2)
d	Depreciation: Depends upon Method Used		(9.22) (9.23)
R	Revenue from Sales		(9.24)

*To obtain before-tax values, set the tax rate (t) to zero.

Example 9.23

For the project given in Example 9.21, the manufacturing costs, excluding depreciation, are \$30 million per year, and the revenues from sales are \$75 million per year. Given the depreciation values calculated in Example 9.21, calculate the following for a 10-year period after start-up of the plant.

a. The after-tax profit (net profit)
b. The after-tax cash flow, assuming a taxation rate of 30%

From Equations (9.27) and (9.28) (all numbers are in $\$10^6$),
After-tax profit $= (75 - 30 - d_k)(1 - 0.3) = 31.5 - (0.7)(d_k)$
After-tax cash flow $= (75 - 30 - d_k)(1 - 0.3) + d_k = 31.5 + (0.3)(d_k)$
A sample calculation for year 1 ($k = 1$) is provided:
From Example 9.21, $d_1^{SL} = 20$, $d_1^{SOYD} = 35$, and $d_1^{DDB} = 42.9$

	SL	SOYD	DDB
Profit after Tax	17.5	7.0	1.47
Cash Flow after Tax	37.5	42.0	44.37

The calculations for years 1 through 10 are plotted in Figure E9.23. From this plot, it can be seen that the cash flow at the start of the project is greatest for the DDB method and lowest for the SL method.

The sum of the profits and cash flows over the 10-year period are \$217 million and \$357 million, respectively. These totals are the same for each of the depreciation schedules used. The difference between the cash flows and the profits is seen to be the depreciation (\$357 – 217 = \$140 million).

Example 9.23 demonstrated how different depreciation schedules affect the after-tax cash flow. The accelerated schedules for depreciation provided the greatest cash flows in the early years. Because money earned in early years has a greater value than money earned in later years, the accelerated schedule of depreciation is the most desirable alternative.

9.9 SUMMARY

In this chapter, the basics of economic analysis required to evaluate project profitability were covered. The material presented in this chapter is founded on the principle that

$$\text{Money} + \text{Time} = \text{More Money}$$

To benefit from this principle, it is necessary to have resources to make an investment and the time to allow the investment to grow. When this principle is

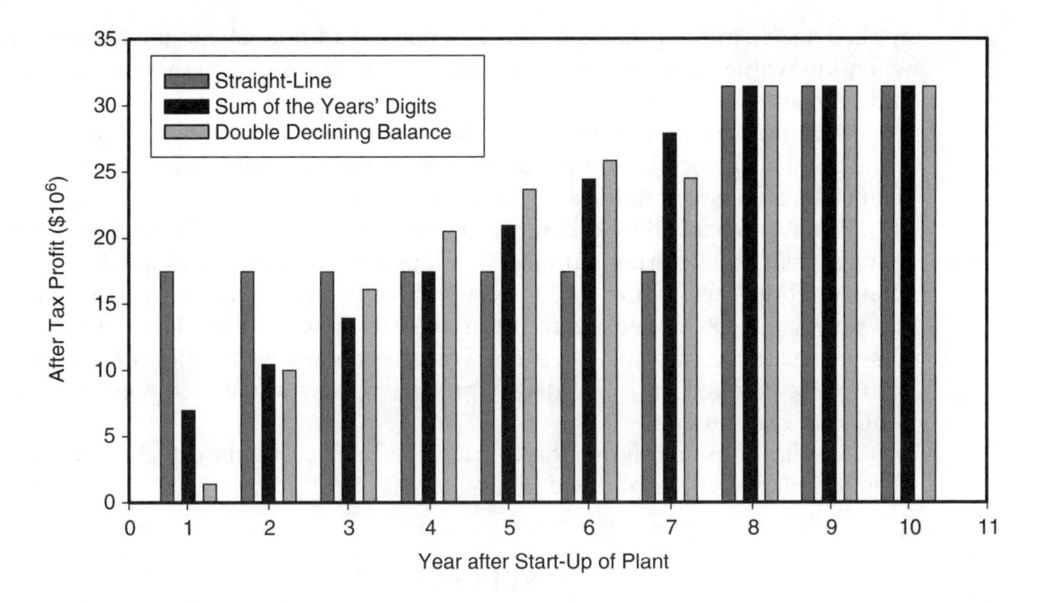

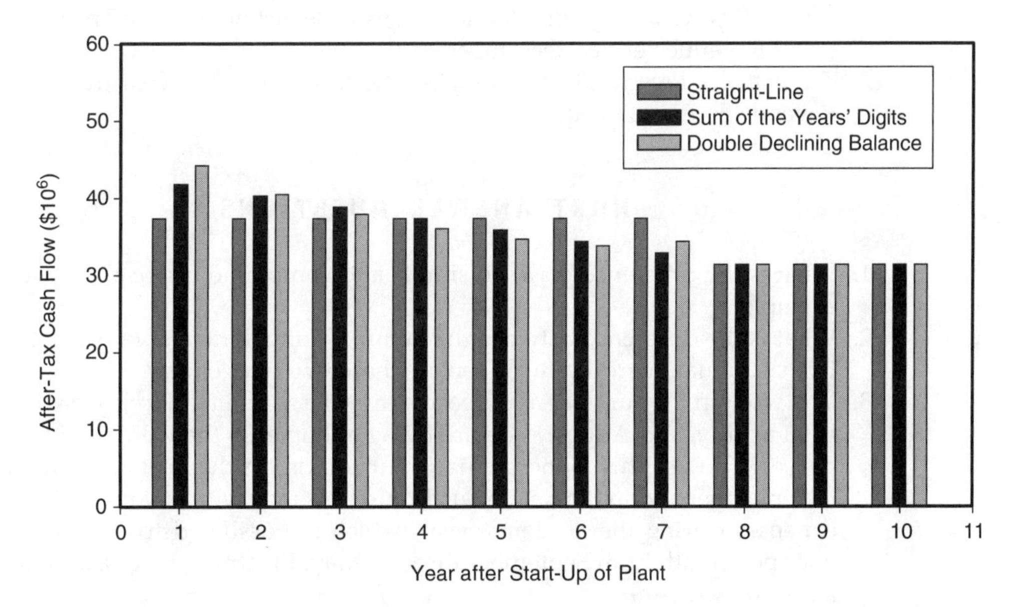

Figure E9.23 Comparison of the After-Tax Profit and Cash Flow Using Different Depreciation Schedules from Example 9.23

applied to chemical processes, the revenue or additional money is generated when low-value materials and services are converted into high-value materials and services.

A central concept identified as the time value of money grows out of this principle. This principle is applied to a wide range of applications, from personal financial management to the analysis of new chemical plants.

The use of cash flow diagrams to visualize the timing of cash flows and to manage cash flows during a project was illustrated. A shorthand notation for the many discount factors involved in economic calculations (factors that account for the time value of money) were introduced to simplify cash flow calculations. Other items necessary for a comprehensive economic evaluation of a chemical plant were covered and included depreciation, taxation, and the evaluation of profit and cash flow.

Applications involving these principles and concepts directly relating to chemical plants will be pursued in Chapter 10.

REFERENCES

1. *How to Depreciate Property,* Publication 946, Department of the Treasury, Internal Revenue Service, December 2000.
2. *Corporations, Property,* Publication 542, Department of the Treasury, Internal Revenue Service, December 2000.

SHORT ANSWER QUESTIONS

1. What is the difference between simple and compound interest? Provide an example.
2. What is the difference between the nominal annual interest rate and the effective annual interest rate? When are these two rates equal?
3. You work in the summer, and receive three large, equal monthly paychecks in June, July, and August. You are living at home, so there are no living expenses. You use this money to pay your tuition for the next academic year in September and in January. You also use this money to pay monthly living expenses during the academic year, which are assumed to be cash flows once per month from September through May. Illustrate these cash flows on a cash flow diagram.
4. Define the term *annuity.*
5. The current value of the Chemical Engineering Plant Cost Index (CEPCI) is 485. If at the same time next year the value has risen to 525, what will be the average inflation rate for the year?
6. Discuss the differences and similarities between interest, inflation, and the time value of money.

7. What term describes the method by which a fixed capital investment can be used to reduce the tax that a company pays?
8. What is depreciation? Explain how it affects the economic analysis of a new chemical process.
9. Why is it advantageous to use an accelerated depreciation schedule?
10. "In general, it is better to depreciate the fixed capital investment as soon as allowable." Give one example of when this statement would not be true.
11. What is the difference between after-tax profit and after-tax cash flow? When are these two quantities the same?

PROBLEMS

12. You need to borrow $1000 for an emergency. You have two alternatives. One is to borrow from a bank. The other is to borrow from your childhood friend Paulie "Walnuts," who is in the "private" financing business. Because you and Paulie go way back, he will give you his preferred rate, which is $5/week until you pay back the entire loan.
    a. What is Paulie's effective annual interest rate?
    b. Your alternative is to borrow from a bank, where the interest is compounded monthly. What nominal interest rate would make you choose the bank over Paulie?
    c. If the bank's interest rate is 7% p.a., compounded monthly, how many months would it be until the interest paid to Paulie equaled the interest paid to the bank?
13. Consider the following three investment schemes:
    9.5% p.a. (nominal rate) compounded daily
    10.0% p.a. (nominal rate) compounded monthly
    10.5% p.a. (nominal rate) compounded quarterly
    a. Which investment scheme is the most profitable, assuming that the initial investment is the same for each case?
    b. What is the effective annual interest rate of the best scheme?
    c. What is the nominal interest rate of the best scheme when compounded continuously?
14. At an investment seminar that I recently attended, I learned about something called "the Rule of 72." According to the person in charge of the seminar, a good estimate for finding how long it takes for an investment to double is given by the following equation:

$$\text{number of years to double investment} = \frac{72}{\text{effective annual interest rate (in \%)}}$$

Using what you know about the time value of money, calculate the error in using the above equation to estimate how long it takes to double an investment made at the following effective annual interest rates: 5%, 7.5%, and 10%.

Express your answers as % error to two significant figures.

Comment on the Rule of 72 and its accuracy for typical financial calculations today.

15. You invested $5000 eight years ago, and you want to determine the value of the investment now, at the end of year 8. During the past eight years, the nominal interest rate has fluctuated as follows.

Year	Nominal Interest Rate (% p.a.)
1	4
2	5
3	7
4	8
5	6
6	4
7	5
8	4

If your investment is compounded daily, how much is it worth today? (Ignore the effect of leap years.)

16. What are the differences between investing $15,000 at 9% p.a. for 15 years when compounded yearly, quarterly, monthly, daily, and continuously? Ignore the effect of leap years.

17. One bank advertises an interest rate on a certificate of deposit (CD) to be 4% compounded daily. Another bank advertises a CD with a 4.1% effective annual rate. In which CD would you invest?

18. You want to begin an investment plan to save for your daughter's college education. Because you believe that your newly born daughter will be a genius (just like you!), you are assuming the cost of an Ivy League education. You plan to put money into an investment account, at the end of each year, for the next 18 years. You believe that you will need about $75,000 per year, each year, 19 through 22 years from now.
    a. Draw a discrete cash flow diagram for this situation.
    b. How much would you have to invest each year to pay for college and have a zero balance at the end of year 21? The effective annual interest rate of your investment is 8%.
    c. What interest rate would be needed if you could invest only $5000/yr?

19. You begin an investment plan by putting $10,000 in an account that you assume will earn 8% annually. For the next 25 years, you add $5000 per year. In anticipation of buying a house with a 15-year mortgage, you expect to need a one-time down payment of $55,000 at the end of year 8. You anticipate being able to make the monthly mortgage payment without affecting the yearly contribution to the savings plan.
    a. Draw a discrete, nondiscounted cash flow diagram for this situation.
    b. Will you have the down payment at the end of year 8?
    c. What will be the value of your savings account at the end of year 25?
20. You begin work on June 1 and work until August 31, and receive pay on the last day of the month. Your expenses for these three months are $1500/mo. At the end of September, you make a $7000 payment, and you make an identical payment at the end of January. Your expenses from September 1 through April 30 are $1000/mo.
    a. Draw a discrete, nondiscounted cash flow diagram for this situation.
    b. If you earn 4% interest, compounded monthly, on the money until it is spent, what monthly salary is required from June 1 through August 31 to break even on May 30?
    c. If you earn 4% interest, compounded monthly, on the money until it is spent and the salary is $4000/mo, how much would you have to earn each month from a different job from September through May to break even on May 31?
    d. What situation is depicted in this problem?
21. The cash flows for a bank account are described by the discrete CFD in Figure P9.21. The bank account has an effective annual interest rate of 4.5%.
    a. Calculate the future value of all cash flows after 15 years.
    b. Calculate the future value of all cash flows after 25 years assuming that there are no more transactions after year 15.

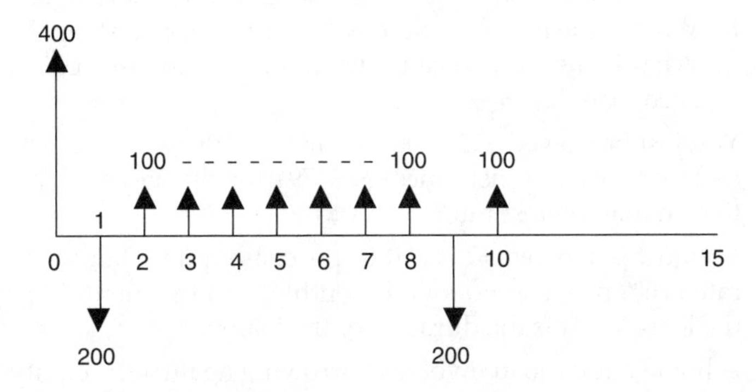

Figure P9.21 Purchased Cost per Unit of Flowrate of a Centrifugal Air Blower (Adapted from Reference [3])

22. You begin to contribute to an investment plan with your company immediately after graduation, when you are 23 years old. Your contribution plus your company's contribution total \$6000/yr. Assume that you work for the same company for 40 years.
    a. What effective annual interest rate is required for you to have \$1 million in 40 years?
    b. Repeat Part (a) for \$2 million.
    c. What is the future value of this investment after 40 years if the effective annual interest rate is 7% p.a.?
23. An IRA is an investment vehicle available to most individuals to save for retirement. The benefit is that the interest is not taxed until you withdraw money, and most retirees are in a lower tax bracket than when they worked. Under current laws, the maximum annual contribution is \$4000 until age 50 and \$5000 thereafter, including the year one turns 50. Suppose that you make the maximum contribution every year for 40 years, beginning at age 25 until age 65.
    a. What is the future value of this retirement fund after 40 years, assuming an effective annual interest rate of 9%?
    b. What effective annual interest rate is required for the future value of this investment to be \$2 million after 40 years?
24. You plan to finance a new car by borrowing \$25,000. The interest rate is 7% p.a., compounded monthly. What is your monthly payment for a three-year loan, a four-year loan, and for a five-year loan?
25. You have just purchased a new car by borrowing \$20,000 for four years. Your monthly payment is \$500. What is your nominal interest rate if compounded monthly?
26. You plan to borrow \$200,000 to purchase a new house. The nominal interest rate fixed at 6.5% p.a., compounded monthly.
    a. What is the monthly payment on a 30-year mortgage?
    b. What is the monthly payment on a 15-year mortgage?
    c. What is the difference in the amount of interest paid over the lifetime of each loan?
27. You just borrowed \$225,000 to purchase a new house. The monthly payment (before taxes and insurance) is \$1791 for the 25-year loan. What is the effective annual interest rate?
28. You just borrowed \$250,000 to purchase a new house. The nominal interest rate is 6% p.a., compounded monthly, and the monthly payment is \$1612 for the loan. What is the duration of the loan?
29. A home-equity loan involves borrowing against the equity in a house. For example, if your house is valued at \$250,000 and you have been paying the mortgage for a sufficient amount of time, you may owe only \$150,000 on the

mortgage. Therefore, your **equity** in the home is $100,000. You can use the home as collateral for a loan, possibly up to $100,000, depending on the bank's policy. Home-equity loans often have shorter durations than regular mortgages. Suppose that you take a home-equity loan of $50,000 for the down payment on a new house in a new location because you have a new job. When you complete the sale of your old house, the home-equity loan will be paid off at the closing. The home-equity loan terms are a 10-year term at 7% p.a., compounded monthly, but you must also pay 0.05% of the original high-equity loan principal each month. What is the monthly payment?

30. Your company is trying to determine whether to spend $500,000 in process improvements. The projected cash flow increases based on the process improvements are as follows.

Year	Annual Increased Cash Flow ($ thousands)
1	25
2	75
3	100
4	125
5	250

The alternative is to do nothing and leave the $500,000 in the investment portfolio earning interest. What interest rate is required in the investment portfolio for the better choice to be to do nothing?

31. It is necessary to evaluate the profitability of proposed improvements to a process prior to obtaining approval to implement changes. For one such process, the capital investment (end of year 0) for the project is $250,000. There is no salvage value. In years 1 and 2, you expect to generate an after-tax revenue from the project of $60,000/yr. In years 3–8, you expect to generate an after-tax revenue of $50,000/yr. Assume that the investments and cash flows are single transactions occurring at the end of the year. Assume an effective annual interest rate of 9%.
    a. Draw a discrete cash flow diagram for this project.
    b. Draw a cumulative, discounted (to year 0) cash flow diagram for this project.
    c. What is the future value of this project at the end of year 8?
    d. Instead of investing in this project, the $250,000 could remain in the company's portfolio. What rate of return on the portfolio is needed to equal the future value of this project at the end of year 8? Would you invest in the project, or leave the money in the portfolio?

32. In Problem 31, the net revenue figures were generated using a taxation rate of 45% and a straight-line depreciation over the eight-year project. Calculate the yearly net revenues if the five-year MACRS depreciation schedule were used.
33. You are evaluating the profitability potential of a process and have the following information. The criterion for profitability is a 15% rate of return over ten operating years. The equipment has zero salvage value at the end of the project.

    Fixed capital investment (including land) in four installments (all values are in millions of dollars as one transaction at the end of the year):

Year 0 land	$10
Year 1 FCI installment 1	$20
Year 2 FCI installment 2	$30
Year 3 FCI installment 3	$20
Start-up capital at end of year 3	$10
Positive cash flow years 4–13	$25

    a. Draw a discrete, discounted cash flow diagram for this process.
    b. Draw a cumulative, discounted cash flow diagram for this process.
    c. What is the present value (at end of year 0) for this process?
    d. What is the future value at the end of year 13?
    e. What would the effective annual interest rate have to be so that the present value (end of year 0) of this investment is zero? (This interest rate is known as the *DCFROR*, and it will be discussed in the next chapter.)
    f. What is your recommendation regarding this process? Explain.
34. What are the MACRS depreciation allowances for recovery periods of four, six, and nine years?
35. For a new process, the land was purchased for $10 million. The fixed capital investment, paid at the end of year 0, is $165 million. The working capital is $15 million, and the salvage value is $15 million. The estimated revenue from years 1 through 10 is $70 million/yr, and the estimated cost of manufacture over the same time period is $25 million/yr. The internal hurdle rate (interest rate) is 14% p.a., before taxes, and the taxation rate is 40%.
    a. Draw a discrete, nondiscounted cash flow diagram for this process.
    b. Determine the yearly depreciation schedule using the five-year MACRS method.
    c. Determine the after-tax profit for each year.
    d. Determine the after-tax cash flow for each year.
    e. Draw a discrete, discounted (to year 0) cash flow diagram for this process.
    f. Draw a cumulative, discounted (to year 0) cash flow diagram for this process.
    g. What is the present value (year 0) of this process?

CHAPTER

10

Profitability Analysis

In this chapter, we will see how to apply the techniques of economic analysis developed in Chapter 9. These techniques will be used to assess the profitability of projects involving both capital expenditures and yearly operating costs. We look at a variety of projects ranging from large multimillion-dollar ventures to much smaller process improvement projects. Several criteria for profitability will be discussed and applied to the evaluation of process and equipment alternatives. We start with the profitability criteria for new large projects.

10.1 A TYPICAL CASH FLOW DIAGRAM FOR A NEW PROJECT

A typical cumulative, after-tax cash flow diagram (CFD) for a new project is illustrated in Figure 10.1. It is convenient to relate profitability criteria to the cumulative CFD rather than the discrete CFD. The timing of the different cash flows are explained below.

In the economic analysis of the project, it is assumed that any new land purchases required are done at the start of the project, that is, at time zero. After the decision has been made to build a new chemical plant or expand an existing facility, the construction phase of the project starts. Depending on the size and scope of the project, this construction may take anywhere from six months to three years to complete. In the example shown in Figure 10.1, a typical value of two years for the time from project initiation to the start-up of the plant has been assumed. Over the two-year construction phase, there is a major capital outlay. This represents the fixed capital expenditures for purchasing and installing the equipment and auxiliary facilities required to run the plant (see Chapter 7). The

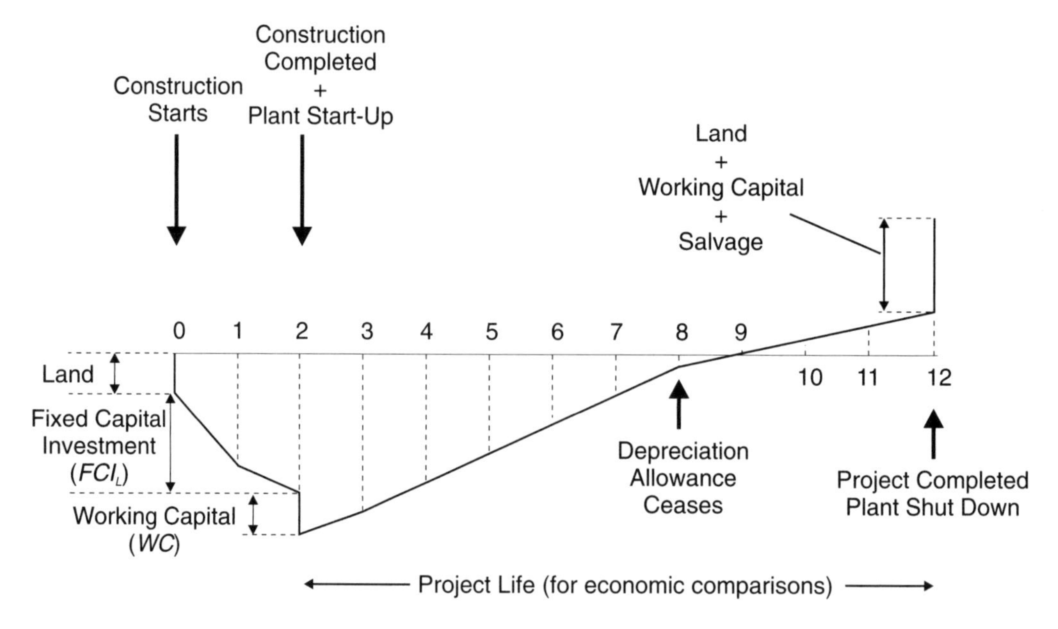

Figure 10.1 A Typical Cumulative Cash Flow Diagram for the Evaluation of a New Project

distribution of this fixed capital investment is usually slightly larger toward the beginning of construction, and this is reflected in Figure 10.1. At the end of the second year, construction is finished and the plant is started. At this point, the additional expenditure for working capital required to float the first few months of operations is shown. This is a one-time expense at the start-up of the plant and will be recovered at the end of the project.

After start-up, the process begins to generate finished products for sale, and the yearly cash flows become positive. This is reflected in the positive slope of the cumulative CFD in Figure 10.1. Usually the revenue for the first year after start-up is less than subsequent years due to "teething" problems in the plant; this is also reflected in Figure 10.1. The cash flows for the early years of operation are larger than those for later years due to the effect of the depreciation allowance discussed in Chapter 9. The time used for depreciation in Figure 10.1 is six years. The time over which the depreciation is allowed depends on the IRS regulations and the method of depreciation used.

In order to evaluate the profitability of a project, a life for the process must be assumed. This is not usually the working life of the equipment, nor is it the time over which depreciation is allowed. It is a specific length of time over which the profitability of different projects are to be compared. Lives of ten, twelve, and fifteen years are commonly used for this purpose. It is necessary to standardize the project life when comparing different projects. This is because profitability is directly related to project life, and comparing projects using different lives biases the results.

Usually, chemical processes have anticipated operating lives much greater than ten years. If much of the equipment in a specific process is not expected to last for a ten-year period, then the operating costs for that project should be adjusted. These operating costs should reflect a much higher maintenance cost to include the periodic replacement of equipment necessary for the process to operate the full ten years. A project life of ten years will be used for the examples in the next section.

From Figure 10.1, we can see a steadily rising cumulative cash flow over the ten operating years of the process, that is, years 2 through 12. At the end of the ten years of operation, that is, at the end of year 12, it is assumed that the plant is closed down and that all the equipment is sold for its salvage or scrap value, that the land is also sold, and that the working capital is recovered. This additional cash flow, received on closing down the plant, is shown by the upward-pointing vertical line in year 12. Remember that in reality, the plant will most likely *not* be closed down; we only assume that it will be in order to perform our economic analysis.

The question that we must now address is how to evaluate the profitability of a new project. Looking at Figure 10.1, we can see that at the end of the project the cumulative CFD is positive. Does this mean that the project will be profitable? The answer to this question depends on whether the value of the income earned during the time the plant operated was smaller or greater than the investment made at the beginning of the project. Therefore, the time value of money must be considered when evaluating profitability. In the following sections, we will look at different ways to evaluate project profitability.

10.2 PROFITABILITY CRITERIA FOR PROJECT EVALUATION

There are three bases used for the evaluation of profitability:

1. Time
2. Cash
3. Interest rate

For each of these bases, discounted or nondiscounted techniques may be considered. The nondiscounted techniques do not take into account the time value of money and are *not* recommended for evaluating new, large projects. Traditionally, however, such methods have been and are still used to evaluate smaller projects, such as process improvement schemes. Examples of both types of methods are presented for all the three bases.

10.2.1 Nondiscounted Profitability Criteria

Four nondiscounted profitability criteria are illustrated in Figure 10.2. The graphical interpretation of these profitability criteria is shown in the figure. Each of the four criteria is explained below.

Time Criterion. The term used for this criterion is the **payback period (PBP),** also known by a variety of other names, such as payout period, payoff period, and cash recovery period. The payback period is defined as follows:

PBP = Time required, after start-up, to recover the fixed capital investment, FCI_L, for the project

The payback period is shown as a length of time on Figure 10.2. Clearly, the shorter the payback period, the more profitable the project.

Cash Criterion. The criterion used here is the **cumulative cash position (CCP),** which is simply the worth of the project at the end of its life. For criteria using cash or monetary value, it is difficult to compare projects with dissimilar fixed capital investments, and sometimes it is more useful to use the **cumulative cash ratio (CCR),** which is defined as

$$CCR = \frac{\text{Sum of All Positive Cash Flows}}{\text{Sum of All Negative Cash Flows}}$$

The definition effectively gives the cumulative cash position normalized by the initial investment. Projects with cumulative cash ratios greater than 1 are

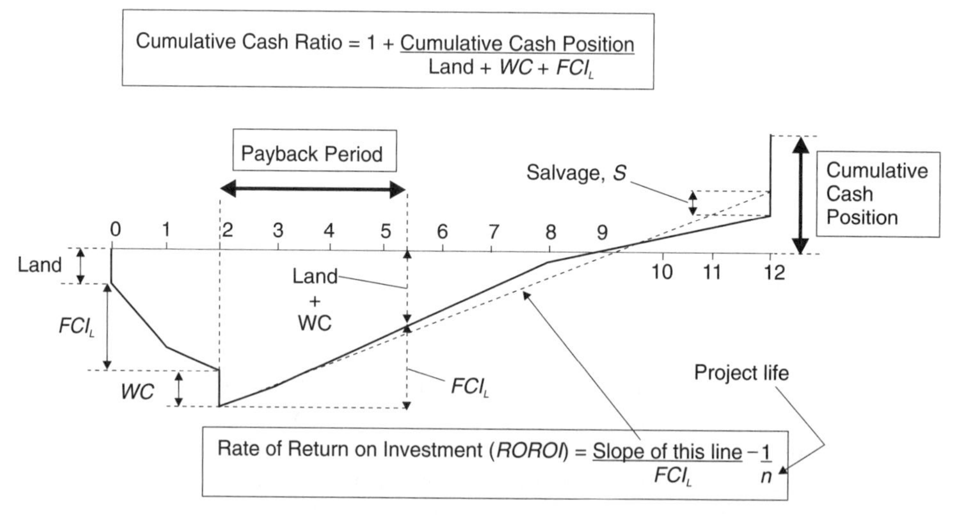

Figure 10.2 Illustration of Nondiscounted Profitability Criteria

potentially profitable, whereas those with ratios less than unity cannot be profitable.

Interest Rate Criterion. The criterion used here is called the **rate of return on investment (ROROI)** and represents the nondiscounted rate at which money is made from our fixed capital investment. The definition is given as

$$ROROI = \frac{\text{Average Annual Net Profit}}{\text{Fixed Capital Investment } (FCI_L)}$$

The annual net profit in this definition is an average over the life of the plant after start-up.

The use of fixed capital investment, FCI_L, in the calculations for payback period and rate of return on investment given above seems reasonable, because this is the capital that must be recovered by project revenue. Many alternative definitions for these terms can be found, and the reader will find that total capital investment ($FCI_L + WC +$ Land) is often used instead of fixed capital investment. When the plant has a salvage value (S), the fixed capital investment minus the salvage value ($FCI_L - S$) could be used instead of FCI_L. However, because the salvage value is usually very small, we prefer to use FCI_L alone. Example 10.1 is a comprehensive profitability analysis calculation using nondiscounted criteria.

Example 10.1

A new chemical plant is going to be built and will require the following capital investments (all figures are in $ million):

Cost of land, L= $10.0
Total fixed capital investment, FCI_L = $150.0
Fixed capital investment during year 1 = $90.0
Fixed capital investment during year 2 = $60.0
Plant start-up at end of year 2
Working capital = $30.0 at end of year 2

The sales revenues and costs of manufacturing are given below:
Yearly sales revenue (after start-up), R = $75.0 per year
Cost of manufacturing excluding depreciation allowance (after start-up), COM_d = $30.0 per year

Taxation rate, t = 45%
Salvage value of plant, S = $10.0
Depreciation: Use 5-year MACRS.
Assume a project life of 10 years.

Calculate each nondiscounted profitability criterion given in this section for this plant.

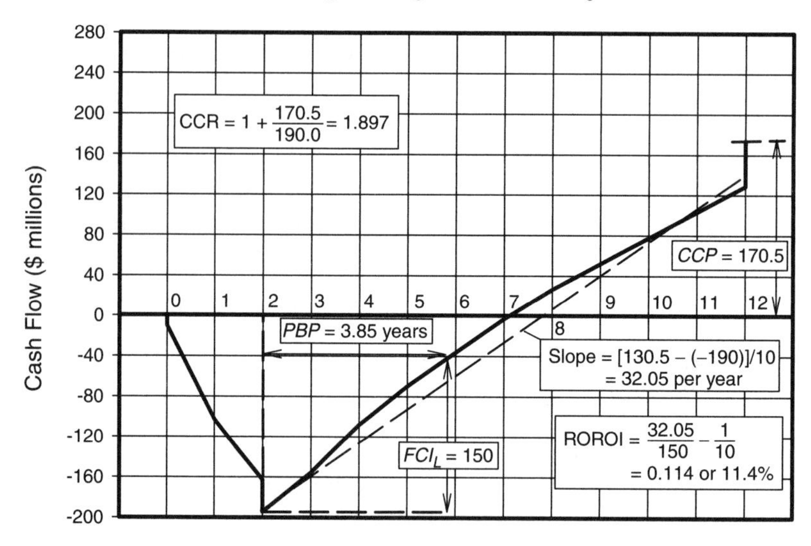

Figure E10.1 Cumulative Cash Flow Diagram for Nondiscounted After-Tax Cash Flows for Example 10.1

The discrete and cumulative nondiscounted cash flows for each year are given in Table E10.1. Using this data, the cumulative cash flow diagram is drawn, as shown in Figure E10.1.

The method of evaluation for each of the criteria is given on Figure E10.1 and in Table E10.1.

Payback Period (*PBP*) = 3.85 years
Cumulative Cash Position (*CCP*) = $ 170.5 × 10^6
Cumulative Cash Ratio (*CCR*) = 1.897
Rate of Return on Investment (*ROROI*) = 11.4%

All these criteria indicate that the project cannot be eliminated as unprofitable. They all fail to take into account the time value of money that is necessary for a thorough measure of profitability. The effects of the time value of money on profitability are considered in the next section.

10.2.2 Discounted Profitability Criteria

The main difference between the nondiscounted and discounted criteria is that for the latter we discount each of the yearly cash flows back to time zero. The resulting discounted cumulative cash flow diagram is then used to evaluate profitability. The three different types of criteria are:

Table E10.1 Nondiscounted After-Tax Cash Flows for Example 10.1 (All Numbers in $\$10^6$)

End of Year (k)	Investment	d_k	$FCI_L - \Sigma d_k$	R	COM_d	$(R-COM-d_k)\times(1-t)+d_k$	Cash Flow	Cumulative Cash Flow
0	(10)*	—	150.00	—	—	—	(10.00)	(10.00)
1	(90)	—	150.00	—	—	—	(90.00)	(100.00)
2	(60 + 30) = (90)	—	150.00	—	—	—	(90.00)	(190.00)
3	—	30.00	120.00	75	30	38.25	38.25	(151.75)
4	—	48.00	72.00	75	30	46.35	46.35	(105.40)
5	—	28.80	43.20	75	30	37.71	37.71	(67.69)
6	—	17.28	25.92	75	30	32.53	32.53	(35.16)
7	—	17.28	8.64	75	30	32.53	32.53	(2.64)
8	—	8.64	0.00	75	30	28.64	28.64	26.00
9	—	—	0.00	75	30	24.75	24.75	50.75
10	—	—	0.00	75	30	24.75	24.75	75.50
11	—	—	0.00	75	30	24.75	24.75	100.25
12	10 + 30 = 40	—	0.00	85	30	30.25	70.25	170.50

*Numbers in () are negative cash flows.

Nondiscounted Profitability Criteria

Payback Period (PBP)

Land + Working Capital = 10 + 30 = $\$40 \times 10^6$—find time after start-up for which cumulative cash flow = $-\$40 \times 10^6$

PBP = 3 + (–67.69 + 40)/(–67.69 + 35.16) = **3.85 years**

Cumulative Cash Position (CCP) and Cumulative Cash Ratio (CCR)

CCP = \$170.50 × 106 and **CCR** = Σ positive cash flows / Σ negative cash flows = (38.25 + 46.35 + 37.71 + ... + 24.75 + 70.25) / (10 + 90 + 90) = **1.897**

Rate of Return on Investment (ROROI)

ROROI = (38.25 + 46.35 + 37.71 + ... + 24.75 + 30.25)/10/150 – 1/10 = **0.114 or 11.4% p.a.**

Time Criterion. The **discounted payback period (*DPBP*)** is defined in a manner similar to the nondiscounted version given above.

$DPBP$ = Time required, after start-up, to recover the fixed capital investment, FCI_L, required for the project, with all cash flows discounted back to time zero

The project with the shortest discounted payback period is the most desirable.

Cash Criterion. The **discounted cumulative cash position,** more commonly known as the **net present value (*NPV*)** or **net present worth (*NPW*)** of the project, is defined as

NPV = Cumulative discounted cash position at the end of the project

Again, the *NPV* of a project is greatly influenced by the level of fixed capital investment, and a better criterion for comparison of projects with different investment levels may be the **present value ratio (*PVR*):**

$$PVR = \frac{\text{Present Value of All Positive Cash Flows}}{\text{Present Value of All Negative Cash Flows}}$$

A present value ratio of unity for a project represents a break-even situation. Values greater than unity indicate profitable processes, whereas those less than unity represent unprofitable projects. Example 10.2 continues Example 10.1 using discounted profitability criteria.

Example 10.2

For the project described in Example 10.1, determine the following discounted profitability criteria:

a. Discounted payback period (*DPBP*)
b. Net present value (*NPV*)
c. Present value ratio (*PVR*)

Assume a discount rate of 0.1 (10% p.a.).

The procedure used is similar to the one used for the nondiscounted evaluation shown in Example 10.1. The discounted cash flows replace actual cash flows. For the discounted case, we must first discount all the cash flows in Table E10.1 back to the beginning of the project (time = 0). We do this simply by multiplying each cash flow by the discount factor $(P/F, i, n)$, where n is the number of years after the start of the project. These discounted cash flows are shown along with the cumulative discounted cash flows in Table E10.2.

The cumulative discounted cash flows are shown on Figure E10.2, and the calculations are given in Table E10.2. From these sources the profitability criteria are given as

a. Discounted payback period (*DPBP*) = 5.94 years

Table E10.2 Discounted Cash Flows for Example 10.2 (All Numbers in Millions of $)

End of Year	Nondiscounted Cash Flow	Discounted Cash Flow	Cumulative Discounted Cash Flow
0	(10.00)	(10)	(10.00)
1	(90.00)	$(90)/1.1 = (81.82)$	(91.82)
2	(90.00)	$(90)/1.1^2 = (74.38)$	(166.20)
3	38.25	$38.25/1.1^3 = 28.74$	(137.46)
4	46.35	$46.35/1.1^4 = 31.66$	(105.80)
5	37.71	$37.71/1.1^5 = 23.41$	(82.39)
6	32.53	$32.53/1.1^6 = 18.36$	(64.03)
7	32.53	$32.53/1.1^7 = 16.69$	(47.34)
8	28.64	$28.64/1.1^8 = 13.36$	(33.98)
9	24.75	$24.75/1.1^9 = 10.50$	(23.48)
10	24.75	$24.75/1.1^{10} = 9.54$	(13.94)
11	24.75	$24.75/1.1^{11} = 8.67$	(5.26)
12	70.25	$70.25/1.1^{12} = 22.38$	17.12

Discounted Profitability Criteria

Discounted Payback Period (DPBP)

Discounted value of land + working capital = $10 + 30/1.12 = \$34.8 \times 10^6$

Find time after start-up when cumulative cash flow = $-\$34.8 \times 10^6$

$DPBP = 5 + (-47.34 + 34.8)/(-47.34 + 33.98) = \mathbf{5.94\ yr}$

Net Present Value (NPV) and Present Value Ratio (PVR)

$NPV = \$17.12 \times 10^6$

PVR = Σ positive discounted cash flows / Σ negative discounted cash flows = (28.74 + 31.36 + 23.41 + ... + 22.38) / (10 + 81.82 + 74.38)

$PVR = (183.31) / (166.2) = 1 + 17.12 / 166.2 = \mathbf{1.10}$

b. Net present value (NPV) = $\$17.12 \times 10^6$

c. Present value ratio (PVR) = 1.10

We can see from these examples that there are significant effects of discounting the cash flows to account for the time value of money. From these results, the following observations may be made.

1. In terms of the time basis, the payback period increases as the discount rate increases. In the above examples, it increased from 3.85 to 5.94 years.
2. In terms of the cash basis, replacing the cash flow with the discounted cash flow decreases the value at the end of the project. In the above examples, it dropped from \$170.5 to \$17.12 million.

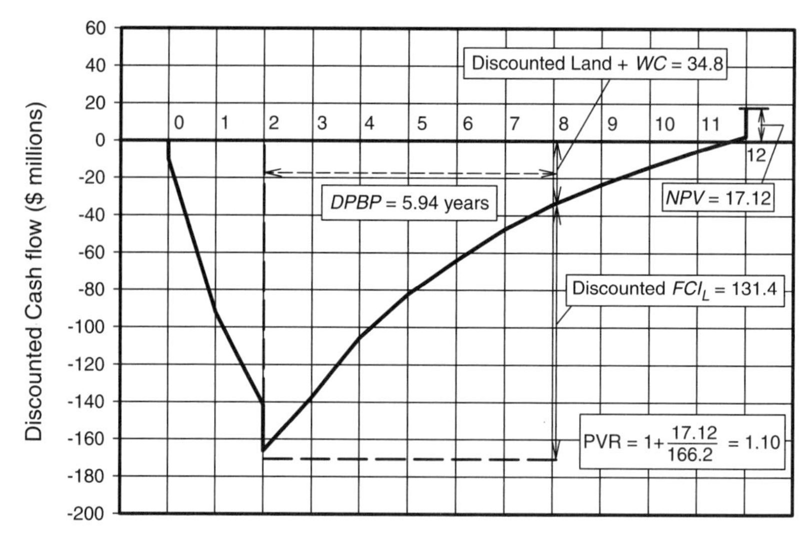

Figure E10.2 Cumulative Cash Flow Diagram for Discounted After-Tax Cash Flows for Example 10.2

3. In terms of the cash ratios, discounting the cash flows gives a lower ratio. In the above examples, the ratio dropped from 1.897 to 1.10.

As the discount rate increases, all of the discounted profitability criteria will be reduced.

Interest Rate Criterion. The **discounted cash flow rate of return (DCFROR)** is defined to be the interest rate at which all the cash flows must be discounted in order for the net present value of the project to be equal to zero. Thus, we can write

$$DCFROR = \text{Interest or discount rate for which the net present value of the project is equal to zero}$$

Therefore, the *DCFROR* represents the highest after-tax interest or discount rate at which the project can just break even.

For the discounted payback period and the net present value calculations, the question arises as to what interest rate should be used to discount the cash flows. This "internal" interest rate is usually determined by corporate management and represents the minimum acceptable rate of return that the company will accept for any new investment. Many factors influence the determination of this discount interest rate, and for our purposes, we will assume that it is always given.

It is worth noting that for evaluation of the discounted cash flow rate of return, no interest rate is required because this is what we calculate. Clearly, if the *DCFROR* is greater than the internal discount rate, then the project is considered to be profitable. Use of *DCFROR* as a profitability criterion is illustrated in Example 10.3.

Example 10.3

For the problem presented in Examples 10.1 and 10.2, determine the discounted cash flow rate of return (*DCFROR*).

The *NPV*s for several different discount rates were calculated and the results are shown in Table E10.3.

The value of the *DCFROR* is found at *NPV* equals 0. Interpolating from Table E10.3 gives:

$$\frac{(DCFROR - 12\%)}{(13\% - 12\%)} = \frac{(0 - 0.77)}{(-6.32 - 0.77)} = 0.109$$

Therefore, $DCFROR = 12 + 1(0.109) = 12.1\%$

An alternate method for obtaining the *DCFROR* is to solve for the value of i in an implicit, nonlinear algebraic expression. This is illustrated in Example 10.4.

Table E10.3 *NPV* for Project in Example 10.1 as a Function of Discount Rate

Interest or Discount Rate	*NPV* ($ million)
0%	170.50
10%	17.12
12%	0.77
13%	–6.32
15%	–18.66
20%	–41.22

Figure 10.3 provides the cumulative discounted cash flow diagram for Example 10.3 for several discount factors. It shows the effect of changing discount factors on the profitability and shape of the curves. It includes a curve for the *DCFROR* found in Example 10.3. For this case, it can be seen that the *NPV* for the project is zero. In Example 10.3, if the acceptable rate of return for our company were set at 20%, then the project would not be considered an acceptable investment. This is indicated by a negative *NPV* for $i = 20\%$.

Each method described above used to gauge the profitability of a project has advantages and disadvantages. For projects having a short life and small

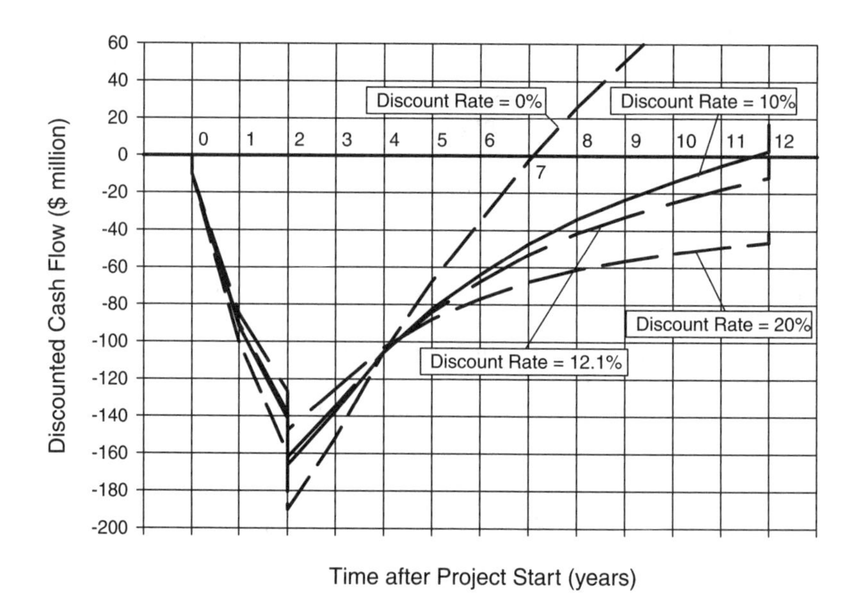

Figure 10.3 Discounted Cumulative Cash Flow Diagrams Using Different Discount Rates for Example 10.3

discount factors, the effect of discounting is small, and nondiscounted criteria may be used to give an accurate measure of profitability. However, it is fair to say that for large projects involving many millions of dollars of capital investment, discounting techniques should always be used.

Because all the above techniques are commonly used in practice, you must be familiar with and be able to use each technique.

10.3 COMPARING SEVERAL LARGE PROJECTS: INCREMENTAL ECONOMIC ANALYSIS

In this section, we compare and select among investment alternatives. When comparing project investments, the *DCFROR* tells us how efficiently we are using our money. The higher the *DCFROR*, the more attractive the individual investment. However, when comparing investment alternatives, it may be better to choose a project that does not have the highest *DCFROR*. The rationale for comparing projects and choosing the most attractive alternative is discussed in this section.

In order to make a valid decision regarding alternative investments (projects), it is necessary to know a baseline rate of return that must be attained in order for an investment to be attractive. A company that is considering whether to invest in a new project always has the option to reject all alternatives offered and invest the cash (or resources) elsewhere. The baseline or benchmark investment rate is related to these alternative investment opportunities, such as investing in the stock market. Incremental economic analysis is illustrated in Examples 10.4 and 10.5.

Example 10.4

Our company is seeking to invest approximately $\$120 \times 10^6$ in new projects. After extensive research and preliminary design work, three projects have emerged as candidates for construction. The minimum acceptable internal discount (interest) rate, after tax, has been set at 10%. The after-tax cash flow information for the three projects using a ten-year operating life is as follows (values in $ million).

	Initial Investment	After-Tax Cash Flow in Year *i*	
		i = 1	*i* = 2–10
Project A	$60	$10	$12
Project B	$120	$22	$22
Project C	$100	$12	$20

For this example it is assumed that the cost of land, working capital, and salvage are zero. Furthermore, it is assumed that the initial investment occurs at time = 0, and the yearly annual cash flows occur at the end of each of the ten years of plant operation.

Determine the following:

a. The *NPV* for each project
b. The *DCFROR* for each project

For Project A we get

$$NPV = -\$60 + (\$10)(P/F, 0.10, 1) + (\$12)(P/A, 0.10, 9)(P/F, 0.10, 1)$$

$$= -\$60 + \frac{(\$10)}{1.1} + (\$12)\frac{1.1^9 - 1}{(0.1)(1.1^9)}\frac{1}{1.1} = \$11.9$$

The *DCFROR* is the value of *i* that results in an $NPV = 0$.

$$NPV = 0 = -\$60 + (\$10)(P/F, i, 1) + (\$12)(P/A, i, 9)(P/F, i, 1)$$

Solving for *i* yields $i = DCFROR = 14.3$ %.

Values obtained for *NPV* and *DCFROR* are as follows:

	***NPV* (*i* = 10%)**	***DCFROR***
Project A	11.9	14.3%
Project B	15.2	12.9%
Project C	15.6	13.3%

Note: Projects A, B, and C are mutually exclusive because we cannot invest in more than one of them, due to our cap of $\$120 \times 10^6$. The analysis that follows is limited to projects of this type. For the case when projects are not mutually exclusive, the analysis becomes somewhat more involved and is not covered here.

Although all the projects in Example 10.4 showed a positive *NPV* and a *DCFROR* of more than 10%, at this point it is not clear how to select the most attractive option with this information. We will see later that the choice of the project with the highest *NPV* will be the most attractive. However, let us consider the following alternative analysis. If Project B is selected, a total of $\$120 \times 10^6$ is invested and yields 12.9%, whereas the selection of Project A yields 14.3% on the $\$60 \times 10^6$ invested. To compare these two options, we would have to consider a situation in which the same amount is invested in both cases. In Project A, this would mean that $\$60 \times 10^6$ is invested in the project and the remaining $\$60 \times 10^6$ is invested elsewhere, whereas in Project B, a total of $\$120 \times 10^6$ is invested in the project.

It is necessary in our analysis that we are sure that the last dollar invested earns at least 10%. To do this we must perform an incremental analysis on the cash flows and establish that at least 10% is made on each additional increment of money invested in the project.

Example 10.5

This is a continuation of Example 10.4.

a. Determine the *NPV* and the *DCFROR* for each increment of investment.

b. Recommend the best option.

Solution

a. Project A to Project C:

Incremental investment is $\$40 \times 10^6 = (\$100 - \$60) \times 10^6$.
Incremental cash flow for $i = 1$ is $\$2 \times 10^6/\text{yr} = (\$12/\text{yr} - \$10/\text{yr}) \times 10^6$.
Incremental cash flow for $i = 2$ to 10 is $\$8 \times 10^6/\text{yr} = (\$20/\text{yr} - \$12/\text{yr}) \times 10^6$.

$$NPV = -\$40 \times 10^6 + (\$2 \times 10^6)(P/F, 0.10, 1) + (\$8 \times 10^6)(P/A, 0.10, 9)(P/F, 0.10, 1)$$
$$NPV = \$3.7 \times 10^6$$

Setting $NPV = 0$ yields $DCFROR = 0.119$ (11.9%).

Project C to Project B:

Incremental investment is $\$20 \times 10^6 = (\$120 - \$100) \times 10^6$.
Incremental cash flow for $i = 1$ is $\$10 \times 10^6/\text{yr} = (\$22/\text{yr} - \$12/\text{yr}) \times 10^6$.
Incremental cash flow for $i = 2$ to 10 is $\$2 \times 10^6/\text{yr} = (\$22/\text{yr} - \$20/\text{yr}) \times 10^6$.

$$NPV = -\$0.4 \times 10^6 \text{ and } DCFROR = 0.094\ (9.4\%)$$

b. It is recommended that we move ahead on Project C.

From Example 10.5, it is clear that the rate of return on the $\$20 \times 10^6$ incremental investment required to go from Project C to Project B did not return the 10% required, and gave a negative *NPV*.

The information from Example 10.4 shows that an overall return on investment of more than 10% is obtained for each of the three projects. However, the correct choice, Project C, also has the highest *NPV* using a discount rate of 10%, and it is this criterion that should be used to compare alternatives.

When comparing mutually exclusive investment alternatives, choose the alternative with the greatest positive net present value.

When carrying out an incremental investment analysis on projects that are mutually exclusive, the following four-step algorithm is recommended:

Step 1: Establish the minimum acceptable rate of return on investment for such projects.
Step 2: Calculate the *NPV* for each project using the interest rate from Step 1.
Step 3: Eliminate all projects with negative *NPV* values.
Step 4: Of the remaining projects, select the project with the highest *NPV*.

10.4 ESTABLISHING ACCEPTABLE RETURNS FROM INVESTMENTS: THE CONCEPT OF RISK

Most comparisons of profitability will involve the rate of return of an investment. Company management usually provides several **benchmarks** or **hurdle rates** for acceptable rates of return that must be used in comparing alternatives.

A company vice president (VP) has been asked to recommend which of the following two alternatives to pursue.

Option 1: A new product is to be produced that has never been made before on a large scale. Pilot plant runs have been made, and the products sent to potential customers. Many of these customers have expressed an interest in the product but need more material to evaluate it fully. The calculated return on the investment for this new plant is 33%.

Option 2: A second plant is to be built in another region of the country to meet increasing demand in the region. The company has a dominant market position for this product. The new facility would be similar to other plants. It would involve more computer control, and attention will be paid to meeting pending changes in environmental regulations. The rate of return is calculated to be 12%.

The recommendation of the VP and the justifications are given below.

Items that favor Option 1 if pursued:

- High return on the investment
- Opens new product possibilities

Items that favor Option 2:

- The market position for Option 2 is well established. The market for the new product has not been fully established.
- The manufacturing costs are well known for Option 2 but are uncertain for the new process because only estimates are available.
- Transportation costs will be less than current values due to the proximity of plant.
- The technology used in Option 2 is mature and well known to us. For the new process there is no guarantee that it will work.

The closing statement from the VP included the following summary:

"We have little choice but to expand our established product line. If we fail to build these new production facilities, our competitors are likely to build a new plant in the region to meet the increasing demand. They could undercut our regional prices, and this would put at risk our market share and dominant market position in the region."

Clearly, the high return on investment for Option 1 was associated with a high risk. This is usually the case. There are often additional business reasons that must be considered prior to making the final decision. The concern for lost market position is a serious one and weighs heavily in any decision. The relatively low return on investment of 12% given in this example would probably not be very attractive had it not been for this concern. It is the job of company management to weigh all of these factors, along with the rate of return, in order to make the final decision.

In this chapter, we often refer to "internal interest/discount rates" or "internal rates of return." This deals with benchmark interest rates that are to be used to make profitability evaluations. There are likely to be different values that reflect dissimilar conditions of risk—that is, the value for mature technology would differ from that for unproven technology. For example, the internal rate of return for mature technology might be set at 12%, whereas that for very new technology might be set at 40%. Using these values the decision by the VP given above seems more reasonable. The analysis of risk is considered in Section 10.7.

10.5 EVALUATION OF EQUIPMENT ALTERNATIVES

Often during the design phases of a project, it will be necessary to evaluate different equipment options. Each alternative piece of equipment performs the same process function. However, the capital cost, operating cost, and equipment life may be different for each, and we must determine which is the best choice using some economic criterion.

Clearly, if there are two pieces of equipment, each with the same expected operating life, that can perform the desired function with the same operating cost, then *common sense* tells us that we should choose the *less expensive alternative!* When the expected life and operating expenses vary, the selection becomes more difficult. Techniques available to make the selection are discussed in this section.

10.5.1 Equipment with the Same Expected Operating Lives

When the operating costs and initial investments are different but the equipment lives are the same, then the choice should be made based on *NPV*. The choice with the least negative *NPV* will be the best choice. Examples 10.6 and 10.7 illustrate evaluation of equipment alternatives.

Example 10.6

In the final design stage of a project, the question has arisen as to whether to use a water-cooled exchanger or an air-cooled exchanger in the overhead condenser loop of a distillation tower. The information available on the two pieces of equipment is provided as follows:

	Initial Investment	Yearly Operating Cost
Air-cooled	\$23,000	\$1,200
Water-cooled	\$12,000	\$3,300

Both pieces of equipment have service lives of 12 years. For an internal rate of return of 8% p.a., which piece of equipment represents the better choice?

The *NPV* for each exchanger is evaluated below.

$$NPV = -\text{Initial Investment} + (\text{Operating Cost})(P/A, 0.08, 12)$$

	NPV
Air-cooled	–\$32,040
Water-cooled	–\$36,870

The air-cooled exchanger represents the better choice.

Despite the higher capital investment for the air-cooled exchanger in Example 10.6, it was the recommended alternative. The lower operating cost more than compensated for the higher initial investment.

10.5.2 Equipment with Different Expected Operating Lives

When process units have different expected operating lives, we must be careful how the best choice is determined. When we talk about expected equipment life, it is assumed that this is less than the expected working life of the plant. Therefore, during the normal operating life of the plant, we can expect to replace the equipment at least once. This requires that different profitability criteria be applied. Three commonly used methods are presented to evaluate this situation. (The effect of inflation is not considered in these methods.) All methods consider both the capital and operating cost in minimizing expenses, thereby maximizing our profits.

Capitalized Cost Method. In this method, a fund is established for each piece of equipment that we wish to compare. This fund provides the amount of cash that would be needed to

a. Purchase the equipment initially
b. Replace it at the end of its life
c. Continue replacing it forever

The size of the initial fund and the logic behind the capitalized cost method are illustrated in Figure 10.4.

From Figure 10.4, we can see that if the equipment replacement cost is P, then the total fund set aside (called the **capitalized cost**) is $P + R$, where R is

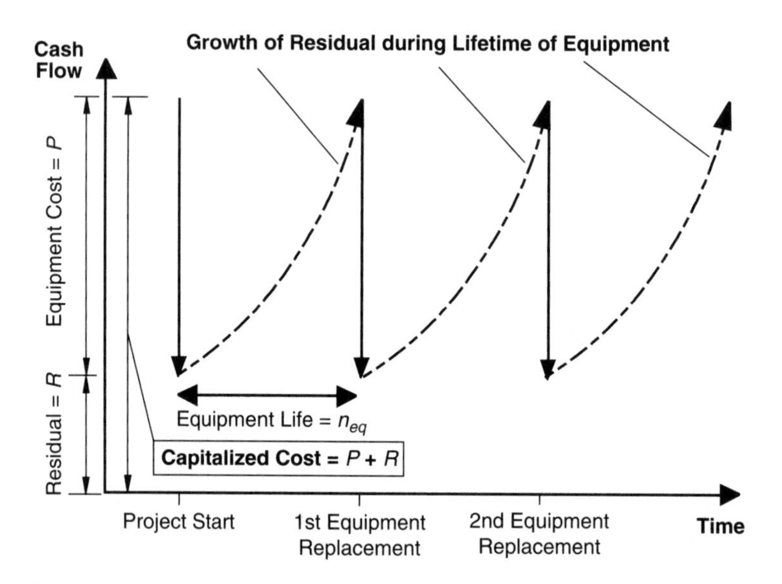

Figure 10.4 An Illustration of the Capitalized Cost Method for the Analysis of Equipment Alternatives

termed the **residual.** The purpose of this residual is to earn sufficient interest during the life of the equipment to pay for its replacement. At the end of the equipment life, n_{eq}, the amount of interest earned is P, the equipment replacement cost. As Figure 10.4 shows, we may continue to replace the equipment every time it wears out. Referring to Figure 10.4, we develop the equation for the capitalized cost defined as $(P + R)$:

$$R(1 + i)^{n_{eq}} - R = P$$

and

$$\text{Capitalized Cost} = P + R = P + \frac{P}{(1 + i)^{n_{eq}} - 1} = P\left[\frac{(1 + i)^{n_{eq}}}{(1 + i)^{n_{eq}} - 1}\right] \quad (10.1)$$

The term in square brackets in Equation (10.1) is commonly referred to as the **capitalized cost factor.**

The capitalized cost obtained from Equation (10.1) does not include the operating cost and is useful in comparing alternatives only when the operating costs of the alternatives are the same. When operating costs vary, it is necessary to capitalize the operating cost. An equivalent capitalized operating cost that converts the operating cost into an equivalent capital cost is added to the capitalized cost calculated from Equation 10.1 to provide the **equivalent capitalized cost (ECC):**

$$\text{Equivalent Capitalized Cost} = \text{Capitalized Cost} + \text{Capitalized Operating Cost}$$

$$\text{Equivalent Capitalized Cost} = \left[\frac{P(1 + i)^{n_{eq}} + YOC(F/A, i, n_{eq})}{(1 + i)^{n_{eq}} - 1}\right] \quad (10.2)$$

This cost considers both the capital cost of equipment and **yearly operating cost (YOC)** needed to compare alternatives. The extra terms in Equation (10.2) represent the effect of taking the yearly cash flows for operating costs from the residual, R.

By using Equation (10.1) or (10.2), we correctly account for the different operating lives of the equipment by calculating an effective capitalized cost for the equipment and operating cost. Example 10.7 illustrates the use of these equations.

Example 10.7

During the design of a new project, we are faced with a decision regarding which type of pump should be used for a corrosive service. Our options are as follows.

	Capital Cost	Operating Cost (per Year)	Equipment Life (Years)
Carbon steel pump	\$8000	\$1800	4
Stainless steel pump	\$16,000	\$1600	7

Assume a discount rate of 8% p.a.
Using Equation (10.2) for the carbon steel pump,

$$\text{Capitalized Cost} = \frac{(8000)(1.08)^4 + (1800)\dfrac{[1.08^4 - 1]}{0.08}}{1.08^4 - 1} = \$52{,}700$$

For the stainless steel pump,

$$\text{Capitalized Cost} = \frac{(16{,}000)(1.08)^7 + (1600)\dfrac{[1.08^7 - 1]}{0.08}}{1.08^7 - 1} = \$58{,}400$$

The carbon steel pump is recommended because it has the lower capitalized cost.

In Example 10.7, the stainless steel pump costs twice as much as the carbon steel pump and, because of its superior resistance to corrosion, will last nearly twice as long. In addition, the operating cost for the stainless steel pump is lower due to lower maintenance costs. In spite of these advantages, the carbon steel pump was still judged to offer a cost advantage.

Equivalent Annual Operating Cost (*EAOC*) Method. In the previous method, we lumped both capital cost and yearly operating costs into a single cash fund or equivalent cash amount. An alternative method is to **amortize** (spread out) the capital cost of the equipment over the operating life to establish a yearly cost. This is added to the operating cost to yield the *EAOC*.

Figure 10.5 illustrates the principles behind this method. From the figure, we can see that the cost of the initial purchase will be spread out over the operating life of the equipment. The *EAOC* is expressed by Equation (10.3).

$$EAOC = (\text{Capital Investment})(A/P, i, n_{eq}) + YOC \tag{10.3}$$

Example 10.8 illustrates this method.

Example 10.8

Compare the stainless steel and carbon steel pumps in Example 10.7 using the *EAOC* method.

For the carbon steel pump,

$$EAOC = \frac{(8000)(0.08)(1.08)^4}{1.08^4 - 1} + 1800 = \$4220 \text{ per year}$$

For the stainless steel pump,

$$EAOC = \frac{(16{,}000)(0.08)(1.08)^7}{1.08^7 - 1} + 1600 = \$4670 \text{ per year}$$

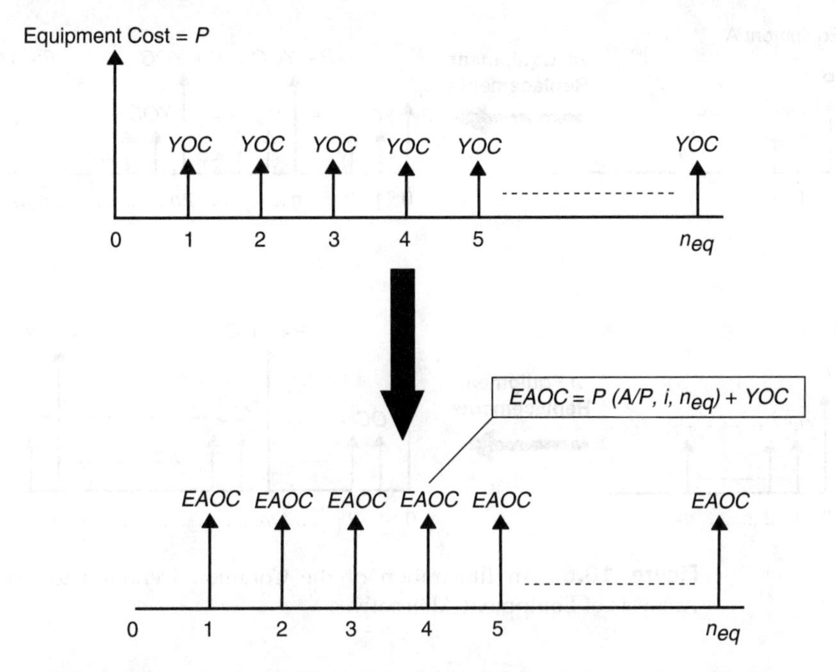

Figure 10.5 Cash Flow Diagrams Illustrating the Concept of Equivalent Annual Operating Cost

The carbon steel pump is shown to be the preferred equipment using the *EAOC* method, as it was in Example 10.7 using the *ECC* method.

Common Denominator Method. Another method for comparing equipment with unequal operating lives is the **common denominator method.** This method is illustrated in Figure 10.6, in which two pieces of equipment with operating lives of n and m years are to be compared. This comparison is done over a period of nm years during which the first piece of equipment will need m replacements and the second will require n replacements. Each piece of equipment has an integer number of replacements, and the time over which the comparison is made is the same for both pieces of equipment. For these reasons the comparison can be made using the net present value of each alternative. In general, an integer number of replacements can be made for both pieces of equipment in a time N, where N is the smallest number into which m and n are both exactly divisible; that is, N is the common denominator. Example 10.9 illustrates this method.

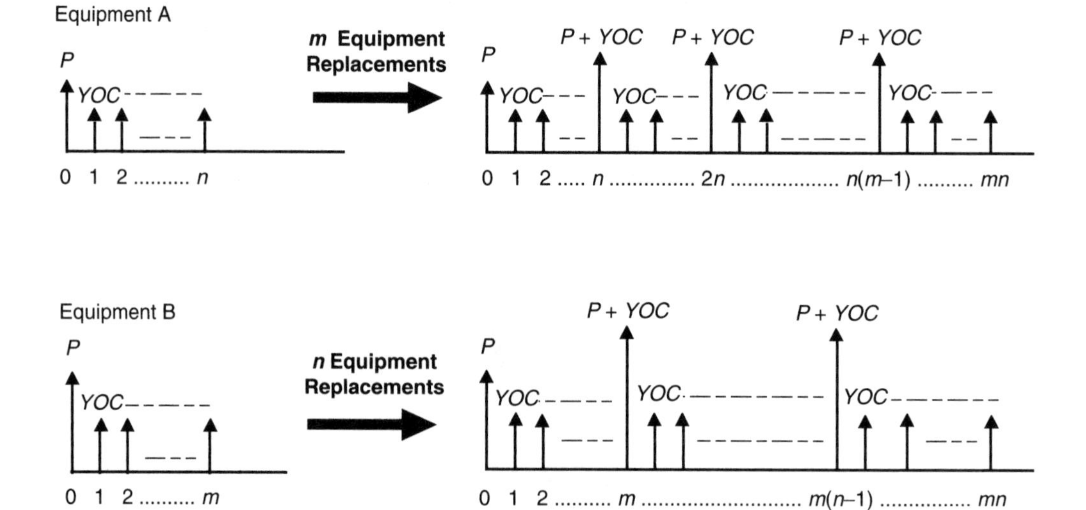

Figure 10.6 An Illustration of the Common Denominator Method for the Analysis of Equipment Alternatives

---

Example 10.9

Compare the two pumps given in Example 10.7 using the common denominator method. The discrete cash flow diagrams for the two pumps are shown in Figure E10.9. The minimum time over which the comparison can be made is 4(7) = 28 years.

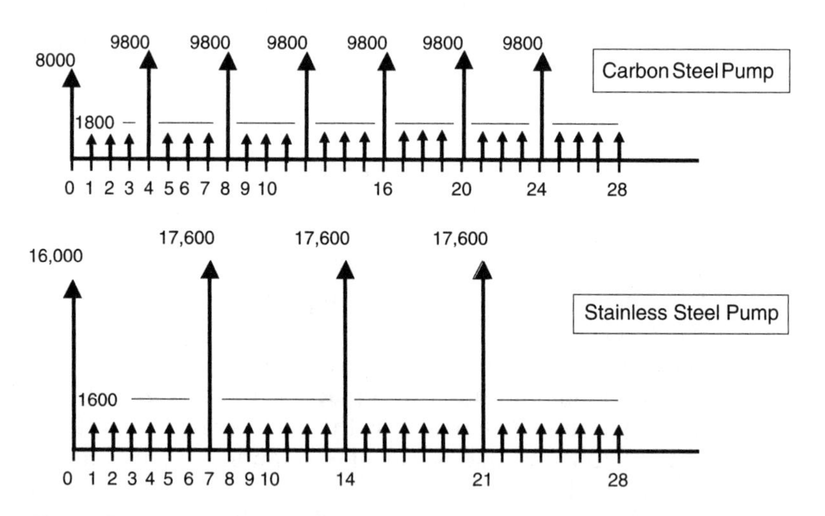

Figure E10.9 Cash Flow Diagrams for the Common Denominator Method Used in Example 10.9

NPV for carbon steel pump:

$$NPV = -(\$8000)(1 + 1.08^{-4} + 1.08^{-8} + 1.08^{-12} + 1.08^{-16} + 1.08^{-20} + 1.08^{-24}) - (\$1800)(P/A, 0.08, 28) = -\$46{,}580$$

NPV for stainless steel pump:

$$NPV = -(\$16{,}000)(1 + 1.08^{-7} + 1.08^{-14} + 1.08^{-21}) - (\$1600)(P/A, 0.08, 28) = -\$51{,}643$$

The carbon steel pump has a less negative *NPV* and is recommended.

As found for the previous two methods, the common denominator method favors the carbon steel pump.

Choice of Methods. Because all three methods of comparison correctly take into account the time value of money, the results of all the methods are equivalent. In most problems, the common denominator method becomes unwieldy. We favor the use of the *EAOC* or the capitalized cost methods for our calculations.

10.6 INCREMENTAL ANALYSIS FOR RETROFITTING FACILITIES

This topic involves profitability criteria used for analyzing situations where a piece of equipment is added to an existing facility. The purpose of adding the equipment is to improve the profitability of the process. Such improvements are often referred to as **retrofitting.** Such retrofits may be extensive—requiring millions of dollars of investment—or small, requiring an investment of only a few thousand dollars.

The decisions involved in retrofitting projects may be of the discrete type, the continuous type, or a combination of both. An example of a discrete decision is whether to add an on-line monitoring and control system to a waste water stream. The decision is a simple yes or no. An example of a continuous decision is to determine what size of heat recovery system should be added to an existing process heater to improve fuel efficiency. This type of decision would involve sizing the optimum heat exchanger, where the variable of interest (heat-exchanger area) is continuous.

Because retrofit projects are carried out on existing operating plants, it becomes necessary to identify all of the costs and savings associated with the retrofit. When comparing alternative schemes, the focus of attention is on the profitability of the incremental investment required. We will consider simple discrete choices in this section. The problem of optimizing a continuous variable is covered in Chapter 14.

The initial step in an incremental analysis of competing alternatives is to identify the potential alternatives to be considered and to specify the increments over which the analysis is to be performed. Our first step is to rank the available alternatives by the magnitude of the capital cost. We will identify the alternatives as $A_1, A_2, \ldots, A_n$. There are n possible alternatives. The first alternative, A_1, which is always available, is the "do nothing" option. It requires no capital cost (and achieves no savings). For each of the available alternatives, the project cost (capital cost), *PC*, and the yearly savings generated (yearly cash flow), *YS*, must be known.

For larger retrofit projects, discounted profitability criteria should be used. The algorithm to compare alternatives using discounted cash flows follows the same four-step method outlined in Section 10.3. For small retrofit projects, nondiscounted criteria may often be sufficiently accurate for comparing alternatives. Both types of criteria are discussed in the next sections.

10.6.1 Nondiscounted Methods for Incremental Analysis

For nondiscounted analyses, two methods are provided below.

1. Rate of Return on Incremental Investment (*ROROII*)

$$ROROII = \frac{\text{Incremental Yearly Savings}}{\text{Incremental Investment}}$$

2. Incremental Payback Period (*IPBP*)

$$IPBP = \frac{\text{Incremental Investment}}{\text{Incremental Yearly Savings}}$$

Examples 10.10–10.12 illustrate the method of comparison of projects using these two criteria.

Example 10.10

A circulating heating loop for an endothermic reactor has been in operation for several years. Due to an oversight in the design phase, a certain portion of the heating loop piping was left uninsulated. The consequence is a significant energy loss. Two types of insulation can be used to reduce the heat loss. They are both available in two thicknesses. The estimated cost of the insulation and the estimated yearly savings in energy costs are given below. (The ranking has been added to the alternatives and is based on increasing project cost).

Ranking (Option #)	Alternative Insulation	Project Cost (*PC*)	Savings Generated by Project (*YS*)
1	No Insulation	0	0
2	B-1″ Thick	$3000	$1400
3	B-2″ Thick	$5000	$1900
4	A-1″ Thick	$6000	$2000
5	A-2″ Thick	$9700	$2400

Assume an acceptable internal rate of return for a nondiscounted profitability analysis to be 15% (0.15).

a. For the four types of insulation determine the rate of return on incremental investment (*ROROII*) and the incremental payback period (*IPBP*).

b. Determine the value of the incremental payback period equivalent to the 15% internal rate of return.

Solution

a. Evaluation of *ROROII* and *IPBP*.

Option #-Option 1	*ROROII*	*IPBP* (years)
2-1	$1400/$3000 = 0.47 (47%)	$3000/$1400 = 2.1
3-1	$1900/$5000 = 0.38 (38%)	$5000/$1900 = 2.6
4-1	$2000/$6000 = 0.33 (33%)	$6000/$2000 = 3.0
5-1	$2400/$9700 = 0.25 (25%)	$9700/$2400 = 4.0

b. $IPBP = 1/(ROROII) = 1/0.15 = 6.67$ yrs

Note that in Part (a) of Example 10.10, the incremental investment and savings are given by the difference between installing the insulation and doing nothing. All of the investments considered in Example 10.10 satisfied the internal benchmark for investment of 15%, which means that the do-nothing option (Option 1) can be discarded. However, which of the remaining options is the best can be determined only using pairwise comparisons.

Example 10.11

Which of the options in Example 10.10 is the best based on the nondiscounted *ROROII* of 15%?

Step 1: Choose Option 2 as the base case, because it has the lowest capital investment.

Step 2: Evaluate incremental investment and incremental savings in going from the base case to the case with the next higher capital investment, Option 3.

Incremental Investment = (\$5000 – \$3000) = \$2000
Incremental Savings = (\$1900/yr – \$1400/yr) = \$ 500/yr
ROROII = 500/2000 = 0.4, or 40% per year

Step 3: Because the result of Step 2 gives an *ROROII* > 15%, we use Option 3 as the base case and compare it with the option with the next higher capital investment, Option 4.

Incremental Investment = (\$6000 – \$5000) = \$1000
Incremental Savings = (\$2000/yr – \$1900/yr) = \$100/yr
ROROII = 100/1000 = 0.1 or 10% per year

Step 4: Because the result of Step 3 gives an *ROROII* < 0.15, we reject Option 4 and compare Option 3 with the option with the next higher capital investment, Option 5.

Incremental Investment = (\$9700 – \$5000) = \$4700
Incremental Savings = (\$2400/yr – \$1900/yr) = \$500/yr
ROROII = 500/4700 = 0.106, or 10.6%.

Step 5: Again, the *ROROII* from Step 4 is less than 15%, and hence we reject Option 5. Because Option 3 (Insulation B-2" thick) is the current base case and no more comparisons remain, we accept Option 3 as the "best option."

It is important to note that in Example 10.11 Options 4 and 5 are rejected even though they give *ROROII* greater that 15% when compared with the do-nothing option (see Example 10.10). The key here is that in going from Option 3 to either Option 4 or 5 the incremental investment loses money, that is, *ROROII* < 15%.

Example 10.12

Repeat the comparison of options in Example 10.10 using a nondiscounted incremental payback period of 6.67 years.

The steps are similar to those used in Example 10.11 and are given below without further explanation.

Step 1: (Option 3 – Option 2) *IPBP* = 2000/500 = 4 years < 6.67
Reject Option 2; Option 3 becomes the base case.

Step 2: (Option 4 – Option 3) *IPBP* = 1000/10 = 10 years > 6.67
Reject Option 4.

Step 3: (Option 5 – Option 3) *IPBP* = 4700/500 = 9.4 years > 6.67
Reject Option 5.

Option 3 is the best option.

10.6.2 Discounted Methods for Incremental Analysis

Incremental analyses taking into account the time value of money should always be used when large capital investments are being considered. Comparisons may be made either by discounting the operating costs to yield an equivalent capital investment or by amortizing the initial investment to give an equivalent annual operating cost. Both techniques are considered in the following section, where the effects of depreciation and taxation are ignored in order to keep the analysis simple. However, it is an easy matter to take these effects into account.

Capital Cost Methods. The incremental net present value (*INPV*) for a project is given by

$$INPV = -PC + YS(P/A, i, n) \tag{10.4}$$

When comparing investment options, a given case option will always be compared with a do-nothing option. Thus, these comparisons may be considered as incremental investments. In order to use *INPV*, it is necessary to know the internal discount rate and the time over which the comparison is to be made. This method is illustrated in Example 10.13.

Example 10.13

Based on the information provided in Example 10.10 for an acceptable internal interest rate of 15% and time n = 5 yrs, determine the most attractive alternative, using the *INPV* criterion to compare options.

For i = 0.15 and n = 5 the value for $(P/A, i, n)$ = 3.352. (See Equation 9.14.)

Equation (10.4) becomes

$$NPV = -PC + 3.352\ YS$$

Option	***INPV***
2-1	1693
3-1	1369
4-1	704
5-1	–1655

From the results above, it is clear that Options 2, 3, and 4 are all potentially profitable as $INPV > 0$. However, the best option is Option 2 because it has the highest *INPV* when compared with the do-nothing case, Option 1. Note that other pairwise comparisons are unnecessary. We simply choose the option that yields the highest *INPV*. The reason that the *INPV* gives the best option directly is because by knowing i and n, each dollar of incremental investment is correctly accounted for in the calculation of *INPV*. Thus, if the incremental investment in going from Option A to Option B is profitable, then the *INPV* will be greater for Option B and vice versa. It should also be pointed out that by using discounting techniques, the best option has changed from Option 3 (in Example 10.11) to Option 2.

Operating Cost Methods. In the previous section, yearly savings were converted to an equivalent present value using the present value of an annuity, and this was measured against the capital cost. An alternative method is to convert all the investments to annual costs using the capital recovery factor and measure them against the yearly savings.

We develop the needed relationship from Equation (10.4), giving

$$INPV/(P/A, i, n) = -PC/(P/A, i, n) + YS$$

It can be seen that the capital recovery factor $(A/P,i,n)$ is the reciprocal of the present worth factor $(P/A,i,n)$. Substituting this relationship and multiplying by –1 gives

$$-(INPV)(A/P, i, n) = (PC)(A/P, i, n) - YS$$

The term on the left is identified as the Equivalent Annual Operating Cost (*EAOC*). Thus, we may write

$$EAOC = (PC)(A/P, i, n) - YS \tag{10.5}$$

When an acceptable rate for i and n is substituted, a negative *EAOC* indicates the investment is acceptable (because a negative cost is the same as a positive savings). Use of EAOC is demonstrated in Example 10.14.

Example 10.14

Repeat Example 10.13 using *EAOC* in place of *NPV*.

For $i = 0.15$ and $n = 5$ the value for $(A/P, i, n) = 1/3.352$.

Equation (10.5) becomes

$EAOC = PC/3.352 - YS$

Option	*EAOC*
2–1	–505
3–1	–408
4–1	–210
5–1	494

The best alternative is Option 2 because it has the most negative *EAOC*.

10.7 EVALUATION OF RISK IN EVALUATING PROFITABILITY

In this section, the concept of risk in the evaluation of profitability is introduced, and the techniques to quantify it are illustrated. Until now, it has been assumed that the financial analysis is essentially deterministic—that is, all factors are known with absolute certainty. Recalling discussions in Chapter 7 regarding the relative error associated with capital cost estimates, it should not be surprising that many of the costs and parameters used in evaluating the profitability of a chemical process are estimates that are subject to error. In fact, nearly all of these factors are subject to change throughout the life of the chemical plant. The question then is not, "Do these parameters change?" but rather, "By how much do they change?" In Table 10.1, due to Humphreys [1], ranges of expected variations for factors that affect the prediction and forecasting of profitability are given. In reviewing the prediction or forecasting of factors affecting the profitability of chemical processes, Humphreys [1] gives the ranges of expected variations shown in Table 10.1.

The most important variable in Table 10.1 is sales volume, with the price of product and raw material being a close second. Clearly, if market forces were such that we were able to sell (and hence produce) only 50% of the originally estimated amount of product, then profitability would be affected greatly. Indeed, the process would quite possibly be unprofitable. The problem is that projections

Table 10.1 Range of Variation of Factors Affecting the Profitability of a Chemical Process

Factor in Profitability Analysis	Probable Variation from Forecasts over 10-year Plant Life, %
Cost of fixed capital investment*	–10 to +25
Construction time	–5 to +50
Start-up costs and time	–10 to +100
Sales volume	–50 to +150
Price of product	–50 to +20
Plant replacement and maintenance costs	–10 to +100
Income tax rate	–5 to +15
Inflation rates	–10 to +100
Interest rates	–50 to + 50
Working capital	–20 to +50
Raw material availability and price	–25 to +50
Salvage value	–100 to +10
Profit	–100 to +10

*For capital cost estimations using CAPCOST, a more realistic range is –20 to +30%

(From *Jelen's Cost and Optimization Engineering,* 3rd ed., edited by K. K. Humphreys (1991), reproduced by permission of The McGraw-Hill Companies, Inc.)

of how the variables will vary over the life of the plant are difficult (and sometime impossible) to estimate. Nevertheless, experienced cost estimators often have a feel for the variability of some of these parameters. In addition, marketing and financial specialists within large companies have expertise in forecasting trends in product demand, product price, and raw material costs. In the next section, the effect that supply and demand have on the sales price of a product is investigated. Following this, methods to quantify risk and to predict the range of profitability that can be expected from a process, when uncertainty exists in some of the profitability parameters listed in Table 10.1, will be discussed.

10.7.1 Forecasting Uncertainty in Chemical Processes

In order to be able to predict the way in which the factors in Table 10.1 vary, it is necessary to take historical data along with information about new developments to formulate a model to predict trends in key economic parameters over the projected life of a process. This prediction process is often referred to as forecasting and is, in general, a very inexact science. The purpose of this section is to introduce some concepts that must be considered when quantifying economic projections. A detailed description of the art of economic forecasting is way beyond the scope of this text. Instead, the basic concepts and factors influencing economic parameters are introduced.

Supply and Demand Concepts in Chemical Markets. Economists use microeconomic theory [2] to describe how changes in the supply of and demand for a given product are affected by changes in the market. Only the most basic supply and demand curves, shown in Figure 10.7, are considered here.

The demand curve (on the left) slopes downward and shows the general trend that as the price for commodity *X* decreases, the demand increases. With very few exceptions, this is always true. Examples of chemical products following this trend are numerous; for example, as the price of gasoline, polyethylene, or fertilizer drops, the demand for these goods increases (all other factors remaining constant). The supply curve (shown on the right) slopes upward and shows the trend that as the price rises, the amount of product X that manufacturers are willing to produce increases. The slope of the supply curve is often positive but may also be negative depending on the product. For most chemical products, it can be assumed that the slope is positive, and with all other factors remaining constant, the quantity supplied increases as the price for the product increases. Unlike physical laws that govern thermodynamics, heat transfer, and so on, these trends are not absolute. Instead, these trends reflect human nature relating to buying and selling of goods.

When market forces are in equilibrium, the supply and demand for a given product are balanced, and the equilibrium price (P_{eq}) is determined by the intersection of the supply and demand curves, as shown in Figure 10.8.

There are many factors that can affect the market for product X. Indeed, a market may not be in equilibrium, and, in this case, the market price must be

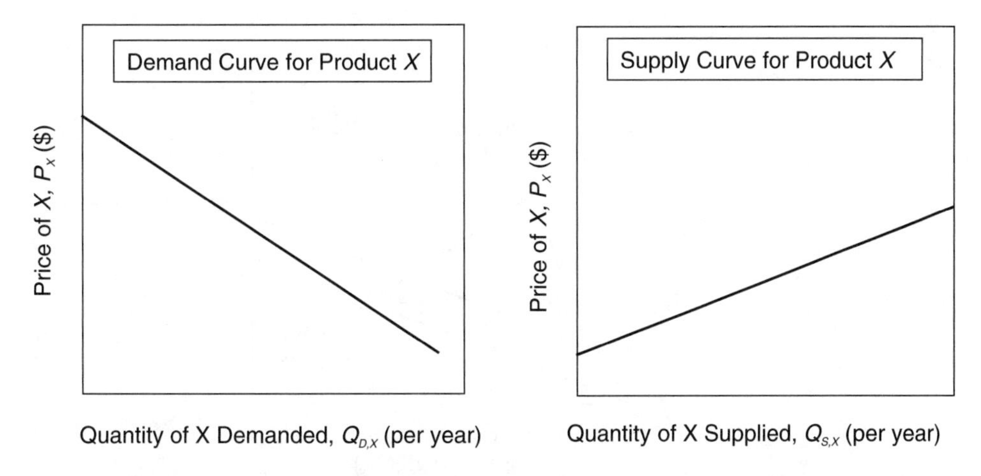

Figure 10.7 Simple Supply and Demand Curves for Product *X*

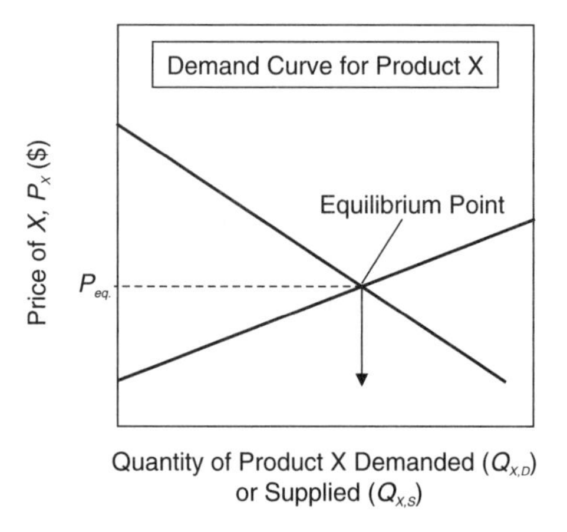

Figure 10.8 Illustration of Market Equilibrium for Product X

determined in terms of rate equations as opposed to equilibrium relationships. However, for the sake of this simplified discussion, it will be assumed that market equilibrium is always reached. If something changes in the market, either the supply or demand curve (or both) will shift, and a new equilibrium point will be reached. As an example, consider the situation when a large new plant that produces X comes on line. Assuming that nothing else in the market changes, the supply curve will be shifted downward and to the right, which will lead to a lower equilibrium price. This situation is illustrated in Figure 10.9. The intersection of the demand curve and the new supply curve gives rise to the new equilibrium price, $P_{eq,2}$, which is lower than the original equilibrium price, $P_{eq,1}$. The magnitude of the decrease in the equilibrium price depends on the magnitude of

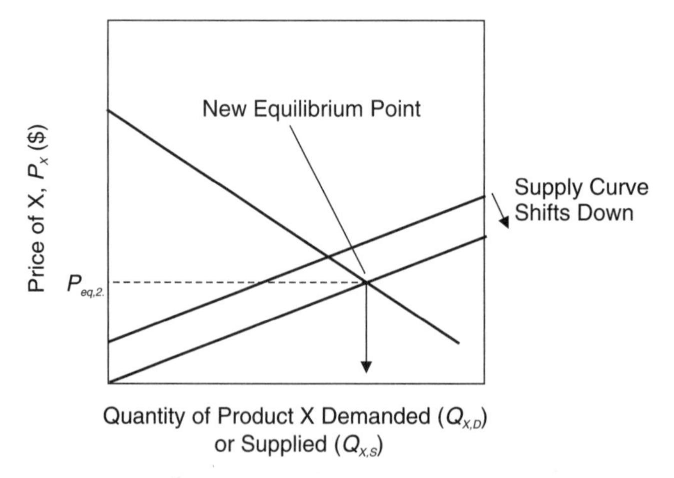

Figure 10.9 Illustration of Market Equilibrium for Product X When a New Plant Comes On-Line

the downward shift in the supply curve. If the new plant is large compared with the total current manufacturing capacity for product X, then the decrease in the equilibrium price will be correspondingly large. If this decrease in price is not taken into account in the economic analysis, the projected profitability of the new project will be overestimated, and the decision to invest might be made when the correct decision would be to abandon the project.

The situation is further complicated when competing products are considered. For example, if product Y can be used as a substitute for product X in some applications, then factors that affect Y will also affect X. It is easy to see that quantifying and predicting changes become very difficult. Torries [3] identifies important factors that affect both the shape and relative location of the supply and demand curves. These factors are listed in Table 10.2.

In order to forecast accurately the prices of a product over a 10- or 15-year project, the factors in Table 10.2 need to be predicted. Clearly, even for the most well-known and stable products, this can be a daunting task. An alternative method to quantifying the individual supply and demand curves is to look at historical data for the product of interest.

The examination of historical data is a convenient way to obtain general trends in pricing. Such data represent the change in equilibrium price for a product with time. Often such data fluctuate widely, and although long-term trends may be apparent, predictions for the next one or two years will often be wildly inaccurate. For example, consider the data for average gasoline prices over the period January 1996 to June 2007, as shown in Figure 10.10. The straight line is a regression through the data and represents the best linear fit of the data. If this were the forecast for gasoline prices over this period, it would be a remarkably good prediction. However, even with this predicting line, significant variations in actual product selling price are noted. The maximum positive and negative deviations are +77¢/gal (+33%) and –54¢/gal (–50%). To illustrate further the effect of these deviations on profitability calculations, consider a new refinery starting

Table 10.2 Factors Affecting the Shape and Relative Location of the Supply and Demand Curves

Factors Influencing Supply	Factors Influencing Demand
Cost and amount of labor	Price of the product
Cost and amount of energy	Price of all substitute products
Cost and amount of raw material	Consumer disposable income
Cost of fixed capital (interest rates) and amount of fixed capital	Consumer tastes
Other miscellaneous factors	Manufacturing technology
	Other miscellaneous factors

(From Torries, T. F., *Evaluating Mineral Projects: Applications and Misconceptions,* by permission of SME, Littleton, CO, 1998; www.smenet.org)

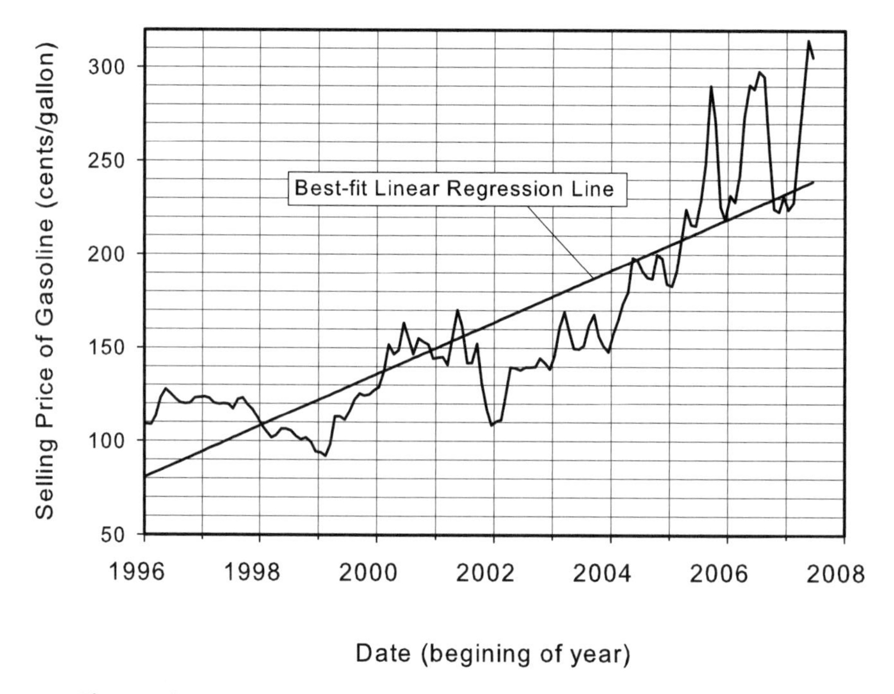

Figure 10.10 Average Price of All Grades of Gasoline over the Period January 1996 to June 2007 (from www.eia.doe.gov)

production in late 2001. For this new plant, the selling price for its major product (gasoline) over the two-year period after start-up is significantly below that predicted by the trend line. If this refinery were contracted to buy crude oil at a price fixed previously, then the profitability of the plant over this initial two-year period would be severely diminished and it would probably lose money.

From the brief discussion given above, it is clear that predicting or forecasting future prices for chemical products is a very inexact and risky business. Perhaps it is best summed up by a quote attributed to "baseball philosopher" Yogi Berra [4]:

"It's tough to make predictions, especially about the future."

In the following section, it will be assumed that such predictions are available and will be used as given, known quantities. The question then is how much meaning can be placed on the results of these predictions when the input data—the basic variability of the parameters—is often poorly known. The answer is that by investigating and looking at how these parameters affect profitability, a better picture of how this variability or uncertainty affects the overall profitability of a project can be

Table 10.3 Values for Uncertain Parameters for the Scenario Analysis (All $ Figures in Millions)

Parameter	Worst Case	Best Case
Revenue, R	–20% = ($75)(0.8) = $60	+5% = ($75)(1.05) = $78.75
Cost of manufacture, COM_d	+10% = ($30)(1.1) = $33	–10% = ($30)(0.9) = $27
Capital investment, FCI_L	+30% = ($150)(1.3) = $195	–20% = ($150)(0.8) = $120

obtained. In general, this type of information is much more useful than the single-point estimate of profitability that has been considered up to this point.

10.7.2 Quantifying Risk

It should be noted that the quantification of risk in no way eliminates uncertainty. Rather, by quantifying it, a better feel can be developed for how a project's profitability may vary. Therefore, more informed and rational decisions regarding whether to build a new plant can be made. However, the ultimate decision to invest in a new chemical process always involves some element of risk.

Scenario Analysis. Returning to Example 10.1 regarding the profitability analysis for a new chemical plant, assume that, as the result of previous experience with similar chemicals and some forecasting of supply and demand for this new product, it is believed that the product price may vary in the range –20% to +5%, the capital investment may vary between –20% and +30%, and the cost of manufacturing may vary in the range –10% to +10%. How can these uncertainties be quantified?

One way to quantify uncertainty is via a **scenario analysis.** In this analysis, the best- and worst-case scenarios are considered and compared with the base case, which has already been calculated. The values for the three parameters for the two cases are given in Table 10.3.

Next, these values are substituted into the spreadsheet shown in Table E10.1, and all the cash flows are discounted back to the start of the project to estimate the *NPV*. The results of these calculations are shown in Table 10.4.

The results in Table 10.4 show that, in the worst-case scenario, the *NPV* is very negative and the project will lose money. In the best-case scenario, the *NPV* is increased over the base case by approximately $45 million. From this result, the

Table 10.4 Net Present Values (*NPV*s) for the Scenario Analysis (All $ Figures in Millions)

Case	Net Present Value
Worst Case	–$59.64
Base Case	$17.12
Best Case	$53.62

decision on whether to go ahead and build the plant is not obvious. On one hand, the process could be highly profitable, but on the other hand, it could lose nearly \$73 million over the course of the ten-year plant life. By taking a very conservative philosophy, the results of the worst-case scenario suggest a decision of "do not invest." However, is the worst-case scenario realistic? Most likely, the worst-case (best-case) scenario is unduly pessimistic (optimistic). Consider each of the three parameters in Table 10.3. It will be assumed that the value of the parameter has an equal chance of being at the high, base-case, or low value. Therefore, in terms of probabilities, the chance of the parameter taking each of these values is 1/3, or 33.3%. Because there are three parameters (R, FCI_L, and COM_d), each of which can take one of three values (high, base case, low), there are $3^3 = 27$ combinations as shown in Table 10.5.

From Table 10.5, it can be seen that Scenario 9 is the worst case and Scenario 19 is the best case. Either of these two cases has a 1 in 27 (or 3.7%) chance of occurring. Based on this result, it is not very likely that either of these scenarios would occur, and hence we should be careful in evaluating the scenario analysis. This is indeed one of the main shortcomings of the scenario analysis [2]. In reviewing Table 10.5, a better measure of the expected profitability might be the weighted average of all 27 possible outcomes. The idea of weighting results based on the likelihood of occurrence is the basis of the probabilistic approach to quantifying risk that will be discussed shortly. However, before looking at that method, it will be instructive to determine the sensitivity of the profitability of the project to changes in important parameters. Sensitivity analysis is covered in the next section.

Sensitivity Analysis. To a great extent, the risk associated with the variability of a given parameter is dependent on the effect that a change in that parameter has on the profitability criterion of interest. For the sake of this discussion, the *NPV* will be used as the measure of profitability. However, this measure could just as easily be the *DCFROR*, *DPEP*, or any other profitability criterion discussed in Section 10.2. If it is assumed that the *NPV* is affected by n parameters $(x_1, x_2, x_3, \ldots, x_n)$, then the first-order sensitivity to parameter x_1 is given in mathematical terms by the following quantity:

$$S_1 = \left[\frac{\partial (NPV)}{\partial x_1}\right]_{x_2 x_3, \ldots, x_n} \tag{10.6}$$

where the partial derivative is taken with respect to x_1, while holding all other parameters constant at their mean value. The sensitivity, S_1, is sometimes called a **sensitivity coefficient.** In general, this quantity is too complicated to obtain via analytical differentiation; hence, it is obtained by changing the parameter by a small amount and observing the subsequent change in the *NPV*, or

$$S_1 \approx \left[\frac{\Delta (NPV)}{\Delta x_1}\right]_{x_2, x_3, \ldots, x_n} \tag{10.7}$$

In Example 10.15, Example 10.1 is revisited to illustrate how the sensitivities of the revenue, cost of manufacturing, and fixed capital investment on the *NPV* are calculated.

Table 10.5 Possible Combinations of Values for Three Parameters

Scenario	R^*	COM_d^*	FCI_L^*	Probability of Occurrence
1	−20%	−10%	−20%	(1/3)(1/3)(1/3) = 1/27
2	−20%	−10%	0%	
3	−20%	−10%	+30%	
4	−20%	0%	−20%	
5	−20%	0%	0%	
6	−20%	0%	+30%	
7	−20%	+10%	−20%	
8	−20%	+10%	0%	
9 (worst)	**220%**	**+10%**	**+30%**	
10	0%	−10%	−20%	
11	0%	−10%	0%	
12	0%	−10%	+30%	
13	0%	0%	−20%	
14 (base)	**0%**	**0%**	**0%**	
15	0%	0%	+30%	
16	0%	+10%	−20%	
17	0%	+10%	0%	
18	0%	+10%	+30%	
19 (best)	**+5%**	**210%**	**-20%**	
20	+5%	−10%	0%	
21	+5%	−10%	+30%	
22	+5%	0%	−20%	
23	+5%	0%	0%	
24	+5%	0%	+30%	
25	+5%	+10%	−20%	
26	+5%	+10%	0%	
27	+5%	+10%	+30%	(1/3)(1/3)(1/3) = 1/27

*0% refers to the base-case value.

Example 10.15

For the chemical process considered in Example 10.1, calculate the sensitivity of R, COM_d, and FCI_L and plot these sensitivities with respect to the NPV.

We consider the effect of a 1% change (½% on either side of the base case) in each parameter on the NPV. These results are shown in Table E10.15.

Table E10.15 Calculations for Sensitivity Analysis for Example 10.1 (All $ Figures Are in Millions)

Parameter	Value	*NPV*	Value	*NPV*	S_i
x_1 (Revenue, R)	+0.5% ($75.375/yr)	$18.17	−0.5% ($74.625/yr)	$16.07	$\frac{(18.17-16.07)}{(75.375-74.625)}=\frac{2.1}{0.75}=2.80$ yr
x_2 (COM_d)	+0.5% ($30.150/yr)	$16.70	−0.5% ($29.850/yr)	$17.54	$\frac{(16.70-17.54)}{(30.15-29.85)}=\frac{-0.84}{0.30}=-2.80$ yr
x_3 (FCI_L)	+0.5% ($150.75)	$16.68	−0.5% ($149.25)	$17.56	$\frac{(16.68-17.56)}{(150.75-149.25)}=\frac{-0.88}{1.50}=-0.59$ yr

The fact that $S_1 = -S_2$ should not be surprising because, in the calculation of yearly cash flows, whenever R appears COM_d is subtracted from it (see Table E10.1). The changes in NPV for percent changes in each parameter are illustrated in Figure E10.15. The slopes

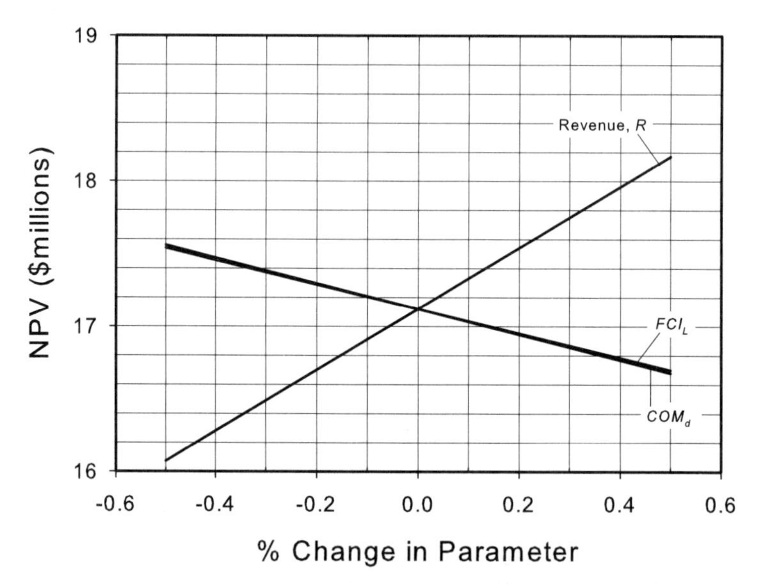

Figure E10.15 Sensitivity Curves for the Parameters Considered in Example 10.15

of the lines are not equal to the sensitivities, because the x-axis is the percent change rather than the actual change in the parameter.

How can the sensitivity values calculated above be used to estimate changes in the profitability criterion of the process? For small changes in the parameters, it may be assumed that the sensitivities are constant and can be added. Therefore, the change in *NPV* can be predicted for a set of changes in the parameters using the following formula:

$$\Delta NPV = S_1\Delta x_1 + S_2\Delta x_2 + \ldots + S_n\Delta x_n \tag{10.8}$$

Example 10.16 illustrates this concept.

Example 10.16

What is the change in the *NPV* for a 2% increase in revenue coupled with a 3% increase in FCI_L?

$$\Delta NPV = S_1\Delta x_1 + S_2\Delta x_2 + S_3\Delta x_3 = (2.80)(0.02)(75) - (2.80)(0) - (0.59)(0.03)(150)$$
$$= \$1.545 \text{ million}$$

A Probabilistic Approach to Quantifying Risk: The Monte-Carlo Method. The basic approach adopted here will involve the following steps.

1. All parameters for which uncertainty is to be quantified are identified.
2. Probability distributions are assigned for all parameters in Step 1.
3. A random number is assigned for each parameter in Step 1.
4. Using the random number from Step 3, the value of the parameter is assigned using the probability distribution (from Step 2) for that parameter.
5. Once values have been assigned to all parameters, these values are used to calculate the profitability (*NPV* or other criterion) of the project.
6. Steps 3, 4, and 5 are repeated many times (for example, 1000).
7. A histogram and cumulative probability curve for the profitability criteria calculated from Step 6 are created.
8. The results of Step 7 are used to analyze the profitability of the project.

The algorithm described in this eight-step process is best illustrated by means of an example. However, before these steps can be completed, it is necessary to review some basic probability theory.

Probability, Probability Distribution, and Cumulative Distribution Functions. A detailed analysis and description of probability theory are beyond the scope of this book. Instead, some of the basic concepts and simple distributions are presented. The interested reader is referred to Resnick [5], Valle-Riestra [6], and Rose [7] for further coverage of this subject.

For any given parameter for which uncertainty exists (and to which some form of distribution will be assigned), the uncertainty must be described via a probability distribution. The simplest distribution to use is a uniform distribution, which is illustrated in Figure 10.11.

From Figure 10.11, the parameter of interest can take on any value between a and b with equal likelihood. Because the uniform distribution is a probability density function, the area under the curve must equal 1, and hence the value of the frequency (y-axis) is equal to $1/(b-a)$. The probability density function can be integrated to give the cumulative probability distribution, which for the uniform distribution is given in Figure 10.12.

Figure 10.12 is interpreted by realizing that the probability of the parameter being less than or equal to x is P. Alternatively, a random, uniformly distributed value of the parameter can be assigned by choosing a random number in the range 0 to 1 (on the y-axis) and reading the corresponding value of the parameter, between a and b, on the x-axis. For example, if the random number chosen is P, then, using Figure 10.12, the corresponding value of the parameter is x. Clearly, the shapes of the density function and the corresponding cumulative distribution influence the values of the parameters that are used in the eight-step algorithm. Which probability density function should be used? Clearly, if frequency occurrence data for a given parameter are available, the distribution can be constructed. However, complete information about the way in which a given parameter will vary is often not available. The minimum data set would be the most likely value (b), and estimates of the highest (c) and lowest (a) values that the parameter could reasonably take. With this information, a triangular probability density function or distribution, shown in Figure 10.13, can be constructed. The

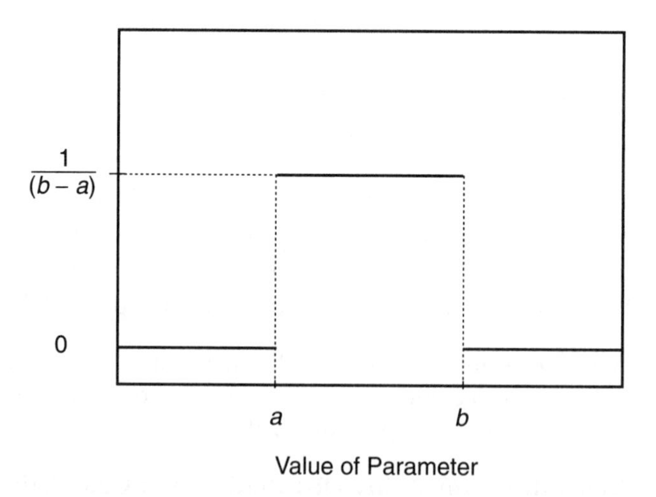

Figure 10.11 Uniform Probability Density Function

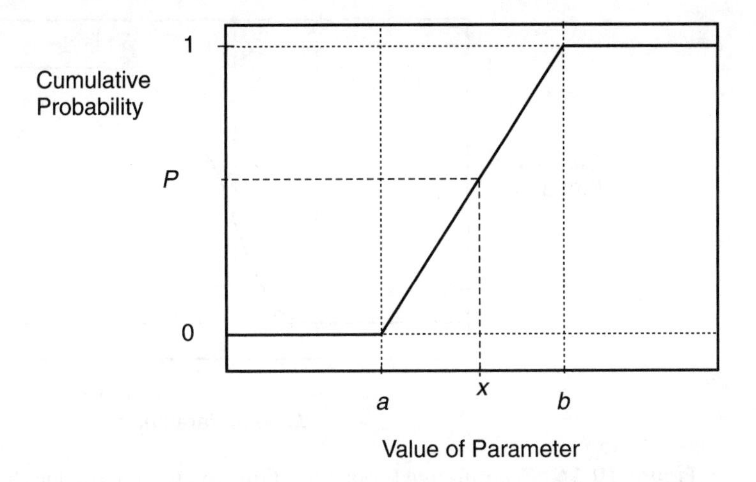

Figure 10.12 Cumulative Probability Distribution for a Uniform Probability Density Function

corresponding cumulative distribution is shown in Figure 10.14. The equations describing these distributions are as follows.

Triangular probability density function:

$$p(x) = \frac{2(x-a)}{(c-a)(b-a)} \quad \text{for } x \leq b$$

$$p(x) = \frac{2(c-x)}{(c-a)(c-b)} \quad \text{for } x > b \tag{10.9}$$

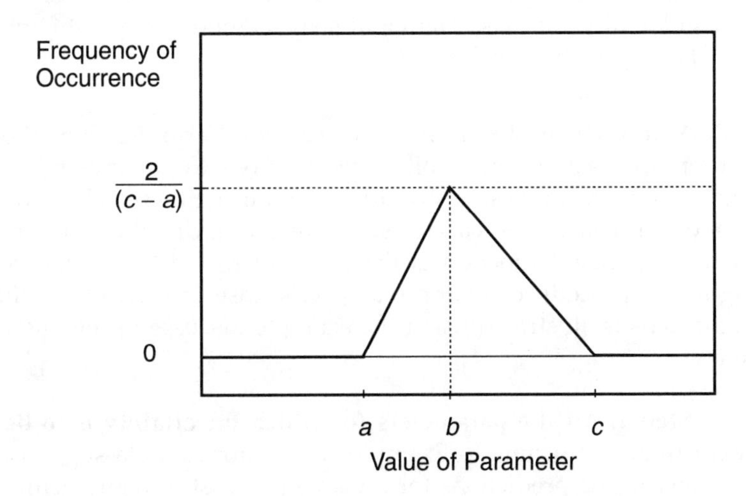

Figure 10.13 Probability Density Function for Triangular Distribution

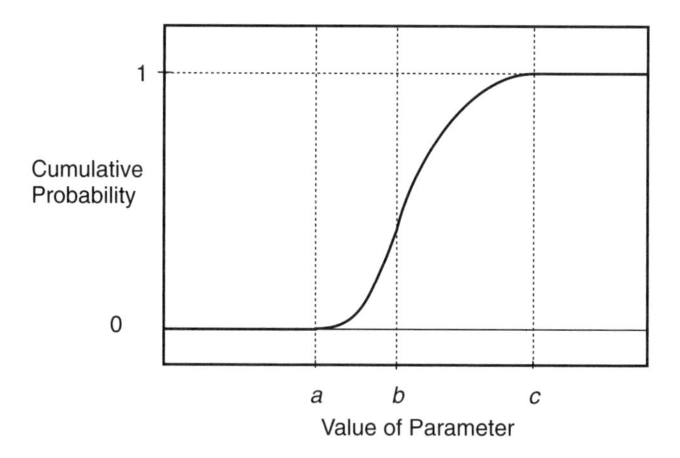

Figure 10.14 Cumulative Probability Function for Triangular Distribution

Triangular cumulative probability function:

$$P(x) = \frac{(x-a)^2}{(c-a)(b-a)} \quad \text{for } x \le b$$

$$P(x) = \frac{(b-a)}{(c-a)} + \frac{(x-b)(2c-x-b)}{(c-a)(c-b)} \quad \text{for } x > b \tag{10.10}$$

Clearly, any probability density function and corresponding cumulative probability distribution could be used to describe the uncertainty in the data. Trapezoidal, normal, lognormal, and so on, are used routinely to describe uncertainty in data. However, for simplicity, the following discussions are confined to triangular distributions. The eight-step method for quantifying uncertainty in profitability analysis is illustrated next.

Monte-Carlo Simulation. The Monte-Carlo (M-C) method is simply the concept of assigning probability distributions to parameters, repeatedly choosing variables from these distributions, and using these values to calculate a function dependent on the variables. The resulting distribution of calculated values of the dependent function is the result of the M-C simulation. Therefore, the eight-step procedure is simply a specific case of the M-C method. Each of the eight steps is illustrated using the example discussed previously in the scenario analysis.

Step 1: All the parameters for which uncertainty is to be quantified are identified. Returning to Example 10.1, historical data suggest that there is uncertainty in the predictions for revenue (R), cost of manufacturing (COM_d), and fixed capital investment (FCI_L).

Step 2: Probability distributions are assigned for all parameters in Step 1. All the uncertainties associated with these parameters are assumed to follow triangular distributions with the properties given in Table 10.6.

Step 3: A random number is assigned for each parameter in Step 1. First, random numbers between 0 and 1 are chosen for each variable. The easiest way to generate random numbers is to use the Rand( ) function in Microsoft's Excel program or a similar spreadsheet. Tables of random numbers are also available in standard math handbooks [8].

Step 4: Using the random number from Step 3, the value of the parameter is assigned using the probability distribution (from Step 2) for that parameter. With the value of the random number equal to the right-hand side of Equation 10.10 and using the corresponding values of a, b, and c, this equation is solved for the value of x. The value of x is the value of the parameter to use in the next step. Table 10.7 illustrates this procedure for R, COM_d, and FCI_L.

To illustrate how the random values for the parameters are obtained, consider the calculation for COM_d. The number 0.6498 was chosen at random from a uniform distribution in the range 0–1 using Microsoft's Excel spreadsheet. This number, along with values of $a = 27$, $b = 30$, and $c = 33$, are then substituted for $P(x)$ in Equation (10.10) to give

$$0.6498 = \frac{(x-27)^2}{(33-27)(30-27)} = \frac{(x-27)^2}{18} \quad \text{for } x \le b$$

$$0.6498 = \frac{(30-27)}{(33-27)} + \frac{(x-30)(2(33)-x-30)}{(33-27)(33-30)} = 0.5 + \frac{(x-30)(36-x)}{18} \quad \text{for } x > b$$

From Equation (10.10), the $P(x)$ value for $x = b$ is given by

$$P(x=30) = \frac{(30-27)^2}{(33-27)(30-27)} = \frac{9}{18} = 0.5$$

Table 10.6 Data for Triangular Distributions for R, COM_d, and FCI_L (All $ Figures Are in Millions)

Parameter	Minimum Value (a)	Most Likely Value (b)	Maximum Value (c)
Revenue, R	\$60.0/yr	\$75.0/yr	\$78.75/yr
Cost of manufacturing, COM_d	\$27.0/yr	\$30.0/yr	\$33.0/yr
Fixed capital investment, FCI_L	\$120.0	\$150.0	\$195.0

Because the value of the random number (0.6498) is greater than 0.5, the form of the equation for $x > b$ must be used. Solving for x yields:

$$0.6498 = 0.5 + \frac{(x - 30)(36 - x)}{18}$$

$$x^2 - 66x + 1082.6964 = 0$$

$$x = \frac{66 \pm \sqrt{(66^2 - (4)(1)(1082.6964)}}{(2)(1)} = 33 \pm 2.51 = 30.49, \textit{ or } 35.51 \text{ (impossible)}$$

Step 5: Once values have been assigned to all parameters, these values are used to calculate the profitability (*NPV* or other criterion) of the project. The spreadsheet given in Table E10.1 was used to determine the *NPV* using the values given in Table 10.7. The *NPV* is also shown in Table 10.7.

Table 10.7 Illustration of the Assignment of Random Values to the Parameters *R*, COM_d, and FCI_L (All $ Figures Are in Millions)

Parameter	Random Number	Random Value of Parameter	*NPV*
Revenue (*R*)	0.3501	\$69.92/yr	
Cost of manufacturing (COM_d)	0.6498	\$30.49/yr	\$–15.60
Fixed capital investment (FCI_L)	0.9257	\$179.16	

Step 6: Steps 3, 4 and 5 are repeated many times (say 1000). For the sake of illustration, Steps 3, 4, and 5 were repeated 20 times to yield 20 values of the *NPV*. These results are summarized in Table 10.8.

Step 7: A histogram or cumulative probability curve is created for the values of the profitability criterion calculated from Step 6. Using the data from Table 10.8, a cumulative probability curve is constructed. To do this, the data are ordered from lowest (–28.20) to highest (28.27), and the cumulative probability of the *NPV* being less than or equal to the value on the x-axis is plotted. The results are shown in Figure 10.15. The dashed line simply connects the 20 data points for

Table 10.8 Results of the 20-Point Monte-Carlo Simulation

Run	Rand (1)	R (\$/yr)	Rand (2)	COM_d (\$/yr)	Rand (3)	FCI_L (\$)	NPV (\$)
1	0.3501	69.92	0.6498	30.49	0.9257	179.16	–15.60
2	0.4063	70.69	0.7859	31.04	0.5531	156.16	–1.45
3	0.8232	75.22	0.3046	29.34	0.7073	163.57	11.59
4	0.9691	77.28	0.6164	30.37	0.8207	170.40	10.45
5	0.4418	71.15	0.2386	29.07	0.7273	164.66	–0.34
6	0.7170	74.20	0.9794	32.39	0.8313	171.14	–4.23
7	0.5626	72.58	0.8368	31.29	0.8891	175.65	–8.84
8	0.9854	77.74	0.1836	28.82	0.8136	169.92	16.34
9	0.8200	75.19	0.7440	30.85	0.5268	155.04	12.31
10	0.6319	73.33	0.1320	28.54	0.3863	149.48	16.84
11	0.1712	66.94	0.9465	32.02	0.0406	129.56	0.99
12	0.4966	71.82	0.3921	29.66	0.5993	158.23	4.34
13	0.2781	68.84	0.1474	28.63	0.7533	166.14	–5.76
14	0.2312	68.06	0.4187	29.75	0.5165	154.60	–4.27
15	0.5039	71.90	0.0042	27.28	0.5681	156.82	12.04
16	0.2184	67.84	0.8629	31.43	0.5107	154.36	–9.44
17	0.7971	74.97	0.3452	29.49	0.0789	133.32	28.27
18	0.2068	67.63	0.7975	31.09	0.9803	186.85	–28.20
19	0.8961	76.05	0.5548	30.17	0.1497	138.35	26.43
20	0.4201	70.87	0.2047	28.92	0.5713	156.96	4.50

this simulation. This line shows several bumps that are due to the small number of simulations. The solid line represents the data for 1000 simulations, and it can be seen that this is curve is essentially smooth. The 1000-point simulation was carried out using the CAPCOST software accompanying the text. The use of the software is addressed at the end of this section.

Step 8: The results of Step 7 are used to analyze the profitability of the project. From Figure 10.15, it can be seen that there is about a 38% chance that the project will not be profitable. The median *NPV* is about \$5 million, and only about 21% of the values calculated lie above \$17.12 million, which is the *NPV* calculated for the base case, using the most likely values of R, COM_d, and FCI_L.

Another way that the data from an M-C analysis can be used is in the comparison of alternatives. For example, consider two competing projects, A and B. A probabilistic analysis of both these projects yields the data shown in Figure 10.16. If only the median profitability is considered, corresponding to a cumulative

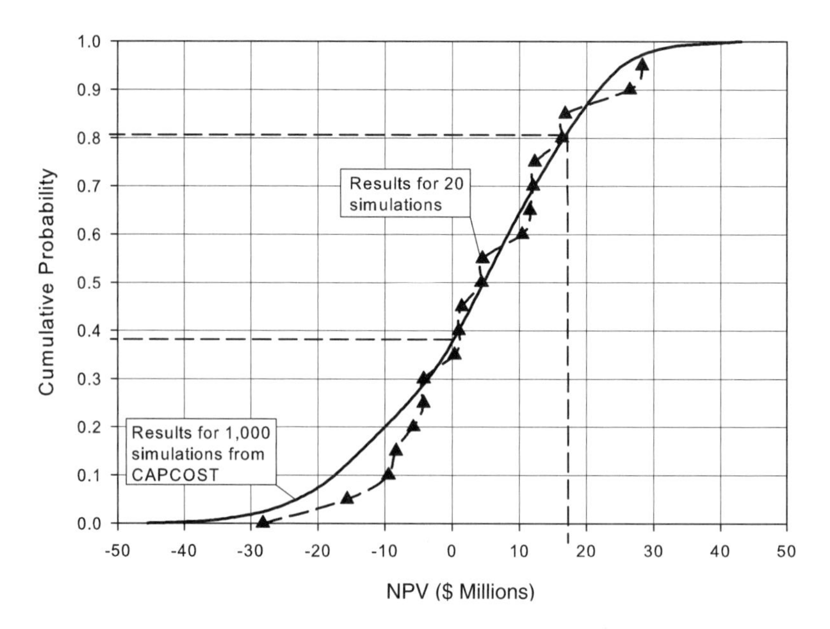

Figure 10.15 Cumulative Probability of *NPV* for Monte-Carlo Simulation

probability of 0.5, then it might be concluded that Project A is better. Indeed, over a wide range of probabilities Alternative A gives a higher *NPV* than Alternative B. However, this type of comparison does not give the whole picture. By looking at the low end of *NPV* predictions, it is found that Project A has a 17% chance of returning a negative *NPV* compared with Project B, which is predicted to have only a 2% chance of giving a negative *NPV*. Clearly, the choice regarding Projects A and B must be made taking into account both the probability of success and the magnitude of the profitability. The Monte-Carlo analysis allows a far more complete financial picture to be painted, and the decisions from such information will be more profound having taken more information into account.

Evaluation of the Risks Associated with Using New Technology. To this point, risks associated with predicting the items listed in Table 10.1 have been considered. For example, predictions for the variations associated with the cost of the plant, the cost of manufacturing, and the revenue generated by the plant were made. Then, by using the M-C technique, the relationship given in Figure 10.15 was generated. For processes using new technology, additional risks will be present, but these risks may be impossible to quantify in terms of the parameters given in Table 10.1. One way to take this additional risk into account is to assign a higher acceptable rate of return for projects using new technology compared with those using mature technologies. The effect of using a higher discount rate is to move the curve in Figure 10.15 to the left. This is illustrated in Figure 10.17. From this figure, it is apparent that if an acceptable rate of return is 15% p.a., then the

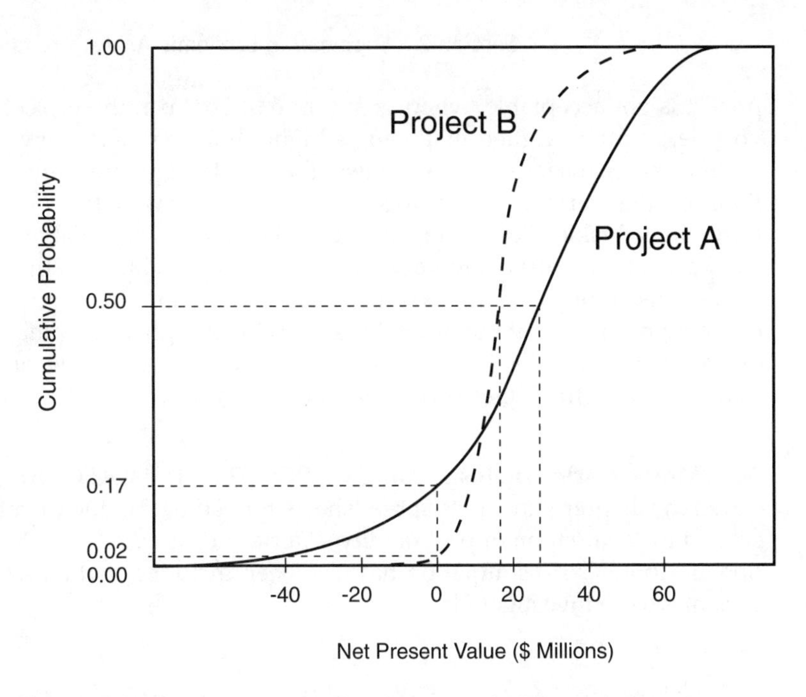

Figure 10.16 A Comparison of the Profitability of Two Projects Showing the *NPV* with Respect to the Estimated Cumulative Probability from a Monte-Carlo Analysis

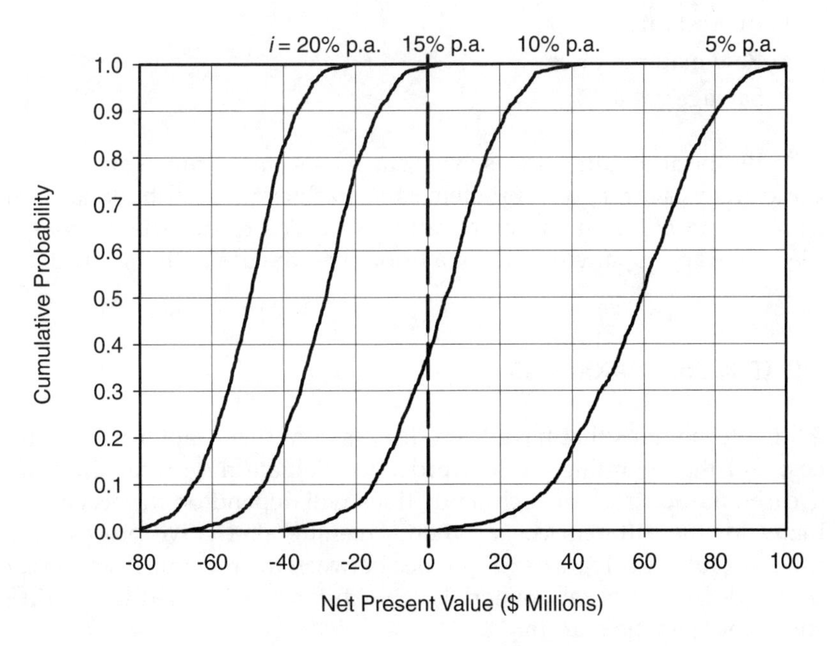

Figure 10.17 The Effect of Interest (Hurdle) Rate on Monte-Carlo Simulations

project is not acceptable, whereas at a rate of 10% p.a. the project looks quite favorable. It can be argued that using a higher hurdle rate for new processes is unnecessary, because, for a new project, there will be greater ranges in the predictions of the variables, and this automatically makes the project using new technology "riskier," as the effect of broader ranges for variables is to flatten the *NPV*-probability curve. However, it may be impossible to estimate the effect of the new technology on, for example, the cost of manufacturing or the acceptance of a new product in the market. By specifying a higher acceptable rate of return on the investment for these projects, the interpretation between projects using new and old technologies is clear and unambiguous.

Monte-Carlo Analysis Using CAPCOST. The CAPCOST program introduced in Chapter 7 includes spreadsheets for estimating the cash flows of a project and the evaluation of profitability criteria such as *NPV* and *DCFROR*. In addition, a Monte-Carlo simulation has also been included that allows the following variables to be investigated:

- FCI_L
- Price of product
- Working capital
- Income tax rate
- Interest rate
- Raw material price
- Salvage value

By specifying the ranges over which these terms are likely to vary, a Monte-Carlo analysis for a given problem can be achieved. Distributions of criteria such as *NPV* and *DPBP* are automatically given. The reader should consult the help file on the accompanying CD for a tutorial on the use of this software.

10.8 PROFIT MARGIN ANALYSIS

All the techniques that have been discussed in this chapter use the fixed capital cost and the operating costs in order to evaluate the profitability of a process. Clearly, the accuracy of such predictions will depend on the accuracy of the estimates for the different costs. When screening alternative processes, it is sometimes useful to evaluate the difference between the revenue from the sale of products and the cost of raw materials. This difference is called the **profit margin** or sometimes just the **margin.**

$$\text{Profit Margin} = \Sigma(\text{Revenue Products}) - \Sigma(\text{Cost of Raw Materials}) \quad (10.11)$$

If the profit margin is negative, the process will not be profitable. This is because no capital cost, utility costs, and other ancillary operating costs have been taken into account. A positive profit margin does not guarantee that the process will be profitable but does suggest that further investigation may be warranted. Therefore, the profit margin is a useful, but limited, tool for the initial screening of process alternatives. This is illustrated in Example 10.17.

Example 10.17

Consider the maleic anhydride process shown in Appendix B.5. Estimate the profit margin for this process using the costs of raw materials and products from Table 8.4.

From Tables 8.4 and B.4 the following flowrates and costs are found:

Cost of benzene = \$0.657/kg
Cost of maleic anhydride = \$1.543/kg
Feed rate of benzene to process (Stream 1, Figure B.4.1) = 3304 kg/h
Product rate of maleic anhydride (Stream 13, Figure B.4.1) = (24.8)(98.058) = 2432 kg/h
Profit Margin = (2432)(1.543) – (3304)(0.657) = \$1581.6/h or \$1581.6/(2432)
= \$0.650/kg of maleic anhydride

Clearly, from an analysis of the profit margin, further investigation of the maleic anhydride process is warranted.

10.9 SUMMARY

In this chapter, the basics of profitability analysis for projects involving large capital expenditures were covered. The concepts of nondiscounted and discounted profitability criteria were introduced, as were the three bases for these criteria: time, money, and interest rate.

How to choose the economically optimum piece of equipment among a group of alternatives using the capitalized cost, equivalent annual operating cost, and the common denominator methods was demonstrated.

The concept of incremental economic analysis was introduced and applied to an example involving large capital budgets and also to a retrofit project. It was shown that both the net present value (*NPV*) and equivalent annual operating cost (*EAOC*) methods were particularly useful when comparing alternatives using discounted cash flows.

Finally, the concept of assigning probabilities to variables in order to quantify risk was discussed. The Monte-Carlo technique was introduced, and its application to simulate the cumulative distribution of net present values of a project was described. The interpretation of results from this technique was presented. Finally, the simulation of risk and the analysis of data using the CAPCOST package was illustrated by an example.

REFERENCES

1. Humphreys, K. K., *Jelen's Cost and Optimization Engineering,* 3rd ed. (New York: McGraw-Hill, 1991).
2. Salvatore, D., *Schaum's Outline of Theory and Problems of Microeconomic Theory,* 3rd ed. (New York: McGraw-Hill, 1992).
3. Torries, T. F., *Evaluating Mineral Projects: Applications and Misconceptions* (Littleton, CO: SME, 1998).
4. http://www.geocities.com/gunjansaraf/yogi.htm.
5. Resnick, W., *Process Analysis and Design for Chemical Engineers* (New York: McGraw-Hill, 1981).
6. Valle-Riestra, J. F., *Project Evaluation in the Chemical Process Industries* (New York: McGraw-Hill, 1983).
7. Rose, L. M., *Engineering Investment Decisions: Planning under Uncertainty* (Amsterdam: Elsevier, 1976).
8. Spiegel, M. R., *Mathematical Handbook of Formulas and Tables,* Schaum's Outline Series (New York: McGraw-Hill, 1968).

SHORT ANSWER QUESTIONS

1. The evaluation of a project requiring a large capital investment has yielded an *NPV* (net present value) of $\$20 \times 10^6$. If the internal hurdle rate for this project was set at 10% p.a., will the *DCFROR* (discounted cash flow rate of return) be greater or less than 10%? Explain.
2. The following are results of a recent evaluation of two projects. Which would you choose? Defend your choice. Your opportunity cost for capital is 15%.

	NPV	*DCFROR*
Project A	10	55%
Project B	10,000	16%

3. Explain the concept of an incremental economic analysis.
4. When comparing two pieces of equipment for a given service, if each piece of equipment has the same life and each costs the same, would the amount of maintenance required for the equipment be an important factor? Why?
5. What economic criterion would you use to choose the best piece of equipment between three alternatives each with a different operating cost, capital cost, and equipment life?
6. Do you agree with the statement, "Monte Carlo simulation enables the design engineer to eliminate risk in economic analysis." Please explain your answer.
7. In evaluating a large project, what are the advantages and disadvantages of probabilistic analysis?

PROBLEMS

For the following problems, unless stated otherwise, you may assume that the cost of land, L, and the salvage value, S, of the plant are both zero.

8. The projected costs for a new plant are given below (all numbers are in $\$10^6$):

   Land Cost = \$7.5
   Fixed Capital Investment = \$120 (\$60 at end of year 1, \$39.60 at end of year 2, and \$20.40 at end of year 3)
   Working Capital = \$35 (at start-up)
   Start-up at end of year 3
   Revenue from sales = \$52
   Cost of manufacturing (without depreciation) = \$18
   Tax rate = 40%
   Depreciation method = Current MACRS over 5 years
   Length of time over which profitability is to be assessed = 10 years after start-up
   Internal rate of return = 9.5% p.a.

   For this project, do the following:
   a. Draw a cumulative (nondiscounted) after-tax cash flow diagram.
   b. From Part (a), calculate the following nondiscounted profitability criteria for the project:
      (i) Cumulative cash position and cumulative cash ratio
      (ii) Payback period
      (iii) Rate of return on investment
   c. Draw a cumulative (discounted) after-tax cash flow diagram.
   d. From Part (c), calculate the following discounted profitability criteria for the project:
      (i) Net present value and net present value ratio
      (ii) Discounted payback period
      (iii) Discounted cash flow rate of return (*DCFROR*)

9. Repeat Problem 10.8 using a straight-line depreciation method over 7 years. Compare the results with those obtained in Problem 10.8. Which depreciation method would you use?

10. The following expenses and revenues have been estimated for a new project:

    Revenues from sales = $\$4.1 \times 10^6$/yr
    Cost of manufacturing (excluding depreciation) = $\$1.9 \times 10^6$/yr
    Taxation rate = 40%
    Fixed Capital Investment = $\$7.7 \times 10^6$
    (two payments of $\$5 \times 10^6$ and $\$2.7 \times 10^6$ at the end of years 1 and 2, respectively)

Start-up at the end of year 2
Working capital = $\$2 \times 10^6$ at the end of year 2
Land cost = $\$0.8 \times 10^6$ at the beginning of the project (time = 0)
Project life (for economic evaluation) = 10 yr after start-up

For this project, estimate the *NPV* of the project assuming an after-tax internal hurdle rate of 11% p.a., using the following depreciation schedules.

a. MACRS method for 5 years
b. Straight-line depreciation with an equipment life (for depreciation) of 9.5 years

Comment on the effect of discounting on the overall profitability of large capital projects.

11. In reviewing current operating processes, the company accountant has provided you with the following information about a small chemical process that was built ten years ago.

Capital Investment = $\$30 \times 10^6$ ($\$10 \times 10^6$ at the end of year 1, $\$15 \times 10^6$ at the end of year 2, $\$5 \times 10^6$ at the end of year 3. Working capital = $\$10 \times 10^6$)

Year after Start-Up	Yearly After-Tax Cash Flow ($\$10^6$/yr)
1	7.015
2	6.206
3	6.295
4	6.852
5	6.859
6	7.218
7	5.954
8	5.459
9	5.789
10	5.898

Over the last ten years, the average (after-tax) return on investment that non-process projects have yielded is 10%.

a. What is the *DCFROR* for this project over the last 12 years? (Ignore land and working capital costs.)
b. In retrospect, was the decision to build this plant a good one?

12. The after-tax cash flows for a new chemical process are shown in Table P10.12. Using these data, calculate the following:

a. Payback period (*PBP*)
b. Cumulative cash position (*CCP*) and cumulative cash ratio (*CCR*)

Table P10.12 Nondiscounted Cash Flow Calculations for Problem 10.12 (All Figures Are in $ Millions)

End of Year	Capital Investment	Depreciation Allowance	Revenue from Sales	Total Annual Costs	Net Profit	Income Tax	Net Profit after Tax	After-Tax Cash Flow
0	(10)*, **	—	—	—	—	—	—	(10)
1	(25)	—	—	—	—	—	—	(25)
2	(20)	—	—	—	—	—	—	(20)
3	(15 + 20)†,‡	—	—	—	—	—	—	(35)
4		8.57	60	50	10	3.80	6.20	14.77
5	—	8.57	120	92	28	10.64	17.36	25.93
6	—	8.57	120	47	73	27.74	45.26	53.83
7	—	8.57	120	50	70	26.80	43.40	51.97
8	—	8.57	120	60	60	22.80	37.20	45.77
9	—	8.57	120	51	69	26.22	42.78	51.35
10	—	—	120	40	80	30.40	49.60	58.17
11	—	—	120	40	80	30.40	49.60	49.60
12		—	120	40	80	30.40	49.60	49.60
13	50	—	120	40	80	30.40	49.60	49.60

*Numbers in ( ) represent negative values.

**Land cost = 0 .

†Plant started up at end of year 3.

‡Working capital = 50.

c. Rate of return on investment (*ROROI*)
d. Discounted payback period (*DPBP*)
e. Net present value (*NPV*)
f. Discounted cash flow rate of return (*DCFROR*)

Use a 10% discount rate for Parts (d) and (e).

13. From the data given in Table P10.12, determine the following information regarding the calculations performed for this analysis.
    a. What taxation rate was assumed?
    b. What was the total fixed capital investment?
    c. What method of depreciation was used?
    d. What was the cost of manufacturing, not including depreciation?
14. Consider the following two new chemical plants, each with an initial fixed capital investment (year 0) of $\$15 \times 10^6$. Their cash flows are as follows.

Year	Process 1 ($ million/yr)	Process 2 ($ million/yr)
1	3.0	5.0
2	8.0	5.0
3	7.0	5.0
4	5.0	5.0
5	2.0	5.0

    a. Calculate the *NPV* of both plants for interest rates of 6% and 18%. Which plant do you recommend? Explain your results.
    b. Calculate the *DCFROR* for each plant. Which plant do you recommend?
    c. Calculate the non-discounted payback period (*PBP*) for each plant. Which plant do you recommend?
    d. Explain any differences in your answers to Parts (a), (b), and (c).
15. In a design, you have the choice of purchasing either of the following batch reactors.

Costs	A	B	C
Material of Construction	CS	SS	Hastalloy
Installed Cost	$15,000	$25,000	$40,000
Equipment Life	3 yrs	5 yrs	7 yrs
Yearly Maintenance Cost	$4000	$3000	$2000

If the internal rate of return for such comparisons is 9% p.a., which of the alternatives is least costly?

16. Two pieces of equipment are being considered for an identical service. The installed costs and yearly operating costs associated with each piece of equipment are as follows.

Costs	A	B
Installed Cost	$5,000	$10,000
Operating Cost	$2000/yr	$1000/yr
Equipment Life	5 years	7 years

a. If the internal hurdle rate for comparison of alternatives is set at 15% p.a., which piece of equipment do you recommend be purchased?
b. Above what internal hurdle rate would you recommend project A?

17. Your company is considering modifying an existing distillation column. A new reboiler and condenser will be required, along with several other peripherals. The equipment lists are as follows.

Option 1

Equipment	Installed Cost ($)	Operating Cost ($/yr)	Equipment Life (yr)
Condenser	50,000	7,000	10
Reboiler	75,000	5,000	15
Reflux Pump	7,500	8,000	10
Reflux Drum	12,500	—	10
Piping	8,000	—	15
Valves	6,500	—	10

Option 2

Equipment	Installed Cost ($)	Operating Cost ($/yr)	Equipment Life (yr)
Condenser	75,000	4,000	15
Reboiler	75,000	5,000	15
Reflux Pump	10,500	5,000	15
Reflux Drum	14,500	—	15
Piping	8,000	—	15
Valves	6,500	—	10

The internal hurdle rate for comparison of investments is set at 12% p.a. Which option do you recommend?

18. Because of corrosion, the feed pump to a batch reactor must be replaced every three years. What is the capitalized cost of the pump?

    Data:

    Purchased cost of pump = $35,000
    Installation cost = 75% of purchased cost
    Internal hurdle rate = 10% p.a.

19. Three alternative pieces of equipment are being considered for solids separation from a liquid slurry.

Equipment Type	Capital Investment ($)	Operating Cost ($/yr)	Service Life (yr)
Rotary Vacuum Filter	15,000	3000	6
Filter Press	10,000	5000	8
Hydrocyclone and Centrifuge	25,000	2000	10

If the internal hurdle rate for this project is 11% p.a., which alternative do you recommend?

20. A change in air pollution control equipment is being considered. A baghouse filter is being considered to replace an existing electrostatic precipitator. Consider the costs and savings for the project in the following data.

    Cost of new baghouse filter = $250,000
    Projected utility savings = $70,000/yr
    Time over which cost comparison should be made = 7 years
    Internal hurdle rate = 7% p.a.

    Should the baghouse filter be purchased and installed?

21. In considering investments in large capital projects, a company is deciding in which of the following projects it will invest.

(All Values in $million)	Project A	Project B	Project C
Capital Required (in year 0)	80	100	120
After-tax, yearly cash flow (yrs 1–10)	11	14	16

The company can always invest its money in long-term bonds that currently yield 5.5% p.a. (after tax). In which, if any, of the projects should the company invest if the capital ceiling for investment is $250 million and a project life of 10 years is assumed? Would you argue to raise the investment ceiling?

22. Because you live in a southern, warm-weather climate, your electricity bill is very large for about eight months of the year due to the need for air conditioning. You have been approached by an agency that would like to assist you in installing solar panels to provide electricity to run your air conditioning. You were also considering adding additional insulation to your house. You also have the option of doing nothing. Which of the following options would you choose based on the given data?
    a. Do nothing.
    b. Install only the solar collector.
    c. Install only the insulation.
    d. Install both the solar collector and the insulation.

    Data (all figures in $thousands)

Purchase and installation cost of solar collector	25
Purchase and installation of insulation	5
Current cooling bill	2.5/yr
Expected savings from insulation alone	0.9/yr
Expected savings from solar collector alone	2.0/yr
Expected savings from insulation and solar collector	2.5/yr
Other maintenance on house	2.0/yr
Assessed value of house (2007)	300
Interest rate of savings if do not spend money	6.5% p.a.
Number of years assumed lifetime of insulation and solar collector	15

    If there were a tax credit for installing the solar collector in the year in which it was installed, how much of a tax credit (in % of initial investment) would be required to make the solar collector alone a worthwhile investment?

23. You have been asked to evaluate several investment opportunities for the biotechnology company for which you work. These potential investments concern a new process to manufacture a new, genetically engineered pharmaceutical. The financial information on the process alternatives are as follows.

Case	Capital Investment ($million)	Yearly, After-Tax Cash Flow ($million/yr)
Base	75	19
Alternative 1	15	3
Alternative 2	25	5
Alternative 3	30	7

    For the alternatives, the capital investment and the yearly after-tax cash flows are incremental to the base case. The assumed plant life is 12 years, and all of the capital investment occurs at time = 0.

a. If an acceptable, nondiscounted rate of return on investment (*ROROI*) is 25% p.a., which is the best option?
b. If an acceptable, after-tax, discounted rate of return is 15% p.a., which option is the best?

24. Your company is considering investing in a process improvement that would require an initial capital investment of $500,000. The projected increases in revenue over the next seven years are as follows.

Year	Incremental Revenue ($thousand/yr)
1	100
2	90
3	90
4	85
5	80
6	80
7	75

a. The company can always leave the capital investment in the stock portfolio, which is projected to yield 8% p.a. over the next seven years. What should the company do?
b. What is the break-even rate of return between the process improvement and doing nothing?
c. If the capital investment could be changed without changing the incremental revenues, what capital investment changes this investment from being profitable (not profitable) to being not profitable (profitable)?

25. During the design of a new process to manufacture nanocomposites, several alternative waste treatment processes are being considered. The base case is the main waste treatment process, and the options are modifications to the base case. Not all options are compatible with each other, and the economic data on the only possible combinations are as follows.

Case	Capital Investment ($million)	Annual, After-Tax Cash flow ($thousand/yr)
Base	5.0	750
Base + option 1	5.1	770
Base + option 1 + option 2	5.3	790
Base + option 3	5.2	782
Base + option 1 + option 3	5.4	810

The nondiscounted, internal hurdle rate for investment is 14% p.a., after tax. Which waste treatment process do you recommend?

26. The installation of a new heat exchanger is proposed for the batch crystallization step in an existing pharmaceutical manufacturing process. The heat exchanger costs $49,600 (installed) and saves $8000/yr in operating costs. Is this a good investment based on a before-tax analysis?

    Data:

    Internal hurdle rate = 12% p.a. (after tax)
    Taxation rate = 40%
    5-year MACRS depreciation used
    Service life of heat exchanger = 12 yrs.

27. A process for the fabrication of microelectronic components has been designed. The required before-tax return on investment is 18% p.a., and the equipment life is assumed to be eight years.

	Base-Case Design	Alternative 1	Alternative 2
Capital Investment ($million)	21	—	—
Additional Investment ($million)	—	2.15	1.35
Product Revenue ($million/yr)	12	12	12
Raw Material Costs ($million/yr)	5.1	5.1	5.1
Other Operating Costs ($million/yr)	0.35	0.27	0.31

    a. Do you recommend construction of the base-case process?
    b. Do you recommend including either of the process alternatives?
    c. Suppose there were another alternative, compatible with either of the previous alternatives, requiring an additional $3.26 million capital investment. What savings in operating cost would be required to make this alternative economically attractive?

28. A new pharmaceutical plant is expected to cost (FCI_L) $25 million, and the revenue (R) from the sale of products is expected to be $10 million/yr for the first four years of operation and $15 million/yr thereafter. The cost of manufacturing without depreciation (COM_d) is projected to be $4 million/yr for the first four years and $6 million/yr thereafter. The cost of land (at the end of year zero) is $3 million, the working capital at start-up (which occurs at the end of year 2) is $3 million, and the fixed capital investment is assumed to be paid out as $15 million at the end of year 1 and $10 million at the end of year 2. Yearly income starts in year 3, and the plant life is ten years after start-up. The before-tax criterion for profitability is 17%. Assume the plant has no salvage value, but that the cost of land and the working capital are recovered at the end of the project life.

a. Draw a labeled, discrete, nondiscounted, cash flow diagram for this project.
b. Draw a labeled, cumulative, discounted (to year zero) cash flow diagram for this project.
c. What is the *NPV* for this project? Do you recommend construction of this plant?

29. How much would you need to save annually to be willing to invest $1.75 million in a process improvement? The internal hurdle rate process for improvements is 18% before taxes over eight years.

30. How much would you be willing to invest in a process improvement to save $2.25 million/yr? The internal hurdle rate for process improvements is 15% before taxes over six years.

31. You can invest $5.1 million in a process improvement to save $0.9 million/yr. What internal hurdle rate would make this an attractive option? The assumed lifetime for these improvements is nine years.

32. Recommend whether your company should invest in the following process. The internal hurdle rate is 17% p.a., before taxes, over an operating lifetime of 12 years. The fixed capital investment is made in two installments: $5 million at time zero and $3 million at the end of year 1. At the end of year 1, $1 million in working capital is required. For the remainder of the project lifetime, $2 million in income is realized.

33. You are considering two possible modifications to an existing microelectronics facility. The criterion for profitability is 18% p.a. over six years. All values are in $million.

	Alternative 1	Alternative 2
Project Cost	2.25	3.45
Yearly Savings	0.65	0.75

a. What do you recommend based on a nondiscounted *ROROII* analysis?
b. What do you recommend based on an *INPV* analysis?
c. Based on the results of Parts (a) and (b), what do you recommend?

34. A new biotech plant is expected to cost (FCI_L) $20 million, with $10 million paid at the beginning of the project and $10 million paid at the end of year 1. There is no land cost because the land is already owned. The annual profit, before taxes, is $4 million and the working capital at start-up (which occurs at the end of year 1) is $1 million, The plant life is 10 years after start-up. The before-tax criterion for profitability is 12%. Assume that the plant has no salvage value, and that the working capital is recovered at the end of the project life.
a. Draw a labeled, discrete, nondiscounted cash flow diagram for this project.

b. Draw a labeled, cumulative, discounted (to year zero) cash flow diagram for this project.
c. What is the *NPV* for this project? Do you recommend construction of this plant?
d. What would the annual profit, before taxes, have to be for the *NPV* to be \$2 million?
e. The plant is built and has been operational for several years. It has been suggested that \$3 million be spent on plant modifications that will save money. Your job is to analyze the suggestion. The criterion for plant modifications is a 16% before-tax return over five years. How much annual savings are required before you would recommend in favor of investing in the modification?

35. You are responsible for equipment selection for a new micro-fiber process. A batch blending tank is required for corrosive service. You are considering three alternatives.

Alternative	Material	Cost (\$thousand)	Maintenance Cost (\$thousand/yr)	Equipment Life (yr)
A	Carbon Steel	5	5.5	3
B	Ni alloy	15	3.5	5
C	Hastalloy	23	2.5	9

The after-tax internal hurdle rate is 14% p.a. Which alternative do you recommend?

36. In an already operational pharmaceutical plant, you are considering three alternative solvent-recovery systems to replace an existing combustion process. The internal hurdle rate is 15% after taxes over eight years. Which alternative do you recommend?

Alternative	Total Module Cost of System (\$thousands)	Natural Gas Savings (\$thousand/yr)	Savings in Solvent Cost (\$thousand/yr)	Net Maintenance Savings (\$thousand/yr)
A	88	15	25	33
B	125	15	45	25
C	250	15	75	18

37. A condenser using refrigerated water is being considered for the recovery of a solvent used in a pharmaceutical coating operation. The amount of solvent

recovered from the gas stream is a function of the temperature to which it is cooled. Three cases, each using a different exchanger and a different amount of refrigerated water, are to be considered. Data for the process cases are given in the table.

Case	Installed Cost of Exchanger ($)	Cost of Refrigerated Water ($/y)	Value of Acetone Recovered ($/y)
A	103,700	15,000	32,500
B	162,400	30,000	65,000
C	216,100	34,500	74,750

a. If the nondiscounted hurdle rate for projects is set at 15% p.a., which case do you recommend?
b. If the discounted hurdle rate is set at 5% p.a. and the project life is set at seven years, which case do you recommend?
c. How does your result from Part (b) change if the project life were changed to 15 years?
d. How does your result from Part (b) change if the project life were changed to 5 years?

38. For Problem 10.8, uncertainties associated with predicting the revenues and cost of manufacturing are estimated to be as follows.

    Revenue: Expected range of variation from base case, low = $45 million, high = $53.5 million

    COM_d: Expected range of variation from base case, low = $16 million, high = $21.5 million

    Using the above information, evaluate the expected distribution of *NPVs* and *DCFRORs* for the project. Would this analysis change your decision compared to that for the base case?

39. Calculate the lowest and highest *NPVs* that are possible for Problem 10.38. Compare these values with the distribution of *NPVs* from Problem 10.38. What are the probabilities of getting an *NPV* within $5 million of these values?

40. Perform a Monte-Carlo analysis on Problem 10.10, using the following ranges for uncertain variables (all figures in $ millions).

	Low	High
FCI_L	6.6	9.3
Revenues	3.5	4.5
Interest rate (%)	9.5	12.0
COM_d	1.7	2.5

What do you conclude? *Hint:* When using the CAPCOST program, set the equation for $COM_d = C_{raw\ materials}$ and input the variability in COM_d as a variability in the cost of raw materials.

41. You are considering buying a house with a mortgage of $250,000. The current interest rates for a mortgage loan for a 15-year period are 7.5% p.a. fixed or 6.75% p.a. variable. Based on historical data, the variation in the variable rate is thought to be from a low of 6.0% to a high of 8.5%, with the most likely value as 7.25%. The variable interest rate is fixed at the beginning of each year. Answer the following questions.
    a. What are the maximum and minimum yearly payments that would be expected if the variable interest option is chosen? For simplicity assume that the compounding period is one year.
    b. Set up a Monte-Carlo simulation by picking 20 random numbers and using these to choose the variable interest rate for each of the 20 years of the loan.
    c. Calculate the yearly payments for each year using the data from (b). *Hint:* You should keep track of the interest paid on the loan and the remaining principal. The remaining principal is used to calculate the new yearly payment with the new yearly interest rate.
42. Perform a Monte-Carlo simulation on Example 10.1 for the following conditions shown. Show that the variation in the *NPV* is the same as shown in Figure 10.15. To which variable is the *NPV* more sensitive?

	Low	High
FCI_L	–20%	35%
Product Price	–25%	10%
Raw Material Price	–10%	25%

Note: Because the Monte-Carlo method is based on the generation of random numbers, no two simulations will be exactly the same. Therefore, you may see some small differences between your results and those shown in Figure 10.15.

43. A product is to be produced in a batch process. The estimated fixed capital investment is $5 million. The estimated raw materials cost is $100,000/batch, the estimated utility costs are $60,000/batch, and the estimated waste treatment cost is $23,000/batch. The revenue is predicted to be $220,000/batch. An initial scheduling scenario suggests that it will be possible to produce 22 batches/yr. The internal hurdle rate is 15% p.a. before taxes over 10 years.
    a. Is this process profitable?
    b. What is the minimum number of batches/yr that would have to be produced to make the process profitable?
    c. If 26 batches/yr is the maximum that can be produced, what is the rate of return on the process?

44. Three different products are manufactured in an existing batch process. The details are as follows.

Product	kg/batch	Product Value ($/kg)	Batches/yr
A	5000	7.5	20
B	2500	6.25	13
C	4000	5.75	16

The annual cost of manufacturing is $0.95 million/yr. The demand for these products is increasing, and the crystallization step has been determined to be the bottleneck to increasing the capacity. It is desired to add 25% capacity to this process. The internal hurdle rate for process improvements is 17% p.a. over five years.

a. If a new batch crystallizer, which allows for a 25% capacity increase, costs $750,000, do you recommend this process improvement?
b. Capital funds are tight, and it has been determined that the maximum investment possible is $600,000, resulting in a smaller, new crystallizer. Using this crystallizer, identical profitability as found in Part (a) has been determined. Determine what capacity increase results from purchasing the smaller crystallizer.
c. Suppose that it is now possible to purchase the $750,000 crystallizer, thereby increasing capacity by 25%. However, the purchase of this crystallizer requires that the *DCFROR* (for this incremental investment) is at least 40% over five years. Is this *DCFROR* reached?

45. A batch process runs on a zero-wait-time schedule (see Chapter 3). It has been determined that a 20% increase in capacity is possible if three equally sized storage tanks are purchased, and the processing schedule is altered appropriately. The cost of manufacture is $1.5 million/yr with annual revenues of $2.75 million/yr. The internal hurdle rate for process improvements is 20% p.a. over six years.

a. If each tank costs $1.7 million, do you recommend the investment?
b. What is the maximum cost per storage tank that will meet the profitability criterion and result in a 20% increase in capacity?
c. Suppose that storage tanks each cost $2.0 million but still result in a 20% increase in capacity. What is the *DCFROR* for the recommended process improvement?

SECTION

3

Synthesis and Optimization of Chemical Processes

In this section, the problems of how to create, simulate, and optimize a process and how to develop a process flow diagram (PFD) are addressed. In order to create a PFD, a considerable amount of information needs to be gathered. This includes reaction kinetics, thermodynamic property data, required purity for products and by-products, types of separations to be used in the process, reactor type, range of conditions for the reaction, and many others. Once this information has been gathered, it must be synthesized into a working process. In order to accommodate the synthesis of information, the chemical engineer relies on solving material balances, energy balances, and equilibrium relationships using a process simulator. The basic data required to perform a simulation of a process are covered, and other aspects of using a process simulator are discussed. Once the PFD has been simulated, the optimization of the process can proceed. In general, process optimization involves both parametric and topological changes, and both these aspects are discussed.

This material is treated in the following chapters.

Chapter 11: Utilizing Experience-Based Principles to Confirm the Suitability of a Process Design

When designing a process, the experienced engineer will often have a good idea as to how large a given piece of equipment needs to be or how many stages a given separation will require. Such information is invaluable in the early stages of design to check the reasonableness of the results of rigorous calculations. To assist the inexperienced engineer, a series of heuristics or guidelines for different equipment is presented in the form of tables.

Chapter 12: Synthesis of the PFD from the Generic BFD

The information required to obtain a base-case process flow diagram is discussed and categorized into the six basic elements of the generic block flow process diagram. The need to obtain reaction kinetics, thermodynamic data, and alternative separation methods is discussed in the context of building a base-case process. Special emphasis is placed on alternative distillation schemes and the sequencing of columns needed for such separations.

Chapter 13: Synthesis of a Process Using a Simulator and Simulator Troubleshooting

The structure of a typical process simulator and the basic process information required to simulate a process are discussed. The various types of equipment that can be simulated, and the differences between alternative modules used to simulate similar process equipment, are reviewed. The importance of choosing the correct thermodynamic package for physical property estimation is emphasized, and strategies to eliminate errors and solve simulation problems are presented.

Chapter 14: Process Optimization

Basic definitions used to describe optimization problems are presented. The need to look at both topological changes in the flowsheet (rearrangement of equipment) and parametric changes (varying temperature, pressure, etc.) is emphasized. Strategies for both types of optimization are included. A new section on batch systems, including batch scheduling and optimal batch cycle times, has been included.

Chapter 15: Pinch Technology

The concepts of pinch technology as applied to the design of new heat and mass exchange networks are presented. The design of the network requiring the minimum utility consumption and the minimum number of exchangers is developed for a given minimum temperature or concentration approach. The temperature- and concentration-interval diagrams, the cascade diagram, and the cumulative enthalpy and cumulative mass exchange diagrams are presented. The identification of the pinch point and its role in the design of the exchanger network are discussed. Finally, the estimation of the heat-exchanger network surface area for a given approach temperature and the effect of operating pressure and materials of construction on the optimum, minimum temperature approach are discussed.

CHAPTER

11

Utilizing Experience-Based Principles to Confirm the Suitability of a Process Design

Experienced chemical engineers possess the skills necessary to perform detailed and accurate calculations for the design, analysis, and operation of equipment and chemical processes. In addition, these engineers will have formulated a number of experienced-based shortcut calculation methods and guidelines useful for the following:

1. Checking new process designs
2. Providing equipment size and performance estimates
3. Helping troubleshoot problems with operating systems
4. Verifying that the results of computer calculations and simulations are reasonable
5. Providing reasonable initial values for input into a process simulator required to achieve program convergence
6. Obtaining approximate costs for process units
7. Developing preliminary process layouts

These shortcut methods are forms of heuristics that are helpful to the practicing engineer. All heuristics are, in the final analysis, fallible and sometimes difficult to justify. They are merely plausible aids or directions toward the solution of a problem [1]. Especially for the heuristics described in this chapter, we need to keep in mind the four characteristics of any heuristic.

1. A heuristic does not guarantee a solution.
2. It may contradict other heuristics.

3. It can reduce the time to solve a problem.
4. Its acceptance depends on the immediate context instead of on an absolute standard.

The fact that one cannot precisely follow all heuristics all the time is to be expected, as it is with any set of technical heuristics. However, despite the limitations of heuristics, they are nevertheless valuable guides for the process engineer.

In Chapter 6, process units and stream conditions that were identified as areas of special concern were analyzed. These areas were highlighted in a series of informational tables. In this chapter, you will complete the analysis of chemical processes by checking the equipment parameters and stream conditions in the PFD for agreement with observations and experiences in similar applications.

The required information to start an analysis is provided in a series of **informational tables** containing shortcut calculation techniques. In this chapter, we demonstrate the use of these resources by checking the conditions given in the basic toluene hydrodealkylation PFD.

11.1 THE ROLE OF EXPERIENCE IN THE DESIGN PROCESS

The following short narrative illustrates a situation that could be encountered early in your career as an engineer.

> *You are given an assignment that involves writing a report that is to be completed and presented in two weeks. You work diligently and feel confident you have come up with a respectable solution. You present the written report personally to your director (boss), who asks you to summarize only your final conclusions. Immediately after providing this information, your boss declares that "your results must be wrong" and returns your report unopened and unread.*
>
> *You return to your desk angry. Your comprehensive and well-written report was not even opened and read. Your boss did not tell you what was wrong, and you did not receive any "partial credit" for all your work. After a while, you cool off and review your report. You find that you had made a "simple" error, causing your answer to be off by an order of magnitude. You correct the error and turn in a revised report.*
>
> *What remains is the nagging question, "How could your boss know you made an error without having reviewed your report or asking any questions?"*

The answer to this nagging question is probably a direct result of your director's experience with a similar problem or knowledge of some guideline that contradicted your answer. The ability of your boss to transfer personal experience to new situations is one reason why he or she was promoted to that position.

It is important to be able to apply knowledge gained through experience to future problems.

11.1.1 Introduction to Technical Heuristics and Shortcut Methods

A **heuristic** is a statement concerning equipment sizes, operating conditions, and equipment performance that reduces the need for calculations. A **shortcut method** replaces the need for extensive calculations in order to evaluate equipment sizes, operating conditions, and equipment performance. These are referred to as "back-of-the-envelope calculations." In this text, we refer to both of these experience-based tools as guidelines or heuristics.

The guidelines provided in this chapter are limited to materials specifically covered in this text (including problems at the end of the chapters). All such material is likely to be familiar to final year B.S. chemical engineering students and new graduates as a result of their education. Upon entering the work force, engineers will develop guidelines that apply specifically to their area of responsibility.

Guidelines and heuristics must be applied with an understanding of their limitations. In most cases, a novice chemical engineer should have sufficient background to apply the rules provided in this text.

The narrative started earlier is now revisited. The assignment remains the same; however, the approach to solving the problem changes.

Before submitting your report, you apply a heuristic that highlights an inconsistency in your initial results. You then review your calculations, find the error, and make corrections before submitting your report. Consider two possible responses to this report.

1. *Your boss accepts the report and notes that the report appears to be excellent and he or she looks forward to reading it.*
2. *Your boss expresses concern and returns the report as before. In this case, you have a reasoned response available. You show that your solution is consistent with the heuristic you used to check your work. With this supporting evidence your boss would have to rethink his or her response and provide you with an explanation regarding his or her concern.*

In either case, your work will have made a good impression.

Guidelines and heuristics are frequently used to make quick estimates during meetings and conferences and are valuable in refreshing one's memory with important information.

11.1.2 Maximizing the Benefits Obtained from Experience

No printed article, lecture, or text is a substitute for the perceptions resulting from experience. An engineer must be capable of transferring knowledge gained from one or more experiences to resolve future problems successfully.

To benefit fully from experience, it is important to make a conscious effort to use each new experience to build a foundation upon which to increase your ability to handle and to solve new problems.

An experienced engineer retains a body of information, made up largely of heuristics and shortcut calculation methods, that is available to help solve new problems.

The process by which an engineer uses information and creates new heuristics consists of three steps. These three steps are predict, authenticate, and reevaluate, and they form the basis of the **PAR** process. The elements of this process are presented in Table 11.1, which illustrates the steps used in the PAR process.

Example 11.1

Evaluate the heat transfer coefficient for water at 93°C (200°F) flowing at 3.05 m/s (10 ft/s) inside a 38 mm (1.5 inch) diameter tube. From previous experience, you know that the heat transfer coefficient for water, at 21°C (70°F) and 1.83 m/s (6 ft/s), in these tubes is 5250 W/m^2°C. Follow the PAR process to establish the heat transfer coefficient at the new conditions.

Step 1—Predict: Assume that the velocity and temperature have no effect.
Predicted Heat Transfer Coefficient = 5250 W/m^2°C

Step 2—Authenticate/Analyze: Using the properties given below we find that the Reynolds number for the water in the tubes is

$$Re = u\rho D_{pipe}/\mu = (1.83)(997.4)(1.5)(0.0254)/(9.8 \times 10^{-4}) = 71 \times 10^3 \rightarrow \text{Turbulent Flow}$$

Table 11.1 PAR Process to Maximize Benefits of Experience: *P*redict, *A*uthenticate, *R*eevaluate

1. **Predict:** This is a precondition of the PAR process. It represents your best prediction of the solution. It often involves making assumptions and applying heuristics based on experience. Calculations should be limited to back-of-the-envelope or shortcut techniques.
2. **Authenticate/Analyze:** In this step, you seek out equations and relationships, do research relative to the problem, and perform the calculations that lead toward a solution. The ability to carry out this activity provides a necessary but not sufficient condition to be an engineer. When possible, information from actual operations is included in order to achieve the best possible solution.
3. **Reevaluate/Rethink:** The best possible solution from Step 2 is compared with the predicted solution in Step 1. When the prediction is not acceptable, it is necessary to correct the reasoning that led to the poor prediction. It becomes necessary to remove, revise, and replace assumptions made in Step 1. This is the critical step in learning from experience.

Use the Sieder-Tate equation [2] to check the prediction:

$$hD/k = (0.023)(Du\rho/\mu)^{0.8}(C_p\mu/k)^{1/3} \quad (11.1)$$

Property	21°C (70°F)	93°C (200°F)	Ratio of (New/Old)
ρ (kg/m^3)	997.4	963.2	0.966
k (W/m°C)	0.604	0.678	1.12
C_p (kJ/kg°C)	4.19	4.20	1.00
μ (kg/m/s)	9.8×10^{-4}	3.06×10^{-4}	0.312

Take the ratio of Equation 11.1 for the two conditions given above, and rearrange and substitute numerical values. Using ′ to identify the new condition at 93°C, we get

$$h'/h = (D/D)^{0.2}(u'/u)^{0.8}(\rho'/\rho)^{0.8}(\mu/\mu')^{0.47}(C_p'/C_p)^{0.33}(k'/k)^{0.67} \quad (11.2)$$

$$= (1)(1.50)(0.973)(1.73)(1.00)(1.08) = 2.725 \quad (11.3)$$

$$h' = (2.725)(5250)\ \text{W/m}^2\text{°C} = 14{,}300\ \text{W/m}^2\text{°C}$$

The initial assumption that the velocity and temperature do not have a significant effect is incorrect. Equation (11.3) reveals a velocity effect of a factor of 1.5 and a viscosity effect of a factor of 1.73. All other factors are close to 1.0.

Step 3—Reevaluate/Rethink: The original assumptions that velocity and temperature had no effect on the heat transfer coefficient have been rejected. Improved assumptions for future predictions are as follows.

1. The temperature effect on viscosity must be evaluated.
2. The effects of temperature on C_p, ρ, and k are negligible.
3. Pipe diameter has a small effect on h (all other things being equal).
4. Results are limited to the range where the Sieder-Tate equation is valid.

With these assumptions, the values for water at 21°C are substituted into Equation (11.2). This creates a useful heuristic for evaluating the heat transfer coefficients for water.

$$h'(\text{W/m}^2\text{°C}) = 125u'^{0.8}/\mu'^{0.47} \text{ for } u'(\text{m/s}),\ \mu'\ (\text{kg/m/s})$$

Although it takes longer to obtain a solution when you start to apply the PAR process, the development of the heuristic and the addition of a more in-depth understanding of the factors that are important offer substantial long-term advantages.

There are hundreds of heuristics covering areas in chemical engineering—some general, and others specific to a given application, process, or material. In the next section, we have gathered a number of these rules that you can use to make predictions to start the PAR analysis.

11.2 PRESENTATION OF TABLES OF TECHNICAL HEURISTICS AND GUIDELINES

We provide a number of these guidelines for you in this section. The information given is limited to operations most frequently encountered in this text. Most of the information was extracted from a collection presented in Walas [3]. In addition, this excellent reference also includes additional guidelines for the following equipment:

1. Conveyors for particulate solids
2. Cooling towers
3. Crystallization from solution
4. Disintegration
5. Drying of solids
6. Evaporators
7. Size separation of particles

The heuristics or rules are contained in a number of tables and apply to operating conditions that are most often encountered. The information provided is used in Example 11.2 and should be used to work problems at the end of the chapter and to check information on any PFD.

Example 11.2

Refer to the information given in Chapter 1 for the toluene hydrodealkylation process, namely, Figure 1.7 and Tables 1.5 and 1.7. Using the information provided in the tables in this chapter, estimate the size of the equipment and other operating parameters for the following units:

a. V-102
b. E-105
c. P-101
d. C-101
e. T-101
f. H-101

Compare your findings with the information given in Chapter 1.

a. V-102 High-Pressure Phase Separator

From Table 11.6, we use the following heuristics:

Rule 3 → Vertical vessel
Rule 4 → L/D between 2.5 and 5 with optimum at 3.0
Rule 5 → Liquid hold-up time is 5 min based on 1/2 volume of vessel
Rule 9 → Gas velocity u is given by

$$u = k\sqrt{\frac{\rho_l}{\rho_v} - 1} \text{ m/s}$$

where $k = 0.0305$ for vessels without mesh entrainers

Rule 12 → Good performance obtained at 30%–100% of u from Rule 9; typical value is 75%

From Table 1.5, we have
Vapor flow = Stream 8 = 9200 kg/h, $P = 23.9$ bar, $T = 38°C$
Liquid flow = Streams 17 + 18 = 11570 kg/h, $P = 2.8$ bar, $T = 38°C$
$\rho_v = 8\ kg/m^3$ and $\rho_l = 850\ kg/m^3$ (estimated from Table 1.7)

From Rule 9, we get $u = 0.0305[850/8 - 1]^{0.5} = 0.313$ m/s
Use $u_{act} = (0.75)(0.313) = 0.23$ m/s
Now mass flowrate of vapor $= u\rho_v\pi D^2/4 = 9200/3600 = 2.56$ kg/s
Solving for D, we get $D = 1.33$ m.

From Rule 5, we have volume of liquid $= 0.5\ L\pi D^2/4 = 0.726L\ m^3$
5 minutes of liquid flow = (5)(60)(11,570)/850/3600 = 1.13 m^3
Equating the two results above, we get $L = 1.56$ m.

From Rule 4, we have L/D should be in range 2.5 to 5. For our case $L/D = 1.56/1.33 = 1.17$
Because this is out of range, we should change to $L = 2.5D = 3.3$ m.
Heuristics from Table 11.6 suggest that V-102 should be a vertical vessel with $D = 1.33$ m, $L = 3.3$ m.

From Table 1.7, we see that the actual V-102 is a vertical vessel with $D = 1.1$ m, $L = 3.5$ m.

We should conclude that the design of V-102 given in Chapter 1 is consistent with the heuristics given in Table 11.6. The small differences in L and D are to be expected in a comparison such as this one.

b. E-105 Product Cooler

From Table 11.11 we use the following heuristics:

Rule 1: Set $F = 0.9$
Rule 6: min. $\Delta T = 10°C$
Rule 7: Water enters at 30°C and leaves at 40°C
Rule 8: $U = 850\ W/m^2{}°C$

We note immediately from Table 1.5 and Figure 1.5 that Rule 6 has been violated because $\Delta T_{min} = 8°C$.

For the moment, ignore this and return to the heuristic analysis:

$\Delta T_{lm} = [(105 - 40) - (38 - 30)]/\ln[(105 - 40)/(38 - 30)] = 27.2°C$
$Q = 1085$ MJ/h = 301 kW (from Table 1.7)
$A = Q/U\Delta T_{lm}F = (301{,}000)/(850)/(27.2)/(0.90) = 14.46\ m^2$
From Rule 9, Table 11.11, this heat exchanger should be a double-pipe or multiple-pipe design.
Comparing our analysis with the information in Table 1.7 we get

Heuristic: Double-pipe design, Area = 14.5 m^2
Table 1.7: Multiple-pipe design, Area = 12 m^2

Again, the heuristic analysis is close to the actual design. The fact that the minimum approach temperature of 10°C has been violated should not cause too much concern, because the actual minimum approach is only 8°C and the heat exchanger is quite small, suggesting that a little extra area (due to a smaller overall temperature driving force) is not very costly.

c. P-101

From Table 11.9, we use the following heuristics:
Rule 1: Power(kW) = (1.67)[Flow(m^3/min)]ΔP(bar)/ε
Rules 4–7: Type of pump based on head

From Figure 1.5 and Tables 1.5 and 1.7, we have
Flowrate (Stream 2) = 13,300 kg/h
Density of fluid = 870 kg/m^3
ΔP = 25.8 – 1.2 = 24.6 bar = 288 m of liquid (head = $\Delta P/\rho g$)
Volumetric flowrate = (13,300)/(60)/(870) = 0.255 m^3/min
Fluid pumping power = (1.67)(0.255)(24.6) = 10.5 kW

From Rules 4–7, pump choices are multistage centrifugal, rotary, and reciprocating. Choose reciprocating to be consistent with Table 1.7. Typical ε = 0.75.

Power (shaft power) = 10.5/0.75 = 14.0 kW → compares with 14.2 kW from Table 1.7.

d. C-101

From Table 11.10, we use the following heuristics:

Rule 2: $W_{rev\ adiab} = mz_1RT_1[(P_2/P_1)^a - 1]/a$

From Table 1.7, we have flow = 6770 kg/h, T_1 = 38°C = 311 K, mw = 8.45, P_1 = 23.9 bar, P_2 = 25.5
k = 1.41 (assume) and a = 0.2908
m = (6770)/(3600)/(8.45) = 0.223 kmol/s
$W_{rev\ adiab} = (223)(1.0)(8.314)(311)\{(25.5/23.9)^{0.2908} - 1)/0.2908 = 37.7$ kW
using a compressor efficiency of 75%
W_{actual} = (37.7)/(0.75) = 50.3 kW → This checks with the shaft power requirement given in Table 1.7.

e. T-101

From Table 11.13, we use the following heuristics:
Rule 5: Optimum reflux in the range of 1.2–1.5 R_{min}
Rule 6: Optimum number of stages approximately $2N_{min}$
Rule 7: $N_{min} = \ln\{[x/(1-x)]_{ovhd}/[x/(1-x)]_{bot}\}/\ln\alpha$
Rule 8: $R_{min} = \{F/D\}/(\alpha - 1)$
Rule 9: Use a safety factor of 10% on number of trays.
Rule 14: L_{max} = 53 m and $L/D < 30$

From Table 11.14, we use the following heuristics:
Rule 2: $F_s = u\rho_v^{0.5} = 1.2 \rightarrow 1.5$ m/s(kg/m^3)$^{0.5}$
Rule 3: ΔP_{tray} = 0.007 bar
Rule 4: ε_{tray} = 60 – 90 %

$x_{ovhd} = 0.9962$, $x_{ovhd} = 0.0308$, $\alpha_{ovhd} = 2.44$, $\alpha_{bot} = 2.13$, $\alpha_{geom\ ave} = (\alpha_{ovhd}\alpha_{bot})^{0.5} = 2.28$
$N_{min} = \ln\{[0.9962/(1-0.9962)]/[0.0308/(1-0.0308)]\}/\ln(2.28) = 10.9$

$R_{min} = \{142.2/105.6\}/(2.28 - 1) = 1.05$
Range of $R = (1.2 \rightarrow 1.5)R_{min} = 1.26 \rightarrow 1.58$

$N_{theoretical} \approx (2)(10.9) = 21.8$
$\varepsilon_{tray} = 0.6$
$N_{actual} \approx (21.6/0.6)(1.1) = 40$ trays

$\rho_v = 6.1\ \text{kg/m}^3$
$u = (1.2 \rightarrow 1.5)/6.1^{0.5} = 0.49 \rightarrow 0.60\ \text{m/s}$
Vapor flowrate (Stream 13) = 22,700 kg/h
Vol. flowrate, $v = 1.03\ \text{m}^3/\text{s}$
$D_{tower} = [4v/\pi u]^{0.5} = [(4)(1.03)/(3.142)/(0.49 \rightarrow 0.60)]^{0.5} = 1.64 - 1.48\ \text{m}$

$\Delta P_{tower} = (N_{actual})(\Delta P_{tray}) = (40)(0.007) = 0.28\ \text{bar}$

A comparison of the actual equipment design and the predictions of the heuristic methods are given below.

	From Tables 1.5 and 1.7 and Figure 1.5	From Heuristics
Tower diameter	1.5 m	1.48 → 1.64 m
Reflux ratio, R	1.75	1.26 →1.58
Number of trays	42	40
Pressure drop, ΔP_{tower}	0.30 bar	0.28 bar

f. H-101

From Table 11.11, we use the following heuristics:
Rule 13: Equal heat transfer in radiant and convective sections
radiant rate = 37.6 kW/m^2, convective rate = 12.5 kW/m^2

$$\text{Duty} = 27{,}040\ \text{MJ/h} = 7511\ \text{kW}$$

Area radiant section = (0.5)(7511)/(37.6) = 99.9 m^2 (106.8 m^2 in Table 1.7)
Area convective section = (0.5)(7511)/(12.5) = 300.4 m^2 (320.2 m^2 in Table 1.7)

From the earlier worked examples, it is clear that the sizing of the equipment in Table 1.7 agrees well with the predictions of the heuristics presented in this chapter. Exact agreement is not to be expected. Instead, the heuristics should be used to check calculations performed using more rigorous methods and to flag any inconsistencies.

11.3 SUMMARY

In this chapter, we have introduced a number of heuristics that allow us to check the reasonableness of the results of engineering calculations. These heuristics or guidelines cannot be used to determine absolutely whether a particular answer is correct or incorrect. However, they are useful guides that allow the engineer to flag possible errors and help focus attention on areas of the process that may require special attention. Several heuristics, provided in the tables at the end of this chapter, were used to check the designs provided in Table 1.5 for the toluene hydrodealkylation process.

LIST OF INFORMATIONAL TABLES

Table 11.2(a) Physical Property Heuristics

	Units	Liquids		Liquids	Gases	Gases	Gases
		Water		Organic Material	Steam	Air	Organic Material
Heat Capacity	kJ/kg°C	4.2		1.0–2.5	2.0	1.0	2.0–4.0
Density	kg/m^3	1000		700–1500		1.29@STP	
Latent heat	kJ/kg	1200–2100		200–1000			
Thermal conductivity	W/m°C	0.55–0.70		0.10–0.20	0.025–0.07	0.025–0.05	0.02–0.06
Viscosity	kg/m s	0°C	1.8×10^{-3}	Wide Range	$10–30 \times 10^{-6}$	$20–50 \times 10^{-6}$	$10–30 \times 10^{-6}$
		50°C	5.7×10^{-4}				
		100°C	2.8×10^{-4}				
		200°C	1.4×10^{-4}				
Prandtl no.		1–15		10–1000	1.0	0.7	0.7–0.8

Table 11.2(b) Typical Physical Property Variations with Temperature and Pressure

	Liquids	Liquids	Gases	Gases
Property	Temperature	Pressure	Temperature	Pressure
Density	$\rho_l \propto (T_c - T)^{0.3}$	Negligible	$\rho_g = MW.P/ZRT$	$\rho_g = MW.P/ZRT$
Viscosity	$\mu_l = Ae^{B/T}$	Negligible	$\mu_g \propto \frac{T^{1.5}}{(T + 1.47T_b)}$	Significant only for $P > 10$ bar
Vapor pressure	$P^* = ae^{b/(T+c)}$	—	—	—

T is temperature (K), T_c is the critical temperature (K), T_b is the normal boiling point (K), ***MW*** is molecular weight, ***P*** is pressure, ***Z*** is compressibility, ***R*** is the gas constant, and P^* is the vapor pressure.

Table 11.3 Capacities of Process Units in Common Usage[a]

Process Unit	Capacity Unit	Max. Value	Min. Value	Comment
Horizontal vessel	Pressure (bar)	400	Vacuum	*L/D* typically 2–5, see Table 11.6
	Temper. (°C)	400[b]	−200	
	Height (m)	10	2	
	Diameter (m)	2	0.3	
	L/D	5	2	
Vertical vessel	Pressure (bar)	400	400	*L/D* typically 2–5, see Table 11.6.
	Temper. (°C)	400[b]	−200	
	Height (m)	10	2	
	Diameter (m)	2	0.3	
	L/D	5	2	
Towers	Pressure (bar)	400	Vacuum	Normal Limits: Diameter / *L/D*
	Temper. (°C)	400[b]	−200	0.5 / 3.0–40[c]
	Height (m)	50	2	1.0 / 2.5–30[c]
	Diameter (m)	4	0.3	2.0 / 1.6–23[c]
	L/D	30	2	4.0 / 1.8–13[c]
Pumps				
Reciprocating	Power[d](kW)	250	< 0.1	
	Pressure (bar)	1000		
Rotary and positive Displacement	Power[d](kW)	150	< 0.1	
	Pressure (bar)	300		
Centrifugal	Power[d](kW)	250	< 0.1	
	Pressure (bar)	300		
Compressors				
Axial, centrifugal + recipr.	Power[d](kW)	8000	50	
Rotary	Power[d](kW)	1000	50	
Drives for Compressors				
Electric	Power[e](kW)	15,000	< 1	
Steam turbine	Power[e](kW)	15,000	100	
Gas turbine	Power[e](kW)	15,000	10	
Internal combustion eng.	Power[e](kW)	15,000	10	
Process heaters	Duty (MJ/h)	500,000	10,000	Duties different for reactive heaters/furnaces.
Heat exchangers	Area (m^2)	1000	10	For area < 10 m^2 use double pipe exchanger.
	Tube Dia. (m)	0.0254	0.019	
	Length (m)	6.5	2.5	
	Pressure (bar)	150	Vacuum	For 150 < *P* < 400 bar need special design.
	Temp. (°C)	400[b]	−200	

[a]Most of the limits for equipment sizes shown here correspond to the limits used in the costing program (CAPCOST) introduced in Chapter 7.
[b]Maximum temperature and pressure are related to the materials of construction and may differ from values shown here.
[c]For 20 < *L/D* < 30 special design may be required. Diameters up to 7 m possible but greater than 4 m must be fabricated on site.
[d]Power values refer to fluid/pumping power.
[e]Power values refer to shaft power.

Table 11.4 Effect of Typical Materials of Construction on Product Color, Corrosion,[a] Abrasion, and Catalytic Effects

Metals		
Material	***Advantages***	***Disadvantages***
Carbon steel	Low cost, readily available, resists abrasion, standard fabrication, resists alkali	Poor resistance to acids and strong alkali, often causes discoloration and contamination
Stainless steel	Resists most acids, reduces discoloration, available with a variety of alloys, abrasion less than mild steel	Not resistant to chlorides, more expensive, fabrication more difficult, alloy materials may have catalytic effects
Monel-Nickel	Little discoloration, contamination, resistant to chlorides	Not resistant to oxidizing environments, expensive
Hasteloy	Improved over Monel-Nickel	More expensive than Monel-Nickel
Other exotic metals	Improves specific properties	Can be very high cost
Nonmetals		
Material	***Advantages***	***Disadvantages***
Glass	Useful in laboratory and batch systems, low diffusion at walls	Fragile, not resistant to high alkali, poor heat transfer, poor abrasion resistance
Plastics	Good at low temperature, large variety to select from with various characteristics, easy to fabricate, seldom discolors, low cost	Poor at high temperature, low strength, not resistant to high-alkali conditions, low heat transfer. Minor catalytic effects possible
Ceramics	Withstands high temperatures, variety of formulations available, modest cost	Poor abrasion properties, high diffusion at walls (in particular hydrogen), low heat transfer, may encourage catalytic reactions

[a]In addition, see Chapter 7 for preliminary selection of materials of construction.

Table 11.5 Heuristics for Drivers and Power Recovery Equipment

1. Efficiency is greater for larger machines. Electric motors are 85%–95%; steam turbines are 42%–78%; gas engines and turbines are 28%–38% efficient (see Figure 8.7).
2. For less than 74.6 kW (100 hp), electric motors are used almost exclusively. They are made for services up to 14,900 kW (20,000 hp).
3. Steam turbines are competitive higher than 76.6 kW (100 hp). They are speed controllable. They are frequently used as spares in case of power failure.
4. Combustion engines and turbines are restricted to mobile and remote locations.
5. Gas expanders for power recovery may be justified at capacities of several hundred horsepower; otherwise any pressure reduction in process is done with throttling valves.
6. The following useful definitions are given:

$$\text{shaft power} = \frac{\text{theoretical power to pump fluid (liquid or gas)}}{\text{efficiency of pump or compressor, } \varepsilon_{sh}}$$

$$\text{drive power} = \frac{\text{shaft power}}{\text{efficiency of drive, } \varepsilon_{dr}}$$

$$\text{overall efficiency} = \varepsilon_{ov} = \varepsilon_{sh}\,\varepsilon_{dr}$$

ε_{dr} values are given in this table and Figure 8.7.

ε_{sh} values are given in Tables 11.9 and 11.10. Usually ε_{sh} are given on PFD.

(Adapted from S. M. Walas, *Chemical Process Equipment: Selection and Design,* Stoneham, MA: Butterworth, 1988.

Table 11.6 Heuristics for Process Vessels (Drums)

1. Drums are relatively small vessels that provide surge capacity or separation of entrained phases.
2. Liquid drums are usually horizontal.
3. Gas-liquid phase separators are usually vertical.
4. Optimum length or diameter = 3, but the range 2.5 to 5 is common.
5. Holdup time is 5 min for half-full reflux drums and gas/liquid separators, 5–10 min for a product feeding another tower.
6. In drums feeding a furnace, 30 min for half-full drum is allowed.
7. Knockout drums placed ahead of compressors should hold no less than 10 times the liquid volume passing per minute.
8. Liquid-liquid separations are designed for settling velocity of 0.085–0.127 cm/s (2–3 in/min).
9. Gas velocity in gas/liquid separators, $u = k\sqrt{\rho_l/\rho_v - 1}$ m/s (ft/sec) $k = 0.11$ (0.35) for systems with mesh deentrainer, and $k = 0.0305$ (0.1) without mesh deentrainer.
10. Entrainment removal of 99% is attained with 10.2–30.5 cm (4–12 in) mesh pad thickness; 15.25 cm (6 in) thickness is popular.
11. For vertical pads, the value of the coefficient in Step 9 is reduced by a factor of 2/3.
12. Good performance can be expected at velocities of 30%–100% of those calculated with the given k; 75% is popular.
13. Disengaging spaces of 15.2–45.7 cm (6–18 in) ahead of the pad and 30.5 cm (12 in) above the pad are suitable.
14. Cyclone separators can be designed for 95% collection at 5 μm particles, but usually only droplets greater than 50 μm need be removed.

(Adapted from S. M. Walas, *Chemical Process Equipment: Selection and Design,* Stoneham, MA: Butterworth, 1988.

Table 11.7 Heuristics for Vessels (Pressure and Storage)

Pressure Vessels

1. Design temperature between −30°C and 345°C is 25°C above maximum operating temperature; higher safety margins are used outside the given temperature range.
2. The design pressure is 10% or 0.69–1.7 bar (10–25 psi) over the max. operating pressure, whichever is greater. The max. operating pressure, in turn, is taken as 1.7 bar (25 psi) above the normal operation.
3. Design pressures of vessels operating at 0–0.69 bar (0–10 psig) and 95–540°C (200–1000°F) are 2.76 barg (40 psig).
4. For vacuum operation, design pressures are 1 barg (15 psig) and full vacuum.
5. Minimum wall thickness for rigidity: 6.4 mm (0.25 in) for 1.07 m (42 in) dia. and less than 8.1 mm (0.32 in) for 1.07–1.52 m (42–60 in) dia., and 11.7 mm (0.38 in) for more than 1.52 m (60 in) dia.
6. Corrosion allowance 8.9 mm (0.35 in) for known corrosive conditions, 3.8 mm (0.15 in) for noncorrosive streams, and 1.5 mm (0.06 in) for steam drums and air receivers.
7. Allowable working stresses are one-fourth of the ultimate strength of the material.
8. Maximum allowable stress depends sharply on temperature.

Temperature: (°F)		−20 to 650	750	850	1000
(°C)		(−30 to 345)	400	455	540
Low alloy steel SA 203	(psi)	18,759	15,650	9950	2500
	(bar)	1290	1070	686	273
Type 302 stainless steel	(psi)	18,750	18,750	15,950	6250
	(bar)	1290	1290	1100	431

Storage Vessels

1. For less than 3.8 m^3 (1000 gal), use vertical tanks on legs.
2. Between 3.8 and 38 m^3 (1000 and 10,000 gal), use horizontal tanks on concrete supports.
3. Beyond 38 m^3 (10,000 gal) use vertical tanks on concrete pads.
4. Liquids subject to breathing losses may be stored in tanks with floating or expansion roofs for conservation.
5. Freeboard is 15% below 1.9 m^3 (500 gal) and 10% above 1.9 m^3 (500 gal) capacity.
6. Thirty-day capacity often is specified for raw materials and products, but depends on connecting transportation equipment schedules.
7. Capacities of storage tanks are at least 1.5 times the size of connecting transportation equipment; for instance, 28.4 m^3 (7500 gal) tanker trucks, 130 m^3 (34,500 gal) rail cars, and virtually unlimited barge and tanker capacities.

(Adapted from S. M. Walas, *Chemical Process Equipment: Selection and Design,* Stoneham, MA: Butterworth, 1988.

Table 11.8 Heuristics for Piping

1. Line velocities (u) and pressure drop (ΔP): (a) For liquid pump discharge: $u = (5 + D/3)$ ft/sec and $\Delta P = 2.0$ psi/100 ft; (b) For liquid pump suction: $u = (1.3 + D/6)$ ft/sec and $\Delta P = 0.4$ psi/100 ft; (c) For steam or gas flow: $u = 20D$ ft/sec and $\Delta P = 0.5$ psi/100 ft, D = diameter of pipe in inches.
2. Gas/steam line velocities = 61 m/s (200 ft/sec), and pressure drop = 0.1 bar/100 m (0.5 psi/100 ft).
3. In preliminary estimates set line pressure drops for an equivalent length of 30 m (100 ft) of pipe between each piece of equipment.
4. Control valves require at least 0.69 bar (10 psi) drop for good control.
5. Globe valves are used for gases, control, and wherever tight shutoff is required. Gate valves for most other services.
6. Screwed fittings are used only on sizes 3.8 cm (1.5 in) or less; otherwise, flanges or welding used.
7. Flanges and fittings are rated for 10, 20, 40, 103, 175 bar (150, 300, 600, 1500, or 2500 psig).
8. Approximate schedule number required = 1000 P/S, where P is the internal pressure in psig and S is the allowable working stress [about 690 bar (10,000 psi)] for A120 carbon steel at 260° (500°F). Schedule 40 is most common.

(Adapted from S. M. Walas, *Chemical Process Equipment: Selection and Design,* Stoneham, MA: Butterworth, 1988.

Table 11.9 Heuristics for Pumps

1. Power for pumping liquids: kW = (1.67)[Flow(m^3/min)][ΔP(bar)]/ε [hp = Flow(gpm) ΔP(psi)/1714/ε] ε = Fractional Efficiency = ε_{sh} (see Table 11.5).
2. Net positive suction head (NPSH) of a pump must be in excess of a certain number, depending upon the kind of pumps and the conditions, if damage is to be avoided. *NPSH* = (pressure at the eye of the impeller − vapor pressure)/(ρg). Common range is 1.2–6.1 m of liquid (4–20 ft).
3. Specific speed N_s = (rpm)(gpm)$^{0.5}$/(head in feet)$^{0.75}$. Pump may be damaged if certain limits on N_s are exceeded, and the efficiency is best in some ranges.
4. Centrifugal pumps: Single stage for 0.057–18.9 m^3/min (15–5000 gpm), 152 m (500 ft) maximum head; multistage for 0.076–41.6 m^3/min (20–11,000 gpm), 1675 m (5500 ft) maximum head. Efficiency 45% at 0.378 m^3/min (100 gpm), 70% at 1.89 m^3/min (500 gpm), 80% at 37.8 m^3/min (10,000 gpm).
5. Axial pumps for 0.076–378 m^3/min (20–100,000 gpm), 12 m (40 ft) head, 65–85% efficiency.
6. Rotary pumps for 0.00378–18.9 m^3/min (1–5000 gpm), 15,200 m (50,000 ft head), 50–80% efficiency.
7. Reciprocating pumps for 0.0378–37.8 m^3/min (10–10,000 gpm), 300 km (1,000,000 ft) head max. Efficiency 70% at 7.46 kW (10 hp), 85% at 37.3 kW (50 hp), and 90% at 373 kW (500 hp).

(Adapted from S. M. Walas, *Chemical Process Equipment: Selection and Design,* Stoneham, MA: Butterworth, 1988.

Table 11.10 Heuristics for Compressors, Fans, Blowers, and Vacuum Pumps

1. Fans are used to raise the pressure about 3% {12 in (30 cm) water}, blowers to raise less than 2.75 barg (40 psig) and compressors to higher pressures, although the blower range is commonly included in the compressor range.
2. Theoretical reversible adiabatic power = $mz_1RT_1[(\{P_2/P_1\}^a - 1)]/a$
   where T_1 is inlet temperature, R = Gas Constant, z_1 = compressibility, m = molar flow rate, $a = (k-1)/k$ and $k = C_p/C_v$.
   Values of R: = 8.314 J/mol K = 1.987 Btu/lbmol R = 0.7302 atm ft^3/lbmol R
3. Outlet temperature for reversible adiabatic process $T_2 = T_1\ (P_2/P_1)^a$.
4. Exit temperatures should not exceed 167–204°C (350–400°F); for diatomic gases (C_p/C_v = 1.4). This corresponds to a compression ratio of about 4.
5. Compression ratio should be about the same in each stage of a multistage unit, ratio = $(P_n/P_1)^{1/n}$, with n stages.
6. Efficiencies of reciprocating compressors: 65% at compression ratios of 1.5, 75% at 2.0, and 80–85% at 3–6.
7. Efficiencies of large centrifugal compressors, 2.83–47.2 m^3/s (6000–100,000 acfm) at suction, are 76–78%.
8. For vacuum pumps use the following:

Reciprocating piston type	Down to 1 Torr
Rotary piston type	Down to 0.001 Torr
Two-lobe rotary type	Down to 0.0001 Torr
Steam jet ejectors	1-stage down to 100 Torr
	3-stage down to 1 Torr
	5-stage down to 0.05 Torr

9. A three-stage ejector needs 100 kg steam/kg air to maintain a pressure of 1 Torr.
10. In-leakage of air to evacuated equipment depends on the absolute pressure, Torr, and the volume of the equipment, V in m^3 (ft^3) according to $W = kV^{2/3}$ kg/h (lb/hr) with k = 0.98 (0.2) when P > 90 Torr, k = 0.39 (0.08) between 3 and 20 Torr, and k = 0.12 (0.025) at less than 1 Torr.

(Adapted from S. M. Walas, *Chemical Process Equipment: Selection and Design,* Stoneham, MA: Butterworth, 1988.

Table 11.11 Heuristics for Heat Exchangers

1. For conservative estimate set $F = 0.9$ for shell-and-tube exchangers with no phase changes, $q = UAF\Delta T_{lm}$. When ΔT at exchanger ends differ greatly then check F, and reconfigure if F is less than 0.85.
2. Standard tubes are 1.9 cm (3/4 in) OD, on a 2.54 cm (1 in) triangle spacing, 4.9 m (16 ft) long.

   A shell 30 cm (1 ft) dia., accommodates 9.3 m^2 (100 ft^2)

   60 cm (2 ft) dia., accommodates 37.2 m^2 (400 ft^2)

   90 cm (3 ft) dia., accommodates 102 m^2 (1100 ft^2)
3. Tube side is for corrosive, fouling, scaling, and high-pressure fluids.
4. Shell side is for viscous and condensing fluids.
5. Pressure drops are 0.1 bar (1.5 psi) for boiling and 0.2–0.62 bar (3–9 psi) for other services.
6. Minimum temperature approach is 10°C (20°F) for fluids and 5°C (10°F) for refrigerants.
7. Cooling water inlet is 30°C (90°F), maximum outlet 45°C (115°F).
8. Heat transfer coefficients for estimating purposes, W/m^2°C (Btu/hr ft^2°F): water to liquid, 850 (150); condensers, 850 (150); liquid to liquid, 280 (50); liquid to gas, 60 (10); gas to gas 30 (5); reboiler 1140 (200). Maximum flux in reboiler 31.5 kW/m^2 (10,000 Btu/hr ft^2).

   When phase changes occur, use a zoned analysis with appropriate coefficient for each zone.
9. Double-pipe exchanger is competitive at duties requiring 9.3–18.6 m^2 (100–200 ft^2).
10. Compact (plate and fin) exchangers have 1150 m^2/m^3 (350 ft^2/ft^3), and about 4 times the heat transfer per cut of shell-and-tube units.
11. Plate and frame exchangers are suited to high-sanitation services and are 25–50% cheaper in stainless steel construction than shell-and-tube units.
12. Air coolers: Tubes are 0.75–1.0 in. OD., total finned surface 15–20 m^2/m^2 (ft^2/ft^2 bare surface), U = 450–570 W/m^2°C (80–100 Btu/hr ft^2 (bare surface) °F). Minimum approach temperature = 22°C (40°F). Fan input power = 1.4–3.6 kW/(MJ/h) [2–5 hp/(1000 Btu/hr)].
13. Fired heaters: radiant rate, 37.6 kW/m^2 (12,000 Btu/hr ft^2); convection rate, 12.5 kW/m^2 (4000 Btu/hr ft^2); cold oil tube velocity = 1.8 m/s (6 ft/sec); approximately equal transfer in the two sections; thermal efficiency 70–90% based on lower heating value; flue gas temperature 140–195°C (250–350°F) above feed inlet; stack gas temperature 345–510°C (650–950°F).

(Adapted from S. M. Walas, *Chemical Process Equipment: Selection and Design*, Stoneham, MA: Butterworth, 1988.

Table 11.12 Heuristics for Thermal Insulation

1. Up to 345°C (650°F), 85% magnesia is used.
2. Up to 870–1040°C (1600–1900°F), a mixture of asbestos and diatomaceous earth is used.
3. Ceramic (refractory) linings at higher temperature.
4. Cryogenic equipment −130°C (−200°F) employs insulation with fine pores of trapped air, e.g., Perlite.
5. Optimal thickness varies with temperature: 1.27 cm (0.5 in) at 95°C (200°F), 2.54 cm (1.0 in) at 200°C (400°F), 3.2 cm (1.25 in) at 315°C (600°F).
6. Under windy conditions 12.1 km/h (7.5 miles/hr), 10–20% greater thickness of insulation is justified.

(Adapted from S. M. Walas, *Chemical Process Equipment: Selection and Design,* Stoneham, MA: Butterworth, 1988.

Table 11.13 Heuristics for Towers (Distillation and Gas Absorption)

1. Distillation is usually the most economical method for separating liquids, superior to extraction, absorption crystallization, or others.
2. For ideal mixtures, relative volatility is the ratio of vapor pressures $\alpha_{12} = P_1^*/P_2^*$.
3. Tower operating pressure is most often determined by the temperature of the condensing media, 38–50°C (100–120°F) if cooling water is used, or by the maximum allowable reboiler temperature to avoid chemical decomposition/degradation.
4. Sequencing of columns for separating multicomponent mixtures:[a]
   a. Perform the easiest separation first, that is, the one least demanding of trays and reflux, and leave the most difficult to the last.
   b. When neither relative volatility nor feed composition varies widely, remove components one by one as overhead products.
   c. When the adjacent ordered components in the feed vary widely in relative volatility, sequence the splits in order of decreasing volatility.
   d. When the concentrations in the feed vary widely but the relative volatilities do not, remove the components in order of decreasing concentration.
5. Economical optimum reflux ratio is in the range of 1.2 to 1.5 times the minimum reflux ratio, R_{min}.
6. The economically optimum number of theoretical trays is near twice the minimum value N_{min}.
7. The minimum number of trays is found with the Fenske-Underwood equation
   $N_{min} = \ln\{[x/(1-x)]_{ovhd}/[x/(1-x)]_{btms}\}/\ln \alpha$.
8. Minimum reflux for binary or pseudobinary mixtures is given by the following when separation is essentially complete ($x_D \approx 1$) and D/F is the ratio of overhead product to feed rate:
   $R_{min}D/F = 1/(\alpha-1)$, when feed is at the bubble point
   $(R_{min}+1)\ D/F = \alpha/(\alpha-1)$, when feed is at the dew point
9. A safety factor of 10% of the number of trays calculated by the best means is advisable.
10. Reflux pumps are made at least 10% oversize.
11. The optimum value of the Kremser absorption factor $A = (L/mV)$ is in the range of 1.25 to 2.0.
12. Reflux drums usually are horizontal, with a liquid holdup of 5 min half-full. A takeoff pot for a second liquid phase, such as water in hydrocarbon systems, is sized for a linear velocity of that phase of 1.3 m/s (0.5 ft/sec), minimum diameter is 0.4 m (16 in).
13. For towers about 0.9 m (3 ft) dia, add 1.2 m (4 ft) at the top for vapor disengagement, and 1.8 m (6 ft) at bottom for liquid level and reboiler return.
14. Limit the tower height to about 53 m (175 ft) max. because of wind load and foundation considerations. An additional criterion is that L/D be less than 30 ($20 < L/D < 30$ often will require special design).

[a]Additional information on sequencing is given in Table 12.2.

(Adapted from S. M. Walas, *Chemical Process Equipment: Selection and Design,* Stoneham, MA: Butterworth, 1988.

Table 11.14 Heuristics for Tray Towers (Distillation and Gas Absorption)

1. For reasons of accessibility, tray spacings are made 0.5–0.6 m (20–24 in).
2. Peak efficiency of trays is at values of the vapor factor $F_s = u\rho^{0.5}$ in the range of 1.2–1.5 m/s $\{kg/m^3\}^{0.5}$ [1–1.2 ft/s $\{lb/ft^3\}^{0.5}$]. This range of F_s establishes the diameter of the tower. Roughly, linear velocities are 0.6 m/s (2 ft/sec) at moderate pressures, and 1.8 m/s (6 ft/sec) in vacuum.
3. Pressure drop per tray is on the order of 7.6 cm (3 in) of water or 0.007 bar (0.1 psi).
4. Tray efficiencies for distillation of light hydrocarbons and aqueous solutions are 60–90%; for gas absorption and stripping, 10–20%.
5. Sieve trays have holes 0.6–0.7 cm (0.25–0.5 in) dia., area being 10% of the active cross section.
6. Valve trays have holes 3.8 cm (1.5 in) dia. each provided with a liftable cap, 130–150 caps/m^2 (12–14 caps/ft^2) of active cross section. Valve trays are usually cheaper than sieve trays.
7. Bubblecap trays are used only when a liquid level must be maintained at low turndown ratio; they can be designed for lower pressure drop than either sieve or valve trays.
8. Weir heights are 5 cm (2 in), weir lengths are about 75% of tray diameter, liquid rate—a maximum of 1.2 m^3/min m of weir (8 gpm/in of weir); multipass arrangements are used at higher liquid rates.

(Adapted from S. M. Walas, *Chemical Process Equipment: Selection and Design*, Stoneham, MA: Butterworth, 1988.

Table 11.15 Heuristics for Packed Towers (Distillation and Gas Absorption)

1. Structured and random packings are suitable for packed towers less than 0.9 m (3 ft) when low pressure drop is required.
2. Replacing trays with packing allows greater throughput and separation in existing tower shells.
3. For gas rates of 14.2 m^3/min (500 ft^3/min), use 2.5 cm (1 in) packing; for 56.6 m^3/min (2000 ft^3/min) or more, use 5 cm (2 in) packing.
4. Ratio of tower diameter to packing diameter should be >15:1.
5. Because of deformability, plastic packing is limited to 3–4 m (10–15 ft) and metal to 6.0–7.6 m (20–25 ft) unsupported depth.
6. Liquid distributors are required every 5–10 tower diameters with pall rings, and at least every 6.5 m (20 ft) for other types of dumped packing.
7. Number of liquid distributors should be >32–55/m^2 (3–5/ft^2) in towers greater than 0.9 m (3 ft) diameter, and more numerous in smaller columns.
8. Packed towers should operate near 70% of flooding (evaluated from Sherwood and Lobo correlation).
9. Height equivalent to theoretical stage (HETS) for vapor-liquid contacting is 0.4–0.56 m (1.3–1.8 ft) for 2.5 cm (1 in) pall rings, and 0.76–0.9 m. (2.5–3.0 ft) for 5 cm (2 in) pall rings.
10. Generalized pressure drops

Generalized pressure drops	Design Pressure Drops (cm of H_2O/m of packing)	Design Pressure Drops (inches of H_2O/ft of packing)
Absorbers and regenerators (nonfoaming systems)	2.1–3.3	0.25–0.40
Absorbers and regenerators	0.8–2.1	0.10–0.25
Atmospheric/pressure stills and fractionators	3.3–6.7	0.40–0.80
Vacuum stills and fractionators	0.8–3.3	0.10–0.40
Maximum value	8.33	1.0

(Adapted from S. M. Walas, *Chemical Process Equipment: Selection and Design,* Stoneham, MA: Butterworth, 1988.

Table 11.16 Heuristics for Liquid-Liquid Extraction

1. The dispersed phase should be the one with the higher volumetric flowrate except in equipment subject to back-mixing, where it should be the one with the smaller volumetric rate. It should be the phase that wets material of construction less well. Because the holdup of continuous phase is greater, that phase should be made up of the less expensive or less hazardous material.
2. There are no known commercial applications of reflux to extraction processes, although the theory is favorable.
3. Mixer-settler arrangements are limited to at most five stages. Mixing is accomplished with rotating impellers or circulation pumps. Settlers are designed on the assumption that droplet sizes are about 150 μm dia. In open vessels, residence times of 30–60 min or superficial velocities of 0.15–0.46 m/min (0.5–1.5 ft/min) are provided in settlers. Extraction stage efficiencies commonly are taken as 80%.
4. Spray towers as tall as 6–12 m (20–40 ft) cannot be depended on to function as more than a single stage.
5. Packed towers are employed when 5–10 stages suffice. Pall rings 2.5–3.8 cm (1–1.5 in) size are best. Dispersed phase loadings should not exceed 10.2 $m^3/min\ m^2$ (25 $gal/min\ ft^2$). HETS of 1.5–3.0 m (5–10 ft) may be realized. The dispersed phase must be redistributed every 1.5–2.1 m (5–7 ft). Packed towers are not satisfactory when the surface tension is more than 10 dyne/cm.
6. Sieve tray towers have holes of only 3–8 mm dia. Velocities through the holes are kept less than 0.24 m/s (0.8 ft/sec) to avoid formation of small drops. Redispersion of either phase at each tray can be designed for. Tray spacings are 15.2 to 60 cm (6 to 24 in). Tray efficiencies are in the range of 20%–30%.
7. Pulsed packed and sieve tray towers may operate at frequencies of 90 cycles/min. and amplitudes of 6–25 mm. In large-diameter towers, HETS of about 1 m have been observed. Surface tensions as high as 30–40 dyne/cm have no adverse effect.
8. Reciprocating tray towers can have holes 1.5 cm (9/16 in) dia., 50–60% open area, stroke length 1.9 cm (0.75 in), 100–150 strokes/min, plate spacing normally 5 cm (2 in) but in the range of 2.5–15 cm (1–6 in). In a 76 cm (30 in) diameter tower, HETS is 50–65 cm (20–25 in) and throughput is 13.7 $m^3/min\ m^2$ (2000 $gal/hr\ ft^2$). Power requirements are much less than that of pulsed towers.
9. Rotating disk contactors or other rotary agitated towers realize HETS in the range of 0.1–0.5 m (0.33–1.64 ft). The especially efficient Kuhni with perforated disks of 40% free cross section has HETS of 0.2 m (0.66 ft) and a capacity of 50 $m^3/m^2\ h$ (164 $ft^3/ft^2\ hr$).

(Adapted from S. M. Walas, *Chemical Process Equipment: Selection and Design,* Stoneham, MA: Butterworth, 1988.

Table 11.17 Heuristics for Reactors

1. The rate of reaction in every instance must be established in the laboratory, and the residence time or space velocity and product distribution eventually must be found from a pilot plant.
2. Dimensions of catalyst particles are 0.1 mm (0.004 in) in fluidized beds, 1 mm in slurry beds, and 2–5 mm (0.078–0.197 in) in fixed beds.
3. The optimum proportions of stirred tank reactors are with liquid level equal to the tank diameter, but at high pressures slimmer proportions are economical.
4. Power input to a homogeneous reaction stirred tank is 0.1–0.3 kW/m^3 (0.5–1.5 hp/1000 gal), but three times this amount when heat is to be transferred.
5. Ideal CSTR (continuous stirred tank reactor) behavior is approached when the mean residence time is 5 to 10 times the length needed to achieve homogeneity, which is accomplished with 500–2000 revolutions of a properly designed stirrer.
6. Batch reactions are conducted in stirred tanks for small daily production rates or when the reaction times are long or when some condition such as feed rate or temperature must be programmed in some way.
7. Relatively slow reactions of liquids and slurries are conducted in continuous stirred tanks. A battery of four or five in series is most economical.
8. Tubular flow reactors are suited to high production rates at short residence times (seconds or minutes) and when substantial heat transfer is needed. Embedded tubes or shell-and-tube construction then is used.
9. In granular catalyst packed reactors, the residence time distribution is often no better than that of a five-stage CSTR battery.
10. For conversion less than about 95% of equilibrium, the performance of a five-stage CSTR battery approaches plug flow.
11. The effect of temperature on chemical reaction rate is to double the rate every 10°C.
12. The rate of reaction in a heterogeneous system is more often controlled by the rate of heat or mass transfer than by the chemical reaction kinetics.
13. The value of a catalyst may be to improve selectivity more than to improve the overall reaction rate.

(Adapted from S. M. Walas, *Chemical Process Equipment: Selection and Design,* Stoneham, MA: Butterworth, 1988.

Table 11.18 Heuristics for Refrigeration and Utility Specifications

1. A ton of refrigeration is the removal of 12,700 kJ/h (12,000 Btu/hr) of heat.
2. At various temperature levels: −18 to −10°C (0 to 50°F), chilled brine and glycol solutions; −45 to −10°C (−50 to −40°F), ammonia, freon, butane; −100 to −45°C (−150 to −50°F) ethane or propane.
3. Compression refrigeration with 38°C (100°F) condenser requires kW/tonne (hp/ton) at various temperature levels; 0.93 (1.24) at −7°C (20°F); 1.31 (1.75) at −18°C (0°F); 2.3 (3.1) at −40°C (−40°F); 3.9 (5.2) at −62°C (−80°F).
4. At less than −62°C (−80°F), cascades of two or three refrigerants are used.
5. In single-stage compression, the compression ratio is limited to 4.
6. In multistage compression, economy is improved with interstage flashing and recycling, so-called economizer operation.
7. Absorption refrigeration: ammonia to −34°C (−30°F), lithium bromide to 7°C (+45°F) is economical when waste steam is available at 0.9 barg (12 psig).
8. Steam: 1–2 barg (15–30 psig), 121–135°C (250–275°F); 10 barg (150 psig), 186°C (366°F); 27.6 barg (400 psig), 231°C (448°F); 41.3 barg (600 psig), 252°C (488°F) or with 55–85°C (100–150°F) superheat.
9. Cooling water: For design of cooling tower use supply at 27–32°C (80–90°F) from cooling tower, return at 45–52°C (115–125°F); return seawater at 43°C (110°F); return tempered water or steam condensate above 52°C (125°F).
10. Cooling air supply at 29–35°C (85–95°F); temperature approach to process, 22°C (40°F).
11. Compressed air 3.1 (45), 10.3 (150), 20.6 (300), or 30.9 barg (450 psi) levels.
12. Instrument air at 3.1 barg (45 psig), −18°C (0°F) dew point.
13. Fuels: gas of 37,200 kJ/m^3 (1000 Btu/SCF) at 0.35–0.69 barg (5–10 psig), or up to 1.73 barg (25 psig) for some types of burners; liquid at 39.8 GJ/m^3 (6 million Btu/bbl).
14. Heat transfer fluids: petroleum oils less than 315°C (600°F), Dowtherms less than 400°C (750°F), fused salts less than 600°C (1100°F), direct fire or electricity above 450°F.
15. Electricity: 0.75–74.7 kW. (1–100 hp), 220–550 V; 149–1864 kW (200–2500 hp), 2300–4000 V.

(Adapted from S. M. Walas, *Chemical Process Equipment: Selection and Design,* Stoneham, MA: Butterworth, 1988.

REFERENCES

1. Koen, B. V., *Definition of the Engineering Method* (Washington, DC: American Society for Engineering Education, 1985).
2. Sieder, E. N., and G. E. Tate, "Heat Transfer and Pressure Drop of Liquids in Tubes," *Ind. Eng. Chem.* 28 (1936): 1429.
3. Walas, S. M., *Chemical Process Equipment: Selection and Design* (Stoneham, MA: Butterworths, 1988).

PROBLEMS

1. For the ethylbenzene process shown in Appendix B, check the design specifications for the following three pieces of equipment against the appropriate heuristics: P-301, V-302, T-302. Comment on any significant differences that you find.
2. For the styrene process shown in Appendix B, check the design specifications for the following three pieces of equipment against the appropriate heuristics: E-401, C-401, T-402. Comment on any significant differences that you find.
3. For the drying oil shown in Appendix B, check the design specifications for the following three pieces of equipment against the appropriate heuristics: V-501, P-501, H-501. Comment on any significant differences that you find.

CHAPTER

12

Synthesis of the PFD from the Generic BFD

The evolutionary procedure to create a full PFD (as presented in Chapter 1) from the generic block flow process diagram (GBFD) (Chapter 2) is described in this chapter. This full PFD truly defines the process in a chemical engineering sense and is the starting point for chemical and other engineers to design the machines, structures, and electrical/electronic components needed to make the chemical engineer's vision a reality.

This crucial step in the design of the chemical plant involves all subareas of chemical engineering: reaction engineering, thermodynamics, process control, unit operations and transport, and material and energy balances. Each is applied to put details into the six general sections of the GBFD—reactor feed preparation, reactor, separator feed preparation, separator, recycle, and environmental control.

In this synthesis, the broader context of the project (e.g., environmental concerns, customer expectations, return on investment) is integrated with the important details such as the type of heat-transfer medium or the number of stages in a column. It is crucial to consider as many alternatives as possible in the early stages to try to avoid becoming trapped in a suboptimal design.

It is a common human trait to resist change more strongly as more effort is expended on a task, design, or product. We describe this as not wanting to abandon our "investment" in the activity.

12.1 INFORMATION NEEDS AND SOURCES

Before the detailed synthesis of the PFD can be completed, one needs basic physical property and kinetics information. We assume here that the very basic chemistry of the desired reaction is known, that is, what main feed materials go to what main product. Before PFD synthesis can begin, the marketing engineers have identified a market need for a specific product, and the chemists have identified at least one way to produce the chemical in the laboratory. Even the marketing and chemistry information, however, will need to be refined. Flowsheet synthesis will uncover the need for more detailed data on the reaction rate, temperature and pressure effects, and market values of products of different purities.

12.1.1 Interactions with Other Engineers and Scientists

Teams of engineers work on the development of the process. For example, the marketing department will find the customer for the plant's product, and product specifications will be identified. Many chemical engineers are employed as marketing engineers, and they will understand that product purity affects product price, often dramatically so. However, the details of this interplay can be determined only by the process design engineers as the PFD is being developed; only through discussions and negotiations with customers can the marketing engineer determine the relationship between product purity and the product value (i.e., maximum selling price) to the customer.

Similarly, there may be more than one chemical pathway to the product. Pathways of greatest interest to the chemical engineer are not necessarily those of greatest interest to the chemist. The abilities to use impure feed materials and to avoid the production of by-products reduce costs but may not be of interest to a chemist. The costs of small-lot, high-purity laboratory reagents may not even qualitatively correlate to those of multiple tank-car, industrial-grade raw materials. Isothermal operation of small laboratory reactors is common but essentially impossible to achieve on a large scale. It is more economical per unit volume to maintain high pressures on the plant scale than it is in the lab. Simple batch operations are common in laboratory work, but, at plant scale, sophisticated optimization of scheduling, ramp rates, cycle sequencing, and choice of operating mode (batch, semibatch, continuous) is vital. Thus, the chemical process design engineer must be in touch with the chemist to make sure that expensive constraints or conditions suggested by laboratory studies are truly needed.

12.1.2 Reaction Kinetics Data

Before reactor design can begin, the kinetics of the main reaction must be known. However, a knowledge of the kinetics of unwanted side reactions is also crucial to the development of PFD **structure** or **topology** (number and position of recycle streams; types, numbers, and locations of separators; batch or continuous operating modes; sterilization operations needed for aseptic operation). Knowledge of detailed reaction pathways, elementary reactions, and unstable reaction intermediates is not required. Rather, the chemical process design engineer needs to know the rate of reaction (main and by-product reactions) as a function of temperature, pressure, and composition. The greater the range of these independent variables, the better the design can be.

For some common homogeneous reactions, kinetics are available [1,2,3]. However, most commercial reactions involve catalysts. The competitive advantage of the company is often the result of a unique catalyst. Thus, kinetics data for catalyzed reactions are not as readily available in open literature but should be available within the company files or must be obtained from experiments. One source of kinetics data for catalytic reactions is the patent literature. The goal of someone writing a patent application, however, is to present as little data as possible about the invention while obtaining the broadest possible protection. This is why patent information is often cryptic. However, this information is often sufficient to develop a base-case PFD. The key data to obtain from the patent are the inlet composition, temperature, pressure, outlet composition, and space time. If the data are for varying compositions, one can develop crude kinetics rate expressions. If the data are for more than one temperature, an activation energy can be determined. These data reduction procedures are described in undergraduate textbooks on reaction engineering [4,5].

Without kinetics data, a preliminary PFD and cost analysis can still be done [6]. In this type of analysis, the differing process configurations and costs for different assumed reaction rates provide estimates of the value of a potential catalyst. If doubling the reaction rate reduces the cost of manufacture by $1 million per year, for example, the value of catalysis research to increase the reaction rate (all other things being equal) is clear. As a guideline, the economic breakpoint is often a catalyst productivity to desired product of ~0.10 kg product per kg catalyst per hour [7]. Another guide is that activation energies are usually between 40 and 200 kJ/mol.

12.1.3 Physical Property Data

In addition to kinetics data, physical property data are required for determining material and energy balances, as well as for sizing of heat exchangers, pumps and compressors, and separation units. These data are, in general, easier to obtain and, when necessary, easier than kinetics data to estimate.

For the material and energy balances, pure-component heat capacity and density data are needed. These are among the most widely measured data and are available on process simulators for more than a thousand substances. (See Chapter 13 for details of process simulators.) There are also reasonably accurate group-contribution techniques for use when no data are available [8]. The enthalpies of mixtures require an accurate equation of state for gases and nonionic liquids. The equations of state available on process simulators are accurate enough for these systems. However, additional heat of solution data are needed for electrolyte solutions, and these data may not be as readily available. For these systems, care should be taken to use accurate experimental data, because estimation techniques are not as well defined.

The design of heat exchangers and the determination of pressure drops across units require thermal conductivity and viscosity data. These data are usually available (often in the databanks of process simulators) and, if unavailable, can be estimated by group-contribution techniques [8].

The most crucial and least available physical property data are for phase equilibrium. Most separators are based on equilibrium stages; thus, these data are usually needed for a process design. For vapor-liquid equilibrium, such as for distillation, either (1) a single equation of state for both phases or (2) a combination of vapor-phase equation of state, pure-component vapor pressure, and liquid-state activity coefficient model is required. The choice of thermodynamics package for process simulators is explained in Section 13.4. The key experimentally determined mixture parameters for either equations of state or activity-coefficient models are called **BIPs** (**binary interaction parameters**), and they have great effect on the design of separation units. A poor estimation of them (e.g., assuming them to be zero!) can lead to severely flawed designs. The solubilities of noncondensables in the liquid phase are also essential but difficult to estimate.

12.2 REACTOR SECTION

For a process with a reactor, often the synthesis of the PFD begins with the reactor section of the GBFD. (See Chapter 20.) A base-case reactor configuration is chosen according to the procedures described in reaction engineering textbooks. This configuration (e.g., plug flow, CSTR, batch, semibatch, adiabatic, isothermal) is used at some base conditions (temperature, pressure, feed composition) and some preliminary base specification (e.g., 60% conversion) to calculate the outlet composition, pressure, and temperature. The goal at this stage is to develop a feasible PFD for the process. Optimization of the PFD can begin only after a suitable base case is developed. If there are obvious choices that improve the process (such as using a fluidized bed instead of a packed bed reactor, or batch operation instead of continuous), these choices are made at this stage; however, these choices should be revisited later.

To enable later optimization, the general effects of varying the feed conditions should be investigated at this point by using the trend prediction approach of Chapter 17. A list of possible reactor configurations should also be developed. These choices often have dramatic effects on the other parts of the GBFD. The earlier these effects are understood, the better the final design will be.

At this stage, the utility needs of the reactor should be considered. If heating or cooling is required, the design of an entire additional system may be required. The choice of heating or cooling medium must be made based on strategies described in Chapter 15, the heuristics presented in Chapter 11, and the costs of these utilities.

The trade-offs of different catalysts, parallel versus series reactors, and conversion versus selectivity should be considered, even though the optimization of these choices occurs after the base case is developed. Again, early identification of alternatives improves later detailed optimization.

Once the base-case reactor configuration is chosen, the duties of the reactor feed preparation and separator feed preparation units are partially determined.

For the reactor, important questions to be considered include the following.

1. *In what phase does the reaction take place (liquid, vapor, mixed, etc.)?* The answer will affect the reactor feed section. For example, it will determine whether a vaporizer or fired heater is required upstream of the reactor when the feed to the plant is liquid.
2. *What are the required temperature and pressure ranges for the reactor?* If the pressure is higher than the feed pressure, pumps or compressors are needed in the reactor feed preparation section. If the required reactor feed temperature is greater than approximately 250°C, a fired heater is probably necessary.
3. *Is the reaction kinetically or equilibrium controlled?* The answer affects both the maximum single-pass conversion and the reactor configuration. The majority of gas- and liquid-phase reactions in the CPI are kinetically controlled. The most notable exceptions are the formation of methanol from synthesis gas, synthesis of ammonia from nitrogen and hydrogen, and the production of hydrogen via the water-gas shift reaction.
4. *Does the reaction require a solid catalyst, or is it homogeneous?* This difference dramatically affects the reactor configuration. For enzymes immobilized on particles, for example, a fluidized bed reactor or packed-bed reactor could be considered, depending on stability of the enzyme and mass-transfer requirements. The immobilization may also impart some temperature stability to the enzyme, which provides additional flexibility in reactor configuration and operating conditions.
5. *Is the main reaction exothermic or endothermic, and is a large amount of heat exchange required?* Again, the reactor configuration is more affected by the heat transfer requirements. For mildly exothermic or endothermic gas-phase reactions, multiple packed beds of catalyst or shell-and-tube reactors (catalyst in tubes) are common. For highly exothermic gas-phase reactions, heat transfer is the dominant concern, and fluidized beds or shell-and-tube reactors with catalyst dilution (with inert particles) are used. For liquid-phase reactions, temperature control can be achieved by pumping the reacting mixture through external heat exchangers (for example, in Figure B.11.1). For some highly exothermic reactions, part of the reacting mixture is vaporized to help regulate the temperature. The vapor is subsequently condensed and returned to the reactor. External jackets and internal heat transfer tubes,

plates, or coils may also be provided for temperature control of liquid-phase reactions. (See Chapters 20 and 21.)

6. *What side reactions occur, and what is the selectivity of the desired reaction?* The formation of unwanted by-products may significantly complicate the separation sequence. This is especially important if these by-products are formed in large quantities and are to be purified for sale. For high-selectivity reactions, it may be more economical to dispose of by-products as waste or to burn them (if they have high heating values), which simplifies the separation section. However, for environmental reasons, great emphasis is placed on producing either salable by-products or none at all.
7. *What is the approximate single-pass conversion?* The final single-pass conversion is determined from detailed parametric optimizations (Chapter 14); however, the range of feasible single-pass conversions affects the structure of the separations section. If extremely high single-pass conversions are possible (e.g., greater than ~98%), it may not be economical to separate and recycle the small amounts of unreacted feed materials. In this case, the feed materials become the impurities in the product, up to the allowable concentration.
8. *For gas-phase oxidations, should the reactor feed be outside the explosive limits?* For example, there are many reactions that involve the partial oxidation of hydrocarbons (see acrylic acid production in Appendix B and phthalic anhydride production in Appendix C [on the CD]). Air or oxygen is fed to a reactor along with hydrocarbons at high temperature. The potential for explosion from rapid, uncontrolled oxidation (ignition) is possible whenever the mixture is within its explosive limits. (Note that the explosive limits widen significantly with increase in temperature.) An inherently safe design would require operation outside these limits. Often, steam is added both as a diluent and to provide thermal ballast for highly exothermic reactions—for example, in the acrylic acid reactor (Figure B.9.1).

12.3 SEPARATOR SECTION

After the reactor section, the separator section should be studied. The composition of the separator feed is that of the reactor effluent, and the goal of the separator section is to produce a product of acceptable purity, a recycle stream of unreacted feed materials, and a stream or streams of by-products. The ideal separator used in the GBFD represents a process target, but it generally represents a process of infinite cost. Therefore, one step is to "de-tune" the separation to a reasonable level. However, before doing that, one must decide what the by-product streams will be. There may be salable by-products, in which case a purity specification is required from the marketing department. For many organic chemical plants, one by-product stream is a mixture of combustible gases or liquids that are then used as fuel. There may also be a waste stream (often a dilute aqueous stream) to be treated downstream; however, this is an increasingly less desirable process feature.

Prior to enactment of current environmental regulations, it was generally thought to be less expensive to treat waste streams with so-called end-of-pipe operations. That is, one produced, concentrated, and disposed of the waste in an acceptable manner. As regulations evolved, the strategy of pollution prevention or green engineering has led to both better environmental performance and reduced costs. More details are given in Chapter 25, but the overall strategy is to minimize wastes at their source or to turn them into salable products.

The separation section then generally accepts one stream from the pre-separation unit and produces product, by-product, and (sometimes) waste streams. In the development of the PFD, one must consider the most inclusive or flexible topology so that choices can be made in the optimization step. Thus, each type of stream should be included in the base case.

Next, the minimum number of simple separation units must be determined. Although there are single units that produce multiple output streams (such as a petroleum refining pipe still with many side draws), most units accept a single inlet stream and produce two outlet streams. For such simple separators, we need at least (N–1) units, where N is the number of outlet streams (products, by-products, and waste). There are two types of questions to answer concerning these units in the separation section: (1) What types of units should be used? and (2) How should the units be sequenced?

12.3.1 General Guidelines for Choosing Separation Operations

There are general guidelines concerning choice of separation unit. Table 12.1 gives a set of rules for the most common choices of separation units on process simulators.

For a base case, it is essential that the separation technique chosen be reasonable, but it is not necessary that it be the best.

For sequencing of the separation units, there is another set of guidelines, given in Table 12.2. In the base case, it is often helpful to consider the same type of separator for each unit. During optimization, one can compare different separator types for the different duties. Again, some separators can do multiple separations in one unit, but these can be found during optimization. Additional heuristics for separation unit sequencing are given in Table 11.13 and in reference [9].

As with all sets of heuristics, these can be mutually contradictory. However, in the initial topology of the separation section, the main goal is to follow as many of these heuristics as possible.

Beyond these general guidelines, beware of azeotropes and multiple phases in equilibrium, especially when water and organics are present. Special techniques are available to deal with these problems, some of which are discussed later. On the other hand, if a single-stage flash will do the separation, do not use a column with reflux.

Table 12.1 Guidelines for Choosing Separation Units

- Use distillation as a first choice for separation of fluids when purity of both products is required.
- Use gas absorption to remove one trace component from a gas stream.
- Consider adsorption to remove trace impurities from gas or liquid streams.
- Consider pressure-swing adsorption to purify gas streams, especially when one of the components has a cryogenic boiling point.
- Consider membranes to separate gases of cryogenic boiling point and relatively low flowrates.
- Choose an alternative to distillation if the boiling points are very close or if the heats of vaporization are very high.
- Consider extraction as a choice to purify a liquid from another liquid.
- Use crystallization to separate two solids or to purify a solid from a liquid solution.
- Use evaporation to concentrate a solution of a solid in a liquid.
- Use centrifugation to concentrate a solid from a slurry.
- Use filtration to remove a solid in almost dry form from a slurry.
- Use screening to separate solids of different particle size.
- Use float/sink to separate solids of different density from a mixture of pure particles.
- Consider reverse osmosis to purify a liquid from a solution of dissolved solids.
- Use leaching to remove a solid from a solid mixture.
- Consider chromatography for final purification of high-value products (such as proteins) from dilute streams.

Table 12.2 Guidelines for Sequencing Separation Units

- Remove the largest product stream first. This makes all of the subsequent separation units smaller.
- For distillation, remove the product with the highest heat of vaporization first, if possible. This reduces the heating/cooling duties of subsequent units.
- Do not recombine separated streams. (This may seem obvious, but it is often disobeyed.)
- Do the easy separations first.
- Do not waste raw materials, and do not overpurify streams based on their uses.
- Remove hazardous or corrosive materials first.
- Use the less expensive, cruder separation technique first (e.g., liquid-liquid extraction before chromatography).

For the separation section, other important questions to be considered include the following.

1. *What are the product specifications for all products?* Product specifications are developed to satisfy customers who will use these products in their own processes. The most common specification is a minimum concentration of the main constituent, such as 99.5 wt%. Maximum impurity levels for specific contaminants may also be specified, as well as requirements for specific physical properties such as color, odor, and specific gravity. A single separation technique may not be sufficient to meet all the required product specifications, as demonstrated in Example 12.1.

Example 12.1

In the production of benzene via the hydrodealkylation of toluene, it is necessary to produce a benzene product stream that contains >99.5 wt% benzene that is water white in color (i.e., absolutely clear). If the feed toluene to the process contains a small amount of color, determine a preliminary separation scheme to produce the desired benzene product.

As a guide, we can look at the toluene hydrodealkylation process shown in Figure 1.5. Because the volatilities of toluene and benzene are significantly different, the main purification step (the separation of benzene from toluene) can be accomplished using distillation, which is consistent with Figure 1.5. However, it has been found that the compound causing the discoloration of the toluene is equally soluble in benzene and toluene, causing the benzene product to be discolored. It is further found by laboratory testing that the benzene product can be decolorized by passing it through a bed of activated carbon. Thus, a second separation step, consisting of an activated carbon adsorber, will be added to the process to decolorize the benzene product.

---

2. *Are any of the products heat sensitive?* If any of the products or by-products are heat sensitive (i.e., they decompose, deactivate, or polymerize at elevated temperatures), the conditions used in the separations section may have to be adjusted, as in Example 12.2.

Example 12.2

It is known that acrylic acid starts to polymerize at 90°C when it is in a concentrated form. Acrylic acid must be separated from acetic acid to produce the required purity product, and the volatilities of both acids are significantly different. This points to distillation as the separation method. The normal boiling points of acrylic acid and acetic acid are 140°C and 118°C, respectively. How should the separation be accomplished to avoid degradation of the acrylic acid product?

The distillation column must be run under vacuum to avoid the problem of acrylic acid degradation. The pressure should be set so that the bottom temperature of the column is less than 90°C. From Figure B.9.1 and Table B.9.1, we see that a column pressure of 0.16 bar at the bottom can accomplish the desired separation without exceeding 90°C.

---

3. *Are any of the products, by-products, or impurities hazardous?* Because separation between components is never perfect, small quantities of toxic or hazardous components may be present in product, fuel, or waste streams. Additional purification or subsequent processing of these streams may be required, depending on their end use.

12.3.2 Sequencing of Distillation Columns for Simple Distillation

Because distillation is still the prevalent separation operation in the chemical industry, it will now be discussed in more detail. **Simple distillation** can be defined as distillation of components without the presence of any thermodynamic anomalies. The most apparent thermodynamic anomaly in distillation systems is an azeotrope. Azeotropic distillation is discussed in the next section. The remainder of this section is for simple distillation.

As stated earlier, as a general guideline, a minimum of N–1 separators are needed to separate N components, and this guideline also applies to distillation systems. Therefore, one distillation column is required to purify both components from a two-component feed. This is the type of problem most often studied in separation classes. To purify a three-component feed into three "pure" components, two distillation columns are required. However, there are two possible sequences, and these are illustrated in Figures 12.1(a) and 12.1(b). Ultimately, the choice of sequence depends upon the economics. However, the results of the economic analysis often follow the guidelines in Table 12.2. For example, if the heavy component (C) is water, it should be removed first due to its high heat of vaporization, so the sequence in Figure 12.1(b) is likely to be more economical for such a situation. This is because the heating and cooling duties in the second column are reduced significantly if the water is removed first. The sequence in Figure 12.1(b) is also likely to be a better choice if component C is present in the largest amount, or if component C is the only corrosive component. This is because in the former case, the second column will be smaller, and in the latter case, the second column may not need the expensive materials of construction needed in the first column.

It must be understood that more than N–1 distillation columns is permissible. There are cases that have been reported where the sequence in Figure 12.1(c) is actually more economical than either of the sequences in Figures 12.1(a) and 12.1(b). This occurs because the sequence in Figure 12.1(c) has lower utility costs that offset the capital cost of the extra column and peripherals [10]. Actually, the sequence represented in Figure 12.1(c) can be accomplished in one column, a partitioned column with a vertical baffle, as illustrated in Figure 12.1(d) [10]. The presence of the vertical baffle makes the column behave like the three columns in Figure 12.1(c). The lesson learned here is that distillation practice can be very different from distillation theory. Textbooks usually state unequivocally that N–1 distillation columns are required for separation of N components. However, the

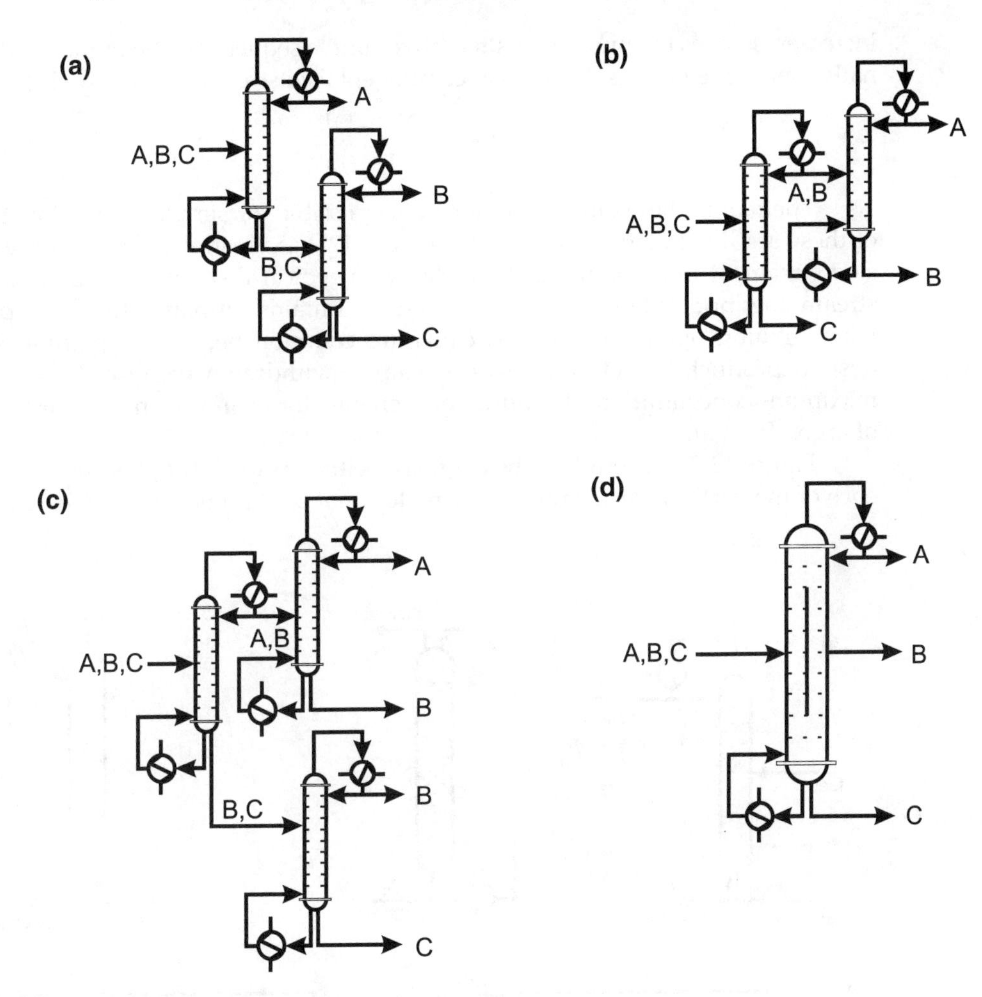

Figure 12.1 Column Arrangements for Simple Distillation of Three-Component Feed

two examples presented here demonstrate that distillation practice does not always conform to theory.

Other distillation column configurations are possible [11–13]. In these references, there are column configurations known as Petlyuk-type columns. It should be noted that even though the configuration in Figure 12.1(c) looks like a Petlyuk–type II column, it is not, due to the presence of a reboiler and a condenser on the first column, units that are not present in the Petlyuk–type II column.

Figures 12.1(a) and (b) show the two theoretical simple sequences possible for separating three components in simple distillation. As the number of components increases, so does the number of alternative simple sequences. The number of alternative simple sequences, S, for an N-component feed stream is given by [14].

$$S = \frac{[2(N-1)]!}{N!(N-1)!} \tag{12.1}$$

There are other column arrangements possible for simple distillation. Some of these are illustrated in Figure 12.2. Figure 12.2(a) illustrates distillation with a side stream. It must be understood that in a typical distillation column, a side stream does not contain a pure component; it contains a mixture. In certain petroleum refining operations, side streams are common because a mixture is the desired product. Sometimes a side stream is withdrawn because it contains a maximum concentration of a third component—for example, in the purification of argon from air.

Figure 12.2(a) should not be confused with Figure 12.1(d), because the presence of the vertical baffle in the latter makes that column behave differently. Fig-

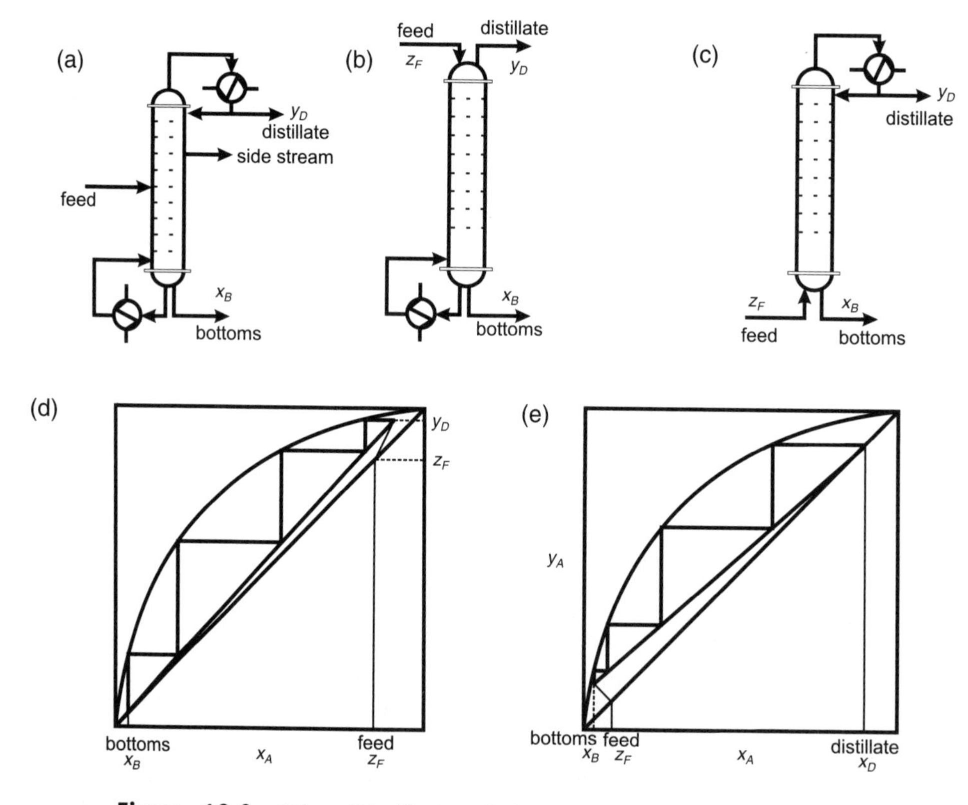

Figure 12.2 Other Distillation Column Arrangements for Simple Distillation

ure 12.2(b) represents a stripping column (sometimes called a reboiled absorber), and Figure 12.2(c) represents a rectifying column. A stripping column is used when the light components are very dilute and when they need not be purified. A rectifying column is used when the heavy components are very dilute and need not be purified. The performance of the stripping and rectifying columns can be understood from their McCabe-Thiele representations, shown in Figures 12.2(d) and 12.2(e), respectively.

12.3.3 Azeotropic Distillation

Distillation involving azeotropes does not conform to the guidelines discussed in Section 12.3.2. This is best illustrated by examining the thermodynamic behavior of a binary homogeneous azeotrope, illustrated in Figure 12.3. Clearly, no McCabe-Thiele construction can be made to produce two pure products from the indicated feed. The heavy component can be purified, but an infinite number of stages above the feed would be required just to approach the azeotropic concentration, (x_{az}, y_{az}), the point where the equilibrium curve crosses the 45° line. There are minimum- and maximum-boiling azeotropes, although minimum-boiling azeotropes (where the azeotrope is at a lower temperature than either pure-component boiling point) are more common. Minimum-boiling azeotropes arise from repulsive forces between molecules. One way of thinking about azeotropes is that the volatility switches. In Figure 12.3, Component A is more volatile when it is present below the azeotropic composition, but Component B is more volatile when Component A is present above the azeotropic composition. Given that simple distillation exploits the difference in volatility between components, it is easy to understand how the switch in volatility makes simple distillation impossible.

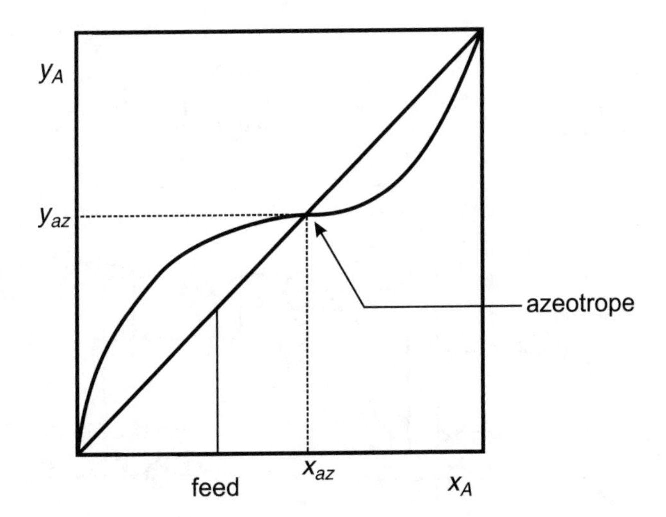

Figure 12.3 *X-Y* Diagram for Component A Forming a Minimum-Boiling, Binary Azeotrope with Component B (Constant Pressure)

In the next section, methods for accomplishing distillation-based separations in the presence of azeotropes for binary systems are discussed. In the subsequent section, methods for azeotropic distillation involving three components are discussed.

Binary Systems. The methods used to distill beyond azeotropes in binary systems are illustrated using McCabe-Thiele diagrams. Four of the more popular methods are discussed here. A more complete discussion is available in any standard separations textbook [15].

The more popular methods for breaking binary azeotropes include the following.

1. One distillation column allows separation close to the azeotrope, and then a different separation method is used to complete the purification (Figure 12.4[a]).
2. If the azeotrope is a binary, heterogeneous azeotrope, that means there is a region where the two components form two mutually immiscible liquid phases (both in equilibrium with the azeotropic vapor composition, which

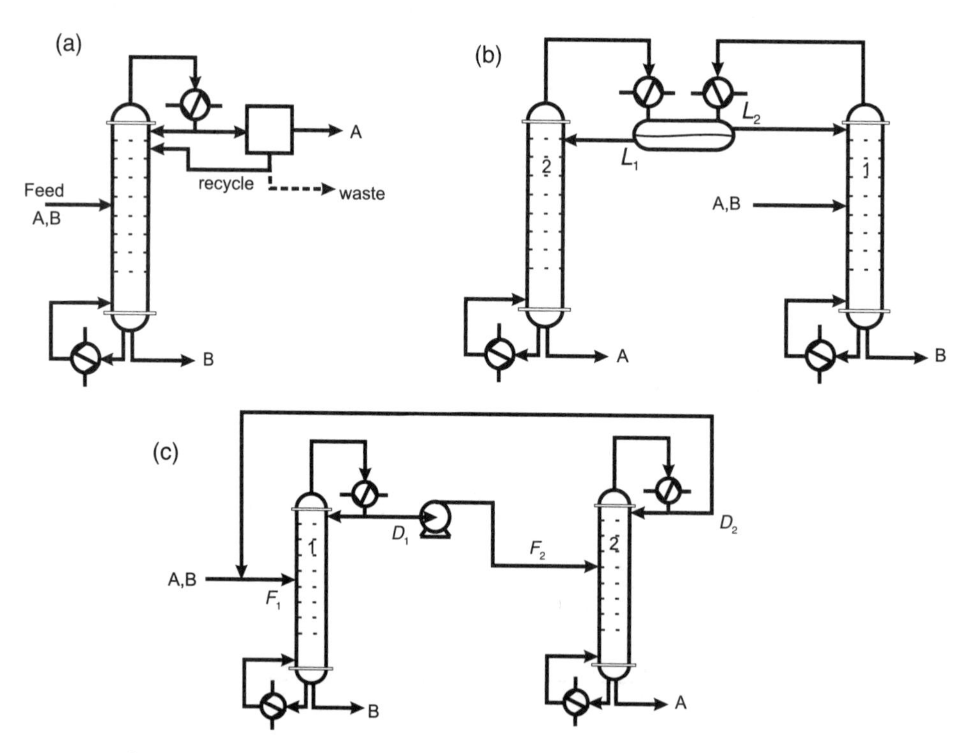

Figure 12.4 Distillation Arrangements to Separate Binary Azeotropes

is between the two liquid compositions), and hence a phase separator and a second column are added. The phase separator provides one phase on the other side of the azeotrope from the feed, so that phase can be purified by distillation in a second column (Figure 12.4[b]).

3. If the azeotrope concentration is pressure sensitive, one column is used to distill close to the azeotrope. One "pure" component is produced from the bottom of this first column. Then the pressure of the distillate is raised so that the azeotropic composition is now below the distillate composition. Then a second column is used to purify the second component. Because the volatility switches as the azeotrope is crossed, the second component is in the bottom stream of the second column (Figure 12.4[c]). This particular sequence assumes that the azeotrope is less concentrated in A at higher pressures.
4. A third component can be added to change the phase behavior. This method is discussed in the next section.

The first alternative is illustrated in Figure 12.4(a). Here, a distillation column is used that will provide "pure" B in the bottoms and a near-azeotropic mixture in the distillate. The McCabe-Thiele construction for this distillation column is illustrated in Figure 12.5(a). Then a different type of separation is used to purify Component A (not shown on the McCabe-Thiele diagram). The impure stream from the second separator should be recycled, if possible; however, treatment as a waste stream is also possible. The recycle stream will most likely be fed to a different tray from the feed stream, because distillation column feeds should always be near the point in the column with the same concentration as the feed stream.

An example of this situation is the ethanol-water system, which has a binary azeotrope in the 90–95 mol% ethanol range (depending on system pressure). A relatively recent method for purifying ethanol beyond the azeotropic composition is **pervaporation** [16]. The following question may arise when considering this method: Why not just use pervaporation (or whatever second separation method is possible) for the entire separation? The answer is that, even with volatile energy prices, the relatively low cost of energy makes distillation a very economical separation method [16]. In most cases, an arrangement like Figure 12.4(a) is far less expensive than using the second separation method alone. This is because separations such as distillation are very economical for producing relatively pure products from roughly equal mixtures. Obtaining ultrapure products from distillation can have unfavorable economics, because large numbers of trays are required for very high purity (think about the McCabe-Thiele construction). Separations like pervaporation (or any membrane separation) have much more favorable economics when removing a dilute component from a relatively pure component. They are also economically unattractive for large processing volumes. Therefore, the combination of distillation and another separation like pervaporation usually provides the economic optimum.

In cases where the two components being distilled form an azeotrope with two immiscible liquid phases, the method illustrated in Figure 12.4(b) can be

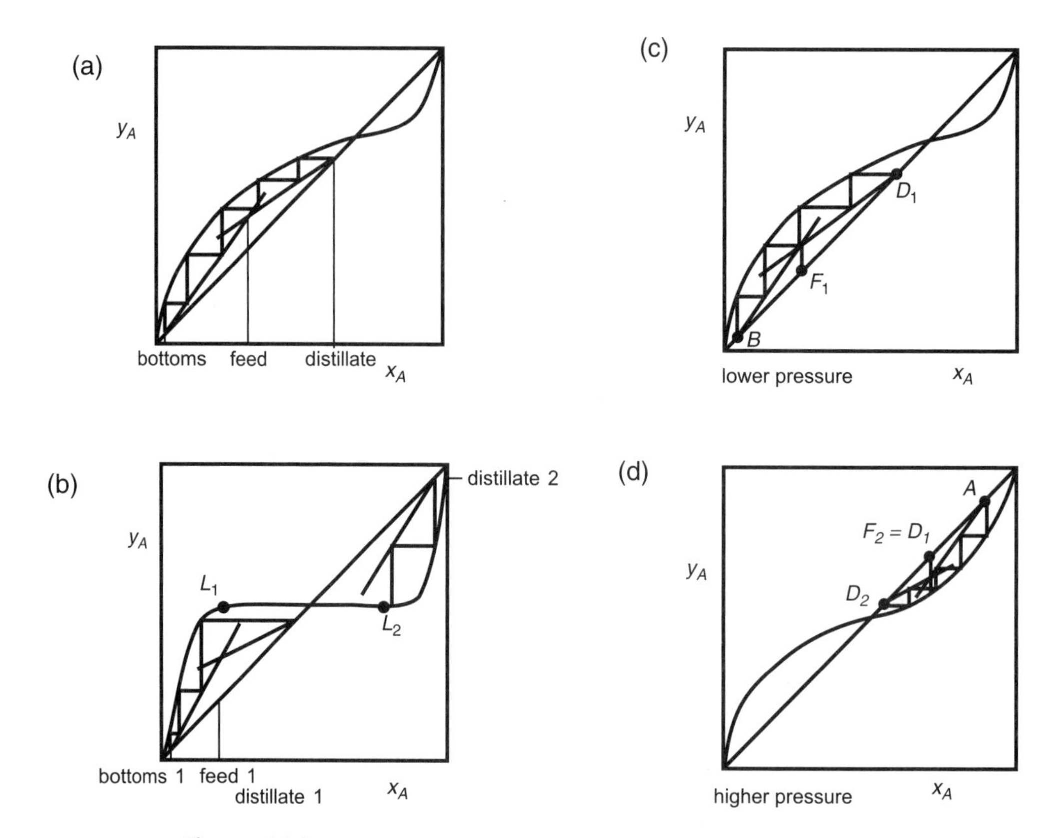

Figure 12.5 McCabe-Thiele Diagrams for Distillation Arrangements in Figure 12.4

used to obtain two "pure" components. The McCabe-Thiele diagram is shown in Figure 12.5(b). A characteristic of the equilibrium in this system is the horizontal segment of the equilibrium curve, which is caused by the phase separation into immiscible phases. The equilibrium between the two phases is illustrated by the ends of the horizontal segment of the equilibrium curve marked by L_1 and L_2. Therefore, in one column, the feed is distilled to near azeotropic conditions, and "pure" component B is in the bottom stream. The impure distillate is condensed and sent to a phase separator. One immiscible phase is on the other side of the azeotrope, and it is sent to a second column to purify component A in the bottom stream. The impure distillate from the second column is condensed and sent to the phase separator. It is important to understand that this method works only for systems exhibiting this type of azeotropic phase behavior.

If a binary azeotrope is pressure sensitive, the method illustrated in Figure 12.4(c) can be used to produce two "pure" products. The McCabe-Thiele construction is illustrated in Figures 12.5(c) and 12.5(d). In this illustration, increasing the system pressure lowers the azeotropic composition of A. The McCabe-Thiele construction in Figure 12.5(d) is at a higher pressure than that in Figure 12.5(c).

Therefore, for the case illustrated, the feed is distilled in one column to produce "pure" B and a near-azeotropic distillate (D_1). This distillate is then pumped to a higher pressure, which lowers the azeotropic composition. At a suitable pressure, the distillate from the first column is now above the azeotropic composition, and a second column is then used to purify A. "Pure" A is the bottoms product of the second column because of the reversal in volatilities caused by the azeotrope. The near-azeotropic distillate (D_2) is recycled to the first column.

A related method for pressure-sensitive azeotropes is to run only one column at vacuum conditions. If the equilibrium behavior is favorable, the azeotrope will be at a mole fraction of A approaching unity. Depending on the desired purity of component A, the maximum possible distillate composition may be sufficient.

It is important to remember that pressure-swing methods are applicable only when the azeotropic composition is highly pressure sensitive. Although there are examples of this behavior, it is actually quite rare.

Azeotropes in Ternary Systems. In binary systems, the McCabe-Thiele method provides a conceptual representation of the distillation process. In ternary systems, there is a method that provides a similar conceptual representation. It is called the **boundary value design method (BVDM),** and it is particularly useful for conceptualizing azeotropic distillation in ternary systems. This method is introduced here; however, the reader seeking a more in-depth treatment of this method and all aspects of azeotropic distillation should consult the definitive reference in the field [17].

In the BVDM, ternary distillation is represented on a right-triangular diagram just as binary distillation is represented on a rectangular plot in the McCabe-Thiele method. Each point on the right-triangular diagram represents the mole fraction of each of the three components on a tray. For example, in Figure 12.6(b), the vertex labeled B is the origin of a rectangular coordinate system. At that point, the mole fractions of A and C are zero, and the mole fraction of B is obtained by subtraction from 1. Hence, the mole fraction of B at the origin is 1. Consider any other point on the diagram, point *p*, as illustrated. The mole fractions of C and A are obtained using the horizontal and vertical coordinates, respectively, based on B as the origin. The mole fraction of B is obtained by subtraction of the mole fractions of A and C from 1.

For a simple ternary distillation process, a curve can be drawn by connecting the compositions on each stage. This is equivalent to the operating line in the McCabe-Thiele method. This is illustrated in Figure 12.6. It is observed that the curves for the rectifying and stripping section intersect. This intersection implies a feasible distillation process, just as intersection of the operating lines for rectifying and stripping implies a feasible binary distillation on the McCabe-Thiele diagram. If the curves do not intersect, then the distillation operation is not feasible.

Each point shown on the curves on the triangular diagram is analogous to the highlighted points on the McCabe-Thiele diagram, where the stepping process intersects the operating lines. In the McCabe-Thiele method, the optimum feed location is determined by the intersection of the operating lines. It is

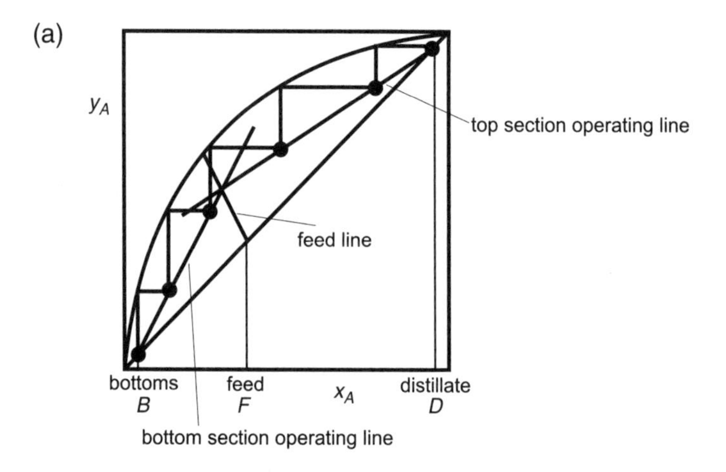

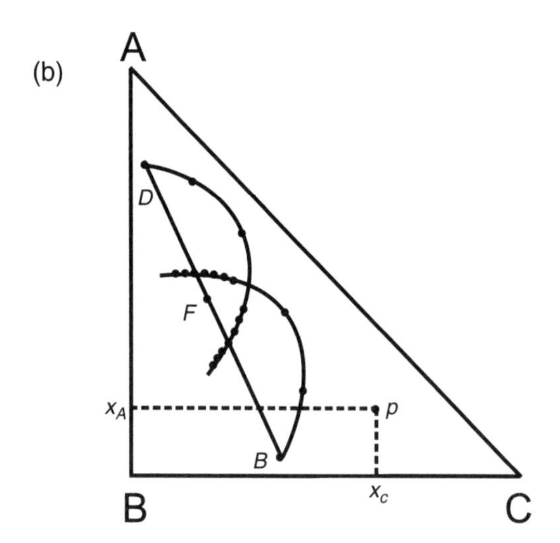

Figure 12.6 Comparison of McCabe-Thiele for Binary Distillation (a) and Triangular Diagram for Ternary Distillation (b)

possible to place the feed in a different location; however, the optimum location minimizes the number of stages required. For simple ternary distillation, the feed tray is fixed at the point of intersection of the rectifying and stripping curves on the triangular diagram. To trace the tray-to-tray compositions, one follows a curve for one section and then switches to the curve for the other section at the intersection of the curves. The closely spaced points not in the range of operation are those analogous to the points on a McCabe-Thiele diagram obtained by not switching operating lines in the stepping process and pinching as the equilibrium curve is approached.

It is also observed that the feed point lies on a straight line connecting the distillate and bottoms product concentrations. This is a consequence of the lever rule and is similar to the representation of mixing and separation processes on triangular diagrams in extraction processes [18,19]. The line connecting points D, F, and B is a representation of the overall material balance on the column. As a consequence of the requirement of the overall material balance, for a column (with only one feed and without side streams) to be feasible, the points D, F, and B must lie on the same straight line.

Instead of using the equilibrium curve, as is done in the McCabe-Thiele diagram, on a triangular diagram, a residue curve is used. A **residue curve** is a plot of the composition of the liquid residue in a single-stage batch equilibrium still with time at a given pressure (Figure 12.7). In a batch still, as the still pot is heated, the more volatile components are boiled off, and the concentration of the less volatile components increases with time in the still pot. The equation used to calculate the residue curve is the unsteady material balance on the still pot for each component *i*.

$$\frac{dx_i}{dt} = (x_i - y_i)\left(\frac{-d\ell nN}{dt}\right) \tag{12.2}$$

$$y_i = K_i x_i \tag{12.3}$$

where N is the total moles of liquid in the pot and the form of K depends on the thermodynamics used to represent the phase equilibrium (i.e., Raoult's Law, equation of state, fugacity, activity coefficient model, etc.). Most process simulation packages have utilities to perform this calculation and plot the result and export the data.

Because the more volatile components are being removed with time, the temperature in the still pot increases with time. The residue curves represent this fact with an arrow in the direction of increasing temperature. It is also true that residue curves never cross. Points on the residue curve map are defined as follows:

Stable node: Arrows on all curves point toward this point (highest temperature).

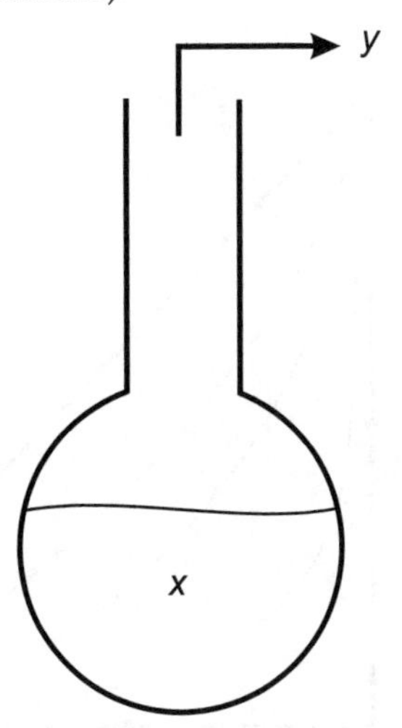

Figure 12.7 Batch Distillation

Unstable node: Arrows on all curves point away from this point (lowest temperature).

Saddle point: Arrows point both toward and away from this point (intermediate temperature).

Figure 12.8 shows the residue curve map for a ternary system without azeotropes. Note that the curves seeming to emanate from the A vertex actually represent initial still pot compositions of nearly pure A with an infinitesimal amount of B (the curve on the A-B line), with infinitesimal amounts of both B and C in differing ratios (the interior curves), or with an infinitesimal amount of C (the A-C line). Each point anywhere on the triangular diagram is at a different temperature. Because the diagram represents liquid compositions, the temperature is the bubble point of the mixture at the given pressure. Therefore, the vertices of the triangular diagram are at the boiling points of the pure components. In Figure 12.8, Component A is the most volatile, and Component C is the least volatile. In the discussion that follows, the convention of decreasing volatility for components A-B-C will be followed.

There are many possible representations of azeotropes on triangular diagrams. Four are shown in Figure 12.9. In Figure 12.9(a), there is a binary, minimum boiling azeotrope between Components A and B. In Figure 12.9(b), there is a binary, minimum boiling azeotrope between Components B and C that boils above pure Component A. In Figure 12.9(c), there is a binary, minimum boiling azeotrope between Components A and C that boils below pure Component A. In Figure 12.9(d), there are two binary minimum boiling azeotropes. One is between Components A and B that boils below pure Component A, and the other between Components A and C that also boils below pure Component A.

There are three key rules to using residue curves to conceptualize distillation processes.

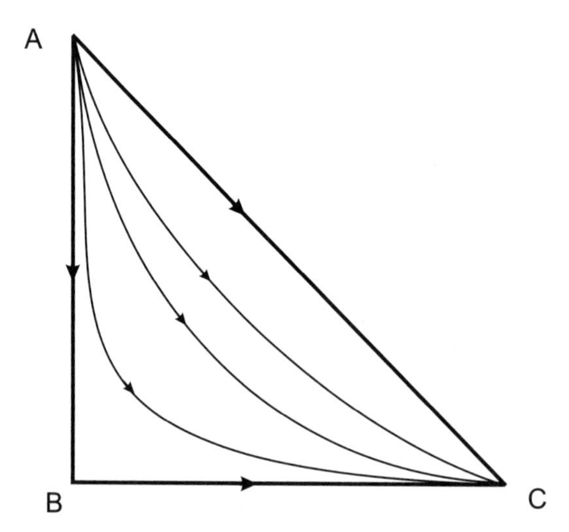

Figure 12.8 Residue Curve Map for Ternary System without Azeotrope

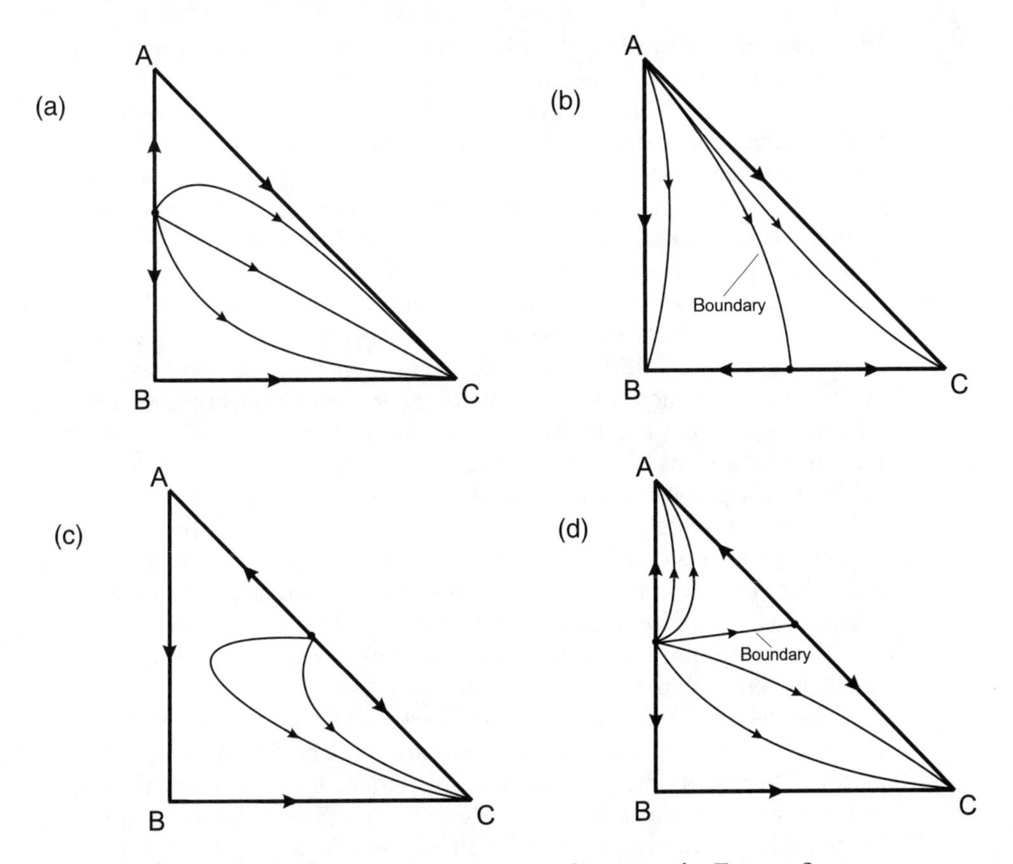

Figure 12.9 Some Possible Azeotropic Situations for Ternary Systems

1. Because the temperature increases from top to bottom in a distillation column, the time variable can be replaced with a height variable in Equation (12.2). This is true only for a packed column; however, the equivalence of a certain column height and a tray (HETP, height equivalent to a theoretical plate) makes generalization to tray columns possible.
2. The residue curve is the composition profile in a continuous, packed distillation column at total reflux.
3. The BVDM (residue curves on triangular diagrams) is useful only for conceptualization, not numerical calculations, in contrast with the McCabe-Thiele method, which can be used for numerical calculations.

Therefore, to represent a feasible distillation process qualitatively, the second point combined with the material balance criterion illustrated in Figure 12.6 (feed, distillate, and bottoms product must lie on same straight line) means that *there must be a straight line connecting the feed, distillate, and bottoms that intersects the same residue curve, with one end at the distillate and the other end at the bottoms.* It must be remembered that there are an infinite number of residue curves on a triangular plot, even though only two or three are actually drawn. Figure 12.10 shows examples of feasible and nonfeasible distillation processes. Figure 12.10(a) is for a system without azeotropes, and Figure 12.10(b) is for a system with a minimum boiling azeotrope between Components A and B.

In the preceding section, a method for breaking binary azeotropes was mentioned that involved adding a third component to break the azeotrope. A key question is how to pick the added component. One answer is to pick an intermediate-boiling component that does not create a new azeotrope and has a residue curve map without any boundaries like Figure 12.9(c). The distillation column sequence and the representation of the sequence on the boundary value diagram are shown in Figure 12.11. The residue curve map suggests a feasible, intermediate-boiling component to break the azeotrope, and also suggests the method for column sequencing and recycles to accomplish the separation. It should be noted that finding an intermediate-boiling component to break an azeotrope can be a difficult task given the narrow boiling point range required. Quite often, high-boiling components are used to break azeotropes. One of the most common examples is the use of ethylene glycol to break the ethanol-water azeotrope. An analysis of this situation is beyond the scope of this discussion. However, it is important to mention that the details of the boundary value design method require that the high-boiling component be added as a separate feed to the column and not be mixed with the process stream feed [17].

On residue curve maps, a **boundary** is defined as the curve that separates two regions within which simple distillation is possible. In Figure 12.9, plots (b) and (d) have boundaries. A boundary separates two regions with residue curves not having the same starting and ending point. Therefore, plots (a) and (c) have no boundary. No simple distillation process in a single column may cross a

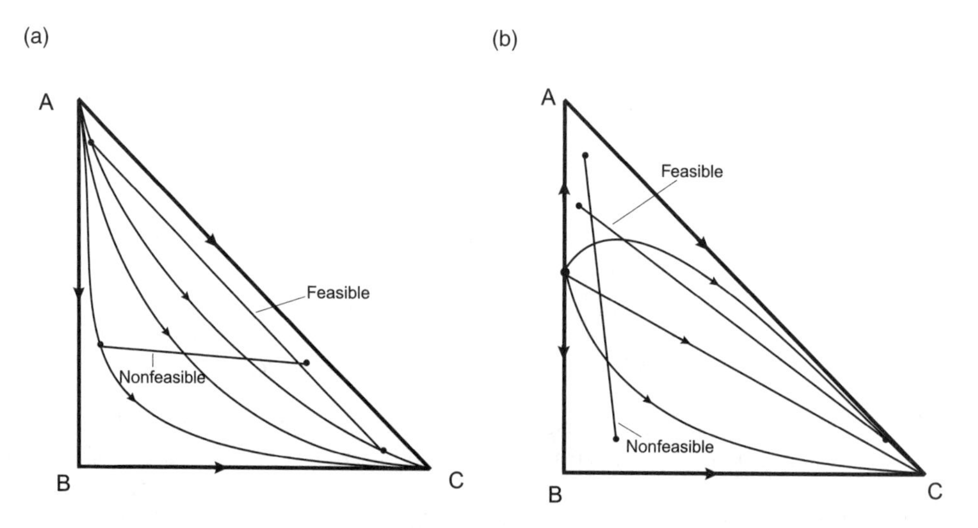

Figure 12.10 Feasible and Nonfeasible Distillation Processes

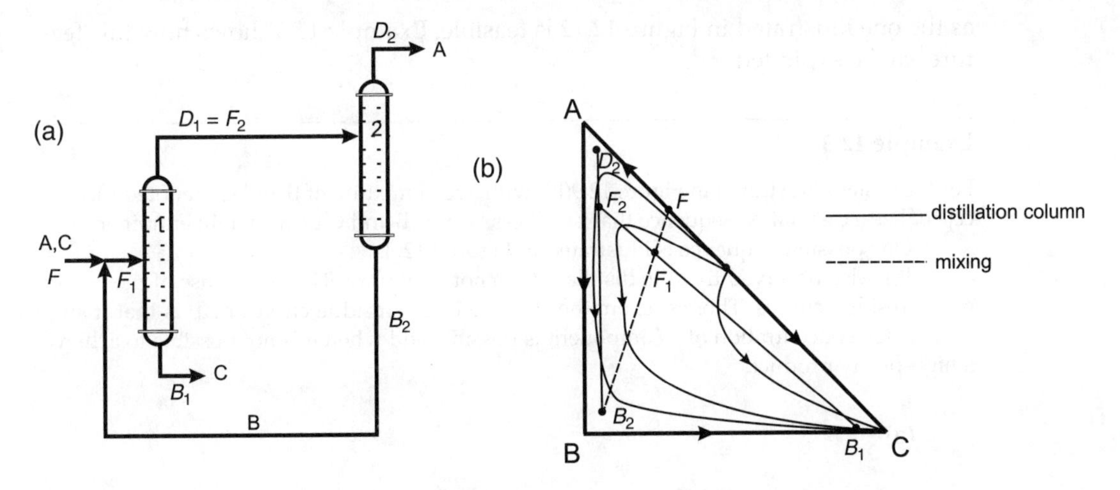

Figure 12.11 Method for Breaking Binary Azeotrope Using Intermediate-Boiling Entrainer

straight boundary. Also, if a boundary is straight, the product streams from a multiple-column arrangement may not cross the boundary.

When the boundary is curved, there are more options for separation sequences. This is because the only requirement is that the distillate and bottoms product from a column be in the same region. The feed can lie in a different region. Also, if a boundary is curved, the product streams from a multiple-column arrangement may lie in a region across the boundary. Therefore, a column such

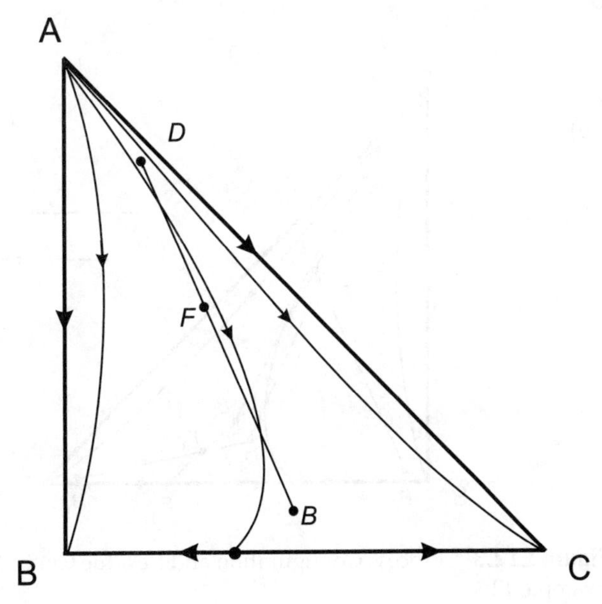

Figure 12.12 Feasible Column Operation with Curved Boundary

as the one illustrated in Figure 12.12 is feasible. Example 12.3 shows how this feature can be exploited.

Example 12.3

For the system illustrated in Figure 12.9(b), with a feed mixture of B and C (denoted F), conceptualize a distillation sequence that produces "pure" B and C using a light entrainer, A.

One possible sequence is illustrated in Figure E12.3.

It is also observed that the distillate D_3 is not very pure. This is because of the layout of the residue curves. Therefore, another feature of the residue curves map is that it suggests when concentration of a component is possible and when it is not possible to achieve a high-purity product.

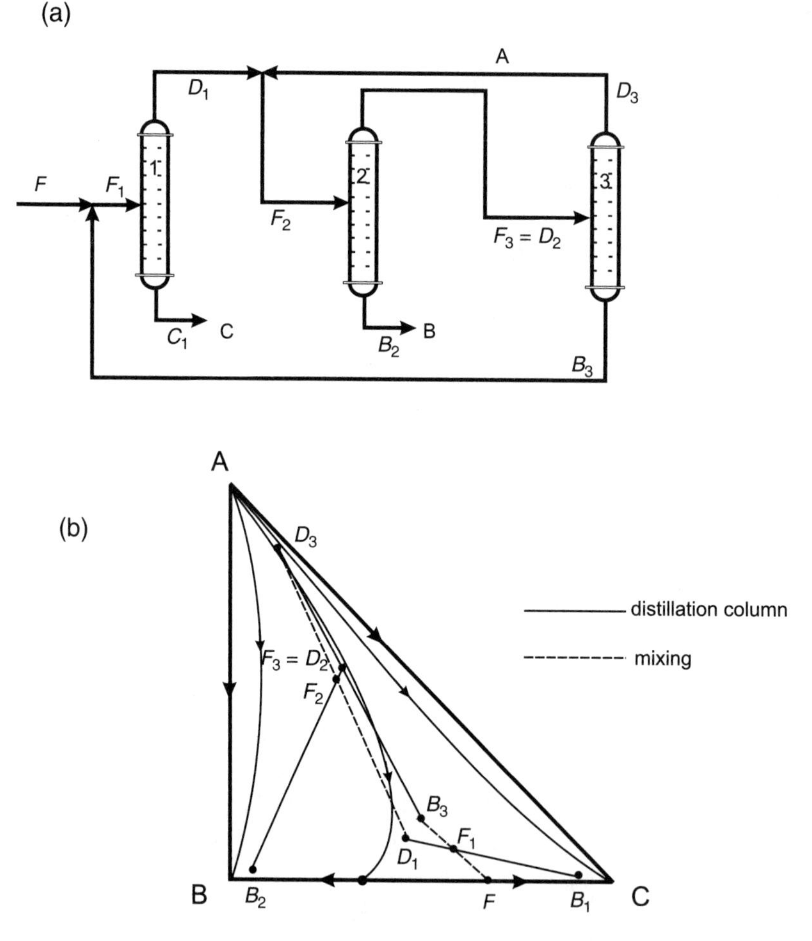

Figure E12.3 Process Configuration and Residue Curve Map for Example 12.3

In summary, the BVDM or residue curve plots can be used to conceptualize feasible distillation sequences. It is particularly useful for azeotropic systems involving three components, either when there are three components in the feed or when an entrainer is added to break a binary azeotrope. This method has only been introduced in this section, so care must be taken when applying this method without additional reading. Reference [17] is suggested for further reading.

12.4 REACTOR FEED PREPARATION AND SEPARATOR FEED PREPARATION SECTIONS

The purpose of these sections is to match the temperature and pressure desired on the inlet streams to the reactor section and to the separation section. If the reactor operates at high temperature (a common occurrence because this increases reaction rate), the reactor feed preparation and separator feed preparation sections are often combined in a single process-process heat exchanger. Such heat integration can be built into the base-case flowsheet, but, if not, it should be caught during the heat-integration step of optimization (see Chapter 15).

Pressure may also need to be increased for the reactor (or, infrequently, decreased), and this requires a pump for liquids or a compressor for gases. When there is a choice, pumps are preferable to compressors, because the operating, capital, and maintenance costs are all lower for pumps. If pressure is reduced between the reactor and the separator sections (or anywhere else in the process), an expander can be considered (for gases), but often it is not economical both because of its high cost and also because it reduces the controllability of the process. A valve allows control at a modest cost, but energy is not recovered.

In these temperature and pressure matching sections, the lowest-cost utility should always be used. For heating the feed to an exothermic reaction, heat integration can be used with the reactor effluent. For low-temperature heating, low-pressure steam or another low-temperature utility is used. For safety reasons, exothermic reactions (when reactor runaway is possible) should be run, with the reactor feed coming in at a temperature high enough to ensure a significant reaction rate. This avoids the buildup of large inventories of unreacted feed materials, which can happen if cold material enters the reactor and quenches the reaction. When sufficient heat is later provided, the entire contents of the reactor could react very rapidly, a process called **ignition.**

When possible, consider operation between 1 and 10 bar. High pressures increase pumping, compression, and capital costs, whereas low pressures tend to increase the size and cost of vessels. The temperature of the feed to the separation unit (at least for the base case) is usually set between the boiling points of the top and bottoms product for distillation, or based on similar considerations for the other separation options.

12.5 RECYCLE SECTION

This section is relatively straightforward. The stream or streams of unreacted raw materials are sent back to the reactor to reduce feed costs, to reduce impurities in the product, or to improve the operation of the process. If the conditions of the recycle streams are close to those of the raw material feed, then the recycle stream should mix with the raw materials prior to the reactor feed preparation section. Otherwise, any heating/cooling or pressure increase/decrease should be done separately. Thus, the recycle stream is combined with the raw material streams when they are all at a similar temperature. Example 12.4 involves a recycle in a biochemical process.

Example 12.4

A recycle is used to return enzyme to the reactor. What particular concerns must be addressed in the recyle section?

The enzyme must be protected from deactivation and from degradation by microbes during the recyle, which might include storage times between batches. The recycle must be maintained in aseptic conditions, at the appropriate temperature and pH.

12.6 ENVIRONMENTAL CONTROL SECTION

As stated previously and in Chapter 24, this section should be eliminated or minimized through pollution prevention and green engineering. However, especially if the contaminant has little or no value if concentrated, there will be relatively dilute waste streams generated and sent to the environmental control section. Here, they are concentrated (by the separation techniques discussed earlier) and then disposed of (by incineration, neutralization, oxidation, burial, or other means). The keys are to concentrate the waste and to make it benign.

12.7 MAJOR PROCESS CONTROL LOOPS

During the initial synthesis of the PFD, the major control loops are developed. These control loops affect more than just one unit of a process. For example, the level control in the condensate tank of a distillation column is necessary for plant operation. On the other hand, a reactor temperature controller that changes the flowrate of molten salt through the cooling tubes of a reactor is a major loop and should be shown on the PFD.

It is through the early development of major control loops that significant design improvements can be made. In the high-temperature exothermic reactor, for example, failure to consider the control loop might lead one to propose an integrated heat-exchanger network that is difficult or impossible to control.

Beyond the importance of the control loops in maintaining steady-state material balance control, assurance of product purity, and safety, they provide focal points for the optimization that will follow the initial PFD synthesis. As described in Chapter 14, the controlled variables are the variables over which we have a choice. We find the best values of these variables through optimization. These loops also provide early clues to the flexibility of the process operation. For example, if the feed to the reactor is cut in half, less heat needs to be removed. Therefore, there must be an increase in the temperature of the cooling medium, which occurs when the coolant flowrate is reduced. Process control can be both very difficult and extremely important in biological processes, as demonstrated in Example 12.5.

Example 12.5

In biological waste treatment, microbes that eat the waste and produce benign products are used. In one class of such processes, called **activated sludge,** the culture is separated from reactor output and recycled. What specfic process control issues arise in such a process?

The culture must maintain sufficient activity throughout the recycle. The recycle conditions may become nutrient poor if nearly all the waste nutrient is consumed, leading to loss of activity. The activity of the culture might be adversely affected by upsets in the feed conditions to the reactor, such as pH extremes or high concentrations of compounds toxic to the culture. Therefore, feed-forward control using measurements of the feed conditions should be considered, as well as control of a reserve of the culture to be used in case an uncontrolled process upset leads to culture death. Immediate, on-line analysis of culture activity is difficult, but off-line measurements can be incorporated into the control scheme.

12.8 FLOW SUMMARY TABLE

The format for the flow summary table is given in Chapter 1. Each of the conditions (temperature, pressure, flowrate, composition, and phase) should be estimated early in the flowsheet development. All are needed to get preliminary costs, for example. Even estimates based on perfect separations can provide sufficient data to estimate the cost of a recycle versus the value of burning the impure, unreacted feed material as a fuel.

Completeness of the estimates, not their accuracy, is important at this stage. For example, an early determination of phase (solid versus liquid versus gas) is needed to help choose a separation scheme or reaction type (see Table 12.1).

12.9 MAJOR EQUIPMENT SUMMARY TABLE

Chapter 1 explains the requirements of a major equipment summary table. In the context of initial PFD development, the process of creating the table forces the process design engineer to question the size (and cost) of various units for which there may well be other options. If, early on, one must specify a large compressor, for example, the process can be changed to a lower pressure or it can be modified to use liquid pumping followed by vaporization rather than vaporization followed by compression. The early consideration of materials of construction provides the clue that normal temperatures and pressures usually result in less-expensive materials.

12.10 SUMMARY

The inclusion of enough detail and the freedom to look at the big picture without the burden of excessive detail are the keys to successful PFD synthesis. One must remain fully aware of the broadest goals of the project while looking for early changes in the structure of the flowsheet that can make significant improvements. Particular attention must be paid to the separations and reactor sections. The formation of azeotropes between components to be separated greatly affects the separation sequence. These must be identified early in the synthesis.

The beginning of the process is the generic block flow process diagram. Although the flow summary table at this stage is based on crude assumptions and the equipment summary table is far from the final equipment specifications, they help keep the chemical engineer cognizant of key choices that need to be made.

REFERENCES

1. *Tables of Chemical Kinetics: Homogeneous Reactions*, National Bureau of Standards, Circular 510, 1951; Supplement 1, 1956; Supplement 2, 1960; Supplement 3, 1961. Now available as NIST Chemical Kinetics Database at www.NIST.gov.
2. Kirk, R. E., and D. F. Othmer, *Encyclopedia of Chemical Technology*, 5th ed. (New York: Wiley, 2004).
3. McKetta, J. J., *Encyclopedia of Chemical Processes and Design* (New York: Marcel Dekker, 1997).
4. Fogler, H. S., *Elements of Chemical Reaction Engineering*, 4th ed. (Upper Saddle River, NJ: Prentice Hall, 2005).
5. Levenspiel, O., *Chemical Reaction Engineering*, 3rd ed. (New York: Wiley, 1999).
6. Viola, J. L., "Estimate Capital Costs via a New, Shortcut Method," in *Modern Cost Engineering: Methods and Data*, Vol. II, ed. J. Matley (New York: McGraw-Hill, 1984), 69 (originally published in *Chemical Engineering*, April 6, 1981).
7. Cropley, J. B., Union Carbide Technical Center, South Charleston, WV, personal communication, 1995.
8. Poling, B. E., J. M. Prausnitz, and J. C. O'Connell, *The Properties of Gases and Liquids*, 5th ed. (New York: McGraw-Hill, 2000).

9. Rudd, D. F., G. J. Powers, and J. J. Siirola, *Process Synthesis* (Englewood Cliffs, NJ: Prentice Hall, 1973).
10. Becker, H., S. Godorr, H. Kreis, and J. Vaughan, "Partitioned Distillation Columns—Why, When and How," *Chemical Engineering* 108, no. 1 (2001): 68–74.
11. Petlyuk, F., V. M. Platonov, and D. M. Slavinskii, "Thermodynamically Optimal Method for Separating Multicomponent Mixtures," *Int. Chem. Eng.* 5 (1965): 555–561.
12. Kaibel, G., "Distillation Columns with Vertical Partitions," *Chem. Engr. Technol.* 10 (1987): 92–98.
13. Agrawal, R., and T. Fidkowski, "More Optimal Arrangements of Fully Thermally Coupled Distillation Columns," *AIChE-J.* 44 (1998): 2565–2568.
14. Seider, W. D., J. D. Seader, and D. R. Lewin, *Product and Process Design Principles: Synthesis, Analysis, and Evaluation*, 2nd ed. (New York: Wiley, 2003).
15. Wankat, P., *Separation Process Engineering*, 2nd ed. (Upper Saddle River, NJ: Prentice-Hall, 2007), ch. 8.
16. Humphrey, J. L., and G. E. Keller II, *Separation Process Technology* (New York: McGraw-Hill, 1997), 99–101, 271–275.
17. Doherty, M. F., and M. F. Malone, *Conceptual Design of Distillation Systems* (New York: McGraw-Hill, 2001).
18. Wankat, P., *Separation Process Engineering*, 2nd ed. (Upper Saddle River, NJ: Prentice-Hall, 2007), ch. 14.
19. Felder, R. M., and R. W. Rousseau, *Elementary Principles of Chemical Processes*, 3rd ed. (New York: Wiley, 2000), ch. 6.

General Reference

Stichlmair, J. G., and J. R. Fair, *Distillation: Principles and Practice* (New York: Wiley-VCH, 1998).

PROBLEMS

1. Choose one of the cases from Appendix B. Identify
   a. All required physical property data
   b. Sources for all data needed but not provided
2. Search the patent literature for kinetics information for one of the processes in Appendix B. Convert the provided data to a form suitable for use on a process simulator.
3. Develop five heuristics for reactor design.
4. For design of an exothermic reactor with cooling, one needs to choose an approach temperature (the nominal temperature difference between the reaction zone and the cooling medium). One engineer claims that the temperature

difference across the tube walls should be as small as possible. Another claims that a large temperature is better for heat transfer.

a. Describe the advantages and disadvantages of these two choices from the point of view of capital costs.
b. Describe the advantages and disadvantages of these two choices from the point of view of operating costs.
c. Describe the advantages and disadvantages of these two choices from the point of view of operability of the process.

5. Are there any safety considerations in the choice of heat transfer driving force (ΔT) in Problem 4? Explain.
6. From the flowsheets in Appendices B and C, identify four examples of each of the following types of recycles described in this chapter:
   a. Recycles that reduce feed costs
   b. Recycles that reduce impurities in products
7. Recycles are also used for the following purposes. Identify four examples of each.
   a. Heat removal
   b. Flowrate control
   c. Improvement of separation
   d. Reduction of utility requirements
8. From your previous courses and experience, identify four process recycle streams that serve different purposes from those identified in 7 (a)–(d).
9. What information is needed about the available utilities before a choice between them can be made for a specific heating duty?
10. Take a flowsheet for one of the processes in Appendix B and identify the reactor feed preparation, reactor, separator feed preparation, separator, recycle, and environmental control sections. Is there any ambiguity concerning the demarcation of these sections? Explain.
11. How many simple distillation columns are required to purify a stream containing four components into four "pure" products? Sketch all possible sequences.
12. How many simple distillation columns are required to purify a stream containing five components into five "pure" products? Sketch all possible sequences.
13. Illustrate a system (PFD and McCabe-Thiele diagrams) to purify two components (A and B) from a binary, homogeneous, minimum-boiling azeotrope that is pressure sensitive. The feed concentration of A is greater than the azeotropic composition at the pressure of the column receiving the feed. The azeotropic composition of A decreases with increasing pressure.
14. In the production of dimethyl carbonate from methanol, it is necessary to separate methanol from formal (also known as methylal—$C_3H_8O_2$). However, an azeotrope exists between these two components. Using a process simulator, examine the equilibrium curves (x-y plots) between 100 kPa and

1000 kPa to determine a strategy for purifying these two components. Compare your results for the following thermodynamic packages for the *K*-value:

a. Ideal
b. NRTL
c. UNIFAC
d. SRK

What do you learn from this comparison?

15. In the production of diethyl ether, it is necessary to purify a stream of equimolar diethyl ether and water that is available at 1500 kPa. Suggest a method for achieving this separation. Use the UNIQUAC model for the *K*-value. Would your answer be different if, for example, only 99 mole % purity were needed instead of 99.9 mole %? What do you learn and how does your answer change if the four models for *K*-values in Problem 14 are used? What do you learn from this comparison?
16. In the production of diethyl ether, assume that it is necessary to purify a stream at 1500 kPa containing 75 mole % diethyl ether, 20 mole % ethanol, and 5 mole % dimethyl ether. Assume that the UNIQUAC model adequately predicts the thermodynamics of this system.
    a. Use the residue curve plotting routine on your simulator to plot the residue curves.
    b. Suggest a method for separating this mixture into three relatively pure components.
    c. How would your answer to Part (b) change if the stream contained 85 mole % diethyl ether, 10 mole % ethanol, and 5 mole % dimethyl ether?
    d. Examine the effect of changing the pressure of the distillation columns. Suggest a much simpler method for achieving the necessary separation.
17. When your parents or grandparents were in college, it was not uncommon to "borrow" some "pure" ethanol (grain alcohol) from the university to add to a party punch. However, because of the azeotrope between ethanol and water, "pure" ethanol is not easy to manufacture. At that time, it was not uncommon for benzene to be added as an entrainer to break the azeotrope. Of course, this means that all "pure" ethanol contained trace amounts of benzene, which was later identified as a carcinogen. By plotting both residue curves and *TPxy* diagrams for the ethanol-water system, suggest a method for purifying ethanol by adding benzene as an entrainer. Assume that the UNIQUAC model applies and that the feed stream of ethanol and water is equimolar and at atmospheric pressure.
18. Does Equation (12.1) give the number of alternative sequences for separation processes other than distillation? If so, give two examples. If not, why not?
19. What constraints on the separation process and on the sequence are assumed in the derivation of Equation (12.1)?
20. Figure 12.1(d) shows a nonsimple separation unit. When one says that the minimum number of simple separation units is $(N–1)$, these nonsimple units

are not considered. Identify four other examples of nonsimple separation units. For each, discuss how it might affect
   a. Capital cost
   b. Operability and safety
   c. Operating costs
21. Does Equation (12.1) apply to batch or semibatch separation processes? Why or why not?
22. The residue curve map of Figure 12.9(a) shows that batch distillation of a mixture at the azeotropic composition for the system (A+B) with a small amount of C added will result in a very pure liquid C residue in the still. Is this a good separation unit choice to obtain "pure" C? Analyze the advantages and disadvantages.
23. On a residue curve map, will the composition of a ternary, minimum-boiling azeotrope always be a stable node, an unstable node, a saddle point, or none of the above?
24. As noted in Section 12.1.1, laboratory studies often involve batch or semibatch reactors for convenience. Develop four heuristics for choosing whether to use batch, semibatch, or continuous reactors for the commercial plant.
25. Semibatch or batch reactors are common in biochemical operations. What characteristics of biological systems, materials, and products lead to the choice of batch over continuous reactors? To what extent do these characteristics also lead to unsteady operations of other GBFD sections:
   a. Reactor feed preparation
   b. Separator feed preparation
   c. Separator
   d. Recycle
   e. Environmental control
26. The unique characteristics of an azeotrope make it ideal for some applications. Find three applications of azeotropic products. How is the production of an azeotropic composition different from the production of "pure" A and B when (A+B) forms an azeotrope? Does the process differ for a minimum-boiling and a maximum-boiling azeotrope?
27. For biological systems (which often have a very narrow range of acceptable temperatures), describe the advantages and disadvantages of large and small ΔT's in reactor temperature control systems.
28. Biological systems often require sterilization as a step before inoculation with the appropriate microbial culture. Sterilization can be considered a reaction process in which organisms are killed, often through cell lysis, using thermal or chemical routes. Consider both batch and continuous sterilization. Which has advantages for sterilizing process equipment? Which has advantages for sterilizing reactor feedstock? (Note: For some biochemical processes, sterilization is neither required not desired, e.g., waste treatment.)

CHAPTER

13

Synthesis of a Process Using a Simulator and Simulator Troubleshooting

The advancement in computer-aided process simulation over the past generation has been nothing short of spectacular. Until the late 1970s, it was rare for a graduating chemical engineer to have any experience in using a chemical process simulator. Most material and energy balances were still done by hand by teams of engineers. The rigorous simulation of multistaged separation equipment and complicated reactors was generally unheard of, and the design of such equipment was achieved by a combination of simplified analyses, shortcut methods, and years of experience. In the present day, however, companies now expect their junior engineers to be conversant with a wide variety of computer programs, especially a process simulator.

To some extent, the knowledge base required to simulate successfully a chemical process will depend on the simulator used. Currently there are several process simulators on the market, for example, CHEMCAD, ASPEN PLUS, HYSYS, PRO/II, and SuperPro Designer. Many of these companies advertise their product in the trade magazines—for example, *Chemical Engineering, Chemical Engineering Progress, Hydrocarbon Processing,* or *The Chemical Engineer*—and on the Internet. A process simulator typically handles batch, semibatch, and continuous processes, although the extent of integration of the batch and continuous processes in a single PFD varies between the various popular simulators. The availability of such powerful software is a great asset to the experienced process engineer, but such sophisticated tools can be potentially dangerous in the hands of the neophyte engineer. The bottom line in doing any process simulation is that you, the engineer, are still responsible for analyzing the results from the computer. The purpose of this chapter is not to act as a primer for one or all of these products. Rather, the general approach to setting up processes is emphasized, and we aim to highlight some of the more common problems that process simulator users encounter and to offer solutions to these problems.

13.1 THE STRUCTURE OF A PROCESS SIMULATOR

The six main features of all process simulators are illustrated in the left-hand column of Figure 13.1. These elements are:

1. **Component Database:** This contains the constants required to calculate the physical properties from the thermodynamic models.
2. **Thermodynamic Model Solver:** A variety of options for vapor-liquid (VLE) and liquid-liquid (LLE) equilibrium, enthalpy calculations, and other thermodynamic property estimations are available.
3. **Flowsheet Builder:** This part of the simulator keeps track of the flow of streams and equipment in the process being simulated. This information can be both input and displayed graphically.
4. **Unit Operation Block Solver:** Computational blocks or modules are available that allow energy and material balances and some design calculations to be performed for a wide variety of process equipment.

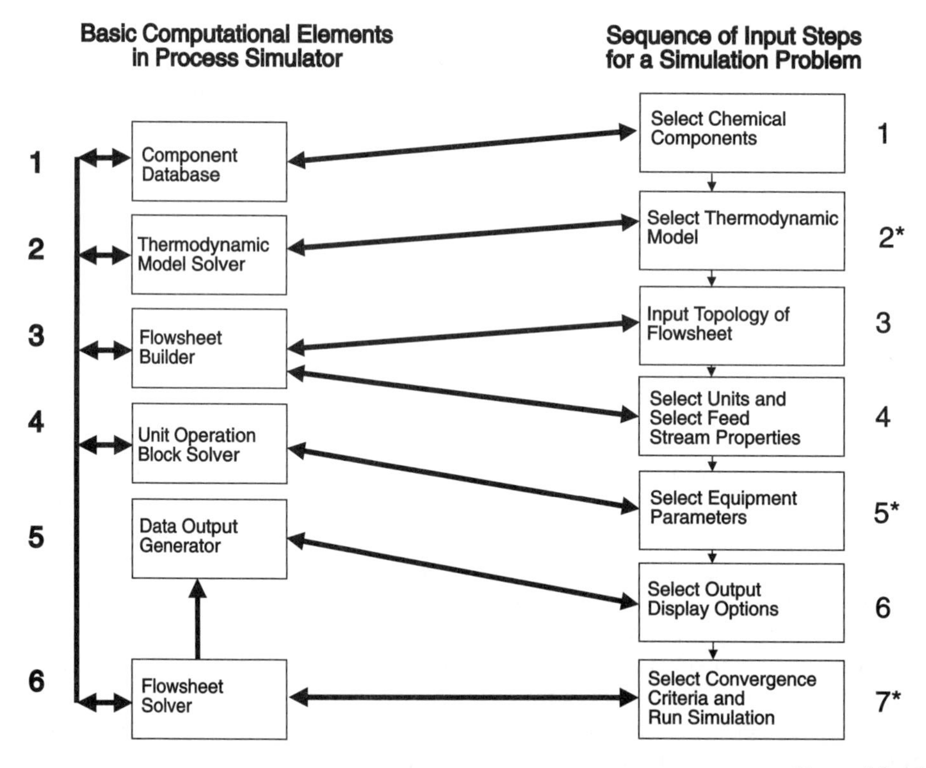

Figure 13.1 Relationship between Basic Computational Elements and Required Input to Solve a Process Simulation Problem

5. **Data Output Generator:** This part of the program serves to customize the results of the simulation in terms of an output report. Often, graphical displays of tower profiles, heating curves, and a variety of other useful process data can be produced.
6. **Flowsheet Solver:** This portion of the simulator controls the sequence of the calculations and the overall convergence of the simulation.

There are several other elements commonly found in process simulators that are not shown in Figure 13.1. For example, there are file control options, the option to use different engineering units, possibly some additional features associated with regressing data for thermodynamic models, and so on. The availability of these other options is dependent on the simulator used and will not be discussed further.

Also shown on the right-hand side of the diagram in Figure 13.1 are the seven general steps to setting up a process simulation problem. The general sequence of events that a user should follow in order to set up a problem on a simulator is as follows:

1. Select all of the chemical components that are required in the process from the component database.
2. Select the thermodynamic models required for the simulation. These may be different for different pieces of equipment. For example, to simulate correctly a liquid-liquid extractor, it is necessary to use a thermodynamic model that can predict liquid-phase activity coefficients and the existence of two liquid phases. However, for a pump in the same process, a less sophisticated model could be used.
3. Select the topology of the flowsheet to be simulated by specifying the input and output streams for each piece of equipment.
4. Select the properties (temperature, pressure, flowrate, vapor fraction, and composition) of the feed streams to the process.
5. Select the equipment specifications (parameters) for each piece of equipment in the process.
6. Select the way in which the results are to be displayed.
7. Select the convergence method and run the simulation.

Step 3 is achieved by constructing the flowsheet using equipment icons and connecting the icons with process streams. Sometimes, it is convenient to carry out this step first.

The interaction between the elements and steps and the general flow of information is shown by the lines on the diagram. Of the seven input steps given above, steps 2, 5, and 7 are the cause of most problems associated with running process simulations. These areas will be covered in more detail in the following sections. However, before these topics are covered, it is worth looking at the basic solution algorithms used in process simulators.

There are basically three types of solution algorithm for process simulators [1]: sequential modular, equation solving (simultaneous nonmodular), and simultaneous modular.

In the **sequential modular** approach, the equations describing the performance of equipment units are grouped together and solved in modules—that is, the process is solved equipment piece by equipment piece. In the equation solving, or **simultaneous nonmodular,** technique, all the relationships for the process are written out together and then the resulting matrix of nonlinear simultaneous equations is solved to yield the solution. This technique is very efficient in terms of computation time but requires a lot of time to set up and is unwieldy. The final technique is the **simultaneous modular** approach, which combines the modularizing of the equations relating to specific equipment with the efficient solution algorithms for the simultaneous equation solving technique.

Of these three types, the sequential modular algorithm is by far the most widely used. In the sequential modular method, each piece of equipment is solved in sequence, starting with the first, followed by the second, and so on. It is assumed that all the input information required to solve each piece of equipment has been provided (see Section 13.2.5). Therefore, the output from a given piece of equipment, along with specific information on the equipment, becomes the input to the next piece of equipment in the process. Clearly, for a process without recycle streams, this method requires only one flowsheet iteration to produce a converged solution. The term **flowsheet iteration** means that each piece of equipment is solved only once. However, there may be many iterations for any one given piece of equipment, and batch units require time–series calculations to match the required scheduling of operations for the given unit. This concept is illustrated in Figure 13.2.

The solution sequence for flowsheets containing recycle streams is more complicated, as shown in Figure 13.3. Figure 13.3(a) shows that the first equipment in the recycle loop (C) has an unknown feed stream (r). Thus, before

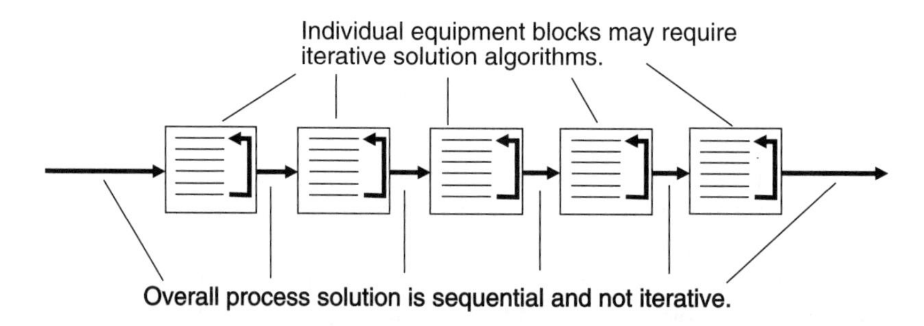

Figure 13.2 Solution Sequence Using Sequential Modular Simulator for a Process Containing No Recycles

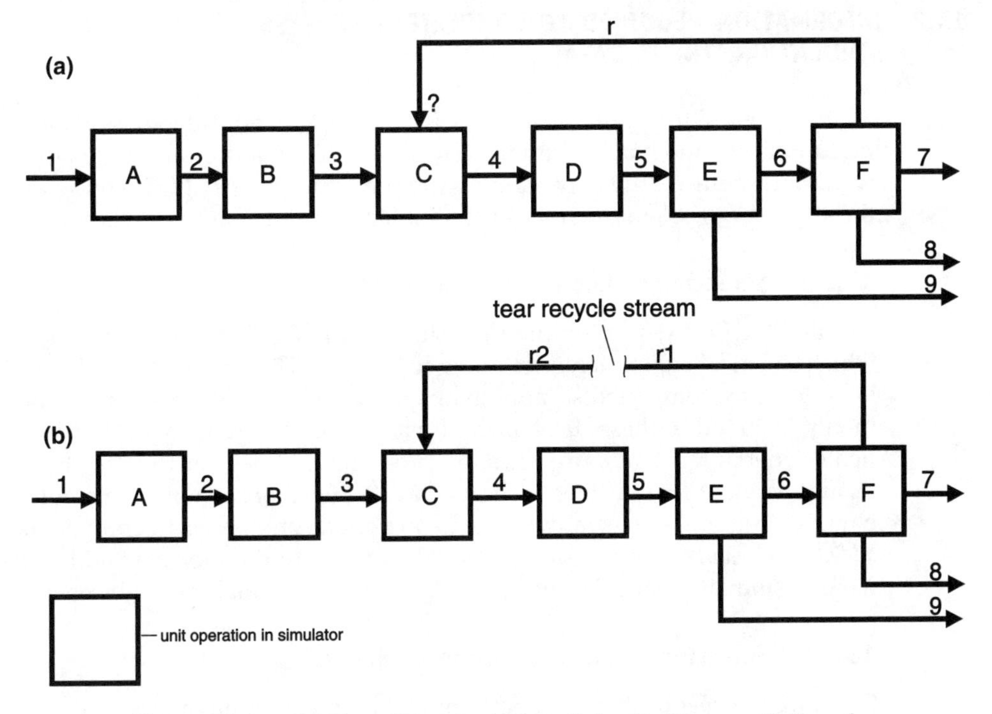

Figure 13.3 The Use of Tear Streams to Solve Problems with Recycles Using the Sequential Modular Algorithm

Equipment C can be solved, some estimate of Stream r must be made. This leads to the concept of tear streams. A **tear stream,** as the name suggests, is a stream that is torn or broken. If the flowsheet in Figure 13.3(b) is considered, with the recycle stream torn, it can be seen, provided information is supplied about Stream r2, the input to Equipment C, that the flowsheet can be solved all the way around to Stream r1 using the sequential modular algorithm. Then compare Streams r1 and r2. If they agree within some specified tolerance, then there is a converged solution. If they do not agree, then Stream r2 is modified and the process simulation is repeated until convergence is obtained. The splitting or tearing of recycle streams allows the sequential modular technique to handle recycles. The convergence criterion and the method by which Stream r2 is modified can be varied, and multivariable successive substitution, Wegstein, and Newton-Raphson techniques [2,3] are all commonly used for the recycle loop convergence. Usually, the simulator will identify the recycle loops and automatically pick streams to tear and a method of convergence. The tearing of streams and method of convergence can also be controlled by the user, but this is not recommended for the novice. Note that heat integration (Chapter 15) introduces recycle streams.

13.2 INFORMATION REQUIRED TO COMPLETE A PROCESS SIMULATION: INPUT DATA

Referring back to Figure 13.1, each input block is considered separately. The input data for the blocks without asterisks (1, 3, 4, and 6) are quite straightforward and require little explanation. The remaining blocks (2, 5, and 7) are often the source of problems, and these are treated in more detail.

13.2.1 Selection of Chemical Components

Usually, the first step in setting up a simulation of a chemical process is to select which chemical components are going to be used. The simulator will have a databank of many components (more than a thousand chemical compounds are commonly included in these databanks). It is important to remember that all components—inerts, reactants, products, by-products, utilities, and waste chemicals—should be identified. If the chemicals that you need are not available in the databank, then there are usually several ways that you can add components *(user-added components)* to your simulation. How to input data for user-added components is simulator specific, and the simulator user manual should be consulted.

13.2.2 Selection of Physical Property Models

Selecting the best physical property model is an extremely important part of any simulation. If the wrong property package or model is used, the simulated results will not be accurate and cannot be trusted. The choice of models is often overlooked by the novice, causing many simulation problems down the road. Simulators use both pure component and mixture properties. These range from molecular weight to activity-coefficient models. Transport properties (viscosity, thermal conductivity, diffusivity), thermodynamic properties (enthalpy, fugacity, *K*-factors, critical constants), and other properties (density, molecular weight, surface tension) are all important.

The physical property options are labeled as "thermo," "fluid package," "property package," or "databank" in common process simulators. There are pure-component and mixture sections, as well as a databank. For temperature-dependent properties, different functional forms are used (from extended Antione equation to polynomial to hyperbolic trigonometric functions). The equation appears on the physical property screen or in the help utility.

For pure-component properties, the simulator has information in its databank for hundreds of compounds. Some simulators offer a choice between DIPPR and proprietary databanks. These are largely the same, but the proprietary databank may contain additional components, petroleum cuts, electrolytes, and so on. DIPPR is the Design Institute for Physical Property Research (a part of AIChE), and sharing of process data across different simulators (e.g., ASPEN Plus, CHEMCAD, HYSYS, PRO/II, SuperPro Designer) can be enhanced by using that databank. (Note that some proprietary databanks may not be supplied in the academic versions of these simulators.) All simulators also have built-in procedures to estimate pure-component properties from group-contribution and other techniques. The details of these techniques are covered in standard chemical engineering thermodynamics texts [4,5,6] and are not described here. However, one must be aware of any such estimations made by the simulator. Any estimation, by definition, increases the uncertainty in the results of the simulation. The entry in the databank for each component should indicate estimations. For example, many long-chain hydrocarbons have no experimental critical point because they decompose at relatively low temperatures. However, because critical temperatures and pressures are needed for most thermodynamic models, they must be estimated. Although these estimations allow the use of equation-of-state and some other models, one must never assume that these are experimental data.

Heat capacities, densities, and critical constants are the most important pure-component data for simulation. The transport and other properties are used in equipment sizing calculations. The techniques used in the simulators are no more accurate than those covered in transport, thermodynamics, unit operations, and separations courses—they are just easier to apply.

Even though simple mass and energy balances cannot be done by the simulator without the above-mentioned pure-component properties, often the most influential decision in a simulation is the choice of a model to predict phase equilibria. Several of the popular simulators have **expert systems** to help the user select the appropriate model for the system. The expert system determines the range (usually with additional user input) of operating temperatures and pressures covered by the simulation and, with data on the components to be used, makes an informed guess of the thermodynamic models that will be best for the process being simulated. The word *expert* should not be taken too seriously! The expert-system choice is only a first guess. Additionally, the model chosen may not be best for a given piece of equipment. A moderately complex simulation will use at least two different thermodynamic packages for different parts of the flowsheet.

Due to the importance of thermodynamic model selection and the many problems that the wrong selection leads to, a separate section (Section 13.4) is dedicated to this subject. An example of how the wrong thermodynamic package can cause serious errors is given in Example 13.1.

Example 13.1

Consider the HCl absorber (T-602) in the separation section of the allyl chloride process, Figure C.3 (Appendix C, on the CD). This equipment is shown in Figure E13.1. The function of the absorber is to contact countercurrently Stream 10a, containing mainly propylene and hydrogen chloride, with water, Stream 11. The HCl is highly soluble in water and is almost completely absorbed to form 32 wt% hydrochloric acid, Stream 12. The gas leaving the top of the absorber, Stream 13, is almost pure propylene, which is cleaned and then recycled.

Table E13.1 shows the results for the two outlet streams from the absorber—Streams 12 and 13—for two simulations, each using a different thermodynamic model for the vapor-liquid equilibrium calculations. The second and third columns in the table show the

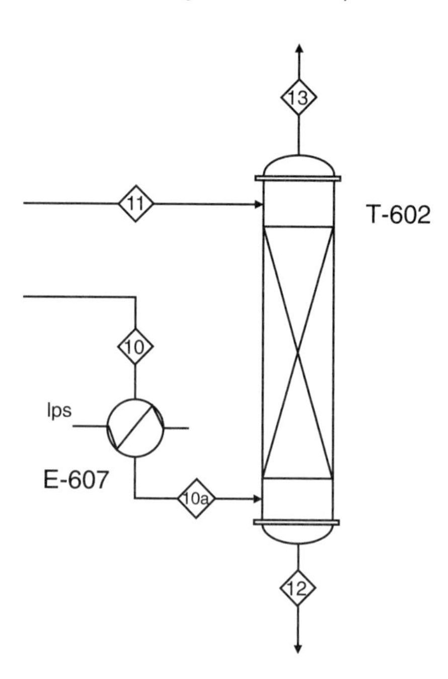

Figure E13.1 HCl Absorber in Allyl Chloride Separation Section (Unit 600), Appendix C

results using the SRK (Soave [7], Redlich and Kwong [8]) model, which is the preferred model for many common components. The fourth and fifth columns show the results using a model that is specially designed to deal with ionic type compounds (HCl) that dissolve in water and then dissociate. The difference in results is remarkable. The HCl-water system is highly non-ideal, and, even though the absorption of an acid gas into aqueous solutions is quite common, the SRK model is not capable of correctly modeling the phase

Table E13.1 Results of Simulation of HCl Absorption Using Two Different Physical Property Models

Phase Component Flows (kmol/h)	Using SRK Model		Using PPAQ* Model	
	Stream 12 Liquid	Stream 13 Vapor	Stream 12 Liquid	Stream 13 Vapor
Propylene	0.05	57.48	—	57.53
Allyl chloride	0.01	—	0.01	—
Hydrogen chloride	0.91	18.78	19.11	0.58
Water	81.37	0.63	81.88	0.12
Total	82.34	76.89	101.00	58.23

*This is a model used in the CHEMCAD simulator especially for HCl-water and similar systems.

behavior of this system. With the SRK model, virtually all the HCl leaves the absorber as a gas. Clearly, if the simulation were done using only the SRK model, the results would be drastically in error. This result is especially disturbing because SRK is the default thermodynamics package in many simulators.

More details of model selection are given in Section 13.4.

The importance of thermodynamic model selection and its impact on the validity of the results of a simulation are discussed at length by Horwitz and Nocera [9], who warn

> **"You absolutely must have confidence in the thermodynamics that you have chosen to represent your chemicals and unit operations. This is your responsibility, not that of the software simulation package. If you relinquish your responsibility to the simulation package, be prepared for dire consequences."**

13.2.3 Input the Topology of the Flowsheet

The most reliable way to input the topology of the process flow diagram is to make a sketch on paper and have this in front of you when you construct the flowsheet on the simulator. Contrary to the rules given in Chapter 1 on the construction of PFDs, every time a stream splits or several streams combine, a simulator equipment module (splitter or mixer) must be included. These "phantom" units were introduced in Chapter 5 and are useful in tracing streams in a PFD as well as being required for the simulator. Most simulators allow the flowsheet topology to be input both graphically and by keyboard. Certain conventions in the numbering of equipment and streams are used by the simulator to keep track of the topology and connectivity of the streams. When using the graphical interface, the streams and equipment are usually numbered sequentially in the order they are added. These can be altered by the user if required. Care must be taken when connecting batch and continuous unit operations, because it is often assumed that "continuous" units approach equilibrium instantaneously.

13.2.4 Select Feed Stream Properties

As discussed in Section 13.1, the sequential modular approach to simulation requires that all feed streams be specified (composition, flowrate, vapor fraction, temperature, and pressure). In addition, estimates of recycle streams should also be made. Although feed properties are usually well defined, some confusion may exist regarding the number and type of variables that must be specified to define completely the feed stream. In general, feed streams will contain n components and consist of one or two phases. For such feeds, a total of $n + 2$ specifications completely defines the stream. This is a consequence of the phase rule. Giving the flowrate (kmol/h, kg/s, etc.) of each component in the feed stream takes care of n of these specifications. The remaining two specifications should also be independent. For example, if the stream is one phase, then giving the temperature and pressure of the stream completely defines the feed. Temperature and pressure also completely define a multicomponent stream having two phases. However, if the feed is a single component and contains two phases, then temperature and pressure are not independent. In this case, the vapor fraction and either the temperature or the pressure must be specified. Vapor fraction can also be used to specify a two-phase multicomponent system, but if used, only temperature or pressure can be used to specify completely the feed. To avoid confusion, it is recommended that vapor fraction (vf) be specified only for saturated vapor ($vf = 1$), saturated liquid ($vf = 0$), and two-phase, single-component ($0 < vf < 1$) streams. All other streams should be specified using the temperature and pressure.

Use the vapor fraction (vf) to define feed streams only for saturated vapor ($vf = 1$), saturated liquid ($vf = 0$), and two-phase, single-component ($0 < vf < 1$) streams.

By giving the temperature, pressure, and vapor fraction for a feed, the stream is overspecified and errors will result.

13.2.5 Select Equipment Parameters

It is worth pointing out that process simulators, with a few exceptions, are structured to solve process material and energy balances, reaction kinetics, reaction equilibrium relationships, phase equilibrium relationships, and equipment performance relationships for equipment in which sufficient process design variables and batch operations scheduling have been specified. For example, consider the design of a liquid-liquid extractor to remove 98% of a component in a feed stream using a given solvent. In general, a process simulator will not be able to solve this design problem directly; that is, it cannot determine the number of equilibrium stages required for this separation. However, if the problem is made into a simulation problem, then it can be solved by a trial-and-error technique. Thus, by specifying the number of stages in the extractor, case studies in which the number of stages are varied can be performed, and this information used to determine the correct number of stages required to obtain the desired recovery of 98%. In other cases, such as a plug flow reactor module, the simulator can solve the design problem directly—that is, calculate the amount of catalyst required to carry out the desired reaction. Therefore, before one starts a process simulation it is important to know what equipment parameters must be specified in order for the process to be simulated.

There are essentially two levels at which a process simulation can be carried out. The first level, Level 1, is one in which the minimum data are supplied in order for the material and energy balances to be obtained. The second level, Level 2, is one in which the simulator is used to do as many of the design calculations as possible. The second level requires more input data than the first. An example of the differences between the two levels is illustrated in Figure 13.4, which shows a heat exchanger in which a process stream is being cooled using cooling water. At the first level, Figure 13.4(a), the only information that is specified is the desired outlet condition of the process stream—for example, pressure and temperature or vapor fraction—if the stream is to leave the exchanger as a two-phase mixture. However, this is enough information for the simulator to calculate the duty of the exchanger and

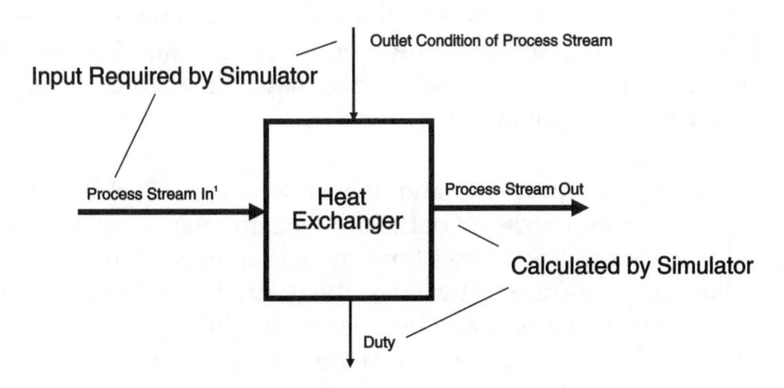

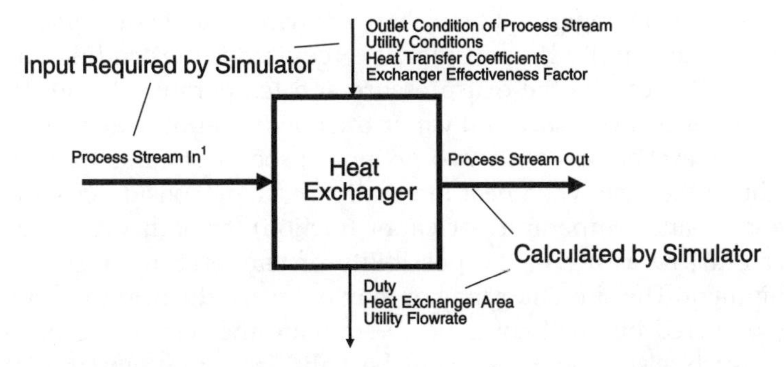

[1] This information may come from the preceding unit operation and thus would be supplied automatically by the simulator.

Figure 13.4 Information Required for Different Levels of Simulation

the properties of the process stream leaving the equipment. At the second level, Figure 13.4(b), additional data are provided: the inlet and desired outlet temperature for the utility stream, the fact that the utility stream is water, the overall heat transfer coefficient, and the heat-exchanger configuration or effectiveness factor, *F*. Using this information, the simulator calculates the heat-exchanger duty, the required cooling water flowrate, and the required heat transfer area.

When attempting to do a simulation on a process for the first time, it is recommended that you provide the minimum data required for a Level 1 simulation. When you have obtained a satisfactory, converged solution, you can go back and provide more data to obtain desired design parameters, that is, a Level 2 solution.

When first simulating a process, input only the data required to perform the material and energy balances for the process.

The structure of the process simulator will determine the exact requirements for the input data, and such information will be available in the user manual for the software or on Help screens. However, for Level 1 simulations, a brief list of typical information is presented below that may help a novice user prepare the input data for a process simulation.

Pumps, Compressors, and Power Recovery Turbines (Expanders). For pumps, the desired pressure of the fluid leaving the pump *or* the desired pressure increase of the fluid as it flows through the pump is all that is required.

For compressors and turbines, the desired pressure of the fluid leaving the device *or* the desired pressure increase of the fluid as it flows through the equipment is required. In addition, the mode of compression or expansion—adiabatic, isothermal, or polytropic—is required.

Heat Exchangers. For exchangers with a single process stream exchanging energy with a utility stream, all that is required is the condition of the exit process stream. This can be the exit pressure and temperature (single-phase exit condition) or the exit pressure and vapor fraction (two-phase exit condition).

For exchangers with two or more process streams exchanging energy (as might be the case when heat integration is being considered), the exit conditions (pressure and temperature or vapor fraction) for both streams are required. The user must be aware of the possibility of temperature crosses in heat-exchange equipment. The simulator may or may not warn the user that a temperature cross has occurred but will continue to simulate the rest of the process. The results from such a simulation will not be valid, and the temperature cross must be remedied before a correct solution can be obtained. Therefore, it is recommended that the user check the temperature profiles for all heat exchangers after the simulation.

Fired Heaters (Furnaces). The same requirements for heat exchangers with a single process fluid apply to fired heaters.

Mixers and Splitters. Mixers and splitters used in process simulators are usually no more than simple tees in pipes. Unless special units must be provided—for example, when the fluids to be mixed are very viscous and in-line mixers might be used—the capital investment of these units can be assumed to be zero.

Mixers represent points where two or more process streams come together. The only required information is an outlet pressure or pressure drop at the mixing point. Usually, the pressure drop associated with the mixing of streams is small, and the pressure drop can be assumed to be equal to zero with little error. If feed streams enter the mixer at different pressures, then the outlet stream is assumed to be at the lowest pressure of the feed streams. This assumption causes little error in the material and energy balance. However, the correct analysis of the pressure profiles in a system where several streams mix is given in Section 19.4.

Splitters represent points at which a process stream splits into two or more streams with different flowrates but identical compositions. The required information is the outlet pressure or pressure drop across the device and the relative flows of the output streams. Usually, there is little pressure drop across a splitter, and all streams leaving the unit are at the same pressure as the single feed stream. In a batch operation, the splitter can be assigned on and off times to divert the inlet flow to various other units on a schedule.

Valves. Either the outlet pressure or pressure drop is required.

Reactors. The way in which reactors are specified depends on a combination of the input information required and the reactor category. Generally there are four categories of reactor: stoichiometric reactor, kinetic (plug flow or CSTR) reactor, equilibrium reactor, and batch reactor. All these reactor configurations require input concerning the thermal mode of operation: adiabatic, isothermal, amount of heat removed or added. Additional information is also required. Each reactor type is considered separately below.

Stoichiometric Reactor: This is the simplest reactor type that can be simulated. The required input data are the number and stoichiometry of the reactions, the temperature and pressure, and the conversion of the limiting reactant. Reactor configuration (plug flow, CSTR) is not required because no estimate of reactor volume is made. Only basic material and energy balances are performed.

Kinetic (Plug Flow and CSTR) Reactor: This reactor type is used to simulate reactions for which kinetics expressions are known. The number and stoichiometry of the reactions are required input data. Kinetics constants (Arrhenius rate constants and Langmuir-Hinshelwood constants, if used) and the form of the rate equation (simple first-order, second-order, Langmuir-Hinshelwood kinetics, etc.) are also required. Reactor configuration (plug flow, CSTR) is required. Options may be available to simulate cooling or heating of reactants in shell-and-tube reactor configurations in order to generate temperature profiles in the reactor.

Equilibrium Reactor: As the name implies, this reactor type is used to simulate reactions that obtain or approach equilibrium conversion. The number and stoichiometry of the reactions and the fractional approach to equilibrium are the required input data. In addition, equilibrium constants as a function of temperature may be required for each reaction or may be calculated directly from information in the database. In this mode, the user has control over which reactions should be considered in the analysis.

Minimum Gibbs Free Energy Reactor: This is another common form of the equilibrium reactor. In the Gibbs reactor, the outlet stream composition is calculated by a free energy minimization technique. Usually data are available from the simulator's databank to do these calculations. The only input data required are the list of components that one anticipates in the output from the reactor. In this mode the equilibrium conversion that would occur for an infinite residence time is calculated.

Batch Reactor: This reactor type is similar to the kinetic reactor (and requires the same kinetics input), except that it is batch. The volume of the reactor is specified. The feeds, outlets flows, and reactor temperature (or heat duty) are scheduled (i.e., they are specified as time series).

As a general rule, one should initially use the least complicated reactor module that will allow the heat and material balance to be established. The reactor module can always be substituted later with a more sophisticated one that allows the desired design calculations to be performed. It should also be noted that a common error made in setting up a reactor module is the use of the wrong component as the limiting reactant when a desired conversion is specified. This is especially true when several simultaneous reactions occur and the limiting component may not be obvious solely from the amounts of components in the feed.

Flash Units. In simulators, the term *flash* refers to the module that performs a single-stage vapor-liquid equilibrium calculation. Material, energy, and phase equilibrium equations are solved for a variety of input parameter specifications. In order to specify completely the condition of the two output streams (liquid and vapor), two parameters must be input. Many combinations are possible—for example, temperature and pressure, temperature and heat load, or pressure and mole ratio of vapor to liquid in exit streams. Often, the flash module is a combination of two pieces of physical equipment, that is, a phase separator and a heat exchanger. These should appear as separate equipment on the PFD. Note that a flash unit can also be specified for batch operation, in which case the unit can serve as a surge or storage vessel.

Distillation Columns. Usually, both rigorous methods (plate-by-plate calculations) and shortcut methods (Fenske and Underwood relationships using key components) are available. In preliminary simulations, it is advisable to use shortcut methods. The advantage of the shortcut methods is that they allow a design calculation (which estimates the number of theoretical plates required for the separation) to be performed. For preliminary design calculations, this is a very useful option and can be used as a starting point for using the more rigorous algorithms, which require that the number of theoretical plates be specified. It should be noted that, in both methods, the calculations for the duties of the reboiler and condenser are carried out in the column modules and are presented in the output for the column. Detailed design of these heat exchangers (area calculations) often cannot be carried out during the column simulation.

Shortcut Module: The required input for the design mode consists of identification of the key components to be separated, specification of the fractional recoveries of each key component in the overhead product, the column pressure and pressure drop, and the ratio of actual to minimum reflux ratio to be used in the column. The simulator will estimate the number of theoretical plates required, the exit stream conditions (bottom and overhead products), optimum feed location, and the reboiler and condenser duties.

If the shortcut method is used in the performance mode, the number of plates must also be specified, but the R/R_{min} is calculated.

Rigorous Module: The number of theoretical plates must be specified, along with the condenser and reboiler type, column pressure and pressure drop, feed tray locations, and side product locations (if side stream products are desired). In addition, the total number of specifications given must be equal to the number of products (top, bottom, and side streams) produced. These product specifications are often a source of problems, and this is illustrated in Example 13.2.

Several rigorous modules may be available in a given simulator. Differences between the modules are the different solution algorithms used and the size and complexity of the problems that can be handled. Tray-to-tray calculations can be handled for several hundred stages in most simulators. In addition, these modules can be used to simulate accurately other equilibrium staged devices, for example, absorbers and strippers.

Batch Distillation: This module is similar to the rigorous module, except that feeds and product draws are on a schedule (not continuous). Therefore, the start and stop times of the feeds and products must be specified, and a time series of tray concentrations and temperatures is generated by the simulator.

Example 13.2

Consider the benzene recovery column in the toluene hydrodealkylation process shown in Figure 1.5. This column is redrawn in Figure E13.2. The purpose of the column is to separate

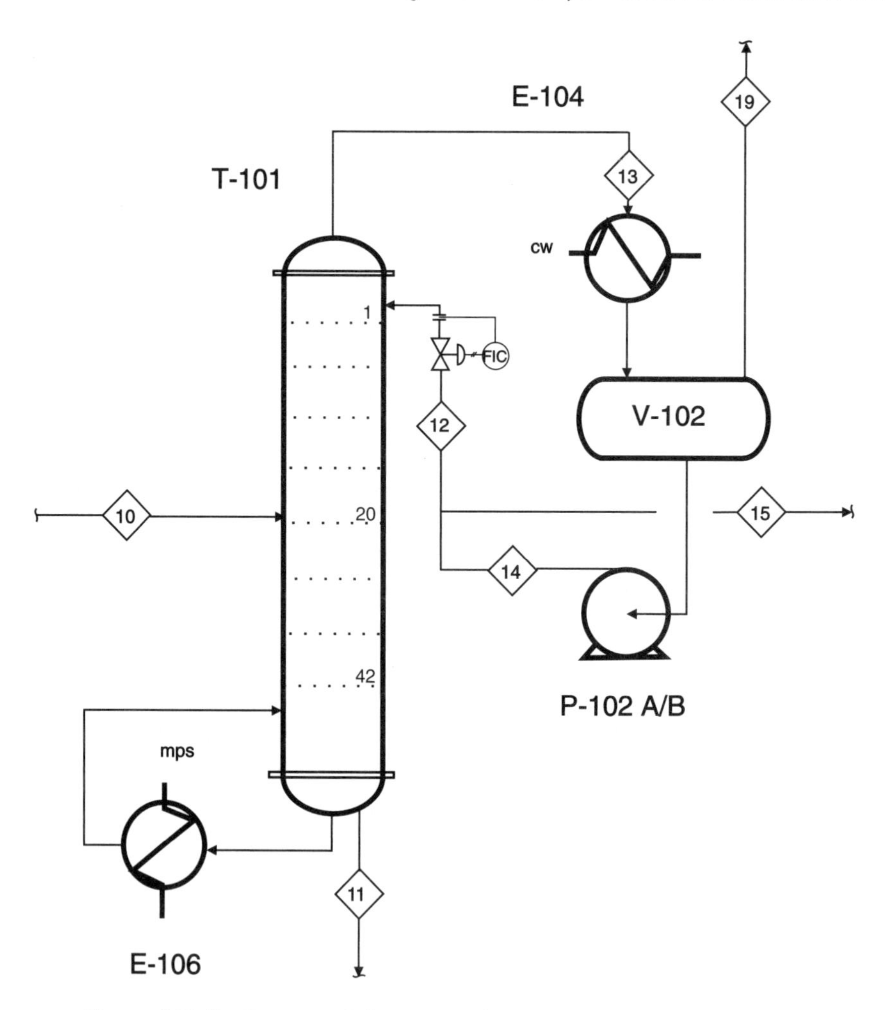

Figure E13.2 Benzene Column in Toluene Hydrodealkylation Process (from Figure 1.5)

Component	Stream 10	Stream 15	Stream 19	Stream 11
Hydrogen	0.02	—	0.02	—
Methane	0.88	—	0.88	—
Benzene	106.3	105.2	—	1.1
Toluene	35.0	0.4	—	34.6

the benzene product from unreacted toluene, which is recycled to the front end of the process. The desired purity of the benzene product is 99.6 mol%. The feed and the top and bottoms product streams are presented in the following table, which is taken from Table 1.5.

There are many ways to specify the parameters needed by the rigorous column algorithm used to simulate this tower.

Two examples are given:

1. The key components for the main separation are identified as benzene and toluene. The composition of the top product is specified to be 99.6 mole% benzene, and the recovery of toluene in the bottoms product is 0.98.
2. The top composition is specified to be 99.6 mol% benzene, and that of the recovery of benzene in the bottoms product is 0.01.

The first specification violates the material balance, whereas the second specification does not. Looking at the first specification, if 98% of the toluene in the feed is recovered in the bottoms product, then 2% or 0.7 kmol/h must leave with the top product. Even if the recovery of benzene in the top product were 100%, this would yield a top composition of 106.3 kmol/h benzene and 0.7 kmol/h toluene. This corresponds to a mole fraction of 0.993. Therefore, the desired mole fraction of 0.996 can never be reached. Thus, by specifying the recovery of toluene in the bottoms product, the specification for the benzene purity is automatically violated.

The second specification shows that both specifications can be achieved without violating the material balance. The top product contains 99% of the feed benzene (105.2 kmol/h) and 0.4 kmol/h toluene, which gives a top composition of 99.6 mol% benzene. The bottoms product contains 1.0% of the feed benzene (1.1 kmol/h) and 34.6 kmol/h of toluene.

When giving the top and bottom specifications for a distillation column, make sure that the specifications do not violate the material balance.

If problems continue to exist, one way to ensure that the simulation will run is to specify the top reflux rate and the boil-up rate (reboiler duty). Although this strategy will not guarantee the desired purities, it will allow a base case to be established. With subsequent manipulation of the reflux and boil-up rates, the desired purities can be obtained.

Absorbers and Strippers. Usually these units are simulated using the rigorous distillation module given above. The main differences in simulating these types of equipment are that condensers and reboilers are not normally used. In addition, there are two feeds to the unit; one feed enters at the top, and the other at the bottom.

Liquid-Liquid Extractors. A rigorous tray-by-tray module is used to simulate this multistaged equipment. It is imperative that the thermodynamic model for this unit be capable of predicting the presence of two liquid phases, each with appropriate liquid-phase activity coefficients.

13.2.6 Selection of Output Display Options

Several options will be available to display the results of a simulation. Often, a report file can be generated and customized to include a wide variety of stream and equipment information. In addition, a simulation flowsheet (not a PFD), T-Q diagrams for heat exchangers, vapor and liquid flows, temperature and composition profiles (tray-by-tray) for multistaged equipment, scheduling charts for batch operations, environmental parameters for exit streams, and a wide variety of phase diagrams for streams can be generated. The user manual should be consulted for the specific options available for the simulator you use.

13.2.7 Selection of Convergence Criteria and Running a Simulation

For equipment requiring iterative solutions, there will be user-selectable convergence and tolerance criteria in the equipment module. There will also be convergence criteria for the whole flowsheet simulation, which can be adjusted by the user.

The two most important criteria are number of iterations and tolerance. These criteria will often have default values set in the simulator. Unless specific problems arise, these default values should be used in your simulations.

If the simulation has not converged, the results do not represent a valid solution and should not be used.

When convergence is not achieved, three common causes are as follows.

1. The problem has been ill posed. This normally means that an equipment specification has been given incorrectly. For example, see the first specification in Example 13.2 for the rigorous column module.
2. The tolerance for the solution has been set too tightly, and convergence cannot be obtained to the desired accuracy no matter how many solution iterations are performed.
3. The number of iterations is not sufficient for convergence. This occurs most often when the flowsheet has many recycle streams. Rerunning the flowsheet simulation with the results from the preceding run may give a converged solution. If convergence is still not obtained, then one way to address this problem is to remove as many recycle streams as pos-

sible. The simulation is then run, and the recycle streams are added back, one by one, using the results from the preceding simulation as the starting point for the new one. This method is discussed in more detail in Section 13.3.

Of the three reasons, the first one is by far the most common.

The most common reason for the failure of a simulation to converge is the use of incorrect or impossible equipment specifications.

13.3 HANDLING RECYCLE STREAMS

Recycle streams are very important and common in process flowsheets. Computationally, they can be difficult to handle and are often the cause for unconverged flowsheet simulations. There are ways in which the problems caused by recycle streams can be minimized. When a flowsheet is simulated for the first time, it is wise to consider carefully any simplifications that may help the convergence of the simulation. Consider the simulation of the DME flowsheet illustrated in Figure B.1.1, Appendix B. This flowsheet is shown schematically in Figure 13.5(a). The DME process is simple, no by-products are formed, the separations are relatively easy, and the methanol can be purified easily prior to being returned to the front end of the process. In attempting to simulate this process for the first time, it is evident that two recycle streams are present. The first is the unreacted methanol that is recycled to the front of the process, upstream of the reactor. The second recycle loop is due to the heat integration scheme used to preheat the reactor feed using the reactor effluent stream. The best way to simulate this flowsheet is to eliminate the recycle streams as shown in Figure 13.5(b). In this figure, two separate heat exchangers have been substituted for the heat integration scheme. These exchangers allow the streams to achieve the same changes in temperature while eliminating the interaction between the two streams. The methanol recycle is eliminated in Figure 13.5(b) by producing a methanol pseudo-output stream. The simulation of the flowsheet given in Figure 13.5(b) is straightforward; it contains no recycle streams and will converge in a single flowsheet iteration. Troubleshooting of the simulation, if input errors are present, is very easy because the flowsheet converges very quickly. Once a converged solution has been obtained, the recycle streams can be added back. For example, the methanol recycle stream would be introduced back into the simulation. The composition of this stream is known from the preceding simulation, and this will be a very good estimate for the recycle stream composition. Once the simulation has been run successfully with the methanol recycle stream, the heat integration around the reactor can be added back and the simulation run again. Although this method may seem unwieldy, it does provide a reliable method for obtaining a converged simulation.

For the DME flowsheet in Figure 13.5, the unreacted methanol that was recycled was almost pure feed material. This means that the estimate of the recycle stream composition, obtained from the once-through simulation using Figure 13.5(b), was very good. When the recycle stream contains significant amounts of by-products, as is the case with the hydrogen recycle stream in Figure 1.5 (Streams 5 and 7), the estimate of the composition using a once-through simulation will be significantly different from the actual recycle stream composition. For such cases, when purification of the recycle stream does not occur, it is best to keep this recycle stream in the flowsheet and eliminate all other recycle streams for the first simulation. Once a converged solution is reached, the other recycle streams can be added back one at a time.

Often, a series of case studies will need to be run using a base-case simulation as a starting point. This is especially true when performing a parametric optimization on the process (see Chapter 14). When performing such case studies, it is wise to make small changes in input parameters in order to obtain a converged simulation. For example, assume that a converged simulation for a reactor module at 350°C has been obtained, and a case study needs to be run at 400°C. When the equipment temperature in the reactor module is changed and the simulation

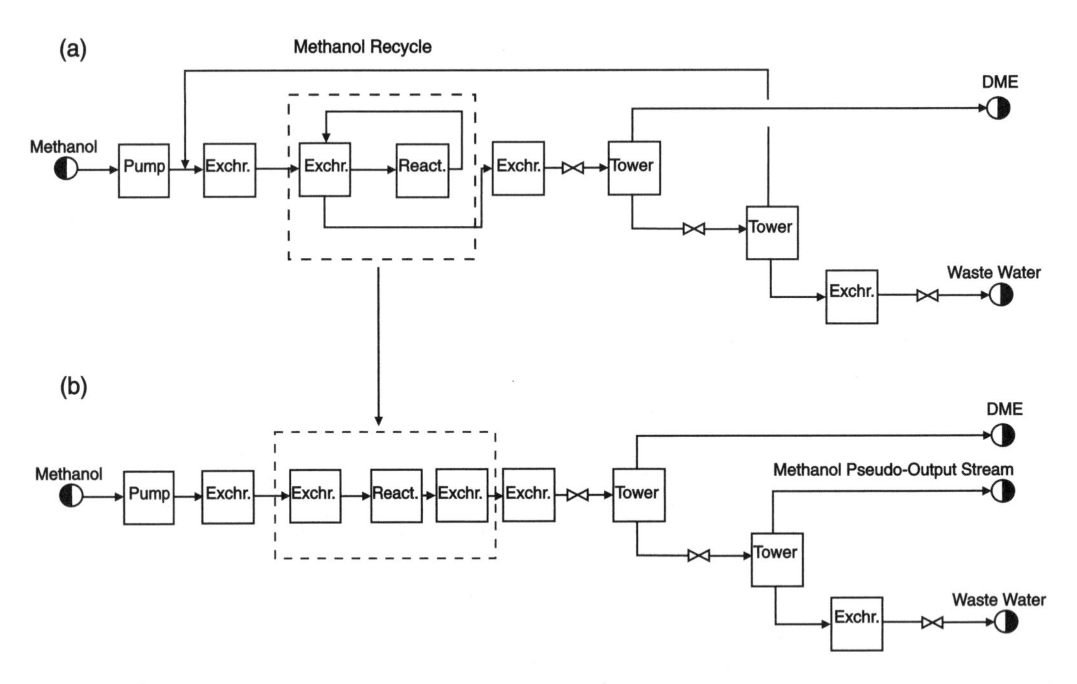

Figure 13.5 Block Flow Diagram for DME Process Showing (a) Recycle Structure and (b) Elimination of Recycles

is rerun, it may be found that the simulation does not converge. If this is the case, then, for example, go back to the base-case run, change the reactor temperature by 25°C, and see whether it converges. If it does, then the input can be changed by another 25°C to give the desired conditions, and so on. The use of small increments or steps when simulating changes in flowsheets often produces a converged simulation when a single large change in input will not.

Often when simulating a process, it is the flowrate of products (not feeds) that is known—for example, production of 60,000 tonne/yr of chemical X, with a purity of 99.9 wt%. Assume that a converged solution has been found in which all the product specifications have been met except that the flowrate of primary product is not at the desired value. For this case, it is a simple matter to multiply all the feeds to the process by a factor to obtain the desired flowrate of the product; that is, just scale the solution up or down by a constant factor and rerun the simulation to get the correct equipment specifications.

For more advanced simulation applications, such as optimizing or simulating existing plants, it may be necessary or useful to use controller modules in the simulation to obtain a desired result. For example, in a recycle loop it might be required that the ratio of two components entering a reactor be set at some fixed value. A controller module could be used to adjust the purge flowrate from the recycle stream to obtain this ratio. The use of controller modules introduces additional recycle loops. The way in which specifications for controllers are given can cause additional convergence problems, and this topic is covered in detail by Schad [10].

13.4 CHOOSING THERMODYNAMIC MODELS

The results of any process simulation are never better than the input data, especially the thermodynamic data.

Everything from the energy balance to the volumetric flowrates to the separation in the equilibrium-stage units depends on accurate thermodynamic data.

If reaction kinetics information is missing, the simulator cannot calculate the conversion from a given reactor volume. Because such a calculation is not possible, only equilibrium reactor modules and those with specified conversions can be used.

Unfortunately, process simulators have default thermodynamics packages, which will—without warning—blindly miscalculate the entire flowsheet.

Only a few, readily available data are required to estimate the parameters in simple thermodynamic models. If the critical temperature and critical pressure are known for each pure component, the parameters for simple, cubic equations of state can be estimated. Even if these critical properties are unknown, they in turn can be estimated from one vapor pressure and one liquid density. Group-contribution models require even less information: merely the chemical structure of the molecule. However, these estimations can never be as accurate as experimental data. In thermodynamics, as elsewhere, *you get only what you pay for—or less!*

Compounding this problem is the development and implementation of expert systems to help choose the thermodynamic model. To date, these features merely offer false hope. Human thermodynamics experts do not recommend them.

A safe choice of thermodynamic model requires knowledge of the system, the calculation options of the simulator, and the margin of error. In this section, guidance on choosing and using a thermodynamic model is given. In an academic setting, the choice of thermodynamic model affects the answers but not the ability of the student to learn how to use a process simulator—a key aspect of this book. Therefore, the examples throughout this book use simplistic thermodynamic models to allow easy simulation. In any real problem, where the simulation will be used to design or troubleshoot a process, the proper choice of thermodynamic model is essential. This section focuses on the key issues in making that choice, in using experimental data, and in determining when additional data are needed.

It has been assumed that the reader understands the basics of chemical engineering thermodynamics as covered in standard textbooks [4,5,6]. As pointed out before, it is extremely important that the chemical engineer performing a process simulation understand the thermodynamics being used. In a course, the instructor can often provide guidance. The help facility of the process simulator provides a refresher on details of the model choices; however, these descriptions do not include the thermodynamics foundation required for complete understanding. If the descriptions in the help facility are more than a refresher, the standard thermodynamics textbooks should be consulted.

If the thermodynamic option used by the process simulator is a mystery, the meaning of the results obtained from the simulation will be equally mysterious.

13.4.1 Pure-Component Properties

Physical properties such as density, viscosity, thermal conductivity, and heat capacity are generally not a serious problem in simulation. The group-contribution methods are reasonably good, and simulator databanks include experimental heat capacity data for more than a thousand substances. Although these correlations have random and systematic errors of several percent, this is close enough for most purposes. (However, they are not sufficient when you are paying for a fluid crossing a boundary based on volumetric flowrate.) As noted in Section 13.2.2, one should always be aware of which properties are estimated and which are from experimental measurements.

13.4.2 Enthalpy

Although the pure-component heat capacities are calculated with acceptable accuracy, the enthalpies of phase changes often are not. Care should be taken in choosing the enthalpy model for a simulation. If the enthalpy of vaporization is an important part of a calculation, simple equations of state should not be used. In fact, the "latent heat" or "ideal" options would be better. If the substance is above or near its critical temperature, equations of state must be used, but the user must beware, especially if polar substances such as water are present, as shown in Example 13.3.

Example 13.3

A gas stream at 3000°F of the following concentration is to be cooled by evaporation of 500 kg/h of water entering at 70°F. Assume atmospheric pressure.

H_2	22.72 kg/h
N_2	272.24
CO	268.40
HCl	26.84

Perform a simulation to determine the final temperature of the cooled gas stream with the default thermodynamic model and with the ideal model.

The default on most process simulators is an equation of state, either PR (Peng-Robinson) or SRK (Soave-Redlich-Kwong). These models give an outlet temperature of 480°F. The ideal model gives an outlet temperature of 348°F. The value calculated by the ideal model is closer to reality in this case because the equations of state do a poor job of estimating the enthalpy of vaporization of water, which is the most important property for the energy balance. The ideal model uses the experimentally determined value of this enthalpy of vaporization (from the databank).

13.4.3 Phase Equilibria

Extreme care must be exercised in choosing a model for phase equilibria (sometimes called the fugacity coefficient, *K*-factor, or fluid model). Whenever possible, phase equilibrium data for the system should be used to regress the parameters in the model, and the deviation between the model predictions and the experimental data should be studied.

There are two general types of fugacity models: equations of state and liquid-state activity-coefficient models. An equation of state is an algebraic equation for the pressure of a mixture as a function of the composition, volume, and temperature. Through standard thermodynamic relationships, the fugacity, enthalpy, and so on for the mixture can be determined. These properties can be calculated for any density; therefore, both liquid and vapor properties, as well as supercritical phenomena, can be determined.

Activity-coefficient models, however, can only be used to calculate liquid-state fugacities and enthalpies of mixing. These models provide algebraic equations for the activity coefficient (γ_i) as a function of composition and temperature. Because the activity coefficient is merely a correction factor for the ideal-solution model (essentially Raoult's Law), it cannot be used for supercritical or "noncondensable" components. (Modifications of these models for these types of systems have been developed, but they are not recommended for the process simulator user without consultation with a thermodynamics expert.)

Equations of state are recommended for simple systems (nonpolar, small molecules) and in regions (especially supercritical conditions for any component in a mixture) where activity-coefficient models are inappropriate. For complex liquid mixtures, activity-coefficient models are preferred, but only if *all* of the binary interaction parameters (BIPs) are available.

Equations of State. The default fugacity model is normally either the SRK (Soave-Redlich-Kwong) or the PR (Peng-Robinson) equation. They (like most popular equations of state) normally use three pure-component parameters per substance and one binary-interaction parameter per binary pair. Although they give qualitatively correct results even in the supercritical region, they are known to be poor predictors of enthalpy changes, and (except for light hydrocarbons) they are not quantitatively accurate for phase equilibria.

The predicted phase equilibrium is a strong function of the binary interaction parameters (BIPs). Process simulators have regression options to determine these parameters from experimental phase-equilibrium data. The fit gives a first-order approximation for the accuracy of the equation of state. This information should always be considered in estimating the accuracy of the simulation. Additional simulations should be run with perturbed model parameters to get a feel for the uncertainty, and the user should realize that even this approach gives an optimistic approximation of the error introduced by the model. If BIPs are provided in the simulator and the user has no evidence that one equation of state is better than another, then a separate, complete simulation should be performed for each of these equations of state. The difference between the simulations is a crude measure of the uncertainty introduced into the simulation by the uncertainty in the models. Again, the inferred uncertainty will be on the low side.

Monte-Carlo simulations (see Section 10.7) can be done with the results of the regression; however, present process simulators are not equipped to perform these directly. A simpler approach is to perform the simulation with a few different values of the BIPs for the equation of state. These values are typically 0.01 to 0.10. Larger values are rare, except in highly asymmetric systems. However, the difference between results calculated with values of, say, 0.01 and 0.02 can be large.

If BIPs are available for only a subset of the binary pairs, caution should be exercised. Assuming the unknown BIPs to be zero can be dangerous. Group-contribution models for estimating BIPs for equations of state can be used with caution.

There are usually six to ten equation-of-state choices, and a few mixing-rule choices. For polar or associating components or for heavy petroleum cuts, the help facility of the simulator should be consulted. Because different choices are available on the different simulators, they will not be covered here.

For most systems containing hydrocarbons and light gases, an equation of state is the best choice. One should initially choose the Peng-Robinson or Soave-Redlich-Kwong equation. (Note that neither the van der Waals nor Redlich-Kwong equation is a standard choice in simulators. These two equations of state were tremendous breakthroughs in fluid property models, but they have long ago been supplanted by other models that give better quantitative results.) VLE (vapor-liquid equilibrium) data for each binary system can then be used with the regression utility to calculate the BIPs for the binary pairs and to plot the resulting model predictions against the experimental data. This regression is done separately for each equation of state. The equation that gives a better fit in the ($PTxy$) region of operation of the unit operation of interest is then used. If phase equilibrium data are available at different temperatures, the temperature-dependent BIP feature of the simulator can be used. In the simulator databank, many BIPs are already regressed and available.

If neither simple equation of state adequately reproduces the experimental data, one of the other equations of state or other mixing rules, or a temperature-dependent BIP, may be needed. These often work better for polar-nonpolar systems. However, running the simulation more than once with different BIPs and with different thermodynamic models to judge the uncertainty of the result is recommended, as shown in Example 13.4. If the difference between the simulations seriously affects the viability of the process, a detailed uncertainty analysis is essential [11]. This is beyond the scope of this book.

Example 13.4

Use both the Peng-Robinson and the Soave-Redlich-Kwong equations of state to calculate the methane vapor molar flowrate from a flash at the following conditions:

Temperature:	225 K
Pressure:	60.78 bar
Feed flowrates:	
Carbon dioxide	6 kmol/h
Hydrogen sulfide	24 kmol/h
Methane	66 kmol/h
Ethane	3 kmol/h
Propane	1 kmol/h

Compare the results for BIPs from the process simulator databank and with the BIPs set to zero.

The following results were obtained using CHEMCAD.

	Databank BIPs	Zero BIPs
Peng-Robinson	52.0 kmol/h	32.5 kmol/h
Soave-Redlich-Kwong	60.1 kmol/h	50.5 kmol/h

The two equations of state give different results, and the effect of setting the BIPs to zero is very significant.

For most chemical systems below the critical region, a liquid-state activity-coefficient model is the better choice.

Liquid-State Activity-Coefficient Models. If the conditions of the unit operation are far from the critical region of the mixture or that of the major component and if experimental data are available for the phase equilibrium of interest (VLE or LLE), then a liquid-state activity-coefficient model is a reasonable choice. Activity coefficients (γ_i) correct for deviations of the liquid phase from ideal solution behavior, as shown in Equation (13.1).

$$\hat{\phi}_i^v y_i P = P_i^* x_i \gamma_i \phi_i^* \exp\left(\frac{1}{RT}\int_{P_i^*}^{P} v_i^l \, dP\right) \tag{13.1}$$

where $\hat{\phi}_i^v$ is the fugacity coefficient of component i in the vapor-phase mixture at system temperature T and pressure P, y_i is the vapor mole fraction of i, P_i^* is the vapor pressure of pure i at T, x_i is the liquid mole fraction of i, ϕ_i^* is the fugacity coefficient of pure i at its vapor pressure at T, and v_i^l is the molar volume of pure liquid i at T.

The roles of the terms in Equation (13.1) are discussed in detail in standard thermodynamics texts. Here, it is sufficient to point out that the two terms closest to the equal sign (on either side of the equal sign) give Raoult's Law and that the most important of the remaining correction terms is usually γ_i, the activity coefficient. Thus, use of an activity-coefficient model requires values for the pure-component vapor pressures at the temperature of the system. There are several important considerations in using activity-coefficient models.

- If no BIPs are available for a given binary system, an activity-coefficient model will give results similar to but not necessarily the same as those for an ideal solution.

- The standard version of the Wilson equation cannot predict liquid-liquid immiscibility.
- The BIPs for various activity-coefficient models can be estimated by UNIFAC. However, caution must be exercised because increased uncertainty is inserted into the model with such estimation.
- Some BIP estimation may be done automatically by the simulator.
- There are no reliable rules for choosing an activity-coefficient model *a priori.* The standard procedure is to check the correlation of experimental data by several such models and then choose the model that gives the best correlation.
- Parameters regressed from VLE data are often unreliable when used for LLE prediction (and vice versa). Therefore, some process simulators provide a choice between two sets of parameter sets.
- Often ternary (and higher) data are not well predicted by activity-coefficient models and BIPs.
- The BIPs are typically highly correlated. This and the empirical nature of these models lead to similar fits to experimental data with very different values of the BIPs.

Some of these considerations are demonstrated in Examples 13.5 and 13.6.

Example 13.5

Use the simulator databank BIPs for NRTL to calculate the vapor-liquid equilibrium for ethanol/water at 1 atm. Compare the results for BIPs set to zero. Regress experimental VLE data [12] to determine NRTL BIPs.

Figure E13.5(a) shows the *Txy* diagrams using the NRTL BIPs from the CHEMCAD databank and for these BIPs set to zero. Note that the latter case results in an ideal solution; thus, the azeotrope is missed. Regressing the experimental data for this system with the simulator regression tool gives the results shown in Figure E13.5(b). Although the BIPs in the databank (–55.1581, 670.441, 0.3031) and those regressed from the data (–104.31, 807.10, 0.28675) are quite different, the VLE calculated is very similar and is close to the experimental data.

Example 13.6

Calculate the LLE for the ternary di-isopropyl-ether/acetic-acid/water using NRTL and the BIPs available for the three binary pairs in the simulator databank. Compare the prediction with ternary LLE data for this system at 24.6°C [13].

See Figure E13.6. The experimental phase envelope (dotted lines) is twice the size of the predicted one (solid lines). This would lead to gross error in extraction calculations. Note that if all the BIPs are set to zero, there is no liquid-liquid immiscibility region.

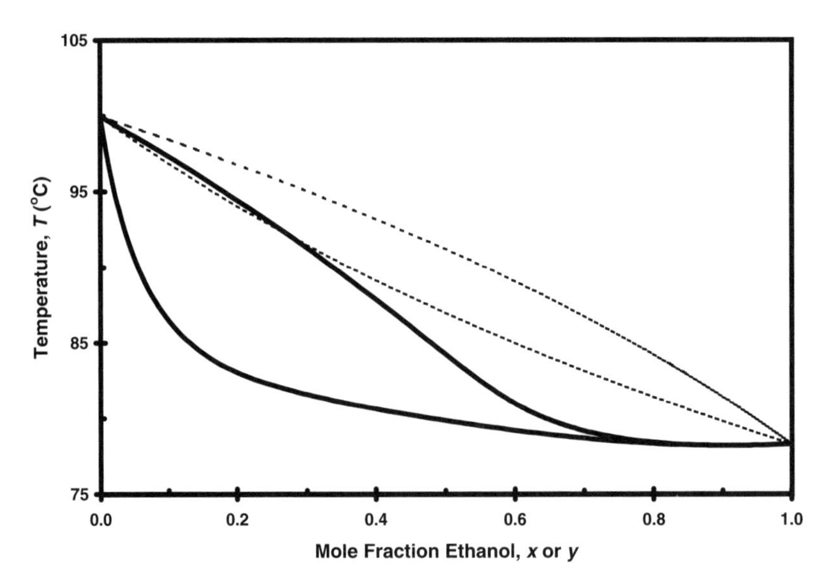

Figure E13.5(a) Vapor-Liquid Equilibrium for Ethanol/Water at 1 atm (*Solid curves are for CHEMCAD databank BIPs for NRTL. Dotted curves are for NRTL BIPs set to zero*).

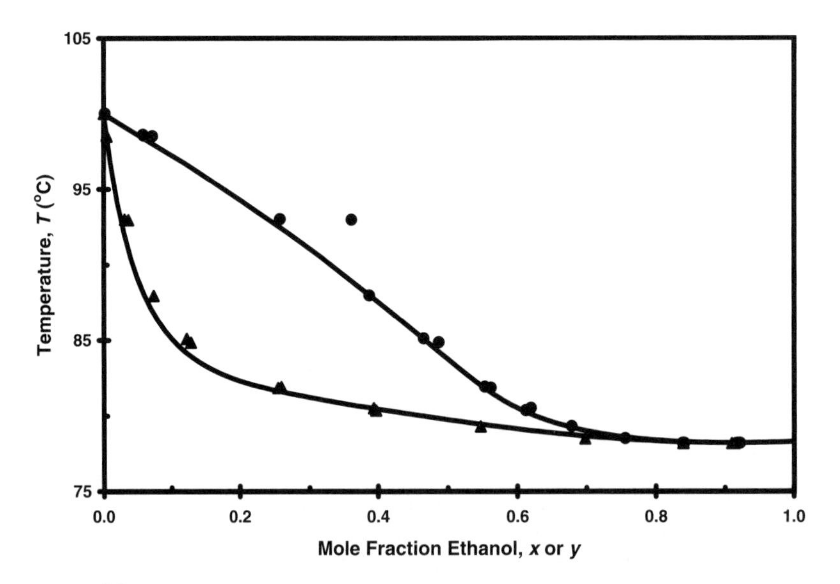

Figure E13.5(b) Vapor-Liquid Equilibrium for Ethanol/Water at 1 atm (*Solid curves are for NRTL BIPs regressed from data points shown*).

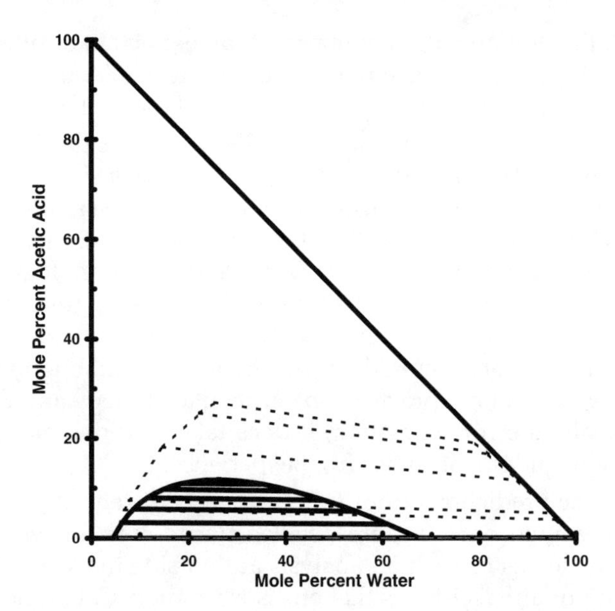

Figure E13.6 Liquid-Liquid Equilibrium for Di-isopropyl-Ether/Acetic-Acid/Water at 24.6°C

However, if the ternary LLE data were used to regress all of the BIPs simultaneously, the fit would be quite good.

The recommended strategy for choosing a liquid-state activity-coefficient model is as follows.

1. The simulator databank is checked for BIPs for all the binary pairs in the system. If these are available, they are most often from the DECHEMA Data Series [12,13], but sometimes they are from different sources. Some simulators provide the literature citation; others do not. Each of the three most common models (Wilson, NRTL, UNIQUAC) has different values for BIPs, and they are not correlated from one model to another. Although BIPs for Wilson may not be available for partially immiscible systems (see above), if a binary pair has BIPs for NRTL or for UNIQUAC, the BIPs for that pair should be available for both models. If they are not, the original data are found and the BIPs are fit for the other model.
2. If phase equilibrium data can be found for the binary pairs with missing BIPs, a regression is done with the simulator to find the missing BIPs.
3. For binary pairs that have no measured phase equilibria (there are many!), the UNIFAC estimation option of the simulator is used to estimate the BIPs. There are two UNIFAC methods: one for VLE and one for LLE. The choice depends on the type of equilibrium for the unit operation of interest.

4. If the unit operation of interest is an extraction or other operation involving LLE or VLLE and ternary data are available, the predictions of the three activity-coefficient models are checked for the ternary LLE.
5. One of the three methods usually shows a significantly better fit than the others. This method, *and the second-best method,* are used for the simulation. Comparing the two results provides a rough sense of the uncertainty of the calculation. Note that the "true" result is certainly not guaranteed to be between the two results. It is possible that this strategy will give a false sense of low uncertainty if the two methods give similar predictions that are far from the experimental data. This uncertainty strategy is used in the same way as is any heuristic from Chapter 11. Although this strategy is an analogue of a long-practiced experimental strategy and the basis for numerical analysis error estimation, it does take considerable engineering judgment. Good judgment comes from experience.
6. If the predictions from the activity-coefficient model chosen do not fit the measured data, a more detailed uncertainty analysis is needed. Although the details of such an analysis are outside the scope of this book, the most important decision is that one is warranted. Calibration of the uncertainty of simulation results is obtained from simulations run with different estimates (high and low, when possible).

The UNIFAC model is never used if experimental data are available for the binary system. UNIFAC is a group-contribution model for determining the BIPs for the UNIQUAC model (and by extension for the NRTL and Wilson models and for equations of state). Only chemical structure data are needed, but the calculations are not very accurate. When determining the numbers of groups within a molecule, one should always start with the largest group. This strategy minimizes the assumptions (and therefore the errors) in the model. Group-contribution models should be used with caution.

For many systems, a model such as UNIFAC may be the only option. If so, even a very crude uncertainty estimate can be difficult. If only one phase equilibrium datum for the system can be found, its deviation from the model prediction is at least some estimate of the uncertainty (as long as that datum was not used to regress parameters).

Using Scarce Data to Calibrate a Thermodynamic Model. Any experimental data on phase equilibria can be used to perform a crude calibration or verification of the model. It need not be the type of data that would be taken in the lab. If one knows the recovery in a column for one set of conditions, for example, and if only one BIP is unknown, then one can easily find the only value of the BIP that will reproduce that datum. Such data are sometimes found in patents.

More Difficult Systems. The above discussions pertain to "easy" systems: (1) small, nonpolar or slightly polar molecules for equations of state and (2) nonelectrolyte, nonpolymeric substances considerably below their critical temperatures for liquid-state activity-coefficient models. Most simulators have some models for electrolytes and for polymers, but these are likely to be even more uncertain than for the easy systems. Again, the key is to find some data, even plant operating data, to verify and to calibrate the models. If the overall recovery from a multistage separation is known, for example, one can simulate the column, using the best-known thermodynamic model, and the deviation between the plant datum and the simulator result is a crude (optimistic) estimate of the uncertainty.

Because most thermodynamic options are semitheoretical models for small, nonpolar molecules, the more difficult systems require another degree of freedom in the model. The most common such modification is to make the parameters temperature dependent. This requires additional data, but there is some theoretical justification for using effective model parameters that vary with temperature.

Hybrid Systems. Often a process includes components as wide ranging as solids at room temperature to supercritical gases. They can include water, strong acids, hydrocarbons, and polymers. Often, no single thermodynamic model can be used reliably to predict the fugacities of such a wide range of components. For these cases, simulators allow for hybrid thermodynamic models. The breadth of hybridization varies from one simulator to another, but all at least allow for some components to be considered immiscible with respect to others. For example, the NRTL model may be used for binary pairs for which both compounds are subcritical, while Henry's Law is used for supercritical (so-called noncondensable) components. Each simulator allows for immiscibility of water and hydrocarbon liquid phases, with the compositions of hydrocarbons in the aqueous phase estimated with Henry's Law and the liquid-liquid equilibrium for water calculated based on ideal solution in the aqueous phase and some chosen model in the hydrocarbon-rich phase. The user should check each of these options, so that it is clear what the simulator is doing. Although Henry's Law does not necessarily mean that the phase is aqueous, the Henry's Law model in a process simulator may be developed only for aqueous systems.

Another kind of hybridization is the choice of auxiliary models for liquid-state activity-coefficient models. One can specify what model to use for the vapor phase and whether to make the Poynting correction. The best choice is to use an equation of state (PR or SRK) for the vapor-phase correction and to use the Poynting correction. Both corrections go smoothly to zero in the low-pressure limit, and neither should add greatly to the computational time for most flowsheets.

The final type of hybridization is the use of different models for different unit operations. Although this appears to be inconsistent at first, it is reality that thermodynamic models are not perfect and that some work much better for LLE than for VLE, some work better for low pressures and others for high pressures, and some work for hydrocarbons but not for aqueous phases. Furthermore, simulators perform calculations for individual units and then pass only component flowrates, temperature, and pressure to the next unit. Thus, consistency is not a problem. Therefore, one should always consider the possibility of using different models for different unit operations. All the simulators allow this, and it is essential for a complex flowsheet. An activity-coefficient model can be used for the liquid-liquid extractor and an equation of state for the flash unit. This hybridization can be extremely important when, for example, some units contain mainly complex organics and other units contain light hydrocarbons and nitrogen.

Other Models. Different simulators have a variety of additional models beyond those mentioned above. For example, some have non-ideal electrolyte thermodynamic models that calculate species equilibria, some have polymer thermodynamic packages, and some allow petroleum cuts to be represented automatically by pseudocomponents. Presently, these packages are less consistent across simulators and are not discussed here. However, the user should always investigate the range of models available for the simulator being used.

13.4.4 Using Thermodynamic Models

In summary, one should use the assumed best thermodynamic model, based on the rough guidelines above. However, one should always resimulate either with another model assumed to be equally good or with different model parameters (see Example 13.7). The appropriate perturbation to apply to the parameters is available from the experimental data regression or from comparison of calculated results with experimental or plant data. Such data should always be sought for the conditions closest to those in the simulation. If the application is liquid-liquid extraction, for example, liquid-liquid equilibrium data (rather than vapor-liquid equilibrium data) should be used in the parameter regression, even though the same activity-coefficient models are used for both liquid-liquid and vapor-liquid equilibria.

The availability of BIPs in the databank of a process simulator must never be interpreted as an indication that the model is of acceptable accuracy.

These parameter values are merely the "best" for some specific objective function, for some specific set of data. The model may not be able to correlate even these data very well with this optimum parameter set. And one should treat a decision to use a BIP equal to zero as equivalent to using an arbitrary value of the BIP. The decision to use zero is, in fact, a decision to use a specific value based on little or no data.

Example 13.7

Consider the DME Tower, T-201, in the DME process in Appendix B. Simulate this unit using a shortcut model with different liquid-state activity-coefficient models to determine the required number of stages for the reflux ratio and the recoveries shown in Appendix B. Include no corrections for heat of mixing. Then perform a rigorous column simulation to check the distillate purity. Compare the results.

The base case uses the UNIFAC model with vapor-phase fugacity correction. The specifications are as follows:

- Column pressures: top, 10.3 bar; bottom, 10.5 bar
- Key components: light, DME; heavy, methanol
- Light key recovery in distillate: 98.93%
- Heavy key recovery in bottoms: 99.08%
- Reflux ratio: 0.3631
- 70% plate efficiency

An initial CHEMCAD simulation confirms the value given in Appendix B of 22 actual stages.

Table E13.7 shows the results obtained from six simulations, all with the same input specifications but with different thermodynamic options. The number of actual stages calculated ranges from 15 to 22; however, the results for two of the simulations (denoted n/a in the table) indicated that the minimum reflux ratio was greater than that specified. Without further information about the ability of the various models to correlate experimental vapor-liquid equilibrium data, a precise solution to the problem is not possible. However, the differences in the results obtained indicate that the choice of thermodynamic model is a crucial one. Of special concern here is the choice of correction of fugacities (denoted w/correction). These corrections are the first and last terms in Equation (13.1). Note that these results were obtained using the CHEMCAD databank BIP values for the NRTL and UNIQUAC models. Different BIP values will yield different results.

To determine which model is best, one finds data on the various binary pairs in the mixture. These data consist of measurements of temperature, pressure, and composition of a liquid in equilibrium with its vapor. Sometimes, the concentration of the vapor is also measured. From the temperature and the liquid-phase composition, the pressure and vapor-phase compositions are calculated with the thermodynamic model. The sum of the squared deviations between the experimental and the calculated pressure and between the experimental and the calculated vapor compositions is the objective function. The decision variables are the adjustable parameters in the thermodynamic model. Through the procedures of Chapter 14, the objective function is minimized and the optimum set of parameters is found. The standard process simulators incorporate a tool to do these regressions.

Of great concern is the purity of the overhead product. This stream is the DME product stream from the process, and the specification is 99.5 wt%. The third column in Table E13.7 shows the purity obtained by rigorous, tray-by-tray simulations of this column. In each case, the number of stages, feed location, column pressures, reflux ratio, and reboiler duty were set at those shown in Appendix B. Again, only the thermodynamic option was varied. In the shortcut calculation, many assumptions are made, including the constancy

Table E13.7 Comparison of Simulations for DME Column, T-201, for Various Thermodynamic Options

Thermodynamic Option	Number of Actual Stages Obtained from Shortcut Method	Distillate Purity Obtained with 22 Stages, Specified Reflux Ratio, and Shortcut Reboiler Duty	Required Reflux Ratio to Obtain DME Purity and Recovery with 22 stages
	Shortcut Method	Rigorous Simulation	Rigorous Simulation
UNIFAC w/correction	22	97.3 wt%	0.52
UNIFAC	15	95.0	0.47
NRTL w/correction	n/a	96.6	0.77
NRTL	21	97.6	0.49
UNIQUAC w/correction	n/a	97.1	0.58
UNIQUAC	18	97.6	0.48

of relative volatilities. For the rigorous calculation, the full power of the thermodynamic package is used in the phase-equilibrium and energy-balance calculations. In this example, the distillate purity specification would be very far off from that calculated with the shortcut method. As expected, the shortcut results are only preliminary to the rigorous column simulation.

The fourth column of Table E13.7 gives the reflux ratio required to meet both the DME purity and recovery specifications. The variation of this ratio from 0.47 to 0.77 is significant, because it is directly related to the required condenser and reboiler duties. If the NRTL w/correction calculation is closest to the truth, the reboiler duty required is 80% greater than that obtained in the baseline shortcut simulation. Ironically, this is the thermodynamics option suggested by the CHEMCAD expert system.

13.5 CASE STUDY: TOLUENE HYDRODEALKYLATION PROCESS

The purpose of this section is to present the input information necessary to make a basic simulation of the toluene hydrodealkylation process presented in Chapter 1. The required input data necessary to obtain a Level 1 simulation is presented in Table 13.1. The corresponding simulator flowsheet is given in Figure 13.6. In Table 13.1, the equipment numbers given in the third column correspond to those used in Figure 13.6. In the first column, the equipment numbers on the toluene hydrodealkylation PFD (Figure 1.5) are given. It should be noted that there is not a one-to-one correspondence between the actual equipment and the simulation modules. For example, three splitters and six mixers are required in the simulation, but these are not identified in the PFD. In addition, several pieces of equipment associated with the benzene purification tower are simulated by a single simulation unit. The numbering of the streams in Figure 13.6 corresponds to that given in Figure 1.5, except when additional stream numbers are required for the simulation. In order to avoid confusion, these extra streams are assigned numbers greater than 90.

In Table 13.2, the specifications for the feed streams are given. For this process, there are only two feed streams—Streams 1 and 3—corresponding to toluene and hydrogen, respectively. In addition, estimates of all the recycle streams should be given prior to beginning the simulation, and these are given in Table 13.2. However, these estimates need not be very accurate. Usually, any estimate is better than no estimate at all.

The data given in Tables 13.1 and 13.2 are sufficient to reproduce the material and energy balances for the toluene hydrodealkylation process. The use of these data to reproduce the flow table in Table 1.5 is left as an example problem at the end of the chapter. As mentioned in Section 13.3, some difficulty may arise when trying to simulate this flowsheet because of the three recycle streams. If you encounter problems in obtaining a converged solution, you should try to eliminate as many recycle streams as possible, run the simulation, and then add the

Table 13.1 Required Input Data for a Level 1 Simulation of Toluene Hydrodealkylation Process

Equipment Number	Simulator Equipment	Simulator Equip. No.	Input Streams		Output Streams		Required Input
TK-101	Mixer	m-1	1	11	90	—	Pressure drop = 0 bar
P-101	Pump	p-1	90	—	2	—	Outlet pressure = 27.0 bar
E-101	Hexch	e-1	92	—	4	—	Outlet stream vapor fraction = 1.0
H-101	Heater	h-1	4	—	6	—	Outlet temperature = 600°C
R-101	Stoic react	r-1	93	—	9	—	Conversion of toluene = 0.75
E-102	Flash	f-1	9	—	8	94	Temperature = 38°C Pressure = 23.9 bar
V-101	Flash	f-1	9	—	8	94	No input required because vessel is associated with flash operation.
V-103	Flash	f-2	94	—	17	18	Temperature = 38°C Pressure = 2.8
E-103	Hexch	e-2	18	—	10	—	Outlet temperature = 90°C
T-101	Shortcut tower	t-1	10	—	19	11	Recovery of benzene in top product = 0.99 Recovery of toluene in top product = 0.01 R/R_{min} = 1.5 Column pressure drop = 0.3 bar
E-104	Shortcut tower	t-1	10	—	19	11	Included in tower simulation
E-106	Shortcut tower	t-1	10	—	19	11	Included in tower simulation
V-102	Shortcut tower	t-1	10	—	19	11	Not required in simulation
P-102	Shortcut tower	t-1	10	—	19	11	Not required in simulation
E-105	Hexch	e-3	95	—	15	—	Outlet temperature = 38°C
C-101	Compr	c-1	97	—	98	—	Outlet pressure = 25.5 bar
	Mixer	m-2	3	5	91	—	Pressure drop = 0 bar
	Mixer	m-3	2	91	92	—	Pressure drop = 0 bar
	Mixer	m-4	6	7	93	—	Pressure drop = 0 bar
	Mixer	m-5	17	96	99	—	Pressure drop = 0 bar
	Mixer	m-6	99	100	16	—	Pressure drop = 0 bar
	Splitter	s-1	8	—	97	96	Pressure drop = 0 bar
	Splitter	s-2	98	—	5	7	Pressure drop = 0 bar
	Splitter	s-3	19	—	100	95	Pressure drop = 0 bar

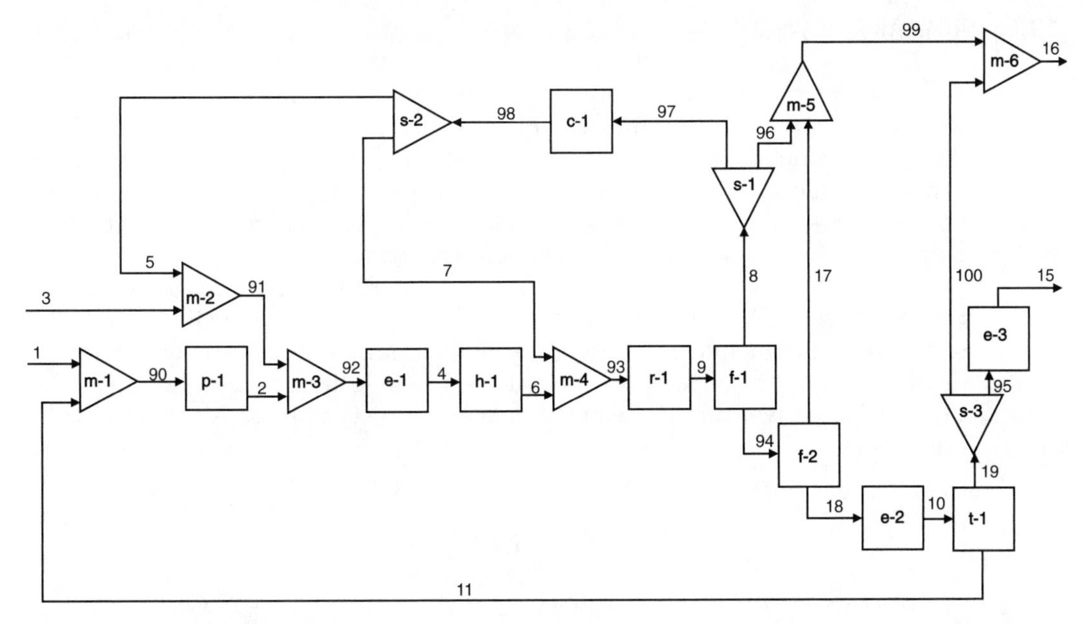

Figure 13.6 Flowsheet Structure Used in the Simulation of the Toluene Hydrodealkylation Process

recycle streams back into the problem one at a time. The thermodynamic models for this simulation should be chosen using the guidelines in Section 13.4 or using the expert system in the simulator that you use. The results given in Chapter 1 for this process were obtained using the SRK models for enthalpy and phase equilibria.

Table 13.2 Feed Stream Properties and Estimates of Recycle Streams

	Stream 1	Stream 3	Stream 11	Stream 5	Stream 7
Temperature (°C)	25.0	25.0	150.0	50.0	50.0
Pressure (bar)	1.9	25.5	2.8	25.5	25.5
Hydrogen (kmol/h)	—	286.0	—	200.0	20.0
Methane (kmol/h)	—	15.0	—	200.0	20.0
Benzene (kmol/h)	—	—	—	—	—
Toluene (kmol/h)	108.7	—	30.0	—	—

13.6 SUMMARY

In this chapter, the general components of a process simulator and the seven types of input required to simulate a process successfully were reviewed. Each of the seven required inputs was covered in detail: selection of chemical components, selection of thermodynamic models, selection of process topology, selection of feed stream properties, selection of equipment parameters, selection of output options, and selection of convergence criteria.

Special attention was paid to the role of recycle streams in obtaining converged solutions, and methods to help convergence were discussed. The selection of thermodynamic models and their importance were discussed in depth. Finally, a case study for the toluene hydrodealkylation process given in Chapter 1 was given and the required data to complete a process simulation were presented.

REFERENCES

1. Westerberg, A. W., H. P. Hutchinson, R. L. Motard, and P. Winter, *Process Flowsheeting* (Cambridge: Cambridge University Press, 1979), ch. 2.
2. Franks, R. G. E., *Modeling and Simulation in Chemical Engineering* (New York: Wiley, 1972), ch. 2.
3. Carnahan, B., H. A. Luther, and J. O. Wilkes, *Applied Numerical Methods* (New York: Wiley, 1969), ch. 5.
4. Elliott, J. R., and C. T. Lira, *Introductory Chemical Engineering Thermodynamics* (Englewood Cliffs, NJ: Prentice Hall, 1999).
5. Sandler, S. I., *Chemical, Biological, and Engineering Thermodynamics*, 4th ed. (New York: Wiley, 2006).
6. Smith, J. M., H. C. Van Ness, and M. M. Abbott, *Introduction to Chemical Engineering Thermodynamics*, 7th ed. (New York: McGraw-Hill, 2005).
7. Soave, G., "Equilibrium Constants from a Modified Redlich-Kwong Equation of State," *Chem. Eng. Sci.* 27 (1972): 1197.
8. Redlich, O., and J. N. S. Kwong, "On the Thermodynamics of Solutions, V: An Equation of State: Fugacities of Gaseous Solutions," *Chem. Rev.* 44 (1949): 233.
9. Horwitz, B. A., and A. J. Nocera, "Are You 'Scotamized' by Your Simulation Software?" *CEP* 92, no. 9 (1996): 68.
10. Schad, R. C., "Don't Let Recycle Streams Stymie Your Simulations," *CEP* 90, no. 12 (1994): 68.
11. Whiting, W. B., "Effects of Uncertainties in Thermodynamic Data and Models on Process Calculations," *J. Chem. Eng. Data* 41 (1996): 935.
12. Gmehling, J., and U. Onken, *Vapor-Liquid Equilibrium Data Collection* (Frankfurt am Main, Germany, DECHEMA, 1977).

13. Sorensen, J. M., and W. Arlt, *Liquid-Liquid Equilibrium Data Collection* (Frankfurt am Main, Germany, DECHEMA, 1979).

PROBLEMS

1. For the toluene HDA process, using the data given in Tables 13.1 and 13.2, simulate the process and compare the results with those given in Chapter 1, Table 1.5. Remember that the number of actual plates is given in Table 1.7, and an efficiency of 0.6 was assumed.
2. For the DME flowsheet given in Appendix B, Figure B.1.1, list the minimum input information required to obtain mass and energy balances for this process. Using the process simulator available to you, simulate the DME process and compare your results to those given in Table B.1.1.
3. For the isopropyl alcohol to acetone process flowsheet given in Appendix B, Figure B.10.1, list the minimum input information required to obtain mass and energy balances for this process. Using the process simulator available to you, simulate the isopropyl alcohol to acetone process, and compare your results to those given in Table B.10.1.
4. Using the results from Problem 1 above and Tables 1.5 and 1.7, compare the results for the simulation of the benzene recovery column, T-101, using a shortcut method and a rigorous method. One way to do this comparison is to use the number of theoretical plates from the shortcut method as an input to the rigorous method. The rigorous method is used to simulate the same separation as the shortcut method, that is, same overhead purity and recovery. The difference in the methods is then reflected by the difference between the reflux required for both methods. Comment on the difference for this nearly ideal system. Remember that there is no need to simulate the whole flowsheet for this problem; just use the input to the column from Table 1.5.
5. In Problem 1 above, you should have simulated the reactor as a stoichiometric reactor with 75% per pass conversion. In order to estimate the volume of the reactor, it is necessary to have kinetics expressions. For the catalytic hydrodealkylation of toluene, assume that the reaction is kinetically controlled with the following kinetics:

$$-r_{tol} = kc_{tol}c_{hyd}^{0.5} \quad \frac{\text{kmol}}{\text{m}^3_{\text{reactor}}\text{s}}$$

where

$$k = 2.833 \times 10^7 e^{-\frac{17814}{T(K)}} \quad \frac{\text{m}^{1.5}}{\text{kmol}^{0.5}\text{s}}$$

With these kinetics, simulate the reactor in Figure 1.5 as a two-stage packed-bed adiabatic reactor with a "cold shot" (Stream 7) injected at the inlet to the second bed. The maximum temperature in the reactor should not exceed 655°C, and this will occur at the exit of both beds; that is, design the system for this maximum outlet temperature for both packed beds. Compare your results with the total volume of the catalyst given in Table 1.7.

6. As noted in Section 13.5, the results provided for the toluene hydrodealkylation process are based on the SRK model for both enthalpy and phase equilibria. Determine the BIPs for this model used by the simulator available to you. If you have access to more than one simulator, compare the BIPs from each. Simulate the benzene column (T-101) using the shortcut simulation module and the specifications given in Table 13.1 and the conditions of feed stream (10) given in Example 13.2. Rerun the simulation with all the BIPs set to zero. Compare the results.
7. Determine what thermodynamic models were used for each of the processes in the appendix. Explain why each was chosen, and give at least one other thermodynamic model that is reasonable and should be tried for each process.
8. For the system DME/methanol/water, determine the BIPs used in the simulator available to you for each of these thermodynamic models: NRTL, Wilson, and UNIQUAC. Simulate T-202 using a shortcut module for each of these models, and compare the number of theoretical stages required for the specified recoveries and $R/R_{min} = 1.5$.
9. Find VLE data in the literature for the system methanol/water. Regress these data to determine the BIPs for the UNIQUAC model. Compare the results of these with the results obtained using the BIPs available in the simulator databank and with the results obtained using the UNIFAC model.
10. Using the Henry's Law model in the simulator available to you, determine the concentration of oxygen (ppm by mass) in water at 25°C and at 35°C if the water is in equilibrium with air. Compare the results obtained to those calculated using the PR model with the BIPs available in the simulator databank.
11. Using the help facility of the simulator available to you, determine how the simulator handles VLE calculations with supercritical components when an activity-coefficient model is specified. (Note that P_i^* in Equation 13.1 is undefined for these components.)
12. For the Ethylbenzene flowsheet given in Appendix B, Figure B.2.1, list the minimum input information required to obtain mass and energy balances for this process. Using the process simulator available to you, simulate the Ethylbenzene process and compare your results to those given in Table B.2.1.
13. For the Styrene flowsheet given in Appendix B, Figure B.3.1, list the minimum input information required to obtain mass and energy balances for this

process. Using the process simulator available to you, simulate the Styrene process and compare your results to those given in Table B.3.1.

14. For the Maleic Anhydride flowsheet given in Appendix B, Figure B.5.1, list the minimum input information required to obtain mass and energy balances for this process. Using the process simulator available to you, simulate the Maleic Anhydride process and compare your results to those given in Table B.5.1.
15. For the Ethylene Oxide flowsheet given in Appendix B, Figure B.6.1, list the minimum input information required to obtain mass and energy balances for this process. Using the process simulator available to you, simulate the Ethylene Oxide process and compare your results to those given in Table B.6.1.
16. For the Formalin flowsheet given in Appendix B, Figure B.7.1, list the minimum input information required to obtain mass and energy balances for this process. Using the process simulator available to you, simulate the Formalin process and compare your results to those given in Table B.7.1.
17. Investigate the batch aspects of the simulator available to you. Remember that these could include reactor, separation, and scheduling modules as well as others.
18. Determine which thermodynamic models were used for each of the processes in Appendix B. Explain why each was chosen, and give at least one other thermodynamic model that is reasonable and should be tried for each process.
19. Using the Henry's Law model in the simulator available to you, determine the concentration of oxygen (ppm by mass) in water at 10°C and 27°C if the water is in equilibrium with air. Compare the results obtained to those calculated using the SRK model with the BIPs available in the simulator databank.
20. As noted in Section 13.5, the results provided for the toluene hydrodealkylation process are based on the SRK model for both enthalpy and phase equilibria. Simulate the benzene column (T-101) with the PR model instead, using the shortcut simulation module and the specifications given in Table 13.1 and the conditions of feed stream (10) given in Example 13.2. Determine the BIPs for the PR model used by the simulator. Rerun the simulation with all the BIPs set to zero. Compare the results.

CHAPTER

14

Process Optimization

Optimization is the process of improving an existing situation, device, or system such as a chemical process. This chapter presents techniques and strategies to

- Set up an optimization problem
- Quantify the value of a potential improvement
- Identify quickly the potential for improvement
- Identify the constraints, barriers, and bottlenecks to improvement
- Choose an appropriate procedure to find the best change
- Evaluate the result of the optimization

This chapter will start with some basic definitions of terms and then investigate several techniques and strategies to perform the optimization of a process.

14.1 BACKGROUND INFORMATION ON OPTIMIZATION

In optimization, various terms are used to simplify discussions and explanations. These are defined below.

Decision variables are those independent variables over which the engineer has some control. These can be continuous variables such as temperature, or discrete (integer) variables such as number of stages in a column. Decision variables are also called **design variables**.

An **objective function** is a mathematical function that, for the best values of the decision variables, reaches a minimum (or a maximum). Thus, the objective function is the measure of value or goodness for the optimization problem. If it is a profit, one searches for its maximum. If it is a cost, one searches for its minimum. There may be more than one objective function for a given optimization problem.

Constraints are limitations on the values of decision variables. These may be linear or nonlinear, and they may involve more than one decision variable. When a constraint is written as an equality involving two or more decision variables, it is called an **equality constraint**. For example, a reaction may require a specific oxygen concentration in the combined feed to the reactor. The mole balance on the oxygen in the reactor feed is an equality constraint. When a constraint is written as an inequality involving one or more decision variables, it is called an **inequality constraint**. For example, the catalyst may operate effectively only below 400°C, or below 20 MPa. An equality constraint effectively reduces the **dimensionality** (the number of truly independent decision variables) of the optimization problem. Inequality constraints reduce (and often bound) the search space of the decision variables.

A **global optimum** is a point at which the objective function is the best for all allowable values of the decision variables. There is no better acceptable solution. A **local optimum** is a point from which no small, allowable change in decision variables in any direction will improve the objective function.

Certain classes of optimization problems are given names. If the objective function is linear in all decision variables and all constraints are linear, the optimization method is called **linear programming**. Linear programming problems are inherently easier than other problems and are generally solved with specialized algorithms. All other optimization problems are called **nonlinear programming**. If the objective function is second order in the decision variables and the constraints are linear, the nonlinear optimization method is called **quadratic programming**. For optimization problems involving both discrete and continuous decision variables, the adjective **mixed-integer** is used. Although these designations are used in the optimization literature, this chapter mainly deals with the general class of problems known as MINLP, **mixed-integer nonlinear programming**.

The use of linear and quadratic programming is limited to a relatively small class of problems. Some examples in which these methods are used include the optimal blending of gasoline and diesel products and the optimal use of manufacturing machinery. Unfortunately, many of the constraints that we run into in chemical processes are not linear, and the variables are often a mixture of continuous and integer. A simple example of a chemical engineering problem is the evaluation of the optimal heat exchanger to use in order to heat a stream from 30°C to 160°C. This simple problem includes continuous variables, such as the area of the heat exchanger and the temperature of the process stream, and integer variables, such as whether to use low-, medium-, or high-pressure steam as the heating medium. Moreover, there are constraints such as the materials of construction that depend, nonlinearly, on factors such as the pressure, temperature, and composition of the process and utility streams. Clearly, most chemical process problems are quite involved, and we must be careful to consider all the constraints when evaluating them.

14.1.1 Common Misconceptions

> **A common misconception is that optimization is a complex, esoteric, mathematical exercise. In fact, optimization is usually a dynamic, creative activity involving brainstorming, exploring alternatives, and asking, "What if . . . ?"**

Although some problems are simple enough to be posed in closed form and to be solved analytically, most real problems are more interesting.

For example, Figure 14.1(a) shows a classical curve of annualized pumping cost versus pipe diameter. The annualized pumping cost includes annualized equipment (pump plus pipe) cost and power (operating) cost. These two components of cost have been calculated based on smooth cost functions and are shown as separate curves on Figure 14.1(a). The combined curve represents the total

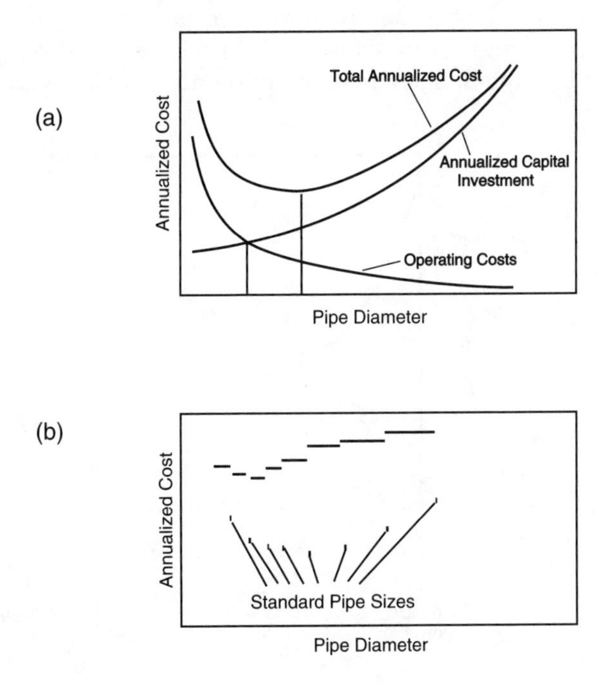

Figure 14.1 Optimization Using Continuous and Discrete Functions of Pipe Diameter

annualized cost. Using these results, one can analytically determine the pipe diameter at which the slope of the combined cost curve is zero and the second derivative is positive. It is important to note that, even with all the assumptions used in this analysis, the minimum annualized cost does not occur at the pipe diameter where the two component curves cross.

The point of intersection of the curves of annual operating cost and of annualized capital cost is not the optimum.

In reality, the cost function (if one were to calculate it) would look something like Figure 14.1(b), because only certain pipe diameters and pump sizes are standard equipment. Other sizes could be produced, but only at much higher costs. Thus, only a few cost evaluations are needed—those at the standard pipe sizes.

Another common misconception is that the optimum will usually be found at a point where the first derivatives of the cost function are zero. Even when the cost function is continuous and smooth, this is seldom the case and should never be assumed. Nearly all problems of any reasonable complexity have optima along at least one constraint. Figure 14.2 shows such a case. Again, both annualized capital costs and operating costs are included. Although there is a point of zero slope (point A), the best design (minimum annual cost) shown is at point B.

One must not assume that the best solution has been found when it merely has been bracketed. Such an assumption not only leads to a false conclusion but

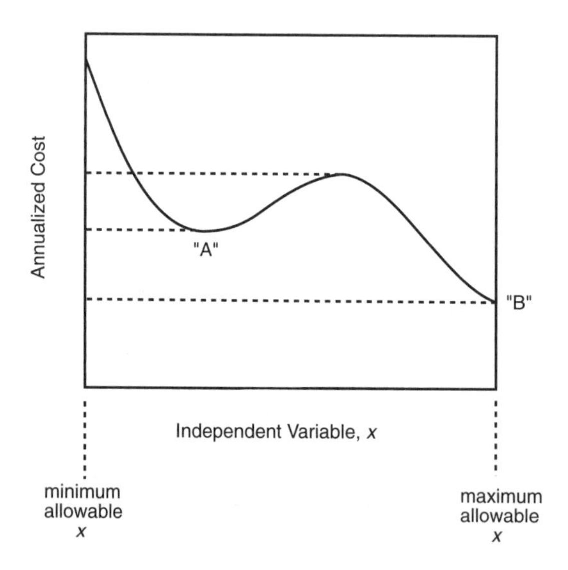

Figure 14.2 Location of the Optimum Value of a Variable

also ignores useful information for estimating the true optimum solution. For example, Figure 14.3(a) shows three points that have been calculated for the optimization of the temperature of a flash unit. If the objective function is continuous and smooth, and if there is indeed only one minimum, then the optimum lies between Points A and C. However, there is no reason to believe that Point B is the optimum, and there are insufficient data to determine whether the optimum lies to the left or to the right of Point B. However, by approximating the objective function simply by a quadratic, for the three points bracketing the optimum, the optimum can be estimated, as shown on the left-hand side of Figure 14.3(b). However, the true curve may look like the right-hand side of Figure 14.3(b). To be certain that the optimum has been located correctly, more points should be evaluated between A and C, using successive quadratic approximations based on the best point achieved plus the closest point on either side of it.

14.1.2 Estimating Problem Difficulty

A key step in any problem-solving strategy is to estimate the effort required to reach the solution. Not only can this estimate provide motivation, but it can also help to redirect resources if the effort (time, money, personnel) required for a

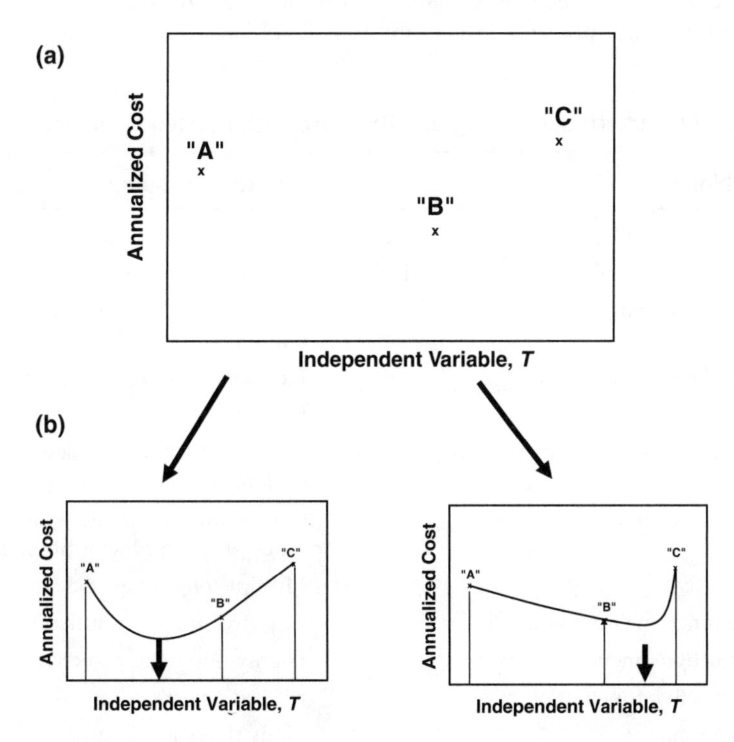

Figure 14.3 Different Locations of Optimum Using Identical Data

"complete" solution is more than the effort available. The first step is to decide whether the problem at hand is an "easy" problem or a "difficult" problem. Table 14.1 lists some characteristics of inherently easy and inherently difficult problems.

For a typical flowsheet, such as the DME (dimethyl ether) PFD in Figure B.1.1 (Appendix B), there are many decision variables. The temperature and pressure of each unit can be varied. The size of each piece of equipment involves decision variables (usually several per unit). The reflux in tower T-201 and the purity of the distillate from T-202 are decision variables. There are many more. Clearly, the simultaneous optimization of all of these decision variables is a difficult problem. However, some subproblems are relatively easy. If Stream 4 (the exit from the methanol preheater) must be at 154°C, for example, the choice of which heat source to use (lps, mps, or hps) is easy. There is only a single decision variable, there are only three discrete choices, and the choice has no direct impact on the rest of the process. The problem becomes more difficult if the temperature of Stream 4 is not constrained.

14.1.3 Top-Down and Bottom-Up Strategies

For any real process of any complexity, the true "global" optimum (at which any change in any decision variable would degrade the system) will not be found. If it could be found with an acceptable level of effort, one would concentrate on

Table 14.1 Characteristics of Easy and Difficult Optimization Problems

Easy Problems	Difficult Problems
Few decision variables	Many decision variables
Independent (uncorrelated) decision variables	Correlated decision variables
Discrete decision variables	Mixed discrete and continuous decision variables
Topological optimization first	Parametric optimization first
Single process units	Multiple, interrelated process units
Separate constraints for each decision variable	Constraints involving several decision variables
Constraints are obvious	Constraints are not obvious or become obvious only after the optimization has begun
Single objective	Multiple objectives
Objective function easy to quantify	Objective function difficult to quantify
Linear objective function	Highly nonlinear objective function
Smooth objective function	Kinked objective function
No local optima	Many deep, local optima

minute changes in decision variables. Rather than become bogged down in such details, experts at optimization tend to look alternately at the big picture and at the details. The overview encourages one to make bold changes in process configuration or variable values, but the closer study is needed to confirm whether the changes are true improvements.

The strategy of looking at the big picture first, followed by the detailed study, is called **top-down**. Looking at the DME flowsheet, such a strategy might lead to one questioning the need for E-201 (Methanol Preheater) and E-203 (DME Cooler). A detailed study of the flowsheet would confirm whether these two units should be eliminated in favor of a larger E-202 (Reactor Cooler). On the other hand, a detailed study of incremental changes in the heat duties of these two heat exchangers would have led to the same solution. Such a strategy is called **bottom-up**. Before investing too much time on detailed calculations, check the big picture. When the big picture is not clear or leads to conflicting alternatives, run the detailed calculations. The key is flexibility.

Details of the top-down and bottom-up approaches are covered in Sections 14.3 and 14.4 under topological and parametric optimization.

14.1.4 Communication of Optimization Results

As will be discussed in Chapter 27 (on the CD), the best strategy for obtaining results is not the best strategy for presenting those results to others. Although the goal of optimization is to find the best solution (and the sensitivity of that solution, as explained in Section 14.6) efficiently, the goal of the presentation is to convince the audience clearly that the solution is the best. Thus, one must explain the ranges of decision variables that were searched, show that the solution is (most likely) not merely a local minimum, and show the degradation in the objective function from moving away from the solution. For example, as discussed in Section 14.4, it is more efficient to change more than one decision variable at a time when searching for the optimum; however, it is better to communicate the validity of the optimum by using families of curves in which any single curve involves the variation of only one decision variable (see Figure 14.6).

14.2 STRATEGIES

In this section, the DME process from Appendix B is used as an example. Stream and unit numbers refer to those in Figure B.1.1.

14.2.1 Base Case

A **base case** is the starting point for optimization. It may be a very simple conceptual flowsheet, it may be a detailed design, or it may be an actual plant whose operation one wishes to improve.

Because the goal of optimization is to *improve* the process, it is essential that one start from a defined process, that is, a base case.

The choice of a base case is straightforward: Choose the best available case. For example, if one has already determined (through prior analysis) that heat integration greatly improves the process, the base case should include the heat integration. The level of detail of the base case is also a crucial decision. At a minimum, the analysis must be detailed enough to provide the calculation of the objective function. If the objective function includes both capital costs and operating costs, the base case analysis must include sizing and costing of equipment, as well as the material and energy balances and utility costs. The analysis must be detailed enough to show the effect of all important decision variables on the objective function. In our example, because the temperature of R-201 affects unit E-202, the base-case analysis must include this effect.

Although the ability to calculate the objective function and how it changes with variations in all decision variables is essential, the base case certainly need not be a completed design. For example, details of T-201 such as tray spacing, weir height, and so on are not important at this point in the problem, and a shortcut rather than a rigorous tower simulation will suffice.

However, it is essential to include enough detail in the base case. Capital costs for individual major pieces of equipment are needed, as are utility costs broken down by type and unit. Creating a calculational model that lumps all costs (even if correctly annualized by the methods shown in Chapter 10) is not very helpful, because specific modifications to the flowsheet are masked. For example, a lumped *EAOC* (equivalent annual operating cost) would not show whether the cost for the reactor, R-201, was relatively small (leading to a conclusion to increase its size) or whether the steam costs for E-201 were especially high (leading to a search for an alternative heat source). Table 14.2 provides a list of useful in-

Table 14.2 Data Required for Base Case (in Addition to PFD and Flow Tables)

Capital	Operating	Material
Installed cost, each equipment	Utility flowrates (each type)	Total cost for each raw material
Installed cost, each category	Utility targets (15.2)	Value of purged or wasted material
Estimated credit for equipment used elsewhere in plant (for existing processes)	Utility costs on $/GJ basis	Total product value
	Estimated uncertainties	Estimated uncertainties
Estimated uncertainties	Other operating costs	

formation that should be available for the base case. If the optimization problem is for an existing plant, the original capital cost is not relevant. However, methods for the estimation of capital costs for new equipment should be available, as should some estimate of the value that any removed equipment has when used in other parts of the total plant.

The scope of the base case (and therefore of the optimization) must be chosen. Usually, the battery limits for the analysis are defined by functional groupings, company organization, location of large surge or storage facilities, or convenience. However, if the scope is too small, important effects will be missed. For example, if the scope of optimization is R-201, the effect of reactor product composition on tower T-201 is missed. On the other hand, if the scope is too broad, one can become overwhelmed by the interactions. One could include, for example, the effect on the internal cost of medium-pressure steam of changes in steam usage. Such an expansion of scope clearly would not be warranted. Typically, column reflux ratio can be optimized separately, but reactor conditions (*T*, *P*, conversion), recycle rates, and separator recoveries should be optimized together.

14.2.2 Objective Functions

The optimization can begin only after the objective function is selected. It must be chosen so that the extreme maximum (or minimum) is the most desired condition. For example, minimizing the *EAOC* is typically considered (rather than the installed capital cost) or maximizing net profit (rather than gross sales). If the objective function is poorly chosen or imprecise, the solution will be worthless.

The most common objective functions have units of dollars. The recurring costs can be discounted to obtain a net present value (*NPV*), or the capital costs can be annualized to obtain an *EAOC*, as discussed in Chapter 10. Usually, this value is for the process unit. However, if a smaller scope is defined for part of the optimization (e.g., T-201 reflux), then the *EAOC* or *NPV* used should not include capital and recurring cost contributions that do not change significantly over the ranges of the varied decision variables. For example, optimization of the reflux for T-201 does not significantly affect the costs for the rest of the flowsheet (other than the ancillary units E-204, E-205, V-201, and P-202). Thus, the raw material costs, reactor costs, and other capital and utility costs should not be included. If they were, the total variation in the objective function might be in the fourth or fifth digit. The total *EAOC* or *NPV* is not accurate to this level, but the changes caused by the tower optimization are.

This focus on the changes to the costs (or savings) rather than on the total costs is called **incremental analysis**. Modification of the design or operation of the plant continues as long as the improvement gained (the **return**) for an incremental investment is greater than our **benchmark** (or hurdle) rate. In Chapter 10, the details of this form of investment analysis are covered. In optimization, the implementation of incremental analysis requires careful choice of the objective function, as described above.

Some objective functions are not directly based on economics. For example, one may be asked to maximize the production of DME from an existing plant or to minimize the contaminant concentration in the waste water stream. However, the objective function should be quantitative. Furthermore, a rational basis for any objective function (monetary or nonmonetary) should be developed. For example, if the goal is to maximize profit (rather than revenues), then maximizing the production of DME may not be desirable. Similarly, if the goal is to cause the least harm to the environment, then minimization of the contaminant concentration (rather than the total contaminant flowrate) may not be the best approach.

14.2.3 Analysis of the Base Costs

The first analysis of the base costs should be to determine targets of an idealized process. The value for the objective function should be determined based on the assumption of equilibrium conversion, no equipment or utility costs, and perfect separations. The profit for such a case is called the **gross profit margin**, or simply the **margin** (see Chapter 10.)

Next, a form of Pareto [1] analysis should be used. This analysis is based on the observation that, for most problems, a large fraction of the objective function (e.g., 80%) is due to a small fraction of the contributing factors (e.g., 20%). For example, Table 14.3 shows the contribution to the *EAOC* for the different types of equipment and for the operating costs and raw materials for the DME process given in Appendix B. Only a few of these categories account for the bulk of the total cost. For our case, as with many chemical processes, the raw material costs swamp all other costs and are treated separately. In this case, the overriding goal

Table 14.3 Ranking of Contributions to the *EAOC* for DME Process

Category	Contribution to *EAOC* ($/yr)
Raw materials	11,215,000
Raw material (at 100% conversion) = $ 11,185,000	
Target savings = $ 30,300	
Medium-pressure steam	695,000
Towers and vessels	210,000
Heat exchangers	170,000
Pumps (including electricity)	160,000
Reactor	70,000
Cooling water	31,000
Waste water treatment	1,000
Total	**12,552,000**

All figures are for 1996 prices.

is to convert as much of the raw materials into product as possible. From this standpoint, the waste of methanol is very low and the net savings that can be made from better use of the methanol amounts to only $30,300/yr. The calculations for this analysis are covered in the next section. However, assuming that methanol is already being purchased from the cheapest source, the focus should not be on reducing the cost of raw materials, because this will not have a significant effect on reducing the cost of producing DME. The single largest operating cost is the medium-pressure steam. A greater improvement in the objective function can be made by lowering steam costs than can be obtained by lowering any other cost category.

Basically, this first analysis of the base case provides a target for optimization and a road map to proceed to the solution. It should be noted that all cost figures for this chapter are based on 1996 prices from the first edition of this text. These numbers will change with time, but the basic approach to the products and main conclusions will remain the same.

14.2.4 Identifying and Prioritizing Key Decision Variables

Based on the Pareto ranking of effects on the objective function, the key decision variables are chosen. For the DME process, the methanol cost is high, but there are assumed to be no by-product reactions. The product DME is assumed to be at its purity specification. The only methanol loss is in the waste water; thus, the purity of the bottoms product of T-202 could be a decision variable. At present, approximately 0.27% of the methanol feed leaves in the waste water. Thus, the target savings attainable can be calculated as $30,300/yr.

Medium-pressure steam is used in three units: E-201, E-204, E-206. The first of these is used to vaporize the reactor feed at 154°C, which is lower than the temperature of low-pressure steam. Because low-pressure steam is less expensive than medium-pressure steam, changing this heat source should be considered. Similarly, the duty of E-204 (a tower reboiler) does not necessarily require medium-pressure steam.

The base case single-pass conversion is 80%. If this can be increased, the costs (capital plus operating) of all units in the recycle loop will be decreased. This includes all the equipment in the flowsheet except P-201 and E-208. The conversion is set by the reactor inlet temperature (T_6) and pressure (P_6) and the volume of the reactor.

Thus, the single-pass conversion has been identified as an important dependent decision variable. With the techniques presented in Section 14.4, the optimum can be determined. The conversion is a secondary (decision) variable. There are an infinite number of temperature, pressure, and reactor-volume combinations for a given conversion. Prioritizing these three primary decision variables requires knowledge of the sensitivity of the objective function to changes in these variables. Although there are elegant mathematical techniques for estimating these sensitivities, the most efficient technique is often to evaluate the objective function at the limits of the variables. This is a standard experimental design technique (two-level factorial design) that is of great help in choosing key decision variables. If the cost changes little when the reactor pressure changes from its upper limit to its lower limit, then another variable (such as temperature) should be chosen.

Another strategy to determine decision variables is to consider how the process is controlled. Any variable that must be controlled is a decision variable. There are alternative control strategies for equipment and for processes, but a well-designed control system reduces the degrees of freedom to zero without overconstraining the process. The controlled variable is a secondary decision variable; the variable manipulated by the final control element is the primary decision variable.

Considering the DME process in Figure B.1.1, the flowrate of methanol is a decision variable, as are the temperature of Stream 5 (controlled by steam pressure in E-201), the temperature of Stream 8 (controlled by the cooling water rate to E-203), the pressure of the reactor (controlled by the valve between Streams 8 and 9), and so on. The two columns have their own control systems, which show that the reflux to the column and the heat input to the reboiler are decision variables. Note that simple (two-product) distillation columns have two decision variables (in addition to the column pressure), but many different combinations are possible. Because the number of decision variables is equal to the number of controlled variables, which is equal to the number of operational specifications, one can look at any process simulator setup for a column and see the myriad choices.

The other type of decision variable is an equipment characteristic. The reactor volume and the number of stages are examples. Each specification that must be made before ordering a piece of equipment is a decision variable. Thus, the area of a heat exchanger and the aspect ratio of a packed-bed reactor are decision variables.

Once the decision variables have been identified and prioritized, the techniques of topological optimization (Section 14.3) and parametric optimization (Section 14.4) can be applied.

14.3 TOPOLOGICAL OPTIMIZATION

As previously discussed in Sections 14.1 and 14.2, there are essentially two types of optimization that a chemical engineer needs to consider. The first is termed topological optimization and deals with the topology or arrangement of process equipment. The second type is parametric optimization, and it is concerned with the operating variables, such as temperature, pressure, and concentration of streams, for a given piece of equipment or process. In this section, topological optimization is discussed. Parametric optimization is addressed in Section 14.4.

14.3.1 Introduction

During the design of a new process unit or the upgrading of an existing unit, topological optimization should, in general, be considered first. The reasons for this are twofold. First, topological changes usually have a large impact on the overall profitability of the plant. Second, parametric optimization is easiest to interpret when the topology of the flowsheet is fixed. It should be noted that combinations of both types of optimization strategies may have to be employed simultaneously, but the major topological changes are still best handled early on in the optimization process.

The questions that a process engineer needs to answer when considering the topology of a process include the following.

1. Can unwanted by-products be eliminated?
2. Can equipment be eliminated or rearranged?
3. Can alternative separation methods or reactor configurations be employed?
4. To what extent can heat integration be improved?

The ordering of these questions corresponds approximately to the order in which they should be addressed when considering a new process. In this section, these questions are answered and examples of processes are given in which such topological rearrangements may be beneficial.

14.3.2 Elimination of Unwanted Nonhazardous By-Products or Hazardous Waste Streams

This is clearly a very important issue and should be addressed early on in the design process. The benefit of obtaining 100% conversion of reactants with a 100% selectivity to the desired product should be clear. Although this goal is never reached in practice, it can be approached through suitable choices of reaction mechanisms, reactor operation, and catalyst. A chemical engineer may not be directly involved in the choice of reaction paths. However, one may be asked to evaluate and optimize designs for using alternative reactions in order to evaluate the optimum scheme.

In many cases, due to side reactions that are suppressed but not eliminated by suitable choice of catalyst and operating conditions, unwanted by-products or waste streams may be produced. The term *unwanted by-product* refers to a stream that cannot be sold for an overall profit. An example of such a by-product would be the production of a fuel stream. In this case, some partial economic credit is obtained from the by-product, but in virtually all cases, this represents an overall loss when compared with the price of the raw materials used to produce it. An example illustrating the cost of purifying a by-product for sale is given in Example 14.1.

Example 14.1

Consider the process given in Figure C.8 in Appendix C (on the CD), for the production of cumene. In this process, benzene and propylene react in a gas-phase catalytic reaction to form cumene, and the cumene reacts further to form p-diisopropyl benzene (DIPB), a by-product sold for its fuel value.

$$C_6H_6 + C_3H_6 \rightarrow \underset{cumene}{C_9H_{12}}$$

$$C_9H_{12} + C_3H_6 \rightarrow \underset{p\text{-diisopropyl benzene}}{C_{12}H_{18}}$$

From the information given in Table C.14, estimate the yearly cost of producing this waste, Stream 14. Assume that the costs of benzene, propylene, and fuel credit are \$0.27/kg, \$0.28/kg, and \$2.5/GJ, respectively.

From Table C.14, the composition of the DIPB waste stream is 2.76 kmol/h DIPB and 0.92 kmol/h cumene. The standard heats of combustion for DIPB and cumene are given below:

Cumene –5.00 GJ/kmol

DIPB –6.82 GJ/kmol

Revenue for sale as fuel = [(2.76) (6.82) + (0.92) (5.00)](8000)(2.5) = \$468,000/yr

Cost of equivalent raw materials:

$$\text{benzene} = (2.76 + 0.92)\,(78)\,(8000)\,(0.27) = \$620{,}000/\text{yr}$$

$$\text{propylene} = (2.76(2) + 0.92)\,(42)\,(8000)\,(0.28) = \$606{,}000/\text{yr}$$

$$\text{net cost of producing DIPB stream} = (620 + 606 - 468) \times 10^3 = \$758{,}000/\text{yr}$$

From Example 14.1, it is clear that the production of DIPB is very costly and impacts negatively the overall profitability of the process. Improved separation of cumene would improve the economics, but even if all the cumene could be recovered, the DIPB would still cost an estimated \$608,000/yr to produce. A better option would be to find ways of minimizing DIPB production. This is investigated further in Example 14.2.

Example 14.2

What process changes could be made to eliminate the production of the DIPB waste stream in Figure C.8 (on the CD)?

There are several possible solutions to eliminate the DIPB stream.

1. Reduce the per-pass conversion of propylene in the reactor. This has the effect of suppressing the DIPB reaction by reducing the cumene concentration in the reactor.
2. Increase the ratio of benzene to propylene in the feed to the reactor. This reduces the propylene concentration in the reactor and hence tends to suppress the DIPB reaction.
3. Obtain a new catalyst for which the DIPB reaction is not favored as much as with the current catalyst.

Each of these remedies has an effect on the process as a whole, and all these effects must be taken into account in order to evaluate correctly the best (most profitable) process. Some of these effects will be discussed in the following sections.

Alternatively, a hazardous waste stream may be produced. In such cases additional costly treatment steps (either on-site or off-site) would be required in order to render the material benign to the environment. The economic penalties for these treatment steps—incineration, neutralization, and so on—are often great and may severely impact the overall economic picture for the process. In addition to the economic penalties that the production of waste streams cause, there are political ramifications that may overshadow the economic considerations. For many companies, the production of hazardous wastes is no longer an acceptable process choice, and alternative reaction routes, which eliminate such waste streams, are aggressively pursued.

14.3.3 Elimination and Rearrangement of Equipment

Both the elimination and rearrangement of equipment can lead to significant improvements in the process economics. Algorithms for these topological changes are under development. However, the approach used here is one based on intuitive reasoning and illustrated by process examples. Both topics are considered separately below. The rearrangement of heat transfer equipment is considered as a separate topic in Chapter 15.

Elimination of Equipment. It is assumed that the starting point for this discussion is a PFD in which all process equipment serves a valid function: that is, the process does not contain any redundant equipment that can be eliminated immediately.

The elimination of a piece of equipment is often the result of a change in operating conditions and can thus be considered the end product of a series of parametric changes. In Example 14.3 the first alternative in Example 14.2 is considered.

Example 14.3

Evaluate the topological changes in the cumene PFD, Figure C.8, which can be made by reducing the per-pass conversion of propylene in the reactor.

As the single-pass conversion of propylene (the limiting reactant) in the reactor is reduced, the DIPB production decreases due to the lowering of the cumene concentration in the reactor. At some point, the second distillation column and the associated equipment can be removed, because all the DIPB produced can leave the process in the cumene product, Stream 13.

Although the second distillation column (T-802), associated equipment (E-805, E-806, V-804, and P-805 A/B), and utility costs may be eliminated, this may not be the most economical alternative. The increase in the size of other equipment in the front end of the process, due to an increase in the recycle flows and the accompanying increase in utility costs, may overwhelm the savings from the elimination of the second tower and associated equipment. However, one should

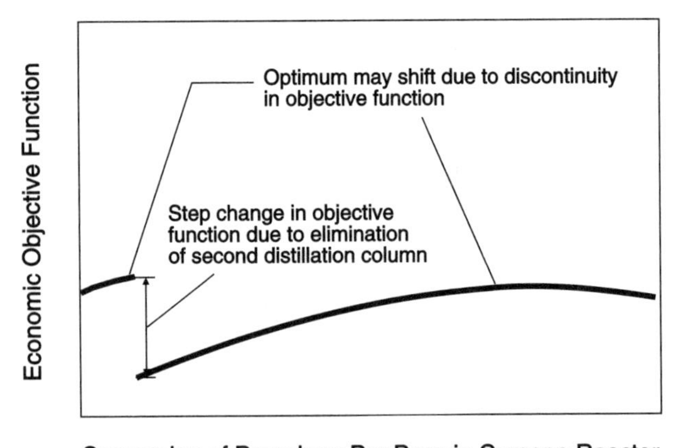

Figure 14.4 The Effect of a Topological Change on a Parametric Optimization

be aware of the consequences of reducing the DIPB production, because the elimination of the tower represents a step change in the profitability function that might otherwise be overlooked. This step change in the objective function is illustrated in Figure 14.4.

Rearrangement of Equipment. There are certain guidelines that should be followed when the sequence of equipment is considered. Some are obvious. For example, one should try to pump a liquid rather than compress a gas; thus, it will always be better to place a pump before a vaporizer rather than a compressor after it. However, other topological changes are somewhat more subtle. The most common examples of equipment rearrangement are associated with the separation section of a process and the integration of heat transfer equipment. In this section the sequencing of separation equipment is the focus, and heat integration is covered in Chapter 15. The separation sequence for the DME process is considered in Example 14.4.

Example 14.4

Consider the dimethyl ether (DME) process shown in Figure B.1.1 (Unit 200) and Tables B.1.1 and B.1.2. The process is quite straightforward and consists of a gas-phase catalytic reaction in which methanol is dehydrated to give DME with no appreciable side reactions.

$$2CH_3OH \rightarrow \underset{DME}{(CH_3)_2O} + H_2O$$

The reactor effluent stream is cooled and sent to two distillation columns. The first column separates DME product from the water and unreacted methanol. The second column separates the methanol, which is recycled, from the water, which is then sent to a waste water treatment facility to remove trace organic compounds.

Is there any economic advantage gained by changing the order of the distillation so that the water is removed first and the DME and methanol are separated in the second column?

There is no simple way to determine whether the separation sequence given in Example 14.4 should be changed. In order to evaluate which alternative is better, a rigorous parametric optimization for both topologies should be made and the configuration with the best economics should be chosen. Although it may not always be possible to determine which sequence is best by inspection only, there are some guidelines that may help determine which sequences are worthy of further consideration. The following guidelines are presented in Tables 11.13 and 12.2:

1. Perform the easiest separation first—that is, the one least demanding of trays and reflux—and leave the most difficult to the last.
2. When neither relative volatility nor feed composition varies widely, remove components one by one as overhead products.
3. When the adjacent ordered components in the feed vary widely in relative volatility, sequence the splits in order of decreasing volatility.
4. When the concentrations in the feed vary widely but the relative volatilities do not, remove the components in order of decreasing concentration.

Example 14.5 applies these guidelines to our DME problem.

Example 14.5

Apply the guidelines for column sequencing to the DME process using the information given in Table B.1.1 and Figure B.1.1.

From Table B.1.1, the composition of the stream leaving the reactor and entering the separation section is as follows.

	Flowrate (kmol/h)	Mole Fraction	Relative Volatility ($P = 10.4$ bar)
DME	130.5	0.398	49.4
MeOH	64.9	0.197	2.2
Water	132.9	0.405	1.0

The relative volatilities for the components are taken from a simulation of the process, using CHEMCAD, with a UNIQUAC/UNIFAC estimation for the vapor-liquid equilibrium (VLE). By applying the guidelines given above, it would appear that the easier separation is the removal of DME, and, according to Rule 1, this should be removed first, which is what is done currently. The other guidelines do not add any additional guidance, and it may be concluded that the current sequence is probably the best. However, no mention is given in the guidelines of special considerations for water. Because water has a very high latent heat of vaporization, the duties of the condensers and reboilers in the columns will be higher than for similar flows of organic materials. By removing the DME first, water must be reboiled in both columns. This suggests that there may be some advantage to removing the water first and then separating the DME and methanol. This case is considered in Problem 14.16 at the end of this chapter. Therefore, this example illustrates a new guideline that should be added to those in Chapters 11 and 12.

Special consideration is necessary when dealing with mixtures of polar compounds or other components that can form azeotropes or give rise to more than one liquid phase. As components are separated from a stream, the remaining mixture of components may fall into regions where two or more liquid phases are present or in which pairs of compounds form azeotropes. This can greatly increase the complexity of a separation and may strongly influence the sequence in which components or products are separated. This topic is considered in Chapter 12, and the reader is referred to this material for a more complete explanation and methods for evaluating azeotropic systems.

14.3.4 Alternative Separation Schemes and Reactor Configurations

Early in the design process, it is important to consider the use of alternative technologies to separate products from unused reactants and waste streams and to evaluate alternative reactor configurations. The topic of alternative reactor configurations is considered in Chapter 20, and only alternative separation technologies are considered here. If one picks up any textbook on mass transfer or unit operations, it is immediately apparent that there exist a myriad of different technologies to separate chemical components. Despite the wide range of separation technologies available to the process engineer, when considering liquid-gas processes, the vast majority of separations are composed of distillation, gas absorption and liquid stripping, and liquid-liquid extraction. According to Humphrey and Keller [2], 90% to 95% of all separations, product recoveries, and purifications in the chemical process industry consist of some form of distillation (including extractive and azeotropic distillation). The relative maturity of distillation technology coupled with its relatively inexpensive energy requirements make distillation the default option for process separations involving liquids and vapors. Many notable exceptions exist, and alternatives to distillation technology may have to be used. Two examples occur if the relative volatilities of two components are close to 1 (for example, less than 1.05) or if excessively high pressures or low temperatures are required to obtain a liquid-vapor mixture. A comprehensive list of alternative separation techniques is not offered in this text. The approach taken here is to emphasize the importance of recognizing when a nontraditional technology might be employed and how it may benefit the process.

Further details of the choice of alternative separation techniques are given in Table 12.1. An example of using alternative separation schemes is given in Example 14.6.

Example 14.6

In the toluene hydrodealkylation process shown in Figure E14.6 (taken from Figure 1.5), the fuel gas leaving the unit, Stream 16, contains a significant amount of hydrogen, a raw material for the process. Currently, there is no separation of Stream 8. The mixture of methane and hydrogen leaving V-102 is split in two, with one portion recycled and the other portion purged as fuel gas. What benefit would there be in separating the hydrogen from the methane in Stream 8 and recycling a hydrogen-rich stream to the reactor? What technology could be used to achieve this separation?

Significant benefits may be derived by sending a hydrogen-rich recycle stream back to the front end of the process. The methane that is currently recycled acts as a diluent in the reactor. By purifying the recycle stream all the equipment in the reactor loop, E-101, H-101, R-101, E-102, and C-101 could be made smaller (because the amount of methane in the feed to each equipment would be reduced), and the utility consumption for the heat-exchange equipment would also be reduced. In addition, the amount of hydrogen feed, Stream 3, would be reduced because far less hydrogen would leave in the fuel gas, Stream 16.

Several technologies exist to purify hydrogen from a stream of light hydrocarbons. It should be noted that distillation is not a viable option for this separation because the temperature at which methane begins to condense from Stream 8 is less than –130°C. The two most likely candidates for this separation are membrane separation, in which hydrogen would preferentially diffuse through a polymer membrane, and pressure swing adsorption, where methane would preferentially adsorb onto a bed of molecular sieve particles. In both of these cases, the potential gains outlined above are offset by the capital cost of the

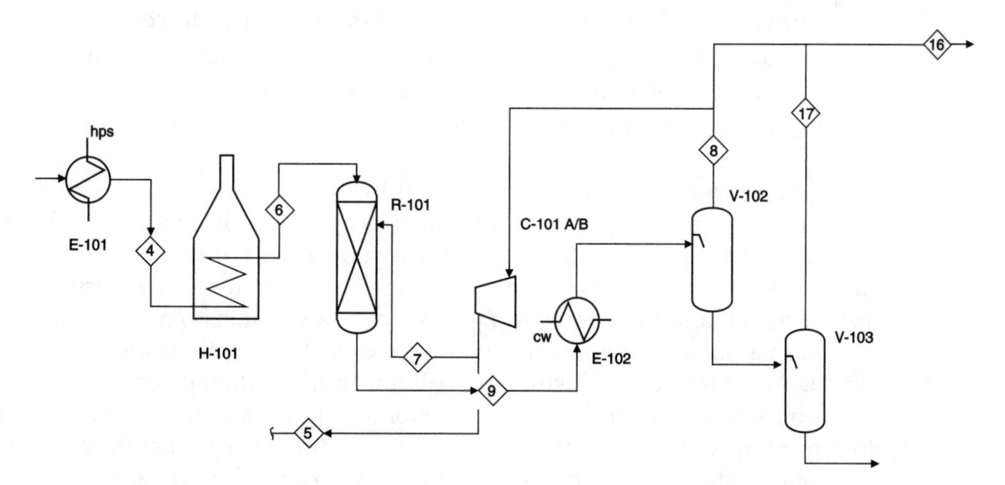

Figure E14.6 Toluene Hydrodealkylation Reactor and Hydrogen Recycle Loop

additional separation equipment and increased compression costs for the membrane case, where the permeate, high in hydrogen, would be obtained at a much lower pressure than the feed gas. In order to evaluate such process alternatives correctly, the economics associated with the new topology would have to be assessed, taking into account all the savings and costs outlined above. The case of implementing a membrane separator is considered in the problem report in Chapter 28 (on the CD).

14.4 PARAMETRIC OPTIMIZATION

In optimizing a chemical process, the key decision variables must be identified early in the optimization procedure. This is necessary in order to reduce the computational effort and time and make the problem tractable. The choice of key decision variables is crucial to the efficiency of the optimization process. An exhaustive list of potential decision variables is not presented here. However, some important variables that should be considered for most processes are listed below.

1. Operating conditions for the reactor—for example, temperature, pressure, concentration of reactants. The temperature range may be restricted by catalyst properties; that is, catalyst may sinter at high temperatures or be inactive at low temperatures.
2. Single-pass conversion in the reactor. The selectivity will be determined by the conditions mentioned in (1) and the single-pass conversion.
3. Recovery of unused reactants.
4. Purge ratios for recycle streams containing inerts.
5. Purity of products (this is often set by external market forces).
6. Reflux ratio and component recovery in columns, and flow of mass separating agents to absorbers, strippers, extractors, and so on.
7. Operating pressure of separators.

Because most chemical processes utilize recycles to recover unused reactants, any changes in operating conditions that occur within a recycle loop will impact all the equipment in the loop. Consequently, the whole flowsheet may have to be resimulated and the economics reworked (capital investments and costs of manufacture) every time a new value of a variable is considered.

For variables that do not lie within a recycle loop, optimization may be simplified. An example is a distillation column in a separation sequence in which two products are purified and sent to storage. The operation of such a column does not impact any part of the process upstream and can therefore be considered independently after the upstream process has been optimized. First, single- and two-variable optimizations for single pieces or small groups of equipment are considered. Then overall process optimization strategies are studied.

14.4.1 Single-Variable Optimization: A Case Study on T-201, the DME Separation Column

When considering the optimization of a distillation column, the variables to be considered are reflux ratio, operating pressure, percent recovery of key components, and purity of the products. For our initial case study, consider that the column pressure and the feed to the column are fixed. Figure B.1.1 and Table B.1.1 present the following information:

Feed, Stream 9	
Temperature	89°C
Pressure	10.4 bar
Vapor fraction	0.148
Molar flows (kmol/h)	
Dimethyl ether	130.5 kmol/h
Methanol	64.9 kmol/h
Water	132.9 kmol/h
Total Flow	**328.3 kmol/h**

In addition, the product specification for the DME is that it be 99.5 wt% pure. Assume that 98.9% of the DME in the feed must be recovered in the final product. Note that for this column, the pressure, reflux ratio, and % recovery of DME are all process variables that can be optimized. Focus first on the reflux ratio; later, relax the constraint on the operating pressure and carry out a two-variable optimization.

In order to proceed with the optimization, choose an objective function. For this example, considering the reflux ratio as the only decision or design variable, the only costs that will be affected by changes in reflux are the capital and operating costs associated with the column. In order to account correctly for both one-time costs and operating costs, an objective function that takes into account the time value of money (see Chapter 10 for the different criteria used for assessing profitability) should be used. Use the before-tax *NPV* as the objective function:

$$\begin{aligned} OBJ &= FCI_{TM} + COM_d(P/A, i, n)(P/F, i, n_{startup}) \\ &= FCI_{TM} + (0.18FCI_{TM} + 1.23C_{UT})(P/A, i, n)(P/F, i, n_{startup}) \end{aligned} \tag{14.1}$$

It should be noted that the value of the cost of manufacturing term without depreciation, COM_d, in Equation (8.2) includes a term with the fixed capital investment based on the total module cost (FCI_{TM}) and the cost of utilities. The other terms are not relevant to the optimization because they do not change with reflux ratio. Using a plant life of ten years after start-up and a 10% internal rate of return and assuming a construction period of one year, Equation (14.1) reduces to

$$OBJ = FCI_{TM} + (0.18FCI_{TM} + 1.23C_{UT})(6.145)(0.909) \\ = 2.005FCI_{TM} + 6.871C_{UT} \tag{14.2}$$

The fixed capital investment term includes the total module costs for T-201, E-204, E-205, V-201, and P-202 A/B. The utility costs include the electricity for P-202 and the heating and cooling utilities for E-204 and E-205, respectively. A series of case studies were run (using the CHEMCAD process simulator) for different reflux ratios for this column, and the equipment costs (evaluated from CAPCOST using mid-1996 prices) and utility costs (also from 1996) are presented in Table 14.4 along with the objective function from Equation (14.2). A plot of the (R/R_{min}) versus the objective function is shown in Figure 14.5. The optimum value of R/R_{min} is seen to be close to 1.12. Moreover, for values greater than 1.1, the objective function changes slowly with R/R_{min}. It should be noted that the results of this univariate search technique are presented as a continuous function of R/R_{min}. However, in reality, the objective function exists only at a set of points on Figure 14.13 (shown by the data symbols on the dotted curve), each point representing a column with an integer number of plates.

14.4.2 Two-Variable Optimization: The Effect of Pressure and Reflux Ratio on T-201, the DME Separation Column

The effect of pressure on the operation of T-201 is considered next. It may be tempting to take the result from 14.4.1, and by holding the value of R/R_{min} constant at 1.12, carry out a univariate search on the pressure. However, the results from such a technique will not yield the optimum pressure or R/R_{min} values. This is because as pressure changes the optimum reflux ratio also changes. In order to optimize correctly this situation, both the pressure and R/R_{min} must be varied and the best combination should be determined. This problem can be approached in many ways. In the following example, pick different pressures and repeat the

Table 14.4 Data for DME Column Optimization, R/R_{min}, versus *OBJ*

R/R_{min}	*FCI* ($\$10^3$)	Steam Cost ($\$10^3$/yr)	Cooling Water Cost ($\$10^3$/yr)	Electrical Cost ($\$10^3$/yr)	Total Utility Cost ($\$10^3$/yr)	*OBJ* from Eq. (19.2) ($103)
1.01	684	72.50	4.04	0.48	77.02	−1911
1.02	509	72.72	4.05	0.48	77.24	−1551
1.03	441	72.96	4.06	0.48	77.50	−1417
1.04	411	73.15	4.07	0.48	77.70	−1358
1.11	354	74.68	4.13	0.49	79.31	−1255
1.27	342	78.16	4.28	0.51	82.95	−1256
1.60	322	85.04	4.54	0.55	90.17	−1265

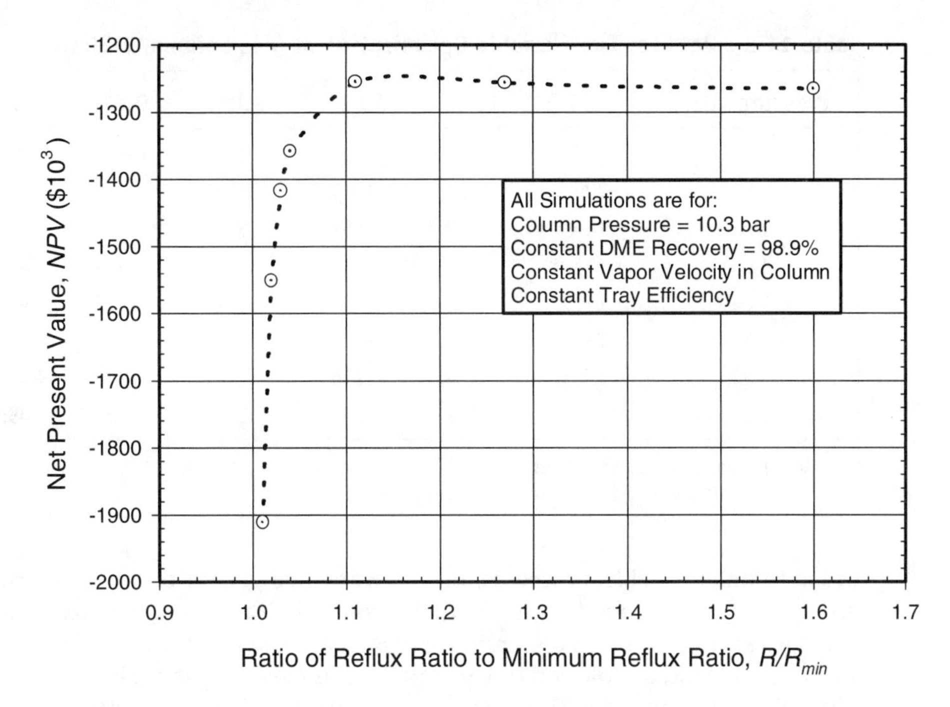

Figure 14.5 Single-Variable Optimization for DME Column, T-201. All figures based on 1996 prices/costs.

procedure used in 14.4.1 for each pressure, and then plot the results. This is not a particularly efficient procedure, but it yields plots that are easy to interpret. This is an example of a bivariate search technique. The data for this problem are presented in Table 14.5 and plotted in Figure 14.6. From the results shown in Figure 14.6, it is clear that the relationship between pressure and *NPV* at the optimum R/R_{min} is highly nonlinear. In addition, the optimum value of R/R_{min} does not remain constant with pressure, as shown by the dotted line in Figure 14.6, although, for this example, $(R/R_{min})_{opt}$ does not change very much with pressure over the range considered here.

It was stated at the beginning of this section that the topology of the distillation column and associated equipment should remain fixed while carrying out the parametric optimization. This may unduly constrain the overall optimization of the process and, for our case, limits the range of pressures over which the optimization may be considered. Moreover, when carrying out this optimization, it was necessary to change some of the utilities used in the process and this significantly impacted the results. Consider the information presented in Table 14.6, in which the utility and fixed capital investment (*FCI*) breakdown is given for the case of $R/R_{min} = 1.27$ (which is close to the optimum value for all the cases). It is clear that several changes in utilities were implemented. Some were required in

Table 14.5 Data for Two-Variable Optimization of DME Column, T-201

Pressure (Bar)	13.5	11.5	10.3	9.0[a]	9.0[b]	7.5
R/R_{min}	*OBJ*	*OBJ*	*OBJ*	*OBJ*	*OBJ*	*OBJ*
1.01	−2052	−1975	−1911	−1890	−1926	−5203
1.02	−1699	−1613	−1551	−1511	−1547	−4847
1.03	−1560	−1474	−1417	−1373	−1409	−4719
1.04	−1499	−1411	−1358	−1310	−1347	−4665
1.11	−1394	−1312	−1255	−1213	−1250	−4616
1.27	−1365	−1288	−1256	−1204	−1243	−4715
1.60	−1385	−1309	−1265	−1216	−1259	−4955

[a]Using lp steam. [b]Using mp steam.
All costs in $1000 using data from 1996, first edition of text.

order to obtain viable processes; others were changed in order to improve the economics. First, consider the switch from medium-pressure steam to low-pressure steam for a process pressure of 9.0 bar. The results are shown in Tables 14.5 and 14.6 and are plotted in Figure 14.7. It is clear that there is a small but significant increase in the *NPV* when low-pressure steam is substituted for medium-pressure steam. The reason for this increase is that the reduction in utility costs (due to less-expensive low-pressure steam) outweighs the increase in equipment

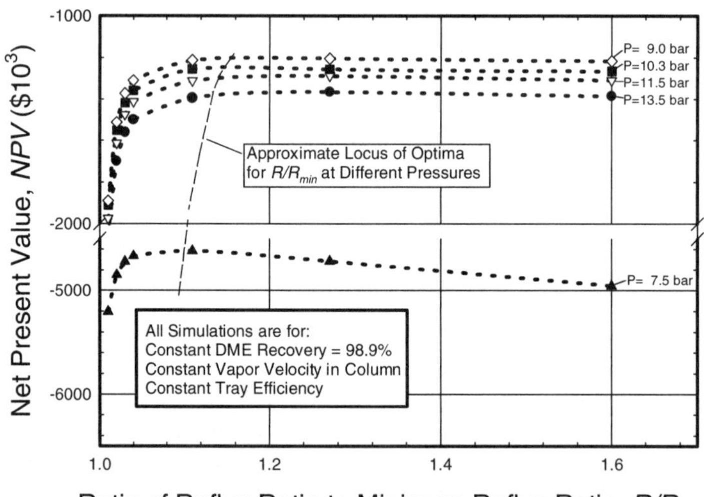

Figure 14.6 Optimization of DME Column, T-201, for Two Variables (Pressure and R/R_{min})

Table 14.6 Breakdown of Costs and Process Information for R/R_{min} = 1.27

Pressure (Bar)	13.5	11.5	10.3	9.0	9.0	7.5
mp steam	89.79	82.44	78.16	—	72.26	—
lp steam	—	—	—	62.80	—	56.67
Cooling water	4.23[a]	4.26[a]	4.28[a]	8.61[b]	8.61[b]	—
Refrig. water	—	—	—	—	—	539.35[c]
Electricity	0.53	0.52	0.51	0.50	0.50	0.49
FCI	357	344	342	354	340	307
NPV	−1339	−1263	−1230	−1178	−1217	−4692
R_{min}	0.3528	0.3214	0.3024	0.2810	0.2810	0.2562
$T_{condenser}$ (°C)	57	50	46	40	40	34
$T_{reboiler}$ (°C)	164	157	153	147	147	140

[a]ΔT = 10°C. [b]ΔT = 5°C. [c]ΔT = 10°C.
All costs in $1,000 or $1,000/yr based on 1996 data.

costs (due to a smaller temperature driving force and higher operating pressure in E-204, the column reboiler). A switch to low-pressure steam is actually possible for all of the operating pressures shown, with the exception of 13.5 bar, because the reboiler temperature for these cases is less than 160°C, the temperature at which low-pressure steam condenses.

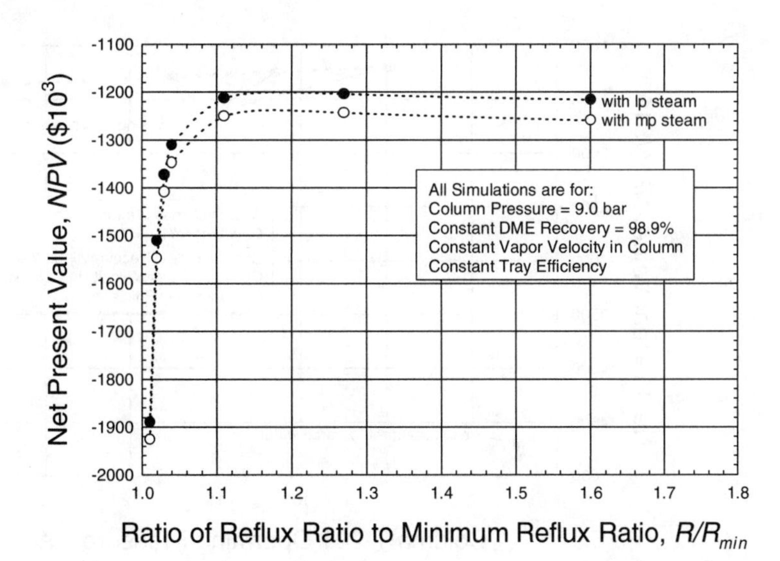

Figure 14.7 The Effect of Changing Utilities in the Reboiler for T-201 at a Column Pressure of 9.0 Bar

Next, consider a change in the cooling water utility. As the operating pressure of the column decreases, the bottom and top temperatures also decrease. One consequence is that low-pressure steam may be used in the reboiler, as considered above. In addition, as the top temperature in the column decreases, a point is reached where a temperature increase of 10°C for the cooling water can no longer be used. This occurs at an operating pressure of 9.0 bar. Cooling water can be used to condense the overhead vapor from the column, but the temperature increase of the utility must be reduced. The results shown are for a change in cooling water temperature of 30°C (in) to 35°C (out). As the operating pressure is reduced still further to 7.5 bar, the column top temperature decreases to 34°C. At this point, the use of cooling water becomes almost impossible; that is, the flow of cooling water and the area of the exchanger would become extremely large, and any slight changes in cooling water temperature would have a large impact on the process. For this reason, refrigerated water is substituted as the cooling utility for this case. The result is a very large increase in the cost of the utilities and a resulting decrease in the objective function (*NPV*) as shown in Figure 14.8.

The example in this section clearly illustrates the need to consider changes in utilities when process conditions change.

During parametric optimization, changes in operating conditions may require corresponding changes in utilities.

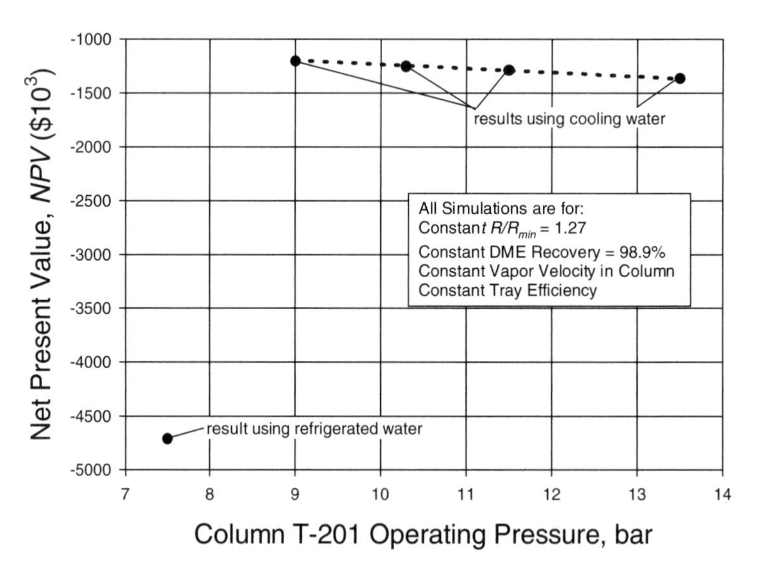

Figure 14.8 The Effect of the Operating Pressure of T-201 on the *NPV* (All Data for $R/R_{min} = 1.27$)

14.4.3 Flowsheet Optimization Using Key Decision Variables

As stated previously, when the decision variables for an optimization lie within a recycle loop, then evaluation of the objective function becomes more complicated. As an example, consider the DME flowsheet given in Appendix B, Figure B.1.1. The number of possible decision variables, for even such a simple flowsheet as this, are numerous. In this example, consider three variables: single-pass conversion in the reactor (x), fractional recovery of the DME product in tower, T-201 (f), and reactor pressure (p). Even with only three variables, it is not clear how many runs should be performed, nor is it clear over what range of the decision variables runs should be performed.

The first step, therefore, in doing such optimizations, should be to choose the range over which the decision variables should be varied. The ranges chosen for this example are

$$0.7 < x < 0.9, 0.983 < f < 0.995, 12.70 \text{ bar} < p < 16.70 \text{ bar}$$

The choice of these ranges is somewhat arbitrary, although it is useful to include maximum and/or minimum constraints as end points for these ranges. For example, it has been found from research in the laboratory that, for the current catalyst, a maximum single-pass conversion of 90% and a maximum operating pressure of 16.70 bar should be used because of equilibrium constraints and to avoid undesirable side reactions. The upper limits for x and p represent upper constraints for these variables, which should not be exceeded. On the other hand, the lower limits for p and x are arbitrary and could be changed if the optimization results warrant looking at lower values. The choice of the range of f values is again arbitrary, although experience tells us that the fractional recovery of product (f) should be close to 1.

The next question to be answered is how many points should be chosen for each variable. If three values for each variable are used, then there would be a total of (3)(3)(3) = 27 simulations to run. A conservative estimate of the length of time to run a simulation, collect the results, evaluate the equipment parameters, price the equipment, estimate the utility costs, and finally evaluate the objective function is, for example, 2 hours per run. This would yield a total time investment of (2)(27) = 54 hours of work! This is a large time investment and would be much larger if, for example, five variables were considered, so it behooves us to minimize the number of simulations carried out. With this in mind, only the end points of the range for each variable are used, namely (2)(2)(2) = 8 simulations will be run. This still represents a significant investment of time but is more reasonable than the previous estimate. Essentially what is proposed is to carry out a 2^k factorial experiment (Neter and Wasserman [3]), where k is the number of decision variables (three for this case). In essence, each evaluation of the objective function (dependent variable) can be considered the result of an experimental run in which the independent variables ($x, f,$ and p) are varied. The test matrix used is given in Table 14.7. The low range for each variable is designated 0, and the high

Table 14.7 Test Matrix and Results for Three Variable Optimizations of DME Process

Run Number	Variable 1 = x 0 = 0.7 1 = 0.9	Variable 2 = f 0 = 0.983 1 = 0.995	Variable 3 = p 0 = 12.7 bar 1 = 16.7 bar	*FCI* ($1000)	Utilities ($1000/yr)	*OBJ* (*NPV*) ($1000)
Base Case	0.8	0.989	14.7	1297	732	−7630
DME000	0.7	0.983	12.7	1378	782	−8136
DME001	0.7	0.983	16.7	1381	872	−8760
DME010	0.7	0.995	12.7	1393	780	−8152
DME011	0.7	0.995	16.7	1440	868	−8851
DME100	0.9	0.983	12.7	1210	539	−6130
DME101	0.9	0.983	16.7	1261	634	−6885
DME110	0.9	0.995	12.7	1232	537	−6160
DME111	0.9	0.995	16.7	1299	634	−6961

All figures based on 1996 data.

range is designated 1. Thus, the run named DME011 represents the process simulation when the conversion is 0.7 (x is at its low value), fractional recovery is 0.995 (f is at its high value), and the pressure is at 16.7 bar (p is at its high value). For each of the eight cases shown in Table 14.7, a simulation was carried out and all equipment and operating costs determined. The results for these cases and the base case are also presented in Table 14.7.

Once the results have been obtained, interpretation and determination of the optimum operating conditions must be carried out. The results in Table 14.7 are shown graphically in Figure 14.9. In this figure, the region over which the optimization has been considered is represented by a cube or box with each corner representing one of the test runs. Thus, the bottom left-hand front corner represents DME000, the top right-hand rear corner represents DME111, and so on. The values of the objective function are shown at the appropriate corners of the box. Also shown is the base-case simulation (for the conditions in Table B.11.1), which lies at the middle of the box (this is due to choosing the ranges for each variable symmetrically about the base-case value). First, analyze these results in terms of general trends using a method known as the analysis of means. Then fit a model for the objective function in terms of the decision variables, and use this to estimate the optimum conditions or, more correctly, to give direction as to where the next test run should be performed. This approach is termed a response surface analysis and is a powerful method to interpret and correlate simulation results.

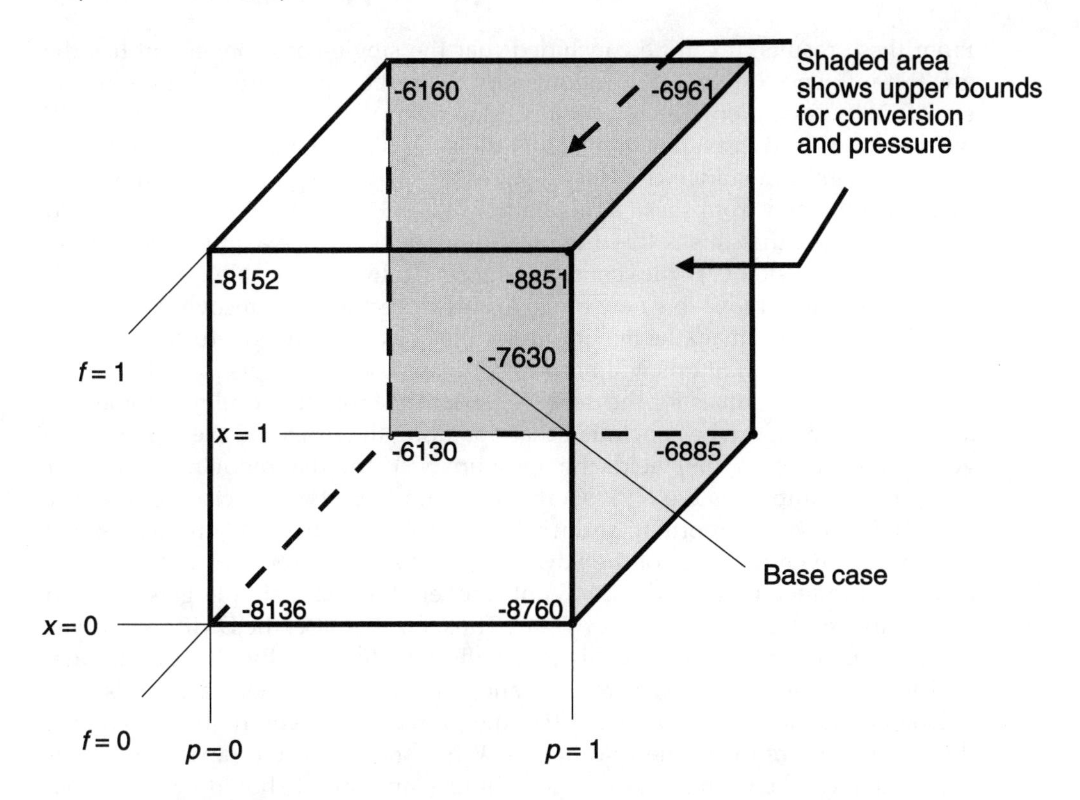

Figure 14.9 Results for Three-Variable Optimization on DME Process; Objective Function Values Are Shown at the Corners of the Test Cube

Sensitivity of Objective Function to Changes in Decision Variables: Analysis of Means. Using the data from Table 14.7 and Figure 14.9, estimate how a change in each variable affects the *NPV*. This is simply done by averaging all the results at a given value of one variable and comparing them with the average value at the other level of that variable. This yields the following results:

$$\text{conversion, } x = 0.7,\ NPV_{avg,x=0} = (-8136 - 8760 - 8152 - 8851)/4 = -8475$$
$$\text{conversion, } x = 0.9,\ NPV_{avg,x=1} = (-6130 - 6885 - 6160 - 6961)/4 = -6534$$

$$\text{recovery, } f = 0.983,\ NPV_{avg,f=0} = (-8136 - 8760 - 6130 - 6885)/4 = -7478$$
$$\text{recovery, } f = 0.995,\ NPV_{avg,f=1} = (-8152 - 8851 - 6160 - 6961)/4 = -7531$$

$$\text{pressure, } p = 12.7 \text{ bar},\ NPV_{avg,p=0} = (-8136 - 8152 - 6130 - 6160)/4 = -7145$$
$$\text{pressure, } p = 16.7 \text{ bar},\ NPV_{avg,p=1} = (-8760 - 8851 - 6885 - 6961)/4 = -7864$$

From these results, it can be concluded that the single-pass conversion has the greatest influence on the *NPV*, followed by the reactor pressure, with the recovery of DME having only a slight effect. Moreover, the results suggest that *NPV* will be maximized (least negative value) by using high conversion, low pressure, and low (within the range chosen) DME recovery. Although the maximum *NPV* that was obtained from these simulations was –$6,130,000 (DME100), it should not be assumed that this is the true maximum, or that the maximum lies within the range of decision variables considered thus far. In fact, the above analysis tells us that we should move to a new range for pressure and DME recovery.

In order to estimate the maximum for the objective function, further simulations at different conditions will need to be done. Use the results from Table 14.7 to choose the next values for the decision variables. From the results in Table 14.7 and Figure 14.9, it can be concluded that operating the reactor at the highest conversion allowable ($x = 0.9$) yields the optimum *NPV* and that reducing the reactor pressure also improves the *NPV*. Compared with these two effects, the recovery of DME is not very important, and it will be ignored for now. It should be noted that for the given topology of the flowsheet, all the DME produced in the reactor leaves as product, because the DME not recovered in T-201 simply gets recycled back to the reactor from the top of T-202. Thus, for this case, the DME recovery is not very important. However, this is usually not the case. By eliminating one-decision variable from our search (f) and noting that another variable is constrained at its maximum value ($x = 0.9$) the problem has been reduced to a one-dimensional search (p). The question is, What should be the next value of p? Although p can be decreased by some arbitrary amount, it should be noted that there may be another constraint as p is lowered. In fact, in Section 14.4.2, a detrimental change in the *NPV* was identified when the pressure in T-201 became so low that refrigerated water was required as the utility for the overhead condenser. Noting the pressure drops through the system and control valves, the lowest pressure at which the reactor can be run and still operate T-201 at 9.0 bar, which is close to the lower limit for using cooling water, is $p = 10.7$ bar. Thus, the next simulation should be carried out at ($x = 0.9, f = 0.983, p = 10.7$).

Modeling the Objective Function in Terms of the Decision Variables. It is useful to be able to estimate *NPV* values for new conditions before actually running the simulations. This can be done by using the results from Table 14.7 to model the *NPV* as a function of x, f, and p. An infinite number of functional forms can be chosen for this model. However, the following form is chosen:

$$NPV = a_0 + a_1x + a_2f + a_3p + a_4xf + a_5xp + a_6fp + a_7xfp \quad (14.3)$$

where $a_0, a_1, \ldots a_7$ are constants that are fit using the data from Table 14.7. This model uses eight arbitrary constants, which will allow the function to predict the *NPV* exactly at each corner point of our experimental design. In addition, this form of model is a simple multivariable polynomial, and the regression techniques to find $a_0, a_1, \ldots a_7$ are well established [3]. For our data, the following form for our objective function is obtained:

$$NPV = -63254 + 55556x + 49448f + 3573p - 44198xf - 3133xp - 3677fp + 3021xfp \quad (14.4)$$

The accuracy of the model is tested by comparing the predictions of Equation (14.4) with all the data in Table 14.7. The results are shown in Table 14.8.

From the results in Table 14.8, it can be seen that the model predicts the eight test runs exactly (differences in the last significant figure are due to round off errors in coefficients in Equation [14.4]). The prediction for the base case is also good, although the model tends to underpredict the *NPV* a little. The expected value of *NPV* can also be predicted for the next simulation ($x = 0.9, f = 0.983, p = 10.7$), and from Equation (14.4), we get $NPV = -5750$. The actual value for the *NPV* for this new simulation is –5947, which is higher than the predicted value by about 3%.

At this point, clearly we are close to the optimum. Whether further simulations are warranted depends on the accuracy of the estimate being used to obtain the costs and the extra effort that must be expended to analyze further simulations. If further refinement of the estimate of the optimum conditions is required, then the model, Equation (14.4), can be modified to include terms in p^2 and used to refine further the grid of decision variables. Alternative approaches for more comprehensive models are considered by Ludlow et al. [4].

One final point should be made. This problem is fairly typical of real design optimizations in that the optimum conditions lie on several of the constraints for the variables. Even though a simple polynomial model has been used to describe

Table 14.8 Predictions of Model for *NPV* in Equation (14.4) with Data from Table 14.7

Test Run	Result from Table 14.7	Prediction from Equation 14.4
Base Case	−7630	−7502
DME000	−8136	−8134
DME001	−8760	−8758
DME010	−8152	−8150
DME011	−8851	−8849
DME100	−6130	−6127
DME101	−6885	−6881
DME110	−6160	−6157
DME111	−6961	−6957

All figures are in $1,000.

the *NPV*, Equation (14.4), which will tend to force the optimum to lie on the constraints, similar results would be obtained if nonlinear models had been used to describe the *NPV*.

14.5 LATTICE SEARCH TECHNIQUES VERSUS RESPONSE SURFACE TECHNIQUES

For problems involving many decision variables, there are three types of parametric optimization techniques.

1. **Analytical Techniques Based on Finding the Location Where Gradients of the Objective Function Are Zero:** These are effective if the objective function is continuous, smooth, and has only a few local extrema. Sometimes the objective function can be approximated as a smooth function, and techniques exist to condition an objective function to reduce the number of extrema. However, for the common, complex problems encountered in flowsheet optimization, analytical techniques are usually ineffective.
2. **Response Surface Strategies Such as Those Discussed in Sections 14.2 and 14.4:** These can be used when the conditions required for analytical techniques are not met. However, one must search the interior of the decision variable space extrema, which can be done by three-level (and higher-level) factorial designs [3,4]. Fractional factorial designs are essential for higher-level and higher-dimensionality problems. As the dimensionality (number of decision variables) and the number of constraints increase, so does the probability that the optimum lies on the boundary of the allowable search space. This makes response surface techniques especially attractive.
3. **Pattern Search Techniques:** These are iterative techniques that are used to proceed from an initial guess toward the optimum, without evaluating derivatives or making assumptions about the shape of the objective function surface. From the initial guesses of the decision variables, a direction is chosen, a move is made, and the objective function is evaluated. If an improvement is detected, further moves are made in that direction. Otherwise, the search continues in a new direction. There are many strategies for choosing the search direction, deciding when to change direction, what direction to change to, and how far to move in each iteration. Crude methods of this type evaluate the objective function at each of the "nearest neighbor" points in each direction. The **Simplex-Nelder-Mead** [5] method uses $(n + 1)$ points when there are n decision variables to determine the next move. This is the smallest number of points that can give information about changes in the objective function in all directions. A related strategy is **simulated annealing** [6], in which all moves (good and bad) have nonzero probability of being accepted. Good moves are always accepted, and bad moves are accepted, but less frequently the larger the degradation in the objective function.

Both the pattern search and the response surface techniques allow the use of discrete decision variables. Topological changes can always be represented by changes in discrete decision variables (such as those defining the flowsheet topology), so these techniques can accommodate both parametric and topological optimization.

The response surface techniques are preferred during the early phases of design. They serve the multiple uses of prioritizing the decision variables, scoping the optimization problem to determine an estimate of the target improvement possible, and performing the parametric and topological optimizations.

14.6 PROCESS FLEXIBILITY AND THE SENSITIVITY OF THE OPTIMUM

Before any optimization result can be understood, one must consider the flexibility of the process and the (related) sensitivity of the result. A process must be able to operate with different feedstocks, under varying weather conditions (especially important for cooling utilities), over a range of catalyst activities, at different production rates, and so on. A process that is optimum for the "design" or "nameplate" conditions may be extremely inefficient for other operating conditions. The common analogy is a car that can operate at only one speed. Clearly, the design of the car could be optimized for this one condition, but the usefulness of such a car is very limited.

The **sensitivity of the optimum** refers to the rate at which the objective function changes with changes in one of the decision variables. Here, there is an apparent contradiction. Decision variables are chosen because the objective function is most sensitive to them, and therefore the process needs to be controlled very close to the optimum values of the decision variables to stay near the optimum. In fact, the objective function evaluations made during the optimization allow one to determine how precise the control must be and, most importantly, to determine the penalty for failure to control the decision variable to within prescribed limits. This is another of the interplays between optimization and process control.

The effects of process sensitivity and flexibility can best be shown with a family of curves such as shown in Figure 14.6. Regardless of the techniques used to *find* the optimum, it is only through these visual representations that the results of optimization can be effectively conveyed to those who will use them. They are essentially performance curves for the process, where the performance is measured as the value of the objective function.

14.7 OPTIMIZATION IN BATCH SYSTEMS

Unlike continuous systems, batch operations do not run under steady-state conditions, and their performance varies with time. As discussed in Chapter 3, the important issues with batch systems are the optimal scheduling of different equipment to produce a variety of products and the determination of optimal cycle times for batch processes. Therefore, the optimization of batch operations often involves determining the "best" processing time for a certain operation, the "best" time at which a certain action should take place, or the "best" distribution of actions over a period of time. The optimization of batch processes is, in itself, a very broad topic and certainly beyond the scope of this section of this chapter. Rather than try to address the many interesting problems in this field, the approach here is to illustrate several important concepts through the use of examples. The interested reader is encouraged to read further into this subject [7–10].

14.7.1 Problem of Scheduling Equipment

Consider a simple case of scheduling a set of equipment used to produce multiple products. For example, let us assume that there are three products of interest (A, B, C) manufactured from the same chemical feedstock F. The production of each of the three chemicals follows a different sequence and requires the use of different equipment (a reactor **R**, a separator **S**, and a precipitator **P**) for different periods of time. This information along with relative feed and product prices are shown in Table 14.9. The times shown in the table reflect the combined time required for filling, operating, emptying, and cleaning each piece of equipment for each product.

Table 14.9 Equipment Time Requirements for Products A, B, and C

Product	Time in Reactor (hr)	Time in Separator (hr)	Time in Precipitator (hr)	Value of Product ($/kg Product)	Cost of Feed ($/kg Product)
A	7	4	0	0.75	0.25
B	15	3	3	1.12	0.27
C	25	4	2	1.41	0.23

The costs of the feeds are different for each product because of differences in process efficiencies. In general, the longer it takes to make a product, the more valuable the product. The question that must be answered is, How should the equipment be scheduled to maximize the revenue from the products?

Another complication in this type of scheduling problem is that the production steps are sequential: R→S→P. This means that when a reactor has finished producing a product, the separator should be available to carry out the next step in the process, and, similarly, the precipitator should be available when the separation has finished. This problem was chosen with the assumption that the separator and precipitator are always available to handle the products from the reactor, because the longest processing time for a given piece of equipment is always shorter than the shortest time for the preceding operation ($4 < 7$, $3 < 4$). For this case, the solution to the problem is simple because the only piece of equipment that must be considered is the reactor; the reaction step is the limiting step in each process.

Problem Formulation. Let the number of batches produced per year for Products A, B, and C be x_1, x_2, and x_3, respectively. Assuming the size of each batch is the same and equal to W kg, then the objective function for the optimization, which is the total profit, is

$$Maximize\ OF = Wx_1(0.75\text{-}0.25) + Wx_2(1.12\text{-}0.27) + Wx_3(1.41 - 0.23)$$

or

$$Maximize\ OF = 0.5\ Wx_1 + 0.85\ Wx_2 + 1.18\ Wx_3 \tag{14.5}$$

The constraints for the problem relate to the total time for processing:

$$\left.\begin{array}{l} x_1 \geq 0 \\ x_2 \geq 0 \\ x_3 \geq 0 \end{array}\right\} \tag{14.6}$$

$$7x_1 + 15x_2 + 25x_3 \leq (365)(24) = 8{,}760 \tag{14.7}$$

Equation (14.7) simply states that the equipment (reactor) is available 24 hours a day for the whole year. The solution of the problem is the maximization of Equation (14.5) subject to the constraints given in Equations (14.6) and (14.7). The solution is trivial, yielding that the optimum strategy is to make only chemical A with $x_1 = 1251$ batches per year and $OF = \$625.7W$

Addition of Market Constraints. Now consider a slightly more realistic version of the problem. This time, consider the situation where market forces tell us that a maximum of 800 batches of A in a year can be sold and that to meet contractual obligations, at least 10 batches of B and 7 batches of C must be produced in a year. These additional constraints are given as

$$\left.\begin{array}{l} x_1 \leq 800 \\ x_2 \geq 10 \\ x_3 \geq 7 \end{array}\right\} \tag{14.8}$$

The resulting solution is $x_1 = 800$, $x_2 = 199$, $x_3 = 7$, and $OF = \$577.4W$. Not surprisingly, the extra constraints have caused the OF to be reduced. This solution tells us that after our contractual obligations have been fulfilled, and as much of Product A has been made as is allowed, as much of Product B as is allowed should be made. In other words, the order of profitable products is A > B > C.

Considering Other Equipment. Now look at the situation where the time taken to process the products through the other equipment becomes a limiting factor. To illustrate this type of problem, the basic problem is modified, as shown in Table 14.10.

Table 14.10 Equipment Time Requirements for Products A, B, and C for the Case When the Time for Separation and/or Precipitation Constrains the Solution

Product	Time in Reactor (hr)	Time in Separator (hr)	Time in Precipitator (hr)	Value of Product ($/kg Product)	Cost of Feed ($/kg Product)
A	7	10	8	0.75	0.25
B	15	7	6	1.12	0.27
C	25	5	2	1.41	0.23

If the complication of sequencing of operations to allow a product to move through the system is ignored, for the moment, then the additional constraints imposed in Table 14.10 can be formulated as

Separator constraint $10x_1 + 7x_2 + 5x_3 \le 8{,}760$ (14.9)

Precipitator constraint $8x_1 + 6x_2 + 2x_3 \le 8{,}760$ (14.10)

Solving the problem including the market constraints given in Equation (14.8) results in $x_1 = 701$, $x_2 = 245$, $x_3 = 7$, and $OF = \$567.2W$. The results show that, with the inclusion of the constraints for the other equipment, it is no longer optimal to make as much of Product A as the market can handle, but, instead, making more of Product B is more profitable. If the values of x_1, x_2, and x_3 are substituted into the equipment time constraints (Equations [14.7], [14.9], and [14.10]) the result is

Reactor usage $7x_1 + 15x_2 + 25x_3 = 8{,}757 \le 8{,}760$

Separator usage $10x_1 + 7x_2 + 5x_3 = 8{,}760 \le 8{,}760$

Precipitator usage $8x_1 + 6x_2 + 2x_3 = 7{,}092 \le 8{,}760$

It should be noted that the values for x_1, x_2, and x_3 have been rounded to the nearest integer, and, in fact, the total times for the reactors and separators are both 8,760 hours. From these results, it is seen that the first two constraints meet the equality, but the constraint for the precipitator shows that this piece of equipment has spare capacity; it is not being used all the time. This spare capacity represents an inefficiency in equipment use and might be worth further investigation; however, some spare capacity is often desirable from an operability point of view. The fact that both the reactor and the separator are used all the time might create a scheduling problem when implementing the "optimal" mix of chemicals predicted by this latest solution. The reason for this is that, in general, when the times for different operations are not the same or not some simple multiple of each other and both pieces of equipment are used all the time, there will be periods of time when one or the other piece of equipment must sit idle. Hence, the true optimum would be less than that obtained by using this approach. This restriction could be eliminated by providing intermediate storage tanks between the different equipment that would allow each piece of equipment to operate all the time, as discussed in Chapter 3, Section 3.5.2. In this case, the optimum solution given for this problem can be realized, but there is an additional hidden cost of buying the intermediate storage tanks. Alternatively, by using multiple single-product campaigns, discussed in Chapter 3, this optimal solution might be obtainable. However, the implementation of single-product campaigns leads to the use of additional product storage and hence more capital investment.

The problem could have been formulated by specifying the constraints in terms of the total number of hours to produce a product. This would be given by

$$\textit{Maximize } OF = 0.5\ Wx_1 + 0.85\ Wx_2 + 1.18\ Wx_3 \tag{14.11}$$

subject to the constraints in Equations (14.6) and (14.8), and

$$(7 + 10 + 8)x_1 + (15 + 7 + 6)x_2 + (25 + 5 + 2)x_3 \leq 8{,}760 \tag{14.12}$$

or

$$25x_1 + 28x_2 + 32x_3 \leq 8{,}760 \tag{14.13}$$

The optimal solution to this problem is $x_1 = 0$, $x_2 = 10$, $x_3 = 265$, and $OF = \$321.2W$. The *OF* is considerably smaller than the previous solution. Moreover, if the total time that the reactor, separator, and precipitator are operating is evaluated, the following results are found:

Reactor usage $7x_1 + 15x_2 + 25x_3 = 6{,}775 \leq 8{,}760$

Separator usage $10x_1 + 7x_2 + 5x_3 = 1{,}395 \leq 8{,}760$

Precipitator usage $8x_1 + 6x_2 + 2x_3 = 590 \leq 8{,}760$

Although this is certainly a valid solution, the separation and precipitation equipment sit idle most of the time. One possible way to sequence the operations for this solution is shown in Figures 14.10(a) and 14.10(b). The sequence for this solution would follow that shown in Figure 14.10(a) for (265)(32) = 8,480 h, with the remaining period of (10)(28) = 280 hours following the sequence shown in Figure 14.10(b). There are many ways to improve the profitability of the scheme shown in Figures 14.10(a) and 14.10(b). For example, it can be seen that a batch run for Product A could be squeezed into the schedule every time a batch of B or C is made. This is illustrated in Figures 14.10(c) and 14.10(d).

By making additional Product A, the value of our *OF* increases from $\$321.4W$ to $321.4W + (265 + 10)(0.5)W = \$458.9W$. Moreover, the equipment usages increase to

Reactor usage $7(275) + 15(10) + 25(265) = 8{,}700 \leq 8{,}760$

Separator usage $10(275) + 7(10) + 5(265) = 4{,}145 \leq 8{,}760$

Precipitator usage $8(275) + 6(10) + 2(265) = 2{,}790 \leq 8{,}760$

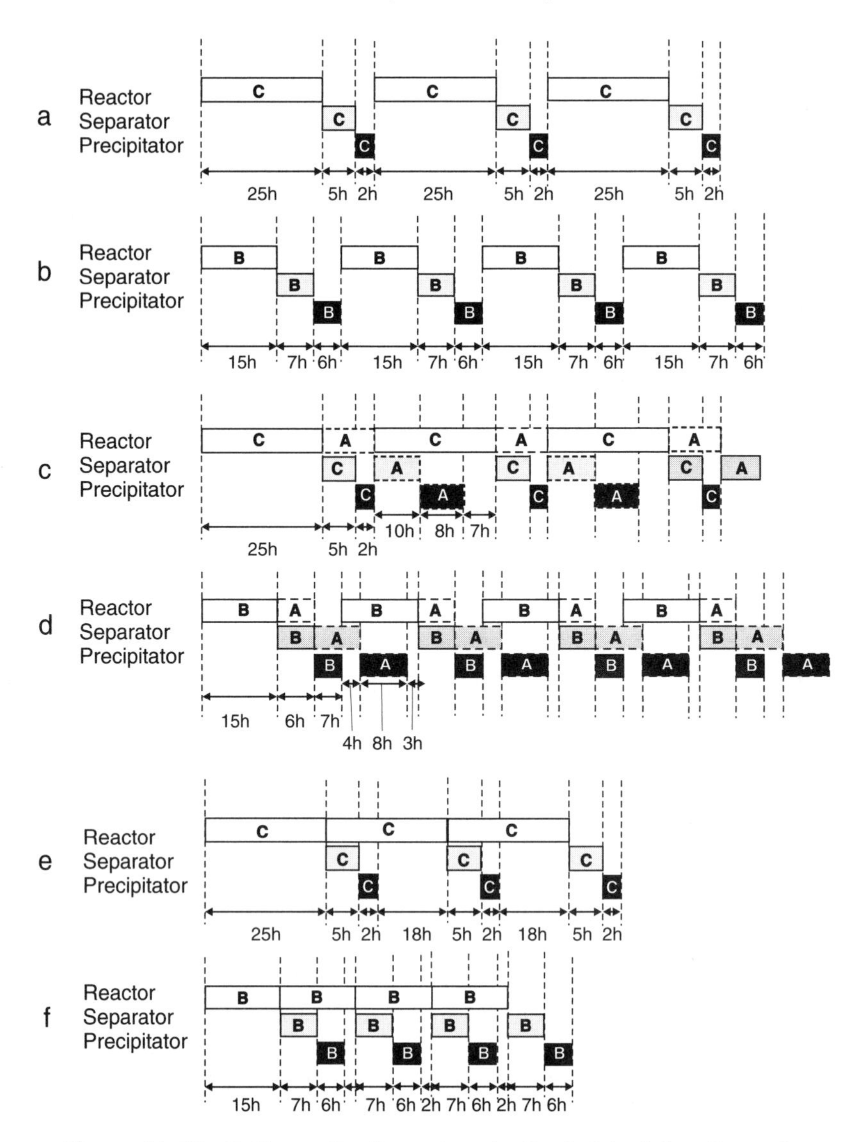

Figure 14.10 Different Configurations for Batch Scheduling Example

An alternative change in the scheduling for this problem is illustrated in Figures 14.10(e) and 14.10(f). In this case, the idle time for the reactor is eliminated by implementing a batch of C or B as soon as the reactor becomes available. This essentially eliminates the reactor downtime. This saves 7 hours per batch of C and 13 hours per batch of B, leaving a period of (7)(265) + (13)(10) = 1985 hours of additional processing time during which more A, B, or C could be made. This same scheme could also be applied to Figure 14.10(d) by moving the sequence of events for producing Product B to the left by 7 hours. This would again eliminate any downtime for the reactor and would free up another (7)(10) = 70 hours per year of time to make other products and additional profit.

Clearly, the number of possible sequences is numerous. In the above examples, reactor downtime could be eliminated, and a constraint was encountered. However, a different configuration could have been used as the starting point, such as producing only A and B, and a different constrained solution that might have been more profitable would have been reached. The optimum configuration is not intuitively obvious, nor is how to go about formulating and optimizing the problem. The problems in this section were all solved using the solver in Microsoft Excel that uses the generalized reduced gradient (GRG2) nonlinear optimization code. However, specialized software is available to help formulate and solve these types of batch scheduling problems. The software most often used in batch scheduling optimization solves mixed-integer, nonlinear programming (MINLP) problems [8]. The topic of batch scheduling has its roots in the field of operations research, which is a shared discipline of industrial and chemical engineers.

A final point of interest involves the degree of consolidation that is possible for batch scheduling problems and the degree that is desirable, which may be different from each other. For the simple example of three chemical products using three pieces of equipment, it was shown how to compress the scheduling so that the reactor is being used all the time. However, this puts additional burden on the operators. For the cases shown in Figures 14.10(a) and 14.10(b), a single operator per shift might be able to cover all the operations (loading feed, monitoring, emptying product, cleaning equipment, etc.) required. However, as operations get squeezed together, more operators are needed, and errors, which are bound to happen on occasion, have a much greater effect on production. The trade-offs between efficient equipment usage and operability must be considered in order to develop robust and profitable operations.

14.7.2 Problem of Optimum Cycle Time

Another problem that arises in batch operations is the question, What is the optimal time to run a particular operation? For example, consider the simple case of a batch reactor producing a product via a first-order, irreversible reaction. The conversion of product follows a simple exponential relationship, as shown in Figure 14.11.

Therefore, the longer the material reacts, the more product is made. However, the rate at which product is formed slows with time, and the average production rate drops. These trends are shown by the curves plotted in Figure 14.11. Both the instantaneous and time-averaged production rate of Product B are seen to decrease with time. Because this is a batch operation, the optimal time to run the reaction must be determined. Because there are costs associated with cleaning and recharging feed to the equipment, these should be taken into account when determining the optimal cycle time for the batch.

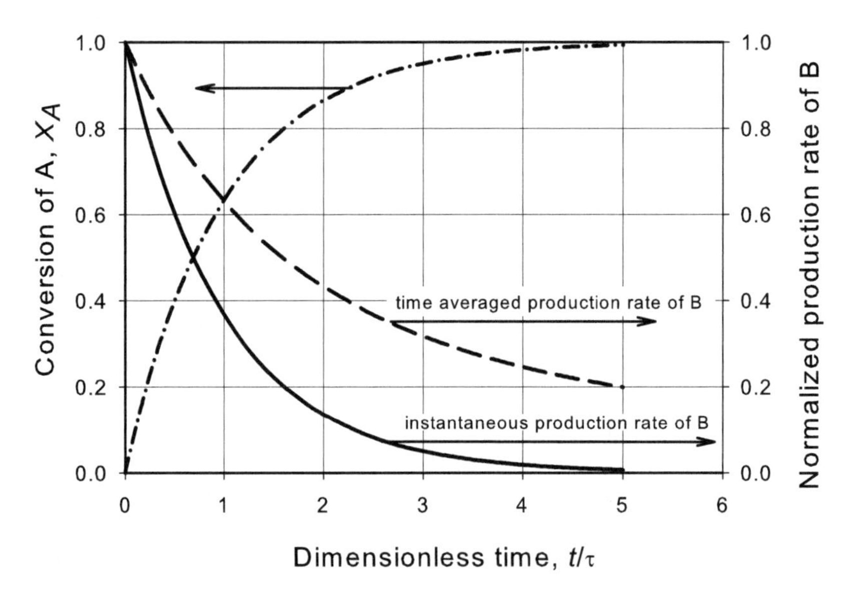

Figure 14.11 Conversion of A and Production of B for the Simple First-Order Reaction A→B as a Function of Dimensionless Time in a Batch Reactor

Formulation of the Problem. Assume that the production of product in a liquid-phase batch reactor, in which A→B, follows a simple first-order, elementary reaction given by

$$N_B = VC_{A0}X_A = VC_{A0}\left[1 - \frac{C_A}{C_{A0}}\right] = VC_{A0}[1 - e^{-kt}] \tag{14.14}$$

where N_B is the moles of Product B formed in the reactor after time = t, V is the volume of the batch reactor, C_{A0} is the concentration of reactant at time = 0, C_A is the concentration of reactant at time = t, and k is the reaction rate constant.

Let the value of the product be $\$x$/kmol, let the value of the reactant be $\$y$/kmol, and let the cost of cleaning and preparing the reactor for the next batch be $\$C_{clean}$. The time to clean and prepare the reactor for the next batch is t_{clean} and the reaction time is t_{react}. The cycle time, θ, is simply the sum of the cleaning and reaction time:

$$\theta = t_{clean} + t_{react} \tag{14.15}$$

The objective function for this problem is the average "profit" over the cycle time, which is given by

$$OF = \frac{xN_B - y(VC_{A0} - N_B) - C_{clean}}{\theta} = \frac{N_B(x + y) - yVC_{A0} - C_{clean}}{t_{react} + t_{clean}} \tag{14.16}$$

The numerator of Equation (14.16) is the profit from Product B minus the cost of unused reactant A minus the cost of cleaning. The denominator is simply the total time for the cycle and thus averages the profit over the cycle time. This formulation of the *OF* assumes that the unused reactant is not recovered and recycled and that the cost of downstream processing (separation of B from A, and purification of B) are not affected by the extent of reaction. Both these assumptions should be relaxed if information about the downstream processes is available.

Substituting Equation (14.14) into Equation (14.16) and noting that $t = t_{react}$ yields

$$OF = \frac{(x + y)VC_{A0}[1 - e^{-kt}] - yVC_{A0} - C_{clean}}{t + t_{clean}} = \frac{A_1 - B_1 e^{-kt}}{C_1 + t} \tag{14.17}$$

where

$A_1 = (x + y)VC_{A0} - yVC_{A0} - C_{clean} = xVC_{A0} - C_{clean}$

$B_1 = (x + y)VC_{A0}$

$C_1 = t_{clean}$

Differentiating the *OF* with respect to t and setting the derivative equal to 0 yields the following expression for $t_{opt} = t_{react,\ optimum}$:

$$(B_1 kC_1 + B_1 kt_{opt} + B_1)e^{-kt_{opt}} - A_1 = 0 \tag{14.18}$$

The positive real root of this equation gives the optimum time to react the batch. An example to illustrate this optimization procedure is given in Example 14.7.

Example 14.7

Determine the optimal batch reaction and cycle time for the first-order, irreversible reaction A→B using the following data and plot the *OF* [Equation (14.17)] as a function of θ.

$V = 5\ m^3$

$C_{A0} = 2\ kmol/m^3$

$k = 0.3173\ h^{-1}$

$t_{clean} = 2$ hr

$x = \$700/kmol$

$y = \$400/kmol$

$C_{clean} = \$1200$

$A_1 = xVC_{A0} - C_{clean} = (700)(5)(2) - 1200 = 5{,}800$

$B_1 = (x + y)VC_{A0} = (400 + 700)(5)(2) = 11{,}000$

$C_1 = t_{clean} = 2$

Substituting into Equation (14.18) yields

$(B_1 kC_1 + B_1 kt_{opt} + B_1)e^{-kt_{opt}} - A_1 =$

$[(11{,}000)(0.3173)(2) + (11{,}000)(0.3173)t_{opt} + 11{,}000]e^{-0.3173t_{opt}} - 5{,}800 = 0$

$(6{,}980 + 3{,}490t_{opt} + 11{,}000)e^{-0.3173t_{opt}} - 5{,}800 = 0$

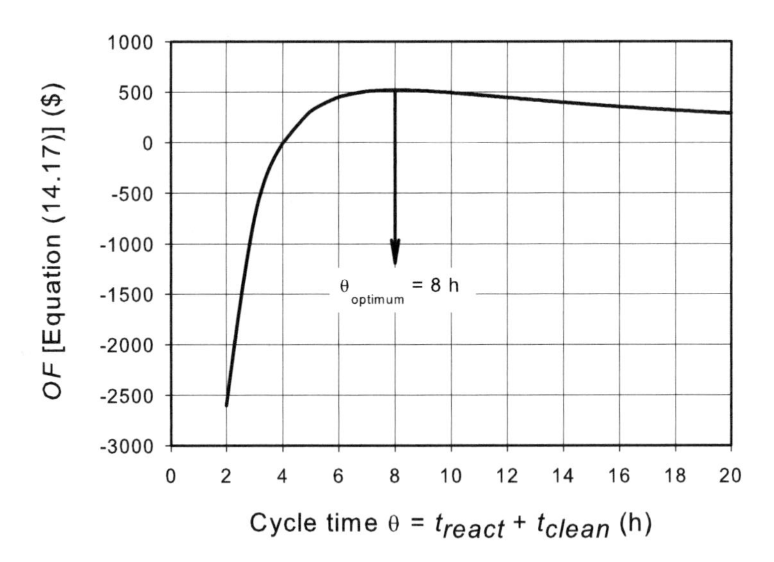

Figure E14.7 Results of Example 14.7 Showing the *OF* as a Function of Cycle Time

From this, $t_{opt} = 6.0$ h, $t_{clean} = 2.0$ h, and $\theta_{opt} = t_{clean} + t_{react} = 2.0 + 6.0 = 8.0$ h

A plot of the *OF* as a function of θ is shown in Figure E14.7.

Similar problems to the optimal cycle time for a batch reaction arise when considering the optimal cycle times for cleaning heat exchangers prone to excessive fouling or for cleaning filters.

14.8 SUMMARY

In this chapter, several definitions commonly used in the area of optimization have been introduced. In addition, the idea of applying a Pareto analysis to the base case was investigated. This type of analysis often reduces the number of decision variables that should be considered and the range over which the optimization should take place.

The differences between parametric and topological optimization were investigated, and strategies were suggested for each type of optimization.

The strategy for optimizing a process flowsheet using response surface concepts from the statistical design of experiments was introduced. The method was then illustrated using the DME flowsheet from Appendix B.

Finally, an introduction to the optimization of batch processes was given. Specifically, methods to determine the optimal mix of products in a multiproduct batch facility and the determination of optimum cycle time were covered.

REFERENCES

1. Juran, J. M., "Pareto, Lorenz, Cournot, Bernoulli, Juran, and Others," *Ind. Qual. Control* 17, no. 4 (1960): 25.
2. Humphrey, J. L., and G. E. Keller II, *Separation Process Technology* (New York: McGraw Hill, 1997), ch. 2.
3. Neter, J., and W. Wasserman, *Applied Linear Statistical Models* (Homewood, IL: Richard D. Irwin, Inc., 1974).
4. Ludlow, D. K., K. H. Schulz, and J. Erjavec, "Teaching Statistical Experimental Design Using a Laboratory Experiment," *Journal of Engineering Education* 84 (1995): 351.
5. Nelder, J. A., and R. Mead, "A Simplex Method for Function Minimization," *Computer Journal* 7 (1965): 308.
6. Kirkpatrick, S., C. D. Gelatt, and M. P. Vecchi, "Optimization by Simulated Annealing," *Science* 220 (1983): 671.
7. Méndez, C. A., J. Cerdá, I. E. Grossmann, I. Harjunkoski, and M. Fahl, "State-of-the-Art Review of Optimization Methods for Short-Term Scheduling of Batch Processes," *Computers & Chemical Engineering* 30 (2006): 913.
8. Biegler, L. T., I. E. Grossman, and A. W. Westerberg, *Systematic Methods of Chemical Process Design* (Upper Saddle River, NJ: Prentice Hall, 1999).
9. Honkomp, S. J., S. Lombardo, O. Rosenand, and J. F. Pekny, "The Curse of Reality: Why Process Scheduling Optimization Problems Are Difficult in Practice," *Computers & Chemical Engineering* 24 (2000): 323.

10 Mauderli, A., and D. W. T. Rippin, "Production Planning and Scheduling for Multi-Purpose Batch Chemical Plants," *Computers & Chemical Engineering* 3 (1979): 199.

SHORT ANSWER QUESTIONS

1. Describe a Pareto analysis. When is it used?
2. What is the difference between parametric optimization and topological optimization? List one example of each.
3. What is an objective function? Give two examples of one.

PROBLEMS

4. In general, when using cooling water (CW) as a utility, the outlet temperature of the water leaving any process exchanger and returning to the cooling tower is limited to about 40°C to avoid excessive fouling in the process exchangers. In a series of laboratory experiments on fouling, the inlet CW temperature

was fixed at 30°C, and the dimensionless fouling ratio was recorded for exit cooling water temperatures in the range of 36°C to 46°C.

Cooling Water Exit Temperature, °C	Dimensionless Fouling Ratio
36	1.11
38	1.05
40	1.00
42	0.86
44	0.74
46	0.64

The fouling ratio is simply the ratio of the time between scheduled cleanings of the exchanger tubes compared to the time required at 40°C. For example, if an exchanger with a CW exit temperature of 40°C requires cleaning once a year (12 months), then an exchanger with a CW exit temperature of 46°C requires cleaning every (0.64)(12) = 7.68 months. The advantage of using higher CW exit temperatures is that less cooling water flow is required; hence, the pumping costs for CW are lower. For a typical heat exchanger in a process with an exit CW temperature of 40°C, the yearly pumping and cleaning costs are $400/yr and $2,400/yr, respectively. Using these costs and the information in the table above, determine the optimal CW exit temperature. Assume that the pumping costs are proportional to the flow of cooling water, that they include an amortized cost for the pump and the cost of electricity to run it, and that the cost of the heat exchanger is essentially unchanged for small differences in the CW exit temperature.

5. Consider the removal of condensable components from the reactor effluent stream of an acetone production process, shown in Figure P14.5. The removal process consists of a cooling water exchanger and a refrigerated water (rw)

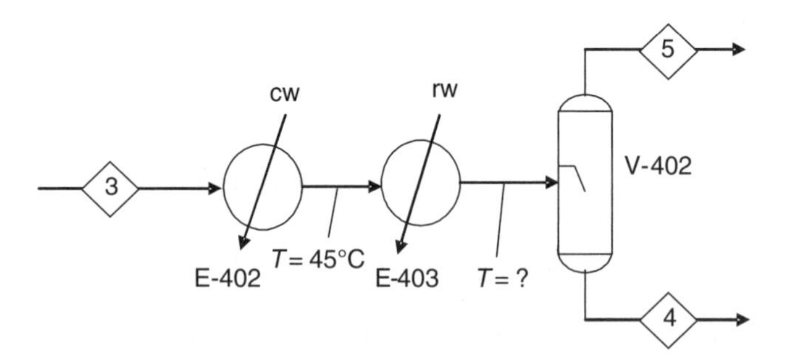

Figure P14.5 Recovery System for Problem 14.5

exchanger, as shown in the figure. For this problem, assume that a downstream acetone stripper will not be used, and all acetone and isopropyl alcohol (IPA) in Stream 5 will be lost and not recovered.

The composition and conditions of Stream 3 are give in the following table. The pressure drop through each exchanger is 0.14 bar. The purpose of the problem is to optimize the recovery of acetone and IPA in Stream 4 by determining the optimal size of heat exchanger E-403. The costs of E-402 and V-402 may be assumed to be fixed. Any acetone lost in Stream 5 should be valued at the equivalent cost of the IPA required to produce it, and the cost of IPA is given in Table 8.4.

Properties of Stream 3 in Problem 14.5

Temperature (°C)	160
Pressure (bar)	1.91
Vapor fraction	1.0
Component mole flow (kmol/h)	
Hydrogen	34.78
Acetone	34.94
Isopropyl alcohol	3.86
Water	19.04

Determine the optimal recovery of IPA and acetone for this system using an objective function of the *EAOC* based on an 8000h operating year. Any simulations should be carried out using a suitable process simulator using the UNIFAC *K*-value option and the latent heat enthalpy option. The overall heat transfer coefficient for E-403 may be taken as 200 W/m^{2}K. Utility costs should be obtained from Chapter 8, and the cost of E-403 is approximated by the following equation:

$$C_{TM}[\$] = 10{,}000\{A[m^2]\}^{0.6} \tag{P14.5}$$

6. Consider the compression of feed air into a process that produces maleic anhydride, shown in Figure P14.6. The air enters the process at a rate of 10 kg/s at atmospheric pressure and 25°C. It is compressed to 3 atm in the first compressor. The compression is 65% efficient based on a reversible, adiabatic process. Prior to entering the second stage of compression (where the air is compressed to 9 atm at the same efficiency), the air flows through a water-cooled heat exchanger where the temperature is cooled to $T_{air,2}$.

   Assume that frictional pressure drops in the connecting pipes and in the exchangers are negligible. For the design of the exchanger, assume that

   $U = 42$ W/m^2°C, $T_{cw,in} = 30$°C, and $T_{cw,out} = 40$°C

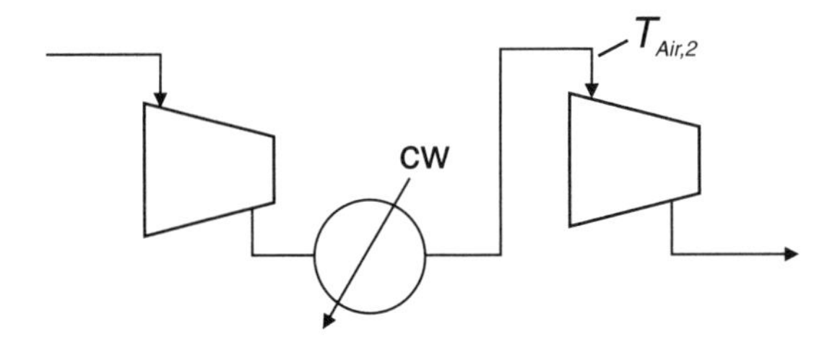

Figure P14.6 Two-Stage Compression for Problem 14.6

For economic calculations, assume the following.

- The fixed capital investment is equal to the total module cost of the compressor (centrifugal), the compressor drive (electric), and heat exchanger (shell-and-tube, floating head). The following equations should be used to evaluate these costs:

$$C_{TM,\,compr}[\$] = 12{,}500\{P[kW]\}^{0.68} \tag{P14.6.1}$$

$$C_{TM,\,drive}[\$] = 600\{P[kW]\}^{0.9} \tag{P14.6.2}$$

$$C_{TM,\,exch}[\$] = 2{,}700\{A[m^2]\}^{0.82} \tag{P14.6.3}$$

- Utility costs should be taken from Table 8.4, and use an electric drive efficiency of 95% and a 65% compressor efficiency.
- *F* factors should be calculated for the heat exchanger, and these will change as $T_{air,2}$ changes (see Chapter 15 for equations for *F*).
- A hurdle rate of 8% and an equipment/project life of 5 years.

Using results from a simulator, determine the optimum size (area) of the heat exchanger and the corresponding value of $T_{air,2}$.

7. When pumping a liquid from one location to another, the diameter of the pipeline plays a critical role in determining the amount of power (electricity) required to pump the liquid. For a given flow of liquid, the amount of power required by the pump to move the liquid between locations decreases as the diameter of the pipe increases. However, the cost of the piping increases with diameter. Therefore, for a given flowrate of liquid and a given distance between locations, there exists an optimal pipe diameter that minimizes the cost of the system comprising the pump and pipe.

   It is required to find the optimal pipe diameter for a system through which 100,000 barrels per day of oil is pumped over a distance of 100 miles.

   The overall cost of the system includes the equipment costs (pump and piping), which are one-time purchases, and the operating cost (electricity for the pump), which occurs all the time. The time value of money must be taken into consideration when defining an objective function. For this project, we want to minimize the equivalent annual operating cost (*EAOC*) of the pump and piping system that is defined in Equation (P14.7.1):

$$EAOC[\$/y] = \sum_{i=1}^{2} PC_i[\$](A/P,i,n)[1/y] + UC[\$/y] \qquad \text{(P14.7.1)}$$

where PC_i is the purchase cost of the pump and the pipe, and UC is the operating (utility) cost for the utilities (electricity to run the pump). Assume that the effective annual interest rate is 9% p.a. and that the length of the project, n, is 15 years.

The relationships between the flowrate (Q) of oil, the pipe diameter (d_{pipe}), and the power required for the pump ($\dot{W}_{pump}$) are

$$\dot{W}_{pump} = \frac{32 f \rho Q^3 L_{pipe}}{\varepsilon_{pump} \pi^2 d_{pipe}^5} \qquad \text{(P14.7.2)}$$

where ρ is the density of the fluid being pumped (930 kg/m^3), and ε_{pump} is the efficiency of the pump (assume this is 85%, or 0.85). Using SI units (m, s, kg), the units of power in Equation (P14.7.2) are Watts (W). Assume that the cost of electricity is \$0.05/kWh, where a kWh is the energy used in one hour by a device consuming 1 kW. Assume that the pump operates 24 h a day 365 days per year.

In Equation (P14.7.2), the term f is the friction factor, which is a function of the flowrate and the pipe diameter and is given by Equation (P14.7.3):

$$\frac{1}{\sqrt{f}} = -4 \log\left[\frac{\varepsilon}{3.7 d_{pipe}} + \left\{\frac{6.81}{\text{Re}}\right\}^{0.9}\right] \qquad \text{(P14.7.3)}$$

where ε is the roughness of the pipe (assume a value of 0.045 mm), $\text{Re} = \text{Reynolds number} = \frac{4Q\rho}{\pi d_{pipe}\mu}$, and μ is the viscosity of the oil (assume μ = 0.01 kg/ms).

The cost of the pump can be estimated using Equation (P14.7.4):

$$PC_{pump} = \$8000(\dot{W}_{pump}[\text{kW}])^{0.6} \qquad \text{(P14.7.4)}$$

The cost of piping is given by Equation (P14.7.5):

$$PC_{pipe}[\$/\text{ft of pipe}] = 10 d_{pipe}[inch] + 2 d_{pipe}^{1.4}[inch] \qquad \text{(P14.7.5)}$$

Present your final results as two plots. The first should show how each term in Equation (P14.7.1) changes with d_{pipe} (x-axis), and the second plot should show the *EAOC* (y-axis) as a function of d_{pipe} (x-axis). Explain the reason for the trends seen in each of these plots.

8. A liquid-phase biological reaction is used to produce an intermediate chemical for use in the pharmaceutical industry. The reaction occurs in a large, well-stirred bioreactor that is illustrated in Figure P14.8. Because this chemical is temperature sensitive, the maximum operating temperature in the reactor is limited to 65°C. The feed material is fed to the reactor through a heat exchanger that can increase the temperature of the reactants (contents of the reactor), which in turn increases the rate of the reaction. This is illustrated in Figure P14.8. The time spent in the bioreactor (space time) must be adjusted in order to obtain the desired conversion of reactant. As the temperature in the reactor increases, so does the reaction rate, and hence the size (and cost) of the reactor required to give the desired conversion decreases. It is desired to determine the optimal temperature at which to maintain the reactor (and to preheat the feed). The costs to be considered are the purchase cost of the reactor and heat exchanger and the operating cost for the energy to heat the feed.

   It is desired to optimize the preheat temperature for a flow of 5000 gal/h. The feed has the properties of water (ρ = 1,000 kg/m^3, C_p = 4.18 kJ/kg°C) and enters the heat exchanger at a temperature of 20°C. Reactor feed is to be heated with a heating medium that is available at a temperature of 65°C and must leave the heat exchanger at 30°C. Therefore, the desired reactor inlet temperature is adjusted by changing the flowrate of the heating medium. The physical properties of the heating medium are ρ = 920 kg/m^3, and C_p = 2.2 kJ/kg°C.

   The reaction rate for this reaction, $-r_A$, is given in terms of the concentration of reactant A (C_A) by the following equation:

$$-r_A = kC_A \qquad \text{(P14.8.1)}$$

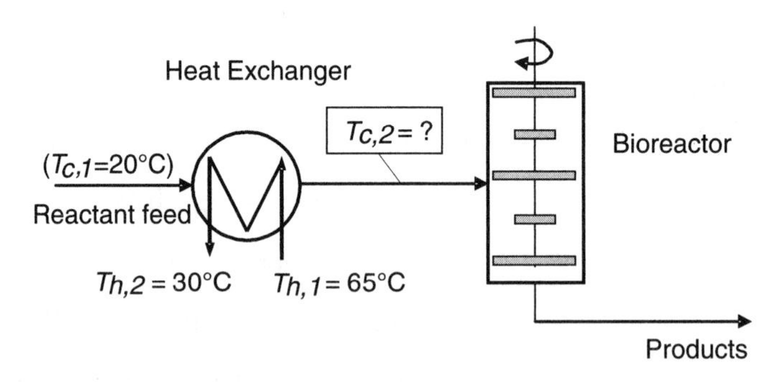

Figure P14.8 Process Flow Diagram of the Feed Preheater and Bioreactor

where

$$k[s^{-1}] = 7\exp\left\{-\frac{3{,}300}{T[K]}\right\} \quad \text{(P14.8.2)}$$

The design equation for the reactor is given by

$$V = \frac{v_o X_A}{k(1 - X_A)} \quad \text{(P14.8.3)}$$

where V is the reactor volume (m^3), v_o is the volumetric flowrate of fluid into the reactor (m^3/s), and X_A is the conversion (assumed to be 80%, or 0.8, for this reaction).

The design equation for the heat exchanger is given by

$$Q = M_c C_{p,c}(T_{c,2} - T_{c,1}) = M_h C_{p,h}(T_{h,1} - T_{h,2}) = UAF\Delta T_{lm} \quad \text{(P14.8.4)}$$

where

$$\Delta T_{lm} = \frac{(T_{h,2} - T_{c,1}) - (T_{h,1} - T_{c,2})}{\ln\dfrac{(T_{h,2} - T_{c,1})}{(T_{h,1} - T_{c,2})}} \quad \text{(P14.8.5)}$$

and

$F = 0.8$ (assume that this is constant for all cases)
U = overall heat transfer coefficient = 400 W/m^2K

The optimal reactor inlet temperature is the one that minimizes the equivalent annual operating costs ($EAOC$) given by

$$EAOC[\$/y] = \sum_{i=1}^{2} PC_i[\$](A/P,i,n)[1/y] + UC[\$/y] \quad \text{(P14.8.6)}$$

where PC_i is the purchase equipment cost for the heat exchanger and reactor and UC is the operating (utility) cost for the heating medium. For this problem use $i = 7\%$ p.a. and $n = 12$ years.

The purchase cost of the reactor is given by

$$PC_{reactor} = \$20{,}000V^{0.85} \quad \text{(P14.8.7)}$$

where V is the volume of the reactor in m^3. The cost of the heat exchanger is

$$PC_{exchanger} = \$12{,}000\{A[m^2]\}^{0.57} \quad \text{(P14.8.8)}$$

where A is the area of the heat exchanger in m^2. The cost of the heating medium is

$$UC[\$/h] = \$5\times10^{-6}Q[kJ/h] \quad \text{(P14.8.9)}$$

Present your final results as two plots. The first should show how each term in Equation (P14.8.6) changes with $T_{c,2}$, and the second plot should show the *EAOC* (y-axis) as a function of $T_{c,2}$ (x-axis). Explain the reason for the trends seen in each of these plots.

9. A typical binary distillation column and related equipment are shown in Figure P14.9.

   When designing distillation columns, it is common to optimize the column using the reflux ratio, R, as the decision variable. The overall cost of building the column and running it must be considered when optimizing the design. The capital investment (cost) for the equipment includes the cost of the column plus the costs of the reboiler, condenser, reflux drum, and reflux pump. For this project we will consider only the costs of the column and the two heat exchangers (reboiler and condenser). The operating costs include the cost of the cooling medium (water) for the condenser and the heating medium (steam) for the reboiler. Because the equipment costs are one-time purchase costs and the operating costs (steam and water) occur all the time, the time

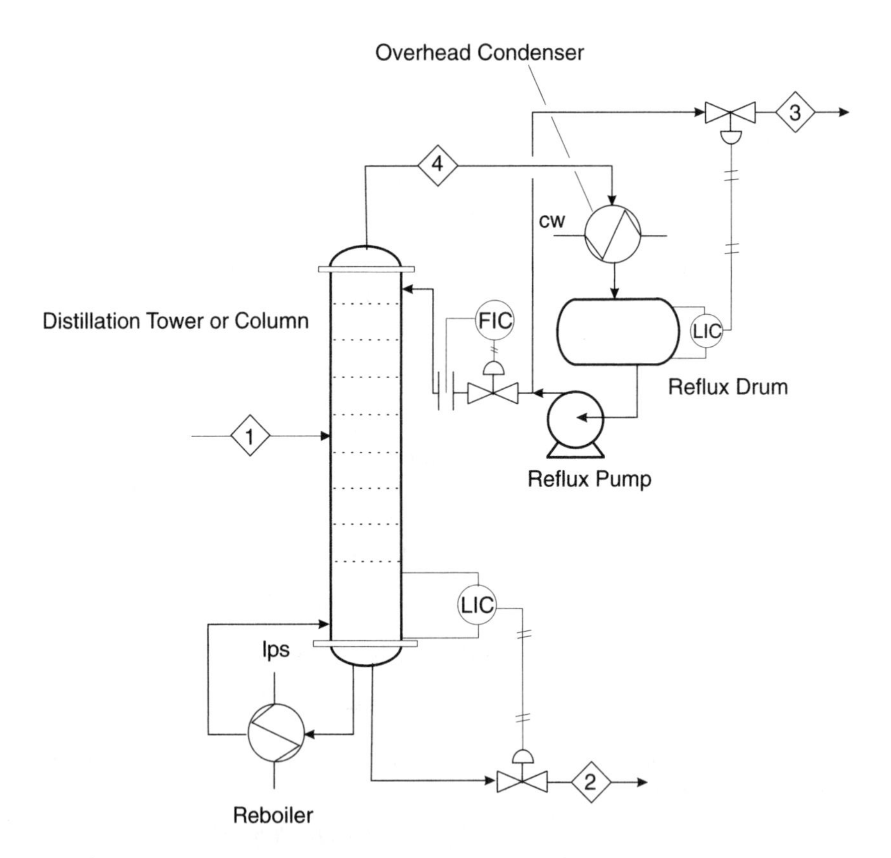

Figure P14.9 Illustration of the Main Pieces of Equipment Used in Distillation

value of money must be taken into consideration when defining an objective function. For this project we want to minimize the equivalent annual operating cost (*EAOC*) of the column that is defined in Equation (P14.9.1):

$$EAOC[\$/y] = \sum_{i=1}^{3} PC_i[\$](A/P,i,n)[1/y] + \sum_{i=1}^{2} UC_i[\$/y] \qquad \text{(P14.9.1)}$$

where PC_i is the purchase cost of the column and heat exchangers, and UC_i is the operating (utility) cost (cooling water and steam). Assume that the effective annual interest rate, i, is 8% p.a. and that the length of the project, n, is 10 years.

It is desired to optimize a distillation column (calculate the value of R that minimizes the *EAOC*) to separate 1000 kmol/h of an equal molar feed of benzene and toluene into a top product containing 99.5 mol% benzene with a recovery of benzene of 98%. The recovery of benzene refers to the ratio of total amount (kmol) of benzene in the top product to that entering in the feed. The column is to operate at a pressure of 1 atm. We define y_i as the mole fraction of benzene in the top stream, and x_i as the mole fraction of benzene in the bottom stream.

The relationship between the reflux ratio, R, and the number of stages or trays in the column, N, is given by the following relationships:

$$N_{min} \quad \frac{\ln\left\{\dfrac{y_i}{x_i}\dfrac{1-x_i}{1-y_i}\right\}}{\ln \alpha} \qquad \text{(P14.9.2)}$$

$$Y = \frac{N - N_{min}}{N + 1} \qquad \text{(P14.9.3)}$$

$$R_{min} = \frac{2}{\alpha - 1} \qquad \text{(P14.9.4)}$$

$$X = \frac{R - R_{min}}{R + 1} \qquad \text{(P14.9.5)}$$

$$Y = 1 - \exp\left\{\left[\frac{1 + 54.4X}{11 + 117.2X}\right]\left[\frac{X - 1}{\sqrt{X}}\right]\right\} \qquad \text{(P14.9.6)}$$

where α is the relative volatility of the benzene with respect to the toluene, which is equal to 2.3 for this system.

The costs of utilities for the overhead condenser (cooling water = CW) and reboiler (steam) are given by

$$UC_{CW}[\$/h] = \$3.54\times10^{-7}V\lambda_V \qquad \text{(P14.9.7)}$$

$$UC_{steam}[\$/h] = 20UC_{CW}[\$/h] \qquad \text{(P14.9.8)}$$

where λ_V is the latent heat of vaporization (kJ/kmol) of benzene at its normal boiling point, and V (kmol/h) is the total vapor flow from the top of the column. The purchase cost of the column is given by

$$PC_{col} = \$10{,}000 V_{col}^{0.85} \quad \text{(P14.9.9)}$$

where $V_{col} = \pi D^2 L/4$ is the volume of the column in m^3 and D[m] and L[m] are the diameter and height of the column, respectively. The diameter and height of the column are given by

$$D[\text{m}] = 0.15(V[\text{kmol/h}])^{0.5} \quad \text{(P14.9.10)}$$

$$L[\text{m}] = N \quad \text{(P14.9.11)}$$

The purchase costs of the condenser and reboiler may be taken as 10% and 20% of the cost of the column, respectively.

Present your final results as two plots. The first should show how each term in Equation (P14.9.1) changes with R/R_{min}, and the second plot should show the $EAOC$ (y-axis) as a function of R/R_{min} (x-axis). Explain the reason for the trends seen in each of these plots.

10. A chemical process recovers heat by transferring energy from a gas stream to a cold brine stream. During the heating of the brine, salts deposit on the heat exchanger surfaces and cause significant fouling. The consequence of this fouling is that the heat transferred to the brine is reduced, causing additional heat, in the form of condensing steam, to be required later on in the process. A portion of a process flow diagram is shown in Figure P14.10.

    It is known that the heat transfer coefficient for a brine/gas heat exchanger changes as the following function of time:

$$U = 75(1 - 0.12t) \text{ W/m}^2{}^\circ\text{C}$$

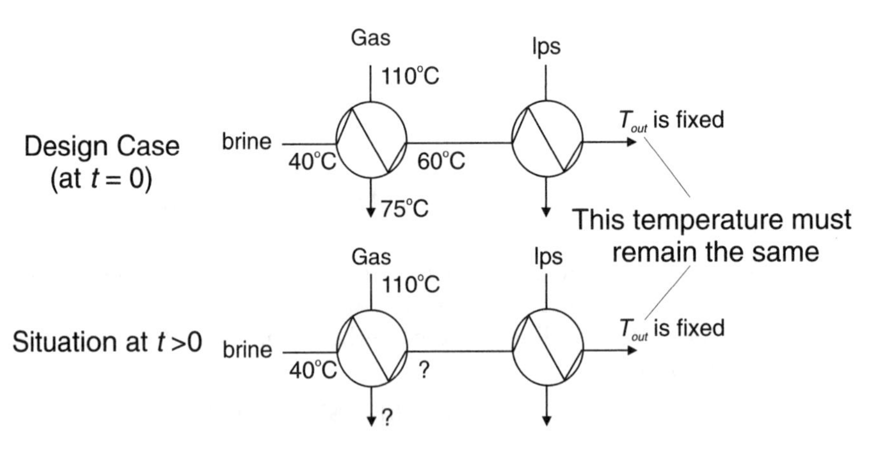

Figure P14.10 Brine Heating Process

where t is the time in months after cleaning the heat exchanger, and U is the overall heat transfer coefficient.

The cost of cleaning the exchanger is $15,000, but there is no downtime for the process. At design conditions (unfouled exchanger surface), the flow of gas (c_p = 800 J/kg°C) is 40,000 kg/h, and the flow of brine (c_p = 4000 J/kg°C) is 14,000 kg/h. The gas enters the heat exchanger at a temperature of 110°C and leaves at 75°C. The brine enters the exchanger at 40°C and leaves at 60°C. The exchanger has an LMTD correction factor F =0.9, which may be assumed not to change during the fouling process. If the flows and inlet temperatures of the streams remain constant, what is the optimal cycle time for cleaning the heat exchanger? Assume that as the brine outlet temperature drops from 60°C due to the fouling of the exchanger surface, additional low-pressure steam at a cost of $8.00/GJ is used to heat the brine back to 60°C.

11. Consider the determination of the optimal cycle time for a batch reactor similar to that covered in Example 14.7. Repeat the analysis using the following parameters:

    $V = 7\ \text{m}^3$
    $C_{A0} = 3\ \text{kmol/m}^3$
    $k = 0.153\ \text{hr}^{-1}$
    $t_{clean} = 1.5\ \text{hr}$
    $x = \$350/\text{kmol}$
    $y = \$200/\text{kmol}$
    $C_{clean} = \$800$

12. Consider the problem given in Example 14.7 for the case when the reaction is 2A→B and the reaction is second order with respect to A. Derive an expression, similar to Equations (14.19) and (14.20), for the optimal cycle time in terms of the relevant parameters.
13. Consider the following batch sequencing problem.

    Two chemicals A and B are to be produced. Each chemical requires the use of a mixer/reactor and a separator unit. The amount of each product made in a batch is the same. The times required to make each chemical along with the value of each chemical are given in the table below.

Process Unit	Process to Make A	Process to Make B
Mixer/Reactor	6 h	8 h
Separator	7 h	6 h
Value of product	$2.3/kg	$4.0/kg
Cost of feed	$0.35/kg	$1.25/kg

Marketing predictions and forecasting suggest that any amount of A can be sold but that a maximum of 250 batches of B can be sold in a year.

a. Ignoring scheduling issues and assuming that both pieces of equipment can be run for all 8760 hours in a year, determine the optimal mix of Products A and B.
b. There is virtually no product storage available in the current plant, and once a batch is made it is sent directly to a tank car and shipped out. Can the answer from Part (a) be accommodated taking into account the scheduling of products without intermediate storage and without using single-product campaigns for A and B (see Chapter 3 for a discussion and an explanation of campaigns)? If the answer is no, then determine the optimum schedule and maximum profit.
c. Would your answers from Parts (a) and (b) change if there were no restriction on the amount of B that could be produced and sold?

14. Consider the problem described in the table below in which Products A, B, and C are made. The amount of each product made per batch is the same.

Product	Time in Reactor (hr)	Time in Separator 1 (hr)	Time in Separator 2 (hr)	Value of Product ($/kg Product)	Cost of Feed ($/kg Product)
A	15	10	8	1.35	0.25
B	7	7	6	0.97	0.27
C	25	5	2	1.41	0.23

At least 50 batches of each product must be produced in a year, and each piece of equipment is available for the same 8000 hours a year.
a. Ignoring scheduling issues, determine the optimal mix of Products A, B, and C.
b. Can the answer from Part (a) be accommodated taking into account the scheduling of products without intermediate storage and without using single-product campaigns? (See Chapter 3 for a discussion and explanation of campaigns.)

15. Perform a parametric optimization for the second distillation column, T-202, in the DME process, Figure B.1.1, using reflux ratio and pressure as the independent variables and *NPV* as the objective function. You should follow the approach used in Section 14.4.2, which considered the first column, T-201, for this process.

16. For the DME process shown in Figure B.1.1, complete a parametric optimization for both columns when the order of separation is reversed. This means that the first column will remove water as a bottoms product and the second column will separate DME from methanol. The approach used in Problem 14.15 should be followed here, with pressure and reflux ratio used as the independent variables and *NPV* as the objective function. Compare the results

from Problem 14.15 with your results from this problem to determine which is the best separation sequence for the DME process.

17. For the DME column (T-201), find the operating pressure at which a switch from medium- to low-pressure steam (for use in the reboiler, E-204) yields an improvement in the *NPV*. Refer to Section 14.4.2 for a discussion of what costs to consider for this problem.

CHAPTER

15

Pinch Technology

15.1 INTRODUCTION

Whenever the design of a system is considered, limits exist that constrain the design. These limits often manifest themselves as mechanical constraints. For example, a distillation of two components that requires 400 equilibrium stages and a tower with a diameter of 20 m would not be attempted, because the construction of such a tower would be virtually impossible with current manufacturing techniques. A combination of towers in series and parallel might be considered but would be very expensive. These mechanical limitations are often (but not always) a result of a constraint in the process design. The example of the distillation column given above is a result of the difficulty in separating two components with similar volatility. When designing heat exchangers and other unit operations, limitations imposed by the first and second laws of thermodynamics constrain what can be done with such equipment. For example, in a heat exchanger, a close approach between hot and cold streams requires a large heat transfer area. Likewise, in a distillation column, as the reflux ratio approaches the minimum value for a given separation, the number of equilibrium stages becomes very large. Whenever the driving forces for heat or mass exchange are small, the equipment needed for transfer becomes large and it is said that the design has a **pinch.** When considering systems of many heat- or mass-exchange devices (called **exchanger networks**), there will exist somewhere in the system a point where the driving force for energy or mass exchange is a minimum. This represents a pinch or pinch point. The successful design of these networks involves defining where the pinch exists and using the information at the pinch point to design the whole network. This design process is designed as pinch technology.

The concepts of pinch technology can be applied to a wide variety of problems in heat and mass transfer. As with other problems encountered in this text, both design and performance cases can be considered. The focus of this chapter is on the implementation of pinch technology to new processes for both heat-exchanger networks (HENs) and mass-exchanger networks (MENs). Retrofitting an existing process for heat or mass conservation is an important but more complicated problem. The optimization of such a retrofit must consider the reuse of existing equipment, and this involves extensive research into the conditions that exist within the process, the suitability of materials of construction to new services, and a host of other issues. By considering the design of a heat- (or mass-) exchange network for existing systems, the solution that minimizes the use of utility streams can be identified and this can be used to guide the retrofit to this minimum utility usage goal.

The approach followed in the remainder of this chapter consists of establishing an algorithm for designing a heat- (mass-) exchanger network that consumes the minimum amount of utilities and requires the **minimum number of exchangers (MUMNE)**. Although this network may not be optimal in an economic sense, it does represent a feasible solution and will often be close to the optimum.

15.2 HEAT INTEGRATION AND NETWORK DESIGN

Even in a preliminary design, some form of heat integration is usually employed. Heat integration has been around in one form or another ever since thermal engineering came into being. Its early use in the process industries was most apparent in the crude preheat trains used in oil refining. In refineries, the thermal energy contained in the various product streams is used to preheat the crude prior to final heating in the fired heater, upstream of the atmospheric column. Because refineries often process large quantities of oil, product streams are also large and contain huge amounts of thermal energy. Even when energy costs were very low, the integration of the energy contained in process streams made good economic sense and was therefore practiced routinely.

The growing importance of heat integration in the chemical process industries (CPI) can be traced to the large increase in the cost of fuel/energy starting in the early 1970s. The formalization of the theory of heat integration and pinch technology has been attributed to several researchers: Linnhoff and Flower [1], Hohmann [2], and Umeda et al. [3]. The approach used here follows that given by Douglas [4], and the interested student is encouraged to study this reference for additional insight into the broader concepts of energy integration.

It was shown in Chapter 2 that, as the PFD evolves, the need to heat and cool process steams becomes apparent. For example, feed usually enters a process from a storage vessel that is maintained at ambient temperature. If the feed is to be reacted at an elevated temperature, then it must be heated. Likewise, after the reaction has taken place, the reactor effluent stream must be purified, which usually requires cooling the stream, and possibly condensing it, prior to separating it. Thus, energy must first be added and then removed from the process. The concept of heat integration, in its simplest form, is to find matches between heat additions and heat removals within the process. In this way, the total utilities that are used to perform these energy transfers can be minimized, or rather optimized. Example 15.1 is presented to give the reader insight into the rationale for heat and energy integration.

Example 15.1

Figure E15.1 shows two configurations for the DME reactor feed and effluent heat exchange system. In both cases the feed enters from the left at 154°C; it is heated to 250°C prior to being fed into the adiabatic catalytic reactor, R-201. The same amount of reaction takes place in both configurations, and the reactor effluent is then cooled to 100°C prior to entering the separation section of the process. The only difference between the two systems is the way in which the heat exchange takes place. In Figure E15.1(a) the feed is heated with high-pressure steam and the effluent is cooled with cooling water. However, this does not make good economic sense. Because heat is generated in the reactor, it would make better sense to use this heat from the reaction to heat the reactor feed. This is what is done in Figure E15.1(b). The reactor effluent is partially cooled by exchanging heat with the cool, incoming feed. Compared with the configuration of Figure E15.1(a), the heat integration saves money in two ways: (1) The cooling water utility is reduced and the high-pressure steam is eliminated, and (2) heat exchanger E-203 is smaller because the duty is reduced, and E-202 is also smaller due to the fact that hps condenses at 254°C, which means the ΔT driving force in the exchanger is very small and the area is large. The economics for the two exchangers and utilities for both cases for a 10% discount rate and a 10-year plant life are summarized in Table E15.1.

Table E15.1 Cost Comparison for Example 15.1

	No Heat Integration	With Heat Integration
Fixed capital investment	\$346,000	\$244,600
Cost of utilities	–\$210,000/yr	–\$36,820/yr
Net present value	–\$1,636,000	–\$471,000

The savings received over the life of the plant by using heat integration are (–471,000 + 1,636,000) = \$1,165,000!

From Example 15.1, it should be apparent that considerable savings are achieved by integrating heat within the process. Rather than try to implement this heat integration on an ad hoc basis, as in the example, a formal way of approaching these types of problems is presented.

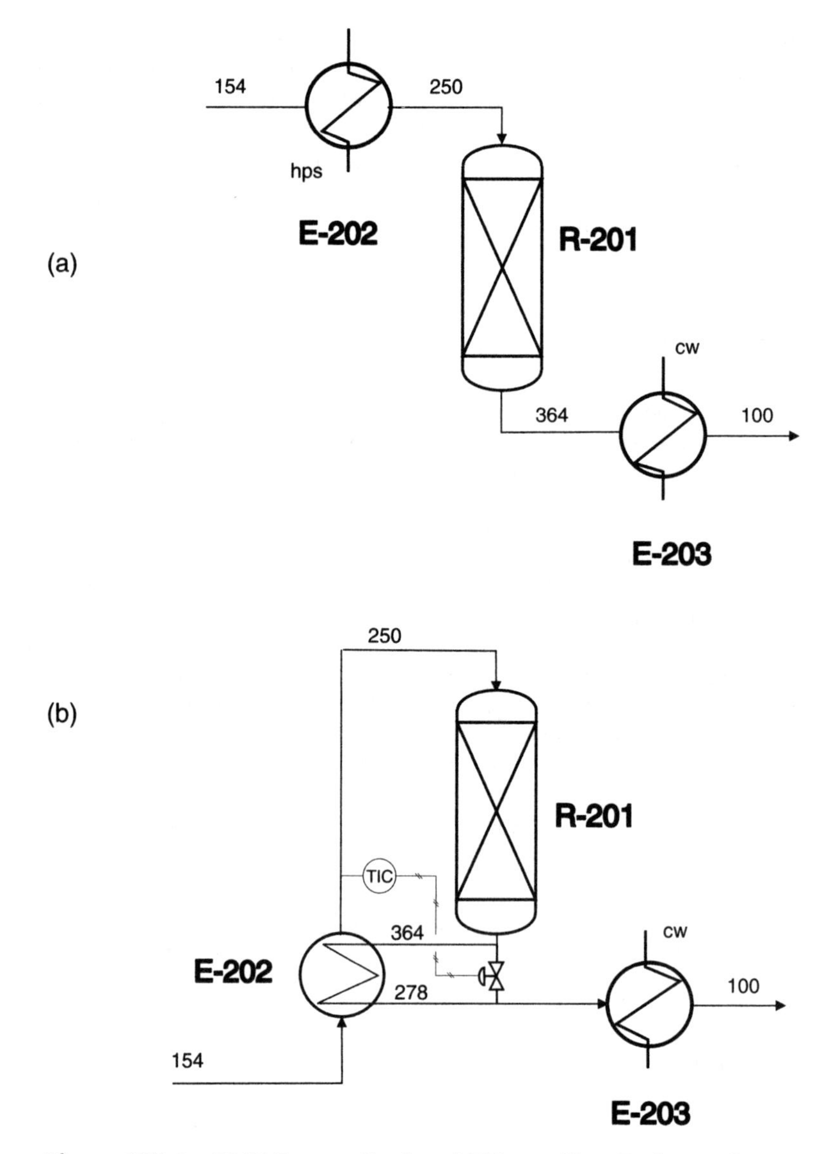

Figure E15.1 DME Reactor Feed and Effluent Heat Exchange System (a) Without Heat Integration, and (b) With Heat Integration

The general algorithm is presented to give the minimum number of exchangers requiring the minimum utility requirements for a given minimum approach temperature. The algorithm to solve the minimum utility (MUMNE) problem consists of the following steps.

1. Choose a minimum approach temperature. This is part of a parametric optimization. See Chapter 14, because for every minimum approach temperature a different solution will be found.
2. Construct a temperature interval diagram.
3. Construct a cascade diagram, and determine the minimum utility requirements and the pinch temperatures.
4. Calculate the minimum number of heat exchangers above and below the pinch.
5. Construct the heat-exchanger network.

It is important to remember that the object of this exercise is to obtain a heat-exchanger network that exchanges the minimum amount of energy between the process streams and the utilities and uses the minimum number of heat exchangers to accomplish this. This network is almost never the optimum economic design. However, it does represent a good starting point for further study and optimization.

Each of the five steps given above is considered in detail as illustrated in Example 15.2.

Example 15.2

In a process, there are a total of six process streams that require heating or cooling. These are listed below along with their thermal and flow data. A stream is referred to as "hot" if it requires cooling, and "cold" if it requires heating. The temperature of the stream is not used to define whether it is "hot" or "cold."

Stream	Condition	Flowrate, $\dot{m}$ (kg/s)	C_p (kJ/kg°C)	$\dot{m}C_p$ (kW/°C)	T_{in} (°C)	T_{out} (°C)	$Q_{available}$ (kW)
1	Hot	10.00	0.8	8.0	300	150	1200
2	Hot	2.50	0.8	2.0	150	50	200
3	Hot	3.00	1.0	3.0	200	50	450
4	Cold	6.25	0.8	5.0	190	290	–500
5	Cold	10.00	0.8	8.0	90	190	–800
6	Cold	4.00	1.0	4.0	40	190	–600
Total							–50

For this system, design the MUMNE network.

It should be noted that the overall heat balance for these streams yields a net enthalpy change of 50 kW. This does not mean that a heat-exchanger network can be designed to exchange heat between the hot and cold streams by receiving only

50 kW from a hot utility. This is because the analysis takes into account only the first law of thermodynamics, that is, an enthalpy balance. In order to design a viable heat-exchanger network, it is also necessary to consider the second law of thermodynamics, which requires that thermal energy (heat) only flow from hot to cold bodies. As a consequence, the utility loads will, in general, be significantly higher than those predicted by a simple overall enthalpy balance.

Step 1: Choose a Minimum Approach Temperature. This represents the smallest temperature difference that two streams leaving or entering a heat exchanger can have. Typical values are from 5°C to 20°C. The value 10°C is chosen for this problem, noting that different results will be obtained by using different temperature approaches. The range of 5°C to 20°C is typical but not cast in concrete. Indeed, any value greater than zero will yield a viable heat-exchanger network. The effect of using different minimum approach temperatures on the economics of the process will be discussed in a later section. But the economic trade-offs are straightforward. As the minimum approach temperature increases, the heat transfer area for the process heat exchangers decreases, but the loads on the hot and cold utilities increase. Therefore, capital investment decreases but the operating costs increase. Methods for estimating the total surface area of the exchanger network and the equivalent annual operating cost of the network are discussed in a later section.

Step 2: Construct a Temperature Interval Diagram. In a temperature interval diagram, all process streams are represented by a vertical line, using the convention that hot streams that require cooling are drawn on the left-hand side and cool streams requiring heating are drawn on the right. The left- and right-hand axes are shifted by the minimum temperature difference chosen for the problem, with the right-hand side being shifted down compared with the left. The temperature interval diagram for Example 15.2 is shown in Figure 15.1. In this figure, each process stream is represented by a vertical line with an arrow at the end indicating the direction of temperature change. Horizontal lines are then drawn through the ends of the lines and divide the diagram into temperature intervals. For our problem, there are four temperature intervals. The net amount of available energy from all the streams in a given temperature interval is given in the right-hand column. The convention of (+) for excess energy and (–) for energy deficit is used. Thus, if the right-hand column contains a positive number for a given temperature interval, this implies that there is more than enough energy in the hot streams to heat the cold streams in that temperature interval. In addition, because the cold streams have been shifted down by the minimum approach temperature, energy can flow from left to right within a given temperature interval without violating the second law of thermodynamics. The summation of the numbers in the right-hand column is the net deficit or surplus enthalpy for all the streams, which for this example is –50 kW.

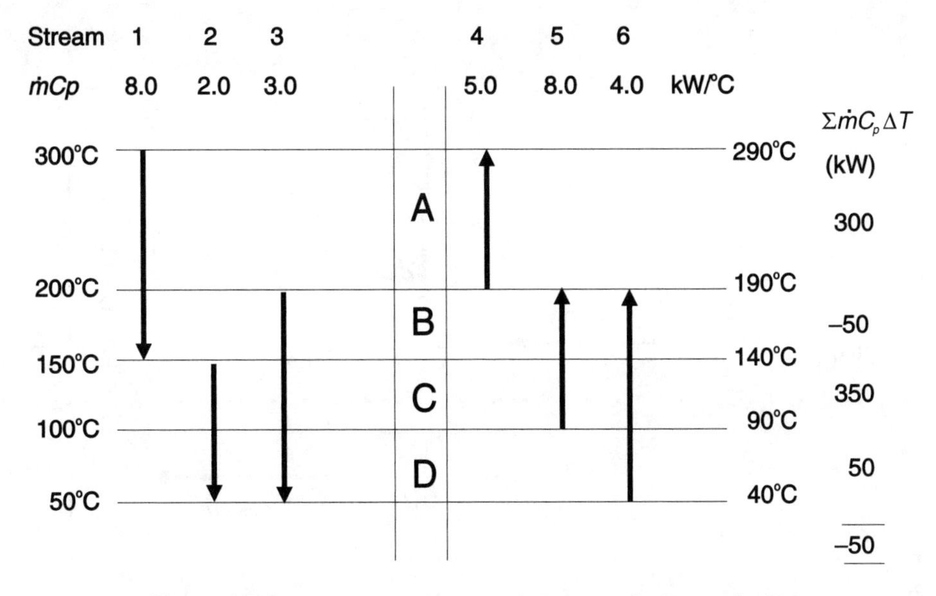

Figure 15.1 Temperature Interval Diagram for Example 15.2

Step 3: Construct a Cascade Diagram. The next step in the MUMNE algorithm involves constructing a cascade diagram like the one illustrated in Figure 15.2. The cascade diagram simply shows the net amount of energy in each temperature interval. Because energy can always be transferred down a temperature gradient, if there is excess energy in a given temperature interval this energy can be **cascaded** down to the next temperature level. It should be noted that for systems in which only thermal energy is being transferred, excess energy cannot be transferred upward to a higher temperature interval. This is a result of the second law of thermodynamics—that transfer of energy up a temperature gradient is possible only if work is done on the system, for example, if a heat pump is used. Energy continues to be cascaded in this manner, and the result is shown in Figure 15.2. From the cascade diagram, it is evident that there is a point in the diagram at which no more energy can be cascaded down, and that energy most often must be supplied from the hot utility to the process. This point is represented by line ab in the diagram. Below this line, the cascading process can continue; but, again, at some point, excess heat must be rejected from the process to the cold utility. The line ab is termed the **pinch zone** or **pinch temperature**. By following the procedure described above, the minimum energy is transferred from the hot utility to the process and from the process to the cold utility. The proof of this is straightforward and is illustrated in Figure 15.3. Consider the situation in Figure 15.2 but now consider an additional transfer of energy, Q, from the hot utility to the process above the pinch. This energy must be cascaded down through the pinch and rejected from the process to the cold utility. This must be true, since a

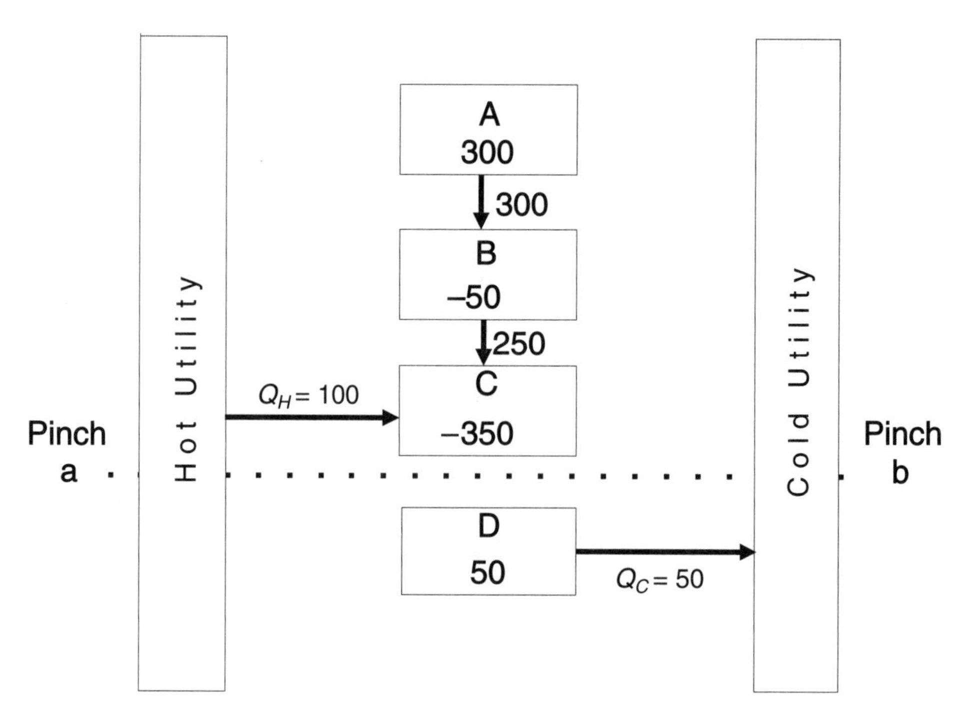

Figure 15.2 The Cascade Diagram for Example 15.2

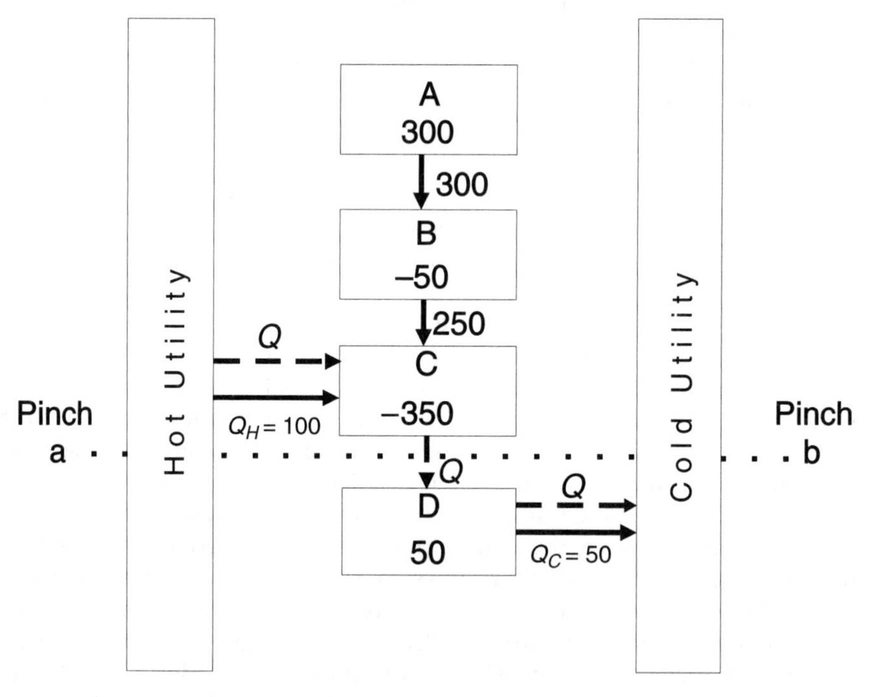

Figure 15.3 The Effect of Transferring Energy through the Pinch

first-law balance on the system requires that the difference between the hot and cold utility demands is constant and equal to the difference in total enthalpies between the hot and cold streams (–50 kW for this example). Therefore, if heat is transferred across the pinch zone, the net result will be that more heat will have to be added from the hot utility and rejected to the cold utility. The criterion of not transferring energy across the pinch zone and cascading energy whenever possible guarantees that the energy requirements to and from the utilities will be minimized. From Figures 15.2 and 15.3, it can be concluded that the minimum utility requirements for this problem are 100 kW from the hot utility and 50 kW to the cold utility, with the net difference equal to –50 kW, which is consistent with the overall enthalpy balance. The pinch temperatures are 100°C and 90°C for the hot and cold streams, respectively.

To minimize the hot and cold utility requirements, energy should not be transferred across the pinch.

At this point, it is important to note that not all problems have a pinch condition. The cascade diagrams for two situations that do not have a pinch are illustrated in Figure 15.4. In Figure 15.4(a), heat is only cascaded downward or rejected to the cold utility. The conditions of the hot and cold streams are such that after cascading energy downward, there is either an excess of energy or the energy is exactly balanced in every temperature interval. In this situation, there is no need to supply energy from the hot utility to the process. In Figure 15.4(b), the opposite situation exists. In this case, heat is only cascaded downward or supplied from the hot utility. The conditions of the hot and cold streams are such that after cascading energy downward, there is either an energy deficit, or the energy is exactly balanced in every temperature interval. In this situation, there is no need to reject energy from the process to the cold utility. Although much of the remaining discussion focuses on processes that have a pinch, the approach for designing the MUMNE network remains essentially the same.

Step 4: Calculate the Minimum Number of Heat Exchangers. Once the pinch temperatures have been found from Step 3, it is necessary to find the minimum number of heat exchangers required to carry out the heat transfer for the minimum utility design. From this point on, the heat transfer problem will be split into two and we will consider above and below the pinch as separate systems.

Above the Pinch. The easiest way to evaluate the minimum number of heat exchangers required is to draw boxes representing the energy in the hot and cold process streams and the hot utility as shown at the top of Figure 15.5. Energy is now transferred from the hot streams and hot utility to the cold streams. These

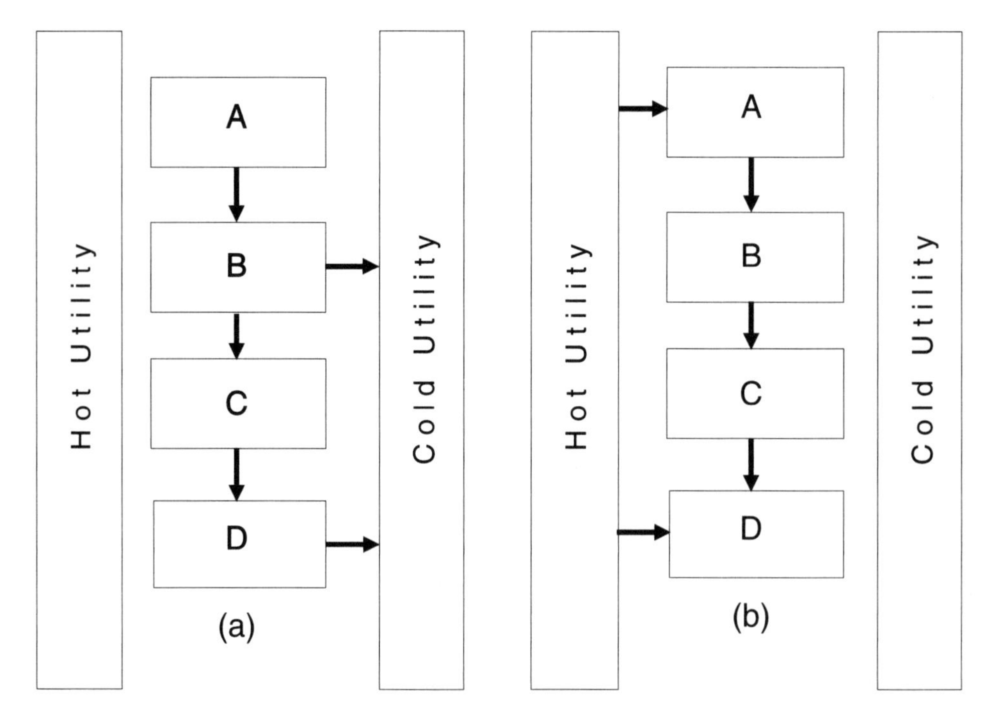

Figure 15.4 Examples of Cascade Diagrams for Problems without a Pinch

energy transfers are indicated by lines, with the amount of energy transferred shown to the side of the lines. Clearly, all the energy in the hot streams and utilities must be transferred to the cold streams. For each line drawn, one heat exchanger is required; thus, by minimizing the number of lines the number of heat exchangers is minimized. It should be pointed out that although the number of heat exchangers equals the number of connecting lines, the lines drawn at this stage may not represent actual heat exchangers. The actual design of exchanger network is covered in Step 5.

Below the Pinch. The same method is used to calculate the minimum number of exchangers below the pinch. The diagrams for above and below the pinch are shown in Figure 15.5, and from this it can be seen that five exchangers are required above the pinch and three below the pinch, or a total of eight heat exchangers for the entire network.

From Figure 15.5, it can be seen that above the pinch, the problem is split into two subproblems. This split is possible because the energy in two of the hot streams (Streams 2 and 3) exactly matches the energy requirement of one of the cold streams (Stream 6). Below the pinch, such a partition of the problem is not

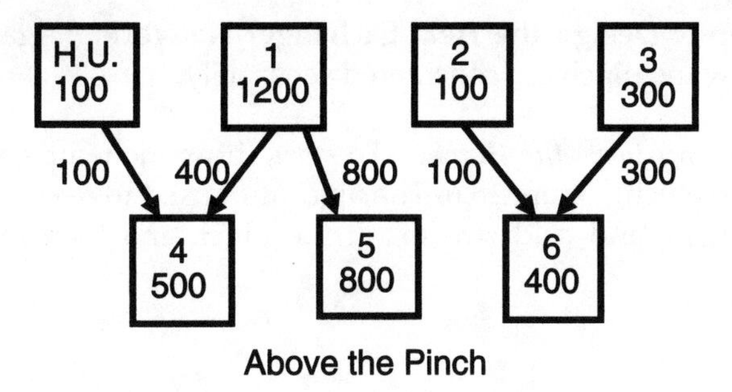

Above the Pinch

Minimum Number of Exchangers, $N_{min,a} = 5$

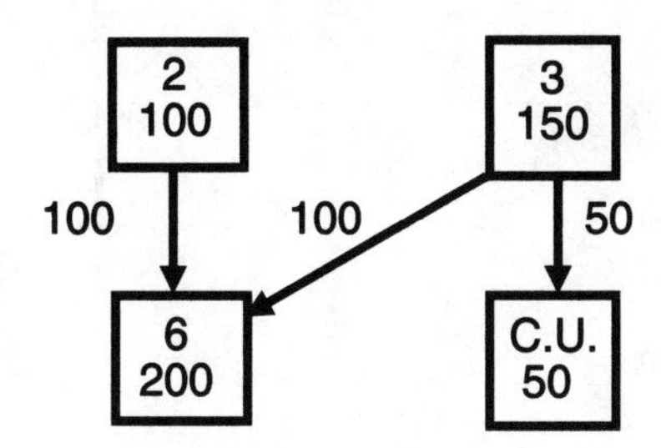

Below the Pinch

Minimum Number of Exchangers, $N_{min,b} = 3$

Figure 15.5 Calculation of Minimum Number of Exchangers for Example 15.2

possible. Such exact matches between groups of hot and cold streams are often not possible, and if such matches do not exist, then for a given problem, above or below the pinch, the following relationship can be written:

$$\text{Min. No. of exchangers} = \text{No. of hot streams} + \text{No. of cold streams} + \text{No. of Utilities} - 1 \quad (15.1)$$

The result in Equation 15.1 is consistent with the result in Figure 15.5 for below the pinch.

For systems without a pinch, the same approach for the minimum number of exchangers as described above can be used, but the problem no longer needs to be split into two parts. For the case considered here, assuming that the problem did not have a pinch (and hence only one utility is required) and that Equation 15.1 is applicable, we get

$$\text{Minimum number of exchangers} = 3 + 3 + 1 - 1 = 6$$

Step 5: Design the Heat-Exchanger Network. Again, the systems above and below the pinch are considered separately.

Design Above the Pinch. To start, draw the temperature interval diagram above the pinch, Figure 15.6. The algorithm used to design the network starts by matching hot and cold streams at the pinch and then moving away from the pinch.

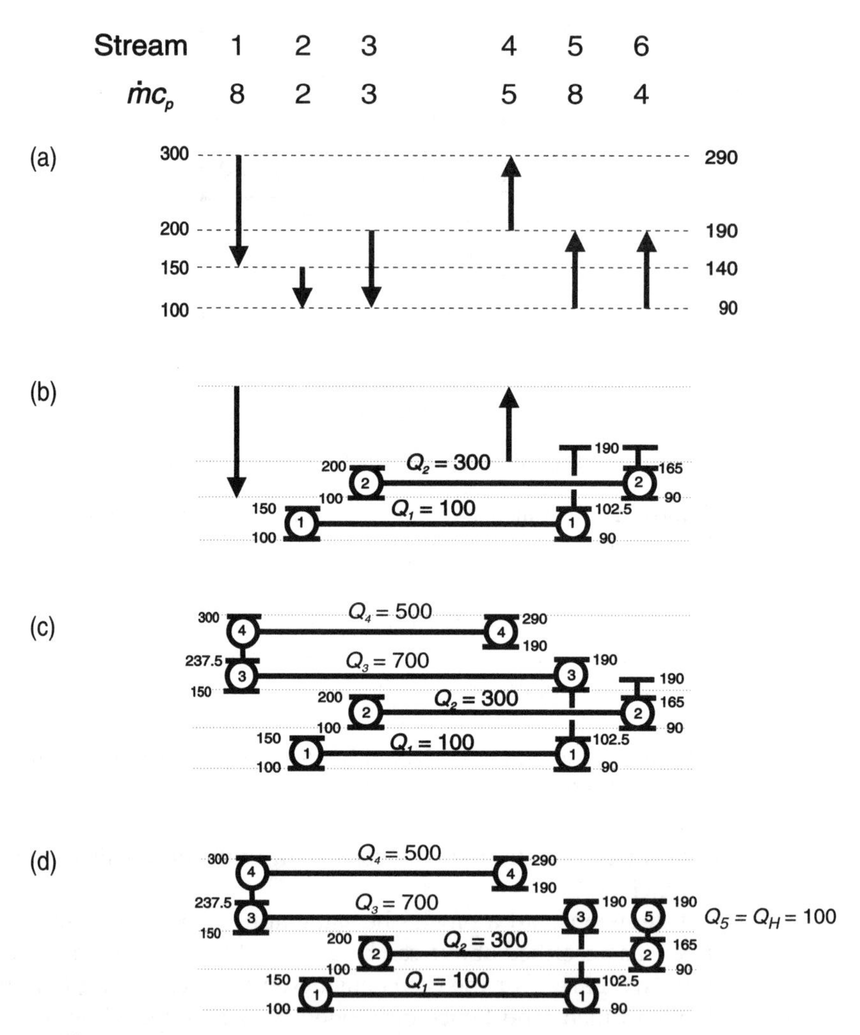

Figure 15.6 Design of Heat-Exchanger Network above the Pinch Zone for Example 15.2

Design at the Pinch. For streams at the pinch, match streams such that $\dot{m}c_{p,hot} \leq \dot{m}c_{p,cold}$. Using this criterion ensures that ΔT_{min} from Step 1 is not violated. Figure 15.6(a) shows that Stream 2 or 3 can be matched with Stream 5 or 6. Note that, for this step, only streams that are present at the pinch temperature are considered. The next step is to transfer heat from the hot streams to the cold streams by placing heat exchangers in the temperature diagram. This step is shown in Figure 15.6(b). There are two possibilities when matching streams at this point: Exchange heat between Streams 2 and 5 and between Streams 3 and 6, or exchange heat between Streams 2 and 6 and between Streams 3 and 5. Only the first combination is shown in Figure 15.6(b). A heat exchanger is represented by two circles connected by a solid line; each circle represents a side (shell or tube) of the exchanger. It is important to exchange as much heat between streams as possible. The temperatures of the streams entering and leaving the exchangers are calculated from an enthalpy balance. For example, consider the enthalpy change of Stream 5 as it passes through Exchanger 1. The total heat transferred is $Q_1 = 100$ kW and $\dot{m}c_{p,5} = 8$; therefore, the temperature change for Stream 5, $\Delta T_5 = 100/8 = 12.5$°C. Thus, the temperature change for Stream 5 through Exchanger 1 is 90°C to 102.5°C, as shown in Figure 15.6(b).

Design Away from the Pinch. The next step is to move away from the pinch and look at the remaining hot and cold streams. There are several ways in which heat may be exchanged from Stream 1 (the only remaining hot stream) and the three cold Streams 4, 5, and 6. The criterion for matching streams at the pinch does not necessarily hold away from the pinch; however, we should make sure that when the network is designed the following constraints are not violated.

1. A minimum approach temperature of 10°C, set in Step 1, is used throughout the design.
2. Only five exchangers are used for the design above the pinch, as calculated in Step 3.
3. Heat is added from the coolest possible source. (This is explained in more detail in the section on problems with multiple utilities.)

The matching of streams and the final design of the network are shown in Figures 15.6(c) and 15.6(d). From these figures, it is clear that there is a design with five heat exchangers, the minimum approach temperature is nowhere less than 10°C, and that heat is added to the process at the lowest temperature consistent with this system, that is, 190°C.

Design Below the Pinch. The approach for the design below the pinch is similar to that described for above the pinch. Start at the pinch and match streams, and then move away from the pinch and match the remaining streams. The temperature interval diagram below the pinch is shown in Figure 15.7(a).

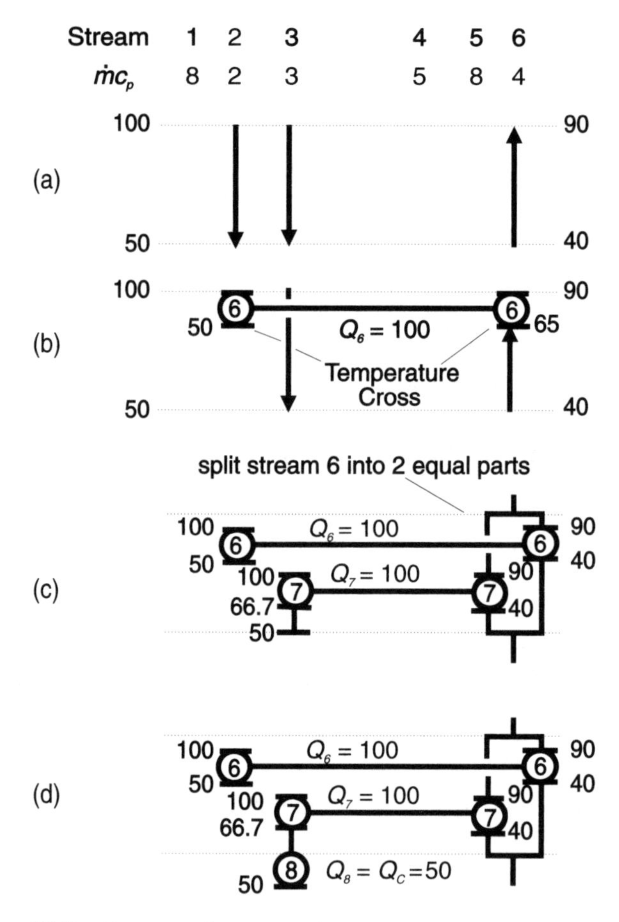

Figure 15.7 Design of Heat-Exchanger Network below the Pinch Zone for Example 15.2

Design at the Pinch. For streams at the pinch, match streams such that $\dot{m}c_{p,hot} \geq \dot{m}c_{p,cold}$. Using this criterion ensures that ΔT_{min} from Step 1 is not violated. Figure 15.7(a) shows that for the three streams at the pinch this criterion cannot be met. This problem can be overcome by splitting the cold Stream 6 into two separate streams. However, before considering this, examine what happens by trying to match streams that violate the above condition. In Figure 15.7(b), Stream 2 is matched with Stream 6. The net result is impossible because it would cause a temperature cross in the exchanger, and that violates the second law of thermodynamics. To maintain the minimum temperature approach set in Step 1, Stream 6 is split into two equal streams, each having an $\dot{m}c_p = 2.0$. These split streams can now be matched with hot Streams 2 and 3 without violating the criterion above.

The net result is shown in Figure 15.7(c), from which it can be seen that the minimum temperature difference is always greater than or equal to 10°C. It should be noted that Stream 6 could be split in a number of ways to yield a viable solution. For example, it could be split into streams with values of 3 and 1.

Design Away from Pinch. The final step is shown in Figure 15.7(d), where the third exchanger is added to transfer the excess heat to the cold utility.

The final heat-exchanger network is shown in Figure 15.8. The exchangers are represented by single circles, with fluid flowing through both sides. This network has the minimum number of heat exchangers, eight, for the minimum utility requirements, $Q_H = 100$ kW and $Q_C = 50$ kW, using a minimum approach temperature, $\Delta T = 10°C$. As mentioned earlier, different results will be obtained by using different minimum approach temperatures.

To this point, the emphasis has been on the topology of the exchanger network, that is, the interaction between the different streams required to give the minimum utility case. In order to complete the design, it is necessary to estimate the heat transfer area of the exchanger network, which in turn allows a capital cost estimate to be made. If heat transfer coefficients for the heat exchangers in Figure 15.8, including appropriate fouling coefficients, are known, then the size

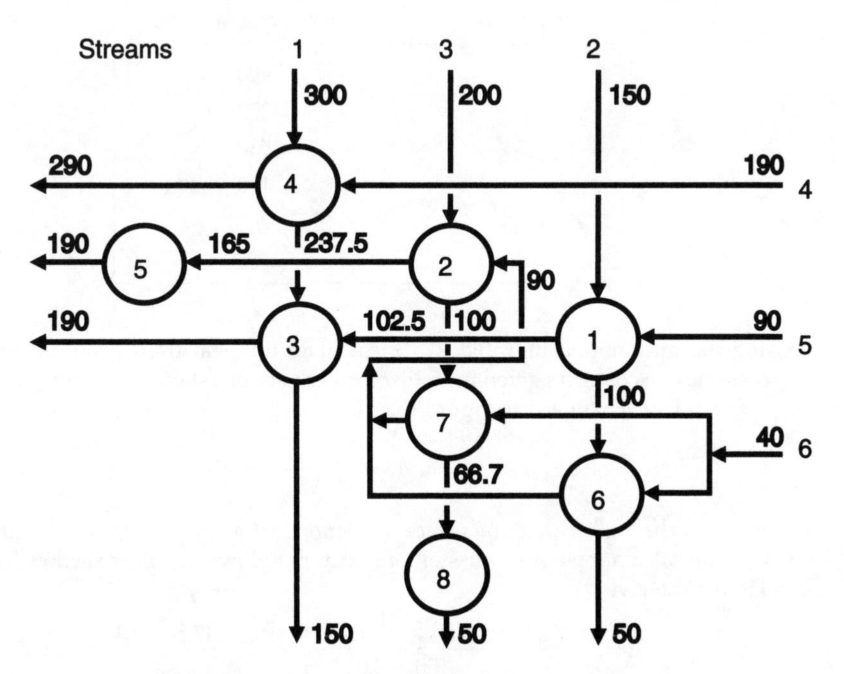

Figure 15.8 MUMNE Network for Example 15.2

of the exchangers can be calculated. Film heat transfer coefficients may be estimated using typical heat transfer coefficients from Table 11.11, previous data on similar streams, published data on similar streams, or values calculated from standard methods. For example, the Seider-Tate equation [5] could be used for the tube side and the Donahue equation [6] for the shell side, along with the fluid properties and typical fluid velocities. The heat transfer areas calculated from such estimates will give a good indication of the heat transfer areas required for the overall exchanger network. Such data are suitable for cost estimates using the CAPCOST program. This procedure is illustrated in Examples 15.3 and 15.4.

Example 15.3

An estimate of the individual film heat transfer coefficients for the hot and cold streams in Example 15.2 are given in Table E15.3. These film coefficients include appropriate fouling factors for the streams.

Table E15.3 Film Heat Transfer Coefficients for Streams in Example 15.2

Stream	Film Heat Transfer Coefficient, W/m²°C
1	400
2	270
3	530
4	100
5	250
6	80

Using the information in Table E15.3, calculate the heat transfer areas for the process-process heat exchangers given in the exchanger network shown in Figure 15.8.

Consider exchanger 1 in Figure 15.6. The design equation is given as

$$A_1 = \frac{Q_1}{U_1 \Delta T_{lm,1} F_1}$$

where F_1 is the log-mean temperature correction factor for exchanger 1. Assume a value of 0.8 for the calculations. A discussion of F factors follows in a later section. The data given in Figure 15.6 give

$$\Delta T_{lm,1} = \frac{(150 - 102.5) - (100 - 90)}{\ln\frac{150 - 102.5}{100 - 90}} = \frac{47.5 - 10}{\ln\frac{47.5}{10}} = 24.1°C$$

$$U_1 = \frac{1}{\frac{1}{270} + \frac{1}{250}} = 129.8 \text{ W/m}^2{}^\circ\text{C}$$

Substituting this information in the design equation gives the following:

$$A_1 = \frac{100 \times 10^3}{(130)(24.1)(0.80)} = 40.0 \text{ m}^2$$

This procedure can be repeated for all the process heat exchangers in the network, for example, exchangers 1, 2, 3, 4, 6, and 7. These results are summarized in the following table.

Exchanger	ΔT_{lm} (°C)	U (W/m²°C)	Q (kW)	F	A (m²)
1	24.1	129.8	100	0.8	40.0
2	20.0	69.5	300	0.8	270.3
3	47.5	153.8	700	0.8	119.7
4	24.1	80.0	500	0.8	324.6
6	10.0	61.7	100	0.8	202.5
7	17.0	69.5	100	0.8	105.8
Total					1063.0

Example 15.4

For Example 15.3, estimate the heat-exchanger area needed for the utility exchangers, that is, exchangers 5 and 8 in Figure 15.8.

Exchanger 5

For this exchanger, the temperature of the cold stream, Stream 6, must be increased from 165°C to 190°C. This requires a hot utility such as steam. Clearly, the correct pressure of steam to use is the lowest one available that gives a temperature driving force of at least 10°C (for this case). A review of Table 8.4 shows that only high-pressure steam is hot enough to heat Stream 6. Assume a film coefficient of 2000 W/m²°C for condensing steam. In addition, assuming that no desuperheating or subcooling occurs, the steam condenses at a constant temperature, and therefore the F correction factor is equal to 1. Hence,

$$\Delta T_{lm,5} = \frac{(255 - 190) - (255 - 165)}{\ln\frac{(255 - 190)}{(255 - 165)}} = \frac{-25}{\ln\frac{65}{90}} = 76.8^\circ\text{C}$$

$$U_5 = \frac{1}{\frac{1}{2000} + \frac{1}{80}} = 76.9\ \mathrm{W/m^2{}^\circ C}$$

$$A_5 = \frac{Q_5}{U_5 \Delta T_{\mathrm{lm},5} F_5} = \frac{100 \times 10^3}{(76.9)(76.8)(1.0)} = 16.9\ \mathrm{m^2}$$

Exchanger 8

For this exchanger, the temperature of the hot stream must be decreased from 66.7°C to 50°C. This can be accomplished using cooling water. Assume that cooling water is available at 30°C and must be returned at 40°C, and a typical film heat transfer coefficient for cooling water (including a scaling or fouling factor) is assumed to be 1000 W/m^2°C. Therefore,

$$\Delta T_{\mathrm{lm},8} = \frac{(66.7 - 40) - (50 - 30)}{\ln \frac{66.7 - 40}{50 - 30}} = \frac{6.7}{\ln 1.335} = 23.2^\circ\mathrm{C}$$

$$U_8 = \frac{1}{\frac{1}{530} + \frac{1}{1000}} = 346\ \mathrm{W/m^2{}^\circ C}$$

$$A_8 = \frac{Q_8}{U_8 \Delta T_{\mathrm{lm},8} F_8} = \frac{50 \times 10^3}{(346)(23.2)(0.8)} = 7.8\ \mathrm{m^2}$$

Total heat transfer area for the entire heat-exchanger network = 7.8 + 16.9 + 1063.0 = 1087.7 m^2.

At this stage, heat-exchanger areas for the network can be combined with the cost of purchasing utilities to give a reasonably complete economic analysis. However, if the effect of the minimum temperature approach is to be investigated, then the complete algorithm outlined in Steps 1–5 and Examples 15.2, 15.3, and 15.4 must be applied repeatedly for different minimum temperature approaches. It should be noted that Steps 1–4 can be readily programmed, but the algorithm for matching streams and exchanging energy outlined in Step 5 is not unique and is not easily computed. Rather than complete the design of the network as given in Step 5, it is possible to estimate approximately the heat transfer area for the network and therefore investigate the effect of changing temperature on the overall cost. The algorithm for estimating the heat transfer area is discussed in the next section.

15.3 COMPOSITE TEMPERATURE-ENTHALPY DIAGRAM

Physically, the pinch zone, previously mentioned, represents a point in the heat-exchanger network at which at least one heat exchanger, or two streams, will have the minimum approach temperature set in Step 1. This point is perhaps more clearly shown in a composite enthalpy-temperature diagram. This diagram is essentially the same as the combination of all the *T-Q* diagrams for all the exchangers in the network. Such a diagram for this example is shown in Figure 15.9 and is constructed by plotting the enthalpy of all the hot streams and all the cold stream as a function of temperature:

Hot Streams

Temperature Interval	Temperature (°C)	Enthalpy of Hot Streams in Temperature Interval (kW)	Cumulative Enthalpy of Hot Streams (kW)
	50	0	0
D			
	100	(2 + 3)(100 – 50) = 250	250
C			
	150	(2 + 3)(150 – 100) = 250	500
B			
	200	(8 + 3) (200 – 150) = 550	1050
A			
	300	(8) (300 – 200) = 800	1850

Cold Streams

Temperature Interval	Temperature (°C)	Enthalpy of Cold Streams in Temperature Interval (kW)	Cumulative Enthalpy of Cold Streams (kW)
	40	0	0
D			
	90	(4)(90 – 40) = 200	200
C			
	140	(8 + 4)(140 – 90) = 600	800
B			
	190	(8 + 4) (190 – 140) = 600	1400
A			
	290	(5) (290 – 190) = 500	1900

These data are plotted in Figure 15.9 as curves 1 and 2, respectively. From Figure 15.1, it can be seen that these two curves cross, which is physically impossible. This situation can be remedied by shifting the cold curve to the right. Curve 3 represents the case when the composite cold stream is shifted 50 kW to the right. For this case, it can seen that there exists a point at which the vertical distance between the hot and cold stream curves is at a minimum. This is the pinch zone. The case shown in curves 1 and 3 corresponds to the situation when the minimum approach temperature between the streams is 10°C, the value chosen in Step 1. Comparing this curve (Curve 3) and the hot stream curve (Curve 1), it can be seen that the horizontal distances between the left- and right-hand ends of the curves represent the

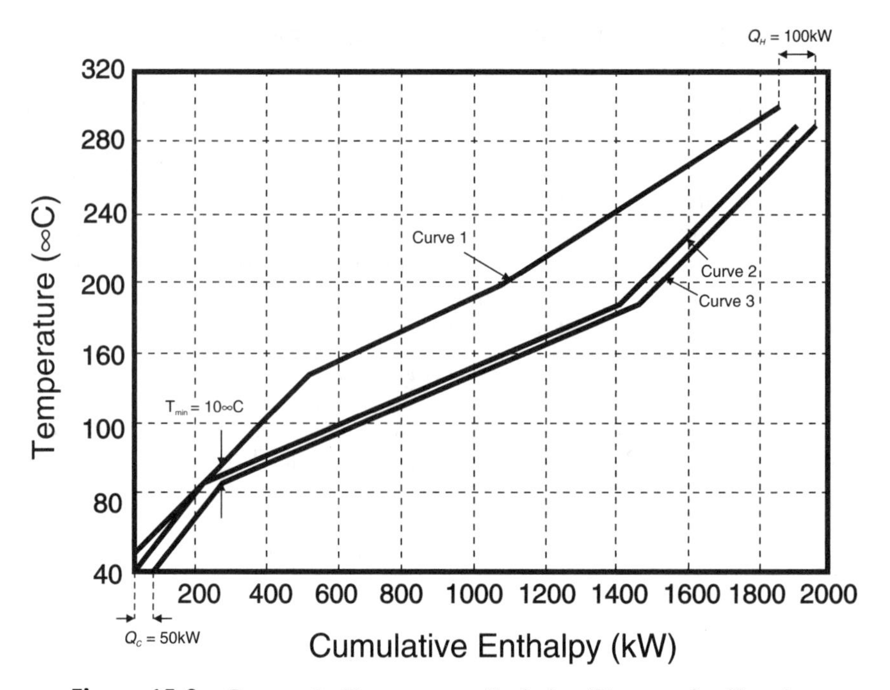

Figure 15.9 Composite Temperature-Enthalpy Diagram for Showing Effect of Changing Minimum Temperature Approach

total duties from the cold and hot utilities, respectively, for the minimum utility case. Thus, the information shown in Figure 15.9 is consistent with the temperature interval diagram and the cascade diagram. In fact, it is exactly the same information but presented in a different way. From Figure 15.9, it should be clear that as the minimum approach temperature decreases so do the minimum utility requirements. However, the temperature driving force within the heat-exchanger network will also decrease, and this will require larger, more expensive heat transfer equipment. Clearly, the optimum network must balance exchanger capital investment against operating (utility) costs.

15.4 COMPOSITE ENTHALPY CURVES FOR SYSTEMS WITHOUT A PINCH

It was shown earlier that some problems do not possess a pinch condition. For these cases, only one utility (hot or cold) is required. This situation is illustrated on the temperature enthalpy diagrams in Figure 15.10. In Figure 15.10(a), the typical heat-exchanger network problem containing a pinch is shown. If the minimum approach temperature is reduced, then the composite enthalpy curve for the cold streams (at the bottom) moves to the left and becomes closer to the hot stream composite enthalpy curve. For the problem shown in Figure 15.10(b),

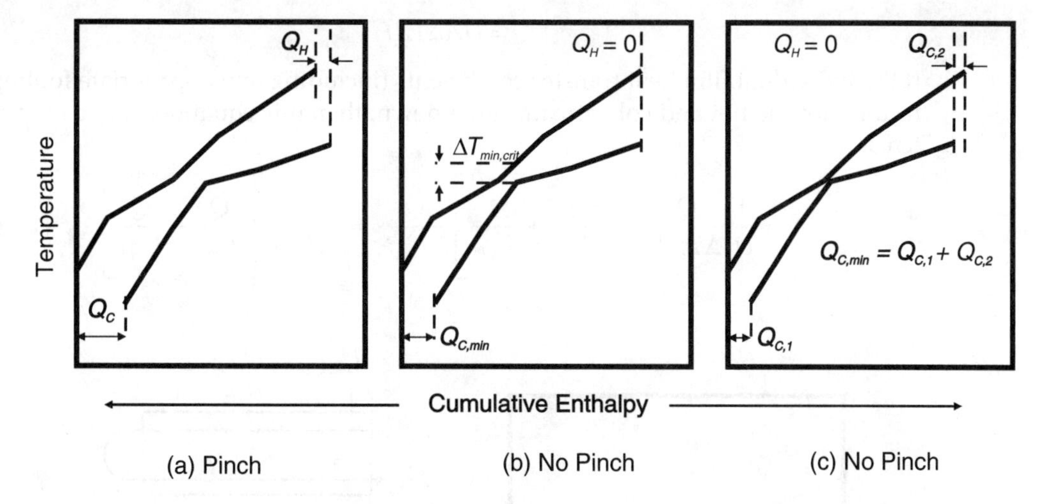

Figure 15.10 Composite Temperature-Enthalpy Diagram Showing a Problem without a Pinch Point

there exists a critical value for the minimum temperature approach, $\Delta T_{min,crit}$, at which the hot utility requirement (Q_H) becomes zero and the cold utility requirement becomes a minimum, $Q_{C,min}$. If the minimum approach temperature is reduced below this critical value, as in Figure 15.10(c), the hot stream composite energy curve overlaps the cold stream curve at both ends. Heat must be rejected to the cold utility at the high temperature end (right-hand side of the figure), as well as at the low temperature end (left-hand side of the figure). However, the combined cold utility duty remains at $Q_{C,min}$. Therefore, below $\Delta T_{min,crit}$, the cold utility duty is constant and the hot utility duty is zero. Above $\Delta T_{min,crit}$, both the hot and cold utility duties increase with increasing ΔT_{min}.

It can be seen that a similar case to that described above exists when the cold stream composite enthalpy curve can overlap the hot stream curve at both ends. For this case, a critical value of ΔT_{min} exists for which the cold utility duty is zero. Above this critical value, both utilities are required, but below it, no cold utility is needed, and the hot utility duty remains constant, but heat must be added to the system at both the hot and cold ends of the diagram.

15.5 USING THE COMPOSITE ENTHALPY CURVE TO ESTIMATE HEAT-EXCHANGER SURFACE AREA

As stated earlier, the composite enthalpy curve can be used to estimate the heat-exchanger surface area of the network. For the case of a single heat exchanger, it is convenient to plot the enthalpy-temperature relationships on a diagram such as Figure 15.11(a). The basic rating equation used for preliminary heat exchanger design is

$$Q = UA\Delta T_{lm}F$$

If the individual film heat transfer coefficients (including any appropriate fouling factors) for the hot and cold streams are known, then this equation can be rewritten as

$$A = \frac{1}{U}\frac{Q}{\Delta T_{\mathrm{lm}}F} = \left[\frac{1}{h_{hot}} + \frac{1}{h_{cold}}\right]\frac{Q}{\Delta T_{\mathrm{lm}}F} = \frac{1}{\Delta T_{\mathrm{lm}}F}\left[\frac{Q}{h_{hot}} + \frac{Q}{h_{cold}}\right] \qquad (15.2)$$

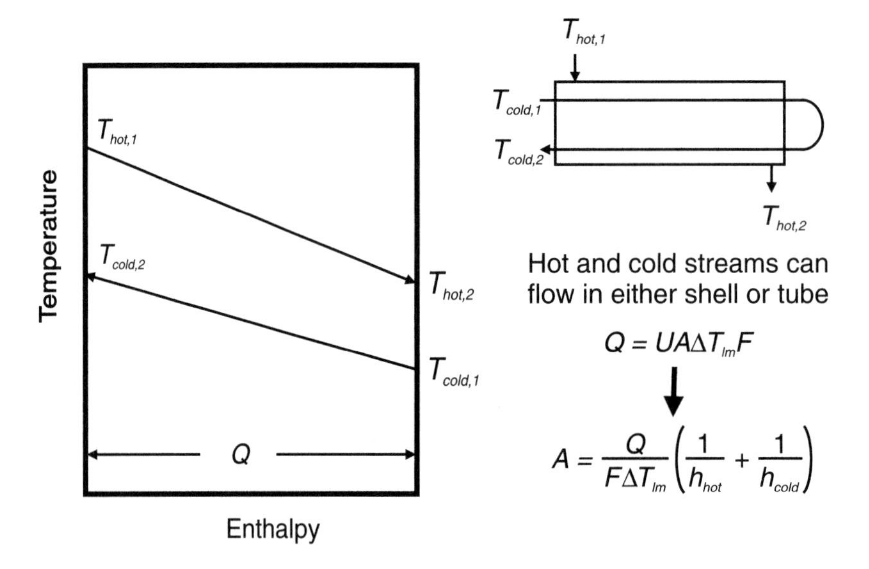

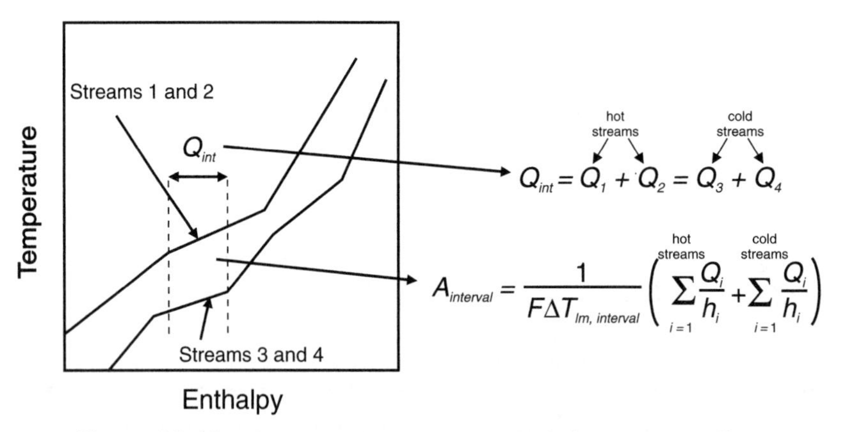

Figure 15.11 Composite Temperature-Enthalpy Diagram Showing a Problem without a Pinch Point

The application of Equation 15.2 was demonstrated in Examples 15.3 and 15.4.

Consider now a portion of the composite enthalpy curve shown in Figure 15.11(b). This curve is made up of two hot streams (1 and 2) and two cold streams (3 and 4). Both the hot streams and both the cold streams have the same temperature change in the interval. However, in general, the enthalpy in either of the hot streams is not the same as the enthalpy in either of the cold streams. Instead, the sum of the enthalpies in the hot streams is equal to that in the two cold streams. This means that different streams cannot be matched in this temperature (enthalpy) interval, because the enthalpy of individual hot and cold streams will be different. However, an estimate of the heat transfer area required to transfer all the energy in the temperature interval can be made by applying Equation 15.2 to both the hot and cold streams in the interval [2,7]. This result may be generalized as follows:

$$A_{\text{interval}} = \frac{1}{\Delta T_{\text{lm,interval}} F} \left[\sum_{i=1}^{\text{hot streams}} \frac{Q_i}{h_i} + \sum_{i=1}^{\text{cold streams}} \frac{Q_i}{h_i} \right] \tag{15.3}$$

where the summation is made for all streams (hot and cold) contained in the interval, and Q_i is the enthalpy change of the stream in the interval.

Clearly, the composite enthalpy diagram can be divided into intervals in which the number of hot and cold streams is constant. The total heat transfer area for the network may be estimated by adding the areas for each interval obtained from Equation 15.3. Ahmad et al. [8] state that the result in Equation 15.3 should predict the minimum area required for a given minimum temperature approach to within about 10%, as long as the film heat transfer coefficients do not differ by a factor of 10 or more. This estimate of accuracy seems to be somewhat optimistic, and a more realistic number might be ±25%. Nevertheless, when looking for the optimum temperature approach, the method used above should give reasonable predictions. Example 15.5 illustrates the methodology to estimate heat transfer areas using this technique.

Example 15.5

For the problem introduced in Example 15.1, and using the film heat transfer coefficients from Examples 15.3 and 15.4, estimate the heat transfer area for the proposed network. Compare the result to that obtained in Example 15.4.

The composite enthalpy diagram for this problem is shown in Figure E15.5. This is the same figure as Figure 15.9, except that temperatures and enthalpies for the different segments have been added. Within each segment, the composite enthalpy curves for the hot and cold streams are straight lines, which implies that the number of streams in each segment is constant because the slope of the line is $1/\Sigma \dot{m} c_p$. In order to generate this information, heat balances must be applied along with the information in the temperature interval diagram, Figure 15.1. For example, consider Segment 1 of Figure E15.5. The hot stream temperature at the border of Segments 1 and 2 must be computed (this is not

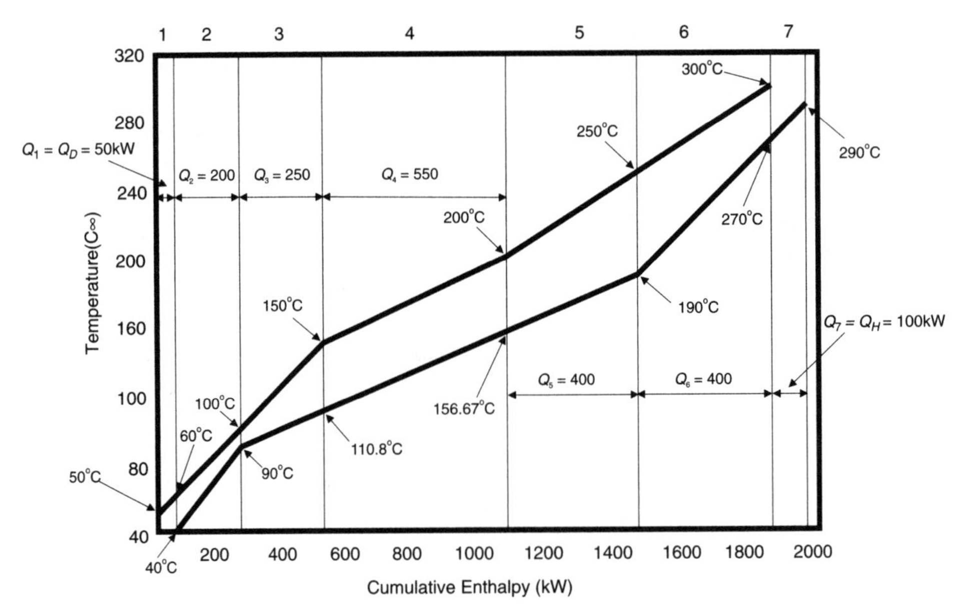

Figure E15.5 Composite Temperature-Enthalpy Diagram for Example 15.5 (Minimum Temperature Approach = 10°C and All Enthalpies Are Given in kW)

shown on the original composite temperature-enthalpy diagram, Figure 15.9). Because Q_1 is the cold utility duty and is equal to 50 kW, the temperature for the composite hot stream is given by the following simple enthalpy balance:

$$\sum_{i=\text{hot streams}} (\dot{m}c_p)_{i,1}\,\Delta T_1 = Q_1$$

Figure 15.1 shows that Streams 2 and 3 lie in the temperature range of Segment 1 (>50°C and <100°C). Therefore,

$$\sum_{i=\text{hot streams}} (\dot{m}Cp)_{i,1}\Delta T_1 = (2000 + 3000)\Delta T_1 = Q_1 = 50{,}000 \text{ W}$$

$$\Delta T_1 = \frac{50{,}000}{5000} = 10°\text{C}$$

Thus, the composite hot stream temperature at the border of Segments 1 and 2, $T_{hot,1\text{-}2} = 50 + 10 = 60°\text{C}$.

In a similar way, the temperature of the composite cold streams between Segments 3 and 4 can be calculated as

$$\sum_{i=cold\ streams} (\dot{m}Cp)_{i,3}\,\Delta T_3 = Q_3$$

From Figure 15.1, Streams 5 and 6 lie in the temperature range (>90°C and <190°C). Therefore, the composite cold stream temperature at the border of Segments 3 and 4 is given by

$$\sum_{i=cold\ streams} (\dot{m}Cp)_{i,3}\Delta T_3 = (8000 + 4000)\Delta T_3 = Q_3 = 250{,}000\ W$$

$$\Delta T_3 \frac{250{,}000}{(8000 + 4000)} = 20.83°C$$

$$T_{cold,3-4} = 90 + 20.83 = 110.83°C$$

Similar calculations must be performed to establish the data given in Figure E15.5.

To estimate the area for each of the segments, Equation 15.3 is applied to each segment. The results of these calculations for the process-process heat exchangers (Segments 2–6) are given in Table E15.5. The total heat transfer area for the heat-exchanger network (ignoring the utility exchangers for now) is 788 m^2.

The area calculated following the five-step algorithm, and computed in Example 15.3, was 1063 m^2. For this case, the method of estimating the heat transfer area of the heat-exchanger network from the composite temperature-enthalpy diagram (Example 15.5) underestimates the area by 26%. Although this value would be unacceptable for final design purposes, this method can be easily programmed, and the effect on the total cost of the network of changing the minimum approach temperature or process changes can be computed easily. The final step in computing the fixed capital investment of the heat-exchanger network is to compute the area and configuration for each of the heat exchangers in the network.

15.6 EFFECTIVENESS FACTOR (*F*) AND THE NUMBER OF SHELLS

To this point, it has been assumed that the log-mean temperature correction factor, *F*, for all exchangers is the same and equal to 0.8. The reason that *F* is not assumed to be equal to unity is that, for heat exchangers in most practical applications, the flows of the hot and cold streams are never purely countercurrent. The most common type of heat exchanger in use in the chemical process industries is the shell-and-tube (S&T) type. These units are typically made as multiples of the basic 1-shell pass, 2-tube pass (1-2) design. When estimating the fixed capital investment associated with the purchase and installation of the heat-exchanger network, the number of 1-2 S&T exchangers is needed in addition to the total surface area of the network.

Table E15.5 Summary of Heat Transfer Area Calculations for Example 15.5

	h kW/m²°C	Temperature Interval Shown in Figure E15.5: 1	2	3	4	5	6	7
$Q_{interval}$, kW		50[a]	200	250	550	400	400	100[b]
$Q_{stream,1}$, kW	0.40		0.0	0.0	400.0	400.0	400.0	
$Q_{stream,2}$, kW	0.27		80.0	100.0	0.0	0.0	0.0	
$Q_{stream,3}$, kW	0.53		120.0	150.0	150.0	0.0	0.0	
$Q_{stream,4}$, kW	0.10		0.0	0.0	0.0	0.0	400.0	
$Q_{stream,5}$, kW	0.25		0.0	166.7	366.7	266.7	0.0	
$Q_{stream,6}$, kW	0.08		200.0	83.3	183.3	133.3	0.0	
$T_{hot,1}$, °C			60	100	150	200	250	
$T_{hot,2}$, °C			100	150	200	250	300	
$T_{cold,1}$, °C			40	90	110.83	156.67	190	
$T_{cold,2}$, °C			90	110.83	156.67	190	270	
ΔT_{lm}, °C			14.43	21.37	41.2	51.22	43.28	
$\Sigma(Q_i/h_i)$			$\frac{80}{0.27}+\frac{120}{0.53}+\frac{200}{0.08}$ $= 3022.7$	$\frac{100}{0.27}+\frac{150}{0.53}+\frac{166.7}{0.25}+\frac{83.3}{0.08}$ $= 2361.4$	$\frac{400}{0.40}+\frac{150}{0.53}+\frac{166.7}{0.25}+\frac{83.3}{0.08}$ $= 5041.1$	$\frac{400}{0.40}+\frac{266.7}{0.25}+\frac{133.3}{0.08}$ $= 3733.1$	$\frac{400}{0.40}+\frac{400}{0.10}$ $= 5000$	
$A_{network}$			$\frac{3022.7}{(14.43)(0.8)} = 261.9$	$\frac{2361.4}{(21.4)(0.8)} = 138.2$	$\frac{5041.1}{(41.2)(0.8)} = 152.9$	$\frac{3733.1}{(51.2)(0.8)} = 91.1$	$\frac{5000}{(43.3)(0.8)} = 144.4$	

[a]This is the duty of the cold utility exchanger.
[b]This is the duty of the hot utility exchanger.
Total Process-Process Exchanger Area for Network = 261.9 + 138.2 +152.9 + 91.1 + 144.4 = 788.5 m²

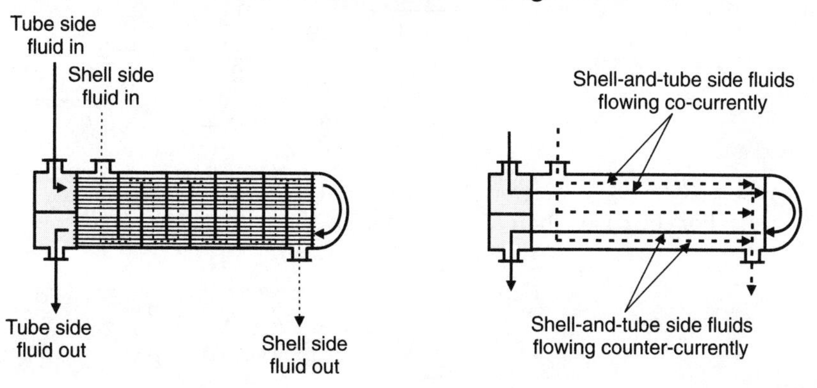

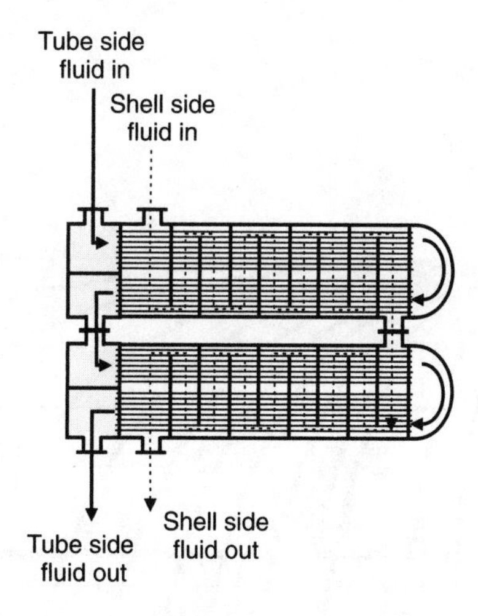

Figure 15.12 Flow Patterns in 1-2 and 2-4 Shell-and-Tube Heat Exchangers

A schematic diagram of a 1-2 S&T exchanger is shown in Figure 15.12. It can be seen that the basic flow patterns of the shell and tube fluids are not countercurrent. Indeed, the flow patterns of the two fluids are very complicated. However, for 1-2 S&T exchangers a reasonable assumption is that the shell-side fluid flows co-currently with the tube-side fluid in one direction and counter-currently in the other direction. Using this idealized model of the fluid flow, an analytical expression may be computed for the F factor (Bowman et al. [9]). This expression is given as

$$F_{1-2,S\&T} = \frac{\sqrt{(R^2+1)}}{(R-1)} \frac{\ln\left[\dfrac{1-P}{1-RP}\right]}{\ln\left[\dfrac{2-P\left(R+1-\sqrt{(R^2+1)}\right)}{2-P\left(R+1+\sqrt{(R^2+1)}\right)}\right]} \tag{15.4}$$

where $R = (\dot{m}C_p)_{cold}/(\dot{m}C_p)_{hot}$ and $P = (t_{c,out} - t_{c,in})/(t_{h,in} - t_{c,in})$.

For the case when $R = 1$, Equation 15.4 reduces to

$$F_{1-2,S\&T} = \frac{P\sqrt{2}}{(1-P)\ln\left[\dfrac{2-2P-P\sqrt{2}}{2-2P+P\sqrt{2}}\right]} \tag{15.5}$$

The relationship between $F_{1\text{-}2,S\&T}$ and P and R is illustrated in Figure 15.13. From this figure it is clear that for a given value of R, the temperature correction factor, F, drops off precipitously as a critical value of P is reached. In fact, for any given value of R, there exists a maximum value of P given by

$$P_{\max} = \frac{2}{R+1+\sqrt{(R^2+1)}} \tag{15.6}$$

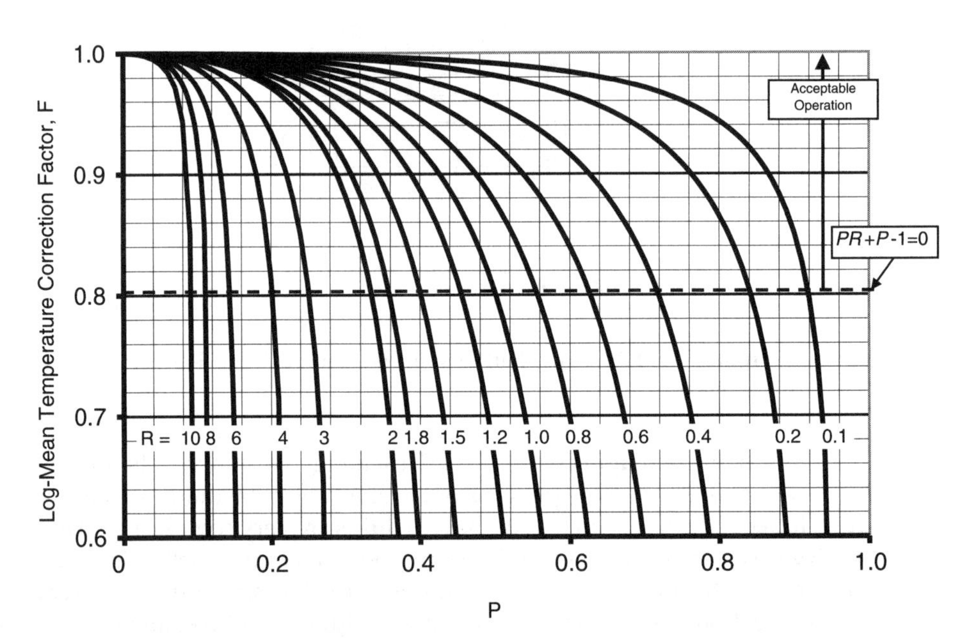

Figure 15.13 Logarithmic-Mean Temperature Correction Factor, F, for a 1-Shell, 2-Tube Pass Heat Exchanger

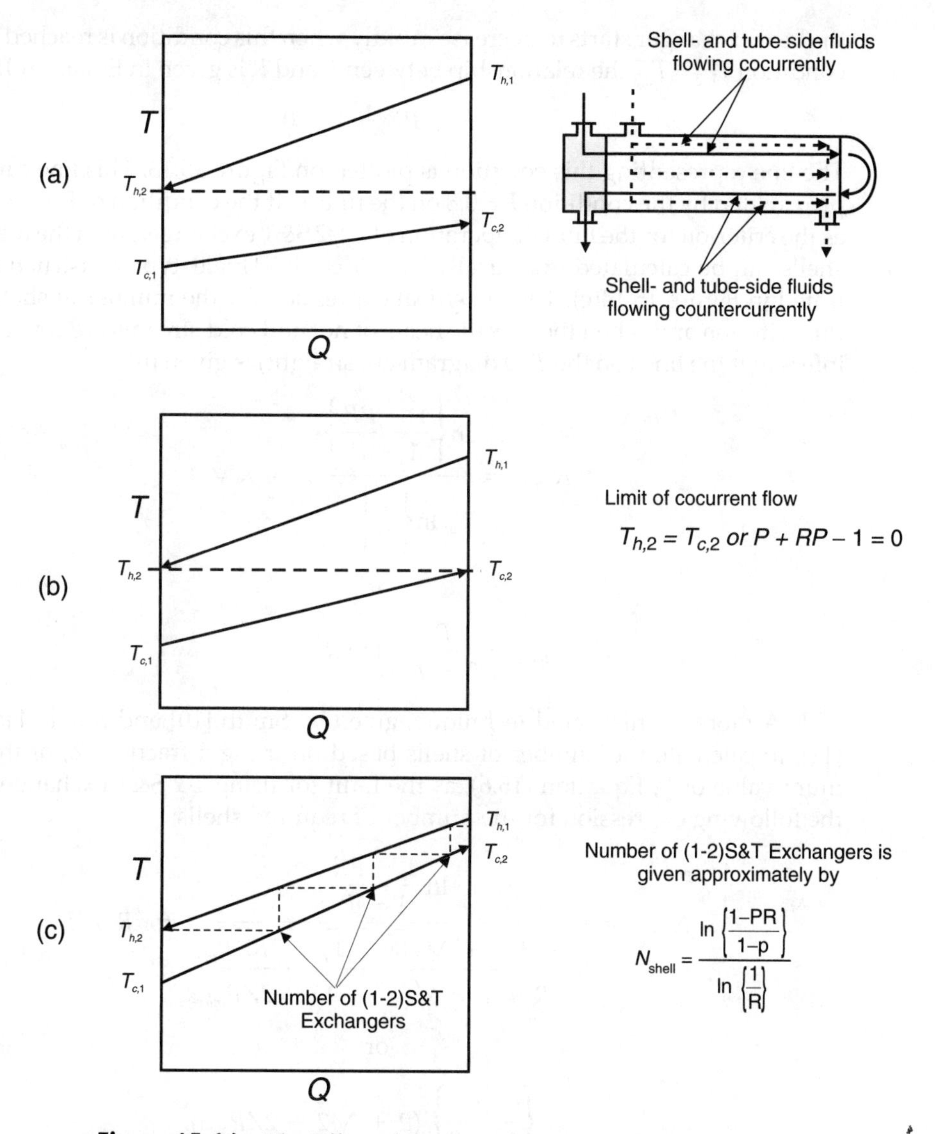

Figure 15.14 The Effect of Temperature Approach on the Number of Shells Required for a Shell-and-Tube Heat Exchanger

This phenomenon can be explained by the fact that as the temperature between the hot exit and cold entrance streams approaches zero, the efficiency of the 1-2 design is reduced. This effect is illustrated in Figure 15.14. For pure cocurrent flow, the highest temperature that the cold stream can be heated to is equal to the hot stream exit temperature. Because half of the flow in a 1-2 S&T exchanger is cocurrent, it is not surprising that the efficiency (temperature correction factor, F) of the heat exchanger starts to decrease rapidly when this condition is reached. For the condition $T_{h,2} = T_{c,2}$ the relationship between P and R is given in Equation 15.7:

$$P + PR - 1 = 0 \tag{15.7}$$

The line representing this equation is plotted on Figure 15.13. This line can be approximated by the condition $F = 0.8$ on the figure. If the condition of $F = 0.8$ is taken as the criterion for the limit of operation of a 1-2 S&T exchanger, then the number of shells can be calculated graphically by a McCabe-Thiele–type construction illustrated in Figure 15.14(c). The analytical expression for the number of shells using this criterion and when the specific heats of hot and cold streams are constant (this infers that the lines on the T-Q diagram are straight) is given by

$$N_{shells} = \frac{\ln\left\{\dfrac{1 - PR}{1 - P}\right\}}{\ln\left\{\dfrac{1}{R}\right\}} \quad \text{for } R \neq 1$$

or (15.8)

$$N_{shells} = \frac{P}{1 - P} \quad \text{for } R = 1$$

A more sophisticated technique, given by Smith [10] and due to Hall et al. [11], to calculate the number of shells based on using a fraction, Z, of the maximum value of P, Equation (15.6), as the limit for using 1-2 S&T exchangers gives the following expression for the number of required shells:

$$N_{shells} = \frac{\ln\dfrac{1 - PR}{1 - P}}{\ln\dfrac{R + 1 + \sqrt{(R^2 + 1)} - 2RZP_{max}}{R + 1 + \sqrt{(R^2 + 1)} - 2ZP_{max}}} \quad \text{for } R \neq 1$$

or (15.9)

$$N_{shells} = \frac{\left\{\dfrac{P}{1 - P}\right\}(2 + \sqrt{2} - 2ZP_{max})}{2ZP_{max}} \quad \text{for } R = 1$$

The number of shells calculated from Equations (15.8) and (15.9) is compared in Table (15.1). The P values are chosen to coincide with the transition from 1 shell to 2 shells using Equation (15.8). According to Smith [7], a limiting value of $F = 0.75$ corresponds to a $Z = 0.9$. Comparing the results in Table 15.1, it is clear that the predictions of Equation (15.8) are not drastically different from those of Equation (15.9). Unless stated otherwise, Equation (15.8) will be used to calculate the number of shells. Example 15.6 demonstrates this procedure.

Table 15.1 Predictions of the Number of Shells Required from Equations (15.8) and (15.9)

R	$P=1/(1+R)$	Eq. (15.8)	Eq. (15.9) $Z=0.8$	Eq. (15.9) $Z=0.9$	Eq. (15.9) $Z=0.95$
0.1	0.91	1	1.71	1.25	1.03
0.2	0.83	1	1.44	1.08	0.92
0.5	0.67	1	1.20	0.93	0.82
0.8	0.56	1	1.14	0.90	0.80
1.0	0.50	1	1.13	0.90	0.80
1.5	0.40	1	1.16	0.91	0.81
2.0	0.33	1	1.20	0.93	0.82
5.0	0.17	1	1.44	1.08	0.92
10.0	0.09	1	1.71	1.25	1.03

Example 15.6

For a heat exchanger that cools a single-phase hot stream from 250°C to 60°C, using a single-phase cooling stream that is heated from 30°C to 200°C, estimate the number of 1-2 S&T heat exchangers required. Assume that the specific heats of the two streams are constant.

The *T-Q* diagram for this heat exchanger is given in Figure E15.6.

$R = (250 - 60)/(200 - 30) = 190/170 = 1.1176$ and $P = (200 - 30)/(250 - 30) = 170/220 = 0.7727$

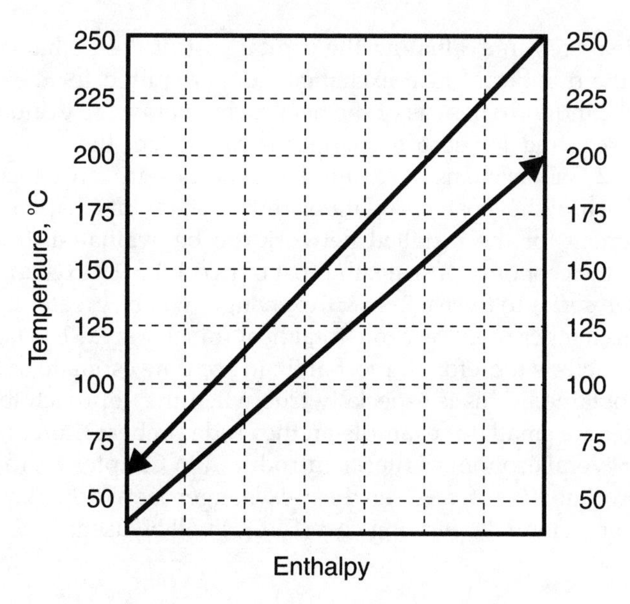

Figure E15.6 Temperature-Enthalpy Diagram for Example 15.6

From Equation (15.8),

$$N_{shells} = \frac{\ln\left\{\dfrac{1 - PR}{1 - P}\right\}}{\ln\left\{\dfrac{1}{R}\right\}} = \frac{\ln\left\{\dfrac{1 - (0.7727)(1.1176)}{1 - 0.7727}\right\}}{\ln\left\{\dfrac{1}{1.1176}\right\}} = 4.59$$

Rounding up, $N_{shells} = 5$.

From Equation (15.9),

$$N_{shells} = \frac{\ln\left[\dfrac{1 - PR}{1 - P}\right]}{\ln\left[\dfrac{R + 1 + \sqrt{(R^2 + 1)} - 2RZ}{R + 1 + \sqrt{(R^2 + 1)} - 2Z}\right]} = \frac{\ln\left[\dfrac{1 - (0.7727)(1.1176)}{1 - 0.7727}\right]}{\ln\left[\dfrac{1.1176 + 1 + \sqrt{(1.1176^2 + 1)} - 2(1.1176)(0.9)}{1.1176 + 1 + \sqrt{(1.1176^2 + 1)} - 2(0.9)}\right]}$$

Rounding up, $N_{shells} = 5$.

Both methods predict the need for an exchanger with five shell passes (and ten tube passes) requiring five 1-2 S&T exchangers stacked on top of each other. This equipment would be referred to as a 5-10 S&T exchanger.

15.7 COMBINING COSTS TO GIVE THE *EAOC* FOR THE NETWORK

The final step in evaluating the capital cost of the exchanger network is to estimate the number of heat-exchanger units required for the network. In Example 15.3, the individual areas of the heat exchangers were evaluated. If the number of shells required for each exchanger is calculated, the total number and areas of each 1-2 S&T exchanger required for the design can be calculated. From these data, using the cost correlations introduced in Chapter 7, the fixed capital investment for the installed network can be evaluated. It should be noted that due to the economy of scale, the cost of two 1-2 S&T exchangers placed with their shells in series to form a 2-4 S&T exchanger will be greater than the equivalent 1-2 S&T exchanger with the same total heat transfer area. The use of the correct number of shells is therefore important if an accurate estimate of the FCI of the HEN is to be obtained. This is especially true when the approach temperatures in an exchanger are small, for example, in the crude preheat train of oil refineries.

Several economic criteria, introduced in Chapter 10, can be used to evaluate the profitability of a given heat-exchanger network. For this discussion, the equivalent annual operating cost (*EAOC*) will be used:

$$EAOC = FCI\frac{i(1 + i)^n}{(1 + i)^n - 1} + \Sigma(\text{utility costs}) \qquad (15.10)$$

where the fixed capital investment, *FCI*, is the total module cost of the exchanger network, the second term on the right-hand side of Equation (15.10) is the yearly cost of the hot and cold utilities, and n is the number of years over which the economic analysis is carried out. For a complete after-tax economic analysis, provision must be made for depreciating the *FCI*. To simplify the analysis, a before-tax hurdle rate, i, can be used to compare cases with different minimum approach temperatures.

When the composite enthalpy curves are used to estimate exchanger areas, there is no simple way of estimating the number and areas of individual exchangers. This is because no specific stream matches are made when estimating the total heat transfer area from the composite enthalpy curves. The simplest method for estimating specific heat transfer areas, which is necessary to calculate the cost of the heat-exchanger network, is to assume that all the heat exchangers have the same area. This approach tends to overestimate somewhat the cost of the network. Thus, the fixed capital investment can be estimated using $(N_{min,a}+N_{min,b})$ times the cost of a heat exchanger with an area of $A_{network}/(N_{min,a}+N_{min,b})$. The *EAOC* can then be estimated using Equation (15.10). Example 15.7 illustrates how the cost of the heat-exchanger network can be estimated.

Example 15.7

Using the results from Example 15.5, estimate the *EAOC* for the heat-exchanger network. Assume that i (before-tax) = 10% p.a. and n = 5 years.

From Example 15.4, the total network heat transfer area (excluding utility exchangers) = 788 m^2.

The minimum number of exchangers was found to be 8.

The fixed capital investment is based on 8 exchangers each with an area of (788/8) = 98.5 m^2.

$$C_{TM,S\&T\ floating\ head\ exchanger} = \$\ 97{,}700$$

The utility costs are estimated as follows.

$$\text{Cooling water utility costs} = (50 \times 10^3)[\text{J/s}]\,(3{,}600)[\text{s/h}](8{,}000)[\text{h/yr}](0.354)[\$/\text{GJ}]\,(1 \times 10^{-9})[\text{GJ/J}] = \$510/\text{yr}$$

$$\text{HPS utility} = (100 \times 10^3)[\text{J/s}]\,(3{,}600)[\text{s/h}](8{,}000)[\text{h/yr}](9.83)[\$/\text{GJ}](1 \times 10^{-9})[\text{GJ/J}] = \$28{,}310/\text{yr}$$

$$EAOC = FCI\frac{(0.1)(1+0.1)^5}{(1+0.1)^5-1} + \$28{,}310/\text{yr} + \$510/\text{yr} = (0.2638)(8)(97{,}000) + \$28{,}820/\text{yr} = \$235{,}000/\text{yr}$$

The procedure illustrated in Example 15.7 can be repeated for different minimum approach temperatures. The basic trends for utility consumption, heat

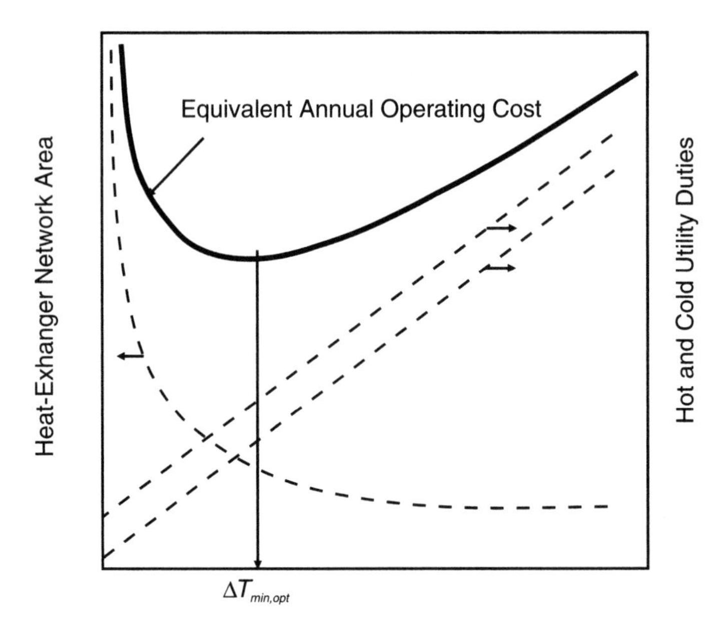

Figure 15.15 Typical Relationships for Heat Transfer Area, Utilities, and *EAOC* for a HEN

transfer area, and *EAOC* as functions of minimum temperature approach are shown in Figure 15.15. It is clear from this figure that there will be an optimum minimum temperature approach for the heat-exchanger network. At the optimum, the increase in utility costs (usage) for an incremental change in minimum temperature approach will be just balanced by the decrease in the equivalent annual operating cost for the smaller heat-exchanger network.

15.8 OTHER CONSIDERATIONS

To this point in the discussion, it has been assumed that the cost of the heat exchanger is simply a function of the heat transfer surface area. From discussions in Chapter 7, it is clear that both the materials of construction and operating pressure are important factors in estimating heat-exchanger costs. These effects must be taken into account when estimating the FCI from the composite temperature-enthalpy diagram. This problem has been addressed by Hall et al. [11] and is explained in the following section.

15.8.1 Materials of Construction and Operating Pressure Issues

Initially, it is assumed that the materials of construction and maximum operating pressures for all heat exchangers are the same. Some streams may require different materials of construction and operating pressures, and, for these streams, the film heat transfer coefficients can be adjusted to correct for the different fixed capital cost associated with exchangers in contact with these streams. For example, assume that all but one of the streams in a HEN are at pressures less than 5 barg and require carbon steel construction. The remaining stream, however, is at a pressure of 25 barg, is corrosive, and requires all exchanger surfaces in contact with it to be made of 304 grade stainless steel. The heat exchangers that contact this fluid will be more expensive than similar sized exchangers that do not contact this fluid. Hall et al. [11] suggest using an effective film heat transfer coefficient for this stream that is lower than the true value. By doing this, the areas predicted for heat exchangers that process this stream increase, causing the costs of the exchangers to also increase. Therefore, using these effective film coefficients allows the same cost equation to be used for all exchangers in the HEN. The form of the cost equations introduced in Chapter 7 (and Appendix A) is not suitable for this analysis. Instead, the simpler form of cost relationship given in Equation (7.2) is used:

$$C_p = KA^n \tag{15.11}$$

The total module cost of the exchanger is given by combining the results of Equations (7.2), (7.15), and (A.4):

$$C_{TM} = 1.18(B_1 + B_2F_MF_P)KA^n \tag{15.12}$$

For the base conditions, operating pressure <5 barg, and carbon steel construction, F_M and F_P in Equation (15.12) are both equal to 1. Let the area of a heat exchanger operating at pressure P and using the correct materials of construction be A, and let the area of an exchanger at the base conditions that has the same total module cost as the other exchanger be A_{bc}. Using Equation (15.12) and equating total module costs gives

$$1.18(B_1 + B_2F_MF_P)KA^n = 1.18(B_1 + B_2)KA_{bc}{}^n$$

$$\frac{A_{bc}}{A} = \left[\frac{B_1 + B_2F_MF_P}{B_1 + B_2}\right]^{\frac{1}{n}} \tag{15.13}$$

From Equation (15.13), it is clear that for equal cost, the heat-exchanger area at base conditions must be greater than the true area by the factor on the right-hand side of the equation. Because the area is inversely proportional to the film heat transfer coefficient, the relationship in Equation (15.14) is obtained:

$$\frac{h}{h_{bc}} = \left[\frac{B_1 + B_2F_MF_P}{B_1 + B_2}\right]^{\frac{1}{n}} \tag{15.14}$$

By modifying heat transfer coefficients for streams requiring non-base-case operating conditions using Equation (15.14), the same cost equation can be used to calculate all exchanger costs in a given network.

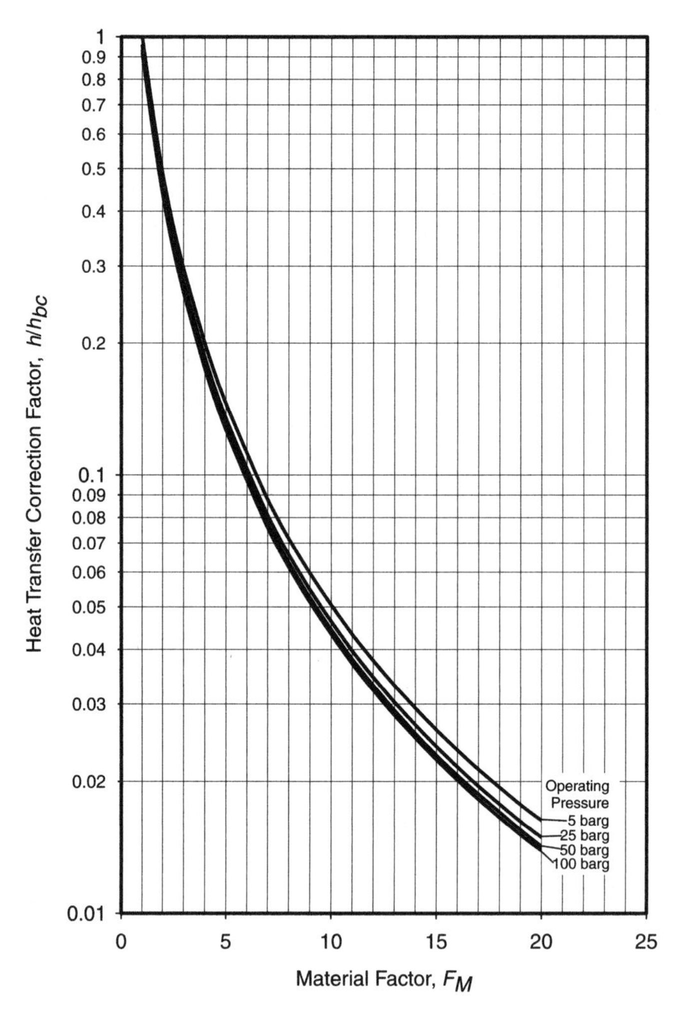

Figure 15.16 Heat Transfer Correction Factor for Shell-and-Tube, Floating-Head Heat Exchangers

The correction factors for a floating-head, shell-and-tube heat exchanger are shown in Figure 15.16. From Appendix A, the values used in Equation (15.14) are $B_1 = 1.63$, $B_2 = 1.66$, and $n = 0.573$. Examples 15.8 and 15.9 illustrate how to use this information.

Example 15.8

For the heat-exchanger network developed in Examples 15.2, 15.3, and 15.4, what heat transfer coefficient should be used to account for the fact that Stream 3 requires stainless steel construction and is at a pressure of 50 barg? Assume that all the streams except Stream 3 are available at pressures <5 bar and require only carbon steel construction.

In order to minimize the investment cost, Stream 3 will always be placed on the tube side of any exchanger through which it passes. From Table A.3, the material factor, F_M, for heat exchangers with carbon steel shells and stainless steel tubes is 1.81. From Figure 15.16, with an operating pressure of 50 barg and F_M of 1.81, the value of h_{bc}/h is 0.517. Therefore, a modified heat transfer coefficient of (530)(0.517) = 274 W/m²°C should be used to estimate heat transfer areas using the cost equation for base-case conditions.

Example 15.9

Estimate the effect of the result of Example 15.8 on the total heat transfer area and cost of the heat-exchanger network calculated in Example 15.4. The adjusted heat transfer areas for the network are shown in Table E15.9.

The results in Table E15.9 show that the effect on the fixed capital investment of the HEN of using stainless steel and an operating pressure of 50 bar for Stream 3 is equivalent to increasing the HEN area from 788 to 848 m² using base-case costs. The fixed capital investment from Example 15.7 was (8)($97,700) = $781,600. The use of nonstandard conditions is based on 8 exchangers each of area (847/8) = 106.0 m², a cost of (8)(103,600) = $828,800. This is an increase of $47,200.

Table E15.9 Heat Transfer Areas for Network Adjusted for Non-Base-Case Conditions for Stream 3

Temperature Interval	$\Sigma(Q_i/h_i)$ (m²°C)	$A_{interval}$ (m²)
2	$\frac{80}{0.27}+\frac{120}{0.274}+\frac{200}{0.08}$ = 3234	$\frac{3241}{(14.43)(0.8)}=280.1$
3	$\frac{100}{0.27}+\frac{150}{0.274}+\frac{166.7}{0.25}+\frac{83.3}{0.08}$ = 2626	$\frac{2626}{(21.4)(0.8)}=153.4$
4	$\frac{400}{0.40}+\frac{150}{0.274}+\frac{366.7}{0.25}+\frac{183.3}{0.08}$ = 5305	$\frac{5305}{(41.2)(0.8)}=161.0$
5	$\frac{400}{0.40}+\frac{266.7}{0.25}+\frac{133.3}{0.08}$ = 3733.1	$\frac{3733}{(43.3)(0.8)}=107.8$
6	$\frac{400}{0.40}+\frac{400}{0.10}$ = 5000	$\frac{5000}{(43.3)(0.8)}=144.3$
Total		847 m²

15.8.2 Problems with Multiple Utilities

For more complicated problems, several pinch temperatures may occur, and energy may have to be added from more than one hot utility and rejected to more than one cold utility. In general, it makes good engineering sense to add heat from the coolest "hot" utility and reject heat to the hottest "cold" utility. For example, if heat must be rejected from the process at a temperature of 190°C, it could be rejected to the cooling water utility. However, it makes more sense—that is, it is more profitable—if this excess energy can be used to make medium-pressure steam, which can be used elsewhere in the plant. Likewise, if heat needs to be added to the process at a temperature of 80°C, it makes little sense to use valuable high-pressure steam (at 250°C); rather, low-pressure steam should be utilized. The use of multiple utilities is required in Problems 15.5 and 15.6 at the end of this chapter.

15.8.3 Handling Streams with Phase Changes

The area of heat integration and pinch technology, in general, is quite broad, and many topics have not been covered here. The MUMNE approach outlined above will give a reasonable first approximation to the optimum heat integration scheme for a given process and is therefore useful in the preliminary design of a process. Although not stated explicitly, the analysis for the MUMNE design assumes that the streams are single phase and that the specific heat capacities of the streams are constant over the temperature range in the process. For streams that undergo a phase change or for which the specific heat capacities are not constant, the analysis becomes more complicated. However, for such streams, dummy streams can be used that have constant heat capacities. This concept is illustrated in Figure 15.17. The left-hand diagram shows a situation where a partial

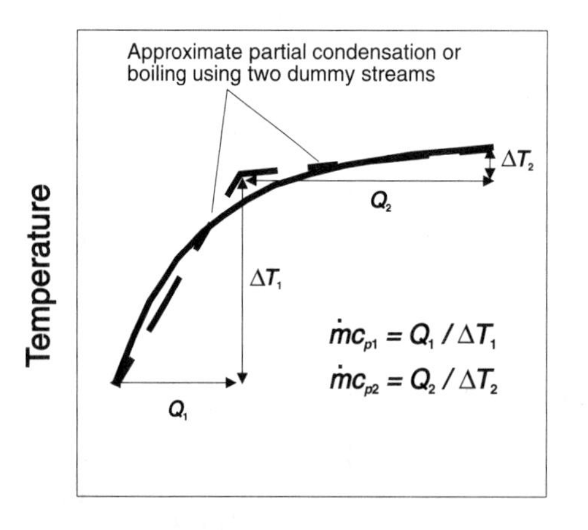

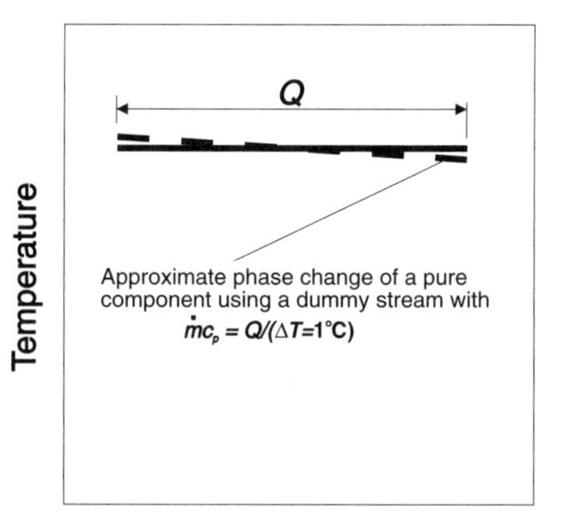

Figure 15.17 Use of Dummy Streams for Phase Changes

phase change is occurring. In this case, the enthalpy change can be approximated by using two dummy streams with constant heat capacities as shown in the diagram. For a single component that changes phase, the *T-Q* diagram is a horizontal line, and for this case, an arbitrarily small ΔT of 1°C is used to get the dummy heat capacity.

For large processes, the hand calculations described above become unwieldy and very time consuming. To help in these calculations, the HENSAD software has been developed and is contained on the CD accompanying the text.

15.9 HEAT-EXCHANGER NETWORK SYNTHESIS ANALYSIS AND DESIGN (HENSAD) PROGRAM

The heat-exchanger network synthesis analysis and design (HENSAD) program was developed to aid in solving the MUMNE problem previously described in this chapter. Once the feed stream information has been entered, temperature interval diagrams, cascade diagrams, and composite temperature-enthalpy plots (corresponding to Steps 1–4 introduced in Section 15.2) are calculated by the program. The design of the optimum exchanger network can also be completed using the software. However, the user must interact with the program to achieve a design that does not violate the minimum approach temperature specified for the problem. Parametric studies using the composite enthalpy diagram can also be carried out. Plots of network exchanger area, utility consumption, and *EAOC* as functions of minimum approach temperature can be made and printed out. The user is referred to a training video (HENSAD.AVI) that accompanies the software, which explains the operation of the program and required user input. Although the software solves comprehensive problems, it is limited by the following assumptions.

- The specific heats of the streams are assumed to be constant.
- Problems with only one hot and one cold utility can be solved.
- For comparing the "optimal" minimum approach temperature, the before-tax *EAOC* is used as the objective function.
- Problems without a pinch may be analyzed, but plots of ΔT_{min} versus heat-exchanger area and *EAOC* are strictly correct only for situations where a pinch occurs.
- Economic analyses are valid for the range of application of the cost correlations.

In order to gain experience with the software, it is recommended that the problem studied in this chapter be entered and checked using the program after the AVI file has been viewed.

15.10 MASS-EXCHANGE NETWORKS

Pinch technology can be applied to the integration of mass-exchange devices in an analogous manner as it is applied to integration of heat exchangers. A mass-exchange device is a separator. The purpose of heat integration is to use energy more efficiently. By matching hot and cold streams, the need for hot and cold utilities is minimized. This has environmental benefits in that utilities require energy, which increases pollution. Energy integration also saves the cost of these utilities, although it must be remembered that the most energy-efficient heat-exchange network is not necessarily the most economical one.

The purpose of mass-exchange networks is to use mass more efficiently. If mass is not used efficiently, the amount of waste created increases, creating more pollution. Without mass integration, raw materials also may not be used very efficiently.

As a simple example of mass integration, consider the acetone process in Appendix B, section B.10. It is observed that process water is needed to scrub the acetone remaining in the vapor stream exiting the vessel, V-1102. It is also observed that there is a wastewater stream exiting the final distillation column, T-1103. This wastewater stream contains trace amounts of acetone and isopropyl alcohol. It should be possible to use some or all of this wastewater stream in the scrubber, T-1101, in place of process water. Because the water now contains some solute, it may be necessary to increase the size of the scrubber. In this example of mass integration, less waste is produced. Furthermore, the expenses of producing pure process water and of treating the wastewater stream are either reduced or eliminated.

The procedure for synthesis of mass-exchange networks is summarized here. Much more detailed treatments are available [12,13], and brief discussions for pollution prevention applications are also available [14,15]. The procedure parallels that for heat-exchange networks. There is a composition-interval diagram (CID) that is analogous to the temperature-interval diagram (TID). There is also a cascade diagram, a similar method for identifying the pinch and the minimum number of exchangers above and below the pinch, a composite mass-exchange diagram, and a final mass-exchange network. In the context of mass-exchange networks, a utility is the separation or addition of a solute to a sink or from a source that is not a process stream. This is similar to the hot and cold utilities in heat-exchange networks, and the analogous terms to *hot* and *cold* utilities are *rich* and *lean* utilities, respectively. Because a hot utility adds energy to a stream, a rich utility adds mass (of a solute) to a stream. Similarly, because a cold utility removes energy from a stream, a lean utility removes mass from a stream (an adsorbent, for example). The term *rich* means a stream that is more concentrated in solute and hence will be losing solute to another process stream by mass transfer. The term *lean* means a stream that is less concentrated in solute and hence will be receiving solute from another process stream by mass transfer. With this analogy between hot and cold and rich and lean utilities understood, then the concept of a MUMNE (minimum utility minimum number of exchangers) network also exists for mass-exchange networks.

However, there are several differences between heat-exchange networks and mass-exchange networks that must be understood.

1. The partition coefficient of the solute between the phases contacted in the mass-exchange devices (separators, probably absorbers, strippers, or liquid-liquid extractors) must be included. This is accomplished by plotting the mole fraction in the rich phase (denoted y) on the left side of the CID, and the mole fraction of the lean phase (denoted x) on the right side of the diagram. In a TID, the temperatures on opposite sides of the same horizontal line differ by the minimum approach temperature. In a CID, the rich- and lean-phase compositions are related by

$$y = m(x + \delta) \tag{15.15}$$

where m is the partition coefficient and δ is the minimum approach composition based on the lean phase. The consequence of the partition coefficient is that the criterion for mass transfer from the rich phase to the lean phase is that $y > mx$. If the partition coefficient, m, were unity, then this criterion would be completely analogous to that for heat transfer, where the criterion for heat transfer from hot to cold streams is $T_H > T_C$. Therefore, if $m < 1$, mass can be transferred from a rich phase with a lower mole fraction than the lean phase.
2. The rules for matching streams at the pinch are different. Above the pinch, the criterion for matching streams is

$$L \geq mR \tag{15.16}$$

where L is the mass flowrate of the lean stream, and R is the mass flowrate of the rich stream. Below the pinch, the criterion for matching streams is

$$L \leq mR \tag{15.17}$$

3. There is also a constraint regarding the number of streams. Above the pinch,

$$N_R \leq N_L \tag{15.18}$$

and below the pinch,

$$N_R \geq N_L \tag{15.19}$$

where N_L is the number of lean streams and N_R is the number of rich streams. If this criterion is not met, this means that stream splitting is required.

The procedure for synthesis of mass-exchange networks is illustrated in Examples 15.10 and 15.11. In these examples, it is assumed that there is only one solute, and it is the only component being transferred. The phases are completely immiscible, that is, no solvent is transferred between phases.

Example 15.10

Consider the following process streams, all containing the same solute. The partition coefficient between both rich phases and the lean phase is 0.5 (i.e., $y = 0.5x$).

	kg/s	y_{in}	y_{out}
$R_1 = 1$	2	0.10	0.03
$R_2 = 2$	4	0.07	0.02
	kg/s	x_{in}	x_{out}
$L = 3$	4	0.00	0.07

Synthesize the MUMNE mass-exchange network for a minimum approach composition of $\delta = 0.01$.

The CID is shown in Figure E15.10(a). It is observed that the values on the left side represent y, the rich stream compositions (mass fractions). The corresponding values on the right side are obtained by solving for x in Equation (15.15). The values far to the right are the total mass transferred in each interval and are obtained from the indicated equation.

The cascade diagram is shown in Figure E15.10(b). Just as in heat-exchange networks, excess mass is cascaded through each interval. In this example, there is no pinch.

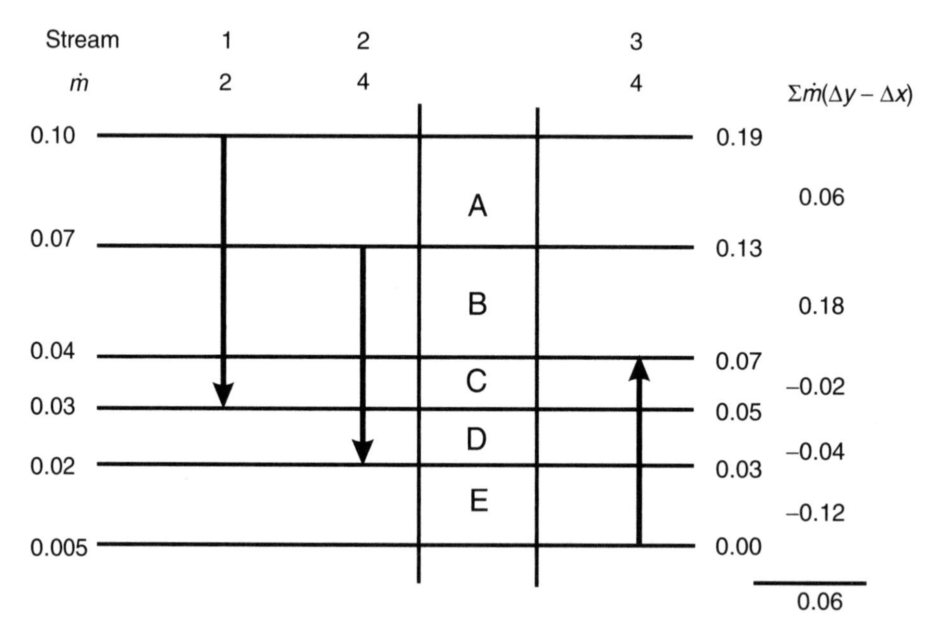

Figure E15.10(a) Composition Interval Diagram for Example 15.10

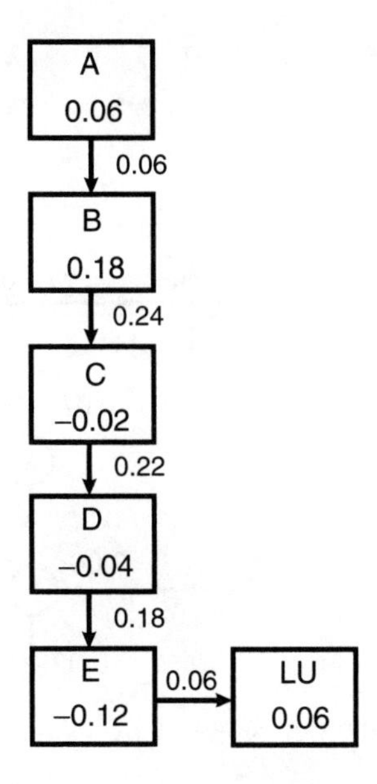

Figure E15.10(b) Cascade Diagram for Example 15.10

At the bottom of the cascade diagram, there is excess mass requiring a lean utility (LU). An LU is a different separation unit, such as an adsorber, that removes solute.

The minimum number of mass exchangers is shown in Figure E15.10(c). The method exactly parallels that for the minimum number of heat exchangers. In this example, there are three mass-exchange devices required.

Figure E15.10(d) shows the MUMNE network. Because there is no pinch, there are no restrictions other than mass balance and concentration gradient.

Figure E15.10(e) shows the composite mass-flow diagram. It is observed that the ordinate is the mass fraction in the lean phase. Therefore, for each rich-phase composition, Equation (15.15) is used to obtain a corresponding lean-phase value. Alternatively, the diagram could have been plotted using the mass fraction in the rich phase. There are two composite lines: one for the rich phase and one for the lean phase. The points needed to plot Figure E15.10(e) are as shown in the table on page 564.

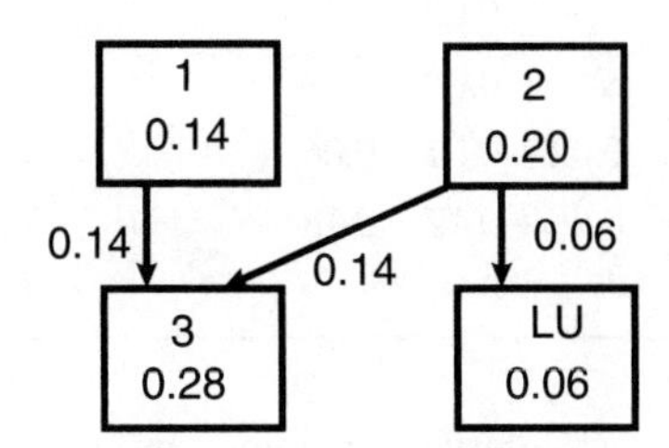

Figure E15.10(c) Minimum Number of Mass Exchangers for Example 15.10

Stream	1	2	3
$\dot{m}$	2	4	4

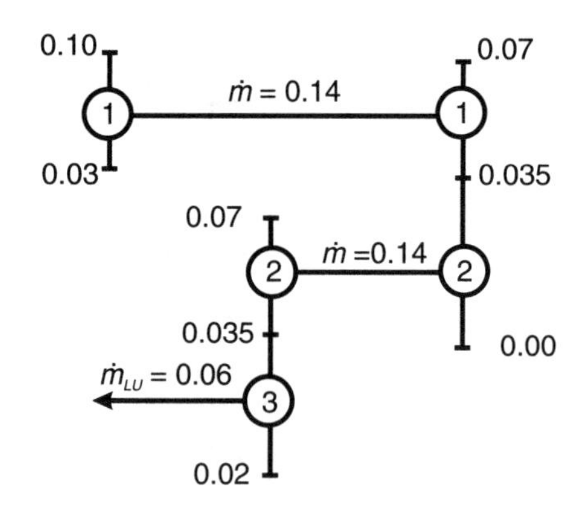

Figure E15.10(d) MUMNE Network for Example 15.10

Lean Streams			
Composition Interval	**Mass Transferred in Interval (kg/s)**	**Cumulative Mass Transferred (kg/s)**	**Mass Fraction in Lean Phase (x)**
	0.00	0.06 (LU)	0.00
E			
	0.12	0.18	0.03
D			
	0.08	0.26	0.05
C			
	0.08	0.34	0.07

Rich Streams			
Composition Interval	**Mass Transferred in Interval (kg/s)**	**Cumulative Mass Transferred (kg/s)**	**Mass Fraction in Lean Phase (x)**
E	0.00	0.00	0.03
D	0.04	0.04	0.05
C	0.02 + 0.04 = 0.06	0.10	0.07
B	0.06 + 0.12 = 0.18	0.28	0.13
A	0.06	0.34	0.19

The mass transfer in each interval is obtained by summing the mass flowrate of each stream multiplied by the composition change across the interval. It should be noted that only the points where the slope of the composite line changes are highlighted in Figure E15.10(e). The

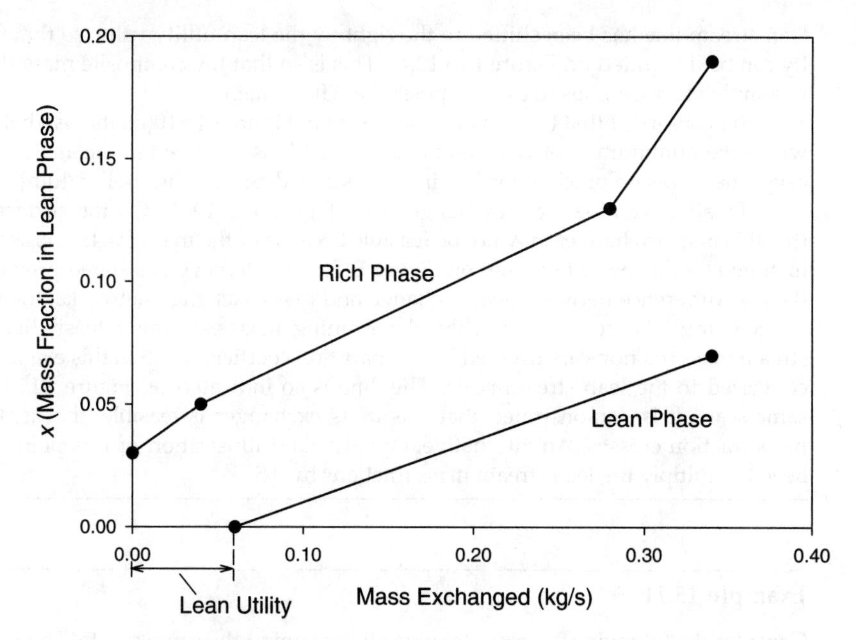

Figure E15.10(e) Composite Mass Flow Diagram for Example 15.10

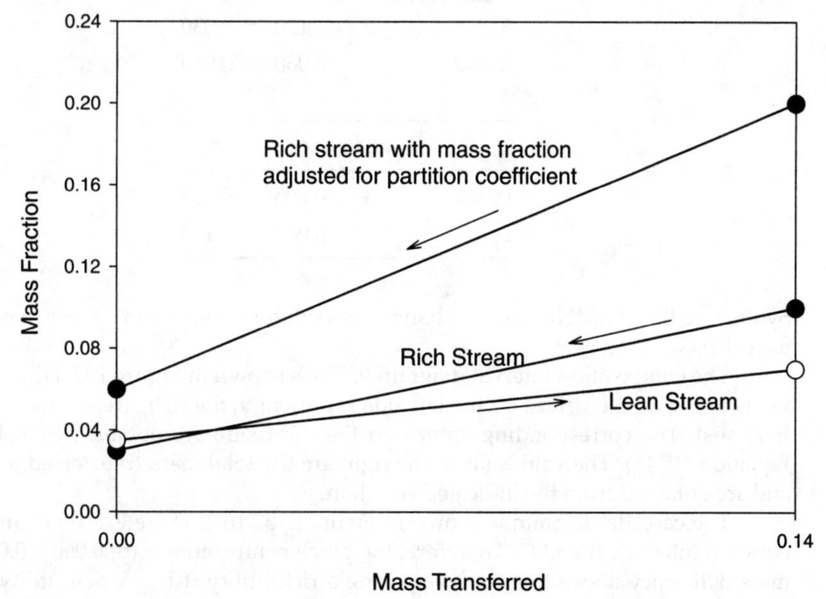

Figure E15.10(f) Illustration of Validity of Mass Exchanger 1 in Example 15.10

lean-stream line has been shifted to the right by the lean utility value so that the lean utility can be illustrated on Figure E15.10(e). This is so that the composite mass flow diagram is completely analogous to the composite heat flow diagram.

It is observed that there is no pinch point in Figure E15.10(e); that is, there is no point where the minimum approach composition is 0.01, as provided in the analysis. This is because there was no pinch identified in the cascade diagram (Figure E15.10[b]).

Finally, examine mass exchanger 1 on Figure E15.10(d). On the surface, it appears that this mass exchanger may not be feasible because of the mass fraction cross, which is illustrated by the lower two lines on Figure E15.10(f). However, as stated previously, under the first difference between heat-exchange and mass-exchange networks, the partition coefficient must be considered when determining mass-exchanger feasibility. If the rich stream mass fractions are divided by the partition coefficient (0.5 in this example), they are converted to the lean stream scale. This line is so indicated on Figure E15.10(f). On the same scale, it is demonstrated that this mass exchanger is feasible; that is, there are no mass fraction crosses. An alternative, equally valid illustration of this point would have been to multiply the lean stream mass fractions by 0.5.

Example 15.11

Consider the following process streams, all containing the same solute. The partition coefficient between both rich phases and the lean phase is 0.667 (i.e., $y = 0.667x$).

	kg/s	y_{in}	y_{out}
$R_1 = 1$	2	0.120	0.030
$R_2 = 2$	1	0.060	0.010

	kg/s	x_{in}	x_{out}
$L_1 = 3$	1	0.005	0.125
$L_2 = 4$	6	0.080	0.125

Synthesize the MUMNE mass-exchange network for a minimum approach composition of $\delta = 0.010$.

The composition interval diagram (CID) is shown in Figure E15.11(a). As in the previous example, the values on the left side represent y, the rich stream compositions (mass fractions). The corresponding values on the right side are obtained by solving for x in Equation (15.15). The values far to the right are the total mass transferred in each interval and are obtained from the indicated equation.

The cascade diagram is shown in Figure E15.11(b). There is a pinch in this example between intervals B and C. Therefore, the pinch compositions are 0.060 – 0.080. There is a mass deficiency above the pinch requiring a rich utility (RU). A rich utility is a separate unit in which pure solute is added to a stream. This may seem unusual. However, it must be remembered that this is not necessarily the optimum economic solution; it is the opti-

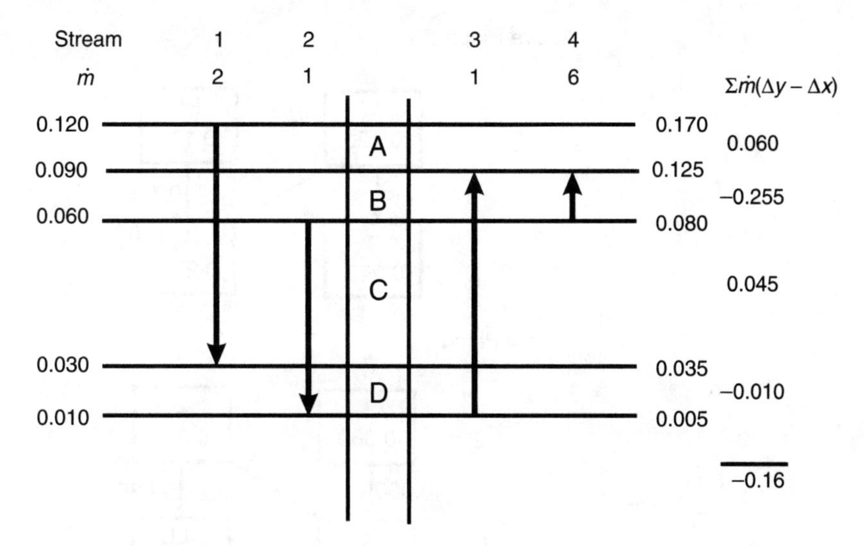

Figure E15.11(a) Composition Interval Diagram for Example 15.11

mum solution for use of mass. There is also a mass excess below the pinch requiring a lean utility (LU). A lean utility is a separation device (such as an adsorber) that removes solute without exchanging solute with other process streams.

The minimum number of mass exchangers above and below the pinch is shown in Figure E15.11(c). In this example, three mass-exchange devices are required both above and below the pinch.

Figure E15.11(d) shows the MUMNE network. It is observed that the criteria in Equations (15.16) through (15.19) are obeyed.

Figure E15.11(e) shows the composite mass-exchange diagram. Once again the ordinate is the mass fraction in the lean phase. The points on Figure E15.11(e) are as shown in the table on page 569.

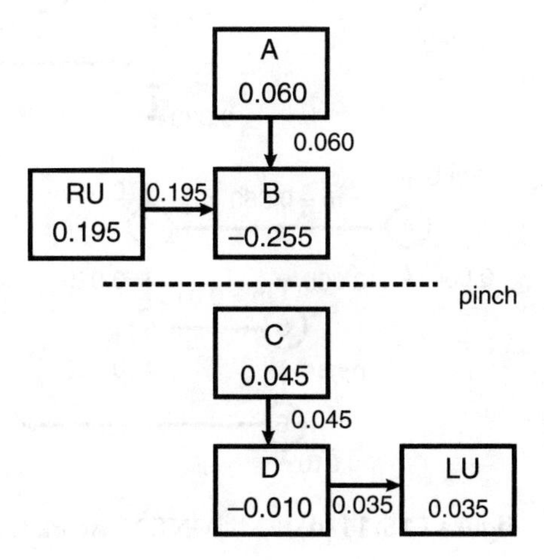

Figure E15.11(b) Cascade Diagram for Example 15.11

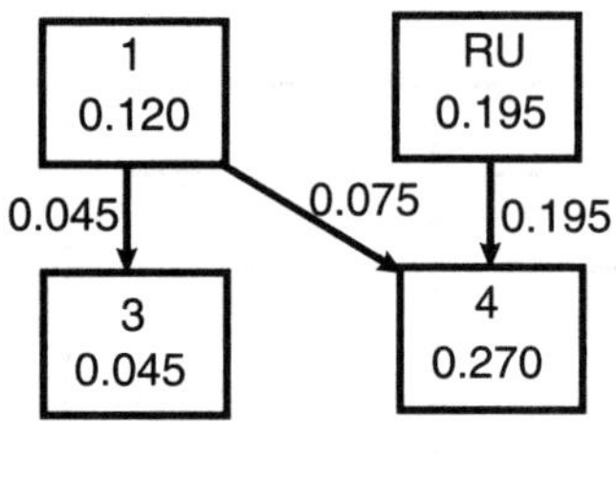

Below Pinch

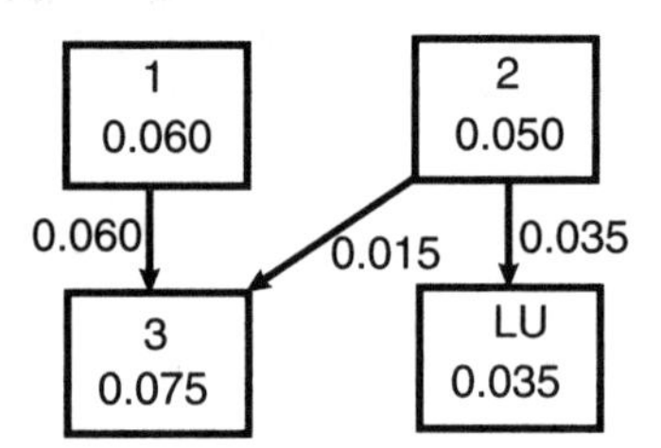

Figure E15.11(c) Minimum Number of Mass Exchangers Above and Below Pinch for Example 15.11

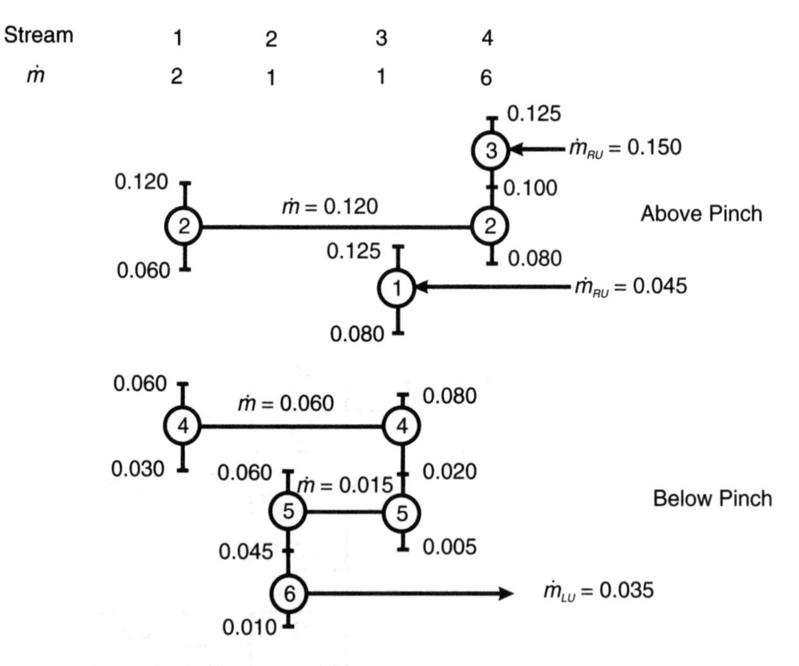

Figure E15.11(d) MUMNE Network for Example 15.11

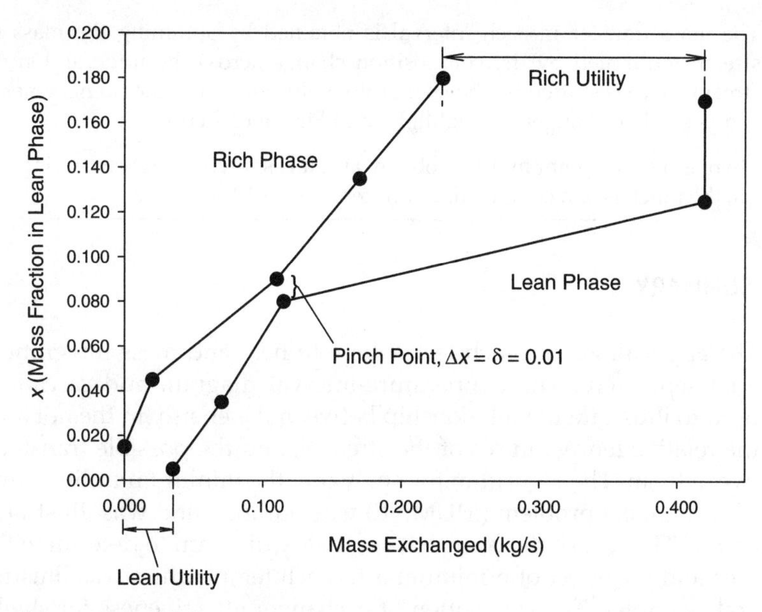

Figure E15.11(e) Composite Mass Flow Diagram for Example 15.11

Rich Streams			
Composition Interval	**Mass Transferred in Interval (kg/s)**	**Cumulative Mass Transferred (kg/s)**	**Mass Fraction in Lean Phase (x)**
E	0.000	0.000	0.015
D	0.020	0.020	0.045
C	0.060 + 0.060 = 0.090	0.110	0.090
B	0.060	0.170	0.135
A	0.060	0.230	0.180

Lean Streams			
Composition Interval	**Mass Transferred in Interval (kg/s)**	**Cumulative Mass Transferred (kg/s)**	**Mass Fraction in Lean Phase (x)**
	0.000	0.035 (LU)	0.005
D	0.035	0.070	0.035
C	0.045	0.115	0.080
B	0.315	0.420	0.125
A	0.000	0.420	0.170

The mass transfer in each interval is obtained by summing the mass flowrate of each stream multiplied by the composition change across the interval. Once again, the lean stream curve is shifted by the lean utility value, and only the points where the slope of the composite line changes are highlighted in Figure E15.11(e).

There is a pinch point in this problem, and it is shown in Figure E15.11(e). As per the criteria, the pinch is at a composition difference of 0.010.

15.11 SUMMARY

The applications of pinch technology to heat and mass integration were considered separately. The temperature-interval diagram and cascade diagram were used to illustrate the relationship between the energy in the hot and cold streams, the relative temperatures of the streams, and the possible transfers of energy between them. The algorithm for analyzing the minimum utility, minimum number of exchangers problem (MUMNE) was outlined and was illustrated using an example. The role of the composite enthalpy diagram to determine the heat transfer area and the effect of minimum approach temperature was illustrated using several examples. The calculation of exchanger effectiveness for shell-and-tube heat exchangers was covered, and a simple expression to determine the number of shells for a given service was introduced. The effects of operating pressure and different materials of construction on the design of the heat exchange network were illustrated with an example. Finally, the HENSAD software was introduced.

Mass-exchanger networks were discussed, and the composition interval diagram and cascade diagrams for these systems were shown to parallel the development for heat-exchange networks. The role of the composite mass flow diagram was introduced, and the MUMNE design algorithm for mass-exchange networks was illustrated with two examples.

REFERENCES

1. Linnhoff, B., and J. R. Flower, "Synthesis of Heat Exchanger Networks: 1. Systematic Generation of Energy Optimal Networks," *AIChE J.* 24 (1978): 633–642.
2. Hohmann, E. C., *Optimum Networks for Heat Exchange*, Ph.D. thesis, University of Southern California, 1971.
3. Umeda, T., Itoh, J., and K. Shiroko, "Heat Exchanger System Synthesis," *Chem. Eng. Prog.* 74, no. 7 (1978): 70–76.
4. Douglas, J. M., *Conceptual Design of Chemical Processes* (New York: McGraw-Hill, 1988).
5. Sieder, E. N., and G. E. Tate, "Heat Transfer and Pressure Drop of Liquids in Tubes," *Ind. Eng. Chem.* 28 (1936): 1429–1435.

6. Donohue, D. A., "Heat Transfer and Pressure Drop in Heat Exchangers," *Ind. Eng. Chem.* 41 (1949): 2499–2511.
7. Townsend, D. W., and B. Linnhoff, "Surface Area Targets for Heat Exchanger Networks," *Annual IchemE* (April 1984).
8. Ahmad, S., B. Linnhoff, and R. Smith, "Cost Optimum Heat Exchanger Networks: II. Targets and Design for Detailed Capital Cost Models," *Comp. Chem. Eng.* 14 (1990): 751–767.
9. Bowman, R. A., A. C. Mueller, and W. M. Nagle, "Mean Temperature Differences in Design," *Trans. ASME* 62 (1940): 283.
10. Smith, R., *Chemical Process Design* (New York: McGraw-Hill, 1995).
11. Hall, S. G., S. Ahmad, and R. Smith, "Capital Cost Target for Heat Exchanger Networks Comprising Mixed Materials of Construction, Pressure Ratings and Exchanger Types," *Computers Chem. Eng.* 14 (1990): 319–335.
12. El-Halwagi, M. M., and V. Manousiouthakis, "Synthesis of Mass Exchange Networks," *AIChE-J* 35 (1989): 1233–1244.
13. El-Halwagi, M. M., *Pollution Prevention through Process Integration. Systematic Design Tools* (San Diego, CA: Academic Press, 1997).
14. Allen, D. T., and D. R. Shonnard, *Green Engineering* (Upper Saddle River, NJ: Prentice Hall PTR, 2001).
15. Allen, D. T., and K. S. Rosselot, *Pollution Prevention for Chemical Processes* (New York: Wiley-Interscience, 1997).

SHORT ANSWER QUESTIONS

1. List all the costs that are affected by changing the minimum temperature approach in a heat-exchanger network (HEN) design.
2. What are the advantages and disadvantages of decreasing the minimum approach temperature in a heat-exchange network?
3. One way to account for different materials of construction (MOCs) in heat-exchanger network design is to "adjust" the film heat transfer coefficients for streams that require MOCs other than carbon steel (CS). How would you adjust (increase or decrease) the heat transfer coefficients for streams requiring (a) MOCs cheaper than CS and (b) MOCs more expensive than CS? Briefly explain your reasoning.
4. In heat-exchanger network designs, there exists an optimum ΔT_{min} for the network. Explain carefully how capital costs and operating costs change as ΔT_{min} is increased.
5. For designing heat exchangers at the pinch, what is the criterion for matching streams above the pinch? Why is such a criterion needed?

6. Sketch a typical composite enthalpy diagram for a heat-exchanger network. On this diagram clearly label the x- and y-axes, and indicate on the diagram the values and location of the overall hot and cold utility demands and the pinch point.
7. What does the cascade diagram illustrate?
8. In general, what is the minimum number of exchangers required to design a heat-exchange network above the pinch that consists of n hot streams and m cold streams?
9. How does one deal with a stream that changes phase in a heat-exchanger network?
10. What is MUMNE? Describe it.

PROBLEMS

11. For a process, the following process streams must be cooled or heated.

Stream No.	$\dot{m}C_p$ $(10^3\text{BTU})/\text{hr}°\text{F}$	Temperature In °F	Temperature Out °F
1	4	600	320
2	4	470	280
3	3	340	580
4	5	300	480

Use the MUMNE algorithm for heat-exchanger networks and a minimum approach temperature of 20°F.

a. Determine the temperature interval diagram.
b. Determine the cascade diagram, the pinch temperatures, and the minimum hot and cold utilities. If there is a choice, for the sake of uniformity, choose the larger values for the pinch temperatures.
c. Determine the minimum number of heat exchangers above and below the pinch.
d. Determine the heat-exchange network above the pinch.
e. Determine the heat-exchange network below the pinch.

12. For a new process, the following process streams must be cooled or heated.

Stream No.	$\dot{m}C_p$ $(10^3\text{BTU})/\text{hr}°\text{F}$	Temperature In °F	Temperature Out °F
1	2	400	320
2	4	300	100
3	3	90	310
4	2	170	310

Use the MUMNE algorithm for heat-exchanger networks and a minimum approach temperature of 10°F.

a. Determine the temperature interval diagram.
b. Determine the cascade diagram, the pinch temperatures, and the minimum hot and cold utilities.
c. Determine the minimum number of heat exchangers above and below the pinch.
d. Determine the heat-exchange network above the pinch.
e. Determine the heat-exchange network below the pinch.

13. For a new process, the following process streams must be cooled or heated.

Stream No.	$\dot{m}C_p$ kW/°C	Temperature In °C	Temperature Out °C
1	3	180	100
2	5	120	80
3	3	70	140
4	2	80	160

Use the MUMNE algorithm for heat-exchanger networks and a minimum approach temperature of 20°C.

a. Determine the temperature interval diagram.
b. Determine the cascade diagram, the pinch temperatures, and the minimum hot and cold utilities.
c. Determine the minimum number of heat exchangers above and below the pinch.
d. Determine the heat-exchange network above the pinch.
e. Determine a heat-exchange network below the pinch that has only one utility exchanger.

14. For a new process, the following process streams must be cooled or heated.

Stream No.	$\dot{m}C_p$ $(10^3\text{BTU})/\text{hr}°\text{F}$	Temperature In °F	Temperature Out °F
1	4	500	360
2	4	430	340
3	3	370	490
4	4	350	440

Use the MUMNE algorithm for heat-exchanger networks and a minimum approach temperature of 10°F.

a. Determine the temperature interval diagram.
b. Determine the cascade diagram, the pinch temperatures, and the minimum hot and cold utilities.
c. Determine the minimum number of heat exchangers above and below the pinch.
d. Determine the heat-exchange network above the pinch.
e. Determine a heat-exchange network below the pinch.

15. In a process design, the following process streams must be cooled or heated.

Stream No.	$\dot{m}C_p$ kW/°C	Temperature In °C	Temperature Out °C
1	3	250	200
2	5	200	40
3	4	30	200
4	3	90	200

Use the MUMNE algorithm for heat-exchanger networks and a minimum approach temperature of 10°C.

a. Determine the temperature interval diagram.
b. Determine the cascade diagram, the pinch temperatures, and the minimum hot and cold utilities.
c. Determine the minimum number of heat exchangers above and below the pinch.
d. Determine a heat-exchange network above the pinch.
e. Determine the heat-exchange network below the pinch.

16. The temperature-interval diagram for a process is shown in Figure P15.16. For this process do the following.
    a. Compute the missing values of $\dot{m}C_p$ for streams 2 and 3, the missing Q values for temperature intervals B, C, and E, and the total of all intervals.
    b. Calculate the minimum hot and cold utility loads for this process subject to a minimum approach temperature of 20°C.
    c. Calculate the pinch temperatures for this process. You may assume that there is only one hot and one cold utility available.
    d. Calculate the minimum number of exchangers needed for the minimum energy case for above and below the pinch.
    e. Design the MUMNE network for the process.

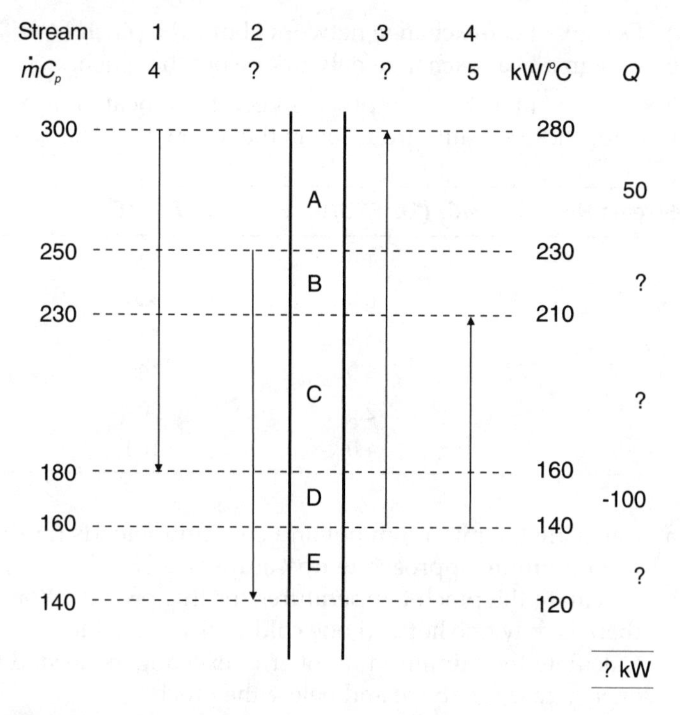

Figure P15.16 Temperature Interval Diagram for Problem 16

17. In a process design, the following process streams must be cooled or heated.

Stream No.	$\dot{m}C_p$ $(10^3$ BTU)/hr°F	Temperature In °F	Temperature Out °F
1	2	350	220
2	4	400	120
3	5	310	380
4	4	100	310

Use the MUMNE algorithm for heat-exchanger networks and a minimum approach temperature of 20°F.

a. Determine the temperature-interval diagram.
b. Determine the cascade diagram, the pinch temperatures, and the minimum hot and cold utilities.
c. Determine the minimum number of heat exchangers above and below the pinch.

d. Design a heat-exchange network above the pinch.
e. Design a heat-exchange network below the pinch.

18. Six streams in a chemical process need to be heated or cooled. The thermal data for these streams are given in the following table.

Stream No.	$\dot{m}C_p$ (10^3BTU/hr°F)	T_{in} (°F)	T_{out} (°F)
1	1.0	620	320
2	6.0	420	120
3	3.0	420	220
4	5.0	400	600
5	2.0	200	300
6	4.0	100	400

a. Calculate the minimum hot and cold utility loads for this process subject to a minimum approach temperature of 20°F.
b. Calculate the pinch temperatures for this process. You may assume that there is only one hot and one cold utility available.
c. Calculate the minimum number of exchangers needed for the minimum energy case for above and below the pinch.
d. Design the MUMNE network for the process.

19. If the streams in Problem 18 have the following film heat transfer coefficients, estimate the optimum minimum approach temperature for this problem.

$h_1 = 75$ BTU/hr/ft²°F
$h_2 = 25$ BTU/hr/ft²°F
$h_3 = 10$ BTU/hr/ft²°F
$h_4 = 45$ BTU/hr/ft²°F
$h_5 = 30$ BTU/hr/ft²°F
$h_6 = 30$ BTU/hr/ft²°F

20. Rework Problem 19, for the case when the streams require special materials of construction with the following material factors:

Stream 1, $F_M = 1.0$
Stream 2, $F_M = 2.0$
Stream 3, $F_M = 3.0$
Stream 4, $F_M = 7.0$
Stream 5, $F_M = 2.5$
Stream 6, $F_M = 1.0$

21. Consider the following process streams, all containing the same solute. The partition coefficient between both rich phases and the lean phase is 0.5 (i.e., $y = 0.5x$).

	kg/s	y_{in}	y_{out}
$R_1 = 1$	2	0.10	0.03
$R_2 = 2$	4	0.07	0.02
	kg/s	x_{in}	x_{out}
$L = 3$	4	0.00	0.06

Synthesize the MUMNE mass-exchange network for a minimum approach composition of $\delta = 0.02$.

22. Ethanol can be made by fermentation of grains such as corn. It can be fermented only up to a mass fraction of about 0.15 because higher concentrations kill the yeast. A big market for ethanol made this way is as an additive to gasoline.

    Suppose that you have 90 kg/h of gasoline, the lean stream, and you would like to increase its concentration of ethanol from zero mass fraction to 0.09. You have two rich ethanol-water streams from fermentation processes. One is at 60 kg/h and a mass fraction of 0.09, and the other is at 30 kg/h and a mass fraction of 0.12. It is desired to reduce the ethanol mass fraction in the rich streams to a mass fraction of 0.02. The partition coefficient between the rich and lean streams is 1.2 (i.e., $y = 1.2x$), where y is the mass fraction in the ethanol-water (rich) stream and x is the mass fraction of ethanol in the gasoline (lean) stream. Synthesize the MUMNE mass-exchange network for a minimum approach composition of $\delta = 0.01$.

23. Consider the following process streams, all containing the same solute. The partition coefficient between both rich phases and the lean phase is 0.5 (i.e., $y = 0.5x$).

	kg/s	y_{in}	y_{out}
$R_1 = 1$	4	0.10	0.06
$R_2 = 2$	6	0.07	0.05
	kg/s	x_{in}	x_{out}
$L_1 = 3$	4	0.10	0.16
$L_2 = 4$	1	0.06	0.12

Synthesize the MUMNE mass-exchange network for a minimum approach composition of $\delta = 0.02$.

24. Consider the following process streams, all containing the same solute. The partition coefficient between both rich phases and the lean phase is 0.667 (i.e., $y = 0.667x$).

	kg/s	y_{in}	y_{out}
$R_1 = 1$	4	0.100	0.060
$R_2 = 2$	6	0.070	0.050
	kg/s	x_{in}	x_{out}
$L_1 = 3$	4	0.065	0.125
$L_2 = 4$	1	0.000	0.080

Synthesize the MUMNE mass-exchange network for a minimum approach composition of $\delta = 0.010$.

25. Consider the following process streams, all containing the same solute. The partition coefficient between both rich phases and the lean phase is 0.667 (i.e., $y = 0.667x$).

	kg/s	y_{in}	y_{out}
$R_1 = 1$	1	0.100	0.060
$R_2 = 2$	6	0.070	0.020
	kg/s	x_{in}	x_{out}
$L_1 = 3$	2	0.020	0.125
$L_2 = 4$	1	0.000	0.080

Synthesize the MUMNE mass-exchange network for a minimum approach composition of $\delta = 0.010$.

SECTION

4

Analysis of Process Performance

In the previous three sections, we focused on problems associated with the design, synthesis, and economics of a new chemical process, where there was freedom to select equipment. In this section, we explore problems associated with an existing chemical process. Two important factors must be understood in dealing with existing equipment.

1. Changes are limited by the performance of the existing equipment.
2. Any changes in operation of the process cannot be considered in isolation. The impact on the total process must always be considered.

Over the ten to thirty years or more a plant is expected to operate, process operations may vary. A plant seldom operates at the original process conditions provided on the design PFD. This is due to the following.

- **Design/Construction:** Installed equipment is often oversized. This reduces risks resulting from inaccuracies in design correlations, uncertainties in material properties, and so on.
- **External Effects:** Feed materials, product specifications and flowrates, environmental regulations, and costs of raw materials and utilities all are likely to change during the life of the process.
- **Replacement of Equipment:** New and improved equipment (or catalysts) may replace existing units in the plant.
- **Changes in Equipment Performance:** In general, equipment effectiveness degrades with age. For example, heat-transfer surfaces foul, packed towers

develop channels, catalysts lose activity, and bearings on pumps and compressors become worn. Plants are shut down periodically for maintenance to restore equipment performance.

To remain competitive, it is necessary to be able to alter process operations in response to changing conditions. Therefore, it is necessary to understand how equipment performs over its complete operating range to quantify the effects of changing process conditions on process performance.

The material provided in this section involves several categories of performance problems.

1. **Predictive Problems:** An examination of the changes that take place for a change in process or equipment input and/or a change in equipment effectiveness.
2. **Diagnostic/Troubleshooting Problems:** If a change in process output (process disturbance or upset) is observed, the cause (change in process input, change in equipment performance) must be identified.
3. **Control Systems Problems:** If a change in process output is undesirable or a change in process input or equipment performance is anticipated, compensating action that can be taken to maintain process output must be identified.
4. **Debottlenecking Problems:** Often, a process change is necessary or desired, such as scale-up (increasing production capacity) or allowance for a change in product or raw material specifications. Identification of the equipment that limits the ability to make the desired change or constrains the change is necessary.

This section introduces the basic principles by which existing equipment and processes can be evaluated, operated, controlled, and subjected to changes in operating conditions. This material is treated in the following chapters.

Chapter 16: Process Input/Output Models
The basic structure of performance problems is considered in the context of an input, an output, and a system.

Chapter 17: Tools for Evaluating Process Performance
Tools needed to analyze performance problems, such as ratios, limiting resistances, and base cases, are presented.

Chapter 18: Performance Curves for Individual Unit Operations
The performance of single pieces of equipment is analyzed for changes in process conditions, flowrates, utility flowrates, and degradation of equipment. It is assumed that the equipment has been designed and built and that the physical parameters of the equipment cannot change.

Chapter 19: Performance of Multiple Unit Operations
The performance of multiple pieces of equipment is analyzed. It is shown how a change in one unit affects the performance of another unit.

Chapter 20: Reactor Performance
Evaluation of the performance of different types of reactors is illustrated. The choice of process conditions to change selectivity is addressed.

Chapter 21: Regulating Process Conditions
Using examples from earlier chapters in this section, it is shown how a deviation in output from a piece of equipment can be controlled by altering an input. This is different from, and complementary to, what is treated in a typical process control class.

Chapter 22: Process Troubleshooting and Debottlenecking
Case studies are presented to introduce the philosophy and methodology for process troubleshooting and debottlenecking.

CHAPTER

16

Process Input/Output Models

Imagine you are in charge of operations for a portion of a chemical plant when you are informed that the pressure of a distillation column has begun to rise slowly. You know that if the pressure continues to rise the structural integrity of the column may be in jeopardy or that a relief valve will open and release valuable product to the stack. Both of these scenarios have serious negative consequences. In order to solve the problem without shutting down the plant, which might be very costly, it is necessary to understand how the distillation column performs in order to diagnose the problem and determine a remedy.

Dealing with the day-to-day performance of a chemical process differs from design of a new process. When designing a new process, there is freedom to choose equipment specifications as long as it produces the desired performance. However, once a piece of equipment has been designed and constructed, it performs in a unique manner. The specific performance of a piece of equipment must be considered when operating such equipment.

When designing equipment, alternative specifications can yield the same operating results. When dealing with existing equipment, day-to-day operation is constrained by the fixed equipment characteristics.

Although it may be tempting to proceed immediately to the analysis of performance of individual pieces of equipment, it is important to develop an intuitive understanding of how equipment performs. This chapter introduces a framework by which individual equipment and complete process performance may be understood. The following chapters develop and use tools for analysis of system performance. Having an intuitive understanding of process and equipment performance is a necessary complement to the ability to do numerous, repetitive, high-speed calculations.

To determine the outputs of a unit operation or chemical process we must know the process inputs and understand the performance for each unit of equipment involved in the process. The relationship between input and output can be described as

$$\text{Output} = f(\text{Input, Unit Performance}) \tag{16.1}$$

Input changes are the driving force for change.

Unit performance defines the characteristics of fixed equipment by which inputs are changed to outputs.

To change process output, one must alter process input and/or equipment performance. Conversely, the cause of a process output disturbance is a change in process input, equipment performance, or both.

Changes in process output result from changes in process inputs and/or equipment performance.

16.1 REPRESENTATION OF PROCESS INPUTS AND OUTPUTS

Figure 16.1 shows us how outputs are connected to inputs through the process system. The process system may consist of a single piece of equipment or several unit operations, with the output from one unit becoming the input to another unit. As a result, a change of input to one unit operation affects all of the down-

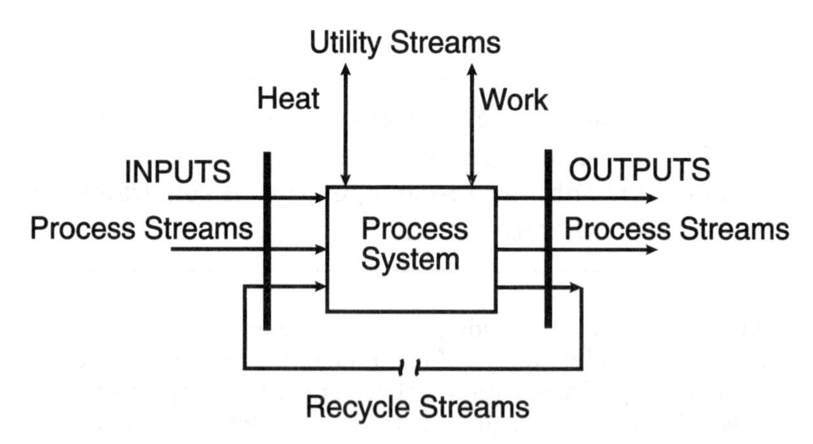

Figure 16.1 Input/Output Diagram for Chemical Process

stream units. For systems containing recycle streams the output from a unit operation returns to affect its input.

Figure 16.1 provides the input/output representation of a chemical process. It shows

a. **Process System** (shown in the center box): This may consist of a single process unit, or a collection of process units (such as a distillation system) that, taken together, performs a specific function or a complete chemical process.
b. **Process Flow Streams:** These are divided into two types.
   i. **Input Streams** (shown on the left): They consist of process flow streams entering a unit operation or process.
   ii. **Output Streams** (shown on the right): They consist of process flow streams exiting a unit operation or process.
c. **Utility Streams:** Utility streams provide for the transfer of energy to or from process streams or process units. They are used to regulate temperatures and pressures required in the process. Utility stream inputs and outputs are shown at the top of Figure 16.1.
   i. **Heat:** Heat is most often provided by the flow of a heating or cooling medium but can be provided by other sources such as electrical energy.
   ii. **Work:** This represents shaft work for a pump or compressor doing work on the fluid.
d. **Recycle Streams:** These streams are internal to the process. They are critical to the operation of the process system. Recycle streams must be identified, and their impact on the process understood. A recycle stream is illustrated by **cutting** or **tearing** the recycle stream (see Chapter 13). The **cut ends** create a set of pseudo-input and pseudo-output streams. They are shown along the bottom of the process system block, connected by a broken line. The pseudo-output stream is fed back to the process as a pseudo-input stream. Recycle streams fall into two categories:
   i. Process recycle streams recover raw materials and/or energy to reduce the cost of raw materials, waste disposal, and energy.
   ii. Equipment recycle streams affect equipment performance. As an example, increasing reflux to a distillation tower improves separations.

Example 16.1 illustrates how to represent process inputs and outputs for individual pieces of equipment.

Example 16.1

Draw an input/output model for

a. A pump
b. A heat exchanger in which a process stream is heated using condensing steam
c. A distillation column

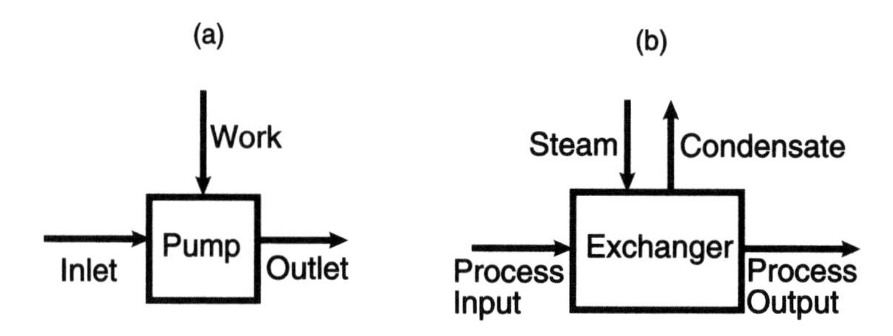

Figure E16.1(a) and (b) Solutions to Example 16.1

Solution

a. The result is shown in Figure E16.1(a). The mass balance on a pump has one input and one output, as shown. The energy balance on a pump (or one's intuitive understanding of how a pump works) indicates that there is energy input to the process stream, as shown.

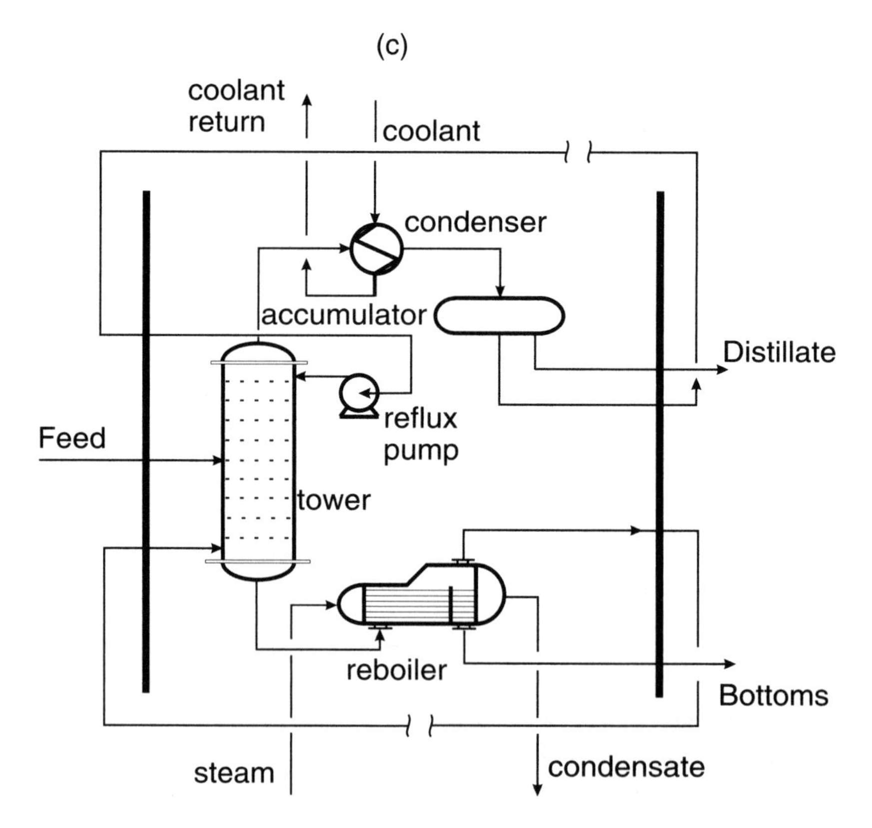

Figure E16.1(c) Solution to Example 16.1(c)

b. The result is shown in Figure E16.1(b). A heat exchanger using a utility has one process stream input and one process stream output. The utility also flows in (steam) and out (condensate) of the heat exchanger.

c. The result is shown in Figure E16.1(c). The process streams are the feed, bottoms, and distillate. The utility streams are steam for the reboiler and cooling water for the condenser. Finally, the reflux from the condenser and the boil-up from the reboiler are represented as recycle streams, which both leave and enter across system boundaries.

Using the input/output diagram in Figure 16.1, two classifications are identified for process analysis problems encountered in this text.

1. **Process Design Analysis:** Input and output streams are fixed. A process system is designed to transform the input into the output.
2. **Process Performance Analysis:** Input and equipment are fixed. The outputs are determined by the process system.

16.2 ANALYSIS OF THE EFFECT OF PROCESS INPUTS ON PROCESS OUTPUTS

An intuitive understanding of the effect of process inputs on process outputs can be obtained by categorizing the relationships used to analyze equipment. We break down these relationships into two groups.

1. **Equipment-Independent Relationships:** These relationships are independent of equipment specifications. Material balance, energy balance, kinetics, and equilibrium relationships are examples of equipment-independent relationships.
2. **Equipment-Dependent Relationships:** These are often called **design equations** and involve equipment specifications. The heat-transfer equation (contains area, A) and the frictional pressure relationship (contains pipe size, D_p, and the equivalent length, L_{eq}) are examples of equipment-dependent relationships.

Unless equipment-dependent relationships are known and used, the impact of a change of process input on the output cannot be determined correctly.

Example 16.2 illustrates equipment-independent and equipment-dependent relationships.

Example 16.2

What relationships are used to analyze the following pieces of equipment? Classify these as equipment independent and equipment dependent.

a. Heat exchanger
b. Adiabatic reactor
c. Multistage extraction

Solution

a. The energy balance—which for a temperature change without phase change is $Q = \dot{m}C_p\Delta T$, and for a phase change is $Q = \dot{m}\lambda$—is an equipment-independent relationship. The design (or performance) equation for a heat exchanger is $Q = UA\Delta T_{lm}F$. It is observed that only the equipment-dependent relationship contains equipment specifications—in this case, the area for heat transfer, A.

b. In a reactor, equipment-independent relationships may include kinetics and/or the definition of equilibrium. If there are no equilibrium limitations, the kinetics may take the form $r = kC_AC_B$, where A and B are the reactants. All kinetic expressions are equipment independent and are only functions of temperature, pressure, and concentration. The energy balance, which is necessary to determine the outlet temperature of the adiabatic reactor, is another relevant equipment-independent relationship. One equipment-dependent relationship is the design equation for the reactor, and it depends upon the type of reactor. For example, for a CSTR, this relationship is $V/F_{Ao} = X/(-r_A)$. It is observed that this relationship contains an equipment specification—in this case, the reactor volume, V. Another equipment-dependent relationship is for the pressure drop through the reactor, which is important in analyzing changes in flowrates.

c. In any multistage separation involving a mass separating agent, the mass balance is the key equipment-independent relationship. For example, it is understood that increasing the solvent rate results in a better separation. Another equipment-independent relationship is the statement of phase equilibrium. An equipment-dependent relationship does not exist as a simple, closed-form equation, except in certain limiting cases. However, an intuitive understanding of the equipment-dependent relationship is possible. For example, it is understood that increasing the number of stages will (usually) improve the separation. For the limiting case of dilute solutions, all the above relationships can be expressed as a single, closed-form equation, the Kremser equation. This will be discussed in Chapter 18. Another equipment-dependent relationship is the efficiency of the equipment, which is dependent on the liquid and vapor flowrates.

16.3 A PROCESS EXAMPLE

We assume that both the equipment-independent and equipment-dependent relationships are understood, and we focus our attention on integrating these relationships to analyze complete processes. In a chemical process, a change in input to one piece of equipment or a change in performance of that piece of equipment affects more than just the output from that piece of equipment. Because the output from the piece of equipment in question usually becomes the input to another piece of equipment, the disturbance caused by the change in the original piece of equipment can propagate through the process.

This is important in two types of problems that are discussed in later chapters. In one type of problem, the effect of a planned disturbance, such as a 10% scale-up in production, can be quantified only by considering the effect of a changed output from one unit on the subsequent unit. Similarly, in a second type of problem, an observed disturbance in the output of a particular unit may trace back to a disturbance in a unit far removed from the observed disturbance.

In order to analyze these problems, the input/output model for an entire process must be understood, as must the equipment-independent and equipment-dependent relationships for each piece of equipment in the process. This is illustrated in Example 16.3.

Example 16.3

Consider the toluene hydrodealkylation process illustrated in Figure 1.5. Draw and label an input/output diagram, similar to Figure 16.1, for this process. Label all process inputs and outputs, all utility streams, and all recycles. Then pick any two adjacent units and draw their input/output diagrams together, showing how the output from one unit affects the adjacent unit.

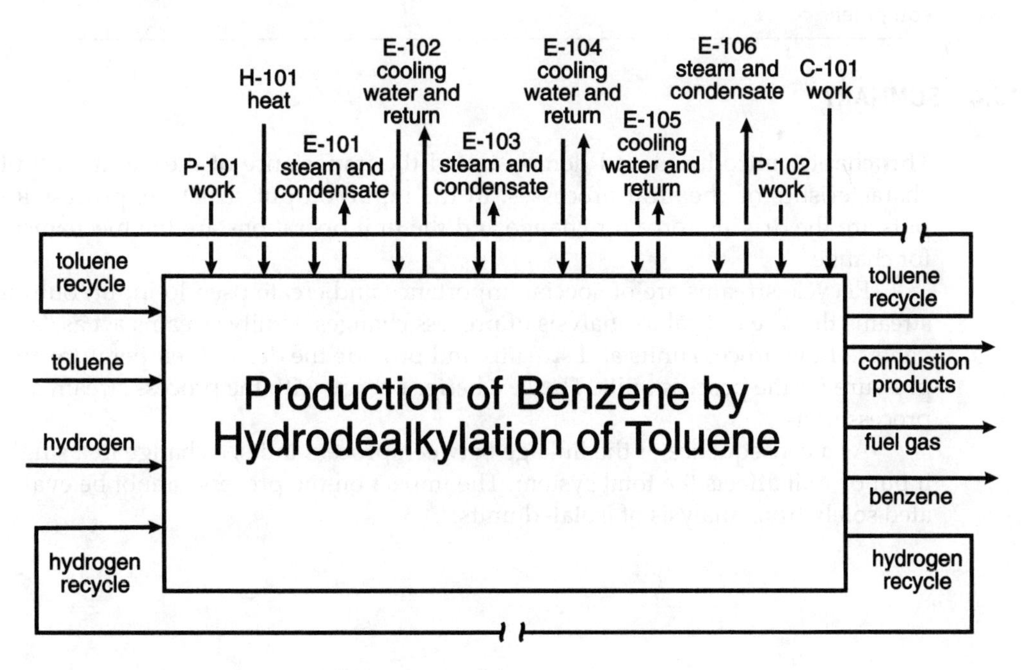

Figure E16.3(a) Solution to Example 16.3

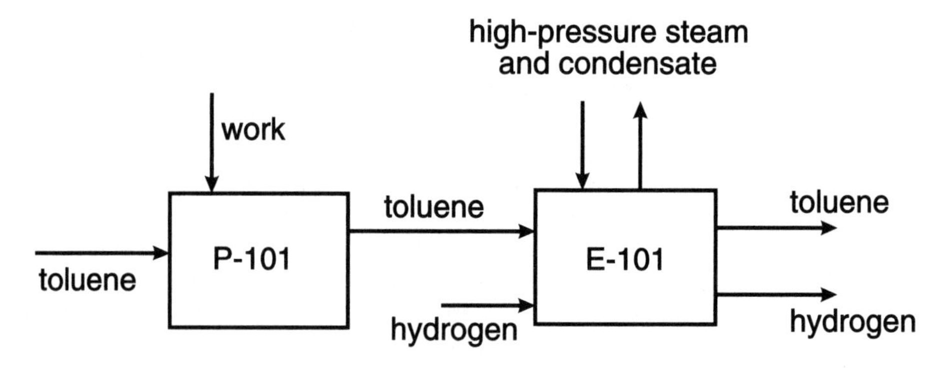

Figure E16.3(b) One Possible Solution to Example 16.3

Figure E16.3(a) is the input/output diagram for the entire process. It is observed that there are two recycles: one for hydrogen and one for toluene, both unreacted reactants. This is a particularly useful representation because all of the inputs and outputs also represent cash flows. By showing all of the utilities, it is less likely that one could be omitted from a cash flow analysis.

Figure E16.3(b) is an input/output diagram for the toluene feed pump (P-101) and the heat exchanger (E-101). This diagram makes it evident that any disturbance in the feed to or the operation of the pump not only affects the pump output but also affects the heat-exchanger output. Clearly, this disturbance would also propagate to other downstream equipment.

16.4 SUMMARY

This chapter introduced and demonstrated the importance of the input/output characteristics of chemical processes. In the input/output structure, process inputs are the driving force for change and the unit operations are the mechanism for change.

Recycle streams are of special importance and create pseudo-input/output streams that are critical to analysis of process changes. Utility streams act as "servants" of the process units and streams and provide the desired temperature and pressure for the process units. They exchange energy with the process stream and process units.

As a consequence of the linkage between process units, a change in a single input or unit affects the total system. The impact on the process cannot be evaluated solely from analysis of isolated units.

PROBLEMS

1. Draw input/output diagrams for the following pieces of equipment:
   a. A fluidized bed
   b. A turbine
   c. A pump
   d. A stripper
   e. An adiabatic batch reactor
   f. A semibatch reactor with heat removal
   g. A heat exchanger between two process streams (no utilities)
2. Write down the equipment-independent and equipment-dependent relationships for each piece of equipment listed in Problem 1.
3. Write down the equipment-independent and equipment-dependent relationships for a distillation column.
4. Draw input/output diagrams for the following processes. Pick any two adjacent pieces of equipment, and draw input/output diagrams for these, clearly showing how the output from one affects the input to the other.
   a. Ethylbenzene, described in Appendix B, Project B.2
   b. Styrene, described in Appendix B, Project B.3
   c. Drying oil, described in Appendix B, Project B.4
   d. Maleic anhydride, described in Appendix B, Project B.5
   e. Ethylene oxide, described in Appendix B, Project B.6
   f. Formalin, described in Appendix B, Project B.7

CHAPTER

17

Tools for Evaluating Process Performance

Several important computational and graphical tools used to analyze plant operations and changes in process design are developed and demonstrated in this chapter. Many of these tools are keyed to actual system performance. Tools presented are used extensively in future chapters.

Intuitive understanding is emphasized over computational complexity. All problems can be solved using a calculator without special functions. Graphical representations are also emphasized. They illustrate the characteristic behavior of a unit or process over a wide range of operations. Graphical presentations reveal critical operating regions that demand careful assessment.

17.1 KEY RELATIONSHIPS

In analyzing equipment performance, there are certain key relationships that are used over and over again. These are shown in Table 17.1. Although these should be familiar to the reader, they are important enough to be reviewed here.

The first entry deals with the pressure drop due to frictional losses in a pipe. The key relationship is the proportionality between pressure drop and the square

Table 17.1 Typical Key Performance Relationships

Situation	Equation	Trends	Comments
Frictional loss for fluid flow	$\Delta P = \frac{2\rho f L_{eq} u^2}{D}$	$\lvert\Delta P\rvert \propto u^2$ $\lvert\Delta P\rvert \propto D^{-5}$ $\lvert\Delta P\rvert \propto L$	Assumes fully developed turbulent flow, i.e., constant friction factor; for laminar flow $\Delta P \propto D^{-4}$.
Heat transfer	Of form $\left(\frac{hD}{k}\right) = c\left(\frac{Du\rho}{\mu}\right)^a\left(\frac{\mu C_p}{k}\right)^b$	$h_i \propto v^{0.8}$ inside closed channels $h_o \propto v^{0.6}$ cross flow outside pipes	Equations given are for no phase change; if phase change, weak flow dependence, but some ΔT dependence.
Kinetics	$r = k\Pi c_i^{a_i}$ $k = k_o e^{-\left(\frac{E}{RT}\right)}$	$\ln k$ vs $1/T$ is linear	As $T\uparrow, k\uparrow$ for ideal gases $P\uparrow, c_i\uparrow$, so $r\uparrow$.
Reactor	mixed flow: $\frac{V}{F_{A0}} = \frac{\tau}{C_{A0}} = \frac{X}{-r_A}$ plug flow: $\frac{V}{F_{A0}} = \frac{\tau}{C_{A0}} = \int_0^X \frac{dX}{-r_A}$	$\tau \propto V$ as $\tau\uparrow$ or $V\uparrow, X\uparrow$	τ is space time, V is reactor volume, X is conversion of limiting reactant, A assumes one reaction and constant volumetric flowrate.
Separator using mass separating agent	Not necessarily described by a single equation	As flow of mass separating agent $\uparrow$, or as number stages or height of packed tower $\uparrow$, degree of separation $\uparrow$	For certain cases, there are limitations to the effect of increasing number of stages or packed tower height.
Distillation	Not necessarily described by a single equation	As reflux ratio $\uparrow$, degree of separation $\uparrow$	Complicated analysis; see Chapter 18.

of the velocity, and the proportionality between pressure drop and length (or equivalent length) of pipe. Although it is intuitive that the pressure drop should increase with length of pipe and with increasing velocity, the quantitative relationship reveals the square dependence on velocity. Similarly, although it is intuitive that a larger pipe diameter should reduce frictional losses, the negative fifth-power dependence is revealed only by the equation describing the actual physical situation. Similarly, the general trends for heat transfer coefficients and for rate constants may be intuitive, but the equation describing the actual physical situation shows that quantitative dependence. For reactors and separators, only the intuitive trends can be shown because the exact, quantitative dependence is either situation specific or not easily quantifiable using a closed-form equation.

The relationships in Table 17.1 will be used extensively in Chapters 18, 19, and 20.

17.2 THINKING WITH EQUATIONS

It is possible to quantify equipment performance without resorting to extensive, detailed calculations. This involves using equations to understand trends. The first step is to identify the equations necessary to quantify a given situation. Wales and Stager [1] have termed this "thinking with equations" and have used the acronym GENI to describe the associated problem-solving strategy. The second step involves predicting trends from equations. These methods are described and then illustrated in an example.

17.2.1 GENI

GENI is a method for solving quantitative problems. The name GENI is an acronym for the four steps in the method.

1. **Goal:** Identify the goal. This is usually the unknown that needs to be calculated.
2. **Equation:** Identify the equation that relates the unknown to known values or properties.
3. **Need:** Identify additional relationships that are needed to solve the equation in Step 2.
4. **Information:** List additional information that is available to determine whether what is needed in Step 3 is known. If the correct information is not known, the need becomes the new goal and the process is repeated.

17.2.2 Predicting Trends

The following method can be used to predict trends from equations known to apply to a given physical situation. In this method, there are four possible modifiers to a term in an equation.

1. ¢: value remains constant.
2. ↑: value increases.
3. ↓: value decreases.
4. ?: value change not known.

Each variable in an expression has one of these identifiers appended. If x is a constant then $x \rightarrow x(¢)$. The symbol in parentheses identifies the effect the variable has on the term; that is, reducing a value in the denominator of a term increases the term.

The application of the techniques described in Sections 17.2.1 and 17.2.2 is illustrated in Example 17.1.

Example 17.1

For a bimolecular, elementary, gas-phase reaction, the rate expression is known to be $-r_A = kC_AC_B$.

a. What is the effect on the reaction rate of increasing the reaction pressure by 10% while maintaining constant temperature?

b. What is the effect on the reaction rate of increasing the reaction temperature by 10% while maintaining constant pressure?

The **goal** is to determine the effect of increasing pressure on the reaction rate. The **equation** containing the unknown value, which is the reaction rate, r_A, is given. We **need** a relationship between pressure and at least one variable in the equation. We have **information,** from the ideal gas law, that tells us that $C_i = P_i/RT$, where i is A or B.

Now that we have the necessary relationships, an intuitive understanding can be obtained by predicting trends.

a. First, let us examine the ideal gas relationship. If the pressure increases with the temperature remaining constant, then the ideal gas relationship can be written as

$$C_i(\uparrow) = y_i(¢)P(\uparrow)/R(¢)T(¢) = P_i(\uparrow)/R(¢)T(¢)$$

Therefore, the concentration increases because the pressure increases while the temperature and the gas constant remain constant. Intuitively, if the pressure increases by 10%, the concentration must increase by 10%.

Moving to the rate expression, because the rate constant is not affected by pressure, the resulting trend is predicted by

$$r_A(\uparrow) = k(¢)C_A(\uparrow)C_B(\uparrow)$$

which means that the reaction rate increases. Intuitively, because each concentration increases by 10%, then the reaction rate changes by a factor of $(1.1)^2 = 1.21$, an increase of 21%.

b. In this case, from the ideal gas relationship,

$$C_i(\downarrow) = y_i(¢)P(¢)/R(¢)T(\uparrow) = P_i(¢)/R(¢)T(\uparrow)$$

From the rate expression, because the rate constant is a function of temperature,

$$r_A(?) = k(\uparrow)C_A(\downarrow)C_B(\downarrow)$$

Because the concentrations decrease and the rate constant increases, it cannot be determined, a priori, whether the rate increases or decreases. If the activation energy in the rate constant were known, a quantitative evaluation could be made.

17.3 BASE-CASE RATIOS

The calculation tool provided in this section combines use of fundamental relationships with plant operating data to form a basis for predicting changes in system behavior.

The ability to predict changes in a process design or in plant operations is improved by anchoring an analysis to a base case. For design changes, we would like to identify a design proven in practice as the base case. For operating plants, actual data are available and are chosen as the base case. It is important to put this base case into perspective. Assuming that there are no instrument malfunctions and the operating data are correct, then these data represent a real operating point at the time the data were taken. As the plant ages, the effectiveness of process units changes and operations are altered to account for these changes. As a consequence, recent data on plant operations should be used in setting up the base case.

Establish predictions of process changes on known operating data, not on design data.

The base-case ratio integrates the "best available" information from the operating plant with design relationships to predict process changes. It is an important and powerful technique with wide application. The base-case ratio, X, is defined as the ratio of a new-case system characteristic, x_2, to the base-case system characteristic, x_1.

$$X \equiv x_2/x_1 \tag{17.1}$$

Using a base-case ratio often reduces the need for knowing actual values of physical and transport properties (physical properties refer to thermodynamic and transport properties of fluids), equipment, and equipment characteristics. The values identified in the ratios fall into three major groups. They are defined below and applied in Example 17.2.

1. **Ratios Related to Equipment Sizes** (L_{eq}, equivalent length; diameter, D; surface area, A): Assuming that the equipment is not modified, these values are constant, the ratios are unity, and these terms cancel out.
2. **Ratios Related to Physical Properties** (such as density, ρ; viscosity, μ): These values can be functions of material composition, temperature, and pressure. Absolute values are not needed, only the functional relationships. Quite often, for small changes in composition, temperature, or pressure, the properties are unchanged, and the ratio is unity and cancels out. An exception to this is gas-phase density.

3. **Ratios Related to Stream Properties:** These usually involve velocity, flowrate, concentration, temperature, and pressure.

> **Using the base-case ratio eliminates the need to know equipment characteristics and reduces the amount of physical property data needed to predict changes in operating systems.**

The base-case ratio is a powerful and straightforward tool to analyze and predict process changes. This is illustrated in Example 17.2.

Example 17.2

It is necessary to scale up production in an existing chemical plant by 25%. It is your job to determine whether a particular pump has sufficient capacity to handle the scale-up. The pump's function is to provide enough pressure to overcome frictional losses between the pump and a reactor.

The relationship for frictional pressure drop is given in Table 17.1. This relationship is now written as the ratio of two base cases as follows:

$$\frac{\Delta P_2}{\Delta P_1} = \frac{2\rho_2 f_2 L_{eq2} u_2^2 D_1}{2\rho_1 f_1 L_{eq1} u_1^2 D_2}$$

Because the pipe has not been changed, the ratios of diameters (D_2/D_1) and lengths (L_{eq2}/L_{eq1}) are unity. Because a pump is used only for liquids, and liquids are (practically) incompressible, the ratio of densities is unity. If the flow is assumed to be fully turbulent, which is usually true for process applications, the friction factor is not a function of Reynolds number. (This fact should be checked for a particular application.) Therefore, the friction factor is constant, and the ratio of friction factors is unity. The above ratio reduces to

$$\frac{\Delta P_2}{\Delta P_1} = \frac{u_2^2}{u_1^2} = \frac{\dot{m}_2^2/A_2^2\rho_2^2}{\dot{m}_1^2/A_1^2\rho_1^2} = \frac{\dot{m}_2^2}{\dot{m}_1^2}$$

where the second equality is obtained by substituting for u_i in numerator and denominator using the mass balance $\dot{m}_i = \rho_i A_i u_i$, canceling the ratio of densities for the same reason as above, and canceling the ratio of cross-sectional areas because the pipe has remained unchanged. Therefore, by assigning the base-case mass flow to have a value of 1, for a 25% scale-up, the new case has a mass flow of 1.25, and the ratio of pressure drops becomes

$$\frac{\Delta P_2}{\Delta P_1} = \left(\frac{\dot{m}_2}{\dot{m}_1}\right)^2 = \left(\frac{1.25}{1}\right)^2 = 1.56$$

Thus, the pump must be able to deliver enough head to overcome 56% additional frictional pressure drop while pumping 25% more material.

It is important to observe that Example 17.2 was solved without knowing any details of the system. The pipe diameter, length, and number of valves and fittings were not known. The liquid being pumped, its temperature, and its density were not known. Yet the use of base-case ratios along with simple assumptions permitted a solution to be obtained. This illustrates the power and simplicity of base-case ratios.

17.4 ANALYSIS OF SYSTEMS USING CONTROLLING RESISTANCES

Design relationships for many operations such as fluid flow, heat transfer, mass transfer, and chemical reactors all involve rate equations of the general form

$$\text{Rate} = \text{Driving Force/Resistance} \tag{17.2}$$

For resistances in series,

$$R_T\,(\text{Total Resistance}) = R_1 + R_2 + R_3 + \cdots + R_N \tag{17.3}$$

For certain situations, one resistance dominates all other resistances. For example, if resistance R_1 dominates, then

$$R_1 >> R_2 + R_3 + \cdots + R_N \tag{17.4}$$

and

$$R_T \approx R_1 \tag{17.5}$$

where R_1 represents the "controlling resistance." Other resistances have little impact on the rate. Only those factors that impact R_1 have a significant impact on the rate.

As an example, an overall heat transfer coefficient for a clean (nonfouling) service, U_o, can be expressed as

$$\frac{1}{U_o} = \frac{1}{h_o} + \frac{D_o \ln\left(\frac{D_o}{D_i}\right)}{2k} + \frac{D_o}{D_i h_i} \tag{17.6}$$

where D_o and D_i are the outer and inner tube radii, respectively, k is the thermal conductivity of the tube material, and h_o and h_i are the outer and inner heat transfer coefficients. We will assume that Equation 17.6 can be simplified by assuming $D_o \approx D_i$, and that the conduction resistance is negligible, so that

$$\frac{1}{U_o} = \frac{1}{h_o} + \frac{1}{h_i} \tag{17.7}$$

Suppose that we wanted to use a base-case ratio of a new overall heat transfer coefficient, U_{o2}, to an original overall heat transfer coefficient, U_{o1}. For a situation when no phase change is occurring on either side of a shell-and-tube heat exchanger, from Equation 17.7, this ratio would be

$$\frac{\left(\frac{1}{U_{o1}}\right)}{\left(\frac{1}{U_{o2}}\right)} = \frac{U_{o2}}{U_{o1}} = \frac{\frac{1}{\alpha u_{i1}^{0.8}} + \frac{1}{\beta u_{o1}^{0.6}}}{\frac{1}{\alpha u_{i2}^{0.8}} + \frac{1}{\beta u_{o2}^{0.6}}} \tag{17.8}$$

where the individual heat transfer coefficients have been expressed as indicated in Table 17.1, and the proportionality factors α and β contain the lead constant and all of the properties contained in the dimensionless groups other than the heat transfer coefficient and the velocity. From Equation 17.8, it would not be possible, as it was in Example 17.2, to determine quantitatively how the overall heat transfer coefficient changes with changing velocity (i.e., mass flowrate) without knowing values for all of the physical properties. However, if it could be assumed that one resistance were dominant, the problem would be greatly simplified. For example, let us assume that a heat transfer fluid in the shell is heating a gas in the tubes. It is likely that the resistance in the tubes would dominate, given the low film heat transfer coefficient for gases. Therefore, the base-case ratio would reduce to

$$\frac{U_{o2}}{U_{o1}} = \frac{h_{i2}}{h_{i1}} = \left(\frac{u_{i2}}{u_{i1}}\right)^{0.8} \tag{17.9}$$

and it would be possible to predict a new heat transfer coefficient without having values for the physical properties.

The reduction of the heat transfer coefficient ratio using a limiting resistance described above is very powerful but must be used with care. It is not valid for all situations. It is most likely valid when one resistance is for a boiling liquid or condensing steam and the other resistance is for a liquid or gas without phase change. In other situations, the base-case ratio can be determined from Equation 17.8 only if the relative magnitude for each resistance is known for the base case, as is illustrated in Example 17.4. It is also important to understand that Equations 17.8 and 17.9 do not represent all possible base-case ratios arising in heat transfer. Each situation must be analyzed individually to ensure that a correct solution is obtained. Problems of this type are given at the end of the chapter.

Situations in which the simplifications leading to Equation 17.9 are valid, along with situations that require a form for the base-case ratio similar to Equation 17.8, will be applied to heat-exchanger performance in Chapter 18. Examples of how Equations 17.8 and 17.9 can be applied are presented in Examples 17.3 and 17.4.

Example 17.3

It is desired to scale down process capacity by 25%. In a particular heat exchanger, process gas in the tubes is heated by condensing steam in the shell. By how much will the overall heat transfer coefficient for the heat exchanger change after scale-down?

Using the subscript 2 for the new, scaled-down operation and the subscript 1 for the original conditions, a scale-down of the mass flowrate of the process stream by 25% means that $u_{i2} = 0.75u_{i1}$. For a process gas heated by condensing steam, it is certain that the heat transfer coefficient for condensing steam will be at least 100 times as large as that for the process gas. Although Equation 17.9 is applicable, it is not derived from Equation 17.8 because of the phase change. The correct derivation is assigned in Problem 17.5 at the end of the chapter. The result is $U_{o2}/U_{o1} = (u_{i2}/u_{i1})^{0.8} = (0.75)^{0.8} = 0.79$, and the overall heat transfer coefficient is reduced by 21%.

Example 17.4

In a similar process as in Example 17.3, there is a heat exchanger between two process streams, neither of which involves a phase change. It is known that the resistances on the shell and tube sides are approximately equal before scale-down. By how much will the overall heat transfer coefficient for the heat exchanger change after scale-down by 25%?

Here, Equation 17.7 is required. Because both original resistances are equal, both original heat transfer coefficients will be denoted h_1. In the tubes, $h_{i2}/h_1 = (u_{i2}/u_{i1})^{0.8}$, and in the shell $h_{o2}/h_1 = (u_{o2}/u_{o1})^{0.6}$. Both velocity ratios are 0.75. Therefore, Equation 17.7 reduces to

$$\frac{U_2}{U_1} = \frac{\frac{1}{h_1} + \frac{1}{h_1}}{\frac{1}{h_1\left(\frac{u_{i2}}{u_{i1}}\right)^{0.8}} + \frac{1}{h_1\left(\frac{u_{o2}}{u_{o1}}\right)^{0.6}}} = \frac{2}{\frac{1}{(0.75)^{0.8}} + \frac{1}{(0.75)^{0.6}}} = 0.82$$

and the new heat transfer coefficient is reduced by 18%.

17.5 GRAPHICAL REPRESENTATIONS

At times, graphical representations are useful descriptions of physical situations. They provide a means to an intuitive understanding of a problem rather than a computational tool. Probably the best known example of this is the McCabe-Thiele diagram for distillation. Even though no one would design a distillation column nowadays using this method, the McCabe-Thiele diagram provides a means for an intuitive understanding of distillation. Another example is the reactor profiles illustrated in Chapter 20.

In this section, three graphical representations that will be used to analyze performance problems are discussed.

17.5.1 The Moody Diagram for Friction Factors

An example of a graphical representation commonly used for illustrative and for computational purposes is the Moody diagram, which gives the friction factor as a function of Reynolds number for varying roughness factors. It is illustrated in Figure 17.1. Although this diagram is often used for numerical calculations, it also provides an intuitive understanding of frictional losses. For example, it is observed that the friction factor increases as the pipe roughness increases. It is also observed that the friction factor becomes constant at high Reynolds numbers, and the dashed line represents the boundary between variable and constant friction factor.

17.5.2 The System Curve for Frictional Losses

In Table 17.1, an equation was given for frictional losses in a pipe. This equation is derived from the mechanical energy balance for constant density.

$$\frac{\Delta P}{\rho g} + \Delta z + \Delta\frac{u^2}{2g} - \frac{W_s}{g} + \frac{F_d}{g} = 0 \tag{17.10}$$

In Equation 17.10, Δ means out – in, and the specific energy dissipation due to drag, F_d, is a positive number. For a constant diameter length of pipe with no pump at constant elevation, $\Delta z = 0$, there is no work, the kinetic energy term is zero, and, using the expression for frictional losses, the result is

$$\Delta P = -\rho F_d = -\frac{2\rho f L_{eq} u^2}{D} \tag{17.11}$$

If fully developed turbulent flow is assumed, the friction factor, f, is constant. For process changes involving the same fluid, the density remains constant, and if the pipe is unchanged, the equivalent length and diameter are unchanged.

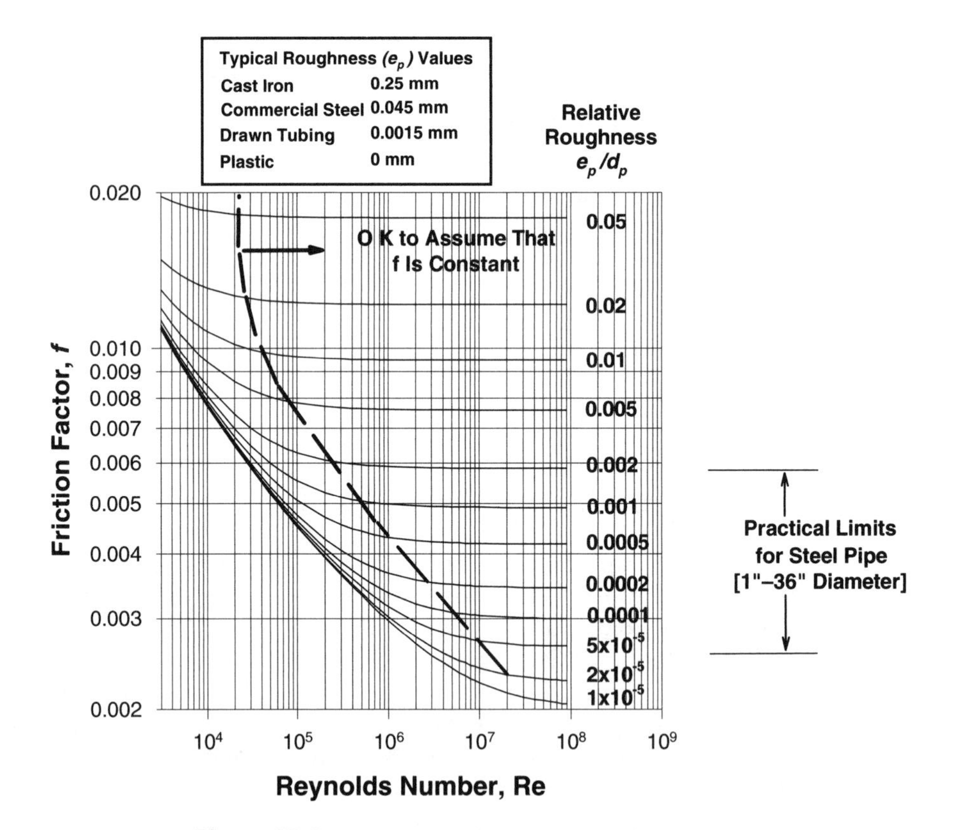

Figure 17.1 Moody Plot for Turbulent Flow in Pipes

Therefore, Equation 17.11 describes how the pressure drop in a length of pipe changes with flowrate or velocity. This can be plotted as ΔP versus velocity, and the result is a parabola passing through the origin. This is called the **system curve** and will be shown in Chapter 18 to be a useful tool in evaluating pump performance.

If, for example, there were an elevation change over the length of pipe in question, there would be an additional term in Equation 17.11, and the result would be a parabola with a nonzero intercept. Examples 17.5 and 17.6 illustrate how a system curve is obtained.

Example 17.5

Develop the system curve for flow of water at approximately 10 kg/s through 100 m of 2 in schedule 40 commercial steel pipe oriented horizontally.

The density of water will be taken as 1000 kg/m^3, and the viscosity of water will be taken as 1 mPa s (0.001 kg/m s). The inside diameter of the pipe is 0.0525 m. The Reynolds number can be determined to be 2.42×10^5. For a roughness factor of 0.001, $f = 0.005$. Equation 17.11 reduces to

$$\Delta P = -19u^2$$

with ΔP in kPa and u in m/s. This is the equation of a parabola, and it is plotted in Figure E17.5. Therefore, from either the equation or the graph, the pressure drop is known for any velocity.

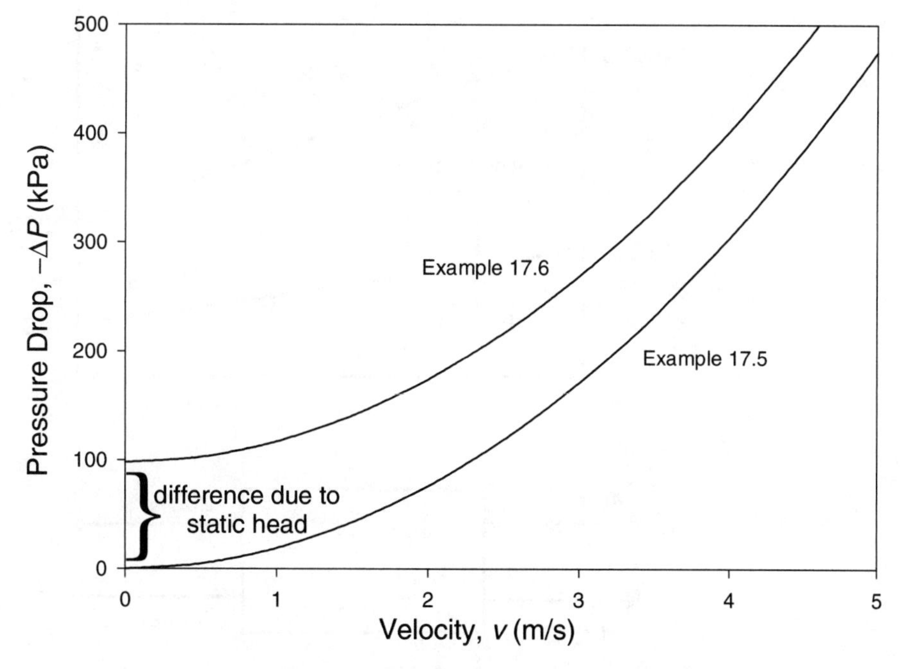

Figure E17.5 System Curves for Examples 17.5 and 17.6

Example 17.6

Repeat Example 17.5 for pipe with a 10 m vertical elevation change, with the flow from lower to higher elevation.

Here, the potential energy term from the mechanical energy balance must be included. The magnitude of this term is 10 m of water, so $\rho g \Delta z = 98$ kPa. Equation 17.10 reduces to

$$\Delta P = -(98 + 19u^2)$$

with ΔP in kPa and u in m/s. This equation is also plotted in Figure E17.5. It is observed that the system curve has the same shape as that in Example 17.5. This means that the frictional component is unchanged. The difference is that the entire curve is shifted up by the constant, static pressure difference.

17.5.3 The T-Q Diagram for Heat Exchangers

Another example of a useful diagram that illustrates the behavior of a piece of equipment is the T-Q diagram for a heat exchanger. Figure 17.2 illustrates a T-Q diagram for the countercurrent heat exchanger with no phase change shown in the figure.

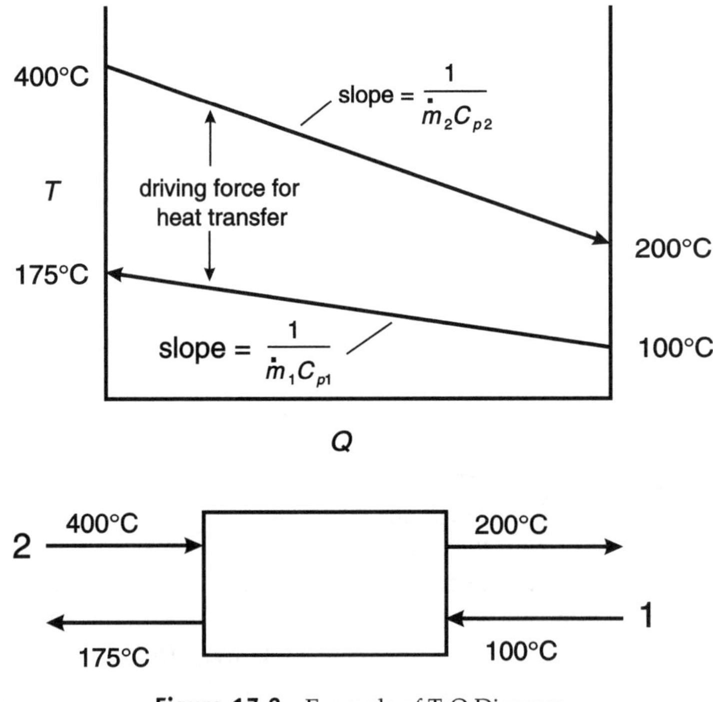

Figure 17.2 Example of T-Q Diagram

A T-Q diagram is a visual representation of the energy balance equation for each stream. Because, for single-phase streams with constant C_p and no pressure effect on the enthalpy,

$$Q = \dot{m}C_p\Delta T \tag{17.12}$$

process streams that undergo no phase change are represented by a line with a slope of $1/(\dot{m}C_p)$. Because, for pure component streams undergoing a phase change

$$Q = \dot{m}\lambda \tag{17.13}$$

process streams that involve constant temperature phase changes (pure component vaporizers and condensers) are represented by a horizontal line.

The temperature differences between the two streams shown on the T-Q diagram provide the actual temperature driving force throughout the exchanger. The greater the temperature separation, the larger the driving force (ΔT_{lm}), and the greater the heat transferred.

The representations described above are for simple situations commonly encountered. However, more complex cases exist. If the heat capacity is not constant, then the line for heat transfer without phase change is a curve. For phase changes involving multicomponent systems, because the bubble and dew points are at different temperatures, the line representing the phase change is not horizontal. For partial condensers and vaporizers, the representation on a T-Q diagram is a curve rather than a straight line.

The T-Q diagram reveals two important truths regarding heat transfer.

1. **Temperature Lines Cannot Cross:** This is an impossible situation. Temperature lines will never cross when dealing with operating equipment. If a temperature cross is encountered in doing a calculation, an error has been made.
2. **Temperature Lines Should Not Approach Each Other Too Closely:** As temperature lines approach each other, the area required for a heat exchanger approaches infinity. The point of closest approach is called the pinch point. When dealing with multiple heat exchangers, the pinch point is a key to heat-exchanger network integration, that is, determining the configuration for most efficient heat transfer between hot and cold streams. This concept was discussed in Chapter 15.

Example 17.7 illustrates the construction of T-Q diagrams.

Example 17.7

Sketch a T-Q diagram for the following situations.

a. A single-phase process stream is heated from 100°C to 200°C by condensation of saturated steam to saturated liquid at 250°C in a countercurrent heat exchanger.

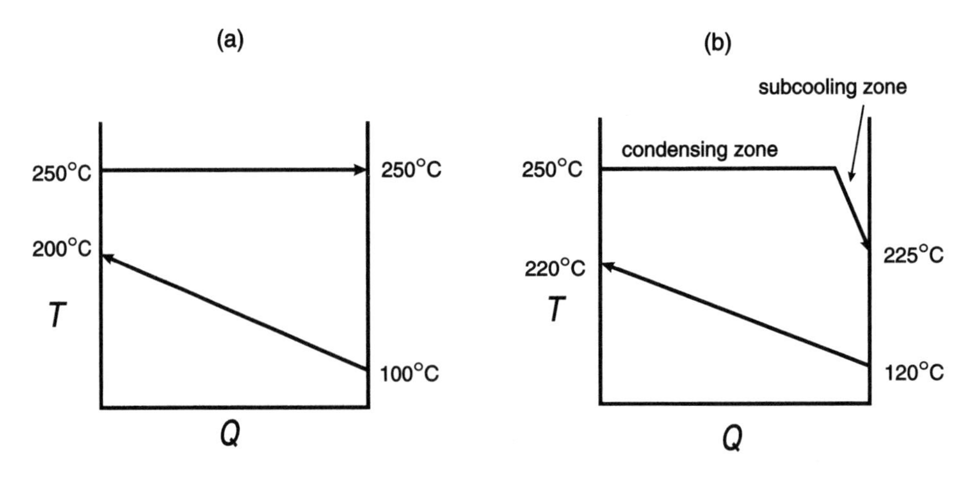

Figure E17.7 Solutions to Example 17.7

b. A single-phase process stream is heated from 120°C to 220°C by condensation of saturated steam at 250°C and subcooling of the condensate to 225°C in a countercurrent heat exchanger.

The solution to Part (a) is shown in Figure E17.7(a). The horizontal line at 250°C is for the steam condensing at constant temperature. The sloped line is for heating of the process stream. The arrow on the sloped line indicates the direction of flow of the process stream, because it is being heated. The arrow on the condensing steam line is opposite because of countercurrent flow.

The solution to Part (b) is shown in Figure E17.7(b). The difference from Part (a) is the subcooling zone. It is important to understand that when there is a pure-component phase change, there must be a horizontal portion of the line. It is incorrect to draw a single straight line between the two end temperatures. In fact, due to the amount of heat associated with phase changes relative to the heat in temperature changes, the horizontal portion associated with the phase change will almost always be the longer segment.

17.6 SUMMARY

In this chapter, several important tools were introduced that are essential in analyzing equipment performance problems. Base-case ratios are the most important tool. They permit comparison of two cases, usually without complex calculations, and often without the need to know physical and transport properties. Under the circumstances for which it is valid, using a limiting resistance simplifies the base-case ratio. Finally, graphical representation of equipment performance is a useful tool in understanding the physical situation. All these techniques require a minimum of calculations. However, they do provide an intuitive understanding of equipment performance that is very rarely achieved from merely doing repetitive, complex computations.

REFERENCE

1. Wales, C. E., and R. A. Stager, *Thinking with Equations* (Morgantown, WV: West Virginia University, 1990).

PROBLEMS

1. It is known that for a certain second-order, elementary, gas-phase reaction, the rate of reaction doubles when the temperature goes from 220°C to 250°C. If the rate of reaction is 5 mole/m^{3}s at the base conditions, which are 10 atm pressure and 225°C, and the feed is pure reactant (no inerts), answer the following questions.
   a. Compared to the base case, by how much does the rate of reaction (at inlet conditions) change if the temperature is increased to 245°C at constant pressure?
   b. Compared to the base case, by how much does the rate of reaction (at inlet conditions) change if the pressure is increased by 10%?
2. A storage tank is connected to a pond (at atmospheric pressure!) by a length of 3 in pipe and two gate valves. From previous operating experience, it has been found that when the tank is at a pressure of 3 atm the flow through the pipe is 12.5 m^3/h, when both gate valves are fully open. If the pressure in the tank increases to 5 atm, what will be the maximum discharge rate from the tank?
3. In a single-pass shell-and-tube heat exchanger, cooling water is used to condense an organic vapor. Under present operating conditions, the heat transfer coefficients are h_o = 2000 W/m^{2}K (turbulent flow of cooling water), h_i = 700 W/m^{2}K. Fouling is negligible, and the tubes are 1.25 in 16 BWG carbon steel. Water flows in the shell. If the cooling water rate were to rise suddenly by 15%, estimate the change in overall heat transfer coefficient. Clearly state all assumptions made.
4. In a shell-and-tube heat exchanger, the overall and individual heat transfer coefficients are U = 300 W/m^{2}K, h_i = 1000 W/m^{2}K, and h_o = 500 W/m^{2}K. Both the shell-side and the tube-side fluids are liquids, and no change of phase occurs. The flow of both fluids is turbulent.
   a. Estimate the new value of U if the flowrate of the shell-side fluid is decreased by 20%.
   b. Estimate the new value of U if the flowrate of the tube-side fluid is decreased by 20%.
   c. Both situations in Parts (a) and (b) exist.

5. Derive the relationship, identical to Equation 17.9, for the situation described in Example 17.3.
6. You are considering two pipes to connect liquid leaving a pump to the entrance to a reactor. For a given mass flowrate of liquid, how much operating cost would you save (or lose) by using 2.5 in schedule 40 pipe rather than 3 in schedule 40 pipe? The pump runs on electricity, and electricity costs $0.06/kWh. The pump efficiency is 80%.
7. For laminar flow of a Newtonian liquid in a pipe, determine the effect on flowrate of the following changes.
   a. The pressure drop triples (everything else remains constant).
   b. The pipe is changed from 1.25 in Schedule 40 to 2 in Schedule 40 (everything else remains constant).
   c. The viscosity of the fluid is increased by 50% (everything else remains constant).
   d. The equivalent length of pipe is decreased by an order of magnitude (everything else remains constant).
   e. (a) and (b)
   f. (a) and (c)
   g. (b) and (d)
8. Sketch T-Q diagrams for the following situations.
   a. The condenser on a distillation column; vapor condenses at 160°C and is subcooled to 130°C, cooling water enters at 30°C and exits at 40°C.
   b. The reboiler on the same distillation column; vapor is reboiled at 240°C, and saturated, high-pressure steam is condensing to saturated water.
   c. Methanol, initially at 30°C and atmospheric pressure, is vaporized to 150°C and atmospheric pressure, and saturated, medium-pressure steam is condensed to saturated water.
9. Air at approximately STP and 15 kg/s flows through 100 m of 8 in commercial steel schedule 40 pipe. Derive an expression for the system curve and sketch the system curve for this situation.
10. In a batch reactor, a first-order, irreversible, liquid-phase reaction, $A \rightarrow B$ occurs. The initial concentration is C_{Ao}.
   a. By what percentage must the reaction time be increased to raise the conversion of A from 0.9 to 0.95?
   b. If the initial conversion is 0.8, what will the final conversion be if the reaction time is increased by 50%?
11. Repeat Problem 10 for a catalytic/enzymatic reaction with a rate expression of the form

$$-r_A = \frac{K_1 C_A}{1 + K_2 C_A}$$

CHAPTER

18

Performance Curves for Individual Unit Operations

As pointed out in the introduction to Section 4, the way in which a process operates will vary significantly throughout its lifetime. Plant operations do not correspond to the conditions specified in the design. This is *not* necessarily a reflection of a poor design. It is a consequence of changes in the process during the life of the plant. There are numerous reasons why a process might not be operated at design conditions. As stated previously, some examples are the following.

- **Design/Construction:** Installed equipment is often oversized. This reduces risks resulting from inaccuracies in design correlations, uncertainties in material properties, and so on.
- **External Effects:** Feed materials, product specifications and flowrates, environmental regulations, and costs of raw materials and utilities all are likely to change during the life of the process.
- **Replacement of Equipment:** New and improved equipment (or catalysts) may replace existing units in the plant.
- **Changes in Equipment Performance:** In general, equipment effectiveness degrades with age. For example, heat transfer surfaces foul, packed towers develop channels, catalysts lose activity, and bearings on pumps and compressors become worn. Plants are shut down periodically for maintenance to restore equipment performance.

With these factors in mind, a good design is one in which operating conditions and equipment performance can be changed throughout the life of the process and plant. This is known as process flexibility. For a company or operation to remain competitive in the marketplace, it must respond to these changes. Therefore, it is essential for us to understand how equipment performs over its entire operating range and to be able to evaluate the effects of changing process conditions on the overall process performance.

Several techniques for evaluating operating systems are presented in Chapters 16 and 17. The purpose of this chapter is to apply these techniques to obtain a solution to a performance problem for a specific piece of equipment. This solution may be a single answer. However, a much more useful solution is a curve or a family of curves that represents the way an existing piece of equipment or system responds to changes in input or equipment variables. These curves are referred to as **performance curves**. They are the basis for predicting the behavior of existing equipment. Performance curves present a range of possible solutions rather than merely a single answer. In principle, performance equations could be used instead of performance curves. However, by representing equipment performance in graphical form, the performance characteristics are easier to visualize, providing a better intuitive understanding. For example, in Figure 17.1, the boundary for fully developed turbulent flow, above which the friction factor is constant, is more effectively represented by the dotted curve on the graph than by the equation of that curve.

Performance curves represent the relationship between process outputs and process inputs.

By plotting the response variables as a function of the input variables, the sensitivity of one to the other becomes immediately obvious. Such sensitivity cannot be inferred from a single (numerical) solution. With the wide availability of spreadsheets and process simulators, the effort expended generating performance curves may be justified.

In order to construct performance and system curves, material and energy balance equations must be used along with the (design) equations relating equipment parameters. However, other constraints may also have to be considered, such as the maximum or minimum system temperature and pressure allowed for the equipment, the maximum velocity of fluid through the equipment to avoid excessive erosion, the maximum or minimum velocity to avoid flooding in packed towers and tray columns, the minimum velocity through a reactor to avoid defluidization, the maximum residence time in a reactor to avoid coking/cracking reactions or by-product formation, the minimum flow through a compressor to avoid surging, or the minimum approach temperature to avoid the condensation of acidic gases inside heat exchangers.

It is possible to construct performance curves for essentially any piece of equipment. In this chapter, performance curves are developed for several specific equipment types. This case-study approach is used to help you develop your own methods for generating performance and system curves for equipment not covered in this text.

18.1 APPLICATIONS TO HEAT TRANSFER

In this section, the steam generator shown in Figure 18.1(a) is analyzed to illustrate the preparation and value of a performance diagram for predicting the response of a heat exchanger to changing conditions. The data shown were obtained from an actual operating system. A steam generator is similar to a shell-and-tube heat exchanger. Figure 18.2 contains drawings of portions of shell-and-tube heat exchangers.

Saturated steam is produced in a kettle-type vaporizer containing long vertical tubes. Heat is provided from a hot light oil stream that enters at 325°C and leaves at 300°C. The effective area for heat transfer is adjusted by changing the level of the boiling liquid in the exchanger. Figure 18.1a provides the current operating conditions, some limited thermodynamic data, data on the vaporizer, and a sketch of the equipment.

The conditions given in Figure 18.1(a) designate the base case. Development of performance curves involves solving the following three equations simultaneously:

$$\text{Heat lost by light oil, } Q = \dot{m}(H_1 - H_2) = \dot{m}C_{p,\,oil}(T_{in} - T_{out}) \quad (18.1)$$

$$\text{Heat gained by water stream, } Q = \dot{m}_s(H_s - H_w) = \dot{m}_s[C_p(T_B - T_w) + \lambda_s] \quad (18.2)$$

$$\text{Heat transferred, } Q = UA\Delta T_{\text{lm}} \quad (18.3)$$

In these equations,

H_1 and H_2 refer to the specific enthalpy, kJ/kg, for the light oil, with the subscript 1 for inlet and 2 for outlet conditions.

H_s and H_w refer to the specific enthalpy, kJ/kg for the steam and water, respectively.

$\dot{m}$ and $\dot{m}_s$ refer to the stream flowrate, kg/h, for the light oil and water/steam stream, respectively.

λ_s is the latent heat of the steam.

The heat transfer equation, Equation 18.3, contains the characteristic factors specific to the heat transfer equipment.

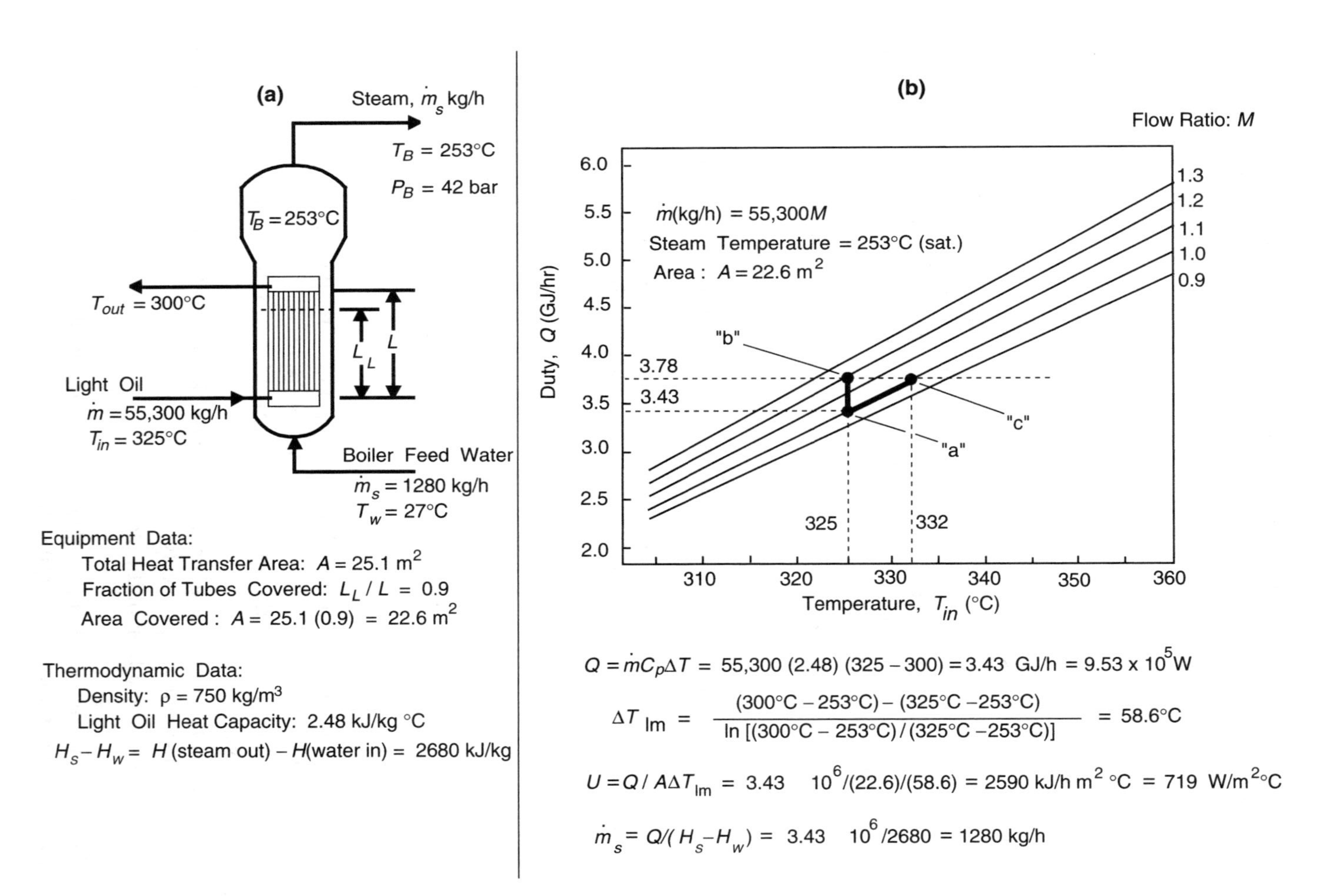

Figure 18.1 Performance Diagram for a Heat-Exchange System

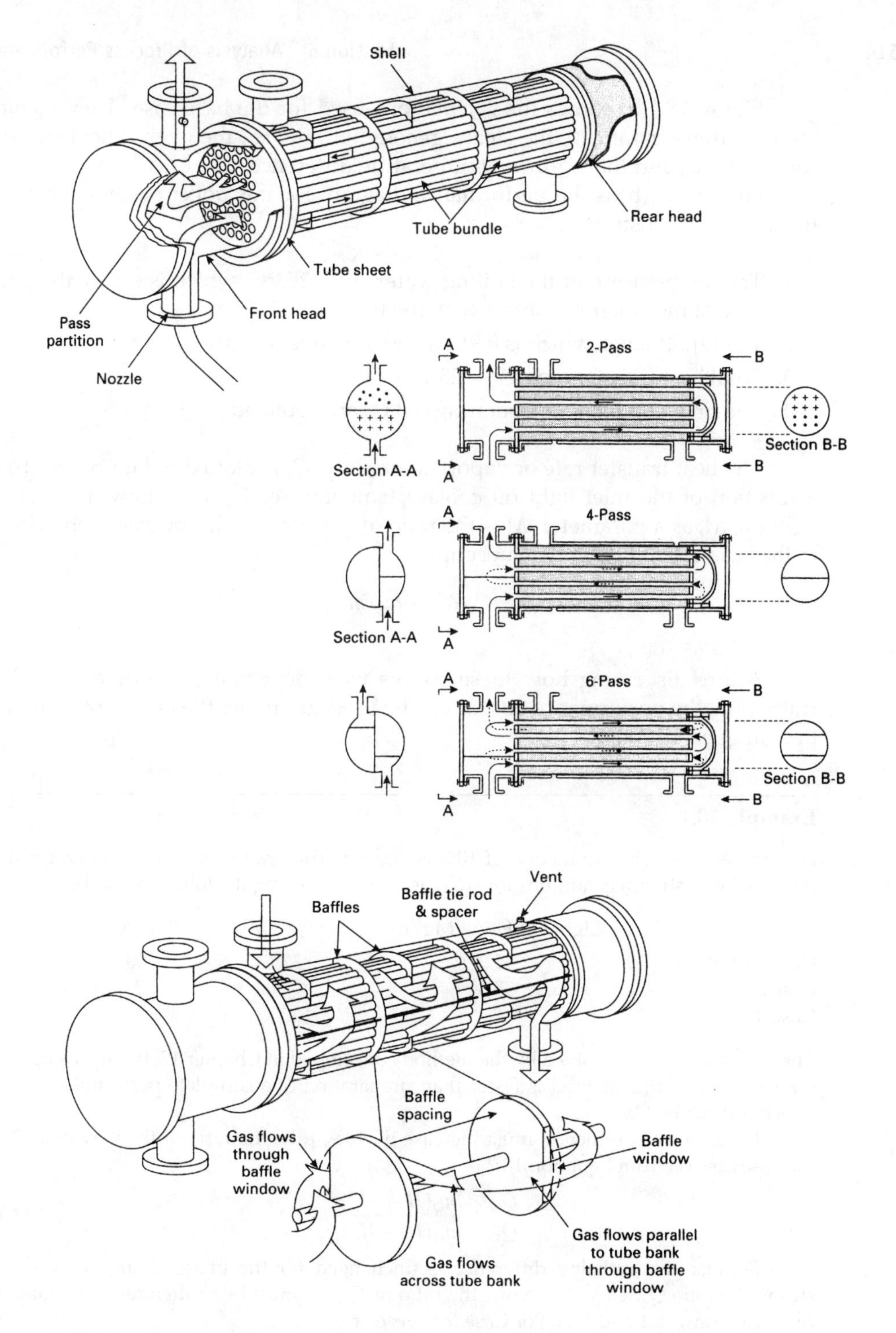

Figure 18.2 Details of a Shell-and-Tube Heat Exchanger (From D. R. Woods, *Process Design and Engineering Practice,* Englewood Cliffs, NJ: Prentice Hall, 1995)

Figure 18.1(b) shows the calculations used for the base case. They include the heat transfer rate, Q, the steam generation rate, $\dot{m}_s$, the overall heat transfer coefficient, U, and the log-mean temperature difference, ΔT_{lm}.

Figure 18.1(b) is the performance diagram for this boiler system subject to the following constraints:

1. The temperature of the boiling water, $T_B = 253°C$, which is set by the pressure of the water in contact with the tubes
2. The liquid level, which is $0.9L$ (where L = tube height)
3. Total heat transfer area, $A_T = 25.1\ m^2$
4. The time at which the operating data were obtained

The heat transfer rate or vaporization duty, Q, is plotted in Figure 18.1(b) as a function of the inlet light oil coolant temperature, T_{in}, with flowrate ratio of light oil, M, as a parameter. M is the ratio of the current oil flowrate (subscript 2) to the base case oil flowrate (subscript 1).

$$M = \dot{m}_2/\dot{m}_1 \tag{18.4}$$

where $\dot{m}_1$= 55,300 kg/h.

Before discussing how these curves were developed, Example 18.1 illustrates how the performance curves can be used to predict the effects of changing operating conditions.

Example 18.1

An increase in steam production of 10% is needed. You are to provide the operator with the new input stream conditions for two cases by completing the following table:

	Light Oil Flow (Mg/h)	T_{in}(°C)	T_{out} (°C)
Current Values	55.3	325	300
Case (a)	55.3	?	?
Case (b)	?	325	?

These problems can be solved by the methods developed in Chapter 17. If only a single answer is desired, this method is faster than preparation of a complete performance curve, which is described later.

If the steam production must increase by 10%, then the ratio of the new case (2) to the base case (1) from Equation 18.2 is

$$\frac{Q_2}{Q_1} = \frac{\dot{m}_{s2}(H_s - H_w)_2}{\dot{m}_{s1}(H_s - H_w)_1} = 1.1 \tag{E18.1a}$$

Because the enthalpy difference is unchanged for the phase change of water to steam, the ratio $Q_2/Q_1 = 1.1$. Now, the ratio of Q_2/Q_1 must be written for the remaining two equations, 18.1 and 18.3. For Case (a), these are

$$\frac{Q_2}{Q_1} = 1.1 = \frac{\dot{m}_2 C_{p2}(T_{in} - T_{out})_2}{\dot{m}_1 C_{p1}(T_{in} - T_{out})_1} = \frac{(T_{in} - T_{out})_2}{25} \tag{E18.1b}$$

$$\frac{Q_2}{Q_1} = 1.1 = \frac{U_2A_2\Delta T_{lm2}}{U_1A_1\Delta T_{lm1}} = \frac{(T_{out} - T_{in})_2}{58.6 \ln\left(\frac{T_{out2} - 253}{T_{in2} - 253}\right)} \frac{U_2}{U_1} \quad \text{(E18.1c)}$$

Because the mass flowrate of the oil does not change, its heat transfer coefficient remains constant. Because a boiling heat transfer coefficient does not change with flowrate, the overall heat transfer coefficient is unchanged. Therefore, $U_1 = U_2$. These assumptions were used in obtaining Equation (E18.1[c]). Solving Equations (E18.1[b]) and (E18.1[c]) simultaneously yields

$$T_{in2} = 332°\text{C}$$

$$T_{out2} = 304.5°\text{C}$$

For Case (b), it is necessary to know more detail about the heat transfer coefficient, because the mass flowrate of the oil changes. For this system, it is known that the boiling heat transfer coefficient (h_o) is two times the oil heat transfer coefficient (h_i). Because only the oil heat transfer coefficient changes with flowrate, assuming negligible fouling and wall resistances, the overall heat transfer coefficient is expressed in terms of the oil heat transfer coefficient as

$$\frac{1}{U_1} = \frac{1}{h_{i1}} + \frac{1}{h_{o1}} = \frac{3}{2h_{i1}} \quad \text{(E18.1d)}$$

$$\frac{1}{U_2} = \frac{1}{h_{i1}M^{0.8}} + \frac{1}{h_{o2}} = \frac{1}{h_{i1}}\left(\frac{1}{M^{0.8}} + 0.5\right) \quad \text{(E18.1e)}$$

where $M = \dot{m}_2/\dot{m}_1$. The base-case ratios now become

$$1.1 = \frac{M(325 - T_{out2})}{25} \quad \text{(E18.1f)}$$

$$1.1 = \frac{3(T_{out2} - 325)}{2(58.6)\left(\frac{1}{M^{0.8}} + 0.5\right) \ln\left(\frac{T_{out2} - 253}{72}\right)} \quad \text{(E18.1g)}$$

Solving these two equations simultaneously yields T_{out2} = 301.5°C and M = 1.17, which means that the flowrate of oil must be increased by 17% to obtain a 10% increase in steam production, and the resulting outlet oil temperature rises slightly.

The same results can be obtained from the performance graph in Figure 18.1(b).

New Steam Production, $\dot{m}_s = 1.1(1280) = 1410$ kg/h

New Exchanger Duty, $Q = \dot{m}_s (H_s - H_w) = 1410(2680) = 3.78 \times 10^6$ kJ/h

Case (a): On Figure 18.1(b), the base-case operating condition is Point "a." From Point "a," follow the constant flow line, M = 1.0, to a vaporizer duty, $Q = 3.78 \times 10^6$ kJ/h. This is line segment "a"–"c," and, from this, we get that the inlet oil temperature, T_{in2} = 332°C.

From Equation 18.1,

$$T_{out2} = T_{in2} - Q/(\dot{m}_2 C_{p,oil}) = 332 - 3.78 \times 10^6/[55{,}300(2.48)] = 305°\text{C}$$

Case (b): On Figure 18.1(b) follow the constant temperature line for T_{in2} = 325°C to the new vaporizer duty, $Q = 3.78 \times 10^6$ kJ/h. This is shown as line segment "a"–"b." From this, we get a value of M = 1.17.

$$\dot{m} = 1.17(55{,}300) = 64{,}700 \text{ kg/h (17\% increase in flow)}$$

From Equation 18.1,

$$T_{out2} = T_{in2} - Q/(\dot{m}_2 C_{p,oil}) = 332 - 3.78 \times 10^6/[64{,}700(2.48)] = 301.5°C$$

In Example 18.1, two sets of operating conditions were calculated that would provide the required increase in steam production. The input oil flowrate, $\dot{m}$, could be increased with the temperature, T_{in}, held constant, or the oil flowrate, $\dot{m}$, could be held constant while the input oil temperature, T_{in}, is increased. These two solutions, along with an infinite number of other solutions, lie on a line of constant heat duty, $Q = 3.78$ GJ/h.

The curves for constant flow ($\dot{m} = 55{,}300M$) in Figure 18.1(b) are straight lines. Equating and rearranging Equations 18.1 and 18.3 produces

$$(T_{in} - T_B)/(T_{out} - T_B) = K \tag{18.5}$$

$$\text{where } K = \exp[UA/(\dot{m}/C_{p,oil})] \tag{18.6}$$

By solving Equation (18.5) for T_{out} and substituting it into Equation (18.1), we obtain

$$Q = \dot{m} C_{p,oil}(T_{in} - T_B)[1 - (1/K)] \tag{18.7}$$

The overall heat transfer coefficient, U, varies with the light oil flowrate, $\dot{m}$. The performance for a constant value of oil flowrate is obtained once the value of K is known.

Equation (18.7) was used to obtain the curves given in Figure 18.1(b). The small amount of heat transferred above the liquid level (where the heat transfer coefficient for the gas phase will be small) was ignored. In addition, it was assumed that the incoming feed water mixes rapidly with the large volume of boiling liquid water in the vaporizer. Thus, the water outside the tubes is essentially constant and equal to the saturation temperature. The simple form of Equation (18.7) is a consequence of this constant temperature. When neither stream involves a phase change, the evaluation is more complicated. One reason is the need to consider the log-mean temperature correction factor for these types of shell-and-tube heat exchangers.

In reality, fouling will affect heat-exchanger performance over time. **Fouling** is a buildup of material on tube surfaces, and it occurs to some extent in all heat exchangers. When a heat exchanger is started up, there will be no fouling; however, with time, trace impurities in the fluids deposit on the heat-exchanger tubes. This is usually more significant for liquids, and among the possible impurities are inorganic salts or microorganisms. The fouling layer provides an additional resistance to heat transfer. For a heat exchanger constructed from material with high thermal conductivity operating with fluids having high heat transfer coefficients, fouling may provide a greater resistance than the convective film resistance. The influence of fouling on heat-exchanger performance is the subject of problems at the end of the chapter.

Which of the two heat transfer surfaces provides the major fouling resistance in the steam generator? The water outside the tubes is designated as boiler feed water. This stream is an expensive source of water. It has been treated extensively to remove trace minerals and hence reduces significantly the fouling of heat transfer surfaces. This suggests that fouling on the outside of the tubes will remain low and the oil stream is the major contributor to fouling.

Example 18.2 suggests ideas for alternative performance of the steam generator. Some of these are the subject of problems at the end of the chapter.

Example 18.2

Assume that neither the oil stream flowrate, $\dot{m}$, nor the inlet temperature, T_{in}, can be changed. What can be done to increase steam production, $\dot{m}_s$?

The heat transfer equation, $Q = UA\Delta T_{lm}$, shows that Q can be increased by

a. **Increasing U:** Clean the tubes.
b. **Increasing A:** Increase the liquid level in the boiler.
c. **Increasing ΔT_{lm}:** Decrease the boiler temperature by lowering the pressure.
d. Combinations of (a), (b), and (c).

18.2 APPLICATION TO FLUID FLOW

In this section, performance curves for centrifugal pumps, reciprocating pumps, and piping networks are discussed. The use of pump and system curves is introduced and the concepts of net positive suction head are discussed.

18.2.1 Pump and System Curves

In this section, performance curves for a centrifugal pump and the flow of a liquid through a pipe network connecting a storage tank to a chemical reactor are presented. Centrifugal pumps are very common in the chemical industry. Figure 18.3 shows the inner workings of a centrifugal pump. It is important to understand that the performance curves presented here are unique to centrifugal pumps. Positive displacement pumps, the other common type of pump used in the chemical industry, have a completely different performance curve and are discussed briefly in Section 18.2.3.

In situations involving the flow of fluid, the equation used to relate pressure changes and flowrate is the mechanical energy balance (or the extended Bernoulli equation):

$$\frac{\Delta P}{\rho g} + \frac{\Delta u^2}{2g} + \Delta z = \frac{W_s}{g} - \frac{F_d}{g} \tag{18.8}$$

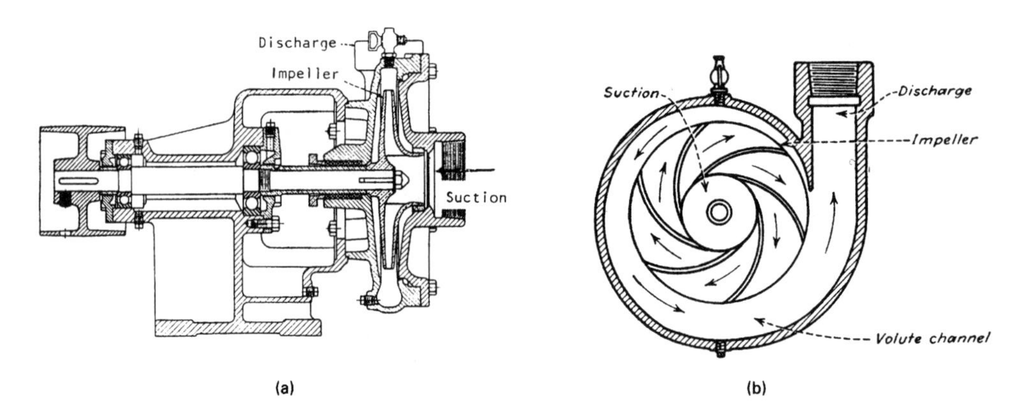

Figure 18.3 Inner Workings of a Centrifugal Pump (From S. Walas, *Chemical Process Equipment. Selection and Design.* Boston, MA: Butterworth, 1988. Reproduced with permission.)

The terms shown on the left-hand side of Equation (18.8) are point properties. They are independent of the path taken by the fluid. The Δ in Equation (18.8) represents the difference between outlet and inlet conditions on the control volume. In contrast, the terms on the right-hand side of the equation are **path properties** that depend on the path followed by the fluid. Work is defined as positive when done on the system. The terms related to the path taken by the fluid are specific to the operating system and form the basis for the performance curves.

Equation (18.8) is written in terms of fluid head. Each term has units of length. (The equation could be written so that each term has units of pressure by multiplying each term by ρg.) **Fluid head** is a way of expressing pressure as an equivalent static pressure of a stationary body of fluid. For example, one atmosphere of pressure is equivalent to about 34 feet of water, because the pressure difference between the top and bottom of a column of 34 feet of water is one atmosphere. Equation 18.8, with each term defined as fluid head, is

$$\Delta h_p + \Delta h_v + \Delta h_z = h_s - h_f \tag{18.9}$$

A centrifugal pump is shown in Figure 18.4 along with its performance diagram. In the performance diagram, the ordinate is in units of pressure head, which is how all of the curves discussed in this section are usually represented. However, it is also possible to represent these curves in pressure units. In this book, examples and problems using both pressure units and head units are included. For this analysis, the change in the velocity head and elevation head between points "1" and "2" are either zero or small, and Equation (18.9) reduces to

$$\Delta h_p = h_s - h_f \tag{18.10}$$

Centrifugal pumps are used in a variety of applications and are available from many manufacturers. Pump impellers may have forward or backward vanes,

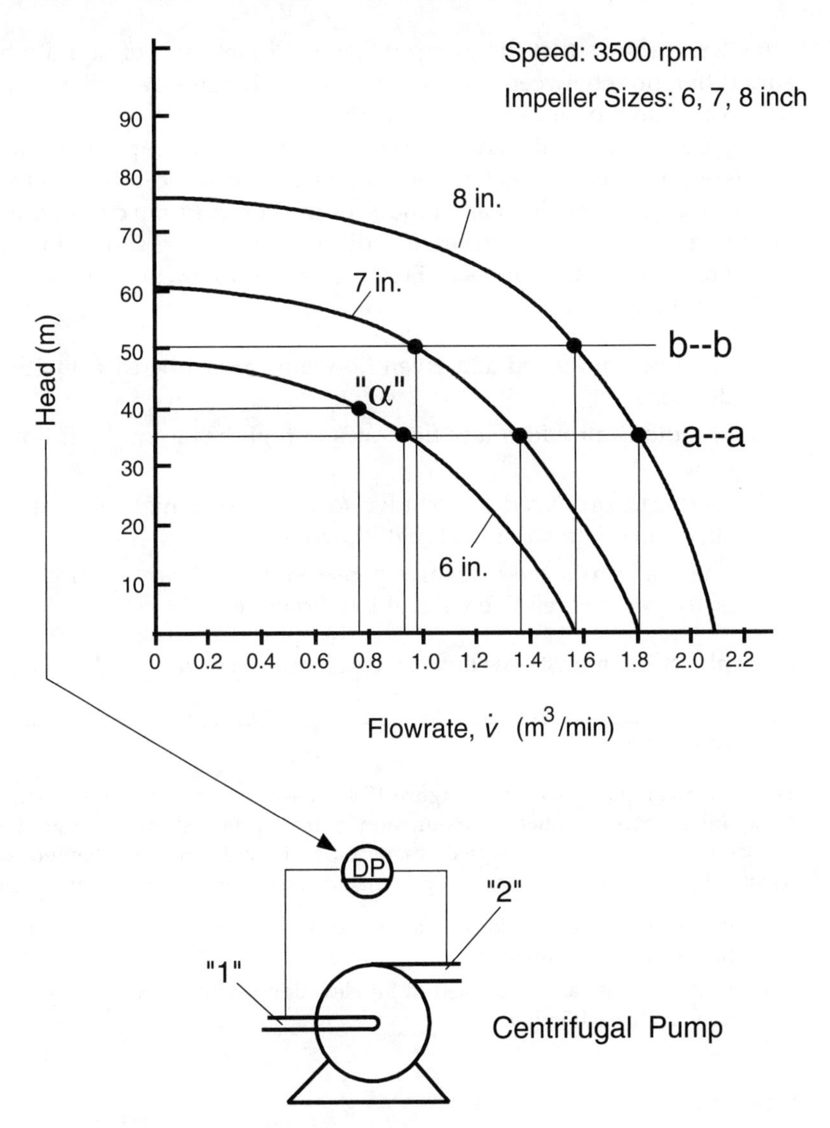

Figure 18.4 Performance Diagram for a Centrifugal Pump

may have enclosed or open configurations, and may operate at different speeds. Their performance characteristics depend on their mechanical design. As a result, experimentally determined performance curves are supplied routinely by the pump manufacturer.

The performance curve, or pump curve, shown in Figure 18.4 is typical of a centrifugal pump handling normal (Newtonian) liquids. It relates the fluid flowrate to the head, h_s. To understand this relationship, consider point "α" shown on the performance curve in Figure 18.4. This point lies on the curve for a 6-inch impeller. It shows that the pump delivers 0.76 m^3/min of liquid when the pressure differential between the suction and discharge side of the pump is equivalent to a column of liquid 40 m in height.

Figure 18.4 contains three curves, each depicting the pump performance with a different impeller diameter. The pump casing can accept any of these three impellers and each impeller has a unique pump curve. Pump curves often display efficiency and horsepower curves in addition to the curves shown in Figure 18.4.

From the pump curves in Figure 18.4, the following trends can be seen for each impeller.

1. The head produced at a given flowrate varies (increases) with the impeller diameter.
2. The pump provides low flowrates at high head and high flowrates at low head.
3. The head produced is sensitive to flowrate at high flowrates; that is, the curve drops off sharply at high flowrates.
4. The head produced is relatively insensitive to flowrate at low flowrates; that is, the curve is relatively flat at low flowrates.

Example 18.3 demonstrates how to read a pump curve.

Example 18.3

The centrifugal pump shown in Figure 18.4 is used to supply water to a storage tank. The pump inlet is at atmospheric pressure and water is pumped up to the storage tank, which is open to atmosphere, via large diameter pipes. Because the pipe diameters are large, the frictional losses in the pipes and any change in fluid velocity can be safely ignored.

a. If the storage tank is located at an elevation of 35 m above the pump, predict the flow using each impeller.

b. If the storage tank is located at an elevation of 50 m above the pump, predict the flow using each impeller.

Solution

a. From Figure 18.4, @Δh_p = 35 m (see line "a—a")
6-in. Impeller: Flow = 0.93 m^3/min
7-in. Impeller: Flow = 1.38 m^3/min
8-in. Impeller: Flow = 1.81 m^3/min

b. From Figure 18.4, @Δh_p = 50 m (see line "b—b")
6-in. Impeller: Flow = 0 m^3/min
7-in. Impeller: Flow = 0.99 m^3/min
8-in. Impeller: Flow = 1.58 m^3/min

In Example 18.3, the flowrate into the tank is restricted to one of three discrete values (depending on the impeller installed in the pump housing). For the system described, once the elevation of the tank and the impeller diameter are chosen, a single unique flowrate is obtained. Other flowrates cannot be obtained for this system.

Example 18.3 demonstrates the need to know the characteristic shape of the performance curve at the current operating point to predict the effect of any change in tank elevation. In Example 18.3, because the 8-in impeller is operating in a region where the flow is not greatly affected by an increase in head, the flowrate was reduced by less than 15%. In contrast, for the 6-in impeller, where a change in head has a large effect, the flowrate is reduced by 100% (no fluid flows).

It is essential to understand the system performance before making predictions or recommendations.

Consider the flow situation shown in Figure 18.5. The figure illustrates a pipe network through which feed material is transported from a storage tank through a heat exchanger to a chemical reactor. From the information provided on Figure 18.5, it is possible to construct the performance curve shown, which is called the **system curve**.

Because the fluid is a liquid, we may assume that the change in velocity head, Δh_v, is small. In addition, the work head $\Delta h_s = 0$, and Equation (18.9) written between points "1" and "2" reduces to

$$\Delta h_p = -h_f - \Delta h_z \qquad (18.11)$$

The pressure head term, Δh_p, and the elevation head term, $-\Delta h_z$, are constant and independent of the flowrate. The friction term, $-h_f$, in Equation (18.11) depends upon

1. The specific system configuration
2. The flowrate of fluid

For this system, for fully developed turbulent flow, the relationship needed to predict the friction term, h_f, as a function of fluid velocity is known. The relationship is given in Table 17.1 and in Example 17.2. This relationship in terms of fluid head is

$$h_{f2} = h_{f1}(u_2/u_1)^2 = h_{f1}(\dot{v}_2/\dot{v}_1)^2 \qquad (18.12)$$

where $\dot{v}$ is the volumetric flow rate, the subscript 2 refers to the new case, and the subscript 1 refers to the base case. Equation (18.12) represents a parabola, which is the

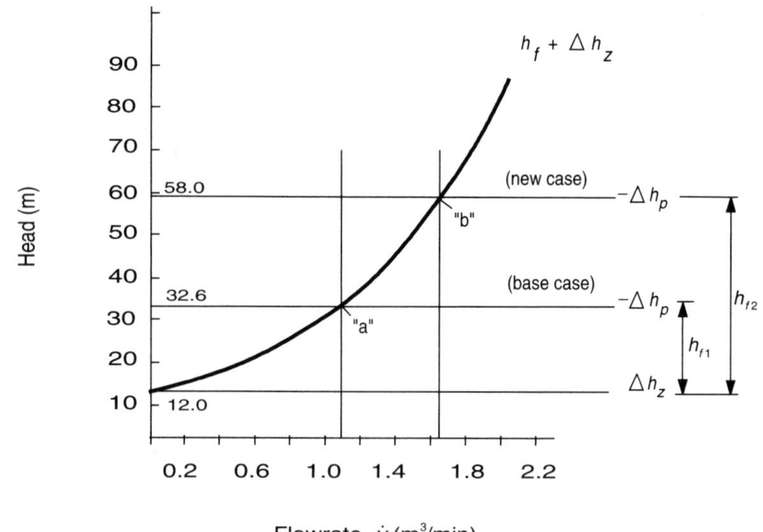

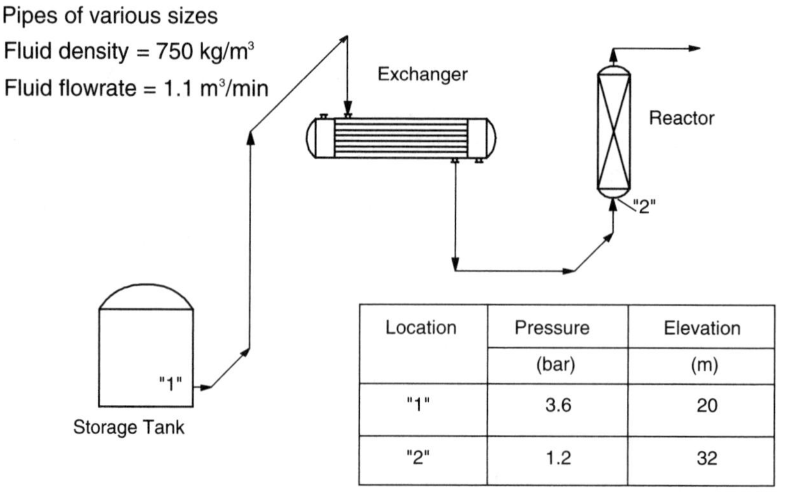

Location	Pressure	Elevation
	(bar)	(m)
"1"	3.6	20
"2"	1.2	32

Figure 18.5 Sketch and Performance Diagram for an Operating Flow System

shape of the system curve in Figure 18.5. In terms of pressure, Equation (18.12) becomes

$$\Delta P_2 = \Delta P_1(\dot{v}_2/\dot{v}_1)^2 \tag{18.13}$$

and it is also possible to plot the system curve in terms of pressure, as is done later in Figure 18.6.

The operating conditions given in Figure 18.5 serve as the base case in Examples 18.4 and 18.5, which illusrate the ratio method discussed previously.

Example 18.4

The fluid flowrate measured for the conditions shown in Figure 18.5 is 1.1 m^3/min.

a. Using the information given on the flow diagram (see Figure 18.5) for the base case, determine the value of h_f.

b. Develop the equation for h_f as a function of the flowrate, $\dot{v}$.

Solution

a. From Equation 18.11,

$$h_f = -\Delta h_z - \Delta h_p$$

$\Delta h_z = z_2 - z_1 = 32 \text{ m} - 20 \text{ m} = 12 \text{ m}$ (Δh_z is drawn as a horizontal line in Figure 18.5)

$$\Delta h_p = (P_2 - P_1)/\rho g = (1.2 - 3.6) \times 10^5/[750(9.81)] = -32.6 \text{ m}$$

$$h_f = -12 \text{ m} + 32.6 \text{ m} = 20.6 \text{ m}$$

b. $h_{f2} = h_{f1}(\dot{v}_2/\dot{v}_1)^2 = (20.6 \text{ m})(\dot{v}_2/1.1)^2 = 17.02\dot{v}_2^2$

Example 18.5

The flow to the reactor in Example 18.4 is increased by 50%. The pressure of the reactor is held constant. Determine the pressure required at the exit of the storage tank.

For a 50% increase, $\dot{v}_2 = 1.5\,\dot{v}_1 = 1.1(1.5) = 1.65$ m^3/min.

The line of constant flow, $\dot{v}$ = 1.65 m^3/min, is shown as the vertical line through Point "b" in Figure 18.5. It intersects the performance line, $h_f + \Delta h_z = 58$ m (Point "b").

$$-\Delta h_p = 58 \text{ m} = (P_1 - P_2)/\rho g = P_1/\rho g - 1.2 \times 10^5/[750(9.81)] = P_1/\rho g - 16.3 \text{ m}$$

$$P_1 = (58 + 16.3)\rho g = 74.3(750)(9.81) = 5 \times 10^5 \text{ Pa} = 5.47 \text{ bar}$$

In this section, two representative types of performance curves were considered. For the pump, the information needed for developing a performance diagram was obtained experimentally and supplied by the manufacturer. In the flow network, a base case was established using actual operating data and the performance diagram, known as the system curve, calculated using the mechanical energy balance.

18.2.2 Regulating Flowrates

In all systems presented in this chapter, input flowrates are the primary variables that are used to change the performance of a system. In fact, in a chemical plant, process regulation is achieved most often by manipulating flowrates, which is accomplished by altering valve settings. If it is necessary to change a temperature, the flow of a heating or cooling medium is adjusted. If it is necessary to change a reflux ratio, a valve is adjusted. In this section how to regulate these input flows to give desired values is considered.

> **Process conditions are usually regulated or modified by adjusting valve settings in the plant.**

Although valves are relatively simple and inexpensive pieces of equipment, they are nevertheless indispensable in any chemical plant that handles liquids or gases.

Figure 18.6 illustrates a fluid system containing three components:

1. A flow system, including piping and three heat exchangers
2. A pump
3. A regulating valve

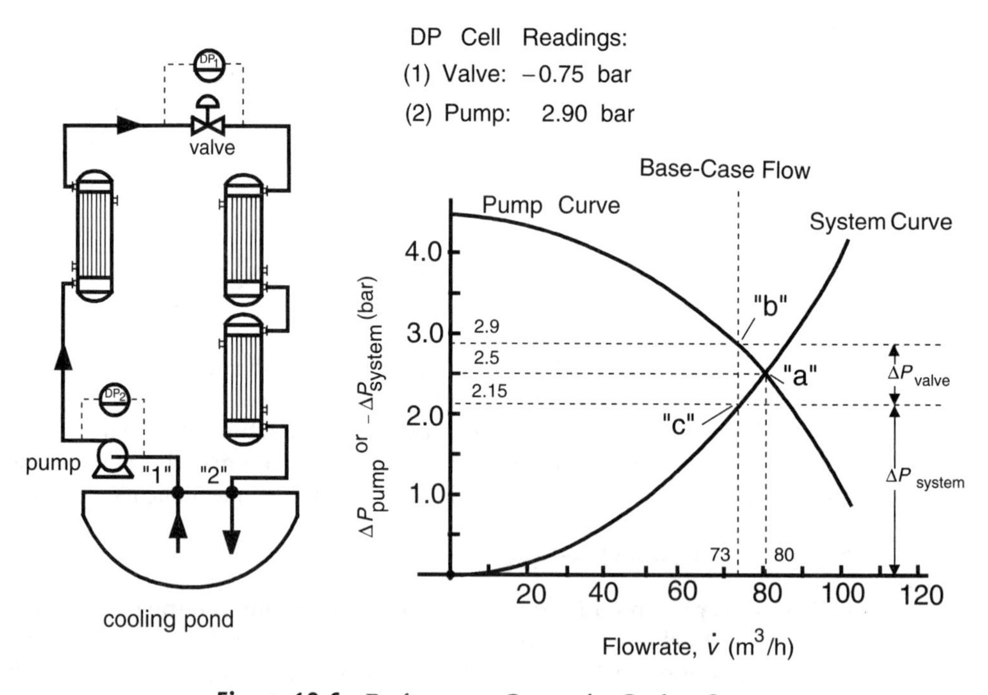

Figure 18.6 Performance Curves for Coolant System

The process shown in Figure 18.6 is described briefly as follows:

A liquid process stream is pumped at a rate of $\dot{v}$ m^3/min from a cooling pond, through a heat exchanger, through a regulating valve and two more heat exchangers connected in series, and returned to the cooling pond. The intake and discharge pipes are at the same elevation. Differential pressure gauges are installed across the pump and the regulating valve. The pump used here has the same characteristics as that shown in Figure 18.4, with a 7-inch impeller.

Conditions shown in Figure 18.6 represent a base case. The mechanical energy balance equation, Equation (18.8), written between Points "1" and "2" in Figure 18.6, gives

$$-(\Delta P_{valve} + \Delta P_{pipes} + \Delta P_{exchangers}) = \Delta P_{pump} \tag{18.14}$$

In Equation (18.14),

ΔP_{valve} represents the pressure drop across the control valve.

ΔP_{pipe} represents the pressure drop due to friction in the piping and pipe fittings between Points "1" and "2."

$\Delta P_{exchangers}$ represents the pressure drop due to friction in the three heat exchangers.

ΔP_{pump} represents the pressure increase produced by the pump.

The frictional drop in pressure through the pipe system and the heat exchangers is given by

$$\Delta P_{system} = \Delta P_{pipes} + \Delta P_{exchangers} \tag{18.15}$$

ΔP_{system} is called the system pressure drop, and Equation (18.14) can be written as

$$-(\Delta P_{valve} + \Delta P_{system}) = \Delta P_{pump} \tag{18.16}$$

The performance curve for this system, shown in Figure 18.6, consists of a plot of pressure against the liquid flowrate, as was done in Figures 18.4 and 18.5. (In this section we use pressure units instead of units of liquid head.)

The system pressure drop is estimated relative to the base case by Equation (18.13).

$$\Delta P_{2,system} = \Delta P_{1,system}(\dot{v}_2/\dot{v}_1)^2 \tag{18.17}$$

$$\Delta P_{2,system}(\text{bar}) = 2.15\ (\dot{v}_2/73)^2 \tag{18.18}$$

This curve has been plotted and labeled as the system curve.

Sufficient information is available to prepare a full performance diagram. In Figure 18.6, for this system, each side of Equation (18.16) is plotted. The pressure produced by the pump, ΔP_{pump}, depends on the volumetric flowrate, $\dot{v}$, as was shown earlier in Figure 18.4. Only the units used to express the pressure and flowrate terms have been changed.

Example 18.6 illustrates how the component terms of the system curve are related to performance parameters.

Example 18.6

Using the pump curve and the differential pressures provided on Figure 18.6, find for the base case

a. The system pressure drop, ΔP_{system}

b. The volumetric flowrate, $\dot{v}$

Solution

a. For the system pressure drop, ΔP_{system}, from Equation (18.16),

$$-\Delta P_{system} = \Delta P_{pump} + \Delta P_{valve} = 2.9 \text{ bar} - 0.75 \text{ bar} = 2.15 \text{ bar}$$

b. For the volumetric flowrate, $\dot{v}$, see Point "b" on Figure 18.6.

$$\text{for } \Delta P_{pump} = 2.9 \text{ bar}, \dot{v} = 73 \text{ m}^3/\text{h}$$

Base-case conditions have been added to Figure 18.6. All pressures for the base case lie along the constant flow line, $\dot{v} = 73$ m^3/h. Point "c" gives the value for the system pressure drop, ΔP_{system}, and Point "b" gives the pressure increase provided by the pump. The difference between these points is the pressure drop over the regulating valve, ΔP_{valve}.

If there were no valve in the line, $\Delta P_{valve} = 0$, the system would operate at Point "a," where the pump curve and the system curve cross. It can be seen from Figure 18.6 that as the pressure across the valve increases, the move is toward the left and lower flowrates. The flow becomes zero when the valve is fully closed. Thus, by manipulating the valve setting, the flow of fluid through the system can be altered.

Example 18.7 illustrates obtaining information from a pump and system curve.

Example 18.7

For the base-case condition shown in Figure 18.6, do the following.

a. Check the pressure drop over the valve against the guideline for control valves (Table 11.8).

b. Determine the percent increase in flow by fully opening the valve.

Solution

a. From guideline, $\Delta P \geq 0.69$ bar (see Table 11.8).
$\Delta P = 0.75$ bar, and therefore the guideline in Table 11.8 is satisfied.

b. With no valve resistance, $\dot{v} = 80\ m^3/h$.
The percent increase in flow = [(80 – 73)/73]100 = 9.6%

The increase in flowrate is limited (less than 10% of the base case) by the pump and system, and only modest increases of flowrate are possible for this system.

18.2.3 Reciprocating or Positive Displacement Pumps

Positive displacement pumps perform differently from centrifugal pumps. They are used to achieve higher pressure increases than centrifugal pumps. Figure 18.7 is a drawing of the inner workings of a positive displacement pump. The performance characteristics are represented on Figure 18.8(a). It can be observed that the flowrate through the pump is almost constant over a wide range of pressure increases. One method for regulating the flow through a positive displacement pump is illustrated in Figure 18.8(b). By carefully regulating the flow of the recycle stream, the pressure rise in the pump is controlled. See Chapter 21 for more details.

18.2.4 Net Positive Suction Head

There is a significant limitation on pump operation called Net Positive Suction Head (NPSH). Its origin is as follows. Although the effect of a pump is to raise the pressure of a liquid, frictional losses at the entrance to the pump, between the feed (suction) pipe and the internal pump mechanism, cause the liquid pressure to drop upon entering the pump. This means that a minimum pressure exists somewhere within the pump. If the feed liquid is saturated or nearly saturated, the liquid can vaporize upon entering due to the pressure drop. This results in

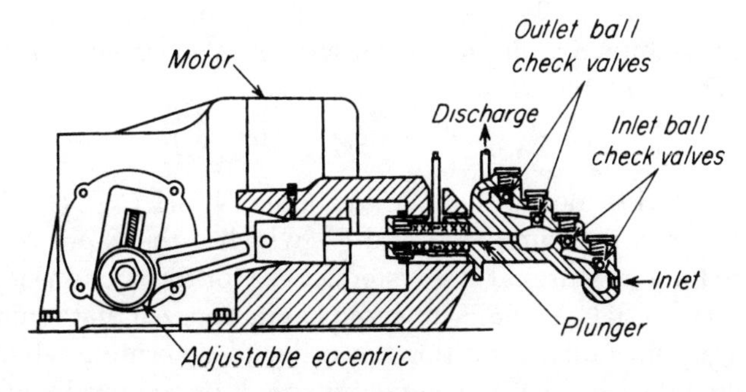

Figure 18.7 Inner Workings of a Positive Displacement Pump (From W. L. McCabe, J. C. Smith, and P. Harriott, *Unit Operations of Chemical Engineering*, 5th ed. New York: McGraw-Hill, 1993, Copyright © 1993 by New York: McGraw-Hill Companies, reproduced with permission of the McGraw-Hill Companies.)

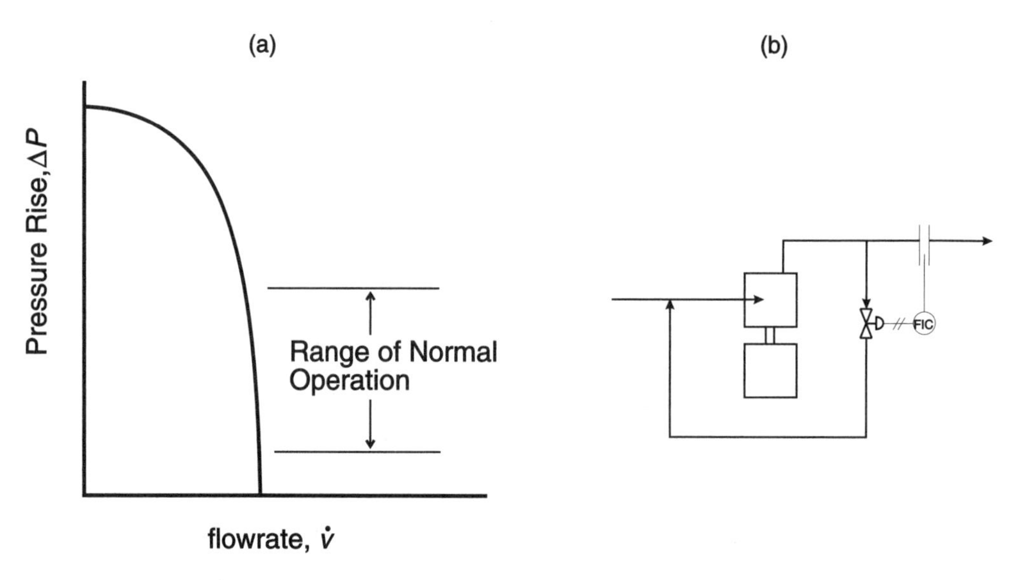

Figure 18.8 Typical Pump Curve for Positive Displacement Pump and Method for Flowrate Regulation

formation of vapor bubbles called **cavitation**. These bubbles rapidly collapse when exposed to the forces created by the pump mechanism. This process usually results in noisy pump operation, and, if it occurs for a period of time, will damage the pump. As a consequence, regulating valves are not normally placed in the suction line to a pump.

Pump manufacturers supply NPSH data with a pump. The required NPSH, denoted $NPSH_R$, is a function of the square of velocity, because it is a frictional loss. Figure 18.9 shows an $NPSH_R$ curve, which defines a region of acceptable pump operation. This is specific to a given liquid. Typical $NPSH_R$ values are in the range of 15–30 kPa (2–4 psi) for small pumps and can reach 150 kPa (22 psi) for larger pumps. On Figure 18.9, there are also curves for $NPSH_A$, the available NPSH. The available $NPSH_A$ is defined as

$$NPSH_A = P_{inlet} - P^* \tag{18.19}$$

Equation (18.19) means that the available NPSH ($NPSH_A$) is the difference between the inlet pressure, P_{inlet}, and P^*, which is the vapor pressure (bubble point pressure for a mixture). It is a system curve for the suction side of a pump. It is required that $NPSH_A \geq NPSH_R$ to avoid cavitation. All that remains is to calculate or know the pump inlet conditions in order to determine whether there is enough available NPSH ($NPSH_A$) to equal or exceed the required NPSH ($NPSH_R$).

For example, consider the exit from a distillation column reboiler, which is saturated liquid. If it is necessary to pump this liquid, cavitation could be a problem. A common solution to this problem is to elevate the column above the pump

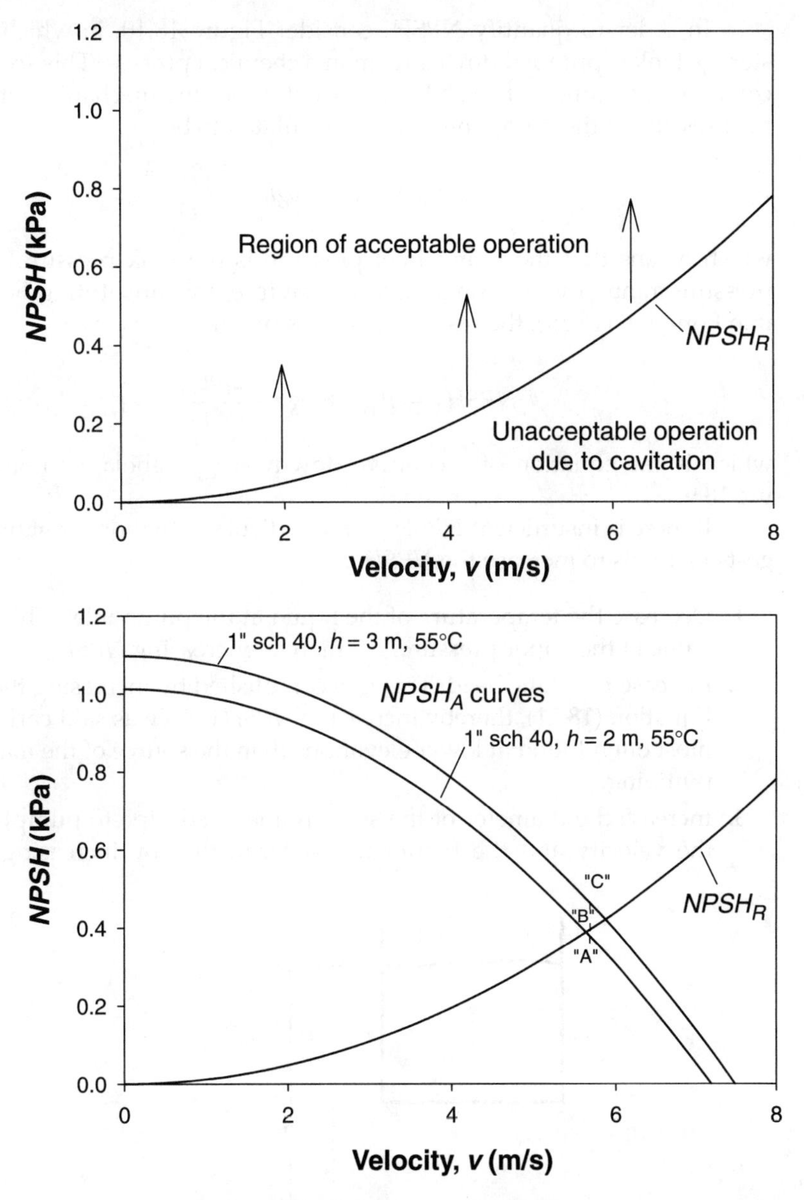

Figure 18.9 $NPSH_A$ and $NPSH_R$ Curves

so that the static pressure increase minus any frictional losses between the column and the pump provides the necessary NPSH. This can be done either by elevating the column above ground level using a metal skirt or by placing the pump in a pit below ground level, although pump pits are usually avoided due to safety concerns arising from accumulation of heavy gases in the pit.

In order to quantify NPSH, consider Figure 18.10, in which material in a storage tank is pumped downstream in a chemical process. This scenario is a very common application of the NPSH concept. From the mechanical energy balance, the pressure at the pump inlet can be calculated to be

$$P_{inlet} = P_{tank} + \rho g h - \frac{2\rho f L_{eq} u^2}{D} \tag{18.20}$$

which means that the pump inlet pressure is the tank pressure plus the static pressure minus the frictional losses. Therefore, by substituting Equation (18.20) into Equation (18.19), the resulting expression for $NPSH_A$ is

$$NPSH_A = P_{tank} + \rho g h - \frac{2\rho f L_{eq} u^2}{D} - P^* \tag{18.21}$$

which is the equation of a concave downward parabola, as illustrated in Figure 18.9.

If there is insufficient $NPSH_A$ for a particular situation, Equation (18.21) suggests methods to increase the $NPSH_A$.

1. Decrease the temperature of the liquid at the pump inlet. This decreases the value of the vapor pressure, P^*, thereby increasing $NPSH_A$.
2. Increase the static head. This is accomplished by increasing the value of h in Equation (18.21), thereby increasing $NPSH_A$. As was said earlier, pumps are most often found at lower elevations than the source of the material they are pumping.
3. Increase the diameter of the suction line (feed pipe to pump). This reduces the velocity and the frictional loss term, thereby increasing $NPSH_A$. It is

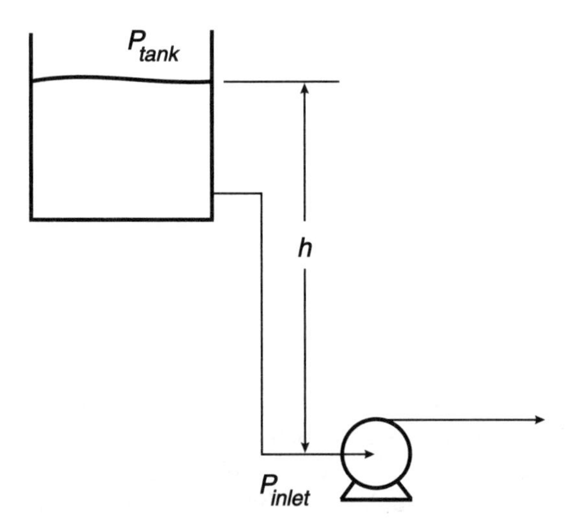

Figure 18.10 Illustration of NPSH for Pumping from Storage Tank

standard practice to have larger-diameter pipes on the suction side of a pump than on the discharge side.

Example 18.8 illustrates how to do NPSH calculations and one of the methods listed above for increasing $NPSH_A$. The other methods are illustrated in a problem at the end of the chapter.

Example 18.8

The feed pump (P-101) on Figure 1.5 pumps toluene from a feed tank (V-101) maintained at atmospheric pressure and 55°C. The pump is located 2 m below the liquid level in the tank, and there is 6 m of equivalent pipe length between the tank and the pump. It has been suggested that 1-in. Schedule 40 commercial steel pipe be used for the suction line. Determine whether this is a suitable choice. If not, suggest methods to avoid pump cavitation.

The following data can be found for toluene: $\ln P^*(\text{bar}) = 10.97 - 4203.06/T(\text{K})$, $\mu = 4.1 \times 10^{-4}$ kg/m s, $\rho = 870$ kg/m^3. For 1 inch Schedule 40 commercial steel pipe, the roughness factor is about 0.001 and the inside diameter is 0.02664 m. From Table 1.5, the flow of toluene is 10,000 kg/h. Therefore, the velocity of toluene in the pipe can be found to be 5.73 m/s. The Reynolds number is 426,000, and, from a friction factor chart, $f = 0.005$. At 55°C, the vapor pressure is found to be 0.172 bar.

From Equation 18.21,

$$NPSH_A = 1.01325 \text{ bar} + 870(9.81)(2)(10^{-5}) \text{ bar} - 2(870)(0.005)(6)(5.73)^2(10^{-5})/(0.02664) \text{ bar} - 0.172 \text{ bar}$$

$$NPSH_A = 0.37 \text{ bar}$$

This is shown as Point "A" on Figure 18.9. At the calculated velocity, Figure 18.9 shows that $NPSH_R$ is 0.40 bar, Point "B." Therefore, there is insufficient $NPSH_A$.

One method for increasing $NPSH_A$ is to increase the height of liquid in the tank. If the height of liquid in the tank is 3 m, with the original 1 in Schedule 40 pipe at the original temperature, $NPSH_A = 0.445$ bar. This is shown as Point "C" on Figure 18.9.

18.2.5 Compressors

The performance of centrifugal compressors is somewhat analogous to that of centrifugal pumps. There is a characteristic performance curve, supplied by the manufacturer, that defines how the outlet pressure varies with flowrate. However, compressor behavior is far more complex than that for pumps because the fluid is compressible.

Figure 18.11 shows the performance curve for a centrifugal compressor. It is immediately observed that the y-axis is the ratio of the outlet pressure to inlet pressure. This is in contrast to pump curves, which have the difference between these two values. Curves for two different rotation speeds are shown. As with pump curves, curves for power and efficiency are often included but are not shown here. Unlike most pumps, the speed is often varied continuously to control the flowrate. This is because the higher power required in a compressor makes it economical to avoid throttling the outlet as in a centrifugal pump.

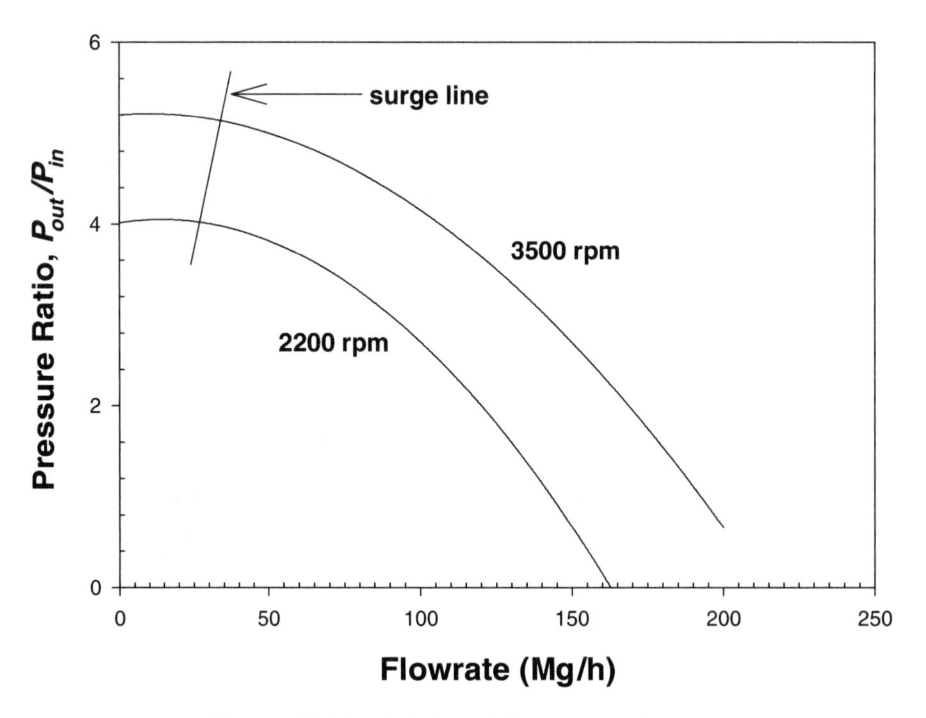

Figure 18.11 Centrifugal Compressor Curves

Centrifugal compressor curves are read just like pump curves. At a given flowrate and rpm there is one pressure ratio. The pressure ratio decreases as flowrate increases. A unique feature of compressor behavior occurs at low flowrates. It is observed that the pressure ratio increases with decreasing flowrate, reaches a maximum, and then decreases with decreasing flowrate. The locus of maxima is called the surge line. For safety reasons, compressors are operated to the right of the **surge line**. The surge line is significant for the following reason. Imagine that we start at a high flowrate and the flowrate is lowered continuously, causing a higher outlet pressure. At some point, the surge line is crossed, lowering the pressure ratio. This means that downstream fluid is at a higher pressure than upstream fluid, causing a backflow. These flow irregularities can severely damage the compressor mechanism, even causing the compressor to vibrate or surge (hence the origin of the term). Severe surging has been known to cause compressors to become detached from the supports keeping them stationary and literally to fly apart, causing great damage. Therefore, the surge line is considered a limiting operating condition, below which operation is prohibited.

Positive displacement compressors also exist and are used to compress low volumes to high pressures. **Centrifugal compressors** are used to compress higher volumes to moderate pressures, and are often staged in order to obtain higher pressures. Figure 18.12 illustrates the inner workings of a compressor.

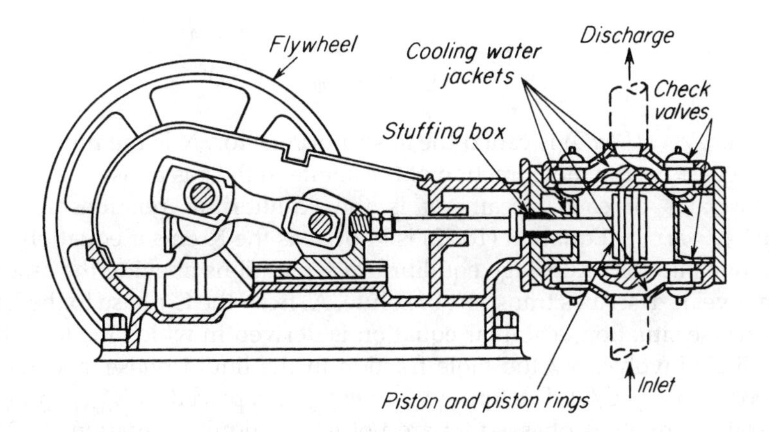

Figure 18.12 Inner Workings of a Positive Displacement Compressor (From W. L. McCabe, J. C. Smith, and P. Harriott, *Unit Operations of Chemical Engineering*, 5th ed. New York: McGraw-Hill, 1993. Copyright © 1993 by McGraw-Hill Companies, reproduced with permission.)

Problems requiring reading compressor curves are given at the end of the chapter.

18.3 APPLICATION TO SEPARATION PROBLEMS

18.3.1 Separations with Mass Separating Agents

Multistage equilibrium separations involve simultaneous solution of material balances and equilibrium relationships for each equilibrium stage. Therefore, there is no simple, closed-form relationship that describes the behavior of these systems for all situations. There are qualitative relationships that are applicable to almost all situations, and these were included in Table 17.1. The key relationships are that a better separation is usually achieved by increasing the number of stages or by increasing solvent flowrate.

Similarly, continuous differential equilibrium separations, which involve simultaneous solution of material balances and mass transfer relationships, do not yield a closed-form solution either. The key qualitative relationships are that better separation is usually achieved by increasing column height or by increasing flowrate.

For the specific assumptions of dilute solutions and a linear equilibrium relationship that results in approximately constant stream flows, an analytical solution is possible for the above situations. For staged separations, the situation is shown in Figure 18.13(a). The result is

$$\frac{y_{A,out} - y^*_{A,out}}{y_{A,in} - y^*_{A,out}} = \frac{1 - A}{1 - A^{N+1}} \tag{18.22}$$

where $A = (L/mG)$ is called the absorption factor, N is the number of equilibrium stages, y_A is the mole fraction of the solute in the gas phase, L and G are the molar flowrates of each stream, m is the equilibrium relationship ($y = mx$), and $y^*_{A,out} = mx_{A,in}$. Equation (18.22) is known as the Kremser equation and is a key relationship for multistage equilibrium separations obeying the assumptions listed above. It describes transport of solute, A, from the G phase to the L phase. For the reverse situation, a similar equation is derived in which the left side of Equation (18.22) involves x_A, the mole fraction in the liquid phase, and the right side involves $S = (mG/L)$. The term $y^*_{A,out} = mx_{A,in}$ is replaced by $x^*_{A,out} = y_{A,in}/m$. If the separation involves phases that are not gas or liquid, Equation (18.22) may still be used by defining L and G appropriately. Figure 18.14 is a plot of Equation (18.22) and contains the performance relationship between key variables for multistage equilibrium separations following the assumptions listed above. Performance curves for a specific staged separation can be generated from the information in Figure 18.14. This is illustrated in Example 18.10. It should be noted that tray performance issues such as flooding, which are specific to a particular column design, are not predicted by the Kremser relationship.

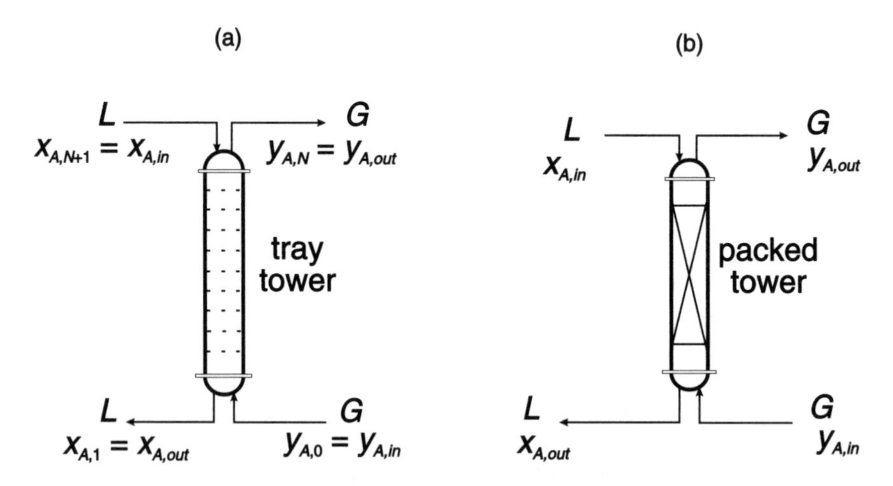

Figure 18.13 Tray and Packed Absorbers

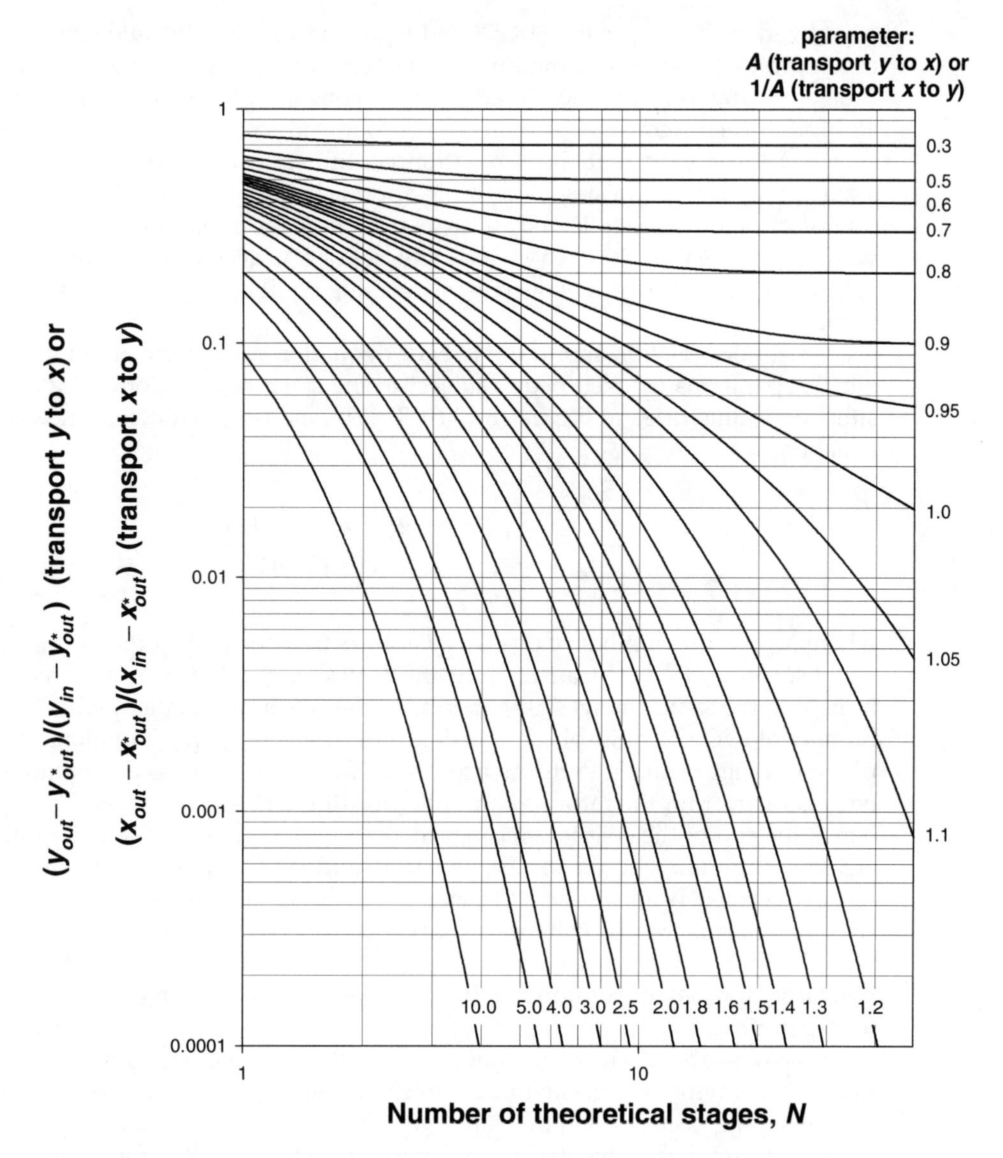

Figure 18.14 Plot of Kremser Equation, Number of Theoretical Stages for Countercurrent Operation, Henry's Law Equilibrium, and Constant A or $1/A$

Even though Equation (18.22) and Figure 18.14 are only valid subject to the assumptions listed above, the qualitative performance of all staged separations can be understood from these relationships. For example, as illustrated in Figure 18.14, if a line of constant A (if $A > 1$) is followed for an increasing number of stages, N, it is seen that the separation continues to improve (lower value of y-axis). For $A < 1$, a best separation is approached asymptotically. If a vertical line of constant N is followed, it is seen that as the separation improves, the value of A increases. A increases if L increases or if G decreases, meaning a larger solvent to feed ratio, or if m decreases, meaning equilibrium more in favor of the L phase.

A similar relationship to the Kremser equation exists for continuous differential separations (packed beds), subject to the same assumptions. The physical situation is illustrated in Figure 18.13(b). This relationship, known as the Colburn equation, is

$$\frac{y_{A,in} - y^*_{A,out}}{y_{A,out} - y^*_{A,out}} = \frac{e^{N_{tOG}[1-(1/A)]} - (1/A)}{1 - (1/A)} \tag{18.23}$$

where N_{tOG} is the number of overall gas-phase transfer units. Equation (18.23) is for absorption, that is, transfer into the L phase. For the reverse direction of transport, the same changes are made as in the Kremser equation, with N_{tOL}, the number of overall liquid-phase transfer units, replacing N_{tOG}. Equation (18.23) is plotted in Figure 18.15. Performance curves for a specific staged separation can be generated from the information in Figure 18.15. This is the subject of a problem at the end of the chapter. Because increasing the number of transfer units increases column height, the same qualitative understanding applicable to all systems is gleaned from Figure 18.15 as was described above. For increasing N_{tOG} (which means increased column height) at constant A, a better separation is observed (for $A > 1$). For $A < 1$, a best separation is approached asymptotically. Similarly, at constant column height (constant N_{tOG}), increasing A results in better separation.

As with the Kremser equation, the Colburn equation does not describe flooding behavior of a packed bed, which is specific to a particular column design. Flooding occurs when the vapor velocity upward through the column is so great that liquid is prevented from flowing downward. When a column is designed, the diameter is chosen so that the vapor velocity is below the flooding limit (typically 75%–80% of the limit). However, if the vapor velocity is increased, flooding can occur. Because flooding is specific to a particular column, it cannot be illustrated on the general performance curve. Care must be taken not to use the Kremser or Colburn graphs to obtain a result that will cause flooding in a particular column or to recommend operation in the flooding zone.

Examples 18.9 and 18.10 illustrate the use of the Kremser and Colburn equations.

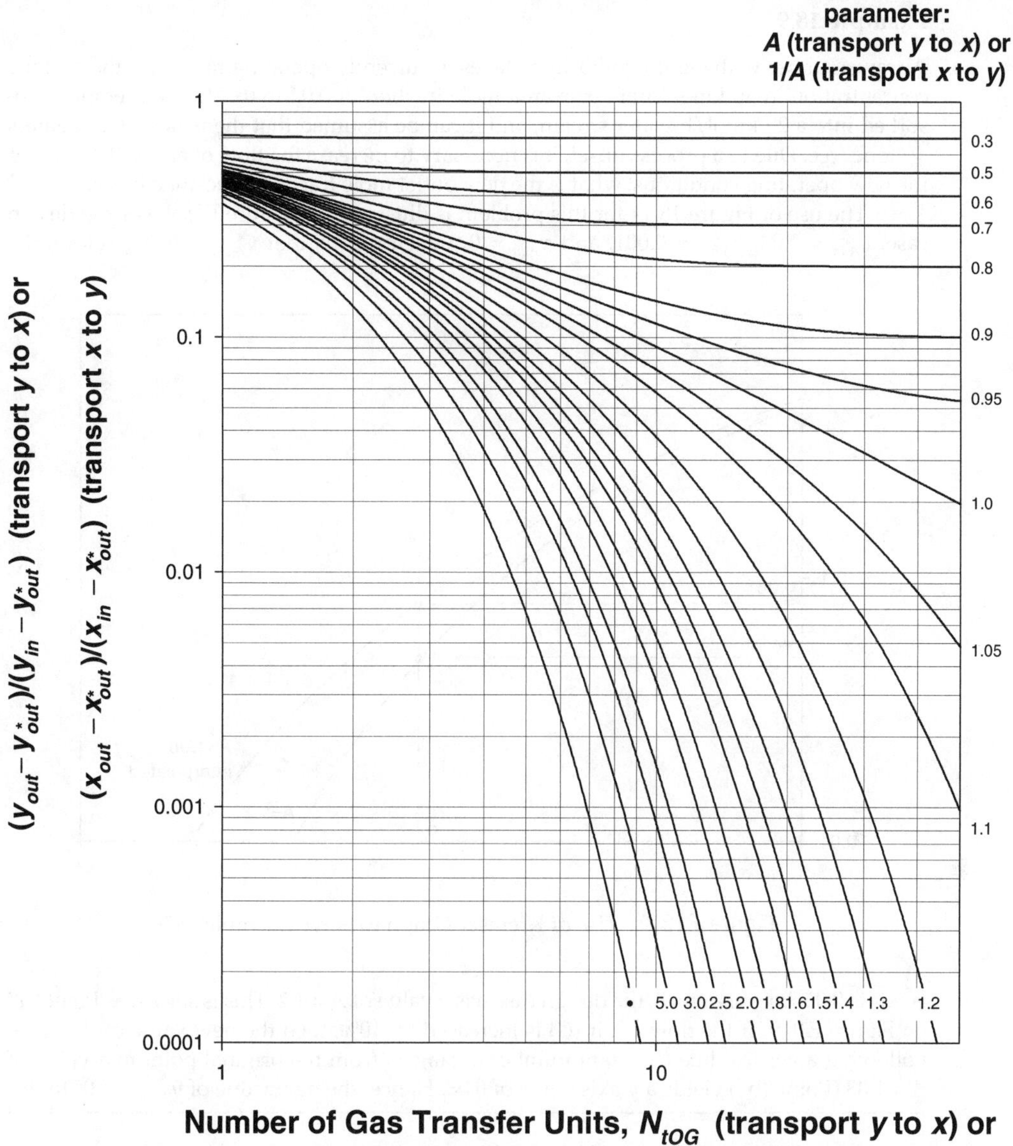

Figure 18.15 Plot of Colburn Equation, Number of Transfer Units for Countercurrent Operation, Henry's Law Equilibrium, and Constant A or $1/A$

Example 18.9

A tray scrubber with eight equilibrium stages is currently operating to reduce the acetone concentration in 40 kmol/h of air from a mole fraction of 0.02 to 0.001. The acetone is absorbed into a 20 kmol/h water stream, and it can be assumed that the water stream enters acetone free. Due to a process upset, it is necessary to increase the flow of air by 10%. Under the new operating conditions, what is the new outlet mole fraction of acetone in air?

The use of Figure 18.14 for this problem is illustrated in Figure E18.9. For the design case, $y_{A,in} = 0.02$, $y_{A,out} = 0.001$, and $x_{A,in} = 0$, which means that $y^*_{A,out} = 0$. Therefore, the

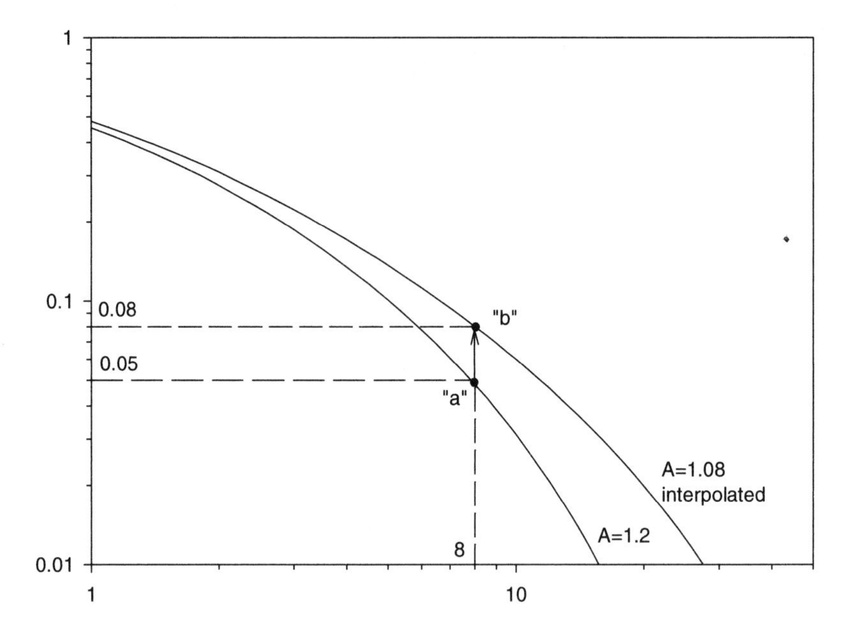

Figure E18.9 Use of Kremser Graph to Solve Example 18.9

y-axis is 0.05. Because $N = 8$, the design case has a value of $A = 1.2$. This is shown as Point "a" on Figure E18.9. If the flow of air (G) is increased by 10%, then the new value of $A = 1.08$. Following a vertical line (constant number of stages) from the original point to a value of $A = 1.08$ (Point "b") yields a y-axis value of 0.08. Hence, the new value of $y_{A,out} = 0.0018$.

Example 18.10

For the situation described in Example 18.9, the inlet air flowrate and the inlet mole fraction of acetone in air may vary during process operation. Prepare a set of performance curves for the liquid rate necessary to maintain the outlet acetone mole fraction for a range of inlet acetone mole fractions. There should be curves for five different values of the inlet gas rate, including the original case and the following percentages of the original case: 80, 90, 110, and 120. The temperature and pressure in the scrubber are assumed to remain constant.

The result is shown on Figure E18.10. The curves are generated as follows. From the original operating point, the value of m can be determined to be 0.417. By moving vertically on a line of constant number of equilibrium stages (8 in this case), values for the y-axis and A can be tabulated. The y-axis value is easily converted into $y_{A,in}$ because $y_{A,out}$ is known and constant. Then, for different values of G, values of L are obtained because m is known and constant.

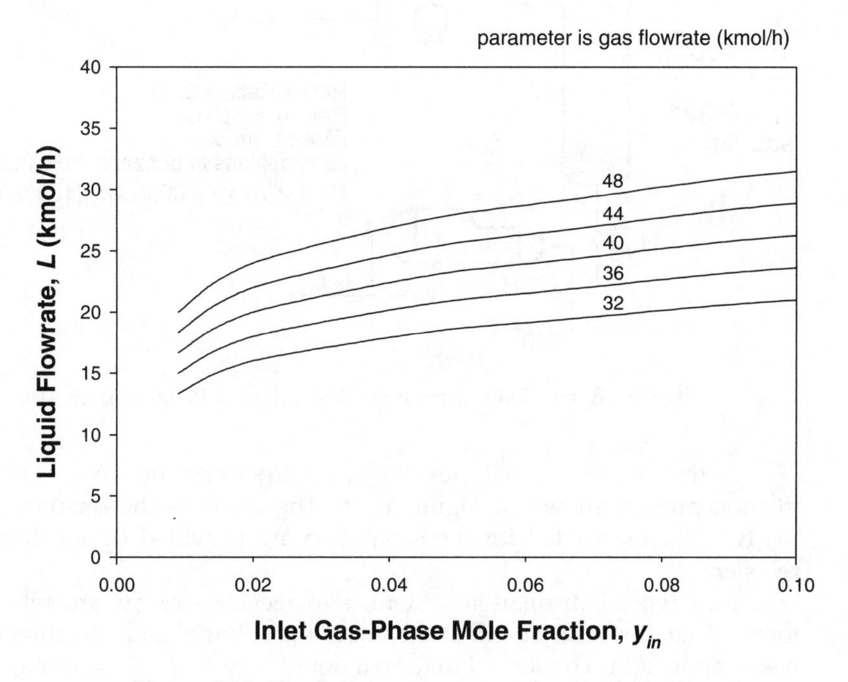

Figure E18.10 Performance Curves for Example 18.10

18.3.2 Distillation

For distillation, there is no universal set of performance curves as for multistage equilibrium separations involving mass separating agents. However, for a given situation, performance curves can be generated. This section demonstrates how this is done.

Figure 18.16 illustrates a multistage distillation system for separating benzene from a mixture of benzene and toluene. Figure 18.17 is a sketch of a typical distillation column. This problem is modified from Bailie and Shaeiwitz [1].

Component separation takes place within the tower. Considering the tower alone, there are three input and two output streams. Separation is achieved by the transfer of components between the two phases. For an operating plant, the size of the tower, the number and type of trays (or height of packing) in the tower, and other tower attributes are fixed. The output streams are established once the input flow streams to the tower are set.

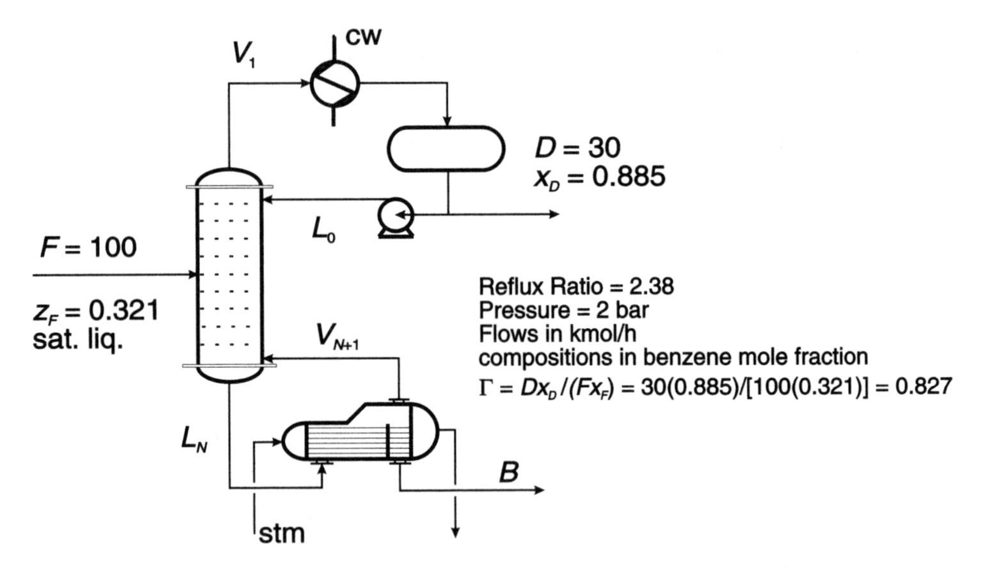

Figure 18.16 Plant Section for Distillation of Benzene from Toluene

In this section we will develop several performance diagrams for the distillation process shown in Figure 18.16. The tower is the separation unit, and the two phases needed for the separation are provided by the condenser and reboiler.

In a typical distillation system, two recycle streams are returned to the tower. A condenser is added at the top of the column, and a fraction of the overhead vapor, V_1, is condensed to form a liquid recycle, L_0. This provides the liquid phase needed in the tower. The remaining fraction is the overhead product, D. A vaporizer (reboiler) is added to the bottom of the column, and a portion of the bottom liquid, L_N, is vaporized and recycled to the tower as stream V_{N+1}. This provides the vapor phase needed in the tower.

Figure 18.16 provides some information on the distillation process taken from an operating plant.

> *Product benzene is separated from a benzene-toluene mixture. The feed stream, F = 100 kmol/h, is a saturated liquid with composition z_F = 0.321 (mole fraction benzene) at a pressure P = 2 bar. The top product consists of a distillate stream, D = 30 kmol/h, with composition x_D = 0.885 and a corresponding benzene recovery, Γ = 0.827. The tower has a diameter of 0.83 m and contains 20 sieve plates with a 0.61 m plate spacing. Feed is introduced on Plate 13, and the tower operates at a reflux ratio of 2.38. No information is provided on the partial reboiler and total condenser. From a benzene balance we obtain B = 70 kmol/h and x_B = 0.079.*

Figure 18.18 provides a McCabe-Thiele diagram for the separation given in Figure 18.16. This was constructed from the equilibrium curve and by matching the concentrations, $x_D = 0.885$, $z_F = 0.321$, and $x_B = 0.079$, the reflux ratio, $R = 2.38$, and the condition of saturated liquid feed with plant operating conditions. This

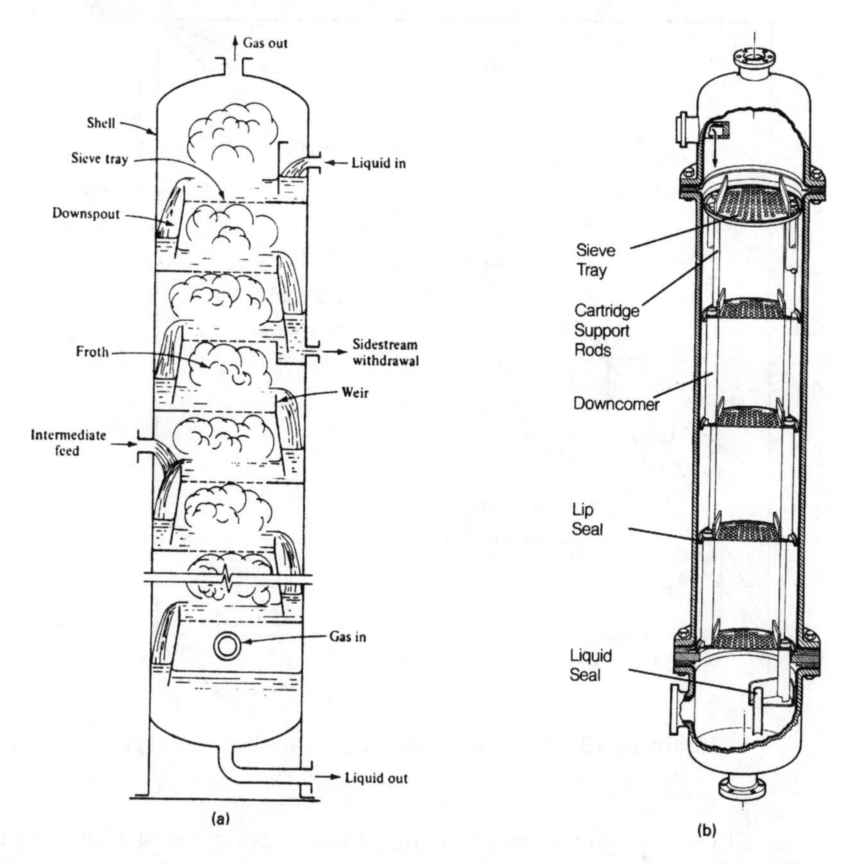

Figure 18.17 Details of Internal Construction of a Distillation Column (From S. Walas, *Chemical Process Equipment: Selection and Design.* Boston, MA: Butterworths, 1988. Reprinted with permission.)

McCabe-Thiele construction yielded 13 theoretical stages (12 trays plus a reboiler). This means that the average plate efficiency was 60% (100[No. of theoretical plates/No. of actual plates] = 100(12/20). The 60% plate efficiency is within the range found for columns separating benzene from toluene. The feed is added on theoretical plate number 8 (13[0.60] ≈ 8). (*Note:* The number 13 appearing in the previous calculation refers to the actual feed location and not to the number of stages.)

The McCabe-Thiele method for evaluating theoretical stages is limited by the following assumptions and constraints.

1. It applies only to binary systems.
2. It assumes "constant molal overflow" and in most situations does not satisfy an overall energy balance.
3. It requires graphical trial-and-error solutions to solve performance problems.

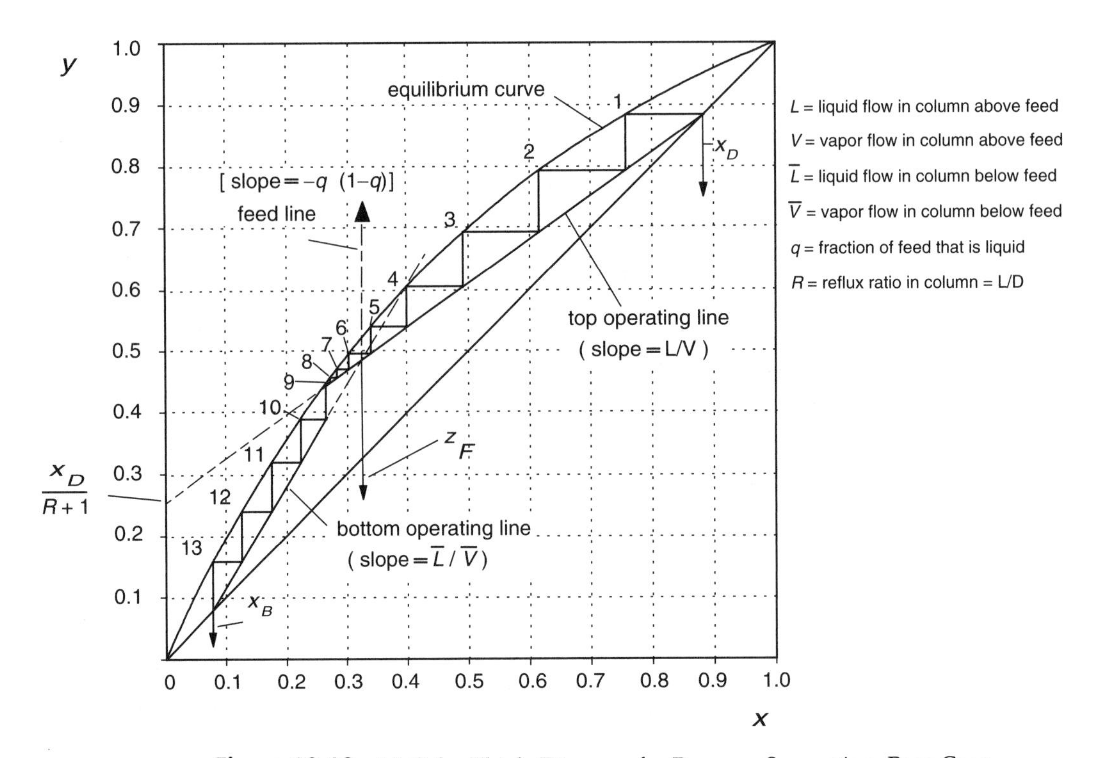

Figure 18.18 McCabe-Thiele Diagram for Benzene Separation: Base Case

Even though the speed of computation using the McCabe-Thiele method is far slower than for process simulators, it nevertheless remains an important analytical tool. Its graphical representation promotes a clarity of understanding that leads to valuable insights. We illustrate this by using the McCabe-Thiele diagram to glean some critical information required to construct a performance curve for the distillation process in Figure 18.18.

Upon studying Figure 18.18 the following factors are revealed.

1. Feed is not being introduced at the optimum location in the column. The optimal location is on Plate 10 ($6/0.60 \approx 10$) and not Plate 13 as currently used.
2. Separation steps near the feed plate are small.

For the current conditions and feed plate location, there are more than the optimum number of stages in the rectifying (top) section. This suggests that the unit may have been designed to process a lower concentration of feed material or to produce a higher-concentration distillate stream than is required currently.

Increasing the slope of the top operating line, L/V, moves the operating lines away from the equilibrium line. This increases the separation of each stage (increases the **step size**). In the special case when the distillate concentration remains the same, the concentration of the more volatile material (in our case, benzene) in the bottom stream will be lowered.

From an inspection of the McCabe-Thiele diagram, the tower appears to be oversized for the present separation. It could process a lower quality of feed (lower benzene concentration), produce a higher-quality product, and improve benzene recovery. These are points that may deserve further investigation and could be overlooked easily without the graphical representation provided by the McCabe-Thiele diagram.

The McCabe-Thiele analysis is difficult to utilize for developing all the information needed in a performance problem. It requires a graphical trial-and-error solution to match the number of stages to a set of values for z_F, x_D, and x_B. Fortunately, modern computer simulators can provide rigorous solutions for a large variety of separators, including distillation towers. The information needed to construct the performance diagrams for the distillation column in this chapter was obtained from a Chemcad simulation. We selected, as a base case, a column with 13 theoretical stages (fed on Tray 8). In our preliminary evaluation, we assume the tray efficiency remained constant. We return to this assumption when we discuss the limitations of these performance curves.

There are three input streams entering the tower shown in Figure 18.16. The flowrate and concentration of these streams, coupled with the performance of the equipment, establish the tower output. The remaining streams in the distillation system can be calculated once these streams are known.

The results of a computer simulation for the base-case conditions are shown in the following table.

	Flowrate (kmol/h)	Mole Fractions	Concentration
Inputs			
F	100.0	z_F	0.321
Outputs from Simulation			
L_0	71.5	x_0	0.885
V_{N+1}	96.4	y_{N+1}	0.079
V_1	101.5	y_1	0.885
L_N	166.4	x_N	0.079

The value for the vapor flowrate changed from 101.5 kmol/h to 96.4 kmol/h moving from the top to the bottom of the column. Even for this "ideal" separation, the assumption of constant molal overflow is not satisfied.

The process simulator provides a rigorous solution by carrying out material balances, energy balances, and equilibrium calculations over each stage. The simulator provides tray-by-tray results that include the composition of liquid and vapor on each tray, temperature, pressure, and K-values for each component on each tray, along with transport properties of both phases. Using data from this simulation, performance diagrams may be obtained. The most important variables are the tower inputs. They cause the outputs to change. Selection of which performance diagrams to prepare depends on the problem being considered. This is illustrated in the following case study.

You are the engineer in charge of the toluene-benzene distillation section in a chemical plant. You have just met with your supervisor and have been informed that, in the future, the feed concentration to this unit will no longer be constant but will vary between 25% and 40% benzene. You have also been told that it is important to maintain a constant distillate flow, D, and concentration, x_D, from this distillation unit to the downstream process.

At this meeting, possibilities for replacing parts of this process, replacing the whole system, and installing a storage system to blend feed and/or product distillate, or a combination of these alternatives, were discussed. The object is to maintain the distillate flow and concentration.

To make a responsible decision on this matter, it will be necessary to assess the effect that changes in feed concentration will have on the operation of the distillation process. Before making a decision, background material will be needed. You have been told to provide a performance diagram that shows the effects of changes in feed concentration on the reflux flowrate. For a preliminary assessment, you are to assume that the reboiler and condenser can meet any new demands. If they are found inadequate, they will be replaced or modified.

In addition, questions have been raised regarding the low benzene recovery in the current operation, and you have been told to consider the impact of using higher benzene recoveries.

For the situation described above, the important variables are the feed concentration, x_F, and the benzene recovery, Γ. The information needed to respond to the above request can be obtained by running simulations at different feed concentrations and reflux flowrates to determine the benzene recovery for each pair of conditions. Reflux is chosen as the dependent variable here because the flow of the reflux stream is the most common way to regulate the performance of a distillation column.

Figure 18.19 gives the performance curves for the reflux, L_0. The performance diagram shows the feed concentration, z_F, on the x-axis with the benzene recovery, Γ, as a parameter. The base case is identified as Point "a" on the diagram.

The performance curves presented show the following trends.

1. **For constant recovery, Γ:** As z_F increases, the reflux decreases. The rate of decrease becomes less as z_F increases.
2. **For a constant feed concentration, z_F:** As Γ increases, the reflux increases. The rate of increase becomes greater as Γ increases.

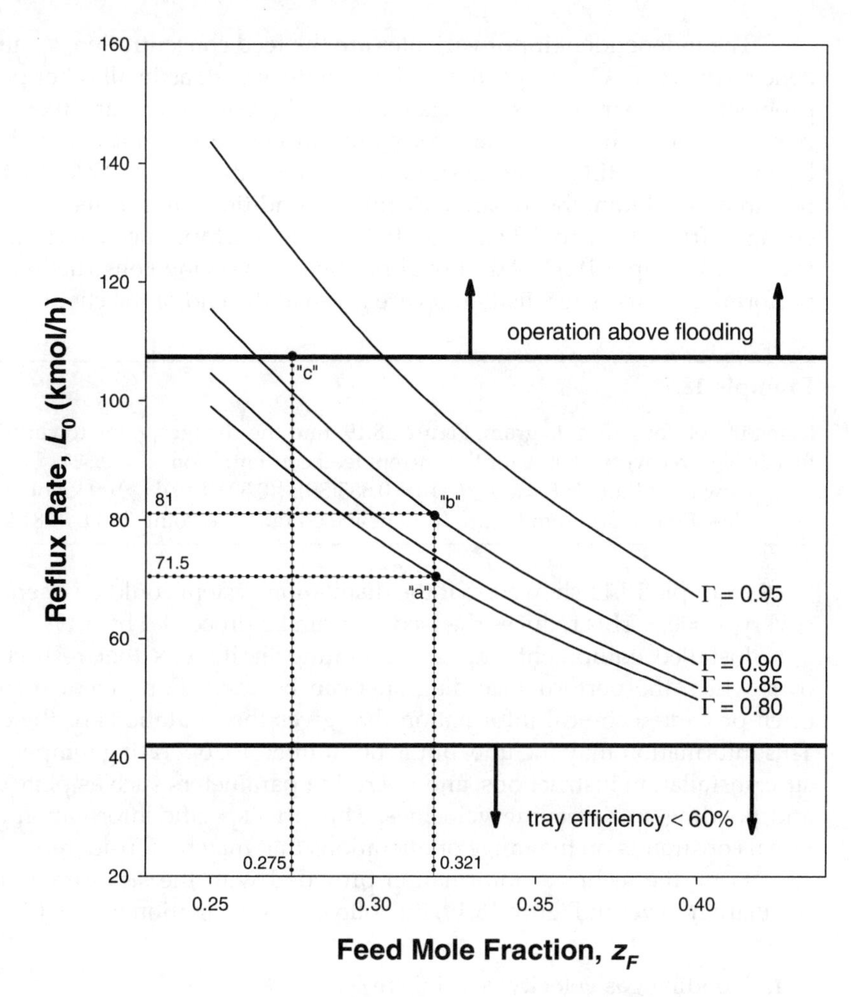

Figure 18.19 Performance Curve for Distillation Tower

The performance diagram given in Figure 18.19 is specific to the existing tower with fixed feed location, saturated liquid feed, and desired distillate conditions. The conditions or variables that are fixed in this problem are as follows:

$D = 30$ kmol/h
$x_D = 0.885$ mole fraction benzene
20 trays (12 theoretical equilibrium stages plus a reboiler)
Feed plate = 13 (theoretical equilibrium Stage 8)
Diameter of tower = 0.83 m
Tray spacing = 0.61 m
Overall tray efficiency = 60%

The independent input variables are the feed concentration, z_F, and the benzene recovery, Γ. One important point to note is that, as in all other performance problems, the parameters associated with the equipment are fixed. A second point to note is that with the information in Figure 18.19, it is possible to work backward and estimate the desired reflux flowrate, for example, that would be required to obtain the desired distillate conditions if the feed concentration changed from 0.321 to 0.295. The utility of the performance curves is best illustrated in Example 18.11. Additional problems involving construction and use of performance curves for distillation are given at the end of the chapter.

Example 18.11

Using the performance diagram, Figure 18.19, find the changes to the tower input needed to achieve a recovery of 90% for the current feed concentration, $x_F = 0.321$.

New Feed Rate, F: $F = x_D D/(z_F \Gamma) = 0.885(30)/[(0.321)(0.9)] = 91.9$ kmol/h

New Reflux, L_0: From Figure 18.19 new feed rate (see Point "b") $L_0 = 81$ kmol/h

---

Example 18.11 showed that the distillation system could be operated at a recovery, $\Gamma = 0.9$. This reduces the feed that can be processed by 8.1%.

Installed equipment imposes operating limitations that restrict the range over which the performance diagrams can be used. Equipment manufacturers often provide technical information that gives the limitations of the equipment. This information may include, but is not limited to, operating temperature, pressure, installation instructions, and operating parameters such as plate efficiencies and flooding and weeping velocities. Thus this specific information adds additional constraints on the range of operations that may be considered.

From the technical information provided with the sieve trays used in the distillation tower in Figure 18.19, the following information was obtained:

1. Flooding gas velocity, $u_f = 1.07$ m/s.
2. Weeping gas velocity, $u_w = 0.35$ m/s.
3. Tray efficiency, $\epsilon = 0.60$ for $2.6 > u[\text{m/s}](\rho_v[\text{kg/m}^3])^{0.5} > 1.2$

Based on this information, two regions on Figure 18.19 are shown. These are "forbidden" or infeasible regions, in which the performance curves are not valid because of the limitations of the specific equipment or the assumptions made (e.g., $\epsilon = 0.60$). For any change to be made, the new condition must be checked to ensure that the new operating point lies within the feasible region for all the performance curves.

The flooding velocity limit is also shown on Figure 18.19. The limit related to weeping lies below that for tray efficiency; that is, the tray efficiency drops below 60% well before the weeping condition is reached. The constraint of operating below a tray efficiency of 60% is shown as the lower infeasible region in Figure 18.19.

Example 18.12 shows how the flooding limit of the equipment, illustrated on the performance graph, identifies another equipment limitation.

Example 18.12

Find the maximum recovery possible from the distillation equipment for a feed concentration, $z_F = 0.275$.

Point "c" on Figure 18.19 provides an estimate of the recovery, $\Gamma = 0.91$.

The constraints associated with the equipment must be included in any analysis of performance.

In our preliminary analysis, we assumed that the tray efficiency was 60% and the reboiler and condenser were adequate. From the manufacturer's data, it was confirmed that the efficiencies were constant over the operating range given above. The limits on the reboiler and the condenser have not been considered in the above analysis and must be checked against the performance limits for these pieces of equipment. In addition, the curves obtained assume that the feed location would not change. However, by using the optimum location, as the feed concentration changes, the performance of the tower could be improved. The use of alternative feed locations requires that feed nozzles be present at different trays. Thus, once again, the constraints of the existing equipment dictate whether this option should be considered further.

18.4 SUMMARY

In this chapter we have discovered how to construct simple performance diagrams for some common individual pieces of equipment. These performance diagrams show how a given piece of equipment responds to changes in input flows. They also can be used to predict what changes in input variables would be required in order to obtain a desired output condition. In some instances, as with valves, pumps, and compressors, performance curves are provided by equipment manufacturers. In other instances, as with the Kremser or Colburn equation, universal performance curves and equations exist subject to specific assumptions. In still other instances, as with heat exchangers or distillation columns, performance curves must be constructed by modeling equipment behavior.

In all real chemical processes, the final element that makes changes in flowrates possible is the regulating valve. Without such valves, it would be impossible for a process to adjust for unforeseen changes in operating conditions. In addition, it would be equally impossible to manipulate a process to give new desired outputs. Regulating valves are relatively simple and inexpensive pieces of equipment, and yet they are absolutely essential in the day-to-day operation of a chemical process. The use of regulating valves was illustrated for a simple flow system including a pump, piping system, process equipment, and a valve.

REFERENCE

1. Bailie, R. C., and J. A. Shaeiwitz, "Performance Problems," *Chem. Eng. Edu.* 28 (1994): 198.

SHORT ANSWER QUESTIONS

1. What is a pump curve? Sketch a typical pump curve, making sure that both axes are clearly labeled. On the same sketch, show a typical system curve. Indicate the typical region of good operating practice. Explain.
2. Comment on the following statement: *For a two-pump system, it is always best to run the two pumps in series rather than in parallel because greater scale-up will be possible.*
3. Explain the meaning of the intersection of the two curves on the diagrams in Figure P18.3.

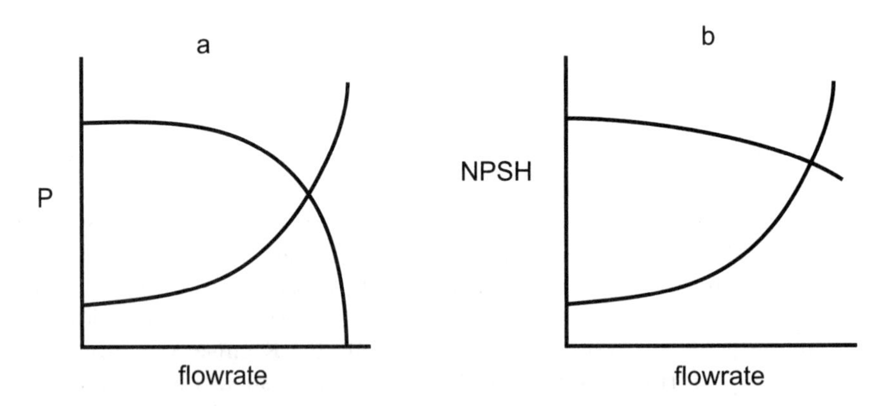

Figure P18.3 Diagrams for Problem 18.3

4. Comment on the following statement: *The film heat transfer coefficient always increases with flowrate to the 0.8 power.*
5. For a given flowrate, it is known that the pump through which the process flows cavitates. The pump has a spare, and a temporary fix to run both pumps simultaneously has been suggested. The pumps can be run either in series or parallel. Which arrangement *might* fix the cavitation problem?
6. A storage tank uses gravity flow to supply a liquid feed material to a holding tank. The pressures and levels in each tank can be considered to be constant. If the viscosity of the fluid is sensitive to temperature (increasing with decreasing temperature), what effect (large or small) will a decrease in temperature have on the flow of the material if the flow is (a) laminar or (b) turbulent? Explain your reasoning.
7. Two columns operate in a chemical plant. The first column operates at a pressure of 10 bar, and the second at a temperature of 1 bar. For process reasons, the pressures in the two columns are both reduced by 0.4 bar. If the internal flows in the columns (mass flows of liquid and vapor) are held constant during this pressure change, which column is more likely to flood?
8. For the separation of a binary mixture in a distillation column, what will be the effect of an increase in column pressure on the following variables?
   a. Tendency to flood at a fixed reflux ratio
   b. Reflux ratio for a given top and bottom purity at a constant number of stages
   c. Number of stages required for a given top and bottom purity at constant reflux ratio
   d. Overhead condenser temperature
9. Explain, using equations where appropriate, why a pump located even with the bottom of a pressurized tank containing a vapor-liquid mixture of a pure component will likely not ever cavitate regardless of the ambient temperature.
10. A storage tank is connected to a pond (at atmospheric pressure!) by a length of 4" pipe and a gate valve. From previous operating experience, it has been found that when the tank is at a pressure of 3 atm the flow through the pipe is 35 m^3/h when the gate valve is fully open. If the pressure in the tank increases to 5 atm, what will be the maximum discharge rate from the tank?
11. Consider the pump and system curves in Figure P18.11 for identical pumps arranged in parallel and series.
    a. If the flowrate is increased from 3 L/s to 4 L/s, which pump arrangement(s) (single pump, two in series, or two in parallel) will give you the desired flow increase?
    b. For your answer(s) to Part (a), will the pump(s) cavitate?

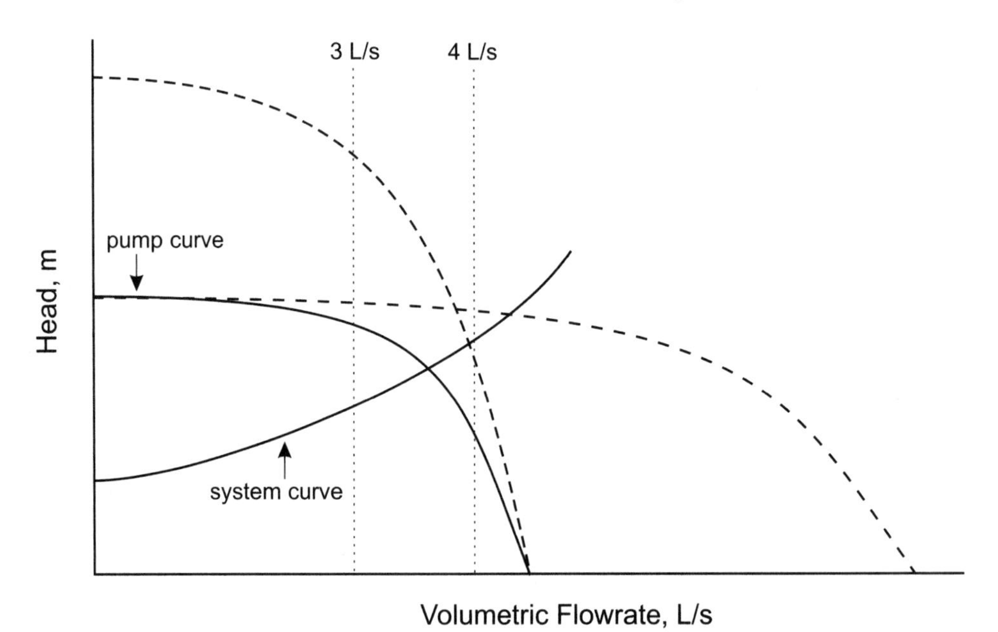

Figure P18.11 Pump and System Curves for Problem 11

PROBLEMS

Background Material for Problems 12–14

For the steam generator illustrated in Figure 18.1(a) and discussed in Section 18.1, the total resistance to heat transfer, $R_T = 1/(UA)$, is given by the following relationship:

$$R_T = R_i + R_o + R_p + R_{f,i} + R_{f,o}$$

In the above equation,

R_T = total resistance to heat transfer.

R_i = convective resistance of the light oil. This resistance is a function of the flowrate, as shown in Chapter 17. Subscript 1 refers to the base case, and subscript 2 refers to the new case.

$$R_{i2} = R_i(\dot{m}_2/\dot{m}_1)^{0.8}$$

R_o = convective resistance of the boiling water. This resistance is not a function of flowrate. It is a weak function of the temperature drop across the film of vapor on the outside of the tubes. For this analysis, it is assumed constant.

R_p = conductive resistance of the tube walls. This resistance is small and is ignored here.

$R_{f,i}$ and $R_{f,o}$ = fouling resistances on the inside and outside of the tubes, respectively. These two resistances are combined into a single fouling term.

$$R_f = R_{f,i} + R_{f,o}$$

For the conditions given above, the total resistance becomes

$$R_T = R_{i1}(\dot{m}_2/\dot{m}_1)^{0.8} + R_o + R_f$$

The only resistance affected by mass flowrate, $\dot{m}_i$, is R_i.

12. Operating data taken from the plant shortly following the cleaning of the reboiler tubes show that

$$T_{in} = 325°C,\ T_{out} = 293°C,\ \dot{m}_s = 1650\ kg/h,\ T_s = 253°C$$

   a. Find the overall heat transfer coefficient.
   b. Estimate the overall resistance to heat transfer.

13. This is a continuation of Problem 12. Estimate the fouling resistance for the base case presented in Figure 18.1(a).
14. This is a continuation of Problem 12. Typical heat transfer coefficients given in Table 11.11 are used to predict the ratio of heat transfer resistances between boiling water and light oil. This ratio is estimated to be approximately 1:2. Estimate the individual resistance and heat transfer coefficient for both the oil and the boiling water.

Background Material for Problems 15–16

The performance curves presented in Figure 18.1 represent those obtained for the heat exchanger used in Problems 15–16.

15. a. Use the guideline from Tables 11.11 and 11.18 to establish "normal" limits on the amount of heat that could be exchanged.
    b. How do these limits explain the fouling of the heat exchanger observed in the operating unit?
    c. Under the constraints identified in Part (a), what is the highest temperature allowed (flowrates limited to range of M values shown)?
16. a. Estimate the relative increase in heat duty that would result from increasing the liquid level from the base case to cover the tubes completely. The input streams and vapor temperature remain the same as in the base case.
    b. Explain why the duty increase is less than the increase in area.

Background Material for Problems 17–19 is given in Section 18.2.

17. For the system presented in Example 18.3 you are asked to appraise the effects of adding a second pump and impeller set identical to the current pump in the system.
    a. Prepare a performance diagram, for the two-pump system and for each impeller size, if the pumps are to be installed in series.
    b. Prepare a performance diagram, for the two-pump system and for each impeller size, if the pumps are to be installed in parallel.
    c. Estimate the pressure achieved for each pump/impeller system in Parts (a) and (b) at flowrates of 400 and 800 gallons/min.
18. Referring to Example 18.3, it has been decided that the storage tank should be kept at 1 bar and the reactor pressure at 1.2 bar. To provide the pressure head required, a pump is placed in line before the heat exchanger. For the 50% increase in flow desired in Example 18.3, answer the following.
    a. What pressure head must be produced by the pump?
    b. Which of the pump/impeller systems given in Figure 18.4 is capable of providing sufficient flow?
    c. At what location in the system would you place the pump? Explain why.
19. A centrifugal pump produces head by accelerating fluid to a velocity approaching the tip velocity of the rotating impeller. The liquid leaving the pump is at a high velocity and is forced into the discharge pipe from the pump, where it comes into contact with relatively slow-moving fluid in the pipe. As the fluid decelerates, the velocity head is converted into pressure head. For a given pump the following relationships can predict performance.

$$h_2/h_1 = [(N_2/N_1)(D_2/D_1)]^2 \text{ and } \dot{v}_2/\dot{v}_1 = [(N_2/N_1)(D_2/D_1)]^3$$

where

N = speed of rotation of the impeller, rpm

h and $\dot{v}$ = head and volumetric flowrate, respectively

2 and 1 = new and base-case conditions, respectively

D = impeller diameter

The speed of rotation of the impeller can be adjusted, even when a constant speed motor is used, by the use of a pulley system as shown in Figure P18.19.

The pump curve given in Figure 18.4 was obtained when the impeller was directly coupled to a constant-speed motor (3500 rpm). Develop performance curves for this pump, with a 7-in diameter impeller, assuming that the pump is driven via a pulley system in which the constant-speed motor (3500 rpm) has a 6-in diameter pulley and the motor impeller is connected to a series of pulleys with diameters of 4 in, 5 in, and 6 in. Construct performance diagrams for each pulley combination.

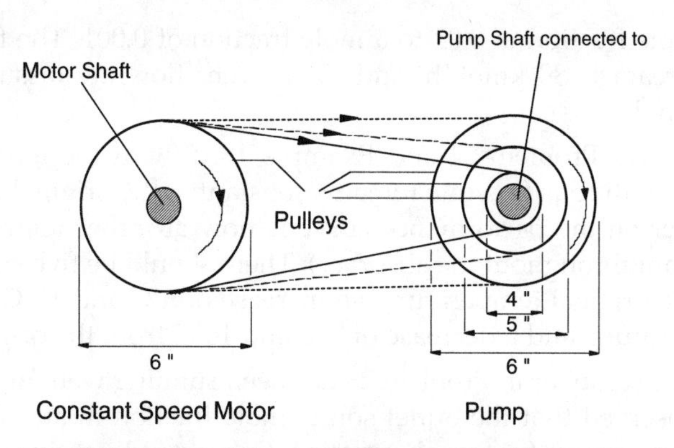

Figure P18.19 Pully System for Problem 19

20. Naphthalene is fed to a phthalic anhydride production process. The feed is available at 208°C and 79 kPa. The flowrate is 18,500 kg/h in 1.5 in Schedule 40 pipe. A pump with NPSH characteristics as plotted in Figure 18.9 is used. Will this pump be suitable for the desired duty? If not, what modifications would be necessary in order to use the existing pump? Be quantitative.
21. Refer to Example 18.8. There are additional methods other than the two presented for increasing the NPSH in order to avoid pump cavitation. For the following situations, sketch the $NPSH_A$ curve on Figure 18.9, keeping head at 2 m and keeping the flowrate constant. Identify the operating point on the $NPSH_A$ curve in order to determine whether $NPSH_A > NPSH_R$.
    a. The diameter of the suction line can be increased. Examine the effect of increasing the diameter of the suction line to 1.25 in Schedule 80.
    b. The temperature of the toluene can be decreased. Examine the effect of decreasing the temperature of the toluene to 40°C using the original pipe.
22. Refer to the compressor curves given in Figure 18.11.
    a. At a flowrate of 125 Mg/h, what is the exit pressure from the compressor operating at 3500 rpm if the inlet pressure is 2 bar?
    b. If the feed pressure is 4 bar and the compressor is operating at 2200 rpm, how can you obtain an outlet pressure of 9 bar at 125 Mg/h?
    c. It is necessary to raise the pressure from 2 bar to 24 bar at 125 Mg/h. How could this be accomplished at each rpm? Which method do you recommend?
23. A packed scrubber with 20 transfer units has been designed to reduce the solute concentration in 80 kmol/h of air from a mole fraction of 0.01 to 0.006. The solute is absorbed into a 20 kmol/h solvent stream, and it can be assumed that the solvent stream enters solute free. If the flow of solvent is decreased by 15%, what is the new outlet mole fraction of solute in air?
24. Prepare a set of performance curves similar to those in Example 18.10 for a packed scrubber with 10 transfer units that removes solute from air from a

mole fraction of 0.02 to a mole fraction of 0.001. The flowrate of the inlet air stream is 80 kmol/h, and the solvent flowrate (assumed solute free) is 40 kmol/h.

25. Repeat Problem 24 and Example 18.10 with the following change: In each case, the gas flowrate remains constant at the original conditions. The parameter on the performance curve is now absorber temperature (assumed constant throughout the absorber). There should be five curves for each absorber: the original temperature, an increase of 5°C and 10°C from the original temperature, and a decrease of 5°C and 10°C from the original temperature.
26. The scrubber in Problem 23 has been running well for several years. It is now observed that the outlet solute mole fraction in air is 0.002. Suggest at least five reasons for this observation. Suggest at least five ways to compensate for this problem in the short term. Evaluate each compensation method as to its suitability.
27. It is necessary to increase the capacity of an existing distillation column by 25%. As a consequence, the amount of liquid condensed in the condenser must increase by 25%. In this condenser, cooling water is available at 30°C and, under present operating conditions, exits the condenser at 45°C, the maximum allowable return temperature without a financial penalty assessed to your process. Condensation takes place at 75°C. You can assume that the limiting resistance is on the cooling water side. Can the existing condenser handle the scale-up without incurring a financial penalty? What is the new outlet temperature of cooling water? By what factor must the cooling water flow change?
28. In the previous problem, it may be necessary to scale up or scale down by as much as 50% from the original operating conditions. Prepare a performance curve similar to the one in Figure 18.1 for operation of the condenser over this range.

Background Material for Problems 29–33

In Section 18.3.2, it was shown that Figure 18.19 could be constructed from results of simulating the distillation column. Table P18.29 shows the results of such a simulation for the problem discussed in Section 18.3.2. The data used in Figure 18.19 are contained in this table. The data in Table P18.29 can be used to construct other performance curves, similar to Figure 18.19. In Problems 29–31, a performance curve should be constructed with the flooding and tray efficiency limits included (given on page 646), as in Figure 18.19.

29. Prepare a performance plot for vapor velocity as a function of feed composition with benzene recovery as a parameter.
30. Prepare a performance plot for boil-up rate as a function of feed composition with benzene recovery as a parameter.

Table P18.29 Performance Data from a Process Simulator for the Benzene-Toluene Tower

Feed Conc z_F	Benz Rcvry Γ	Feed F kmol/h	Bottm Prod B kmol/h	Reflux Ratio R	Ovhd Vapor V_1 kmol/h	Boil-Up V_{N+1} kmol/h	Liquid Reflux L_0 kmol/h	Cond Duty Q_c GJ/h	Rebl Duty Q_r GJ/h	Bottm Temp °C	Vapor Vel at Top u m/s
0.25	0.80	132.8	102.8	3.30	129.0	122.4	99.0	−3.93	3.87	133.4	1.01
0.30	0.80	110.6	80.6	2.57	107.1	101.7	77.1	−3.26	3.21	132.6	0.84
0.35	0.80	94.8	64.8	2.05	91.5	86.9	61.5	−2.79	2.75	131.8	0.73
0.40	0.80	83.05	53.0	1.66	79.8	75.7	49.8	−2.43	2.39	130.7	0.62
0.25	0.85	124.9	94.9	3.49	134.7	127.8	104.7	−4.10	4.04	134.1	1.05
0.30	0.85	104.1	74.1	2.70	111.0	105.4	81.0	−3.38	3.33	133.5	0.87
0.35	0.85	89.3	59.3	2.14	94.3	89.5	64.3	−2.87	2.83	132.8	0.74
0.40	0.85	78.1	48.1	1.73	81.8	77.6	51.8	−2.49	2.45	132.0	0.64
0.25	0.90	118.0	88.0	3.85	145.4	138.0	115.4	−4.42	4.36	134.9	1.14
0.30	0.90	98.3	68.3	2.96	118.8	112.7	88.8	−3.56	3.61	134.4	0.93
0.35	0.90	84.3	54.3	2.33	99.9	94.8	69.9	−3.04	3.00	133.9	0.78
0.40	0.90	73.8	43.8	1.86	85.5	81.4	55.8	−2.61	2.57	133.3	0.67
0.25	0.95	111.8	81.8	4.78	173.5	164.5	143.5	−5.26	5.20	135.6	1.36
0.30	0.95	93.2	63.2	3.64	139.2	132.1	109.2	−4.23	4.17	135.4	1.09
0.35	0.95	79.9	49.9	2.83	115.0	109.2	85.0	−3.50	3.45	135.1	0.90
0.40	0.95	69.9	39.9	2.24	97.1	92.0	67.1	−2.95	2.91	134.8	0.76

D = 30 kmol/h, x_D = 0.885, P_{bot} = 2 bar, overall tray efficiency = 60%.
Assume ideal system for evaluation of K values and other thermodynamic data.

31. Prepare a performance plot for reboiler duty as a function of feed composition with benzene recovery as a parameter.
32. For the distillation problem outlined in Section 18.3.2,
    a. Determine the values of F, B, and x_B required when the concentration of the feed becomes $z_F = 0.25$. Assume that the values of D and x_D are to remain the same (D = 30 kmol/h and x_D = 0.885) and that the recovery is to remain 90%.
    b. Report any risks associated with the solution.
    c. What are the values of L_0 and V_{N+1} for this case?

33. The information given in Figure 18.18 showed that the benzene recovery column was not operating at maximum benzene recovery.
    a. Estimate the maximum value of benzene recovery for the base case ($z_F = 0.321$).
    b. Give several reasons for operating the current column at less than maximum benzene recovery.
    c. Plot a performance curve that shows estimated values for F, L_0, and V_1 at maximum benzene recovery as a function of feed composition.

Background Information for Problems 34 and 35

In preparing the performance diagrams for the benzene-toluene system in Section 18.3.2, it was noted that under current operation, the feed stream was not introduced at the optimum location. The performance curves, Figure 18.19, were constructed for this nonoptimal feed location.

34. For the same feed, distillate, and bottoms shown in Figure 18.16, determine the following.
    a. The boil-up, V_{N+1}, and reflux, L_0, required if the feed is now introduced at the optimum location.
    b. Using typical utility costs from Chapter 8, estimate the yearly savings ($/yr) obtained by changing the feed location.
35. Assume that the distillation column in Figure 18.16 is to be operated at the same value of L_0, D, and B and the same feed as the base case.
    a. Calculate x_D, x_B, and the benzene recovery, Γ, obtained by introducing the feed at the optimum location.
    b. Assuming that the benzene lost in the bottom is valued at $0.20/lb, calculate the savings resulting from locating the feed at the optimum location.
36. A heat exchanger was put into service approximately one year ago. The design conditions are that process gas is cooled from 100°C to 50°C, with cooling water entering at 30°C and exiting at 40°C. Initially, the heat exchanger operated satisfactorily, meaning the temperatures for the cooling water and process streams were the same as the design conditions given above. However, over time, it has been observed that, due to impurities in the gas stream, a fouling layer of dirt has built up on the outside of the tubes. This dirt layer has caused the temperatures through the heat exchanger to change, and, in order to maintain the process gas outlet temperature at 50°C with the same design gas flowrate, the mass flow of water has had to be increased to 150% of the design flow, with a corresponding change in cooling water outlet temperature.

    Assuming that, for the design case, all the resistance to heat transfer is on the process-gas side, estimate the fouling heat transfer coefficient for the current operation as a fraction or multiple of the gas-phase heat transfer coefficient.

37. In a process that produces a temperature-sensitive product, the final step is to cool the product from 70°C to 35°C prior to sending it to storage. This cooling is achieved using cooling water that is available at 30°C and exiting at 40°C. A shell-and-tube heat exchanger is used with cooling water in the tubes. The process-side resistance is dominant. It is desired to scale down this process. Answer the following questions.
    a. If the flowrate of the process fluid decreases to 70% of its design value, by how much must the cooling water flowrate change to maintain the desired exit temperature of the process fluid of 35°C?
    b. What is the exit temperature of the cooling water for this case?
38. A desuperheater permits the temperature of saturated steam entering a heat exchanger to be controlled. This is accomplished by having a valve that changes the pressure of the source steam to a known value. When the steam pressure is lowered, the steam becomes superheated. If the correct amount of bfw is sprayed into the stream, it can be resaturated, and at the lower pressure, the steam is at a lower temperature.

    In a particular heat exchanger, condensing steam in the shell at 160°C is used to heat 2 kg/s of a process fluid from 50°C to 100°C. The condensing coefficient is 5000 W/m^2K, and the process-side coefficient is 200 W/m^2K. It is required that the process fluid flowrate be increased by 15%. What steam temperature and flowrate are needed to maintain the outlet process temperature at 100°C?

CHAPTER

19

Performance of Multiple Unit Operations

In Chapter 18, techniques for evaluating the performance of individual unit operations were introduced. The use of performance curves was illustrated. Performance curves describe the behavior of a unit over a wide range of inputs. The techniques developed in Chapter 17 were also used to obtain results for a single set of input conditions, which is one point on a performance curve.

In a chemical process, units do not operate in isolation. If the input to one unit is changed, its output changes. However, this output is usually the input to another unit. In certain cases, what is usually considered to be a single unit is actually multiple units operating together. A distillation column is an example of this situation, because the reboiler and condenser are part of the distillation unit. Any change in the column or in either of the heat exchangers will affect the performance of the other two. Therefore, it is necessary to understand the interrelationship between performance of multiple units and to gain experience in analyzing these types of problems.

Given the complexity of the performance of multiple units, performance curves will not be generated. The techniques developed in Chapters 16 through 18 will be used to solve for one point on a performance curve. As in Chapter 18, the case-study approach is used. Four examples of multiple units are illustrated in order to help you understand the approach to analyzing performance problems involving multiple units.

19.1 ANALYSIS OF A REACTOR WITH HEAT TRANSFER

Many industrially significant reactions are exothermic. Therefore, in a reactor, it is necessary to remove the heat generated by the heat of reaction. There are several ways to accomplish heat removal. For relatively low heats of reaction, staged adiabatic packed beds (packed with catalyst) with intercooling may be used.

Here, no heat is removed in the reaction section of the equipment. Instead, the process fluid is allowed to heat up in the reactor and heat is removed in heat exchangers between short beds of catalyst (reactor sections). For larger exothermic heats of reaction, a shell-and-tube configuration, much like a heat exchanger, is often used. Here, the reaction usually occurs in the tubes, which are packed with catalyst. As the heat of reaction increases, tube diameter is decreased in order to increase the heat transfer area. For extremely large heats of reaction, fluidized beds with internal heat transfer surfaces are often used, due to the constant temperature of the fluid bed and the relative stability of such reactors because of the large thermal mass of well-mixed solid particles.

In all three of these cases, the performance of the reactor and the heat exchanger are coupled. It is not correct to analyze one without the other. In this section, an example of the performance of a shell-and-tube-type reactor (tube bundle containing catalyst immersed in a pool of boiling water) with heat removal is used to illustrate the interrelationship between reactor performance and heat transfer. In Chapter 20, other examples of reactor performance are presented.

A shell-and-tube reactor, as illustrated in Figure 19.1, is used for the following reaction to produce cumene from benzene and propylene.

$$\underset{benzene}{C_6H_6} + \underset{propylene}{C_3H_6} \rightarrow \underset{cumene}{C_9H_{12}}$$

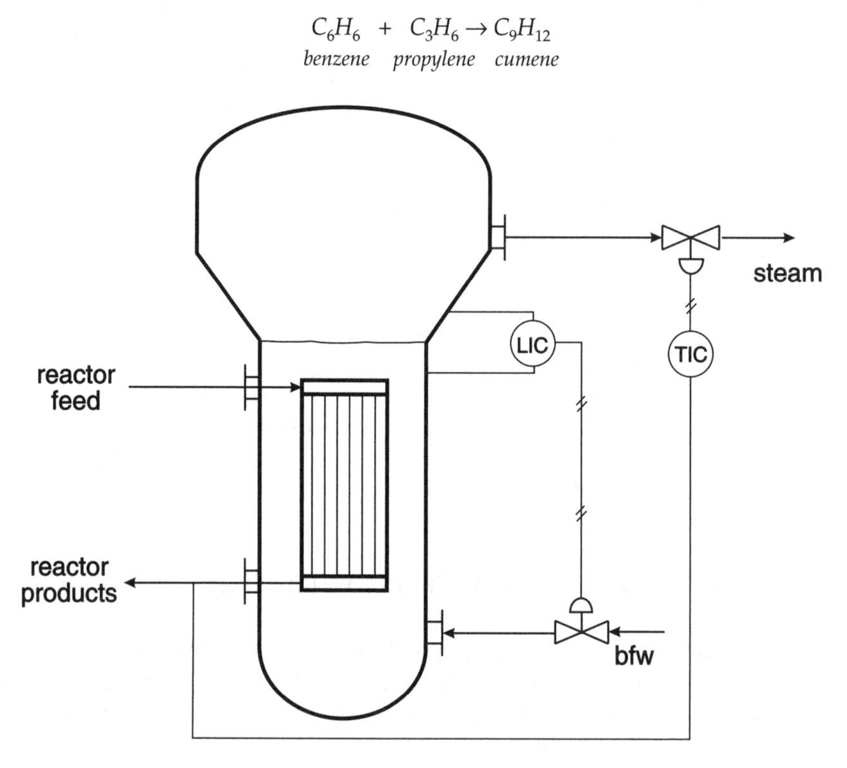

Figure 19.1 Shell-and-Tube-Type Packed-Bed Reactor

The reaction kinetics are as follows:

$$r = kc_b c_p \quad \text{mol/L s}$$

$$k = 3500 \exp(-13.28\ (\text{kcal/mol})/RT)$$

Normally, the reaction occurs at 350°C and 3000 kPa (the reaction is not really isothermal; see item 3 below). Heat evolved by the reaction is removed by producing high-pressure steam (4237 kPa, 254°C) from boiler feed water in the reactor shell. Propylene is the limiting reactant, with benzene present in excess. The feed propylene is a raw cut, and contains 5 wt% propane impurity. Recently, the propane supplier has been having difficulty meeting specifications, and the propane impurity exceeds 5 wt%. The reduced feed concentration of propylene is causing a decrease in cumene production. In order to maintain the desired cumene production rate, it has been suggested that the reaction temperature be raised to compensate.

The following assumptions are made in order to simplify the problem:

1. There are no other impurities except for the propane.
2. The pressure drop in the reactor is negligible relative to the operating pressure.
3. The temperature profile in the reactor is flat; that is, the temperature is 350°C everywhere in the reactor. This is the most serious assumption, and it is clearly not correct. There will most likely be a hot spot near the reactor entrance. Figure 19.2 illustrates the temperature profiles for the reactor and for the steam. The dotted line is the anticipated temperature profile. By assuming

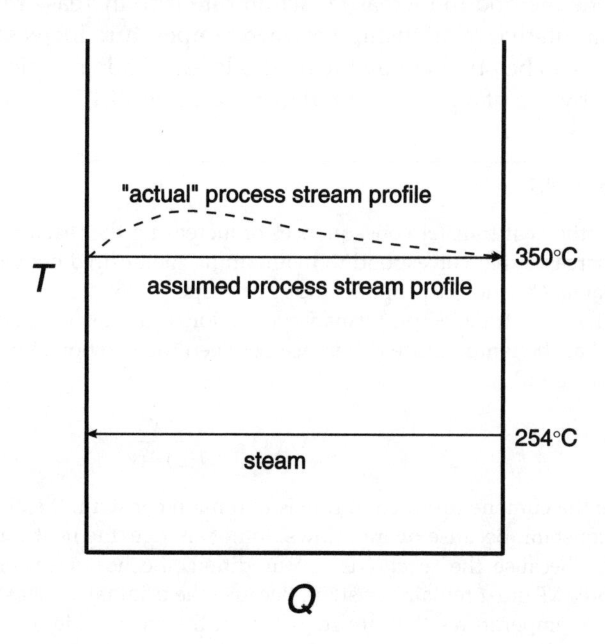

Figure 19.2 Temperature Profiles in Cumene Reactor

a constant temperature, calculations are simplified, and the concept illustrated in this case study can still be understood.

Before proceeding with a detailed analysis, it is instructive to understand why a temperature increase was suggested as compensation for decreased propylene feed concentration. If the total feed rate of impure propylene is consistently maintained, the feed rate of propylene decreases with increasing propane impurity. Because propylene is the limiting reactant, cumene production is related to propylene feed by the following relationship.

$$F_c = F_{po} X \tag{19.1}$$

where

F_c = molar rate of cumene produced
F_{po} = molar rate of propylene feed
X = conversion of propylene

If the molar feed rate of propylene, F_{po}, is reduced, cumene production will decrease unless the conversion, X, is increased. In the reactor, the overall flow of material is constant, because propane replaces propylene; therefore, the space time remains constant. From the reaction expression, a reduced concentration of propylene results in a reduced reaction rate. Intuitively, a reduced reaction rate at constant space time results in reduced conversion. Therefore, from Equation (19.1), the cumene production rate is decreased because of both a decreased reactant feed rate and a decreased conversion.

One method to increase reaction rate is to increase the reactor temperature. The quantitative relationship between temperature increase and cumene production rate can be obtained by the methods learned in a typical reaction engineering class or by use of a process simulator. Example 19.1 illustrates this method.

Example 19.1

Analyze the heat transfer consequences of increasing the reactor temperature by 10°C and 20°C, increases that correspond to maintaining the desired cumene production rate at approximately 6% and 7% propane impurity, respectively.

The heat transfer performance equation must also be obeyed. Predicting trends shows that the temperature difference between the reaction side and steam side must remain constant:

$$\Delta T(¢) = \frac{Q(¢)}{U(¢)A(¢)} \tag{E19.1a}$$

Because the cumene production rate is to remain constant, the heat load on the reactor remains constant. Because overall flows do not change, the heat transfer coefficient remains constant. Because the reactor is not modified, the heat transfer area remains constant. Therefore, ΔT must remain constant. Because the original ΔT was 96°C, the following table gives the temperatures that are required on the steam side. In order to produce higher-temperature steam, the boiler feed water pressure, and therefore steam pressure, must be increased on the shell side. This is because the temperature at which boiler feed water vaporizes to saturated steam increases with pressure.

Temperature Increase (°C)	Steam Temperature (°C)	Steam Pressure (kPa)
0	254	4237
10	264	5002
20	274	5853

For a 10°C increase in reactor temperature, the pressure on the shell side must increase by almost 20%, and for a 20°C increase in reactor temperature, the pressure on the shell side must increase by about 38%.

Without considering the heat transfer analysis, it would have been easy to suggest the temperature increases be implemented. However, they could not be accomplished without raising the shell-side temperature and pressure. Because the reactor is also a heat exchanger, the performance equation for a heat exchanger must be used in the analysis. Therefore, in order to increase the reactor temperature, the steam temperature, and consequently the steam pressure, must be increased. The unanswered question is whether the reactor can withstand the indicated pressure increases. It is unlikely that a vessel designed for a high pressure like 4237 kPa would be overdesigned to handle a 20% increase in pressure, and a design for 38% increase would be extremely uncommon.

It is instructive to continue this analysis and determine whether there are other ways to increase the conversion in the reactor to compensate for the reduced feed, and this is illustrated in Example 19.2.

Example 19.2

Suggest another method for increasing the conversion in the reactor so that the cumene production rate remains constant with reduced propylene feed concentration.

The question to be answered is what can be adjusted, other than temperature, that will affect the reaction rate and/or the conversion. In the reaction rate expression, the primary temperature effect is in the Arrhenius expression. For gas-phase reactions, assuming ideal gas behavior, the concentrations $c_i = P_i/RT$. The increase in temperature discussed in Example 19.1 will actually reduce the concentrations; however, the exponential increase in the reaction rate constant is the primary effect. It is also observed that increasing the pressure increases the concentration, thereby increasing the reaction rate. There is an additional effect. This can be seen from the equation of a plug flow reactor.

$$\frac{\tau}{c_{Ao}} = \int_0^X \frac{dX}{-r_A}$$

Because the mass flowrate $\dot{m} = \rho A v$, and increasing the pressure increases the density, at constant mass flowrate, the velocity drops. Therefore, the residence time, τ, in the reactor increases, which results in increased cumene conversion (upper limit of integral increases) and increased cumene production. Of course, pressure limitations of the equipment in the process or in the tube side of the reactor may limit the pressure increase possible.

In this case study, the most likely solution to the problem is some combination of temperature and pressure increase. If this is insufficient, it may be necessary to try to increase the feed rates of the reactants. However, consistent with the lesson of this case study, the performance of the pumping equipment may limit possible feed rate increases.

In summary, this case study introduces the concept that performance of any unit cannot be considered in isolation. For the reactor, it seems straightforward to do an analysis involving only reaction kinetics and conclude that a temperature increase is a solution to the problem of reduced feed. However, it is now clear that because the reactor is also a heat exchanger, both reactor and heat exchanger analyses must be done simultaneously in order to obtain a correct result. In order to get the correct numerical result, the correct temperature profile on the reaction side must be used. It is obtained by using a process simulator or by solving the differential equations for conversion and temperature simultaneously. The correct temperature profile, as illustrated in Figure 19.2, arises because of the reaction rate, $r \propto e^{-E/RT}$. Therefore, as the temperature increases, r increases by a large amount, generating even more heat, and the peak in the temperature profile may increase significantly, causing hot spots to develop. Hot spots can damage catalyst or promote undesired side reactions. Chapter 20 treats these issues in more detail.

19.2 PERFORMANCE OF A DISTILLATION COLUMN

A distillation column requires a reboiler to add the energy necessary to accomplish the separation. A condenser is also required to reject heat to the surroundings. The performance of a distillation column cannot be analyzed without consideration of the performance of both heat exchangers. This is illustrated in the following case study.

The distillation column illustrated in Figure 19.3 is used to separate benzene and toluene. It contains 35 sieve trays, with the feed on tray 18. The relevant flows are given in Table 19.1. Your assignment is to recommend changes in the tower operation to handle a 50% reduction in feed. Overhead composition must be maintained at 0.996 mole fraction benzene. Cooling water is used in the condenser, entering at 30°C and exiting at 45°C. Medium-pressure steam (185°C, 1135 kPa) is used in the reboiler.

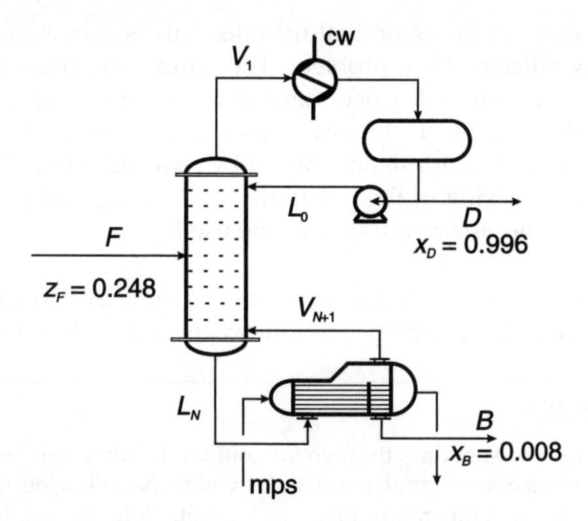

Figure 19.3 Distillation of Benzene from Toluene

The operating conditions of the tower before reduction in feed are in Table 19.1.

Table 19.1 Operating Conditions for Process in Figure 19.3

Input/Output	Flow (kmol/h)	Mole Fraction Benzene	Temperature (°C)
Inputs			
Feed, F	141.3	0.248	90
Reflux, L_0	130.7	0.996	112.7
Boil-up, V_{N+1}	189.5	0.008	145.3
Outputs			
Distillate, D	34.3	0.996	112.7
Still bottoms, L_N	296.5	0.008	145.3
Bottoms product, B	107.0	0.008	145.3

Among the several possible operating strategies for accomplishing the necessary scale-down are the following.

1. Scale down all flows by 50%. This is possible only if the original operation is not near the lower velocity limit that initiates weeping or reduced tray efficiency. If 50% reduction is possible without weeping, reduced tray efficiency, or poor heat-exchanger performance, this is an attractive option.

2. Operate at the same boil-up rate. This is necessary if weeping or reduced tray efficiency is a problem. The reflux ratio must be increased in order to maintain the reflux necessary to maintain the same liquid and vapor flows in the column. In this case, weeping, lower tray efficiency, or poor heat-exchanger performance caused by reduced internal flows is not a problem. The downside of this alternative is that a purer product will be produced and unnecessary utilities will be used.

In Example 19.3, the analysis will be done by assuming that it is possible to scale down all flows by 50% without weeping or reduced tray efficiency.

Example 19.3

Estimate the pressure drop through the column. To what weir height does this correspond?

Interpolation of tabulated data [1] yields the following relationships for the vapor pressures in the temperature range of interest. Note that the temperatures predicted by these interpolated relationships differ slightly from those predicted by the simulator results in Table 19.1.

$$\ln P^* \text{ (kPa)} = 15.1492 - \frac{3706.84}{T(\text{K})} \quad \text{benzene}$$

$$\ln P^* \text{ (kPa)} = 15.3877 - \frac{4131.14}{T(\text{K})} \quad \text{toluene}$$

It is assumed that the bottom is pure toluene and the distillate is pure benzene. This is a good assumption for estimating top and bottom pressures given the mole fractions specified for the distillate and bottoms. Therefore, at the bottom temperature of 141.7°C, the vapor pressure of toluene, and hence the pressure at the bottom of the column, is 227.2 kPa. At the top temperature of 104.2°C, the vapor pressure of benzene, and hence the pressure at the top of the column, is 204.8 kPa.

Because there are 35 trays, the pressure drop per tray is

$$\Delta P = (227.2 - 204.8)/35 = 0.64 \text{ kPa/tray}$$

If it assumed that the weir height is the major contribution to the pressure drop on a tray, then

$$\Delta P = \rho g h$$

where h is the weir height. Assuming an average density of 800 kg/m^3, then

$$h = (640 \text{ Pa})/[(800 \text{ kg/m}^3)(9.8 \text{ m/s})] = 0.08 \text{ m} \approx 3 \text{ in}$$

This is a typical weir height and is consistent with the assumption that the weir height is dominant.

The pressure drop is assumed to remain constant after the scale-down because the weir height is not changed. In practice, there is an additional contribution to the pressure drop due to gas flow through the tray orifices. This would change if column flows changed, but the pressure drop through the orifices is small and should be a minor effect. Examples 19.4 and 19.5 illustrate how the performance of the reboiler and condenser affect the performance of the distillation column.

Example 19.4

Analyze the reboiler to determine how its performance is altered at 50% scale-down.

Figure E19.4 shows the T-Q diagram for this situation. Because the amounts of heat transferred for the base and new cases are different, the Q values must be normalized by the total heat transferred in order for these profiles to be plotted on the same scale. The solid lines are the original case (subscript 1). For the new, scaled-down case (subscript 2), a ratio of the energy balance on the reboiled stream for the two cases yields

$$\frac{Q_2}{Q_1} = \frac{\dot{m}_2\lambda}{\dot{m}_1\lambda} \tag{E19.4a}$$

If it is assumed that the latent heat is unchanged for small temperature changes, then $Q_2/Q_1 = 0.5$, because at 50% scale-down, the ratio of the mass flowrates in the reboiler is 0.5. The ratio of the heat transfer equations yields

$$\frac{Q_2}{Q_1} = 0.5 = \frac{U_2 A \Delta T_2}{U_1 A \Delta T_1} \approx \frac{\Delta T_2}{43.3} \tag{E19.4b}$$

In Equation (E19.4b), it is assumed that the overall heat transfer coefficient is constant. We assume here that $U \neq f(T)$. This assumption should be checked, because for boiling heat transfer coefficients with large temperature differences, the boiling heat transfer coefficient may be a strong function of temperature difference.

From Equation (E19.4b), it is seen that $\Delta T_2 = 21.7°C$. Therefore, for the reboiler to operate at 50% scale-down, with the steam side maintained constant, the boil-up temperature must be 163.4°C. This is shown as the dotted line in Figure E19.4.

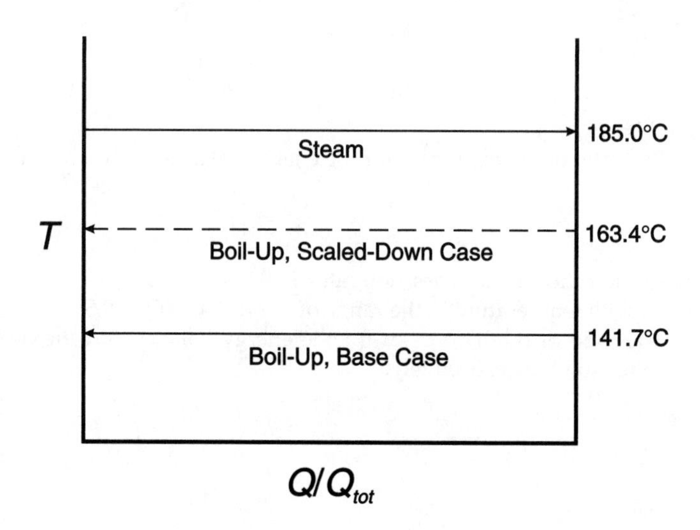

Figure E19.4 Temperature Profiles in Reboiler for Example 19.4

From the vapor pressure expression for toluene, the pressure at the bottom of the column is now 372.5 kPa. It would have to be determined whether the existing column could withstand this greatly increased pressure.

Example 19.4 shows that the reboiler operation requires that the pressure of the boil-up stream be increased. This is not the only possible alternative. All that is necessary is that the temperature difference be 21.7°C. This could be accomplished by reducing the temperature of the steam to 163.4°C. In most chemical plants, steam is available at discrete pressures, and the steam used here is typical of medium-pressure steam. Low-pressure steam would be at too low a temperature to work in the scaled-down column. However, the pressure and temperature of medium-pressure steam could be reduced if there were a throttling valve in the steam feed line to reduce the steam pressure. The resulting steam would be superheated, so desuperheating would also be necessary. This could be accomplished by spraying water into the superheated steam. A change in a utility stream is almost always preferred to a change in a process stream. The condenser is now analyzed in Examle 19.5.

Example 19.5

Using the results of Example 19.4, analyze the condenser for the scaled-down case.

Under the assumption that the pressure drop for the column remains at 22.4 kPa, the pressure in the condenser is now 350.0 kPa. From the vapor pressure expression for benzene, the new temperature in the condenser is 126.0°C. The remainder of the analysis is similar to Example 18.1. The T-Q diagram is illustrated in Figure E19.5. Because an organic is condensing, it is assumed that the resistance on the cooling water side is approximately equal to the resistance on the condensing side. Therefore,

$$\frac{U_2}{U_1} = \frac{\dfrac{1}{h_{o1}} + \dfrac{1}{h_{i1}}}{\dfrac{1}{h_{o2}} + \dfrac{1}{h_{i2}}} = \frac{\dfrac{2}{h_{o1}}}{\dfrac{1}{h_{o1}} + \dfrac{1}{h_{o1}M^{0.8}}} = \frac{2}{1 + \dfrac{1}{M^{0.8}}} \quad \text{(E19.5a)}$$

where $M = \dot{m}_2/\dot{m}_1$.

The ratio of the base cases for the energy balance on the condensing stream is

$$\frac{Q_2}{Q_1} = \frac{\dot{m}_2\lambda}{\dot{m}_1\lambda} \quad \text{(E19.5b)}$$

Because the ratio of the mass flowrates is 0.5, and assuming that the latent heat is unchanged with temperature in the range of interest, $Q_2/Q_1 = 0.5$.

The ratio of the base cases for the energy balance and the heat exchanger performance equation are, respectively,

$$0.5 = M\frac{T - 30}{15} \quad \text{(E19.5c)}$$

$$0.5 = \frac{2(T - 30)}{(66.42)\left(1 + \dfrac{1}{M^{0.8}}\right)\ln\left(\dfrac{96}{126 - T}\right)} \quad \text{(E19.5d)}$$

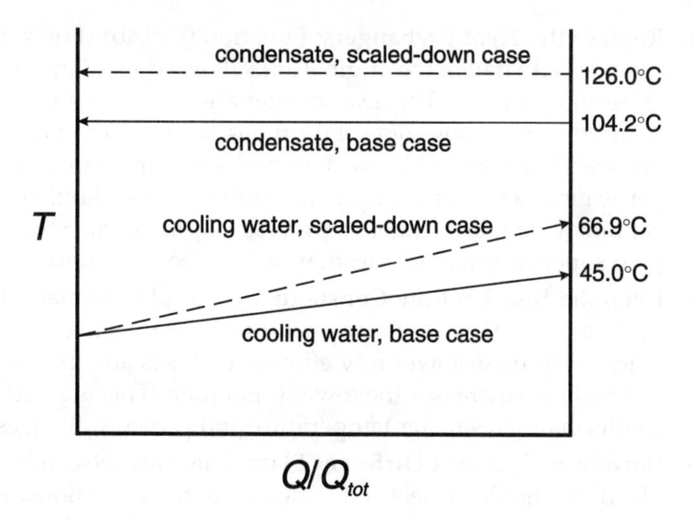

Figure E19.5 Temperature Profiles in Condenser for Example 19.5

Solution of Equations (E19.5c) and (E19.5d) yields

$$M = 0.203$$
$$T = 66.9°\text{C}$$

Therefore, the cooling water rate must be reduced to 20% of the original rate, and the outlet water temperature is increased by about 22°C. This would cause increased fouling problems on the cooling water side. Therefore, it may be better to use a higher reflux ratio rather than reduce the flows in the column.

Observation indicates that in order to operate the distillation column at 50% scale-down with process flows reduced by 50%, a higher pressure is required and the cooling water will be returned at a significantly higher temperature than before scale-down. The operating pressure of the column is determined by the performance of the reboiler and condenser.

Examples 19.4 and 19.5 reveal a process bottleneck at the reboiler that must be resolved in order to accomplish the scale-down. In the process of changing operating conditions (capacity) in a plant, a point will be reached where the changes cannot be increased or reduced any further. This is called a **bottleneck**. A bottleneck usually results when a piece of equipment (usually a single piece of equipment) cannot handle additional change. In this problem, one bottleneck is the high cooling water return temperature, which would almost certainly cause excessive fouling in a short period of time. Another potential bottleneck is tray weeping due to the greatly reduced vapor velocity. In addition to the solution presented here, there are a variety of other possible adjustments, or debottlenecking strategies, which follow with a short explanation for each. Another debottlenecking problem is presented in Chapter 22.

1. **Replace the Heat Exchangers:** Equation (E19.4b) shows that a new heat exchanger with half of the original area allows operation at the original temperature and pressure. The heat transfer area of the existing exchanger could be reduced by plugging some of the tubes, but this modification would require a process shutdown. This involves both equipment downtime and capital expense for a new exchanger. Your intuitive sense should question the need to get a new heat exchanger to process less material. You can be assured that your supervisor would question such a recommendation.
2. **Keep the Boil-Up Rate Constant:** This would maintain the same vapor velocity in the tower. If we are operating near the lower velocity limit that initiates weeping or lower tray efficiency, this is an attractive option. The constant boil-up increases the tower separation. This is an attractive option that results in much smaller temperature and pressure changes.
3. **Introduce Feed on a Different Plate:** This must be combined with Option 4. The plate should be selected to decrease the separation and increase the bottoms concentration of the lower boiling fraction. This lowers the process temperature, increases the ΔT for heat transfer and the reboiler duty. This may or may not be simple to accomplish depending on whether the tower is piped to have alternate feed plates. As in Option 2, the reflux and coolant inputs will change.
4. **Recycle Bottoms Stream and Mix with Feed:** This must be combined with Option 3. By lowering the concentration of the feed, the concentration of the low boiler in the bottoms can be increased, the temperature lowered, and the reboiler duty increased. This represents a modification of process configuration and introduces a new (recycle) input stream into the process.

The adjustments outlined were based on the following.

1. Replace the exchanger (Option 1).
2. Modify utility stream (discussed after Example 19.4).
3. Modify process stream (Option 2).
4. Modify equipment (Option 3 or Option 1, if plugging of tubes to reduce area is chosen).
5. Modify process configuration (Option 4).
6. Modify operating conditions T and P (as in Examples 19.4 and 19.5).

We have shown that there are a number of paths that remove the bottleneck that exists at the reboiler. Adjustments 1 and 2 required no changes in process inputs. All others required modifications of several input streams to maintain the output quality.

It should be noted that, at 50% scale-down, it is possible that weeping or low tray efficiency may be observed (see Section 18.3.2 and Figure 18.19). This must be considered before recommending such a scale-down. Furthermore, reduction of the cooling water flowrate in the condenser, combined with increased cooling water temperatures, could cause fouling problems, as pointed out in Example 19.5. Care must taken to understand the consequences of process modifications.

During plant operations, when seeking the best operating conditions, incremental changes can be made and the effects observed. This is termed "evolutionary operation" [2]. Amoco has reported a significant increase in its annual income from evolutionary operation [3]. In the design case, changes recommended are necessarily conservative to ensure that equipment purchased and installed is adequate to meet the conditions when the plant is operated.

19.3 PERFORMANCE OF A HEATING LOOP

In Section 19.2, we observed how the performance of a distillation column was affected by the performance of the reboiler and condenser. It was possible to analyze the reboiler, column, and condenser sequentially in order to solve the scale-down problem. In this section, we examine the performance of a reactor in which the large, exothermic heat of reaction is removed by a heat transfer fluid. It will be seen that the bottleneck for scale-up of this process is the performance of the heat removal loop. The analysis of this problem is more complex than for the distillation column, because all units involved must be analyzed simultaneously in order to solve the problem.

The problem to be analyzed is illustrated in Figure 19.4. It is a part of the allyl chloride problem discussed in more detail in Appendix C, on the accompanying CD.

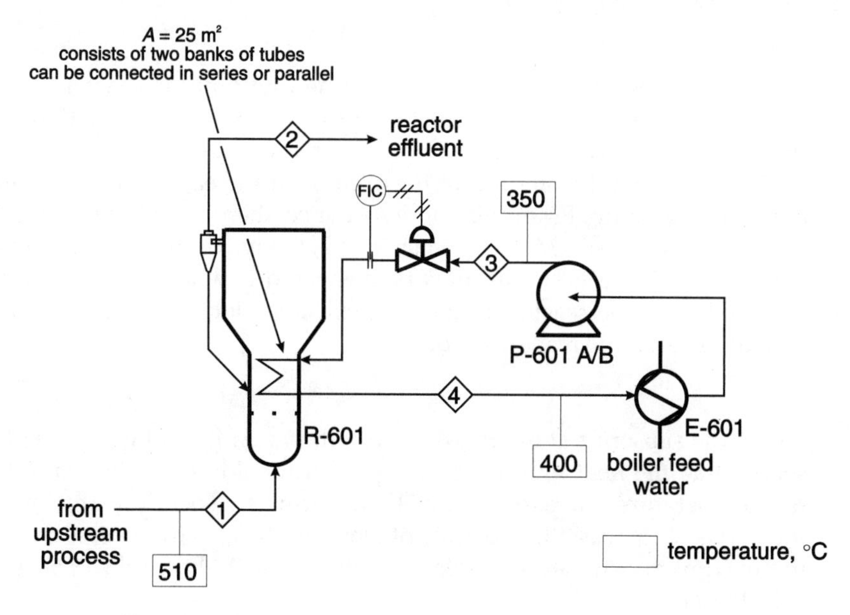

Figure 19.4 Dowtherm A Loop for Allyl Chloride Reactor

Due to an emergency, an unscheduled shutown at an another allyl chloride production plant operated by your company, it is necessary to increase production temporarily at your plant. Your job is to determine the maximum scale-up possible for the allyl chloride production reactor. The reactor is a fluidized bed, with a cyclone for solids recovery, operating isothermally at 510°C. The reactor is operating at two times the minimum fluidization velocity. Based on the cyclone design and the solids handling system, it is known that an increase of at least 100% of process gas flow can be accommodated. The reaction proceeds to completion as long as the reactor temperature remains at 510°C. A temperature increase destroys the catalyst, and a temperature decrease results in incomplete reaction and undesired side products. The reaction is exothermic and heat is removed by Dowtherm A circulating in the loop. The maximum operating temperature for Dowtherm A is 400°C, and its minimum operating pressure at this temperature is 138 psig. There are two heat-exchanger units in the reactor, and they are currently operating in series. The heat transfer resistance on the reactor side is four times that on the Dowtherm A side. The inside of the reactor is always at 510°C due to the well-mixed nature of the fluidized solids. There is a spare pump. High-pressure steam is made from boiler feed water in the heat exchanger. The boiler feed water is available at 90°C, and has been pumped to 600 psig by a pump that is not shown. The pool of liquid being vaporized can be assumed to be at the vaporization temperature, 254°C. All resistance in the heat exchanger is assumed to be on the Dowtherm A side. At normal operating conditions, Dowtherm A circulates at 85 gal/min in the loop. The temperatures of each stream are shown on Figure 19.4. The pump curve for the pump in the loop is shown on Figure 19.5.

An energy balance on the process fluid in the reactor yields

$$Q_R = \dot{n}_{Cl}\Delta H_{rxn} \tag{19.2}$$

where $\dot{n}_{Cl}$ is the molar flowrate of chlorine (the limiting reactant), and the subscript R refers to the reactor. The heat of reaction term in Equation (19.2) is on a unit mole basis. Because the heat of reaction is constant during scale-up, Equation (19.2) shows that the amount of heat removed increases proportionally with the amount of scale-up. From this energy balance, there is nothing to suggest that the reactor cannot be scaled up by a factor of 2, the maximum allowed by the reactor. However, the question that must be answered is whether twice the heat can be removed from the reactor by the heat exchange loop. An energy balance on the Dowtherm A in the reactor yields

$$Q_R = \dot{m}_D C_{pD}(T_4 - T_3) \tag{19.3}$$

where the subscript D refers to the Dowtherm A. From Equation (19.3), it is observed that the heat removal rate is proportional to both the mass flowrate and the temperature increase of the Dowtherm A. Therefore, the most obvious method is to increase the flowrate of Dowtherm A, so it is necessary to determine the maximum flowrate possible for Dowtherm A. Example 19.6 illustrates this calculation.

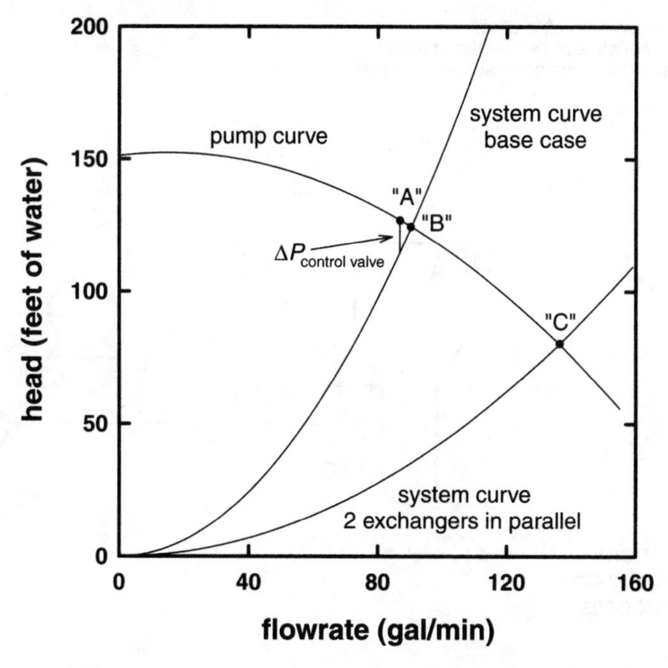

Figure 19.5 Performance of Dowtherm A Loop

Example 19.6

For the pump curve provided in Figure 19.5, determine the maximum possible flowrate for Dowtherm A in the loop. The pump operates at only one speed, described by the curve given in the figure. The pressure drops between streams are shown in Figure E19.6.

The pump operates only at conditions shown on the pump curve. The pressure drop in the system and regulating valves must equal the pressure delivered by the pump. The base case solution gives one point on this line (Point "a"). Of this total pressure drop, 15 feet of water is from the control valve, which can be adjusted independently. The remaining pressure drop, 110 feet of water (125 – 15 feet of water), is the **system pressure drop** and is dependent on the flowrate through the system. The system pressure drop can be obtained from the equation for frictional pressure losses.

$$\Delta P = \frac{2\rho f L_{eq} v^2}{D} \quad \text{(E19.6a)}$$

Taking the ratio of Equation (19.6a) for the scaled-up case (subscript 2) to the base case (subscript 1) for constant L, D, and the assumed constant friction factor (high Reynolds number, fully developed turbulent flow) yields

$$\frac{\Delta P_2}{110} = \frac{v_2^2}{v_1^2} = \frac{\dot{m}_2^2}{\dot{m}_1^2} \quad \text{(E19.6b)}$$

where M is the scale-up factor, the ratio of the scaled-up Dowtherm A flowrate to that for the base case. Equation (19.6b) is plotted with the pump curve on Figure 19.5. The

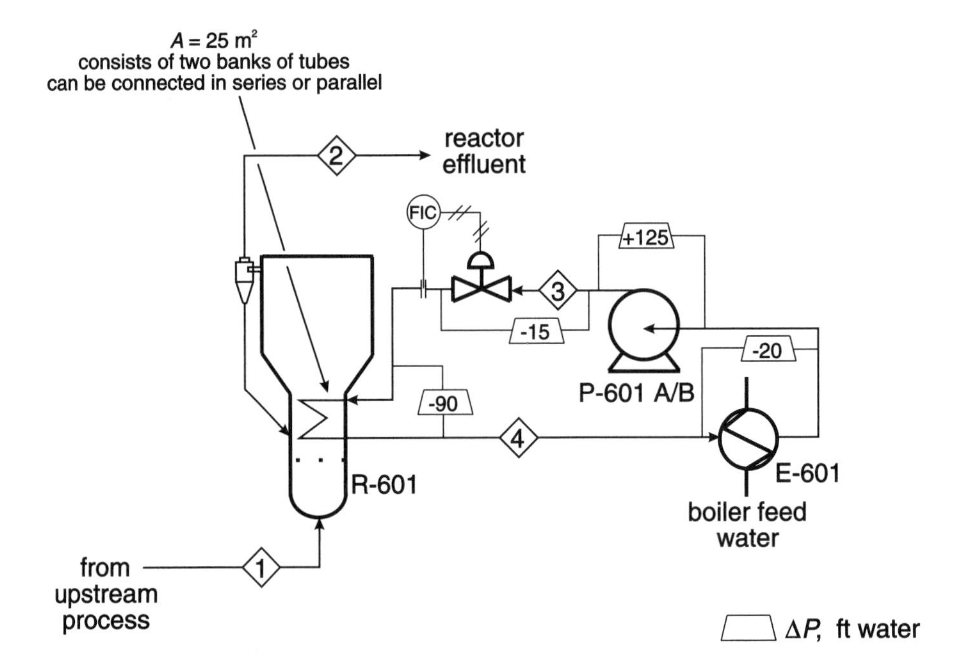

Figure E19.6 Pressure Drops in Dowtherm A Loop for Allyl Chloride Reactor

maximum flowrate occurs when the valve is wide open. When the valve is wide open, there is no (or very little) pressure drop across the valve. This is shown as Point "b" and occurs at 89 gal/min.

Example 19.6 shows that the maximum scale-up is limited to about 5% by the Dowtherm A flowrate in the loop. A bottleneck has been identified. In order for additional scale-up to be possible, a method for increasing the maximum Dowtherm A flowrate must be found. Several options are possible.

1. Run only one pump, but operate the reactor heat exchangers in parallel.
2. Run both pumps (the operating pump and the spare) in parallel, with the reactor heat exchangers in series.
3. Run both pumps in series, with the reactor heat exchangers in series.
4. Run both pumps (series or parallel), and operate the reactor heat exchangers in parallel.

Option 1 is illustrated in Example 19.7. The remaining options are the subject of a problem at the end of the chapter.

Example 19.7

Determine the maximum possible Dowtherm A flowrate for one pump operating with the two reactor heat exchangers operating in parallel.

From Equation (E19.6a), it is seen that

$$\Delta P \propto L_{eq} v^2 \quad \text{(E19.7a)}$$

The equivalent length in the heat exchanger is reduced by half for operation in parallel. Half the flowrate goes through each heat exchanger, which means that the Dowtherm A has half the original velocity. Therefore, the pressure drop through the reactor heat exchangers drops by a factor of 8 and becomes 11.25 feet of water. The base-case pressure drop is now 31.25 feet of water. Therefore, a new system curve is obtained. Its equation is

$$\Delta P = 31.25\, M^2 \quad \text{(E19.7b)}$$

The new system curve along with the pump curve are plotted on Figure 19.5. Point "c" represents the maximum possible flowrate of Dowtherm A, 135 gal/min. This represents a 58.8% scale-up.

Therefore, if only the original pump is used with the reactor heat exchangers in parallel, the Dowtherm A circulation rate can be increased by almost 50%. For Options 2 through 4, additional increases are possible. It should be noted that liquid velocities exceeding 10 ft/sec are not recommended due to potential erosion problems. Therefore, if the original case was designed for the maximum recommended velocity, operation at a 50% higher velocity for an extended period of time is not a good idea. We will proceed with this problem under the assumption that the original design permits 50% scale-up without exceeding the maximum recommended velocity.

Because the flowrate of the Dowtherm A can be scaled up by 50% does not mean that the heat removal rate can be scaled up by the same factor. Analysis of the heat transfer in both heat exchangers is required. For the reactor, the energy balances were given by Equations (19.2) and (19.3), for the process side and the Dowtherm A side, respectively. The performance equation for the reactor heat exchanger is

$$Q_R = U_R A_R \Delta T_{\text{lm}} \quad (19.4)$$

For the heat exchanger producing steam, there are two additional equations. One is the energy balance on the boiler feed water to steam side.

$$Q_h = \dot{m}_s (C_{p,bfw} \Delta T + \lambda) \quad (19.5)$$

where $\dot{m}_S$ is the mass flowrate of steam, ΔT is the temperature difference between the boiler feed water inlet and the vaporization temperature, and λ is the latent heat of vaporization of the steam. The second equation is the performance equation for the heat exchanger.

$$Q_h = U_h A_h \Delta T_{lm} \quad (19.6)$$

The energy balance on the Dowtherm A side is identical (with the temperature difference reversed) to Equation (19.3). It is assumed that all heat removed from the reactor is transferred to make steam ($Q_R = Q_h$).

Therefore, there are five independent equations that describe the performance of the loop. The unknowns are T_3, T_4, $Q_R = Q_h$, $\dot{m}_p$, $\dot{m}_D$, and $\dot{m}_s$. For a given value of $\dot{m}_D$, the mass flowrate of Dowtherm A, a unique solution exists. The method for obtaining this solution is illustrated in Example 19.8.

Example 19.8

From a heat transfer analysis, determine the maximum scale-up possible for the allyl chloride reactor for the original case and for the case of one pump with the reactor heat exchangers operating in parallel.

Ratios of the scaled-up case (subscript 2) to the base case (subscript 1) will be used. Q_2/Q_1 is the ratio of the heat transfer rates, M is the ratio of the mass flowrates of Dowtherm A, and M_s is the ratio of the mass flowrates of steam. Only three of the five equations are coupled. Base-case ratios of Equations (19.3), (19.4), and (19.6) must be solved for T_3, T_4, and Q_2/Q_1. Then M_s and M are determined from Equations (19.5) and (19.2), respectively. The value of Q_2/Q_1 defines the level of scale-up possible.

The base-case ratios for Equations (19.3), (19.4), and (19.6) are

$$\frac{Q_2}{Q_1} = \frac{M(T_4 - T_3)}{50} \tag{E19.8a}$$

$$\frac{Q_2}{Q_1} = \frac{5(T_4 - T_3)}{133.44\left(4 + \frac{1}{M^{0.8}}\right)\ln\left(\frac{510 - T_3}{510 - T_4}\right)} \tag{E19.8b}$$

$$\frac{Q_2}{Q_1} = \frac{M^{0.8}(T_4 - T_3)}{119.26\ln\left(\frac{T_4 - 254}{T_3 - 254}\right)} \tag{E19.8c}$$

For the case of maximum scale-up with the original heat-exchanger configuration, $M = 89/85 = 1.05$. The solutions are

$$T_3 = 348.7°\text{C}$$
$$T_4 = 397.4°\text{C}$$
$$Q_2/Q_1 = 1.02$$

From Equation (19.2), because $Q_2/Q_1 = 1.02$, and because the heat of reaction is constant, the process flowrate increases by 2%. So only 2% scale-up is possible.

For one pump with the reactor heat exchangers in parallel, the mass flowrate in each reactor heat exchanger is half of the total flow. Therefore, Equation (E19.8b) becomes

$$\frac{Q_2}{Q_1} = \frac{5(T_4 - T_3)}{133.44\left(4 + \left(\frac{2}{M}\right)^{0.8}\right)\ln\left(\frac{510 - T_3}{510 - T_4}\right)} \tag{E19.8d}$$

For $M = 1.588$, solution of Equations (E19.8a), (E19.8c), and (E19.8d) yields

$$T_3 = 331.7°\text{C}$$
$$T_4 = 367.8°\text{C}$$
$$Q_2/Q_1 = 1.15$$

Therefore, 15% scale-up is possible.

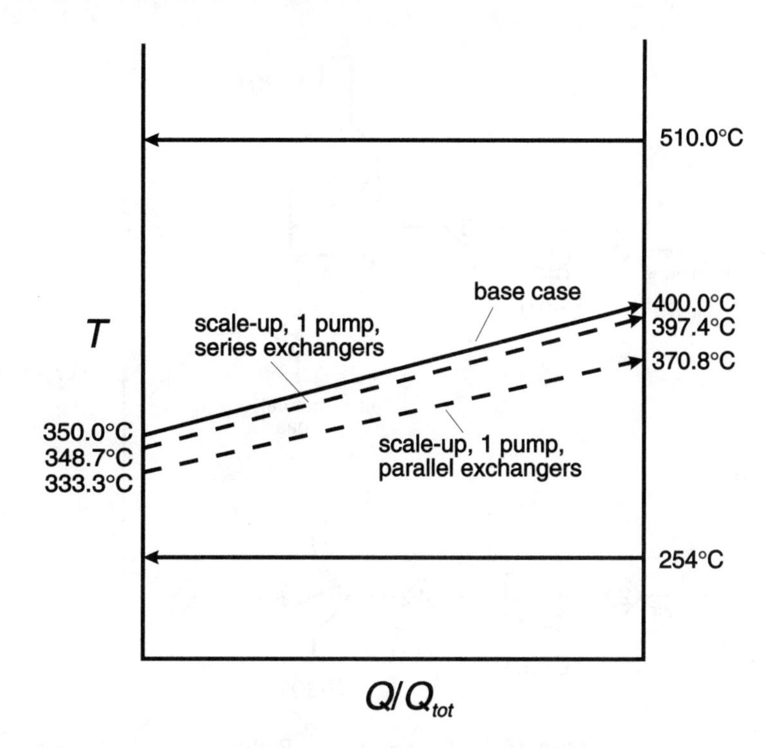

Figure 19.6 Temperature Profiles for Exchangers in Dowtherm A Loop

Performance of the heat exchange loop is illustrated on the T-Q diagram shown in Figure 19.6. The lines for the reactor and the steam-generating heat exchanger are unchanged. The line for the Dowtherm A changes slope as the mass flowrate, the reactor heat-exchanger configuration, or the number and configuration of pumps change.

This complex problem illustrates an important feature of chemical processes. Often the bottleneck to solving a problem is elsewhere in the process. In this problem, the bottleneck to reactor scale-up is not in the reactor itself, but in the heat removal loop. Nevertheless, several alternatives are available to increase the Dowtherm A flowrate and scale up the process.

19.4 PERFORMANCE OF THE FEED SECTION TO A PROCESS

A very common feature of chemical processes is the mixing of reactant feeds prior to entering a reactor. When two streams mix, they are at the same pressure. The consequences of this are illustrated by the following scenario.

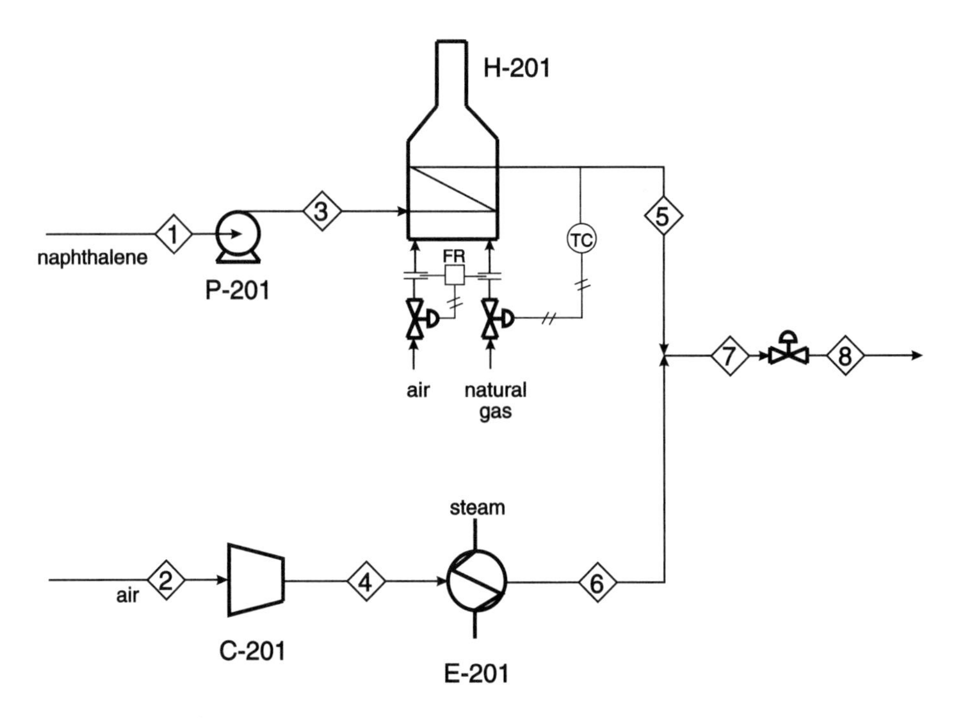

Figure 19.7 Feed Section to Phthalic Anhydride Process

Phthalic anhydride can be produced by reacting naphthalene and oxygen. The feed section to a phthalic anhydride process is shown in Figure 19.7. The mixed feed enters a fluidized bed reactor operating at 5 times the minimum fluidization velocity. A stream table is given in Table 19.2. It is assumed that all frictional pressure losses are associated with equipment and that frictional losses in the piping are negligible. It is temporarily necessary to scale down production by 50%. Your job is to determine how to scale down the process and to determine the new flows and pressures.

Table 19.2 Partial Stream Table for Feed Section in Figure 19.7

	Stream							
	1	2	3	4	5	6	7	8
P (kPa)	80.00	101.33	343.00	268.00	243.00	243.00	243.00	200.00
Phase	L	V	L	V	V	V	V	V
Naphthalene (Mg/h)	12.82	—	12.82	—	12.82	—	12.82	12.82
Air (Mg/h)	—	151.47	—	151.47	—	151.47	151.47	151.47

It is necessary to have pump and compressor curves in order to do the required calculations. In this example, we will use equations for the pump curves. These equations can be obtained by fitting a polynomial to the curves provided by pump manufacturers. As discussed in Chapter 18, pump curves are usually expressed as pressure head versus volumetric flowrate. This is so that they can be used for a liquid of any density. In this section, pressure head and volumetric flowrate have been converted to absolute pressure and mass flowrate using the density of the fluids involved. Pump P-201 operates at only one speed, and an equation for the pump curve is

$$\Delta P(\text{kPa}) = 500 + 4.663\dot{m} - 1.805\dot{m}^2 \quad \dot{m} \leq 16.00 \text{ Mg/h} \tag{19.7}$$

Compressor C-201 operates at only one speed, and the equation for the compressor curve is

$$\frac{P_{out}}{P_{in}} = 5.201 + 2.662 \times 10^{-3}\,\dot{m} - 1.358 \times 10^{-4}\,\dot{m}^2 + 4.506 \times 10^{-8}\dot{m}^3 \quad \dot{m} \leq 200 \text{ Mg/h} \tag{19.8}$$

From Figure 19.7, it is seen that there is only one valve in the feed section, after the mixing point. Therefore, the only way to reduce the production of phthalic anhydride is to close the valve to the point at which the naphthalene feed is reduced by 50%. Example 19.9 illustrates the consequences of reducing the naphthalene feed rate by 50%.

Example 19.9

For a reduction in naphthalene feed by 50%, determine the pressures and flows of all streams after the scale-down.

Because it is known that the flowrate of naphthalene has been reduced by 50%, the new outlet pressure from P-201 can be calculated from Equation (19.7). The feed pressure remains at 80 kPa. At a naphthalene flow of 6.41 Mg/h, Equation (19.7) gives a pressure increase of 455.73 kPa, so P_3 = 535.73 kPa. Because the flowrate has decreased by a factor of 2, the pressure drop in the fired heater decreases by a factor of 4 (see Equation [E19.7a]). Therefore, P_5 = 510.73 kPa. Therefore, the pressure of Stream 6 must be 510.73 kPa. The flowrate of air can now be calculated from the compressor curve equation.

There are two unknowns in the compressor curve equation: the compressor outlet pressure and the mass flowrate. Therefore, a second equation is needed. The second equation is obtained from a base-case ratio for the pressure drop across the heat exchanger. The two equations are

$$\frac{P_4}{101.33} = 5.201 + 2.662 \times 10^{-3}\dot{m}_{2,new}^2 - 1.358 \times 10^{-4}\dot{m}_{2,new}^2 + 4.506 \times 10^{-8}\dot{m}_{2,new}^3 \tag{E19.8a}$$

$$P_4 - 510.73 = 25\left(\frac{\dot{m}_{2,new}}{151.47}\right)^2 \quad \text{(E19.8b)}$$

The solution is

$$P_4 = 512.84 \text{ kPa}$$

$$\dot{m}_2 = 43.80 \text{ Mg/h}$$

The stream table for the scaled-down case is given in Table 19.3. Although it is not precisely true, for lack of additional information, it has been assumed that the pressure of Stream 8 remains constant.

It is observed that the flowrate of air is reduced by far more than 50% in the scaled-down case. This is because of the combination of the compressor curve and the new pressure of Streams 5 and 6 after the naphthalene flowrate is scaled down by 50%. The total flowrate of Stream 8 is now 50.21 Mg/h, which is 30.6% of the original flowrate to the reactor. Given that the reactor was operating at five times minimum fluidization, the reactor is now in danger of not being fluidized adequately. Because the phthalic anhydride reaction is very exothermic, a loss of fluidization could result in poor heat transfer, which might result in a runaway reaction. The conclusion is that it is not recommended to operate at these scaled-down conditions.

The question is how the air flowrate can be scaled down by 50% to maintain the same ratio of naphthalene to air as in the original case. The answer is in valve placement. Because of the requirement that the pressures at the mixing point be equal, with only one valve after the mixing point, there is only one possible flowrate of air corresponding to a 50% reduction in naphthalene flowrate. Effectively, there is no control of the air flowrate. A chemical process would not be designed as in Figure 19.7. The most common design is illustrated in Figure 19.8. With valves in both feed streams, the flowrates of each stream can be controlled independently. Design of control systems is discussed in more detail in Chapter 21.

Table 19.3 Partial Stream Table for Scaled-Down Feed Section in Figure 19.7

	Stream							
	1	**2**	**3**	**4**	**5**	**6**	**7**	**8**
P (kPa)	80.00	101.33	535.73	512.84	510.73	510.73	510.73	200.00
Phase	L	V	L	V	V	V	V	V
Naphthalene (Mg/h)	6.41	—	6.41	—	6.41	—	6.41	6.41
Air (Mg/h)	—	43.80	—	43.80	—	43.80	43.80	43.80

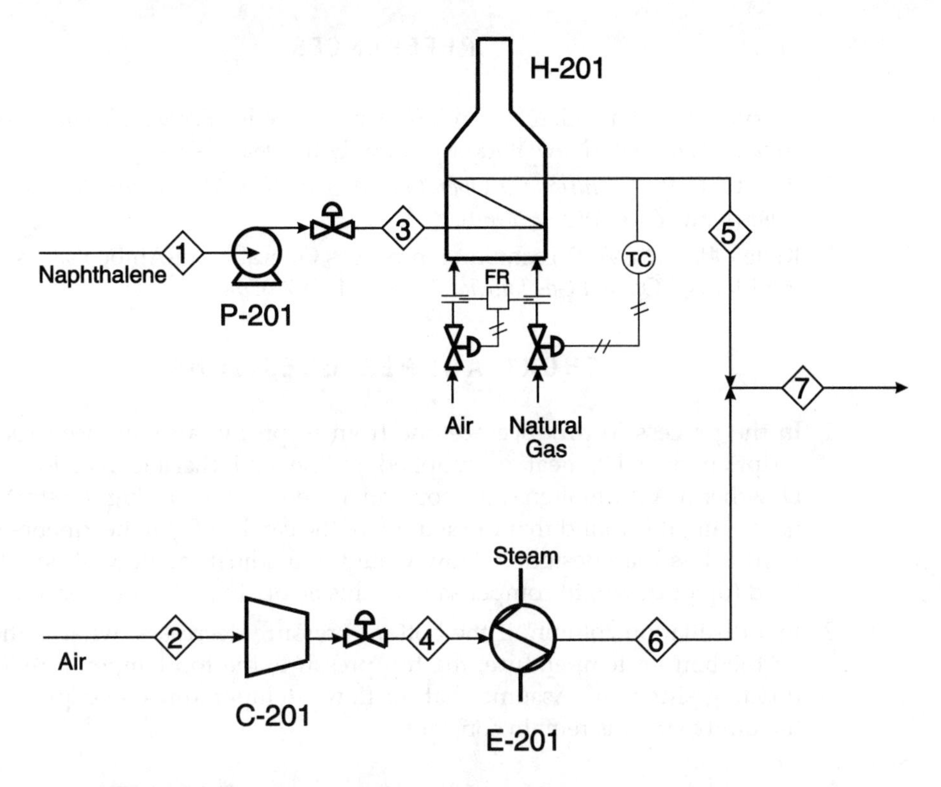

Figure 19.8 Feed Section to Phthalic Anhydride Process with Better Valve Placement than Shown in Figure 19.7

19.5 SUMMARY

In this chapter, we have demonstrated that performance of existing equipment is affected by other equipment. The input/output models discussed in Chapter 16 suggested this outcome. If the input to a unit is altered, its output is altered. Because the output from one unit is the input to the next, the interaction observed in the case studies in this chapter are expected. It is always important to remember the interaction between equipment performance in a chemical process.

In terms of obtaining numerical solutions for performance problems involving multiple units, we have seen examples such as the reactor heat exchanger and the distillation column, in which the adjacent units are analyzed sequentially. We have also seen examples such as the heat-exchange loop and the feed section in which simultaneous solution of the relationships for multiple units was required. The exact set of calculations necessary and the difficulty of these calculations are specific to each problem encountered. However, the tools developed in Chapter 17, such as determining trends, base case ratios, and T-Q diagrams, are essential to obtaining desired solutions.

REFERENCES

1. Perry, R. H., D. W. Green, and J. O. Maloney, eds., *Perry's Chemical Engineers' Handbook,* 6th ed. (New York: McGraw-Hill, 1984), 3-50–3-63.
2. Box, G. E. P., *Evolutionary Operation: A Statistical Method for Process Improvement* (New York: Wiley, 1969).
3. Kelley, P. E., "EVOP Technique Improves Operation of Amoco's Gas Producing Plants," *Oil and Gas Journal* 71, no. 44 (1973): 94.

SHORT ANSWER QUESTIONS

1. In the process to produce acetone from isopropyl alcohol (see process description on CD), heat is supplied to the endothermic reaction using a Dowtherm A (or molten salt) loop and a fired heater. During the start-up of a new plant, it is found that the activity of the catalyst (k_0 in the kinetics expression) is less than designed. How would you adjust the flow of heat transfer fluid (up or down) to compensate for this error? Explain your reasoning.
2. In a distillation column, if the bottom pressure decreases, what is the effect on the bottom temperature, the top pressure, the top temperature, and the flooding situation? Assume that all flows, temperatures, and pressures for the utility streams remain constant.

PROBLEMS

3. It is possible to generate performance curves for the reactor in Section 19.1. One type of performance curve would have the rate of cumene production (kmol/h) on the y-axis, and the percent propane impurity in the propylene on the x-axis. There would be lines for different operating temperatures all at constant pressure.

   Assume that the feed to the reactor is 108 kmol/h propylene, 8 kmol/h propane, and 203 kmol/h benzene. The pressure is kept constant at 3000 kPa. Prepare performance curves on the same graph for temperatures from 350°C to 400°C in 10°C intervals. Superimpose on this graph the maximum allowable conditions, which correspond to a steam-side pressure of 4800 kPa. A process simulator should be used to generate points on the performance curves. Other data: reactor volume = 7.89 m^3, heat transfer area A = 436 m^2, overall heat transfer coefficient = 65 W/m^{2}°C.
4. Repeat Problem 3 for varying reaction pressure at a constant temperature of 350°C. The pressure range is 3000 kPa to 3300 kPa, in intervals of 50 kPa.
5. For the distillation performance problem in Section 19.2, assume that scale-down occurs while maintaining constant boil-up rate. Determine the conditions in the reboiler, column, and condenser for these operating conditions. A process simulator should be used for the distillation column.

6. The benzene-toluene distillation column in Section 19.2 must temporarily handle a 25% increase in throughput while maintaining the same outlet concentrations. Determine the operating conditions required if the reflux ratio remains constant. What other factors must be considered in order to determine whether the column can handle this amount of increased throughput?
7. Suggest alternative changes in process conditions or addition of new equipment in the distillation column in Section 19.2 that would allow 50% scale-down. Suggest as many alternatives as you can think of, and discuss the advantages and disadvantages of each.
8. Consider the Dowtherm A loop described in Section 19.3. For the options involving multiple pumps either in series or parallel, determine the configuration that provides the maximum scale-up capacity. What are the temperatures of Dowtherm A, T_3, and T_4, and the maximum percent scale-up?
9. Repeat the analysis in Section 19.3 for the case when cooling water is used in the heat exchanger in place of boiler feed water. The cooling water enters at 30°C and must exit at 45°C.
10. Repeat the analysis in Problem 8 using cooling water as described in Problem 9.
11. Suggest alternative solutions, not involving increasing the Dowtherm A flowrate, to obtain the maximum scale-up possible in the Dowtherm A loop in Section 19.3. Suggest as many alternatives as you can think of, and discuss the advantages and disadvantages of each.
12. Consider the molten salt loop for removal of the heat of reaction in the production of phthalic anhydride described in Appendix C (on the CD) and illustrated in Figure C.5. It is necessary to scale down phthalic anhydride production by 50%. Estimate the flow of molten salt for the scaled-down case and the temperatures of molten salt entering and exiting the reactor. You may assume that sensible heat effects (energy necessary to heat reactants to reaction temperature) are negligible.
13. Consider the situation illustrated in Figure P19.13. Due to downstream considerations, P_8 is always maintained at 200 kPa. It is known that $P_1 = 100$ kPa

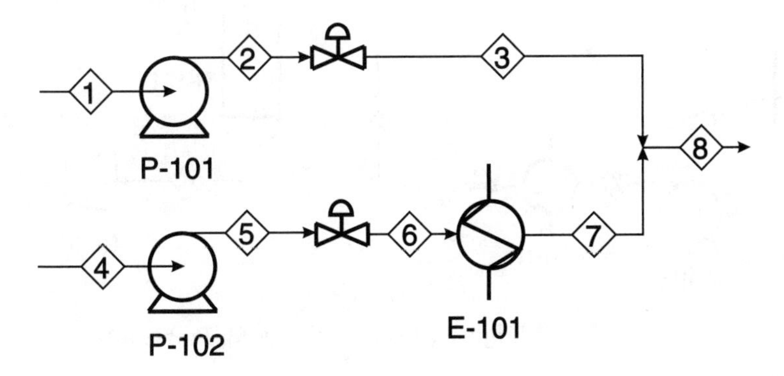

Figure P19.13 Process for Problem 13

and $P_4 = 100$ kPa, and because the feeds come from storage tanks maintained at constant pressure, they are always constant. It is also known that $P_3 = 225$ kPa, $P_5 = 375$ kPa, $P_6 = 250$ kPa, $P_7 = 225$ kPa, and $\dot{m}_1 = 15{,}500$ kg/h. The pump curves are given by the following equations, with ΔP in kPa and $\dot{m}$ in Mg/h.

$$\text{Pump 1} \quad \Delta P = 340 - 0.913\dot{m} + 0.0535\dot{m}^2 - 0.0101\dot{m}^3$$

$$\text{Pump 2} \quad \Delta P = 312 - 1.924\dot{m} + 0.0302\dot{m}^2 - 0.01124\dot{m}^3$$

a. Calculate $\dot{m}_4$ and ΔP_{23} for this situation.
b. Sketch the pump and system curves for this situation as illustrated in Section 18.2.1. Identify the pressure drop across the valves on the sketch.

14. Refer to Figure P19.14 and the data provided with the figure. The compressor (C-101) exhausts to 101 kPa. Tank TK-102 is controlled to be always 101 kPa. Assume all flows are turbulent.
    a. If we close the valve, the flowrate from the compressor exhaust (Stream 7) is 100 Mg/h. What is the pressure in the storage tank (TK-101)?
    b. If the pressure of TK-101 is kept constant at the pressure calculated in Part (a), what is the maximum flowrate of liquid (Stream 2) when the valve is wide open (i.e., no pressure drop across the valve)?
    c. What is the flowrate of the compressor exhaust (Stream 7) when the valve is wide open?

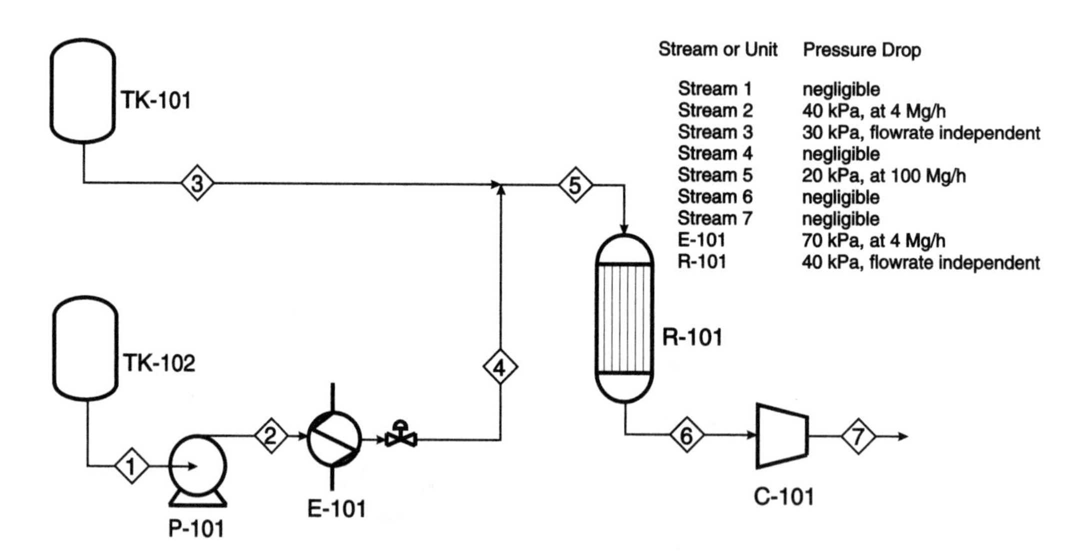

Stream or Unit	Pressure Drop
Stream 1	negligible
Stream 2	40 kPa, at 4 Mg/h
Stream 3	30 kPa, flowrate independent
Stream 4	negligible
Stream 5	20 kPa, at 100 Mg/h
Stream 6	negligible
Stream 7	negligible
E-101	70 kPa, at 4 Mg/h
R-101	40 kPa, flowrate independent

Figure P19.14 Process for Problem 14

C-101 compressor curve:

$$\frac{P_{out}}{P_{in}} = 4.015 + 5.264 \times 10^{-3}\dot{m} - 1.838 \times 10^{-4}\dot{m}^2$$

P-101 pump curve:

$$\Delta P = 500 + 4.662\dot{m} - 1.805\dot{m}^2$$

where P is in kPa and $\dot{m}$ is in Mg/h.

15. Consider a process in which a fluid is pumped from a storage tank through a heat exchanger, where it is heated and then pumped into another storage tank. The process flow diagram for this process is shown in Figure P19.15(a), and the pump and system curves are shown in Figure P19.15(b).

    Currently, the system is operating with the control valve (cv-1) wide open, and the flow through the pump is measured at 100 gallons per minute (gpm). The elevations and pressures of both tanks are equal, and the pressure drop across the heat exchanger is 5 psi, with the remaining 5 psi pressure drop occurring in the piping. It has been noted that the current velocity through the heat exchanger is very high and must be reduced by half. It is proposed to add a bypass line with a valve around the heat exchanger. The valve will be adjusted so that the flow through the heat exchanger is 50% of the current flow. The control valve, cv-1, will remain wide open. For the proposed change, find the following.
    a. The pressure drop across the heat exchanger.
    b. The total flowrate of fluid passing through the pump.

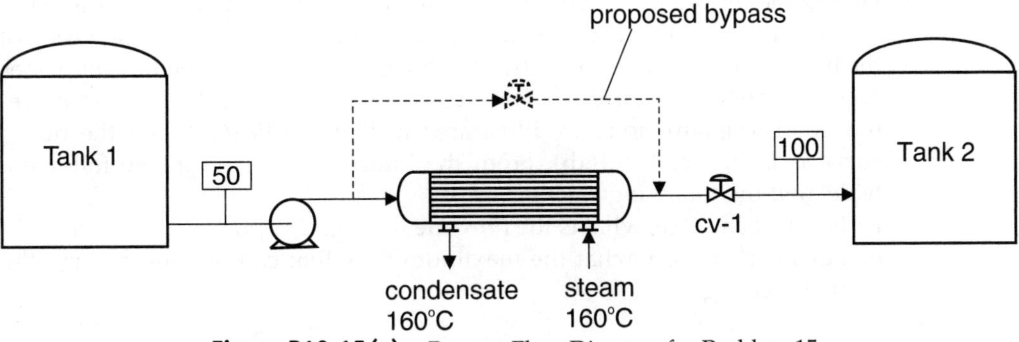

Figure P19.15(a) Process Flow Diagram for Problem 15

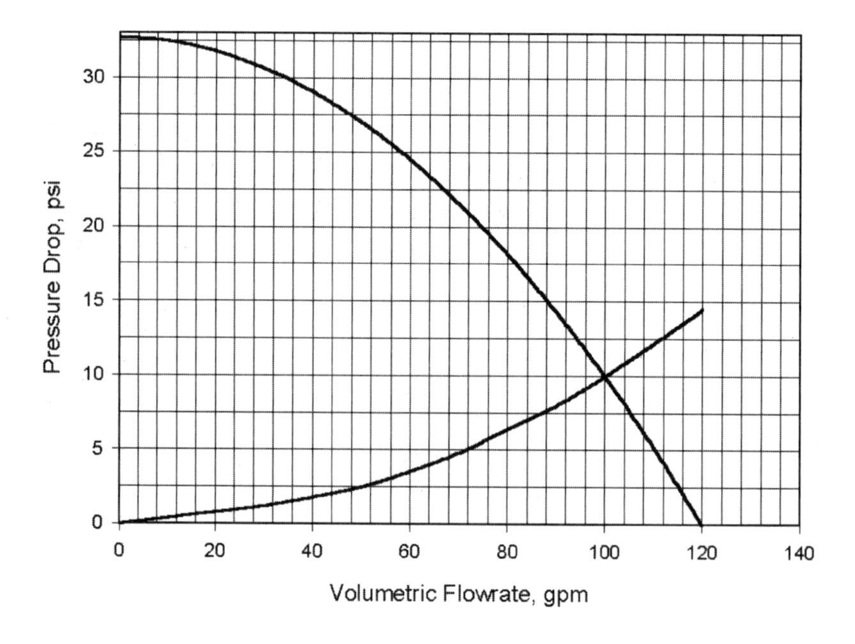

Figure P19.15(b) Pump Curve and System Curve for Problem 15

c. Assume that all the resistance to heat transfer is on the process fluid side and that the temperature of the condensing steam remains constant. Determine the temprature of the fluid entering Tank 2.

16. During the retrofit of a chemical plant, it was decided to place valves and bypass lines around two exchangers placed in series. The flows through each of the bypass lines are controlled by the valves so that 40% of the flow bypasses each exchanger. Because of safety concerns, the bypass lines must be removed. These situations are illustrated in Figure P19.16(a), and the pump curve is in Figure P19.16(b). From the information given in the following table, find the following:
    a. For the base case, what is the flowrate through the pump?
    b. For the new case, what the maximum flow that can be sent through the two exchangers?

Pressure Profile for Base Case

Location	Pressure (kPa)
1	300
2	290
3	250
4	240
5	200
6	170

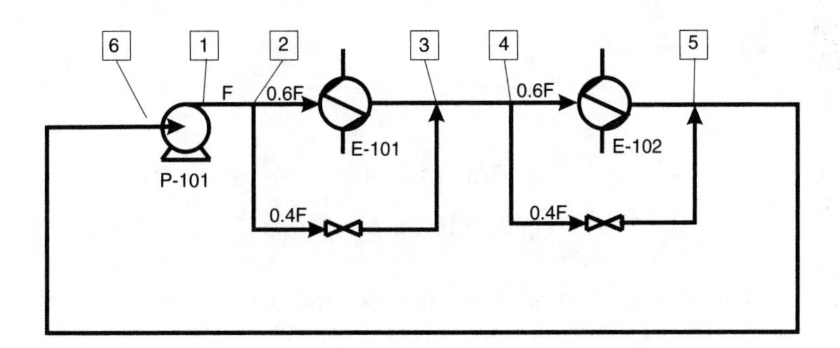

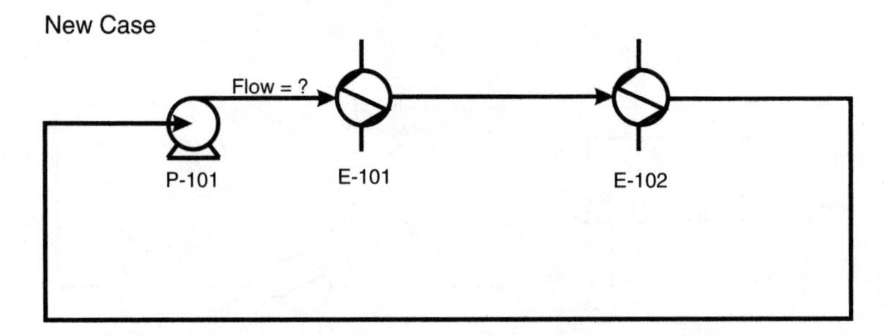

Figure P19.16(a) Process Flow Diagram for Problem 16

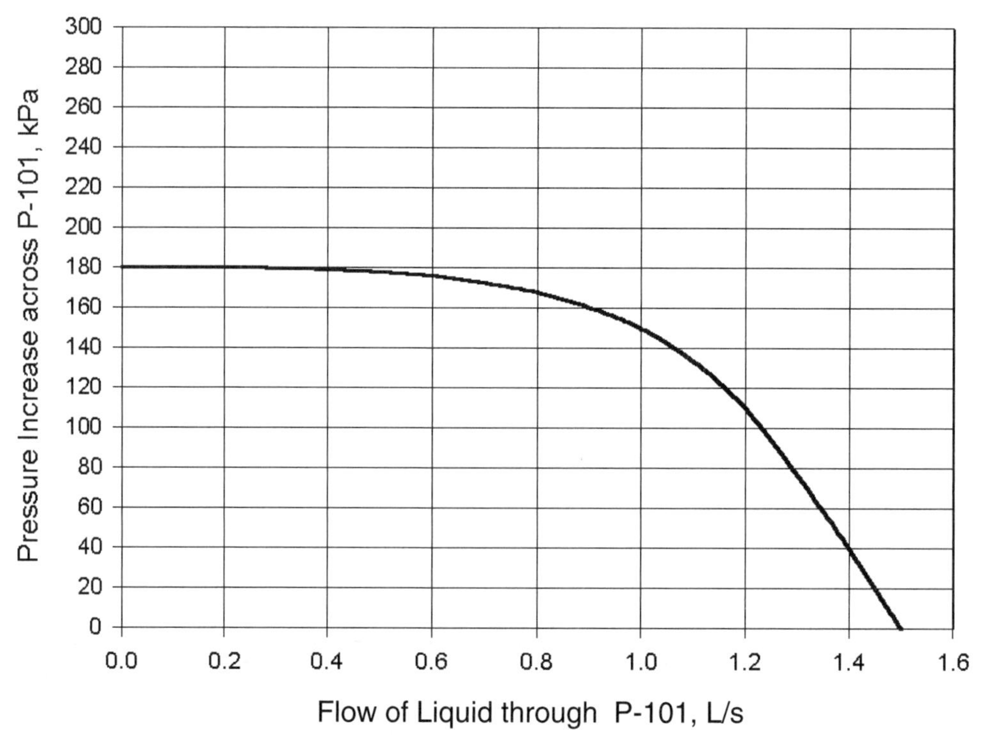

Figure P19.16(b) Pump Curve for P-101

17. The feed system for a distillation process is shown in Figure P19.17(a). The feed to be distilled is an organic liquid stored in V-301 under a nitrogen blanket at a constant pressure of 21 psi. The liquid is pumped through a pump,

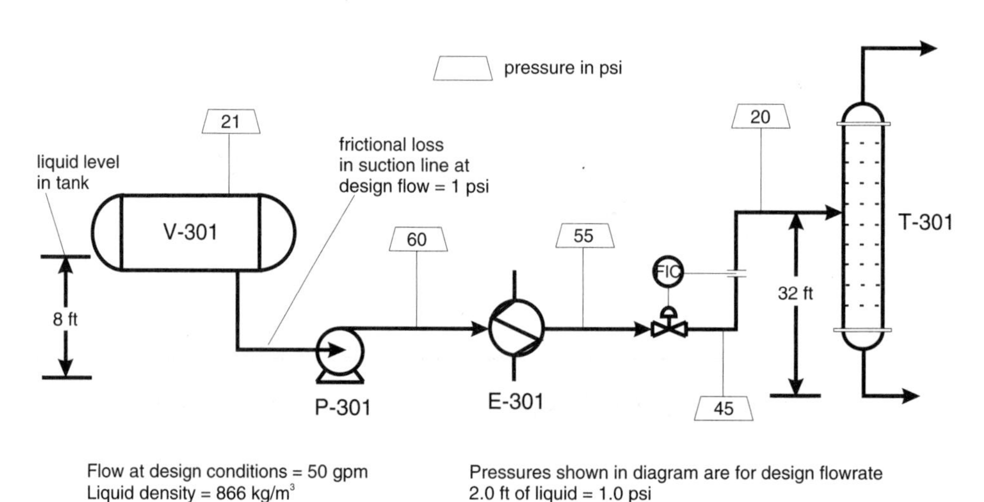

Figure P19.17(a) Process Flow Diagram for Problem 17

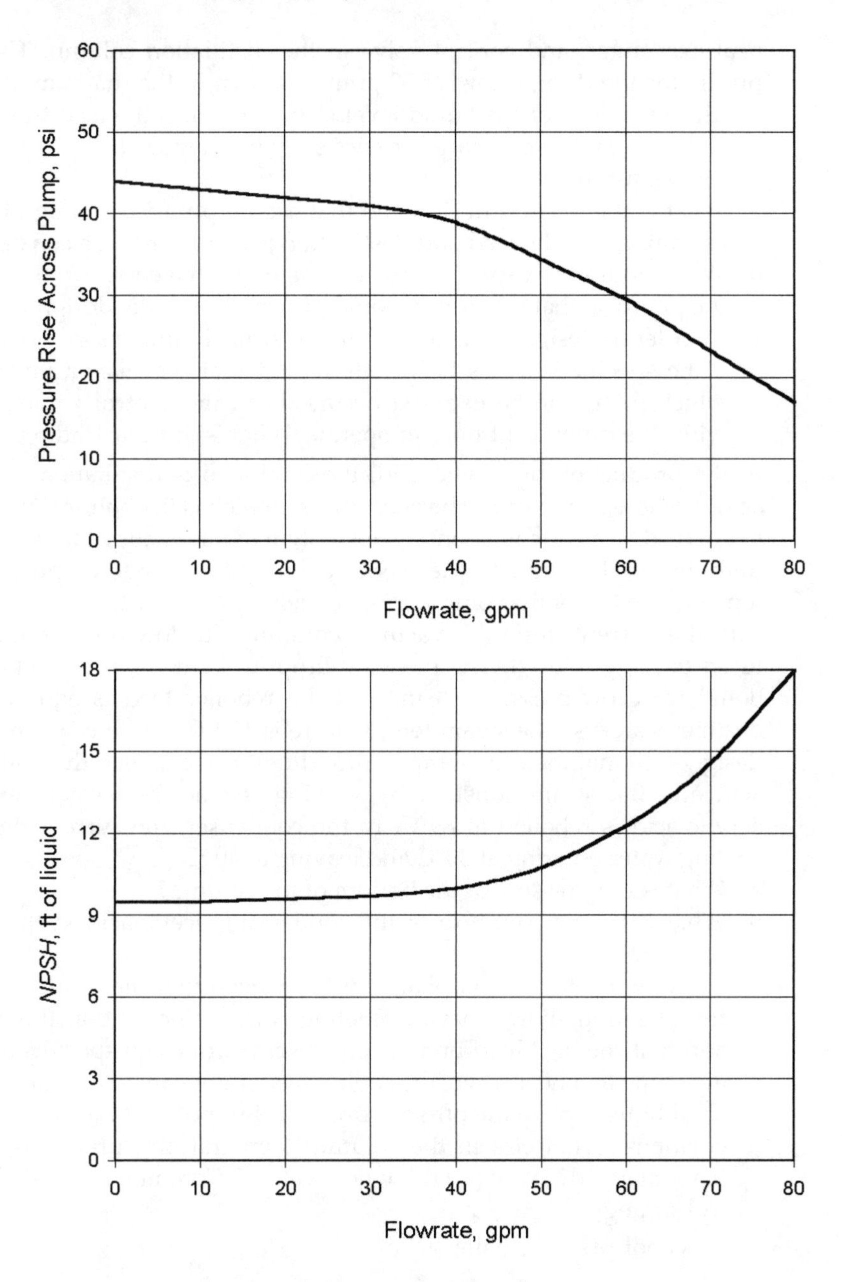

Figure P19.17(b) Pump Characteristics for P-301

heat exchanger, and control valve to the distillation column. The pressure profile for the design flow of 50 gpm is shown in the diagram. Also shown are the elevations of the liquid level in the tank and the feed tray in the column. Using the pump characteristics shown in Figure P19.17(b), answer the following questions.

a. What is the maximum flowrate that can be pumped through this system assuming that the feed and destination pressures remain constant?
b. What would the vapor pressure of the liquid have to be at the entrance to the pump so that the pump would just cavitate at the design flow?
c. In order to design the heat exchanger, a maximum design pressure needs to be specified. It has been determined that the maximum pressure to which E-301 can be exposed occurs when the control valve fails closed while the pump continues to operate. What is this maximum pressure?

18. In the production of acrylic acid, the final step is distillation of an acrylic acid/acetic acid mixture. The acrylic acid, which is the bottoms product, may not exceed 90°C due to spontaneous polymerization above that temperature. Both the acrylic acid and the acetic acid may be considered pure at the bottom and the top of the column, respectively.

    In the current design, a vacuum column with low-pressure-drop, structured packing is used. The pressure drop in the column is 9.5 kPa. The reboiler uses low-pressure steam, but the reboiler feed is equipped with a desuperheater, so the steam temperature is 120°C. There is a control system designed to maintain the temperature difference between the boiling acrylic acid and the steam constant by varying the desuperheater settings. The acrylic acid is reboiled at 89°C. In the condenser, the current design is for cooling water entering at 30°C and leaving at 40°C.

    a. What is the pressure at the bottom of the column?
    b. What is the temperature of the condensing acetic acid at the top of the column?
    c. Based on customer complaints, it is believed that the product is off spec. Investigation shows that the cooling water is flowing at the design rate and that the heat load on the condenser is at design specifications. However, due to a heat wave, the cooling water appears to be entering at 35°C. Could this explain the off-spec product? Support your argument with calculations. What else in the column (temperatures, pressures, etc.) must have changed? What are these new values? Comment on the reboiler control strategy.

    Vapor pressure data:

$$\text{For acrylic acid} \quad \ln P^* \,(kPa) = 17.31 - \frac{5264.4}{T(K)}$$

$$\text{For acetic acid} \quad \ln P^* \,(kPa) = 17.54 - \frac{4981.5}{T(K)}$$

19. The distillate stream from a distillation column (flowrate 35 m^3/h, density of water) is recycled. The destination of the recycle stream is at a location 3 m below the fluid level in the reflux drum. First, the recycle passes through a pump to supply sufficient head to overcome frictional losses (20 kPa) and reach the pressure of the stream with which it is mixed (250 kPa above distillate stream). Then it is heated from 90°C to 120°C using low-pressure steam at 140°C. Assume that all resistance is on the process-stream side of the exchanger (tube side).

    It is desired to scale up the entire plant capacity by 15%. It has been determined that the distillation column can be modified to maintain the recycle stream temperature at 90°C. It has also been determined that, with the required changes in the feed section of the process, the minimum temperature of the recycle stream must be between 119°C and 121°C to satisfy liquid-phase reactor inlet conditions after mixing with fresh feed.

    a. Can the temperature criterion be satisfied?
    b. Can the flows in this portion of the process be scaled up by 15%? The pump curve is in Figure P19.19.

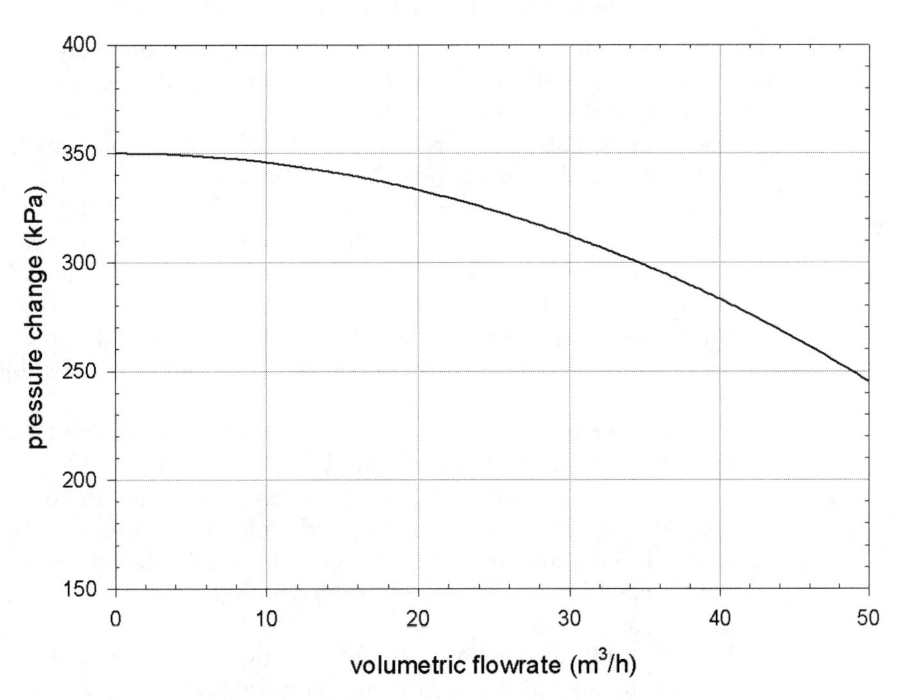

Figure P19.19 Pump Curve for Problem 19

c. With a scale-up of 15% desired, what feature must be included in this distillation column so that column operation can be modified to maintain the recycle stream at 90°C?
d. It is suggested that because a lower temperature of the recycle stream would reduce the reaction rate, if needed, the pumps in the feed section could be used to increase the reactor pressure and hence the reaction rate. What is your opinion of this suggestion?

20. A process heat exchanger has a bypass around it that is normally closed, thus forcing all the process flow through the exchanger. The process fluid (vapor) flows through the shell side of an exchanger, and cooling water flows through the tubes. You may assume that all the heat transfer resistance is on the shell side. During a planned change in operations, it has been determined that additional gas flow will be generated, resulting in an increase of 70% above the current flow. However, the pressure drop across the exchanger would be excessive if all of this flow passed through the exchanger. Therefore, the bypass will be opened so that the pressure drop across the process exchanger increases by only 69%. For this situation answer the following questions.
   a. By how much does the flow through the heat exchanger increase?
   b. What fraction of the new flow will flow through the bypass?
   c. What will be the new overall heat transfer coefficient?

21. You are trying to evaluate whether an existing, idle distillation column can be used for a separation for which it was not originally designed. Answer the following questions about this column.
   a. The reflux pump (there are two in parallel) must pump fluid from the reflux drum to the top of the column. The fluid density is 775 kg/m^3. The reflux drum is 6 m above the pump inlet. The top of the tower is 42 m above the pump discharge. Frictional pressure drop for the entire lengths of pipe is calculated to be 130 kPa. The desired flowrate is 7 L/s. The pump curve and $NPSH_R$ curves are in Figure P19.21. Determine whether this pump can be used for this job, and, if so, what configuration is necessary. The frictional pressure drop seems high. Suggest a possible explanation.
   b. The reboiler is connected to the low-pressure steam line (160°C). However, there is also a desuperheater. The area of the exchanger is 300 m^2, and the heat transfer coefficients have been estimated to be 5000 W/m^{2}K and 3000 W/m^{2}K for the boiling side and the condensing side, respectively. The boiling point of the process fluid is 87°C, and the required heat load is 6000 kW. What is the saturated steam temperature required at the desuperheater exit?
   c. List at least three items you would investigate regarding the ability of this column to do the desired separation. For each item, discuss what you would look for or calculate to determine the suitability of the column for this duty.

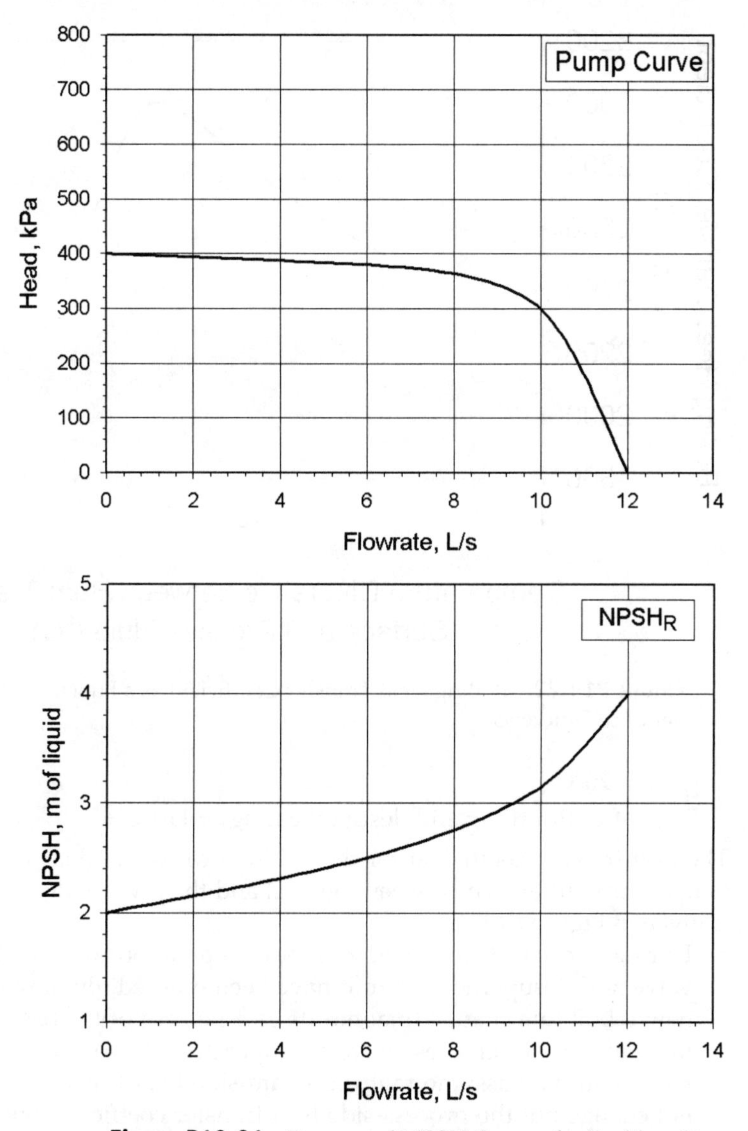

Figure P19.21 Pump and NPSH Curves for Problem 21

22. During the design of a new separation system, a reboiler using throttled and desuperheated low-pressure steam (5 barg, T_{sat} = 160°C) was specified. The details of this design are given below:

Duty = 5,000 kW
Overall heat transfer coefficient = 1500 W/m²°C
Steam-side coefficient = process-side coefficient

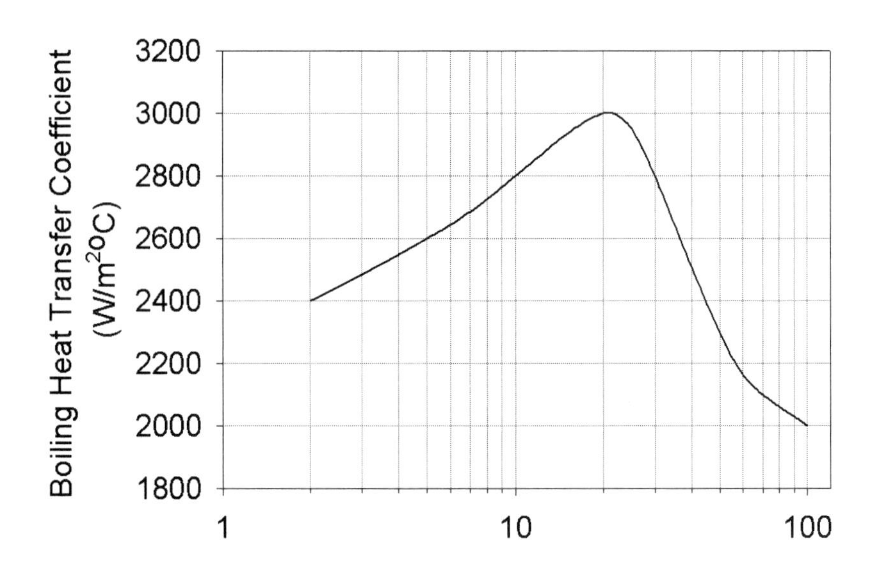

Figure P19.22 Boiling Heat Transfer Coefficient as a Function of Temperature Difference

$T_{process} = 70°C$

T_{steam} (after throttling and desuperheating) = 110°C

The heat transfer coefficient for the boiling process fluid as a function of the temperature difference between the wall and the process fluid for this design is given in Figure P19.22.

a. During the construction of this new separation system, the throttling valve and desuperheating unit have been omitted. By how much will the new reboil rate change (magnitude as a percent of the base case), assuming that the column pressure and temperature do not change?
   Hint: You may assume that the steam-side heat transfer coefficient does not change but the process-side heat transfer coefficient will change. *This means that the wall temperature will also change, and the solution to the problem will require trial and error.*
b. If the column were allowed to operate without the desuperheater and throttling units and no adjustments or changes were made to the column operation, how would the pressure in the column change? This answer should be consistent with your answer to Part (a).

23. The bottoms product from a distillation column is pumped through two heat exchangers to recover energy and is then sent to a storage tank, which is at atmospheric pressure. The system is shown in Figure P19.23(a).

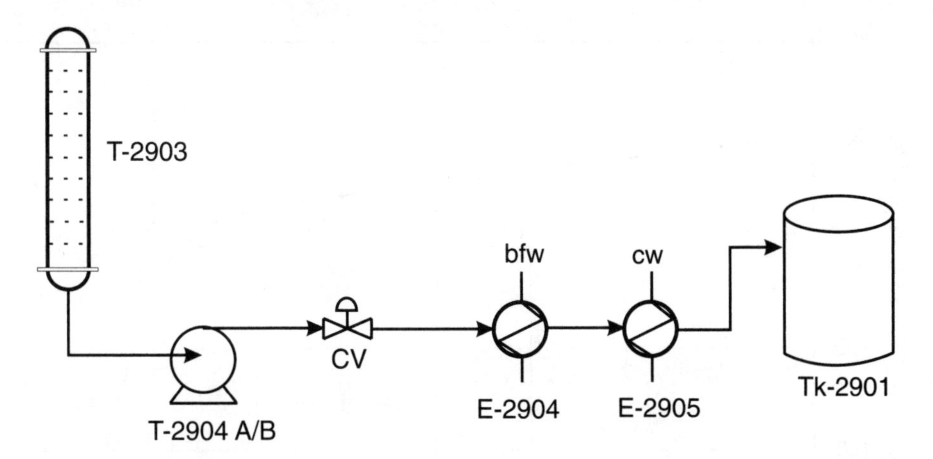

Figure P19.23(a) Process Flow Diagram for Problem 23

The pump curve for P-2904A/B is shown in Figure P19.23(b). At design conditions, the flow through the pump is 5 liters/s, the pressure in T-2903 is 200 kPa, and the frictional loss (for the process fluid) through the piping and heat exchangers, but excluding the control valve, is 250 kPa. You may assume that the difference in height between the liquid level in T-2903 and Tk-2901 is negligible. For this system, answer the following questions.

a. What is the pressure drop across the control valve at design conditions?
b. Under design conditions, the 250 kPa frictional pressure drop is distributed as shown in Table P19.23.

Table P19.23 Pressure Drops in Pipe Sections

Equipment or Pipe	Pressure Drop at Design Flowrate (kPa)
Suction piping	5
Piping from P-2904 to CV	10
Piping from P-2904 to CV	10
E-2904	50
Piping from E-2904 to E-2905	25
E-2905	50
Piping from E-2905 to Tk-2901	100

c. What is the pressure of the liquid entering E-2904 and E-2905 at a scaled-up condition, which is 150% of the base-case flowrate?

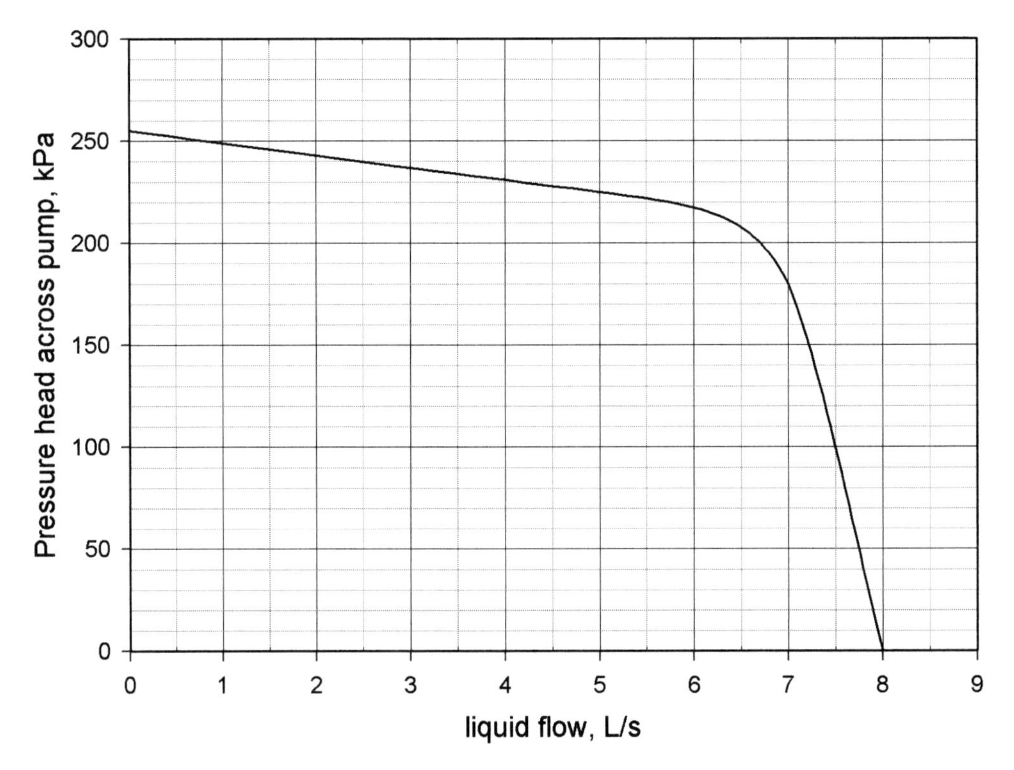

Figure P19.23(b) Pump Curve for Problem 23

24. Two heat exchangers of equal area have been designed to heat a certain process fluid from 50°C to 100°C using condensing saturated steam at 150°C. The designs of the two exchangers are different and result in different pressure drops and different overall heat transfer coefficients. Pilot-test results for the two heat exchangers operating individually are given in Table P19.24 for a single set of operating conditions.

    In the plant, these two exchangers are now to be piped in parallel, and at the current operating conditions, the pressure drop across both heat exchangers is measured at 52 kPa.

    Assuming that the physical properties and inlet temperature of the process stream (in the tube side) are the same as given in Table P19.24 and that condensing steam is used in the shell side, answer the following questions.

    a. What is the mass flowrate of process fluid through Exchanger 1?
    b. What is the mass flowrate of process fluid through Exchanger 2?
    c. What is the new overall heat transfer coefficient for Exchanger 1?
    d. What is the new overall heat transfer coefficient for Exchanger 2?

Table P19.24 Heat-Exchanger Pilot-Test Results

Process Variable	Exchanger 1	Exchanger 2
Mass flow of process liquid (tube side)	2.0 kg/s	1.5 kg/s
Inlet temperature of liquid	50°C	50°C
Outlet temperature of liquid	100°C	100°C
Process liquid C_p	1000 J/kgK	1000 J/kgK
Steam temperature (shell side)	150°C	150°C
ΔT_{lm}	72.13°C	72.13°C
Area	6.93 m^2	6.93 m^2
Condensing steam heat transfer coefficient	1000 W/m^2K	1000 W/m^2K
U	200 W/m^2K	150 W/m^2K
Tube-side pressure drop	40 kPa	25 kPa

e. If the process exit stream from Exchanger 1 is to be maintained at 100°C, what must be the temperature of the condensing steam that is fed to this exchanger?

f. If the process exit stream from Exchanger 2 is to be maintained at 100°C, what must be the temperature of the condensing steam that is fed to this exchanger?

g. Under another set of conditions, if the two exchangers were piped in parallel and the flow measured through Exchanger 1 was 3.1 kg/s, what would the flow be through Exchanger 2?

CHAPTER

20

Reactor Performance

Chemical reactors are used to produce high-value chemicals from lower value chemicals. Reactor performance depends on the complex interaction between four effects, as is illustrated in Figure 20.1. In order to understand fully the performance of a chemical reactor in the context of a chemical process, all four effects must be considered.

1. **Reaction Kinetics and Thermodynamics:** The influence of extensive variables (pressure, temperature, concentration) on reactor performance is defined by reaction kinetics and reaction equilibrium. These extensive variables affect the reaction rate and determine the extent to which reactants can be converted into products in a given reactor or the size of reactor needed to achieve a given conversion. Additionally, catalysts are used to increase the rate of reaction. Thermodynamics sets a theoretical limit on the extent to which reactants can be converted into products and cannot be changed by catalysts.
2. **Reactor Parameters:** These include the reactor volume, space time (reactor volume/inlet volumetric flowrate), and reactor configuration. For given kinetics, thermodynamics, reactor and heat transfer configuration, and space time, the reactor volume needed to achieve a given conversion of reactants is determined. This is the design problem. For a fixed reactor volume, the conversion is affected by the temperature, pressure, space time, catalyst, and reactor and heat transfer configuration. This is the performance problem.

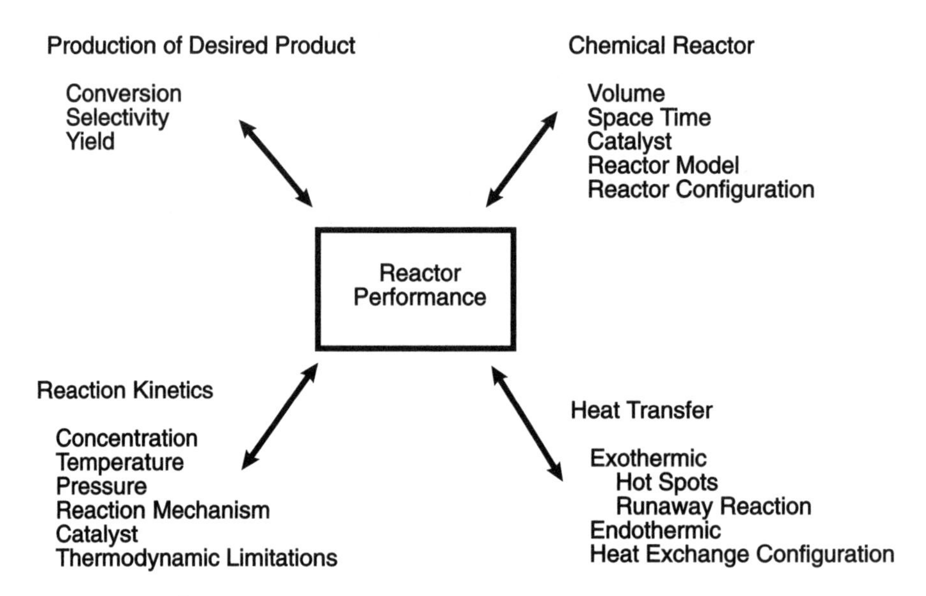

Figure 20.1 Phenomena Affecting Reactor Performance

3. **Production of Desired Product:** *Conversion*, *selectivity*, and *yield* are terms that quantify the amount of reactants reacted to form desired products. Reactor performance is expressed in terms of these parameters. For a fixed reactor volume, these parameters are functions of the temperature, pressure, reactor and heat transfer configuration, and space time.
4. **Heat Transfer in Reactor:** This important effect is often overlooked. Energy is released or consumed in chemical reactions. The rate of chemical reaction is highly temperature dependent. The key point to consider is the interaction between chemical kinetics and heat transfer (see Example 19.1). For exothermic reactions, the heat of reaction must be removed efficiently to avoid large temperature increases that can damage catalyst and to prevent runaway reaction. For endothermic reactions, heat must be supplied efficiently so that the reaction can proceed. The rate of heat transfer depends on reactor configuration (including the heat transfer configuration), the properties of the reacting stream, the properties of the heat transfer medium, and the temperature driving force.

In this chapter, the basic concepts needed for prediction of reactor performance are reviewed. Multiple reaction systems, reactor models, and heat transfer considerations are emphasized. Case studies are presented based on reactors found in the processes used as examples throughout the text and in the appendixes.

20.1 PRODUCTION OF DESIRED PRODUCT

Consider the following reaction scheme:

$$aA + bB \xrightarrow[rxn\,1]{k_1} pP \xrightarrow[rxn\,2]{k_2} uU$$
$$\beta B \xrightarrow[rxn\,3]{k_3} vV \qquad (20.1)$$

Equation (20.1) shows three reactions involving five species A, B, P, U, and V, with stoichiometric coefficients a, b, p, u, and v, respectively. It is assumed that P is the desired product and that U and V are unwanted by-products. The reaction scheme in Equation (20.1) can be used to illustrate the effects commonly observed in actual reaction kinetics.

1. A **single reaction** produces desired product. Here $k_2 = k_3 = 0$, and only the first reaction proceeds. Acetone production (Appendix B, Figure B.9.1) is an example of this situation. Hydrodealkylation of toluene to benzene behaves like this because the catalyst suppresses the undesired reactions sufficiently so that single-reaction behavior is approached.
2. **Parallel** (competing) reactions produce desired products and unwanted by-products. Here $k_2 = 0$, and no U is formed. Species B reacts to form either P or V. In the phthalic anhydride reaction sequence (Appendix C, Figure C.5), the reaction of o-xylene to form either phthalic anhydride, maleic anhydride, or combustion products (Appendix C, Table C.10, Reactions 1, 3, and 4) is an example of a parallel reaction.
3. **Series** (sequential) reactions produce desired products and unwanted by-products. Here $k_3 = 0$, and no V is formed. Species A reacts to form desired product P, which further reacts to form unwanted by-product U. In the phthalic anhydride reaction sequence (Appendix C, Table C.10), the reaction of o-xylene to phthalic anhydride to combustion products (Reactions 1 and 2) is an example of a series reaction.
4. **Series** and **parallel** reactions produce desired products and unwanted by-products. Here, all three reactions in Equation (20.1) occur to form desired product P and unwanted by-products U and V. The entire phthalic anhydride sequence (Appendix C, Table C.10) as well as the cumene production reaction (Appendix C, Table C.17) are examples of series or parallel reactions.

For reaction schemes like the one shown in Equation (20.1), there are three key definitions used to quantify production of desired product.

The term **conversion** quantifies the amount of reactant reacted. The single-pass conversion (or reactor conversion) is

$$\text{single-pass conversion} = \frac{\text{reactant consumed in reactor}}{\text{reactant fed to reactor}} \tag{20.2}$$

and is generally reported in terms of the limiting reactant. This is different from the overall process conversion, which is defined as

$$\text{overall conversion} = X = \frac{\text{reactant consumed in process}}{\text{reactant fed to process}} \tag{20.3}$$

Most processes recover and recycle unreacted material to provide a high overall conversion.

High reactor conversions are neither necessary nor desirable for optimum reactor performance. At low reactor conversions, high overall conversions can be achieved with increased recycle.

The term **selectivity** quantifies the conversion to desired product.

$$\text{selectivity} = \eta = \frac{\text{rate of production of desired product}}{\text{rate of production undesired by products}} \tag{20.4}$$

A high selectivity is always desirable.

Competition from undesired reactions limits conversion to the desired product.

Another term used to quantify production of the desired product is **yield**, defined as

$$\text{yield} = \frac{\text{moles of reactant reacted to produce desired product}}{\text{moles of limiting reactant reacted}} \tag{20.5}$$

The terms defined in this section will be used in examples and the case studies presented later in this chapter.

20.2 REACTION KINETICS AND THERMODYNAMICS

The kinetics of a reaction quantify the rate at which the reaction proceeds. When designing a new reactor for a given conversion, a faster reaction requires a smaller volume reactor. When analyzing an existing reactor of fixed volume, a faster reaction means increased conversion. As stated previously, thermodynamics provides limits to the conversions obtainable from a chemical reaction.

20.2.1 Reaction Kinetics

The reaction rate, r_i, is defined as

$$r_i = \frac{1}{V}\frac{dN_i}{dt} = \frac{\text{moles of } i \text{ formed}}{(\text{volume of reactor})\,(\text{time})} \tag{20.6}$$

The reaction rate is an intensive property. This means that the reaction rate depends only on state variables such as temperature, concentration, and pressure, and not on the total mass of material present.

For solid catalyzed reactions, the reaction rate is often defined based on the mass of catalyst present, W:

$$r_i = \frac{1}{W}\frac{dN_i}{dt}\rho_b = \frac{1}{V}\frac{dN_i}{dt} \tag{20.7}$$

where ρ_b is the bulk catalyst density (mass catalyst/volume reactor). The density of solid catalyst is defined as ρ_{cat} (mass catalyst particle/volume catalyst particle). So the bulk density of the catalyst, ρ_b, is defined as

$$\rho_b = (1 - \varepsilon)\rho_{cat} \tag{20.8}$$

Here, ε is the void fraction in the reactor; so $(1 - \varepsilon)$ is volume of catalyst/volume of reactor.

If a reaction is an elementary step, the kinetic expression can be obtained directly from the reaction stoichiometry. For example, in Equation (20.1), if the first reaction is an elementary step, the rate expression is

$$-r_A = k_1 c_A^{\alpha} c_B^{\beta} \tag{20.9}$$

For catalytic reactions, the rate expressions are often more complicated because the balanced equation is not an elementary step. Instead, the rate expression can be obtained by an understanding of the details of the reaction mechanism. The resulting rate expressions are often of the form

$$r_i = \frac{k_1 \prod_{i=1}^{n} c_i^{\alpha_i}}{\left[1 + \sum_{j=1}^{m} K_j c_j\right]^{\gamma}} \tag{20.10}$$

Equation (20.10) describes a form of Langmuir-Hinshelwood kinetics. The constants (k_1 and K_j) in Equation (20.10) are catalyst specific. The constants in Equation (20.10) must be obtained by fitting reaction data.

In heterogeneous catalytic reacting systems, reactions take place on the surface of the catalyst. Most of this surface area is internal to the catalyst pellet or particle. The series of resistances that can govern the rate of catalytic chemical reaction are as follows:

1. Mass film diffusion of reactant from bulk fluid to external surface of catalyst
2. Mass diffusion of reactant from pore mouth to internal surface of catalyst
3. Adsorption of reactant on catalyst surface
4. Chemical reaction on catalyst surface
5. Desorption of product from catalyst surface
6. Mass diffusion of product from internal surface of catalyst to pore mouth
7. Mass diffusion from pore mouth to bulk fluid

Each step offers a resistance to chemical reaction. Reactors often operate in a region where only one or two resistances control the rate. For a good catalyst, the intrinsic rates are so high that internal diffusion resistances are usually controlling.

For solid catalyst systems, reactor performance is usually controlled by resistances to mass transfer.

The temperature dependence of the rate constants in Equations (20.9) and (20.10) is given by the Arrhenius equation

$$k_i = k_o e^{-\frac{E}{RT}} \tag{20.11}$$

where k_o is called the **pre-exponential factor**, and E is the **activation energy** (units of energy/mol, always positive). Equation (20.11) reflects the significant temperature dependence of the reaction rate. For gas-phase reactions, the concentrations can be expressed or estimated from the ideal gas law, so $c_i = P_i/RT$. This is the origin of the pressure dependence of gas-phase reactions. As pressure increases, so does concentration, and so does the reaction rate. The temperature dependence of the Arrhenius equation usually dominates the opposite temperature effect on the concentration.

As temperature increases, the reaction rate always increases, and usually significantly.

20.2.2 Thermodynamic Limitations

Thermodynamics provides limits on the conversion obtainable from a chemical reaction. For an equilibrium reaction, the equilibrium conversion may not be exceeded.

Thermodynamics sets limits on possible conversions in a reacting system.

The limitations placed on conversion by thermodynamic equilibrium are best illustrated by Example 20.1.

Example 20.1

Methanol can be produced from syngas by the following reaction:

$$CO + 2H_2 = CH_3OH$$

For the case when no inerts are present and for stoichiometric feed, the equilibrium expression has been determined to be

$$K = \frac{X(3-2X)^2}{4(1-X)^3 P^2} = 4.8 \times 10^{-13} \exp(11{,}458/T) \qquad \text{(E20.1)}$$

where X is the equilibrium conversion, P is the pressure in atmospheres, and T is the temperature in Kelvin. Construct a plot of equilibrium conversion versus temperature for four different pressures: 15 atm, 30 atm, 50 atm, and 100 atm. Interpret the significance of the results.

The plot is shown on Figure E20.1. By following any of the four curves from low to high temperature, it is observed that the equilibrium conversion decreases with increasing temperature at constant pressure. This is a consequence of LeChatelier's principle, because the methanol formation reaction is exothermic. By following a vertical line from low to high pressure, it is observed that the equilibrium conversion increases with increasing pressure at constant temperature. This is also a consequence of LeChatelier's principle.

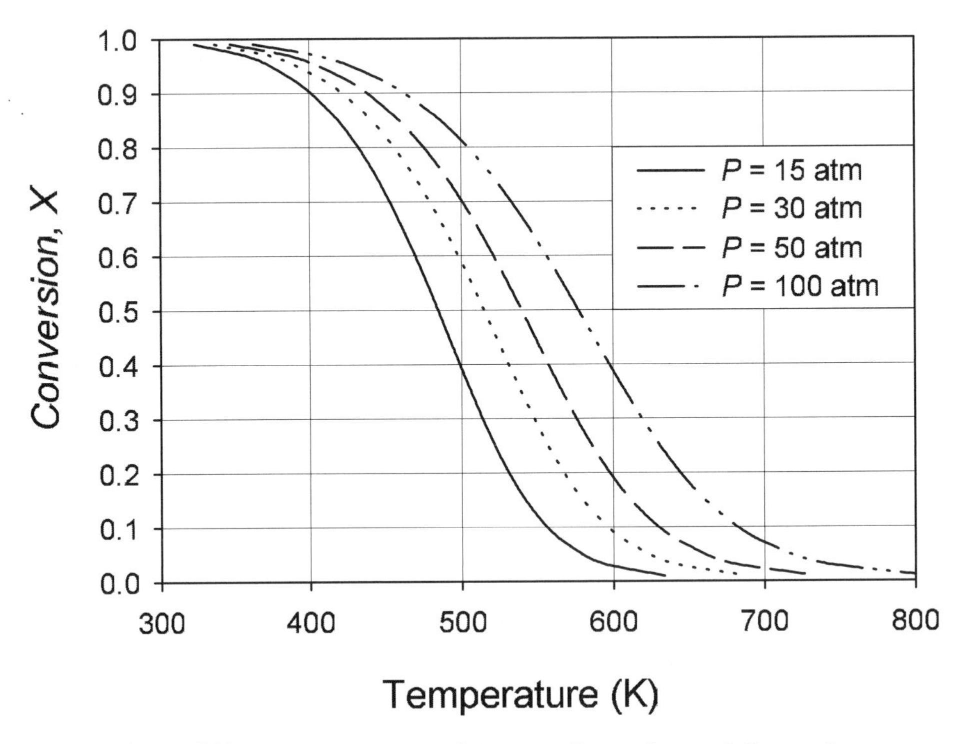

Figure E20.1 Temperature and Pressure Dependence of Conversion for Methanol from Syngas

Because there are fewer moles on the right-hand side of the reaction, increased conversion is favored at high pressures.

From thermodynamic considerations alone, it appears that this reaction should be run at low temperatures in order to achieve maximum conversion. However, as discussed in Section 20.2.1, because the rate of reaction is a strong function of temperature, this reaction is usually run at higher temperatures, with low single-pass conversion, in order to take advantage of the faster kinetics. As discussed in Section 20.1, despite a low single-pass conversion in the reactor, a large overall conversion is still achievable by recycling unreacted reactants.

20.3 THE CHEMICAL REACTOR

For continuous flow reactors, the two ideal models are the plug flow reactor and the continuous stirred tank reactor. In the plug flow reactor (PFR), which is basically a pipe within which the reaction occurs, the concentration, pressure, and temperature change from point to point. The performance equation is

$$\frac{V}{F_{Ao}} = \frac{\tau}{c_{Ao}} = \int_0^{X_A} \frac{dX_A}{-r_A} \tag{20.12}$$

In the continuous stirred tank reactor (CSTR), the reactor is assumed to be well mixed, and all properties are assumed to be uniform within the reactor. The performance equation is

$$\frac{V}{F_{Ao}} = \frac{\tau}{c_{Ao}} = \frac{X_A}{-r_A} \tag{20.13}$$

where

V = volume of the reactor
F_{Ao} = molar flow of limiting reactant A
τ = space time (reactor volume/inlet volumetric flowrate)
c_{Ao} = inlet concentration of A
X_A = conversion of A
r_A = rate of reaction of A

Equations (20.12) and (20.13) must be solved for a given rate expression to get the actual relationship between the indicated parameters. However, several generalizations are possible. The reactor volume and conversion change in the same direction. As one increases, so does the other; as one decreases, so does the other. As the reaction rate increases, the volume required decreases, and vice versa. As the reaction rate increases, the conversion increases, and vice versa. It has already been discussed how temperature, pressure, and concentration affect the reaction rate. Their effect on conversion and reactor volume is derived from their effect on reaction rate. Example 20.2 illustrates these qualitative generalizations.

Example 20.2

Discuss qualitatively the effect of temperature and pressure on the conversion, space time, and conversion in a chemical reactor.

Gas-Phase Reaction: From the Arrhenius equation in Equation (20.11), the reaction rate constant increases exponentially with temperature. From the ideal gas law, the concentration decreases less drastically. From Equations (20.12) and (20.13), at constant pressure, if the reaction rate increases, at constant conversion, a smaller reactor volume (and space time) is required. At constant reactor volume and space time, the conversion increases.

From a typical reaction rate expression such as Equation (20.9), as the pressure increases, so do the concentration and the reaction rate. Therefore, the same trends are observed as with temperature. However, the quantitative effect is not as great, because the concentration dependence on pressure is not exponential.

Liquid-Phase Reaction: For a liquid-phase reaction, temperature and pressure do not affect concentration. Therefore, there is no effect of pressure on reactor performance for a liquid-phase reaction. Because the temperature effect is determined by the Arrhenius equation only, the reaction rate increases exponentially with temperature.

The PFR is a hypothetical system. PFRs are elementary models for packed-bed reactors. The PFR model excludes the effects of axial mixing resulting from molecular movement and fluid turbulence while assuming radial mixing is complete. This mixing takes place between adjacent fluid elements of approximately the same conversion. Reactors approximating PFRs are often built using many small-diameter tubes (less than 3 in) of significant length (20 to 30 ft) operated at high fluid velocity with small space times. This minimizes axial fluid mixing, limits radial temperature profiles, and provides needed heat transfer area. Tubes are arranged in a bundle characteristic of those found in many heat exchangers. If heat exchange is not desired in the reaction zone, a single or series of larger-diameter packed beds can be used.

The CSTR is also a hypothetical system in which there is perfect mixing so that temperature, pressure, concentration, and reaction rate are constant over the reactor volume. Reactors approximating CSTRs are used for liquid-phase reactions. This represents a theoretical limit because perfect mixing can only be approached. Transit time for fluid elements varies. The exit stream is at the same temperature, pressure, and conversion as the reactor contents. Feed is mixed with the reactor contents that have a high conversion. As a result, the CSTR requires a higher volume than a PFR when operated isothermally at the same temperature and conversion for simple, elementary reactions.

Reactors are modeled using different combinations and arrangements of CSTRs and PFRs. Intermediate degrees of mixing are obtained that can be used to model various real reactor systems. Some of these models are shown in Figure 20.2 and described below.

System A is a PFR for an exothermic reaction. The concentration and temperature vary from point to point in the reactor. This model is used to simulate the cumene reactor in Appendix C.

System B is a CSTR. The temperature and concentration are constant within the reactor.

System C consists of a PFR and CSTR in parallel. By increasing the ratio of CSTR feed rate to PFR feed rate, the degree of mixing is increased. By increasing the ratio of PFR feed rate to CSTR feed rate, dead space (regions that are not well mixed) in a CSTR can be simulated.

System D consists of a series of CSTRs in series. The degree of mixing is less than that in a single CSTR. Although each reactor is completely mixed, there is no mixing between each section. In the limit of an infinite number of CSTRs in series, PFR behavior is approached.

System E consists of an isothermal PFR with a bypass stream. In a fluidized bed, solid catalyst is circulated within the bed; however, the fluid moves through the bed essentially in plug flow. Maldistribution of fluid, due to the formation of bubbles and voids, and channeling, may cause some fluid to bypass the catalyst. This model is used in catalytic fluidized bed applications or in any other application where the solid does not react. A portion of the reactants forms an emulsion phase with little mixing (plug flow). The remaining

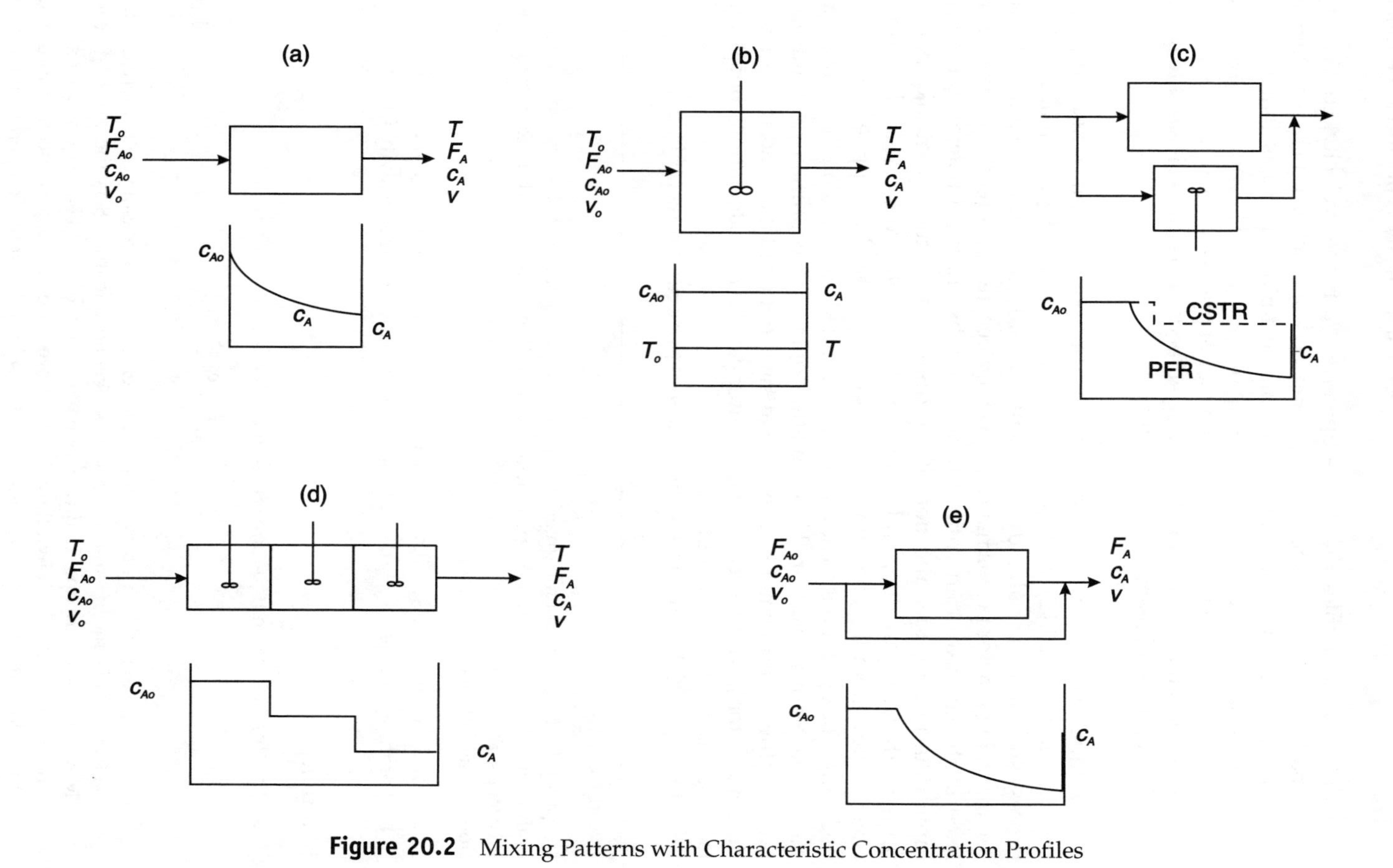

Figure 20.2 Mixing Patterns with Characteristic Concentration Profiles

fluid forms bubbles that move upward at a higher velocity, contact only a small portion of the solid catalyst, and do not react as much as in the emulsion phase. An important feature of the fluidized bed is that it operates isothermally as a result of the mixing of the solids. The model shown in System E is a crude approximation of what happens in a real fluidized bed, but it predicts the trends found in fluidized beds. This model is used to simulate the acrylic acid reactor in Appendix B.

As the reactor model becomes more complex, it becomes more difficult to solve Equations (20.12) and (20.13) to predict reactor performance. Because process simulators are already programmed to do this, they are a logical choice for analyzing reactor performance, especially because real reactors must be simulated in the context of an entire process. For the chemical reactor, the required input data include identification of the dominant reactions that take place, the form of the reaction rate, and values of the kinetic constants. All these may change with operating conditions, and a given set of values applies only to a limited range of operations. They must be obtained experimentally. If the experimental data are flawed or the simulation operating conditions are outside the range of the experimental studies, the answer obtained can lead to poor prediction of reactor performance. The computer can do the calculations but is unable to recognize the reliability of the input.

Example 20.3 summarizes the concepts presented thus far in this chapter.

Example 20.3

Consider the reaction scheme given in Equation (20.1) where P is the desired product, with both U and V as undesired by-products. Assume that Equation (20.1) represents elementary steps, that $a = b = p = u = v = 1$, that $\beta = 2$, and that the activation energies for the reactions are as follows: $E_1 > E_2 > E_3$.

a. For the case where $k_2 = 0$, what conditions maximize the selectivity for P?

b. For the case where $k_3 = 0$, what conditions maximize the selectivity for P?

Solution

a. For this case, from Equation (20.4), the selectivity is written as

$$\eta = \frac{r_P}{r_V} = \frac{k_1 c_A c_B}{k_3 c_B^2} = \frac{k_1 c_A}{k_3 c_B} \qquad \text{(E20.3a)}$$

There are several ways to maximize the selectivity. Increasing c_A/c_B increases the selectivity. This means that excess A is needed and that B is the limiting reactant. Many reactions are operated with one reactant in excess. The reason is usually to improve selectivity as shown here. Because pressure affects all concentrations equally, it is seen that pressure does not affect the selectivity here. Temperature has its most significant effect on the rate constant. Because the activation energy for rxn 1 is larger than that for rxn 3, k_1 is more strongly affected by temperature changes than is

k_3. Therefore, increasing the temperature increases the selectivity. In summary, higher temperatures and excess A maximize the selectivity for P.

b. For this case, the selectivity is written as

$$\eta = \frac{r_P}{r_U} = \frac{k_1 c_A c_B - k_2 c_P}{k_2 c_P} = \frac{k_1 c_A c_B}{k_2 c_P} - 1 \tag{E20.3b}$$

Because pressure appears to the second power in the numerator and to the first power in the denominator (it is in each concentration term), increasing the pressure increases the selectivity. Because the activation energy for rxn 1 is larger, increasing the temperature increases the selectivity. Increasing both reactant concentrations increases the selectivity, but increasing the concentration of Component P decreases the selectivity. The question is how the concentration of Component P can be kept to a minimum. The answer is to run the reaction at low conversions (small space time, small reactor volumes). Quantitatively, the selectivity in Equation (20.3b) is maximized. Intuitively, because this is a series reaction with the desired product intermediate in the series, a low conversion maximizes the intermediate product and minimizes the undesired product. This can be illustrated by the concentration profiles obtained by assuming that these reactions take place in a PFR, as shown in Figure E20.3. It is seen that the ratio of

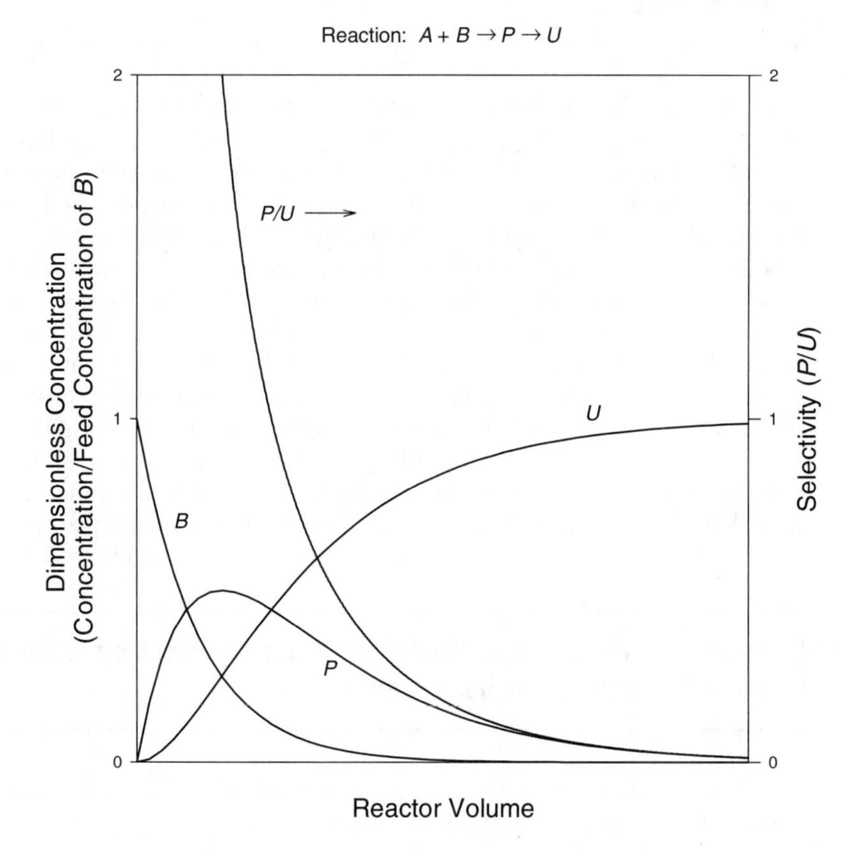

Figure E20.3 Concentration Profiles in a PFR for Series Reaction

P/U is at a maximum at low reactor volumes, which corresponds to low conversion. Therefore, increasing the temperature and pressure both increases the selectivity for Component P. Running at low conversion probably does more to increases the selectivity than can be accomplished by manipulating temperature and pressure alone. However, there is a trade-off between selectivity and overall profitability, because low conversion per pass means very large recycles and larger equipment.

20.4 HEAT TRANSFER IN THE CHEMICAL REACTOR

Chemical reactions are either exothermic (release heat) or endothermic (absorb heat). Therefore, heat transfer in reactor systems is extremely important. For exothermic reactions, heat may have to be removed so that the temperature in the reactor does not increase above safe limits. For endothermic reactions, heat may have to be added so that the reaction will occur at an acceptable rate.

Reactor performance is often limited by the ability to add or remove heat.

Most industrially significant chemical reactions are exothermic. Highly exothermic reactions, especially in a reactor behaving like a PFR, can be dangerous if the rate of heat removal is not sufficient. For a rate expression such as Equation (20.9), the concentration of reactants is largest at the entrance to the reactor. The reaction at the entrance generates heat if the reaction is exothermic. The concomitant increase in temperature increases the reaction rate even further, which increases the temperature further, and so forth. Therefore, it is possible to have large temperature increases, called **hot spots.** The large temperature increase can be offset by removing heat from the reactor. This can cause radial temperature gradients, with severe hot spots near the center of the tube. Hot spots can damage and deactivate catalyst particles. In the extreme case, if the rate of heat removal is not sufficient to offset the rate of heat production by reaction, the temperature may increase rapidly causing damage to the reactor and its contents or an explosion. This is called **runaway reaction**. Reactor temperature profiles illustrating the situation leading to runaway reaction are illustrated in Figure 20.3(b).

Beware of exothermic reactions! Runaway reaction is possible if there is insufficient heat removal.

For endothermic reactions, heat must be added to the reactor. **Cold zones** are possible based upon the same logic as for the hot spots for exothermic reactions.

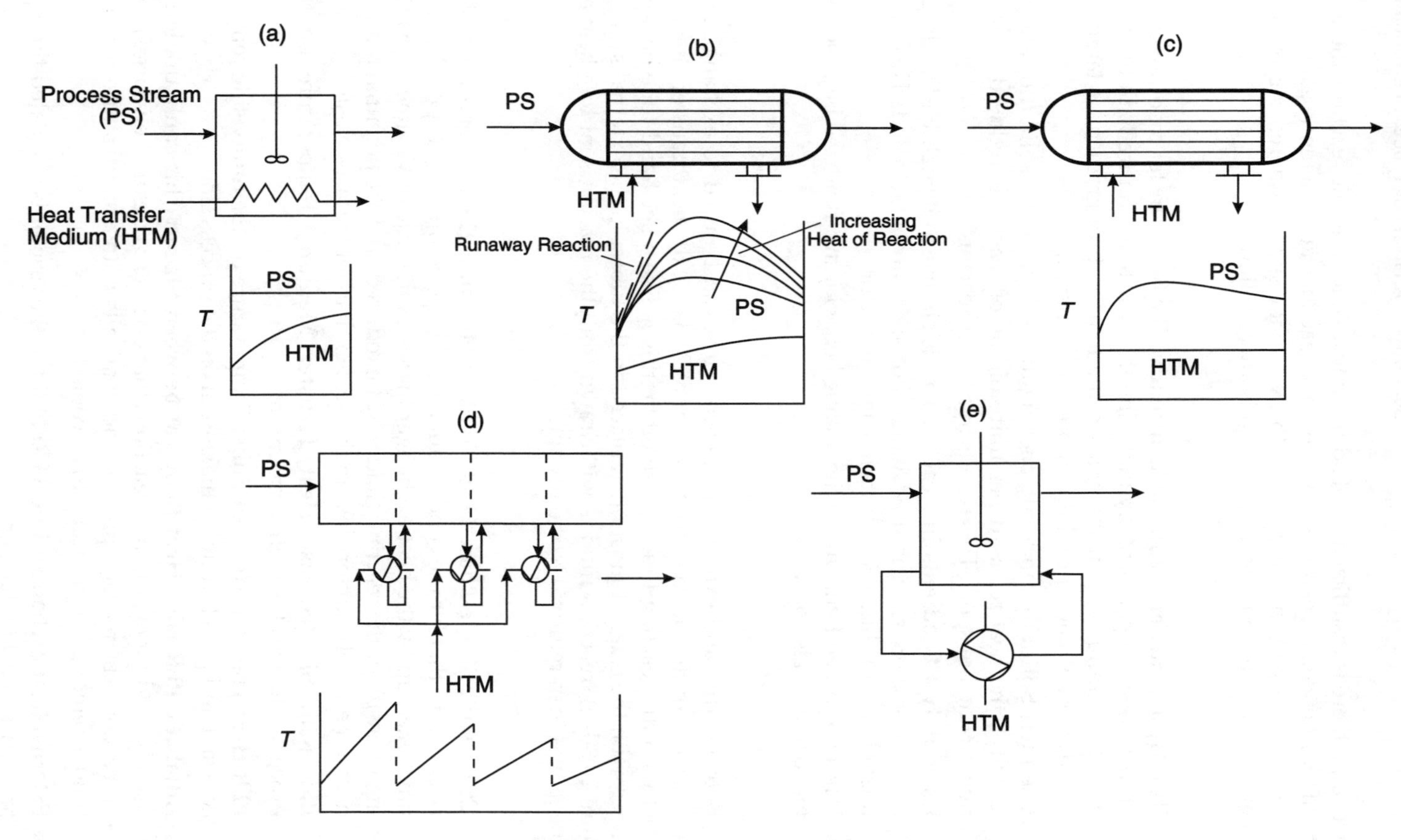

Figure 20.3 Alternative Heat-Exchange Systems for Reactors Shown for Exothermic Reaction: a–c, Internal Heat Exchanger; d–e, External Heat Exchanger

However, if there is insufficient heat addition to an endothermic reaction, the reaction will be quenched. Therefore, endothermic reactions are inherently safe.

There are many methods used to avoid hot spots in a reactor with an exothermic reaction. Some of these methods are as follows.

1. **Run Heat Transfer Medium Cocurrently:** The heat transfer medium enters at its lowest temperature. By matching this low temperature with the highest temperature zone in the reactor, the temperature gradient is largest at the reactor entrance, providing more heat removal.
2. **Use Inert Solid:** By randomly packing the bed with inert solid (no catalytic activity), the heat released per unit volume of reactor is reduced, thereby minimizing hot spots. This results in a larger reactor.
3. **Use Catalyst Gradients in Reactor:** This is similar to item 2. Here, a larger fraction of inert solid is used where hot spots are anticipated. This minimizes hot spots but also results in a larger reactor.
4. **Use a Fluidized Bed:** Due to the mixing behavior in a fluidized bed, isothermal operation is approached.

Both internal and external heat exchangers can be used to exchange heat with a reactor. In the internal systems (Figure 20.3a–c), the heat transfer area is in contact with the reacting fluid. Quite often, this is like a shell-and-tube heat exchanger, with the catalyst most often placed in the tubes. In the external systems (Figure 20.3d–e), reaction fluid is withdrawn from the reactor, sent to a heat exchanger, and then returned to the reactor.

a. **CSTR with Single-Phase Heat Transfer Medium:** The temperature profile is similar to that for the condensation of a single component vapor. The volume taken up by the heat exchanger increases the reactor volume. The mixing intensity of the reacting fluid may be reduced, and the potential for local nonisothermal behavior close to the heat exchanger surfaces exists.
b. **PFR with Single-Phase Heat Transfer Medium:** Temperature on the process side goes through a maximum.
c. **PFR Heat Transfer Medium Boils:** Temperature on the utility side remains constant and goes through a maximum on the process side.
d. **Adiabatic PFR with Heat Removal between Stages:** This configuration is usually used for exothermic reactions that do not have extremely large heats of reaction. The reaction proceeds adiabatically with a temperature rise, and fluid is removed periodically to be cooled.
e. **External heat exchanger in CSTR:** Fluid in the reactor is circulated through an external heat exchanger.

In order to solve a problem involving a chemical reactor with heat transfer, the following equations must be solved:

1. Reactor performance equation (such as Equations [20.12] and [20.13])
2. Energy balance on reaction side
3. Energy balance on heat transfer medium side
4. Heat transfer performance equation ($Q = UA\Delta T_{lm}$)
5. Pressure drop in packed bed (Ergun equation)

Therefore, as the heat transfer configuration becomes more complex, it becomes very difficult to solve the necessary equations to predict reactor performance. Because process simulators are already programmed to do this, they are a logical choice for analyzing reactor performance with heat transfer.

For a chemical reactor with heat transfer, in addition to the input discussed earlier, the required input data must include the heat transfer coefficient, the heat transfer area, the length and diameter of tubes, and the number of tubes. The output obtained is the temperature, pressure, and concentration of each component at each point in the reactor.

Heterogeneous catalytic reactions are even more difficult to simulate. The major resistance to reaction can change from chemical reaction to external diffusion or to pore diffusion within a single reactor. Stream temperatures are not necessarily the temperatures of the catalyst surface where the reaction takes place. Large temperature gradients may exist in the radial direction and within the catalyst particles. Although they do not consider these important factors, process simulators can often be used to obtain approximate solutions.

20.5 REACTOR SYSTEM CASE STUDIES

This section presents case studies involving reactors from process flow diagrams presented throughout this text. The process changes presented illustrate typical problems a chemical engineer working in production may face on a routine basis.

All studies presented contain three parts:

Part A: A statement of the problem.

Part B: A solution to the specific problem, which includes discussion of the strategy used in the solution.

Part C: A discussion of the significance of the solution to a wider range of applications. Figures showing process response to important process variables are developed and used to reveal trends that yield insight into the problem. These performance curves provide better understanding of reactor behavior. They are also used as a guide to develop alternative problem solutions.

The availability of computers and computer software permits process simulation. Once the process is simulated, variables are easily varied to discover important performance characteristics that can be plotted over a wide range of operating conditions.

20.5.1 Replacement of Catalytic Reactor in Benzene Process

Problem. The catalyst in our benzene reactor has lost activity and must be replaced. The benzene process presented in Chapter 1 (Figure 1.5) is used to produce benzene. The reaction is carried out in an adiabatic PFR. Because of the cost of periodic catalyst replacement, it has been suggested that the catalytic reactor system could be replaced with a noncatalytic adiabatic PFR system that might be less costly. The new system is to match the input/output conditions of the current system so that the remainder of the plant will operate with little, if any, modification. You are asked to investigate the merits of this suggestion by identifying major cost items. Table 20.1 provides reactor information based on experimental studies on the toluene/benzene reaction.

Figure 20.4(a) provides a flow diagram of the current reactor taken from Figure 1.5. A-A identifies the boundaries of the reactor system to be replaced. Flowrates and stream temperature and pressure at these boundaries are given.

The dominant chemical reaction at plant operating conditions is

$$\underset{toluene}{C_7H_8} + \underset{hydrogen}{H_2} = \underset{benzene}{C_6H_6} + \underset{methane}{CH_4}$$

Reactor conversion, $X_{toluene} = 1 - (36.1/144.0) = 0.749$. The amount of by-product formed by side reaction is negligible; however, the small amounts of carbon formed can deposit on the catalyst surface.

Table 20.1 Findings from Experimental Studies on Dealkylation of Toluene [1]

1. Reaction rate (mol/liter reactor s)	$-r_{tol} = 3.0 \times 10^{10} \exp(-25{,}164/T)\, c_{tol} c_H^{0.5}$
2. Temperature range	700–950°C
3. Carbon formation	None observed < 850 °C with ratio of H_2/Tol > 1.5
4. Selectivity	Benzene/Unwanted > 19/1 with toluene conversion < 0.75
5. H_2/toluene	>2
6. By-product yield W = mass of all by-products/mass feed	W(Mass % of feed) $= 1.0\times10^8 (\theta/b) \exp(-25{,}667/T)$ where θ is the residence time (in seconds) and b is the H_2/toluene ratio
7. Observed components in output	Hydrogen, methane, toluene, various diphenyls
T in K, c in mol/liter.	

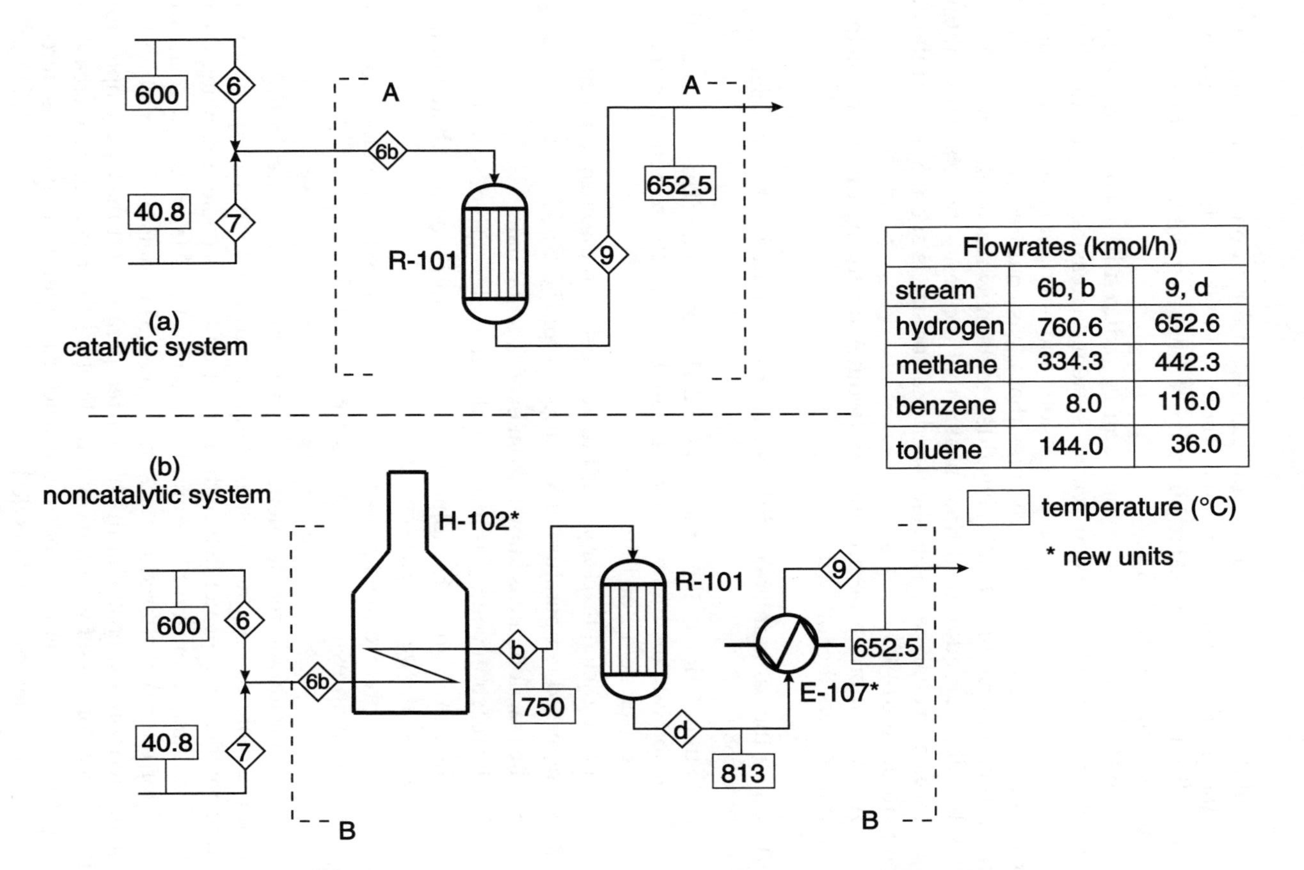

Flowrates (kmol/h)		
stream	6b, b	9, d
hydrogen	760.6	652.6
methane	334.3	442.3
benzene	8.0	116.0
toluene	144.0	36.0

Figure 20.4 Reactor Systems for Production of Benzene from Toluene

Solution. The reactor volume calculated for a noncatalytic adiabatic PFR using the kinetics expression given in Table 20.1 that met the conditions shown in Figure 20.4(a) is 257 m^3 (7800 ft^3). This is six times as great as the current catalytic reactor (41.5 m^3; see Table 1.7).

Intuitively, the reactor volume can be reduced by increasing the pressure and/or temperature in the new reactor system. Increasing the pressure increases the concentrations in the rate expression, and increasing the temperature increases the rate constant. This requires that a gas compressor and/or a heat exchanger (additional fired heater) be added to the new reactor system.

Installing an additional fired heater that increases the feed temperature to 750°C was also analyzed. The calculated reactor size is about 5 m^3 and the outlet temperature is 814°C.

An evaluation of the preliminary findings for the noncatalytic reactor configuration shows the following.

1. Using the same inlet conditions as for the catalytic reactor (inlet temperature 588°C),
   a. Extrapolation of the data in Table 20.1 is uncertain. This is because the reaction rate expression is extrapolated outside the region of experimental data (see item 1 of Table 20.1).
   b. The predicted reactor size is much larger than the current reactor.
2. At increased temperature (inlet temperature 750°C),
   a. The reaction rate is known from experimental data.
   b. The reaction volume is much smaller than the current reactor.
   c. The temperature, H_2/toluene ratio, and conversion are all in a range that gives high benzene selectivity.
3. Increasing reactor feed pressure: This is considered an expensive option due to gas compression costs and was not evaluated.

The noncatalytic replacement system is shown in Figure 20.4(b) as region B-B. It is a more complicated system. Higher feed temperature requires the addition of another fired heater before the reactor and a gas cooler following the reactor. (Heat integration might be another alternative.) Major capital costs are required for the reactor, fired heater, and cooler. Materials of construction for the reactor are more expensive at the higher temperature (at the exit temperature of 814°C, a reactor vessel would glow red). Fuel for the fired heater increases operating cost. The economically optimum reactor temperature that considered these items could be determined if needed.

Unless the cost of catalyst is extremely high, replacement of the catalytic reactor with a noncatalytic system does not appear to be an attractive option.

Discussion. It is observed that the low-temperature reactor (at the temperature of the catalytic reaction) was very large, a result obtained by assuming that it is valid to extrapolate the data in Table 20.1 outside the stated temperature range. Assuming the data to be accurate in the low temperature range, reasons a large reactor was obtained can be investigated.

Several important factors considered in selecting a chemical reactor are revealed in this case study. Normally the advantages of a catalytic over a noncatalytic reactor are reduction in operating temperature and pressure and improvement in product selectivity.

In this case study, the catalyst reduced the temperature and pressure requirements, as expected. However, high selectivity for the desired product could be obtained from both the catalytic and noncatalytic reactors.

Carbon is a common by-product formed during organic reactions carried out in high-temperature reactors. It is most significant in catalytic reactors where carbon forms on the catalyst surface and reduces activity. High hydrogen concentrations discourage carbon formation. This helps explain the high H_2-to-toluene ratio of 5:1 used in the catalytic system compared to the lower value of 2.0 required for the noncatalytic reactor.

Equilibrium constants for the reactions involving the components found in the laboratory studies on the benzene reaction are published [2]. The primary by-product at high temperatures is diphenyl, formed by the following reaction.

$$\underset{benzene}{2C_6H_6} = \underset{diphenyl}{C_{12}H_{10}} + \underset{hydrogen}{H_2}$$

At equilibrium, excess H_2 increases the toluene conversion, decreases diphenyl conversion, and improves benzene selectivity. This trend is assumed to hold for nonequilibrium conditions, which suggests that high excess H_2 be used.

Increasing the reaction temperature decreases the equilibrium constant of the benzene reaction, because the main reaction is exothermic and increases the equilibrium constant of the diphenyl reaction because the side reaction is endothermic. As a result, the selectivity decreases and the raw material costs increase.

In our system, any diphenyl present in the reactor effluent, Stream 9, would be largely removed in the separation section and recycled with the toluene to the reactor. The concentration of diphenyl in the reactor feed, Stream b, would be increased. When sufficient diphenyl is introduced in Stream b to provide the equilibrium concentration, little diphenyl is formed in the reactor. This retains a high benzene selectivity at higher reactor temperatures.

For a new design, it may be desired to consider scenarios to increase the reaction rate for the noncatalytic reaction. Figure 20.5 shows the reaction rate, temperature, and toluene mole fraction plotted against the reactor volume for the original feed temperature. This figure shows that low reaction rates exist at the

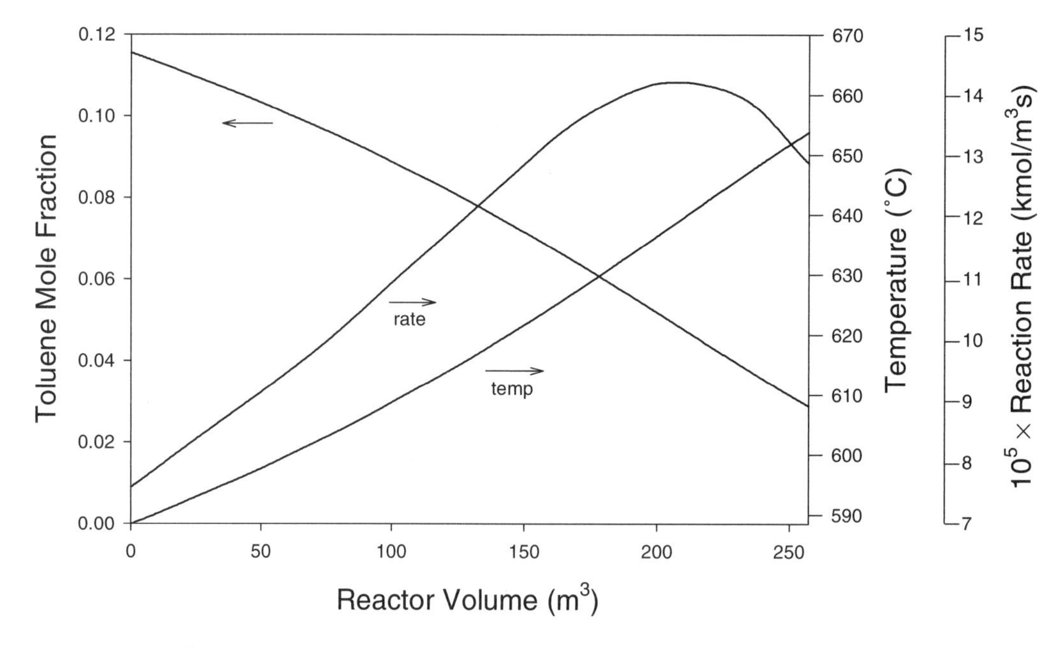

Figure 20.5 Behavior of Benzene PFR at Original Feed Temperature (585ºC)

feed end of the reactor, with rapidly falling reaction rates at the reactor exit. The reactor temperature increases over the volume of the reactor. Figure 20.5 identifies steps to take that would improve reactor performance. Increased performance can be obtained by

a. Elimination of the low reaction rate region at the beginning of the reactor. The low reaction rates are due to the low reaction rate constant, k, at low temperature. Increasing the feed temperature would increase reaction rate in this section.

b. Avoidance of operating in the region at the end of the reactor where the reaction rate falls rapidly. The falling reaction rate results from the lowering of toluene concentrations at high conversions. This region can be avoided by operating at lower reactor conversion (and higher recycle rates of unreacted feed).

This case study demonstrates that a dramatic decrease in the reactor volume required for a given conversion results from an increase in feed temperature.

One could consider replacing the current reactor with an adiabatic CSTR. The reaction rate throughout the reactor would be constant at reactor exit conditions. This reaction rate is shown in Figure 20.5. The average reaction rate for the PFR is less than that of the CSTR, and the volume for a CSTR is less than that of the PFR. The elimination of the low-temperature region in the PFR overcomes the effect of lower toluene concentration in the CSTR.

The conclusion of this case study is that replacing the reactor with a non-catalytic PFR is not an attractive option. This may not be the conclusion if a new plant were to be built. In a new plant, the excess hydrogen could be substantially reduced. This would increase the toluene concentration and reduce the amount of recycle. Lower reactor conversions could be considered. Higher temperatures, higher toluene concentrations, and lower conversions would all contribute to a smaller reactor.

20.5.2 Replacement of Cumene Catalyst

Problem. Your company produces cumene. At your plant, the shell-and-tube, packed-bed reactor shown in Appendix C, Figure C.8, is used. At other plants owned by your company, proprietary fluidized bed reactors providing low gas bypass are used. The remainder of the process is identical at all locations. Your supplier has informed you that it has developed a new cumene catalyst that will improve fluidized bed reactor performance. The new catalyst is supported on the same inert material as the old catalyst. Therefore, the fluidizing properties are identical to that of the old catalyst. You are assigned the task of verifying claims of improved performance and identifying changes in operating conditions that would maximize the improvement this new catalyst could provide.

Figure 20.6 provides the operating conditions for this base case. In addition to the general process information provided with the process flow diagram (see cumene flow diagram, Figure C.8, in Appendix C), the following specific background information is provided for your plant.

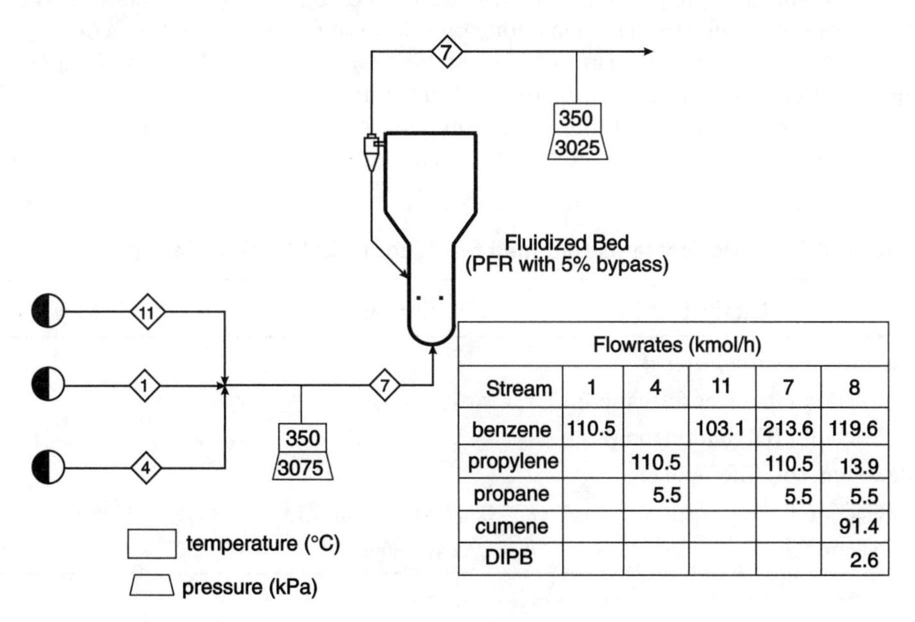

Flowrates (kmol/h)					
Stream	1	4	11	7	8
benzene	110.5		103.1	213.6	119.6
propylene		110.5		110.5	13.9
propane		5.5		5.5	5.5
cumene					91.4
DIPB					2.6

Figure 20.6 Base-Case Flow Diagram of Cumene Reactor System

In the first stage of a plant start-up, the plant is operated without the recycle stream, Stream 11. Benzene normally provided by this recycle stream is added with the feed stream. In the initial plant start-up, the plant operated for a substantial period without the recycle. During this time, the plant was operated at a range of temperatures, pressures, and excess benzene rates. These data were used to determine effective kinetics that are presented in Table 20.2. The active reactor volume is 7.88 m^3. It was found that the fluidized bed could be modeled accurately over a narrow range of superficial velocities as a PFR with 5% bypass.

The information from the open system (no recycle) was used to establish a base case. At each plant start-up following a catalyst change, the plant is run at base-case conditions. If the same reactor performance is not obtained, the catalyst and flow pattern in the reactor are examined. Problems of reactor performance are resolved before continuing plant start-up.

Solution. The dominant reactions involved within the cumene plant are

$$\underset{\textit{propylene}}{C_3H_6} + \underset{\textit{benzene}}{C_6H_6} \xrightarrow{k_1} \underset{\textit{cumene}}{C_9H_{12}} \qquad \text{Reaction 1}$$

$$\underset{\textit{propylene}}{C_3H_6} + \underset{\textit{cumene}}{C_9H_{12}} \xrightarrow{k_2} \underset{\textit{p-diisopropyl benzene}}{C_{12}H_{18}} \qquad \text{Reaction 2}$$

The kinetics for the catalyst are provided in Table 20.2.

In this analysis, to ensure a low bypass ratio (5%), the superficial gas velocity was held constant at the base-case conditions. For higher reactor temperatures, the molar flowrate to the reactor was decreased slightly so as to maintain the same superficial velocity. This was done by retaining the base-case flows of Stream 11 and Stream 4, and lowering the excess benzene Stream 1. The fluidized bed was assumed to behave as an isothermal PFR with a 5% bypass stream. This is shown in Figure 20.7. Some of the simulation results are shown in Figures 20.8 and 20.9. Figure 20.8 presents the production of cumene obtained for the current and new cata-

Table 20.2 Kinetic Constants Obtained for Hypothetical Cumene Catalysts

Catalyst	Current	New
Reaction Rate (mol/liter s)		
Reaction 1 (cumene formation)	$r_1 = k_1 c_p c_b$	$r_1 = k_1 c_p c_b$
Reaction 2 (DIPB formation)	$r_2 = k_2 c_p c_c$	$r_2 = k_2 c_p c_c$
Rate Constant (liter/mol s)		
Reaction 1	$k_1 = 3500\exp(-6680/T)$	$k_1 = 2.8 \times 10^7\exp(-12{,}530/T)$
Reaction 2	$k_2 = 290\exp(-6680/T)$	$k_2 = 2.32 \times 10^9\exp(-17{,}650/T)$

T has the units of K, liters refers to liters of reactor.

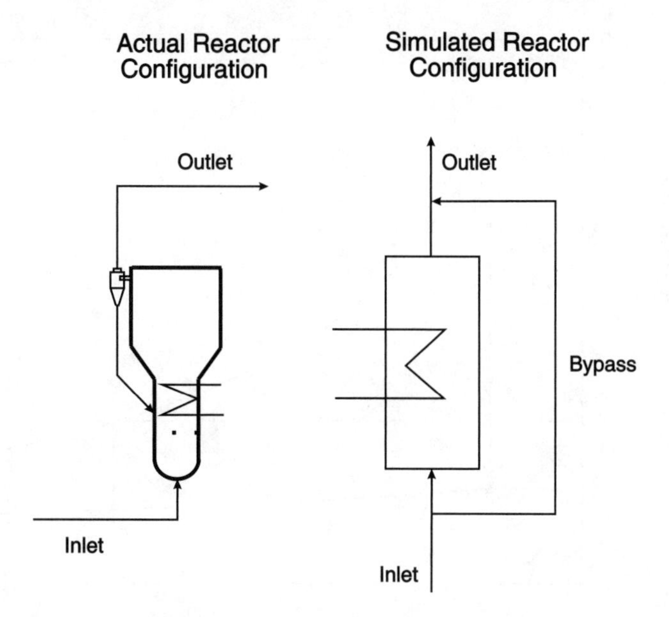

Figure 20.7 Fluidized Bed Reactor Showing Configuration (5% Bypass) Used for Simulation

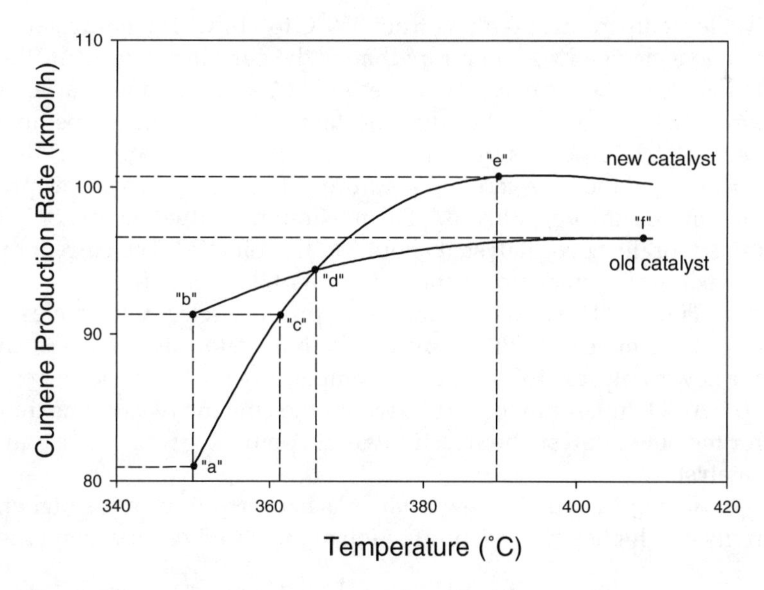

Figure 20.8 Cumene Production Rates at Constant Reactor Volume and Superficial Velocity

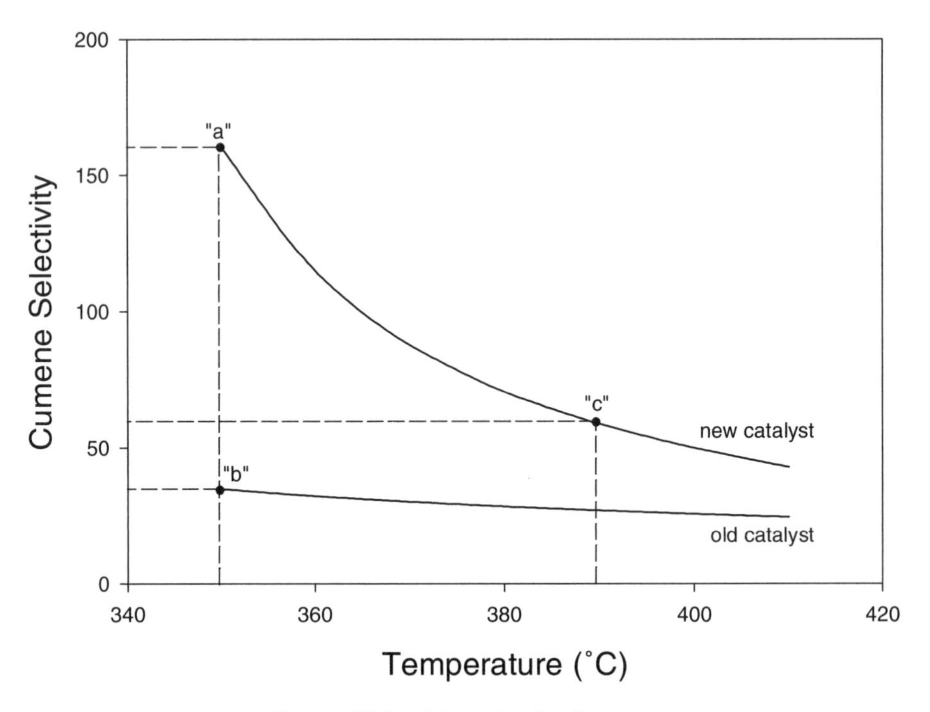

Figure 20.9 Selectivity for Cumene

lyst for temperatures ranging from 350°C to 410°C. The new catalyst is more sensitive to a change in temperature than is the current catalyst. At the base-case temperature, the new catalyst produces only 81 kmol/h, point "a," compared with 91 kmol/h, point "b," for the current catalyst. At a reactor temperature of 361°C, the new catalyst produces the same amount of cumene, point "c," as the old catalyst operating at the base-case temperature. At 367°C, both systems produce the same amount of cumene, point "d." The maximum production rate of 101 kmol/h is obtained for the new catalyst at about 390°C, point "e." The current catalyst provides a maximum cumene production of 96 kmol/h, point "f."

Figure 20.9 shows the trends in product selectivity (moles of cumene produced per mole of DIPB produced) for both catalysts. At 350°C, the selectivity of the new catalyst is 160, point "a," compared with 35 for the current catalyst, point "b." At a temperature of 390°C, where the cumene reaches maximum production for the new catalyst, the selectivity is 59, point "c," well above that for the current catalyst.

In conclusion, the new catalyst is expected to provide higher product selectivity and higher cumene production at increased reactor temperature.

Discussion. It was found in the simulation of the dominant reactions in the production of cumene using the new catalyst that high temperatures increase the formation of the undesirable product, DIPB. This is expected based on the relative activation energies of the two reactions. The higher activation energy for the DIPB reaction leads to a more rapid increase in the rate of reaction of DIPB with temperature compared with cumene. Therefore, it is important to avoid hot spots in the reactor to avoid DIPB production. From a reaction point of view, isothermal reactors represent a desirable goal. This goal is obtained with a CSTR but is not easy to obtain in a PFR with a high heat of reaction. The fluidized bed system can approach isothermal behavior because of rapid mixing of the solid particles making up the bed. For highly exothermic reactions, the fluidized bed is inherently safe, hot spots do not develop, and heat can be removed from the bed.

However, fluid bypassing takes place in a fluidized bed. This adversely affects performance. This is because the bypassed fluid fraction does not come into contact with the catalyst and so a portion of the feed does not react. In the case study given above, the bypass fraction had a low value for a fluidized bed, 5%. As a result, the highest conversion that can be achieved is 95%.

If the products from the reactor are easily separated, the unreacted reactants can be recovered and recycled to achieve high overall conversion. From a kinetic point of view, the formation of DIPB in the product could be effectively eliminated by operating at low reactor conversions. This can be seen from the reaction rate expression, which shows that the formation of DIPB is directly proportional to the concentration of cumene. If cumene concentration is kept low, the amount of DIPB formed is small.

Separation of the reactor products containing propane or propylene presents a problem. Propylene and propane have similar volatilities and are difficult to separate. If propylene and propane are not separated, there are three options to be evaluated.

1. Propylene-propane mix is discarded as waste. Fuel value may be recovered.
2. Total recycle of propylene-propane mix. This is not possible. Because the propane does not react, it must be purged from system.
3. Partial recycle of propylene-propane mix. The propane in the feed would increase. This decreases the concentrations of the reactants and reduces the production of cumene. The superficial velocity in the reactor would increase. This reduces residence time and the fraction bypassed in the fluidized bed.

In the case study, the constant superficial velocity was maintained at higher temperatures by decreasing the excess benzene. The total concentration of gas in the reactor was reduced. This tends to decrease the reaction rate. In addition, the reaction equations show that the lower benzene concentration reduces the cumene reaction rate and has no impact on the DIPB reaction. As a result of maintaining a constant required superficial velocity by lowering the excess benzene, the selectivity is reduced.

An alternative approach to maintain the superficial velocity at increased temperature and retain the excess benzene flow is to increase the system pressure. This should provide higher reaction rates at a given temperature, resulting in higher conversions, and would not effect the selectivity.

The conclusions given regarding the catalyst replacement assumed that the reaction took place in a region where chemical reaction controlled. When chemical reaction controls, the effect of temperature is large. Figure 20.8 indicates there was little effect of temperature on cumene production using the current catalyst, but temperature had major impact on cumene production for the new catalyst. The low sensitivity in the base case suggests that the reaction is not chemical reaction controlled. Often when dealing with catalytic reactions, one finds that due to high catalyst activity, the reaction may be diffusion controlled. The activation energy for diffusion is significantly lower than that for chemical reactions. Therefore, the effective activation energy for a diffusion-controlled reaction is also significantly lower. This means that intrinsic reaction kinetics supplied by manufacturers may not be directly applicable in process simulation. The reader is referred to standard reaction engineering texts.

For catalytic reactions, always consider mass transfer effects. If present, they reduce the temperature sensitivity of the reaction.

20.5.3 Increasing Acetone Production

Problem. You are responsible for the operation of the acetone production unit in your plant. The acetone produced is used internally elsewhere in your plant. A need to increase internal production by as much as 50% is anticipated. If this full increase cannot be met from your acetone unit you will have to purchase additional acetone. You have been requested to evaluate how much increase in capacity can be obtained using our current chemical reactor.

Acetone is produced by the endothermic reaction

$$\underset{\textit{isopropyl alcohol}}{(CH_3)_2CHOH} \rightarrow \underset{\textit{acetone}}{(CH_3)_2CO} + \underset{\textit{hydrogen}}{H_2}$$

There are no significant side reactions. The reaction is endothermic and takes place in the gas phase on a solid catalyst. The reaction kinetics are given by

$$-r_{IPA} = kc_{IPA} \quad \text{kmol/m}^3 \text{ reactor s}$$

where

$$k = [3.156 \times 10^5 \text{ m}^3 \text{ gas/(m}^3 \text{ reactor s)}]\exp[-8702/T]$$

where T is in Kelvin. The process flowsheet for our plant is given in Appendix B, Figure B.3. Figure 20.10 shows the reactor portion of this process along with reactor specifications obtained from Table B.9.1. The process uses a PFR with the heat transfer medium, HTM, on the shell side. Because there is concern about operation of the downstream separation section, the IPA reaction conversion is held constant at 90% to ensure a constant feed composition to the separation section even though the flowrate of this stream may change.

Solution. The conversion of an endothermic reaction is most often limited by the amount of heat that can be provided to the reaction. Current operating conditions define the base case. Figure 20.11(a) shows the temperature profile in the reactor for the reacting stream and HTM for the base case. There is a small temperature driving force for heat transfer over most of the reactor. Only in a small region near the reactor entrance is there a large temperature difference

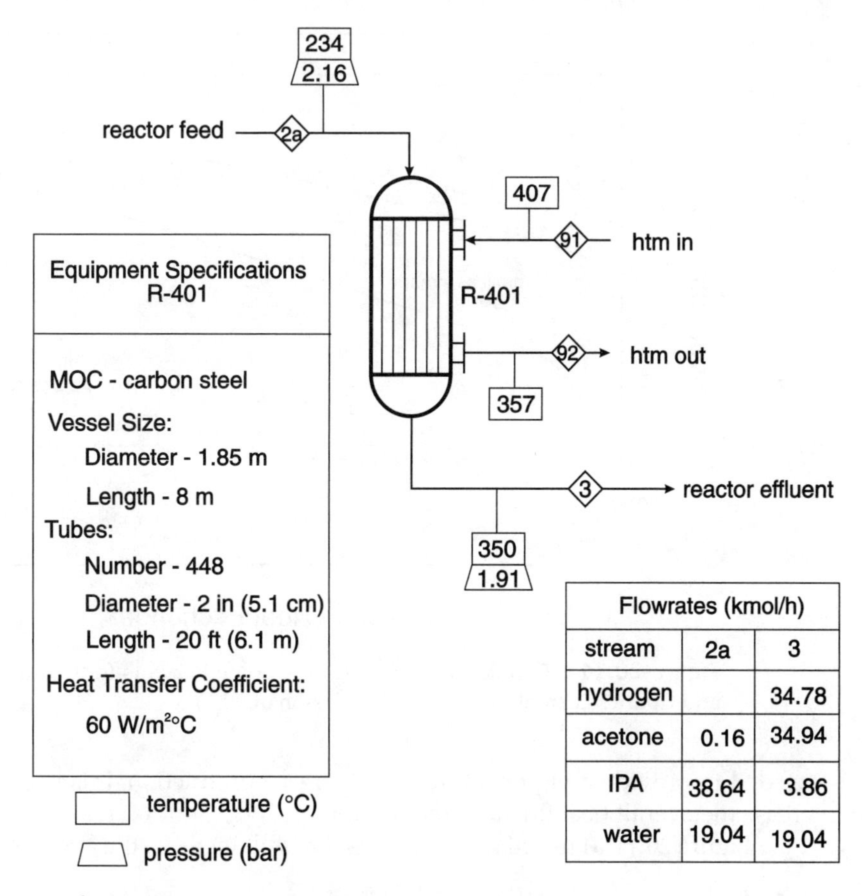

Flowrates (kmol/h)		
stream	2a	3
hydrogen		34.78
acetone	0.16	34.94
IPA	38.64	3.86
water	19.04	19.04

Figure 20.10 Reaction Section of Acetone Process (From Figure B.9.1 and Table B.9.1)

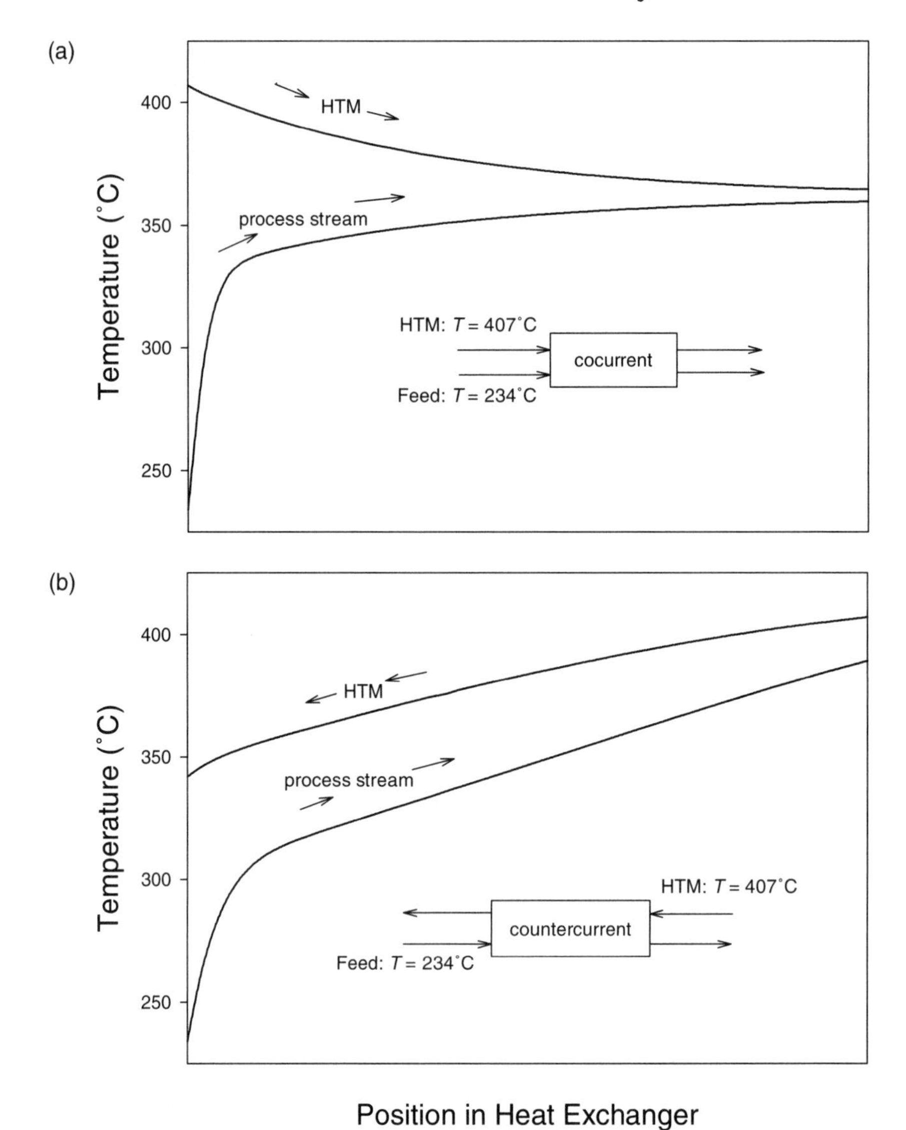

Figure 20.11 Temperature Profiles in Acetone Reactors for Cocurrent and Countercurrent Flow (IPA Conversion of 90%)

needed to provide a high heat flux. To increase production, it is necessary to increase the overall heat flux into the reactor.

Figure 20.11(a) reveals four options that will increase the average heat flux.

1. Increase the temperature of the HTM.
2. Decrease the temperature of the reaction stream.

3. Operate with the HTM introduced countercurrent to the reacting stream.
4. Heat the feed stream to a higher temperature.

An evaluation of these options follows.

Option 1, Increase HTM Temperature: Figure 20.12 presents results obtained for increased acetone production resulting from a change in HTM inlet temperature (flowrate of HTM held constant). Increasing the HTM by 50°C increased acetone production by 48%. This would require a 48% increase in heat duty and a 50°C increase in HTM temperature from the fired heater, H-401.

Option 2, Decrease Reactor Temperature: Lowering reactor temperatures decreases the reaction rate (lower reaction rate constant, k), reducing acetone production. This option has no merit and is rejected.

Option 3, Countercurrent Flow: Figure 20.11(b) shows the temperature profile obtained by introducing the HTM countercurrent to the reaction stream. Compared with the base case (cocurrent flow), it provides a larger, more constant temperature driving force (and heat flux). This change provides a 13% increase in acetone production. The heat duty of the fired heater, H-401, must be increased by 13% to provide this increase in duty.

Option 4, Heat Feed Stream to Higher Temperature: This provides energy in the form of sensible heat that adds to the energy transferred from the

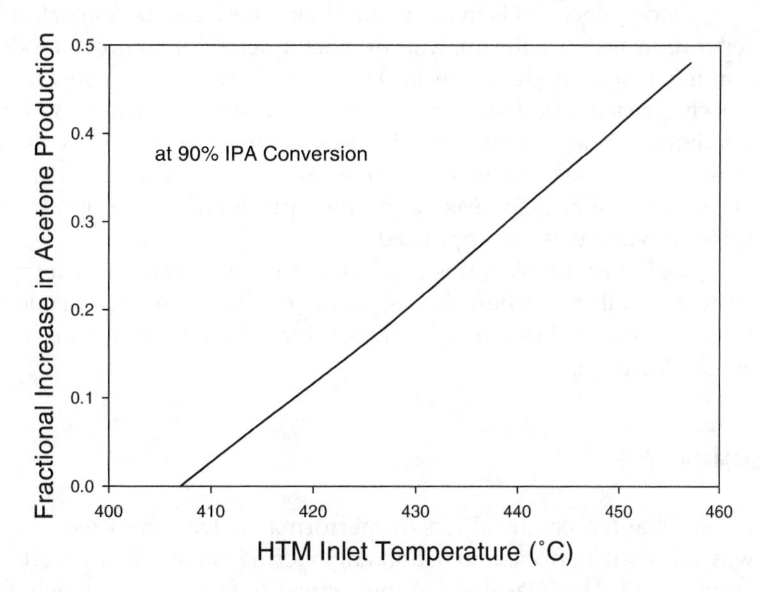

Figure 20.12 Increase in Acetone Production Resulting from Increase in HTM Inlet Temperature (HTM Flow Constant)

HTM. For an increase in feed temperature of 100°C, the acetone production increases by 14%. This would require an additional fired heater for the reactor feed stream.

This investigation showed there are a number of options that can effect a significant increase in acetone production. A 50% increase in capacity can be obtained from the reactor. All options require an increase in fired heater capacity.

Discussion. It has been shown that there are a number of options that provide increased acetone production. Except for Option 2, the options provided can be combined. The heat transfer configuration, Option 3, can be combined with an increase in HTM temperature, Option 1. In addition, the reactor feed temperature, Option 4, can be increased. In most situations, there is more than one option to consider to make a desired change in performance.

All options to increase acetone production require an increase in capacity of the fired heater (or addition of a new fired heater). Therefore, the fired heater is likely to represent a bottleneck to increased production. Fired heaters are expensive, and it is not likely that the current heater was oversized enough to provide a 50% increase in heat duty. Until this bottleneck is removed, the additional acetone production required cannot be obtained.

Once the fired heater bottleneck is resolved, the separation system must be able to process the large increase in throughput. It is unlikely that the separation system can process the 50% increase. Therefore, if the fired heater bottleneck is removed, the separation system becomes the bottleneck.

Once a decision is made to eliminate these two bottlenecks (fired heater and separation system), the analysis of reactor behavior could be modified to consider a number of secondary effects. The case study did not consider the effect of increasing reactor flowrates on the heat transfer coefficient (gas-phase resistance dominates). The impact of the increased pressure drop over the reactor was not examined. These options were not necessary to confirm that the reactor is capable of providing a 50% increase in acetone production. The option of increasing the HTM flowrate was not appraised.

Endothermic reactions are inherently safe systems in terms of undergoing an uncontrolled reaction that is characteristic of an exothermic reactions. If the heat source is withdrawn, the temperature of the reactor drops and the rate of reaction decreases.

20.6 SUMMARY

In this chapter, chemical reactor performance was analyzed differently from the way it is usually done in reaction engineering classes. Key qualitative trends were emphasized. This was done in the context of Figure 20.1, which illustrated the interrelationship between product production, reaction kinetics, reactor behavior, and heat transfer limitations. The case studies presented reinforce the interrelationship between these factors.

REFERENCES

1. Zimmerman, C. C., and R. York, "Thermal Demethylation of Toluene," *Ind. Eng. Chem. Proc. Des. Dev.* 3 (1964): 254.
2. Rase, H. F., *Chemical Reactor Design for Process Plants,* Vol. 2, Case Studies and Design Data (New York: Wiley-Interscience, 1977), 38.

SHORT ANSWER QUESTIONS

1. Explain why very high conversions (>99%) are often difficult to achieve in practice, even for plug flow reactors and nonequilibrium constrained reactions.

2. Comment on the following statement: *For an exothermic reaction, if the temperature is increased at constant pressure with a fixed reactor size, the conversion decreases.*

3. It is known that for a certain third-order, gas-phase reaction, the rate of reaction doubles when the temperature goes from 250°C to 270°C. The form of the rate equation is

$$-r_A = k_o e^{-\left(\frac{E}{RT}\right)} c_A c_B^2$$

   If the rate of reaction is 10 moles/m^3s at the base conditions, which are 20 atm pressure and 250°C, and an equal molar flow of A and B are in the feed (no inerts), answer the following questions.
   a. Compared to the base case, by how much does the rate of reaction (at inlet conditions) change if the temperature is increased to 260°C?
   b. Compared to the base case, by how much does the rate of reaction (at inlet conditions) change if the pressure is increased by 15%?
4. For the reactor system shown in Figure P20.4, answer the following.
   a. Is the reaction exothermic or endothermic?
   b. Give two reasons why this reaction is being run at a high temperature.

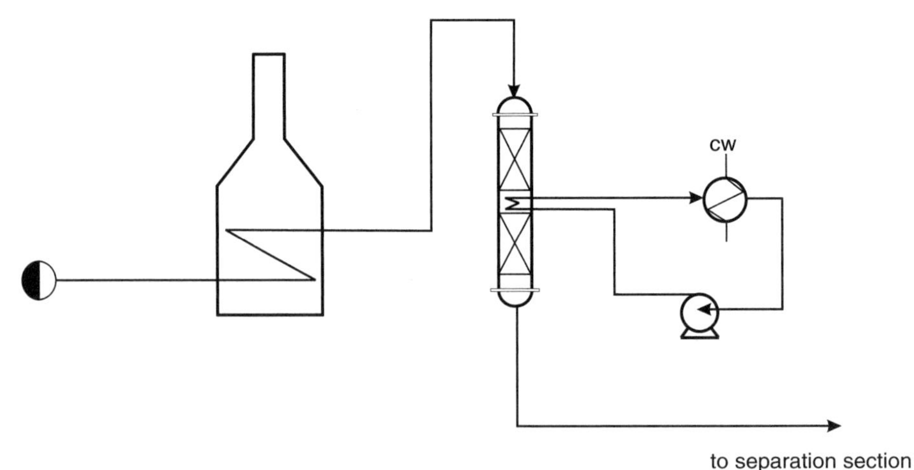

Figure P20.4

PROBLEMS

Note: Information for problems referenced as Appendix C is provided on the accompanying CD.

5. For the hypothetical, endothermic, gas-phase reaction

$$A + B = C$$

sketch a plot similar to Figure E20.1. Discuss the relationship between equilibrium conversion, temperature, and pressure.

6. For the situation in Example 20.3, if all of the reactions take place, what conditions maximize the selectivity for P?

7. Examine the reaction network and the reaction kinetics for drying oil production in Appendix C, Table C.10. What conditions will maximize the selectivity for drying oil? Sketch concentration profiles for the case when this reaction is run in a packed bed modeled as a PFR.

8. Examine the reaction network and the reaction kinetics for ethylene oxide production in Appendix C, Table C.17. What conditions will maximize the selectivity for ethylene oxide? Sketch concentration profiles for the case when this reaction is run in a packed bed modeled as a PFR.

9. Sketch the T-Q diagram for an endothermic reaction in a PFR for
   a. Countercurrent flow of HTM (heat transfer medium)
   b. Cocurrent flow of HTM

   What are the consequences on reactor size of choosing each configuration? Explain your answer by examining the trends (see Chapter 17).

10. In the production of cumene from propylene, the following elementary, vapor-phase, irreversible reaction takes place:

$$\underset{\text{propylene}}{C_3H_6} + \underset{\text{benzene}}{C_6H_6} \xrightarrow{k_1} \underset{\text{cumene}}{C_9H_{12}}$$

The reaction rate is given by

$$r_1 = k_1 c_p c_b \text{ mol/(g cat) sec and } k_1 = 3.5\times10^4 \exp\left(\frac{-12{,}530}{T(K)}\right)$$

The feed to the reactor consists of an equal ratio of benzene and propylene. The reaction takes place in a fluidized bed reactor operating at 300°C and 3 MPa pressure. For this problem, the fluidized bed may be assumed to be a constant-temperature CSTR reactor.

At design conditions, you may assume that the side reaction does not take place to any great extent and the conversion 68%.

It is desired to scale up production by 25%. All flows to the reactor will increase by 25% at the same feed concentration. Determine the following.

a. What is the single-pass conversion if the process conditions and amount of catalyst remain unchanged?
b. What percent change in catalyst would be required to achieve the scale-up assuming that the pressure, temperature, and conversion were held constant?
c. Estimate how much the temperature would have to be changed (without changes in catalyst amount, operating pressure, or conversion) to achieve the desired scale-up.
d. By how much would the pressure have to be changed (without changes in catalyst amount, operating temperature, or conversion) to achieve the desired scale-up?

11. Consider a liquid-phase reaction occurring in a constant-volume, isothermal, batch reactor.
a. For a first-order decomposition, what is the ratio of the time to reach 75% conversion to the time to reach 50% conversion?
b. For a first-order decomposition, what is the ratio of the time to reach 90% conversion to the time to reach 50% conversion?
c. Repeat Parts (a) and (b) for a second-order reaction between reactants initially in equimolar quantities.
d. Explain the results of Parts (a)–(c).

For Problems 12–26, it is recommended that a process simulator be used.

Problems 12–16 investigate the performance of the reactor section for a non-catalytic process for the hydrodealkylation of toluene to produce benzene. Reactor R-101 in Figure 20.4(b) operated at the flow conditions shown represents a base case (inlet pressure is 25 bar). Kinetics are given in Table 20.1. Ignore reactor pressure drop. Reactor volume for base case is 4.76 m^3.

12. Investigate the effect of increasing reactor volume on toluene conversion. Plot the results.
13. Investigate the effect of a variation of hydrogen feed rate (all other parameters held constant at base-case values).
    a. Plot toluene conversion versus percent excess oxygen.
    b. Determine the maximum conversion possible.
14. Investigate a two-reactor system to replace the single adiabatic PFR in the base case. The two-reactor system consists of an adiabatic CSTR followed by an adiabatic PFR. For the base-case toluene conversion (0.75),
    a. Determine the size of the CSTR and PFR that provides the minimum total reactor volume.
    b. Determine the fraction of the total conversion obtained in the CSTR from Part (a).
15. Investigate the effect of increasing the feed rate on reactor performance.
    a. Prepare a performance curve of the toluene conversion versus feed rate.
    b. Prepare a performance curve of benzene generated versus feed rate.
16. Investigate the effect of feed temperature on reactor performance. Prepare a performance curve for toluene conversion versus feed temperature.

Problems 17–20 investigate the performance of the cumene reactor (fluidized bed), described in Section 20.5.2, that results from replacing the catalyst. The replacement catalyst kinetics are given in Table 20.2. The base case selected is the currently operating reactor (with an active volume of 7.49 m^3). Process output for the base case is given in the flow table in Figure 20.6.

17. a. Derive an equation for the ratio

    $$\eta = \text{rate of cumene formation}/\text{rate of DIPB formation}$$

    Rewrite the equation in the form $\eta = A/B - 1$.
    b. Present the equation for η given above in terms of the following variables: T, P, k_o, E_{act}, and mole fractions.
    c. Using the equation from Part (b), determine the trend resulting from a change in the following variables. After each item, provide ↑ (increases), ↓ (decreases), ¢ (constant) or ? (cannot be determined).

    Increasing system temperature
    Increasing system pressure
    Increasing benzene mole fraction
    Increasing propylene mole fraction
    Increasing propane mole fraction
    Increasing cumene mole fraction
    Increasing DIPB mole fraction
18. Replace the current catalyst with the new catalyst (maintain same feed rate and conditions) and
    a. Provide a flow table (similar to that in Figure 20.6).

b. Compare cumene production with the new to the old catalyst.
c. Compare DIPB production with the new to the old catalyst.

19. Evaluate the effectiveness of the new catalyst at 350°C that results from increasing the reactor volume.
a. Plot cumene production versus reactor volume.
b. Plot selectivity (defined as cumene generated/DIPB generated) versus volume.

20. In Figures 20.8 and 20.9, the volumetric flowrate to the reactor was held constant at increased temperatures by reducing the hydrogen in the feed. This altered the feed concentration. Investigate the alternative of reducing the total feed rate (at the same composition) to maintain constant volumetric flowrate.
a. Plot the cumene production rate versus temperature, and compare to Figure 20.8.
b. Plot the selectivity (cumene generated/DIPB generated) versus temperature, and compare to Figure 20.9.

Problems 21–26 investigate the performance of the acetone reactor system. The reaction kinetics are presented in Section 20.5.3. The reactor and internal heat exchanger are fixed. For the base case, the active volume of the reactor is 5.1 m^3 at the operating conditions given in Figure 20.10. Unless otherwise stated, the heat transfer coefficient and reactor pressure drop are to be assumed constant. The feed temperature, pressure, and component concentrations remain constant. For these calculations, assume the utility fluid to be n-heptadecane (at pressures more than 30 bar). At the base case, the utility flow is 58 kmol/h.

21. Reconfigure the heating system to operate in countercurrent fashion. Retain the same utility stream input and IPA conversion (90%).
a. Determine the change in process feed rate.
b. Determine the percent change in process feed rate.

22. Increase the utility feed rate by 50%, and maintain the base-case process feed stream.
a. Determine the conversion obtained.
b. Determine the percent change in the acetone produced.

23. Increase the utility feed rate by 50%. Retain the IPA conversion of 0.9.
a. Determine the process feed rate.
b. Determine the percent change in acetone produced.

24. Increase the utility feed temperature by 50°C. Retain the IPA conversion at 0.9.
a. Determine the process feed rate.
b. Determine the percent change in acetone production.

25. Prepare a performance curve for the change in utility flowrate necessary to maintain an IPA conversion of 0.9 at increased process feed rates.

26. Assume the individual film coefficients change according to the relationship

$$h_i \propto \text{velocity}^{0.8}$$

   a. Estimate the effect of process flow changes on the overall heat transfer coefficient.
   b. Determine the process flowrate that provides an IPA conversion of 0.9 resulting from an increase in utility temperature of 50°C.
   c. Does the assumption of constant overall heat transfer coefficient result in a higher or lower estimate of the process feed rate? If you worked Problem 25, you have this value for comparison.

CHAPTER

21

Regulating Process Conditions

Over the life of a plant, it is important for operating conditions to be regulated in order to obtain stable operations and to produce quality products efficiently and economically. In this chapter, we consider regulation; a subset of this is process control. Regulation establishes the strategy by which the process can be controlled. Regulation also involves the dynamic response (transient behavior) of the process to changes in operating variables. The latter will not be considered here as it is covered in the typical undergraduate process control course.

The regulation of process operations involves an understanding of two facts:

In most situations, processes are regulated, either directly or indirectly, by the manipulation of the flowrates of utility and process streams.

Changes in flowrates are achieved by opening or closing valves.

In order to decouple the effects of changes in process units, adjustments are most often made to the flow of the utility streams. Utility streams are generally supplied via large pipes called **headers**. Changes in flowrates from these headers have little effect on the other utility flows, and hence, they can be changed independently of each other. The one notable exception to this concept is when the flowrate of a process stream is to be controlled. In this case, the process stream clearly must be controlled or regulated directly by a valve placed in the process line.

In this chapter, we consider that utility streams are available in unlimited amounts at discrete temperatures and pressures. This is consistent with normal plant operations, and a typical (although not exhaustive) list of utilities is given in Table 8.4.

This chapter looks at several aspects of regulation that are important for the successful control of processes. The following topics are covered:

The characteristics of regulating valves
The regulation of flowrates and pressures
The measurement of process variables
Some common control strategies
Exchange of heat and work between process streams and utility streams
Case studies of a reactor and a distillation column

Before we look at these topics, however, we start out with a simple regulation problem from the overhead section of a distillation column.

21.1 A SIMPLE REGULATION PROBLEM

Consider the flow of liquid from an overhead condenser on a distillation column to a reflux drum, as illustrated in Figure 21.1(a). This is a section of the process flow diagram for the DME process shown in Appendix B. The liquid condensate flows from the heat exchanger, E-205, to the reflux drum, V-201. From the drum, the liquid flows to a pump, P-202, from which a portion (Stream 16) is returned to the distillation column, T-201, and the remainder (Stream 10) is sent to product storage. Let us assume that the amount returned to the column as reflux is set by a control valve, shown in the diagram. The amount of reflux is fixed in order to maintain the correct internal flows in the column and hence the product purity. Consider what happens if there is an upset and there is an increase in the amount of liquid being sent to the reflux drum. If no additional control strategy is employed, then the level of liquid in the reflux drum will start to increase, and at some point the drum may flood, causing liquid to back up into the overhead condenser and causing the condenser to malfunction. Clearly, this is an undesirable situation. The question is, How do we control the situation? The answer is illustrated in Figure 21.1(b). A control valve is placed in the product line, and a level indicator and controller are placed on the reflux drum. As the level in the drum starts to increase, a signal is sent from the controller to open the control valve. This allows more flow through the valve and causes the level in the drum to drop. Although this may seem like a simple example, it illustrates an important principle of process control, namely, that a major objective of any control scheme is to maintain a steady-state material balance.

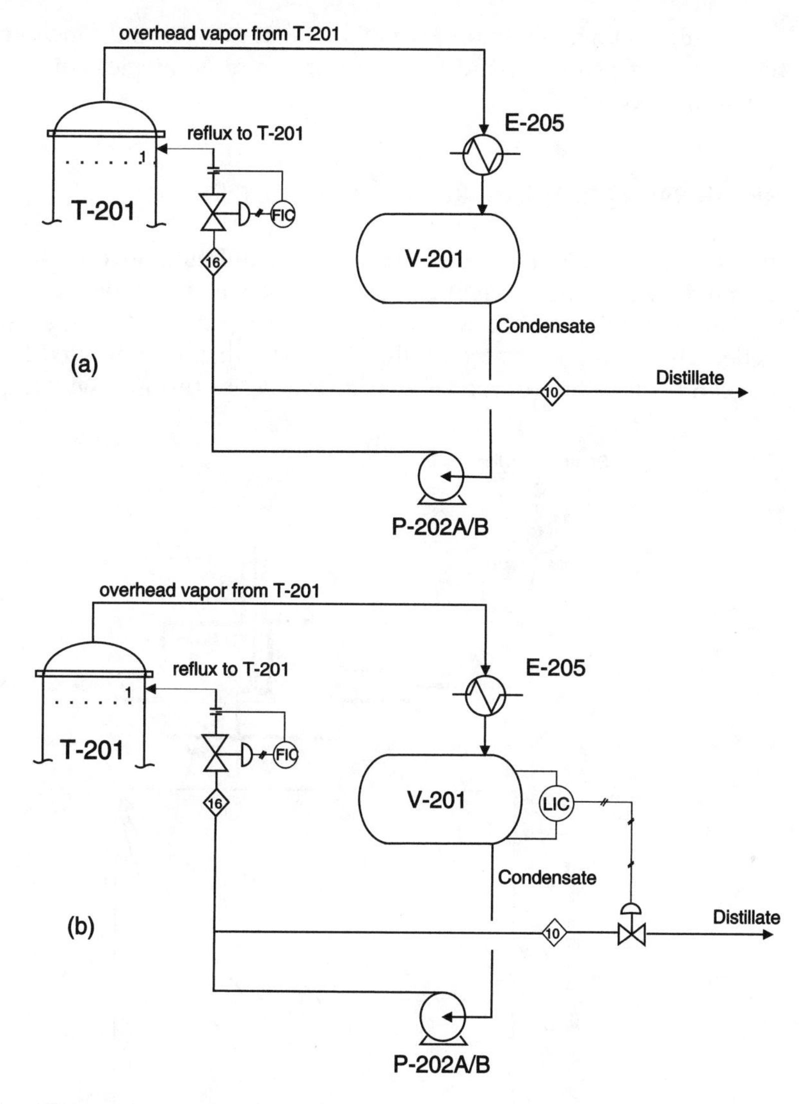

Figure 21.1 Basic Material Balance Control for DME Overhead Product

When controlling a process, it is important to ensure that the steady-state material balance is maintained, that is, to avoid the accumulation (positive or negative) of material in the process.

In the following sections, we will look at the role and functions of valves in the control of processes and the types of control strategies commonly found in chemical processes.

21.2 THE CHARACTERISTICS OF REGULATING VALVES

Figure 21.2(a) is a crude representation of a regulating or control valve. Fluid enters on the left of the valve. It flows under the valve seat, where it changes direction and flows upward between the valve seat and the disk. It again changes direction and leaves the valve on the right. The disk is connected to a valve stem that can be adjusted (in the vertical direction) by turning the valve handle. The

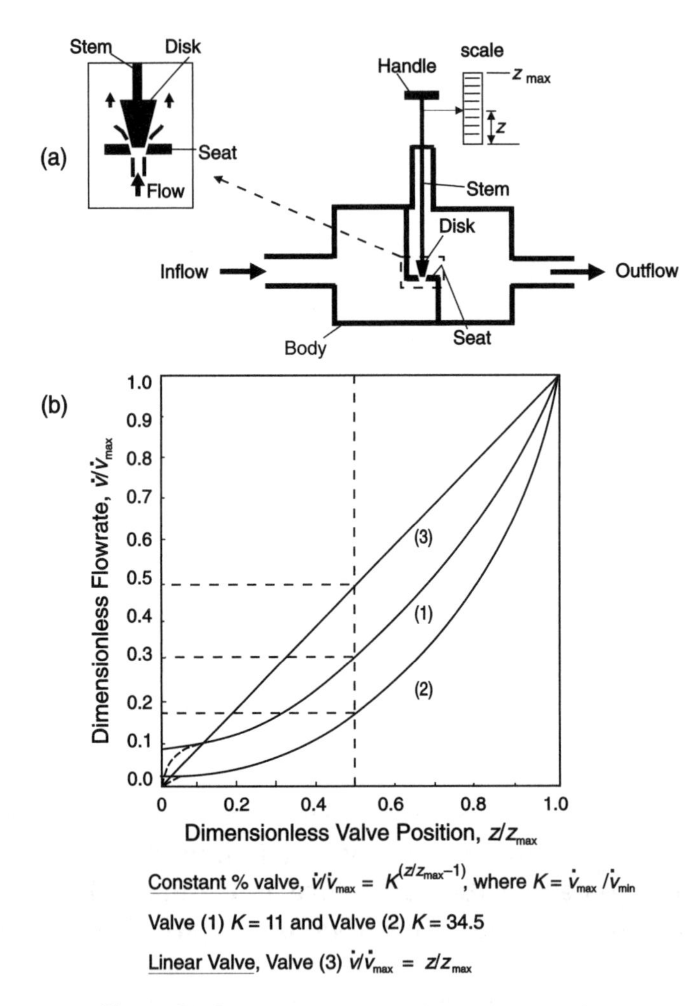

Figure 21.2 Characteristics of Regulating Valves

position of the disk is given on the linear scale. The enlarged section shows the critical region involving the disk and the valve seat. The liquid flows through the annular space between the disk and the seat. As the disk is lowered, the area of this annulus decreases. The relationship between flowrate and valve position depends upon the shape of both the valve disk and the valve seat. It should be noted that the direction of flow through the valve could also be from the right to the left. In this case, the flow of the fluid pushes down on the disk and pushes it toward the seat. This configuration is preferred if the fluid is at a relatively high pressure or there is a large pressure drop across the valve.

Figure 21.2(b) shows the pressure drop versus flowrate characteristics of regulating valves. To make a specific change in pressure drop across the valve, the valve must be opened or closed, as appropriate. In order to evaluate the desired valve position, it is necessary to know the performance diagram or characteristic for the valve.

Figure 21.3 is a mechanical drawing of the cross section of a globe valve. Globe valves are often used for regulation purposes. There are two general valve types used for regulation.

Linear Valves. In a **linear valve**, the flowrate is proportional to the valve position. This is given by the relationship $\dot{v} = kz$, where $\dot{v}$ is the volumetric flow and z is the vertical position of the disk or valve stem.

Constant or Equal Percentage Valves. For an **equal percentage valve**, an equal change in the valve position causes an equal percentage change in flowrate. The valve equation for the constant percentage valve is given by $\dot{v}/\dot{v}_{max} = K^{(z/z_{max}-1)}$. We note that a true constant percentage valve does not completely stop the flow. However, in reality, the valve does stop the flow when fully closed, and

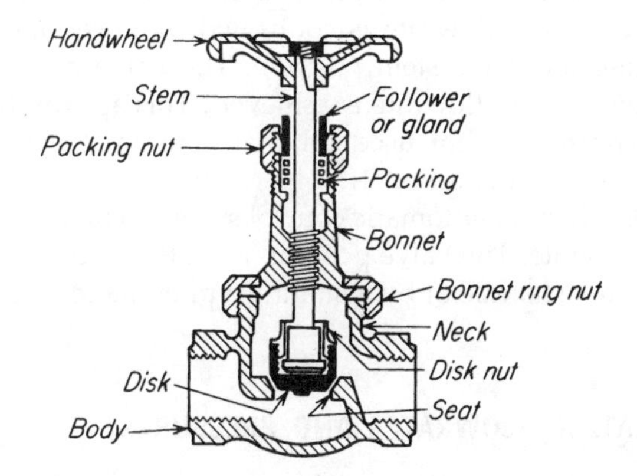

Figure 21.3 Cross Section of a Globe Valve (From W. L. McCabe, J. C. Smith, and P. Harriott, *Unit Operations of Chemical Engineering,* 5th ed. (New York: McGraw-Hill, 1993), copyright © 1993 by McGraw-Hill Companies, reproduced with permission of the McGraw-Hill Companies)

the constant percentage characteristic does not apply to very low flowrates, as indicated by the dashed lines at the lower end of the curves in Figure 21.2(b).

Both of these valve characteristics are described by the equations shown on Figure 21.2(b). Note that these equations strictly apply only for a constant pressure drop across the regulating valve. Example 21.1 illustrates the effect of using different valves.

Example 21.1

A valve on a coolant stream to a heat exchanger is operating with the valve stem position at 50% of full scale. You are asked to find what the maximum flowrate would be if the valve were opened all of the way (at the same pressure drop). What is an appropriate response given the information in Figure 21.2(b)?

The flow would increase by 100% or more. The exact increase depends upon the design of the valve.

From Figure 21.2(b) we see that a linear valve (Curve 3 on Figure 21.2(b)) has the smallest increase, from $\dot{v}/\dot{v}_{max} = 0.5$ to $\dot{v}/\dot{v}_{max} = 1$ (100% increase). For constant percent valve 1 (Curve 1), the increase is from $\dot{v}/\dot{v}_{max} = 0.3$ to $\dot{v}/\dot{v}_{max} = 1.0$ (233%), and for constant percent valve 2 (Curve 2), the increase is from $\dot{v}/\dot{v}_{max} = 0.17$ to $\dot{v}/\dot{v}_{max} = 1.0$ (488%). The flow increases in the range of 100% to 488%. The problem of predicting the flow from the valve position is even more complex because there usually exists some form of hysteresis. Thus, the pressure drop over the valve changes not only with flowrate of fluid through the valve but also with the direction of change, that is, whether the flow was last increased or decreased.

Most valves installed are constant percentage valves, and there is no generally accepted standard design. Such valves offer fast response at high flowrates and fine control at low flowrates. The prediction of the disk position needed to give the required flowrate would be difficult using the performance diagram in Figure 21.2. In practice, the flowrate is controlled by observing a measured flowrate while changing the valve stem position. The valve position continues to be adjusted until the desired flowrate is achieved. This approach forms the basis of the **feedback control system** used for many automatic flow control schemes and discussed further in Section 21.5.

In practice, automatic control systems change valve positions to obtain a desired flowrate. The valve position is modified by installing a servomotor in place of the valve handle or by installing a pneumatic diaphragm on the valve stem.

21.3 REGULATING FLOWRATES AND PRESSURES

The rate equation describing the flowrate of a stream is given by

$$\text{Flowrate} = \frac{\text{Driving force for flow}}{\text{Resistance to flow}} \tag{21.1}$$

The driving force for flow is proportional to ΔP (pressure head), and the resistance to flow is proportional to friction. The resistance to flow can be varied by opening or closing valves placed in the flow path.

Figure 21.4 shows a pipe installed between a high-pressure header, containing a liquid, and a low-pressure header. The pressure difference, ΔP, between

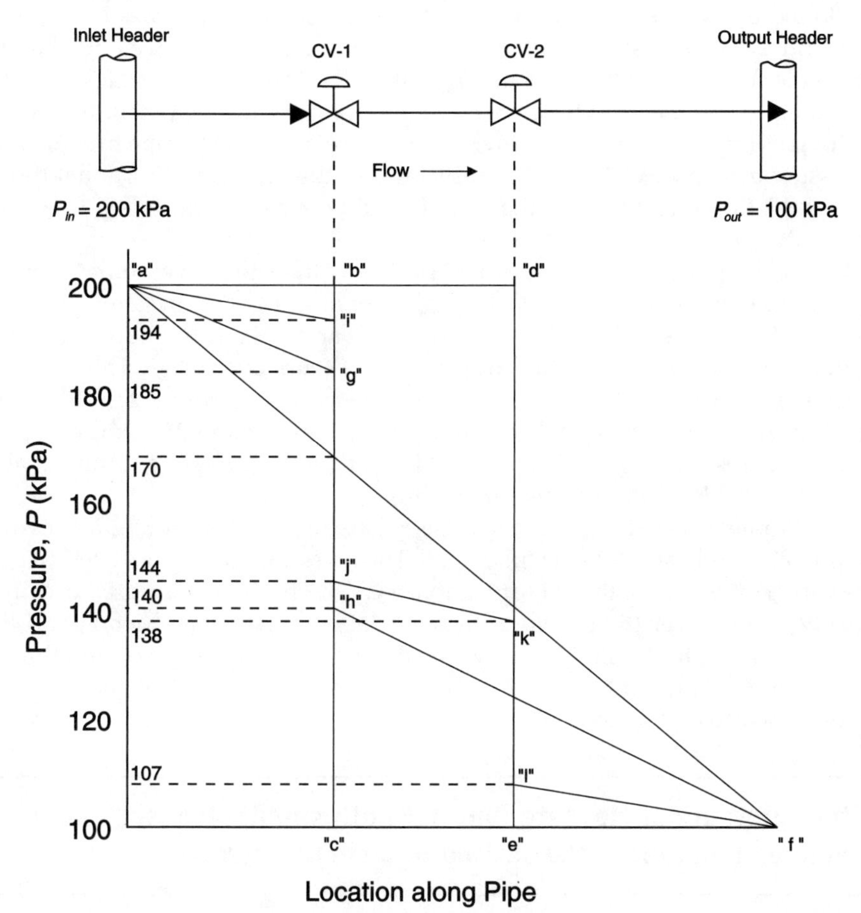

Profile	CV-1	CV-2	Flow (kmol/h)	ΔP (friction) (kPa)	P (before CV-1) (kPa)	ΔP over CV-1 (kPa)	ΔP over CV-2 (kPa)
1: a–f	open	open	100	100	170	0	0
2: a–b–c–f	closed	open	0	0	200	100	0
3: a–d–e–f	open	closed	0	0	200	0	100
4: a–g–h–f	partially open	open	74	55	185	45	0
5: a–i–j–k–l–f	partially open	partially open	44	19	194	50	31

Figure 21.4 Pressure Profiles in a Pipe Containing Two Valves

these two headers is the driving force for flow. Two valves are shown in the line, and when these valves are fully open, they offer little resistance to flow. The resistance in the transport line is due to frictional losses in the pipe. The flowrate is at a maximum value when the valves are fully open. When either valve in the line is fully closed, the resistance to flow is infinite, and the flowrate is zero. The valves may be adjusted to provide flows between these limits.

Plotted below the diagram in Figure 21.4 are pressure profiles for various valve settings. The pressure (in kPa) is plotted on the *y*-axis, and we note that for the example shown P_{in} = 200 kPa and P_{out} = 100 kPa. The *x*-axis indicates the relative location of the valves in the process. In addition, Figure 21.4 includes a table that provides information on the flowrate, pressure drop due to pipe friction, and the pressure before valve CV-1. The resistance in the pipe is proportional to the square of the flowrate, which is the case for fully developed turbulent flow (see Chapter 17).

Profile 1 (a–f) shows the pressure profile with both valves fully open. It gives the maximum flowrate, 100 kmol/h, possible for this system. Profile 2 (a–b–c–f) is for the case when CV-1 is fully closed and CV-2 is fully open. For this case, the flow is zero, and all the pressure drop occurs over CV-1. The pressure upstream of CV-1 is P_{in} (200 kPa), and the downstream pressure is P_{out} (100 kPa). Profile 3 (a–d–e–f) is the case when CV-2 is fully closed and CV-1 is fully open. The pressure upstream of CV-2 is P_{in} (200 kPa), the downstream pressure is also equal to P_{in} (200 kPa), and the flowrate is zero.

For Profile 4 (a–g–h–f), valve CV-1 is partially open, providing a pressure drop of 45 kPa, and valve CV-2 is fully open. The pressure drop across CV-1 ($\Delta P_{g\text{-}h}$) can be varied by changing the valve position of CV-1. The greater the pressure drop across CV-1, the lower the pressure drop available to overcome friction and the lower the flowrate. In Profile 4, the flow is reduced to 74% of the maximum flow.

For every setting of CV-1 (with CV-2 fully open), a unique value for pressure and flowrate is obtained.

Either pressure or flowrate, but not both simultaneously, can be regulated by altering the setting of a single valve.

Two valves are required to regulate simultaneously both the pressure and flowrate of a stream. The total system resistance (pipe and valves) determines the flowrate. The ratio of valve resistances establishes the pressure profile through the process. This is shown in Profile 5 (a–i–j–k–l–f), where 50% of the available pressure drop is taken over CV-1 and 31% is taken over CV-2. The resistance ratio, for the two valves, is 31/50 = 1.61, the flow is 44% of the maximum flow, and the pressure upstream of CV-1 is 194 kPa. To illustrate this concept further, consider Example 21.2.

Example 21.2

Consider the flow diagram in Figure 21.4. At design conditions, we have 70% of the total available pressure drop across the two control valves, and the flowrate of fluid at these conditions is given as $100\ [30/100]^{0.5} = 54.8$ kmol/h. If we consider a point, "z," midway between the two valves, over what range can the pressure at point "z" be varied at the design flowrate?

To solve this problem, we consider the two extreme cases: (1) CV-1 regulates the flow and CV-2 is fully open, and (2) CV-2 regulates the flow and CV-1 is fully open. Both these situations are illustrated in Figure E21.2. From the diagram, we can see that the pressure at point "z" may vary between 185 kPa (CV-1 fully open) and 115 kPa (CV-2 fully open).

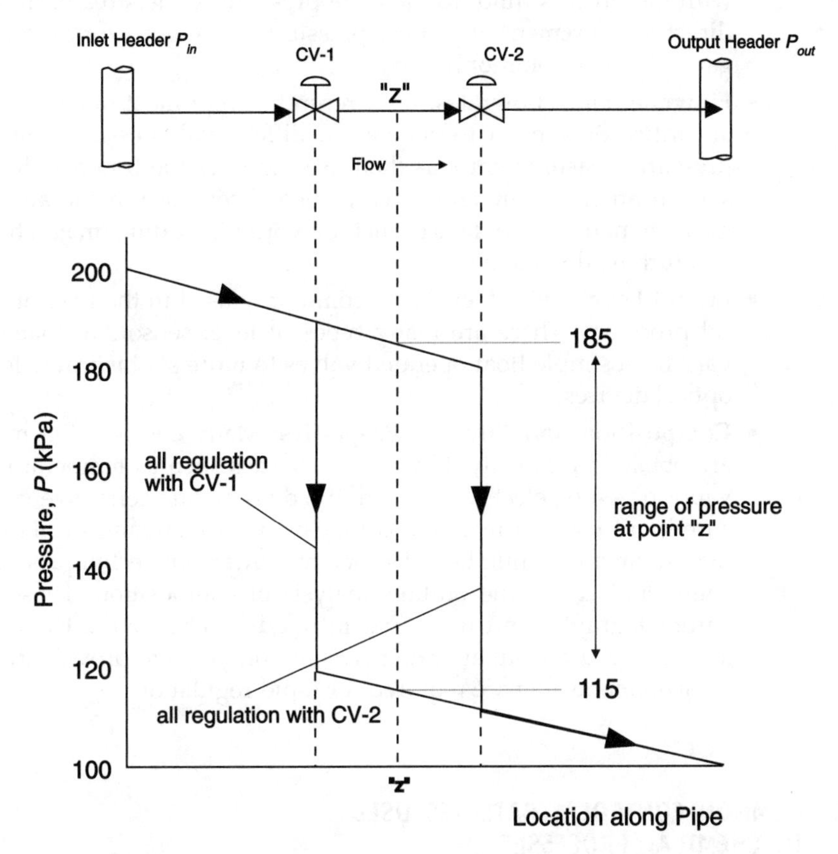

Figure E21.2 Range of Pressure at Point "z" for Design Flow

Note that all possible combinations of partially open valves give pressures at point "z" between these two limits.

21.4 THE MEASUREMENT OF PROCESS VARIABLES

The process variables that are most commonly measured and used to regulate process performance are as follows.

- **Temperature:** Several instruments are available that provide continuous measurement of temperature—for example, thermocouples, thermometers, thermopiles, and resistance thermometric devices.
- **Pressure:** A variety of sensors are available to measure the pressure of a process stream. Many sensors use the deflection of a diaphragm, in contact with the process fluid, to infer the pressure of the stream. In addition, the direct measurement of process pressure by gauges—for example, Bourdon gauge—is still commonly used.
- **Flowrate:** Fluid flowrates, until recently, were most often measured using an orifice or venturi to generate a differential pressure. This differential pressure measurement was then used to infer the flowrate. More recently, several other instruments have gained acceptance in the area of flowrate measurement. These devices include vortex shedding, magnetic, ultrasonic, and turbine flowmeters.
- **Liquid Level:** Liquid levels are commonly used in the regulation of chemical processes. There are many types of level sensors available, and these vary from simple float operated valves to more sophisticated load cells and optical devices.
- **Composition and Physical Properties:** Many composition measurements are obtained indirectly. Physical properties such as temperature, viscosity, vapor pressure, electric conductivity, density, and refractive index are measured and used to infer the composition of a stream, in place of a direct measurement. A number of other measurement techniques have become commonplace for the on-line analysis of composition. These include gas chromatography and mass and infrared spectrometry. These instruments are very accurate but are expensive and often fail to provide the continuous measurements that are required for rapid regulation.

21.5 COMMON CONTROL STRATEGIES USED IN CHEMICAL PROCESSES

There are many strategies used to control process variables in an operating plant. For more details, see Anderson [1] and Shinskey [2]. In this section, we consider only feedback, feed-forward, and cascade systems.

21.5.1 Feedback Control and Regulation

The process variable to be controlled is measured and compared with its desired value (set-point value). If the process variable is not at the set-point value, then appropriate control action is taken.

Advantage. The cause of the change in the output variable need not be identified for corrective action to be taken. Corrections continue until the set-point value is achieved.

Disadvantage. No action is taken until after an error has propagated through the process and the error in the process variable has been measured. If there are large process lag times, then significant control problems may occur.

Example 21.3 illustrates the use of feedback control in a chemical process.

Example 21.3

Identify all the feedback control loops in the process flow diagram for the production of DME, Figure B.1, and explain the control action of each. *Note:* There are other control valves on the utility streams, but these are shown only on the P&ID and not considered here.

All the control loops associated with the control valves shown on the PFD exhibit feedback control strategies. These are shown individually in Figure E21.3. Each control action is explained below.

Figure E21.3(a): In this figure, the object of the controller is to control the flowrate of Stream 4, which is sent through two heat exchangers and then into reactor R-201. Because P-201 is a positive displacement pump, the flow is controlled by varying the amount of liquid bypassed from the exit to the inlet of the pump (see Section 18.2.3). The signal to adjust the setting of the valve in the bypass line is obtained from an orifice placed in Stream 3. Thus, if the flow sensed by the pressure cell (not shown) across the orifice is below the set-point value, a signal is sent to the bypass valve to close the valve slightly. This has the effect of reducing the amount of liquid bypassed around the pump, and that increases the flow of Stream 2 and hence increases the flow of Stream 3. If the flow of Stream 3 lies above the set point, then the opposite control action is initiated.

Figure E21.3(b): The flowrate of the bottoms product in both distillation columns is adjusted to maintain a constant level of liquid in the bottom of the tower. The control strategies for the bottoms of both T-201 and T-202 are identical. Assume that the level in the bottom of the column drops below its set point. A signal from the level sensor would be sent to close slightly the valve on the bottoms product stream. This would reduce the flow of bottoms product and result in an increase in liquid inventory at the bottom of the column, causing the liquid level to rise. If the liquid level were to increase above the set-point value, then the opposite control action would be initiated.

Figure E21.3(c): The control strategy at the top of both columns (T-201 and T-202) is illustrated in Figure E21.3(c). The control action was explained previously in Example 21.1, where the reflux stream is held constant by a control valve and the liquid level in the reflux drum is held constant by adjusting the flow of the overhead product.

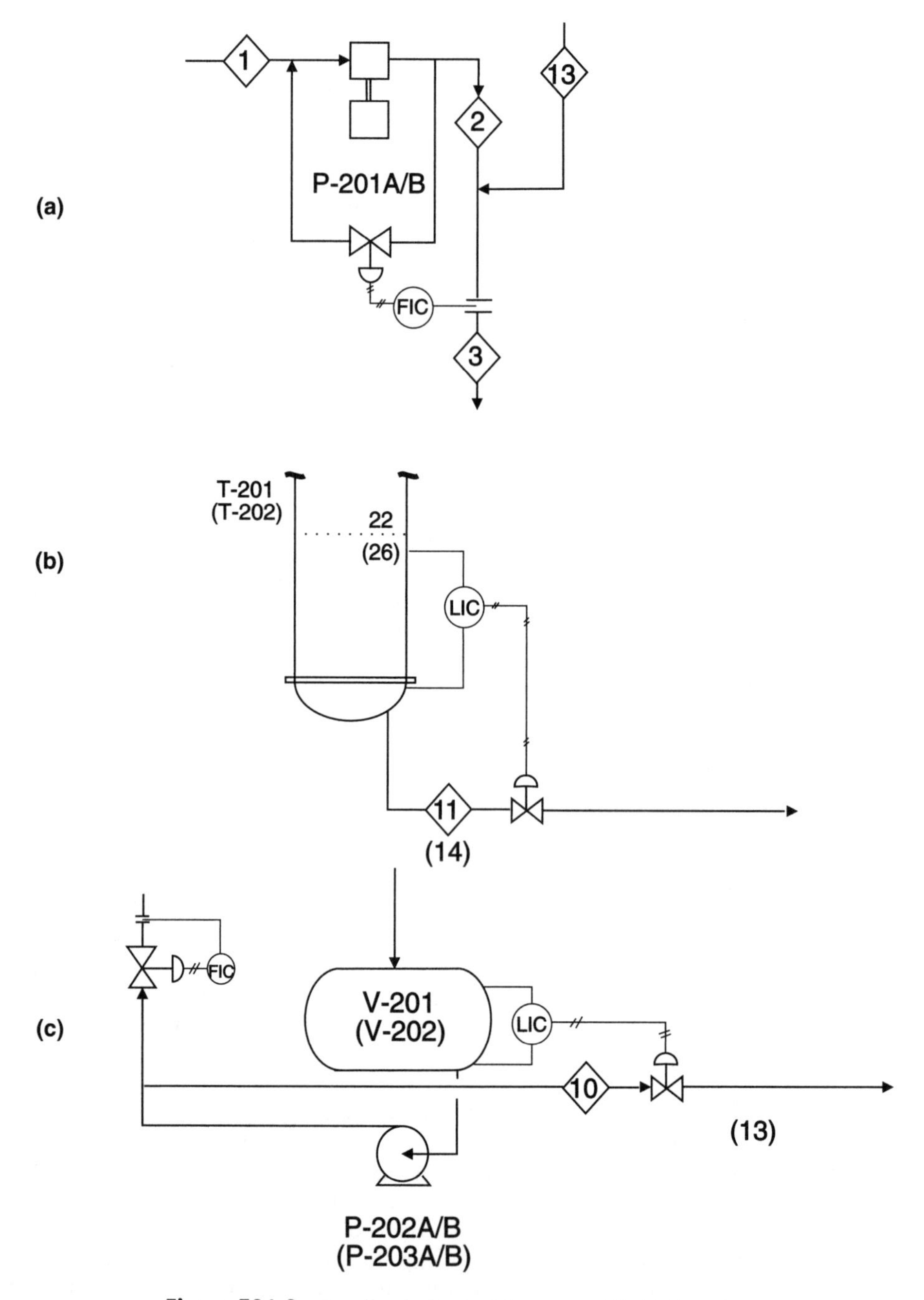

Figure E21.3 Feedback Control Loops on DME Flowsheet

It should be noted that at present there is no product quality (purity) control on either column. This is addressed under cascade control below.

21.5.2 Feed-Forward Control and Regulation

Process input variables are measured and used to provide the appropriate control action.

Advantages. Changes in the output process variable are predicted, and adjustments are made before any deviation from the desired output takes place. This is useful, especially when there are large lag times in the process.

Disadvantages. It is necessary to identify all factors likely to cause a change in the output variable and to describe the process by a model. The regulator must perform the calculations needed to predict the response of the output variable. The output variable being regulated is not used directly in the control algorithm. If the control algorithm is not accurate and/or the cause of the deviation is not identified, then the process output variable will not be at the desired value. The accuracy and effectiveness of the control scheme are directly linked to the accuracy of the model used to describe the process.

Example 21.4 illustrates how feedback control, feed-forward control, or a combination of both can be used to control a process unit.

Example 21.4

Consider the process illustrated in Figure E21.4. The object is to cool, in a heat exchanger, a process stream (Stream 1) to a desired temperature (Stream 2) using cooling water supplied from a utility header. In Figure E21.4(a), a feedback control scheme is illustrated. The control strategy is to measure the temperature of the process stream leaving the heat exchanger (Stream 2) and to adjust the flow of cooling water to obtain the desired temperature. Thus, if the temperature of Stream 2 were to be greater than the desired set-point value, then the control action would be to increase the flow of cooling water and vice versa.

A feed-forward control scheme for this process is illustrated in Figure E21.4(b). Here, both the flowrate and temperature of the input process stream and the inlet cooling water temperature are measured. A calculation is then made that predicts the flowrate of cooling water needed to satisfy the energy balance and the exchanger performance equations as given in Chapter 18. Using the subscripts c and p to refer to the cooling water and process streams, respectively, we get

$$Q = \dot{m}_c C_{p,c}(T_b - T_a) \tag{E21.4a}$$

$$Q = \dot{m}_p C_{p,p}(T_1 - T_2) \tag{E21.4b}$$

$$Q = UA\frac{(T_1 - T_b) - (T_2 - T_a)}{\ln\dfrac{(T_1 - T_b)}{(T_2 - T_a)}} \tag{E21.4c}$$

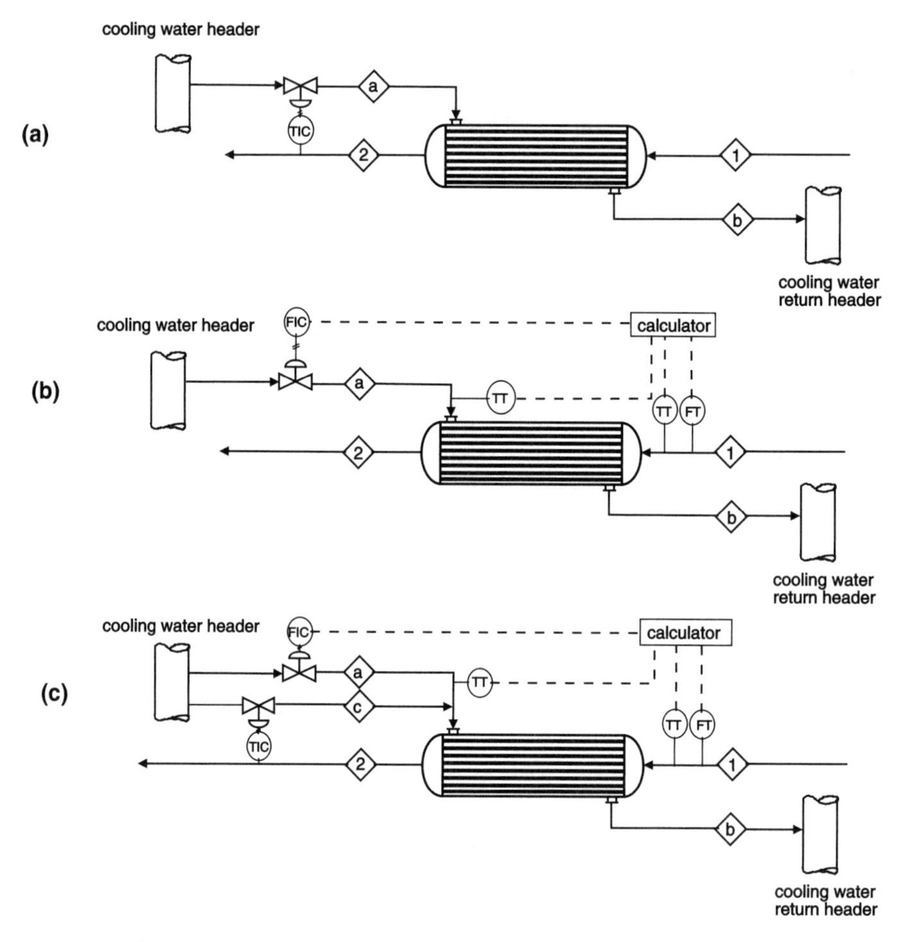

Figure E21.4 Control Strategies for Cooling a Process Stream

From Equations (E21.4a)–(E21.4c), we can categorize the variables as unknown, measured, known, and specified according to the following table.

Unknown	Measured	Known	Specified
Q, $\dot{m}_c$, T_b	$\dot{m}_p$, T_a, T_1	$C_{p,c}$, $C_{p,p}$, U, A	T_2

Therefore, for a given value of the exit process temperature (T_2), Equations (E21.4a), (E21.4b), and (E21.4c) can be solved to give the unknowns in the table. Thus, by measuring the inlet process stream flow and temperature and the inlet cooling water temperature, the model given by Equations (E21.4a)–(E21.4c) can be solved to yield the desired value of the cooling water flowrate. The valve on the cooling water inlet can then be adjusted to give this value. There are some assumptions associated with this model that will affect the performance of this control scheme. The first assumption is that both streams flow countercurrently, and hence the exchanger effectiveness factor, F, is assumed to be equal to 1. If the value of F is not close to unity, then the model given above will not be accurate. However, the model can be modified by including another equation to calculate F based on the process variables. A more serious assumption is that the overall heat transfer coefficient, U, has a constant value. In real processes, this is seldom the case, either because of flowrate changes or due to fouling of heat-exchange surfaces occurring throughout the life of the process. As the heat transfer surfaces foul or flowrates change, the value of U changes and the predictions of the above model will deviate increasingly from the true value of T_2. This illustrates one of the major disadvantages of feed-forward control strategies, in that the efficacy of the control strategy is only as good as the predictions of the equations used to model the process.

An alternative scheme to pure feed-forward control is to use a combination of feedback and feed-forward control strategies. This is illustrated in Figure E21.4(c) and discussed below.

In feed-forward control, the process must be modeled using material and energy balances and the performance equations for the equipment.

The effectiveness of feed-forward control is strongly influenced by the accuracy of the process model used.

21.5.3 Combination Feedback and Feed-Forward Control

By correctly combining these two control strategies, the advantages of both methods can be realized. The feed-forward strategy uses measured input variables to predict changes necessary to regulate the output. The feedback regulator then measures the regulated output variable and makes additional changes to ensure that the process output is at the set point.

Advantages. The advantages of both feedback and feed-forward regulation are achieved. The feed-forward regulator makes the major corrections before any deviation in process output takes place. The feedback regulator ensures that the final set-point value is achieved.

Disadvantages. The control algorithm and hardware to achieve this form of control are somewhat more complicated than either individual strategy.

Example 21.5 considers the operation of the combination of feedback and feed-forward control.

Example 21.5

We return to Example 21.4 and look at the operation of the combined control strategy illustrated in Figure E21.4(c). This system is essentially a combination of Figures E21.4(a) and E21.4(b). The feed-forward scheme of Figure E21.4(b) has been retained but is used to regulate only a fraction of the cooling water required, say 80%. In addition, a feedback control loop has been added. This feedback loop consists of a second parallel cooling water feed stream that will be adjusted so as to maintain the desired set-point value for the temperature of Stream 2. The amount of cooling water regulated by the feedback loop is the remaining 20% required to satisfy the regulation. This combined system will respond quickly to changes in process flowrate or inlet temperatures via the feed-forward control loop. In addition, a second cooling water stream is adjusted, based on the feedback control loop, to control exactly the process outlet temperature.

21.5.4 Cascade Regulation

The **cascade** system uses controllers connected in series. The first control element in the system measures a process variable and alters the set point of the next control loop.

Advantages. This system reduces lags and allows finer control.

Disadvantages. It is more complicated than single-loop designs.

Example 21.6 illustrates cascade regulation.

Example 21.6

Consider the control of the top-product purity in distillation column T-201 in the DME process. As shown in Example 21.3, the material balance control is achieved by a flow controller on the reflux stream and a level controller on the top product stream. From previous operating experience, it has been found that the top-product purity can be measured accurately by monitoring the refractive index (RI) of the liquid product from the top of the column, Stream 10. Using a cascade control system, indicate how the control scheme at the top of T-201 should be modified to include the regulation of the top-product composition.

The solution to the problem is illustrated in Figure E21.6. An additional control loop has been added to the top of the column that consists of a refractive index measuring element that sends a signal to the flow controller placed on the reflux stream (Stream 16). The purpose of this signal is to change the set point of the flow controller. Consider a situation where the purity of the top product has fallen below its desired value. This change may be due to a number of reasons; for example, the flowrate or purity of the feed to the column may have changed. However, the reason for the change is not important at this stage. The decrease in purity will be detected by the change in refractive index of the liquid product, Stream 10, and a signal will be sent from the RI sensor to increase the reflux flowrate of Stream 16. The flow of Stream 16 will increase, which will result in an improved separa-

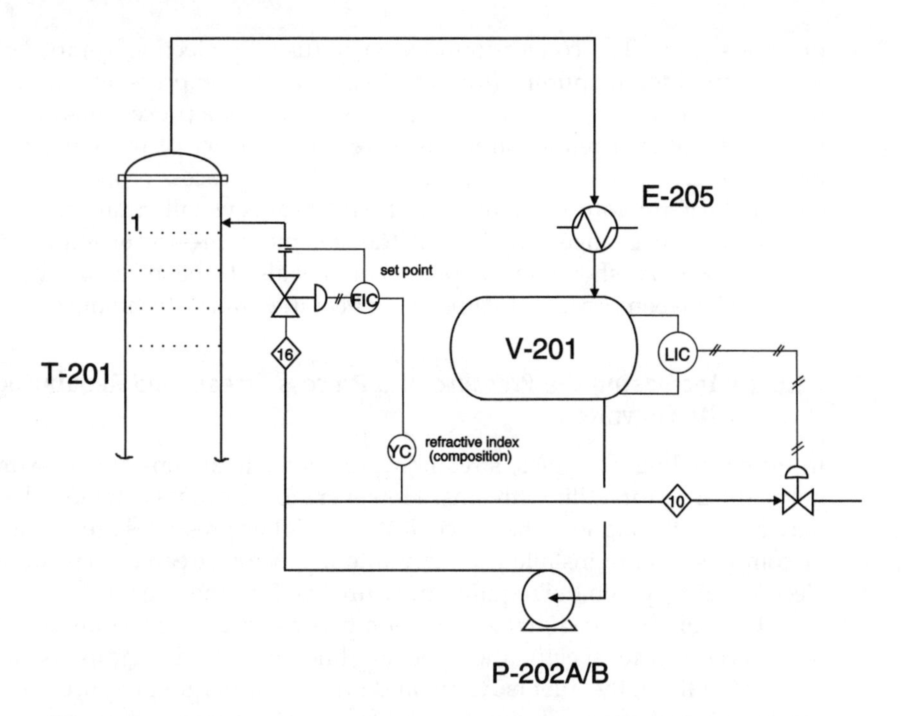

Figure E21.6 Cascade Control for DME Overhead Product Purity

tion in the column and an increase in purity of the top product. For an increase in product purity, the reverse control action will occur.

These control strategies do not represent a comprehensive list of possibilities. The range of strategies is very large. This is due, in part, to the flexibility that software has added to the control field by allowing sophisticated process models to be used in simulating processes, as well as the use of more complicated control algorithms.

No mention has been made of the types of controller (proportional, differential, integral, etc.) and strategies to tune these controllers. Such information is covered in traditional process control texts, such as Stephanopolous [3], Coughanowr [4], Smith and Corripio [5], Seborg et al. [6], and Marlin [7].

21.6 EXCHANGING HEAT AND WORK BETWEEN PROCESS AND UTILITY STREAMS

In order to obtain the correct temperatures and pressures for process streams, it is often necessary to exchange energy, in the forms of heat and work, with utility streams. In order to increase the pressure of a process stream, work is done on the process stream. The conversion of energy (usually electrical) into the pressure head is provided by pumps (for liquid streams) or compressors (for gas streams) that are inserted directly into the process stream. For a process gas stream undergoing a significant reduction in pressure, the recovery of work using a gas turbine may have a significant impact (favorable) on process economics. In order to change the temperature of a process stream, heat is usually transferred to or from a process by exchange with a heat transfer medium—for example, steam and cooling water. As shown in Chapter 15, it may also be beneficial for heat to be exchanged between process streams using heat integration techniques.

21.6.1 Increasing the Pressure of a Process Stream and Regulating Its Flowrate

In the preceding examples, streams were available at constant pressure. This is normally true for utility streams. However, for a process stream, the pressure may change throughout the process. When higher pressures are needed, pumps or compressors are installed directly into a process stream. These units convert electrical energy into the required pressure head for the stream.

It should be noted that when a pump or compressor is required in a process, it is necessary to specify the type of fluid and fluid conditions, the design flowrate of fluid, the inlet (suction) and outlet (discharge) pressures at the design conditions, and an overdesign (safety) factor. The pump or compressor conforming to these specifications must be able to operate at and somewhat above the design conditions. However, in order to regulate the flow of the process streams, as will be required throughout the life of the process, it is necessary to implement a control or regulation system. Some typical methods of regulating the flow of process streams are outlined below and shown in Figure 21.5.

Figure 21.5(a) shows a centrifugal pump system that increases the pressure of a process stream. The position of the control valve, CV-1, is at the discharge side of the pump. The flowrate of Stream 2 is changed in order to regulate the flow of the stream passing through the pump. The operation of the pump and the interaction of the pump curve, the system curve, and the control valve were discussed in the presentation of pump performance curves in Chapter 18. In reviewing this material, it is important to remember that the maximum flowrate of the fluid is determined by the intersection of the pump and system curves. Regulation of flowrate is possible only for flows less than this maximum value, that is, to the left of the intersection of the pump and system curves, point "a" in Figure 18.6.

Figure 21.5(b) shows a positive displacement pump increasing the pressure of a process stream. The positive displacement pump can be considered to be a constant-flow, variable-head device; that is, it will deliver the same flowrate of fluid at a wide range of discharge pressures. Thus we may write:

$$\dot{v}_2 + \dot{v}_3 = \dot{v}_1 = \text{a constant} \tag{21.2}$$

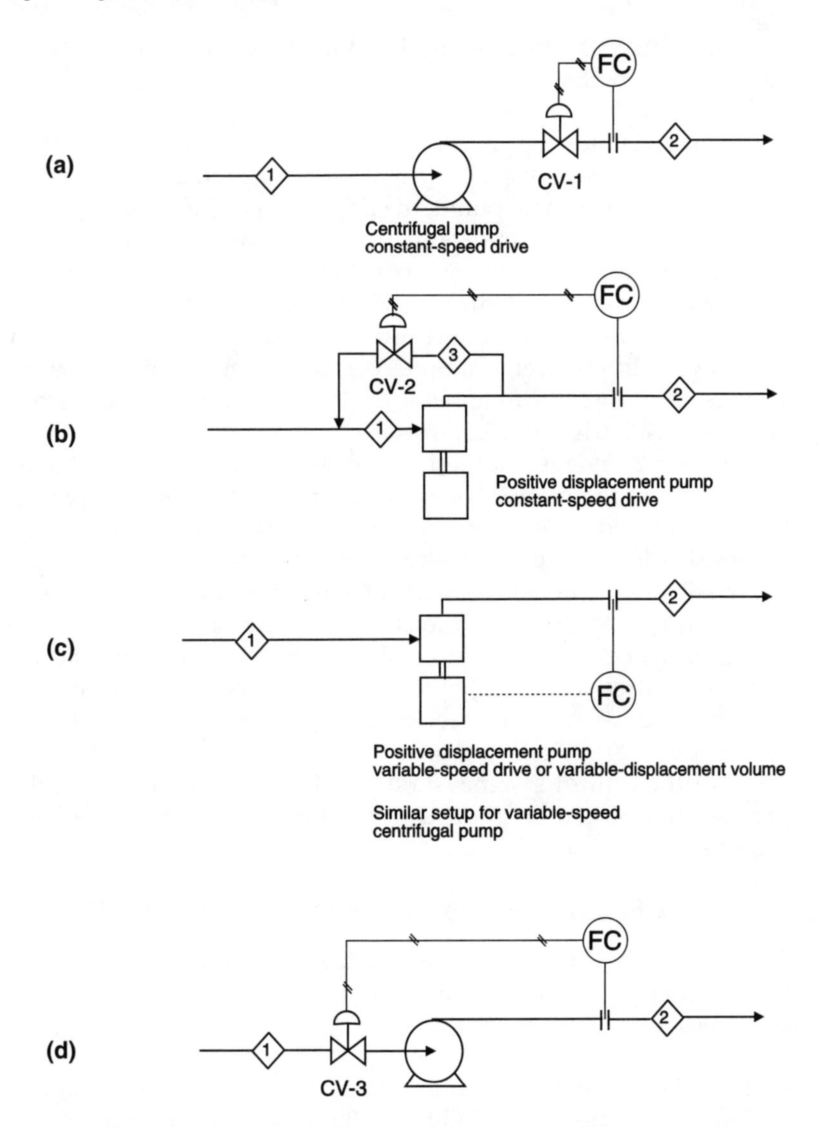

Figure 21.5 Flowrate Feedback Control Schemes for Pumping Liquids

For this reason, in order to regulate the flow of the process stream, Stream 2, a portion of the output stream from the pump must be recycled. In Figure 21.5(b) this is accomplished by returning Stream 3 to the suction side of the pump. By altering the position of valve CV-2, the recycle stream flowrate is altered, and because the flow of liquid through the pump is almost constant, this also provides flow regulation for the main process stream, Stream 2. It should be noted that one should never throttle the output stream of a positive displacement pump because the pump curve is almost vertical and the flow will change little with throttling, although the discharge pressure will increase drastically (see Section 18.2.3, Figure 18.8).

Figure 21.5(c) uses a variable-speed drive—or variable displacement volume, for a positive displacement pump—to regulate the flow of the process stream. For a centrifugal pump, the impeller speed is regulated to provide the required flow, at the desired pressure, for the process stream. The advantage of using this type of control strategy is that there is no wasted energy due to throttling in a control valve. However, variable-speed controls and motors are expensive and less efficient. Therefore, this type of control scheme is usually cost effective only for gas blowers, compressors, and large liquid pumps, because a large savings in utilities is required to offset the large capital investment for the variable-speed drives and controls.

Figure 21.5(d) regulates the input flow with a valve, CV-3, installed in the suction line to the pump. The position of CV-3 is altered to provide the desired flowrate of fluid. This may work well for gases but is seldom used for liquid pumps. For liquid streams, the reduction in pressure at the pump inlet increases the possibility of cavitation, that is, $NPSH_A$ is reduced drastically. The causes of cavitation and NPSH calculations were covered in Section 18.2.4. On the other hand, for the case of gas compression using centrifugal machines, the throttling of the inlet stream is essential for start-up purposes. Thus, this method is often the preferred control strategy for flow regulation in centrifugal blowers and compressors.

All the control systems shown in Figure 21.5 for the regulation of liquid process streams can be applied to compressor systems to regulate the flow of gaseous process streams.

21.6.2 Exchanging Heat between Process Streams and Utilities

The amount of heat added to or removed from a process stream is usually altered by changing the flow or pressure of utility streams. The primary utility streams in a chemical plant are as follows.

1. Cooling water is used to remove heat from a process stream. The heat transferred to the coolant adds to the sensible heat (enthalpy) of the coolant stream (temperature increases).
2. Air can be used to remove heat from a process stream and is often used in product coolers and overhead condensers. The heat transferred to the coolant adds to the sensible heat (enthalpy) of the coolant stream (temperature increases).
3. Steam condensation is used to add heat to a process stream. The heat transferred from the steam decreases the enthalpy of the steam (steam condenses).

4. Boiler feed water (BFW) is used to remove heat from a process stream to make steam that can be used elsewhere in the process. The heat transferred to the bfw increases the enthalpy of the bfw stream (BFW vaporizes).

When using cooling water or air in which there is no change of phase of the utility stream, regulation is generally performed by changing the flowrate of the cooling stream. For example, see Figure E21.4 and Example 21.4. In this section, we concentrate on systems in which the utility stream undergoes a change of phase. The focus is on boiler feed water and steam. However, the results apply equally well to other heat-transfer media that undergo a phase change. It is further assumed that utility streams are taken from and returned to headers and that the process streams are a single phase. The control of reboilers and condensers in a distillation column is covered in the case study on distillation given in the following section.

There are many ways to regulate systems using steam. We do not try to present an exhaustive list of possible schemes but concentrate on several of the more common techniques. Figures 21.6 and 21.7 show several systems used to control the transfer of heat between a utility stream and a process stream. The systems on the left-hand side of the figures add heat to the process stream by condensing steam. Systems on the right-hand side remove heat from the process stream by vaporizing BFW. In either case, there are two phases present in equilibrium on the utility side of the exchanger. The temperature and pressure are related by the vapor pressure relationship for water. If steam from the header is significantly

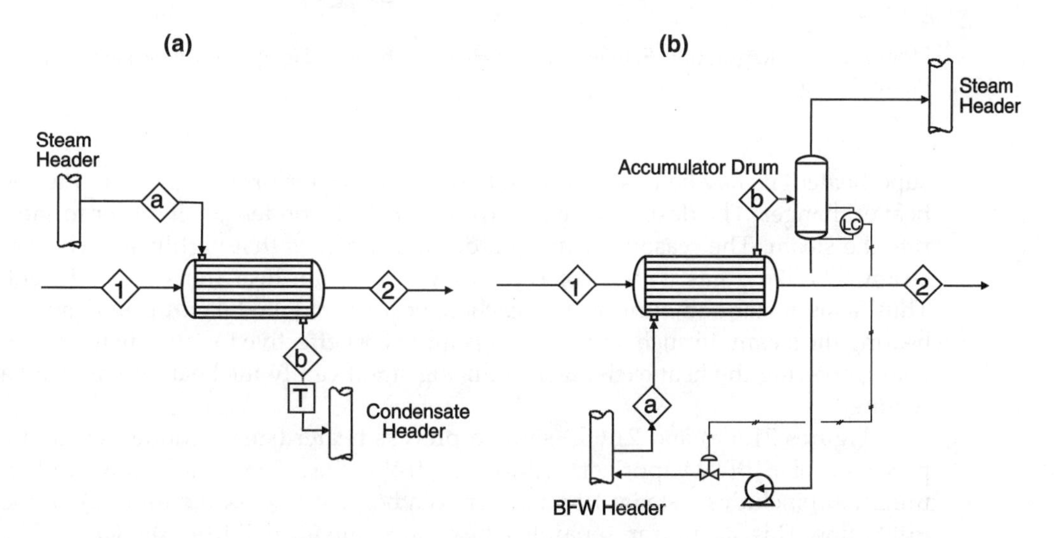

Figure 21.6 Unregulated Heat Exchanger Using Utility Streams with a Phase Change: (a) Process Heater and (b) Process Cooler

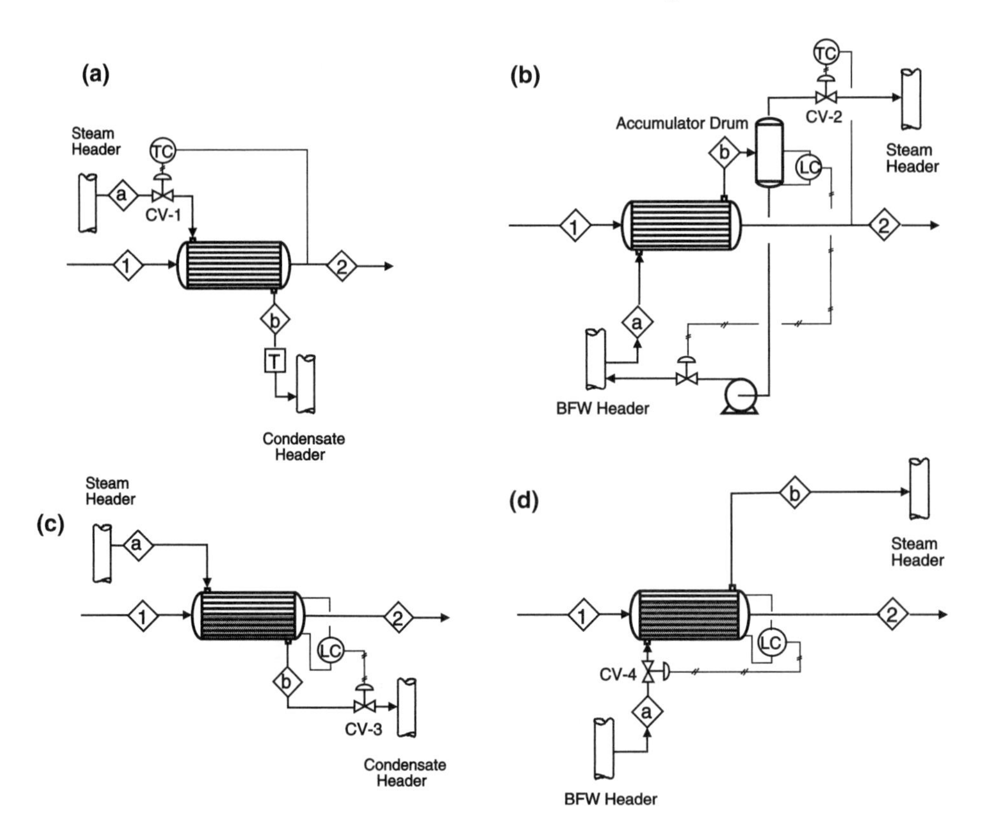

Figure 21.7 Regulation Schemes for a Heat Exchanger Using Utility Streams with a Phase Change

superheated, it may be passed through a desuperheater prior to being fed to the heat exchanger. The desuperheater adds just enough condensate in order to saturate the steam. The reason for using a desuperheater is that highly superheated steam acts like a gas with a correspondingly low film heat transfer coefficient. Thus, a significant amount of heat-exchanger area may be taken up in desuperheating the steam. In such situations, it is more cost effective to saturate the steam before entering the heat exchanger, reducing significantly the heat exchange area required.

Figures 21.6(a) and 21.6(b) show a process heater (steam condenser) and a process cooler (BFW vaporizer) without control valves. The heater shown in Figure 21.6(a) includes a **steam trap** (shown as a box with a *T* in the middle) on the utility line. This steam trap separates the condensate formed from the vapor. The condensate is collected by the trap in the bottom of the exchanger and discharged intermittently to the condensate header. The steam flowrate, Stream a, is established by the rate at which heat is exchanged. This is different from normal utility flow regulation. As the steam condenses, there is a large decrease in the specific volume of the utility stream. Thus, steam is continually "sucked" from the header into the heat exchanger. For this system, no regulation of process temperature is attempted.

The process cooler shown in Figure 21.6(b) vaporizes BFW. In this case, the flow of BFW is more than sufficient to provide for the vaporization taking place. The exit stream is a two-phase mixture of steam and BFW. The BFW is separated from the vapor in an accumulation drum (phase separator) and recycled back to the inlet of the exchanger. Again, for this system, no regulation of process temperature is attempted.

Although no control schemes are implemented in the systems in Figures 21.6(a) and 21.6(b), to some extent, both systems are self-regulating. The reason for this self-regulating property is that the major resistance to heat transfer is on the process side (film heat transfer coefficients for condensing steam and vaporizing BFW are very high). Consider an increase in the flowrate of the process stream in Figure 21.6(a). As the flow increases, so does the process-side heat transfer coefficient, and this increase is proportional to the (process flowrate)$^{0.8-0.6}$. See Table 17.1 for an explanation of the range of exponents. Because the major resistance to heat transfer is on the process side, this suggests that the overall heat transfer coefficient changes by the same amount. Thus, for a 10% increase in process flowrate, the overall heat transfer coefficient increases by 6% to 8%. Because the change in the overall heat transfer coefficient does not quite match the change in process flowrate, the exit temperature of the process stream, Stream 2, must drop slightly. Although the temperature of the exit stream drops, it does not drop by much due to the partial regulation provided by the increase in heat transfer coefficient.

In discussing the operation of this heat exchanger, no mention was made of the change in flowrate of the steam that would be required to provide the increase in heat-exchanger duty caused by the increased process flowrate. As the exchanger duty increases, more steam will condense, and this increased demand for steam does have an effect on the operation. It should be noted that steam is supplied from the header to the exchanger in the quantity that is required; that is, it is determined by the energy balance. As the steam flowrate increases, the frictional pressure losses in the supply pipe from the header to the exchanger increase. This causes the pressure on the steam side of the heat exchanger to decrease. Thus, the utility side of the heat exchanger essentially "floats" on the header pressure, with an appropriate pressure loss to account for flow through the supply line. As the pressure of the steam side of the exchanger drops, this causes the temperature at which the steam condenses to drop slightly, thus reducing the temperature driving force for heat exchange. However, compared with the change in heat transfer coefficient, this change is minor.

In order to regulate the temperature of the process stream leaving the heat exchanger, it is necessary to control the heat duty. In order to do this, we must be able to regulate a variable on the right-hand side of the heat-exchanger performance equation:

$$Q = UA\Delta T_{lm} \quad (21.3)$$

This can be achieved by doing one or more of the following.

1. Regulate the temperature driving force (ΔT_{lm}) between the process fluid and the utility.
2. Adjust the overall heat transfer coefficient (U) for the heat exchanger.
3. Change the area (A) for heat exchange.

We consider two alternatives for regulating the temperature of the process exit stream.

Regulate the Temperature Driving Force between the Process Fluid and the Utility. The systems shown in Figures 21.7(a) and 21.7(b) include a valve on the utility line. This valve is used to change the pressure of the utility stream in the exchanger. Because there are two phases present, the pressure establishes the temperature at which the phase change takes place. Reducing pressure reduces temperature according to the vapor pressure-temperature relationship for water, for example, from Antoine's equation.

To increase the heat duty for the process heater, Figure 21.7(a), the temperature driving force is increased by increasing the pressure. This is achieved by opening CV-1 to decrease the frictional resistance to flow. Because CV-1 is located on the input side of the heat exchanger, opening the valve increases the pressure of the steam in the exchanger, and the temperature at which condensation occurs will also increase. To increase the process cooler heat duty, Figure 21.7(b), the temperature driving force is increased by opening CV-2, which decreases the resistance to flow. CV-2 is located on the discharge side of the utility, that is, between the exchanger and the steam header. In this case, the exchanger once again floats on the steam header pressure. As CV-2 is opened the pressure on the utility side of the exchanger decreases, and this reduces the temperature at which the BFW boils and increases the temperature driving force for heat transfer.

Adjust the Overall Heat Transfer Coefficient for the Heat Exchanger by Adjusting the Area Exposed to Each Phase. The systems shown in Figures 21.7(c) and 21.7(d) operate such that the interface between the liquid and vapor on the utility side of the exchanger covers some of the tubes. For this situation, a fraction of the heat exchange area is immersed in the liquid region, with the remaining fraction in the vapor phase region. The heat transfer coefficients and the temperature driving forces for these two regions differ significantly. Thus, by adjusting the level of the liquid-vapor interface in the exchanger it is possible to regulate the duty.

For the process heater, Figure 21.7(c), decreasing the level increases the heat duty. The reason for this is that in the vapor-phase region the utility-side film coefficient and temperature driving force are large because steam is condensing.

For the liquid-phase region, both the film coefficient and temperature driving force are low because the condensate is being subcooled. Therefore, as the liquid level decreases, more heat transfer area is exposed to condensing steam, the overall average value of *UA* increases, and the duty increases.

For the process cooler, Figure 21.7(d), increasing the liquid level increases the heat duty. The reason for this is that in the vapor-phase region, the utility-side film coefficient and temperature driving force are low because steam is being superheated. For the liquid-phase region, both the film coefficient and temperature driving force are high because the BFW is boiling. Therefore, as the liquid level increases, more heat transfer area is exposed to boiling water, the overall average value of *UA* increases, and the duty increases.

21.6.3 Exchanging Heat between Process Streams

The integration of heat between process streams is often economically advantageous, and a strategy to do this is explained in Chapter 15. When heat integration is implemented in a process, it is likely that heat exchange will occur between two process streams. The regulation of the flow of these process streams cannot effectively be used to control the heat transfer, because the flow of the two streams will be directly coupled. One method to control the exit temperature in such an exchanger is to bypass some portion of one or both of the process streams around the exchanger. By altering the amount of the stream that bypasses the equipment, temperature regulation is possible. This concept is considered further in Problem 21.4.

21.7 CASE STUDIES

In this section two operations that require more complex control schemes are studied in detail. Again, we emphasize that these are not the only ways to control these units, but are typical schemes that might be employed. Many alternatives exist, and in the course of a career in process engineering, you will encounter a wide variety of regulation systems.

21.7.1 The Cumene Reactor, R-801

Consider the reactor system for the production of cumene shown in Appendix C, Figure C.8 and redrawn in Figure 21.8. The reactor feed, Stream 7, comes from a fired heater, where it is heated to the temperature required in the catalytic reactor. The reaction takes place in a bank of parallel catalyst-filled tubes and is mildly exothermic. In order to regulate the temperature of the reacting mixture and catalyst, the tubes are immersed in boiler feed water. The heat is removed by vaporizing the water to make high-pressure steam, which is then sent to the high-pressure steam header.

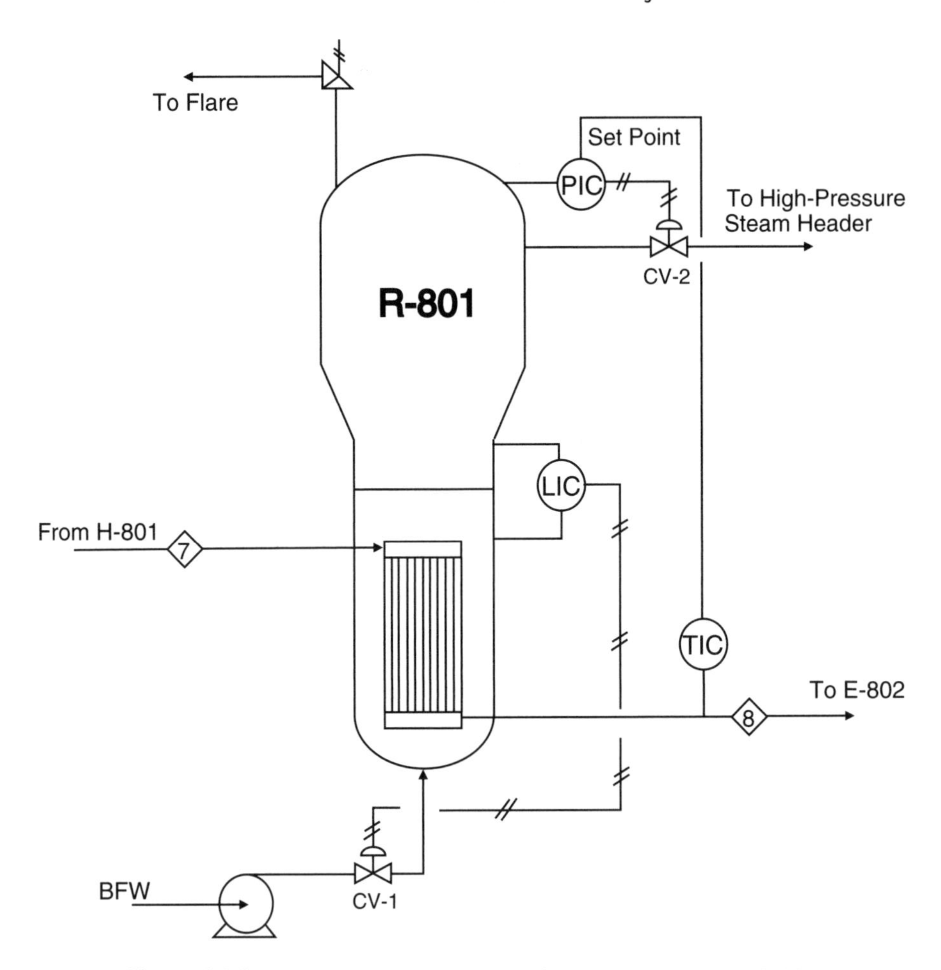

Figure 21.8 Basic Regulation Scheme for Cumene Reactor, R-801

The first issue is what should be controlled in the reactor and why. This is not a simple question to answer because many things must be considered. Some of the important considerations for exothermic catalytic reactions were discussed in Chapter 20. Here, we assume that the exit temperature from the reactor, Stream 8, is the variable that must be controlled. If the reactor is designed such that the temperature increases monotonically from the inlet to the outlet, then the exit stream is the hottest point in the reactor. More commonly, there will be a temperature bump or warm (hot) spot somewhere within the reactor, and the exit temperature will be cooler than at the temperature bump. In this case, we could use a series of in-bed thermocouples to measure the temperature profile within the catalyst tubes and use the maximum temperature as our controlled variable. In either case, it is important to control the reactor temperature, because the reaction rate and catalyst activity are both affected strongly by temperature.

The first part of the control scheme is a feedback material balance control on the boiler feed water. Control valve CV-1 is adjusted using a signal from the level controller mounted on the side of the reactor. This scheme ensures that the bank of catalyst tubes is always totally immersed in the BFW. Regulation of the BFW level below the top of the tubes, as discussed in the previous section, is inadvisable for this reactor, because the poor heat transfer coefficient for the portion of the tubes above the liquid level might cause a hot spot to occur at the entrance of the reactor. The second part of the control scheme uses cascade control to regulate the reactor exit temperature. The temperature at which the BFW boils in the reactor is regulated by the setting of CV-2. The logic is the same as described for the situation in Figure 21.7(b): that the pressure of the steam in the reactor is set by the steam header pressure (fixed) and the pressure drop across CV-2 (adjustable). The temperature of Stream 8 is monitored and used to adjust the set point on CV-2 as required.

Consider a situation in which the reactor exit temperature is seen to drop slowly. This situation may be caused by a number of reasons—for example, the catalyst slowly loses its activity and has the undesirable effect of reducing the conversion in the reactor with the possibility of a reduction in cumene production. With the control scheme shown in Figure 21.8, the system would respond in the following manner. As the temperature of Stream 8 drops, the set point on CV-2 would be adjusted to close the valve and increase the pressure of the steam in the reactor. This would have the effect of increasing the temperature at which the BFW boiled and would reduce the temperature driving force for heat transfer. The net effect would be to reduce the amount of heat removed from the reactor, causing the exit stream temperature (and reactor temperature profile) to increase and also causing the amount of reaction to increase. If the reactor exit temperature were to increase for some reason, then the control strategy would be the opposite of that described above. The increase in temperature of Stream 8 would be sensed by the temperature controller, which would then adjust downward the set-point pressure for the reactor. This would cause CV-2 to open, reducing the pressure and temperature of the steam in the reactor. This, in turn, would increase the temperature driving force for heat transfer and increase the amount of heat removed from the reactor, causing the exit stream temperature to decrease and also causing the amount of reaction to decrease. Clearly, for exothermic reactions, the control of temperature is imperative to safe reactor operation. For such equipment, several safety features would be incorporated into the overall control scheme. These safety features are extremely important and are treated separately in Chapter 24.

21.7.2 A Basic Control System for a Binary Distillation Column

The purpose of a control system for a distillation column is to provide stable operation and to produce products with the desired purity. Figure 21.9 shows a control scheme typically used to regulate a binary distillation system. Flowrates,

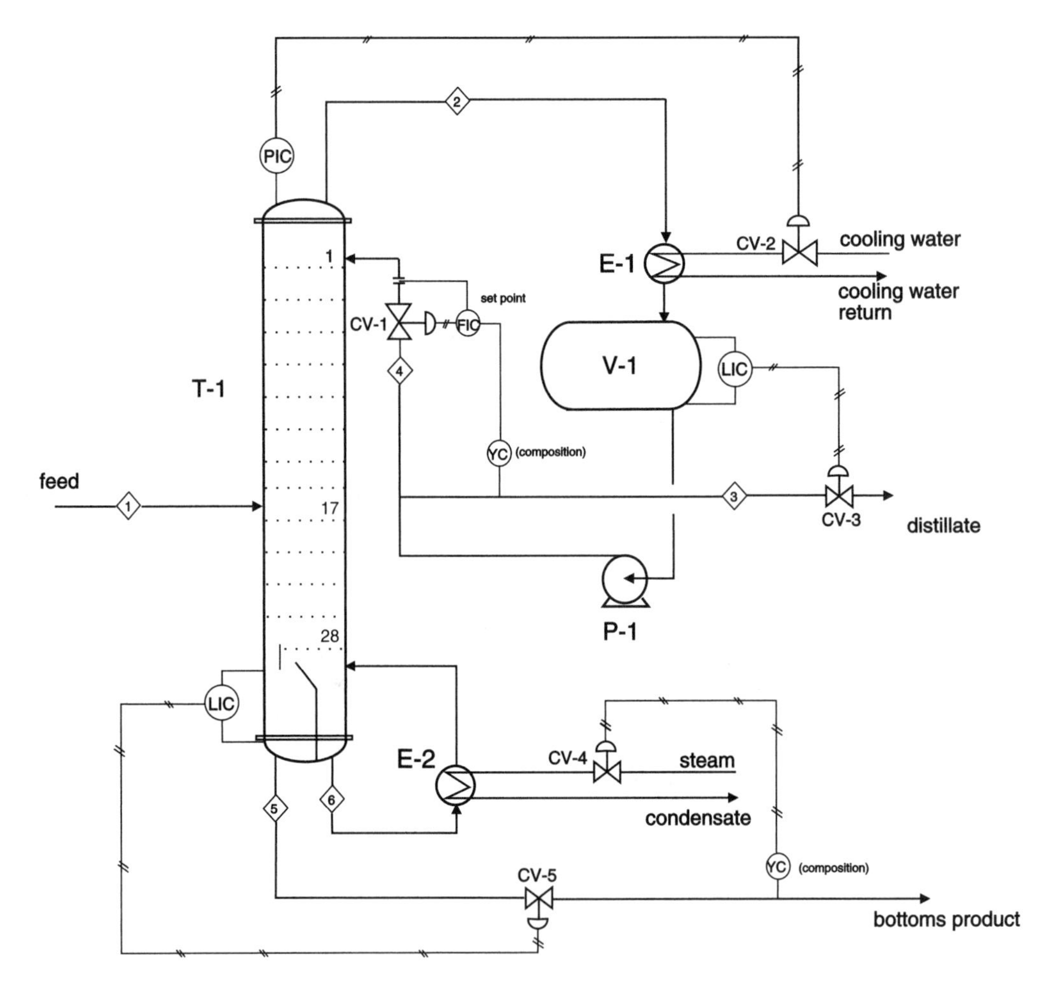

Figure 21.9 Typical Basic Control Scheme for a Binary Distillation Column

pressure, stream composition, and liquid levels are regulated. The five control variables to be regulated are shown (column pressure, composition of distillate, composition of bottoms product, liquid level in the bottom of the column, and the liquid level in the reflux drum).

Shown in Figure 21.9 are five feedback control loops and one cascade control loop used to regulate the five control variables.

- Valve CV-1 regulates the liquid reflux, Stream 4, in order to maintain the desired distillate composition. In addition, a composition measurement is made on the overhead product stream, Stream 3, and this is used to change the set point of CV-1.

- Valve CV-2 regulates the condenser heat duty (by changing the flowrate of cooling water) in order to maintain a constant column pressure.
- Valve CV-3 regulates the distillate flowrate, Stream 3, in order to maintain a constant liquid level in the reflux drum, V-1.
- Valve CV-4 regulates the reboiler heat duty (by changing the pressure of the stream flowing into E-2) in order to maintain a constant distillate bottoms composition.
- Valve CV-5 regulates the bottoms product flowrate, Stream 5, in order to maintain a constant liquid level in the bottom of the column.

Consider the changes that result from a decrease in the concentration of the light component in the feed—assuming the feed flowrate remains unchanged—and the regulation that the control system provides. It is observed for this system that as a lower amount of light material is fed to the column, the purity of the distillate and amount of overhead vapor begin to drop. The concentration detector located on Stream 3 will detect this change and will send a signal to change the set point of CV-1. Valve CV-1 will begin to open in order to increase the reflux rate. Simultaneously, the level of liquid in V-1 will start to drop. The level control will sense this change and start to close CV-3, thus reducing the flow of overhead product, Stream 3. In addition, as the amount of overhead vapor, Stream 2, flowing to E-1 also drops, the column pressure will start to decrease slightly and the cooling water flowrate, regulated via CV-2, will be reduced. At the bottom of the column, the heavier material will start to accumulate. For this system, it is observed that this will result in an increase in the liquid level in the bottom of the column and an increase in the purity of the bottoms product. The level controller at the bottom of the column will sense the increase in liquid level and open valve CV-5, increasing the bottoms product flowrate. Finally, as the composition detector on Stream 5 senses the increase in purity of the bottoms product, it will send a signal to CV-4 to close, reducing the boil-up.

The action of the control scheme to the change in feed composition essentially regulated five streams in the following manner:

1. Stream 3↓
2. Stream 5↑
3. Stream 4↑
4. Cooling water to condenser (E-1)↓
5. Steam to reboiler (E-2)↓

Actions 1, 2, 4, and 5 are consistent with satisfying the material and energy balances for the distillation system. Action 3 is required to adjust the purity of the products.

The regulation scheme shown in Figure 21.9 is only one of many possible systems that can be used to regulate a binary distillation column. If there is a sudden change in a process variable, this system may become unstable, and careful tuning of the controllers is necessary to avoid such problems. It should also be pointed out that the material balance for the column is automatically satisfied by using the liquid levels in the column and reflux drum to control the product flowrates. The same is not true for the energy balance. For example, if valve CV-2 opens rapidly while valve CV-4 opens slowly, then a disparity in the energy balance will occur. Less vapor is produced in the reboiler than is condensed in the condenser, and the pressure drops. Fortunately, the distillation column tends to be self-regulating in response to pressure. If pressure decreases, the temperature decreases because of saturation conditions in the column. This increases the driving force for heat transfer in the reboiler and decreases the driving force for heat transfer in the condenser. This increases the boil-up and decreases the condensation of overhead vapor. Therefore, this results in an increase in the system pressure that tends to correct for the disparity in the energy balance.

For distillation columns with side products, the control strategy is more complicated. In general, for each additional variable that is to be regulated or controlled, an additional control loop must be added.

It is worth noting that most major column upsets arise from a change in the conditions of the feed stream to the column. These changes are most often caused by process upsets occurring upstream of the distillation column. The effects of sudden changes in column feed conditions can be significantly reduced by installing a surge tank upstream of the column. This tank acts as a buffer or capacitor by storing and mixing feed of differing composition. The overall effect is to dampen the amplitude of concentration fluctuations and reduce the impact of sudden changes on the column operation. The use of surge tanks is not restricted to distillation columns only. In fact, any place where a significant inventory of liquid is stored acts to dampen changes in feed conditions to the downstream units.

A surge tank reduces the effects of sudden changes in feed conditions.

21.7.3 A More Sophisticated Control System for a Binary Distillation Column

Figure 21.10 shows a control system for a binary distillation process, adapted from Skrokov [8]. The distillation unit is similar to the one shown in Figure 21.9. Valves CV-1, CV-2, CV-3, CV-4, and CV-5 are in the same location and perform the same function in both systems.

However, for the system shown in Figure 21.10, the concentrations and flows of the input stream, exit streams, and reflux stream are measured and sent to the monitor/analyzer. This unit performs energy and material balances and

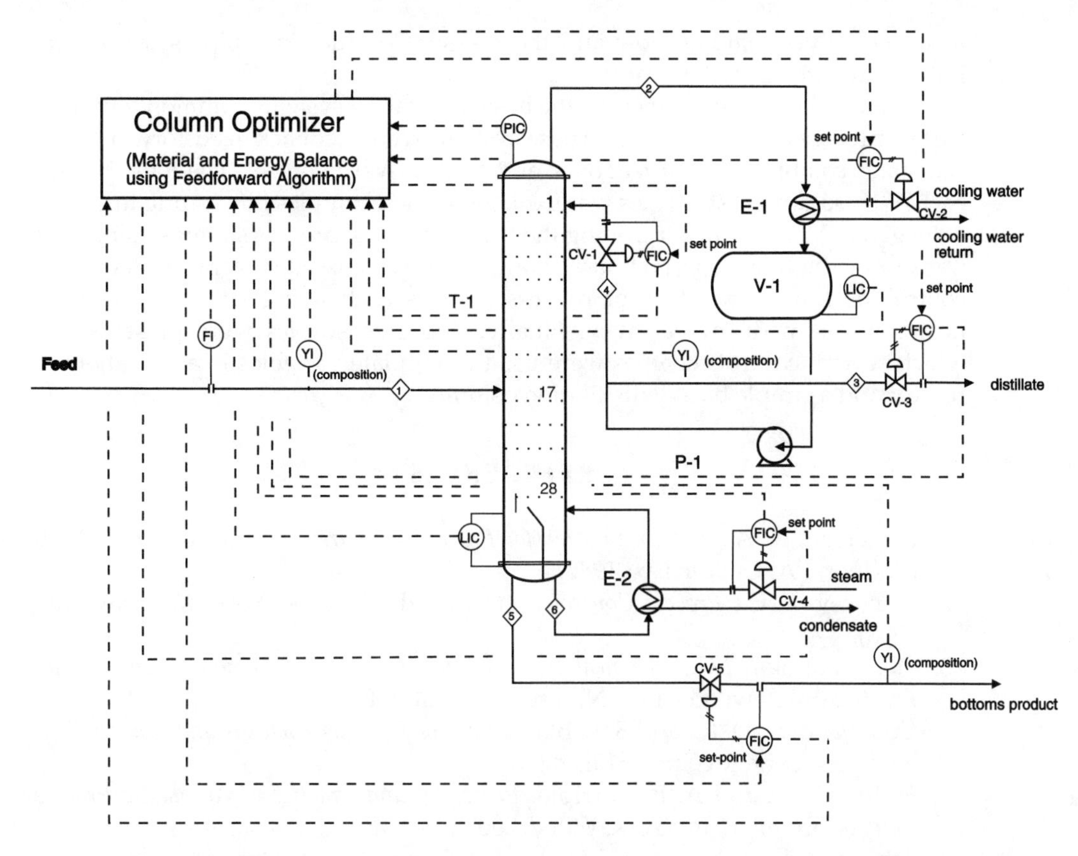

Figure 21.10 An Advanced Control Scheme for a Binary Distillation Column (Adapted from M. R. Skrokov, *Mini- and Microcomputer Control in Industrial Processes: Handbook of Systems and Applications Strategies* [New York: Van Nostrand Reinhold, 1980]. Reproduced by permission.)

uses performance relationships to evaluate new set points for the recycle, distillate, and bottom streams. It is a feed-forward system that predicts and makes changes in operations based on the predictions of a detailed process model.

Modern control capabilities involving sophisticated computer hardware and software also require a high degree of chemical engineering expertise in order to develop a system that optimizes the performance of process units.

21.8 SUMMARY

In controlling any chemical process, the final control element is almost always a valve that regulates the flow of a process or, more commonly, a utility stream. The basic construction and operation of a control valve were reviewed. The way in which valves regulate flow and the pressure profiles in a pipe system were illustrated with several examples.

In this chapter, we reviewed the basic regulation systems commonly used in simple process control schemes. These systems were feedback, feed-forward, cascade, and combinations of feedback and feed-forward control. The logic behind each of these control strategies was explained, and examples illustrating this logic were given. Methods of controlling the temperature of process streams using heat transfer media that undergo phase changes were discussed. Several strategies for controlling the temperature of process streams exchanging energy with these heat transfer media were given. Finally, two case studies were presented in which several control schemes were used to regulate variables in an exothermic reactor and a simple binary distillation column.

REFERENCES

1. Anderson, N. A., *Instrumentation for Process Measurement and Control,* 3rd ed. (Radnor, PA: Chilton Co., 1980).
2. Shinskey, F. G., *Process Control Systems,* 3rd ed. (New York: McGraw-Hill, 1988).
3. Stephanopolous, G., *Chemical Process Control: An Introduction to Theory and Practice* (Englewood Cliffs, NJ: Prentice Hall, 1984).
4. Coughanowr, D. R., and S. LeBlanc, *Process Systems Analysis and Control,* 3rd ed. (New York: McGraw-Hill, 2005).
5. Smith, C. A., and A. B. Corripio, *Principles and Practice of Automatic Process Control,* 3rd ed. (New York: Wiley, 2005).
6. Seborg, D. E., T. Edgar, and D. A. Mellichamp, *Process Dynamics and Control,* 2nd ed. (New York: Wiley, 2003).
7. Marlin, T. E., *Process Control: Designing Process and Control Systems for Dynamic Performance,* 2nd ed. (New York: McGraw-Hill, 2000).
8. Skrokov, M. R., *Mini- and Microcomputer Control in Industrial Processes: Handbook of Systems and Application Strategies* (New York: Van Nostrand Reinhold, 1980).

PROBLEMS

1. Consider an alternative reactor configuration for the cumene reactor, R-801, discussed in Section 21.7.1. The alternative is shown in Figure P21.1 and consists of a series of adiabatic catalytic packed beds through which the reactants pass. As the reactants and products pass through each bed, they are cooled by a cold shot of inert diluent—in this case, propane—before entering the next bed. The catalyst must be protected from excessive temperatures that cause it to sinter and to lose activity rapidly. For each of the control

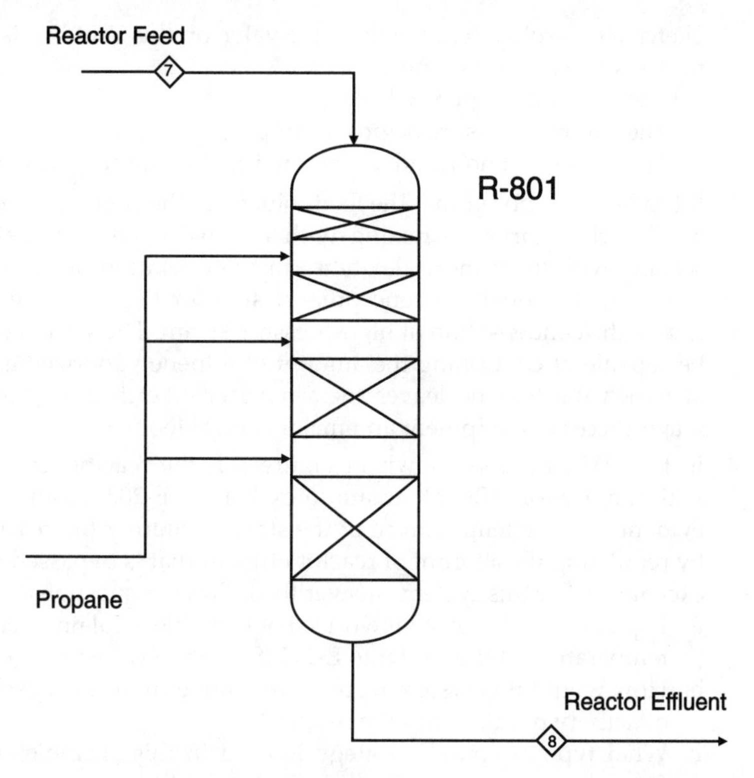

Figure P21.1 Alternative Configuration for Cumene Reactor from Problem 21.1

strategies given below, sketch the control loops that must be added to the system and discuss how the regulation is achieved—for example, "that is, as x increases, valve y opens, causing . . ."

a. Use the exit temperature from a given bed to adjust the amount of diluent fed to that bed.
b. Use the exit temperature from the previous bed to adjust the amount of diluent fed to the next bed.
c. To what types of control strategy do Parts (a) and (b) conform? Which scheme do you think is better for this type of reactor? Explain your reasoning.
d. Devise a control strategy that incorporates both the ideas used in Parts (a) and (b).

2. Consider the benzene purification column, T-101, in Figure 1.5 for the toluene hydrodealkylation process. The column feed contains noncondensables, mainly methane and a small amount of hydrogen, that must be vented from the system. The vent is taken off the top of the reflux drum, V-104.

Sketch a control system in which the valve on this vent line is used to control the pressure of the column.

Explain what happens when

a. The column pressure begins to drop
b. The amount of noncondensables fed to the column suddenly increases

3. It has been proposed that the feed toluene to the process shown in Figure 1.5 be vaporized prior to mixing with the hydrogen. It is believed that this scheme will allow more flexibility in the operation of the plant. You have been asked to devise a conceptual design for this vaporizer. The source of heat is the condensation of high-pressure steam. The system designed should be capable of controlling the amount of toluene vaporized and the pressure at which the toluene leaves the vaporizer. Sketch a diagram showing the major pieces of equipment and major control loops.
4. In the DME process shown in Figure B.1, the reactor feed exchanges heat with the reactor effluent stream in exchanger E-202. From the diagram, it is evident that the temperature of the stream entering the reactor is controlled by regulating the amount of reactor effluent that is bypassed around the heat exchanger. For this system, answer the following questions.
   a. Explain how this system works. For example, explain what happens if the temperature of the feed into E-202 (Stream 4) were to increase or decrease.
   b. How would this system respond to fouling in the heat exchanger or a loss of activity of catalyst in the reactor?
   c. What type of control strategy is used in this example—feedback, feed-forward, and so on?
   d. Design a control system that would regulate the exit temperature of the reactor (Stream 6) rather than the inlet stream temperature.
5. For the benzene column (T-101) in Figure 1.5, do the following.
   a. Implement a control scheme that will regulate the purity of the benzene product (Stream 15). You may assume that the purity of this stream can be evaluated by an on-line refractive index monitor placed on Stream 15.
   b. What is the type of control system that you have designed in Part (a) called?
6. Consider the feed section of the phthalic anhydride process in Appendix C, Figure C.5. A single control valve is used to control the flow of the combined vaporized naphthalene and air streams, Stream 7. In this process it is important to regulate both the flowrate of Stream 7 and the relative amounts of air and naphthalene in Stream 7. Does the current control scheme allow this type of regulation to occur? If not, devise a control scheme that will allow both these variables to be controlled independently.
7. Describe which form of control strategy (feed-forward, feedback, cascade, or a combination of these) best describes how a responsible person drives an automobile.

   What would be the consequence of using a feedback control strategy alone?

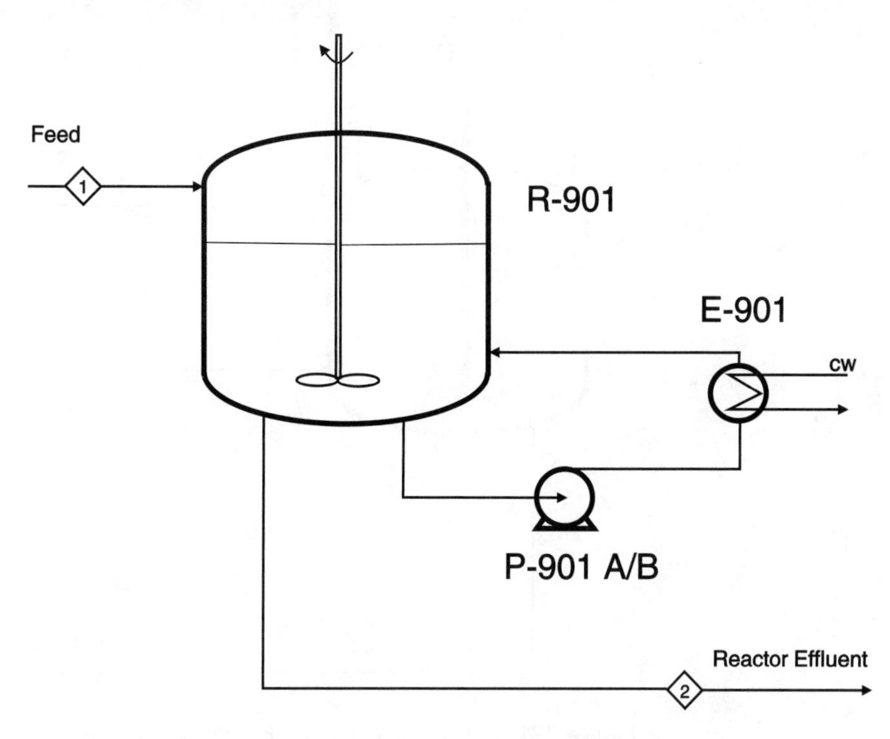

Figure P21.8 CSTR Configuration for Problem 21.8

8. Figure P21.8 shows a CSTR, R-901, carrying out a liquid-phase exothermic reaction. The feed, Stream 1, comes from an upstream unit, and its flowrate is known to vary. The heat of reaction is removed by circulating a portion of the contents of the reactor through an external heat exchanger. Cooling water is used as the cooling utility in E-901.

   For this reactor scheme, devise a system to implement the following control actions:

   a. Regulate the temperature of the reactor
   b. Regulate the inventory of the reactor, that is, keep a constant liquid level

   Explain how your control system would compensate for a change in the flowrate of Stream 1.

   What additional control scheme would be needed to maintain a constant conversion in the reactor, assuming that constant conversion can be achieved by ensuring constant residence time in the reactor?

9. Figure P21.9 shows a CSTR carrying out a liquid-phase, exothermic reaction. The heat released partially vaporizes the contents of the reactor. This vapor is condensed in an external condenser, and the condensate is returned to the reactor. The reactor operates at the boiling point (*T* and *P*) of the mixture in the CSTR. The vapor formed consists of reactants and products.

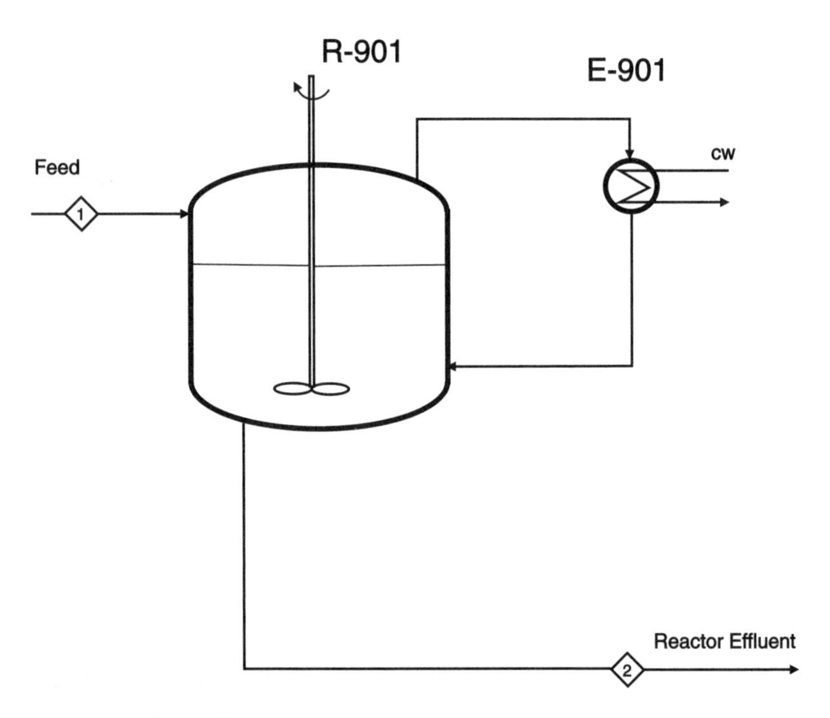

Figure P21.9 CSTR Configuration for Problem 21.9

For this system, devise a system to regulate the following variables:

a. Temperature in reactor
b. Residence time in the reactor

Explain how your system would respond to the following changes:

c. Increase in flowrate of feed
d. Fouling of the tubes (on cw side) of the condenser.

10. Consider the feed section of the toluene process given in Figure 1.5. The flow of fresh toluene, Stream 1, is regulated by monitoring the level in the toluene storage tank, V-101. If the level is seen to drop, then the valve on Stream 1 opens and vice versa. This is an example of a feedback material balance control loop. For this system, implement a feed-forward control scheme in which the flow of the recycle, Stream 11, is measured and used to control the flow of Stream 1. Do you foresee any potential problems with this control strategy?

CHAPTER

22

Process Troubleshooting and Debottlenecking

Imagine that you are responsible for a chemical process unit. The pressure in a chemical reactor begins to increase. You are concerned about material failure and explosion. What do you do? For a case such as this with potential catastrophic consequences, it may be necessary to shut the process down. However, process shutdown and start-up are very costly, and if a safe alternative were available, you would certainly want to consider it as an option. In another scenario, what would you do if it had been observed that the purity of product from your unit had been decreasing continuously for several days, and customers had begun to complain of poor product quality and have threatened to cancel lucrative contracts?

The situations described above may be classified as **process troubleshooting** problems. Once a plant is built and operating, it is anticipated that it will operate for a number of years (10–30 years). During this time, there will be times when the plant displays unusual behavior. This unusual behavior may represent a problem or a symptom of a problem that has not yet become apparent. The procedure for identifying the root cause of unusual behavior is part of troubleshooting. The other part is to provide guidance as to what action should be taken to correct the problem or, in the case of a symptom that has not yet resulted in a problem to prevent a problem from developing. Problems that affect process performance represent financial losses and potential safety hazards, so these problems must be quickly identified and resolved. The key to smooth plant operations is preventive action based on correct diagnosis of early symptoms. Troubleshooting problems associated with process start-up are beyond the scope of this text.

Even during a period of successful operation, the process does not operate at a steady state. Distillation units operate differently in summer than in winter, as well as between night and day, as a result of internal reflux changes resulting from heat losses or gains from the tower. Feed materials fluctuate, the temperature of cooling water changes, catalyst decays, heat exchangers foul, and so on. The control system responds to these changes and alters utility flows to maintain process streams at close to normal operating conditions.

The key to solving troubleshooting problems is to make use of the information regarding the process taken during periods of successful operation. Based on operating experience, the range over which changes can take place without a significant effect on the performance of the process is learned. Consequently, there is no single base case to represent process behavior as there was in Chapter 18. When comparing current operations to normal operations, we check to determine whether current operation lies within the range of normal operations. The range of normal process operation provides the base case.

Three steps can be identified to troubleshoot a process.

1. **Treat the Symptoms:** In this situation, the observed problem is addressed without investigation of the root cause. If the reactor pressure is increasing, find a way to relieve the pressure. This is a short-term solution. Because the root cause of the pressure increase has not been identified and addressed, the pressure may increase again. However, the immediate problem (an explosion) has been avoided, and there is now time to seek the root cause of the problem.
2. **Identify the Cause of the Problem:** Eventually, the cause of the problem should be diagnosed. This is particularly true if the problem recurs or if it is safety related. Because this may take time, the symptoms must continue to be treated.
3. **Fix the Problem:** Ultimately, the problem should be fixed.

Process troubleshooting involves solving open-ended problems for which there are likely to be several possible solutions. It is necessary for the engineer faced with such a problem to consider many identifiable solutions. Failure to consider a sufficient number of possible solutions may result in missing the actual solution.

Now, consider a different situation. It is necessary to determine how much scale-up is possible for the process for which you are responsible. You determine that one process unit can be scaled up only by 10%, whereas all other process units can be scaled up by at least 15%. The process unit that can be scaled up only by 10% is called a bottleneck. Elimination of this bottleneck is called **debottlenecking** and involves determining how to remove obstacles limiting process changes.

To put troubleshooting and debottlenecking problems in context, consider the input/output model shown in Figure 22.1. This model was first introduced in Chapter 16 and was used to define the design and performance problems. The input/output model can also be used to define troubleshooting and debottlenecking problems. Troubleshooting problems involve identification and correc-

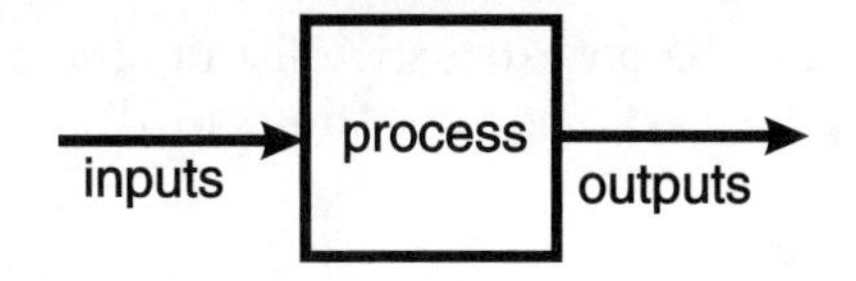

Type of Problem	Known	Unknown
design	inputs and outputs	process
performance	inputs and process	outputs
troubleshooting	observed Δoutput	causative Δinputs and/or Δprocess
debottlenecking	desired Δoutput or Δinput	portion of process limiting magnitude of change

Figure 22.1 Input/Output Relationships for Various Types of Problems

tion of the change in inputs and/or the process responsible for observed changes in outputs. Debottlenecking problems involve identification and modification of the portion of a process limiting the ability to change inputs or outputs.

In this chapter, a recommended methodology for attacking process troubleshooting problems is presented. Then five case studies of increasing complexity are presented that will serve to enhance the reader's skills in attacking troubleshooting problems. Finally, an example of a debottlenecking problem is presented.

22.1 RECOMMENDED METHODOLOGY

When solving complex problems such as troubleshooting and debottlenecking, having a reliable, personal problem-solving strategy is important. Problem-solving strategies are discussed briefly here. More details can be found elsewhere [1].

22.1.1 Elements of Problem-Solving Strategies

Three elements of successful problem-solving strategies are recommended here for attacking process troubleshooting problems (including debottlenecking problems).

This is not meant to be an exhaustive list of strategies; it is simply some of the strategies that can be used. The three strategies discussed here are

1. Brainstorming
2. Using known or observed data plus your understanding of equipment behavior
3. Considering the unexpected

Each is considered separately. They can also be combined into a general methodology for solving open-ended problems.

Brainstorming involves generating an extensive list of possible ideas. This need not be a formal process; it can be done informally and rapidly. In fact, the natural response to a troubleshooting situation is to think immediately and rapidly of several possible causes. If the product is not meeting specifications, think of all of the reasons that can be considered as potential causes, no matter how remote the possibility. Then brainstorm possible solutions. The main rule of brainstorming is that there are no bad ideas. The goal is to generate as many ideas as is possible. After brainstorming, only then is it time to evaluate all items critically to generate the most likely causes and solutions. Although brainstorming is usually a group activity, often it must be done individually. For a large, long-term problem, time can be taken to brainstorm in a group. For an everyday problem, especially an emergency situation, one should train oneself to brainstorm automatically. Brainstorming is a component of most problem-solving strategies.

When brainstorming a troubleshooting problem, consider all ideas, no matter how unusual they may seem.

When troubleshooting a chemical process, an **understanding of equipment behavior** should be used to narrow the list of possibilities. For example, in a staged separation using a mass separating agent, the Kremser equation (see Section 18.3.1) quantifies the relationship between process variables. The most important of these relationships for a variety of unit operations were summarized in Table 17.1. Use of these relationships will be illustrated in the case studies presented later in this chapter.

When there is a problem with process operation, the cause of the problem must be identified. The problem may or may not be located at the unit where poor operation is observed. For example, the output from the separator may be off spec due to lower-quality product exiting the reactor rather than due to poor separator operation. In troubleshooting plant problems, a vast amount of data exists that can be used to help identify the problem. These data would likely include current and historical operating conditions. In addition, the plant P&IDs would show where to look for additional current operating data. Any methodology for troubleshooting must consider this information, and any solution must be consistent with the operating data. Often, solving the problem is facilitated by selection of the data that will lead to identification of the problem.

Although a knowledge of equipment behavior is essential to solving troubleshooting problems, it is important that all alternatives, no matter how **unexpected**, be considered. If the pressure drop in a tray tower is increasing, a knowledge of equipment function suggests that there may be loading or flooding, so an increased liquid or vapor flowrate is a possible cause. However, was the possibility that someone left a toolbox in the downcomer during a recent maintenance shutdown also considered? The lesson here is to expect the unexpected! Consider all possibilities no matter how remote they may seem.

Given the open-ended nature of troubleshooting problems, their solution may best be attacked by creative problem-solving strategies. One such strategy, presented by Fogler and LeBlanc [1], is discussed here in the context of process troubleshooting. (Another similar strategy is presented in Chapter 23.) Their problem-solving strategy involves five steps:

1. Define
2. Generate
3. Decide
4. Implement
5. Evaluate

First, the correct problem must be defined. If the problem is incorrectly defined, it is likely that an incorrect solution will be found. If the product is not meeting specifications, this is the problem. However, if after further investigation, it is found that the stream leaving the reactor is not at design conditions but the reactor feed is at design conditions, it may be necessary to redefine the problem to be incorrect reactor performance.

Once the problem is defined, ideas must be generated. This is identical to brainstorming. It is important to generate as many ideas as possible. It is poor problem-solving strategy to focus on one possible solution or to assume that there is only one possible solution. For process troubleshooting, ideas may need to be generated both for the cause of the problem and for remedies to the problem.

Once ideas have been generated, the next step is to decide how to proceed. This is where knowledge of equipment can be used to select the most likely items from the brainstorming list to implement first. The next step is to implement the chosen solution.

Once the chosen solution method is implemented, it is necessary to evaluate the chosen solution. Is it working? If not, why not? Should another solution be implemented? If several solutions have been attempted, none of which appears to be solving the problem, this may be the time to think about whether there is an unexpected solution that can solve the problem.

22.1.2 Application to Troubleshooting Problems

A troubleshooting strategy is given in Table 22.1. It involves five steps. This sequence of steps is shown to parallel the problem-solving strategy of Fogler and LeBlanc discussed in Section 22.1.1.

Table 22.1 Strategy for Troubleshooting Existing Plants

Phase 1	Check out primary suspects. a. Verify the identified problem or symptom. b. Check input to the process. c. If only one unit is involved, check operating conditions of unit. d. Check for fully open or closed control valves.
Phase 2	Identify the unit operation producing the problem or symptom. a. System size is reduced systematically until the unit operation that is the source of the problem or symptom is determined. b. Inputs to each system are checked.
Phase 3	Perform a detailed analysis of unit operation uncovered in Phase 2 to determine and to verify the root cause of the problem or symptom.
Phase 4	Report your diagnosis of the root cause of the problem, and recommend action to remove the problem or symptom.
Phase 5	Report significant observations uncovered during the analysis that may be important to your organization.

Phase 1. Screen the whole process for the most common causes of problems or symptoms of problems in the process. This might involve brainstorming done informally as part of your thought process.

It is important to **define** the correct problem, or to determine whether one even exists. One common situation is false identification of a problem or symptom. An instrument could have been read incorrectly or could be broken, the analytical analysis (online or in lab) may not be correct, reagents may have been prepared incorrectly, and so on. In this situation, there is no problem with the process; it is a false indication of a problem. Never accept the initial problem identification without verification.

Once it has been determined that a problem really exists, the suspects should be screened. This is the **generate** step. A common cause of problems results from changes in process inputs. The adage "garbage in, garbage out" is universal and applies to chemical processes. Component flows into the process must be verified and compared to those for normal operations. If a problem appears as a result of process inputs, go to Phase 3. If a problem is known to involve a single unit (which may be true only for academic problems), consider possible unit malfunctions (some are listed under Phase 2). In this case, you should also go directly to Phase 3.

A third common cause results from limitations of the control systems on the utility streams used to maintain the temperature and pressure of process streams. If any control valves are found to be fully open or fully closed, there is a high probability that the desired control is not being achieved. All of the utility control valves are not normally shown on the PFD, and it is necessary to review the P&ID. All control valves should be checked. This information is used as input in later phases to identify the cause of the problem observed.

Phase 2. Locate the unit operation that is producing the problem or symptom. This is part of the **decide** step. The process is divided into subsections. If there is no obvious choice for selecting subsections, the process sections identified in the block flow diagram represent a reasonable starting point. Each system analyzed contains the stream identified as having a problem or symptom. Analyze system inputs. This identifies the subsection containing the cause of the problem. The subsection size is reduced, and the inputs are again analyzed. This continues until the unit operation producing the problem is identified. Then the operation of the identified unit should be checked. Ask key questions about each unit. Is there evidence of heat-exchanger fouling? Is the reflux ratio on the distillation column within its normal range? Are the temperature and pressure of the reactor at normal conditions?

Phase 3. Determine the root cause of the problem or symptom by a detailed analysis of the unit operation identified in Phase 1 or 2. This is the remainder of the **decide** step. Normal operation is used as a base case. This is the first place where the utility flows are analyzed. This involves using heuristics (Chapter 11), operating conditions of special concern (Chapter 6), calculation tools, and other material presented in Chapters 16–18.

Phase 4. Present all available evidence that establishes the root cause identified in Phase 3 as valid. Also, present any evidence that may not support the argument. Recommend action to be taken to correct the problem or treat the symptom. This is the **implement** step.

Phase 5. Present an evaluation of any significant observations that resulted from your analysis that could impact this or other processes within your company. This is the **evaluate** step. If you identify process improvements or potential for future problems, you are acting professionally. This is true especially if the suggestions are related to the environment or safety of personnel.

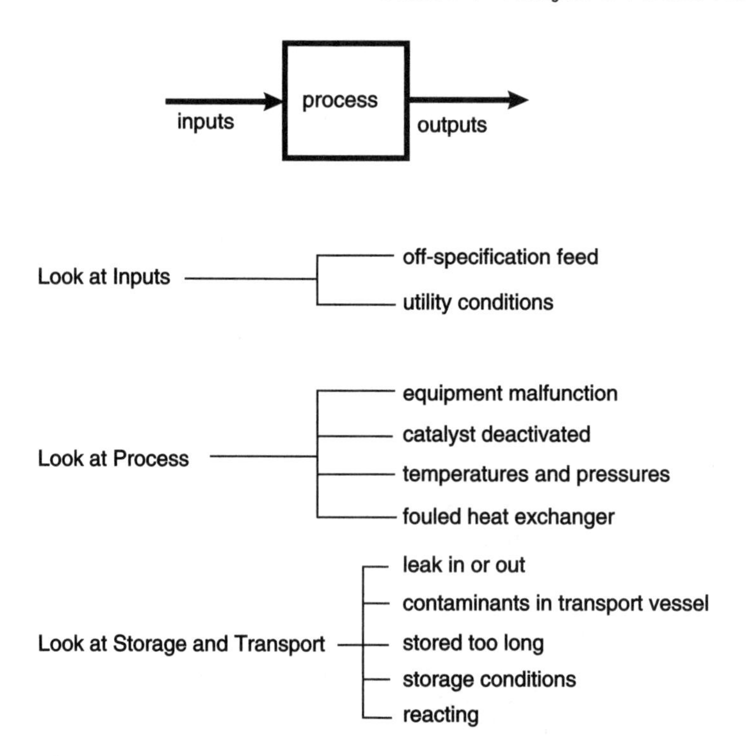

Figure 22.2 Suggestions for Solving Troubleshooting Problems

In summary, the cause of a change in output originates with a change in input and/or a change in process operation. Some possibilities are illustrated in Figure 22.2. This list is not meant to be exhaustive. It illustrates some possible causes for changes in output. The strategy discussed above is illustrated in the problems that follow.

22.2 TROUBLESHOOTING INDIVIDUAL UNITS

The first two case studies presented involve troubleshooting individual pieces of equipment. Although real troubleshooting situations usually involve an entire process, the necessary skills can be developed on simpler problems such as the two presented in this section.

22.2.1 Troubleshooting a Packed-Bed Absorber

The first troubleshooting problem involves a packed-bed absorber. The absorber has been designed to remove a contaminant from an air purge stream and has been operating for some time as designed. Then it is observed that the outlet air contains more contaminant than it should. A similar problem involving a tray absorber is given at the end of chapter.

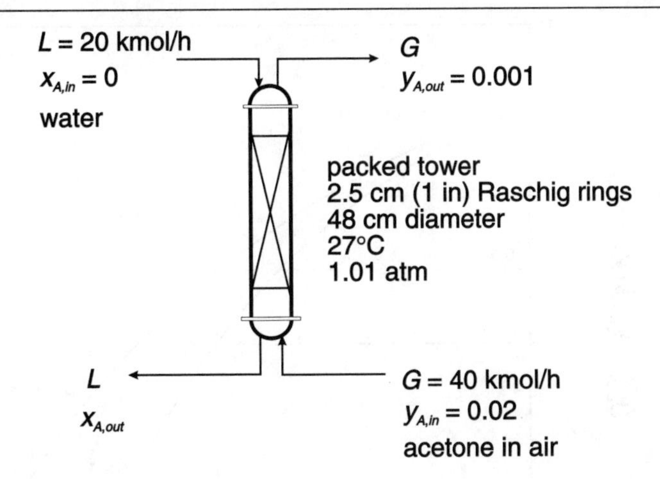

Figure 22.3 Packed Absorber for Troubleshooting Case Study

A packed absorber (Figure 22.3) has been designed to reduce the acetone concentration in 40 kmol/h of air from a mole fraction of 0.02 to 0.001. Acetone is absorbed into pure water at 20 kmol/h. Acetone is recovered from the effluent liquid, and the water, which is assumed pure, is recycled to the absorption unit. After a period of successful operation, it is observed that the exit acetone mole fraction in air is now 0.002.

The column is packed with 2.5 cm (1 in) Raschig rings and has a 48 cm diameter, which was obtained by designing for 75% of flooding. The column is assumed to operate isothermally at 27°C, and the nominal pressure is 1 atm. Raoult's law is assumed, and the partition coefficient for acetone, $m = y/x = P^*/P$, where $\ln P^* = 10.92 - 3598/T(\text{K})$, has been determined from tabulated data [2]. At 27°C and 1.01 atm, $m = 0.337$.

This problem, which involves dilute solutions, can be analyzed using the Colburn graph, Figure 18.15. On this graph, the interrelationship between the number of transfer units, N_{tOG}, the absorption factor, A ($A = L/mG$), and the mole fraction is defined. The base-case point can be located. The y-axis is at a value of 0.05, and $A = 1.48$. This gives $N_{tOG} = 6.2$. This is shown as point "a" on Figure 22.4.

The first step is to verify that the acetone concentration has indeed increased. This might involve having an operator or technician make flow and concentration measurements on the effluent stream in question. We will assume that the increased concentration at the normal flowrate has been verified. The next step in Phase 1 is to check process inputs and process operation (because only one unit is known to be involved) for potential causes for the observed change in output concentration. Changes in input are evaluated first. If the inputs are not found to be the cause of the problem, process operation is then investigated. Example 22.1 shows seven possible causes ("suspects") for the observed change in output concentration.

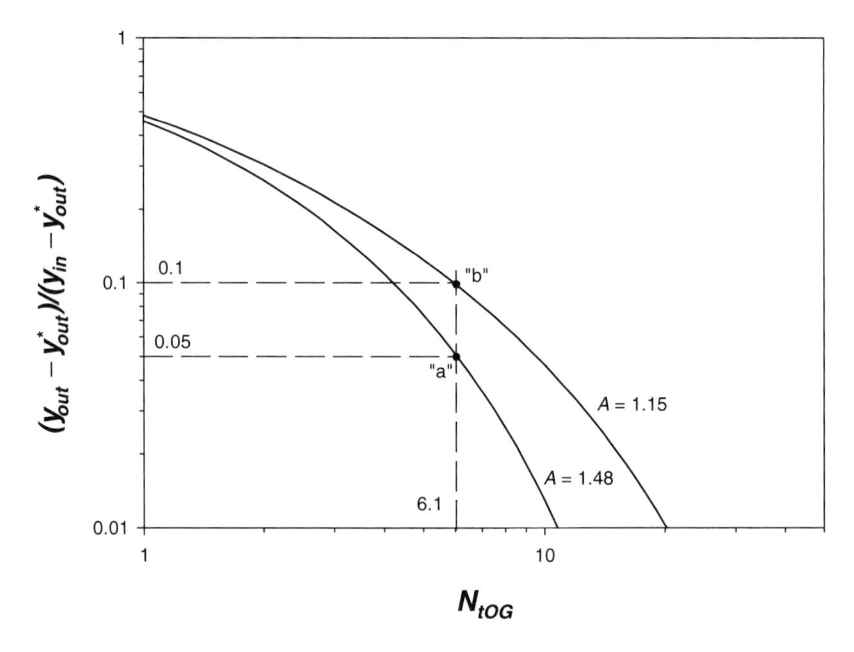

Figure 22.4 Use of Colburn Graph to Solve Example 22.1

Example 22.1

Generate a list of causes for the observed change in absorber output. First, examine input changes, and then examine process operation changes.

The knowledge of packed-bed absorber performance illustrated on the Colburn graph is a good starting point. Any or all of these parameters could be different from design conditions.

Potential input problems include the following.

1. **Increased Flowrate of Gas To Be Treated:** A disturbance in A is a possible cause. Intuitively, if it is the gas rate, G, that has been disturbed, an increased gas rate would be the cause of an increased acetone concentration in the exit gas stream. This would result in a decrease in A.
2. **Decreased Flowrate of Water:** This would also result in a decrease in A.
3. **The Water Does Not Enter Acetone Free:** If the entering water contains acetone, intuitively, less acetone can be absorbed.
4. **There Is More Acetone in the Feed:** If the mole fraction of acetone in the feed were increased, the same fractional removal of acetone from air would result in a higher acetone mole fraction in the exit air stream.

These four items would be checked first to determine whether they are the cause of the faulty absorber performance. If they were not found to be the cause, the process operation would be investigated. Potential process operation problems include the following.

5. **A Decrease in Column Pressure:** Intuitively, a decrease in column pressure favors the vapor phase. Mathematically, because $A = L/mG$, and $m = P^*/P$, a decrease in P increases m, causing a decrease in A.
6. **An Increase in Column Temperature:** Intuitively, an increase in column temperature favors the vapor phase. Mathematically, because P^* increases with temperature, so does m, causing a decrease in A.
7. **There Is Channeling in the Packed Bed:** If there were channeling in the packed bed, not all of the available (and designed for) mass transfer area would be used. This could cause faulty absorber performance.

To reiterate, the process inputs would be checked. If they were not found to be the problem, then process operation would be investigated. The next step would be to check for fully open or closed control valves. In this problem, it is assumed that control valves are not involved. Phase 2 is also not applicable because there is only one unit operation involved in this problem.

Phase 3 involves a detailed analysis of the suspects. In Example 22.1, items 1, 2, 5, and 6 can be represented on the Colburn graph. For these items, diagnosis is that the absorption factor has decreased, which moves the operating point for the column vertically at constant $N_{tOG} = 6.2$ to point "b." The new absorption factor is 1.15. The problem could be in any (or all) of the parameters of the absorption factor. L could have decreased to 17,391 mol/h, or G could have increased to 46,000 mol/h. Alternatively, the value of m could have changed to 0.388, meaning that the temperature of the column increased to about 30°C (30.5°C), or that the column pressure decreased to 0.87 atm. However, because the air stream discharges to the atmosphere, a decrease in column pressure below 1 atm is not possible.

There is an alternative diagnosis, however, which is item 3 in Example 22.1. The operating point can remain fixed in the original position (point "a"), but the outlet acetone concentration in air increased due to the presence of acetone in the water fed to the column. This makes the second term in the numerator and denominator of the y-axis nonzero. Solution for the inlet acetone concentration in water yields a mole fraction of 0.00312.

Thus far, five possible causes for the observed increase in outlet acetone concentration in air have been identified. There are certainly additional possible causes, some of which are associated with equipment operation such as liquid distribution, channeling, and fouling, which could also contribute to the observed performance decrease. One of these is identified as item 6 in Example 22.1.

At this point, there is not enough information given to complete Phase 3 to identify the root cause of the problem. The next step would be to measure the input flows and concentrations, column temperature, column pressure, and column pressure drop (a measure of channeling). This should allow identification and verification of the root cause of the problem.

If the root cause of the reduced performance of the absorber is understood, then possible methods of compensation are straightforward (Phase 4). Here, it is assumed that compensation cannot be achieved by altering the cause of the disturbance; that is, if the cause is an increased gas rate, then the gas rate cannot be lowered. However, in the context of a chemical process, the possibility of reversing the disturbance should be investigated.

If the gas rate is too high, the liquid rate can be increased to compensate. However, flooding could be a problem, especially if both gas and liquid rates are increased. The Colburn graph does not account for flooding, which is specific to a given packed column.

If the liquid rate is too low, it is unlikely that the gas rate can be decreased without scaling down the entire process. A better choice might be to decrease the temperature of the absorber to 22°C, increase the pressure in the absorber to 1.15 atm, or make a combination of changes in temperature and pressure, in order to make the absorption equilibrium more favorable and bring the absorption factor back to a value of 1.48.

If a temperature increase is the problem and altering flowrates is not desirable due to flooding considerations, one possible compensation is to alter the pressure. Another might be to decrease the temperature of the water used to remove the acetone from the air. Increasing the liquid rate moves the column toward flooding, but a small increase should not be a serious problem.

Finally, if the cause of the disturbance is acetone in the water, compensation can be accomplished by decreasing the temperature, increasing the liquid rate, or increasing the pressure. For this situation, the stripping column used to remove scrubbed acetone from the water stream should also be investigated, because the cause of the faulty absorber performance may lie in an adjacent piece of equipment.

In general, adjusting the temperature or pressure is probably the best method of compensation, because flooding is not an issue. Of course, there can be multiple causes of the disturbance, and compensation can be achieved by adjusting two variables by smaller amounts rather than by adjusting only one variable.

As an example of Phase 5, suppose that the root cause was determined to be an increase in acetone content of the scrubber liquor (water). It is necessary for you to notify anyone else using the same water supply of the acetone contamination. Similarly, suppose that channeling in the absorber was identified as the problem. It is necessary to report the operating conditions that caused the channeling to other parts of the company using the same packing material.

To review, this relatively simple problem illustrates how there can be multiple possible causes and multiple solutions to a troubleshooting problem. The strategy presented here for solving troubleshooting problems was illustrated.

22.2.2 Troubleshooting the Cumene Process Feed Section

This problem deals with the feed section to the cumene process in Appendix C, Figure C.8. This problem is actually part of the problem presented in Appendix C, and the process flowsheet along with calculations associated with the pumps can be found there. The problem is restated here.

A problem has recently arisen regarding the feed pumps to the cumene process. A maintenance check showed that P-802 (propylene feed pump) needed a new bearing, and a new one was installed. Premature bearing failure in this pump, often associated with cavitation, has occurred several times. The latest problem occurred during a recent warm spell, when the ambient temperature reached 110°F. The same maintenance check showed that P-801 (benzene feed pump) was fine. The ambient temperature has now returned to an average of 70°F, and both pumps seem to be working fine. Suggest a diagnosis and a method for compensation for the problem with P-802. In the process description in Appendix C, it is stated that propylene is stored in a tank as a saturated vapor/liquid mixture with liquid drawn from the tank as feed, and that liquid benzene is stored in a tank (most likely with an inert vapor blanket such as nitrogen) at atmospheric pressure.

In this case, the problem is immediately verified; the bearing had to be replaced. A check of the pump input might identify a contaminant that could have caused the bearing to deteriorate. We will assume that no such contaminant was found. Because the unit operation producing the problem has already been identified, we skip to Phase 3 to identify the root cause of the problem.

An intuitive understanding of pump operation suggests that the primary suspect is cavitation of P-802. This is consistent with the need for replacement of a bearing because cavitation can damage the internals of a pump. This may also be consistent with the recent warm spell, because the available Net Positive Suction Head, $NPSH_A$, often decreases at higher temperatures due to increasing vapor pressure. Recalling the discussion in Section 18.2.4,

$$NPSH_A = P_{inlet} - P^* \tag{22.1}$$

Because the vapor pressure P^* increases with increasing temperature, it seems logical that the available NPSH decreased during the warm spell, causing cavitation that damaged the pump bearing. Example 22.2 shows the results of an analysis of the propylene feed pump.

Example 22.2

Calculate the $NPSH_A$ for the propylene feed pump under normal ambient conditions (70°F) and for the warm spell (110°F).

Equation (22.1) can be rewritten as in Section 18.2.4, with an expression for P_{inlet} substituted.

$$NPSH_A = P_{tank} + \Delta P_{static} + \Delta P_{friction} - P^* = P_{tank} + \rho g h - \frac{2\rho f L_{eq} u^2}{D} - P^* \tag{E22.1}$$

For the case of propylene being stored as a vapor/liquid mixture, the pressure in the tank is equal to the vapor pressure. As a consequence, $NPSH_A$ does not change with temperature because $P_{tank} = P^*$. Therefore, the $NPSH_A = 7$ feet of liquid at the **level alarm low** (LAL—the tank level at which an alarm goes off warning of too low a liquid level in the tank) in the tank at all temperatures, as per the calculations shown in Appendix C.

The lesson learned from Example 22.2 is that although a belief in the understanding of how equipment works can help you focus in on a solution to a troubleshooting problem, the tendency to focus on the first solution or on only one solution can lead to an erroneous solution.

Resist the temptation to focus on only one solution or the first solution that comes to mind.

The reason why the pump malfunctioned is still not clear. If cavitation were the reason, the $NPSH_A$ expression should be investigated further. It is now known, for this situation, that

$$NPSH_A = \Delta P_{static} + \Delta P_{friction} \tag{22.2}$$

so we are seeking reasons why this expression may have decreased in value. There may be reasons for the pump malfunction not associated with the $NPSH_A$. Therefore, five additional possibilities that should be analyzed (Phase 3) for causing the pump malfunction are as follows.

1. The static pressure has decreased. Perhaps the level in the tank has fallen below the low alarm level and the alarm has failed. Therefore, these items should be checked.
2. The frictional pressure has increased. Perhaps the flowrate from the tank has increased, causing an increase in frictional losses. The flowrate monitors and the settings on the control valve after the pump should be checked. This scenario would have additional consequences, because an increased propylene flowrate would have an effect on reactor conversion and possibly on product production rate and purity. Therefore, the flowrate and purity of the reactor effluent should also be checked.
3. The bearing in the pump wore out with age. Just because pump cavitation often causes damage to pump internals such as bearings does not mean that it was the cause of this pump's malfunction. Given the analysis above, it is possible, indeed likely, that the pump malfunction observed here was simply caused by a worn bearing.
4. Mechanical problems in the pump. The bearing could be wearing out prematurely due to poor shaft alignment within the pump.
5. There is a manufacturing defect within the pump. If the pump shaft or other internals were defective (incorrect size, for example), the bearing might wear out prematurely.

The recommended action (Phase 4) will depend upon which (if any) of the five causes of bearing failure listed above is identified to be the problem. If problems are found with the pump (item 4 and/or 5), it would be appropriate for you to notify others in your company using the same pump or similar pumps from the same manufacturer of the problem you have had with your pump.

22.3 TROUBLESHOOTING MULTIPLE UNITS

In this section, two troubleshooting case studies are presented that involve multiple unit operations. One of the lessons of these case studies is that the symptoms of a problem are not necessarily observed at the source of the problem. In these two case studies, only one solution is discussed. Generation of alternative possibilities is the subject of problems at the end of the chapter.

22.3.1 Troubleshooting Off-Specification Acrylic Acid Product

This problem concerns an acrylic acid production process similar to the one in Appendix B, Figure B.10.1. The process flow diagram, stream flows, and equipment specifications are presented there.

At another acrylic acid plant owned by your company, process shutdown for modifications and improvements has recently been completed, and the process has been started up once again. Customers have begun to complain that the acrylic acid product does not appear to be meeting specifications. They have observed that the acrylic acid has a yellowish color, which is different from the clear liquid they had previously received. Their tests also found that the viscosity of the acrylic acid has increased.

The following process modifications were completed during the recent shutdown.

1. A new catalyst that is supposed to minimize side reactions was installed in the reactor. The reactor specifications were not changed.
2. A new solvent is now being used in the extraction unit. It is less expensive than the previous solvent, and the performance of the extraction unit is supposed to be unchanged. No modifications were made to the extraction equipment.
3. As a cost-cutting measure, refrigerated water (entering at 10°C) has been replaced by cooling water (entering at 30°C) in the acrylic acid purification column. The column has 25 actual trays with 2.25 in weirs, a total condenser, and a partial reboiler.

The following operating restrictions are also known. For the new catalyst, the operating range is between 250°C and 350°C and between 1 bar and 5 bar. Once the acrylic acid has been produced and condensed into the liquid phase, the temperature is to be maintained below 90°C to avoid polymerization of the acrylic acid.

The first step is to verify the problem. Let us assume that you had a technician take a sample of acrylic acid product, and the yellowish color and increased viscosity were both verified. This means that the problem is not in the shipping and storage steps, but is within the plant. Therefore, the problem is yours, not the customer's. Further, let us assume that all process inputs have been checked and found to be within normal operating conditions, and that a check has found no control valves to be fully open or closed.

There are several possible causes for the off-specification acrylic acid product. The most likely causes are changes that may have occurred during shutdown. Perhaps the new catalyst is not performing as designed. Perhaps the new solvent is contaminating the product. The reactor effluent and the extractor effluent streams should be checked for contamination. Suppose that has been done, and everything has been found to be within normal conditions. Therefore, the most likely cause involves the temperature in the acrylic acid distillation column. In this column, acrylic acid and acetic acid are separated. It will be assumed that this is a simple binary distillation.

There must have been a reason for using refrigerated water in the distillation column prior to the recent shutdown. This can be understood using an analysis similar to the one in Section 19.2. From tabulated data [2], the following vapor pressure expressions can be obtained for acrylic acid and acetic acid:

$$\text{acrylic acid} \qquad \ln P^* \text{ (mm Hg)} = 19.776 - \frac{5450.06}{T(\text{K})} \tag{22.3}$$

$$\text{acetic acid} \qquad \ln P^* \text{ (mm Hg)} = 18.829 - \frac{4786.41}{T(\text{K})} \tag{22.4}$$

Because acrylic acid is the heavier component, the bottom of the column must remain below 90°C to avoid undesired acrylic acid polymerization. From Equation (22.3), assuming that pure, saturated acrylic acid leaves the bottom of the column, the pressure at the bottom of the column is 118 mm Hg, which is 5.26 ft of liquid assuming that liquid acrylic acid and liquid acetic acid have the same density as liquid water. The pressure drop per tray will be approximated by the height of liquid on the tray. The height of liquid on each tray will be approximated by the weir height. Because there are 25 trays, the pressure drop in the column is

$$25 \text{ trays}[(2.25/12)\text{ft liquid/tray}] = 4.69 \text{ ft liquid}$$

Therefore, the pressure at the top of the column is 5.26 – 4.69 = 0.57 ft liquid = 12.8 mm Hg. From Equation (22.4), assuming pure, saturated acetic acid at the top of the column, $T = 21°\text{C}$. Because cooling water enters at 30°C, it is not possible to condense acetic acid at 21°C with cooling water. This is why refrigerated water was used in the original design.

The above discussion suggests that the switch to cooling water could be one reason for the off-specification acrylic acid product. With cooling water available at 30°C, it is not possible for the top of the column to be less than 30°C, which places a lower bound on the top pressure of the column. Because the pressure drop in the column is fixed by the weir height on the trays, there is also a lower bound on bottom pressure in the column, which places a lower bound on the bottom temperature, which is greater than 90°C, thereby promoting polymerization.

Apparently, this column has no control room pressure reading (perhaps the instrument is out of order). Therefore, an operator would be sent to measure the pressure at the top of the column.

There is one simple remedy. Refrigerated water should be used in the condenser. If this is not desirable for the long term, one could consider the incremental economics of modifying the trays to have lower weirs, which could happen only at the next plant shutdown. It is also necessary for you to report this problem to any other plants within your company making acrylic acid so that they can avoid (or correct) problems arising from the use of cooling water instead of refrigerated water.

It is observed that rough calculations involving certain reasonable, simplifying assumptions were used to obtain an approximate result very quickly. It is neither necessary nor desirable to do detailed calculations when screening alternatives in a troubleshooting problem.

When screening alternatives, rough calculations using reasonable approximations are more useful than detailed simulations.

The approximations that were made to facilitate a rapid calculation for the distillation column were that the top and bottoms products were pure and that the height of liquid on the trays equaled the weir height. Although these are not exactly true, detailed calculations would show that these are good approximations.

22.3.2 Troubleshooting Steam Release in Cumene Reactor

This problem involves the reactor in a cumene process similar to the one in Appendix C, Figure C.8. The process flow diagram is presented in Appendix C.

Our company has been testing a new cumene catalyst at a facility producing the identical amount of cumene as the one in Appendix C. The new catalyst completely suppresses the undesired DIPB formation reaction. However, the reaction rate for the desired reaction is lower. Therefore, when the new catalyst is used in the existing reactor, a single-pass conversion of only 50% is obtained. This new catalyst, which is less expensive than the previous catalyst, is known to have a higher initial activity that decays rapidly to constant activity. During the initial activity period, a 33% increase in cumene production has been observed. The operating parameters of the benzene distillation column have been altered, the recycle benzene stream has been increased, the DIPB column has been taken off-line, and the plant has been producing cumene successfully.

The reactor for cumene production, which is shown in Figure 22.5, is of the shell-and-tube design, with catalyst in the tubes and boiler feed water vaporized to form high-pressure steam in the shell. The pipe to the steam header is 2 in schedule 40 and contains 32.5 m of equivalent pipe length after the regulating valve. Under normal operating

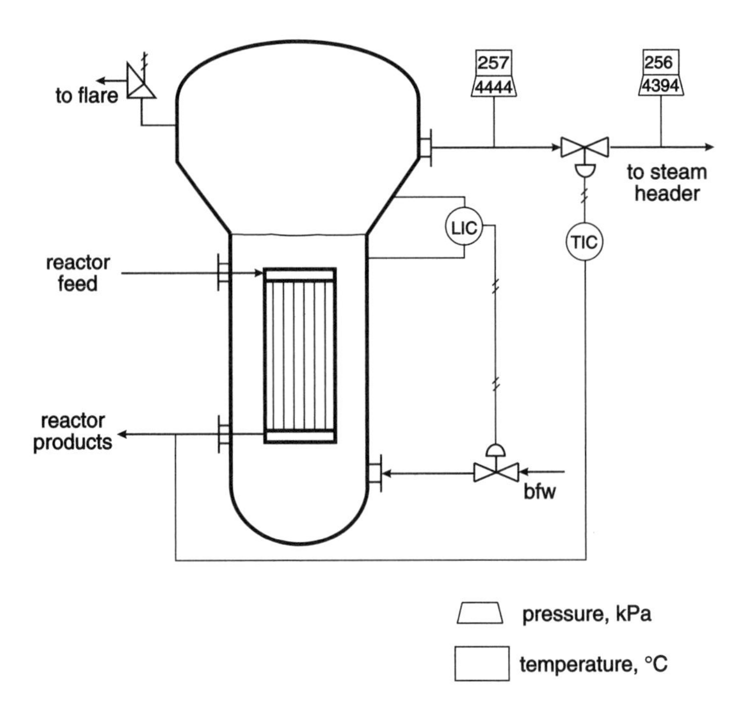

Figure 22.5 Cumene Reactor in Case Study

conditions, the pressure drop across the regulating valve is 50 kPa. There is a pressure-relief valve on the shell side of the reactor rated at 4500 kPa. The tubes are completely submerged in the boiler feed water. As part of the test of the new catalyst, the plant has been shut down briefly every three months for the past year, and the catalyst has been replaced so that the spent catalyst could be studied.

After start-up subsequent to each shutdown, the pressure-relief valve on the shell side of the reactor has opened periodically for a few days. Because only steam was released and it is not in a part of the plant where anyone could be harmed, the problem was attributed to start-up transients and was ignored. A recent safety audit has resulted in the suggestion that the cause of this problem be identified and corrected.

The problem is verified by the observation of a steam release. Let us assume that you have checked all process inputs, the operation of all other units, and control valves and you have decided that the problem must be at the reactor. A check of the reactor feed and the boiler feed water input reveals normal conditions. Therefore, the problem must be with reactor operation, which points to the higher-activity catalyst.

First, it is necessary to calculate some operating parameters for the design conditions. Under normal operating conditions, from Appendix C, the heat generated in the reactor is 9800 MJ/h. The amount of steam formed can be calculated.

$$Q = \dot{m}_{stm}[C_p(T_{stm} - T_{bfw}) + \lambda_{stm}] \tag{22.5}$$

Solving for the mass flow of steam yields

$$\dot{m}_{stm} = \frac{9{,}800{,}000 \text{ kJ/h}}{[4.35 \text{ kJ/kg°C}(256 - 90°\text{C}) + 1700 \text{ kJ/kg}]} = 4046 \text{ kg/h}$$

Here, the heat capacity of water is taken at an average temperature of 173°C, and the temperature of steam formed in the shell corresponds to the pressure in the shell calculated below. Because high-pressure steam is made, the pressure of steam downstream of the reactor is that of high-pressure steam, 4237 kPa. Under design conditions, the pressure of steam in the reactor exceeds this value to account for the pressure drop in the pipe leading to the steam header and across the regulating valve. The density of the steam varies with the pressure. It will be assumed that the average steam conditions between the reactor shell and the steam header are 255°C and 4300 kPa. At these conditions, the density of steam is about 17.6 kg/m^3. For 2 in schedule 40 pipe ($D = 0.05250$ m), at the given mass flow and density of steam, the velocity in the pipe is 28.3 m/s, so the Reynolds number is

$$\text{Re} = \frac{(0.0525 \text{ m})(28.3 \text{ m/s})(17.6 \text{ kg/m}^3)}{1.88 \times 10^{-5} \text{ kg/m s}} = 1.39 \times 10^6$$

For commercial steel pipe, $e/d = 4.6 \times 10^{-5}$ m/0.0525 m = 0.0009. From Figure 17.1, the friction factor $f = 0.009$. Therefore, the pressure drop in the pipe leading to the steam header is

$$\Delta P = \frac{2f\rho L_{eq}u^2}{D} = \frac{2(0.009)(17.6 \text{ kg/m}^3)(32.5 \text{ m})(28.3 \text{ m/s})^2}{0.0525 \text{ m}} = 157 \text{ kPa}$$

The pressure in the reactor shell, under normal conditions, is the pressure at the steam header plus the pressure drop in the pipe plus the pressure drop across the valve, which is

$$P_{rxr} = 4237 + 157 + 50 = 4444 \text{ kPa}$$

The situation immediately after catalyst replacement can be analyzed using base-case ratios. If the activity of the new catalyst is higher than the old catalyst, then the reaction rate is increased. Because the only reaction involved is for cumene production, 33% more steam is produced. Therefore, the steam velocity in the pipe leading to the steam header is increased by 33%. Because the flow is fully turbulent, the friction factor remains constant, and, assuming unchanged density,

$$\frac{\Delta P_2}{\Delta P_1} = \left(\frac{v_2}{v_1}\right)^2 = 1.33^2 = 1.77 \tag{22.6}$$

where subscript 2 refers to new conditions and subscript 1 refers to design conditions. Therefore,

$$\Delta P_2 = 157 \text{ kPa}(1.77) = 278 \text{ kPa}$$

If it is assumed that the control system has responded by opening the regulating valve completely, so there is no pressure drop across the valve, then the pressure in the reactor shell is

$$P_{rxr} = 4237 + 278 = 4515 \text{ kPa}$$

which exceeds the rating for the pressure-relief valve, causing it to open and release steam.

Now that the problem has been identified, the next question is how to compensate for this problem. It is likely that necessary changes can be made only during the next shutdown, which is not a problem because there are no releases for most of an operating cycle. One alteration would be to reset the pressure-relief valve to a higher pressure. Caution is warranted here because the 4500 kPa limit was originally chosen for a reason. The pressure ratings for the materials of construction and reactor design should be checked very carefully to determine their limits.

Another simple solution would be to replace the line leading to the steam header with larger-diameter pipe. The next larger size, 2.5 in schedule 40, has a diameter $D = 0.06271$ m. Using a relationship from Table 17.1, a base-case ratio at the original flowrate is

$$\frac{\Delta P_2}{\Delta P_1} = \left(\frac{D_1}{D_2}\right)^5 = \left(\frac{0.05250}{0.06271}\right)^5 = 0.41$$

So the pressure drop, under normal operating conditions with the new pipe, is

$$\Delta P_2 = 157 \text{ kPa}(0.41) = 64.4 \text{ kPa}$$

and the pressure drop with the new pipe with increased catalyst activity is

$$\Delta P_2 = 64.4 \text{ kPa}(1.77) = 114 \text{ kPa}$$

and the pressure in the reactor is

$$P_{rxr} = 4237 + 114 = 4351 \text{ kPa}$$

With the larger-diameter pipe, it is possible for there to be a pressure change across the regulating valve without exceeding the cutoff pressure of the relief valve.

Finally, you should report these results and explanations clearly so that proper modifications can be made in similar plants prior to switching to the new catalyst.

22.4 A PROCESS TROUBLESHOOTING PROBLEM

In this section, a troubleshooting problem involving an entire process is presented [3]. It is based on the production of cumene problem presented in Appendix C, Figure C.8. The process flowsheet and stream flow table are included in Appendix C.

Lately, Unit 800 has not been operating within standard conditions. We have recently switched suppliers of propylene; however, our contract guarantees that the new propylene feed will contain less than 5 wt% propane.

Upon examining present operating conditions, we have made the following observations.

1. Production of cumene has dropped by about 8%, and the reflux in T-801 was increased by approximately 8% in order to maintain 99 wt% purity. The flows of benzene (Stream 1) and propylene (Stream 2) have remained the same. Pressure in the storage tanks (not shown on flowsheet) has not changed appreciably when measured at the same ambient temperature.
2. The amount of fuel gas being produced has increased significantly and is estimated to be 78% greater than before. Additionally, it has been observed that the pressure control valve on the fuel gas line (Stream 9) leading from V-802 is now fully open, whereas previously it was controlling the flow.
3. The benzene recycle Stream 11 has increased by about 5%, and the temperature of Stream 3 into P-201 has increased by about 3°C.
4. Production of steam in the reactor has fallen by about 6%.
5. Catalyst in the reactor was changed six months ago, and previous operating history (over the last ten years) indicates that no significant drop in catalyst activity should have occurred over this time period.
6. p-diisopropyl benzene (p-DIPB) production, Stream 14, has dropped by about 20%.

Suggest possible causes and potential remedies for the observed problems.

The reactions are as follows:

$$\underset{propylene}{C_3H_6} + \underset{benzene}{C_6H_6} \rightarrow \underset{cumene}{C_9H_{12}} \tag{22.7}$$

$$\underset{propylene}{C_3H_6} + \underset{cumene}{C_9H_{12}} \rightarrow \underset{p\text{-diisopropyl benzene}}{C_{12}H_{18}} \tag{22.8}$$

Assume that all of the above symptoms have been verified. The next step would be to check process inputs, which might reveal off-specification feed. This might immediately identify the problem. A check of the control valves verifies that only the valve in Stream 9 is fully open, as stated in Observation 2. If there were no problems with the feed, the next step would be to check the individual units to determine which ones were not operating within normal limits. As part of Phase 2, let us analyze the six observations above to determine what they suggest.

An analysis of the six observations above suggests the following.

1. Observation 1 suggests that either less cumene is being produced in the reactor or that significant cumene is being lost as fuel gas. If it is assumed that the feeds are unchanged, these are the only possibilities.

2. Observation 2 suggests that the fuel gas rate has increased significantly. Components of fuel gas could be cumene, unreacted propylene and benzene, propane, and p-DIPB.
3. Observation 3 suggests that additional benzene is being processed in the distillation column. The temperature increase could be due to the increased concentration of benzene relative to propane and propylene at the top of T-801.
4. Observation 4 suggests that less cumene is being formed in the reactor. This is the opposite situation to that in the case study presented in the previous section.
5. Observation 5 suggests that catalyst deactivation should not be a problem. This does not necessarily guarantee that catalyst deactivation is not a problem.
6. Observation 6 suggests that the selectivity for the desired reaction, Equation (22.7), has increased.

Among the possible causes of some of these observations are the following.

1. The propylene feed contains propane impurity in excess of 5 wt%. Even though the new supplier claims that the propylene meets specifications, it is possible that the propylene is off specification. This would be verified in Phase 1. If the propylene feed contains undetected excess propane, it is likely that the feed rate would be unchanged. (This could also be verified.) Therefore, the concentration of propylene in the reactor feed would decrease, thereby decreasing the reaction rate. Conversion in the reactor would then decrease. An examination of Figure 22.6, the approximate reactor profiles, shows that a decrease in reactor conversion would have a larger percentage effect on the p-DIPB, because its concentration is lower. Examination of the kinetics, which are based on the assumption that Equations (22.7) and (22.8) are elementary steps, also supports this diagnosis. If the concentration of propylene is decreased, the rates of Equations (22.7) and (22.8) decrease. This decreases the concentration of cumene in the reactor, which causes the rate of Equation (22.8) to decrease even further, reducing the p-DIPB concentration more than the cumene concentration. If this reactor scenario were true, then there would be additional propane leaving the reactor. This would increase the flow of fuel gas, which is mostly propane and propylene, and increase the recycle, which is mostly benzene. It is seen that this scenario is consistent with all six observations.
2. The catalyst is defective and has begun to deactivate. This could be checked by analyzing samples of the reactor input and reactor output. If the catalyst is defective and has begun to deactivate, all reaction rates would decrease.

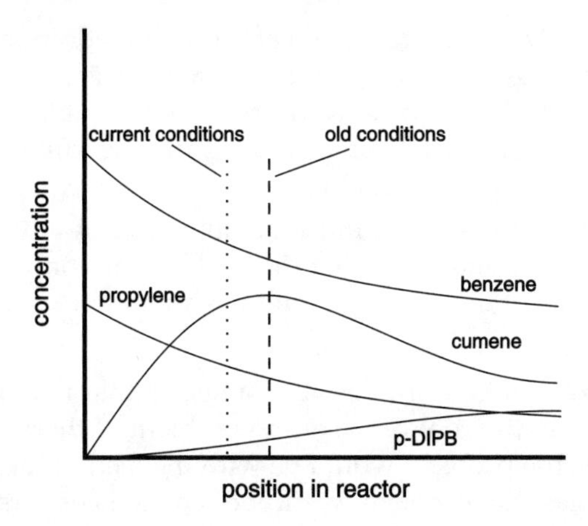

Figure 22.6 Concentration Profiles in Reactor for Case Study

This would result in less cumene and p-DIPB and more unreacted benzene and propylene. This scenario is consistent with five of the six observations. Because the possibility of bad catalyst must be acknowledged, this scenario is also possible.

3. Reactor temperature has decreased. This could be checked in the control room, but the possibility of a faulty thermocouple or indicator should also be considered if the "correct" temperature was observed. A decrease in reactor temperature would result in decreased reaction rates. Qualitatively, the results are similar to those for deactivated catalyst in item 2.
4. Reactor pressure has decreased. This could be checked in the control room, but the possibility of a faulty pressure transducer or indicator should also be considered if a "correct" pressure was observed. A decrease in reactor pressure would result in a decrease in all gas-phase concentrations and a decrease in all reaction rates. Qualitatively, the results are similar to items 2 and 3.
5. The flow controllers on Streams 1 and/or 2 have failed. Suppose that the benzene flow were increased or the propylene flow were decreased. The benzene concentration would increase, which would increase the excess benzene and increase the selectivity for the desired reaction. This is consistent with the last observation above. If this were not observed in the control room, the possibility of faulty instrumentation should be considered.
6. The temperature and/or pressure in V-802 is not at specification. This could be checked, but a "correct" reading could be due to faulty instrumentation. If the temperature and/or pressure in the flash vessel were incorrect, the desired separation would not be accomplished. If the temperature were too

high and/or the pressure were too low, additional fuel gas would be produced causing less cumene to leave as product and less benzene in the recycle. If the opposite were true, additional feed to the distillation column would be produced, which could cause flooding in the column and compromise the desired separation.

7. The problem stems from a combination of any or all of the above items. There is no guarantee that the problem has only one cause. Therefore, combinations of the above possibilities should be considered.

An examination of items 1–7 suggests that taking a sample of the fuel gas would identify the root cause of the problem. If there were too much propane in the fuel gas, the problem would be with the feed. If there were too much benzene in the fuel gas, the problem would be reduced conversion in the reactor. If there were too much or too little cumene in the fuel gas, the problem would be that the temperature and/or pressure of the flash were incorrect.

The next question is how to remedy the problem. It is not always necessary to identify the cause of a problem in order to begin to compensate. However, it is necessary to identify the problem to ensure that it does not recur. It is also necessary to be certain that what appears to be a remedy is not itself a problem. For example, if the propylene feed contains excess propane or if the catalyst was deactivated, it would still be necessary to find a way to compensate temporarily until new feed and/or catalyst could be obtained. Also, detailed quantitative solutions may not be necessary as long as the qualitative trends are understood.

In order to suggest remedies, it will be assumed that the cause of the observed process upset is in the process feed and in the reactor (items 1 and 2). Possible remedies include, but are not limited to, the following.

1. *Increase the temperature in the reactor.* Intuitively, this seems like a reasonable possibility. However, Chapters 19 and 20 should be consulted for the limitations associated with increasing reactor temperature. Increasing the reactor feed temperature can be accomplished by increasing the air and natural gas flows to the fired heater, but the reactor operation is limited by heat transfer. Increasing the reaction temperature increases the reaction rate exponentially, so a large temperature increase should not be needed. However, increasing the inlet temperature does not guarantee that the reactor operation will change appreciably. This situation was illustrated in Example 19.1 and is an example of heat transfer limitations of reactor performance, as discussed in Section 20.4. One consequence of increasing reactor temperature is that the temperature of the boiling water used to remove the heat of reaction must increase. This requires an increased steam pressure. It must be determined whether the materials of construction of the reactor can withstand

the required pressure increase. The temperature limitations of the catalyst support and of catalyst activity must also be considered. Increasing the temperature high enough to damage the catalyst is an example of the remedy causing another problem.

2. *Increase the pressure on the process side in the reactor.* Increasing the pressure in the reactor can be accomplished by closing the valve after the reactor. The increase in pressure increases the reaction rate by increasing the concentration. The effect is not as significant as for temperature, because temperature increases the reaction rate exponentially.
3. *Increase the flow of propylene feed.* Increasing the propylene feed can be accomplished only by using the spare pump (P-802B) in series or in parallel with the operating pump (P-802A), due to limitations in pumping capacity shown on the pump and system curve plot, which is shown in Appendix C. In principle, the flow of propylene can be increased enough so that the specified cumene production rate is achieved in the reactor. However, it must be determined how each piece of equipment will perform when subjected to the increased capacity and concentration changes caused by this remedy. The required calculations are performance problems like those discussed in Chapters 18 and 19.

There are several lessons to be learned from this process troubleshooting problem. As discussed earlier in this chapter, it is important not to focus on one possible solution to the exclusion of others. It is important to consider as many alternatives as possible. Several possible causes were presented here for the observed process upsets. Without detailed measurements and/or simulations, which take more time to perform, the other possibilities could not be ruled out. They would cause the same qualitative trends, but different quantitative values for the upset parameters.

It is also important to observe that a problem in the reactor was manifested in process locations far removed from the reactor: the fuel gas, the benzene recycle, the cumene production rate, and the p-DIPB production rate.

The cause of an observed process upset may be located in a different part of the process.

Finally, if the problem was identified to be impure propylene feed or defective catalyst, it would be necessary to report this problem to other plants in your company using the same propylene feed supplier or the same catalyst.

22.5 DEBOTTLENECKING PROBLEMS

In the course of operating a chemical process, it may become necessary to modify operating conditions. Possible changes include scale-up, scale-down, handling a new feed composition, and so on. A bottleneck is defined as the part of the process that limits the desired change. Troubleshooting the bottleneck is called debottlenecking. Debottlenecking is a sequential process. A bottleneck is identified and removed. The next bottleneck is then identified and removed, and so on. The removal of these bottlenecks does not usually involve a large capital investment.

The performance of a heating loop problem discussed in Section 19.3 was actually a debottlenecking problem. In this problem, described in detail in Appendix C, it is necessary to scale up production of our allyl chloride facility (see Appendix C, Figure C.4) due to an unscheduled shutdown at a similar facility owned by our company. The problem is to determine the maximum level of scale-up possible for our allyl chloride facility. Therefore, it is necessary to determine the maximum scale-up possible for the reactor. (*Note:* Some conditions in the problem in the appendix are different from those in Section 19.3.)

For the problem in Section 19.3, it was determined that the pump in the Dowtherm A loop could handle only 5% increased flow, which allowed only 2% increased reactor operation. Within the reactor portion of the allyl chloride process, the pump is identified as the bottleneck. From the perspective of the entire process, assuming that all other units can be scaled up by more than 5%, the Dowtherm A loop is the bottleneck. In Section 19.3, it was shown that by operating the two reactor heat-exchange coils in parallel, the reactor could be scaled up by 15%. Altering configuration of the reactor heat-exchange coils is the act of debottlenecking the process. For this problem, it can be shown that additional scale-up in the reactor and Dowtherm A loop is possible by operating the pump and the spare either in series or in parallel (Problem 19.6). Suppose that the maximum scale-up were now found to be 25% (not the correct answer to Problem 19.6). What if one of the distillation columns downstream could handle only 20% increased throughput before flooding? Then the distillation column would become the new bottleneck. You would now focus on methods for debottlenecking the distillation column. If this were possible, then another unit would become the bottleneck, and so forth. Therefore, debottlenecking is a progressive problem in which bottlenecks are removed from the process one at a time. Eventually, the maximum possible change, where a bottleneck cannot be removed, will be reached. At this point, a decision would have to be made whether a significant capital investment should be made in order to increase further the maximum scale-up. When significant process modifications involving new equipment are required, the procedure is called **retrofitting**.

22.6 SUMMARY

In summary, process troubleshooting problems and debottlenecking problems, such as the ones described here, are very realistic problems in terms of what the process engineer will experience. Unlike comprehensive design problems, the troubleshooting problems rely on simple, approximate calculations along with an intuitive understanding of a chemical process rather than repetitive, complex calculations. In order to solve troubleshooting and debottlenecking problems, it is important to develop both an intuitive feel for chemical processes and the ability to do approximate calculations to complement your ability to do repetitive, detailed calculations.

REFERENCES

1. Fogler, H. S., and S. E. LeBlanc, *Strategies for Creative Problem Solving,* 2nd ed. (Upper Saddle River, NJ: Prentice-Hall, 2008), 20.
2. Perry, R. H., D. W. Green, and J. O. Maloney, eds., *Perry's Chemical Engineers' Handbook,* 6th ed. (New York: McGraw-Hill, 1984), 3–50.
3. Shaeiwitz, J. A., and R. Turton, "A Process Troubleshooting Problem," *1996 Annual ASEE Conference Proceedings,* Session 3213.

PROBLEMS

1. For the absorber problem in Section 22.2.1, it is necessary to adjust process operation temporarily to handle a 20% increase in gas to be treated.
   a. Can this be accomplished by increasing the liquid rate by 20%?
   b. Suggest at least two additional methods for handling the increase. Be quantitative.
2. A five-equilibrium-stage, tray absorber is used for the acetone separation described in Section 22.2.1. It is now observed that the outlet mole fraction of acetone in air is 0.002.
   a. Suggest at least six individual causes for the faulty absorber performance.
   b. For the causes listed in Part (a) that are represented on the Kremser graph, determine the exact value of the parameter (i.e., what flowrate would cause the observed outlet mole fraction?).
   c. For each cause listed in Part (b), suggest at least three compensation methods. Be quantitative.

3. For the situation in Problem 2, how would you handle the following temporary process upsets? Be quantitative. Suggest at least three alternatives for each situation.
   a. The gas rate must increase by 10%.
   b. The feed mole fraction of acetone must increase to 0.025.
   c. The outlet mole fraction of acetone in air must be reduced to 0.00075.
4. Suggest additional alternatives for the off-specification acrylic acid in the case study in Section 22.3.1. Analyze the alternatives quantitatively.
5. Our acrylic acid facility has been designed to operate successfully using cooling water in the distillation column condenser. However, after a recent warm spell in which the temperature exceeded 100°F for a week, our customers complained that our acrylic acid product had the same yellowish color and increased viscosity observed in our other plant, as described in Section 22.3.1. Suggest possible causes and remedies for this situation.
6. Suggest additional alternatives for the steam release in the cumene reactor in the case study in Section 22.3.2. Analyze the alternatives quantitatively.
7. During the start-up of a chemical plant, one of the final steps before introducing the process chemicals is a steam-out procedure. Essentially, this step involves filling all the equipment with low-pressure steam and leaving it for a period of time in order to clean the equipment. During one such steam-out, in a plant in Wisconsin, a vessel was accidentally isolated from other equipment and left to stand overnight. Upon inspection the following morning, it was found that this vessel had ruptured.

   The company responsible for the design of the plant claimed that the design was not at fault and cited a similar situation that occurred at a plant in Southern California in which no damage to the vessel was seen to occur.

   From the above information can you explain what happened? What would you suggest be done in the future in order to ensure that this problem does not reoccur?
8. During the hydrogenation of a certain plant-derived oil, the fresh feed is pumped from a vessel to the process unit. The process is illustrated in Figure P22.8(a).

   Because the oil is very viscous, it is heated with steam that passes through a heating coil located in the vessel. The present system uses 50 psig saturated steam to heat the oil. Due to unforeseen circumstances, the 50 psig steam supply will be down for maintenance for about a week. It has been suggested that a temporary connection from the high-pressure steam line (600 psig, saturated) be made via a regulator (to reduce the pressure to 50 psig) to supply steam to the steam coil as shown in Figure P22.8(b).

   Do you foresee any problems with the recommendation regarding the high-pressure steam? If so, what recommendations do you suggest?

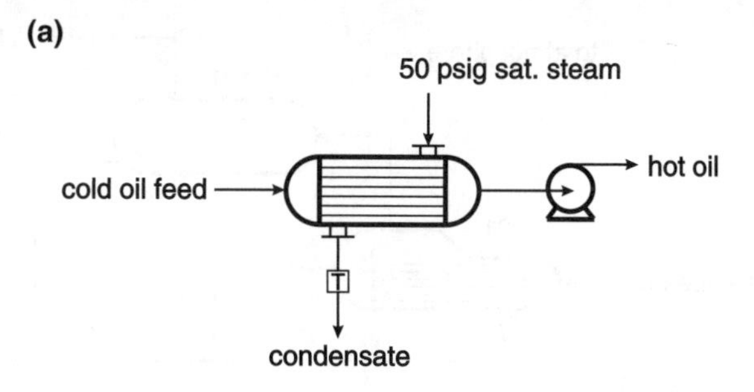

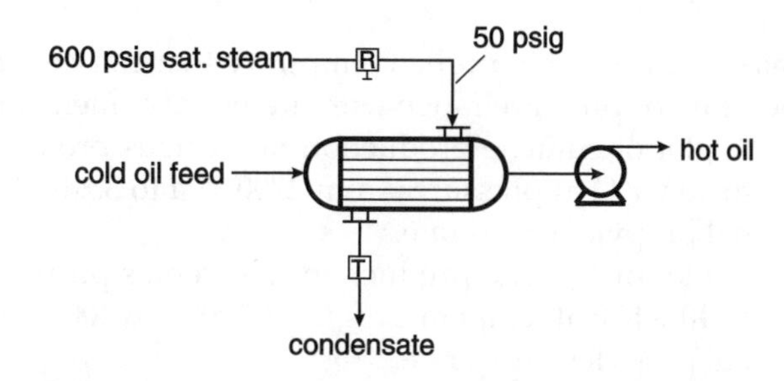

Figure P22.8 Heat-Exchanger Configurations for Problem 22.8

9. During the start-up and operation of a new plant (see Figure P22.9) the pressure-relief/safety valve on top of the steam drum (V-101) of a waste heat boiler (E-101) has opened and low-pressure steam is escaping to the atmosphere through the open valve.

   Upon questioning the operators from another unit, you discover that similar incidents have occurred during the initial operation of other process units. However, it appears that the situation remedies itself after a few months.

   What could be causing this phenomenon to occur? If the situation does not remedy itself, what permanent solution (if any) would you suggest to fix the problem?

10. For the cumene troubleshooting problem in Section 22.4, it has been determined that off-specification propylene will have to be used for the next several months. To assist in handling this feed, prepare the following performance graphs for use in determining what reactor temperature and pressure are required to compensate for increased propane impurity. Determine the

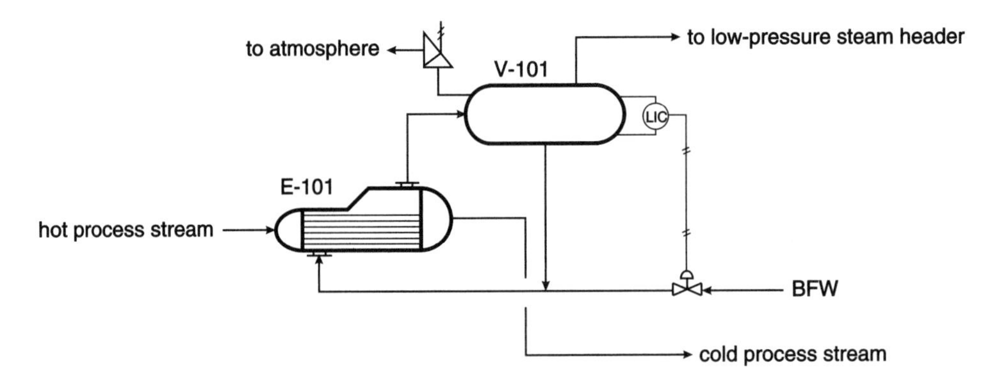

Figure P22.9 PFD for Problem 22.9

maximum possible propane impurity that can be handled in the given temperature or pressure range without loss of cumene production capability.

a. A plot of cumene production rate versus propane impurity (from 5 wt% to 10 wt%) at pressures from 2700 kPa to 3300 kPa at 400°C and the original propylene feed rate.
b. A plot of cumene production rate versus propane impurity (from 5 wt% to 10 wt%) at temperatures from 300°C to 400°C at 3300 kPa and the original propylene feed rate.

11. In the cumene problem in Section 22.4, assume that one possible remedy involves increased benzene recycle with a corresponding decrease in fresh benzene feed. What are some potential consequences on the benzene feed pump, P-201? Support your answer with calculations.

12. You are in charge of a process to manufacture a polymer in which acetone is used as a solvent. The last step is a drying oven in which residual acetone is removed from the polymer into an air stream. The air is fed to the absorber in Example 22.1. All flows, mole fractions, and physical property data in Example 22.1 are assumed to hold.

    It is necessary to scale up polymer production as much as possible. Upstream, there is a spare reactor, dryer, and peripherals, because the plant has been operating below capacity for many years. However, the acetone scrubber was designed and installed recently for the current production capacity. Therefore, it is anticipated that the acetone scrubber will be the process bottleneck. What limit does the acetone scrubber place on scale-up? How would you debottleneck this situation?

13. The phthalic acid production scale-down project (Appendix C, Project 3) involves determining a method for scale-down of phthalic acid production by 50%. The feed section of this process was discussed in Section 19.4.

    a. Identify potential bottlenecks to the 50% scale-down.

b. Quantify the answer to Part (a); that is, determine the primary bottleneck to 50% scale-down.
c. Debottleneck your answer to Part (b).
d. Repeat Parts (b) and (c) until 50% scale-down or a maximum possible scale-down is achieved.

14. For the cumene problem in Section 22.4, it has been determined that the feed is off specification. Until we can contract with a new propylene supplier we will have to use the propylene from the current supplier. Tests have shown that we can expect this propylene to have between 5 wt% and 10 wt% propane impurity. It has been decided to try to maintain the design cumene production rate by increasing the propylene feed rate so that a constant, design amount of propylene enters the reactor. Identify the bottlenecks to the proposed process change. Debottleneck this situation.

15. A saturated liquid is stored in a tank in equilibrium with its vapor. The liquid is pumped to a heat exchanger, where the liquid is vaporized. Frictional losses in the suction line to pump are negligible; however, frictional losses after the pump are such that the liquid is saturated upon entering the heat exchanger. The heat exchanger vaporizes the saturated liquid to saturated vapor while condensing saturated steam to saturated liquid.

    In the base case, the liquid vaporizes at 130°C, and the steam used is passed through a desuperheater to create saturated steam at 155°C from saturated, low-pressure steam at 160°C. It is now necessary to increase the flowrate of the vaporized stream. Pump and system curves are provided in Figures P22.15a and P22.15b. Assume that the pressure entering the vaporizer remains constant.

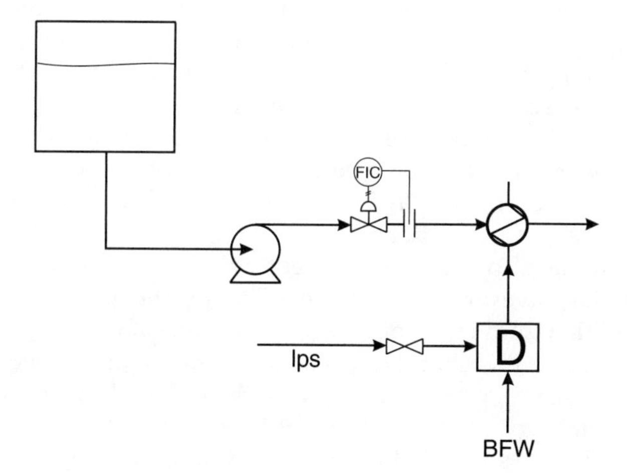

Figure P22.15a Process Flow Diagram for Problem 15

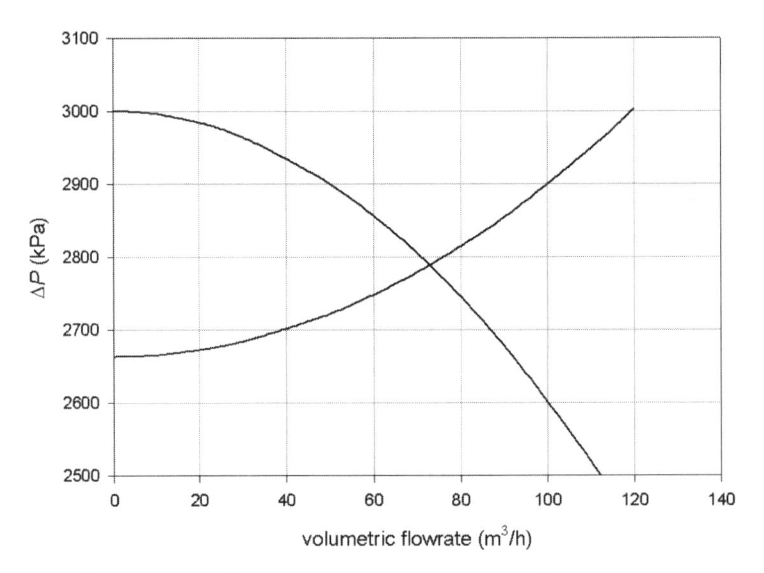

Figure P22.15b Pump and System Curves for Problem 15

a. If the base-case flowrate is 50 m^3/h, identify the bottleneck to process flowrate scale-up. What is the maximum possible flowrate increase?
b. With scale-up, the frictional losses in the suction line to the pump increase. Is the pump in danger of cavitating? Explain your answer.

For Problems 16–17, it may be useful to refer to the following document: http://www.che.cemr.wvu.edu/publications/projects/drying_oil/do12.pdf.

16. Refer to the drying oil process in Appendix B. The following problems were encountered after the drying oil facility was started up in early August.
    a. Pump P-501 A/B has been noisy since start-up, and the noise (a high-pitched whine) continues to get louder.
    b. Problems have been observed in T-501. Column performance has been below specification. Specifically, the flowrate of Dowtherm A through the reboiler, E-502, has had to be increased in order to keep the products at design specifications.
    c. It has also been noted recently that production rates have fallen during the graveyard shift (12 p.m. – 8 a.m.). The head of operations believed that this was due to some of the operators not paying attention to the plant during this shift. However, after talking to all concerned, it appears that there might be a problem with the feed of ACO from the storage tank (not shown on the PFD) to the feed vessel, V-501.
    d. The pressure-relief valve on E-506 has been open from the start of production, and steam has been venting to the atmosphere. This problem was more severe right after start-up; however, it is still occurring.

Suggest causes for these problems, identify the most likely causes, and suggest potential remedies.

17. Refer to the drying oil process in Appendix B. Due to an unfavorable business climate for the product, it is possible that production will have to be cut by 20%. Suggest at least three equipment-related issues that might have to be addressed to implement the production cut.

For Problem 18, it may be useful to refer to the following document: http://www.che.cemr.wvu.edu/publications/projects/formalin/formalin12.pdf.

18. Refer to the formalin process in Appendix B.

    You have recently joined a chemical company. Among the chemicals that this company produces is methanol, mostly for internal consumption. A major use of internal methanol is to produce formalin, which is a 37 wt% solution of formaldehyde in water. Formaldehyde and urea are used to make urea-formaldehyde resins that are subsequently used as adhesives and binders for particleboard and plywood.

    We recently have received a memorandum from a customer who buys our formalin indicating that there are periodic problems, mostly in the summer, with the formalin they are receiving from us. They claim that the formalin contains unacceptably high amounts of formic acid relative to formaldehyde. It seems that this alters the urea-formaldehyde resin properties such that the particle board and plywood subsequently manufactured do not have the appropriate tensile and load strength. They are threatening legal action unless this problem is rectified immediately.

    Storage of formaldehyde/water mixtures is tricky. At high temperatures, undesirable polymerization of formaldehyde is inhibited, but formic acid formation is favored. At low temperatures, acid formation is inhibited, but polymerization is favored. There are stabilizers that inhibit polymerization, but they are incompatible with resin formation. Methanol, at concentrations between 5 and 15 wt %, can also inhibit polymerizaton, but no separation equipment for methanol currently exists on site, and methanol greater than 1 wt % also causes defective resin production. With ≤ 1 wt% methanol, the storage tank contents must be maintained between 35°C and 45°C.

    You have toured the plant, checked the logbooks, spoken to engineers and operators, and have gathered the following information.

    - The catalyst is replaced once per year, and for about five days after start-up, a pressure-relief valve in the reactor (R-801) releases.
    - During the same time period, BFW consumption in R-801 was not seen to increase beyond normal, day-to-day fluctuations of ±1%, and the reactor outlet temperature was not seen to increase significantly.
    - Excess acidity occurs mainly in the summer.
    - All existing equipment is made of carbon steel.
    - One year ago, an unscheduled, emergency shutdown was required to replace a control system between the tower (T-802) and the subsequent

pump (P-803 A/B). Due to availability and time constraints, 1-in schedule 40 pipe was used in place of 1.5-in schedule 40 pipe.

- During hot-weather periods, the inlet temperature of cooling water has been observed to increase by as much as 5°C.
- During these hot weather periods, the pressure in the distillation column has been observed to increase slightly, but operation over a long period of time shows no leakage or safety problems associated with this change in pressure.
- Pump (P-803 A/B) is located at a height of 5 m below the tower (T-802) exit. There is a total of 30 equivalent meters of pipe, elbows, and valves in the line.
- An inspection of the pressure-relief valve on the reactor (R-801) indicates that it is rated at 1115 kPag. It will open at or above that pressure.
- Over long periods of time after catalyst replacement, the pressure of medium-pressure steam leaving R-801 is 1110 kPag.
- During hot weather, the volume of off gas leaving the absorber increases by 1%. However, the increase in formaldehyde is about 22%, and the increase in water is about 3%.
- Pump (P-803 A/B) constantly makes noise, and each individual pump has been replaced within the last year.

Suggest potential causes for the observed problems, and suggest possible remedies.

19. Delayed coking is an oil refinery process that takes a heavy oil fraction and heats it up to approximately 900°F (480°C) in a furnace. Upon leaving the furnace, the oil is fed to one of two (or more) coking drums, where it is allowed to sit and "cook" for up to 24 hours. During the cooking (coking) process, the oil cracks and gives off a variety of gaseous and liquid products that are processed further. The material remaining in the drums is a solid coke that is subsequently "cut out" of the drum using a very high-pressure water jet. This coke is then shipped out as a saleable product. The process is shown Figure P22.19.

    The control scheme on the furnace is set to regulate the flow of fuel gas (fg) based on the measured temperature of the exit oil stream (Stream 2). A ratio flow controller adjusts the flow rate of air so that a constant air/fuel ratio is maintained. An oxygen analyzer measures the excess oxygen in the flue gas and adjusts the set point of the ratio controller to maintain the desired air/fuel ratio. A flow alarm is also present to detect a low flow condition in Stream 1 and activates a solenoid shutoff valve to the fuel gas.

    A few days ago there was a process upset in which the oil flow to the fired heater was momentarily interrupted. This interruption triggered the shutoff valve, which closed the fuel gas line. The flow of Stream 1 resumed its normal rate after about 30 seconds, at which time the fuel gas solenoid valve opened and the flow of fuel gas resumed. During the upset, the flow of air changed in proportion to the fuel gas flow.

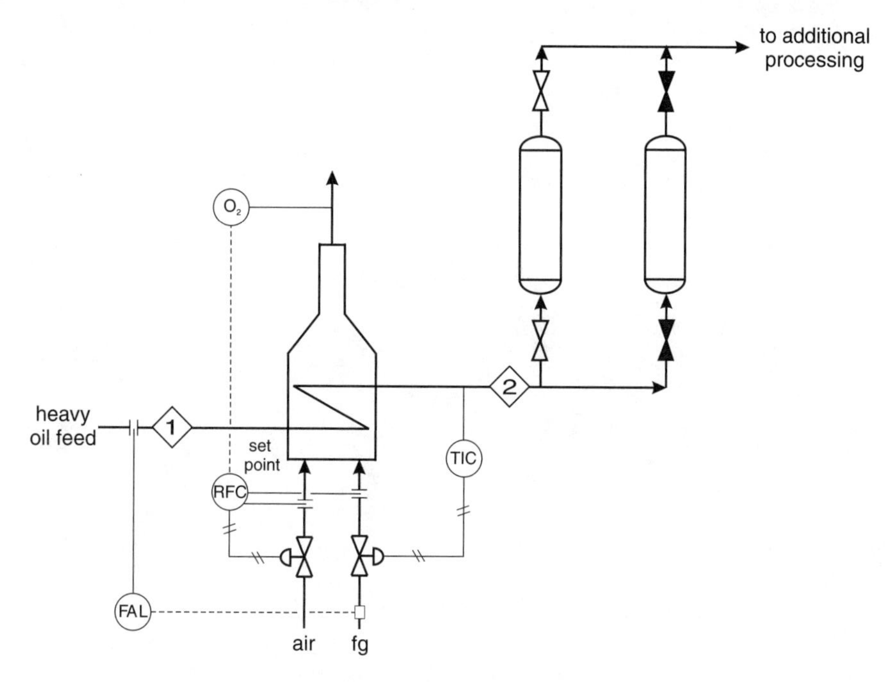

Figure P22.19 PFD for Problem 19

A short time after this process upset, the following conditions were observed. For each scenario given below, give an explanation of what may have happened to cause the observed changes.

a. The flow of fuel gas to the furnace has slowly increased, and the air flowrate has increased proportionally to maintain the same air/fuel ratio. No other changes have been observed. What could be causing the problem? What do you recommend we do?
b. The flow of fuel gas to the furnace has increased, the air flowrate has increased proportionally to maintain the same air/fuel ratio, and the flow of products and coke production rate appear to be slightly lower than before (the upset). What could be causing the problem? What do you recommend we do?
c. The flow of fuel gas to the furnace has increased, the air flowrate has also increased, and the ratio of air/fuel has increased slightly. The flow of products and coke production rate are lower than before (the upset). What could be causing the problem? What do you recommend we do?

SECTION

5

The Impact of Chemical Engineering Design on Society

Throughout this book, the focus has been on the role that a chemical engineer plays in the analysis, synthesis, and design of chemical processes. In this section, that role is put in the context of the profession of chemical engineering, which is defined by the American Institute of Chemical Engineers as "the profession in which a knowledge of mathematics, chemistry, and other natural science gained by study, experience, and practice is applied with judgment to develop economic ways of using materials and energy for the benefit of mankind."

Far from being the "soft" side of engineering, the topics in this section are very much the crucial steps between a design in the mind of the engineer and the realization of a new or improved process or product in the service of humanity. In the past 20 years, the importance of the impact of chemical engineering design on society in terms of how the chemical process industry and chemical engineers are viewed has increased dramatically. No longer is being a "good engineer" defined by one's ability to solve complicated process problems correctly. Instead, a company's reputation and success are often measured in terms of its track record in operating environmentally friendly processes, its ability to find innovative solutions to existing problems, and its ability to develop new products. Chemical engineers are responsible for all these initiatives, and the success of modern engineers is based in large part on their contributions to these areas.

In this section, focus is placed on the engineer's ability to

- Gain the necessary experience and knowledge
- Earn the trust of society

- Ensure that the process and products are safe
- Protect the environment
- Continue to look for new processes that are less harmful to the environment and that pose less risk to society
- Work with other engineers and professionals to bring new chemical engineering based products to the marketplace

Chapter 23: Ethics and Professionalism
Engineering problem solving within an ethical framework, legal responsibilities, and professional registration are developed. Ethics case studies and the content of the Fundamentals of Engineering (FE) and Professional Engineering (PE) examinations are included.

Chapter 24: Health, Safety, and the Environment
Methods of analyzing risk are provided, and the basic types of health, safety, and environmental regulations are explained, with references to government databases. Pollution prevention strategies and the assessment of plant safety are introduced through their relationships to **hazard and operability (HAZOP)** studies, the Dow Fire and Explosion Index, and the Dow Chemical Hazards Index.

Chapter 25: Green Engineering
A brief introduction to green engineering design methods—including the pollution prevention hierarchy, green chemistry principles, and flowsheet analysis for pollution prevention—is discussed. Examples of the economics of pollution prevention activities using the profitability methods in Chapter 10 and life cycle analysis are presented. Brief descriptions of environmental risk assessment, analysis of releases, and the software available to do this analysis are given.

CHAPTER

23

Ethics and Professionalism

Engineering has been described as "the strategy for causing the best change in a poorly understood or uncertain situation within the available resources" [1]. The realm of ethics and professionalism entails very real, poorly understood problems that are as challenging as any technical problems an engineer will face. This chapter presents heuristics, objective functions (i.e., ways of defining what is *best*), and constraint identification strategies that are especially useful in solving such problems.

As described in Chapter 11, all heuristics are, in the final analysis, fallible and incapable of justification. They are merely plausible aids or directions toward the solution of a problem [1]. Especially for the heuristics described in this chapter, we need to keep in mind the four characteristics of any heuristic.

1. A heuristic does not guarantee a solution.
2. It may contradict other heuristics.
3. It can reduce the time to solve a problem.
4. Its acceptance depends on the immediate context instead of on an absolute standard.

The fact that one cannot precisely follow all ethical heuristics all the time is to be expected, as it is with any set of technical heuristics.

The purpose of this chapter is to help develop strategies to make the best choice when faced with an ethical problem. The set of strategies developed will be different for each reader. A general overview of engineering ethics is presented, and a series of ethics scenarios is introduced. The authors have found that the best way to develop the facility to deal with ethical problems is, after reading the overview, to discuss these scenarios in small groups. Each group presents its solution to the class. Even with seemingly straightforward situations, the solutions can be diverse. A class discussion of the different solutions is followed by a reflection on what was discussed. Then we go on to the next problem. Many such scenarios are given at the end of this chapter and in books and articles [2, 3, 4, 5, 6, 7].

23.1 ETHICS

Whenever chemical engineers develop products, design processes and equipment, manage process operations, communicate with other engineers and non-engineers, develop markets and sell products, lead other engineers, interact with clients, represent their firms to the government or to the public—in short whenever chemical engineers do anything that impacts the lives of anyone—their choices of action are based on ethics. Even when faced with two different equations, one equation is selected, based in part on ethical values. Does the less precise equation include a safety factor that lowers the risk to our employer, employees, or the public? Should we spend more time to do more rigorous calculations, costing the firm more money but providing a better answer to the client? How do we decide?

In each of these circumstances, engineers apply their own moral standards, mindful of the legal requirements, using their personal code of ethics to make the decision. To help in the development of a personal code of ethics that will provide a framework for making these decisions, we begin by identifying the three types of reasons for ethical behavior:

- Moral
- Legal
- Ethical

Although nearly all people share *some* fundamental moral ideals, each engineer has his or her own distinct set of moral principles. Typically, these principles are shaped by religion, conscience, and especially early childhood family experiences. The basic framework by which one decides what is right and what is wrong is generally very well developed by the time one reads an engineering text. Thus, this text will not take up the moral dimensions of engineering ethics other than to stress the importance of continually reminding oneself to be true to one's moral values as one works through ethical problems.

A few legal aspects of solving ethical problems are covered in Section 23.3, but the full legal consequences of engineering decisions are far beyond the scope of this book. The legal system (which includes government regulations) is a collection of rules of conduct for a society to assist orderly transactions between people. Chemical engineers should seek skilled legal advice whenever these rules, or the consequences for not following them, are unclear.

The aspect of ethical decision making covered here is that commonly referred to as engineering ethics. There are generally accepted codes of conduct for engineers, although, as will become clear, they are too broad to be used alone as prescriptions for engineering choices in difficult situations. **Engineering ethics** is the system of principles and strategies that engineers use to solve complex problems involving other people's lives. It includes aspects of moral principles and legal responsibilities, as well as recognized codes of ethics and generally accepted norms of engineering and business behavior.

23.1.1 Moral Autonomy

Inasmuch as all engineers do not share a single set of moral principles by which to make ethical decisions, it is fully expected that different readers will make different decisions, especially in complex situations. The goal of this chapter is not *uniformity* of decisions by all engineers but *autonomy* of each engineer to make the right decision. In this context, the right decision can be identified by the use of a heuristic. The right decision is one that is

- Consistent with the engineer's moral principles
- Consistent with the generally accepted codes of engineering conduct
- Consistent with obligations that the engineer has accepted
- Consistent with the law
- Consistent with the applicable code of ethics

But, most importantly, *the right decision is one that the engineer can live with.* Of course, it is always possible that one person's decision would not be acceptable to another person.

The ability to make one's own ethical decisions is known as *moral autonomy*.

Moral autonomy does not require that you be able to look back and always be confident that the choice made was the best of all possible choices. Although this is the goal, it is a moving target. Rather, you are exercising **moral autonomy** if you are in control of your decision, if you make the choice based on a reasonable analysis of the potential consequences consistent with your moral, legal, and ethical beliefs, rights, duties, and obligations. If you do not understand your moral principles, if you have no strategy for ethically analyzing a situation, or if you defer your own ethical responsibilities to others, you cannot claim to be exercising moral autonomy.

The goal of this chapter is to help you develop moral autonomy. Previewing the kinds of ethical problems that you are likely to have to resolve is the most powerful tool you can use to learn about and to develop your moral autonomy.

23.1.2 Rehearsal

When learning any new skill, one usually practices or rehearses. To learn to apply the ideal gas law in process calculations, one does end-of-chapter problems in thermodynamics. Understanding the theory behind the ideal gas law is no guarantee that one will be able to solve applied problems. Some people can and some people cannot. But few would argue that one can solve a problem as quickly, as easily, or as correctly the first time as the *n*th time one solves it. Such is the power of rehearsal.

Rehearsal becomes more important when decisions must be made quickly, extremely accurately, or under great stress. And great stress often accompanies ethical problems. Take a look at Example 23.1, for instance.

Example 23.1

The Falsified Data [2, Reprinted by special permission from *Chemical Engineering*, May 5, 1980, and September 22, 1980, Copyright © 1980 by McGraw-Hill, Inc., New York, NY 10020]

Jay's boss is an acknowledged expert in the field of catalysis. Jay is the leader of a group that has been charged with developing a new catalyst system, and the search has narrowed to two possibilities, Catalyst A and Catalyst B.

The boss is certain that the best choice is A, but directs that tests be run on both, "just for the record." Owing to inexperienced help, the tests take longer than expected, and the results show that B is the preferred material. The engineers question the validity of the tests, but, because of the project's timetable, there is no time to repeat the series. So the boss directs Jay to work the math backward and come up with phony data to substantiate the choice of Catalyst A, a choice that all the engineers in the group, including Jay, fully agree with. Jay writes the report.

In this simple scenario, there is a great deal of stress. If Jay had never thought about what to do in such a situation, it is highly unlikely that he would make a decision that he could live with. It is much more likely that Jay would look back on the event and wish he could go back and make a different choice. Rehearsal gives us the opportunity to do just that. The first time we see the situation, we would like to be in a low-risk environment. If we make the wrong decision, it does not matter. The second time, we can make the better choice.

In early education and in skills training at any level, the goal of rehearsal is to work through a scenario that is likely to occur in the future and to develop the best response we can. In advanced professional education, and especially in engineering education, the goal of rehearsal is to work through a scenario representative of a broad range of situations that are likely to occur in the future and to develop a strategy for responding to the broad range of problems, many of which cannot be imagined. For ethical decision making, this strategy must be powerful, adaptive, and personal. Throughout your career, new ethical problems will arise. The key is to rehearse frequently, using example and homework problems and conceiving your own representative scenarios.

23.1.3 Reflection in Action

One of the characteristics of successful professionals in a variety of fields is frequent postmortem analysis. This self-imposed study of events that have occurred in one's professional life is called "reflection in action" [8]. After an engineering event has occurred in which ethical decisions were made, one sits down (individually or in a small group) and reviews the case, analyzing the facts, the missing information, the constraints, the unnecessary perceived constraints, the options considered, the options not considered, and the strategy used to arrive at the decision.

As demonstrated in Example 23.2, there are many reasons why reflection in action is so powerful, but we focus on two.

- It forces one to analyze the strengths and weaknesses of one's own strategy.
- It provides continual opportunities and encouragement for rehearsal.

Example 23.2

Reflection on the Falsified Data

Let us assume that Example 23.1 has occurred. We can use any standard problem-solving strategy [9] to reconstruct the scenario. Here we have used the McMaster five-step strategy [10].

1. **Define:** Was the problem well defined by the participants? Was the real problem that the experimental results were unexpected? Was the real problem that things were now more uncertain than before? There are other possible definitions of the problem, but if one defines the problem as how to deal with obviously flawed data, one may miss the entire point. The problem would be thereby unnecessarily over-constrained. What is the real problem? If the characters had defined the problem better, would they have reached a different conclusion? Would it have been a better conclusion?
2. **Explore:** What alternatives were explored? Are there other alternatives, such as requesting additional time to rerun the experiments, alerting the users of the data to their possible inaccuracy, writing a detailed analysis of the theoretical and experimental discrepancies, debriefing the technician who performed the experiments to determine whether errors in technique could cause the discrepancy? Would a more

careful exploration of the alternatives have been helpful in this case? What brainstorming techniques might have been helpful?

3. **Plan:** Did the participants develop an adequate plan? What would (should) you have done differently?
4. **Do It:** Did the participants execute the plan well?
5. **Look Back:** Here is the play within the play. Would it have been helpful if the participants had reflected on past experiences? Would it have been helpful if the participants had projected what could happen as a consequence of their decision and had analyzed the expected effectiveness of their approach?

Because the outcome of the "Falsified Data" case is unknown, the *Plan* and *Do It* stages cannot be fully analyzed. When the full case is reflected upon, we could, and would, do just that.

At the end of any rehearsal or reflection, one (or the group) should develop a list of heuristics to use in future ethical problem solving. These could be the heuristics that had been used effectively (in the case of a reflection) or new heuristics that can be used (in the case of a rehearsal). For Example 23.2, the following heuristics might be identified (use the three blank lines to develop your own heuristics).

- Use a traditional problem-solving strategy for solving ethical problems.
- Consider the possibility that inexperienced people can be right.
- Debrief people fully before assuming facts about their actions.
- Consider what will happen if a specific decision is based on a false assumption.
- Be honest.
- Be concerned about the welfare of your company.
- Be concerned about the welfare of your employees.
- Do not let other people make ethical decisions for you.
-
-
-

23.1.4 Mobile Truth

It is quite natural for people to assume that groups to which they belong are right, and other groups are wrong. This basic response gives rise to loyalty, strong familial pride and duty, willful obedience, and leadership. The strength of collective action depends on this response, which, in terms of ethical decision making in engineering, manifests itself in **mobile truth**.

Let us say that your AIChE student chapter and the IEEE student chapter are playing volleyball. If the referee makes a difficult call, you are apt to find the call right if it favors your team and wrong if it favors the other. You are not being unethical, but you are perhaps being unreasonable. Similarly, when you start to work for an organization, you begin to develop attitudes toward it that are similar to the bonding that occurs within families, nationalities, and schools. The faster these bonds develop, the faster you will be accepted. The stronger these bonds are, the more loyal you will become. Example 23.3 demonstrates mobile truth.

Example 23.3

If you work for Company A, which produces polyethylene, you know that Type-A polyethylene is the best. Because you are offered a promotion, higher pay, and desirable relocation, you move to Company B. Almost immediately, Type-B polyethylene becomes the best. It could be because you are such a great chemical engineer, but it is probably because of mobile truth. The world did not change, but your frame of reference and, especially, your loyalty, did.

> **When your affinity to a group clouds your ethical decision making, you are being affected by *mobile truth*.**

The point here is to learn to recognize mobile truth and to filter it out of ethical decision making. Obviously, a chemical engineer has ethical duties and obligations to an organization, but one must be ever-vigilant to identify the intrusion of mobile truth into the ethical decision-making process. The point is not just to make the process more "objective" or "fair," but rather to try to see the situation from the point of view of those outside the organization. Here are a few heuristics for avoiding being misled by mobile truth.

- Ask yourself whether your decision would be different if you worked for another part of the company or for another company.
- Imagine that you live just outside the plant fence.
- Imagine that you work for the Environmental Protection Agency.
- Ask for the opinion of someone else in your organization. Explain the situation, suggesting that the facts pertain to a different organization. The response is likely to be less affected by mobile truth.

More heuristics can be developed through rehearsal and reflection.

Related to the concept of mobile truth is the concept of **postrationalization**. Again, it is quite natural and normal for people to try to justify their actions (and the actions of their colleagues and organizations), whether or not they are defensible. Because complex ethical problems can be analyzed and evaluated in so many different ways, it is often easy to fool oneself into thinking that one has acted ethically. But one must keep in mind the frame of reference or point of view of others (outside your organization). If you want to be sure that what you have done or are about to do is ethical, imagine how someone outside the organization would view the decision. If there is a difference, chances are you were post-rationalizing (or prerationalizing).

23.1.5 Nonprofessional Responsibilities

Each chemical engineer has personal responsibilities: to family, to friends, to oneself. These responsibilities, like professional responsibilities, will change dramatically throughout one's life. The obvious example is one's family situation, which is likely to grow and change over a 40-year career. In general, it is easier to make ethical choices when they affect only oneself. Choices that affect one's nuclear family are especially difficult. Some choices might cost you your job, make you a social outcast, compel you to move from your home, even estrange you from your family.

Generally, you can mitigate consequences by

- Thinking early about the effects of your decisions on your family
- Taking into account likely changes in your family situation
- Most important, talking with your family about your decision

The choice of when to discuss a professional ethics problem with one's family (and with which members of the family) is certainly a difficult one. Considering what a dramatic effect your decisions can have on your family, it would be wise to consider a hypothetical situation before a real ethical problem arises. Think about various options of whom to tell what and when. Rehearsal can be done with one's family, with a peer group, or by oneself. Whichever you choose, it should make any potential conflict easier to resolve to everyone's satisfaction.

The crucial impact of ethical decisions on one's family is generally not fully appreciated until one is married or becomes a parent. In using rehearsal to prepare for difficult ethical situations, one should anticipate potential changes in one's family situation that might affect the decision. In a typical senior chemical engineering class, most students are single without children. However, there may be students with a significant range of circumstances. Some might be married without any children; several might be parents. Additionally, some may intend to be the only or the main potential wage earner, whereas others may expect to work or earn less than their husbands or wives. Whenever possible, it is better to form nonhomogeneous groups for rehearsal of ethical scenarios. The addition of a married student to an otherwise single group usually changes the discussion substantially.

While discussing ethical scenarios, financial and other concerns also must be considered. Prospective chemical engineers often say, "I just wouldn't work for a company that would do that," or "I'd quit and get another job before I'd agree to that." But these comments are looked upon as naive by most who have made tough ethical choices, especially whistle-blowers, as is discussed in Section 23.1.8. Often, chemical engineers who no longer have young children or who have saved well for retirement express more willingness to do what they feel is ethically correct. It is unclear whether this expressed willingness translates into action. Some choose to live more frugally, and others choose jobs that require fewer or less difficult ethical choices. Those who become overcommitted—for example, financially—and then feel compelled to make the choice to perform an act that they consider ethically wrong, seldom would claim to be morally autonomous, let alone happy, as seen in Example 23.4.

Example 23.4

A Question of Integrity [3, Reprinted by special permission from *Chemical Engineering*, March 2, 1987, and September 28, 1987, Copyright © 1987 by McGraw-Hill, Inc., New York, NY 10020]

Under the Toxic Substances Control Act (U.S. Public Law 94-469), all chemicals in general use, or that had been in use, were required to be placed on an inventory list in 1979. Some chemicals were omitted from the initial list through oversight. To produce a chemical not on the list, or a new chemical, a manufacturer must submit a premanufacturing marketing notification to the EPA.

(The purpose of notification is to allow EPA 90 days to review a chemical, to ensure that its production, distribution, and use will not be detrimental to human health and the environment. The agency has the authority to place controls as necessary. Heavy penalties can be, and have been, assessed against violators of this regulation.)

Chris supervises a unit that has begun making a "new" chemical (one not on the list) and becomes aware that a premanufacturing notification has not been filed.

If Chris blows the whistle, Chris's career with the company could be over, despite laws to protect whistle-blowers. If Chris does nothing, workers may be exposed to the chemical without safeguards, and be harmed by it. And, if the company is caught, Chris's professional reputation could be stained, especially if it could be shown that the inaction resulted in harm to workers.

Should Chris discuss this ethical problem with her family?
What financial or other hardships might result if she reports the situation?
What obligations does she have to her family in this regard?
What obligations does she have to the community?
What obligations does she have to her employer?

23.1.6 Duties and Obligations

Chemical engineers have certain duties by virtue of their positions, and they acquire other obligations in a number of ways—for example, by accepting assignments, joining professional organizations, and through their family choices.

Throughout ethical problem solving, one needs to remind oneself of all of the duties and obligations to which one has agreed.

According to the National Science Foundation, less than 1% of the U.S. population are engineers. Of these, fewer than one in ten are chemical engineers. The American Institute of Chemical Engineers estimates that only 0.05% of U.S. citizens are chemical engineers. Clearly, the vast majority of people are not and do not think like chemical engineers. Few will understand what chemical engineers do, but all will be affected by their actions. Keeping this fact in mind helps to put the awesome responsibilities of chemical engineers in perspective.

One's duties and obligations form the basis for some additional important heuristics in ethical problem solving.

- Remind yourself of relevant duties and obligations that you have accepted.
- Remind yourself of otherwise relevant duties and obligations that you have *not* accepted.
- When accepted duties and obligations are necessarily in conflict, rank these responsibilities.
- If you choose a solution that violates an obligation, discuss the decision with those to whom the obligation was made and determine the consequences.

23.1.7 Codes of Ethics

Codes of ethics are formal obligations that persons accept when they join organizations or when they are allowed to enter a profession. In chemical engineering, there are three main types of codes of ethics: employer, technical society, and government. The employer-based codes of ethics are usually incorporated into the codes of business conduct that one agrees to upon employment with a particular firm. These are covered in Section 23.4. The government-based codes are the professional engineer rules, regulations, and laws that exist in all states and territories of the United States and their counterparts throughout the world. These are covered in Section 23.2.

The most important technical society code for U.S. chemical engineers is the "Code of Ethics' of the American Institute of Chemical Engineers. Similar codes have been adopted by other engineering societies. When you sign your application for membership in AIChE, you agree to abide by this code (Figure 23.1).

The code can be divided into three parts. The first part identifies to whom the code applies (members of AIChE) and the purpose of the code (to uphold and advance the integrity, honor, and dignity of the engineering profession). Note that this purpose is based on the historical concept that each member of a group (AIChE in this case) has a responsibility to help maintain the "good name" of all members of the group through professional behaviors that tend to increase the trust that society has in the profession. This is a powerful concept in ethics be-

AIChE Code of Ethics

(Revised January 17, 2003)

Members of the American Institute of Chemical Engineers shall uphold and advance the integrity, honor and dignity of the engineering profession by: being honest and impartial and serving with fidelity their employers, their clients, and the public; striving to increase the competence and prestige of the engineering profession; and using their knowledge and skill for the enhancement of human welfare. To achieve these goals, members shall

- Hold paramount the safety, health and welfare of the public and protect the environment in performance of their professional duties.
- Formally advise their employers or clients (and consider further disclosure, if warranted) if they perceive that a consequence of their duties will adversely affect the present or future health or safety of their colleagues or the public.
- Accept responsibility for their actions, seek and heed critical review of their work and offer objective criticism of the work of others.
- Issue statements or present information only in an objective and truthful manner.
- Act in professional matters for each employer or client as faithful agents or trustees, avoiding conflicts of interest and never breaching confidentiality.
- Treat fairly and respectfully all colleagues and co-workers, recognizing their unique contributions and capabilities.
- Perform professional services only in areas of their competence.
- Build their professional reputations on the merits of their services.
- Continue their professional development throughout their careers, and provide opportunities for the professional development of those under their supervision.
- Never tolerate harassment.
- Conduct themselves in a fair, honorable and respectful manner.

Figure 23.1 AIChE Code of Ethics (Reprinted with permission of AIChE)

cause it acknowledges that the unethical behavior of one member can create trust issues between the society and all members of the profession. Thus, all members have a vested interest in the ethical behavior of all members.

The second part is the list of three goals that identify the duties that chemical engineers have (being honest, impartial, loyal), to whom (employers, clients, the public, the profession), and why (enhancement of human welfare). It also points out that practicing chemical engineering ethically involves improving the profession (striving to increase competence). This section serves at least two important purposes. First, as a chemical engineer reads the goals, he or she realizes the very broad responsibilities to the society that may not be apparent in day-to-day work. A decision made by a chemical engineer might save thousands of lives by providing fertilizer to grow much needed food, or a decision might kill scores of people in a catastrophic release of toxic materials. These goals are a powerful tool in helping chemical engineers to do the right thing. Second, the list of goals points out quite clearly that not all goals can be met simultaneously all of the time. In fact, one frequently encounters ethical dilemmas, in which *no* choice is a perfect choice, completely satisfying all of our moral, legal, and ethical responsibilities.

The third part of the code of ethics concerns 11 relatively specific responsibilities that chemical engineers have. Many ethical problems can be attacked by referring to these responsibilities. While reiterating and clarifying the responsibilities of chemical engineers for the safety and health of the society at large, they speak to the responsibilities to clients, employers, employees, and the profession itself. Of particular note are two principles that are sometimes overlooked by beginning engineers. Both refer to the need for continuing education. The seventh principle states that chemical engineers shall "perform professional services only in their areas of competence." It may be obvious that a chemical engineer should not be doing electrical engineering work without having had significant education in the relevant area of electrical engineering, but, even within chemical engineering, there are many areas that are not taught in even the most rigorous B.S.Ch.E. program. Therefore, throughout one's career, one is required by the code to evaluate one's own chemical engineering capabilities continuously and to acquire any needed education (through reading, working with more experienced engineers, consulting experts, or taking courses) before accepting an assignment in any area of chemical engineering. This principle states that the chemical engineer, not some outside governmental agency, must take final responsibility for professional competence. A governmental agency may certify you as a professional engineer after you pass some day-long examinations and a multiyear internship, but you are still the responsible party for your own chemical engineering competence.

The ninth principle states that chemical engineers shall "continue their professional development throughout their careers, and provide opportunities for the professional development of those under their supervision." Many young engineers may think that one's chemical engineering education ends after the B.S., the M.S., or surely after the Ph.D., but the code makes it clear that this is not the case. In the past 40 years, tremendous strides have been made in chemical engineering, and we certainly expect at least as great a change in the next 40 years. Thus, one is required to keep up with the latest advances through such activities as participating in in-house (i.e., within the company) training programs, taking continuing education courses offered by AIChE, universities, and other organizations, reading technical journals, consulting with experts, and so on. The facts that one learns during the B.S. experience are overshadowed by the strategies that one learns to attack problems. And some of the greatest of these problems are to evaluate one's knowledge, to decide what new material needs to be learned, and to develop and implement a plan to acquire that new knowledge.

But why are these principles "ethical" considerations? When one claims to be a chemical engineer, society in general (and employers, clients, and employees in particular) puts trust in that individual. To earn that trust, one must be competent; however, most people do not have the background to judge the competence of a chemical engineer. Therefore, it is an ethical responsibility for the chemical engineer to practice within a scope of competency that can be defined completely only by that chemical engineer. And, as technology expands, a chemical engineer's capability must expand just to maintain a given scope of competency.

In addition to the AIChE code of ethics, two other codes are frequently used in chemical engineering. One is called the "Engineers' Creed":

> *As a professional engineer, I dedicate my professional knowledge and skill to the advancement and betterment of human welfare. I pledge to give the utmost of performance, to participate in none but honest enterprise, to live and work according to the laws of man and the highest standards of professional conduct, to place service before profit, honor and standing of the profession before personal advantage, and the public welfare above all other considerations. In humility and with need for divine guidance, I make this pledge.*

Some states incorporate this creed into the code of ethics for professional engineers, and some engineering colleges ask engineering graduates to recite this creed at the commencement ceremonies. The Engineers' Creed is a very much more general and "moral" (as opposed to ethical) obligation than is the AIChE code of ethics. The creed can serve as a bonding and an inspirational pledge.

The third common code of ethics is that of the National Society for Professional Engineers (Figure 23.2). This code is more detailed and more specific than is the AIChE "Code of Ethics." It not only includes the canons and principles that are the total of the AIChE Code, but it also prescribes rather specific actions to take in specific circumstances. The NSPE code applies to those who join the organization and to those licensed to practice engineering in states where this code is included in the professional engineers' code. However, it is instructive for any chemical engineer to read the NSPE code periodically, as a reminder of some of the ethical problems that arise in the profession.

23.1.8 Whistle-Blowing [12]

When a chemical engineer notices behavior that is possibly or potentially unethical, the question is, What action should he or she take? As noted in the AIChE code of ethics and in the NSPE code of ethics, there are specific avenues for action in the form of heuristics. For example, the AIChE code requires members to "formally advise their employers or clients (*and consider further disclosure, if warranted*) if they perceive that a consequence of their duties will adversely affect the present or future health or safety of their colleagues or the public." The code also requires chemical engineers to "offer objective criticism of the work of others." The code states that chemical engineers shall "issue statements or present information only in an objective and truthful manner." It is clear that one has the responsibility to tell those who engage one's professional services when there is a problem or potential problem. It is clear that one should be truthful. What is not so clear is what a chemical engineer should do if, after such disclosure, the situation persists.

Code of Ethics for Engineers

Preamble

Engineering is an important and learned profession. As members of this profession, engineers are expected to exhibit the highest standards of honesty and integrity. Engineering has a direct and vital impact on the quality of life for all people. Accordingly, the services provided by engineers require honesty, impartiality, fairness, and equity, and must be dedicated to the protection of the public health, safety, and welfare. Engineers must perform under a standard of professional behavior that requires adherence to the highest principles of ethical conduct.

I. Fundamental Canons

Engineers, in the fulfillment of their professional duties, shall:

1. Hold paramount the safety, health, and welfare of the public.
2. Perform services only in areas of their competence.
3. Issue public statements only in an objective and truthful manner.
4. Act for each employer or client as faithful agents or trustees.
5. Avoid deceptive acts.
6. Conduct themselves honorably, responsibly, ethically, and lawfully so as to enhance the honor, reputation, and usefulness of the profession.

II. Rules of Practice

1. Engineers shall hold paramount the safety, health, and welfare of the public.
   a. If engineers' judgment is overruled under circumstances that endanger life or property, they shall notify their employer or client and such other authority as may be appropriate.
   b. Engineers shall approve only those engineering documents that are in conformity with applicable standards.
   c. Engineers shall not reveal facts, data, or information without the prior consent of the client or employer except as authorized or required by law or this Code.
   d. Engineers shall not permit the use of their name or associate in business ventures with any person or firm that they believe is engaged in fraudulent or dishonest enterprise.
   e. Engineers shall not aid or abet the unlawful practice of engineering by a person or firm.
   f. Engineers having knowledge of any alleged violation of this Code shall report thereon to appropriate professional bodies and, when relevant, also to public authorities, and cooperate with the proper authorities in furnishing such information or assistance as may be required.
2. Engineers shall perform services only in the areas of their competence.
   a. Engineers shall undertake assignments only when qualified by education or experience in the specific technical fields involved.
   b. Engineers shall not affix their signatures to any plans or documents dealing with subject matter in which they lack competence, nor to any plan or document not prepared under their direction and control.
   c. Engineers may accept assignments and assume responsibility for coordination of an entire project and sign and seal the engineering documents for the entire project, provided that each technical segment is signed and sealed only by the qualified engineers who prepared the segment.
3. Engineers shall issue public statements only in an objective and truthful manner.
   a. Engineers shall be objective and truthful in professional reports, statements, or testimony. They shall include all relevant and pertinent information in such reports, statements, or testimony, which should bear the date indicating when it was current.
   b. Engineers may express publicly technical opinions that are founded upon knowledge of the facts and competence in the subject matter.
   c. Engineers shall issue no statements, criticisms, or arguments on technical matters that are inspired or paid for by interested parties, unless they have prefaced their comments by explicitly identifying the interested parties on whose behalf they are speaking, and by revealing the existence of any interest the engineers may have in the matters.
4. Engineers shall act for each employer or client as faithful agents or trustees.
   a. Engineers shall disclose all known or potential conflicts of interest that could influence or appear to influence their judgment or the quality of their services.
   b. Engineers shall not accept compensation, financial or otherwise, from more than one party for services on the same project, or for services pertaining to the same project, unless the circumstances are fully disclosed and agreed to by all interested parties.
   c. Engineers shall not solicit or accept financial or other valuable consideration, directly or indirectly, from outside agents in connection with the work for which they are responsible.
   d. Engineers in public service as members, advisors, or employees of a governmental or quasi-governmental body or department shall not participate in decisions with respect to services solicited or provided by them or their organizations in private or public engineering practice.
   e. Engineers shall not solicit or accept a contract from a governmental body on which a principal or officer of their organization serves as a member.
5. Engineers shall avoid deceptive acts.
   a. Engineers shall not falsify their qualifications or permit misrepresentation of their or their associates' qualifications. They shall not misrepresent or exaggerate their responsibility in or for the subject matter of prior assignments. Brochures or other presentations incident to the solicitation of employment shall not misrepresent pertinent facts concerning employers, employees, associates, joint venturers, or past accomplishments.
   b. Engineers shall not offer, give, solicit, or receive, either directly or indirectly, any contribution to influence the award of a contract by public authority, or which may be reasonably construed by the public as having the effect or intent of influencing the awarding of a contract. They shall not offer any gift or other valuable consideration in order to secure work. They shall not pay a commission, percentage, or brokerage fee in order to secure work, except to a bona fide employee or bona fide established commercial or marketing agencies retained by them.

III. Professional Obligations

1. Engineers shall be guided in all their relations by the highest standards of honesty and integrity.
   a. Engineers shall acknowledge their errors and shall not distort or alter the facts.
   b. Engineers shall advise their clients or employers when they believe a project will not be successful.
   c. Engineers shall not accept outside employment to the detriment of their regular work or interest. Before accepting any outside engineering employment, they will notify their employers.
   d. Engineers shall not attempt to attract an engineer from another employer by false or misleading pretenses.
   e. Engineers shall not promote their own interest at the expense of the dignity and integrity of the profession.
2. Engineers shall at all times strive to serve the public interest.
   a. Engineers are encouraged to participate in civic affairs; career guidance for youths; and work for the advancement of the safety, health, and well-being of their community.
   b. Engineers shall not complete, sign, or seal plans and/or specifications that are not in conformity with applicable engineering standards. If the client or employer insists on such unprofessional conduct, they shall notify the proper authorities and withdraw from further service on the project.
   c. Engineers are encouraged to extend public knowledge and appreciation of engineering and its achievements.
   d. Engineers are encouraged to adhere to the principles of sustainable development[1] in order to protect the environment for future generations.

Figure 23.2 NSPE Code of Ethics [11] (Reprinted with Permission of NSPE)

3. Engineers shall avoid all conduct or practice that deceives the public.
   a. Engineers shall avoid the use of statements containing a material misrepresentation of fact or omitting a material fact.
   b. Consistent with the foregoing, engineers may advertise for recruitment of personnel.
   c. Consistent with the foregoing, engineers may prepare articles for the lay or technical press, but such articles shall not imply credit to the author for work performed by others.
4. Engineers shall not disclose, without consent, confidential information concerning the business affairs or technical processes of any present or former client or employer, or public body on which they serve.
   a. Engineers shall not, without the consent of all interested parties, promote or arrange for new employment or practice in connection with a specific project for which the engineer has gained particular and specialized knowledge.
   b. Engineers shall not, without the consent of all interested parties, participate in or represent an adversary interest in connection with a specific project or proceeding in which the engineer has gained particular specialized knowledge on behalf of a former client or employer.
5. Engineers shall not be influenced in their professional duties by conflicting interests.
   a. Engineers shall not accept financial or other considerations, including free engineering designs, from material or equipment suppliers for specifying their product.
   b. Engineers shall not accept commissions or allowances, directly or indirectly, from contractors or other parties dealing with clients or employers of the engineer in connection with work for which the engineer is responsible.
6. Engineers shall not attempt to obtain employment or advancement or professional engagements by untruthfully criticizing other engineers, or by other improper or questionable methods.
   a. Engineers shall not request, propose, or accept a commission on a contingent basis under circumstances in which their judgment may be compromised.
   b. Engineers in salaried positions shall accept part-time engineering work only to the extent consistent with policies of the employer and in accordance with ethical considerations.
   c. Engineers shall not, without consent, use equipment, supplies, laboratory, or office facilities of an employer to carry on outside private practice.
7. Engineers shall not attempt to injure, maliciously or falsely, directly or indirectly, the professional reputation, prospects, practice, or employment of other engineers. Engineers who believe others are guilty of unethical or illegal practice shall present such information to the proper authority for action.
   a. Engineers in private practice shall not review the work of another engineer for the same client, except with the knowledge of such engineer, or unless the connection of such engineer with the work has been terminated.
   b. Engineers in governmental, industrial, or educational employ are entitled to review and evaluate the work of other engineers when so required by their employment duties.
   c. Engineers in sales or industrial employ are entitled to make engineering comparisons of represented products with products of other suppliers.
8. Engineers shall accept personal responsibility for their professional activities, provided, however, that engineers may seek indemnification for services arising out of their practice for other than gross negligence, where the engineer's interests cannot otherwise be protected.
   a. Engineers shall conform with state registration laws in the practice of engineering.
   b. Engineers shall not use association with a nonengineer, a corporation, or partnership as a "cloak" for unethical acts.
9. Engineers shall give credit for engineering work to those to whom credit is due, and will recognize the proprietary interests of others.
   a. Engineers shall, whenever possible, name the person or persons who may be individually responsible for designs, inventions, writings, or other accomplishments.
   b. Engineers using designs supplied by a client recognize that the designs remain the property of the client and may not be duplicated by the engineer for others without express permission.
   c. Engineers, before undertaking work for others in connection with which the engineer may make improvements, plans, designs, inventions, or other records that may justify copyrights or patents, should enter into a positive agreement regarding ownership.
   d. Engineers' designs, data, records, and notes referring exclusively to an employer's work are the employer's property. The employer should indemnify the engineer for use of the information for any purpose other than the original purpose.
   e. Engineers shall continue their professional development throughout their careers and should keep current in their specialty fields by engaging in professional practice, participating in continuing education courses, reading in the technical literature, and attending professional meetings and seminars.

Footnote 1 "Sustainable development" is the challenge of meeting human needs for natural resources, industrial products, energy, food, transportation, shelter, and effective waste management while conserving and protecting environmental quality and the natural resource base essential for future development.

As Revised July 2007

"By order of the United States District Court for the District of Columbia, former Section 11(c) of the NSPE Code of Ethics prohibiting competitive bidding, and all policy statements, opinions, rulings or other guidelines interpreting its scope, have been rescinded as unlawfully interfering with the legal right of engineers, protected under the antitrust laws, to provide price information to prospective clients; accordingly, nothing contained in the NSPE Code of Ethics, policy statements, opinions, rulings or other guidelines prohibits the submission of price quotations or competitive bids for engineering services at any time or in any amount."

Statement by NSPE Executive Committee

In order to correct misunderstandings which have been indicated in some instances since the issuance of the Supreme Court decision and the entry of the Final Judgment, it is noted that in its decision of April 25, 1978, the Supreme Court of the United States declared: "The Sherman Act does not require competitive bidding."

It is further noted that as made clear in the Supreme Court decision:

1. Engineers and firms may individually refuse to bid for engineering services.
2. Clients are not required to seek bids for engineering services.
3. Federal, state, and local laws governing procedures to procure engineering services are not affected, and remain in full force and effect.
4. State societies and local chapters are free to actively and aggressively seek legislation for professional selection and negotiation procedures by public agencies.
5. State registration board rules of professional conduct, including rules prohibiting competitive bidding for engineering services, are not affected and remain in full force and effect. State registration boards with authority to adopt rules of professional conduct may adopt rules governing procedures to obtain engineering services.
6. As noted by the Supreme Court, "nothing in the judgment prevents NSPE and its members from attempting to influence governmental action . . ."

Note: In regard to the question of application of the Code to corporations vis-a-vis real persons, business form or type should not negate nor influence conformance of individuals to the Code. The Code deals with professional services, which services must be performed by real persons. Real persons in turn establish and implement policies within business structures. The Code is clearly written to apply to the Engineer, and it is incumbent on members of NSPE to endeavor to live up to its provisions. This applies to all pertinent sections of the Code.

1420 King Street
Alexandria, Virginia 22314-2794
703/684-2800 • Fax:703/836-4875
www.nspe.org
Publication date as revised: July 2007 • Publication #1102

Figure 23.2 *(Continued)*

When is further disclosure "warranted"? What is further disclosure? Does a chemical engineer have a responsibility to disclose further, if warranted? What are the likely consequences of further disclosure? These questions are likely to be the most difficult and most important of one's professional career.

Deciding that further disclosure is warranted and then making that disclosure is called **whistle-blowing**. Most instances of whistle-blowing share the following characteristics.

1. Whistle-blowing rarely results in correction of the specific situation; however, it sometimes changes the prevailing strategy for decision making and thus reduces the chance of a further occurrence.
2. Whistle-blowing often brings about severe personal and professional problems for the whistle-blower.

Given the above characteristics, why would any chemical engineer be a whistle-blower? All three codes of ethics mentioned in Section 23.1.7 require chemical engineers to dedicate their skills to the public welfare, and whistle-blowing has led to improved automobile safety, safer nuclear and chemical plants, better control of toxic wastes, reduced government waste, safety improvements in NASA launches, and other laudable results. Whistle-blowers themselves have stated that they could not stand the stress of nondisclosure, which they viewed as an abdication of their ethical responsibilities. In other words, they could be morally autonomous only by whistle-blowing.

Many laws have been created that protect some whistle-blowers in some circumstances. Federal employees are probably the most protected. There is a specific conduit for an employee, former employee, or applicant for federal employment to make a whistle-blowing disclosure, with the identity of the whistle-blower protected to some extent. And no federal employee may engage in reprisal for whistle-blowing—"take, fail to take, or threaten to take or fail to take a personnel action with respect to any employee or applicant because of any disclosure of information by the employee or applicant that he or she reasonably believes evidences a violation of a law, rule or regulation; gross mismanagement; gross waste of funds; abuse of authority; and substantial and specific danger to public health and safety" (U.S. Code § 2302(b)). Employees in the private sector are protected by other law and regulations, many of which are summarized on the Department of Labor Web site (www.dol.gov/compliance/laws/comp-whistleblower.htm). Many whistle-blower laws and regulations exist in other countries and at state and local levels. In each case, there is a limited period (30 days is not uncommon) during which to file a complaint of retaliation. And even though the professional engineers' code of ethics requires whistle-blowing in some states, protection of the whistle-blower is not ensured. Many states have "employment at will" laws, giving any employer the right to fire an employee for any (or no) reason, and this can make it difficult to prove reprisal. Engineers in private practice (as sole proprietors or partners) may not be protected by many of these laws.

One may go through one's entire chemical engineering career without whistle-blowing, but it is unlikely that one would go through a 40-year career without having to face the question of whether to blow the whistle. Thus, to be able to have the moral autonomy to make the decision, one must consider what one should do.

Before deciding to blow the whistle, one needs to examine four key questions.

1. **What Should One Do to Solve the Problem without Whistle-Blowing?** The assumption here is that a resolution of the problem through normal channels is likely to be more effective, more timely, and less stressful than whistle-blowing would be. The AIChE and NSPE codes require that one attempt these avenues, if they are likely to be successful. Furthermore, most whistle-blowers (successful or not) indicate that whistle-blowing should be a last resort.
2. **Is Whistle-Blowing Likely to Solve the Problem?** One needs to be reasonably certain that there is a problem and that the disclosure outside normal channels will resolve it. Although merely exposing a problem may make you feel virtuous, such action would not be considered whistle-blowing unless the problem were, in the view of the whistle-blower, serious, and unless the disclosure were likely to effect a change for the better.
3. **What Whistle-Blowing Actions Should One Take?** Some whistle-blowing occurs within an organization, the most common being going over your boss's head. Other whistle-blowing involves disclosures (anonymous or attributed) to the news media. Many levels of action between these two examples are possible. One must consider the action that is most likely to effect the change that one wants, consistent with the risks that one is willing to take. The goal is to change the situation for the better, not just to expose it.
4. **What Are the Potential Consequences to One's Personal and Professional Life?** The consequences of disclosure are often serious. Loss of one's job is a clear possibility. Identification as a whistle-blower can derail or end one's career; finding another job may be very difficult. If the disclosure results in a plant closing, loss of jobs, or financial loss to the company or to other property owners, those affected may be very angry with the whistle-blower. One's family may encounter cruel treatment from those who consider the whistle-blower to be the problem. On the other hand, nondisclosure could result in danger to employees or to the general public, financial damage to shareholders or to taxpayers, or charges of a cover-up. Most whistle-blowers recommend early discussion of any potential whistle-blowing action with one's family. Nearly all wish they had legal protection from retribution. Some of these whistle-blowing issues are demonstrated in Example 23.5.

Example 23.5

Insider Information [3, Reprinted by special permission from *Chemical Engineering*, March 2, 1987, and September 28, 1987, Copyright © 1987 by McGraw-Hill, Inc., New York, NY 10020]

One day, Lee, a process engineer in an acrylonitrile plant, runs into a former classmate at a technical society luncheon. The friend has recently taken a job as a regional compliance officer for OSHA and reveals, after several drinks, that there will be an unannounced inspection of Lee's plant. In a telephone conversation a few days later, the friend mentions that the inspection will occur on the following Tuesday.

Lee believes that unsafe practices are too often tolerated in the plant, especially in the way that toxic chemicals are handled. However, although there have been many small spills, no serious accidents have occurred in the plant during the past few years.

What should Lee do? Let us suggest a problem-solving strategy for this scenario, leaving the details to a class discussion.

1. **Define:** Determine what the problem is and what the ethically desired final outcome should be. Reread the relevant codes of ethics. What are Lee's obligations according to the AIChE code of ethics? According to the NSPE code of ethics? Are there conflicting obligations?
2. **Explore:** Brainstorm for solutions in a small group. Consider internal (within the company) actions, external actions, and nonaction.
3. **Plan:** Rank the possible actions according to their likelihood to bring about the ethically desired result. Determine the most effective order for these actions. Which should be done first? Under what circumstances should you go to the next step? When would you involve your family and others? Remember that internal actions, if effective, are often the faster method to reach the desired solution.
4. **Do It:** Imagine that you followed through with your plan.
5. **Look Back:** Consider the consequences of the proposed actions. Do the likely consequences of the actions change the ranking or the ordering? Predict the final outcome of the action plan. Evaluate the outcome based on your moral, ethical, and legal responsibilities.

23.1.9 Ethical Dilemmas

Some ethical problems can be solved easily. In fact, one does this every day. The goal of this chapter is to help the reader to prepare for the difficult choices, the answers to the *ethical dilemmas*. Some problems (such as the examples in this chapter and the problems at the end) have no "perfect" solution. Whistle-blowing, for example, may satisfy one ethical obligation while violating another. The same can be said of not blowing the whistle. A simple example that many would judge an ethical dilemma is Example 23.6.

Example 23.6

Initial Employment [6]

Robin is a senior seeking employment. In January, Robin is offered a job by Company X for $4000/month and given 10 days to accept the offer. Robin accepts this offer. Two weeks later, Robin receives an offer of $4500/month and a more exciting position from Company Y. What should Robin do?

At first many may think the answer is obvious. Some will wonder whether this example is even an ethical dilemma. Yet if you ask five people what their responses would be, you are likely to get more than one answer. What are the student's obligations to Company X after accepting employment but before becoming an employee? Some companies have rescinded job offers after they have been accepted. Does this fact change Robin's obligations? Should Robin's family responsibilities affect the solution?

When faced with an ethical dilemma, *one must necessarily rank one's ethical obligations.* This is perhaps the important heuristic in solving ethical dilemmas, but here are others.

- Rank order your ethical obligations again after brainstorming for solutions.
- Admit that you may not be able to satisfy all of your obligations, but then *try* to satisfy them all.
- Combine the individual actions identified in the brainstorming into action plans. Evaluate each entire plan (rather than individual actions) for its consequences and effectiveness.
- At some point in the decision-making process, involve those who are most affected by the consequences but who are not active participants in the solution (often family, friends, trusted colleagues).

23.1.10 Additional Ethics Heuristics

Many heuristics for the solution of ethical problems have been presented in Section 23.1. A crucial heuristic is *always try to develop new heuristics.* In this vein, we offer more.

- Acquire all the information you can about the situation. The problem may be more serious than it first appears, or there may be no real problem at all.
- Be honest and open. This is especially important when dealing with those who are predisposed to distrust you.
- Acknowledge the concerns of others, whether or not you share their concerns.
- Remember that one is only as ethical as one can afford to be. One can enhance one's ability to afford to do the right thing by knowing the consequences and by balancing the responsibilities that one accepts with the rest of one's life.

For other heuristics and for cases to use for rehearsal, please see the references listed at the end of this chapter and the Internet-based resources given in the following section.

23.1.11 Other Resources

There are a myriad of resources on engineering ethics. The following are a few of the most helpful.

Engineering Ethics Center at the National Academy of Engineering. This center, founded in 2007, incorporates the Online Ethics Center (www.onlineethics.org), which has many engineering ethics resources, including the following.

- **Ethics Cases:** More than 100 engineering ethics cases that have been developed by organizations (such as NSPE), by universities, and others can be accessed through this Web site, as well as dozens of essays on engineering ethics. These include famous accidents with significant engineering ethics issues, such as Three-Mile Island and the space shuttle *Challenger*.
- **Moral Exemplars:** These are accounts of engineers and scientists showing leadership in very difficult and important ethical situations.
- **Guidelines to Professional Employment for Engineers and Scientists:** These guidelines are endorsed by AIChE, ASME, ASCE, IEEE, and NSPE and cover issues specifically related to recruitment, employment, professional development, and termination and transfer. They identify responsibilities of employers and or employees.
- **Educational Resources:** This section of the Web site provides typical assignments, syllabi, strategies for exploring professional ethics, and assessment procedures.

Engineering Ethics at TAMU. This Web site at http://ethics.tamu.edu was developed at Texas A&M University through a National Science Foundation grant and provides the ethics case studies mentioned above, as well as links to current news media articles on real ethics issues.

National Institute for Engineering Ethics (NIEE). This Web site (http://www.niee.org) is maintained by the Murdough Center for Engineering Professionalism at Texas Tech University. This institute was initially created by NSPE, and it developed and distributes the excellent engineering ethics videos titled "Gilbane Gold" (1989) and "Incident at Morales" (2003). The NSPE ethics cases since 1976 are also available. Additional resources available through this web site include the following.

- **Applied Ethics in Professional Practice Case of the Month.** Each month, a new ethics case is presented. One can respond with a suggested action, and the results of these responses are available on the site the next month.
- **Ethics Resource Guide.** An extensive bibliography of 400 books, articles, and videos on engineering ethics is given.
- **NAFTA Ethics.** This is a report describing the efforts under the North American Free Trade Agreement to develop a code of ethics for engineering throughout Canada, Mexico, and the United States. The "Principles of Ethical Conduct in Engineering Practice under the North American Free Trade

Agreement" are presented. These principles include the aspects of the AIChE code of ethics and go further in two specific areas. The NAFTA principles specifically recognize the responsibility of all engineers to strive for efficient use and conservation of the world's resources and energy and to examine the environmental impact of their actions. It also specifies that whistle-blowing (although that term is not used) is a responsibility of all engineers and that the order of disclosure should be to employer, client, public agencies, and the public.

American Chemical Society. This technical society is for chemists and chemical engineers, although chemists outnumber chemical engineers by a wide margin. The ACS "Chemical Professional's Code of Conduct" is given in Figure 23.3. This code is "for the guidance of Society members" and is not a formal agreement when one accepts membership, as is the case for AIChE. It also focuses more on scientific integrity and less on professional responsibility to the public than do the engineering codes of ethics.

National Society of Professional Engineers. The Web site of NSPE (http://www.nspe.org) contains one of the most extensive collections of items related to engineering ethics. In addition to a list of articles, books, and Web sites on engineering ethics, resources include the following.

- **Engineering Ethics Cases:** Since the 1950s, NSPE's Board of Ethical Review has accepted factual cases—from engineers, public officials, and the public—involving engineering ethics issues and has rendered opinions. From 1958 through 2006, the board published 500 opinions. The cases from 1976 on are available through the NIEE Web site. The rest are available through the NSPE Web site, which also includes a full table of contents for all cases and a cross-reference for cases against the applicable sections of the NSPE code of ethics.
- **Gilbane Gold:** This 24-minute video presents an excellent case study involving a chemical/environmental engineer, an electrical engineer, a manager, a mechanical/industrial engineer, a public agency, two engineering consultants, and the news media. A study guide, PowerPoint presentation, and the script can be downloaded free from the NIEE Web site. Produced in 1989.
- **Incident at Morales:** This 36-minute DVD presents a complex case involving a new chemical product, moving production from the United States. to Mexico, environmental, technical, safety, and financial concerns, and the need to make decisions quickly. A study guide, PowerPoint presentation, and the script can be downloaded free from the NIEE Web site. Produced in 2003, re-released in 2005 with subtitles in 13 languages.

Chemical Professional's Code of Conduct

The American Chemical Society expects its members to adhere to the highest ethical standards. Indeed, the Federal Charter of the Society (1937) explicitly lists among its objectives "the improvement of the qualifications and usefulness of chemists through high standards of professional ethics, education and attainments..." The chemical professional has obligations to the public, to colleagues, and to science.

"The Chemist's Creed," was approved by the ACS Council in 1965. The principles of The Chemist's Code of Conduct were prepared by the Council Committee on Professional Relations, approved by the Council (March 16, 1994), and replaced "The Chemist's Creed." They were adopted by the Board of Directors (June 3, 1994) for the guidance of Society members in various professional dealings, especially those involving conflicts of interest. The Chemist's Code of Conduct was updated and replaced by The Chemical Professional's Code of Conduct to better reflect the changing times and current trends of the Society. It was approved by Council on March 28, 2007 and adopted by the Board of Directors on June 2, 2007.

Their Responsibilities:

- ***The Public***
  Chemical professionals have a responsibility to serve the public interest and safety and to further advance the knowledge of science. They should actively be concerned with the health and safety of co-workers, consumers and the community. Public comments on scientific matters should be made with care and accuracy, without unsubstantiated, exaggerated, or premature statements.
- ***The Science of Chemistry***
  Chemical professionals should seek to advance chemical science, understand the limitations of their knowledge, and respect the truth. They should ensure that their scientific contributions, and those of their collaborators, are thorough, accurate, and unbiased in design, implementation, and presentation.
- ***The Profession***
  Chemical professionals should strive to remain current with developments in their field, share ideas and information, keep accurate and complete laboratory records, maintain integrity in all conduct and publications, and give due credit to the contributions of others. Conflicts of interest and scientific misconduct, such as fabrication, falsification, and plagiarism, are incompatible with this Code.
- ***Their Employer***
  Chemical professionals should promote and protect the legitimate interests of their employers, perform work honestly and competently, fulfill obligations, and safeguard proprietary and confidential business information.
- ***Their Employees***
  Chemical professionals, as employers, should treat subordinates with respect for their professionalism and concern for their well-being, without bias. Employers should provide them with a safe, congenial working environment, fair compensation, opportunities for advancement, and proper acknowledgment of their scientific contributions.

Figure 23.3 Chemical Professional's Code of Conduct (Reprinted with Permission of the American Chemical Society)

- ***Students***
  Chemical professionals should regard the tutelage of students as a trust conferred by society for the promotion of the students' learning and professional development. Each student should be treated fairly, respectfully, and without exploitation.
- ***Associates***
  Chemical professionals should treat associates with respect, regardless of the level of their formal education, encourage them, learn with them, share ideas honestly, and give credit for their contributions.
- ***Their Clients***
  Chemical professionals should serve clients faithfully and incorruptibly, respect confidentiality, advise honestly, and charge fairly.
- ***The Environment***
  Chemical professionals should strive to understand and anticipate the environmental consequences of their work. They have a responsibility to minimize pollution and to protect the environment.

Figure 23.3 *(Continued)*

- **Ethics Test:** This 25-item true-false examination is an excellent way to examine one's knowledge of the NSPE code of ethics. Answers, with section references, are provided.

23.2 PROFESSIONAL REGISTRATION

To be recognized as a chemical engineer and to offer such services to the public, one must be a licensed (registered) professional engineer. Each state (and other U.S. jurisdictions) has its own Board of Professional Engineers and its own regulations on licensure and practice. However, engineers registered in one state can generally become registered in any other state: one needs only to certify the experience, testing, and other requirements to the new state and pay a fee.

Although the professional engineer laws bar those not registered from offering their engineering services to the public, most states have a "corporate exemption." This exemption excludes from the licensure requirement engineers who offer their engineering services only to their employer and not to any outside firm or individual. If the firm offers engineering services, a licensed professional engineer must still sign, seal, and take responsibility for all such services. Stretched to the limit, this would mean that only the chief engineer of the firm would need to be a professional engineer. However, because the seal of the professional engineer on the engineering report or drawing certifies that the details of work have been checked by that engineer, in practice, all principal engineers in the firm need to be licensed. In the language of the law, anyone in "responsible charge" of engineering work must be licensed.

There are many reasons to become a licensed professional engineer, even though many chemical engineers historically have not been.

1. One cannot be in "responsible charge" of engineering work that will be offered to other firms or to the public (or indeed call oneself an "engineer" in some states) without a PE license. Effectively, this means that chemical engineers who work for engineering design or engineering and construction firms need to be licensed early in their careers. Chemical engineers in many other kinds of firms will need to be licensed before they can rise above a certain technical level. And one cannot be a consultant or an expert witness without a license.
2. The corporate exemption described above is under review in many states and could be eliminated. This is a serious career concern, because most states have already eliminated the "grandfather clauses" that previously allowed someone with many years of significant experience in the field to be granted a PE license without the normal examinations. If the corporate exemption is rescinded, the status of engineers who are licensed will be enhanced, whereas other engineers will need to study for and pass the examinations and to document the appropriate years of engineering experience just to retain their status.
3. In the past, many states exempted their own government employees from PE requirements; however, there is a strong move to eliminate all such exemptions. In fact, many states now *require* all government employees who have the word "engineer" in their job title to be PEs. There is also a move to require engineering design professors to be registered, which is already the case in 11 states.
4. Professional engineer registration is an indication to potential employers, as well as to the general public, of one's competence in the field.

There are two steps to becoming registered: engineer-in-training and professional engineer.

23.2.1 Engineer-in-Training

The first level of certification is as an engineer-in-training (EIT). As an EIT (also known as an engineering intern), one may begin to acquire engineering experience as an employee, but one may not offer services as an engineer directly to the public (or to firms other than one's employer). To obtain EIT certification, one must take and pass the Fundamentals of Engineering (FE) exam, sometimes referred to as the EIT exam. This is an eight-hour, closed-book, multiple-choice examination on a wide range of technical subjects typically covered in engineering programs. The exam is given twice a year on a Saturday (in October and in April), on the same day in all U.S. jurisdictions.

To sit for the FE exam, one must have completed or be about to complete a B.S. in an engineering field from a department that is accredited by the Accreditation Board for Engineering and Technology (ABET). Most U.S. chemical engineering departments are so accredited. Half of the states still allow a person with a nonengineering degree (or a nonaccredited engineering degree) to take the exam after acquiring substantial experience. Applicants with engineering degrees from other countries can sit for the exam if ABET certifies that the curriculum they followed is "substantially equivalent" to a U.S. ABET-accredited program. All applicants must pay a fee to the state board, and references must be provided to certify that one is of good character. The address of the State Board for Professional Engineers can be obtained from any chemical engineering department or on the Internet at http://www.ncees.org.

The examination is best taken while one is still an undergraduate. Some states allow one to take the examination only during the last semester before graduation, and some allow it to be taken any time during the senior year or earlier. The examination is written by the National Council of Examiners for Engineering and Surveying (NCEES, Clemson, SC 29633, http://www.ncees.org), from whom sample examinations and study materials can be obtained. A 230-page *FE Supplied-Reference Handbook* is provided at the examination. Applicants cannot bring any materials to the examination except a calculator that meets certain requirements. The *FE Supplied-Reference Handbook* is available for free downloading from the NCEES Web site, and it should be studied extensively before the examination date.

The morning session of the FE exam consists of 120 multiple-choice questions (4 choices per question) on general engineering, math, and science topics, as shown in Table 23.1. Each correct answer is worth 1 point, and no points are deducted for wrong answers. Therefore, all questions should be answered.

The afternoon session of the FE exam consists of 60 multiple-choice questions (four choices per question) on the specific engineering field chosen. There are separate afternoon exams for chemical, civil, electrical, environmental, industrial, and mechanical engineering. There is also a "general engineering" afternoon exam, with more difficult questions in the same subjects as in the morning section. Table 23.2 shows the breakdown of topics for the afternoon chemical engineering exam. Each correct answer is worth 2 points, and no points are deducted for wrong answers. As in the morning section, all questions should be answered.

Over the years, the definitions of these subject categories have changed. The latest information is always available on the NCEES Web site.

Many engineering schools provide refresher courses for the morning session during the six weeks or so before the exam date. In addition to this refresher or in case no such opportunity is available, a good strategy for preparing for the examination is the following.

1. At least six weeks before the exam, form a study group of people who are planning to take the FE exam on the same day.
2. Study all the chapters of the *FE Supplied-Reference Handbook*. There are two reasons to do this. First, this is the only reference material available during

Table 23.1 Subject Categories for the Morning Session of the FE exam

Subject	Average Number of Questions	Typical Preparation Courses in Chemical Engineering Curriculum
Chemistry	10.8	General chemistry
Computers	8.4	Introduction to programming (any structured language) plus spreadsheets
Electricity and Magnetism	10.8	Introduction to circuit theory or general physics plus AC power and Op amps
Engineering Economics	9.6	Section 2 of this book
Engineering Probability & Statistics	8.4	Introduction to probability and statistics
Ethics and Business Practices	8.4	Chapter 23 of this book and the Code of Professional Conduct given in the *FE Supplied-Reference Handbook*
Fluid Mechanics	8.4	Fluid mechanics or introduction to transport phenomena
Material Properties	8.4	Introduction to materials science
Mathematics	18.0	Multivariate calculus, differential equations, linear algebra, analytic geometry
Strength of Materials	8.4	Strength of materials
Engineering Mechanics	12.0	Statics and dynamics, general physics
Thermodynamics	8.4	Physical chemistry plus cycles or general engineering thermodynamics

the exam. Second, it shows the level of depth, breadth, and detail that is involved in the examination questions. Note that most of what is in the civil engineering chapter is not relevant to the morning or the ChE afternoon sessions, but about half of the electrical and computer engineering chapter is basic electricity and magnetism that is found on the morning session of the exam. In the environmental engineering chapter, the session on dispersion modeling is included in the afternoon ChE exam. In the industrial engineering chapter, be sure to study the statistical quality control and the ANOVA sections. And in the mechanical engineering chapter, study the fluid mechanics and heat transfer sections. Become familiar with the location of the material in the book and the use of the table of contents and the index. If any parts of these chapters are unfamiliar, study those topics using prior course notes, textbooks, and FE study books. This part of the preparation should take about two weeks.

3. If any of the topics covered in the chapters are unfamiliar because they were not covered in your curriculum, consult the course instructor for appropri-

Table 23.2 Subject Categories for the Chemical Engineering (Afternoon) Session of the FE exam

Subject	Average Number of Questions
Chemistry	6
Chemical Reaction Engineering	6
Chemical Engineering Thermodynamics	6
Computer Usage in Chemical Engineering	3
Fluid Dynamics	6
Heat Transfer	6
Mass Transfer	6
Material/Energy Balances	9
Process Control	3
Process Design and Economic Optimization	6
Safety, Health, and Environmental	3

ate introductory textbooks and study guides. The group should study these topics together. In most cases, this should take about a week.

4. Purchase sample exams either through the NCEES Web site or through one of the many commercially available FE study books. The study group will need approximately 240 morning questions and 60 afternoon chemical engineering questions.
5. Randomly choose 15 morning questions for each member of the study group. Each member takes one exam in 30 minutes, using only the *FE Supplied-Reference Handbook* and an approved calculator (see NCEES Web site). After the exam is taken, look at the correct answers and then share experiences among the group members, both in terms of test-taking strategy and the questions themselves. Then do the same for the afternoon ChE examination, using 8 questions in 32 minutes.
6. After studying the topics of the questions answered incorrectly in Step 5, a quarter-scale or half-scale version of the exam should be attempted, about a week later. Again, the experiences of all group members should be shared.
7. After studying any topics related to questions answered incorrectly on this exam, a final, full-length practice exam should be taken under real FE exam conditions. In all of these postexam study sessions, it is important to study the general topic area of concern, and not just the specific question.
8. On the night before the exam, a good night's rest is essential. On the day of the exam, arrival at the exam site early is a must.

Scores on the FE examination are reported to the state board 12 weeks after the test date. Shortly after that, the results are usually released to the applicant by the state board. The notification usually indicates only whether the applicant passed or failed. The nominal passing score is approximately 168 points. However, the NCEES recalibrates the passing score after the examination date because some examinations turn out to be slightly more or less difficult than anticipated. The procedure used to perform this recalibration is standard across nearly all standardized tests. It involves experimental questions and a procedure called equating. The bottom line is, if one routinely scores more than 168 out of 240 points on full-scale practice FE exams, one is probably ready to take the real exam. At the spring 2007 examination, 85% of all applicants taking the chemical engineering afternoon session of the FE exam for the first time passed, higher than in any other field. Only 46% of applicants retaking the chemical engineering exam passed. These are also the typical passing rates for the last four years. Rules for retaking the exam vary from state to state.

Many engineers who take the FE exam after they graduate wish they had taken it earlier. The examination is geared toward the material covered in B.S. engineering programs, and many engineers become more specialized during their careers. Those engineers would require more preparation for the FE examination. Also, the designation of engineer-in-training on one's credentials certainly makes a candidate more attractive to a potential employer. Finally, the next step in the registration process requires specific engineering experience, and many states count only experience that is gained after registration as an engineer-in-training. Thus, if one decides to become a PE at a later date, one may have to wait four years more to become registered. Significant opportunities may be lost in the meantime.

23.2.2 Registered Professional Engineer

After becoming an engineer-in-training, one must acquire considerable engineering experience before one can take the Principles and Practice examination (often called the PE exam). In most states, four years of responsible work must be certified by engineers for whom the applicant has worked or engineers who have reviewed the applicant's work in detail and can attest to its quality. Generally, only registered professional engineers may certify experience. Character references are also required.

The Chemical Engineering Principles and Practice exam is an eight-hour, open-book, multiple-choice examination. There are 40 questions in the morning session and 40 questions in the afternoon session. Each correct answer is worth 1 point. As with the FE exam, each question has four possible answers, and there is no deduction for incorrect answers. A separate exam is given for each field of engineering. The Chemical Engineering exam is given twice a year (April and October) on the Friday just before the date the FE exam is given. The nominal passing score is 48 points; however, as with the FE exam, recalibration to ensure uni-

Table 23.3 Subject Categories for the Chemical Engineering PE exam

Subject	Average Number of Questions
Mass and Energy Balances	19
Heat Transfer	13
Fluids	14
Mass Transfer	10
Kinetics	9
Plant Design and Operation	15

form difficulty in passing the examination is performed. On the April 2007 Chemical Engineering PE exam, 81% of applicants taking the exam for the first time passed. Only 27% of those retaking the exam passed. The results are sent to the state board 12 weeks after the exam is taken and are then sent to the applicant by the state board after that.

The difficulty of questions on the PE exam is roughly equivalent to those on a final exam in the respective course. The questions are design oriented. The topics covered are shown in Table 23.3. Note that thermodynamics questions are included in both the mass and the energy balances subjects and in the mass transfer subject. Preparing for the PE exam generally takes more effort than preparing for the FE exam. Full-length, hard-copy practice exams are available from NCEES and various publishers. Many applicants also take a refresher course, available through AIChE local sections and commercial services. In addition to the national Chemical Engineering PE exam, most states require a short law exam on the legal requirements for engineering practice in that state.

To renew a PE license, one must demonstrate continuing professional competency by documenting 15 professional development hours (PDHs) per year, although some states allow averaging over two years. Typical activities that can count (under specific rules) for PDHs are taking a course, attending seminars, publishing articles or books, and receiving a patent.

A PE license can be revoked by the state board for violation of the engineering code of ethics for that state. Such revocation becomes a matter of public record.

23.3 LEGAL LIABILITY [13]

Chemical engineers often encounter legal liability for their work. Legal advice cannot be provided in this book, but some of the situations in which a chemical engineer should seek legal counsel are noted. Perhaps the best advice is to take a short course in legal issues for engineers early in your career and seek the advice of a licensed attorney if there are any doubts about your legal rights, obligations, or liabilities.

When applying for registration as an engineer-in-training or a professional engineer, each person is given a copy of the appropriate state laws and regulations and he or she is required to certify that it is understood. These should be studied in detail. Beyond these specific licensing requirements, a chemical engineer must deal with government regulations, contracts, and issues of civil and even criminal liability.

For example, throughout a chemical engineering career, one will deal with government agencies such as the Environmental Protection Agency (EPA), Occupational Safety and Health Administration (OSHA), Department of Transportation (DOT), and others. Federal and state laws give these agencies authority to regulate industry and commerce by enacting various regulations. For federal agencies, these regulations are first published in the *Federal Register* and then periodically cataloged in the *Code of Federal Regulations* (CFR). These documents are available in most large libraries and in all law libraries as well as on-line (www.gpoaccess.gov). It is worthwhile to look at some of these regulations while you are still in school, to become aware of the scope, detail, and format of these documents. The regulations cover, for example, maximum concentrations for wastewater discharges, approved process safety procedures, and requirements for transportation of hazardous materials. They are generally written by technically trained professionals (often engineers), although the wording will have been checked by the legal staff. Thus, it is expected that competent chemical engineers can and will read, understand, and follow these regulations.

Sometimes the engineer's responsibilities are greater than they might first appear. For example, OSHA sets maximum permissible employee exposure levels for many substances, but adherence to these may not be enough, because the **general duty clause** of the OSHA Act of 1970 has been interpreted as requiring an employer to avoid exposing employees to any hazards that are known or that *should be known* to the employer. If a substance is known to cause harmful effects (e.g., such a study has been published or the company has proprietary knowledge of such effects), the employer must control employee exposure even though OSHA has no standard for that substance.

Contracts are another form of legally binding document, in which two or more parties agree to accept one or more obligations. Before signing any written contract or agreeing to any oral contract, one should understand one's obligations and what consideration (e.g., payment) is promised in return from the other party. Upon obtaining employment, one may be asked to sign an employment contract in which consideration (salary) is offered in return for certain chemical engineering services to the firm, and for adherence to specific codes of conduct (see Section 23.4). If any aspect of a contract is unclear, one should obtain legal counsel. If one is signing the contract on behalf of the firm, the firm's attorney is consulted. If one is signing as an individual, a private attorney is consulted.

Whenever there is any indication that either party may not live up to its obligations, appropriate counsel should be consulted.

Beyond the legal issues of contracts, civil legal actions (known as **torts**) sometimes involve engineers. These suits are brought when some action or lack of action by the defendant is alleged to have caused injury (physical, financial, or emotional) to the plaintiff. Both compensatory damages (to pay for correction of the injury) and punitive damages (to punish the party that did the injury) can be recovered by the plaintiff. Again, good advice here is to consult legal counsel whenever such circumstances arise or are anticipated to arise.

Finally, engineers can face criminal prosecution for actions such as falsifying records submitted to federal regulatory agencies or willfully subjecting employees to hazardous environments. Although such actions are rare, the penalties are severe.

23.4 BUSINESS CODES OF CONDUCT [14, 15, 16]

Most firms have formal codes of conduct that must be adhered to as a condition of employment. Often one is asked to sign these on the first day of work. As with any contract, it is important to read and to understand all of the details and to then fulfill all of the obligations undertaken. If an employee does not adhere to the code, he or she may be fired.

In consideration for employment, a chemical engineer (or any employee) accepts what are called **fiduciary responsibilities**. This means that trust has been placed in the engineer to act faithfully for the good of the firm. This is a general legal principle and is covered in the AIChE code of ethics.

Related to fiduciary responsibilities is the avoidance of conflicts of interest. The focus here is to avoid circumstances where you have, or appear to have, contradictory obligations. For example, if you own stock in a valve manufacturer, it would be a conflict of interest if you were to order valves from that company for your present employer, if there were another supplier of equal-quality valves at a lower price. In general—and especially in government service, where the regulations are quite strict—one must avoid not only actual conflicts of interest but also apparent conflicts of interest. The test is that, if a reasonable person could reasonably assume there to be a conflict of interest, one has an apparent conflict of interest.

It is assumed that much information about employees will become known to the company. In a typical business code of conduct, one agrees to keep personnel information confidential. That is, one agrees not to release such information to anyone outside the firm and to release it only to those within the firm who have a clear need to know. Strict adherence to such an agreement is essential.

In the business code of conduct or in a proprietary secrecy or patent agreement, the employee agrees that certain knowledge gained and discoveries made through employment are the property of the firm and may not be divulged to others. It is crucial that engineers read and understand this agreement in detail before signing. Much of the value of a firm resides in **proprietary** knowledge, that is, knowledge that is not shared with others. Breaching a secrecy agreement can be very costly to a firm, and the engineer could be in serious legal trouble. Similarly, a firm hires a chemical engineer to do work that often results in patentable discoveries. If such discoveries are made on company time, with company property, or with proprietary knowledge, the patent agreement will require disclosure to the firm and assignment of patent rights to the firm. If one does make discoveries completely outside the realm of employment, one should keep extremely careful and complete notes to avoid patent ownership difficulties.

If the firm has significant international operations, the code of conduct generally covers the conduct of employees representing the firm abroad. Because business customs vary widely from country to country, this can be an especially important subject. Before leaving the country on assignment, it is crucial to obtain as complete knowledge as possible both about the customs and laws of the other country and about any changes in business procedure that are authorized by the firm. The U.S. government forbids representatives of U.S. corporations from engaging in certain business practices, even if they are customary in that country.

A final aspect of business conduct that must be considered, and that might be included in the code of conduct, is employee relations. Specific guidelines for hiring and firing of employees must be followed to avoid legal difficulties. Whenever you are in a position to hire or fire, you must be proactive to learn the appropriate company procedures. There are specific legal requirements about what information you cannot request from job applicants and what information you can use to make employment decisions. Also, most business codes of conduct include requirements to avoid any discriminatory or harassing behavior. Good advice here, as mentioned above regarding conflicts of interest, is to avoid even the appearance of discrimination or sexual harassment.

23.5 SUMMARY

In this chapter, we explained the role of ethics in the chemical engineering profession and provided strategies for making ethical decisions. We also described the framework for registration of professional engineers in the United States, including strategies for preparing for the licensing examination.

REFERENCES

1. Koen, B. V., *Discussion of the Method: Conducting the Engineer's Approach to Problem Solving* (Oxford, UK: Oxford University Press, 2003).

2. Kohn, P. M., and R. V. Hughson, "Engineering Ethics," *Chemical Engineering*, May 5, 1980, 97–103. Responses in Hughson, R. V., and P. M. Kohn, "Ethics," *Chemical Engineering*, September 22, 1980, 132–147.
3. Matley, J., and R. Greene, "Ethics of Health, Safety and Environment: What's 'Right,'" *Chemical Engineering*, March 2, 1987, 40–46. Responses in Matley, J., R. Greene, and C. McCauley, "Ethics of Health, Safety, and Environment: CE Readers Say What's 'Right,'" *Chemical Engineering*, September 28, 1987, 108–120.
4. Mascone, C. F., A. G. Santaquilani, and C. Butcher, "Engineering Ethics: What Are the Right Choices?" *Chemical Engineering Progress* (April 1991): 61–64. Responses in Mascone, C. F., A. G. Santaquilani, and C. Butcher, "Engineering Ethics: How ChEs Respond," *Chemical Engineering Progress* (October 1991): 73–82.
5. Rosenzweig, M., and C. Butcher, "Can You Use That Knowledge?" *Chemical Engineering Progress* (April 1992): 76–80. Responses in Rosenzweig, M., and C. Butcher, "Should You Use That Knowledge?" *Chemical Engineering Progress* (October 1992): 7 and 85–92.
6. Woods, D. R., *Financial Decision Making in the Process Industries* (Englewood Cliffs, NJ: Prentice Hall, 1975).
7. Woods, D. R., "Teaching Professional Ethics," *Chemical Engineering Education* 18, no. 3 (1984): 106.
8. Schön, D. A., *The Reflective Practitioner: How Professionals Think in Action* (New York: Basic Books, 1983); Schön, D. A., *Educating the Reflective Practitioner* (San Francisco: Jossey-Bass, 1987).
9. Fogler, H. S., and S. E. LeBlanc, *Strategies for Creative Problem Solving* (Englewood Cliffs, NJ: Prentice Hall, 1995).
10. Woods, D. R., *A Strategy for Problem Solving*, 3rd ed., Department of Chemical Engineering, McMaster University, Hamilton, Ontario, 1985.
11. *The NSPE Ethics Reference Guide* (Alexandria, VA: National Society of Professional Engineers, 2006).
12. Martin, M. W., and R. Schinzinger, *Ethics in Engineering*, 4th ed. (New York: McGraw-Hill, 2004); Schinzinger, R., and M. W. Martin, *Introduction to Engineering Ethics* (New York: McGraw-Hill, 2000).
13. Heines, M. H., and K. B. Dow, "Proprietary Information: What Are Your Rights and Responsibilities?" *Chemical Engineering Progress* (July 1994): 78–84.
14. National Academy of Sciences, National Academy of Engineering, and Institute of Medicine, *On Being a Scientist: Responsible Conduct in Research*, 2nd ed. (Washington, DC: National Academy Press, 1995).
15. *Guidelines to Professional Employment for Engineers and Scientists*, Online Ethics Center for Engineering, National Academy of Engineering, 2006 (www.onlineethics.org).
16. Unger, S. H., *Controlling Technology: Ethics and the Responsible Engineer,* 2nd ed. (New York: Wiley, 1994).

PROBLEMS

1. **Proprietary Information** [6]: I am a senior. Last summer, I had an excellent job designing a new type of heat exchanger that the company was developing. When I returned to school, my professor asked me to use my knowledge on a design project, and to explain heat-exchanger design to other seniors in the design group. I really feel that I have something exciting to share with my fellow classmates. Is it ethical for me to share my experiences with my colleagues?
2. **Medical School** [6]: I am a senior in chemical engineering, but I plan to go to medical school next year. I have not been accepted yet. I want a good summer job, and yet, if my application for medical school is turned down, I would want the job to be permanent. Jobs are hard to get, and the interviewers with whom I have talked so far will offer me permanent employment but not summer employment. I have three interviews left. What do I tell these interviewers?
3. **Sophomore** [6]: I am a sophomore. None of my classmates has a summer job. Jobs are hard to get, and I need money to pay for school next year. I have been offered a well-paying job with lots of engineering experience to work on the production of propellants/explosives for antipersonnel mines. Personally, I strongly believe that this product should not be manufactured. Should I accept the job? Should I tell the company of my personal beliefs?
4. **Not a Hazard as Defined** [3, Reprinted by special permission from *Chemical Engineering*, March 2, 1987, and September 28, 1987, Copyright © 1987 by McGraw-Hill, Inc., New York, NY 10020]: In a unit where grain is steeped, sulfur dioxide is added directly to grain and water. Operators have long complained about sulfur dioxide fumes, citing runny noses, teary eyes, coughing, and headaches. The concentration has been checked many times and has always measured lower than OSHA specifications. Management's stance has always been, "Don't spend money to fix something if it isn't broken." A few employees have quit, citing allergies and other medical problems. Operators in the area have requested an engineering study to remedy the situation. Chris is given the job of investigating whether the exhaust fan should be replaced with an expensive ventilation system. What are Chris's obligations? Can they all be met? What creative strategies can you suggest to Chris to deal with this situation?
5. **Improving a Reaction** [5, Reprinted with permission of AIChE]: Look up this article, and write a one-page response to the case. Then look up the reader responses. Write a one-page reflection on the case.
6. **The Falsified Data Strike Back** [2, Reprinted by special permission from *Chemical Engineering*, May 5, 1980, and September 22, 1980, Copyright © 1980 by McGraw-Hill, Inc., New York, NY 10020]: In the case described in Section 23.1.2, Jay has written the report to suit the boss, and the company has gone ahead with an ambitious commercialization program for Catalyst A. Jay has

been put in charge of the pilot plant where development work is being done on the project. To allay personal doubts, Jay runs some clandestine tests on the two catalysts. To Jay's astonishment and dismay, the tests determine that even though Catalyst A works better under most conditions (as everyone had expected), at the operating conditions specified in the firm's process design, Catalyst B is indeed considerably superior. What should Jay do?

7. Contact the Board of Registration for Professional Engineers in your area. Determine what code of ethics is required and what are the specific requirements for registration.
8. In a small group (approximately four students), obtain examples of codes of conduct from different companies and compare their features.
9. Develop five heuristics (not mentioned in this chapter) for ethical problem solving.
10. Develop a scenario of an ethical dilemma, and, in a small group, use a problem-solving strategy to analyze it.
11. Find the Process Safety Management regulation in the *Code of Federal Regulations* (29CFR1900.119), and write a two-page synopsis of it.
12. View the (24-minute) video "Gilbane Gold," National Society for Professional Engineers, 1989. Evaluate the actions taken by the plant manager, the production engineer, the environmental engineer, and the two consulting engineers. Write a group report detailing this evaluation and referring to sections of the NSPE code of ethics where appropriate.
13. View the (two-hour) movie *Acceptable Risks,* ABC-TV productions, 1986. Evaluate the actions taken by the headquarters management, the plant manager, the production engineer, and the maintenance engineer. Write a group report detailing this evaluation and referring to sections of the AIChE code of ethics where appropriate.
14. Study the student code of conduct (sometimes referred to as the honor code, the definition of academic dishonesty, or the students' rights and responsibilities) at your institution. Develop a hypothetical (but plausible) case concerning engineering students. Present the case to the class for a 30-minute discussion.
15. The NCEES code of professional conduct (in the *FE Supplied-Reference Handbook*) was developed as a model code to be accepted by or modified by the individual state boards of professional engineers. Compare it with the AIChE code of ethics and with the NSPE code of ethics.

CHAPTER

24

Health, Safety, and the Environment

One goal of chemical engineering is to produce goods and services that enhance the quality of life. Chemical engineers have been at the forefront of efforts to improve health (e.g., pharmaceuticals), safety (e.g., shatterproof polymer glasses), and the environment (e.g., catalytic converters). Moreover, throughout the product life cycle, chemical engineers are concerned with potential harm to the health and safety of people and damage to the environment. This chapter focuses on assessment of these potential dangers. Chapter 25 focuses on proactive strategies to avoid them.

Although many of the chemical processing industries regard improvements of health, safety, and the environment (HSE) as one general function, the U.S. government has separated the field into distinct categories, based on the varying rights afforded to different classes and for historical purposes. The health and safety of employees, for example, is regulated by the Occupational Safety and Health Administration (OSHA), whereas the health of nonemployees and the environment are regulated by the Environmental Protection Agency (EPA). Other activities are regulated by the Department of Transportation (DOT) or the Department of Energy (DOE), among other agencies.

An exact description of applicable laws, rules, and regulations would be out of date before any textbook could be printed; rather, this chapter describes the types of regulations that are relevant to chemical engineering and provides guidance on where to find the current (and proposed) regulations. The focus is on the general concepts and strategies of risk assessment and reduction that transcend the details of regulations.

24.1 RISK ASSESSMENT

There are many ways to view risks to a person's health or safety. Some people will knowingly accept tremendous risk (such as in rock climbing or smoking), some expose themselves to risk perhaps without knowing (such as by sunbathing), and some refuse to be exposed at all to certain risks (such as extremely low concentrations of pesticides in food). Generally, people are much more accepting of risks that they choose (voluntary exposures) than to risks that are forced upon them (involuntary exposures). Also, people are generally much more accepting of risks that they understand than they are of risks that they do not understand. For example, the population at large tends to accept that human activity often degrades rivers and lakes with biological and chemical pollution, but they do not accept any measurable radioactive emissions from power plants.

24.1.1 Accident Statistics

Engineers quantify risks to provide a rational basis for deciding what activities should be undertaken and which risks are worth the benefits provided. These decisions are frequently made by nonengineers; thus, it is essential that the measures that engineers use be understandable to the public. For this reason, several measures have been established.

The **OSHA incidence rate** is the number of injuries and illnesses per 200,000 hours of exposure. Any injury or work-related illness that results in a "lost workday" is counted in the ratio. Thus, minor injuries that can be treated with first aid are not counted, but counted injuries range all the way to death. The 200,000 hours is roughly equivalent to 100 worker years. The OSHA incidence rate illustrates two features of risk measures: The details of the accident are not included, and these measures can be used for non-work-related exposures.

The **fatal accident rate (FAR)** is the number of fatalities per 10^8 hours of exposure. Only fatalities are counted, and the 100,000,000 hours is roughly equivalent to 1000 worker lifetimes. Although only deaths are counted, in many fields there is a strong correlation between numbers of injuries and numbers of deaths. Thus, the reasonable presumption is that a decrease in the FAR will also decrease the OSHA incidence rate.

For some exposures, the available data are insufficient to determine the total time of exposure. For these cases, the **fatality rate** can be used. This measure is the number of fatalities per year divided by the total population at risk. For example, the exposure of individual smokers is extremely variable because of differences in type of cigarette, number of inhalations per cigarette, and so on. However, one can determine the fatality rate (~0.005), which gives a rough estimate of the chance that a smoker will die this year from smoking-related causes of 0.5%.

Table 24.1 shows some representative numbers of these three risk measurements. One of the possibly surprising observations is that the chemical process industries are relatively safe (for workers) compared with many other industries.

Table 24.1 Comparison of Three Risk Measurements [1]

Activity	OSHA Incident Rate (Injuries and Deaths per 200,000 h)	Fatal Accident Rate (Deaths per 100,000,000 h)	Fatality Rate (Deaths per Person per Year)
Working in chemical industry	0.49	4	
Staying at home		3	
Working in steel industry	1.54	8	
Working in construction	3.88	67	
Traveling by car		57	170×10^{-6}
Rock climbing		4000	40×10^{-6}
Smoking (1 pack per day)			5000×10^{-6}
Being struck by lightning			0.1×10^{-6}

There are many reasons for this, ranging from the high level of remote sensing and operation, which separates workers from the most dangerous parts of the plant, to the historical concern for the hazards of industrial chemicals. Regardless of the reasons, chemical engineers continue to try to improve the safety of chemical plants.

Another observation is that there are differences in relative rankings if one uses the different risk measurements. Perhaps most striking is that the FAR for employees of the chemical industry is only slightly higher than the FAR for people staying in their homes. However, this is a potentially misleading comparison. The portion of the population that works is on average more healthy than the portion that does not. Also, most people take risks at home that they would not be allowed to take at work. Finally, a worker is probably more tired and prone to accidents at home after a hard day at work.

The main use of these statistics is not to compare one activity with another (unless they can be substituted for one another) but to monitor improvements to the health and safety of workers and others achieved by process modifications.

24.1.2 Worst-Case Scenarios

Another measure of risk is to imagine the worst possible consequence of an operation. Such a study is called a **worst-case scenario** study, and it is required by some government agencies. These studies have some drawbacks, but they can be very useful in identifying ways to avoid serious accidents. A related strategy that is routinely used throughout the industry for identifying potential hazards is called HAZOP and is discussed in Section 24.4.

The development of worst-case scenarios is certainly subjective, but government agencies and organizations develop guidelines for this task. For example, there is certainly a chance that an asteroid will impact Earth directly in the middle of the chemical plant. Should this be the worst-case scenario? Most people would argue that this takes the worst-case scenario study beyond reason, but there are no clear-cut rules. The subjective rules that have been developed contain definitions such as the worst accident that might reasonably be assumed possible over the life of the facility. Different people would define "possible" (or "probable") in different ways. Is sabotage by an employee possible? Is the simultaneous failure of three independent safety systems probable? Certainly the events of September 11, 2001, have indicated that a terrorist-attack scenario is not impossible. Sometimes probabilities of occurrence can be estimated, but often they cannot.

By EPA definition (40CFR68.3), "*Worst-case release* means the release of the largest quantity of a regulated substance from a vessel or process line failure that results in the greatest distance to an endpoint defined in § 68.22(a)." This end point is a specified concentration of a toxic substance, an overpressure of 1 psi for explosions, a radiant heat flux of 5 kW/m^2 for 40 seconds for a fire, or the lower flammability limit for a flammable substance, whichever is appropriate. Thus, the worst case is defined by the size of the area adversely affected by the release. There are also definitions of how much of the material in a tank (often 100%) should be considered over what period of time (often 10 minutes) in the scenario.

All this is to say that worst-case scenario analyses can be extremely helpful, but they are difficult to perform and potentially difficult to understand. They are most useful when specific guidelines are followed and when they are used to enhance safety by developing safeguards against the accident scenarios developed. For the risk management plan described in Section 24.2.2, worst-case scenario analyses are required by regulation.

These guidelines are constantly changing, so they are not described in detail here. The most current guidelines should be obtained from the EPA, OSHA, or other agencies.

24.1.3 The Role of the Chemical Engineer

As the professional with the best knowledge of the risks of a chemical processing operation, the chemical engineer has a responsibility to communicate those risks to employers, employees, clients, and the public.

It can be very difficult to explain FARs in deaths per 10^8 hours to the general, nontechnical public. However, the consequences of failure to explain rationally and honestly to the public the risks and the steps taken to reduce those risks are tremendous. The ethical role of the chemical engineer in this communication is discussed in Chapter 23. Beyond those responsibilities, the damage from poor communication can destroy an industry. For example, the nuclear power industry in the United States has been destroyed in large part by the failure to communicate risks to the general public in a way that it could understand. When even relatively minor accidents occur, many feel that they were misled by engineers who seemed to have said that there were no risks, no chance of an accident.

24.2 REGULATIONS AND AGENCIES

Rules and regulations arise both from governmental agencies and from nongovernmental organizations (such as AIChE). These rules and regulations change constantly, and the most up-to-date rules should always be determined. The following sections describe the kinds of rules and regulations promulgated (put into effect by formal public announcement) by various organizations. The actual agency should be contacted directly for the latest rules.

Federal government rules and regulations are published in the *Federal Register* (FR), which is issued daily. Notices of proposed or pending regulations are also given in the FR. Periodically, the regulations of a specific type are collated and published in the *Code of Federal Regulations* (CFR). Both the FR and the CFR are available in large libraries and in all law libraries, as well as on the Internet. In Table 24.2 the Internet address for the FR and CFR as of the date this book is written is provided. Because Internet addresses change frequently, we suggest use of a search engine to find new URLs if the links given here are broken. However, note that only server addresses ending in ".gov" are official U.S. government Web sites.

In addition to the direct government sources, numerous private firms collate federal regulations (more quickly than does the government), and they sell their compendia, often tailored to the needs of the customer.

State and local government regulations are also available in hard copy and increasingly in on-line form. However, it is best to contact these agencies directly to be sure that one has all the relevant regulations. Nongovernmental organizations can be contacted directly for their rules. *However, sources of information on the Internet are notoriously inaccurate and out of date.* Unless it is an official government Web site (with the date of the latest update), beware.

Table 24.2 Internet Addresses for Federal Agencies

Code of Federal Regulations (CFR)	http://www.gpoaccess.gov/cfr
Federal Register (FR) and U.S. Code	http://www.gpoaccess.gov
DOT regulations (FR and 49 CFR)	http://hazmat.dot.gov/regs/rules.htm
EPA regulations (FR and 40 CFR)	http://www.epa.gov/epahome/rules.html
MSHA regulations (FR and 30 CFR)	http://www.msha.gov/regsinfo.htm
NIOSH databases	http://www.cdc.gov/niosh/database.html
OSHA regulations (FR and 29 CFR)	http://www.osha.gov

24.2.1 OSHA and NIOSH

In general, the Occupational Safety and Health Administration promulgates regulations having to do with worker safety and health in industries other than mining (Mine Safety and Health Administration, MSHA, serves a similar role there). The National Institute for Occupational Safety and Health (NIOSH) is a federal research organization that provides information to employees, employers, and OSHA to help assess health and safety risks. Regulations are not promulgated by NIOSH, although it does certify various analytical techniques and equipment (such as respirators).

One major law and four major regulations in this area are the OSHA Act [2], Hazard Communication (29CFR1910.1200), Air Contaminants (29CFR1910.1000), Occupational Exposure to Hazardous Chemicals in Laboratories (29CFR1910.1450), and Process Safety Management of Highly Hazardous Chemicals (29CFR1910.119).

OSHA Act and Air Contaminants Standard. The original act of Congress that set up OSHA, the Occupational Safety and Health Act of 1970 [2], specified certain regulations and standards and required OSHA to promulgate others. Perhaps of most importance is the so-called **general duty clause** of the OSHA Act stating that "each employer shall furnish to each of his employees employment and a place of employment which are free from recognized hazards that are causing or are likely to cause death or serious physical harm to his employees" [Reference 2, Section 5.a.1]. This clause has been interpreted to mean that an employer must avoid exposing employees to hazards that *should have been known* to the employer, whether or not that hazard is specifically regulated by OSHA. Thus, the responsibility of researching the literature to see whether anyone has identified a hazard is placed on the employer. Most chemical engineers are employees, and yet they often represent the employer and therefore assume the responsibilities under the general duty clause.

To search for published data on chemical hazards, a good place to start is the Integrated Risk Information System (IRIS) and the Hazardous Substances Data Bank (HSDB) at http://toxnet.nlm.nih.gov.

Some specific regulations ensued from the OSHA Act, notably for the chemical industry exposure limits for certain substances in breathing air for employees (29CFR1910.1000 Air Contaminants). These limits are called **permissible exposure limits (PEL)** and are often given in parts per million (by volume). Related limits are updated yearly by the American Conference of Governmental and Industrial Hygienists (ACGIH, a nongovernmental association) and are called **threshold limit values (TLV)** [3]. In fact, the original OSHA regulations merely made these TLVs the official government limits. NIOSH also publishes **recommended exposure limits (REL)**, but these are not legally binding regulations. Each of these exposure limits is based on the assumption that a typical worker can be exposed to that concentration of the substance for eight hours a day, five days a week, for a working lifetime, without ill effects. These assumptions are often based on sketchy data extrapolated from animal studies. All these limits (PEL, TLV, REL) are given in the convenient *NIOSH Pocket Guide to Chemical Hazards* [4], available in printed form from NIOSH for a small charge or available free as an HTML download from the NIOSH Web site. Respirator requirements and a description of the health effects of exposure are also given.

The exposure limits are based on a **time-weighted average (TWA)** over an eight-hour shift, which means that higher concentrations are allowable, as long as the average over the shift is not greater than the PEL. When even short-term exposure to higher levels is harmful, there is a separate **short-term exposure limit (STEL)**, which is a 15-minute time-weighted average concentration that must never be exceeded, or an OSHA **ceiling concentration**, which is an instantaneous concentration that must never be exceeded. Also, maximum concentrations from which one could escape within 30 minutes without experiencing escape impairing or irreversible health effects are identified as **immediately dangerous to life and health (IDLH)** concentrations. The IDLH limit is used under conditions in which a respirator is normally used. In the event of a respirator failure, the employee might not be able to escape if the concentration is greater than the IDLH. These limits are also given in the NIOSH handbook [4].

Hazard Communication Standard. This regulation is also known as the **Worker Right to Know** or simply **HazCom**. The reference is 29CFR1910.1200. It requires that the employer train all employees so that the employees understand the hazards of the substances that they are handling or are exposed to or potentially exposed to. It is explicitly stated that the employer has not met this standard merely by giving the employee the hazard information orally or in written form. The employer must make sure that the employee *understands* the hazards; thus, much effort in training goes into satisfying this requirement. Two very obvious results and requirements of this regulation are proper labeling of containers and availability of **material safety data sheets (MSDS)**. Table 24.3 lists typical

Table 24.3 Typical Sections of a Material Safety Data Sheet (MSDS)

1. Material and manufacturer identification
2. Hazardous ingredients/identity information
3. Physical/Chemical characteristics
4. Fire and explosion hazard data
5. Reactivity data
6. Health hazard data
7. Precautions for safe handling and use
8. Control measures

From OSHA Form 1218-0072.

major sections of an MSDS, but the precise format of an MSDS is not presently defined by regulation, although some of the minimum requirements are, and these are also listed in Table 24.3. The MSDS must list the substances, their known hazards, and procedures for proper handling of the material and for proper actions in an emergency. Although only hazardous materials must have these MSDSs, they are available for almost all substances and mixtures in commerce. They are available directly from the supplier or manufacturer or on the Internet. It is recommended to use the Internet only as a preliminary source for MSDSs. One should get an MSDS directly from more than one manufacturer if possible, because errors in MSDSs do occur and because different MSDSs will contain different data.

Unfortunately, often MSDSs are difficult to read quickly. They are filled with much useful information, but they are often several pages long. In addition, different chemical suppliers use different formats. (OSHA now recommends the ANSI 16-section format.) They must be studied extensively *before* any hazardous situation is encountered. A group of international agencies (UN Environment Programme, International Labour Office, and the World Health Organization in cooperation with NIOSH and the European Community) has developed a simpler, standard, two-page data card that can be understood quickly and used in an emergency. In addition to the required MSDSs, the ready availability of this **international chemical safety card** wherever a hazardous material is used, stored, or transported is highly recommended. They can be found at http://www.cdc.gov/niosh/ipcs/ipcscard.html.

On June 1, 2007, the European Community put into force a new regulation called **Registration, Evaluation, Authorisation and Restriction of Chemicals (REACH)**. Although it specifically applies to manufacturers within the European Union and manufacturers that import into the EU, it is becoming the de facto worldwide standard for regulation of chemical products. All firms that handle one metric tonne or more per year of any chemical product or produce any chemical product that is new to commerce are covered. For many existing chemical products, new data will need to be measured to determine hazards more accurately. Details of the regulation are available at http://echa.europa.eu.

Minimum Requirements for MSDS (29CFR1910.1200[g]).

There are numerous facts that may be given in an MSDS; however, the minimum information required is listed below.

1. An MSDS is required for each "hazardous chemical," including those specifically listed in 29CFR1910 (subpart Z), any material assigned a TLV, or any material determined to be cancer causing, corrosive, toxic, an irritant, a sensitizer, or one that has damaging effects on specific body organs.
2. Written in English.
3. Identity used on label.

4. Chemical name and common name of all ingredients that are hazardous and that are present in ≥1% concentration or that could be released in harmful concentrations.
5. Chemical name and common name of all ingredients that are carcinogens and that are present in ≥0.1% concentration or that could be released in harmful concentrations.
6. Physical and chemical characteristics of the hazardous chemical (such as vapor pressure, flash point).
7. Physical hazards of the hazardous chemical, including the potential for fire, explosion, and reactivity.
8. Health hazards of the hazardous chemical, including signs and symptoms of exposure, and any medical conditions that are generally recognized as being aggravated by exposure to the chemical.
9. Primary routes of entry.
10. OSHA permissible exposure limit, ACGIH TLV, and any other exposure limit used or recommended by the chemical manufacturer, importer, or employer.
11. Whether the hazardous chemical is listed in the National Toxicology Program (NTP) *Annual Report on Carcinogens* (latest edition) or has been found to be a potential carcinogen in the International Agency for Research on Cancer (IARC) Monographs (latest editions), or by OSHA.
12. Any generally applicable precautions for safe handling and use that are known to the chemical manufacturer, importer, or employer preparing the MSDS, including appropriate hygienic practices, protective measures during repair and maintenance of contaminated equipment, and procedures for cleanup of spills and leaks.
13. Any generally applicable control measures that are known to the chemical manufacturer, importer, or employer preparing the MSDS, such as appropriate engineering controls, work practices, or personal protective equipment.
14. Emergency and first aid procedures.
15. Date of preparation of the MSDS or the last change to it.
16. Name, address, and telephone number of the chemical manufacturer, importer, employer, or other responsible party preparing or distributing the material safety data sheet that can provide additional information on the hazardous chemical and appropriate emergency procedures, if necessary.

Process Safety Management of Highly Hazardous Chemicals (29CFR1910.119). This OSHA regulation applies to essentially all of the chemical processing industries and requires action in 13 different types of activities as shown in Table 24.4. Note, however, that transportation of hazardous materials is regulated by the U.S. Department of Transportation under 49CFR100 through

Table 24.4 Process Safety Management of Highly Hazardous Chemicals (29CFR1910.119)

1. Employee participation
2. Process safety information
3. Process hazards analysis
4. Operating procedures
5. Training
6. Contractors
7. Pre-start-up safety review
8. Mechanical integrity
9. Hot work permit
10. Management of change
11. Incident investigation
12. Emergency planning and response
13. Compliance safety audit

49CFR185 (http://hazmat.dot.gov/regs/rules.htm). Similarly, the safe operation of chemical laboratories is regulated by OSHA through 29CFR1910.1450, which recognizes that laboratories and production facilities present different kinds of hazards.

When OSHA promulgated the Process Safety Management Regulation in 1992, it used the already existing rules of the American Institute of Chemical Engineers' *Guidelines for Technical Management of Chemical Process Safety* [5] and the American Petroleum Institute's Recommended Practices 750 [6] as guides. In fact, as is often the case, OSHA essentially gave the force of law to these voluntary nongovernmental standards.

Process safety management (PSM) embraces nearly the entire safety enterprise of a chemical process organization. It requires employee training, written operating procedures, specific quality in the engineering design of components and systems, very specific procedures for some activities, investigation and reporting of accidents that do occur, and an internal audit of the safety enterprise of the company. Following is a description of each of the 13 components.

1. **Employee Participation:** The employer must actively involve the employees in the development and implementation of the safety program. Employees are more likely to understand the hazards and to follow the established safety procedures when they are involved early and continuously in the development of the safety program. This item was added to the earlier API and AIChE standards.

2. **Process Safety Information:** The employer must research the materials, process, and operation to determine the potential hazards and keep in an immediately accessible form all safety information. This includes all MSDSs, as well as information on the process itself, such as up-to-date process flow diagrams and P&IDs.
3. **Process Hazards Analysis:** Before a process is started up and periodically thereafter (typically every three to five years or whenever significant modifications are made), a detailed study must be made of the process to determine potential hazards and to correct them. There are several approved procedures, and an organization can opt to use an alternative procedure if it can be shown to be as effective. In fact, most of the chemical processing industry uses the HAZOP technique, which is described in Section 24.4. This technique is a modified brainstorming process in which potential hazards are identified, their consequences are determined, and an action to deal with the hazard is identified.
4. **Operating Procedures:** Written operating procedures must be available to operators, and any deviations in the plant operation from these procedures must be noted. These procedures must include start-up, shutdown, and emergency response to process upset.
5. **Training:** The employer must train all employees in the hazards present and the procedures for mitigating them.
6. **Contractors:** The employer is responsible for the safe conduct of any contractors. Although each contractor is responsible for the safe conduct of the contractor's employees, the owner or operator of the plant who enters into a contract with the contractor remains liable for the safe operation of the contractor. This is an OSHA addition to the earlier API and AIChE standards.
7. **Pre-start-up Safety Review:** The regulation specifically requires that there be a review of the safety aspects of the process before any processing occurs on the site. The review must be documented, and any deviations of the plant as built from the design specifications must be addressed.
8. **Mechanical Integrity:** Vessels and other equipment must meet existing codes and be inspected during manufacture and after installation. Appropriate procedures for maintenance must be developed and followed.
9. **Hot Work Permit:** This is a very specific procedure by which a wide range of people in the plant are notified before **hot work,** such as welding, can occur. Many chemical plants use flammable materials, and everyone in the area needs to be informed so that no flammable vapors are released during the operation.
10. **Management of Change:** During accident investigations in the chemical process industries, it has often been found that severe incidents (involving deaths and massive destruction) occurred because equipment, processes, or procedures were changed from the original design without careful study of

the consequences. Thus, the OSHA regulation requires companies to have in place a system by which any modification is reviewed by all of the appropriate people. For example, any change in the reactor design must be reviewed not only by the design engineer but also by the process engineer who can evaluate how the overall process is affected. The maintenance leader must also make sure that the modification does not adversely affect the maintenance schedule or the ability of workers to get to or to maintain the equipment.

11. **Incident Investigation:** When there is a hazardous process upset, it must be investigated and a written report must be developed indicating the details of the incident, the probable cause, and the steps taken to avoid future incidents.
12. **Emergency Planning and Response:** There must be a written plan, and employees must be trained to respond to possible emergency situations.
13. **Compliance Safety Audit:** Periodically, all of the elements of the safety system (including items 1–12 above) must be audited to make sure that the approved procedures are being followed and that they are effective.

One item that is included in the industry codes but not in the PSM regulation is the entry of workers into **confined spaces**. This situation—in which the environment of the space or the difficulty of egress from the space could create a hazard—is encountered frequently in the chemical process industries. OSHA regulation 29CFR1910.146 covers the required permitting procedures to ensure that workers are protected in confined spaces.

24.2.2 Environmental Protection Agency (EPA)

The role of the EPA is to protect the environment from the effects of human activity. Although this is a very broad role, in the context of the chemical processing industries it usually relates to emissions of harmful or potentially harmful materials from the plant site to the outside by air or by water. There are three classes of such emissions: (1) planned emissions, (2) fugitive emissions, and (3) emergency emissions. This section describes some of the present regulations for these classes of emissions. There are many more regulations that are not mentioned here. In any facility, one must keep constantly aware of new and modified regulations through research, use of an environmental compliance consulting firm, or communication with the local, state, and federal environmental protection agencies.

Planned Emissions. Any process plant will have emissions. These may be harmful to the environment, benign, or, in rare cases, beneficial. In any case, a permit is usually required before construction or operation of the plant. Significant modifications to the plant (especially if they change the design emissions) will likely require a modification to the permit or a new permit.

These emissions permits are normally obtained through the state environmental protection agency, but federal regulations must be met. In some regions, states, or localities, the requirements for the permit may be significantly more stringent than the federal EPA regulations. One must contact the local agencies. However, searching the EPA databases described in Section 24.2 can provide a good preliminary overview.

Permits frequently require an extensive **environmental impact statement (EIS)** detailing the present environment and any potential disturbances that the planned activity could produce. For process plants, these EISs are typically written by a team of chemical engineers, biologists, and others, and they deal not only with planned emissions but also with potential process upsets and emergencies. In this regard, the worst-case scenario mentioned in Section 24.1.2 is used.

Permitting is based on assessment of potential degradation of the environment, and thus both the level of emissions from the plant and the present level of contamination in the local environment are considered. There are National Ambient Air Quality Standards (NAAQS) for a few materials and National Emissions Standards for Hazardous Air Pollutants (NESHAP) and New Source Performance Standards (NSPS) for these and others. Major sources (defined as those plants that emit more than some annual threshold quantity such as 25 tons of hazardous air pollutants) must meet the most stringent emissions criteria and require more permits. Similar standards are applied for water quality and for discharges into the water. Many of these regulations are based on the Clean Air Act and Clean Water Act, among others.

Beyond the effects on the environment after emissions are fully dispersed in the air or the water, there can be acute, short-term effects on nearby populations. Often chemical engineers perform dispersion studies to determine the range and longevity of the plume of harmful materials that flows from the point of discharge into the air or water.

The focus of the Clean Air Act Amendments of 1990, Title I, is the release of **volatile organic compounds (VOCs)**, which are precursors to the photochemical production of ozone (smog), especially in areas that have not attained NAAQS. The definition of a VOC is any organic compound with an appreciable vapor pressure at 25°C. **Hazardous air pollutants (HAP)** are also regulated through Title III of the act.

An important part of planned emissions are so-called **fugitive emissions**. These are losses from seals in rotating equipment (e.g., pumps, agitators, compressors), losses through connections between equipment (e.g., piping connections, valves), and other losses that result from incomplete isolation of the interior of the process from the atmosphere (e.g., tanks). Although substantial progress has been made over the last two decades to reduce fugitive emissions through improved equipment design, fugitive emissions are still substantial and are, in some cases, the major source of all emissions from a process plant.

Emergency Releases. Process upsets can create catastrophic releases of hazardous materials, and regulations require that there be an effective plan to deal with these occurrences and that the consequences for affected populations not be too serious. As mentioned earlier, worst-case scenarios and dispersion modeling are used to make this assessment.

One such regulation is the Emergency Planning and Community Right to Know Act (EPCRA) of 1986, also known as SARA, Title III. This regulation requires plants to provide the local community with information about potentially hazardous or toxic materials or processes. Further, the plant must work with the local community to develop effective emergency procedures that will be implemented automatically in the event of an accident. A local emergency planning committee is formed of members of the local government, emergency response organizations, and plant personnel. Also, releases of certain hazardous substances must be reported to the EPA and a compilation of these releases made available to the public through the Toxic Release Inventory System.

The EPA provides querying and mapping functions for its databases through the EPA Envirofacts Data Warehouse at http://www.epa.gov/enviro/html/qmr.html. Included are the Toxic Release Inventory and databases on Superfund sites, drinking water, water discharge, hazardous waste, UV index, and air releases.

Through the DOT, regulations require all over-the-road transport vehicles to carry a manifest of hazardous materials that is immediately available to all emergency personnel in the event of an accident. Also, the DOT and the U.S. Coast Guard regulate the conditions under which hazardous cargo can be transported. For example, these regulations frequently require stabilizing additives to prevent runaway polymerization.

The National Response Center (1-800-424-8802) is operated by the U.S. Coast Guard "to serve as the sole national point of contact for reporting all oil, chemical, radiological, biological, and etiological discharges into the environment anywhere in the United States and its territories." When a call is received, the information is relayed to the National Response Team as well as to various government agencies that maintain incident databases.

Many other EPA regulations that are beyond the scope of this book impact the operation of chemical processing facilities, including the Resource Conservation and Recovery Act (RCRA); the Comprehensive Environmental Response, Compensation, and Liability Act known as Superfund (CERCLA); the Superfund Amendments and Reauthorization Act (SARA); and the Toxic Substances Control Act (TSCA).

EPA Risk Management Program. The Clean Air Act Amendments of 1990 also "require the owner or operator of stationary sources at which a regulated substance is present to prepare and implement a Risk Management Plan (RMP) and provide emergency response in order to protect human health and the environment" [40 CFR 68]. The final rule was implemented in 1999. As with the OSHA Act, there is a general duty clause in this regulation specifying that owners and operators of plants have "a general duty . . . to identify hazards which may result from such releases using appropriate hazard assessment techniques, to design and maintain a safe facility taking such steps as necessary to prevent releases, and to minimize the consequences of accidental releases which do occur"

[Reference 7, Section 112(r)(1)]. The RMPs must be registered with the EPA, they must be made public, and they must be periodically updated. The risk management program, which includes as a subset the risk management plan, must include three elements:

- Hazard assessment
- Prevention
- Emergency response

This program is coordinated with OSHA's process safety management (PSM). In fact, compliance with the PSM standard is considered equivalent to the prevention part of the RMP. The following overview of the risk management program pertains to all plants covered under the PSM standard, which includes most plants in the chemical process industries.

The **hazard assessment** must include a worst-case analysis, an analysis of non-worst-case accidental releases, and a five-year accident history. The **worst-case release scenario** is defined by the EPA [40 CFR 68] as

> *the release of the largest quantity of a regulated substance from a vessel or process line failure, including administrative controls and passive mitigation that limit the total quantity involved or the release rate. For most gases, the worst-case release scenario assumes that the quantity is released in 10 minutes. For liquids, the scenario assumes an instantaneous spill; the release rate to the air is the volatilization rate from a pool 1 cm deep unless passive mitigation systems contain the substance in a smaller area. For flammables, the worst case assumes an instantaneous release and a vapor cloud explosion.*

The EPA rule specifies default values of wind speed, atmospheric stability class, and other parameters for the development of the offsite consequence analysis of worst-case scenarios. It also specifies the end point for the consequence analysis, based on the calculated concentration of toxic materials, the overpressure (1 psi) from vapor cloud explosions, and the radiant heat exposure for flammable releases (5 kW/m^2 for 40 seconds).

The prevention program is identical to the PSM standard, except that the emergency planning and response item is covered under a separate category in the RMP.

The emergency response program portion of the risk management plan is coordinated with other federal regulations. For example, compliance with the OSHA Hazardous Waste and Emergency Operations (HAZWOPER) rule (29 CFR 1910.120), the emergency planning and response portion of the PSM standard, and EPCRA will satisfy this requirement in the RMP regulation.

The RMP must designate a qualified person or position with overall responsibility for the program, as well as show the lines of authority or responsibility for implementation of the plan.

Overall, then, the only additional RMP requirement for plants already covered by the OSHA process safety management regulation is the hazard assessment (including offsite consequence analyses of worst-case and non-worst-case accidental release scenarios). This hazard assessment must not be confused with the process hazard analysis (PHA). The hazard assessment is a study of what will happen in the event of an accidental release and usually includes, for example, air dispersion simulations. The PHA (e.g., HAZOP) studies the hazards present in the process and seeks to minimize them through redesign or modifications to operating procedures.

24.2.3 Nongovernmental Organizations

Many professional societies and industry associations develop voluntary standards, and these are often accepted by government agencies and thereby are given the force of law. Examples of such organizations and their standards are as follows:

- American Petroleum Institute (API), Recommended Practices 750
- American Institute of Chemical Engineers (AIChE)
  Center for Chemical Process Safety (CCPS)
  Design Institute for Emergency Relief Systems (DIERS)
- American National Standards Institute (ANSI)
- American Society for Testing and Materials (ASTM)
- National Fire Protection Association (NFPA), fire diamond
- American Conference of Governmental Industrial Hygienists (ACGIH), TLVs
- American Chemistry Council, Responsible Care program
- Synthetic Organic Chemicals Manufacturers Association (SOCMA)
- American Society of Mechanical Engineers (ASME), boiler and pressure vessel code

The Responsible Care program is a chemical industry initiative started in 1988. All the members of the American Chemistry Council (about 140 companies) as well as several other industry organizations in the United States and in 50 other countries agree to operate according to this health, safey, and environment code. The details are given on the ACC Web site (http://www.americanchemistry.com). Its key areas include

1. Environmental impact
2. Employee, product, and process safety
3. Energy
4. Chemical industry security
5. Product stewardship: Managing product safety and public communications
6. Accountability through management system certification
7. Contribution to the economy

24.3 FIRES AND EXPLOSIONS

The most common hazards on many chemical plant sites are fires and explosions. Whenever a fuel, an oxidizer, and an ignition source are present, such a hazard exists. Detailed analyses of these hazards and their consequences are covered in other books [1,8]. Here, the terminology of the field is introduced.

24.3.1 Terminology

Combustion is the very rapid oxidation of a fuel. Most fuels oxidize slowly at room temperature. As a fuel is heated, it oxidizes more rapidly. If the heating source is removed, the fuel cools, and its oxidation returns to its normal rate for room temperature.

However, if a certain temperature (the **auto-ignition temperature**) is exceeded, the heat liberated by the oxidation is sufficient to sustain the temperature, even if the external heating source is removed. Thus, above the auto-ignition temperature, the reaction zone will expand into other areas having appropriate mixtures of fuel and oxygen. The minimum energy required to heat a small region to the auto-ignition temperature is called the **ignition energy** and is often exceedingly small.

A gaseous mixture of fuel and air will ignite only if it is within certain concentration limits. The **lower flammability** (or **explosive) limit** (LFL or LEL) is the minimum concentration of fuel that will support combustion and is somewhat below the stoichiometric concentration. The maximum concentration of fuel that will support combustion is called the **upper flammability** (or **explosive) limit** (UFL or UEL). Above the UFL, the mixture is too "rich"—that is, it does not contain enough oxygen. These two limits (UFL and LFL) straddle the stoichiometric concentration for complete combustion of the fuel. It is because of the convenient upper flammability limit of gasoline that this fuel is not more dangerous than it is. (See Problem 24.8.) Any mixture within the flammability limits should be avoided or very carefully controlled.

The **flash point** of a liquid is the temperature at which the vapor in equilibrium with the standard atmosphere above a pool of the liquid is at the LFL. Thus, a low flash point indicates a potential flammability problem if the liquid is spilled. Diesel fuel, for example, is much safer than is gasoline because diesel has a higher flash point. Regulations for transportation and use of gasoline are therefore much more stringent than they are for diesel. Flash point can be measured by the open-cup or the closed-cup method. In the **open-cup** method, an open container of the liquid is heated while a flare-up of the vapor is intentionally attempted with an ignition source. The temperature at which flare-up occurs is the flash point. In the **closed-cup** method, the liquid is placed in a closed container and allowed to come to equilibrium with air at standard pressure. Ignition is attempted at increasing temperatures. Although the open-cup and closed-cup flash points for many materials are very close, materials that vaporize slowly and disperse in the atmosphere quickly can have much higher open-cup flash points than their closed-cup flash points. Although the MSDS will give the flash point, one must notice which type of flash point is being reported.

Explosions are very rapid combustions in which the pressure waves formed propagate the combustion. The combustion creates a local pressure increase, which heats the flammable mixture to its auto-ignition temperature. This secondary combustion causes the pressure wave to propagate through the mixture. This traveling pressure pulse is called a **shock wave**. Often, a strong wind accompanies the shock wave. The combination of shock wave and wind, called a **blast wave**, causes much of the damage from explosions. When the shock wave speed is less than the speed of sound in the ambient atmosphere, the explosion is called a **deflagration**. When the speed is greater than the speed of sound, the explosion is called a **detonation**. Detonations can cause considerably more damage from the combination of blast wave, overpressure, and concussion. The damage from an overpressure of only 1 psi on structures can be extensive. Such a pressure differential on a typical door, for example, would result in considerably more than one ton of pressure on the door—enough to break most locks.

Of special concern when flammable gaseous mixtures (or dispersions of combustible dusts in air) are present is the so-called **vapor cloud explosion (VCE)**. If there is a natural gas leak, for example, the cloud (mostly methane) will spread and mix with air. The cloud, parts of which are within the flammable limits, can be quite large. If it ignites, the deflagration will cause a shock wave perpendicular to the ground that can cause great damage, often flattening buildings. When a liquid stored above its ambient boiling point suddenly comes in contact with the atmosphere (through a rupture in the tank, for example), the rapid release and expansion of the vapor can cause a massive shock wave. This phenomenon is called a **boiling-liquid expanding-vapor explosion (BLEVE)**. The failure of a steam drum, for example, can cause a BLEVE. When the BLEVE is of a flammable substance, the resulting cloud can explode. This combination of BLEVE and VCE is one of the most destructive forces in chemical plant accidents. The classic example is a propane tank that ruptures when it becomes overheated in a normal fire (a BLEVE). The propane is stored as a liquid under pressure. As the tremendous quantity of propane that vaporizes mixes rapidly with the atmosphere, it creates a massive VCE.

Runaway reactions are confined, exothermic reactions that go from their normal operating temperatures to greater than the ignition temperature; that is, they liberate more heat than can be dissipated. Thus, the temperature increases, increasing the reaction rate. Although there may be a steady state at a higher temperature (as there is in combustion), often the limits of mechanical integrity of the reaction vessel are reached before that point, causing catastrophic failure of the vessel. Such a failure can cause direct injuries, release toxic material, cause a fire, or lead to a BLEVE and/or a VCE.

To reduce the chance of a runaway condition during process upsets, the temperature difference between the reacting mixture and the cooling medium should be kept small. This may seem counterintuitive. However, consider a case where the temperature driving force is 1°C. If the temperature of the reacting mixture increases by 1°C during a process upset, the driving force for cooling has doubled! If the heat-exchange system had been designed for a 10°C driving force, that same upset would result in only a 10% increase in cooling. In systems with a chance of runaway, increased heat transfer area is the cost of an inherently safer system.

A common scenario for an accident involving an exothermic reaction is the **loss of coolant accident (LOCA)**. Unless the cooling system is backed up to the extent that it is essentially 100% reliable, one must consider this scenario in designing the vessels and the pressure-relief systems.

24.3.2 Pressure-Relief Systems

During a severe process upset, the pressure and/or temperature limits of integrity for vessels can be approached. To avoid an uncontrolled, catastrophic release of the contents or the destruction of the vessel, **pressure-relief systems** are installed. Usually, these are relief valves on vessels or process lines that open automatically at a certain pressure. Downstream, they are connected to **flares** (for flammable or toxic materials), **scrubbers** (for toxic materials), or a **stack** directed away from workers (for materials such as steam that present physical hazards). The design of the pressure-relief system is especially important, because the worst-case scenario must be considered, which is sometimes the simultaneous failure of multiple relief systems, as was the case for the Bhopal tragedy in 1984.

The design of such systems is complicated by several factors. The devices are designed to operate under unsteady conditions. Therefore, a dynamic simulation is required. Also, the flow through the relief system may be single-phase or two-phase flow. For two-phase flow, not only are the calculations more difficult, but also more factors affect the pressure drop, such as whether the line is horizontal or vertical.

In addition to the relief valves (which are called **safety valves**, **relief valves**, **pressure-relief valves**, or **pop valves** depending on service), rupture disks are used to open the process to the discharge system. **Rupture disks** are specially manufactured disks that are installed in a line, similar to the metal blanks used between flanges to close a line permanently. However, the disks are designed to fail rapidly at a set pressure. Ideally, the rupture disk allows no flow when the pressure is less than the set pressure, and it ruptures immediately, offering no resistance to flow, when the pressure hits the set point.

24.4 PROCESS HAZARD ANALYSIS

Under the "Process Hazard Analysis" requirement of the Process Safety Management of Highly Hazardous Chemicals regulation (29 CFR 1910.119), employers must complete such an analysis of all covered processes using one or more of the following techniques:

- What-If
- Checklist
- What-If/checklist
- Failure mode and effects analysis (FMEA)
- Fault-tree analysis (FTA)
- Hazards and operability study (HAZOP)

or "an appropriate equivalent methodology." The OSHA regulation specifically refers to the AIChE Center for Chemical Process Safety for details of process hazard analysis methods [9], which is an excellent source for details of these techniques.

The **what-if** technique involves a group of engineers and others going through the flowsheet and operating procedures methodically and considering what would happen if something were not as expected. For example, What if the reactor where not at the specified temperature? The answers to these hypothetical situations can uncover potential problems. This process hazard analysis technique is normally used only for simple, small-scale processes, such as laboratory experiments. For more complicated processes, the more rigorous HAZOP technique is used. This technique, which is described in the next section, is a formalized version of what-if.

Checklists have been developed by various companies for their specific processes. These lists can include hundreds of items [1,9]. Checklists are very specific and focused; they do not typically lead to the identification of safety problems that have never been encountered. Therefore, checklists (which are focused on areas of known concern) are often used in combination with what-if techniques (which are focused on thinking "outside the box").

The **FMEA** was invented by NASA in the 1960s. The underlying principle is that failures of individual components cannot be avoided, but these component failures must not cause a catastrophic failure of the system. Therefore, the analysis begins by identifying the various ways that each individual component can fail (a failure mode). Then the effect of these failures (individually and in combination) is studied. FMEA is thus a bottom-up approach that leads to identification of critical combinations of component failures that can cause some catastrophic failure. The result is usually an attempt to improve the reliability of specific components or to design protective redundancy into the system. In principle, FMEA requires the prediction and consideration of all failure modes of all components—a very large task for a complex system.

FTA is based on the premise that many of the component failure modes that would be studied in the FMEA technique would not contribute to any system failure. FTA is a top-down analysis of the system failures. First, the catastrophic system failures to be avoided are identified. Then contributing failures of subsystems and individual components are considered. FTA is widely used in the nuclear power industry, where catastrophic system failures are clearly defined.

In both the FTA and FMEA analyses, large logic diagrams are created to show the connections between low-level failures and higher-level failures. If a combination of failures is required to create a higher-level failure, the connection is denoted as an **AND gate**. If any one of several failures can create a high-level failure, the connection is denoted as an **OR gate**. There can be many levels of failures, dozens of systems failures, and several failure modes for each of thousands of components. This logic diagram leads directly to the probability of system failure if the reliability of the individual components is known.

24.4.1 HAZOP

The most widely used process hazards analysis technique in the chemical process industries is HAZOP. Unlike FTA and FMEA, the **HAZOP** technique is an outside-the-box technique. It is a modified brainstorming technique for identifying and resolving process hazards by considering seemingly unusual occurrences. Although it is a bottom-up technique, it is more efficient than the FMEA because it involves early dismissal of component failure modes that are of no consequence to system operation and focuses early on the more probable failure scenarios. A HAZOP is especially useful in identifying human factors that can contribute to system failures. For example, a HAZOP based on a sabotage scenario could consider failure modes not apt to be uncovered by the FMEA.

HAZOP consists of asking questions about possible deviations that could occur in the process (or part of a process) under consideration. A HAZOP is always done in a group, and the regulation requires that the team have "expertise in engineering and process operations," have "experience and knowledge specific to the process being evaluated," and be "knowledgeable" about the HAZOP methodology. As with any brainstorming process, the ideas and suggestions can come very quickly, and there must be an identified **scribe** ready with appropriate software to capture them. Various software packages are available to speed this process and to offer additional triggers in the brainstorming process.

The first step in a HAZOP is to identify the normal operating condition or purpose of the process or unit. This is called the **intention**. Next, a **guide word** is used to identify a possible **deviation** in the process. For example, the intention may be to keep the temperature in a vessel constant. The guide words are

- None, no, *or* not
- More of
- Less of
- More than *or* As well as
- Part of
- Reverse
- Other than

In our example, there may be no coolant flow. Once such a possible deviation is identified, the team notes any possible *causes* of the deviation. If there are any safety *consequences* of the deviation, those are noted. Suppose the coolant flow ceased because of a pump failure. The consequence may be a runaway reaction. The *action* to be taken is assigned by the HAZOP team. In this case, the action might be assigning the process engineer to investigate a backup pumping system. The team then goes on to the next possible deviation, until all reasonable deviations have been considered. The team does not solve the safety problem during the HAZOP; its job is to identify the problem and to assign its resolution to a specific person. An example of part of a HAZOP for the feed heater (H-101) of the hydrodealkylation of toluene process (Figure 1.5) is shown in Table 24.5. Several features of a HAZOP are shown. Several of the items are dismissed for "no probable cause." Others are redundant. A few are outside-the-box deviations that could lead to important safeguards for rare events. Also, the result of the HAZOP is a list of action items. These action items are not themselves decisions to change the process. They are decisions to study potential changes.

The OSHA process safety management regulation requires that the actions assigned be taken in a timely manner and that all process hazard analyses be updated at least every five years.

24.4.2 Dow Fire & Explosion Index and Chemical Exposure Index

The Dow Fire & Explosion Index was developed by Dow Chemical in the 1960s and is today used by many companies to identify high-risk systems. It is a form of process hazards analysis that focuses on fires or explosions, but it also goes beyond the identification of these hazards to quantifying the probable loss from a resulting fire or explosion.

The format of this index is similar in many ways to an income tax form, as seen in Figures 24.1 and 24.2, which are for a fictitious reactor in a polymer plant (provided by Dow Chemical, Inc., May 1998, at a workshop for faculty). The details of these procedures are given in the official guide, available from AIChE [10].

First, a specific process unit is selected. The components in the unit have the greatest impact on the hazard, so a "material factor" is determined. This factor is a measure of the energy released during a fire or explosion involving a specific material and varies from 1 (e.g., for sulfur dioxide) to 40 (e.g., for nitromethane). Material factors for 328 materials (substances or defined mixtures) are given in the official guide, as well as procedures for determining the factor for any other material based on flammability and reactivity. In the case of the hypothetical example, the reactor contains several materials, but the material factor for butadiene is the highest, so it is used as the base.

Various corrections are made for type of reaction, facility, and materials handling to arrive at a **general process hazards factor**. In the case of Figure 24.1, there is a polymerization (exothermic) penalty, and there is a **drainage and spill control** penalty because there is only a 20-minute supply of fire water available.

Table 24.5 HAZOP for the Feed Heater of the Hydrodealkylation (HDA) Process

Process Unit: H-101, Feed Heater, Figure 1.3

Intention: To provide feed to the reactor (Stream 6) at 600°C

Guide Word	Deviation	Cause	Consequence	Action
No	No flow (Stream 4)	Blockage in line	Fluid in H-101 overheats	Consider an interlock on fuel gas flow.
⇓	No O_2 in combustion products	Rich fuel:air mixture	Unburned fuel and CO in combustion products	None. O_2 analyzer with self-checking circuit controls ratio reliably.
⇓	⇓	O_2 analyzer malfunction	Potentially rich fuel:air mixture	⇓
⇓	No flow (Stream 6)	Heat tubes burst	Explosion	Interlock with sudden pressure drop alarm and shutdown.
⇓	No benzene in Stream 6	C-101 not working	Hydrogen:Toluene ratio off to R-101 and loss of hydrogen to fuel gas	Maintain spare compressor C-101.
⇓	No fuel gas flow	Supply pipe rupture	Cold shot to R-101, quenching reaction	Interlock with process shutdown.
⇓	No flame	Momentary loss of fuel gas	Explosive mixture	Automatic flame detection with reignition cycle.
More of	More flow in Stream 4	Surge of C-101	Unstable operation	Alarm.
⇓	Higher temperature	Sudden reduction in Stream 6 flowrate	Reactor overheats	Consider an interlock on fuel gas flow.
⇓	Higher pressure	Dowstream blockage	Tube failure	Pressure-relief system on tubes.
⇓	Higher temperature in heater	Increased temperature of Stream 4	Flame becomes erratic	Robust demister design.

(continued)

Table 24.5 HAZOP for the Feed Heater of the Hydrodealkylation (HDA) Process (*Continued*)

Guide Word	Deviation	Cause	Consequence	Action
⇓	⇓	Loss of furnace control	Higher temperature in Stream 6	Interlock on furnace controls.
⇓	Higher concentration of O_2 in exhaust	Lean fuel:air ratio	Waste of fuel	
⇓	⇓	No flame	Explosive mixture	Automatic flame detection with reignition cycle.
Less of	Lower temperature	Flameout	Cold shot to R-101, quenching reaction	Include automatic flame detection with reignition cycle.
⇓	Less flow in Stream 6 than in Stream 4	Heat tubes burst	Explosion	Interlock with differential flow alarm and shutdown.
⇓	Less pressure in tubes	Burst pipe downstream	Explosive and toxic release	Alarm on low pressure or low flow.
⇓	Fuel gas flow	Process upset	Cold shot to R-101	Automatic supervisory control of plant.
⇓	Less atmospheric pressure	Storm	No consequence	
⇓	⇓	Tornado	Destruction of plant	Monitor severe weather.
As well as	Liquid drops in fuel gas	Failure of V-101 demister	Flame becomes erratic	Robust demister design.
⇓	⇓	Failure of V-103 demister	⇓	⇓
⇓	Water in fuel gas	No probable cause		
⇓	Benzene in fuel gas	Unlike overflow of V-102		

(*continued*)

Table 24.5 HAZOP for the Feed Heater of the Hydrodealkylation (HDA) Process (*Continued*)

Guide Word	Deviation	Cause	Consequence	Action
Part of	Low toluene in Stream 6	P-101 not working	No reaction in R-101	Install low-flow alarm on Stream 2.
⇓	Low benzene in Stream 6	C-101 not working	Hydrogen:Toluene ratio off to R-101 and loss of hydrogen to fuel gas	Maintain spare compressor C-101.
Reverse	Decrease in Stream 6 temperature through H-101	No probable cause		
⇓	Reversal of flow (from Stream 6 to 4)	No probable cause		
Other than	Impurities in Stream 6	Impurities in feed or overheating in tubes	Impurities in product and/or catalyst deactivation	Monitor concentrations and H-101 temperatures.
⇓	Toluene in Stream 4 replaced by other hydrocarbon	Wrong connection to TK-101 by sabotage	Explosion and loss of product	Redundant management controls on storage facilities.
⇓	Fuel gas replaced by liquid fuel	Wrong connection during plant modification	Explosion	Better management of change procedures.

FIRE & EXPLOSION INDEX

AREA/COUNTRY	DIVISION	LOCATION	DATE
North America	North Central	Your State	05/18/98

SITE	MANUFACTURING UNIT	PROCESS UNIT
No Loss	My Polymer	Reactor

PREPARED BY:	APPROVED BY: (Superintendent)	BUILDING
Pat Smith	Robin Doe	A-103

REVIEWED BY: (Management)	REVIEWED BY: (Technology Center)	REVIEWED BY: (Safety & Loss Prevention)
Jo Big	Chris Wright	Lee Safe

MATERIALS IN PROCESS UNIT

Cyclohexane, Isopentane, DFI, Isoprene, Strene, Butadiene, IPA, Phosphoric Acid

STATE OF OPERATION	BASIC MATERIAL(S) FOR MATERIAL FACTOR
__DESIGN __STARTUP _X_NORMAL OPERATION __SHUTDOWN	Butadiene

	Penalty Factor Range	Penalty Factor Used(1)
Material Factor (See Table 1 or Appendices A or B) Note requirements when unit temperature over 140°F (60°C)		24
1. General Process Hazards		
Base Factor	1.00	1.00
A. Exothermic Chemical Reactions	0.30 to 1.25	0.50
B. Endothermic processes	0.20 to 0.40	
C. Material Handling and Transfer	0.25 to 1.05	
D. Enclosed or Indoor Process Units	0.25 to 0.90	
E. Access	0.20 to 0.35	
F. Drainage and Spill Control ________ gal or cu.m.	0.25 to 0.50	0.25
General Process Hazards Factor (F_1)		1.75
2. Special Process Hazards		
Base Factor	1.00	1.00
A. Toxic Material(s) $N_H = 2$	0.20 to 0.80	0.40
B. Sub-Atmospheric Pressure (<500 mm Hg)	0.50	
C. Operation In or Near Flammable Range __ Inerted __ Not Inerted		
1. Tank Farms Storage Flammable Liquids	0.50	
2. Process Upset or Purge Failure	0.30	0.30
3. Always in Flammable Range	0.80	
D. Dust Explosion (See Table 3)	0.25 to 2.00	
E. Pressure (See Figure 2) Operating Pressure _15_ psig ~~or kPa gauge~~; Relief Setting _90_ psig ~~or kPa gauge~~		0.06
F. Low Temperature	0.20 to 0.30	
G. Quantity of Flammable/Unstable Material: Quantity _59.6K_ lb ~~or kg~~; H_c = _19.0K_ BTU/lb ~~or kcal/kg~~		
1. Liquids or Gases in Process (See Figure 3) 1.132×10^9 BTU		1.56
2. Liquids or Gases in Storage (See Figure 4)		
3. Combustible Solids in Storage, Dust in Process (See Figure 5)		
H. Corrosion and Erosion	0.10 to 0.75	0.10
I. Leakage--Joints and Packing	0.10 to 1.50	0.10
J. Use of Fired Equipment (See Figure 6)		
K. Hot Oil Heat Exchange System (See Table 5)	0.15 to 1.15	
L. Rotating Equipment	0.50	0.50
Special Process Hazards Factor (F_2)		4.02
Process Unit Hazards Factor ($F_1 \times F_2$) = F_3		7.04
Fire and Explosion Index (F_3 x MF = F&EI)		169

(1) For no penalty use 0.00.

Figure 24.1 Dow Fire & Explosion Index (Form reproduced with permission of the American Institute of Chemical Engineers. Copyright © 1994 AIChE. All rights reserved. Example used by permission of SACHE (www.aiche.org/sache), a component of the Center for Chemical Process Safety providing instructional materials to member chemical engineering departments throughout the world.)

LOSS CONTROL CREDIT FACTORS

1. Process Control Credit Factor (C_1)

Feature	Credit Factor Range	Credit Factor Used(2)	Feature	Credit Factor Range	Credit Factor Used(2)
a. Emergency Power	0.98	0.98	f. Inert Gas	0.94 to 0.96	0.96
b. Cooling	0.97 to 0.99	0.97	g. Operating Instructions/Procedures	0.91 to 0.99	0.91
c. Explosion Control	0.84 to 0.98	1.00	h. Reactive Chemical Review	0.91 to 0.98	1.00
d. Emergency Shutdown	0.96 to 0.99	0.96	I. Other Process Hazard Analysis	0.91 to 0.98	0.94
e. Computer Control	0.93 to 0.99	0.93			

C_1 Value(3) **0.70**

2. Material Isolation Credit Factor (C_2)

Feature	Credit Factor Range	Credit Factor Used(2)	Feature	Credit Factor Range	Credit Factor Used(2)
a. Remote Control Valves	0.96 to 0.98	0.96	c. Drainage	0.91 to 0.97	0.91
b. Dump/Blowdown	0.96 to 0.98	1.00	d. Interlock	0.98	0.98

C_2 Value(3) **0.86**

3. Fire Protection Credit Factor (C_3)

Feature	Credit Factor Range	Credit Factor Used(2)	Feature	Credit Factor Range	Credit Factor Used(2)
a. Leak Detection	0.94 to 0.98	0.94	f. Water Curtains	0.97 to 0.98	1.00
b. Structural Steel	0.95 to 0.98	1.00	g. Foam	0.92 to 0.97	1.00
c. Fire Water Supply	0.94 to 0.97	0.94	h. Hand Extinguishers/Monitors	0.93 to 0.98	1.00
d. Special Systems	0.91	1.00	I. Cable Protection	0.94 to 0.98	1.00
e. Sprinkler Systems	0.74 to 0.97	0.97			

C_3 Value(3) **0.86**

Loss Control Credit Factor = C_1 x C_2 xC_3(3) = **0.51** (Enter on line 7 below)

PROCESS UNIT RISK ANALYSIS SUMMARY

1. Fire & Explosion Index (F&EI) (See Front)	169	
2. Radius of Exposure (Fig. 7)	142 ft ~~or m~~	
3. Area of Exposure	63,192 ft^2 ~~or m^2~~	
4. Value of Area of Exposure		$MM 5.00
5. Damage Factor (Fig. 8)	0.88	
6. Base Maximum Probable Property Damage -- (Base MPPD [4 x 5]		$MM 4.38
7. Loss Control Credit Factor (See Above)	0.51	
8. Actual Maximum Probable Property Damage -- (Actual MPPD) [6 x 7]		$MM 2.24
9. Maximum Probable Days Outage (MPDO) (Fig. 9)	30 days	
10. Business Interruption -- (BI)		$MM 4.74

(2) For no credit factor enter 1.00. (3) Product of all factors used.

Refer to *Fire & Explosion Index Hazard Classification Guide* for details.

Figure 24.2 Loss Control Credit Factors (Form reproduced with permission of the American Institute of Chemical Engineers. Copyright © 1994 AIChE. All rights reserved. Example used by permission of SACHE (www.aiche.org/sache), a component of the Center for Chemical Process Safety providing instructional materials to member chemical engineering departments throughout the world.)

Then a **special process hazards** factor is calculated based on extreme process conditions, storage of large quantities of hazardous materials, corrosion and erosion, fired equipment, and so on. In the example, there is a toxic penalty of 0.4 for butadiene. The operation is nitrogen padded but operates in the flammable range, so a penalty of 0.3 applies. There is a 0.06 penalty based on high-pressure operation with the specified relief setting. The flammable material penalty is based on the total heat of combustion for all materials in the unit (not only the butadiene). Corrosion has been estimated at less than 0.127 mm/yr, and minor leakage at pumps occurs, which lead to the respective penalties. There is an agitator, which accounts for the rotating equipment penalty.

From these three hazards indicators (materials factor, general process hazards factor, and special process hazards factor), the Fire & Explosion Index (F&EI) is calculated, 169 in the example. Table 24.6 shows the qualitative level of hazard for various values of the F&EI. Our example is in the "severe" hazard class. Although the F&EI is useful in identifying process units where hazardous conditions exist, it does not estimate the damage that might result from such an event.

The second part of the analysis (Figure 24.2) involves estimating the probable damage if the hazard leads to a catastrophic event. The damage from a fire or explosion depends on the area affected by the event, the value of the equipment destroyed, and the loss of production while the equipment damage is being repaired. All three of these are estimated by detailed algorithms described in the guide. Three categories of mitigating factors are considered: process control, material isolation, and fire protection. Typical values are used in the example. These allow the calculation of a **loss control credit factor**, which is used to correct the damage estimates.

The final part of the analysis (bottom of Figure 24.2) is the calculation of probable loss of property and loss of business if a fire or explosion were to occur. The area likely to be damaged is estimated from the F&EI. The value of the equipment in this area ($5 million in the example) is used to estimate the likely property loss, which is a function of loss control credits. The business interruption loss is estimated based on (1) probable days of outage and (2) annual fixed costs plus before-tax profit.

The Dow Chemical Hazards Index is a somewhat similar index that provides an estimate of the hazard from accidental atmospheric release of toxic sub-

Table 24.6 Dow Fire & Explosion Index [10]

Fire & Explosion Index	Qualitative Hazard Level
1–60	Light
61–96	Moderate
97–127	Intermediate
128–158	Heavy
159–	Severe

stances. A central factor in this analysis is the CEI, which is proportional to the square root of the ratio (toxic release flowrate):(threshold limit value). The details are available in the official guide [11].

A Dow "risk analysis package" consists of the analyses developed with the Dow F&EI and the Dow Chemical Exposure Index plus reports of loss prevention measures. This risk analysis package is used by industrial insurance carriers to predict the likelihood and size of loss from catastrophic events.

24.5 CHEMICAL SAFETY AND HAZARD INVESTIGATION BOARD

The Clear Air Act Amendments of 1990 (Reference 7, Section 112[r][6]) created the Chemical Safety and Hazard Investigation Board to "investigate, determine and report to the public in writing the fact, conditions and circumstances and the cause or probable cause of any accidental release resulting in fatality, serious injury or substantial property damages." The board is an independent scientific investigatory agency with no regulatory or enforcement duties. This board investigates chemical accidents, but it does not investigate all such accidents. Investigations are prioritized according to the likelihood that they would reduce further such accidents, either through enhanced knowledge of the causes or through stimulating a regulatory agency to consider further actions. Although the board was authorized in 1990, it was not funded until 1998. Since that time, it has produced numerous investigation reports, which are available on its Web site (http://www.chemsafety.gov).

The board receives initial incident reports through the National Response Center, EPA, OSHA, and DOT. From these sources, it has created a Chemical Incident Reports Center, where data on releases (whether investigated by the board or not) can be searched through the Web site.

As with other governmental safety and accident investigation boards, the Chemical Safety and Hazard Investigation Board seeks to determine root causes of accidents, with a focus on avoiding future accidents and not on assigning blame.

24.6 INHERENTLY SAFE DESIGN

Although safety controls can be added to existing processes, a more effective and efficient strategy is called **inherently safe design** [12]. The idea is to streamline the process to eliminate hazards, even if there is a major process upset. This strategy is based on a hierarchy of six approaches to process plant safety.

1. **Substitution:** One avoids using or producing hazardous materials on the plant site. If the hazardous material is an intermediate product, for example, alternative chemical reaction pathways might be used. In other words, the most inherently safe strategy is to avoid the use of hazardous materials.

2. **Intensification:** One attempts to use less of the hazardous materials. In terms of a hazardous intermediate, the two processes could be more closely coupled, reducing or eliminating the inventory of the intermediate. The inventories of hazardous feeds or products can be reduced by enhanced scheduling techniques such as **just-in-time (JIT)** manufacturing [13].
3. **Attenuation:** Reducing, or attenuating, the hazards of materials can often be effected by lowering the temperature or adding stabilizing additives. Any attempt to use materials under less-hazardous conditions inherently reduces the potential consequences of a leak.
4. **Containment:** If the hazardous materials cannot be eliminated, they at least should be stored in vessels with mechanical integrity beyond any reasonably expected temperature or pressure excursion. This is an old but effective strategy to avoid leaks. However, it is not as inherently safe as substitution, intensification, or attenuation.
5. **Control:** If a leak of hazardous material does occur, there should be safety systems that reduce the effects. For example, chemical facilities often have emergency isolation of the site from the normal storm sewers, and large tanks for flammable liquids are surrounded by dikes that prevent any leaks from spreading to other areas of the plant. Scrubbing systems and relief systems in general are in this category. They are essential, because they allow a controlled, safe release of hazardous materials, rather than an uncontrolled, catastrophic release from a vessel rupture.
6. **Survival:** If leaks of hazardous materials do occur and they are not contained or controlled, the personnel (and the equipment) must be protected. This lowest level of the hierarchy includes firefighting, gas masks, and so on. Although essential to the total safety of the plant, the greater the reliance on survival of leaks rather than elimination of leaks, the less inherently safe the facility.

24.7 SUMMARY

This chapter describes the overall framework of health, safety, and environmental activities in the chemical process industries. The specific regulations change constantly, and the cognizant agencies must be consulted for the current rules.

24.8 GLOSSARY

ACC: American Chemistry Council

ACGIH: American Congress of Governmental Industrial Hygienists

AIChE: American Institute of Chemical Engineers

ANSI: American National Standards Institute

API: American Petroleum Institute

ASME: American Society of Mechanical Engineers

ASTM: American Society for Testing and Materials

BLEVE: boiling-liquid expanding-vapor explosion

CAA: Clean Air Act

CCPS: Center for Chemical Process Safety

CERCLA: Comprehensive Environmental Response, Compensation, and Liability Act

CFR: *Code of Federal Regulations*

CMA: Chemical Manufacturers' Association; former name of the American Chemistry Council

DIERS: Design Institute for Emergency Relief Systems

DOT: Department of Transportation

EIS: environmental impact study (or statement)

EPA: Environmental Protection Agency

EPCRA: Emergency Planning and Community Right to Know Act, also known as SARA, Title III

F&EI: Dow Fire & Explosion Index

FAR: fatal accident rate

FMEA: failure modes and effects analysis

FR: *Federal Register*

FTA: fault-tree analysis

HAP: hazardous air pollutant

HAZOP: hazard and operability study

IDLH: immediately dangerous to life and health

LEL: lower explosive limit

LFL: lower flammability Limit

LOCA: loss of coolant accident

MSDS: material safety data sheet

MSHA: Mine Safety and Health Agency

NAAQS: National Ambient Air Quality Standards

NESHAP: National Emissions Standards for Hazardous Air Pollutants

NFPA: National Fire Protection Association

NIOSH: National Institute for Occupational Health

NIOSHTIC: NIOSH Technical Information Center

NSPS: New Source Performance Standards

OSHA: Occupational Safety and Health Agency

PEL: permissible exposure limit

PHA: process hazard analysis

PSM: process safety management

RCRA: Resource Conservation and Recovery Act

REL: recommended exposure limit

RMP: risk management program (or plan)

SARA: Superfund Amendments and Reauthorization Act

SOCMA: Synthetic Organic Chemical Manufacturers' Association

STEL: short-term exposure limit

TLV: threshold limit values

TSCA: Toxic Substances Control Act

TWA: time-weighted average

UEL: upper explosive limit

UFL: upper flammability limit

VCE: vapor cloud explosion

VOC: volatile organic compound

REFERENCES

1. Crowl, D. A., and J. F. Louvar, *Chemical Process Safety: Fundamentals with Applications,* 2d ed. (Englewood Cliffs, NJ: Prentice Hall, 2002).
2. Occupational Safety and Health Act of 1970, Public Law 91-596, 29 U.S. Code §651 et seq., December 29, 1970.

3. *2007 TLVs® and BEIs®: Threshold Limit Values for Chemical Substances and Physical Agents and Biological Exposure Indices* (Cincinnati, OH: American Conference of Governmental Industrial Hygienists, 2007).
4. *NIOSH Pocket Guide to Chemical Hazards,* National Institute for Occupational Safety and Health, Cincinnati, OH, printed version: 2005. (Updated and downloadable at http://www.cdc.gov/niosh/npg.)
5. *Guidelines for Technical Management of Chemical Process Safety* (New York: American Institute of Chemical Engineers, 1989).
6. *API Recommended Practices 750* (Washington, DC: American Petroleum Institute, 1990).
7. Clean Air Act Amendments of 1990, Public Law 101-549, 42 U.S. Code §7401 et seq., November 15, 1990.
8. Bodurtha, F. T., *Industrial Explosion Prevention and Protection* (New York: McGraw-Hill: 1980).
9. *Guidelines for Hazard Evaluation Procedures,* 2nd ed. with worked examples (New York: Center for Chemical Process Safety of the American Institute for Chemical Engineers, 1992).
10. *Dow's Fire & Explosion Index Hazard Classification Guide,* 7th ed. (New York: American Institute of Chemical Engineers, 1994).
11. *Dow's Chemical Exposure Index Guide,* 2nd ed. (New York: American Institute of Chemical Engineers, 1998).
12. Kletz, T. A., *Cheaper, Safer Plants or Wealth and Safety at Work: Notes on Inherently Safer and Simpler Plants* (Rugby, England: Institution of Chemical Engineers, 1984).
13. Hall, R. W., *Attaining Manufacturing Excellence: Just-in-Time, Total Quality, Total People Involvement* (Homewood, IL: Business One Irwin, 1987).

PROBLEMS

1. You work for a chemical company with 30,000 employees. If your company has a typical safety record for the chemical industry, what is your best estimate of how many of your employees
   a. Will succumb to a fatal accident while on the job this year?
   b. Will be injured but not killed on the job?
2. Locate the nearest steam plant on campus.
   a. Develop two possible accident scenarios that should be considered when searching for worst-case scenarios.
   b. List the safeguards that have been built into the system to mitigate some (or all) of these effects.
3. Find an MSDS for each of the components listed in the HDA process.

4. Summarize all regulations (safety, environmental, transportation) that you can find for benzene.
5. Assume that the unit operations lab in your department must meet the process safety management standard. Choose two of the thirteen components of PSM, and write a critical analysis of these aspects of lab operation.
6. Some paints have a closed-cup flash point near room temperature, whereas they have no measurable open-cup flash point.
   a. Explain this apparent paradox.
   b. Which is the more useful flash point when using a paint?
7. Assume gasoline to be 87 vol% iso-octane (2,2,4 tri-methyl pentane) and 13 vol% n-heptane. At room temperature, the air above a pool of gasoline will become saturated.
   a. Is the air-gasoline mixture within its flammable limits?
   b. Could there be any location where the mixture will be within its flammable limits? Explain.
8. Perform a HAZOP on R-101 of the HDA process. Be sure to perform the analysis in a team.
9. Benzene in gasoline is now limited by regulation to a maximum of 0.62 vol%. Is the benzene concentration in the air above a pool of gasoline greater than or less than the PEL? Show the effect of temperature.
10. Calculate the Dow Fire & Explosion Index for the HDA Reactor 101. Compare this with the F&EI for tank TK-101.
11. Use the various resources from this chapter to find health, safety, and environmental information about a chemical process plant in your area.
12. Obtain an environmental impact study for a facility in your area. Prepare a synopsis and lead a class discussion.

CHAPTER

25

Green Engineering

Green engineering can be defined as engineering for the environment. In terms of chemical engineering and chemical process design, green engineering means design for reduction of emissions, design to eliminate particularly hazardous chemicals, design to minimize the use of natural resources, and design to minimize energy usage.

Several books have been published on green engineering and pollution prevention [1,2,3]; therefore, only a summary of the bare essentials is possible in one chapter. The reader desiring a more in-depth treatment should consult these references.

In this chapter, environmental regulations, particularly the Pollution Prevention Act of 1990, are reviewed. Methods for understanding and estimating the behavior of chemicals in the environment are surveyed. Then the concept of green chemistry is defined and discussed briefly. Next, methods for pollution prevention at every stage of a design are discussed. The economics of pollution prevention are discussed. Finally, life cycle analysis, a study of the environmental consequences of manufacturing, using, and disposing of a product, is introduced.

There are several Web-based resources for pollution prevention and green engineering. The Environmental Protection Agency (EPA) Web site (http://www.epa.gov) is one, as is its green engineering program Web site (http://www.epa.gov/oppt/greenengineering).

25.1 ENVIRONMENTAL REGULATIONS

The Pollution Prevention Act of 1990 has one major requirement. Companies are required to report on their pollution prevention activities. It establishes a waste management hierarchy, which, from most to least desirable, is as follows.

1. **Source Reduction:** Source reduction means that the process is modified so that less waste and/or less-hazardous waste is generated. The best way to accomplish this is by modification of the chemistry in the reactor. If hazardous waste is not made, there is no chance of its becoming a pollutant.
2. **In-Process Recycle:** This is the recycle of unreacted feed so that it will not become a waste product. It has already been established several places in this book that this is also sound economic policy, because raw materials are almost always the largest operating cost in a chemical process. Therefore, in-process recycle, along with other pollution prevention concepts, has economic as well as environmental advantages.
3. **On-Site Recycle:** An example of on-site recycle is to convert waste generated in a reactor to a useful product in another reactor.
4. **Off-Site Recycle:** Off-site recycle is the separation of waste, the transfer of the waste off-site, followed by its conversion to a useful product at another facility. The definition of another facility may include a different part of the same chemical plant.
5. **Waste Treatment:** Waste treatment involves separation of waste generated in the process followed by treatment to make it less hazardous.
6. **Secure Disposal:** Secure disposal involves separation of waste generated in the process followed by disposing of it in a secure facility such as a landfill.
7. **Release to Environment:** Here, waste generated in the process is separated and released to the environment.

Although items 1–6 could each be considered pollution prevention in its own way, the Pollution Prevention Act of 1990 defines only the first two as being pollution prevention. However, most engineers consider anything that converts waste into a useful product to be pollution prevention.

The hierarchy is clear. The top priority is not to produce waste or to minimize the production of waste, which we understand to be a difficult task. If waste products are made, they should be converted to useful products. If this cannot occur, they should be rendered less hazardous. Only as last resorts should there be disposal to a landfill (which could ultimately become a release to the environment) or direct release to the environment.

There are other laws regulating different aspects of the environment. As a whole, they are the lower bounds for environmental protection and human safety for the chemical industry. Many companies go beyond these minima (see, for example, "Responsible Care," in Section 24.2.3.). Nine current, major laws are summarized in Table 25.1, and many of these laws have subsequent amendments. The text of all of these laws is available at http://www.epa.gov/lawsregs/laws/index.html.

Table 25.1 Summary of Environmental Laws

Law	Year	Summary
Occupational Safety and Health Act (OSH Act)	1970	Created OSHA; provides regulations protecting worker safety on the job; requires development of MSDS sheets, training of employees on safe handling of chemicals.
Clean Air Act (CAA)	1970	Establishes air quality standards for criteria pollutants; defines criteria pollutants as CO, Pb, NO_x, SO_2, ozone, and particulates.
Clean Water Act (CWA)	1972	Requires permits for all discharges; permit holders must monitor discharges.
Federal Insecticide, Fungicide, and Rodenticide Act (FIFRA)	1947 (amended 1972)	Requires that pesticides be registered with EPA. Manufacturer must prove product efficacy and that it is not harmful to humans.
Toxic Substances Control Act (TSCA)	1976	Chemical manufacturers must report processing information for all chemicals; premanufacturing notices required for all new chemicals.
Resource Conservation and Recovery Act (RCRA)	1976	Requires maintenance of records of hazardous waste generated and its ultimate fate; those transporting such waste also must maintain similar records.
Comprehensive Environmental Response, Compensation, and Liability Act (CERCLA)	1980	Amended by the Superfund Amendments and Reauthorization Act (SARA) 1986; defines "Superfund" sites; parties responsible for cleanup identified; assigns economic liability for cleanup; makes such liability retroactive.
Emergency Planning and Community Right to Know Act (EPCRA)	1986	Before this act, it was common for the community not to know what went on in a chemical plant; requires facilities to work with local agencies to develop plans for dealing with accidental release; also requires reporting of waste stored, transferred off-site, and released.
Pollution Prevention Act (PPA)	1990	Establishes pollution prevention hierarchy; only mandated provision is to report pollution prevention activities.

25.2 ENVIRONMENTAL FATE OF CHEMICALS

Methods have been developed to estimate the fate of chemicals if they have been released to the environment. This is done by defining properties of a molecule that are a measure of the molecule's behavior in the environment. These estimation techniques are based on group contribution methods used to estimate thermodynamic properties [4]. In group contribution methods, a physical property is estimated by breaking the molecule down into its component parts—for example, the number of carbons, the number of alcohol (or ether, ester, etc.) groups, and so on. There are correction factors quite often for branching, aromatic structure, and so on, and there are also methods based on counting the number and type of bonds. The physical property is estimated by adding the contributions of all component parts of the molecule. Most of the group contribution values have been obtained empirically by correlating measured values with a molecule's group structure. These methods are beyond the scope of this textbook, and the interested reader can find more detailed descriptions elsewhere [1]. Software (EPIWINSuite—downloadable at http://www.epa.gov/oppt/exposure/pubs/episuite.htm) has been developed that does the group contribution calculations for any molecule [1].

In pollution prevention, the properties calculated by group contribution methods are estimates of how a molecule behaves in the environment. These properties are best used for comparison between molecules. Table 25.2 lists selected properties that can be estimated using group contribution methods that have been used to estimate environmental fate.

Example 25.1 illustrates the use of one of these properties to estimate environmental fate.

Example 25.1

Methyl tertiary butyl ether (MTBE, CH_3-O-$C(CH_3)_3$) is an oxygenated gasoline additive. Oxygenated gasoline additives are mandated by law to reduce air pollution. Oxygenated fuel additives burn cleaner—for example, producing CO_2 rather than CO. However, because there is inevitably some leakage from underground gasoline tanks, the potential exists for the components of gasoline to enter the water table, ultimately contaminating drinking water. This is a problem for water treatment facilities, and it is an even more significant problem for those using well water.

For this calculation, gasoline is assumed to be 2,2,4-trimethyl pentane ($CH_3C(CH_3)_2$ $CH_2CH(CH_3)CH_3$) (also known as isooctane). The table summarizes several of the properties of these two compounds, calculated using the EPISuite software.

Property	MTBE	Gasoline
Boiling point (°C)	47.04	81.44
$\log_{10}$ of octanol/water partition coefficient	1.43	4.09
Water solubility (mg/L)	1.98×10^4	9.91
Henry's law coefficient in water (atm m^3/mol)	2.02×10^{-3}	3.01
$\log_{10}$ soil sorption coefficient	0.721	2.44

The observation is that MTBE has much higher water solubility than gasoline and a much lower soil sorption coefficient. Therefore, MTBE in leaking from underground tanks

Table 25.2 Physical Properties that Influence Environmental Phase Partitioning [2]

Property	Definition	Significance in Estimating Environmental Fate and Risks
Melting point (T_m)	Temperature at which solid and liquid co-exist at equilibrium	Sometimes used as a correlating parameter in estimating other properties for compounds that are solids at ambient or near-ambient conditions
Boiling point (T_b)	Temperature at which the vapor pressure of a compound equals atmospheric pressure; normal boiling point	Characterizes the partitioning between gas and liquid phases; frequently used as a correlating variable in estimating other properties
Vapor pressure (P^*)	Partial pressure exerted by a vapor when the vapor is in equilibrium with its liquid	Characterizes the partitioning between gas and liquid phases
Henry's law constant (H)	Equilibrium ratio of the concentration of a compound in the gas phase to the concentration of the compound in a dilute aqueous solution (sometimes reported as atm-m^3/mol)	Characterizes the partitioning between gas and aqueous phases
Octanol-water partition coefficient (K_{ow})	Equilibrium ratio of the concentration of a compound in octanol to the concentration of the compound in water	Characterizes the partitioning between hydrophilic and hydrophobic phases in the environment and the human body; frequently used as a correlating variable in estimating other properties
Water solubility (S)	Equilibrium solubility in mol/L	Characterizes the partitioning between hydrophilic and hydrophobic phases in the environment
Soil sorption coefficient (K_{oc})	Equilibrium ratio of the mass of a compound adsorbed per unit weight of organic carbon in a soil (in μg/g organic carbon) to the concentration of the compound in a liquid phase (in μg/mL)	Characterizes the partitioning between solid and liquid phases in soil, which in turn determines mobility in soils; frequently estimated based on octanol-water partition coefficient, and water solubility
Bioconcentration factor (BCF)	Ratio of a chemical's concentration in the tissue of an aquatic organism to its concentration in water (reported as L/kg)	Characterizes the magnification of concentrations through the food chain

preferentially enters groundwater, whereas gasoline leaking from underground tanks preferentially adsorbs to the soil.

25.3 GREEN CHEMISTRY

When a chemical is manufactured, a raw material is used. In the chemical industry, this raw material can usually be traced back to a depletable natural resource such as crude oil. Furthermore, when a chemical is manufactured, there are usually side reactions that produce undesired contaminants, which become hazardous waste if they cannot be converted to useful products. The processes illustrated in this book generally ignore all except the most important undesired contaminants; however, in real applications, there are usually trace amounts of several undesired contaminants that must be separated and eventually treated. Another possible source of waste is the need to use hazardous solvents or catalysts to promote a reaction.

Three issues in green chemistry are a search for the following:

- Alternative feedstocks
- Green solvents
- New synthesis pathways

One component of green chemistry is to use alternative feedstocks that either improve the environmental performance of a process or that do not deplete nonrenewable resources as much as the traditional process. For example, it has been suggested that adipic acid, a feedstock for nylon, can be manufactured from glucose, a renewable resource, by using a microbial biocatalyst [5]. The traditional method uses benzene, a known carcinogen, which can trace its origin back to nonrenewable fossil fuels such as oil and coal.

Green solvents are another component of green chemistry. In principle, the group contribution methods described in Section 25.2 could be used to design a solvent providing the desired performance and having properties that are less toxic to humans and less hazardous to the environment. The applicability of such a solvent would also depend on how it is manufactured. If the only way to make a green solvent is by a process more hazardous than the process that uses the original solvent, then there is no advantage to using the green solvent. The expense of the green solvent is also an issue.

The use of new synthesis pathways that avoid hazardous intermediates and/or hazardous by-products is also a component of green chemistry. For example, an elimination reaction ($AB \rightarrow A + B$), where A is the desired product and B is a waste product, is less desirable than an addition reaction ($A + B \rightarrow C$) because no waste products are made. Another example is the elimination of unwanted side reactions, such as the production of p-diisopropyl benzene in the cumene process shown on the CD-ROM accompanying this book. New synthesis pathways will most likely be obtained only by development of new catalysts.

Clearly, green chemistry is in its infancy. The concepts are clear, but the implementation is years away. Example 25.2 relates to green chemistry.

Example 25.2

Phosgene ($COCl_2$) is one of the most toxic chemicals known. It was used as mustard gas in World War I. Currently, phosgene is used as a raw material in the manufacture of pesticides and urethanes. It has been suggested that dimethyl carbonate (DMC, CH_3-O-(C=O)-O-CH_3) can serve as a reactant in many of the same processes that use phosgene [6]. This is an example of an alternative feedstock.

However, the traditional method for manufacturing DMC is from phosgene and methanol.

Phosgene manufacture $CO + Cl_2 \rightarrow COCl_2$

DMC manufacture $COCl_2 + 2CH_3OH \rightarrow CH_3\text{-O-(C=O)-O-}CH_3 + 2HCl$

Patents have been issued on a process to manufacture DMC directly from CO and O_2 [7]. The reaction is

$$CO + O_2 + 2CH_3OH \rightarrow CH_3\text{-O-(C=O)-O-}CH_3 + H_2O$$

This new process illustrates two of the principles of green chemistry listed above. First of all, DMC is manufactured without using the hazardous phosgene feed. In fact, the phosgene is not even produced in the new pathway. Second, the DMC can be a raw material substitution for phosgene in the manufacture of pesticides and urethanes. Furthermore, the first principle of waste management hierarchy is demonstrated in that the by-product HCl is not produced (although, if market conditions were appropriate, the HCl produced could be purified and sold as a commodity).

25.4 POLLUTION PREVENTION DURING PROCESS DESIGN

The most important issues when trying to prevent pollution during process design are to try to minimize generation of waste products in the reactor, to try to design separation systems for maximum recovery and minimum energy usage, to minimize effluent streams containing waste, and to minimize leaks, particularly during storage and transfer operations.

Because raw materials are almost always the largest operating cost, their efficient use is necessary for economical process operation. As every chemical engineer learned in the material and energy balances class, unreacted raw materials are separated and recycled. This is also green engineering. If the raw materials were not recycled, their fate would either be subsequent reaction (the same effect as recycle), emission, or, for organic chemicals, combustion to produce energy. Clearly, emission is undesirable, both from an economic and a green engineering perspective. Combustion is also undesirable from both perspectives. From a green engineering viewpoint, combustion generates carbon dioxide, a greenhouse gas, and possibly trace amounts of other pollutants. From an economic viewpoint, using a valuable chemical for fuel is a bad idea. Where supplies are available, natural gas is commonly used as fuel to produce energy (for process steam, for example). Any chemical more valuable than methane is a poor choice as a fuel. In general, the more carbons in an organic chemical, the higher its value. Therefore, in terms of raw material usage, green engineering and economics coincide (or have similar goals).

Heat integration is another example in which green engineering and economics coincide. Fuel must be burned to create a heat source. It was shown in Section 8.3.2 that there are electricity costs associated with producing cooling water. Most electricity is obtained from burning fossil fuels. Therefore, the production of both heating and cooling utilities produces carbon dioxide, and both increase manufacturing costs. Heat integration can be implemented to minimize utility (e.g., cooling water, steam) usage (MUMNE method; see Chapter 15), so heat integration is both green engineering and economically beneficial.

It was mentioned previously that most chemical reactions produce undesired side products. Research in green chemistry aims to find catalysts and/or reaction pathways that reduce or eliminate these undesired side products. However, until this research bears fruit, the chemical engineer must try to reduce production of undesired side products. This requires maximizing selectivity for the desired product. Maximizing selectivity is also sound process economics, because it minimizes the fraction of raw materials not converted to the desired product. Methods for maximizing selectivity include optimizing the reactor temperature, optimizing the ratio of reactants, and optimizing the method of delivery of reactants to the reactor. Some of these methods are discussed in Chapter 20, and a more detailed discussion is available elsewhere [8].

Separation processes contribute to the problem of waste management because no separation is perfect. There are always trace contaminants in any "pure" stream. The goal in pollution prevention is to minimize these trace contaminants. For example, if absorption is used, a low-volatility solvent should be chosen to minimize contamination of the effluent gas stream. Distillation does not introduce additional contaminants, but it requires heating and cooling, both of which produce carbon dioxide and increase manufacturing costs. Therefore, distillation sequences should be designed for optimum energy usage. If solvents are required (e.g., absorption, extraction), they must be recovered. Organic solvents are expensive, so their recycle is tantamount to recycle of raw materials; if they were not recycled they would be burned at great expense and with carbon dioxide emission. In some processes (ethylene oxide, formalin—see Appendix B) water is used as a solvent. Water can even be used as an inert in a reactor (styrene—see Appendix B). Water that comes in contact with organics must be sent to wastewater treatment, which is expensive and consumes energy. Therefore, recycle of all solvents, including water, is both green engineering and economically beneficial. Solvent recycle is a simple example of mass integration, which is discussed in Chapter 15.

One potential source of pollution that does not typically receive much attention is the emissions generated during the loading and unloading of storage tanks. Consider the following scenario: A storage tank contains a somewhat volatile liquid. Therefore, there will be vapor in equilibrium above the liquid. The tank is to be filled by pumping liquid into the bottom of the tank. Where does the vapor go as the liquid level rises? An unsatisfactory solution to this problem is to vent the vapor to the atmosphere. A greener solution is to collect the vapor as it leaves the tank and recycle it to the tanker truck or rail car providing the liquid to fill the tank. This is commonly done when filling underground gasoline tanks. In some areas, gasoline pumps used to fill passenger cars have this capability. Emissions during normal tank operation (not during loading) are also a pollution prevention issue. Different types of tanks have different problems. A fixed-roof tank has the problem described above during loading and also must have a vent to prevent overpressurization when, for example, the ambient temperature increases. Placing a separator on the vent can minimize emissions, as can a vapor recompression/cooling system. A floating-roof tank, in which the roof floats on top of the liquid, minimizes the vapor problem described above, but it creates the problem of thin films of liquid remaining on the inside vertical surface of the tank, which is exposed to the atmosphere when the liquid level in the tank decreases. The inevitable result is that storage tanks leak to the atmosphere, and the best that can be done is to minimize these leaks. Additional methods for minimizing these leaks can be found [1,3]. TANKS software can be used to estimate losses from tanks, and it is available for download at http://www.epa.gov/ttn/chief/software/tanks/index.html.

The problem of leaking equipment is not unique to storage tanks. Flanges and valves also leak. Reflux drums and reactors usually have relief valves, and they also might leak. These are called **fugitive emissions** and occur because of leaks in valves, pumps, flanges, and so on. Fugitive emissions are especially prevalent in moving equipment (e.g., pumps and valves), where there must be a seal between the process side and the outside that allows movement of a shaft. Typically, these seals are made of a packing material that offers a tortuous, high-pressure-drop path between the inside and outside. Some leakage occurs, and often, for liquid systems, the small leakage provides lubrication for the shaft. For such a packing seal, the lower the fugitive emissions, the higher the frictional losses from the shaft rotation.

As plants are designed for lower "stack" emissions, the fugitive emissions are increasingly the dominant emission type. Thus, the emissions from thousands of valves, pumps, flanges, and so on must be estimated and reported. There are several procedures to make these estimates [9], from crude assumptions that all valves leak at the same rate to experiments in which working valves are bagged in plastic and the fugitive emission rate directly measured. Fugitive emissions *can* be measured for existing plants, but they *must* be estimated for new plants in the design stage. The Clean Air Act defines a plant to be a "major source" if it emits 25 ton/yr of any combination of hazardous air pollutants or 10 ton/yr of any single hazardous air pollutant. Plants defined as major sources can get permits to operate only by demonstrating **best available control technology (BACT)**, which is tantamount to a negotiation with the EPA. The estimates for fugitive emissions virtually guarantee that any new chemical plant in the design stage will be considered to be a major source. Once a plant is constructed, careful monitoring is needed to ensure that fugitive emissions are minimized.

25.5 ANALYSIS OF A PFD FOR POLLUTION PERFORMANCE AND ENVIRONMENTAL PERFORMANCE

After a process design has been conceptualized and a preliminary PFD developed, further analysis is possible to reduce pollution. The PFD should be studied to make certain that all possible recycle opportunities have been exploited. For processes with multiple streams containing the same solute, mass integration is possible using the method of mass-exchange networks (MENs) described in Chapter 15. More detail on the MEN method and other methods for analysis to maximize solute recovery is available [10]. Example 25.3 demonstrates this type of analysis of a PFD.

Example 25.3

Examine the PFD for the styrene process in Appendix B, project B.3. Suggest methods for pollution prevention.

One of the most obvious methods is to maximize the selectivity for styrene in the reactor. The ethylbenzene is already being recycled; however, minimizing the amount of ethylbenzene in the benzene/toluene effluent stream is both sound economics and pollution prevention. The fate of the benzene/toluene stream is also important. Because ethylbenzene is made from benzene, the benzene can be separated and returned to the ethylbenzene plant, which is almost always at the same facility. This is sound pollution prevention, although it might not qualify as pollution prevention according to the Pollution Prevention Act, because it does not constitute in-process recycle.

Another possibility is to recycle the wastewater stream to make the low-pressure steam used in the feed. On the surface, this appears to be a good idea for pollution prevention. However, there is a problem. There will certainly be trace amounts of organics in the wastewater stream. In a fired heater, the temperature is so high that the tubes, in which the process stream flows, are undoubtedly glowing red hot. These high temperatures can cause the organics to carbonize, that is, form carbon deposits on the tubes with hydrogen gas being released into the process stream. The hydrogen gas is not a problem because it is made in the reactor, and there are already provisions in the process for its separation. However, the carbon deposits will foul the fired heater tubes, reducing the heat transfer performance. Therefore, if recycling the wastewater stream is desired, a separation step such as carbon adsorption should be included to remove the trace organics from that stream.

Finally, there are models that permit analysis of the fate of chemicals in the atmosphere that might be released from a chemical process. The most popular model is the Mackay Level III model [1,11]. In this model, air, water, soil, and sediment are considered to be separate, well-mixed compartments in equilibrium with each other. Additional inputs and outputs are allowed to model air flow, river flow, and so on, for certain of these compartments. The equilibrium between compartments is modeled using fugacity. Therefore, for an emission into, for example, air, by knowing the physical properties of the chemicals involved (calculated from the group contribution method described in Section 25.2), the ultimate fate of all chemicals emitted can be predicted.

25.6 AN EXAMPLE OF THE ECONOMICS OF POLLUTION PREVENTION

The economics of pollution prevention can be analyzed using the incremental economic analysis discussed in Chapter 10. First, let us assume that pollution prevention activities require an additional capital investment. This incremental capital investment can be that required for modifications to an existing plant or the incremental capital investment required to include pollution prevention technology in a new design. The incremental net present value is

$$INPV = -PC + YS(P/A,i,n) \tag{10.4}$$

This means that the incremental net present value for the pollution prevention improvement, *INPV*, can be calculated from the incremental project cost: the additional capital investment needed for pollution prevention technology, *PC*, and the yearly savings resulting from the pollution prevention improvement, *YS*. The question is how to estimate *YS*.

As an example, consider the Love Canal near Buffalo, New York [12]. Hooker Chemical Company followed all existing regulations when burying toxic waste drums in a clay dome. Years later, the local municipality decided, against the strong objections of Hooker, to build homes and a school near that disposal site. During construction, the clay dome was disturbed, and the toxic chemicals began to leak. However, this leak was not discovered until health problems were observed years later. Subsequently, CERCLA (see Table 25.1) was enacted. Despite having followed all existing regulations at the time and having repeatedly warned the municipality not to disturb the site, it was Hooker, and not the municipality or any of the construction companies, that was held liable for the cleanup and for civil actions. This is an example of the meaning of "retroactive liability" in CERCLA.

The lesson from this true story is that it is not always possible to predict the yearly savings from pollution prevention improvements with any certainty. This is one motivation for the Monte-Carlo uncertainty analysis discussed in Chapter 10, because this method allows these types of uncertainties to be factored in to an economic analysis. Although it may look as if there are no yearly savings, estimating future liability with any certainty is difficult, if not impossible. Because the government has already established a precedent that it will change the rules and impose retroactive liability, a company doing business legally today may find itself liable for damages from laws not yet enacted. The conclusion is that pollution prevention activities are not only good environmental policy but also probably excellent long-term economic policy.

25.7 LIFE CYCLE ANALYSIS

A **life cycle analysis** or **life cycle assessment (LCA)** is a detailed technical study of the "environmental consequences of a product, production process, package, or activity (done) holistically, across its entire life cycle" [13]. The time frame of such an analysis is often termed "cradle-to-grave."

The first step in any life cycle analysis is to define the boundaries of the analysis. If a vessel is used in the process, are the manufacture and possible disposal of the vessel within the boundaries? Is the production of the steel for the vessel included? Should the environmental effects of the mining of the iron ore be included? Or should the analysis begin with the on-site installation of the equipment? The choice of boundaries defines the scope of the analysis. A broader scope requires greater effort but leads to a more complete understanding of the environmental impact of a process.

The remainder of the LCA consists of three stages.

1. **Inventory Analysis:** This is a quantitative study of the material and energy inputs and the air, water, and solid waste outputs of the entire life of the product or process.
2. **Impact Analysis:** The environmental consequences of both the inputs and the outputs of the inventory analysis are enumerated in this part of the LCA.
3. **Improvement Analysis:** The opportunities for improving the environmental consequences through modifications to the product or process are included in this section.

The life cycle itself can be divided into five stages [3]:

1. Raw materials acquisition
2. Material (of construction) manufacture
3. Product manufacture
4. Product use and reuse
5. Product and equipment disposal

Some models combine material manufacture and product manufacture into one stage [2]. At every stage in the life cycle, energy and material inputs are considered, as well as waste emissions. Therefore, a very "clean" process to manufacture a chemical may be considered less desirable, from an LCA standpoint, if a raw material (or a raw material for a raw material) is manufactured using a process that is very energy intensive or that produces large amounts of waste.

Simple life cycle analyses are available [14]. They are most useful when parallel LCAs are done on competing products or processes, because they allow one to evaluate objectively the overall environmental impact of the choices. In this case, it is especially important that all parallel LCAs use the same boundaries. Example 25.4 demonstrates a very simplified life cycle analysis.

Example 25.4

Consider polyethylene, which can be made into plastic wrap (e.g., Glad Wrap) or plastic containers such as milk jugs. You work in a manufacturing facility that produces a variety of polyethylenes (e.g., low density and high density). Trace the life cycle of the products of your facility qualitatively.

Polyethylene is manufactured from very pure ethylene. Ethylene is a by-product of oil refining, and hence the ultimate source of polyethylene is a fossil fuel natural resource. Therefore, all of the energy and waste disposal issues associated with oil refining (and even with oil production) are part of the polyethylene life cycle. Ethylene is usually purified from other light hydrocarbons; therefore, their fate is also part of this life cycle. This is usually not a problem, because methane can be used as a fuel and propane and propylene also have uses and are not disposed of. In preliminary separation steps, ethylene and ethane are in the same stream, but ultimately, they must be separated from each other. Because ethane and ethylene are close boilers and high-purity ethylene is needed, their separation by distillation is difficult, requiring large columns with large reflux ratios. The large reflux ratios increase the energy consumption, so this is also part of the polyethylene life cycle.

The ultimate fate of the plastic wrap and plastic containers is also part of the polyethylene life cycle. If used containers are recycled and refilled, this is a beneficial part of the life cycle. If either the containers or the wrap could be collected and remanufactured into new containers or wrap, this would also be part of the life cycle. If the containers, for example, could be converted back into ethylene as part of a recycling program, this would also be part of the life cycle. If the polyethylene, in whatever form, ends up in a landfill, where it does not degrade, this is the ultimate part of the life cycle.

Clearly, a quantitative life cycle analysis can easily become very involved. This example is simplified and does not include all aspects of the life cycle. For example, the materials used to manufacture the process equipment were not considered to be within the system boundaries. You can imagine how complex a life cycle analysis can get if all aspects are included.

25.8 SUMMARY

Green engineering is important because it is the right thing to do from an environmental standpoint, because it is likely to be good long-term economic policy, and because of the realities of the political climate involved in regulation of the chemical industry. This chapter has provided only a flavor of green engineering or pollution prevention. It was designed to give the reader a perspective on the issues associated with pollution prevention. It should be considered as a beginning, not an end. Further study is necessary to understand how to apply the principles introduced in this chapter to an actual chemical process design.

REFERENCES

1. Allen, D. T., and D. R. Shonnard, *Green Engineering* (Upper Saddle River, NJ: Prentice Hall, 2001).
2. Bishop, P. L., *Pollution Prevention: Fundamentals and Practices* (New York: McGraw Hill, 2000).
3. Allen, D. T., and K. S. Rosselot, *Pollution Prevention for Chemical Processes* (New York: Wiley-Interscience, 1997).
4. Sandler, S. I., *Chemical and Engineering Thermodynamics,* 3rd ed. (New York: Wiley, 1999), 424.
5. Draths, K. M., and J. W. Frost, "Microbial Biocatalysts: Synthesis of Adipic Acid from D-Glucose," in *Benign by Design: Alternative Synthetic Design for Pollution Prevention,* ed. P. T. Anastas and C. A. Farris (Washington, DC: ACS Symposium Series #577, American Chemical Society, 1994).
6. Rivetti, F., U. Romano, and D. Delledonne, "Dimethylcarbonate and Its Production Technology," in *Benign by Design: Alternative Synthetic Design for Pollution Prevention,* ed. P. T. Anastas and C. A. Farris (Washington, DC: ACS Symposium Series #577, American Chemical Society, 1994).
7. Kricsfalussy, Z., H. Steude, H. Waldmann, K. Hallenberger, W. Wagner, and H.-J. Traenckner, "Process for Preparing Dimethyl Carbonate," U.S. Patent #5523452, June 4, 1996.
8. Levenspiel, O., *Chemical Reaction Engineering,* 3rd ed. (New York: Wiley, 1999), chs. 7, 8.
9. *Emission Factors for Equipment Leaks of VOC and HAP,* EPA-450/3-86-002 (Washington, DC: U.S. Environmental Protection Agency, 1986); *Improving Air Quality: Guidance for Estimating Fugitive Emissions from Equipment* (Washington, DC: Chemical Manufacturers Association, 1989).
10. El-Halwagi, M. M., *Pollution Prevention through Process Integration* (San Diego, CA: Academic Press, 1997).
11. Mackay, D., and S. Paterson, "Evaluating the Multimedia Fate of Organic Chemicals: A Level III Fugacity Model," *Environmental Science and Technology* 25, no. 3 (1991): 427–436.
12. http://ublib.buffalo.edu/libraries/projects/lovecanal/.
13. *Life Cycle Assessment: Inventory Guidelines and Principles,* EPA/600/SR-92/245 (Washington, DC: U.S. Environmental Protection Agency, 1993).
14. Allen, D. T., N. Bakshani, and K. S. Rosselot, *Pollution Prevention: Homework & Design Problems for Engineering Curricula* (New York: Center for Waste Reduction Technologies, American Institute of Chemical Engineers, 1992).

PROBLEMS

For the following processes in Appendix B, suggest modifications that would make the process greener.

1. The ethylbenzene process
2. The styrene process
3. The maleic anhydride process
4. The ethylene oxide process
5. The formalin process
6. The ethylbenzene process
7. The DME process
8. The acetone process

SECTION

6

Interpersonal and Communication Skills

A chemical engineer is very much a member of a tiny technical elite of modern society. In the United States, less than 1% of the population is trained in engineering and sciences. Chemical engineers account for only 0.05% of the population, and yet they design and manage the plants that produce such essential goods as pharmaceuticals, plastics, paper, fertilizers and pesticides, fuels, synthetic fabrics, and clean water. Chemical engineers use jargon and perform calculations that are, at best, mysterious to most people. Their success depends not only on performing the calculations correctly but also on convincing the public that they can add value to the quality of life. In addition, to communicate clearly with the public as a whole, chemical engineers work most often as part of a team of professionals. The interaction and communication between team members are absolutely crucial to the success of any given project.

The following chapters deal with the issues of team building and effective communications.

Chapter 26: Teamwork
The essential elements of team building and teamwork are described. The choice of group members, initial organization determination, roles within and outside the group, group management, team building, team member roles and responsibilities, and team self-evaluation are described. References to the most accessible team-building literature and examples of problems typically encountered by poorly functioning teams are included.

Chapter 27: Written and Oral Communication
Through a focus on audience analysis, strategies for improving the effectiveness of both written and oral presentations are explained. Commonly

accepted (but frequently broken) formatting rules for figures and tables are covered, as are hints to effective use of communication software. (Available on the accompanying CD.)

Chapter 28: A Report Writing Case Study
Following a sample student design report, this chapter offers models of both strong and weak written communication in several formats: memoranda, visual aides, and short design reports. A checklist of common errors is provided. (Available on the accompanying CD.)

CHAPTER

26

Teamwork

Chemical engineers work both with and for other chemical engineers, other types of engineers, scientists, and nontechnical people—often all at the same time. In short, chemical engineers do not work alone.

Teamwork is not just working with or for other people; it is working together as a unit to accomplish goals in a way that each member of the team accomplishes more than that person could accomplish alone. The concept of teams and teamwork is an old and powerful one.

Some would say that teamwork cannot be learned or that it is easy. In this chapter, both of these misconceptions are addressed.

Developing the skill set for effective teamwork is an essential part of being a chemical engineer. As with the rest of engineering, although it is not easy, it can be learned.

26.1 GROUPS

A team is a subset of a group. Therefore, this chapter starts by addressing groups and group work.

Any collection of people working on a common project can be considered a group. Two students working together in the chemistry laboratory course compose a group. Five students working on a design project are a group. A chemical engineer, a civil engineer, and a project manager working to get a water treatment plant on line act as a group. In each of these examples, and in myriad others inside and outside the engineering world, collections of people form groups to get a job done. One certainly hopes that the group is more *effective* than any of the people working alone. However, that could not justify the existence of the group (although some groups only barely meet this criterion). To be an effective group, the group output should be better than the total output from all the individuals working alone. The concept that explains this increased *efficiency* is called **synergy**. It is related to the concept of economy of scale discussed in Chapter 7.

When people form groups, various organizational behaviors occur. Some of these behaviors are productive, some are nonproductive, and some are counterproductive. However, none of these organizational behaviors is unchangeable. Section 26.1.3 describes some of the organizational behaviors of groups and provides strategies to improve the efficiency and effectiveness of the group. The next two sections discuss characteristics of effective groups and offer a general strategy for improving group work.

26.1.1 Characteristics of Effective Groups

An effective group produces an output that is better than the total of outputs from the individual group members working alone. There are three important keys to achieving this effectiveness:

1. Task differentiation
2. Work environment
3. Coordination

Task Differentiation. In any group, members do different tasks. However, imagine a group in which each member did the same job, and then someone chose the best result to go forward. The group would be terribly inefficient. In fact, such a group would not even satisfy the first criterion of a group: that the group be *more effective* than any of the people working alone. In this example, the output of the group is only as good as the output of a single person working alone. However, this example suggests one of the keys to effective task differentiation:

Each job should be done by the group member who can do it best.

This is a heuristic. As with all heuristics, it cannot always be applied (nor should it), but it is a very helpful guide. For this heuristic, it is very important that the special abilities of each group member, vis-à-vis the project tasks, be known.

Of course, a single group member might be best at a large proportion of the tasks, and that would lead to unequal workload, which would be a problem. Therefore, a second heuristic is used.

Each group member should contribute equally to the group outcome.

The overall group outcome should be the best that it can be (i.e., it should be optimized). This heuristic typically overrides the previous two heuristics for highly effective groups. It could easily be that the best outcome occurs when one (or a few) members do more than their share of work and/or when individual tasks are done by members who are not experts in that task. As in all of engineering, optimization is important.

Other aspects of task differentiation may be important. For example, it may be optimal to train a member to do a task that another already does well. It may be helpful to have more than one member do the same task and then compare answers. (This is a classic error-checking strategy.)

Too much task differentiation can create problems. The project outcome should never be dependent on any one person. For example, if a group member becomes sick, someone must be able to step in and do the job. This can be accomplished in two ways. Either all tasks have assigned backups (understudies), or all members develop and maintain competencies in a broad range of tasks. The former is a quick fix. The latter is more stable and leads to greater flexibility.

Work Environment. People are more productive when the environment that they work in is safe, supportive, and challenging. It is incumbent on all group members to create and maintain this environment.

Safety in this context often refers to the safety of one's job. One of the most damaging events in a work environment is notice that someone is being reassigned or even fired. People work best when they feel somewhat safe, but they also need to be challenged.

When the job is demanding, one tends to rise to the occasion. Clearly then, there is an anxiety level, between safe and overly demanding, that optimizes one's performance. Because this optimum anxiety level is affected by many factors, it is difficult to predict. To be a good group member, one needs to know one's own optimum anxiety level and to give feedback to the group when one is outside the optimal range.

The supportive aspect of the group work environment refers to help that other group members give when one's efforts for the group do not meet expectations. Group members who are supported are much more likely to stretch for greater productivity or to be more creative. Both of these activities give rise to both higher overall outcomes and higher numbers of failures. Without a supportive environment, people take the easy path, which leads to few failures but only mediocre success.

There are many heuristics to help group members improve the work environment and thereby increase group effectiveness. Here are some.

- Tell the group when you are unable to meet expectations.
- Be constructive and accept constructive criticism.
- Assume that negative feedback was meant to help you improve.
- Give positive reinforcement to group members.
- Provide assistance to group members in achieving their professional goals.
- Give credit to other group members when talking with group members and when talking with those outside the group.
- Thank group members for taking risks.
- Accept group responsibility for failures.
- Avoid competition within the group.
- Focus competition outside the group.
- Offer assistance when group members have nonwork problems.
- Try to obtain support for the group, not for individual group members.
- Immediately cover for a group member who is having trouble.

Coordination. Synergy requires a coordinated effort by the group members. This coordination involves breaking down the project into components, assigning these components to group members, ensuring communication of results, and assembling the results into the final product. As in any problem solving, these steps are taken many times and often out of the order just described. An essential aspect of coordination is flexibility. If a project component is initially assigned to what turns out to be the wrong group member, it must be reassigned, without prejudice. If the chosen communication scheme breaks down, it must be changed. Example 26.1 describes such a scenario.

Example 26.1

In a typical design assignment, there are many aspects of the problem: problem definition, alternatives generation, research, communication with the client, technical analysis, economic analysis, error checking, report generation, presentation to the client, and others. Coordination in such a project might initially involve meeting together to discuss the problem, choosing a leader, determining a schedule, and specifying a division of duties. These activities are generally serial to the flow of the project. Then various aspects of the research can be done in parallel. One group member studies the given flowsheet and puts it on the process simulator. Another member researches processes commonly used for the application. Another member meets with the client to gather any additional information.

Still another studies the economics of the process and identifies opportunities. Then the group convenes to share the information, to verify the problem statement in light of the information obtained, and to brainstorm alternatives. Members then go off to work on their assigned tasks, with frequent communication with the other group members. Then the group reassembles to plan the final output and the format for delivery to the client.

Obviously, many other steps are taken by the group, but, in each step, coordination is important. If two group members do the same job, effort is wasted (although, as noted in the previous section, this situation may provide some error checking). If something remains undone because it is not explicitly in any one member's "job description," the project may not get done on time. It may not get done at all. Here are some heuristics for group coordination.

- Always know what everyone else's assignment is.
- Share information when it conceivably could be relevant to the work of another group member.
- Ask the group, "Are we missing anything?"
- When a gap in assignments is noticed, inform the group and offer a modification to solve the coordination problem.
- Role-play the client/group interaction.
- When information is shared, make sure you understand, even if it is not directly within the sphere of your assignment.
- Focus on your assignment as a *contribution* to the group effort, not as a *division* of what is and is not your responsibility.
- Be flexible.

26.1.2 Assessing and Improving the Effectiveness of a Group

Throughout any engineering activity, self-assessment is important. For a calculational homework problem, checking the answer before submission is standard procedure. On an examination, it can mean the difference between a good grade and a poor one. Group work is assessed in a similar way. The final answers (e.g., the details of the final report) must be checked by the entire group. Beyond that, the effectiveness of the group itself must be assessed repeatedly. For this assessment, the form in Figure 26.1 can be used. This form was developed by Dr. C. J. Coronella at University of Nevada, Reno (2001).

The assessment is short and therefore can be used at frequent intervals. The form should be used after the first meeting, halfway through the project, and at the end of the project. In addition, it should be used at least every week for longer projects. At this stage, some of the items on the form may not yet apply. Note that the form is parallel to many of the heuristics given above. Immediately after the group members fill out the form, a summary of all the forms should be shared with all. A discussion should immediately follow about the results, with the only

Group Assessment

Name

Please enter a score from 0 (worst) to 10 (best) in each row, for yourself and for your team members. If you enter uniform scores across the board, I know you haven't taken this assessment seriously. A 10 should be an unusual score.

Enter your initials in the first column and those of the members of your group in the remaining columns.	(me)				
Leadership 10 Very strong leader, provided direction, inspired and encouraged others 5 Willing follower, took directions easily 0 Frustrated the group, blocked progress, criticized others					
Cooperation 10 Worked readily with others, outstanding contributor, anticipated requests 5 Cooperated with occasional prompting 0 Rarely contributed or cooperated, worked alone, had to be coerced					
Initiative 10 Produced good ideas which helped others, "went the extra mile" 5 Accepted other's ideas and improved on them 0 Criticized other's ideas, never contributed original ideas					
Attitude 10 Positive, enthusiastic, encouraging others to work better 5 Neutral, worked with the group without either enthusiasm or grumbling 0 Negative, complained about the project, worked unwillingly					
Effort 10 Worked hard on assigned tasks, independently and cooperatively 5 Worked reasonably hard, also socialized a lot. Occasional prodding needed. 0 Didn't contribute much at all, tasks were unsatisfactory.					
Preparation and conduct of Lab work 10 Enthusiastic, came prepared to meetings and to lab. 5 Came to all meetings, somewhat prepared. Often helpful. 0 Missed lab, was unprepared, we'd have accomplished more without him/her.					
Writing of report 10 Enthusiastic, contributed substantially to production 5 Contributed everything that was asked. Work needed revision 0 Didn't contribute to the production					
Preparation and delivery of presentation 10 Enthusiastic, contributed substantially to production 5 Contributed everything that was asked. Work needed revision 0 Didn't contribute.					
Computation 10 Helped others to understand and use computer tools 5 Good, but not incredible, use of computers 0 Uninterested or unable to use computer tools effectively					
Grade You have **100** points to divide among your team members. Distribute the points in an equitable manner, where each score reflects effort and contribution, (which aren't always the same.)					
Assessment of the group 10 Best group I've ever worked with; the project was fun as a result 5 Group sometimes worked well together, with occasional problems 0 Worst group I've ever worked with (or not); this was a miserable experience					

Σ Who in your group took most of the leadership for this last assignment?

Σ Is your group functioning better, worse, or just differently than at the beginning of the semester? Explain.

Figure 26.1 Group Assessment Form (Used by permission of Dr. C. J. Coronella, University of Nevada, Reno, copyright 2001.)

goal being to make changes to improve the working of the group. All group members have a vested interest in improving the group effectiveness. Each member should try to take the feedback as constructive. The group result will be better, the group members will be less stressed, and the positive contributions of each member will be clear to all, as seen in Example 26.2.

Example 26.2

By using the form in Figure 26.1, the group finds that Jo believes that the other group members are putting in little effort. Robin thinks that effort was fairly consistent across the group but that the attitude of Sam and Lee were negative. Chris received low scores in leadership from everyone except Sam. After discussion of these results, it was determined (and accepted by all group members) that Chris needed to be more positive about group successes in group meetings. All of the group members began to understand the effect of morale on the effectiveness of the group. Jo found out that Robin, Sam, and Pat actually spent much effort in researching process alternatives that did not end up providing improvements. Because these alternatives were not part of the ongoing design work, Jo was unaware of this essential effort. It was decided that roles for and results from all group members would be discussed in the group meetings.

Sometimes major problems occur in groups that make them highly ineffective and inefficient, such as withdrawal of a group member from the communal effort or competition for credit between group members. To be resolved, these problems must come to the attention of the group. If the matter is very personal, the group leader may need to deal with it directly. Otherwise, the group as a whole can seek to solve the problem, but only after the problem is identified through assessment. The following section describes typical group problems and offers often successful solutions.

26.1.3 Organizational Behaviors and Strategies

Throughout the group work, behaviors of members of the group will sometimes be conducive to group effectiveness and sometimes not. The following are some typical organizational behaviors. For more complete lists and strategies, see [1, 2, 3].

Friction. The group is like an engineering system, and the unappreciated dissipation of otherwise useful energy is friction. This friction comes from a variety of sources and must be reduced or eliminated as quickly as possible so that the energies of the group members are focused on the project. Here are some examples.

- **Apparently Unequal Work Distribution:** Members of the group who believe they have more to do than others operate at lower efficiency. They have a responsibility to mention this perception to the group leader, who has the responsibility either to explain why the distribution is fair or to equalize the workload if it is unfair. Maintaining silence about this situation increases stress. Sharing perceptions with only a subset of the group makes the matter worse and harder to correct.

- **Low Motivation, Low Morale:** People work harder when they are motivated. More important, they also work smarter. For example, motivated group members fill gaps between individual assignments when they notice them. Group members can become demotivated when they do not feel that their work is that important, when they do not care whether the group succeeds, when they are overwhelmed by group and nongroup pressures, and so on. Individual group members should try to be aware of their own low motivation; however, it is often more noticeable to other group members. In either case, this is a problem to be solved with the help of the group. The group leader should talk with the individual involved, try to determine the underlying cause of the low motivation, and make reassignments or offer other help. The extent to which the entire group is involved depends on whether the members have already sensed the problem. If so, the resolution should be described to them. A related source of friction is low morale. Groups work best when they have an emotional stake in the group result. If good group work increases one's sense of well-being, one works harder and smarter. High morale can be maintained by frequent positive feedback to all group members, individually and jointly. Giving credit to group members whenever it might be deserved is a good policy.
- **Lack of Concurrence with Group Decisions:** This is especially troublesome with groups in which the members are inexperienced in effective group work. First, the opinions of the members need to be aired. Then the group leader should lead a discussion of how the various options will advance the group project. If there are multiple options that appear to advance the group work nearly equally, the group leader makes the decision. The group leader should explain the value of continuing with the project, rather than worrying about the decision. The leader also has responsibility to explain the process by which decisions are made so that all the group members can accept the decision and move on, even if they do not believe the decision to be the best one.
- **Overlapping or Gapping Work Assignments:** All group members have the responsibility to do their assigned work and to fill in the noticeable gaps in assignments between group members. If these gaps are too great, they require reassignment of work. The group leader should be alert to this possibility at all times, and the group members must inform the leader when gaps appear. Some overlap between work assignments should be expected and appreciated. Group members should try not to take offense when another member is assigned some of the same work.

- **Uncomfortable Group Interactions:** Some people may be uncomfortable interacting with other group members for any number of reasons. Although less common today, one historical example is the discomfort that some men have felt while working for a woman. Another common problem in undergraduate design groups is the existence of cliques and "outsiders." These types of friction can be the most disruptive to group work, and they must be dealt with directly and quickly. Group interactions must be based on trust, respect, and responsibility. Reminding all group members of this principle of group work may mitigate the friction. However, in the context of a course design group, faculty help should be requested if needed.
- **Assigning Blame:** Sometimes, a decision made by the group is later found to be a bad one. Unfortunately, too often this leads to one group member blaming another, which results in decreased efficiency for all. When problems arise, it is best to view them as group problems. "It's not his fault or her fault; it's our fault." This viewpoint encourages a group solution that will be more durable than any individual's solution.
- **Disengagement:** For maximum effectiveness and efficiency, all group members must be engaged in the group work process. When a group member is disengaged, often the first symptom is a statement such as, "Just give me my work assignment and let me go." The group members should offer encouragement by indicating the value of the individual's work to the group result and by explaining that the group needs to work as a unit to avoid extra effort. Disengagement often follows from low motivation and/or low morale.

Leadership. The most important thing about the leadership of the group is that there be an identified leader. Leaderless groups waste time discussing unimportant decisions, and all decisions seem to be temporary—open to additional time-consuming discussions at a later date. The role of the leader is to focus discussion, develop the structure of the group, pace the group members, and make decisions that advance the group and are acceptable to the group members. It is extremely important that all group members accept that the leader has these responsibilities to avoid friction from control issues. In response, the goal of the leader should be to improve the effectiveness of the group effort; the goal should not be to exert control. If the leader is flexible and yet consistent in decision making, friction can be minimized and more group member time will be spent effectively on task. Effective leaders ask for feedback on their leadership and use that feedback to improve. Some first-time leaders make too many plans without group input and make too many demands without sufficient explicit appreciation of the work done by the group members. Effective leaders guide their groups to develop plans that all group members appreciate and find that their group members are eager to do even more than what is assigned. Every group leader should read a complete book about developing the many unique skills required [4].

Organization. In a small group, the organizational structure is simple—only the leader and everyone else. For larger groups (ten or more) a three (or more) tiered structure may be required. However, such a structure necessarily reduces the ratio of "workers" to "management," which is apt to reduce group effectiveness. Therefore, the temptation to create multiple levels of organization should be resisted. If two group members research patents, there is no need to identify which person is "in charge" of patents. In large groups, the subgroups are best organized according to capabilities rather than according to fixed tasks. For example, a subgroup might be formed of five people who have the capability to find information in the literature and to analyze its value. The task assigned to this subgroup would vary during a design assignment, from patent searching, to alternatives research, to economic research. However, some of the members of this subgroup may be needed in another subgroup halfway through the project to work on process simulation. Thus, the organizational structure should be focused on both flexibility and capability (not task) functionality.

Choosing Group Members. Two types of characteristics are important in choosing group members. Obviously, the group needs members who are competent in certain technical fields—for example, chemical engineers, chemists, and mechanical engineers. The technical expertise may need to be more specific, such as catalysis, simulation, thermodynamics, and so on. However, beyond their technical (or task-oriented) expertise, group members must have complementary process-oriented skills. "Process" here means the group process, or how group members interact. Many engineering students are currently evaluated by the Myers-Briggs Type Indicator (MBTI) or another tool designed to give feedback on one's preferred style of learning and interacting with others on intellectual tasks. A well-constructed group will contain members with a breadth of learning and working styles. For example, a group of only introverts may ignore the concerns of the client, and a group of only intuitive members may fail to recommend much-needed experimental work in the final report. Crucial to the success of a group with a spectrum of styles is respect for all styles by all members of the group. Therefore, the best strategy for selecting group members depends on the sophistication of the members. If they already understand and value learning styles other than their own, the broad-spectrum approach is appropriate. If the concept of learning styles is new to the group (or the group leader!), it is better to choose a more coherent group—people who generally approach problem solving in a similar (but not identical) way. However, the latter approach will nearly always reduce the creativity of the group, even as it reduces friction. It should be used only until the group members can get to a learning styles workshop, which is available through most universities, technical societies, and large corporations.

One indicator of different styles that is always available in a course is grade-point average. Students with straight A's are different from students who earn B's or C's—and this difference often has more to do with learning styles and working styles than anything else. Therefore, it is advisable to choose group members so that the group is heterogeneous on this metric also. Grouping all *A* students together in one group and all *B* students in another is an ineffective strategy that has led to numerous group failures.

Another consideration in choosing group members is friendship. Groups based on friendship are frequently unsuccessful on technical projects. In any reasonably complex project, there will be disagreements as to the best route to the final report. Friends may take these disagreements personally, in a way that people who view their relationship as purely professional may not. Professional relationships based on respect allow a more direct and honest dialogue, whereas personal relationships involve the added dimension of predicting the emotional response and modifying communication based on that prediction. In addition, other group members may hold back communication if they perceive themselves to be outside a clique of other group members. All these concerns have led to dysfunctional groups in design courses, with resulting poor project performance. Overall, it is easier to be efficient and effective working with a colleague than with a friend.

An extreme case of this type of dysfunctional group behavior can occur when there are prior, current, or potential romantic relationships between group members. It is best to avoid such conflicts in choosing group members. The more high performing the group, the more damaging romantic complications can be. When conflicts of this type arise, they can become intense and lead to low morale, disengagement from group activity, and polarization of the entire group.

Throughout one's career, one works in groups with a wide variety of people. Sometimes another group member will be one's polar opposite. For example, one member prefers to take in all the information, think about it, and give a solution without detailing the steps. Another prefers to go step-by-step and always keeps a neat calculations notebook. Group members need to interact on a basis of respect for each other's unique styles. Interactions based on respect and professionalism make use of this diversity to enhance creativity, to improve communication with a wide variety of clients, and to help all group members develop a broader array of capabilities.

When groups are chosen by someone outside the group (a professor in a course or a manager in industry), one hopes that the above heuristics are used. Whether they are or not, self-selection of group members is the exception; outside selection is the norm.

Roles and Responsibilities. The glue that holds the group together, as well as the lubricant that reduces the friction, is the clear understanding of roles and responsibilities of group members. The leader has the role of making decisions based on the best input from the group; however, this entails a responsibility to make those decisions even when it becomes unpopular to do so. Another leader responsibility is to monitor the group process continually to keep everyone on track, to make sure that nothing falls between the cracks, and to reduce unproductive tensions between group members proactively if possible (but retroactively if necessary).

Each group member should know all group members' roles on the project. If a role is unclear, it is the responsibility of the group member to obtain clarification from the leader. All group members have the responsibility for the final project results, jointly and severally. If the report is poor, it reflects poorly on all group members; if the report is good, all receive credit. If any one group member is unaware of the group results, it casts doubt on the entire group effort.

Mobile Truth, Groupthink, Mob Effect. Earlier, group friction was described. Although this characterization covers most disruptive problems encountered by groups, another problem is "mobile truth" and related issues. As described in Chapter 23, when your affinity to a group clouds your ethical decision making, you are being affected by mobile truth. However, related situations arise frequently in group work that do not seem exactly like ethical problems. For example, group cohesion can lead to an autocatalytic agreement on group design decisions once they are proposed. If a tray column is chosen over a packed column by one group member, other group members may suspend their own judgment to add their vote. As more members agree, it becomes less likely that *any* other group member will disagree. The disturbing fact is that the same scenario could describe the same group, with the same available data, if the first person chose the packed column over the tray column. Thus, a random suggestion becomes the group choice without sufficient analysis. This phenomenon, known as **groupthink**, is well known to observers of groups but is often difficult for groups to see themselves. All group members must guard against it. One heuristic is to ask whether there is a better way whenever all the group members appear to be in agreement. It is potentially disastrous to know that you are right (by unanimous agreement!) when you are clearly (to those outside the group) wrong.

26.2 GROUP EVOLUTION

As groups develop, they go through stages. A widely used categorization of these stages devised by Tuckman [5] is used in this section. As with problem-solving strategies, the order of the stages is the typical case, but active groups of experienced group members tend to recycle back to earlier stages to optimize the group work.

26.2.1 Forming

The first stage is **group formation**. This stage may include the actual choosing of the group members, but often the members are chosen by someone external to the group. The group members get to know each other, including their technical abilities and preferred working styles. They share contact information to allow sharing of results outside normally scheduled group meetings. The leader is identified by the group members or externally, and the leader establishes an organizational structure. Rules and procedures for the group work are established. The charter or mission of the group is established. A **charter** is a statement of scope, task, and responsibility determined externally and given to the group. A **mission** is the same type of statement, but it is developed by the group itself as is common in what are referred to as self-directed work teams. These teams are given more autonomy, along with the responsibility to plan and manage their performance. They are expected to self-assess both their task results and group process activities. Example 26.3 is an example of forming.

Example 26.3

A group of six students in a design class is formed to design a portable fuel cell that can compete with existing gasoline-fueled generators for electrical power in remote locations. They introduce each other and share e-mail addresses and cell phone numbers. The members describe the way they do their engineering homework and study for exams as a low-key way to share their working styles. They discuss the charter presented to them by the course instructor. The group leader (chosen by the course instructor) sets up procedures, assigns tasks, and schedules daily meetings.

26.2.2 Storming

Creating a well-functioning group often requires a period of **storming** or flailing about. Group members test one another. They become aware of differences between themselves and focus on them as barriers to success. They disagree and debate issues—sometimes very minor issues. They question the mission of the group or the meaning of the problem posed to the group. They question the leadership and organizational structure developed in the forming stage. Although this is a normal stage in the development of the group, it can derail the entire project. Therefore, experienced group members recognize that reining in the discord is essential after the storming becomes counterproductive. The ability to guide this stage requires practice and self-reflection after each such situation. One heuristic for beginners: Just before the discord starts to digress into personal rather than professional attacks, point out that *storming* is going on and that it is time to move on to the next stage, as shown in Example 26.4.

Example 26.4

The group begins to complain about the wording of the charter, and they openly question why Chris was chosen as leader by the course instructor. Jo asks why they must meet every day; Lee wants to meet whenever the need arises. Robin wants to go to the course instructor to settle all these issues, but Sam thinks the group should function on its own. Pat says, "Let's just stop this storming and get on with it."

26.2.3 Norming

Norms are rules and procedures (some stated, some not) that guide group interactions. During this stage, norms are developed and accepted by the group members. The diversity of backgrounds, viewpoints, and experiences of the group members is accepted. Disagreement between group members on technical issues and on group process issues is accepted, but the group moves on. Individuals take on and accept their assigned roles and develop a sense of shared responsibility for the group effort. At this stage, detailed scheduling may be most effective. The group develops a Gantt chart or a PERT chart, if the project is complex enough to benefit (most are). The Gantt chart is a matrix of tasks (rows) versus time (columns). An open bar in the appropriate row going from the start of effort on the task to its conclusion is shown on the schedule. The bar is filled in proportion to the completed results of the task. At any time, the Gantt chart thus indicates which tasks are ahead of schedule and which are behind schedule. Various notations are made to show linkages between tasks, milestones, and slack time, but this is all beyond the scope of this text. Please see the excellent article by Dewar [6] for a clear and very useful description of this technique, which is the most common scheduling technique in engineering. The **critical path method (CPM)** and the **PERT technique (program evaluation and review technique)** can be used to determine the set of activities that must be performed in series without delays to minimize the total project time. These incorporate a flowsheet to show linkages between tasks and are especially useful for complex projects. See the excellent description by Smith [7], which also includes helpful software for the CPM. The PERT technique also allows for including uncertainty in the projected time required for individual tasks. Example 26.5 is an example of a scheduling meeting.

Example 26.5

Chris has read an article about Gantt charts and organizes a meeting to create one. It has been decided that all group meetings must have an agenda, so Chris provides one. At the meeting, Lee has many suggestions of tasks, but Jo is nearly silent. Once Lee stops talking, Jo asks, "Do we really need to do task number 7? It doesn't seem to be linked to our overall goal." Robin believes that task 7 is essential, Pat and Sam are not sure, and a short discussion ensues. Finally, the entire group realizes that they need to link each task to the goal and that the diversity of opinions just witnessed helped them to reach a better solution. It is decided that, at each meeting, one group member will be assigned to observe discussions and to indicate when enough time has been spent on that issue. If a consensus has not been reached, Chris will make a decision. It is decided that communication with the course instructor will be primarily through Pat, who will report all such communication at each meeting.

The Gantt chart created by the group is given in Figure E26.5, updated for four weeks after this meeting. Note that the group is a week behind in research and technical analysis but a week ahead in economic analysis. It becomes clear from this diagram that

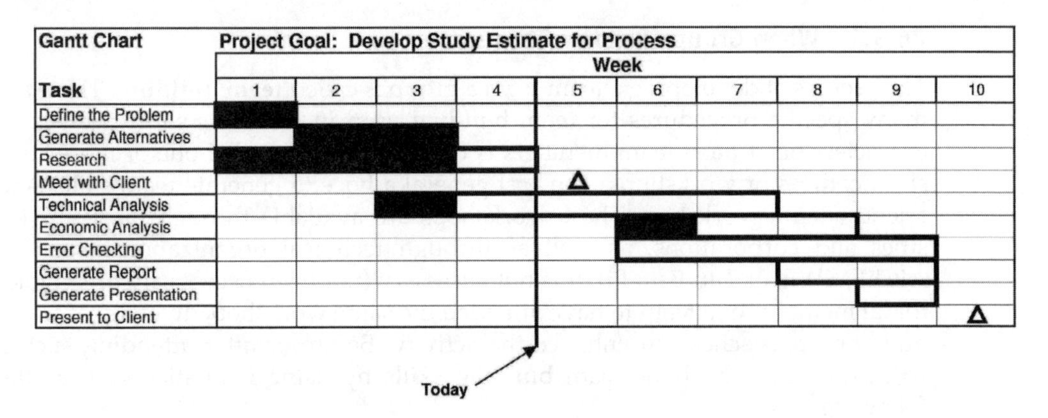

Figure E26.5 Gantt Chart for Example 26.5

the effort on economic analysis should be redirected immediately to research and technical analysis to catch up, especially because a client meeting occurs next week.

26.2.4 Performing

During the **performing** stage, most of the group on-task work is done. The preceding stages allow the development of the group processing that is essential to the smooth functioning of the group on the task. There is agreement on the project schedule, assigned tasks, and leadership. Results are shared with the group, and the project progresses to its conclusion. Diversity of analyses, conclusions, and recommendations by different group members is regarded as an advantage in developing creative solutions, optimizing group processes, and assessing the group results. The performing stage is the goal of the earlier stages, and results-oriented professionals such as engineers often try to jump right into this stage, without valuing the earlier stages. Although experienced group members can shorten the time in the earlier stages, they especially understand the need for developing an effective group culture, which is the goal of the forming, storming, and norming stages. It is essential for all groups to spend time and exert effort to develop this culture.

26.3 TEAMS AND TEAMWORK

A team is the ultimate group. As people become comfortable in group work, when they become efficient in group processing, and when they can maximize the effectiveness of the group to develop the best group result, their groups become teams. Thus, two characteristics that differentiate groups and teams are morale and task performance [2].

26.3.1 When Groups Become Teams

The process of developing a team from a group is called **team building**. There are many specific procedures for team building used in countless short courses. A characteristic of most team members is that each has attended one or more such short courses or workshops. The best general advice to generate team skills is to attend such a workshop. These workshops are available through most universities and corporations, as well as through technical organizations (such as AIChE, SWE, or Tau Beta Pi). It is not necessary (nor even necessarily preferable) for all members of a team to have attended the same workshop; diversity of team-building approaches can enhance the activity. Before or after attending such a workshop, one can hone team building skills by using heuristics such as the following.

- Actively and consciously follow the four stages of group development.
- Find common ground with each team member.
- Find value in each team member's contribution.
- Create a team identity.
- Appreciate diversity as a trigger in brainstorming, as a facilitator in filling gaps, and as a mirror to simulate the response of the client.
- Accept all criticism as constructive.
- Present all criticism in a constructive way.
- Read through a chapter on teamwork while thinking about the working of your team. Choose one or two aspects of teamwork to work on as you work on the present task.
- Joke with team members. Show concern when other team members are having trouble.
- Scan the team for team members who may feel left out. Actively bring them into the core of the teamwork.
- Share your analyses and opinions, but strive for the best decisions for the team, not for yourself.
- Have and show concern for the development of other team members.
- Ask frequently, "Are we a team yet?"

When everyone in the group not only wants to but also works to improve the performance of every *other* member of the group, a team has been created.

26.3.2 Unique Characteristics of Teams

Teams can be identified (internally and externally) by several key characteristics. These improve morale, effectiveness, efficiency, and stability. Some of these are as follows.

- **Concern for Each Other:** Team members help each other on their assignments in such a way that both team members become better problem solvers. A team member in difficulty asks for help, and other team members automatically provide it. The helping team member tries to teach the other how to avoid the difficulty in the future. When there is a crisis for one team member (professional or personal), the other team members immediately step in to help.
- **Leaving No One Behind:** In teams, all team members are assigned essential tasks, tasks that if not done properly will negatively impact the team result. A goal (shared and individual) of team members is to help each team member improve. All team members share in the successes and failures of the team. Individual team members are not blamed for failures; the team reflects on the failure and develops procedures to avoid it in the future. Team members congratulate each other (within the team and outside the team) all the time.
- **Constructive Criticism:** Team members feel comfortable discussing each other's performance. They reflect on team activities and discuss ways for all team members to improve. There is more criticism and self-reflection in a team than in other groups, but there are few hurt feelings.
- **Consensus, Not Voting:** Voting is an impersonal attempt to involve all group members in a decision. Although a decision is made, the procedure creates two (or more) subgroups that have identified their differences. **Consensus building** is a process designed to produce a result by having team members (1) identify common ground and (2) actively keep the best interests of the team (rather than the individual) in mind. Consensus building is definitely not a vote followed by the "losers" giving in to the majority, and it is not a way to find the "least objectionable" or "least common denominator" solution. Consensus building begins with a discussion to determine the features of a solution that are acceptable to all team members. Then the solution is built by identifying divergent ideas and parsing them all (as a team) according to which is best for the team result, without concern for whose ideas they were or how a decision will affect the individual team member. A common misconception is that consensus means that all members believe the decision reached to be the best one. Rather, consensus means that all members consent to the decision. Some may still think that it is not best, but all agree to accept the decision as the group decision and move on. Groups vote. Teams build consensus. Consensus generally results in better decisions and maintains team cohesiveness so that the team can move on without delay.

- **Trust:** Team members share information freely because they have high levels of trust in the team and all its members. This enhanced information flow speeds the teamwork, and it avoids "missing something" along the way. This trust also makes it easier for team members to accept decisions made by individual team members.
- **Sacrifice:** Team members work to advance the team toward its goals. This often means working longer or harder when team progress or the inability of other team members requires it. Team members check their egos at the door. They recognize that a good team result will reflect on them, that helping other team members will improve their own productivity, and that they can depend on other team members to help them.
- **Leadership:** In a high-functioning team, most or all of the team members have the capability to be the leader. They all accept the leader chosen, they try to help the leader improve leadership skills, and they accept the leader's decisions. When decisions are made between alternatives that are nearly equally good, the leader makes those decisions. In other cases, consensus building is led by the leader. The leader is the outside contact for the team, but the leader always gives credit to the entire team.
- **Automatic Compromise:** Team members often have divergent views. As soon as a team member judges another's view as more beneficial to the team, the team member automatically compromises. Team members compromise; group members negotiate. Compromising is essentially acknowledging another's view as better than one's own. Negotiation is coming to an agreement by surrendering one's position to reach a central position. One irrational result of negotiating is that the final position is influenced by the original position. If one artificially hardens one's initial position, the final outcome will tend to be closer to one's views. Thus, negotiation favors exaggeration of differences, which itself causes group friction.
- **Doing More than Is Expected:** Group members never say, "That's not my job." Team members never think it. Team members always tend toward doing more than their assignment. This decreases the chance that the team result will be adversely affected by errors in planning or work assignments.
- **Time on Task:** Of the time spent, teams generally spend more time on task because they do not waste time complaining or trying to promote themselves over other team members.

26.4 MISCONCEPTIONS

Teamwork and team building are recognized as essential components in engineering. In fact, they are required of all accredited degree-granting engineering programs in the United States. As stated at the beginning of this chapter, chemical engineers work in teams (or groups) in all aspects of their work. This fact is sometimes misinterpreted in a way that leads to dysfunctional behaviors. Two of the main misconceptions arising in chemical engineering courses are addressed in this section.

26.4.1 Team Exams

"If engineers always work in teams, isn't the most appropriate engineering examination a team exam?" No. Examinations test the ability of an engineering student to do the work of a team member. This is clearly an essential skill for an engineer, and it cannot be evaluated by giving a typical engineering examination to students working as a group. The value of such an examination would be to evaluate the *group*, but not the individual group member.

In practice, there is often only one engineer on a team. When there are multiple engineers, often there is only one chemical engineer. Without development of and evaluation of individual engineering abilities, the chemical engineer would be unprepared to work in a group, let alone on a team.

An extremely effective strategy for developing teamwork skills is to study in a team and be evaluated as an individual.

26.4.2 Overreliance on Team Members

In a group, there will be members of varying expertise and ability. As noted earlier, task differentiation is a characteristic of group work, but no group output should be totally dependent on any one member. A common misconception is that individual members can focus only on a narrow field of expertise because other group members will do the rest. Without task overlap, flexibility, and member backups, a slight disturbance can destroy the group momentum. A chemical engineering student may mistakenly conclude that an understanding of thermodynamics or differential equations is unnecessary because other group members can be relied on. The fallacy of this argument becomes clear when one remembers that there may be only one engineer in the group. On a team, broad and deep knowledge and understanding of basic engineering concepts are even more essential because a team member is expected to fill in for another team member—even for a nonengineer. Thus, teamwork requires a broader and higher level of individual abilities, not a narrower or lower level.

26.5 LEARNING IN TEAMS

Teamwork can occur explicitly as a part of a course, in which case it is called **cooperative** or **collaborative learning**. The team can be formed outside the course structure but with the goal of improving the performance of all team members, in which case it is called a **study group**. In either case, the functioning of the group will be enhanced as it advances toward being a team, as described in Section 26.3.

Learning to be an engineer is a challenging, time-consuming, and complex project—the ideal situation for teamwork.

High achievers and low achievers both improve their learning when they study in teams. In fact, study groups function better for all members when there is a diversity of learners in the group. Although some study groups are formed for specific courses and then disbanded, a more effective scheme is to develop a study group across courses for students in the same curriculum.

The following heuristics apply especially to teams set up with learning as the goal.

- State as a goal that each team member will understand all the material, not that the "team" will understand the material.
- Accept all approaches offered by any team member that lead to a correct answer or help to explain the material.
- Realize that explaining information already understood to another team member will enhance learning for both people.
- Congratulate all team members when everyone understands the material.
- Celebrate high exam scores by any member of the team.
- If a team member does poorly on an exam, make it a team goal to improve that team member's learning.
- Verify learning of each team member for a given learning objective.
- Allow each team member time to do the problem individually. Then compare problem-solving strategies as well as answers.
- Plan "time on task" for the entire team, and share contact information so that individual team members can get assistance when doing assignments alone.
- Test each member of the team; that is, the team assesses individual performance.

Instructors often examine teams by choosing which team member will answer any given question. Because each member is responsible for understanding all of the material studied by the team, all team members receive the grade earned by the single team member chosen. This strategy is consistent with the basis for team learning, and it reinforces the concept that a learning team can claim success only when all team members have acquired the individual knowledge and understanding to be successful.

26.6 OTHER READING

In this section, several excellent teamwork resources are listed. Each is easily accessible and can provide more in-depth information and ideas than possible in this chapter. A good team resource library should include each of the references.

Teamwork from Start to Finish [1] by Fran Rees is a fun read (200 pages, paperback) that allows the reader to develop a sense of teamwork quickly. Although it does not go into depth in the various aspects of teamwork, it is an excellent foundation that can be read at leisure. This book (and the other references) is typical of the best teamwork literature. The style is not dense as in a typical engineering textbook. It is meant for a general audience and especially for the standard one-day workshop on teamwork that nearly all U.S. engineers are sent to by their employers.

Although *Problem-Based Learning* [2] by Donald Woods is focused on using teams in learning, its scope includes everything from building a team to being an effective chairperson and from coping creatively with conflict to self-assessment. It has a wealth of essential information on teamwork, and an annotated subject index where one can find definitions of terms as well as key page numbers. Dr. Woods is a chemical engineering professor at McMaster University. His engineering viewpoint is obvious from the directness of his explanations of what works and from his many tables, forms, figures, and mind maps. There are problems at the end of each chapter to help the reader process and internalize the information presented.

Kimball Fisher and colleagues present the 100 most common problems that teams encounter and several suggested solutions for each in *Tips for Teams* [3]. This is one of the most popular and accessible of the books that offer advice on how to avoid and to correct conflicts in teamwork.

Although all of these resource books have sections or chapters on leadership skills, anyone who ever wants to be a team leader should read at least one book focused on developing these specific skills. *Leading Self-Directed Work Teams* [4] by Kimball Fisher is an excellent choice.

Every team needs a plan, and the most common planning technique is the Gantt chart. The lively and short article by Jeff Dewar [6] demonstrates the technique and shows why it is so powerful. It is all the reference needed to develop and use Gantt charts.

Project Management and Teamwork [7] by Karl Smith of the University of Minnesota teaches engineering students to work effectively and efficiently in project teams. Case studies of student work demonstrate the various aspects of project work, including the project life cycle, project scoping, monitoring, documentation, and scheduling. The explanation of the critical path method is excellent, as are the chapters on teamwork skills.

The Team Handbook [8] by Peter Scholtes and colleagues is considered by many in industry as a teamwork bible. It has been used in one-day and two-day workshops by companies since the late 1980s, and its format is geared to that application. It is about 200 pages and spiral-bound, with plenty of blank space on every page for notes. The many forms, lists, and guidelines are meant to be used in a team setting to identify problems, understand group behaviors, and move the group toward effective teamwork. The viewpoint of the authors is a common management strategy, continuous quality improvement, that demands teamwork.

Donald Woods has been a world leader in problem solving for decades. Many of the resource materials from his McMaster Problem Solving (MPS) program are available from him through the Web [9, 10]. The modules most useful in teamwork are 28 and 53. They are straightforward, hands-on guides in team building, with strategies that can be used immediately.

The U.S. National Aeronautics and Space Administration (NASA) provides a library Web site with lists of articles, books, and Internet resources on a wide range of topics. Two of the lists dealing with teamwork are listed along with their URLs [11]. These are excellent bibliographies.

H. Scott Fogler and Steven LeBlanc's *Strategies for Creative Problem Solving* [12] provides a wide array of exercises for use in team building. Through focusing on the stages of problem solving and the various proven heuristics for implementing them, these authors provide teamwork strategies that can be implemented immediately.

26.7 SUMMARY

Good engineers are team animals. One of their finely honed *individual* skills is the ability to be an excellent *team* member.

In this introductory chapter about teamwork, the features of groups that make them effective and efficient were introduced. Heuristics were given to enhance teamwork by reducing the friction that can build in groups. The four stages of group evolution were described and examples given. The concept of a team as the ultimate group was presented, and misconceptions about individual responsibility versus team responsibility were discussed. Beyond this chapter, there is a wealth of literature on the subject. Some of the most useful resources were described.

REFERENCES

1. Rees, F., *Teamwork from Start to Finish: 10 Steps to Results* (San Francisco: Jossey-Bass, 1997).
2. Woods, D. R., *Problem-Based Learning: How to Gain the Most from PBL* (Hamilton, Ontario, Canada: McMaster University Bookstore, 1994).
3. Fisher, K., W. Belgard, and S. Rayner, *Tips for Teams: A Ready Reference for Solving Common Team Problems* (New York: McGraw-Hill, 1995).
4. Fisher, K., *Leading Self-Directed Work Teams: A Guide to Developing New Team Leadership Skills*, 2nd ed. (New York: McGraw-Hill, 1999).
5. Tuckman, B. W., "Development Sequence in Small Groups," *Psychological Bulletin* 63 (1965): 384.
6. Dewar, J. D., "If You Don't Know Where You're Going, How Will You Know When You Get There?" *CHEMTECH* 19, no. 4 (1989): 214–217.

7. Smith, K. A., *Teamwork and Project Management*, 3rd ed. (New York: McGraw-Hill, 2007).
8. Scholtes, P. R., B. L. Joiner, and B. J. Streibel, *The Team Handbook*, 3rd ed. (Madison, WI: Oriel, 2003).
9. Woods, D. R., *Group Skills*, available at http://www.chemeng.mcmaster.ca/innov1.htm as MPS-28, 1998.
10. Woods, D. R., *Team Building: How to Develop and Evaluate Individual Effectiveness in Teams*, available at http://www.chemeng.mcmaster.ca/innov1.htm as MPS-53, 1998.
11. NASA Headquarters Library Bibliographies, *Interpersonal Relations and Group Dynamics*, 2007 (http://www.hq.nasa.gov/office/hqlibrary/ppm/ppm29.htm), and *Teams and Teamwork*, 2007 (http://www.hq.nasa.gov/office/hqlibrary/ppm/ppm5.htm).
12. Fogler, H. S., and S. E. LeBlanc, *Strategies for Creative Problem Solving*, 2nd ed. (Upper Saddle River, NJ: Prentice Hall, 2008).

PROBLEMS

These problems are not like normal homework problems to be done after just reading the chapter. Instead, they are assignments that should be done in the context of an actual group activity to improve group process. Although a group of four or five students is envisioned, the exercises will work with somewhat larger or smaller groups. They are often used during class sessions so that the instructor can provide immediate feedback.

1. In turn, each group member describes the last homework assignment submitted, sharing the strategy used to solve the problem. The assignment chosen should have been done individually. After each member's strategy has been described, the group should discuss the value in each strategy. The goal is not to rank the strategies but to develop respect for each strategy through constructive discussion. After the discussion, the group answers the question, "What have we learned from this exercise?"
2. Each group member describes the most valuable course taken so far and why that course was so valuable. Then the group brainstorms criteria for determining the value of a course. After developing a list of twenty criteria in two minutes, the group creates a set of three criteria to use in assessing courses by developing a consensus rather than voting.
3. As a group, develop a set of norms for a study group. Address such issues as who will be the leader, when and where the group should meet, how much preparation is expected of each member before each meeting, the agenda for a typical meeting, and how the success of the group will be assessed.

4. Within the context of a group design project, meet to discuss and plan task differentiation. When the plan is complete, answer the following questions.
   a. What happens if one of the group members becomes ill and cannot complete the assigned task? (In turns, do this for each member.)
   b. For which tasks is there overlap? Why?
   c. What percent of the effort is being done by each group member?
5. For the last group project completed, each group member fills out the form in Figure 26.1. One group member then collates the responses in a spreadsheet, and the group discusses how the group performed on the project. Finally, the group prepares a list of ways to improve the group work in future projects.
6. Repeat Problem 1 for the last examination taken. Be sure to discuss timing as well as problem-solving strategies used.
7. Each group member describes an example of friction in a prior group situation, and the group does the following:
   a. Determines which of the categories of friction from Section 26.1.3 best describes the example.
   b. Develops a suggestion for how the situation could have been handled better.
8. For your most productive group experience, analyze whether it was a "team," using the characteristics of teams given in Section 26.3.2.
9. Each group member chooses several chapters of references [1], [3], [4], or [7]. Read them and report to the group how the strategies presented can help improve group effectiveness and efficiency.
10. Choose the most difficult course you are likely to face next semester, and create a Gantt chart for succeeding in it. (If you are finishing this semester, choose finding a job or making the best impression in the job you have already chosen.)

APPENDIX

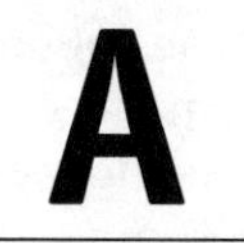

Cost Equations and Curves for the CAPCOST Program

The purpose of this appendix is to present the equations and figures that describe the relationships used in the capital equipment-costing program CAPCOST introduced in Chapter 7 and used throughout the text. The program is based on the module factor approach to costing that was originally introduced by Guthrie [1, 2] and modified by Ulrich [3].

A.1 PURCHASED EQUIPMENT COSTS

All the data for the purchased cost of equipment for the second edition of this book were obtained from a survey of equipment manufacturers during the period May to September of 2001, so an average value of the CEPCI of 397 over this period should be used when accounting for inflation.

Additional process equipment has been added to the third edition and is listed below:

- Conveyors
- Crystallizers
- Dryers
- Dust Collectors
- Filters
- Mixers
- Reactors
- Screens

The purchased costs for these types of equipment were obtained in 2003 but the costs given here have been normalized to 2001. For this new equipment, bare module factors were not available, nor were pressure factors or materials of construction factors. In general, these units are generally bought as a package, and installation in the plant is not expensive. The bare module factors for these units are taken to be the field installation factors given by Guthrie [1, 2].

Data for the purchased cost of the equipment, at ambient operating pressure and using carbon steel construction, C_p^o, were fitted to the following equation:

$$\log_{10} C_p^o = K_1 + K_2 \log_{10}(A) + K_3[\log_{10}(A)]^2 \quad (A.1)$$

where A is the capacity or size parameter for the equipment. The data for K_1, K_2, and K_3, along with the maximum and minimum values used in the correlation, are given in Table A.1. These data are also presented in the form of graphs in Figures A.1–A.17. It should be noted that in these figures, the data are plotted as C_p^o/A as a function of size attribute, A. This form of the graph clearly illustrates the decreasing cost per unit of capacity as the size of the equipment increases.

Data from the R-Books software marketed by Richardson Engineering Services, Inc. [4], were used as a basis for several of the graphs and correlations; acknowledgment is given in the appropriate figures.

(text continues on p. 941)

Table A.1 Equipment Cost Data to Be Used with Equation A.1

Equipment Type	Equipment Description	K_1	K_2	K_3	Capacity, Units	Min Size	Max Size
Blenders	Kneader	5.0141	−0.4133	0.3224	Volume, m^2	0.14	3
	Ribbon	4.1366	−0.4928	0.0070	Volume, m^2	0.7	11
	Rotary	4.1366	−0.4928	0.0070	Volume, m^2	0.7	11
Centrifuges	Auto batch separator	4.7681	−0.0260	0.0240	Diameter, m	0.5	1.7
	Centrifugal separator	4.3612	−0.1236	−0.0049	Diameter, m	0.5	1
	Oscillating screen	4.8600	−0.6660	0.1063	Diameter, m	0.5	1.1
	Solid bowl w/o motor	4.9697	0.1689	0.0038	Diameter, m	0.3	2
Compressors	Centrifugal, axial, and reciprocating	2.2897	1.3604	−0.1027	Fluid power, kW	450	3000
	Rotary	5.0355	−1.8002	0.8253	Fluid power, kW	18	950
Conveyors	Apron	3.9255	−0.4961	0.1506	Area, m^2	1.0	15
	Belt	4.0637	−0.7416	0.1550	Area, m^2	0.5	325
	Pneumatic	4.6616	−0.6795	0.0638	Area, m^2	0.75	65
	Screw	3.6062	−0.7341	0.1982	Area, m^2	0.5	30
Crystallizers	Batch	4.5097	−0.8269	0.1344	Volume, m^3	1.5	30
Drives	Gas turbine	−21.7702	13.2175	−1.5279	Shaft power, kW	7500	23,000
	Intern comb. engine	2.7635	0.8574	−0.0098	Shaft power, kW	10	10,000
	Steam turbine	2.6259	1.4398	−0.1776	Shaft power, kW	70	7500
	Electric—explosion-proof	2.4604	1.4191	−0.1798	Shaft power, kW	75	2600
	Electric—totally enclosed	1.9560	1.7142	−0.2282	Shaft power, kW	75	2600
	Electric—open/drip-proof	2.9508	1.0688	−0.1315	Shaft power, kW	75	2600
Dryers	Drum	4.5472	−0.7269	0.1340	Area, m^2	0.5	50
	Rotary, gas fired	3.5645	0.1118	−0.0777	Area, m^2	5	100
	Tray	3.6951	−0.4558	−0.1248	Area, m^2	1.8	20

(continued)

Table A.1 Equipment Cost Data to Be Used with Equation A.1 (*Continued*)

Equipment Type	Equipment Description	K_1	K_2	K_3	Capacity, Units	Min Size	Max Size
Dust Collectors	Baghouse	4.5007	−0.5818	0.0813	Volume, m^3	0.08	350
	Cyclone scrubbers	3.6298	−0.4991	0.0411	Volume, m^3	0.06	200
	Electrostatic precipitator	3.6298	−0.4991	0.0411	Volume, m^3	0.06	200
	Venturi scrubber	3.6298	−0.4991	0.0411	Volume, m^3	0.06	200
Evaporators	Forced circulation (pumped)	5.0238	0.3475	0.0703	Area, m^2	5	1000
	Falling film	3.9119	0.8627	−0.0088	Area, m^2	50	500
	Agitated film (scraped wall)	5.0000	0.1490	−0.0134	Area, m^2	0.5	5
	Short tube	5.2366	−0.6572	0.3500	Area, m^2	10	100
	Long tube	4.6420	0.3698	0.0025	Area, m^2	100	10,000
Fans	Centrifugal radial	3.5391	−0.3533	0.4477	Gas flowrate, m^3/s	1	100
	Backward curve	3.3471	−0.0734	0.3090	Gas flowrate, m^3/s	1	100
	Axial vane	3.1761	−0.1373	0.3414	Gas flowrate, m^3/s	1	100
	Axial tube	3.0414	−0.3375	0.4722	Gas flowrate, m^3/s	1	100
Filters	Bent	5.1055	−0.5001	0.0001	Area, m^2	0.9	115
	Cartridge	3.2107	−0.2403	0.0027	Area, m^2	15	200
	Disc and drum	4.8123	−0.7142	0.0420	Area, m^2	0.9	300
	Gravity	4.2756	−0.6480	0.0714	Area, m^2	0.5	80
	Leaf	3.8187	−0.3765	0.0176	Area, m^2	0.6	235
	Pan	4.8123	−0.7142	0.0420	Area, m^2	0.9	300
	Plate and frame	4.2756	−0.6480	0.0714	Area, m^2	0.5	80
	Table	5.1055	−0.5001	0.0001	Area, m^2	0.9	115
	Tube	5.1055	−0.5001	0.0001	Area, m^2	0.9	115
Furnaces	Reformer furnace	3.0680	0.6597	0.0194	Duty, kW	3000	100,000
	Pyrolysis furnace	2.3859	0.9721	−0.0206	Duty, kW	3000	100,000
	Nonreactive fired heater	7.3488	−1.1666	0.2028	Duty, kW	1000	100,000

(continued)

Table A.1 Equipment Cost Data to Be Used with Equation A.1 (*Continued*)

Equipment Type	Equipment Description	K_1	K_2	K_3	Capacity, Units	Min Size	Max Size
Heat exchangers	Scraped wall	3.7803	0.8569	0.0349	Area, m^2	2	20
	Teflon tube	3.8062	0.8924	−0.1671	Area, m^2	1	10
	Bayonet	4.2768	−0.0495	0.1431	Area, m^2	10	1000
	Floating head	4.8306	−0.8509	0.3187	Area, m^2	10	1000
	Fixed tube	4.3247	−0.3030	0.1634	Area, m^2	10	1000
	U-tube	4.1884	−0.2503	0.1974	Area, m^2	10	1000
	Kettle reboiler	4.4646	−0.5277	0.3955	Area, m^2	10	100
	Double pipe	3.3444	0.2745	−0.0472	Area, m^2	1	10
	Multiple pipe	2.7652	0.7282	0.0783	Area, m^2	10	100
	Flat plate	4.6656	−0.1557	0.1547	Area, m^2	10	1000
	Spiral plate	4.6561	−0.2947	0.2207	Area, m^2	1	100
	Air cooler	4.0336	0.2341	0.0497	Area, m^2	10	10000
	Spiral tube	3.9912	0.0668	0.2430	Area, m^2	1	100
Heaters	Diphenyl heater	2.2628	0.8581	0.0003	Duty, kW	650	10750
	Molten salt heater	1.1979	1.4782	−0.0958	Duty, kW	650	10750
	Hot water heater	2.0829	0.9074	−0.0243	Duty, kW	650	10750
	Steam boiler	6.9617	−1.4800	0.3161	Duty, kW	1200	9400
Mixers	Impeller	3.8511	−0.2991	−0.0003	Power, kW	5	150
	Propeller	4.3207	−0.9641	0.1346	Power, kW	5	500
	Turbine	3.4092	−0.5104	0.0030	Power, kW	5	150
Packing	Loose (for towers)	2.4493	0.9744	0.0055	Volume, m^3	0.03	628
Process vessels	Horizontal	3.5565	0.3776	0.0905	Volume, m^3	0.1	628
	Vertical	3.4974	0.4485	0.1074	Volume, m^3	0.3	520
Pumps	Reciprocating	3.8696	0.3161	0.1220	Shaft power, kW	0.1	200
	Positive displacement	3.4771	0.1350	0.1438	Shaft power, kW	1	100
	Centrifugal	3.3892	0.0536	0.1538	Shaft power, kW	1	300

(continued)

Table A.1 Equipment Cost Data to Be Used with Equation A.1 (*Continued*)

Equipment Type	Equipment Description	K_1	K_2	K_3	Capacity, Units	Min Size	Max Size
Reactors	Autoclave	4.5587	−0.7014	0.0020	Volume, m^3	1	15
	Fermenter	4.1052	−0.4680	−0.0005	Volume, m^3	0.1	35
	Inoculum tank	3.7957	−0.5407	0.0160	Volume, m^3	0.07	1
	Jacketed agitated	4.1052	−0.4680	−0.0005	Volume, m^3	0.1	35
	Jacketed nonagitated	3.3496	−0.2765	0.0025	Volume, m^3	5	45
	Mixer/settler	4.7116	−0.5521	0.0004	Volume, m^3	0.04	60
Screens	DSM	3.8050	−0.4144	0.2120	Area, m^2	0.3	6
	Rotary	4.0485	−0.8882	0.3260	Area, m^2	0.3	15
	Stationary	3.8219	0.0368	−0.6050	Area, m^2	2	11
	Vibrating	4.0485	−0.8882	0.3260	Area, m^2	0.3	15
Towers	Tray and packed	3.4974	0.4485	0.1074	Volume, m^3	0.3	520
Tanks	API—fixed roof	4.8509	−0.3973	0.1445	Volume, m^3	90	30000
	API—floating roof	5.9567	−0.7585	0.1749	Volume, m^3	1000	40000
Trays	Sieve	2.9949	0.4465	0.3961	Area, m^2	0.07	12.30
	Valve	3.3322	0.4838	0.3434	Area, m^2	0.70	10.50
	Demisters	3.2353	0.4838	0.3434	Area, m^2	0.70	10.50
Turbines	Axial gas turbines	2.7051	1.4398	−0.1776	Fluid power, kW	100	4000
	Radial gas/liquid expanders	2.2476	1.4965	−0.1618	Fluid power, kW	100	1500
Vaporizers	Internal coils/jackets	4.0000	0.4321	0.1700	Volume, m^3	1	100
	Jacketed vessels	3.8751	0.3328	0.1901	Volume, m^3	1	100

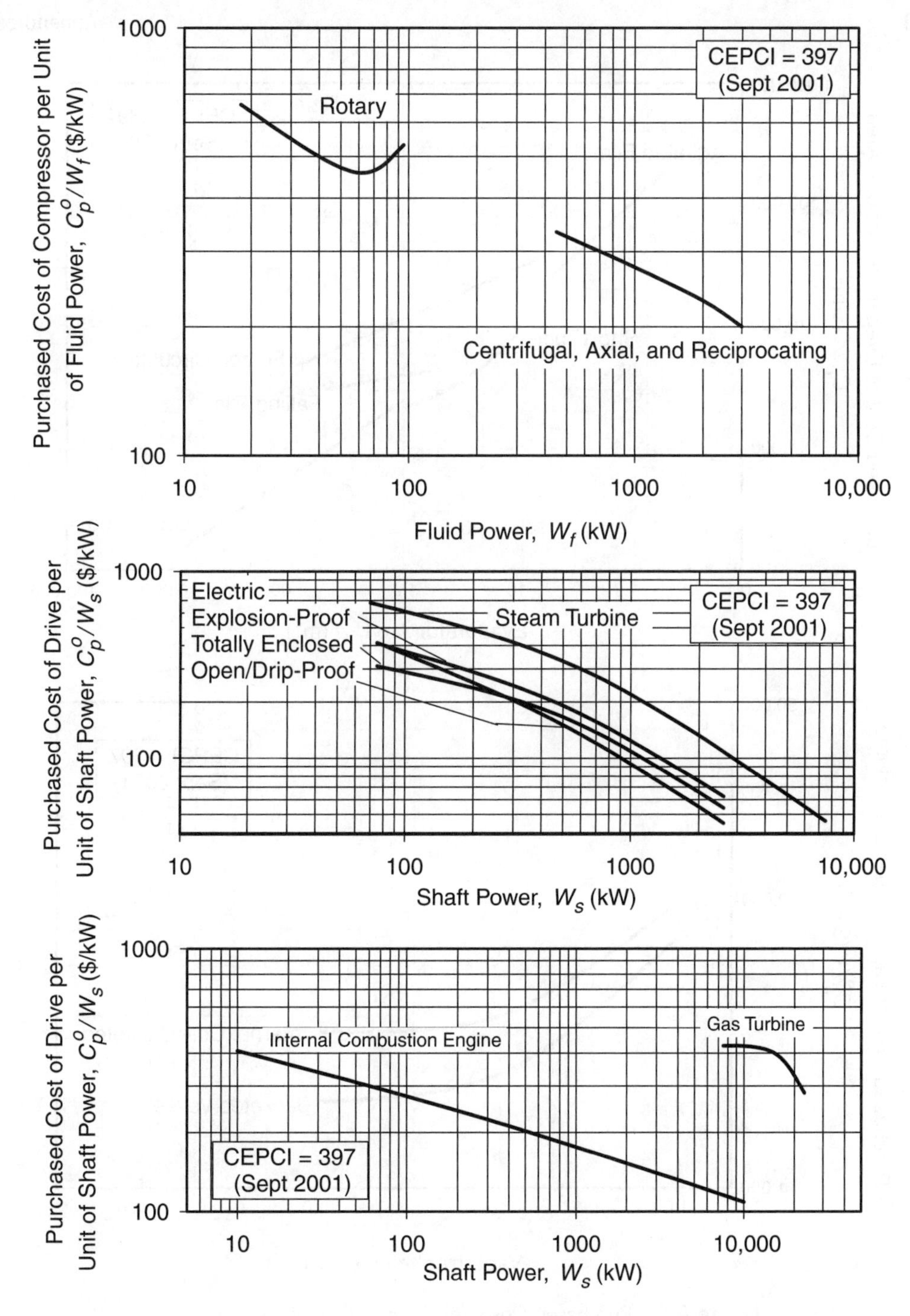

Figure A.1 Purchased Costs for Compressors and Drives (Cost Data for Compressors and Drives Taken from R-Books Software by Richardson Engineering Services, Inc. [4])

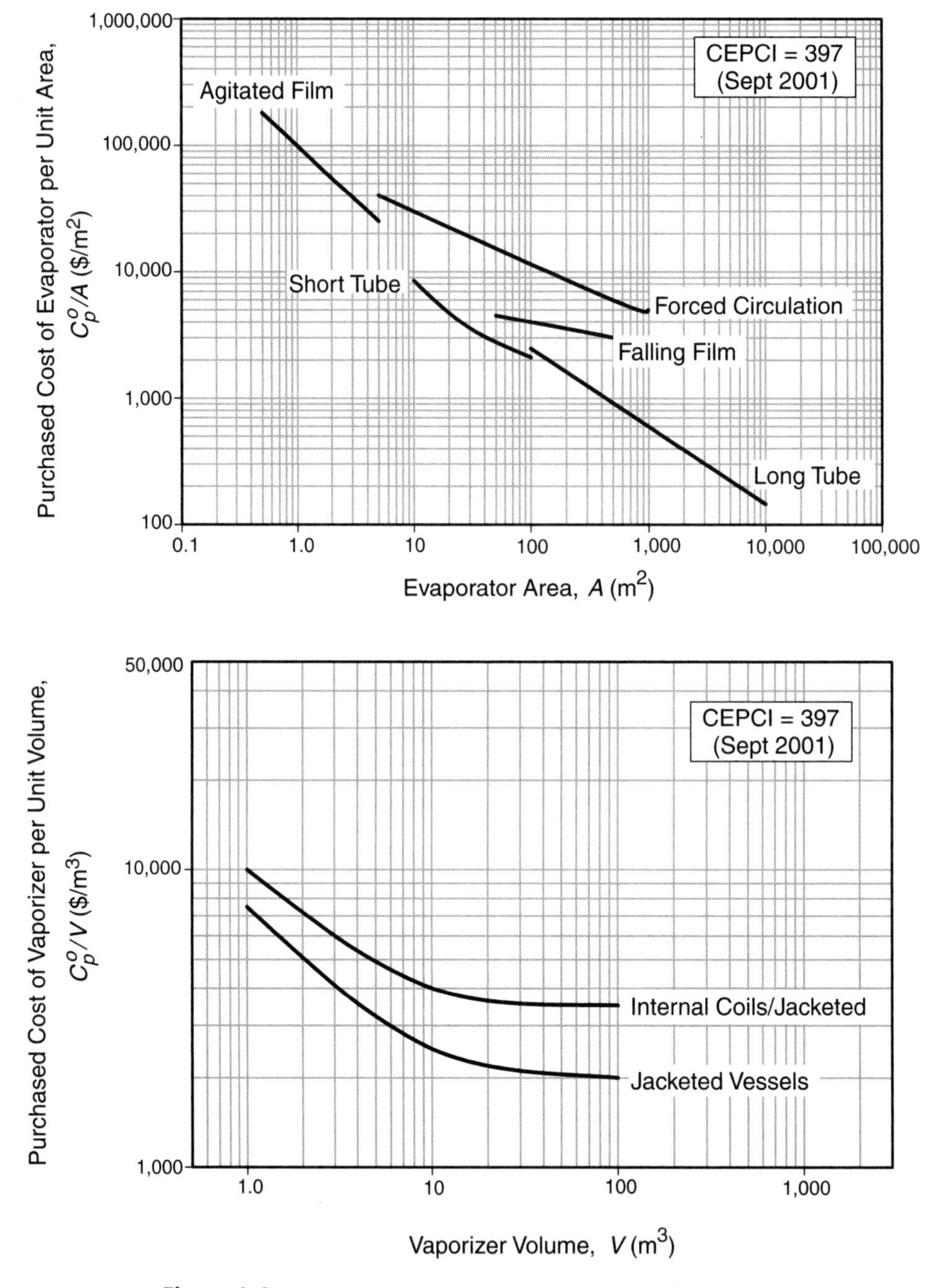

Figure A.2 Purchased Costs for Evaporators and Vaporizers

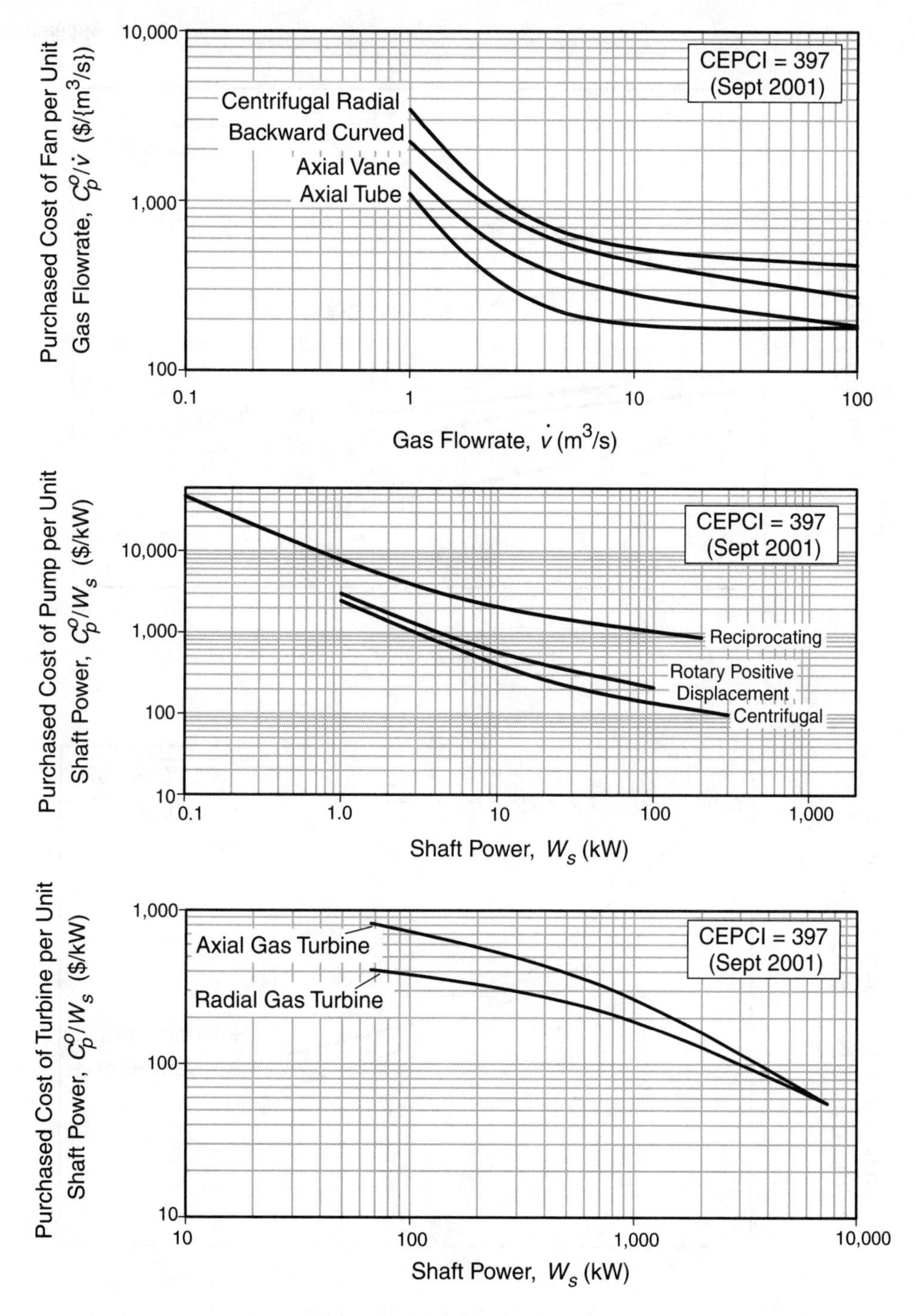

Figure A.3 Purchased Costs for Fans, Pumps, and Power Recovery Equipment (Cost Data for Fans Taken from R-Books Software by Richardson Engineering services [4])

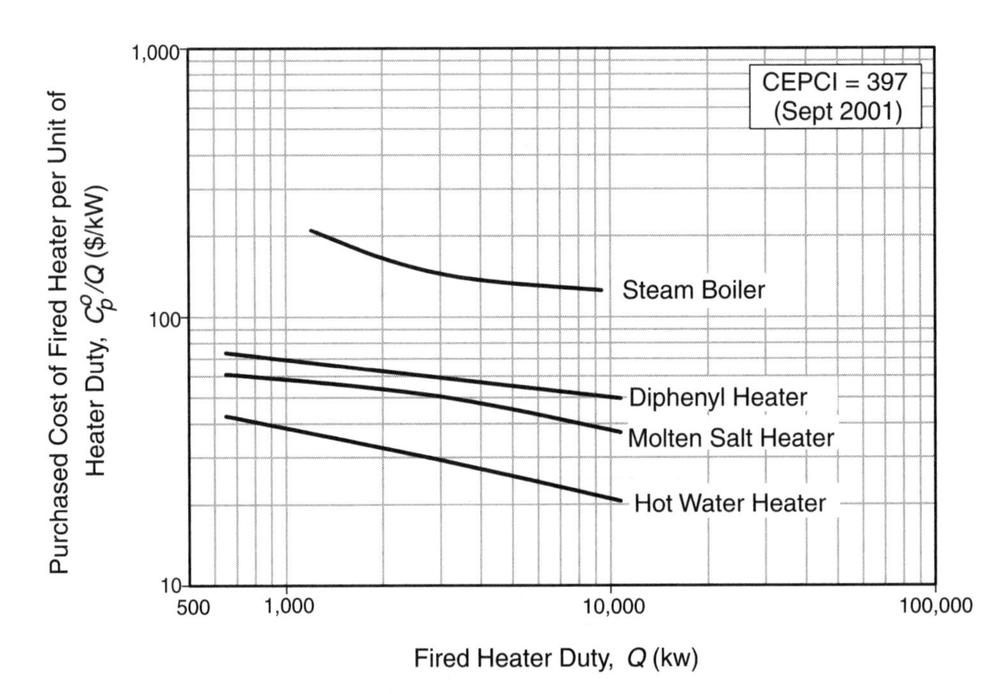

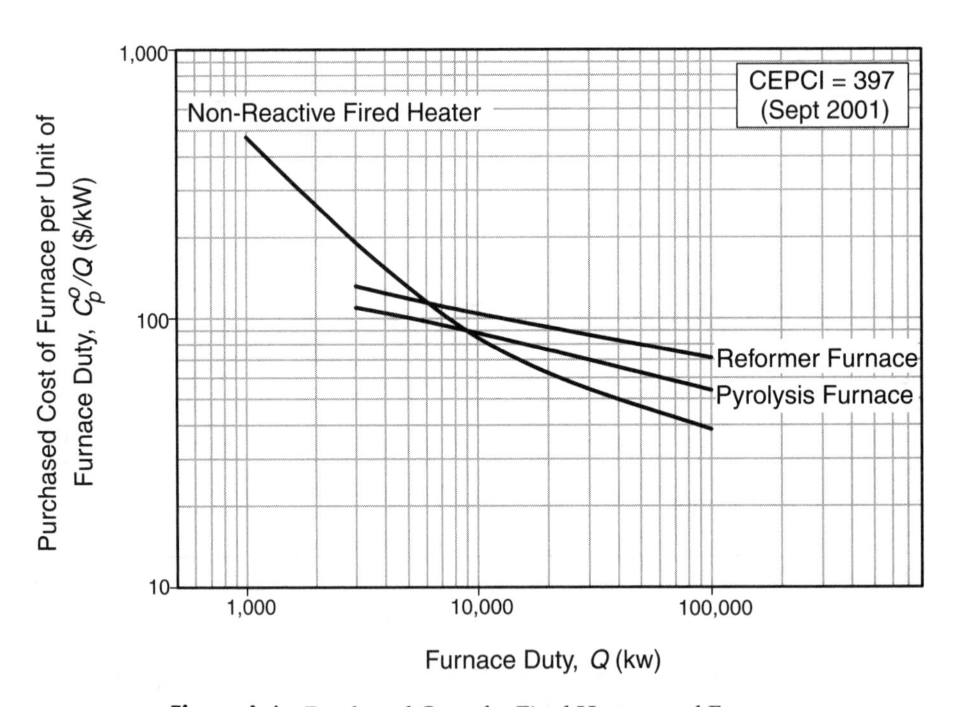

Figure A.4 Purchased Costs for Fired Heaters and Furnaces

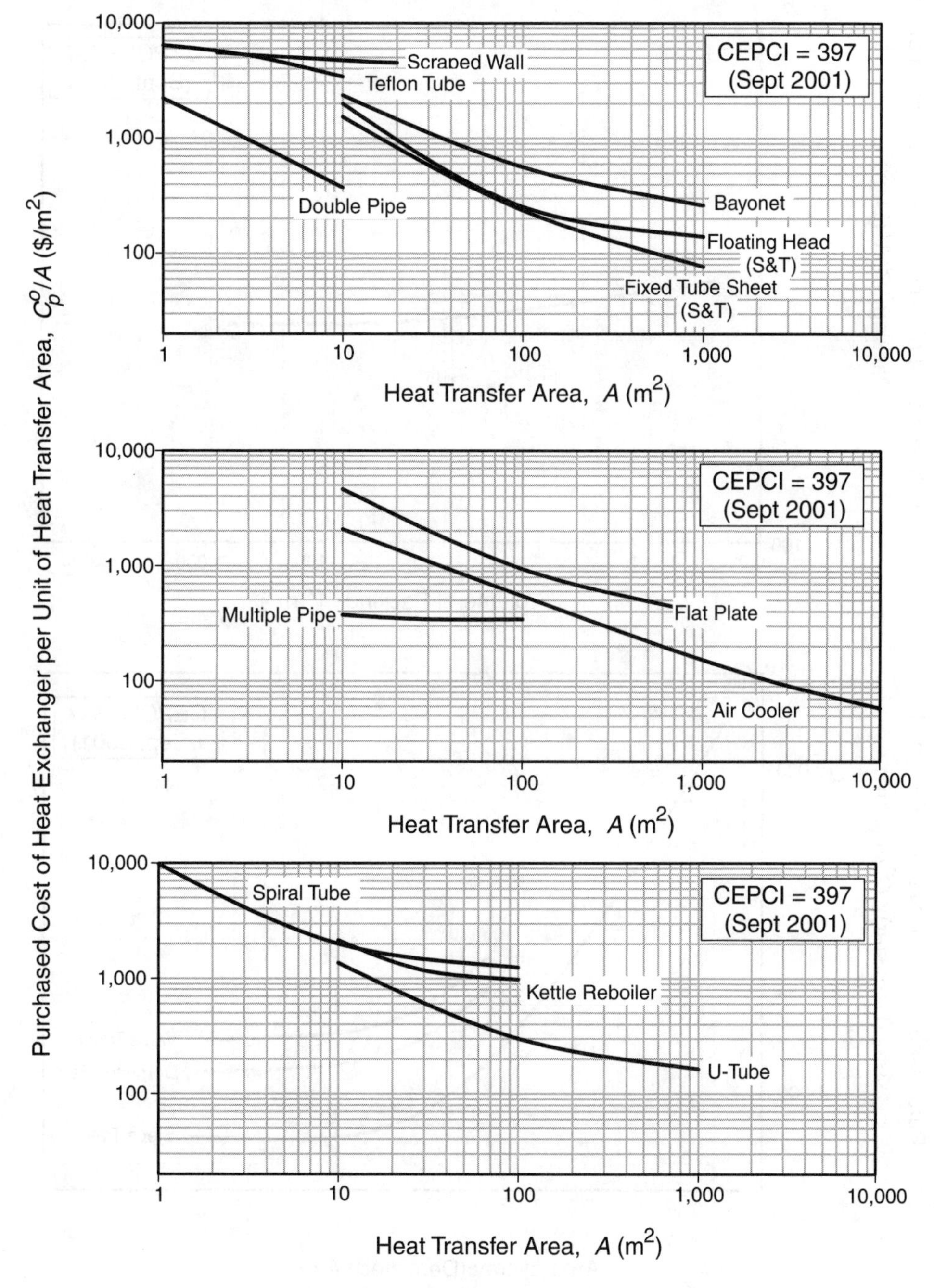

Figure A.5 Purchased Costs for Heat Exchangers

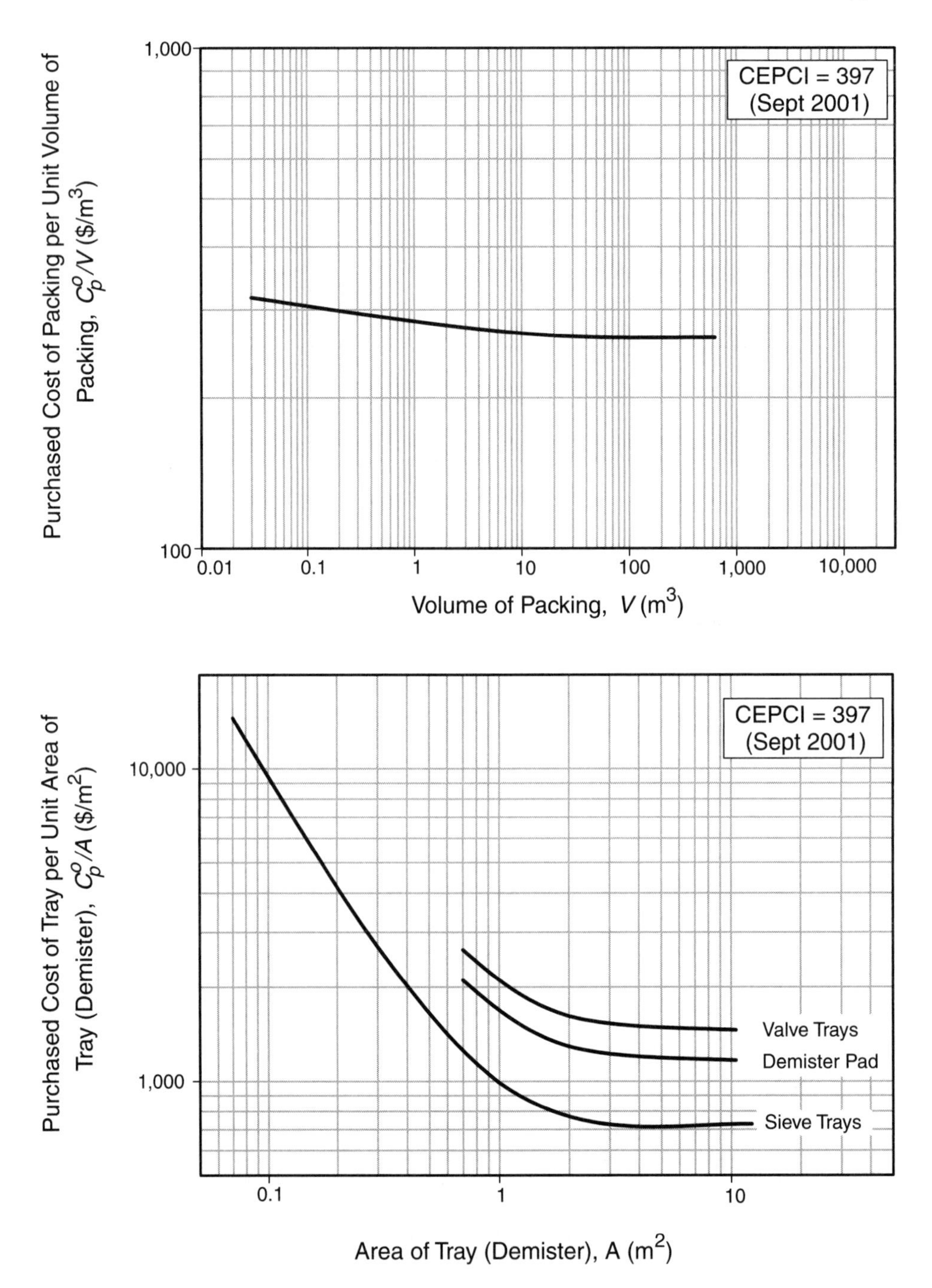

Figure A.6 Purchased Costs for Packing, Trays, and Demisters

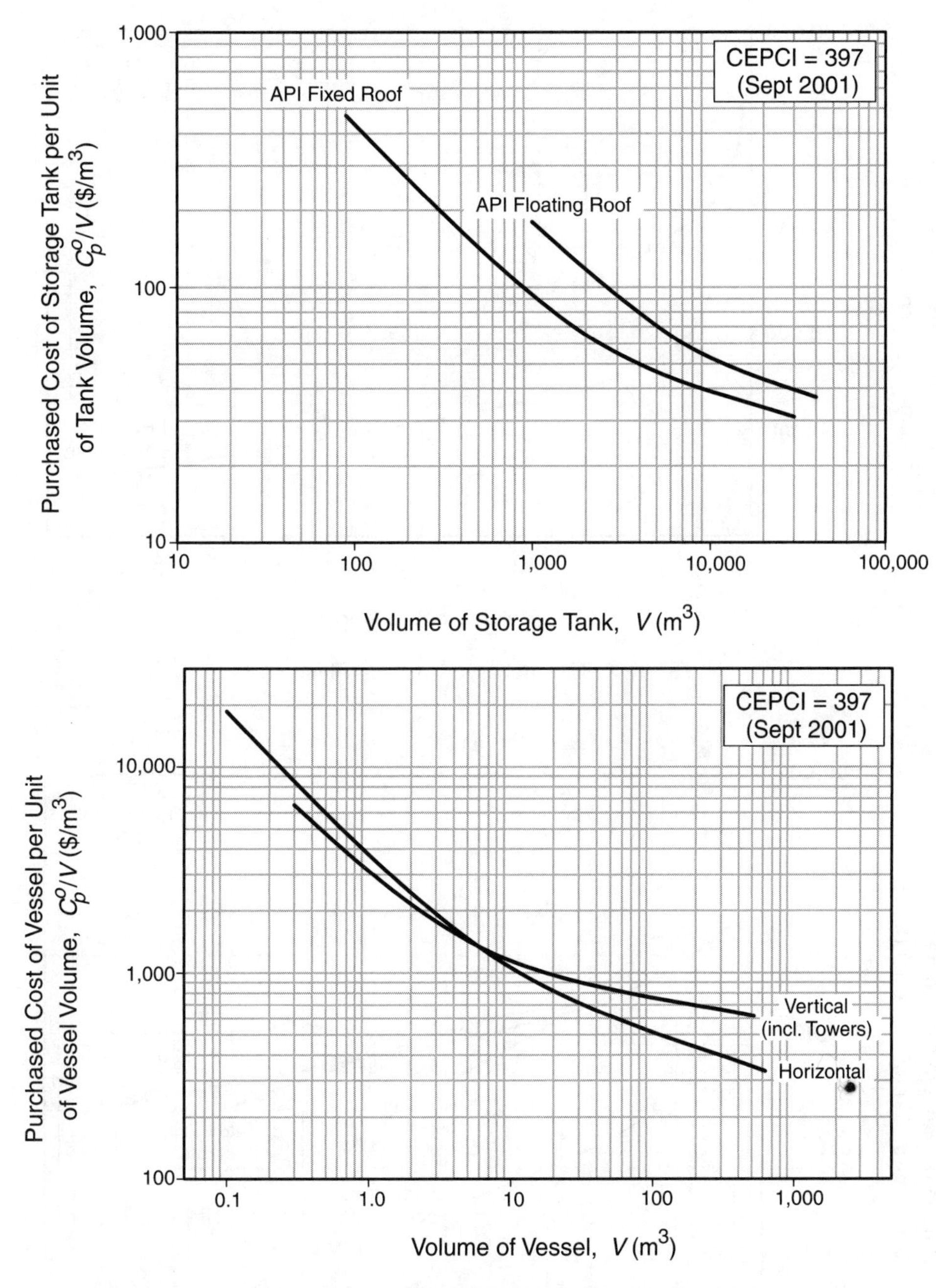

Figure A.7 Purchased Costs of Storage Tank and Process Vessels. (Data for Storage Tanks Taken from R-Books Software by Richardson Engineering Services [4])

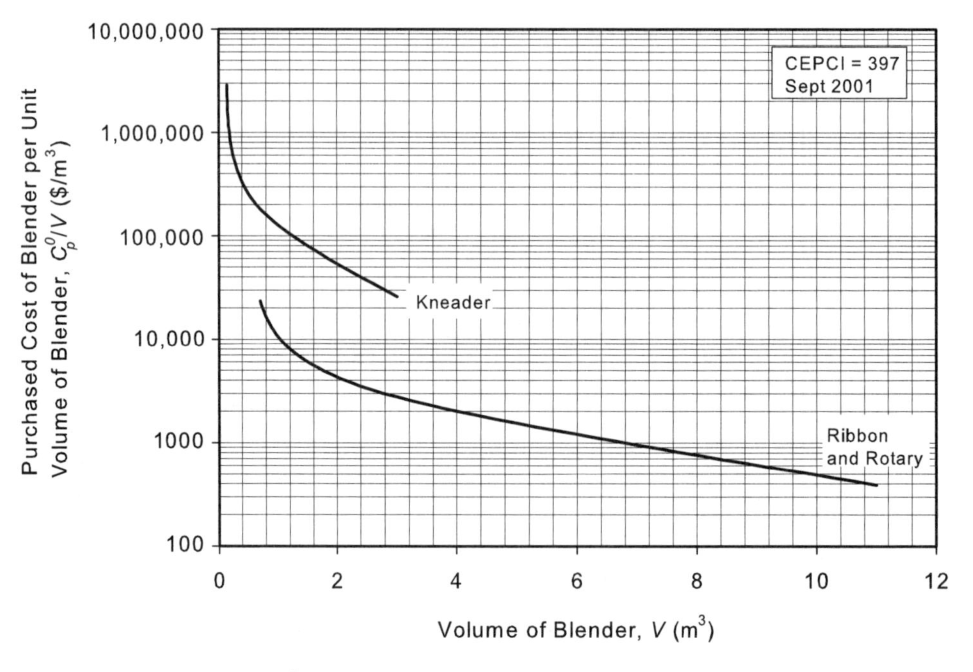

Figure A.8 Purchased Costs for Blenders

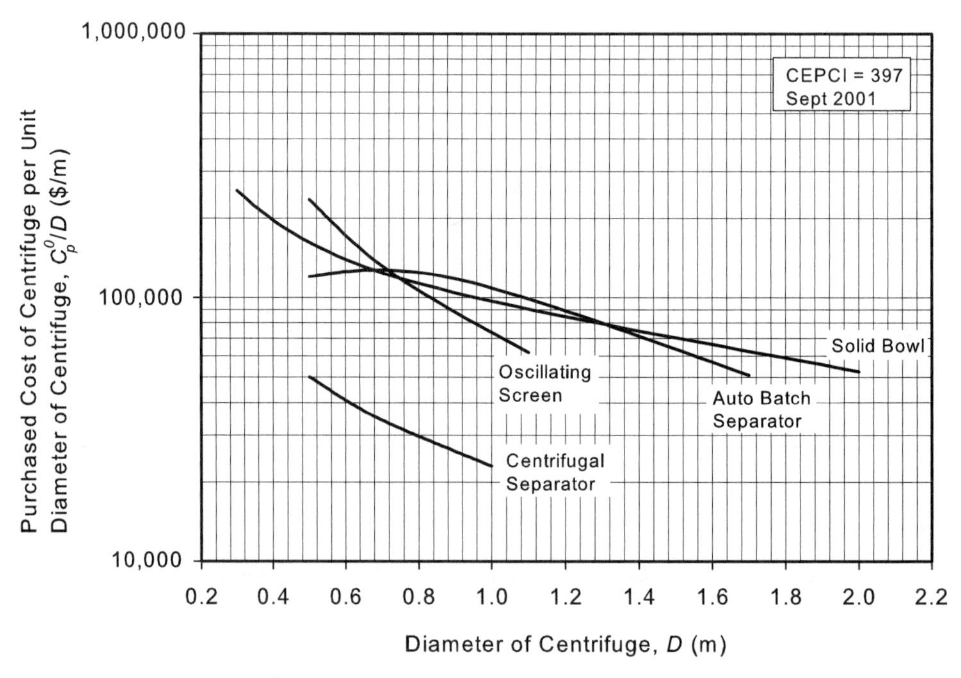

Figure A.9 Purchased Costs of Centrifuges

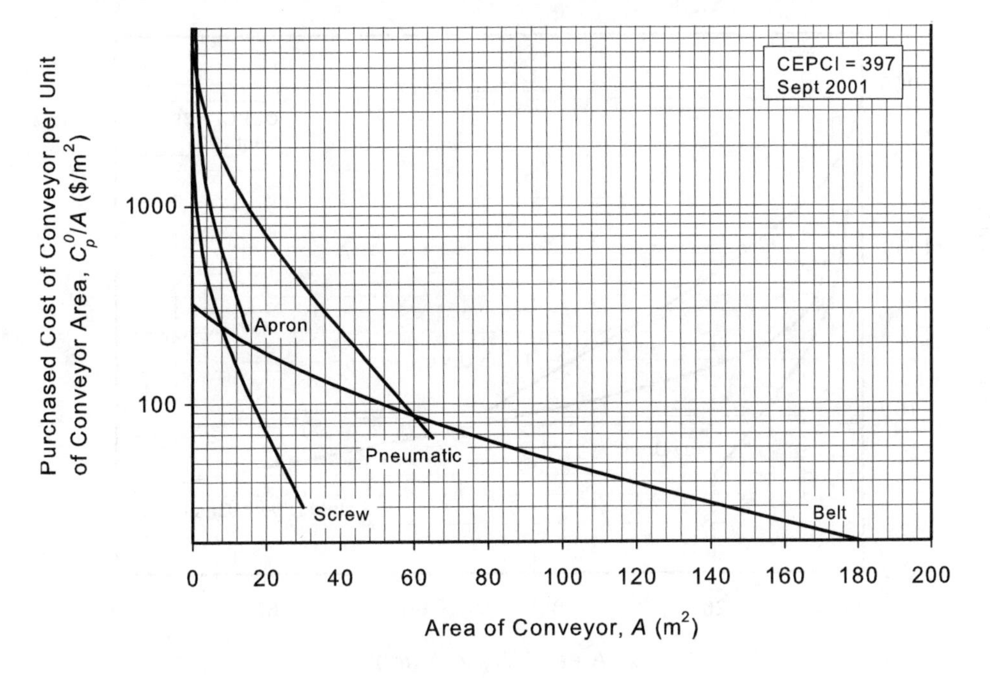

Figure A.10 Purchased Costs for Conveyors

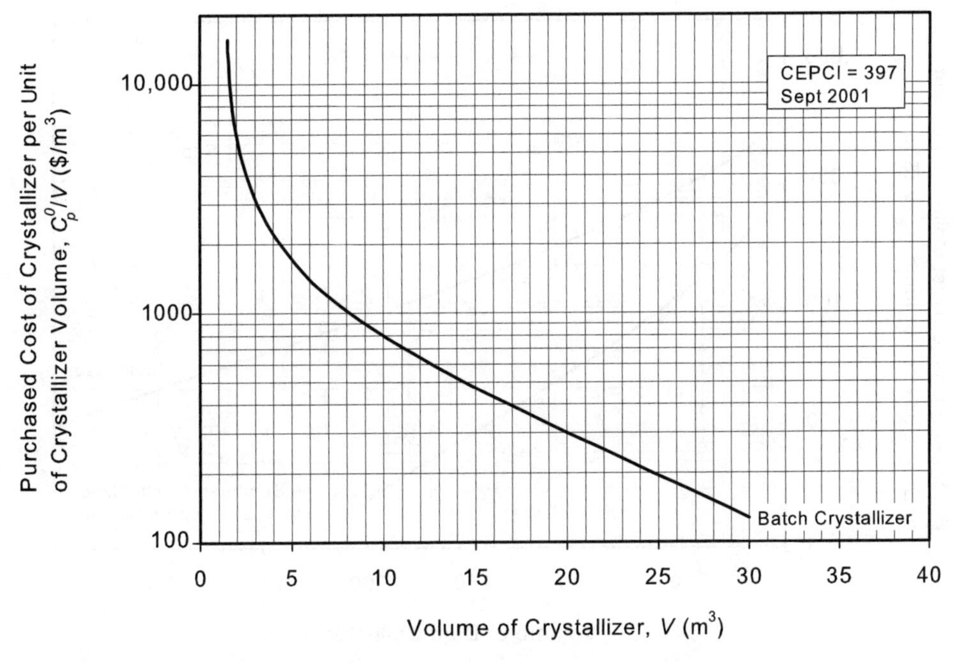

Figure A.11 Purchased Costs for Crystallizers

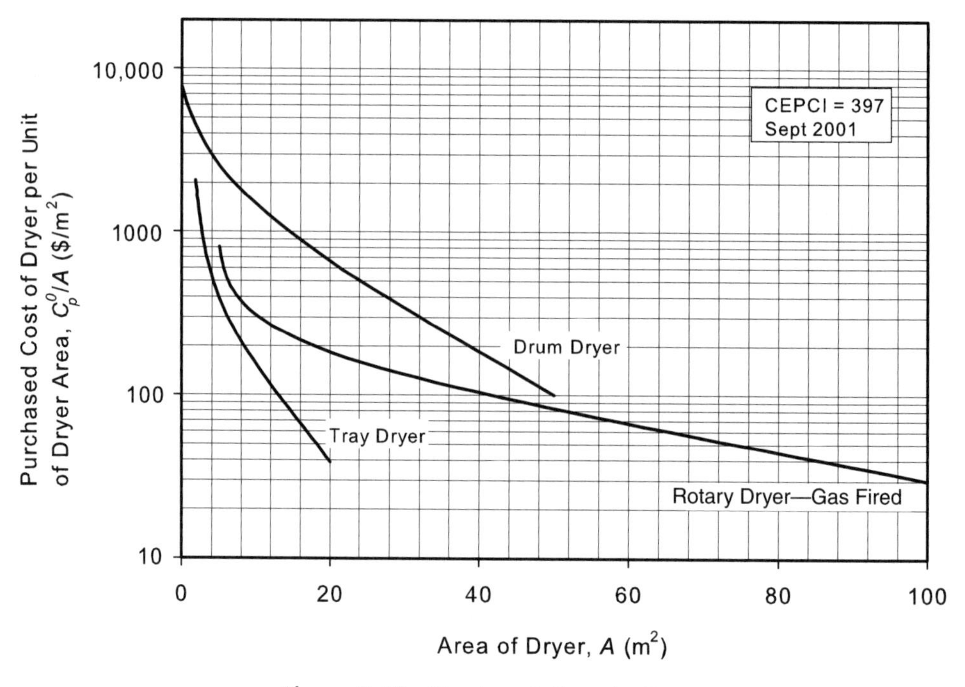

Figure A.12 Purchased Costs for Dryers

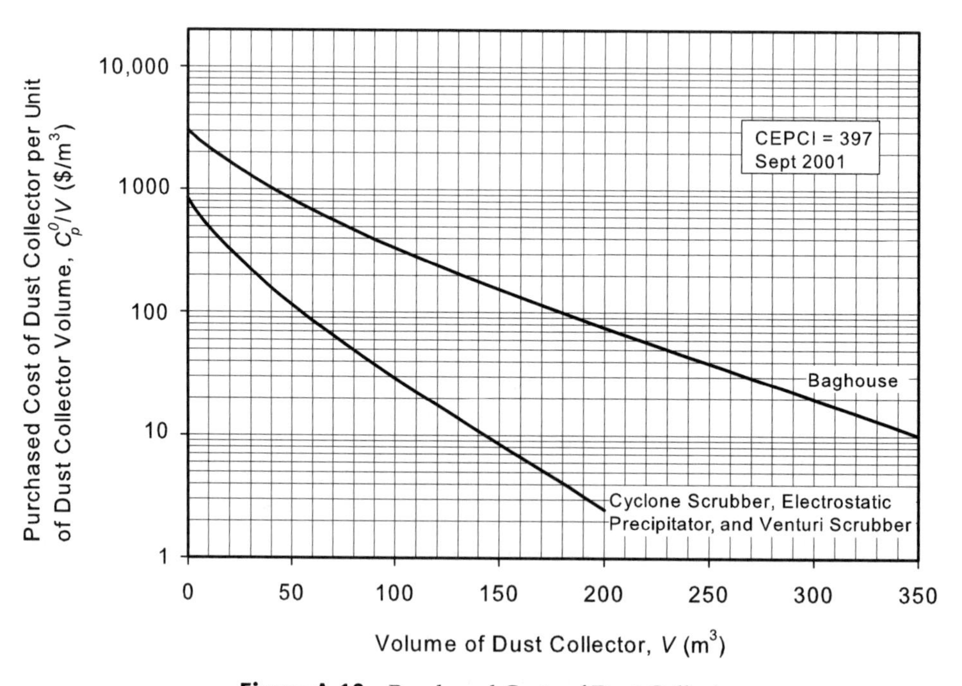

Figure A.13 Purchased Costs of Dust Collectors

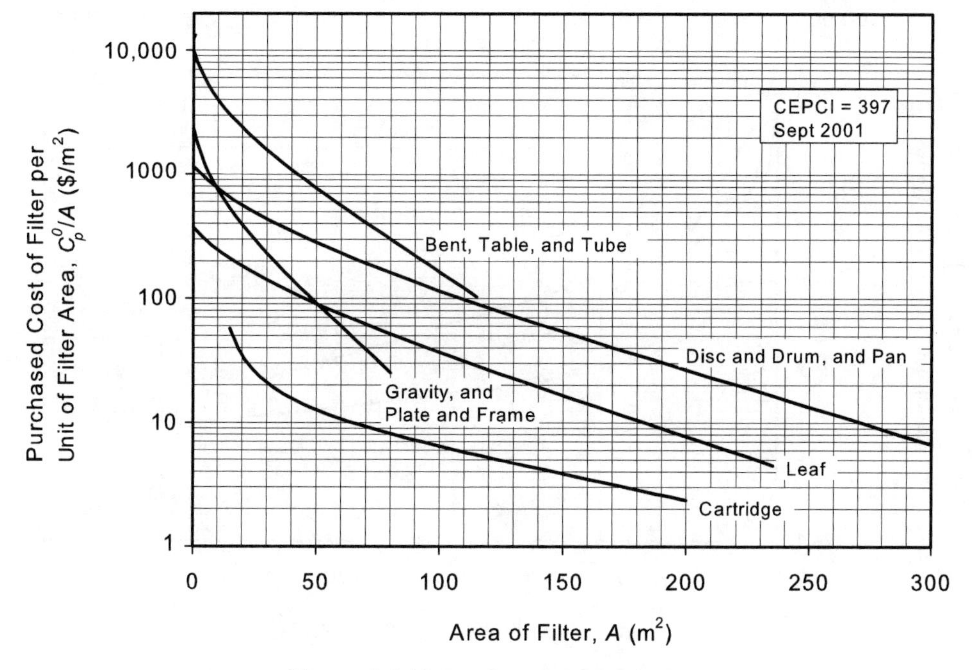

Figure A.14 Purchased Costs of Filters

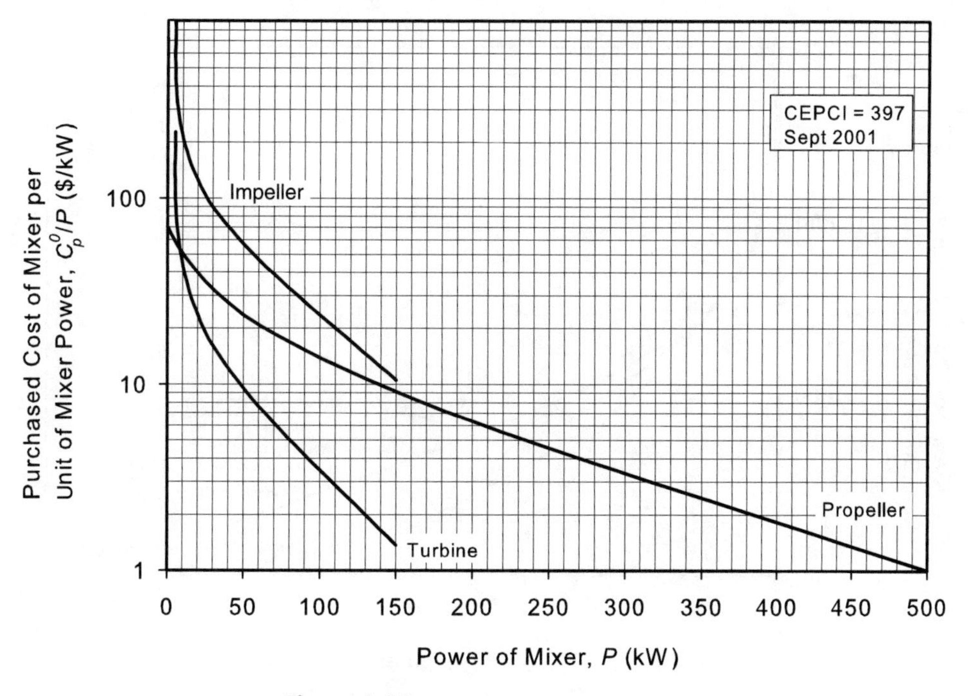

Figure A.15 Purchased Costs of Mixers

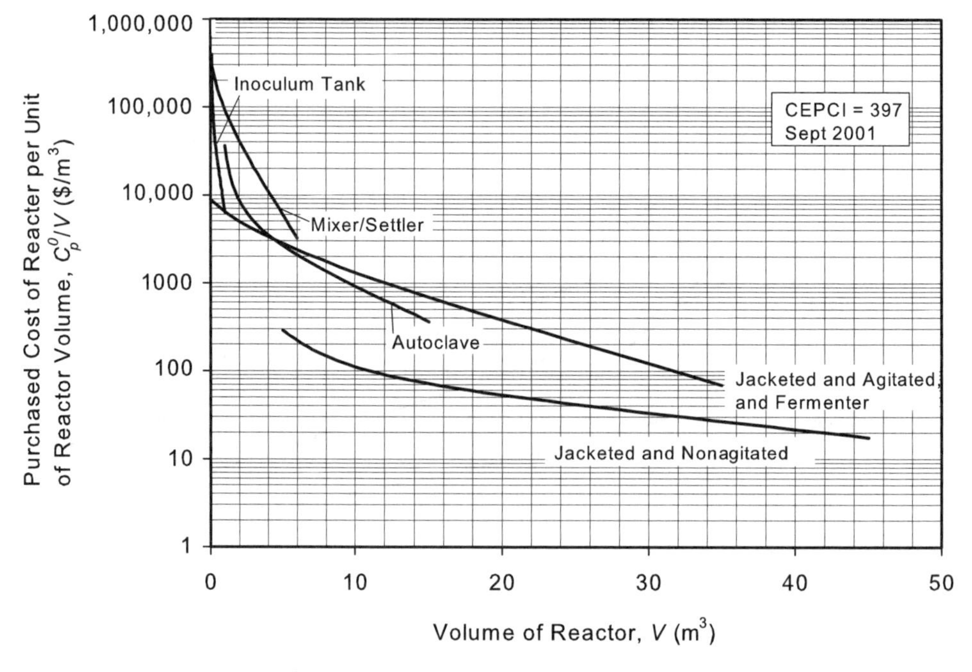

Figure A.16 Purchased Costs of Reactors

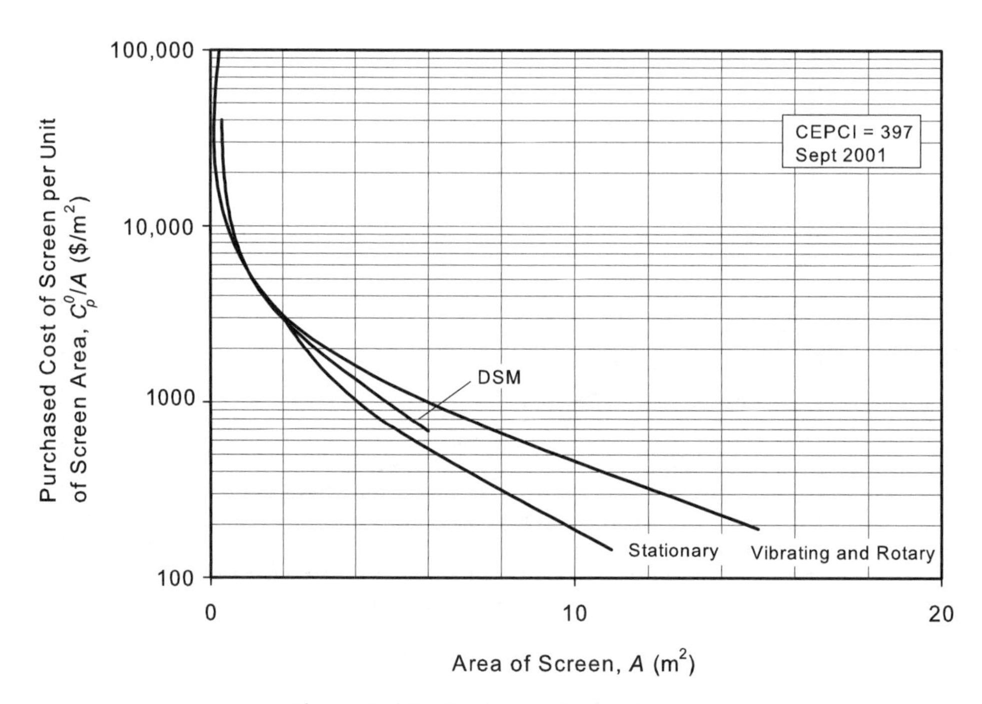

Figure A.17 Purchased Costs of Screens

A.2 PRESSURE FACTORS

As was pointed out in Chapter 7, the costs of equipment increase with increasing operating pressure. In this section, the method of accounting for changes in operating pressure through the use of pressure factors is covered.

A.2.1 Pressure Factors for Process Vessels

The pressure factor for horizontal and vertical process (pressurized) vessels of diameter D meters and operating at a pressure of P barg is based on the ASME code for pressure vessel design [5]. At base material conditions using a maximum allowable stress for carbon steel, S, of 944 bar, a weld efficiency, E, of 0.9, a minimum allowable vessel thickness of 0.0063 m (1/4 inch), and a corrosion allowance, CA, of 0.00315 m (1/8 inch) gives the following expression:

$$F_{P,vessel} = \frac{\dfrac{(P+1)D}{2[850-0.6(P+1)]} + 0.00315}{0.0063} \quad \text{for } t_{vessel} > 0.0063 \text{ m} \tag{A.2}$$

If $F_{P,\ vessel}$ is less than 1 (corresponding to $t_{vessel} < 0.0063$ m), then $F_{P,\ vessel} = 1$. For pressures less than –0.5 barg, $F_{P,\ vessel} = 1.25$. It should be noted that Equation (A.2) is strictly true for the case when the thickness of the vessel wall is less than ¼ D; for vessels in the range D = 0.3 to 4.0 m, this occurs at pressures of approximately 320 barg.

A.2.2 Pressure Factors for Other Process Equipment

The pressure factors, F_P, for the remaining process equipment are given by the following general form:

$$\log_{10} F_P = C_1 + C_2 \log_{10} P + C_3 (\log_{10} P)^2 \tag{A.3}$$

The units of pressure, P, are bar gauge or barg (1 bar = 0.0 barg) unless stated otherwise. The pressure factors are always greater than unity. The values of constants in Equation (A.3) for different equipment are given in Table A.2, and also shown are the ranges of pressures over which the correlations are valid. The values for the constants given in Table A.2 were regressed from data in Guthrie [1, 2] and Ulrich [3]. Extrapolation outside this range of pressures should be done with extreme caution. Some equipment does not have pressure ratings and therefore has values of C_1–C_3 equal to zero. If cost estimates are required for these units at high pressures and the equipment cost is affected by pressure, then the correlations should again be used with caution.

Table A.2 Pressure Factors for Process Equipment (Correlated from Data in Guthrie [1, 2], and Ulrich [3])

Equipment Type	Equipment Description	C_1	C_2	C_3	Pressure Range (barg)
Compressors	Centrifugal, axial, rotary, and reciprocating	0	0	0	–
Drives	Gas turbine	0	0	0	–
	Intern. comb. engine	0	0	0	–
	Steam turbine	0	0	0	–
	Electric—explosion-proof	0	0	0	–
	Electric—totally enclosed	0	0	0	–
	Electric—open/drip-proof	0	0	0	–
Evaporators	Forced circulation (pumped), falling film, agitated film (scraped wall), short tube, and long tube	0	0	0	$P<10$
		0.1578	−0.2992	0.1413	$10<P<150$
Fans*	Centrifugal radial, and centrifugal backward curve	0	0	0	$\Delta P<1$kPa
		0	0.20899	−0.0328	$1<\Delta P<16$kPa
	Axial vane and axial tube	0	0	0	$\Delta P<1$kPa
		0	0.20899	−0.0328	$1<\Delta P<4$kPa
Furnaces	Reformer furnace	0	0	0	$P<10$
		0.1405	−0.2698	0.1293	$10<P<200$
	Pyrolysis furnace	0	0	0	$P<10$
		0.1017	−0.1957	0.09403	$10<P<200$
	Nonreactive fired heater	0	0	0	$P<10$
		0.1347	−0.2368	0.1021	$10<P<200$
Heat exchangers	Scraped wall	0	0	0	$P<40$
		0.6072	−0.9120	0.3327	$40<P<100$
		13.1467	−12.6574	3.0705	$100<P<300$
	Teflon tube	0	0	0	$P<15$

(*continued*)

Table A.2 Pressure Factors for Process Equipment (Correlated from Data in Guthrie [1, 2], and Ulrich [3]) (*Continued*)

Equipment Type	Equipment Description	C_1	C_2	C_3	Pressure Range (barg)
	Bayonet, fixed tube sheet, floating head, kettle reboiler, and U-tube (both shell and tube)	0	0	0	P<5
		0.03881	−0.11272	0.08183	5<P<140
	Bayonet, fixed tube sheet, floating head, kettle reboiler, and U-tube (tube only)	0	0	0	P<5
		−0.00164	−0.00627	0.0123	5<P<140
	Double pipe and multiple pipe	0	0	0	P<40
		0.6072	−0.9120	0.3327	40<P<100
		13.1467	−12.6574	3.0705	100<P<300
	Flat plate and spiral plate	0	0	0	P<19
	Air cooler	0	0	0	P<10
		−0.1250	0.15361	−0.02861	10<P<100
	Spiral tube (both shell and tube)	0	0	0	P<150
		−0.4045	0.1859	0	150<P<400
	Spiral tube (tube only)	0	0	0	P<150
		−0.2115	0.09717	0	150<P<400
Heaters	Diphenyl heater, molten salt heater, and hot water heater	0	0	0	P<2
		−0.01633	0.056875	−0.00876	2<P<200
	Steam boiler	0	0	0	P<20
		2.594072	−4.23476	1.722404	20<P<40
Packing	Loose (for towers)	0	0	0	–
Process vessels	Horizontal and vertical			†	
Pumps	Reciprocating	0	0	0	P<10
		-0.245382	0.259016	−0.01363	10<P<100
	Positive displacement	0	0	0	P<10
		−0.245382	0.259016	−0.01363	10<P<100
	Centrifugal	0	0	0	P<10
		−0.3935	0.3957	−0.00226	10<P<100

(*continued*)

Table A.2 Pressure Factors for Process Equipment (Correlated from Data in Guthrie [1, 2], and Ulrich [3]) *(Continued)*

Equipment Type	Equipment Description	C_1	C_2	C_3	Pressure Range (barg)
Towers	Tray and packed			†	
Tanks	API—fixed roof	0	0	0	P<0.07
	API—floating roof	0	0	0	P<0.07
Trays	Sieve	0	0	0	–
	Valve	0	0	0	–
	Demisters	0	0	0	–
Turbines	Axial gas turbines	0	0	0	–
	Radial gas/liquid expanders	0	0	0	–
Vaporizers	Internal coils / jackets and jacket vessels	0	0	0	P<5
		−0.16742	0.13428	0.15058	5<P<320

*Pressure factors for fans are written in terms of the pressure rise across the fan, ΔP, where ΔP is measured in kPa.
†See Equation (A.2).

A.3 MATERIAL FACTORS AND BARE MODULE FACTORS

As was pointed out in Chapter 7, the costs of equipment change with changes in the material of construction. In this section, the method of accounting for different materials of construction is covered.

A.3.1 Bare Module and Material Factors for Heat Exchangers, Process Vessels, and Pumps

The material factors, F_M, for heat exchangers, process vessels, and pumps are given in Figure A.18, with the appropriate identification number listed in Table A.3. The bare module factors for this equipment are given by the following equation:

$$C_{BM} = C_p^o F_{BM} = C_p^o(B_1 + B_2 F_M F_P) \tag{A.4}$$

The values of the constants B_1 and B_2 are given in Table A.4. The bare module cost for ambient pressure and carbon steel construction, C_{BM}^o, and the bare module factor for the equipment at these conditions, F_{BM}^o, are found by setting F_M and F_P equal to unity. The data given in Tables A.3 and A.4 and Figure A.18 are average values from the following references: Guthrie [1, 2], Ulrich [3], Navarrete [6], Perry et al. [7], and Peters and Timmerhaus [8].

(*text continues on p. 949*)

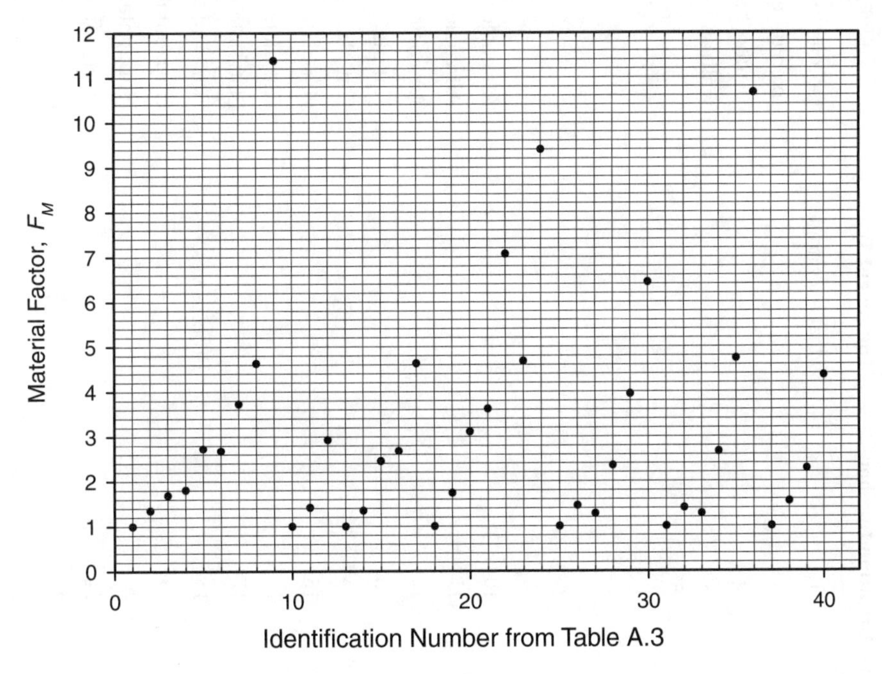

Figure A.18 Material Factors for Equipment in Table A.3 (Averaged Data from References [1, 2, 3, 6, 7, and 8])

Table A.3 Identification Numbers for Material Factors for Heat Exchangers, Process Vessels, and Pumps to Be Used with Figure A.18 (*Continued*)

Identification Number	Equipment Type	Equipment Description	Material of Construction
1	Heat exchanger	Double pipe, multiple pipe,	CS-shell/CS-tube
2		fixed tube sheet, floating head,	CS-shell/Cu-tube
3		U-tube, bayonet, kettle reboiler, scraped	Cu-shell/Cu-tube
4		wall, and spiral tube	CS-shell/SS-tube
5			SS-shell/SS-tube
6			CS-shell/Ni alloy tube
7			Ni alloy, shell/Ni alloy-tube
8			CS-shell/Ti-tube
9			Ti-shell/Ti-tube
10		Air cooler	CS tube
11		Air cooler	Al tube
12		Air cooler	SS tube
13		Flat plate and spiral plate	CS (in contact with fluid)
14		Flat plate and spiral plate	Cu (in contact with fluid)
15		Flat plate and spiral plate	SS (in contact with fluid)
16		Flat plate and spiral plate	Ni alloy (in contact with fluid)
17		Flat plate and spiral plate	Ti (in contact with fluid)
18	Process vessels	Horizontal, vertical (including towers)	CS
19		Horizontal, vertical (including towers)	SS clad
20		Horizontal, vertical (including towers)	SS
21		Horizontal, vertical (including towers)	Ni alloy clad
22		Horizontal, vertical (including towers)	Ni alloy
23		Horizontal, vertical (including towers)	Ti clad
24		Horizontal, vertical (including towers)	Ti

(*continued*)

Table A.3 Identification Numbers for Material Factors for Heat Exchangers, Process Vessels, and Pumps to Be Used with Figure A.18 (*Continued*)

Identification Number	Equipment Type	Equipment Description	Material of Construction
25	Pumps	Reciprocating	Cast iron
26		Reciprocating	Carbon steel
27		Reciprocating	Cu alloy
28		Reciprocating	SS
29		Reciprocating	Ni alloy
30		Reciprocating	Ti
31		Positive displacement	Cast iron
32		Positive displacement	Carbon steel
33		Positive displacement	Cu alloy
34		Positive displacement	SS
35		Positive displacement	Ni alloy
36		Positive displacement	Ti
37		Centrifugal	Cast iron
38		Centrifugal	Carbon steel
39		Centrifugal	SS
40		Centrifugal	Ni alloy

Table A.4 Constants for Bare Module Factor to Be Used in Equation A.4 (Correlated from Data in Guthrie [1, 2] and Ulrich [3])

Equipment Type	Equipment Description	B_1	B_2
Heat exchangers	Double pipe, multiple pipe, scraped wall, and spiral tube	1.74	1.55
	Fixed tube sheet, floating head, U-tube, bayonet, kettle reboiler, and Teflon tube	1.63	1.66
	Air cooler, spiral plate, and flat plate	0.96	1.21
Process vessels	Horizontal	1.49	1.52
	Vertical (including towers)	2.25	1.82
Pumps	Reciprocating	1.89	1.35
	Positive displacement	1.89	1.35
	Centrifugal	1.89	1.35

Table A.5 Equations for Bare Module Cost for Equipment Not Covered by Tables A.3 and A.4

Equipment Type	Equation for Bare Module Cost
Compressors and blowers without drives	$C_{BM} = C_p^o F_{BM}$
Drives for compressors and blowers	$C_{BM} = C_p^o F_{BM}$
Evaporators and vaporizers	$C_{BM} = C_p^o F_{BM} F_P$
Fans with electric drives	$C_{BM} = C_p^o F_{BM} F_P$
Fired heaters and furnaces	$C_{BM} = C_p^o F_{BM} F_P F_T$ F_T is the superheat correction factor for steam boilers ($F_T = 1$ for other heaters and furnaces) and is given by $F_T = 1 + 0.00184\Delta T - 0.00000335(\Delta T)^2$ where ΔT is the amount of superheat in °C.
Power recovery equipment	$C_{BM} = C_p^o F_{BM}$
Sieve trays, valve trays, and demister pads	$C_{BM} = C_p^o N F_{BM} F_q$ Where N is the number of trays and F_q is a quantity factor for trays only given by $\log_{10} F_q = 0.4771 + 0.08516 \log_{10} N - 0.3473 (\log_{10} N)^2$ for $N<20$ $F_q = 1$ for $N \geq 20$
Tower packing	$C_{BM} = C_p^o F_{BM}$

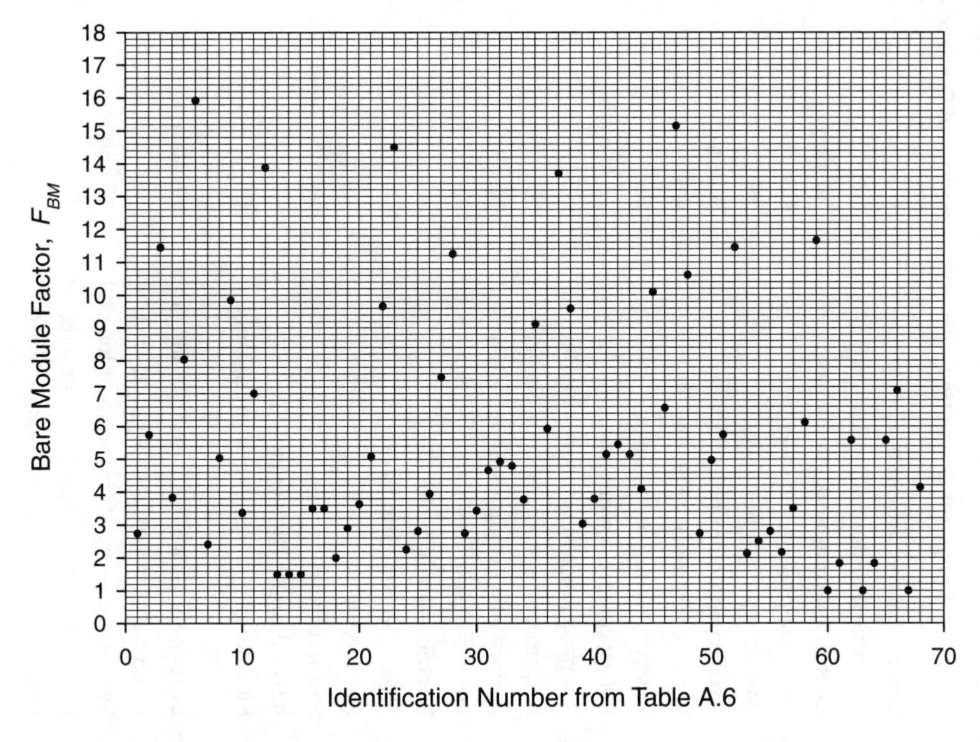

Figure A.19 Bare Module Factors for Equipment in Table A.6 (Average Data from References [1, 2, 3, 6, 7, and 8])

A.3.2 Bare Module and Material Factors for the Remaining Process Equipment

For the remaining equipment, the bare module costs are related to the material and pressure factors by equations different from Equation (A.4). The form of these equations is given in Table A.5. The bare module factors that correspond to the equations in Table A.5 are given in Figure A.19 using the identification numbers listed in Table A.6. Again, the data used to construct Figure A.19 are compiled from average values taken from Guthrie [1, 2], Ulrich [3], Navarrete [6], Perry et al. [7], and Peters and Timmerhaus [8]. In addition, bare module factors for the equipment added to the third edition of the book (conveyors, crystallizers, dryers, dust collectors, filters, mixers, reactors, and screens) are given separately in Table A.7.

Table A.6 Identification of Material Factors for Equipment Listed in Table A.5 to Be Used with Figure A.19

Identification Number	Equipment Type	Equipment Description	Material of Construction
1	Compressors/blowers	Centrifugal compressor or blower	CS
2		Centrifugal compressor or blower	SS
3		Centrifugal compressor or blower	Ni alloy
4		Axial compressor or blower	CS
5		Axial compressor or blower	SS
6		Axial compressor or blower	Ni alloy
7		Rotary compressor or blower	CS
8		Rotary compressor or blower	SS
9		Rotary compressor or blower	Ni alloy
10		Reciprocating compressor or blower	CS
11		Reciprocating compressor or blower	SS
12		Reciprocating compressor or blower	Ni alloy
13	Drives for compressors and blowers	Electric—explosionproof	—
14		Electric—totally enclosed	—
15		Electric—open/dripproof	—
16		Gas turbine	—
17		Steam turbine	—
18		Internal combustion engine	—
19	Evaporators and vaporizers	Evaporator—forced circ, short or long tube	CS
20		Evaporator—forced circ, short or long tube	Cu alloy
21		Evaporator—forced circ, short or long tube	SS
22		Evaporator—forced circ, short or long tube	Ni alloy
23		Evaporator—forced circ, short or long tube	Ti
24		Evaporator—falling film, scraped-wall	CS

(continued)

Table A.6 Identification of Material Factors for Equipment Listed in Table A.5 to Be Used with Figure A.19 (*Continued*)

Identification Number	Equipment Type	Equipment Description	Material of Construction
25		Evaporator—falling film, scraped-wall	Cu alloy
26		Evaporator—falling film, scraped-wall	SS
27		Evaporator—falling film, scraped-wall	Ni alloy
28		Evaporator—falling film, scraped-wall	Ti
29		Vaporizer—jacketed vessel	CS
30		Vaporizer—jacketed vessel	Cu
31		Vaporizer—jacketed vessel	Glass lined/SS coils
32		Vaporizer—jacketed vessel	Glass lined/Ni coils
33		Vaporizer—jacketed vessel	SS
34		Vaporizer—jacketed vessel	SS clad
35		Vaporizer—jacketed vessel	Ni alloy
36		Vaporizer—jacketed vessel	Ni alloy clad
37		Vaporizer—jacketed vessel	Ti
38		Vaporizer—jacketed vessel	Ti clad
39		Vaporizer—jacketed vessel + internal coil	CS
40		Vaporizer—jacketed vessel + internal coil	Cu
41		Vaporizer—jacketed vessel + internal coil	Glass lined/SS coils
42		Vaporizer—jacketed vessel + internal coil	Glass lined/Ni coils
43		Vaporizer—jacketed vessel + internal coil	SS
44		Vaporizer—jacketed vessel + internal coil	SS clad
45		Vaporizer—jacketed vessel + internal coil	Ni alloy
46		Vaporizer—jacketed vessel + internal coil	Ni alloy clad
47		Vaporizer—jacketed vessel + internal coil	Ti
48		Vaporizer—jacketed vessel + internal coil	Ti clad
49	Fans	Fan with electric drive	CS

(*continued*)

Table A.6 Identification of Material Factors for Equipment Listed in Table A.5 to Be Used with Figure A.19 (*Continued*)

Identification Number	Equipment Type	Equipment Description	Material of Construction
50		Fan with electric drive	Fiberglass
51		Fan with electric drive	SS
52		Fan with electric drive	Ni alloy
53	Fired heaters and furnaces	Tube for furnaces and nonreactive process heater	CS
54		Tube for furnaces and nonreactive process heater	Alloy steel
55		Tube for furnaces and nonreactive process heater	SS
56		Thermal fluid heater—hot water, molten salt, or diphenyl-based oil	—
57	Power recovery equipment	Turbines	CS
58		Turbines	SS
59		Turbines	Ni alloy
60	Trays and demister pads	Sieve and valve trays	CS
61		Sieve and valve trays	SS
62		Sieve and valve trays	Ni alloy
63		Demister pad	SS
64		Demister pad	Fluorocarbon
65		Demister pad	Ni alloy
66	Tower packing	Packing	Metal (304SS)
67		Packing	Polyethylene
68		Packing	Ceramic

Figure A.7 Bare Module Factors for Conveyors, Crystallizers, Dryers, Dust Collectors, Filters, Mixers, Reactors, and Screens

Equipment Type	Equipment Description	Bare Module Factor, FBM
Blenders	Kneader	1.12*
	Ribbon	1.12*
	Rotary	1.12
Centrifuges	Auto batch separator	1.57*
	Centrifugal separator	1.57
	Oscillating screen	1.57*
	Solid bowl w/o motor	1.27
Conveyors	Apron	1.20
	Belt	1.25
	Pneumatic	1.25*
	Screw	1.10
Crystallizers	Batch	1.60
Dryers	Drum	1.60
	Rotary, gas fired	1.25
	Tray	1.25
Dust Collectors	Baghouse	2.86*
	Cyclone scrubbers	2.86*
	Electrostatic precipitator	2.86*
	Venturi scrubber	2.86*
Filters	Bent	1.65*
	Cartridge	1.65*
	Disc and drum	1.65*
	Gravity	1.65*
	Leaf	1.65
	Pan	1.65*
	Plate and frame	1.80
	Table	1.65*
	Tube	1.65*
Mixers	Impeller	1.38*
	Propeller	1.38
	Turbine	1.38
Reactors	Autoclave	4.0*
	Fermenter	4.0*
	Inoculum tank	4.0*
	Jacketed agitated	4.0*
	Jacketed nonagitated	4.0*
	Mixer/settler	4.0*
Screens	DSM	1.34*
	Rotary	1.34*
	Stationary	1.34*
	Vibrating	1.34

When possible, bare module factors are taken to be equal to the Field Installation Factors from Guthrie [2]. Items marked * are estimates.

REFERENCES

1. Guthrie, K. M., "Data and Techniques for Preliminary Capital Cost Estimating," *Chem. Eng.*, March 24, 1969, 114–142.
2. Guthrie, K. M., *Process Plant Estimating Evaluation and Control* (Solana Beach, CA: Craftsman Book Co., 1974).
3. Ulrich, G. D., *A Guide to Chemical Engineering Process Design and Economics* (New York: Wiley, 1984).
4. R-Books software marketed by Richardson Engineering Services, Inc., January 2001.
5. Section VIII, *ASME Boiler and Pressure Vessel Code,* ASME Boiler and Pressure Vessel Committee, ASME, New York, 2000.
6. Navarrete, P. F., *Planning, Estimating, and Control of Chemical Construction Projects* (New York: Marcel Dekker, Inc., 1995).
7. Perry, R. H., D. W. Green, and J. O. Maloney, eds., *Perry's Chemical Engineering Handbook,* 7th ed. (New York: McGraw-Hill, 1997).
8. Peters, M. S., and K. D. Timmerhaus, *Plant Design and Economics for Chemical Engineers,* 4th ed. (New York: McGraw-Hill, 1991).

APPENDIX

B

Information for the Preliminary Design of Eleven Chemical Processes

The purpose of the process designs contained in this appendix is to provide the reader with a preliminary description of several common chemical processes. The designs provided are the result of preliminary simulation using the CHEMCAD process simulation software and often contain simplifying assumptions such as ideal column behavior (shortcut method using the Underwood Gilliland method) and in some cases the use of ideal thermodynamics models (k-value = ideal gas, enthalpy = ideal). These designs are used throughout the book in the end-of-chapter problems and provide a starting point for detailed design. The authors recognize that there are additional complicating factors, such as non-ideal phase equilibrium behavior (such as azeotrope formation and phase separation), feed stream impurities, different catalyst selectivity, side reaction formation, and so on. The presence of any one of these factors may give rise to significant changes from the preliminary designs shown here. Thus, the student, if asked to perform a detailed process design of these (or other) processes, should take the current designs as only a starting point and should be prepared to do further research into the process to ensure that a more accurate and deeper understanding of the factors involved is obtained.

Following is a list of the sections and projects discussed in this appendix.

B.1 DIMETHYL ETHER (DME) PRODUCTION, UNIT 200

DME is used primarily as an aerosol propellant. It is miscible with most organic solvents, has a high solubility in water, and is completely miscible in water and 6% ethanol [1]. Recently, the use of DME as a fuel additive for diesel engines has been investigated due to its high volatility (desirable for cold starting) and high cetane number. The production of DME is via the catalytic dehydration of methanol over an acid zeolite catalyst. The main reaction is

$$\underset{methanol}{2CH_3OH} \rightarrow \underset{DME}{(CH_3)_2O} + H_2O \tag{B.1.1}$$

In the temperature range of normal operation, there are no significant side reactions.

B.1.1 Process Description

A preliminary process flow diagram for a DME process is shown in Figure B.1.1, in which 50,000 metric tons per year of 99.5 wt% purity DME product is produced. Due to the simplicity of the process, an operating factor greater than 0.95 (8,375 h/yr) is used.

Fresh methanol, Stream 1, is combined with recycled reactant, Stream 14, and vaporized prior to being sent to a fixed-bed reactor operating between 250°C and 370°C. The single-pass conversion of methanol in the reactor is 80%. The reactor effluent, Stream 7, is then cooled prior to being sent to the first of two distillation columns: T-201 and T-202. DME product is taken overhead from the first column. The second column separates the water from the unused methanol. The methanol is recycled back to the front end of the process, and the water is sent to waste water treatment to remove trace amounts of organic compounds.

Stream summaries, utility summaries, and equipment summaries are presented in Tables B.1.1–B.1.3.

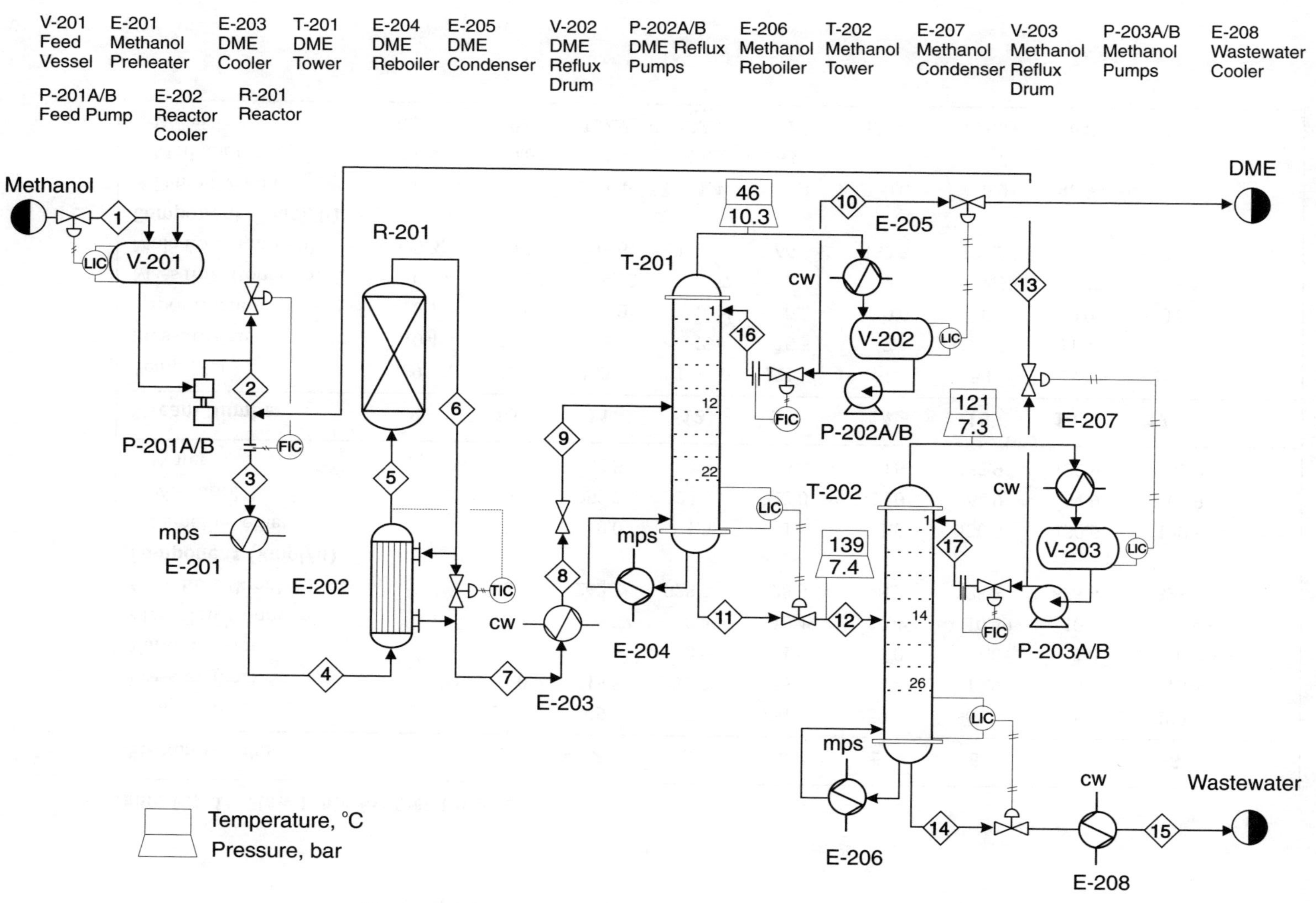

Figure B.1.1 Unit 200: Dimethyl Ether Process Flow Diagram

Table B.1.1 Flow Table for DME Unit 200

Stream Number	1	2	3	4	5	6	7	8
Temp (°C)	25	25	45	154	250	364	278	100
Pressure (bar)	1.0	15.5	15.2	15.1	14.7	13.9	13.8	13.4
Vapor fraction	0.0	0.0	0.0	1.0	1.0	1.0	1.0	0.0798
Mass flow (tonne/h)	8.37	8.37	10.49	10.49	10.49	10.49	10.49	10.49
Mole flow (kmol/h)	262.2	262.2	328.3	328.3	328.3	328.3	328.3	328.3
Component (kmol/h)								
Dimethyl ether	0.0	0.0	1.5	1.5	1.5	130.5	130.5	130.5
Methanol	259.7	259.7	323.0	323.0	323.0	64.9	64.9	64.9
Water	2.5	2.5	3.8	3.8	3.8	132.9	132.9	132.9

Stream Number	9	10	11	12	13	14	15	16	17
Temp (°C)	89	46	153	139	121	167	50	46	121
Pressure (bar)	10.4	11.4	10.5	7.4	15.5	7.6	1.2	11.4	7.3
Vapor fraction	0.148	0.0	0.0	0.04	0.0	0.0	0.0	0.0	0.0
Mass flow (tonne/h)	10.49	5.97	4.52	4.52	2.13	2.39	2.39	2.17	3.62
Mole flow (kmol/h)	328.3	129.7	198.6	198.6	66.3	132.3	132.3	47.1	113.0
Component (kmol/h)									
Dimethyl ether	130.5	129.1	1.4	1.4	1.4	0.0	0.0	46.9	2.4
Methanol	64.9	0.6	64.3	64.3	63.6	0.7	0.7	0.2	108.4
Water	132.9	0.0	132.9	132.9	1.3	131.6	131.6	0.0	2.2

Table B.1.2 Utility Summary Table for Unit 200

E-201	E-203	E-204	E-205	E-206	E-207	E-208
mps 7220 kg/h	cw 297,100 kg/h	mps 1250 kg/h	cw 75,120 kg/h	mps 2900 kg/h	cw 142,600 kg/h	cw 28,700 kg/h

Table B.1.3 Major Equipment Summary for Unit 200

Heat Exchangers

E-201
$A = 99.4\ m^2$
Floating head, carbon steel, shell-and-tube design
Process stream in tubes
$Q = 14,400\ MJ/h$
Maximum pressure rating of 15 bar

E-202
$A = 171.0\ m^2$
Floating head, carbon steel, shell-and-tube design
Process stream in tubes and shell
$Q = 2,030\ MJ/h$
Maximum pressure rating of 15 bar

E-203
$A = 101.8\ m^2$
Floating head, carbon steel, shell-and-tube design
Process stream in shell
$Q = 12,420\ MJ/h$
Maximum pressure rating of 14 bar

E-204
$A = 22.0\ m^2$
Floating head, carbon steel, shell-and-tube design
Process stream in shell
$Q = 2,490\ MJ/h$
Maximum pressure rating of 11 bar

E-205
$A = 100.6\ m^2$
Fixed head, carbon steel, shell-and-tube design
Process stream in shell
$Q = 3,140\ MJ/h$
Maximum pressure rating of 10 bar

E-206
$A = 83.0\ m^2$
Floating head, carbon steel, shell-and-tube design
Process stream in shell
$Q = 5,790\ MJ/h$
Maximum pressure rating of 11 bar

E-207
$A = 22.7 m^2$
Floating head, carbon steel, shell-and-tube design
Process stream in shell
$Q = 5,960\ MJ/h$
Maximum pressure rating of 7 bar

E-208
$A = 22.8\ m^2$
Floating head, carbon steel, shell-and-tube design
Process stream in shell
$Q = 1,200\ MJ/h$
Maximum pressure rating of 8 bar

Pumps

P-201 A/B
Reciprocating/electric drive
Carbon steel
Power = 7.2 kW (actual)
60% efficient
Pressure out = 15.5 bar

P-202 A/B
Centrifugal/electric drive
Carbon steel
Power = 1.0 kW (actual)
40% efficient
Pressure out = 11.4 bar

P-202 A/B
Centrifugal/electric drive
Carbon steel
Power = 5.2 kW (actual)
40% efficient
Pressure out = 16 bar

(continued)

Table B.1.3 Major Equipment Summary for Unit 200 (*Continued*)

Reactor **R-201** Carbon steel Packed-bed section 7.2 m high filled with catalyst Diameter = 0.72 m Height = 10 m Maximum pressure rating of 14.7 bar	
Towers **T-201** Carbon steel 22 SS sieve trays plus reboiler and condenser 24-in tray spacing Column height = 15.8 m Diameter = 0.79 m Maximum pressure rating of 10.6 bar	**T-202** Carbon steel 26 SS sieve trays plus reboiler and condenser 18-in tray spacing Column height = 14.9 m Diameter = 0.87 m Maximum pressure rating of 7.3 bar
Vessels **V-201** Horizontal Carbon steel Length = 3.42 m Diameter = 1.14 m Maximum pressure rating of 1.1 bar **V-202** Horizontal Carbon steel Length = 2.89 m Diameter = 0.98 m Maximum pressure rating of 10.3 bar	**V-203** Horizontal Carbon steel Length = 2.53 m Diameter = 0.85 m Maximum pressure rating of 7.3 bar

B.1.2 Reaction Kinetics

The reaction taking place is mildly exothermic with a standard heat of reaction, $\Delta H_{reac}(25°C) = -11{,}770$ kJ/kmol. The equilibrium constant for this reaction at three different temperatures is given below:

T	K_p
473 K (200°C)	92.6
573 K (300°C)	52.0
673 K (400°C)	34.7

The corresponding equilibrium conversions for pure methanol feed over the above temperature range is greater than 99%. Thus this reaction is kinetically controlled at the conditions used in this process.

The reaction takes place on an amorphous alumina catalyst treated with 10.2% silica. There are no significant side reactions at less than 400°C. At greater than 250°C the rate equation is given by Bondiera and Naccache [2] as:

$$-r_{methanol} = k_0 \exp\left[\frac{-E_0}{RT}\right] p_{methanol} \tag{B.1.2}$$

where $k_0 = 1.21 \times 10^6$ kmol/(m^{3}cat.h.kPa), $E_0 = 80.48$ kJ/mol, and $p_{methanol}$ = partial pressure of methanol (kPa)

Significant catalyst deactivation occurs at temperatures greater than 400°C, and the reactor should be designed so that this temperature is not exceeded anywhere in the reactor. The design given in Figure B.1.1 uses a single packed bed of catalyst, which operates adiabatically. The temperature exotherm occurring in the reactor of 118°C is probably on the high side and gives an exit temperature of 368°C. However, the single-pass conversion is quite high (80%), and the low reactant concentration at the exit of the reactor tends to limit the possibility of a runaway.

In practice the catalyst bed might be split into two sections, with an intercooler between the two beds. This has an overall effect of increasing the volume (and cost) of the reactor and should be investigated if catalyst damage is expected at temperatures lower than 400°C. In-reactor cooling (shell-and-tube design) and cold quenching by splitting the feed and feeding at different points in the reactor could also be investigated as viable alternative reactor configurations.

B.1.3 Simulation (CHEMCAD) Hints

The DME-water binary system exhibits two liquid phases when the DME concentration is in the 34% to 93% range [2]. However, upon addition of 7% or more alcohol, the mixture becomes completely miscible over the complete range of DME concentration. In order to ensure that this non-ideal behavior is simulated correctly, it is recommended that binary vapor-liquid equilibrium (VLE) data for the three pairs of components be used in order to regress binary interaction parameters (BIPs) for a UNIQUAC/UNIFAC thermodynamics model. If VLE data for the binary pairs are not used, then UNIFAC can be used to estimate BIPs, but these should be used only as preliminary estimates. As with all non-ideal systems, there is no substitute for designing separation equipment using data regressed from actual (experimental) VLE.

B.1.4 References

1. "DuPont Talks about Its DME Propellant," *Aerosol Age* (May and June 1982).
2. Bondiera, J., and C. Naccache, "Kinetics of Methanol Dehydration in Dealuminated H-mordenite: Model with Acid and Basic Active Centres," *Applied Catalysis* 69 (1991): 139–148.

B.2 ETHYLBENZENE PRODUCTION, UNIT 300

The majority of ethylbenzene (EB) processes produce EB for internal consumption within a coupled process that produces styrene monomer. The facility described here produces 80,000 tonne/yr of 99.8 mol% ethylbenzene that is totally consumed by an on-site styrene facility. As with most EB/styrene facilities, there is significant heat integration between the two plants. In order to decouple the operation of the two plants, the energy integration is achieved by the generation and consumption of steam within the two processes. The EB reaction is exothermic, so steam is produced, and the styrene reaction is endothermic, so energy is transferred in the form of steam.

B.2.1 Process Description [1, 2]

The PFD for the EB process is shown in Figure B.2.1. A refinery cut of benzene is fed from storage to an on-site process vessel (V-301), where it is mixed with the recycled benzene. From V-301, it is pumped to a reaction pressure of approximately 2,000 kPa (20 atm) and sent to a fired heater (H-301) to bring it to reaction temperature (approximately 400°C). The preheated benzene is mixed with feed ethylene just prior to entering the first stage of a reactor system consisting of three adiabatic packed-bed reactors (R-301 to R-303), with interstage feed addition and cooling. Reaction occurs in the gas phase and is exothermic. The hot, partially converted reactor effluent leaves the first packed bed, is mixed with more feed ethylene, and is fed to E-301, where the stream is cooled to 380°C prior to passing to the second reactor (R-302), where further reaction takes place. High-pressure steam is produced in E-301, and this steam is subsequently used in the styrene unit. The effluent stream from R-302 is similarly mixed with feed ethylene and is cooled in E-302 (with generation of high-pressure steam) prior to entering the third and final packed-bed reactor, R-303. The effluent stream leaving the reactor contains products, by-products, unreacted benzene, and small amounts of unreacted ethylene and other noncondensable gases. The reactor effluent is cooled in two waste-heat boilers (E-303 and E-304), in which high-pressure and low-pressure steam, respectively, is generated. This steam is also consumed in the styrene unit. The two-phase mixture leaving E-304 is sent to a trim cooler (E-305), where the stream is cooled to 80°C, and then to a two-phase separator (V-302), where the light gases are separated and, because of the high ethylene conversion, are sent overhead as fuel gas to be consumed in the fired heater. The condensed liquid is then sent to the benzene tower, T-301, where the unreacted benzene is separated as the overhead product and returned to the front end of the process. The bottoms product from the first column is sent to T-302, where product EB (at 99.8 mol% and containing less than 2 ppm diethylbenzene [DEB]) is taken as the top product and is sent directly to the styrene unit. The bottoms product from T-302 contains all the DEB and trace amounts of higher ethylbenzenes. This stream is mixed with recycle benzene and passes through the fired heater (H-301) prior to being sent to a fourth packed-bed reactor (R-304), in which the excess benzene is reacted with the DEB to produce EB and unreacted benzene. The effluent from this reactor is mixed with the liquid stream entering the waste-heat boiler (E-303).

Stream summary tables, utility summary tables, and major equipment specifications are shown in Tables B.2.1–B.2.3.

B.2.2 Reaction Kinetics

The production of EB takes place via the direct addition reaction between ethylene and benzene:

$$\underset{benzene}{C_6H_6} + \underset{ethylene}{C_2H_4} \rightarrow \underset{ethylbenzene}{C_6H_5C_2H_5} \tag{B.2.1}$$

The reaction between EB and ethylene to produce DEB also takes place:

$$\underset{ethylbenzene}{C_6H_5C_2H_5} + \underset{ethylene}{C_2H_4} \rightarrow \underset{diethylbenzene}{C_6H_4(C_2H_5)_2} \tag{B.2.2}$$

Additional reactions between DEB and ethylene yielding triethylbenzene (and higher) are also possible. However, in order to minimize these additional reactions, the molar ratio of benzene to ethylene is kept high, at approximately 8:1. The production of DEB is undesirable, and its value as a side product is low. In addition, even small amounts of DEB in EB cause significant processing problems in the downstream styrene process. Therefore, the maximum amount of DEB in EB is specified as 2 ppm. In order to maximize the production of the desired EB, the DEB is separated and returned to a separate reactor in which excess benzene is added to produce EB via the following equilibrium reaction:

$$\underset{diethylbenzene}{C_6H_4(C_2H_5)_2} + \underset{benzene}{C_6H_6} \rightleftharpoons \underset{ethylbenzene}{2C_6H_5C_2H_5} \tag{B.2.3}$$

The incoming benzene contains a small amount of toluene impurity. The toluene reacts with ethylene to form ethyl benzene and propylene:

$$\underset{toluene}{C_6H_5CH_3} + 2C_2H_4 \rightarrow \underset{ethylbenzene}{C_6H_5C_2H_5} + \underset{propylene}{C_3H_6} \tag{B.2.4}$$

The reaction kinetics derived for a new catalyst are given as

$$-r_i = k_{o,i}e^{-Ei/RT}C^a_{ethylene}C^b_{EB}C^c_{toluene}C^d_{benzene}C^e_{DEB} \tag{B.2.5}$$

where i is the reaction number above (B.2.i), and the following relationships pertain.

i	E_i kcal/kmol	$k_{o,i}$	a	b	c	d	e
1	22,500	1.00×10^6	1	0	0	1	0
2	22,500	6.00×10^5	1	1	0	0	0
3	25,000	7.80×10^6	0	0	0	1	1
4	20,000	3.80×10^8	2	0	1	0	0

The units of r_i are kmol/s/m^3-reactor, the units of C_i are kmol/m^3-gas, and the units of $k_{o,i}$ vary depending upon the form of the equation.

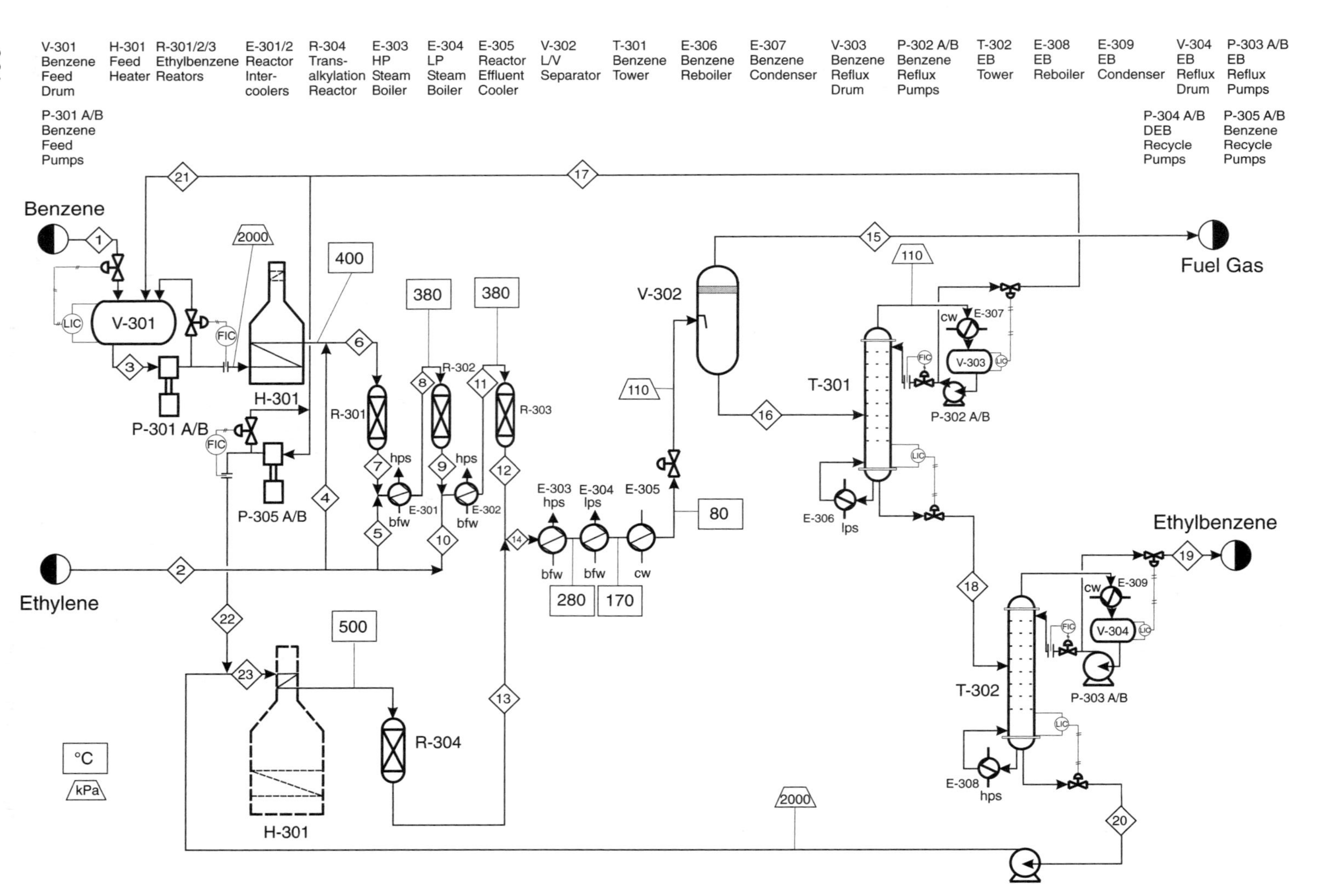

Figure B.2.1 Unit 300: Ethylbenzene Process Flow Diagram

Table B.2.1 Stream Table for Unit 300

Stream Number	1	2	3	4	5	6
Temp (°C)	25.0	25.0	58.5	25.0	25.0	383.3
Pres kPa	110.0	2000.0	110.0	2000.0	2000.0	1985.0
Vapor mole fraction	0.0	1.0	0.0	1.0	1.0	1.0
Total kmol/h	99.0	100.0	229.2	30.0	35.0	259.2
Total kg/h	7761.3	2819.5	17952.2	845.9	986.8	18797.9
Flowrates in kmol/h						
Ethylene	0.00	93.00	0.00	27.90	32.55	27.90
Ethane	0.00	7.00	0.00	2.10	2.45	2.10
Propylene	0.00	0.00	0.00	0.00	0.00	0.00
Benzene	97.00	0.00	226.51	0.00	0.00	226.51
Toluene	2.00	0.00	2.00	0.00	0.00	2.00
Ethylbenzene	0.00	0.00	0.70	0.00	0.00	0.70
1,4-DiEthBenzene	0.00	0.00	0.00	0.00	0.00	0.00

Stream Number	7	8	9	10	11	12
Temp (°C)	444.1	380.0	453.4	25.0	380.0	449.2
Pres kPa	1970.0	1960.0	1945.0	2000.0	1935.0	1920.0
Vapor mole fraction	1.0	1.0	1.0	1.0	1.0	1.0
Total kmol/h	234.0	269.0	236.4	35.0	271.4	238.7
Total kg/h	18797.9	19784.7	19784.7	986.8	20771.5	20771.5
Flowrates in kmol/h						
Ethylene	0.85	33.40	0.62	32.55	33.17	0.54
Ethane	2.10	4.55	4.55	2.45	7.00	7.00
Propylene	1.83	1.81	2.00	0.00	2.00	2.00
Benzene	203.91	203.91	174.96	0.00	174.96	148.34
Toluene	0.19	0.19	0.0026	0.00	0.0026	0.00
Ethylbenzene	24.28	24.28	49.95	0.00	49.95	70.57
1,4-DiEthBenzene	0.87	0.87	4.29	0.00	4.29	10.30

(continued)

Table B.2.1 Stream Table for Unit 300 (*Continued*)

Stream Number	13	14	15	16	17	18
Temp (°C)	497.9	458.1	73.6	73.6	81.4	145.4
Pres kPa	1988.0	1920.0	110.0	110.0	105.0	120.0
Vapor mole fraction	1.0	1.0	1.0	0.0	0.0	0.0
Total kmol/h	51.3	290.0	18.6	271.4	170.2	101.1
Total kg/h	4616.5	25387.9	1042.0	24345.9	13321.5	11024.5
Flowrates in kmol/h						
Ethylene	0.00	0.54	0.54	0.00	0.00	0.00
Ethane	0.00	7.00	7.00	0.00	0.00	0.00
Propylene	0.00	2.00	2.00	0.00	0.00	0.00
Benzene	29.50	177.85	8.38	169.46	169.23	0.17
Toluene	0.00	0.00	0.00	0.00	0.00	0.00
Ethylbenzene	21.69	92.25	0.71	91.54	0.92	90.63
1,4-DiEthBenzene	0.071	10.37	0.013	10.35	0.00	10.35

Stream Number	19	20	21	22	23
Temp (°C)	139.0	191.1	82.6	82.6	121.4
Pres kPa	110.0	140.0	2000.0	2000.0	2000.0
Vapor mole fraction	0.0	0.0	0.0	0.0	0.0
Total kmol/h	89.9	11.3	130.2	40.0	51.3
Total kg/h	9538.6	1485.9	10190.9	3130.6	4616.5
Flowrates in kmol/h					
Ethylene	0.00	0.00	0.00	0.00	0.00
Ethane	0.00	0.00	0.00	0.00	0.00
Propylene	0.00	0.00	0.00	0.00	0.00
Benzene	0.17	0.00	129.51	39.78	39.78
Toluene	0.00	0.00	0.00	0.00	0.00
Ethylbenzene	89.72	0.91	0.70	0.22	1.12
1,4-DiEthBenzene	0.0001	10.35	0.00	0.00	10.35

Table B.2.2 Utility Summary for Unit 300

Stream Name	bfw to E-301	bfw to E-302	bfw to E-303	bfw to E-304	cw to E-305
Flowrate (kg/h)	851	1,121	4,341	5,424	118,300
Stream Name	**lps to E-306**	**cw to tE-307**	**hps to E-308***	**cw to E-309**	
Flowrate (kg/h)	4,362	174,100	3,124	125,900	

*Throttled and desuperheated at exchanger

Table B.2.3 Major Equipment Summary for Unit 300

Heat Exchangers

E-301
$A = 62.6\ m^2$
1-2 exchanger, floating head, stainless steel
Process stream in tubes
$Q = 1,967$ MJ/h
Maximum pressure rating of 2,200 kPa

E-302
$A = 80.1\ m^2$
1-2 exchanger, floating head, stainless steel
Process stream in tubes
$Q = 2,592$ MJ/h
Maximum pressure rating of 2,200 kPa

E-303
$A = 546\ m^2$
1-2 exchanger, floating head, stainless steel
Process stream in tubes
$Q = 10,080$ MJ/h
Maximum pressure rating of 2,200 kPa

E-304
$A = 1,567\ m^2$
1-2 exchanger, fixed head, carbon steel
Process stream in tubes
$Q = 12,367$ MJ/h
Maximum pressure rating of 2,200 kPa

E-305
$A = 348\ m^2$
1-2 exchanger, floating head, carbon steel
Process stream in shell
$Q = 4,943$ MJ/h
Maximum pressure rating of 2,200 kPa

E-306
$A = 57.8\ m^2$
1-2 exchanger, fixed head, carbon steel
Process stream in shell
$Q = 9,109$ MJ/h
Maximum pressure rating of 2,200 kPa

E-307
$A = 54.6\ m^2$
1-2 exchanger, floating head, carbon steel
Process stream in shell
$Q = 7,276$ MJ/h
Maximum pressure rating of 200 kPa

E-308
$A = 22.6\ m^2$
1-2 exchanger, fixed head, carbon steel
Process stream in shell
$Q = 5,281$ MJ/h
Maximum pressure rating of 200 kPa

(continued)

Table B.2.3 Major Equipment Summary for Unit 300 (*Continued*)

Heat Exchangers (*Continued*)

E-309
$A = 17.5\ m^2$
1-2 exchanger, floating head, carbon steel
Process stream in shell
$Q = 5{,}262$ MJ/h
Maximum pressure rating of 200 kPa

Fired Heater

H-301
Required heat load = 22,376 MJ/h
Design (maximum) heat load = 35,000 MJ/h
Tubes = Stainless steel
75% thermal efficiency
Maximum pressure rating of 2,200 kPa

Pumps

P-301 A/B
Positive displacement/electric drive
Carbon steel
Actual power = 15 kW
Efficiency 75%

P-302 A/B
Centrifugal/electric drive
Carbon steel
Actual power = 1 kW
Efficiency 75%

P-303 A/B
Centrifugal/electric drive
Carbon steel
Actual power = 1 kW
Efficiency 75%

P-304 A/B
Centrifugal/electric drive
Carbon steel
Actual power = 1.4 kW
Efficiency 80%

P-305 A/B
Positive displacement/electric drive
Carbon steel
Actual power = 2.7 kW
Efficiency 75%

Reactors

R-301
316 stainless steel packed bed, ZSM-5 molecular sieve catalyst
$V = 20\ m^3$
11 m long, 1.72 m diameter
Maximum pressure rating of 2,200 kPa
Maximum allowable catalyst temperature = 500°C

R-302
316 stainless steel packed bed, ZSM-5 molecular sieve catalyst
$V = 25\ m^3$
12 m long, 1.85 m diameter
Maximum pressure rating of 2,200 kPa
Maximum allowable catalyst temperature = 500°C

R-303
316 stainless steel packed bed, ZSM-5 molecular sieve catalyst
$V = 30\ m^3$
12 m long, 1.97 m diameter
Maximum pressure rating of 2,200 kPa
Maximum allowable catalyst temperature = 500°C

R-304
316 stainless steel packed bed, ZSM-5 molecular sieve catalyst
$V = 1.67\ m^3$
5 m long, 0.95 m diameter
Maximum pressure rating of 2,200 kPa
MMximum allowable catalyst temperature 525°C

(*continued*)

Table B.2.3 Major Equipment Summary for Unit 300 (*Continued*)

Towers	
T-301 Carbon steel 45 SS sieve trays plus reboiler and total condenser 42% efficient trays Feed on tray 19 Additional feed ports on trays 14 and 24 Reflux ratio = 0.3874 24-in tray spacing Column height = 27.45 m Diameter = 1.7 m Maximum pressure rating of 300 kPa	**T-302** Carbon steel 76 SS sieve trays plus reboiler and total condenser 45% efficient trays Feed on tray 56 Additional feed ports on trays 50 and 62 Reflux ratio = 0.6608 15-in tray spacing Column height = 28.96 m Diameter = 1.5 m Maximum pressure rating of 300 kPa
Vessels	
V-301 Carbon steel Horizontal L/D = 5 V = 7 m^3 Maximum operating pressure = 250 kPa	**V-303** Carbon steel Horizontal L/D = 3 V = 7.7 m^3 Maximum operating pressure = 300 kPa
V-302 Carbon steel with SS demister Vertical L/D = 3 V = 10 m^3 Maximum operating pressure = 250 kPa	**V-304** Carbon steel Horizontal L/D = 3 V = 6.2 m^3 Maximum operating pressure = 300 kPa

B.2.3 Simulation (CHEMCAD) Hints

A CHEMCAD simulation is the basis for the design. The thermodynamics models used were K-val = UNIFAC and Enthalpy = Latent Heat.

It should be noted that in the simulation a component separator was placed after the high-pressure flash drum (V-302) in order to remove noncondensables from Stream 16 prior to entering T-301. This is done in order to avoid problems in simulating this tower. In practice, the noncondensables would be removed from the overhead reflux drum, V-303, after entering T-301.

As a first approach, both towers were simulated as Shortcut columns in the main simulation, but subsequently each was simulated separately using the rigorous TOWER module. Once the rigorous TOWER simulations were completed, they were substituted back into the main flowsheet and the simulation was run again to converge. A similar approach is recommended. The rigorous TOWER module provides accurate design and simulation data and should be used to assess column operation, but using the shortcut simulations in the initial trials speeds up overall conversion of the flowsheet.

B.2.4 References

1. William J. Cannella, "Xylenes and Ethylbenzene," *Kirk-Othmer Encyclopedia of Chemical Technology*, on-line version (New York: John Wiley & Sons, 2006).
2. "Ethylbenzene," *Encyclopedia of Chemical Processing and Design*, vol. 20, J. J. McKetta, ed. (New York: Marcel Dekker, Inc., 1984), 77–88.

B.3 STYRENE PRODUCTION, UNIT 400

Styrene is the monomer used to make polystyrene, which has a multitude of uses, the most common of which are in packaging and insulated Styrofoam beverage cups. Styrene is produced by the dehydrogenation of ethylbenzene. Ethylbenzene is formed by reacting ethylene and benzene. There is very little ethylbenzene sold commercially, because most ethylbenzene manufacturers convert it directly into styrene.

B.3.1 Process Description [1, 2]

The process flow diagram is shown in Figure B.3.1. Ethylbenzene feed is mixed with recycled ethylbenzene, heated, and then mixed with high-temperature, superheated steam. Steam is an inert in the reaction, which drives the equilibrium shown in Equation (B.3.1) to the right by reducing the concentrations of all components. Because styrene formation is highly endothermic, the superheated steam also provides energy to drive the reaction. Decomposition of ethylbenzene to benzene and ethylene, and hydrodealkylation to give methane and toluene, are unwanted side reactions shown in Equations (B.3.2) and (B.3.3). The reactants then enter two adiabatic packed beds with interheating. The products are cooled, producing steam from the high-temperature reactor effluent. The cooled product stream is sent to a three-phase separator, in which light gases (hydrogen, methane, ethylene), organic liquid, and water exit in separate streams. The hydrogen stream is further purified as a source of hydrogen elsewhere in the plant. The benzene/toluene stream is currently returned as a feed stream to the petrochemical facility. The organic stream containing the desired product is distilled once to remove the benzene and toluene and distilled again to separate unreacted ethylbenzene for recycle from the styrene product.

$$\underset{ethylbenzene}{C_6H_5C_2H_5} \underset{2}{\overset{1}{\rightleftharpoons}} \underset{styrene}{C_6H_5C_2H_3} + \underset{hydrogen}{H_2} \tag{B.3.1}$$

$$\underset{ethylbenzene}{C_6H_5C_2H_5} \xrightarrow{3} \underset{benzene}{C_6H_6} + \underset{ethylene}{C_2H_4} \tag{B.3.2}$$

$$\underset{ethylbenzene}{C_6H_5C_2H_5} + \underset{hydrogen}{H_2} \xrightarrow{4} \underset{toluene}{C_6H_5CH_3} + \underset{methane}{CH_4} \tag{B.3.3}$$

The styrene product can spontaneously polymerize at higher temperatures. Because product styrene is sent directly to the polymerization unit, experience suggests that as long as its temperature is maintained at less than 125°C, there is no spontaneous polymerization problem. Because this is less than styrene's normal boiling point, and because low pressure pushes the equilibrium in Equation (B.3.1) to the right, much of this process is run at vacuum.

Stream tables, utility summaries, and major equipment summaries are given in Tables B.3.1, B.3.2, and B.3.3, respectively.

B.3.2 Reaction Kinetics

The styrene reaction may be equilibrium limited, and the equilibrium constant is given as Equation (B.3.4).

$$K = \left(\frac{y_{sty} y_{hyd} P}{y_{eb}}\right) \tag{B.3.4}$$

$$\ln K = 15.5408 - \frac{14852.6}{T}$$

where T is in K and P is in bar.

The equilibrium calculation is given as:

$$C_6H_5C_2H_5 \overset{1}{\underset{2}{\rightleftharpoons}} C_6H_5C_2H_3 + H_2$$

$$\begin{matrix} 1 & 0 & 0 \\ 1-x & x & x \end{matrix}$$

total = $N + 1 + x$ includes N moles of inert steam

$$K = \frac{x^2 P}{(1-x)(N+1+x)} \tag{B.3.5}$$

where P is in bar.

Equation (B.3.5) can be used to generate data for equilibrium conversion, x, versus P, T, and N.

The kinetic equations are adapted from Snyder and Subramaniam [3]. Subscripts on r refer to reactions in Equations (B.3.1)–(B.3.3), and the positive activation energy can arise from nonelementary kinetics; it is thought that perhaps these kinetics are an elementary approximation to nonelementary kinetics.

$$r_1 = 4.24 \times 10^6 \exp\left(-\frac{21708}{RT}\right) p_{eb} \tag{B.3.6}$$

$$r_2 = 0.755 \exp\left(\frac{7804}{RT}\right) p_{sty} p_{hyd} \tag{B.3.7}$$

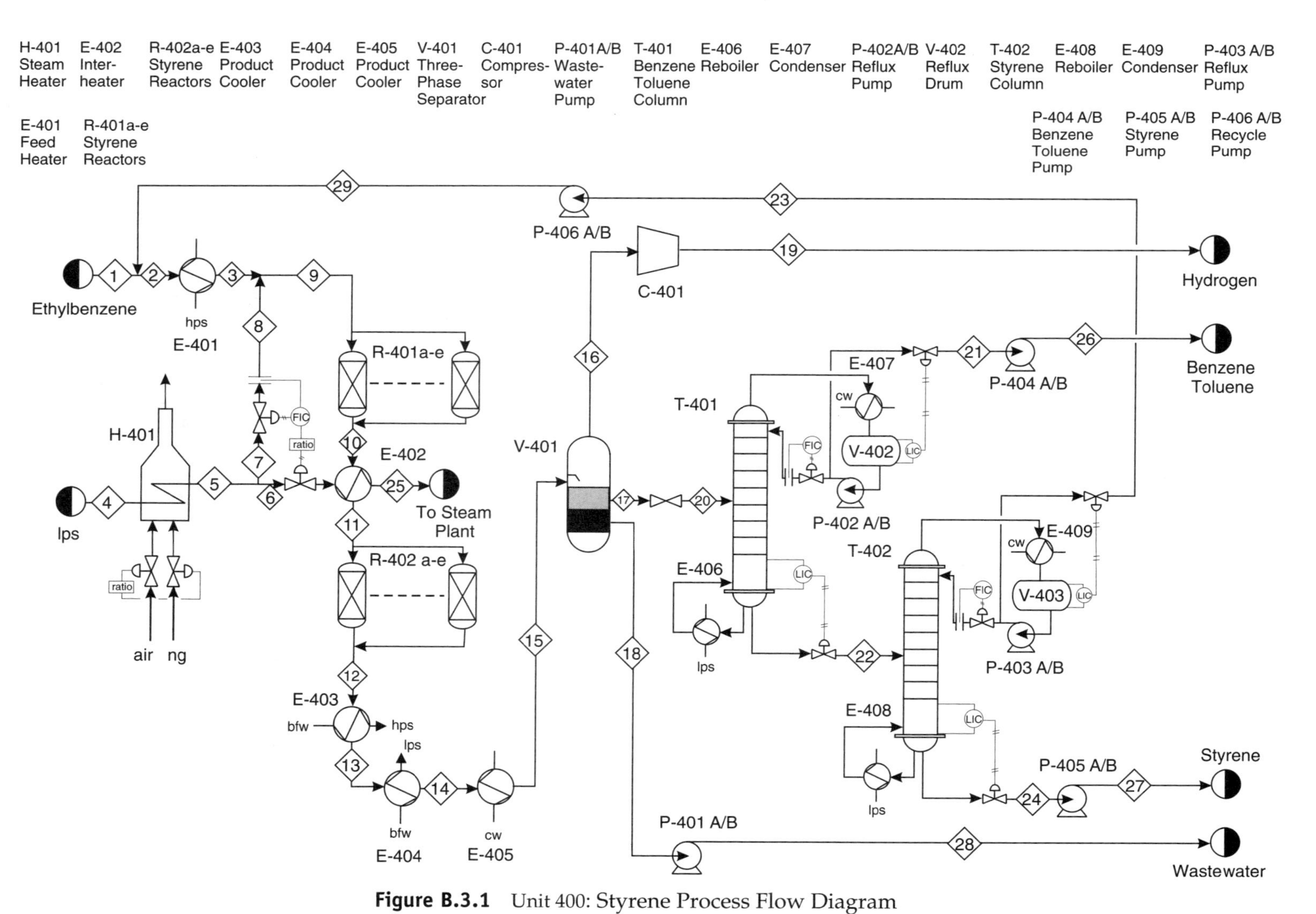

Figure B.3.1 Unit 400: Styrene Process Flow Diagram

Table B.3.1 Stream Tables for Unit 400

Stream Number	1	2	3	4	5	6
Temperature (°C)	16	116	225	159	800	800
Pressure (kPa)	20	200	180	600	565	565
Vapor mole fraction	0	0	1	1	1	1
Total flow (kg/h)	19,417	54,890	54,890	227,784	227,784	83,664
Total flow (kmol/h)	183.6	517.8	517.8	12,644	12,644	4644
Component flows						
Water	0.0	0.0	0.0	12,644	12,644	4644
Ethylbenzene	180	512.7	512.7	0.0	0.0	0.0
Styrene	0.0	1.2	1.2	0.0	0.0	0.0
Hydrogen	0.0	0.0	0.0	0.0	0.0	0.0
Benzene	1.8	1.8	1.8	0.0	0.0	0.0
Toluene	1.8	2.13	2.13	0.0	0.0	0.0
Ethylene	0.0	0.0	0.0	0.0	0.0	0.0
Methane	0.0	0.0	0.0	0.0	0.0	0.0
Stream Number	**7**	**8**	**9**	**10**	**11**	**12**
Temperature (°C)	800	799	632	609	650	640
Pressure (kPa)	565	180	170	160	145	135
Vapor mole fraction	1	1	1	1	1	1
Total flow (kg/h)	44,120	144,120	199,010	199,010	199,010	199,010
Total flow (kmol/h)	8000	8000	8517.6	8614.7	8614.7	8662.7
Component flows						
Water	8000	8000	8000	8000	8000	8000
Ethylbenzene	0.0	0.0	512.7	399.1	399.1	336.36
Styrene	0.0	0.0	1.2	86.8	86.8	120.67
Hydrogen	0.0	0.0	0.0	69.0	69.0	88.1
Benzene	0.0	0.0	1.8	13.1	13.1	27.5
Toluene	0.0	0.0	2.13	18.7	18.7	33.3
Ethylene	0.0	0.0	0.0	11.3	11.3	25.7
Methane	0.0	0.0	0.0	16.6	16.6	31.1

(*continued*)

Table B.3.1 Stream Tables for Unit 400 (*Continued*)

Stream Number	13	14	15	16	17	18
Temperature (°C)	270	180	65	65	65	65
Pressure (kPa)	120	105	90	75	75	75
Vapor mole fraction	1	1	0.025	1	0	0
Total flow (kg/h)	199,010	199,010	199,010	2682	53,493	142,715
Total flow (kmol/h)	8662.7	8662.7	8662.7	216.3	517.8	7928.6
Component flows						
Water	8000	8000	8000	71.4	0.0	7928.6
Ethylbenzene	336.36	336.36	336.36	0.0	336.36	0.0
Styrene	120.67	120.67	120.67	0.0	120.67	0.0
Hydrogen	88.1	88.1	88.1	88.1	0.0	0.0
Benzene	27.5	27.5	27.5	0.0	27.5	0.0
Toluene	33.3	33.3	33.3	0.0	33.3	0.0
Ethylene	25.7	25.7	25.7	25.7	0.0	0.0
Methane	31.1	31.1	31.1	31.1	0 0.0	0.0

Stream Number	19	20	21	22	23	24
Temperature (°C)	197	65	66	119.5	105	124.5
Pressure (kPa)	240	60	40	60	210	60
Vapor mole fraction	1	0	0	0	0	0
Total flow (kg/h)	2682	53,493	5548	47,905	35,473	12,432
Total flow (kmol/h)	216.3	517.8	63.9	453.9	334.2	119.7
Component flows						
Water	71.4	0.0	0.0	0.0	0.0	0.0
Ethylbenzene	0.0	336.36	3.36	333.0	332.66	0.34
Styrene	0.0	120.67	0.1	120.53	1.20	119.3
Hydrogen	88.1	0.0	0.0	0.0	0.0	0.0
Benzene	0.0	27.5	27.5	0.0	0.0	0.0
Toluene	0.0	33.3	32.9	0.33	0.33	0.0
Ethylene	25.7	0.0	0.0	0.0	0.0	0.0
Methane	31.1	0.0	0.0	0.0	0.0	0.0

(*continued*)

Table B.3.1 Stream Tables for Unit 400 (*Continued*)

Stream No.	25	26	27	28	29
Temperature (°C)	0.0	66	124.5	65	105
Pressure (kPa)	565	200	200	200	210
Vapor mole fraction	1	0.0	0.0	0.0	0.0
Total flow (kg/h)	83,664	548	12,432	14,2715	35,473
Total flow (kmol/h)	4644	63.9	119.7	7928.6	334.2
Component flows					
Water	4644	0.0	0.0	7928.6	0.0
Ethylbenzene	0.0	3.36	0.34	0.0	332.66
Styrene	0.0	0.1	119.3	0.0	1.20
Hydrogen	0.0	0.0	0.0	0.0	0.0
Benzene	0.0	27.5	0.0	0.0	0.0
Toluene	0.0	32.9	0.0	0.0	0.33
Ethylene	0.0	0.0	0.0	0.0	0.0
Methane	0.0	0.0	0.0	0.0	0.0

Table B.3.2 Utility Summary for Unit 400

E-401	E-403	E-404	E-405
hps	bfw → hps	bfw → lps	cw
17,566 kg/h	67,256 kg/h	14,790 kg/h	9,455,376 kg/h
E-406	**E-407**	**E-408**	**E-409**
lps	cw	lps	cw
15,878 kg/h	671,941 kg/h	163,505 kg/h	8,173,446 kg/h

Table B.3.3 Major Equipment Summary for Unit 400

Compressors and Drives	
C-401 Carbon steel W = 364.2 kW 80% adiabatic efficiency	**D-401A/B** (not shown on PFD) Electric/explosion proof W = 380 kW 98% efficiency

Heat Exchangers*	
E-401 Carbon steel $A = 541\ m^2$ Boiling in shell, condensing in tubes 1 shell – 2 tube passes Q = 29695 MJ/h	**E-406** Carbon steel $A = 76.7\ m^2$ Boiling in shell, condensing in tubes 1 shell – 2 tube passes Q = 331.2 GJ/h
E-402 316 stainless steel $A = 456\ m^2$ Steam in shell, process fluid in tubes 1 shell – 2 tube passes Q = 19237 MJ/h	**E-407** Carbon steel $A = 127\ m^2$ Condensing in shell, cw in tubes 1 shell – 2 tube passes Q = 281.1 GJ/h
E-403 316 stainless steel $A = 2006\ m^2$ Boiling in shell, process fluid in tubes 1 shell – 2 tube passes Q = 162.8 GJ/h	**E-408** Carbon steel $A = 902\ m^2$ Boiling in shell, condensing in tubes 1 shell – 2 tube passes Q = 341.1 GJ/h
E-404 Carbon steel $A = 2133\ m^2$ Boiling in shell, process fluid in tubes 1 shell – 2 tube passes Q = 35.16 GJ/h	**E-409** Carbon steel $A = 680\ m^2$ Condensing in shell, cw in tubes 1 shell – 2 tube passes Q = 342.0 GJ/h
E-405 Carbon steel $A = 2904\ m^2$ cw in shell, process fluid in tubes 1 shell – 2 tube passes Q = 395.9 GJ/h	

Fired Heater	
H-401 Fired heater – refractory Carbon steel tubes Design Q = 88.26 MW Max Q = 100 MW	

(continued)

*See Figure B.3.1 and Table B.3.1 for shell-and-tube side pressures.

Table B.3.3 Major Equipment Summary for Unit 400 (*Continued*)

Pumps	
P-401A/B Centrifugal/electric drive Carbon steel W = 6.4 kW (actual) 80% efficient	**P-404 A/B** Centrifugal/electric drive Carbon steel W = 0.38 kW (actual) 80% efficient
P-402 A/B Centrifugal/electric drive Carbon steel W = 2.0 kW (actual) 80% efficient	**P-405 A/B** Centrifugal/electric drive Carbon steel W = 0.75 kW (actual) 80% efficient
P-403 A/B Centrifugal/electric drive Carbon steel W = 2.0 kW (actual) 80% efficient	**P-406 A/B** Centrifugal/electric drive Carbon steel W = 2.65 kW (actual) 80% efficient
Reactors	
R-401 a-e 316 stainless steel, packed bed Cylindrical catalyst pellet (1.6 mm × 3.2 mm) Void fraction = 0.4 V = 1000 m^3 5 reactors in parallel at 200 m^3 20 m tall, 3.6 m diameter	**R-402 a-e** 316 stainless steel, packed bed Cylindrical catalyst pellet (1.6 mm × 3.2 mm) Void fraction = 0.4 V = 700 m^3 5 reactors in parallel at 140 m^3 20 m tall, 3.0 m diameter
Towers	
T-401 Carbon steel 46 SS sieve trays 55% efficient Feed on tray 26 Reflux ratio = 11.4 24-in tray spacing, 2-in weirs Column height = 92 ft = 28 m Top diameter = 3.6 m Bottom diameter = 2.7 m Tapered column	**T-402** Carbon steel Total condenser (E-409) Feed at location equivalent to tray 36 Reflux ratio = 25.8 Structured packing C_f = 1 Diameter = 4.1 m *HETP* = 0.3 m Height = 34.5 m
Vessels	
V-401 Carbon steel L/D = 3 V = 34.7 m^3	**V-403** Horizontal Carbon Steel L/D = 3 V = 5 m^3
V-402 Horizontal Carbon Steel L/D = 3 V = 5 m^3	

$$r_3 = 7.21 \times 10^8 \exp\left(-\frac{49675}{RT}\right)p_{eb} \quad \text{(B.3.8)}$$

$$r_4 = 1723 \exp\left(-\frac{26857}{RT}\right)p_{eb}\, p_{hyd} \quad \text{(B.3.9)}$$

where p is in kPa, T is in K, R = 1.987 cal/mol K, and r_i is in mol/m^3-reactor s.

You should assume that the catalyst has a bulk catalyst of 1282 kg/m^3, an effective diameter of 25 mm, and a void fraction = 0.4.

B.3.3 Simulation (CHEMCAD) Hints

Results for the simulation given here were obtained using SRK as the k-value and enthalpy options in the thermodynamics package.

B.3.4 References

1. Shiou-Shan Chen, "Styrene," *Kirk-Othmer Encyclopedia of Chemical Technology*, on-line version (New York: John Wiley & Sons, 2006).
2. "Styrene," *Encyclopedia of Chemical Processing and Design*, vol. 55, J. J. McKetta, ed. (New York: Marcel Dekker, Inc., 1984), 197–217.
3. Snyder, J. D., and B. Subramaniam, "A Novel Reverse Flow Strategy for Ethylbenzene Dehydrogenation in a Packed-Bed Reactor," *Chem. Engr. Sci.* 49 (1994): 5585–5601.

B.4 DRYING OIL PRODUCTION, UNIT 500

Drying oils are used as additives to paints and varnishes to aid in the drying process when these products are applied to surfaces. The facility manufactures drying oil (DO) from acetylated castor oil (ACO). Both of these compounds are mixtures. However, for simulation purposes, acetylated castor oil is modeled as palmitic (hexadecanoic) acid ($C_{15}H_{31}COOH$) and drying oil is modeled as 1-tetradecene ($C_{14}H_{28}$). In an undesired side reaction, a gum can be formed, which is modeled as 1-octacosene ($C_{28}H_{56}$).

B.4.1 Process Description

The process flow diagram is shown in Figure B.4.1. ACO is fed from a holding tank where it is mixed with recycled ACO. The ACO is heated to reaction temperature in H-501. The reaction does not require a catalyst, since it is initiated at high temperatures. The reactor, R-501, is simply a vessel with inert packing to promote radial mixing. The reaction is quenched in E-501. Any gum that has been formed is removed by filtration. There are two holding vessels, V-502 A/B. One of them is used to hold reaction products, while the other one feeds the filter (not shown). This allows a continuous flow of material into Stream 7. In T-501 the ACO is

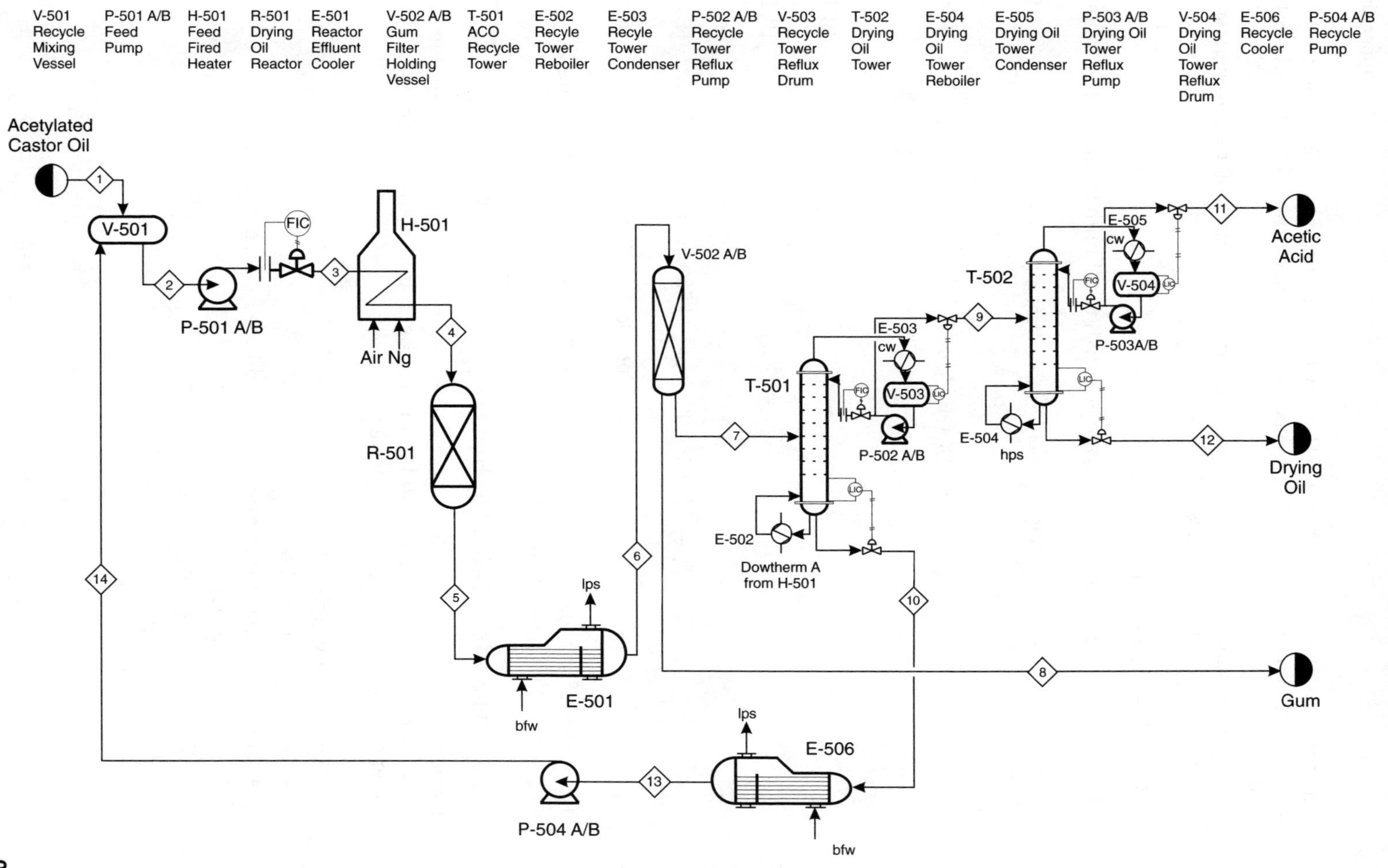

Figure B.4.1 Unit 500: Drying Oil Process Flow Diagram

separated and recycled, and in T-502, the DO is purified from the acetic acid. The contents of Streams 11 and 12 are cooled (exchangers not shown) and sent to storage.

Stream summary tables, utility summary tables, and major equipment specifications are shown in Tables B.4.1–B.4.3.

B.4.2 Reaction Kinetics

The reactions and reaction kinetics are adapted from Smith [1] and are as follows.

$$\underset{ACO}{C_{15}H_{31}COOH(l)} \xrightarrow{k_1} \underset{acetic\ acid}{CH_3COOH(g)} + \underset{DO}{C_{14}H_{28}(l)} \tag{B.4.1}$$

$$\underset{DO}{2C_{14}H_{28}(l)} \xrightarrow{k_2} \underset{gum}{C_{28}H_{56}(s)} \tag{B.4.2}$$

where

$$-r_1 = k_1 C_{ACO} \tag{B.4.3}$$

$$-r_2 = k_2 C_{DO}^2 \tag{B.4.4}$$

and

$$k_1 = 5.538 \times 10^{13} \exp(-44{,}500/RT) \tag{B.4.5}$$

$$k_2 = 1.55 \times 10^{26} \exp(-88{,}000/RT) \tag{B.4.6}$$

The units of reaction rate, r_i, are kmol/m^{3}s, and the activation energy is in cal/mol (which is equivalent to kcal/kmol).

B.4.3 Simulation (CHEMCAD) Hints

If you want to simulate this process and 1-octacosene is not a compound in your simulator's database, you can add gum as a compound to the simulator databank using the following physical properties:

- molecular weight = 392
- boiling point = 431.6°C
- For the group contribution method add the following groups
  - 1 $-CH_3$ group
  - 25 $>CH_2$ groups
  - 1 $=CH_2$ group
  - 1 $=CH-$ group

Table B.4.1 Stream Tables for Unit 500

Stream Number	1	2	3	4
Temp (°C)	25.00	151.03	151.13	380.00
Pres (kPa)	110.00	105.00	230.00	195.00
Vapor mole fraction	0.00	0.00	0.00	0.00
Flowrate (kg/h)	1628.70	10,703.10	10,703.10	10,703.10
Flowrate (kmol/h)	6.35	41.75	41.75	41.75
Component flowrates (kmol/h)				
Acetic acid	0.00	0.00	0.00	0.00
1-Tetradecene (Drying Oil)	0.00	0.064	0.064	0.064
Hexadecanoic Acid (ACO)	6.35	41.69	41.69	41.69
Gum	0.00	0.00	0.00	0.00

Stream Number	5	6	7	8
Temp (°C)	342.81	175.00	175.00	175.00
Pres (kPa)	183.00	148.00	136.00	136.00
Vapor mole fraction	0.39	0.00	0.00	0.00
Flowrate (kg/h)	10,703.10	10,703.10	10,703.08	0.02
Flowrate (kmol/h)	48.07	48.07	48.07	4.61×10^{-5}
Component flowrates (kmol/h)				
Acetic acid	6.32	6.32	6.32	0.00
1-Tetradecene (Drying Oil)	6.38	6.38	6.38	0.00
Hexadecanoic Acid (ACO)	35.38	35.38	35.38	0.00
Gum	4.61×10^{-5}	4.61×10^{-5}	0.00000	4.61×10^{-5}

Stream Number	9	10	11	12
Temp (°C)	107.96	344.75	119.19	252.83
Pres (kPa)	125.00	90.00	105.00	125.00
Vapor mole fraction	0.00	0.00	0.00	0.00
Flowrate (kg/h)	1628.68	9074.40	378.64	1250.04
Flowrate (kmol/h)	12.67	35.40	6.29	6.38
Component flowrates (kmol/h)				
Acetic acid	6.32	0.00	6.28	0.03
1-Tetradecene (Drying Oil)	6.32	0.06	0.01	6.31
Hexadecanoic Acid (ACO)	0.04	35.34	0.00	0.04
Gum	0.00	0.00	0.00	0.00

(continued)

Table B.4.1 Stream Tables for Unit 500 (*Continued*)

Stream Number	13	14
Temp (°C)	170.00	170.03
Pres (kPa)	65.00	110.00
Vapor mole fraction	0.00	0.00
Flowrate (kg/h)	9074.40	9074.40
Flowrate (kmol/h)	35.40	35.40
Component flowrates (kmol/h)		
Acetic acid	0.00	0.00
1-Tetradecene (Drying Oil)	0.06	0.06
Hexadecanoic Acid (ACO)	35.34	35.34
Gum	0.00	0.00

Table B.4.2 Utility Flow Summary for Unit 500

E-501	E-502	E-503	E-504	E-505	E-506
bfw→lps	Dowtherm A	cw	hps	cw	bfw→lps
2664 kg/h	126,540 kg/h	24,624 kg/h	425 kg/h	5508 kg/h	2088 kg/h

Table B.4.3 Major Equipment Summary for Unit 500

Fired Heater

H-501

Total heat duty required = 13219 MJ/h = 3672 kW

Design capacity = 4000 kW

Carbon steel tubes

85% thermal efficiency

Heat Exchangers

E-501

$A = 26.2\ m^2$

1-2 exchanger, floating head, stainless steel

Process stream in tubes

$Q = 6329$ MJ/h

E-502

$A = 57.5\ m^2$

1-2 exchanger, floating head, stainless steel

Process stream in shell

$Q = 5569$ MJ/h

E-503

$A = 2.95\ m^2$

1-2 exchanger, floating head, stainless steel

Process stream in shell

$Q = 1029$ MJ/h

E-504

$A = 64.8\ m^2$

1-2 exchanger, floating head, stainless steel

Process stream in shell

$Q = 719$ MJ/h

E-505

$A = 0.58\ m^2$

1-2 exchanger, floating head, stainless steel

Process stream in shell

$Q = 230$ MJ/h

E-506

$A = 919\ m^2$

1-4 exchanger, floating head, stainless steel

Process stream in tubes

$Q = 4962$ MJ/h

(*continued*)

Table B.4.3 Major Equipment Summary for Unit 500 (*Continued*)

Pumps

P-501 A/B
Centrifugal/electric drive
Carbon steel
Power = 0.9 kW (actual)
80% efficient
$NPSH_R$ at design flow = 14 ft of liquid

P-502 A/B
Centrifugal/electric drive
Stainless steel
Power = 1 kW (actual)
80% efficient

P-503 A/B
Centrifugal/electric drive
Stainless steel
Power = 0.8 kW (actual)
80% efficient

P-504 A/B
Stainless steel/electric drive
Power = 0.3 kW (actual)
80% efficient
$NPSH_R$ at design flow = 12 ft of liquid

Reactor

R-501
Stainless steel vessel
$V = 1.15\ m^3$
5.3 m long, 0.53 m diameter

Towers

T-501
Stainless steel
56 sieve trays plus reboiler and condenser
25% efficient trays
Total condenser
Feed on tray = 32
Reflux ratio = 0.15
12-in tray spacing, 2.2-in weirs
Column height = 17 m
Diameter = 2.1 m below feed and 0.65 m above feed

T-502
Stainless steel
35 sieve trays plus reboiler and condenser
52% efficient trays
Total condenser
Feed on tray = 23
Reflux ratio = 0.52
12-in tray spacing, 2.8-in weirs
Column height = 11 m
Diameter = 0.45 m

Vessels

V-501
Horizontal
Carbon steel
L/D = 3
$V = 2.3\ m^3$

V-502
Vertical
Stainless steel
L/D = 5
$V = 3\ m^3$

V-503
Horizontal
Stainless steel
L/D = 3
$V = 2.3\ m^3$

V-504
Horizontal
Carbon steel
L/D = 3
$V = 0.3\ m^3$

B.4.4 Reference

1. Smith, J. M., *Chemical Engineering Kinetics*, 3rd ed. (New York: John Wiley & Sons, 1981), 224–228.

B.5 PRODUCTION OF MALEIC ANHYDRIDE FROM BENZENE, UNIT 600

Currently, the preferred route to maleic anhydride in the United States is via isobutene in fluidized-bed reactors. However, an alternative route via benzene may be carried out using a shell-and-tube reactor, with catalyst in the tubes and a cooling medium being circulated through the shell.

B.5.1 Process Description

A process flow diagram for the reactor section of the maleic anhydride process is shown in Figure B.5.1. Benzene is vaporized in E-601, mixed with compressed air, and then heated in a fired heater, H-601, prior to being sent to a packed-bed catalytic reactor, R-601, where the following reactions take place.

$$\underset{\text{benzene}}{C_6H_6} + 4.5O_2 \xrightarrow{k_1} \underset{\text{maleic anhydride}}{C_4H_2O_3} + 2CO_2 + 2H_2O \qquad \text{(B.5.1)}$$

$$\underset{\text{benzene}}{C_6H_6} + 7.5O_2 \xrightarrow{k_2} 6CO_2 + 3H_2O \qquad \text{(B.5.2)}$$

$$\underset{\text{maleic anhydride}}{C_4H_2O_3} + 3O_2 \xrightarrow{k_3} 4CO_2 + H_2O \qquad \text{(B.5.3)}$$

$$\underset{\text{benzene}}{C_6H_6} + 1.5O_2 \xrightarrow{k_4} \underset{\text{quinone}}{C_6H_4O_2} + 2H_2O \qquad \text{(B.5.4)}$$

All the reactions are highly exothermic. For this reason, the ratio of air to benzene entering the reactor is kept very high. A typical inlet concentration (Stream 6) of approximately 1.5 vol% of benzene in air is used. Cooling is achieved by circulating molten salt (a mixture of sodium nitrite and sodium nitrate) cocurrently through the shell of the reactor and across the tubes containing the catalyst and reactant gases. This molten salt is cooled in two external exchangers—E-602 and E-607—prior to returning to the reactor.

The reactor effluent, Stream 7—containing small amounts of unreacted benzene, maleic anhydride, quinone, and combustion products—is cooled in E-603 and then sent to an absorber column, T-601, which has both a reboiler and condenser. In T-601, the vapor feed is contacted with recycled heavy organic solvent (dibutyl phthalate), Stream 9. This solvent absorbs the maleic anhydride, quinone, and small amounts of water. Any water in the solvent leaving the bottom of the absorber, T-601, reacts with the maleic anhydride to form maleic acid, which must be removed and purified from the maleic anhydride. The bottoms product from the absorber is sent to a separation tower, T-602, where the dibutyl phthalate is recovered as the bottoms product, Stream 14, and recycled back to the absorber. A small amount of fresh solvent, Stream 10, is added to account for losses. The overhead product from T-602, Stream 13, is sent to the maleic acid column, T-603, where 95 mol% maleic acid is removed as the bottoms product.

The overhead stream is taken to the quinone column, T-604, where 99 mol% quinone is taken as the top product and 99.9 mol% maleic anhydride is removed as the bottoms product. These last two purification columns are not shown in Figure B.5.1 and are not included in the current analysis.

Stream summaries, utility summaries, and equipment summaries are presented in Tables B.5.1–B.5.3.

B.5.2 Reaction Kinetics

The reactions and reaction kinetics [3] given in Equations (B.5.1)–(B.5.4) are given by the expression

$$-r_i = k_i C_{benzene} \text{ or } -r_3 = k_3 C_{maleic\ anhydride} \quad \text{(B.5.5)}$$

where

$$k_1 = 7.7 \times 10^6 \exp(-25{,}143/RT) \quad \text{(B.5.6)}$$

$$k_2 = 6.31 \times 10^7 \exp(-29{,}850/RT) \quad \text{(B.5.7)}$$

$$k_3 = 2.33 \times 10^4 \exp(-21{,}429/RT) \quad \text{(B.5.8)}$$

$$k_4 = 7.20 \times 10^5 \exp(-27{,}149/RT) \quad \text{(B.5.9)}$$

The units of reaction rate, r_i, are kmol/m^3(reactor)s, the activation energy is given in cal/mol (which is equivalent to kcal/kmol), the units of k_i are m^3(gas)/m^3 (reactor)s, and the units of concentration are kmol/m^3(gas).

The catalyst is a mixture of vanadium and molybdenum oxides on an inert support. Typical inlet reaction temperatures are in the range of 350°C to 400°C. The catalyst is placed in 25 mm diameter tubes that are 3.2 m long. The catalyst pellet diameter is 5 mm. The maximum temperature that the catalyst can be exposed to without causing irreversible damage (sintering) is 650°C. The packed-bed reactor should be costed as a shell-and-tube exchanger. The heat transfer area should be calculated based on the total external area of the catalyst-filled tubes required from the simulation. Because of the high temperatures involved, both the shell and the tube material should be stainless steel. An overall heat transfer coefficient for the reactor should be set as 100 W/m^2°C. (This is the value specified in the simulation.)

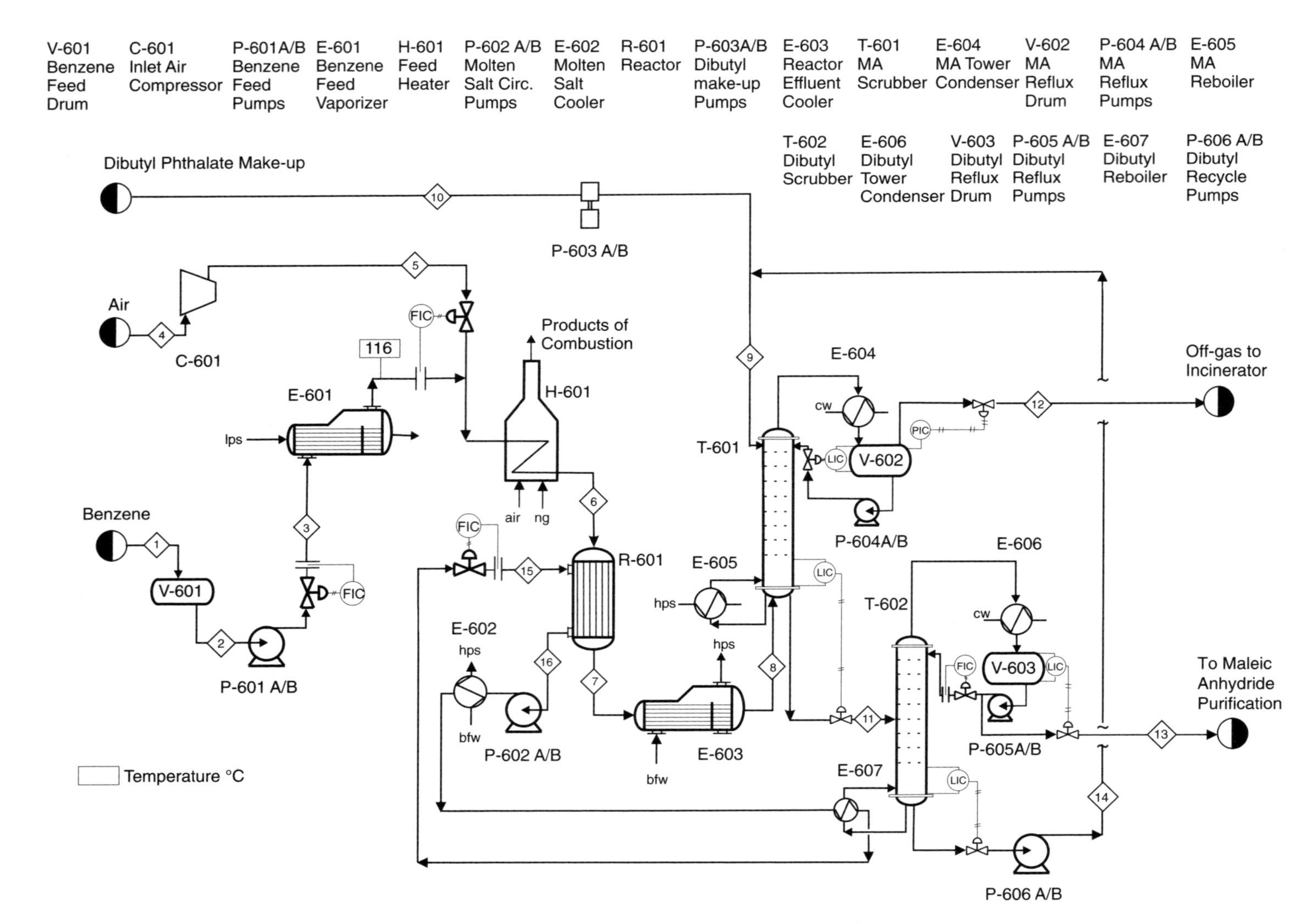

Figure B.5.1 Unit 600: Maleic Anhydride Process Flow Diagram

Table B.5.1 Flow Summary for Unit 600

Stream Number	1	2	3	4	5	6	7	8
Temp (°C)	30	30	30	30	170	460	608	270
Pres kPa	101	101	280	101	250	235	220	215
Total kmol/h	42.3	42.3	42.3	2790.0	2790.0	2832.3	2825.2	2825.3
Total kg/h	3304	3304	3304	80,490	80,490	83,794	83,794	83,794
Flowrates in kmol/h								
Maleic anhydride	0.0	0.0	0.0	0.0	0.0	0.0	26.3	26.3
Dibutyl phthalate	0.0	0.0	0.0	0.0	0.0	0.0	0.0	0.0
Nitrogen	0.0	0.0	0.0	2205.0	2205.0	2205.0	2205.0	2205.0
Water	0.0	0.0	0.0	0.0	0.0	0.0	91.2	91.2
Oxygen	0.0	0.0	0.0	585.0	585.0	585.0	370.9	370.9
Benzene	42.3	42.3	42.3	0.0	0.0	42.3	2.6	2.6
Quinone	0.0	0.0	0.0	0.0	0.0	0.0	0.7	0.7
Carbon dioxide	0.0	0.0	0.0	0.0	0.0	0.0	128.3	128.3
Maleic acid	0.0	0.0	0.0	0.0	0.0	0.0	0.0	0.0
Sodium nitrite	0.0	0.0	0.0	0.0	0.0	0.0	0.0	0.0
Sodium nitrate	0.0	0.0	0.0	0.0	0.0	0.0	0.0	0.0

Stream Number	9	10	11	12	13	14	15	16
Temp (°C)	330	320	194	84	195	330	419	562
Pres kPa	82	100	82	75	80	82	200	200
Total kmol/h	500.1	0.1	526.2	2797.9	26.2	500.0	5000.0	5000.0
Total kg/h	139,191.6	30.6	141,866	81,225	2597	139,269	391,925	391,925
Flowrates in kmol/h								
Maleic anhydride	0.0	0.0	0.0	0.0	24.8	0.0	0.0	0.0
Dibutyl phthalate	0.0	0.0	0.0	0.0	0.0	500.4	0.0	0.0
Nitrogen	0.0	0.0	0.0	0.0	0.0	0.0	0.0	0.0
Water	0.0	0.0	0.0	0.0	0.0	0.0	0.0	0.0
Oxygen	0.0	0.0	0.0	0.0	0.0	0.0	0.0	0.0
Benzene	0.0	0.0	0.0	0.0	0.0	0.0	0.0	0.0
Quinone	0.0	0.0	0.0	0.0	0.4	0.0	0.0	0.0
Carbon dioxide	0.0	0.0	0.0	0.0	0.0	0.0	0.0	0.0
Maleic acid	0.0	0.0	0.0	0.0	1.0	.005	0.0	0.0
Sodium nitrite	0.0	0.0	0.0	0.0	0.0	0.0	2065.6	2065.6
Sodium nitrate	0.0	0.0	0.0	0.0	0.0	0.0	2934.4	2934.4

Table B.5.2 Utility Summaries for Unit 600

E-601	E-602	E-603	E-604	E-605	E-606
lps	bfw → hps	bfw → hps	cw	hps	cw
1750 MJ/h	16,700 MJ/h	31,400 MJ/h	86,900 MJ/h	19,150 MJ/h	3050 MJ/h
841 kg/h	7295 kg/h	13,717 kg/h	2.08×10^6 kg/h	11,280 kg/h	73,000 kg/h

Table B.5.3 Equipment Summary for Unit 600

Compressor and Drives

C-601
Centrifugal/electric drive
Carbon steel
Discharge pressure = 250 kPa
Efficiency = 65%
Power (shaft) = 3,108 kW
MOC carbon steel

D-601A/B (not shown on PFD)
Electric/explosionproof
W = 3200 kW
98% efficient

Fired Heater

H-601
Total (process) heat duty required = 26,800 MJ/h
Design capacity = 32,000 kW
Carbon steel tubes
85% thermal efficiency
Design pressure = 300 kPa

Heat Exchangers

E-601
$A = 14.6$ m^2
1-2 exchanger, floating head, stainless steel
Process stream in tubes
$Q = 1{,}750$ MJ/h
Design pressure = 600 kPa

E-602
$A = 61.6$ m^2
1-2 exchanger, floating head, stainless steel
Process stream in shell
$Q = 16{,}700$ MJ/h
Design pressure = 4100 kPa

E-603
$A = 1{,}760$ m^2
1-2 exchanger, floating head, stainless steel
Process stream in shell
$Q = 31{,}400$ MJ/h
Design pressure = 4100 kPa

E-604
$A = 1{,}088$ m^2
1-2 exchanger, fixed head, stainless steel
Process stream in tubes
$Q = 86{,}900$ MJ/h
Design pressure = 300 kPa

E-605
$A = 131$ m^2
1-2 exchanger, floating head, stainless steel
Process stream in shell
$Q = 19{,}150$ MJ/h
Design pressure = 4100 kPa

E-606
$A = 11.7$ m^2
1-2 exchanger, floating head, stainless steel
Process stream in shell
$Q = 3{,}050$ MJ/h
Design pressure = 300 kPa

E-607
$A = 192$ m^2
1-2 exchanger, floating head, stainless steel
Molten salt in tubes
$Q = 55{,}600$ MJ/h
Design pressure = 4100 kPa

(continued)

Table B.5.3 Equipment Summary for Unit 600 (*Continued*)

Pumps	
P-601 A/B Centrifugal/electric drive Carbon steel Power = 0.3 kW (actual) 65% efficient Design pressure = 300 kPa	**P-604 A/B** Centrifugal/electric drive Stainless steel Power = 6.75 kW (actual) 65% efficient Design pressure = 200 kPa
P-602 A/B Centrifugal/electric drive Stainless steel Power = 3.8 kW (actual) 65% efficient Design pressure = 300 kPa	**P-605 A/B** Centrifugal/electric drive Stainless steel Power = 0.7 kW (actual) 65% efficient Design pressure = 400 kPa
P-603 A/B Reciprocating/electric drive Stainless steel Power = 0.1 kW (actual) 65% efficient Design pressure = 200 kPa	**P-606 A/B** Centrifugal/electric drive Stainless steel Power = 2.4 kW (actual) 65% efficient Design pressure = 150 kPa
Reactor	
R-601 Shell-and-tube vertical design Stainless steel L = 7.0 m D = 3.8 m 12,100 1-in diameter, 6.4 m length catalyst-filled tubes Design pressure = 300 kPa	
Towers	
T-601 Stainless steel 14 sieve trays plus reboiler and condenser 50% efficient trays Partial condenser Feeds on trays 1 and 14 Reflux ratio = 0.189 24-in tray spacing, 2.2-in weirs Column height = 10 m Diameter = 4.2 m Design pressure = 110 kPa	**T-602** Stainless steel 42 sieve trays plus reboiler and condenser 65% efficient trays Total condenser Feed on tray 27 Reflux ratio = 1.24 15-in tray spacing, 1.5-in weirs Column height = 18 m Diameter = 1.05 m Design pressure = 110 kPa

(*continued*)

Table B.5.3 Equipment Summary for Unit 600 (*Continued*)

Vessels	
V-601	**V-603**
Horizontal	Horizontal
Carbon steel	Stainless steel
L = 3.50 m	L = 3.90 m
D = 1.17 m	D = 1.30 m
Design pressure = 110 kPa	Design pressure = 110 kPa
V-602	
Horizontal	
Stainless steel	
L = 13.2 m	
D = 4.4 m	
Design pressure = 110 kPa	

B.5.3 Simulation (CHEMCAD) Hints

The CHEMCAD simulation used to generate the PFD shown in Figure B.5.1 has several simplifications that are valid for this system. The removal of trace amounts of noncondensables is achieved after the absorber using a component separator, which avoids problems with column convergence downstream. The formation of maleic acid is simulated by using a stoichiometric reactor and setting the conversion of water to 1.

Tower T-601, the maleic anhydride scrubber, is simulated using the rigorous tower simulator. Tower T-602, the dibutyl phthalate tower, is simulated using the Shortcut column module. Currently, we do not have any experimental vapor pressure data for the components in this simulation. It appears that the vapor pressures of the components differ widely, and no azeotropes are known at this time. For this reason, the ideal vapor pressure *K*-value option and the latent heat enthalpy option are used.

In order to simulate the temperature spike in the reactor, the reactor is simulated as a cocurrent, packed-bed kinetic reactor, with a molten salt stream as the utility. This configuration provides a greater temperature differential at the front end of the reactor, where the reaction rate is highest. Countercurrent flow could be investigated as an alternative. The kinetics given above are used in the simulation. Dimensions of the reactor tubes are given in Section B.5.2.

B.5.4 References

1. Felthouse, T. R., J. C. Burnett, B. Horrell, M. J. Mummey, and Y-J Kuo, "Maleic Anhydride, Maleic Acid, and Fumaric Acid," *Kirk-Othmer Encyclopedia of Chemical Technology*, on-line version (New York: John Wiley & Sons, 2001).
2. "Maleic Acid and Anhydride," *Encyclopedia of Chemical Processing and Design*, vol. 29, J. J. McKetta, ed. (New York: Marcel Dekker, Inc., 1984), 35–55.

3. Wohlfahrt, Emig G., "Compare Maleic Anhydride Routes," *Hydrocarbon Processing* (June 1980): 83–90.

B.6 ETHYLENE OXIDE PRODUCTION, UNIT 700

Ethylene oxide is a chemical used to make ethylene glycol (the primary ingredient in antifreeze). It is also used to make polyethylene-oxide, and both the low-molecular-weight and high-molecular-weight polymers have many applications including as detergent additives. Because ethylene oxide is so reactive, it has many other uses as a reactant. However, because of its reactivity, danger of explosion, and toxicity, it is rarely shipped outside the manufacturing facility but instead is often pumped directly to a nearby consumer.

B.6.1 Process Description [1, 2]

The process flow diagram is shown in Figure B.6.1. Ethylene feed (via pipeline from a neighboring plant) is mixed with recycled ethylene and mixed with compressed and dried air (drying step not shown), heated, and then fed to the first reactor. The reaction is exothermic, and medium-pressure steam is made in the reactor shell. Conversion in the reactor is kept low to enhance selectivity for the desired product. The reactor effluent is cooled, compressed, and sent to a scrubber, where ethylene oxide is absorbed by water. The vapor from the scrubber is heated, throttled, and sent to a second reactor, followed by a second series of cooling, compression, and scrubbing. A fraction of the unreacted vapor stream is purged, with the remainder recycled to recover unreacted ethylene. The combined aqueous product streams are mixed, cooled, throttled, and distilled to produce the desired product. The required purity specification is 99.5 wt% ethylene oxide.

Stream summary tables, utility summary tables, and major equipment specifications are shown in Tables B.6.1–B.6.3.

B.6.2 Reaction Kinetics

The pertinent reactions (adapted from Stoukides and Pavlou [3]) are as follows.

$$C_2H_4 + 0.5\,O_2 \rightarrow C_2H_4O \tag{B.6.1}$$

$$C_2H_4 + 3\,O_2 \rightarrow 2CO_2 + 2H_2O \tag{B.6.2}$$

$$C_2H_4O + 2.5\,O_2 \rightarrow 2CO_2 + 2H_2O \tag{B.6.3}$$

The kinetic expressions are, respectively,

$$r_1 = \frac{1.96\exp(-2400/RT)p_{ethylene}}{1 + 0.00098\exp(11{,}200/RT)p_{ethylene}} \tag{B.6.4}$$

$$r_2 = \frac{0.0936\exp(-6400/RT)p_{ethylene}}{1 + 0.00098\exp(11{,}200/RT)p_{ethylene}} \tag{B.6.5}$$

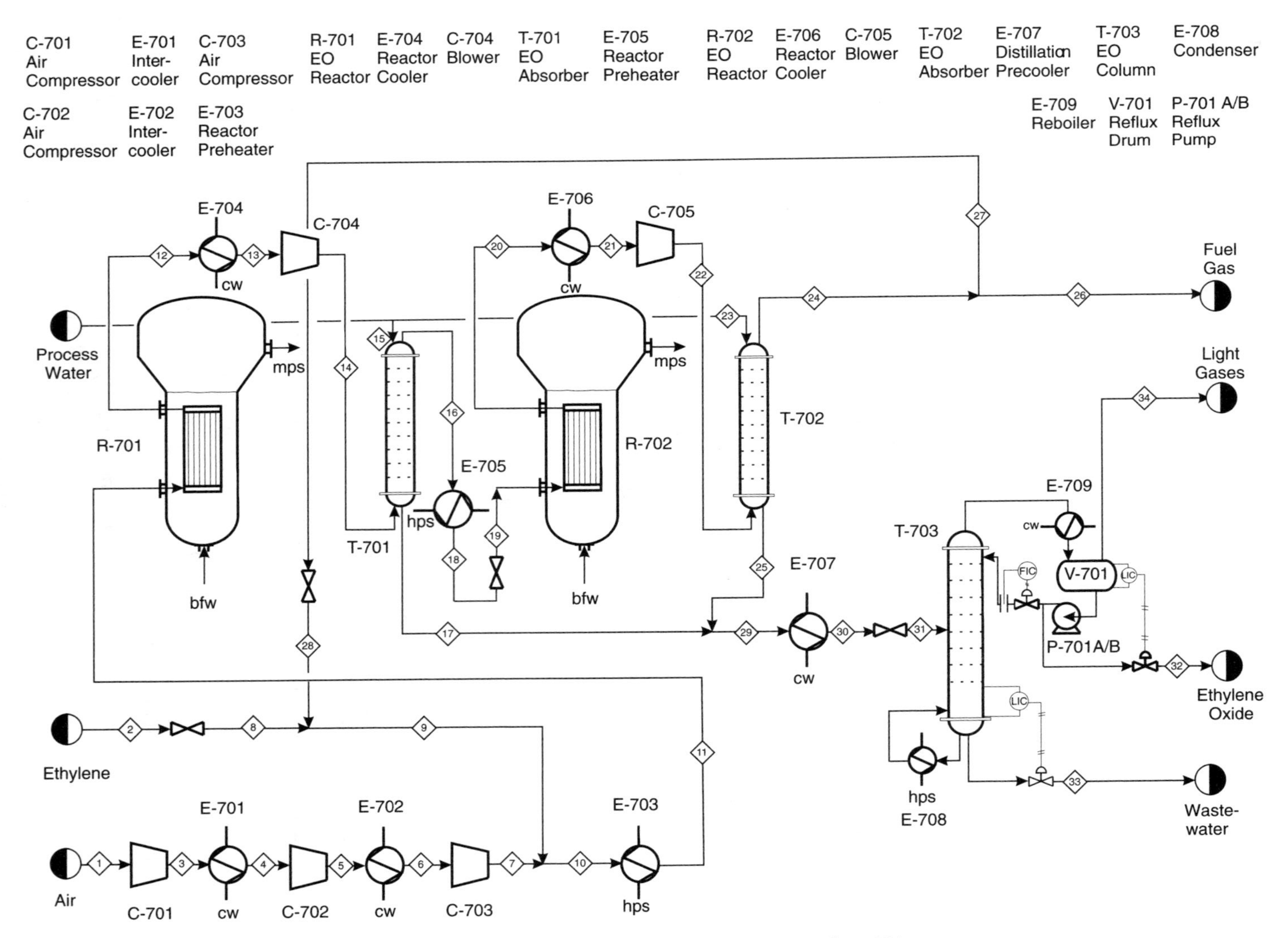

Figure B.6.1 Unit 700: Ethylene Oxide Process Flow Diagram

$$r_3 = \frac{0.42768 \exp(-6200/RT) p^2_{ethylene\ oxide}}{1 + 0.000033 \exp(21{,}200/RT) p^2_{ethylene\ oxide}} \quad \text{(B.6.6)}$$

The units for the reaction rates are moles/m^3 s. The pressure unit is bar. The activation energy numerator is in cal/mol. The catalyst used for this reaction is silver on an inert support. The support consists of 7.5 mm diameter spheres that have a bulk density of 1250 kg/m^3 and a void fraction of 0.4.

Table B.6.1 Stream Table for Unit 700

Stream Number	**1**	**2**	**3**	**4**
Temp (°C)	25.00	25.00	159.19	45.00
Pres (bar)	1.01325	50.00	3.00	2.70
Vapor mole fraction	1.00	1.00	1.00	1.00
Flowrate (kg/h)	500,000	20,000	500,000	500,000
Flowrate (kmol/h)	17,381.45	712.91	17,381.45	17,381.45
Component flowrates (kmol/h)				
Ethylene	0.0	712.91	0.0	0.0
Ethylene oxide	0.0	0.0	0.0	0.0
Carbon dioxide	0.0	0.0	0.0	0.0
Oxygen	3281.35	0.0	3281.35	3281.35
Nitrogen	14,100.09	0.0	14,100.09	14,100.09
Water	0.0	0.0	0.0	0.0
Stream Number	**5**	**6**	**7**	**8**
Temp (°C)	206.11	45.00	195.21	–6.30
Pres (bar)	9.00	8.70	27.00	27.00
Vapor mole fraction	1.00	1.00	1.00	1.00
Flowrate (kg/h)	500,000	500,000	500,000	20,000
Flowrate (kmol/h)	17,381.45	17,381.45	17,381.45	712.91
Component flowrates (kmol/h)				
Ethylene	0.0	0.0	0.0	712.91
Ethylene oxide	0.0	0.0	0.0	0.0
Carbon dioxide	0.0	0.0	0.0	0.0
Oxygen	3281.35	3281.35	3281.35	0.0
Nitrogen	14,100.09	14,100.09	14,100.09	0.0
Water	0.0	0.0	0.0	0.0

(continued)

Table B.6.1 Stream Table for Unit 700 (*Continued*)

Stream Number	9	10	11	12
Temp (°C)	26.34	106.74	240.00	240.00
Pres (bar)	27.00	26.80	26.50	25.75
Vapor mole fraction	1.00	1.00	1.00	1.00
Flowrate (kg/h)	524,042	1,023,980	1,023,980	1,023,979
Flowrate (kmol/h)	18,260.29	35,639.59	35,639.59	35,539.42
Component flowrates (kmol/h)				
Ethylene	1047.95	1047.91	1047.91	838.67
Ethylene oxide	6.48	6.47	6.47	206.79
Carbon dioxide	31.71	31.71	31.71	49.56
Oxygen	3050.14	6331.12	6331.12	6204.19
Nitrogen	14,093.02	28,191.39	28,191.39	28,191.39
Water	30.99	30.98	30.98	48.82

Stream Number	13	14	15	16
Temp (°C)	45.00	63.72	25.00	30.30
Pres (bar)	25.45	30.15	30.00	30.00
Vapor mole fraction	1.00	1.00	0.00	1.00
Flowrate (kg/h)	1,023,980	1,023,980	360,300	1,015,669
Flowrate (kmol/h)	35,539	35,539	20,000	35,358
Component flowrates (kmol/h)				
Ethylene	838.67	838.67	0.0	837.96
Ethylene oxide	206.79	206.79	0.0	15.45
Carbon dioxide	49.56	49.56	0.0	49.56
Oxygen	6204.19	6204.19	0.0	6202.74
Nitrogen	28,191.39	28,191.39	0.0	28,188.72
Water	48.82	48.82	20,000	63.24

(*continued*)

Table B.6.1 Stream Table for Unit 700 (*Continued*)

Stream Number	**17**	**18**	**19**	**20**
Temp (°C)	51.92	240.0	239.9	240.0
Pres (bar)	30.00	29.70	26.50	25.75
Vapor mole fraction	0.00	1.00	1.00	1.00
Flowrate (kg/h)	368,611	1,015,669	1,015,669	1,015,669
Flowrate (kmol/h)	20,181.77	35,357.65	35357.66	35,277.47
Component flowrates (kmol/h)				
Ethylene	0.70	837.96	837.96	670.64
Ethylene oxide	191.34	15.45	15.45	175.83
Carbon dioxide	0.01	49.55	49.55	63.44
Oxygen	1.45	6202.74	6202.74	6101.72
Nitrogen	2.68	28,188.72	28,188.72	28,188.72
Water	19,985.58	63.24	63.24	77.13

Stream Number	**21**	**22**	**23**	**24**
Temp (°C)	45.00	63.78	25.00	30.0851
Pres bar	25.45	30.15	30.00	30.00
Vapor mole fraction	1.00	1.00	0.00	1.00
Total kg/h	1,015,669	1,015,669	60,300	1,008,084
Total kmol/h	35,277.47	35,277.47	20,000	35094.76
Flowrates in kmol/h				
Ethylene	670.64	670.64	0.0	670.08
Ethylene oxide	175.83	175.83	0.0	12.96
Carbon dioxide	63.44	63.44	0.0	63.43
Oxygen	6101.72	6101.72	0.0	6100.28
Nitrogen	28,188.72	28,188.72	0.0	28,186.04
Water	77.13	77.13	20,000	61.96

(continued)

Table B.6.1 Stream Table for Unit 700 (*Continued*)

Stream Number	25	26	27	28
Temp (°C)	52.26	30.09	30.09	29.48
Pres (bar)	30.00	30.00	30.00	27.00
Vapor mole fraction	0.00	1.00	1.00	1.00
Flowrate (kg/h)	367,885	504,042	504,042	504,042
Flowrate (kmol/h)	20,182.72	17,547.38	17,547.38	17,547.38
Component flowrates (kmol/h)				
Ethylene	0.57	335.04	335.04	335.04
Ethylene oxide	162.88	6.48	6.48	6.48
Carbon dioxide	0.01	31.71	31.71	31.71
Oxygen	1.43	3050.14	3050.14	3050.14
Nitrogen	2.68	14,093.02	14,093.02	14,093.02
Water	20,015.15	30.99	30.99	30.99

Stream Number	29	30	31	32
Temp (°C)	52.08	45.00	45.02	86.40
Pres (bar)	30.00	29.70	10.00	10.00
Vapor mole fraction	0.00	0.00	0.00	0.00
Flowrate (kg/h)	736,497	736,497	736,218	15,514
Flowrate (kmol/h)	40,364.48	40,364.48	40,354.95	352.39
Component flowrates (kmol/h)				
Ethylene	1.27	1.27	1.27	0.0
Ethylene oxide	354.22	354.22	354.22	352.04
Carbon dioxide	0.02	0.02	0.02	0.0
Oxygen	2.89	2.89	2.89	0.0
Nitrogen	5.35	5.35	5.35	0.0
Water	40,000.74	40,000.74	40,000.74	0.35

(*continued*)

Table B.6.1 Stream Table for Unit 700 (*Continued*)

Stream Number	33	34
Temp (°C)	182.30	86.40
Pres (bar)	10.50	10.00
Vapor mole fraction	0.00	1.00
Flowrate (kg/h)	720,703	278.78
Flowrate (kmol/h)	40,002.57	9.53
Component flowrates (kmol/h)		
Ethylene	0.0	1.27
Ethylene oxide	2.18	0.0
Carbon dioxide	0.0	0.02
Oxygen	0.0	2.88
Nitrogen	0.0	5.35
Water	40,000.39	0.0

Table B.6.2 Utility Summary for Unit 700

E-701	E-702	E-703	E-704
cw	cw	hps	cw
1,397,870 kg/h	1,988,578 kg/h	87,162 kg/h	5,009,727 kg/h
E-705	**E-706**	**E-707**	**E-708**
hps	cw	cw	hps
135,789 kg/h	4,950,860 kg/h	513,697 kg/h	258,975 kg/h
E-709	**R-701**	**R-702**	
cw	bfw→mps	bfw→mps	
29,609 kg/h	13,673 kg/h	10,813 kg/h	

Table B.6.3 Major Equipment Summary for Unit 700

Compressors*

C-701
Carbon steel
Centrifugal
Power = 19 MW
80% adiabatic efficiency

C-702
Carbon steel
Centrifugal
Power = 23 MW
80% adiabatic efficiency

C-703
Carbon steel
Centrifugal
Power = 21.5 MW
80% adiabatic efficiency

C-704
Carbon steel
Centrifugal
Power = 5.5 MW
80% adiabatic efficiency

C-705
Carbon steel
Centrifugal
Power = 5.5 MW
80% adiabatic efficiency

Heat Exchangers

E-701
$A = 5553\ m^2$
1-2 exchanger, floating head, carbon steel
Process stream in tubes
$Q = 58{,}487$ MJ/h

E-702
$A = 6255\ m^2$
1-2 exchanger, floating head, carbon steel
Process stream in tubes
$Q = 83{,}202$ MJ/h

E-703
$A = 12{,}062\ m^2$
1-2 exchanger, floating head, carbon steel
Process stream in tubes
$Q = 147{,}566$ MJ/h

E-704
$A = 14{,}110\ m^2$
1-2 exchanger, floating head, carbon steel
Process stream in tubes
$Q = 209{,}607$ MJ/h

E-705
$A = 14{,}052\ m^2$
1-2 exchanger, floating head, carbon steel
Process stream in tubes
$Q = 229{,}890$ MJ/h

E-706
$A = 13{,}945\ m^2$
1-2 exchanger, floating head, carbon steel
Process stream in tubes
$Q = 207{,}144$ MJ/h

E-707
$A = 1478\ m^2$
1-2 exchanger, floating head, carbon steel
Process stream in tubes
$Q = 21{,}493$ MJ/h

E-708
$A = 566\ m^2$
1-2 exchanger, floating head, stainless steel
Process stream condenses in shell
$Q = 43{,}844$ MJ/h

E-709
$A = 154\ m^2$
1-2 exchanger, floating head, stainless steel
Process stream boils in shell
$Q = 14{,}212$ MJ/h

(continued)

*Note that all compressors have electric-explosion-proof drives with a backup. These units are designated D-701 A/B through D-705 A/B but are not shown on the PFD.

Table B.6.3 Major Equipment Summary for Unit 700 (*Continued*)

Pump
P-701 A/B
Centrifugal/electric drive
Stainless steel
Power = 4 kW (actual)
75% efficient

Reactors

R-701
Carbon steel, shell-and-tube packed bed
Spherical catalyst pellet, 9 mm diameter
Void fraction = 0.4
$V = 202\ m^3$
10 m tall, 7.38 cm diameter tubes
4722 tubes
100% filled with active catalyst
$Q = 33{,}101$ MJ/h
mps made in shell

R-702
Carbon steel, shell-and-tube packed bed
Spherical catalyst pellet, 9 mm diameter
Void fraction = 0.4
$V = 202\ m^3$
10 m tall, 9.33 cm diameter tubes
2954 tubes
100% filled with active catalyst
$Q = 26{,}179$ MJ/h
mps made in shell

Towers

T-701
Carbon steel
20 SS sieve trays
25% efficient trays
Feeds on trays 1 and 20
24-in tray spacing, 3-in weirs
Column height = 12.2 m
Diameter = 5.6 m

T-702
Carbon steel
20 SS sieve trays
25% efficient trays
Feeds on trays 1 and 20
24-in tray spacing, 3-in weirs
Column height = 12.2 m
Diameter = 5.6 m

T-703
Stainless steel
70 SS sieve trays plus reboiler and condenser
33% efficient trays
Total condenser (E-709)
Feed on tray 36
Reflux ratio = 0.89
12-in tray spacing, 3-in weirs
Column height = 43 m
Diameter = 8.0 m

Vessel
V-701
Stainless steel
Horizontal
L/D = 3.0
$V = 12.7\ m^3$

B.6.3 Simulation (CHEMCAD) Hints

The following thermodynamics packages are strongly recommended for simulation of this process.

- **K-values:** Use a global model of PSRK but use UNIFAC as a local model for T-701 and T-702.
- **Enthalpy:** Use SRK.

B.6.4 References

1. Dever, J. P., K. F. George, W. C. Hoffman, and H. Soo, "Ethylene Oxide," *Kirk-Othmer Encyclopedia of Chemical Technology*, on-line version (New York: John Wiley & Sons, 2004).
2. "Ethylene Oxide," *Encyclopedia of Chemical Processing and Design*, vol. 20, J. J. McKetta, ed. (New York: Marcel Dekker, Inc., 1984), 274–318.
3. Stoukides, M., and S. Pavlou, "Ethylene Oxidation on Silver Catalysts: Effect of Ethylene Oxide and of External Transfer Limitations," *Chem. Eng. Commun.* 44 (1986): 53–74.

B.7 FORMALIN PRODUCTION, UNIT 800

Formalin is a 37 wt% solution of formaldehyde in water. Formaldehyde and urea are used to make urea-formaldehyde resins that subsequently are used as adhesives and binders for particle board and plywood.

B.7.1 Process Description [1, 2]

Unit 800 produces formalin (37 wt% formaldehyde in water) from methanol using the silver catalyst process. Figure B.7.1 illustrates the process.

Air is compressed and preheated, fresh and recycled methanol is pumped and preheated, and these two streams are mixed to provide reactor feed. The feed mixture is about 39 mole % methanol in air, which is greater than the upper flammability limit for methanol. (For methanol, UFL = 36 mole %; LFL = 6 mole %.) In the reactor, the following two reactions occur.

$$\underset{methanol}{CH_3OH} + \frac{1}{2}O_2 \rightarrow \underset{formaldehyde}{HCHO} + H_2O \qquad \Delta H_{rxn} = -37.3 \text{ kcal/mole} \qquad \text{(B.7.1)}$$

$$\underset{methanol}{CH_3OH} \rightarrow \underset{formaldehyde}{HCHO} + H_2 \qquad \Delta H_{rxn} = 20.3 \text{ kcal/mole} \qquad \text{(B.7.2)}$$

The reactor is a unique configuration, in which the silver catalyst is in the form of wire gauze, suspended above a heat exchanger tube bank. Because the net reaction is very exothermic, the heat generated in the adiabatic reactor section must be removed quickly, hence the close proximity of the heat exchanger tubes. The heat exchanger resembles a pool boiler, with a pool of water on the shell side. If the temperature of the effluent is too high, the set point on the steam pressure line is lowered to increase the vaporization of boiler feed water (BFW). In general, the liquid-level controller on the BFW is adjusted to keep the tube bundle fully immersed. The reactor effluent enters an absorber in which most of the methanol and formaldehyde are absorbed into water, with most of the remaining light gases purged into the off-gas stream. The methanol, formaldehyde, and water enter a distillation column, in which the methanol overhead is recycled; the bottoms product is a formaldehyde/water mixture that contains ≤1 wt% methanol as an inhibitor. This mixture is cooled and sent to a storage tank, which is sized at four days' capacity. This storage tank is essential, because some of the downstream processes are batch. The composition in the storage tank exceeds 37 wt% formaldehyde, so the appropriate amount of water is added when the downstream process draws from the storage tank. This is not shown in the PFD (Figure B.7.1).

Storage of formaldehyde/water mixtures is tricky. At high temperatures, undesirable polymerization of formaldehyde is inhibited, but formic acid formation is favored. At low temperatures, acid formation is inhibited, but polymerization is favored. There are stabilizers that inhibit polymerization, but they are incompatible with resin formation. Methanol, at concentrations between 5 wt% and 15 wt%, can also inhibit polymerizaton, but no separation equipment for methanol currently exists on site, and methanol greater than 1 wt% also causes defective resin production. With ≤1 wt% methanol, the storage tank contents must be maintained between 35°C and 45°C.

Stream summary tables, utility summary tables, and major equipment specifications are shown in Tables B.7.1–B.7.3.

B.7.2 Reaction Kinetics

Due to the very high temperature and large surface area of the wire gauze, the reaction may be considered to be instantaneous.

B.7.3 Simulation (CHEMCAD) Hints

Solutions of formaldehyde and water are very non-ideal. Individually, the volatilities are, from most volatile to least volatile, formaldehyde, methanol, and water. However, formaldehyde associates with water so that when this three-component mixture is distilled, methanol is the light key and water is the heavy key. The formaldehyde will "follow" the water. The ESDK K-value package in CHEMCAD simulates this appropriately and was used for the simulation presented here. Latent heat should be used for enthalpy calculations. The expert system will recommend these choices. Alternatively, the data provided in Table B.7.4 can be used directly or to fit an appropriate non-ideal VLE model.

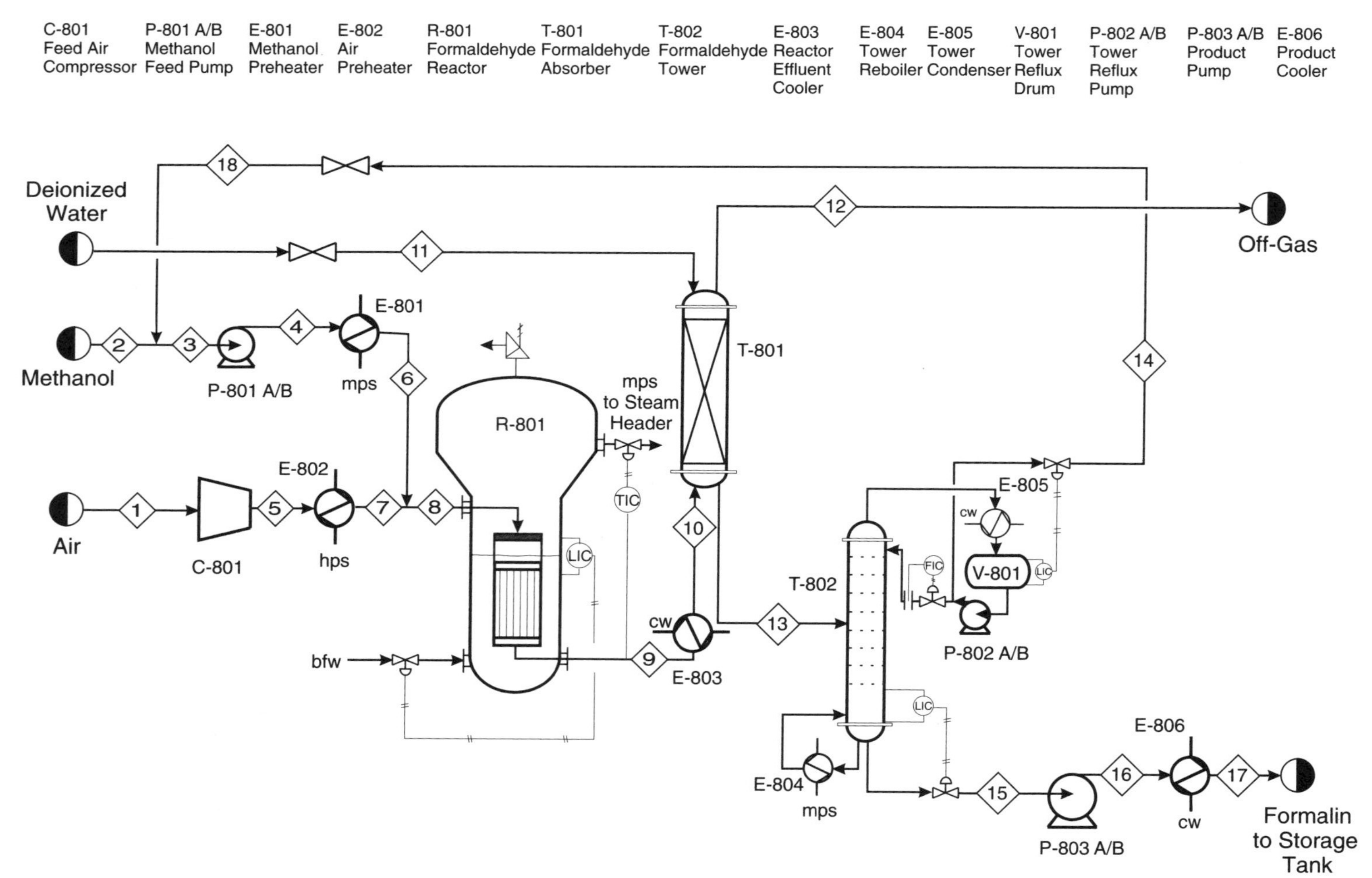

Figure B.7.1 Unit 800: Formalin Process Flow Diagram

Table B.7.1 Stream Tables for Unit 800

Stream Number	1	2	3	4	5	6
Temp (°C)	25.00	30.00	40.66	40.78	183.01	150.00
Pres kPa	101.325	120.00	101.325	300.00	300.00	265.00
Vapor fraction	1.0	0.0	0.0	0.00	1.0	1.0
Total kg/h	4210.54	2464.75	3120.31	3120.31	145.94	99.92
Total kmol/h	145.94	76.92	99.92	99.92	4210.54	3120.31
Component kmol/h						
Methanol	0.0	76.92	94.11	94.11	0.0	94.12
Oxygen	30.66	0.0	0.0	0.0	30.66	0.0
Formaldehyde	0.0	0.0	0.0	0.0	0.0	0.0
Water	0.0	0.0	5.81	5.81	0.0	0.0
Hydrogen	0.0	0.0	0.0	0.0	0.0	0.0
Nitrogen	115.28	0.0	0.0	0.0	115.28	0.0

Stream Number	7	8	9	10	11	12
Temp (°C)	200.00	171.94	200.00	100.00	30.00	84.57
Pres kPa	265.00	255.00	185.00	150.00	150.00	140.00
Vapor fraction	1.0	1.0	1.0	1.0	0.0	1.0
Total kmol/h	145.94	245.86	7330.82	7330.82	2576.15	5354.21
Total kg/h	4210.54	7330.85	278.03	278.03	143.00	224.16
Component kmol/h						
Methanol	0.0	94.12	31.45	31.45	0.0	13.35
Oxygen	30.66	30.66	0.15	0.15	0.0	0.15
Formaldehyde	0.0	0.0	62.67	62.67	0.0	0.04
Water	0.0	5.81	66.82	66.82	143.00	93.68
Hydrogen	0.0	0.0	1.66	1.66	0.0	1.66
Nitrogen	115.28	115.28	115.28	115.28	0.0	115.28

(continued)

Table B.7.1 Stream Tables for Unit 800 (*Continued*)

Stream Number	13	14	15	16	17	18
Temp (°C)	89.85	75.46	106.64	106.71	35.00	73.36
Pres kPa	150.00	130.00	150.00	350.00	315.00	120.00
Vapor mole fraction	0.0	0.0	0.0	0.0	0.0	0.0
Total kg/h	4552.75	655.56	3897.06	3897.06	3897.06	655.56
Total kmol/h	196.87	23.00	173.86	173.86	173.86	23.00
Component kmol/h						
Methanol	18.10	17.19	0.90	0.90	0.90	17.19
Oxygen	0.00	0.00	0.00	0.00	0.00	0.00
Formaldehyde	62.63	0.00	62.63	62.63	62.63	0.00
Water	116.14	5.81	110.33	110.33	110.33	5.81
Hydrogen	0.00	0.00	0.00	0.00	0.00	0.00
Nitrogen	0.00	0.00	0.00	0.00	0.00	0.00

Table B.7.2 Utility Stream Flow Summary for Unit 800

E-801	E-802	E-803	E-804
mps 2063 kg/h	hps 45.43 kg/h	cw 23,500 kg/h	mps 18,949 kg/h
E-805	**E-806**	**R-801**	
cw 775,717 kg/h	cw 27,957 kg/h	bfw → mps 3723 kg/h	

Table B.7.3 Major Equipment Summary for Unit 800

Compressor
C-801
Carbon steel
Centrifugal
Power = 183 kW (shaft)
70% efficient

D-801 A/B (not shown on PFD)
Electric/explosionproof
W = 195 kW
95% efficient

(*continued*)

Table B.7.3 Major Equipment Summary for Unit 800 (*Continued*)

Heat Exchangers

E-801
$A = 405\ m^2$
1-2 exchanger, floating head, carbon steel
Process stream in shell
$Q = 4111$ MJ/h
Maximum pressure rating of 350 kPa

E-802
$A = 4.62\ m^2$
1-2 exchanger, floating head, carbon steel
Process stream in tubes
$Q = 76.75$ MJ/h
Maximum pressure rating of 350 kPa

E-803
$A = 28.16\ m^2$
1-2 exchanger, floating head, carbon steel
Process stream in shell
$Q = 983.23$ MJ/h
Maximum pressure rating of 350 kPa

E-804
$A = 37.3\ m^2$
1-2 exchanger, kettle reboiler, stainless steel
Process stream in shell
$Q = 37{,}755$ MJ/h
Maximum pressure rating of 250 kPa

E-805
$A = 269\ m^2$
1-2 exchanger, floating head, stainless steel
Process stream in shell
$Q = 32{,}456$ MJ/h
Maximum pressure rating of 250 kPa

E-806
$A = 41\ m^2$
1-2 exchanger, floating head, stainless steel
Process stream in tubes
$Q = 1169.7$ MJ/h
Maximum pressure rating of 400 kPa

Reactors

R-801, Heat-Exchanger Portion
$A = 140.44\ m^2$
Counterflow exchanger, floating head, carbon steel
Process stream in tubes
$Q = 8{,}928$ MJ/h
Maximum pressure rating of 350 kPa

R-801, Reactor Portion
Thin layers of silver wire gauze suspended above heat exchanger tube bank

Pumps

P-801 A/B
Centrifugal/electric drive
Carbon steel
Power = 0.3 kW
80% efficient

P-802 A/B
Centrifugal/electric drive
Carbon steel
Power = 1.7 kW
80% efficient

P-803 A/B
Centrifugal/electric drive
Stainless steel
Power = 0.5 kW
75% efficient

(*continued*)

Table B.7.3 Major Equipment Summary for Unit 800 (*Continued*)

Towers	
T-801	**T-802**
Carbon steel	Stainless steel
10 m of packing	31 SS sieve trays plus reboiler and partial condenser
2-in ceramic Berl Saddles	70% efficient trays
20 theoretical stages	Feed on tray 18
1.00 kPa/m pressure drop	Reflux ratio = 37.34
Diameter = 0.86 m	0.6096 m tray spacing, 0.091 m weirs
Packing factor = 45	Column height = 19 m
Maximum pressure rating of 300 kPa	Diameter = 2.5 m
	Maximum pressure rating of 200 kPa
Vessel	
V-801	
Horizontal	
Stainless steel	
L/D = 4.0	
Volume = 4.2 m^3	

When simulating an entire process, we recommend first using the Shortcut distillation column within the process for the methanol-water/formaldehyde distillation. A rigorous column solver should then be used as a separate item to simulate the column based on the results obtained from the shortcut column. However, due to the non-ideality of the thermodynamics, the actual column sim-

Table B.7.4 *K*-values for Formaldehyde/Water/Methanol System [2]

P(psia) =14.696	Chemical Component		
T (°C)	Formaldehyde	Water	Methanol
0.1	0.123	1.000	0.273
67.1	0.266	0.491	1.094
72.1	0.336	0.394	1.435
74.8	0.374	0.453	1.598
84.6	0.546	0.607	2.559
97.6	0.693	1.105	2.589
99.9	0.730	1.198	2.595
150.1	1.220	2.460	3.004

ulation using the rigorous column will probably require many more stages than predicted by the shortcut simulation, possibly twice the number. Once the parameters for the rigorous column have been established, the Shortcut column can be replaced by the rigorous column and the simulation rerun to get a converged simulation.

B.7.4 References

1. Gerberich, H. R., and G. C. Seaman, "Formaldehyde," *Kirk-Othmer Encyclopedia of Chemical Technology*, on-line version (New York: John Wiley & Sons, 2004).
2. "Formaldehyde," *Encyclopedia of Chemical Processing and Design*, vol. 23, J. J. McKetta, ed. (New York: Marcel Dekker, Inc., 1984), 350–371.
3. Gmehling, J., U. Onken, and W. Arlt, *Vapor-Liquid Equilibrium Data Collection*, Chemistry Data Series (Aqueous-Organic Systems, Supplement 1), vol. 1, part 1a, DECHEMA, 1981, 474–475.

B.8 BATCH PRODUCTION OF L-PHENYLALANINE AND L-ASPARTIC ACID, UNIT 900

Phenlyalanine and L-aspartic acid are amino acids. When they bond together, the corresponding di-peptide methyl ester is aspartame, known by the brand name Nutra-Sweet or Equal. Production of both amino acids can be accomplished via fermentation of genetically altered bacteria. Production rates of 1,000 and 1,250 tonnes/yr of L-aspartic acid and L-phenylalanine are desired.

B.8.1 Process Description

To accomplish a fermentation process, bacteria must grow in the presence of appropriate nutrients that facilitate the production of the desired product. In a processing context, the fermentation reactor must first be primed with the bacteria and the nutrients. The nutrient feed includes the reactant that the bacteria metabolize to produce the desired amino acid. Air is also sparged into the fermenter as a source of oxygen. All of these feeds are passed through sterilization filters prior to entering the reactor. The bacteria are then allowed to multiply, and the desired product, an amino acid in this case, is produced. In this process, both products are extracellular. After the desired production level of the amino acid is reached, the fermentation broth pH is lowered by addition of sulfuric acid, the bacteria are removed from the fermentation broth by filtration, and the product stream is sent to a holding tank. The addition of acid titrates the amino acid, making it positively charged. The addition of acid is done only for phenylalanine, because L-aspartic acid bypasses the ion exchange column and is crystallized directly via precipitation from solution.

In this process, both amino acids are produced in the same facility. Because fermentation is involved and production levels are low compared with typical commodity chemicals, batch processes are involved. In batch processes, the key variable is the batch time, or the length of time that the unit is allowed to run. For example, in a batch reactor, the batch time is analogous to the space time in a continuous reactor.

The separation sequence is a continuous process, which is accomplished by a continuous feed from the holding tank. This is not uncommon in batch facilities, because many separation processes are more easily accomplished in the continuous mode. The separation sequence for the two amino acids differs slightly. Phenylalanine is isolated using ion exchange followed by crystallization; in contrast, L-aspartic acid is crystallized directly from the filtered fermentation broth. For phenylalanine, it is adsorbed on the ion exchange resin and subsequently eluted using a basic solution. The addition of base neutralizes the positive charge to promote desorption from the ion exchange resin. For both amino acids, filtration follows crystallization. The product is then sent to storage. The process is shown in Figure B.8.1.

The use of batch processing requires batch scheduling of the type discussed in Chapter 3, which allows use of the same equipment to manufacture both amino acids in the same facility. In this description, only the PFD, reactor calculations, and general descriptions of the separation units are presented. The design of individual equipment, the utility consumption, and the production schedule for the plant are left as exercises for the student. A description of a process to produce four amino acids (including the two amino acids in this process) in the same facility is available at http://www.che.cemr.wvu.edu/publications/projects/large_proj/batch-production_of_amino_acids.pdf. This process description includes possible batch schedules for both the reactors and the separation section.

B.8.2 Reaction Kinetics

L-Aspartic Acid. The reaction of fumaric acid to form L-aspartic acid is an enzymatic conversion carried out using the aspartase activity of bacteria *Escherichia Coli (E. Coli)* cells according to the following reaction.

$$\underset{\text{fumaric acid}}{C_4H_4O_4} + NH_3 \overset{\text{aspartase}}{\leftrightarrows} \underset{\text{L-aspartic acid}}{C_4H_7NO_4} \qquad \text{(B.8.1)}$$

The bacteria cells are suspended in a matrix polyacrylamide gel, and the reacting species must diffuse in and out of the matrix. The diffusivities of the substrate (fumaric acid) and product (L-aspartic acid) in the gel decrease as their concentrations increase due to the tendency of the gel to shrink at low pH.

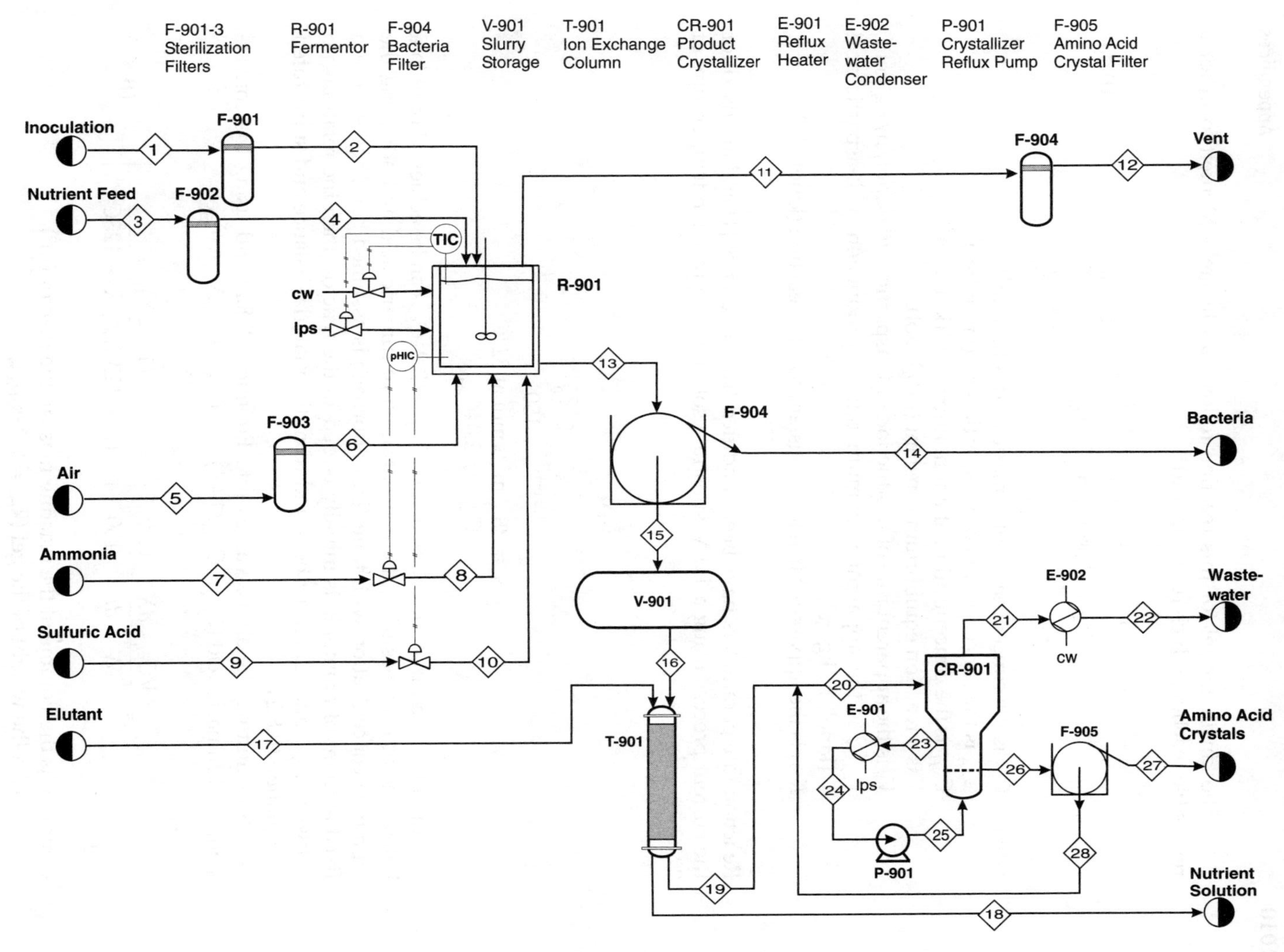

Figure B.8.1 Unit 900: Amino Acid Process Flow Diagram

The kinetic model for this reaction follows a Michaelis-Menten form for a reversible reaction, which rearranges to

$$\frac{1}{V_a} = \frac{1}{V_{a,\max}} + \frac{K_{a,m}}{V_{a,\max}\left[C_{FA} - \dfrac{C_{AA}}{KC_{NH_4^+}}\right]} \tag{B.8.2}$$

where C_{FA} is the concentration of fumaric acid (kmol/m^3).
C_{AA} is the concentration of L-aspartic acid (kmol/m^3).
$C_{NH_4^+}$ is the concentration of ammonium ions (kmol/m^3).
K is a reaction equilibrium constant (m^3/kmol).
V_a is the apparent rate of production of L-aspartic acid (kmol/h/kg-gel).
$V_{a,\max}$ is the apparent maximum rate of production of L-aspartic acid (kmol/h/kg-gel).
$K_{a,M}$ is the apparent Michaelis constant for the reaction (kmol/m^3).

Reaction rate parameters have been modified from reference [1] and are used for the current process using a 1.0 M substrate solution at a reaction temperature of 32°C.

$$K_{a,M} = 0.68C_{FA,0}^{1.04}$$

$$V_{a,\max} = \frac{C_{FA,0}^{0.77}}{150}$$

$$K = 88.7 \text{ m}^3/\text{kmol at } 32°\text{C}$$

$$C_{NH_4^+} = 2.04C_{FA}$$

It should be noted that the relationship $C_{NH_4^+} = 2.04C_{FA}$ can be achieved only in a batch reactor by measuring the concentration of fumaric acid and adjusting the ammonia concentration with time. This approach is assumed here; however, if a fixed amount of ammonia is initially added to the reactor, then the relationship between $C_{NH_4^+}$ and C_{FA} must be found from the material balance and substituted in Equation (B.8.2).

Substituting the above values into Equation (B.8.2) and using the conversion, X, of fumaric acid ($C_{FA} = C_{FA,0}(1-X)$ and $C_{AA} = C_{FA,0}X$), we get

$$V_a = \frac{VC_{FA,0}}{W}\frac{dX}{dt} = \frac{C_{FA,0}^{0.77}}{150}\left[\frac{(1-2X)}{1-(2+123C_{FA,0}^{1.04})X + 123C_{FA,0}^{1.04}}\right] \tag{B.8.3}$$

where V is the volume of the reacting mass in the reactor (m^3).
W is the weight of the gel (kg) = $V(1-\varepsilon)\rho_{bead}$.
ε is the void fraction of beads in the reacting mass.
ρ_{bead} is the bead density (kg/m^3).

Substituting into Equation (B.8.3) we have

$$\frac{dX}{dt} = \frac{(1-\varepsilon)\rho_{bead}}{C_{FA,0}^{0.23}150}\left[\frac{(1-2X)}{(1+123C_{FA,0}^{1.04}) - (2+123C_{FA,0}^{1.04})X}\right] \quad \text{(B.8.4)}$$

For the specified initial concentration of fumaric acid of 1.0M = 1 kmol/m^3 and with $\rho_{bead} \sim$ 1,000 kg/m^3 and assuming a voidage of 0.5, Equation (B.8.4) simplifies to

$$\frac{dX}{dt} = 3.33\left[\frac{(1-2X)}{124-125X}\right] \quad \text{(B.8.5)}$$

Separating variables and integrating Equation (B.8.5) yield the conversion as a function of batch reaction time. This relationship is shown in Figure B.8.2.

Preliminary Sizing of Reactor R-901. For a conversion of 45% (90% of equilibrium), a reaction time of approximately 30 hours is required. Assuming that an additional 5 hours is required for filling, cleaning, and heating, the total reactor step time is 35 hours.

Using a reactor size of 37.9 m^3 (10,000 gal) and assuming a 90% fill volume and a voidage of 0.5, the amount of fumaric acid fed to the batch is (37.9)(0.9)(0.5) = 17.04 m^3, or 17.04 kmol (17.04 × 116 = 1977 kg). The amount of L-aspartic acid produced = (17.04)(0.45) = 7.67 kmol = (7.67)(133) = 1020 kg.

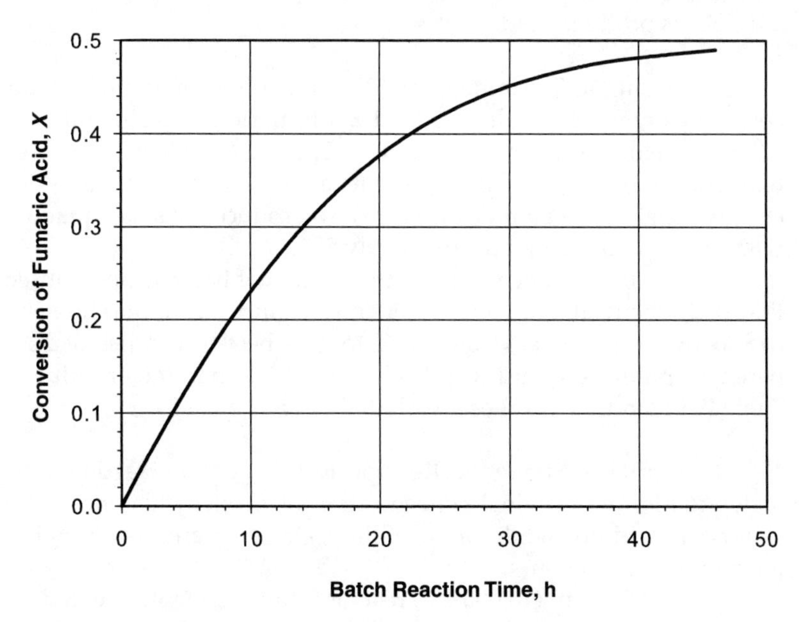

Figure B.8.2 Conversion of Fumaric Acid to L-Aspartic Acid as a Function of Reaction Time

Production rate of L-aspartic acid from a 10,000 gallon reactor is 1020 kg/batch using a batch time of 35 hr.

L-Phenylalanine. L-phenylalanine is produced via fermentation using a mutant *Brevibacterium lactofermentum* 2256 (ATCC No. 13869) known as No. 123 [2]. The rate equations for biomass (bacteria, X), substrate (mainly glucose, *S*), and product (L-phenylalanine, *P*) are described by Monod kinetics.

$$\frac{dX}{dt} = \frac{\mu_m S}{K_s + S} X \tag{B.8.6}$$

$$\frac{dS}{dt} = -\frac{1}{Y_s}\frac{\mu_m S}{K_s + S} X \tag{B.8.7}$$

$$\frac{dP}{dt} = -\frac{Y_{ps}}{Y_{xs}}\frac{\mu_m S}{K_s + S} X \tag{B.8.8}$$

where

X is the concentration of bacteria (kg/m^3).
S is the concentration of substrate (glucose) (kg/m^3).
P is the concentration of product (L-phenylalanine) (kg/m^3).
μ_m is the maximum specific growth rate (hr^{-1}).
K_s is the Monod constant (kg/m^3).
Y_{XS} is biomass yield.
Y_{PS} is product yield.

According to Tsuchida et al. [2], for a culture medium containing 13% glucose, 1% ammonium sulfate, and 1.2% fumaric acid (plus other trace nutrients etc.) the yield of L-phenylalanine was 21.7 mg/ml after 72 hr of cultivation at a temperature of 31.5°C. This represents a yield of approximately 16.7% from glucose by weight. Other amino acids are also produced in small quantities (<5 kg/m^3), with Lysine making up approximately 50%.

To obtain a kinetic model of the growth of bacteria and subsequent production of L-phenylalanine and depletion of glucose, the parameters in Equations (B.8.6)–(B.8.8) were back calculated to give best-fit profiles of *X*, *S*, and *P* compared to published values ([2], Figure 4). The parameter values are shown in Table B.8.1, and the profiles are plotted in Figure B.8.3.

Preliminary Sizing of Reactor R-901. For a reaction time of approximately 60 hours, the final concentration of L-phenylalanine is 21.0 kg/m^3. Assuming an additional 5 hours for filling, cleaning, and heating gives a total reactor step time of 65 hours.

Using a 37.9 m^3 (10,000 gal) reactor for the fermentation and assuming that a 90% fill volume is used, the volume of reactants is (37.9)(0.9) = 34.11 m^3.

Table B.8.1 Best-Fit Parameters for Monod Kinetics Using *Brevibacterium lactofermentum* 2256 (ATCC No. 13869) Strain No. 123

Parameter	Value
μ_m	0.25 hr^{-1}
K_s	105.4 kg/m^3
Y_{xs}	0.07
Y_{ps}	0.167
X_0	0.0114 kg/m^3
S_0	130.0 kg/m^3
P_0	0.0 kg/m^3

The amount of L-phenylalanine produced in a 60 h batch reaction is (34.11)(21.0) = 716 kg.

Production rate of L-aspartic acid from a 10,000 gallon reactor is 716 kg/batch, with a reactor step time of 65 hr.

Preliminary Information on Other Equipment. As mentioned in the process description, the production of both L-aspartic acid and L-phenylalanine follows similar paths. A brief discussion of the unit operations involved with the separation and purification of the final products is in order, because these operations are not typical of the unit operations covered in this text. Size information is not included for the equipment described next; however, estimates of processing times, where applicable, are given.

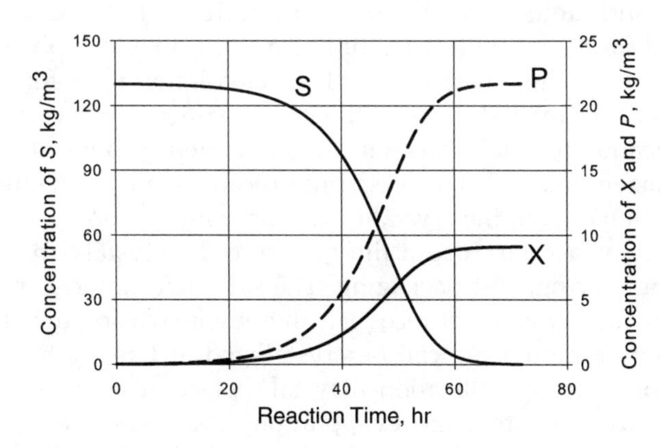

Figure B.8.3 Concentration Profiles as a Function of Reactor Time

Filtration of Bacteria. After reaction, the bacteria must be filtered from the mother liquor prior to storage. The bacteria tend to give rise to slimy filter cakes, and the filtration of such material is best accomplished using a rotary drum filter utilizing a precoat. Typical precoating materials are dolomite, perlite, and cellulose, and these are applied to the drum in a two-stage process prior to filtration. The precoating process involves depositing a layer of the precoat material (5–15 cm thick) on the drum prior to the filtration operation. Once the precoat has been applied, the filtration starts, and the biomass forms a thin layer on the precoat. This layer of biomass is continuously removed, along with a thin layer of the precoat material, using a sharp-edged "Doctor" blade. Additional information is given at http://www.solidliquid-separation.com [3]. Without doing detailed calculations, it is difficult to determine the time required for the precoat stage and filtration stages. For this project you may assume that these steps take 25 h and 5 h, respectively.

Intermediate Storage. The fermentation broth (free of solids, or biomass) leaving the filters is stored in an intermediate storage tank prior to being sent to the ion exchange column (for L-phenylalanine) and on to the crystallizer. The use of an intermediate storage vessel allows the remainder of the process to operate as a continuous process.

Ion Exchange Column. The ion exchange columns operate as semibatch processes. Hydrochloric acid is added to the L-phenylalanine-containing solution and is passed through freshly regenerated ion exchange resin such as DOW's Dowex Marathon C [4]. The resin captures the positively charged amino acid. Once the bed is full, it is backwashed with a basic solution of ammonium or sodium hydroxide, which breaks the resin amino acid bond. The resin is subsequently washed free of the hydroxide, and the cycle starts again.

Continuous Crystallizer and Filtration. Draft tube baffle crystallizers can be used for the crystallization of L-aspartic acid and L-phenylalanine. These crystallizers offer the advantages of high circulation rates for efficient mixing. Fines removal is facilitated through the use of baffles, and a certain amount of product classification (crystal size control) is obtained through the elutriating leg. Batch crystallizers could also be used, but product quality and efficiency suffer. The saturated liquid from the crystallizer, containing amino acid crystals, is sent to a filter (such as a rotary drum filter), where the crystals are removed and sent for washing, drying, and packaging. The saturated liquid is returned to the crystallizer for further product recovery, thereby increasing the efficiency of the operation. Both amino acids can be crystallized at temperatures greater than 100°C. Therefore, the crystallization may take place at ambient pressure by removing the excess water through evaporation. The solubilities of L-aspartic acid and L-phenylalanine at 100°C are 67 g/liter and 100 g/liter, respectively.

B.8.3 References

1. Takamatch, T., K. Yamashita, and A. Sumi, "Kinetics of Production of L-aspartic Acid by Aspartase of Immobilized E. Coli Cells," *Japanese Journal of Fermentation Technology* 58, no. 2 (1980): 129–133.
2. Tsuchida, T., K. Kubota, Y. Morinaga, H. Matsui, H. Enei, and F. Yoshinga, "Production of L-phenylalanine by a Mutant of *Brevibacterium lactofermentum* 2256," *Agric. Bio. Chem.* 51, no. 8 (1987): 2095–2101.
3. http://www.solidliquid-separation.com.
4. DOWEX Marathon C data sheet, http://www.dow.com/PublishedLiterature/dh_0082/0901b80380082af5.pdf?filepath=liquidseps/pdfs/noreg/177-01593.pdf&fromPage=GetDoc.

B.9 ACRYLIC ACID PRODUCTION VIA THE CATALYTIC PARTIAL OXIDATION OF PROPYLENE [1, 2, 3, 4, 5]

Acrylic acid (AA) is used as a precursor for a wide variety of chemicals in the polymer and textile industries. There are several chemical pathways to produce AA, but the most common one is via the partial oxidation of propylene. The usual mechanism for producing AA utilizes a two-step process in which propylene is first oxidized to acrolein and then further oxidized to AA. Each reaction step usually takes place over a separate catalyst and at different operating conditions. The reaction stoichiometry is given below:

$$C_3H_6 + O_2 \rightarrow \underset{\text{Acrolein}}{C_3H_4O} + H_2O$$

$$C_3H_4O + \frac{1}{2}O_2 \rightarrow \underset{\text{Acrylic Acid}}{C_3H_4O_2}$$

Several side reactions may occur, most resulting in the oxidation of reactants and products. Some typical side reactions are given below:

$$C_3H_4O + \frac{7}{2}O_2 \rightarrow 3CO_2 + 2H_2O$$

$$C_3H_4O + \frac{3}{2}O_2 \rightarrow \underset{\text{Acetic Acid}}{C_2H_4O_2} + CO_2$$

$$C_3H_6 + \frac{9}{2}O_2 \rightarrow 3CO_2 + 3H_2O$$

Therefore, the typical process setup consists of a two-reactor system, with each reactor containing a separate catalyst and operating at conditions so as to maximize the production of AA. The first reactor typically operates at a higher temperature than the second.

As with any reaction involving the partial oxidation of a fuel-like feed material (propylene), considerable attention must be paid to the composition of hydrocarbons and oxygen in the feed stream. In the current design, a fluidized-bed reactor is used, which provides essentially isothermal conditions in the reactor and, with the addition of large amounts of steam, allows safe and stable operation. The second safety concern is associated with the highly exothermic polymerization of AA, which occurs in two ways. First, if this material is stored without appropriate additives, then free radical initiation of the polymerization can occur. This potentially disastrous situation is discussed by Kurland and Bryant [1]. Second, AA dimerizes when in high concentrations at temperatures greater than 90°C, and thus much of the separation sequence must be operated under high vacuum in order to keep the bottom temperatures in the columns below this temperature.

B.9.1 Process Description

The process shown in Figure B.9.1 produces 50,000 metric tons per year of 99.9% by mole AA product. The number of operating hours is taken to be 8000/yr, and the process is somewhat simplified because there is only one reactor [5]. It is assumed that both reactions take place on a single catalyst to yield AA and by-

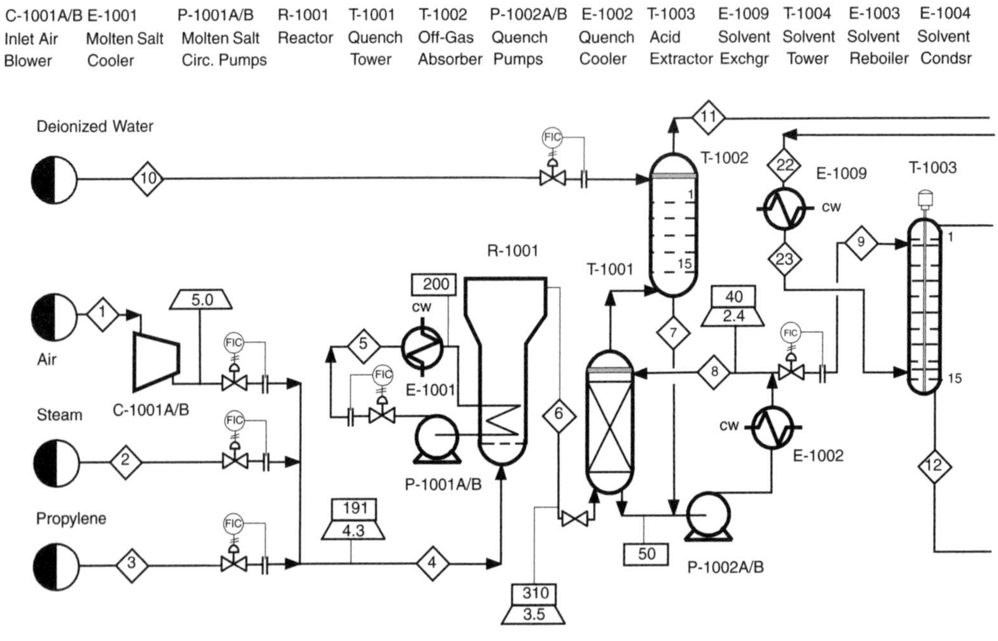

Figure B.9.1 Unit 1000: Production of Acrylic Acid from Propylene PFD (The point where Streams 1 and 2 are mixed with Stream 3 to form Stream 4 actually occurs within Reacor R-1001.)

products. It is imperative to cool the products of reaction quickly to avoid further oxidation reactions, and this is achieved by rapidly quenching the reactor effluent with a cool recycle, Stream 8, of dilute aqueous AA in T-1001. Additional recovery of AA and acetic acid (a by-product) is achieved in the absorber, T-1002. The stream leaving the absorption section is a dilute aqueous acid, Stream 9. This is sent to a liquid-liquid extractor, T-1003, to remove preferentially the acid fraction from the water prior to purification. There are many possible solvents that can be used as the organic phase in the separation; high solubility for AA and low solubility for water are desirable. Some examples include ethyl acrylate, ethyl acetate, xylene, diisobutyl ketone, methyl isobutyl ketone, and diisopropyl ether (DIPE), which is used here. The organic phase from T-1003 is sent to a solvent recovery column, T-1004, where the diisopropyl ether (and some water) is recovered overhead and returned to the extractor. The bottom stream from this column, Stream 14, contains virtually all the AA and acetic acid in Stream 9. This is sent to the acid purification column, T-1005, where 95% by mole acetic acid by-product is produced overhead, and 99.9 % by mole AA is produced as a bottoms product and cooled prior to being sent to storage.

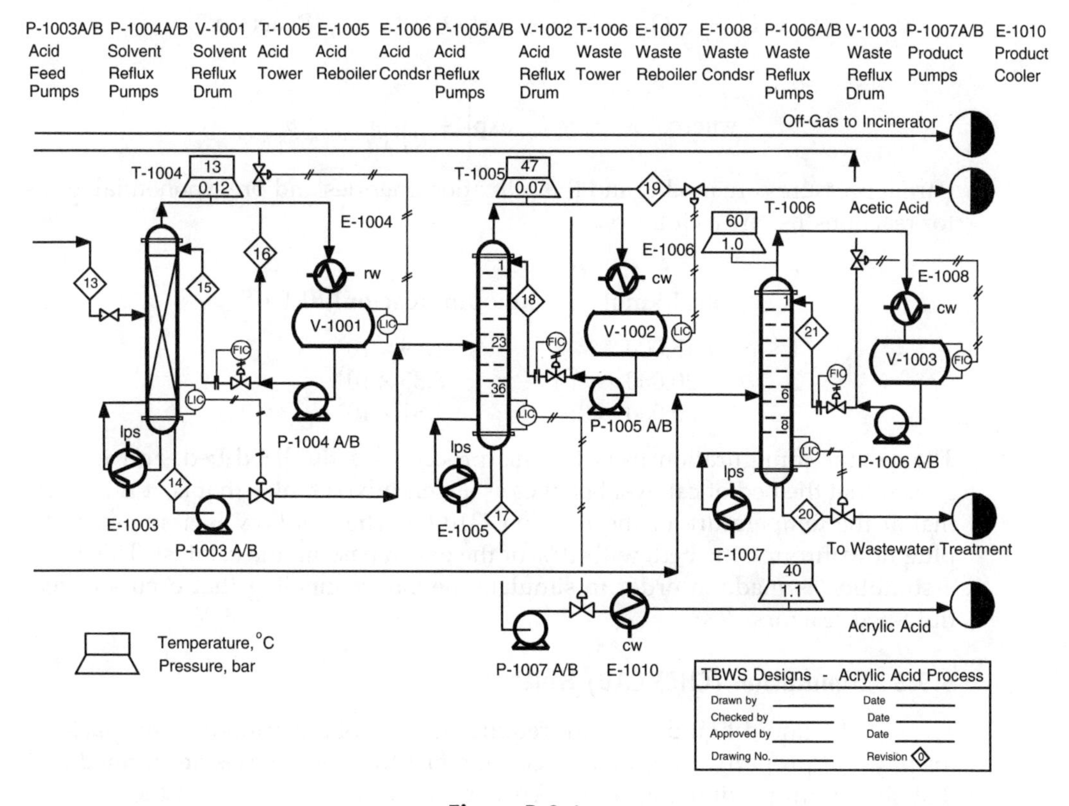

Figure B.9.1 *(Continued)*

The aqueous phase from the extractor, Stream 12, is sent to a wastewater column, T-1006, where a small amount of DIPE is recovered overhead and returned to the extractor. The bottoms product, containing water and trace quantities of solvent and acid, is sent to wastewater treatment. Process stream information and preliminary equipment summaries are given in Tables B.9.1 and B.9.2, respectively. A utility summary is also provided in Table B.9.3.

B.9.2 Reaction Kinetics and Reactor Configuration

The reactions taking place are kinetically controlled at the conditions used in the process; that is, equilibrium lies far to the right. The reaction kinetics for the catalyst used in this process are given below:

$$C_3H_6 + \frac{3}{2}O_2 \rightarrow \underset{Acrylic\ Acid}{C_3H_4O_2} + H_2O \qquad \text{Reaction 1}$$

$$C_3H_6 + \frac{5}{2}O_2 \rightarrow \underset{Acetic\ Acid}{C_2H_4O_2} + CO_2 + H_2O \qquad \text{Reaction 2}$$

$$C_3H_6 + \frac{9}{2}O_2 \rightarrow 3CO_2 + 3H_2O \qquad \text{Reaction 3}$$

$$\text{where} \qquad -r_i = k_{o,i}\exp\left[-\frac{E_i}{RT}\right]p_{propylene}\,p_{oxygen}$$

Partial pressures are in kPa, and the activation energies and preexponential terms for reactions 1–3 are as follows.

i	E_i kcal/kmol	$k_{o,i}$ kmol/m³ reactor h/(kPa)²
1	15,000	1.59×10^5
2	20,000	8.83×10^5
3	25,000	1.81×10^8

The reactor configuration used for this process is a fluidized bed, and it is assumed that the bed of catalyst behaves as a well-mixed tank—that is, it is isothermal at the temperature of the reaction (310°C). The gas flow is assumed to be plug flow through the bed, with 10% of the gas bypassing the catalyst. This latter assumption is made in order to simulate the gas channeling that occurs in real fluid-bed reactors.

B.9.3 Simulation (CHEMCAD) Hints

The use of a liquid-liquid extractor requires the use of a thermodynamic package (or physical property data) that reflects the fact that two phases are formed and that significant partitioning of the AA and acetic acid occurs, with the majority

Table B.9.1 Flow Summary Table for Acrylic Acid Process in Figure B.9.1

Stream Number	1	2	3	4	5	6	7	8	9
Temperature (°C)	25	159	25	191	250	310	63	40	40
Pressure (bar)	1.0	6.0	11.5	4.3	3.0	3.5	1.4	2.4	2.4
Vapor fraction	1.0	1.0	1.0	1.0	0.0	1.0	0.0	0.0	0.0
Mass flow (tonne/h)	39.05	17.88	5.34	62.27	1075.0	62.27	3.08	1895.	27.46
Mole flow (kmol/h)	1362.9	992.3	127.0	2482.2	0.00	2444.0	148.5	85200.0	1249.6
Component mole flow (kmol/h)					HiTec Molten Salt				
Propylene	0.00	0.00	127.0	127.0	0.00	14.7	0.00	0.00	0.00
Nitrogen	1056.7	0.00	0.00	1056.7	0.00	1056.7	0.00	0.00	0.00
Oxygen	280.9	0.00	0.00	280.9	0.00	51.9	0.00	0.00	0.00
Carbon dioxide	0.00	0.00	0.00	0.00	0.00	60.5	0.00	0.00	0.00
Water	25.3	992.3	0.00	1017.6	0.00	1165.9	140.9	78870	1156.7
Acetic acid	0.00	0.00	0.00	0.00	0.00	6.54	0.65	415	6.08
Acrylic acid	0.00	0.00	0.00	0.00	0.00	87.79	6.99	5915	86.81
Solvent (diisopropyl ether)	0.00	0.00	0.00	0.00	0.00	0.00	0.00	0.00	0.00

(continued)

Table B.9.1 Flow Summary Table for Acrylic Acid Process in Figure B.9.1 (*Continued*)

Stream Number	10	11	12	13	14	15	16	17	18
Temperature (°C)	25	48	40	40	90	13	13	89	47
Pressure (bar)	5.0	1.0	2.4	2.4	0.19	0.12	3.0	0.16	0.07
Vapor fraction	0.0	1.0	0.0	0.0	0.0	0.0	0.0	0.0	0.0
Mass flow (tonne/h)	2.54	37.35	20.87	143.0	6.63	155.3	136.4	6.26	5.28
Mole flow (kmol/h)	141.0	1335.4	1156.9	1591.2	93.19	1705.7	1498.0	86.85	90.49
Component mole flow (kmol/h)									
Propylene	0.00	14.7	0.00	0.00	0.00	0.00	0.00	0.00	0.00
Nitrogen	0.00	1056.7	0.00	0.00	0.00	0.00	0.00	0.00	0.00
Oxygen	0.00	51.9	0.00	0.00	0.00	0.00	0.00	0.00	0.00
Carbon dioxide	0.00	60.5	0.00	0.00	0.00	0.00	0.00	0.00	0.00
Water	141.0	150.2	1156.6	198.8	0.30	226.0	198.5	0.00	4.28
Acetic acid	0.00	0.46	0.03	6.08	6.08	0.00	0.00	0.05	86.07
Acrylic acid	0.00	0.98	0.00	86.81	86.81	0.00	0.00	86.80	0.14
Solvent (diisopropyl ether)	0.00	0.00	0.30	1299.5	0.00	1479.7	1299.5	0.00	0.00

(*continued*)

Table B.9.1 (*Continued*)

Stream Number	**19**	**20**	**21**	**22**	**23**
Temperature (°C)	47	102	60	13	40
Pressure (bar)	1.1	1.1	1.0	3.0	2.8
Vapor fraction	0.0	0.0	0.0	0.0	0.0
Mass flow (tonne/h)	0.37	20.84	37.37	136.4	136.4
Mole flow (kmol/h)	6.34	1156.43	470.2	1498.5	1498.5
Component mole flow (kmol/h)					
Propylene	0.00	0.00	0.00	0.00	0.00
Nitrogen	0.00	0.00	0.00	0.00	0.00
Oxygen	0.00	0.00	0.00	0.00	0.00
Carbon dioxide	0.00	0.00	0.00	0.00	0.00
Water	0.30	1156.4	126.8	198.7	198.7
Acetic acid	6.03	0.03	0.00	0.00	0.00
Acrylic acid	0.01	0.00	0.00	0.00	0.00
Solvent (diisopropyl ether)	0.00	0.00	343.4	1299.8	1299.8

Table B.9.2 Preliminary Equipment Summary Table for Acrylic Acid Process

Equipment	T-1001	T-1002	T-1003	T-1004	T-1005	T-1006	R-1001*
MOC	Stainless Steel	Stainless Steel	Stainless Steel	Stainless Steel	Carbon Steel	Stainless Steel	Carbon Steel Fluidized Bed
Diameter (m)	5.3	3.5	2.2	7.5	2.4	2.3	3.6
Height/length (m)	12	11	9.5	34	25	7.0	10
Orientation	Vertical	Vertical	Vertical	Vertical	Vertical	Vertical	Vertical
Internals	10m of High-efficiency Packing Polyethylene	15 Sieve Trays + Demister Stainless Steel	15 Perforated Plates + Mixer Stainless Steel	31 m of High-efficiency Structured Packing Stainless Steel	36 Sieve Plates Stainless Steel	8 Sieve Plates Stainless Steel	1451 Heat Transfer Tubes (1420 m^2) Filled with molten salt
Pressure (barg)	1.4	1.0	1.4	–1.0	–1.0	0	3.0

(continued)

*Installed cost of reactor (mid-1996) = $\$ 2 \times 10^5 [\text{Area } (m^2)]^{0.5}$.

Table B.9.2 ***(Continued)***

Equipment	P-1001 A/B	P-1002 A/B	P-1003 A/B	P-1004 A/B	P-1005 A/B	P-1006 A/B	P-1007 A/B
MOC	Carbon Steel	Stainless Steel	Stainless Steel	Stainless Steel	Carbon Steel	Carbon Steel	Stainless Steel
Power (shaft) (kW)	32.3	106.2	0.9	51.3	1.2	9.0	1.0
Efficiency	75%	75%	40%	75%	40%	60%	40%
Type/drive	Centrifugal/ Electric	Centrifugal/ Electric	Centrifugal/ Electric	Centrifugal/ Electric	Centrifugal/ Electric	Centrifugal/ Electric	Centrifugal/ Electric
Temperature (°C)	250	50	90	13	47	60	89
Pressure In (bar)	2.0	1.4	0.19	0.12	0.07	1.0	0.16
Pressure Out (bar)	3.6	2.9	2.05	4.62	3.31	4.62	2.46

(*continued*)

Table B.9.2 Preliminary Equipment Summary Table for Acrylic Acid Process (*Continued*)

Equipment	E-1001	E-1002	E-1003	E-1004	E-1005	E-1006	E-1007
Type	Floating Head	Fixed TS	Floating Head Reboiler	Floating Head Condenser	Floating Head Reboiler	Fixed TS Condenser	Floating Head Reboiler
Duty (MJ/h)	83,400	70,300	101,000	108,300	2230	2280	21,200
Area (m^2)	160	2550	891	7710	19.7	73.3	187
Shell							
Max temp. (°C)	40	40	160	10	160	40	160
Pressure (barg)	4.0	4.0	5.0	4.0	5.0	4.0	5.0
Phase	L	L	Cond. Steam	L	Cond. Steam	L	Cond. Steam
MOC	Carbon Steel	Carbon Steel	Carbon Steel	Carbon Steel	Carbon Steel	Carbon Steel	Carbon Steel
Tube side							
Max temp(°C)	250	50	90	13	90	47	102
Pressure (barg)	2.0	2.0	–0.81	–0.88	–0.84	–0.93	0.1
MOC	Carbon Steel	Stainless Steel	Stainless Steel	Stainless Steel	Carbon Steel	Carbon Steel	Stainless Steel
Phase	L	L	Boiling Liq.	Cond. Vapor	Boiling Liq.	Cond. Vapor	Boiling Liq.

(*continued*)

Table B.9.2 Preliminary Equipment Summary Table for Acrylic Acid Process (*Continued*)

Equipment	E-1008	E-1009	E-1010
Type	Fixed TS Condenser	Floating Head	Floating Head
Duty (MJ/h)	15,800	8,000	698
Area (m^2)	210	19.7	10.3
Shell side			
Max temp. (°C)	40	160	40
Pressure (barg)	4.0	5.0	4.0
Phase	L	Cond. Steam	L
MOC	Carbon Steel	Carbon Steel	Carbon Steel
Tube side			
Max temp. (°C)	60	40	89
Pressure (barg)	0.0	2.0	1.4
MOC	Carbon Steel	Carbon Steel	Stainless Steel
Phase	Cond. Vapor	L	L

(continued)

Equipment	C-1001 A/B	V-1001	V-1002	V-1003
MOC	Carbon Steel	Stainless Steel	Carbon Steel	Carbon Steel
Power (shaft) (kW)	2260	—	—	—
Efficiency	77%	—	—	—
Type/drive	Centrifugal Centrifugal –2 Stage/Electric	—	—	—
Temperature (°C)	25	—	—	—
Pressure in (bar)	1.0	—	—	—
Pressure out (bar)	5.0	—	—	—
Pressure (barg)	—	–0.88	–0.93	0.0
Diameter (m)	—	2.4	1.0	1.5
Height/length (m)	—	7.2	2.5	4.5
Orientation	—	Horizontal	Horizontal	Horizontal
Internals	—	—	—	—

(continued)

Table B.9.3 Utility Summary for Unit 1000

Utility	cw	cw	lps	rw	lps	cw	lps	cw	lps	cw
Equipment	E-1001	E-1002	E-1003	E-1004	E-1005	E-1006	E-1007	E-1008	E-1009	E-1010
Flow (tonne/h)	1995.0	1682.0	48.5	5182.0	1.07	54.5	10.19	378.0	3.85	16.7

going to the organic phase (in this case DIPE). Distribution coefficients for the organic acids in water and DIPE as well as mutual solubility data for water/DIPE are desirable. The process given in Figure B.2 was simulated using a UNIFAC thermodynamics package and the latent heat enthalpy option on CHEMCAD and should give reasonable results for preliminary process design. Much of the process background material and process configuration was taken from the 1986 AIChE student contest problem in Reference [5]. The kinetics presented above are fictitious but should give reasonable preliminary estimates of reactor size.

B.9.4 References

1. Kurland, J. J., and D. B. Bryant, "Shipboard Polymerization of Acrylic Acid," *Plant Operations Progress* 6, no. 4 (1987): 203–207.
2. *Kirk-Othmer Encyclopedia of Chemical Technology,* 3rd ed., vol. 1 (New York: John Wiley and Son, 1978), 330–354.
3. *Encyclopedia of Chemical Processing and Design,* ed. J. J. McKetta and W. A. Cunningham, vol. 1 (New York: Marcel Dekker, 1976), 402–428.
4. Sakuyama, S., T. Ohara, N. Shimizu, and K. Kubota, "A New Oxidation Process for Acrylic Acid from Propylene," *Chemical Technology* (June 1973): 350.
5. "1986 Student Contest Problem," *The AIChE Student Annual 1986,* ed. B. Van Wie, and R. A. Wills (AIChE, 1986), 52–82.

B.10 PRODUCTION OF ACETONE VIA THE DEHYDROGENATION OF ISOPROPYL ALCOHOL (IPA) [1, 2, 3, 4], UNIT 1100

The prevalent process for the production of acetone is as a by-product of the manufacture of phenol. Benzene is alkylated to cumene, which is further oxidized to cumene hydroperoxide and finally cleaved to yield phenol and acetone. However, the process shown in Figure B.10.1 and discussed here uses isopropyl alcohol (IPA) as the raw material. This is a viable commercial alternative, and a few

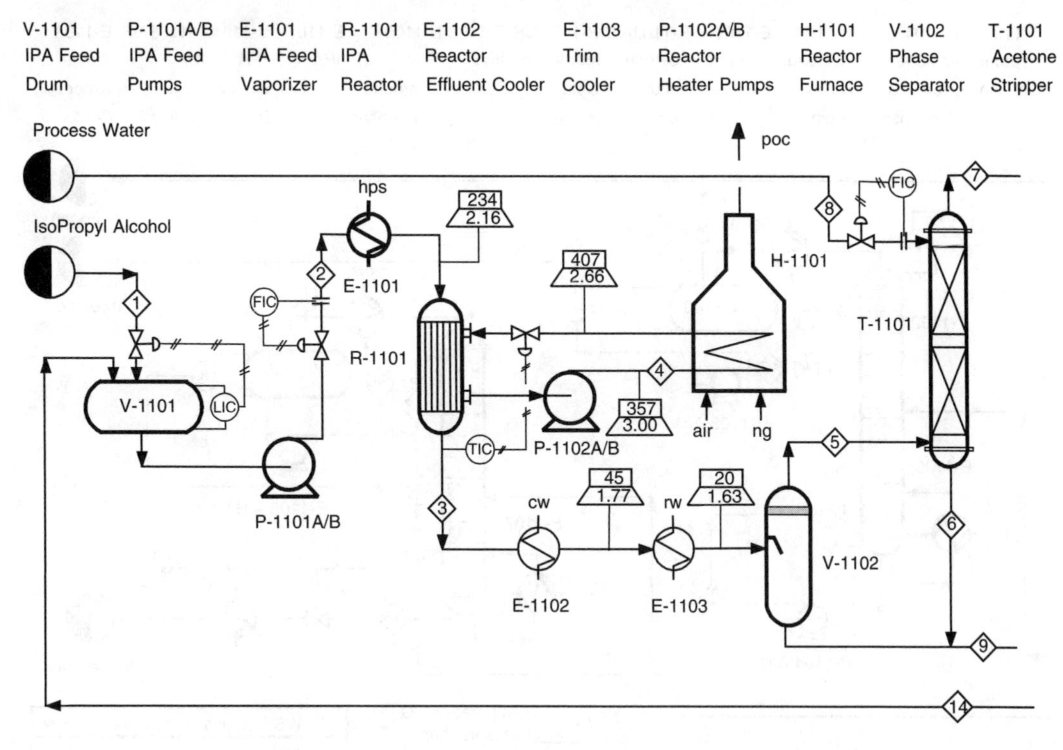

Figure B.10.1 Unit 1100: Production of Acetone from Isopropyl Alcohol PFD

plants continue to operate using this process. The primary advantage of this process is that the acetone produced is free from trace aromatic compounds, particularly benzene. For this reason, acetone produced from IPA may be favored by the pharmaceutical industry due to the very tight restrictions placed on solvents by the Food and Drug Administration (FDA). The reaction to produce acetone from IPA is as follows.

$$\underset{\textit{Isopropyl Alcohol}}{(CH_3)_2CHOH} \rightarrow \underset{\textit{Acetone}}{(CH_3)_2CO} + H_2$$

The reaction conditions are typically 2 bar and 350°C, giving single-pass conversions of 85%–92%.

B.10.1 Process Description

Referring to Figure B.10.1, an azeotropic mixture of isopropyl alcohol and water (88 wt% IPA) is fed into a surge vessel (V-1101), where it is mixed with the recycled unreacted IPA/water mixture, Stream 14. This material is then pumped and

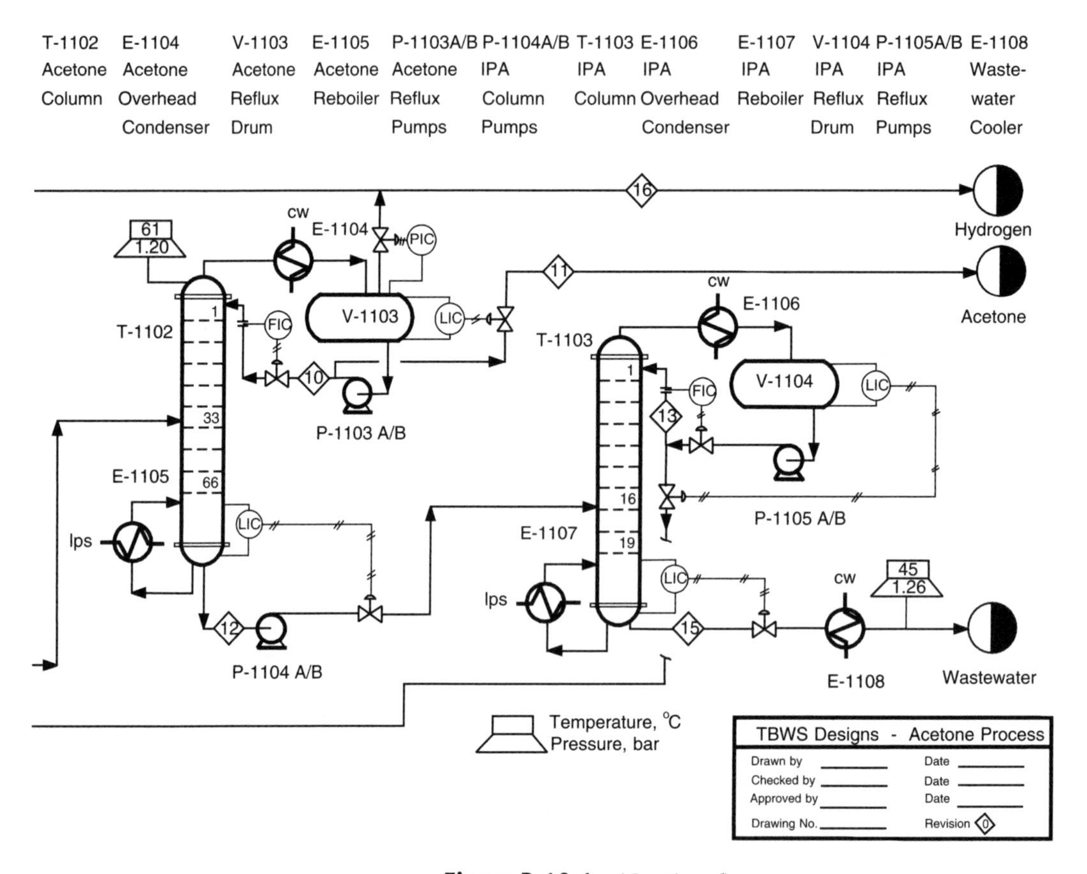

Figure B.10.1 *(Continued)*

vaporized prior to entering the reactor. Heat is provided for the endothermic reaction using a circulating stream of molten salt, Stream 4. The reactor effluent, containing acetone, hydrogen, water, and unreacted IPA, is cooled in two exchangers prior to entering the phase separator (V-1102). The vapor leaving the separator is scrubbed with water to recover additional acetone, and then this liquid is combined with the liquid from the separator and sent to the separations section. Two towers are used to separate the acetone product (99.9 mole %) and to remove the excess water from the unused IPA, which is then recycled back to the front end of the process as an azeotropic mixture. Stream summaries, preliminary equipment and utility summaries are given in Tables B.10.1, B.10.2 and B.10.3, respectively.

Table B.10.1 Flow Table for Acetone Process in Figure B.10.1

Stream Number	**1**	**2**	**3**	**4**	**5**	**6**	**7**	**8**
Temperature (°C)	25	32	350	357	20	27	33	25
Pressure (bar)	1.01	2.30	1.91	3.0	1.63	1.63	1.50	2.0
Vapor fraction	0.0	0.0	1.0	0.0	1.0	0.0	1.0	0.0
Mass flow (tonne/h)	2.40	2.67	2.67	35.1	0.34	0.46	0.24	0.36
Mole flow (kmol/h)	51.96	57.84	92.62	0.00	39.74	21.14	38.60	20.00
Component mole flow (kmol/h)				Molten Salt				
Hydrogen	0.00	0.00	34.78	0.00	34.78	0.00	34.78	0.00
Acetone	0.00	0.16	34.94	0.00	4.44	1.93	2.51	0.00
Isopropyl alcohol	34.82	38.64	3.86	0.00	0.12	0.10	0.02	0.00
Water	17.14	19.04	19.04	0.00	0.40	19.11	1.29	20.00

(continued)

Table B.10.1 Flow Table for Acetone Process in Figure B.10.1 (*Continued*)

Stream Number	**9**	**10**	**11**	**12**	**13**	**14**	**15**	**16**
Temperature (°C)	22	61	61	90	83	83	109	33
Pressure (bar)	1.63	1.5	1.5	1.4	1.2	1.2	1.4	1.2
Vapor fraction	0.0	0.0	0.0	0.0	0.0	0.0	0.0	1.0
Mass flow (tonne/h)	2.79	4.22	1.88	0.92	8.23	0.27	0.65	0.24
Mole flow (kmol/h)	74.02	72.51	32.29	41.73	177.18	5.88	35.85	38.60
Component mole flow (kmol/h)								
Hydrogen	0.00	0.00	0.00	0.00	0.00	0.00	0.00	34.78
Acetone	32.43	72.46	32.27	0.16	4.82	0.16	0.00	2.51
Isopropyl alcohol	3.84	0.05	0.02	3.82	115.10	3.82	0.00	0.02
Water	37.75	0.00	0.00	37.75	57.26	1.90	35.85	1.29

Table B.10.2 Preliminary Equipment Summary Table for Acetone Process

Equipment	P-1101 A/B	P-1102 A/B	P-1103 A/B	P-1104 A/B	P-1105 A/B	V-1101	V-1102
MOC	Carbon Steel	Carbon Steel	Carbon Steel	Carbon Steel	Carbon Steel	Carbon Steel	Carbon Steel
Power (shaft) (kW)	0.43	2.53	1.75	0.06	1.45	0.00	0.00
Efficiency	40%	50%	40%	40%	40%	0.00	0.00
Type/drive	Centrifugal/ Electric	Centrifugal/ Electric	Centrifugal/ Electric	Centrifugal/ Electric	Centrifugal/ Electric	0.00	0.00
Temperature (°C)	25	400	61	90	83	0.00	0.00
Pressure in (bar)	1.13	1.83	1.41	1.93	1.42	0.00	0.00
Pressure out (bar)	3.00	3.00	4.48	2.78	3.25	0.00	0.00
Diameter (m)	0.00	0.00	0.00	0.00	0.00	0.80	0.75
Height/length (m)	0.00	0.00	0.00	0.00	0.00	2.40	2.25
Orientation	0.00	0.00	0.00	0.00	0.00	Horizontal	Vertical
Internals	0.00	0.00	0.00	0.00	0.00	0.00	SS Demister
Pressure (barg)	0.00	0.00	0.00	0.00	0.00	0.0	0.63

(*continued*)

Table B.10.2 Preliminary Equipment Summary Table for Acetone Process (*Continued*)

Equipment	V-1103	V-1104	T-1101	T-1102	T-1103	H-1101	R-1101
MOC	Carbon Steel	Carbon Steel	Carbon Steel	Carbon Steel	Carbon Steel	Carbon Steel	Carbon Steel
Diameter (m)	0.83	0.93	0.33	1.25	1.36	0.00	1.85
Height/length (m)	2.50	2.80	3.20	37.0	18.6	0.00	8.0
Orientation	Horizontal	Horizontal	Vertical	Vertical	Vertical	0.00	Vertical
Internals	0.00	0.00	2.5 m of Packing 1″ Ceramic Rashig Rings	66 SS Sieve Plates @ 18″ Spacing	19 SS Sieve Plates @ 24″ Spacing	0.00	448 2″ Diameter, 20′ Long Catalyst Filled Tubes
Pressure (barg)	0.2	0.2	1.0	0.4	0.4	0.00	2.0
Type	0.00	0.00	0.00	0.00	0.00	Fired Heater	0.00
Duty (MJ/h)	0.00	0.00	0.00	0.00	0.00	2,730	0.00
Area radiant (m^2)	0.00	0.00	0.00	0.00	0.00	10.1	0.00
Area convective (m^2)	0.00	0.00	0.00	0.00	0.00	30.4	0.00
Tube pressure (barg)	0.00	0.00	0.00	0.00	0.00	2.0	0.00

(*continued*)

Table B.10.2 ***(Continued)***

Equipment	E-1101	E-1102	E-1103	E-1104	E-1105	E-1106	E-1107	E-1108
Type	Float. Head Vaporizer	Float. Head Partial Cond.	Float. Head Partial Cond.	Fixed TS Condenser	Float. Head Reboiler	Fixed TS Condenser	Float. Head Reboiler	Double Pipe
Duty (MJ/h)	3,550	3,260	563	3,095	3,500	7,340	7,390	174
Area (m^2)	70.3	77.6	8.5	39.1	30.9	50.2	65.1	1.6
Shell side								
Max. temp (°C)	234	350	45	61	90	83	109	109
Pressure (barg)	1.0	1.0	1.0	0.2	0.4	0.2	0.4	0.4
Phase	Boiling Liq.	Cond. Vapor	Cond. Vapor	Cond. Vapor	Boiling Liq.	Cond. Vapor	Boiling Liq.	L
MOC	Carbon Steel	Carbon Steel	Carbon Steel	Carbon Steel	Carbon Steel	Carbon Steel	Carbon Steel	Carbon Steel
Tube side								
Max. temp (°C)	254	40	15	40	160	40	160	40
Pressure (barg)	41.0	3.0	3.0	3.0	5.0	3.0	5.0	3.0
Phase	Cond. Steam	L	L	L	Cond. Steam	L	Cond. Steam	L
MOC	Carbon Steel	Carbon Steel	Carbon Steel	Carbon Steel	Carbon Steel	Carbon Steel	Carbon Steel	Carbon Steel

Table B.10.3 Summary Table Unit 1100

Utility	hps	cw	rw	cw	lps	cw	lps	cw
Equipment	E-1101	E-1102	E-1103	E-1104	E-1105	E-1106	E-1107	E-1108
Flow (tonne/h)	2.09	77.90	13.50	74.00	1.68	176.00	3.55	4.16

B.10.2 Reaction Kinetics

The reaction to form acetone from isopropyl alcohol (isopropanol) is endothermic, with a standard heat of reaction of 62.9 kJ/mol. The reaction is kinetically controlled and occurs in the vapor phase over a catalyst. The reaction kinetics for this reaction are first order with respect to the concentration of alcohol and can be estimated from the following equation [3,4]:

$$-r_{IPA} = k_0 \exp\left[-\frac{E_a}{RT}\right] C_{IPA} \quad \frac{\text{kmol}}{\text{m}^3\text{reactor s}}$$

where $E_a = 72.38\text{MJ/kmol}$, $k_0 = 3.51 \times 10^5 \dfrac{\text{m}^3\text{gas}}{\text{m}^3\text{reactor s}}$, $C_{IPA} = \dfrac{\text{kmol}}{\text{m}^3\text{gas}}$

In practice, several side reactions can occur to a small extent. Thus, trace quantities of propylene, diisopropyl ether, acetaldehyde, and other hydrocarbons and oxides of carbon can be formed [1]. The noncondensables are removed with the hydrogen, and the aldehydes and ethers may be removed with acid washing or adsorption. These side reactions are not accounted for in this preliminary design.

For the design presented in Figure B.10.1, the reactor was simulated with catalyst in 2-in (50.4 mm) diameter tubes, each 20 feet (6.096 m) long, and with a cocurrent flow of a heat transfer medium on the shell side of the shell-and-tube reactor. The resulting arrangement gives a 90% conversion of IPA per pass.

B.10.3 Simulation (CHEMCAD) Hints

Isopropyl alcohol and water form a minimum boiling point azeotrope at 88 wt% isopropyl alcohol and 12 wt% water. Vapor-liquid equilibrium (VLE) data are available from several sources and can be used to back-calculate binary interaction parameters or liquid-phase activity coefficients. The process presented in Figure B.3 and Table B.6 was simulated using the UNIQUAC VLE thermodynamics package and the latent heat enthalpy option in the CHEMCAD simulator. This package correctly predicts the formation of the azeotrope at 88 wt% alcohol.

B.10.4 References

1. *Kirk-Othmer Encyclopedia of Chemical Technology,* 3d ed., vol. 1 (New York: John Wiley & Sons, 1976), 179–191.

2. *Shreve's Chemical Process Industries,* 5th ed., ed. G. T. Austin (New York: McGraw-Hill, 1984), 764.
3. *Encyclopedia of Chemical Processing and Design,* ed. J. J. McKetta and W. A. Cunningham, vol. 1 (New York: Marcel Dekker, 1976), 314–362.
4. Sheely, C. Q., *Kinetics of Catalytic Dehydrogenation of Isopropanol,* Ph.D. Thesis, University of Illinois, 1963.

B.11 PRODUCTION OF HEPTENES FROM PROPYLENE AND BUTENES [1], UNIT 1200

This background information for this process is taken from Chauvel et al. [1]. This example is an illustration of a preliminary estimate of a process to convert a mixture of C_3 and C_4 unsaturated hydrocarbons to 1-heptene and other unsaturated products. The market for the 1-heptene product would be as a high-octane blending agent for gasoline or in the production of plasticizers. Based on preliminary market estimates, a production capacity of 20,000 metric tons per year of 1-heptene using 8000 operating hours/yr was set. This process differs from the other examples in Appendix B in several ways. First, the raw materials to the process contain a wide variety of chemicals. This is typical for oil refinery and some petrochemical operations. Second, no specific kinetic equations are given for the reactions. Instead, the results of laboratory tests using the desired catalyst at different conditions and using different feed materials are used to guide the process engineer to an optimum, or close to an optimum, reactor configuration. The flowsheet in Figure B.11.1 and stream, equipment summary, and utility summary tables, Tables B.11.1–B.11.3, have been developed using such information. It should be noted that a preliminary economic analysis, and hence the feasibility of the process, can be determined without this information, as long as yield and conversion data are available and the reactor configuration can be estimated.

B.11.1 Process Description

Two liquid feed streams containing propylene and butene and a stream of catalyst slurried with 1-hexene are mixed at a pressure of approximately 8 bar prior to being sent to the reactor. The reactor consists of five essentially well-mixed sections, with similar concentrations in each section. Heat removal is achieved by using pump-arounds from each stage through external heat exchangers. The reactor effluent is partially vaporized before being fed to the first of three distillation columns. The first column (T-1201) removes the unreacted C_3 and C_4 components, which are used subsequently as fuel (Stream 7) or sent to LPG storage (Stream 6). The next column (T-1202) separates the 1-hexene product overhead (Stream 10) and sends the bottoms stream to the final column (T-1203). In T-1203, the main 1-heptene product (Stream 13) is taken overhead, and the C_8 and heavier

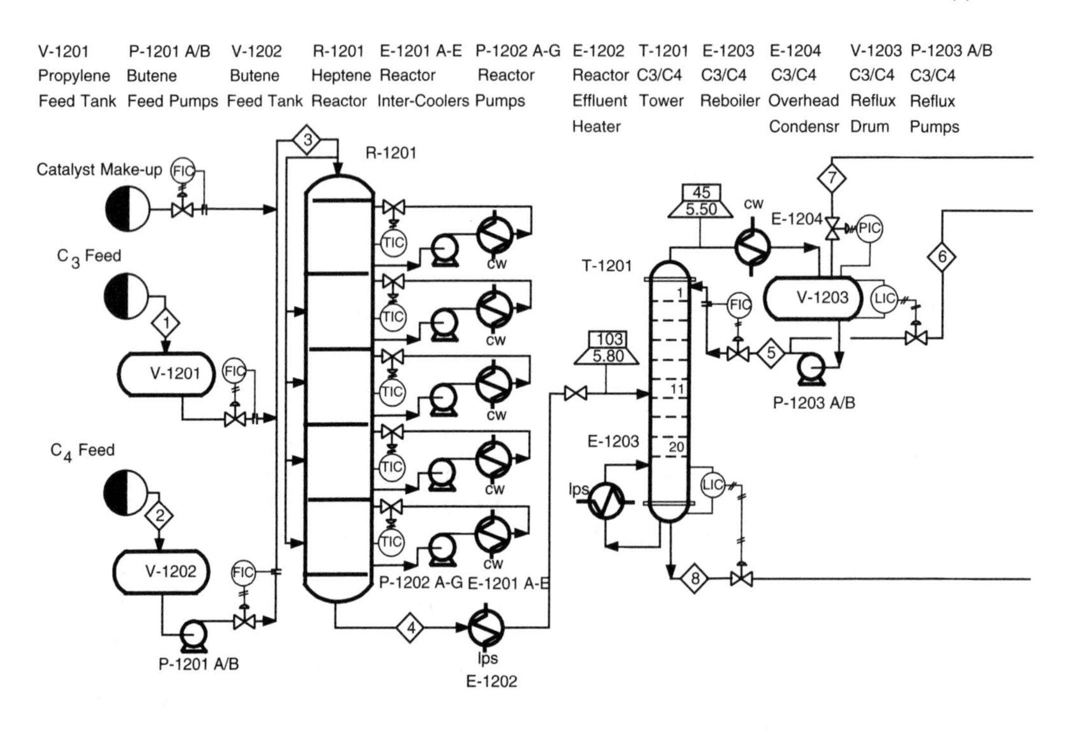

Figure B.11.1 Unit 1200: Production of Heptenes from Propylene and Butenes PFD

compounds are taken as the bottoms product (Stream 14). The bottoms product is processed off-site to remove the heavy material and to recover spent catalyst.

B.11.2 Reaction Kinetics

The process given in Figure B.11.1 is based on the liquid-phase catalytic co-dimerization of C_3 and C_4 olefins using an organometallic catalyst. This catalyst is slurried with a small volume of the hexenes product and fed to the reactor with the feed streams. The volume of the catalyst stream is small compared with the other streams and is not included in the material balance given in Table B.11.1. In 1976 (CEPCI = 183), consumption of catalyst amounted to \$9.5/1000 kg of 1-heptene product [1].

The primary reactions that take place are as follows:

$$C_3H_6 + C_3H_6 \rightarrow \underset{\textit{1-Hexene}}{C_6H_{12}}$$

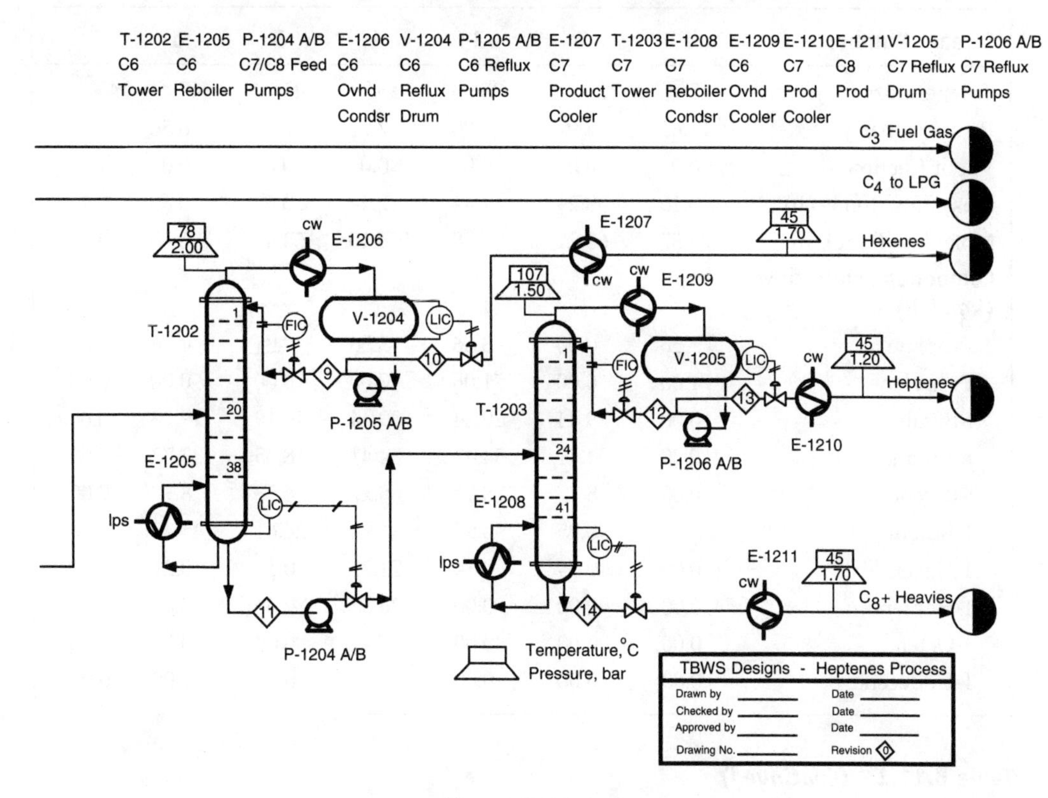

Figure B.11.1 (Continued)

$$C_3H_6 + C_4H_8 \rightarrow C_7H_{14}$$

1-Heptene

$$C_4H_8 + C_4H_8 \rightarrow C_8H_{16}$$

1-Octene

$$C_3H_6 + 2C_4H_8 \rightarrow C_{11}H_{22}$$

1-Undecene

In order to maximize the selectivity of the heptene reaction, several reactor configurations were considered [1]. The reactor configuration that maximized the heptene production, in a minimum volume, was found to be a plug flow reactor in which the butene feed was introduced at one end and the propylene stream was injected along the side of the reactor. However, due to other considerations

Table B.11.1 Flow Summary Table for Heptene Process in Figure B.11.1

Stream Number	1	2	3	4	5	6	7
Temperature (°C)	25	25	26	45	45	45	45
Pressure (bar)	11.62	2.97	8.00	7.7	7.50	6.50	5.00
Vapor fraction	0.0	0.0	0.0	0.0	0.0	0.0	1.0
Mass flow (tonne/h)	3.15	9.29	12.44	12.44	3.68	6.66	0.13
Mole flow (kmol/h)	74.57	163.21	237.78	178.10	64.41	116.45	3.00
Component mole flow (kmol/h)							
Propane	3.56	0.00	3.56	3.56	0.31	0.56	3.00
Propylene	71.06	0.00	71.06	0.00	0.00	0.00	0.00
i-Butane	0.00	29.44	29.44	29.44	16.19	29.28	0.00
n-Butane	0.00	34.41	34.41	34.41	18.65	33.72	0.00
i-Butene	0.00	8.27	8.27	8.27	4.53	8.19	0.00
1-Butene	0.00	90.95	90.95	44.94	24.61	44.49	0.00
1-Hexene	0.00	0.14	0.14	21.21	0.12	0.21	0.00
1-Heptene	0.00	0.00	0.00	26.53	0.00	0.00	0.00
1-Octene	0.00	0.00	0.00	7.41	0.00	0.00	0.00
1-Undecene	0.00	0.00	0.00	2.34	0.00	0.00	0.00

Table B.11.1 *(Continued)*

Stream Number	8	9	10	11	12	13	14
Temperature (°C)	151	78	78	135	107	107	154
Pressure (bar)	5.80	4.47	4.47	2.50	4.00	4.00	2.00
Vapor fraction	0.0	0.0	0.0	0.0	0.0	0.0	0.0
Mass flow (tonne/h)	5.64	5.79	1.86	3.79	4.30	2.53	1.26
Mole flow (kmol/h)	58.65	69.84	22.44	36.22	43.78	25.76	10.46
Component mole flow (kmol/h)							
Propane	0.00	0.00	0.00	0.00	0.00	0.00	0.00
Propylene	0.00	0.00	0.00	0.00	0.00	0.00	0.00
i-Butane	0.16	0.50	0.16	0.00	0.00	0.00	0.00
n-Butane	0.69	2.15	0.69	0.00	0.00	0.00	0.00
i-Butene	0.08	0.25	0.08	0.00	0.00	0.00	0.00
1-Butene	0.45	1.40	0.45	0.00	0.00	0.00	0.00
1-Hexene	21.00	64.70	20.79	0.21	0.36	0.21	0.00
1-Heptene	26.52	0.84	0.27	26.26	43.28	25.47	0.79
1-Octene	7.41	0.00	0.00	7.41	0.14	0.08	7.33
1-Undecene	2.34	0.00	0.00	2.34	0.00	0.00	2.34

Table B.11.2 Preliminary Equipment Summary Table for Heptene Process Unit 1200

Equipment	P-1201 A/B	P-1202 A-G*	P-1203 A/B	P-1204 A/B	P-1205 A/B	P-1206 A/B
MOC	Carbon Steel	Carbon Steel	Carbon Steel	Carbon Steel	Carbon Steel	Carbon Steel
Power (shaft) (kW)	6.75	5.13	2.75	0.66	2.15	1.93
Efficiency	40%	70%	40%	40%	40%	40%
Type/drive	Centrifugal/ Electric	Centrifugal/ Electric	Centrifugal/ Electric	Centrifugal/ Electric	Centrifugal/ Electric	Centrifugal/ Electric
Temperature (°C)	25	45	45	151	78	107
Pressure In (bar)	2.97	8.00	5.50	2.50	2.00	1.50
Pressure Out (bar)	9.00	9.00	7.55	4.00	4.47	4.00

*7 identical pumps: 5 operating + 2 spares.

Table B.11.2 Preliminary Equipment Summary Table for Heptene Process Unit 1200 (*Continued*)

Equipment	V-1201	V-1202	V-1203	V-1204	V-1205
MOC	Carbon Steel	Carbon Steel	Carbon Steel	Carbon Steel	Carbon Steel
Diameter (m)	1.40	1.90	1.10	0.95	0.92
Height/length (m)	4.20	5.7	3.30	2.85	2.75
Orientation	Horizontal	Horizontal	Horizontal	Horizontal	Horizontal
Internals	—	—	—	—	—
Pressure (barg)	20.0	5.0	4.5	1.0	0.5

Equipment	R-1201	T-1201	T-1202	T-1203
MOC	Carbon Steel	Carbon Steel	Carbon Steel	Carbon Steel
Diameter (m)	3.00	1.05	1.10	0.90
Height/length (m)	13.0	20.7	26.0	27.3
Orientation	Vertical	Vertical	Vertical	Vertical
Internals	Reactor Split into 5 Equal Sections	20 SS Sieve Plates @ 24″ Spacing	38 SS Sieve Plates @ 18″ Spacing	41 SS Sieve Plates @ 18″ Spacing
Pressure (barg)	7.0	5.0	1.5	1.0

(*continued*)

Table B.11.2 (*Continued*)

Equipment	E-1201 A-E*	E-1202	E-1203	E-1204	E-1205	E-1206
Type	Fixed TS	Float. Head Partial Vap.	Float. Head Reboiler	Fixed TS Condenser	Float. Head Reboiler	Fixed TS Condenser
Duty (MJ/h)	846	3827	1251	3577	2184	2630
Area (m^2)	61.4	33.8	32.1	128.5	21.1	20.0
Shell side						
Max. temp (°C)	45	160	160	45	160	78
Pressure (barg)	8.0	5.0	5.0	4.0	5.0	1.0
Phase	L	Cond. Steam	Cond. Steam	Cond. Vapor	Cond. Steam	Cond. Vapor
MOC	Carbon Steel	Carbon Steel	Carbon Steel	Carbon Steel	Carbon Steel	Carbon Steel
Tube side						
Max. temp (°C)	40	103	151	40	135	40
Pressure (barg)	3.0	6.7	4.8	3.0	1.5	3.0
Phase	L	L+V	Boiling Liq.	L	Boiling Liq.	L
MOC	Carbon Steel	Carbon Steel	Carbon Steel	Carbon Steel	Carbon Steel	Carbon Steel

*Area and duty given for one exchanger; five identical exchangers are needed.

Table B.11.2 Preliminary Equipment Summary Table for Heptene Process Unit 1200 (*Continued*)

Equipment	E-1207	E-1208	E-1209	E-1210	E-1211
Type	Double Pipe	Float. Head Reboiler	Fixed TS Condenser	Double Pipe	Double Pipe
Duty (MJ/h)	146	2026	2146	372	330
Area (m^2)	2.1	75.3	9.7	3.9	2.4
Shell side					
Max. temp (°C)	78	160	107	107	154
Pressure (barg)	1.0	5.0	0.5	0.3	0.7
Phase	L	Cond. Steam	Cond. Vapor	L	L
MOC	Carbon Steel	Carbon Steel	Carbon Steel	Carbon Steel	Carbon Steel
Tube side					
Max. temp (°C)	40	154	40	40	40
Pressure (barg)	3.0	1.0	3.0	3.0	3.0
Phase	L	Boiling Liq.	L	L	L
MOC	Carbon Steel	Carbon Steel	Carbon Steel	Carbon Steel	Carbon Steel

Table B.11.3 Utility Summary for Unit 1200

Utility	cw	lps	lps	cw	lps	cw	cw	lps	cw	cw	cw
Equipment	E-1201 A-E	E-1202	E-1203	E-1204	E-1205	E-1206	E-1207	E-1208	E-1209	E-1210	E-1211
Temperature In (°C)	30	160	160	30	160	30	30	160	30	30	30
Temperature Out (°C)	40	160	160	40	160	40	40	160	40	40	40
Flow (tonne/h)	20.20*	1.84	0.60	85.50	1.05	62.90	3.49	0.97	51.30	8.90	7.89

*Flow of cooling water shown for one exchanger only.

such as reactor complexity, it was finally decided to use a reactor with five equal stages in which the concentration in each stage is maintained approximately the same. Heat removal and mixing in each stage are accomplished by withdrawing a stream of material and pumping it through an external heat exchanger and back into the same stage of the reactor. The liquid cascades downward from stage to stage by means of liquid downcomers. The inside of the reactor can thus be considered similar to a five-plate distillation column (without vapor flow). The distribution of the feeds into the different stages is not shown in Figure B.4, and the dimensions of the reactor are taken directly from Chauvel et al. [1].

B.11.3 Simulation (CHEMCAD) Hints

All the hydrocarbon components used in the simulation can be considered to be well behaved, i.e., no azeotrope formation. The simulations were carried out using the SRK VLE and enthalpy packages using the CHEMCAD simulator.

B.11.4 Reference

1. Chauvel, A., P. Leprince, Y. Barthel, C. Raimbault, and J-P Arlie, *Manual of Economic Analysis of Chemical Processes,* trans. R. Miller and E. B. Miller (New York: McGraw-Hill, 1976), 207–228.

Index

D

N

CD-ROM Warranty

Prentice Hall warrants the enclosed CD-ROM to be free of defects in materials and faulty workmanship under normal use for a period of ninety days after purchase (when purchased new). If a defect is discovered in the CD-ROM during this warranty period, a replacement CD-ROM can be obtained at no charge by sending the defective CD-ROM, postage prepaid, with proof of purchase to:

Disc Exchange
Prentice Hall
Pearson Technology Group
75 Arlington Street, Suite 300
Boston, MA 02116
Email: AWPro@aw.com

Prentice Hall makes no warranty or representation, either expressed or implied, with respect to this software, its quality, performance, merchantability, or fitness for a particular purpose. In no event will Prentice Hall, its distributors, or dealers be liable for direct, indirect, special, incidental, or consequential damages arising out of the use or inability to use the software. The exclusion of implied warranties is not permitted in some states. Therefore, the above exclusion may not apply to you. This warranty provides you with specific legal rights. There may be other rights that you may have that vary from state to state. The contents of this CD-ROM are intended for personal use only.

More information and updates are available at:
informit.com/ph